CALCULUS

CALCULUS

Gilbert Strang
Massachusetts Institute of Technology

WELLESLEY-CAMBRIDGE PRESS
Box 812060 Wellesley MA 02482

A Manual for Instructors and a Student Study Guide are also available from Wellesley-Cambridge Press.

Strang, Gilbert
   Calculus, Second edition
   Includes index.
   ISBN 978-0-9802327-4-5
   1. Calculus, Second edition   I. Title.
   QA303.S8839 2010 515.20

Other texts by Gilbert Strang from Wellesley-Cambridge Press

Introduction to Linear Algebra, Fourth edition, ISBN 978-0-9802327-1-4.

Introduction to Linear Algebra, Fourth Intl edition, ISBN 978-0-9802327-2-1.

Computational Science and Engineering, ISBN 978-0-9614088-1-7.

Introduction to Applied Mathematics, ISBN 978-0-9614088-0-0.

An Analysis of the Finite Element Method, Second edition (2008), Gilbert Strang and George Fix, ISBN 978-0-9802327-0-7.

Algorithms for Global Positioning, with Kai Borre (2011), ISBN 978-0-9802327-3-8.

Wavelets and Filter Banks, with Truong Nguyen, ISBN 978-0-9614088-7-9.

Highlights of Calculus, (2011), ISBN 978-0-9802327-5-2.

Wellesley-Cambridge Press   (not Wellesley College or Cambridge University)
Box 812060
Wellesley MA 02482 USA
(781) 431-8488 FAX (617) 253-4358
www.wellesleycambridge.com

LaTeX text preparation by Valutone Software Solutions Pvt. Ltd., **www.valutone.co.in**

Cover art © Jasper Johns / Licensed by VAGA, New York, NY.
Catenary (Manet-Degas) 1999.

Cover preparation by Gail Corbett and Lois Sellers (**birchdesignassociates.com**)

Wellesley-Cambridge Press books can also be ordered from SIAM

Distribution outside North America is now by Cambridge University Press

Video lectures: *Highlights of Calculus* are freely available on **ocw.mit.edu**. See also iTunes U and YouTube and **math.mit.edu/calculus**.

Preface

My goal is to help you learn calculus. It is a beautiful subject and its central ideas are not so hard. Everything comes from the relation between two different functions. Here are two important examples:

Function (1) The *distance* a car travels   Function (2) Its *speed*
Function (1) The *height* of a graph   Function (2) Its *slope*

Function (2) is telling us how quickly Function (1) is changing. The distance will change quickly or slowly based on the speed. The height changes quickly or slowly based on the slope. You see the same in climbing — Function (1) can be the height of a mountain and Function (2) is its steepness. The height and the distance in Function (1) are "running totals" that add up the changes that come from Function (2).

The clearest example is when the speed is **CONSTANT**. The distance is steadily increasing. If you travel at 50 miles per hour, or at 80 kilometers per hour, then after 3 hours you know the distance traveled. I can write the answer by multiplying 3 times 50 or 3 times 80. I can write the formula using algebra, which allows any constant speed s and any time of travel t:

The distance f at constant speed s in travel time t is $f = s$ times t.

We don't need calculus when the speed is constant or the slope is constant:

If $s =$ slope and $x =$ distance across, the distance up is $y = s$ times x.

Those rules find Function (1) from Function (2). We can also find Function (2) from Function (1). To know the speed or the slope, divide instead of multiplying:

$$\text{speed } s = \frac{\text{distance } f}{\text{travel time } t} \qquad \text{slope } s = \frac{\text{distance up}}{\text{distance across}}$$

From the distance, we find the speed. This is *Differential Calculus*. Knowing the speed s, we find the distance f. This is *Integral Calculus*.

Algebra is enough for this example of constant speed. But when s is continually changing, and we speed up or slow down, then multiplication and division are not enough! **A new idea is needed and that idea is the heart of calculus.**

Differential Calculus finds Function (2) from Function (1). We recover the speedometer information from knowing the trip distance at all times.

Integral Calculus goes the other way. The "integral" adds up small pieces, to get the total distance traveled. That integration brings back Function (1).

Function (1) is $f(t)$ or $y(x)$ (2) Its "derivative" s is df/dt or dy/dx

---

The derivative in Function (2) is the "*rate of change*" of Function (1). The book will explain the meaning of these symbols df/dt and dy/dx for the derivative.

CHANGING SPEED AND CHANGING SLOPE

Let me take a first step into the real problem of calculus, when s is not constant. Now Function (1) will not have a straight line graph. The speed and the slope of the graph will change, but only every hour. From the numbers you can see the pattern:

Distances   0   1   4   9   16   Subtract the distances to get Function (2)

Speeds   1   3   5   7   Add up the speeds to get Function (1)

Going from (1) to (2) we are subtracting, as in $4-1=3$ and $9-4=5$. Those differences 3 and 5 are the speeds in the second hour and third hour.

Going from (2) to (1) we are adding, as in $1+3+5=9$. The trip meter adds up the distances from hours 1 and 2 and 3. Addition is the opposite of subtraction.

The essential point of calculus is to see this same pattern in "continuous time." It's not enough to look at the total or the change every hour or every minute. The distance and speed can be **changing at every instant**. In that case addition and subtraction are not enough. The central idea of calculus is **continuous change**.

There are so many pairs like (1) and (2)—not just cars and graphs and mountains. This is what makes calculus important. The functions are changing continuously—not just in finite steps. This is what makes calculus different from arithmetic and algebra.

IMPORTANT FUNCTIONS

Let me repeat the right name for the step from (1) to (2). When we know the distance or the height or the function $f(x)$, calculus can find the speed (*velocity*) and the slope and *the derivative*. That is **differential calculus**, going from Function (1) to Function (2). It will take time to find the slopes (*the derivatives*) for the examples we need.

I finally realized that the list of truly essential functions is not extremely long! I now include only five special choices for $y(x)$ or $f(t)$ on my basic list:

Important functions x^n $\sin x$ and $\cos x$ e^x and $\ln x$

For those distance functions, the speed (the slope) is continually changing. If we divide the total distance by the total time, we only know one number: the *average speed*. What calculus finds is the *speed at each separate moment*—the whole history of speed from the whole history of distance.

Your car has a speedometer to tell the derivative. It has a trip meter to tell the total mileage. They have the same information, recorded in different ways. From a record of the speeds we could recover a lost trip meter and vice versa. *One black box is enough, we could recover the other one* :

The derivative (speedometer) tells how the distance is changing.

The integral (odometer) adds up the changes to find the distance.

This is the "Big Picture" and the details are in this book. We need examples, we need formulas, we need rules, and especially we want and need applications — how to use this subject that you are learning.

VIDEO LECTURES

May I mention something very new? As this second edition goes to press, I am preparing video lectures to explain the "Highlights of Calculus." If all goes well, these videos will be freely available to everyone. (The first five were just released in 2010.) My hope is to help students and teachers everywhere.

At this moment I cannot know all the ways the videos may be used (my website **math.mit.edu/~gs** will try to keep up to date). *OpenCourseWare* on **ocw.mit.edu** is the focus of this plan. That site has notes and reading materials for almost 2000 MIT courses. For large classes there are also videos. The linear algebra lectures have had 2 million viewers (*amazing*). Those videos for the 18.06 course can help for review (at any hour!). There is also the next course 18.085 on Computational Science and Engineering. Now we are preparing this series of short calculus lectures.

One plan is to include these calculus videos on "Highlights for High School" within **ocw.mit.edu**. The highlights are shorter than the full multi-semester subject 18.01-18.02. I started with the idea that it must be so easy to get lost in a calculus textbook of 1000 pages. (This book is not so long, but still there is a lot to learn.) The video lectures go far beyond the page you are reading now, to help you capture the main ideas of calculus.

I so much hope that those videos are useful. As I prepare them, I think again about what is most important. Of course I make notes. The notes follow exactly the sequence of videos. They offer practice exercises on each topic (this book has many more problem sets). Probably those notes will turn into a new and quite short book called "*Highlights of Calculus*." That will be a supplement to this book and other textbooks, and a guide to the videos.

THE EXPONENTIAL $y = e^x$

There is one special function that I want to mention separately. It is created by calculus (not seen in algebra because a limit is involved). This is the function $y = e^x$, and one big question for the new Section 0.3 is how to construct it.

My answer now is not one of the usual four or five ways. All those definitions of e^x seem rather indirect and subtle. Instead I will try a direct attack that only uses the powers x^n. At that point of the book, the functions $y = 1, x, \ldots, x^n$ have known slopes. We are looking for *the magic function $y = e^x$ that equals its own slope*.

Then Function (1) equals Function (2)! See what you think of this approach.

Contents

Contents

Supporting Materials

The **Manual for Instructors** contains solutions to all exercises, with extended comments on the even-numbered problems. There are also notes on each chapter for classroom use. The manual comes to faculty on request, and is sent automatically on adoption. It may be ordered by readers who are using the text for self-teaching.

The **Student Study Guide** provides additional worked examples, which are coordinated section by section with the text. It also contains supplementary drill problems and completed solutions to all "read-through questions." Selected even-numbered problems from each section are solved.

Chapter 1 of the Study Guide is reproduced at the end of this text. The full Guide can be ordered from Wellesley-Cambridge Press, Box 812060, Wellesley MA 02482.

WELLESLEY-CAMBRIDGE PRESS

Wellesley-Cambridge is a small press. Our books are listed on the copyright page, and *Introduction to Linear Algebra* is especially popular. The printer and bookstore do their part equally for all publishers. When the author stays involved, the book is ready a year sooner. ***Most important, the connection between author and reader is not broken***. And we can keep book prices reasonable for students—an important goal.

Since our press was founded in 1985, hundreds of readers have sent encouragement. We feel part of a larger effort, which certainly lies behind this book. So does the need, widely expressed, for a fresh approach to the teaching of calculus. I hope that the pleasure of writing it will be felt by the reader.

Because the press has few spokesmen, the book has to speak for itself. We depend on teachers who like it to say so. If the direction is right, support will come. What counts in the end is moving steadily forward (with flashes of insight). All we can do is our best—and mathematics deserves it.

To the Student

I hope you will learn calculus from this book. On this page I will admit that I even hope for more. If you find that the explanations are clear, and also the *purpose* is clear—that means you see not only equations but ideas. Then the book was worth writing, and the course is a success.

I am trying to say that this subject is alive. As long as there are problems to solve, mathematics will keep growing. It is not wrapped up inside some giant formula! We want to know the chance of winning the Florida lottery (Chapter 2). You can see the meaning of an electrocardiogram (Chapter 3), and where we are on the population **S**-curve. There is no reason to pretend that mathematics has all the answers. This book is really the *life story of an idea*—which isn't finished.

Most of mathematics is about patterns and functions. There is a pattern in the graphs of Section 1.6 (which I don't understand). There is a pattern in the slopes of x^2 and x^3 and x^n (which you will understand). They change as x changes. Every function contains a whole history of growth or decay. The pattern is sometimes clear and sometimes hidden—the goal is to find it.

The reader will understand that calculus is not all sweetness and light. There is work to do. You absolutely have to solve problems—and think carefully. As one of my students said, "*My God, I have to read the words.*" I guess that's true, and I hope the words come to the point. The book will try not to waste your time. Its object is not to "cover" material, but to uncover and explain it. In the end, teaching a subject comes down to teaching a person.

It is not easy to stay inspired for a year—probably impossible. But mathematics is more active and cheerful than most people know. This book was written in a happy spirit, with a serious purpose—I hope you enjoy it.

A SMALL REWARD

Some mistakes may have crept into the solutions. I still have the ten dimes that George Thomas offered in 1952, for correcting ten errors. This reward is hereby increased to $\$e$. It should be e^t but that could grow exponentially. ***More important*** : *All suggestions are welcome*. Please write about any part of the book or the videos.

Derivatives

Sum: $\frac{d}{dx}(u+v) = \frac{du}{dx} + \frac{dv}{dx}$

Product: $\frac{d}{dx}(uv) = u\frac{dv}{dx} + v\frac{du}{dx}$

Quotient: $\frac{d}{dx}\left(\frac{u}{v}\right) = \frac{v\,du/dx - u\,dv/dx}{v^2}$

Power: $\frac{d}{dx}(u^n) = nu^{n-1}\frac{du}{dx}$

Chain: $\frac{d}{dx}z(y(x)) = \frac{dz}{dy}\frac{dy}{dx}$

Inverse: $\frac{dx}{dy} = \frac{1}{dy/dx}$

$\frac{d}{dx}\sin x = \cos x$

$\frac{d}{dx}\cos x = -\sin x$

$\frac{d}{dx}\tan x = \sec^2 x$

$\frac{d}{dx}\cot x = -\csc^2 x$

$\frac{d}{dx}\sec x = \sec x \tan x$

$\frac{d}{dx}\csc x = -\csc x \cot x$

$\frac{d}{dx}e^{cx} = ce^{cx}$

$\frac{d}{dx}b^x = b^x \ln b$

$\frac{d}{dx}\ln x = \frac{1}{x}$

$\frac{d}{dx}\sin^{-1} x = \frac{1}{\sqrt{1-x^2}}$

$\frac{d}{dx}\tan^{-1} x = \frac{1}{1+x^2}$

$\frac{d}{dx}\sec^{-1} x = \frac{1}{|x|\sqrt{x^2-1}}$

Limits and Continuity

$\frac{\sin x}{x} \to 1 \quad \frac{1-\cos x}{x} \to 0 \quad \frac{1-\cos x}{x^2} \to \frac{1}{2}$

$a_n \to 0: |a_n| < \varepsilon$ for all $n > N$

$a_n \to L: |a_n - L| < \varepsilon$ for all $n > N$

$f(x) \to L: |f(x) - L| < \varepsilon$ for $0 < |x-a| < \delta$

$f(x) \to f(a)$: Continuous at a if $L = f(a)$

$\frac{f(x)-f(a)}{x-a} \to f'(a)$: Derivative at a

$\frac{f(x)-f(a)}{x-a} = f'(c)$: Mean Value Theorem

$\frac{f(x+\Delta x)-f(x)}{\Delta x} \to f'(x)$: Derivative at x

$\frac{f(x+\Delta x)-f(x-\Delta x)}{2\Delta x} \to f'(x)$: Centered

$\lim \frac{f(x)}{g(x)} = \lim \frac{f'(x)}{g'(x)}$ l'Hôpital's Rule for $\frac{0}{0}$

Maximum and Minimum

Critical: $f'(x) = 0$ or no f' or endpoint

Minimum $\qquad f'(x) = 0$ and $f''(x) > 0$

Maximum $\qquad f'(x) = 0$ and $f''(x) < 0$

Inflection point $\qquad\qquad f''(x) = 0$

Newton's Method $x_{n+1} = x_n - \frac{f(x_n)}{f'(x_n)}$

Iteration $x_{n+1} = F(x_n)$ attracted to

fixed point $x^* = F(x^*)$ if $|F'(x^*)| < 1$

Stationary in 2D: $\partial f/\partial x = 0, \partial f/\partial y = 0$

Minimum $\qquad f_{xx} > 0 \quad f_{xx}f_{yy} > f_{xy}^2$

Maximum $\qquad f_{xx} < 0 \quad f_{xx}f_{yy} > f_{xy}^2$

Saddle point $\qquad\qquad f_{xx}f_{yy} < f_{xy}^2$

Newton in 2D $\begin{cases} g + g_x\Delta x + g_y\Delta y = 0 \\ h + h_x\Delta x + h_y\Delta y = 0 \end{cases}$

Algebra

$\frac{x/a}{y/b} = \frac{bx}{ay} \qquad x^{-n} = \frac{1}{x^n} \qquad \sqrt[n]{x} = x^{1/n}$

$(x^2)(x^3) = x^5 \quad (x^2)^3 = x^6 \quad x^2/x^3 = x^{-1}$

$ax^2 + bx + c = 0$ has roots $x = \frac{-b \pm \sqrt{b^2 - 4ac}}{2a}$

$x^2 + 2Bx + C = 0$ has roots $x = -B \pm \sqrt{B^2 - C}$

Completing square $ax^2 + bx + c = a\left(x + \frac{b}{2a}\right)^2 + c - \frac{b^2}{4a}$

Partial fractions $\frac{cx+d}{(x-a)(x-b)} = \frac{A}{x-a} + \frac{B}{x-b}$

Mistakes $\quad \frac{a}{b+c} \neq \frac{a}{b} + \frac{a}{c} \quad \sqrt{x^2+a^2} \neq x+a$

Fundamental Theorem of Calculus

$\frac{d}{dx}\int_a^x v(t)dt = v(x) \quad \int_a^b \frac{df}{dx}dx = f(b) - f(a)$

$\frac{d}{dx}\int_{a(x)}^{b(x)} v(t)dt = v(b(x))\frac{db}{dx} - v(a(x))\frac{da}{dx}$

$\int_0^b y(x)dx = \lim_{\Delta x \to 0} \Delta x[y(\Delta x) + y(2\Delta x) + \cdots + y(b)]$

Circle, Line, and Plane

$x = r\cos\omega t$, $y = r\sin\omega t$, speed ωr

$y = mx + b$ or $y - y_0 = m(x - x_0)$

Plane $ax + by + cz = d$ or

$a(x - x_0) + b(y - y_0) + c(z - z_0) = 0$

Normal vector $a\mathbf{i} + b\mathbf{j} + c\mathbf{k}$

Distance to $(0,0,0)$: $|d|/\sqrt{a^2 + b^2 + c^2}$

Line $(x, y, z) = (x_0, y_0, z_0) + t(v_1, v_2, v_3)$

No parameter: $\frac{x-x_0}{v_1} = \frac{y-y_0}{v_2} = \frac{z-z_0}{v_3}$

Projection: $\mathbf{p} = \frac{\mathbf{b}\cdot\mathbf{a}}{\mathbf{a}\cdot\mathbf{a}}\mathbf{a}$, $|\mathbf{p}| = |\mathbf{b}|\cos\theta$

Trigonometric Identities

$\sin^2 x + \cos^2 x = 1$

$\tan^2 x + 1 = \sec^2 x$ (divide by $\cos^2 x$)

$1 + \cot^2 x = \csc^2 x$ (divide by $\sin^2 x$)

$\sin 2x = 2\sin x \cos x$ (double angle)

$\cos 2x = \cos^2 x - \sin^2 x = 2\cos^2 x - 1 = 1 - 2\sin^2 x$

$\sin(s \pm t) = \sin s \cos t \pm \cos s \sin t$ (Addition

$\cos(s \pm t) = \cos s \cos t \mp \sin s \sin t$ formulas)

$\tan(s + t) = (\tan s + \tan t)/(1 - \tan s \tan t)$

$c^2 = a^2 + b^2 - 2ab\cos\theta$ (Law of cosines)

$a/\sin A = b/\sin B = c/\sin C$ (Law of sines)

$a\cos\theta + b\sin\theta = \sqrt{a^2+b^2}\cos(\theta - \tan^{-1}\frac{b}{a})$

$\cos(-x) = \cos x$ and $\sin(-x) = -\sin x$

$\sin(\frac{\pi}{2} \pm x) = \cos x$ and $\cos(\frac{\pi}{2} \pm x) = \mp\sin x$

$\sin(\pi \pm x) = \mp\sin x$ and $\cos(\pi \pm x) = -\cos x$

Trigonometric Integrals

$\int \sin^2 x \, dx = \frac{x - \sin x \cos x}{2} = \int \frac{1 - \cos 2x}{2} dx = \frac{x}{2} - \frac{\sin 2x}{4}$

$\int \cos^2 x \, dx = \frac{x + \sin x \cos x}{2} = \int \frac{1 + \cos 2x}{2} dx = \frac{x}{2} + \frac{\sin 2x}{4}$

$\int \tan^2 x \, dx = \tan x - x$

$\int \cot^2 x \, dx = -\cot x - x$

$\int \sin^n x \, dx = -\frac{\sin^{n-1} x \cos x}{n} + \frac{n-1}{n}\int \sin^{n-2} x \, dx$

$\int \cos^n x \, dx = +\frac{\cos^{n-1} x \sin x}{n} + \frac{n-1}{n}\int \cos^{n-2} x \, dx$

$\int \tan^n x \, dx = \frac{\tan^{n-1} x}{n-1} - \int \tan^{n-2} x \, dx$

$\int \sec^n x \, dx = \frac{\sec^{n-2} x \tan x}{n-1} + \frac{n-2}{n-1}\int \sec^{n-2} x \, dx$

$\int \tan x \, dx = -\ln|\cos x|$

$\int \cot x \, dx = \ln|\sin x|$

$\int \sec x \, dx = \ln|\sec x + \tan x|$

$\int \csc x \, dx = \ln|\csc x - \cot x| = -\ln|\csc x + \cot x|$

$\int \sec^3 x \, dx = \frac{1}{2}\sec x \tan x + \frac{1}{2}\ln|\sec x + \tan x|$

$\int \sin px \sin qx \, dx = \frac{\sin(p-q)x}{2(p-q)} - \frac{\sin(p+q)x}{2(p+q)}$

$\int \cos px \cos qx \, dx = \frac{\sin(p-q)x}{2(p-q)} + \frac{\sin(p+q)x}{2(p+q)}$

$\int \sin px \cos qx \, dx = -\frac{\cos(p-q)x}{2(p-q)} - \frac{\cos(p+q)}{2(p+q)}$

Integration by Parts

$\int \ln x \, dx = x\ln x - x$

$\int x^n \ln x \, dx = \frac{x^{n+1}\ln x}{n+1} - \frac{x^{n+1}}{(n+1)^2}$

$\int x^n e^x \, dx = x^n e^x - n\int x^{n-1} e^x dx$

$\int e^{cx} \sin kx \, dx = \frac{e^{cx}}{c^2+k^2}(c\sin kx - k\cos kx)$

$\int e^{cx} \cos kx \, dx = \frac{e^{cx}}{c^2+k^2}(c\cos kx + k\sin kx)$

$\int x\sin x \, dx = \sin x - x\cos x$

$\int x\cos x \, dx = \cos x + x\sin x$

$\int x^n \sin x \, dx = -x^n \cos x + n\int x^{n-1}\cos x \, dx$

$\int x^n \cos x \, dx = +x^n \sin x - n\int x^{n-1}\sin x \, dx$

$\int \sin^{-1} x \, dx = x\sin^{-1} x + \sqrt{1-x^2}$

$\int \tan^{-1} x \, dx = x\tan^{-1} x - \frac{1}{2}\ln(1+x^2)$

Integrals with x^2 and a^2 and $D = b^2 - 4ac$

$\int \frac{dx}{x^2+a^2} = \frac{1}{a}\tan^{-1}\frac{x}{a}$

$\int \frac{dx}{a^2-x^2} = \frac{1}{2a}\ln|\frac{x+a}{x-a}| = \frac{1}{a}\tanh^{-1}\frac{x}{a}$

$\int \frac{dx}{\sqrt{x^2+a^2}} = \ln|x + \sqrt{x^2+a^2}|$

$\int \frac{dx}{\sqrt{a^2-x^2}} = \sin^{-1}\frac{x}{a}$

$\int \sqrt{x^2+a^2}\,dx = \frac{x}{2}\sqrt{x^2+a^2} + \frac{a^2}{2}\ln|x + \sqrt{x^2+a^2}|$

$\int \sqrt{a^2-x^2}\,dx = \frac{x}{2}\sqrt{a^2-x^2} + \frac{a^2}{2}\sin^{-1}\frac{x}{a}$

$\int \frac{dx}{x\sqrt{x^2-a^2}} = \frac{1}{a}\cos^{-1}\frac{a}{x}$

$\int \frac{dx}{x\sqrt{x^2+a^2}} = \frac{1}{a}\ln|\frac{\sqrt{x^2+a^2}-a}{x}|$

$\int \frac{dx}{ax^2+bx+c} = \frac{1}{\sqrt{D}}\ln|\frac{2ax+b-\sqrt{D}}{2ax+b+\sqrt{D}}|, D > 0$
$\qquad = \frac{2}{\sqrt{-D}}\tan^{-1}\frac{2ax+b}{\sqrt{-D}}, D < 0$
$\qquad = \frac{-2}{2ax+b}, D = 0$

$\int \frac{dx}{\sqrt{ax^2+bx+c}} = \frac{1}{\sqrt{a}}\ln|2ax+b+2\sqrt{a}\sqrt{ax^2+bx+c}|$
$\qquad = \frac{1}{\sqrt{-a}}\sin^{-1}\frac{-2ax-b}{\sqrt{D}}, a < 0$

Definite Integrals

$\int_0^\infty x^n e^{-x} dx = n! = \Gamma(n+1)$

$\int_0^\infty e^{-a^2 x^2} dx = \sqrt{\pi}/2a$

$\int_0^1 x^m (1-x)^n dx = \frac{m! n!}{(m+n+1)!}$

$\int_0^\infty \frac{\sin^2 x}{x^2} dx = \int_0^\infty \frac{\sin x}{x} dx = \frac{\pi}{2}$

$\int_0^{\pi/2} \sin^n x \, dx = \int_0^{\pi/2} \cos^n x \, dx =$

$\frac{1}{2}\frac{3}{4}\cdots\frac{n-1}{n}(\frac{\pi}{2})$ or $\frac{2}{3}\frac{4}{5}\cdots\frac{n-1}{n}$

n even $\qquad$ n odd > 1

CHAPTER 0

Highlights of Calculus

0.1 Distance and Speed // Height and Slope

Calculus is about functions. I use that word "functions" in the first sentence, because we can't go forward without it. Like all other words, we learn this one in two different ways: We *define* the word and we *use* the word.

I believe that seeing examples of functions, and using the word to explain those examples, is a fast and powerful way to learn. I will start with three examples:

Linear function	$y(x) = 2x$
Squaring function	$y(x) = x^2$
Exponential function	$y(x) = 2^x$

The first point is that those are not the same! Their formulas involve 2 and x in very different ways. When I draw their graphs (this is a good way to understand functions) you see that all three are increasing when x is positive. The slopes are positive.

When the input x increases (moving to the right), the output y also increases (the graph goes upward). The three functions increase at different *rates*.

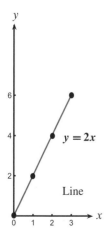

$y = 2x$

Line

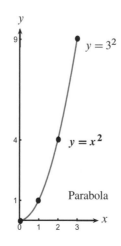

$y = 3^2$

$y = x^2$

Parabola

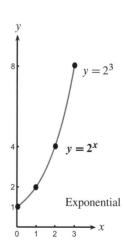

$y = 2^3$

$y = 2^x$

Exponential

Near the start at $x = 0$, the first function increases the fastest. But the others soon catch up. All three graphs reach the same height $y = 4$ when $x = 2$. Beyond that point the second graph $y = x^2$ pulls ahead. At $x = 3$ the squaring function reaches $y = 3^2 = 9$, while the height of the third graph is only $y = 2^3 = 8$.

Don't be deceived, *the exponential will win*. It pulls even at $x = 4$, because 4^2 and 2^4 are both 16. Then $y = 2^x$ moves ahead of $y = x^2$ and it stays ahead. When you reach $x = 10$, the third graph will have $y = 2^{10} = 1024$ compared to $y = 10^2 = 100$.

The graphs themselves are a *straight line* and a *parabola* and an *exponential*. The straight line has constant growth rate. The parabola has increasing growth rate. The exponential curve has exponentially increasing growth rate. I emphasize these because calculus is all about growth rates.

The whole point of differential calculus is to discover the growth rate of a function, and to use that information. So there are actually **two functions** in play at the same time—*the original function and its growth rate*. Before I go further down this all-important road, let me give a working definition of a function $y(x)$:

A function has inputs x and outputs $y(x)$. To each x it assigns one y.

The inputs x come from the "domain" of the function. In our graphs the domain contained all numbers $x \geqslant 0$. The outputs y form the "range" of the function. The ranges for the first two functions $y = 2x$ and $y = x^2$ contained all numbers $y \geqslant 0$. But the range for $y = 2^x$ is limited to $y \geqslant 1$ when the domain is $x \geqslant 0$.

Since these examples are so important, let me also allow x to be *negative*. The three graphs are shown below. Strictly speaking, these are new functions! Their domains have been extended to *all real numbers x*. Notice that the three ranges are also different:

The range of $y = 2x$ is all real numbers y

The range of $y = x^2$ is all nonnegative numbers $y \geqslant 0$

The range of $y = 2^x$ is all positive numbers $y > 0$

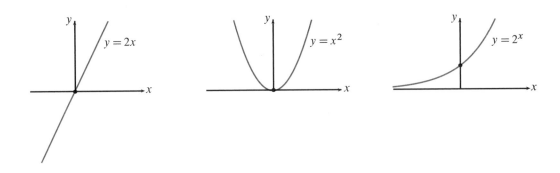

One more note about the idea of a function, and then calculus can begin. We have seen the three most popular ways to describe a function:

1. Give a *formula* to find y from x. Example: $y(x) = 2x$.
2. Give a *graph* that shows x (distance across) and y (distance up).
3. Give the *input-output pairs* (x in the domain and y in the range).

In a high-level definition, the "function" is the set of all the input-output pairs. We could also say: The function is the rule that assigns an output y in the range to every input x in the domain.

This shows something that we see for other words too. Logically, the definition should come first. Practically, we understand the definition better after we know examples that use the word. Probably that is the way we learn other words and also the way we will learn calculus. Examples show the general idea, and the definition is more precise. Together, we get it right.

The first words in this book were *Calculus is about functions*. Now I have to update that.

PAIRS OF FUNCTIONS

Calculus is about pairs of functions. Call them Function (1) and Function (2). Our graphs of $y = 2x$ and $y = x^2$ and $y = 2^x$ were intended to be examples of Function (1). Then we discussed the growth rates of those three examples. **The growth rate of Function (1) is Function (2).** This is our first task—to find the growth rate of a function. Differential calculus starts with a formula for Function (1) and aims to produce a formula for Function (2).

Let me say right away how calculus operates. There are two ways to compute how quickly y changes when x changes:

Method 1 (*Limits*): Write $\dfrac{\text{Change in } y}{\text{Change in } x} = \dfrac{\Delta y}{\Delta x}$. Take the limit of this ratio as $\Delta x \to 0$.

Method 2 (*Rules*): Follow a rule to produce new growth rates from known rates.

For each new function $y(x)$, look to see if it can be produced from known functions—obeying one of the rules. An important part of learning calculus is to see different ways of producing new functions from old. Then we follow the rules for the growth rate.

Suppose the new function is *not* produced from known functions (2^x is not produced from $2x$ or x^2). Then we have to find its growth rate directly. By "directly" I mean that we compute a limit which is Function (2). This book will explain what a "limit" means and how to compute it.

Here we begin with examples—almost always the best way. I will state the growth rates "dy/dx" for the three functions we are working with:

Function (1)	$y = 2x$	$y = x^2$	$y = 2^x$
Function (2)	$\dfrac{dy}{dx} = 2$	$\dfrac{dy}{dx} = 2x$	$\dfrac{dy}{dx} = 2^x (\ln 2)$

The linear function $y = 2x$ has constant growth rate $dy/dx = 2$. This section will take that first and easiest step. It is our opportunity to connect the growth rate to the **slope of the graph**. The ratio of *up* to *across* is $2x/x$ which is 2.

Section 0.2 takes the next step. The squaring function $y = x^2$ has linear growth rate $dy/dx = 2x$. (This requires the idea of a limit—so fundamental to calculus.) Then we can introduce our first two rules:

Constant factor   The growth rate of $Cy(x)$ is C times the growth rate of $y(x)$.

Sum of functions   The growth rate of $y_1 + y_2$ is the sum of the two growth rates.

The first rule says that $y = 5x^2$ has growth rate $10x$. The factor $C = 5$ multiplies the growth rate $2x$. The second rule says that $y_1 + y_2 = 5x^2 + 2x$ has growth rate $10x + 2$. Notice how we immediately took $5x^2$ as a function y_1 with a known growth rate. Together, the two rules give the growth rate for any "linear combination" of y_1 and y_2:

$$\textbf{The growth rate of } \; C_1 y_1 + C_2 y_2 \; \textbf{ is that same combination } C_1 \frac{dy_1}{dx} + C_2 \frac{dy_2}{dx}.$$

In words, the step from Function (1) to Function (2) is *linear*. The slope of $y = x^2 - x$ is $dy/dx = 2x - 1$. This rule is simple but so important.

Finally, Section 0.3 will present the exponential functions $y = 2^x$ and $y = e^x$. Our first job is their meaning—what is "2 *to the power* π"? We understand $2^3 = 8$ and $2^4 = 16$, but how can we multiply 2 by itself π times?

When we meet e^x, we are seeing the great creation of calculus. This is a function with the remarkable property that $dy/dx = y$. **The slope equals the function**. This requires the amazing number e, which was never seen in algebra—because it only appears when you take the right limit.

So these first sections compute growth rates for three essential functions. We are ready for $y = 2x$.

THE SLOPE OF A GRAPH

The slope is distance up divided by distance across. I am thinking now about the graph of a function $y(x)$. The "distance across" is the change $x_2 - x_1$ in the inputs, from x_1 to x_2. The "distance up" is the change $y_2 - y_1$ in the outputs, from y_1 to y_2. The slope is large and the graph is steep when $y_2 - y_1$ is much larger than $x_2 - x_1$. *Change in y divided by change in x* matches our ordinary meaning of the word slope:

$$\textbf{Average slope } = \frac{\text{change in } y}{\text{change in } x} = \frac{y_2 - y_1}{x_2 - x_1} = \frac{\Delta y}{\Delta x}. \tag{1}$$

I introduced the very useful Greek letter Δ (delta), as a symbol for *change*. We take a step of length Δx to go from x_1 to x_2. For the height $y(x)$ on the graph, that produces a step $\Delta y = y_2 - y_1$. The ratio of Δy to Δx, up divided by across, is the average slope between x_1 and x_2. The slope is the steepness.

Important point: I had to say "average" because the slope could be changing as we go. The graph of $y = x^2$ shows an increasing slope. Between $x_1 = 1$ and $x_2 = 2$, what is the average slope for $y = x^2$? *Divide Δy by Δx*:

$$y_1 = 1 \text{ at } x_1 = 1 \qquad \textbf{Average slope } = \frac{4-1}{2-1} = \frac{\Delta y}{\Delta x} = 3.$$
$$y_2 = 4 \text{ at } x_2 = 2$$

Between $x_1 = 0$ and $x_2 = 2$, we get a different answer (not 3). This graph of x^2 shows the problem of calculus, to deal with changes in slope and changes in speed.

The graph of $y = 2x$ has constant slope. The ratio of Δy to Δx, distance up to distance across, is always 2:

$$\boxed{\textbf{Constant slope} \qquad \frac{\Delta y}{\Delta x} = \frac{y_2 - y_1}{x_2 - x_1} = \frac{2x_2 - 2x_1}{x_2 - x_1} = 2.}$$

The mathematics is easy, which gives me a chance to emphasize the words and the ideas:

Function (1) = **Height** of the graph   Function (2) = **Slope** of the graph

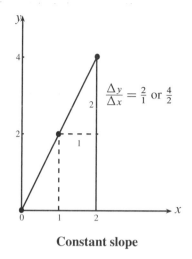

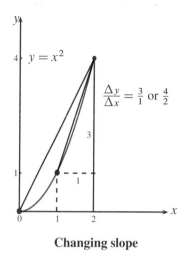

Constant slope   **Changing slope**

When Function (1) is $y = Cx$, the ratio $\Delta y / \Delta x$ is always C. A linear function has a constant slope. And those same functions can come from driving a car at constant speed:

Function (1) = **Distance** traveled $= Ct$ Function (2) = **Speed** of the car $= C$

For a graph of Function (1), its rate of change is the **slope**. When Function (1) measures distance traveled, its rate of change is the **speed** (or **velocity**). When Function (1) measures our height, its rate of change is our **growth rate**.

The first point is that *functions are everywhere*. For calculus, those functions come in pairs. *Function (2) is the rate of change of Function (1).*

The second point is that Function (1) and Function (2) are measured in different units. That is natural:

$$\left(\text{Speed in } \frac{\textbf{miles}}{\textbf{hour}}\right) \text{ multiplies } \left(\text{Time in } \textbf{hours}\right) \text{ to give } \left(\text{Distance in } \textbf{miles}\right)$$

$$\left(\text{Growth rate in } \frac{\textbf{inches}}{\textbf{year}}\right) \text{ multiplies } \left(\text{Time in } \textbf{years}\right) \text{ to give } \left(\text{Height in } \textbf{inches}\right)$$

When time is in seconds and distance is in meters, then speed is automatically in meters per second. We can choose two units, and they decide the third. Function (2) always involves a division: Δy is divided by Δx or distance is divided by time.

The delicate and tricky part of calculus is coming next. We want the *slope at one point* and the *speed at one instant*. What is the rate of change in *zero time* ?

The distance across is $\Delta x = 0$ at a point. The distance up is $\Delta y = 0$. **Formally, their ratio is $\frac{0}{0}$.** It is the inspiration of calculus to give this a useful meaning.

Big Picture

Calculus connects Function (1) with Function (2) = **rate of change** of (1)

Function (1)   Distance traveled $f(t)$ Function (2)   Speed $s(t) = df/dt$

Function (1)   Height of graph $y(x)$ Function (2)   Slope $s(x) = dy/dx$

Function (2) tells how quickly Function (1) is changing

KEY   Constant speed $s = \dfrac{\text{Distance } f}{\text{Time } t}$ Constant slope $s = \dfrac{\text{Distance up}}{\text{Distance across}}$

> Graphs of (1) and (2)
>
> f = increasing distance
>
> s = constant speed

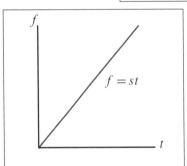 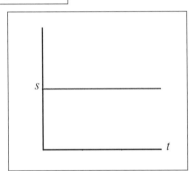

Slope of f-graph $= \dfrac{\text{up}}{\text{across}} = \dfrac{st}{t} = s$

Area under s-graph $=$ area of rectangle $= st = f$

> Now run the car backwards.
>
> Speed is negative
>
> Distance goes down to 0
>
> Area "under" $s(t)$ is zero

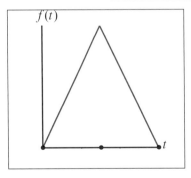 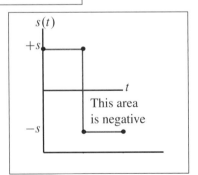

This area is negative

Example with increasing speed   Then distance has steeper slope

$$f = 10t^2$$
$$s = 20t$$

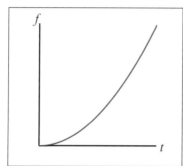
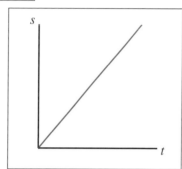

When speed is changing, algebra is not enough $s = \dfrac{f}{t}$ is wrong

Still true that area under s = triangle area = $\dfrac{1}{2}(t)(20t) = 10t^2 = f$

Still true that s = slope of $f = \dfrac{df}{dt}$ = "derivative" of f

---

When f is increasing, the slope s is **positive**

When f is decreasing, the slope s is **negative**

When f is at its maximum or minimum, the slope s is **zero**

The graphs of any $f(t)$ and $f(t) + 10$ have the same slope at every t

To recover f = Function (1) from $\dfrac{df}{dt}$, good to know a starting height $f(0)$

Practice Questions

1. Draw a graph of $f(t)$ that goes up and down and up again.

Then draw a reasonable graph of its slope.

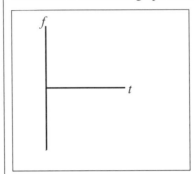 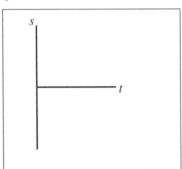

2. The world population $f(t)$ increased slowly at first, now quickly, then slowly again (we hope and expect). Maybe a limit ≈ 12 or 14 billion.

Draw a graph for $f(t)$ and its slope $s(t) = \dfrac{df}{dt}$

3. Suppose $f(t) = 2t$ for $t \leq 1$ and then $f(t) = 3t + 2$ for $t \geq 1$

Describe the slope graph $s(t)$. Compare its area out to $t = 3$ with $f(3)$

---

4. Draw a graph of $f(t) = \cos t$. Then sketch a graph of its slope. At what angles t is the slope zero (slope $= 0$ when $f(t)$ is "flat").

5. Suppose the graph of $f(t)$ is shaped like the capital letter **W**. Describe the graph of its slope $s(t) = \dfrac{df}{dt}$. What is the total area under the graph of s ?

6. A train goes a distance f at constant speed s. Inside the train, a passenger walks forward a distance F at walking speed S. What distance does the passenger go ? At what speed ? (Measure distance from the train station)

0.2 The Changing Slope of $y = x^2$ and $y = x^n$

The second of our three examples is $y = x^2$. Now the slope is changing as we move up the curve. The average slope is still $\Delta y / \Delta x$, but that is not our final goal. We have to answer the crucial questions of differential calculus:

What is the meaning of "slope at a point" and how can we compute it ?

My video lecture on *Big Picture*: *Derivatives* also faces those questions. Every student of calculus soon reaches this same problem. What is the meaning of "rate of change" when we are at a single moment in time, and nothing actually changes in that moment ? Good question.

The answers will come in two steps. Algebra produces $\Delta y / \Delta x$, and then calculus finds dy/dx. Those steps dy and dx are infinitesimally short, so formally we are looking at $0/0$. Trying to define dy and dx and $0/0$ is not wise, and I won't do it. The successful plan is to realize that the ratio of Δy to Δx is clearly defined, and those two numbers can become very small. *If that ratio $\Delta y / \Delta x$ approaches a limit, we have a perfect answer*:

$$\text{The slope at } x \text{ is the limit of} \qquad \frac{\Delta y}{\Delta x} = \frac{y(x + \Delta x) - y(x)}{\Delta x}.$$

The distance across, from x to $x + \Delta x$, is just Δx. The distance up is from $y(x)$ to $y(x + \Delta x)$. Let me show how algebra leads directly to $\Delta y / \Delta x$ when $y = x^2$:

$$\frac{\Delta y}{\Delta x} = \frac{(x + \Delta x)^2 - x^2}{\Delta x} = \frac{x^2 + 2x\Delta x + (\Delta x)^2 - x^2}{\Delta x} = 2x + \Delta x.$$

Notice that calculation! The "leading terms" x^2 and $-x^2$ cancel. The important term here is $2x\Delta x$. This "first-order term" is responsible for most of Δy. The "second-order term" in this example is $(\Delta x)^2$. *After we divide by Δx, this term is still small.* That part $(\Delta x)^2 / \Delta x$ will disappear as the step size Δx goes to zero.

That limiting process $\Delta x \to 0$ produces the slope dy/dx at a point. The first-order term survives in dy/dx and higher-order terms disappear.

$$\text{Slope at a point} \qquad \frac{dy}{dx} = \text{limit of } \frac{\Delta y}{\Delta x} = \text{limit of } 2x + \Delta x = 2x.$$

Algebra produced $\Delta y / \Delta x$. In the limit, calculus gave us dy/dx. Look at the graph, to see the geometry of those steps. The ratio up/across $= \Delta y / \Delta x$ is the slope between two points on the graph. *The two points come together in the limit.* Then $\Delta y / \Delta x$ approaches the slope dy/dx at a **single point**.

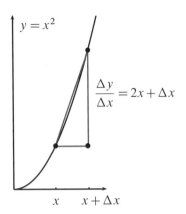 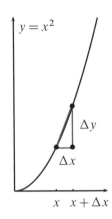 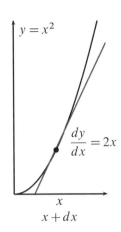

The color lines connecting points on the first two graphs are called "chords." They approach the color line on the third graph, which touches at only *one* point. This is the "**tangent line**" to the curve. Here is the idea of differential calculus:

$$\textbf{Slope of tangent line} = \textbf{Slope of curve} = \text{ Function (2)} = \frac{dy}{dx} = \textbf{2}\textbf{\textit{x}}.$$

To find the equation for this tangent line, return to algebra. Choose any specific value x_0. Above that position on the x axis, the graph is at height $y_0 = x_0^2$. The slope of the tangent line at that point of the graph is $dy/dx = 2x_0$. **We want the equation for the line through that point with that slope**.

> **Equation for the tangent line** $\qquad y - y_0 = (2x_0)(x - x_0)$ $\qquad$ (1)

At the point where $x = x_0$ and $y = y_0$, this equation becomes $0 = 0$. The equation is satisfied and the point is on the line. Furthermore the slope of the line matches the slope $2x_0$ of the curve. You see that directly if you divide both sides by $x - x_0$:

$$\textbf{Tangent line} \qquad \frac{\text{up}}{\text{across}} = \frac{y - y_0}{x - x_0} = 2x_0 \text{ is the correct slope } \frac{dy}{dx}.$$

Let me say this again. The curve $y = x^2$ is bending, the tangent line is straight. This line stays as close to the curve as possible, near the point where they touch. The tangent line gives a *linear* approximation to the nonlinear function $y = x^2$:

> **Linear approximation** $\qquad y \approx y_0 + (2x_0)(x - x_0) = y_0 + \frac{dy}{dx}(x - x_0)$ $\quad$ (2)

I only moved y_0 to the right side of equation (1). Then I used the symbol $\approx$ for "approximately equal" because the symbol $=$ would be wrong: The curve bends.

Important for the future: This bending comes from the **second derivative** of $y = x^2$.

THE SECOND DERIVATIVE

The first derivative is the slope $dy/dx = 2x$. **The second derivative is the slope of the slope.** By good luck we found the slope of $2x$ in the previous section (easy to do, it is just the constant 2). Notice the symbol d^2y/dx^2 for the slope of the slope:

Second derivative $\dfrac{d^2y}{dx^2}$ The slope of $\dfrac{dy}{dx} = 2x$ is $\dfrac{d^2y}{dx^2} = 2.$ (3)

In ordinary language, the first derivative dy/dx tells how fast the function $y(x)$ is changing. The second derivative tells whether we are *speeding up or slowing down*. The example $y = x^2$ is certainly speeding up, since the graph is getting steeper. The curve is bending and the tangent line is steepening.

Think also about $y = x^2$ on the left side (the negative side) of $x = 0$. The graph is coming down to zero. Its slope is certainly negative. But the curve is still bending upwards! The algebra agrees with this picture: The slope $dy/dx = 2x$ is *negative* on the left side of $x = 0$, but the second derivative $d^2y/dx^2 = 2$ is still positive.

An economist or an investor watches all three of those numbers: $y(x)$ tells where the economy is, and dy/dx tells which way it is going (short term, close to the tangent line). But it is d^2y/dx^2 that reveals the longer term prediction. I am writing these words near the end of the economic downturn (I hope). I am sorry that dy/dx has been negative but happy that d^2y/dx^2 has recently been positive.

DISTANCE AND SPEED AND ACCELERATION

An excellent example of $y(x)$ and dy/dx and d^2y/dx^2 comes from driving a car. The function y is the *distance traveled*. Its rate of change (first derivative) is the *speed*. The rate of change of the speed (second derivative) is the *acceleration*. If you are pressing on the gas pedal, all three will be positive. If you are pressing on the brake, the distance and speed are probably still positive but the acceleration is negative: The speed is dropping. If the car is *in reverse* and you are *braking*, what then?

> The speed is negative (going backwards)
> The speed is increasing (less negative)
> The acceleration is positive (increasing speed).

The video lecture mentions that car makers don't know calculus. The distance meter on the dashboard does not go back toward zero (in reverse gear it should). The speedometer does not go below zero (it should). There is no meter at all (on my car) for acceleration. Spaceships do have accelerometers, and probably aircraft too.

We often hear that an astronaut or a test pilot is subjected to a high number of g's. The ordinary acceleration in free fall is one g, from the gravity of the Earth. An airplane in a dive and a rocket at takeoff will have a high second derivative—the rocket may be hardly moving but it is accelerating like mad.

One more very useful point about this example of motion. *The natural letter to use is not x but t.* The distance is a function of **time**. The slope of a graph is up/across, but now the right ratio is (change of distance) divided by (change in time):

Average speed between t and $t + \Delta t$ $\dfrac{\Delta y}{\Delta t} = \dfrac{y(t + \Delta t) - y(t)}{\Delta t}$

Speed at t itself (instant speed) $\dfrac{dy}{dt} = $ limit of $\dfrac{\Delta y}{\Delta t}$ as $\Delta t \to 0$

The words "rate of change" and "rate of growth" suggest t. The word "slope" suggests x. But calculus doesn't worry much about the letters we use. If we graph the distance traveled as a function of time, then the x axis (across) becomes the t axis. And the slope of that graph becomes the speed (velocity is the best word).

Here is something not often seen in calculus books—the **second difference**. We know the first difference $\Delta y = y(t + \Delta t) - y(t)$. It is the change in y. The second difference $\mathbf{\Delta^2 y}$ **is the change in** $\mathbf{\Delta y}$:

Second difference
$$\Delta^2 y = (y(t + \Delta t) - y(t)) - (y(t) - y(t - \Delta t)) \qquad \frac{\Delta^2 y}{(\Delta t)^2} \to \frac{d^2 y}{dt^2} \quad (4)$$

$\Delta^2 y$ simplifies to $y(t + \Delta t) - 2y(t) + y(t - \Delta t)$. We divide by $(\Delta t)^2$ to approximate the acceleration. In the limit as $\Delta t \to 0$, this ratio $\Delta^2 y / (\Delta t)^2$ becomes the **second derivative** $\mathbf{d^2 y / dt^2}$.

THE SLOPE OF $y = x^n$

The slope of $y = x^2$ is $dy/dx = 2x$. Now I want to compute the slopes of $y = x^3$ and $y = x^4$ and all succeeding powers $y = x^n$. The rate of increase of x^n will be found again in Section 2.2. But there are two reasons to discover these special derivatives early:

1. Their pattern is simple: **The slope of each power** $\mathbf{y = x^n}$ **is** $\dfrac{dy}{dx} = nx^{n-1}$.

2. The next section can then introduce $y = e^x$. This amazing function has $\dfrac{dy}{dx} = y$.

Of course $y = x^2$ fits into this pattern for x^n. The exponent drops by 1 from $n = 2$ to $n - 1 = 1$. Also $n = 2$ multiplies that lower power to give $nx^{n-1} = 2x$.

The slope of $\mathbf{y = x^3}$ ***is*** $\mathbf{dy/dx = 3x^2}$. Watch how $3x^2$ appears in $\Delta y / \Delta x$:

$$\frac{\Delta y}{\Delta x} = \frac{(x + \Delta x)^3 - x^3}{\Delta x} = \frac{x^3 + 3x^2 \,\Delta x + 3x(\Delta x)^2 + (\Delta x)^3 - x^3}{\Delta x}. \qquad (5)$$

Cancel x^3 with $-x^3$. Then divide by Δx:

Average slope
$$\frac{\Delta y}{\Delta x} = 3x^2 + 3x \Delta x + (\Delta x)^2.$$

When the step length Δx goes to zero, the limit value dy/dx is $3x^2$. This is nx^{n-1}.

To establish this pattern for $n = 4, 5, 6, \dots$ the only hard part is $(x + \Delta x)^n$. When n was 3, we multiplied this out in equation (5) above. The result will always start with x^n. We claim that the next term (the "first-order term" in Δy) will be $nx^{n-1} \Delta x$. When we divide this part of Δy by Δx, we have the answer we want—the correct derivative nx^{n-1} of $y(x) = x^n$.

How to see that term $nx^{n-1} \Delta x$? Our multiplications showed that $2x\Delta x$ and $3x^2 \Delta x$ are correct for $n = 2$ and 3. Then we can reach $n = 4$ from $n = 3$:

$$(x + \Delta x)^4 = (x + \Delta x)^3 \text{ times } (x + \Delta x)$$
$$= (x^3 + 3x^2 \Delta x + \cdots) \text{ times } (x + \Delta x)$$

That multiplication produces x^4 and $4x^3 \Delta x$, exactly what we want. We can go from each n to the next one in the same way (this is called "induction"). The derivatives of all the powers $x^4, x^5, \dots, x^n$ are $4x^3, 5x^4, \dots, nx^{n-1}$.

Section 2.2 of the book shows you a slightly different proof of this formula. And the video lecture on the *Product Rule* explains one more way. Look at x^{n+1} as the product of x^n times x, and use the rule for the slope of y_1 times y_2:

Product Rule   Slope of $y_1 y_2 = y_2$ (slope of y_1) $+ y_1$ (slope of y_2)   (6)

With $y_1 = x^n$ and $y_2 = x$, the slope of $y_1 y_2 = x^{n+1}$ comes out right:

$$x(\text{slope of } x^n) + x^n(\text{slope of } x) = x(nx^{n-1}) + x^n(1) = (n+1)x^n. \qquad (7)$$

Again we can increase n one step at a time. Soon comes the truly valuable fact that this derivative formula is correct for *all powers* $y = x^n$. The exponent n can be negative, or a fraction, or any number at all. The slope dy/dx is always nx^{n-1}.

By combining different powers of x, you know the slope of every "polynomial." An example is $y = x + x^2/2 + x^3/3$. Compute dy/dx one term at a time, as the Sum Rule allows:

$$\frac{d}{dx}\left(x + \frac{x^2}{2} + \frac{x^3}{3}\right) = 1 + x + x^2.$$

The slope of the slope is $d^2y/dx^2 = 1 + 2x$. The fourth derivative is zero!

Function (1)   tells us the height y above each point x

The problem is to find the "instant slope" at x

This slope $s(x)$ is written $\dfrac{dy}{dx}$ It is **Function (2)**

KEY: $\dfrac{\boldsymbol{\Delta y}}{\boldsymbol{\Delta x}} = \dfrac{y(x + \Delta x) - y(x)}{\Delta x} = \dfrac{\text{up}}{\text{across}}$ approaches $\dfrac{dy}{dx}$ as $\Delta x \to 0$

Compute the **instant slope** $\dfrac{dy}{dx}$ for the function $y = x^3$

First find the average slope between x and $x + \Delta x$

Average slope $= \dfrac{\Delta y}{\Delta x} = \dfrac{(x + \Delta x)^3 - x^3}{\Delta x}$

Write $(x + \Delta x)^3 = x^3 + 3x^2\Delta x + 3x(\Delta x)^2 + (\Delta x)^3$

Subtract x^3 and divide by Δx

$\dfrac{\Delta y}{\Delta x} = \dfrac{3x^2\Delta x + 3x(\Delta x)^2 + (\Delta x)^3}{\Delta x} = 3x^2 + 3x\Delta x + (\Delta x)^2$

When $\Delta x \to 0$, this becomes $\dfrac{dy}{dx} = 3x^2$ $\dfrac{d}{dx}(x^n) = nx^{n-1}$

$y = Cx^n$ has slope Cnx^{n-1} The slope of $y = 7x^2$ is $\dfrac{dy}{dx} = 14x$

Multiply y by $C \to$ Multiply Δy by $C \to$ Multiply $\dfrac{dy}{dx}$ by C

Neat Fact: The slope of $y = \sin x$ is $\dfrac{dy}{dx} = \cos x$

The graphs show this is reasonable

Slope at the start is 1 (to find later)

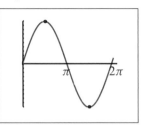

$y = \sin x$

slope $= \cos x$

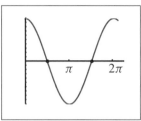

Sine curve climbing $\rightarrow$ Cosine curve > 0

Top of sine curve (flat) $\rightarrow$ Cosine is zero at the first bullet

Sine curve falling $\rightarrow$ Cosine curve < 0 between bullets

Bottom of sine curve (flat) $\rightarrow$ Cosine back to zero at the second bullet

---

Practice Questions

1. For $y = 2x^3$, what is the average slope $= \dfrac{\Delta y}{\Delta x}$ from $x = 1$ to $x = 2$?

2. What is the instant slope of $y = 2x^3$ at $x = 1$?   What is $\dfrac{d^2 y}{dx^2}$?

3. $y = x^n$ has $\dfrac{dy}{dx} = nx^{n-1}$. What is $\dfrac{dy}{dx}$ when $y(x) = \dfrac{1}{x} = x^{-1}$?

4. For $y = x^{-1}$, what is the average slope $\dfrac{\Delta y}{\Delta x}$ from $x = \dfrac{1}{2}$ to $x = 1$?

5. What is the instant slope of $y = x^{-1}$ at $x = \dfrac{1}{2}$?

---

6. Suppose the graph of $y(x)$ climbs up to its maximum at $x = 1$

Then it goes downward for $x > 1$

6A. What is the sign of $\dfrac{dy}{dx}$ for $x < 1$ and then for $x > 1$?

6B. What is the instant slope at $x = 1$?

7 If $y = \sin x$, write an expression for $\dfrac{\Delta y}{\Delta x}$ at any point x.

We see later that this $\dfrac{\Delta y}{\Delta x}$ approaches $\cos x$

0.3 The Exponential $y = e^x$

The great function that calculus creates is the exponential $y = e^x$. There are different ways to reach this function, and Section 6.2 of this textbook mentions five ways. Here I will describe the approach to e^x that I now like best. It uses the derivative of x^n, the first thing we learn.

In all approaches, a "limiting step" will be involved. So the amazing number $e = 2.7\ldots$ is not seen in algebra (e is not a fraction). The question is where to take that limiting step, and my answer starts with this truly remarkable fact: When $y = e^x$ is **Function (1)**, **it is also Function (2)**.

The exponential function $y = e^x$ solves the equation $\dfrac{dy}{dx} = y$.

The function equals its slope. This is a first example of a **differential equation**—connecting an unknown function y with its own derivatives. Fortunately $dy/dx = y$ is the most important differential equation—a model that other equations try to follow.

I will add one more requirement, to eliminate solutions like $y = 2e^x$ and $y = 8e^x$. When $y = e^x$ solves our equation, all other functions Ce^x solve it too. ($C = 2$ and $C = 8$ will appear on both sides of $dy/dx = y$, and they cancel.) At $x = 0$, e^0 will be the "zeroth power" of the positive number e. *All zeroth powers are* 1. So we want $y = e^x$ to equal 1 when $x = 0$:

$y = e^x$ is the solution of $\dfrac{dy}{dx} = y$ that starts from $y = 1$ at $x = 0$.

Before solving $dy/dx = y$, look at what this equation means. When y starts from 1 at $x = 0$, its slope is also 1. So y increases. Therefore dy/dx also increases, staying equal to y. So y increases faster. The graph gets steeper as the function climbs higher. This is what "growing exponentially" means.

INTRODUCING e^x

Exponential growth is quite ordinary and reasonable. When a bank pays interest on your money, the interest is proportional to the amount you have. After the interest is added, you have more. The new interest is based on the higher amount. Your wealth is growing "geometrically," one step at a time.

At the end of this section on e^x, I will come back to *continuous* compounding—interest is added at every instant instead of every year. That word "continuous" signals that we need calculus. There is a limiting step, from every year or month or day or second to every instant. You don't get infinite interest, you do get exponentially increasing interest.

I will also describe other ways to introduce e^x. This is an important question with many answers! I like equation (1) below, because we know the derivative of each power x^n. If you take their derivatives in equation (1), you get back the same e^x: *amazing*. So that sum solves $dy/dx = y$, starting from $y = 1$ as we wanted.

The difficulty is that the sum involves every power x^n: an *infinite series*. When I go step by step, you will see that those powers are all needed. For this infinite series, I am asking you to believe that everything works. *We can add the series to get e^x*, and we can add all derivatives to see that the slope of e^x is e^x.

For me, the advantage of using only the powers x^n is overwhelming.

CONSTRUCTING $y = e^x$

I will solve $dy/dx = y$ a step at a time. At the start, $y = 1$ means that $dy/dx = 1$:

Start $\begin{array}{l} y = 1 \\ dy/dx = 1 \end{array}$ **Change** y $\begin{array}{l} y = 1+x \\ dy/dx = 1 \end{array}$ **Change** $\dfrac{dy}{dx}$ $\begin{array}{l} y = 1+x \\ dy/dx = 1+x \end{array}$

After the first change, $y = 1+x$ has the correct derivative $dy/dx = 1$. But then I had to change dy/dx to keep it equal to y. And I can't stop there:

$\begin{array}{l} y = 1+x \\ dy/dx = 1+x \end{array}$ **Update** y **to** $1+x+\dfrac{1}{2}x^2$ **Then update** $\dfrac{dy}{dx}$ **to** $1+x+\dfrac{1}{2}x^2$

The extra $\frac{1}{2}x^2$ gave the correct x in the slope. Then $\frac{1}{2}x^2$ also had to go into dy/dx, to keep it equal to y. Now we need a new term with this derivative $\frac{1}{2}x^2$.

The term that gives $\frac{1}{2}x^2$ has x^3 divided by 6. The derivative of x^n is nx^{n-1}, so I *must divide by* n (to cancel correctly). Then the derivative of $x^3/6$ is $3x^2/6 = \frac{1}{2}x^2$ as we wanted. After that comes x^4 divided by 24:

$$\frac{x^3}{6} = \frac{x^3}{(3)(2)(1)} \quad \text{has slope} \quad \frac{x^2}{(2)(1)}$$

$$\frac{x^4}{24} = \frac{x^4}{(4)(3)(2)(1)} \quad \text{has slope} \quad \frac{4x^3}{(4)(3)(2)(1)} = \frac{x^3}{6}.$$

The pattern becomes more clear. The x^n term is divided by n *factorial*, which is $n! = (n)(n-1)\ldots(1)$. The first five factorials are $1, 2, 6, 24, 120$. **The derivative of that term** $x^n/n!$ **is the previous term** $x^{n-1}/(n-1)!$ (because the n's cancel). As long as we don't stop, this sum of infinitely many terms does achieve $dy/dx = y$:

$$\boxed{\; y(x) = e^x = 1 + x + \tfrac{1}{2}x^2 + \tfrac{1}{6}x^3 + \cdots + \tfrac{1}{n!}x^n + \cdots \qquad (1)}$$

If we substitute $x = 10$ into this series, do the infinitely many terms add to a finite number e^{10}? *Yes.* The numbers $n!$ grow much faster than 10^n (or any other x^n). So the terms $x^n/n!$ in this "exponential series" become extremely small as $n \to \infty$. Analysis shows that the sum of the series (which is $y = e^x$) does achieve $dy/dx = y$.

Note 1   Let me just remember a series that you know, $1 + \frac{1}{2} + \frac{1}{4} + \frac{1}{8} + \cdots = 2$. If I replace $\frac{1}{2}$ by x, this becomes the *geometric series* $1 + x + x^2 + x^3 + \cdots$ and it adds up to $1/(1-x)$. This is the most important series in mathematics, but it runs into a problem at $x = 1$: the infinite sum $1 + 1 + 1 + 1 + \cdots$ doesn't "converge."

I emphasize that the series for e^x is always safe, because the powers x^n are divided by the rapidly growing numbers $n! = n$ *factorial*. This is a great example to meet, long before you learn more about convergence and divergence.

Note 2   Here is another way to look at that series for e^x. Start with x^n and take its derivative n times. First get nx^{n-1} and then $n(n-1)x^{n-2}$. Finally the nth derivative is $n(n-1)(n-2)\ldots(1)x^0$, which is n *factorial*. When we divide by that number, **the** n**th derivative of** $x^n/n!$ **is equal to 1**.

Now look at e^x. All its derivatives are still e^x. They all equal 1 at $x = 0$. *The series is matching every derivative of the function* e^x *at the starting point* $x = 0$.

Set $x = 1$ in the exponential series. This tells us the amazing number $e^1 = e$:

> **The number e** $\qquad e = 1 + 1 + \frac{1}{2} + \frac{1}{6} + \frac{1}{24} + \frac{1}{120} + \cdots$ $\qquad$ (2)

The first three terms add to 2.5. The first five terms almost reach 2.71. *We never reach* 2.72. With quite a few terms (how many?) you can pass 2.71828. It is certain that e is not a fraction. It never appears in algebra, but it is the key number for calculus.

MULTIPLYING BY ADDING EXPONENTS

We write e^2 in the same way that we write 3^2. *Is it true that e times e equals e^2?* Up to now, e and e^2 come from setting $x = 1$ and $x = 2$ in the infinite series. The wonderful fact is that for every x, the series produces the "xth power of the number e." When $x = -1$, we get e^{-1} which is $1/e$:

Set $x = -1$ $\qquad e^{-1} = \frac{1}{e} = 1 - 1 + \frac{1}{2} - \frac{1}{6} + \frac{1}{24} - \frac{1}{120} + \cdots$

If we multiply that series for $1/e$ by the series for e, we get 1.

The best way is to go straight for all multiplications of e^x times any power e^X. The rule of adding exponents says that the answer is e^{x+X}. The series must say this too! When $x = 1$ and $X = -1$, this rule produces e^0 from e^1 times e^{-1}.

> **Add the exponents** $\qquad (e^x)(e^X) = e^{x+X}$ $\qquad$ (3)

We only know e^x and e^X from the infinite series. For this all-important rule, we can multiply those series and recognize the answer as the series for e^{x+X}. Make a start:

Multiply each term $\qquad e^x = 1 + x + \frac{1}{2}x^2 + \frac{1}{6}x^3 + \cdots$

e^x times e^X

Hoping for $\qquad e^X = 1 + X + \frac{1}{2}X^2 + \frac{1}{6}X^3 + \cdots$

e^{x+X} $\qquad\qquad (e^x)(e^X) = 1 + x + X \quad + \frac{1}{2}x^2 + xX + \frac{1}{2}X^2 + \cdots$ $\qquad$ (4)

Certainly you see $x + X$. Do you see $\frac{1}{2}(x + X)^2$ in equation (4)? No problem:

$$\frac{1}{2}(x + X)^2 = \frac{1}{2}(x^2 + 2xX + X^2) \quad \text{matches the "second degree" terms.}$$

The step to third degree takes a little longer, but it also succeeds:

$$\frac{1}{6}(x + X)^3 = \frac{1}{6}x^3 + \frac{3}{6}x^2 X + \frac{3}{6}xX^2 + \frac{1}{6}X^3 \quad \text{matches the next terms in (4).}$$

For high powers of $x + X$ we need the *binomial theorem* (or a healthy trust that mathematics comes out right). When e^x multiplies e^X, the coefficient of $x^n X^m$ will be $1/n!$ times $1/m!$. Now look for that same term in the series for e^{x+X}:

$$\frac{(x + X)^{n+m}}{(n+m)!} \quad \text{includes} \quad \frac{x^n X^m}{(n+m)!} \quad \text{times} \quad \frac{(n+m)!}{n! \, m!} \quad \text{which gives} \quad \frac{x^n X^m}{n! \, m!}. \qquad (5)$$

That binomial number $(n+m)!/n!\,m!$ is known to successful gamblers. It counts the number of ways to choose n aces out of $n+m$ aces. Out of 4 aces, you could choose 2 aces in $4!/2!2!=6$ ways. To a mathematician, there are 6 ways to choose 2 x's out of $xxxx$. This number 6 will be the coefficient of $x^2 X^2$ in $(x+X)^4$.

That 6 shows up in the fourth degree term. It is divided by 4! (to produce $1/4$). When e^x multiplies e^X, $\frac{1}{2}x^2$ multiplies $\frac{1}{2}X^2$ (which also produces $1/4$). All terms are good, but we are not going there—we accept $(e^x)(e^X)=e^{x+X}$ as now confirmed.

Note A different way to see this rule for $(e^x)(e^X)$ is based on $dy/dx=y$. Starting from $y=1$ at $x=0$, follow this equation. At the point x, you reach $y=e^x$. Now go an additional distance X to arrive at e^{x+X}.

Notice that the additional part starts from e^x (instead of starting from 1). That starting value e^x will multiply e^X in the additional part. So e^x times e^X must be the same as e^{x+X}. (This is a "differential equations proof" that the exponents are added. Personally, I was happy to multiply the series and match the terms.)

The rule immediately gives e^x times e^x. The answer is $e^{x+x}=e^{2x}$. If we multiply again by e^x, we find $(e^x)^3$. This is equal to $e^{2x+x}=e^{3x}$. We are finding a new rule for all powers $(e^x)^n=(e^x)(e^x)\cdots(e^x)$:

Multiply exponents	$(e^x)^n = e^{nx}$ (6)

This is easy to see for $n=1,2,3,\ldots$ and then $n=-1,-2,-3,\ldots$ It remains true for all numbers x and n.

That last sentence about "all numbers" is important! Calculus cannot develop properly without working with all exponents (not just whole numbers or fractions). The infinite series (1) defines e^x for every x and we are on our way. Here is the graph: **Function (1) = Function (2) = e^x = exp(x).**

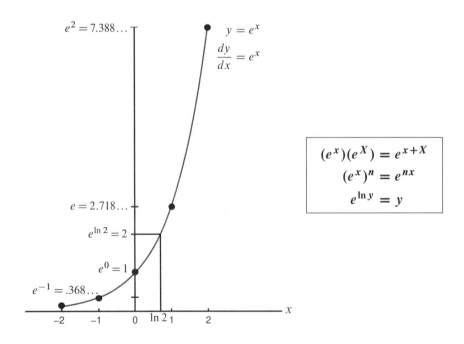

$e^2 = 7.388\ldots$

$y = e^x$

$\dfrac{dy}{dx} = e^x$

$e = 2.718\ldots$

$e^{\ln 2} = 2$

$e^0 = 1$

$e^{-1} = .368\ldots$

$-2 \quad -1 \quad 0 \quad \ln 2 \; 1 \quad\quad 2$

x

$(e^x)(e^X) = e^{x+X}$
$(e^x)^n = e^{nx}$
$e^{\ln y} = y$

THE EXPONENTIALS 2^x AND b^x

We know that $2^3 = 8$ and $2^4 = 16$. But what is the meaning of 2^π ? One way to get close to that number is to replace π by 3.14 which is $314/100$. As long as we have a fraction in the exponent, we can live without calculus:

Fractional power $2^{314/100} = $ 314th power of the 100th root $2^{1/100}$.

But this is only "close" to 2^π. And in calculus, we will want the slope of the curve $y = 2^x$. The good way is to connect 2^x with e^x, whose slope we know (it is e^x again). So we need to connect 2 with e.

The key number is the **logarithm of 2**. This is written "ln 2" and it is the power of e that produces 2. It is specially marked on the graph of e^x:

Natural logarithm of 2 $e^{\ln 2} = 2$

This number ln 2 is about $7/10$. A calculator knows it with much higher accuracy. In the graph of $y = e^x$, the number ln 2 on the x axis produces $y = 2$ on the y axis.

This is an example where we want the output $y = 2$ and we ask for the input $x = \ln 2$. That is the opposite of knowing x and asking for y. "The logarithm $x = \ln y$ is the *inverse* of the exponential $y = e^x$." This idea will be explained in Section 4.3 and in two video lectures—inverse functions are not always simple.

Now 2^x has a meaning for every x. When we have the number ln 2, meeting the requirement $2 = e^{\ln 2}$, we can take the xth power of both sides:

Powers of 2 from powers of e $2 = e^{\ln 2}$ and $2^x = e^{x \ln 2}$.   (7)

All powers of e are defined by the infinite series. The new function 2^x also grows exponentially, but not as fast as e^x (because 2 is smaller than e). Probably $y = 2^x$ could have the same graph as e^x, if I stretched out the x axis. That stretching multiplies the slope by the constant factor ln 2. Here is the algebra:

Slope of $y = 2^x$ $\dfrac{d}{dx} 2^x = \dfrac{d}{dx} e^{x \ln 2} = (\ln 2) e^{x \ln 2} = (\textbf{ln 2}) 2^x$.

For any positive number b, the same approach leads to the function $y = b^x$. First, find the natural logarithm ln b. This is the number (positive or negative) so that $b = e^{\ln b}$. Then take the xth power of both sides:

> **Connect b to e** $b = e^{\ln b}$ and $b^x = e^{x \ln b}$ and $\dfrac{d}{dx} b^x = (\ln b) b^x$ (8)

When b is e (the perfect choice), $\ln b = \ln e = 1$. When b is e^n, then $\ln b = \ln e^n = n$. "*The logarithm is the exponent*." Thanks to the series that defines e^x for every x, that exponent can be any number at all.

Allow me to mention Euler's Great Formula $e^{ix} = \cos x + i \sin x$. The exponent ix has become an **imaginary number**. (You know that $i^2 = -1$.) If we faithfully use $\cos x + i \sin x$ at 90° and 180° (where $x = \pi/2$ and $x = \pi$), we arrive at these amazing facts:

> **Imaginary exponents** $e^{i\pi/2} = i$ and $e^{i\pi} = -1$.   (9)

Those equations are not imaginary, they come from the great series for e^x.

CONTINUOUS COMPOUNDING OF INTEREST

There is a different and important way to reach e and e^x (not by an infinite series). We solve the key equation $dy/dx = y$ in small steps. As these steps approach zero (a limit is always involved !) the small-step solution becomes the exact $y = e^x$.

I can explain this idea in two different languages. Each step multiplies by $1 + \Delta x$:

1. *Compound interest.* After each step Δx, the interest is added to y. Then the next step begins with a larger amount, and y increases exponentially.

2. *Finite differences.* The continuous dy/dx is replaced by small steps $\Delta Y/\Delta x$:

$$\frac{dy}{dx} = y \text{ changes to } \frac{Y(x + \Delta x) - Y(x)}{\Delta x} = Y(x) \text{ with } Y(0) = 1. \qquad (10)$$

This is Euler's method of approximation. $Y(x)$ approaches $y(x)$ as $\Delta x \to 0$.

Let me compute compound interest when 1 year is divided into 12 months, and then 365 days. The interest rate is 100% and you start with $Y(0) = \$1$. If you only get interest once, at the end of the year, then you have $Y(1) = \$2$.

If interest is added every month, you now get $\frac{1}{12}$ of 100% each time (12 times). So Y is multiplied each month by $1 + \frac{1}{12}$. (The bank adds $\frac{1}{12}$ for every 1 you have.) Do this 12 times and the final value \$2 is improved to \$2.61:

After 12 months $\qquad Y(1) = \left(1 + \frac{1}{12}\right)^{12} = \$2.61$

Now add interest every day. $Y(0) = \$1$ is multiplied 365 times by $1 + \frac{1}{365}$:

After 365 days $\qquad Y(1) = \left(1 + \frac{1}{365}\right)^{365} = \$2.71 \ (\text{close to } e)$

Very few banks use minutes, and nobody divides the year into $N = 31,536,000$ seconds. It would add less than a penny to \$2.71. But many banks are willing to use *continuous compounding*, the limit as $N \to \infty$. After one year you have \$$e$:

Another limit gives e $\qquad \left(1 + \frac{1}{N}\right)^N \to e = 2.718\ldots \text{ as } N \to \infty \qquad (11)$

You could invest at the 100% rate for x years. Now each of the N steps is for x/N years. Again the bank multiplies at every step by $1 + \frac{x}{N}$. The 1 keeps what you have, the x/N adds the interest in that step. After N steps you are close to e^x:

A formula for e^x $\qquad \left(1 + \frac{x}{N}\right)^N \to e^x \text{ as } N \to \infty \qquad (12)$

Finally, I will change the interest rate to a. Go for x years at the interest rate a. The differential equation changes from $dy/dx = y$ to $dy/dx = ay$. The exponential function still solves it, but now that solution is $y = e^{ax}$:

Change the rate to a $\qquad \dfrac{dy}{dx} = ay \text{ is solved by } y(x) = e^{ax} \qquad (13)$

You can write down the series $e^{ax} = 1 + ax + \frac{1}{2}(ax)^2 + \cdots$ and take its derivative:

$$\frac{d}{dx}(e^{ax}) = a + a^2 x + \cdots = a(1 + ax + \cdots) = ae^{ax} \qquad (14)$$

The derivative of e^{ax} brings down the extra factor a. So $y = e^{ax}$ solves $dy/dx = ay$.

The Exponential $y = e^x$

Looking for a function $y(x)$ that equals its own derivative $\dfrac{dy}{dx}$

A differential equation! We start at $x = 0$ with $y = 1$

Infinite Series $y(x) = 1 + x + \dfrac{x^2}{2!} + \dfrac{x^3}{3!} + \cdots + \left(\dfrac{x^n}{n!}\right) + \cdots$

Take derivative $\dfrac{dy}{dx} = 0 + 1 + x + \dfrac{x^2}{2!} + \cdots + \left(\dfrac{x^{n-1}}{(n-1)!}\right) + \cdots$

Term by term $\dfrac{dy}{dx}$ agrees with y Limit step = add up this series

$n! = (n)(n-1)\cdots(1)$ grows much faster than x^n so the terms get very small

At $x = 1$ the number $y(1)$ is called e. Set $x = 1$ in the series to find e

$$e = 1 + 1 + \frac{1}{2} + \frac{1}{6} + \frac{1}{24} + \cdots = 2.71828\ldots$$

GOAL Show that $y(x)$ agrees with e^x for all x Series gives powers of e

Check that the series follows the rule to add exponents as in $e^2 e^3 = e^5$

Directly multiply series e^x times e^X to get e^{x+X}

$\left(1 + x + \dfrac{1}{2}x^2\right)$ times $\left(1 + X + \dfrac{1}{2}X^2\right)$ produces the right start for e^{x+X}

$1 + (x + X) + \dfrac{1}{2}(x + X)^2 + \cdots$ HIGHER TERMS ALSO WORK

The series gives us e^x for EVERY x, not just whole numbers

CHECK $\dfrac{de^x}{dx} = \lim \dfrac{e^{x+\Delta x} - e^x}{\Delta x} = e^x \left(\lim \dfrac{e^{\Delta x} - 1}{\Delta x}\right) = e^x$ YES!

SECOND KEY RULE $(e^x)^n = e^{nx}$ for every x and n

Another approach to e^x uses multiplication instead of an infinite sum

Start with \$1. Earn interest every day at yearly rate x

Multiply 365 times by $\left(1 + \dfrac{x}{365}\right)$. End the year with $\$\left(1 + \dfrac{x}{365}\right)^{365}$

Now pay n times in the year. End the year with $\left(1 + \dfrac{x}{n}\right)^n \to \$\,e^x$ as $n \to \infty$

We are solving $\dfrac{\Delta Y}{\Delta x} = Y$ in n short steps Δx.   The limit solves $\dfrac{dy}{dx} = y$.

Practice Questions

1. What is the derivative of $\dfrac{x^{10}}{10!}$?   What is the derivative of $\dfrac{x^9}{9!}$?

2. How to see that $\dfrac{x^n}{n!}$ gets small as $n \to \infty$?

Start with $\dfrac{x}{1}$ and $\dfrac{x^2}{2}$, possibly big. But we multiply by $\dfrac{x}{3}, \dfrac{x}{4}, \cdots$ which gets small.

3. Why is $\dfrac{1}{e^x}$ the same as e^{-x}?   Use equation (3) and also use (6).

4. Why is $e^{-1} = 1 - 1 + \dfrac{1}{2} - \dfrac{1}{6} + \cdots$ between $\dfrac{1}{3}$ and $\dfrac{1}{2}$? Then $2 < e < 3$.

5. Can you solve $\dfrac{dy}{dx} = y$ starting from $y = 3$ at $x = 0$?

Why is $y = 3e^x$ the right answer?   Notice how 3, multiplies e^x.

6. Can you solve $\dfrac{dy}{dx} = 5y$ starting from $y = 1$ at $x = 0$?

Why is $y = e^{5x}$ the right answer?   Notice 5 in the exponent!

7. Why does $\dfrac{e^{\Delta x} - 1}{\Delta x}$ approach 1 as Δx gets smaller?   Use the $e^{\Delta x}$ series.

8. Draw the graph of $x = \ln y$, just by flipping the graph of $y = e^x$ across the $45°$ line $y = x$. Remember that y stays positive but $x = \ln y$ can be negative.

9. What is the exact sum of $1 + \ln 2 + \dfrac{1}{2}(\ln 2)^2 + \dfrac{1}{3!}(\ln 2)^3 + \cdots$?

10. If you replace $\ln 2$ by 0.7, what is the sum of those four terms?

11. From Euler's Great Formula $e^{ix} = \cos x + i \sin x$, what number is $e^{2\pi i}$?

12. How close is $\left(1 + \dfrac{1}{10}\right)^{10}$ to e?

13. What is the limit of $\left(1 + \dfrac{1}{N}\right)^{2N}$ as $N \to \infty$?

0.4 Video Summaries and Practice Problems

This section is to help readers who also look at the **Highlights of Calculus** video lectures. The first five videos are just released on `ocw.mit.edu` as I write these words. Sections 0.1–0.2–0.3 discussed the content of three lectures in full detail. The summaries and practice problems for the other two will come first in this section:

   4. Maximum and Minimum and Second Derivative

   5. Big Picture of Integrals

That Lecture 5 is a taste of *Integral Calculus*. A second set of video lectures goes deeper into *Differential Calculus*—the rules for computing and using derivatives.

   This second set is right now with the video editors, to zoom in when I write on the blackboard and zoom out for the big picture. I just borrowed a video camera from MIT's OpenCourseWare and set it up in an empty room. I am not good at looking at the audience anyway, so it was easier with nobody watching !

   I hope it will be helpful to print here the summaries and practice problems that are planned to accompany those videos. Here are the topics:

   6. Derivative of the Sine and Cosine

   7. Product and Quotient Rules

   8. Chain Rule for the Slope of $f(g(x))$

   9. Inverse Functions and Logarithms

   10. Growth Rates and Log Graphs

   11. Linear Approximation and Newton's Method

   12. Differential Equations of Growth

   13. Differential Equations of Motion

   14. Power Series and Euler's Formula

   15. Six Functions, Six Rules, Six Theorems

That last lecture summarizes the theory of differential calculus. The other lectures explain the steps. Here are the first lines written for the max-min video.

Maximum and Minimum and Second Derivative

> To find the maximum and minimum values of a function $y(x)$
>
> Solve $\dfrac{dy}{dx} = 0$ to find points x^* where **slope = zero**
>
> Test each x^* for a possible minimum or maximum
>
> Example $y(x) = x^3 - 12x$ $\dfrac{dy}{dx} = 3x^2 - 12$ Solve $3x^2 = 12$
>
> The slope is $\dfrac{dy}{dx} = 0$ at $\boldsymbol{x^* = 2}$ and $\boldsymbol{x^* = -2}$
>
> At those points $y(2) = 8 - 24 = -16 = \mathbf{min}$ and $y(-2) = -8 + 24 = 16 = \mathbf{max}$

$x^* = 2$ is a minimum   Look at $\dfrac{d}{dx}\left(\dfrac{dy}{dx}\right) =$ **second derivative**

$\dfrac{d^2y}{dx^2} =$ derivative of $3x^2 - 12$.   This second derivative is $6x$.

$\dfrac{d^2y}{dx^2} > 0$ $\dfrac{dy}{dx}$ increases   Slope goes from down to up at $x^* = 2$

The bending is upwards and this x^* is a **minimum**

$\dfrac{d^2y}{dx^2} < 0$ $\dfrac{dy}{dx}$ decreases   Slope goes from up to down at $x^* = -2$

The bending is downwards and x^* is a **maximum**

Find the maximum of $y(x) = \sin x + \cos x$ using $\dfrac{dy}{dx} = \cos x - \sin x$

The slope is zero when $\cos x = \sin x$ at $x^* = 45$ degrees $= \dfrac{\pi}{4}$ radians

That point x^* has $y = \sin \dfrac{\pi}{4} + \cos \dfrac{\pi}{4} = \dfrac{\sqrt{2}}{2} + \dfrac{\sqrt{2}}{2} = \sqrt{2}$

The second derivative is $\dfrac{d^2y}{dx^2} = -\sin x - \cos x$

At $x^* = \dfrac{\pi}{4}$ this is < 0 y is bending down x^* is a **maximum**

$\dfrac{d^2y}{dx^2} > 0$ when the curve bends up $\dfrac{d^2y}{dx^2} < 0$ when the curve bends down

Direction of bending changes at a **point of inflection** where $\boldsymbol{\dfrac{d^2y}{dx^2} = 0}$

Which x^* gives the minimum of $y = (x-1)^2 + (x-2)^2 + (x-6)^2$?

You can write $y = (x^2 - 2x + 1) + (x^2 - 4x + 4) + (x^2 - 12x + 36)$

The slope is $\dfrac{dy}{dx} = 2x - 2 + 2x - 4 + 2x - 12 = 0$ at the minimum point x^*

Then $6x^* = 18$ and $x^* = 3$ Minimum point is the average of $1, 2, 6$

Key for **max / min** word problems is to choose a suitable meaning for x

Practice Questions

1. Which x^* gives the minimum of $y(x) = x^2 + 2x$? Solve $\dfrac{dy}{dx} = 0$.

2. Find $\dfrac{d^2y}{dx^2}$ for $y(x) = x^2 + 2x$. This is > 0 so the parabola bends up.

3. Find the maximum height of $y(x) = 2 + 6x - x^2$. Solve $\dfrac{dy}{dx} = 0$.

4. Find $\dfrac{d^2y}{dx^2}$ to show that this parabola bends down.

5. For $y(x) = x^4 - 2x^2$ show that $\dfrac{dy}{dx} = 0$ at $x = -1, 0, 1$.

   Find $y(-1)$, $y(0)$, $y(1)$.   Check max versus min by the sign of d^2y/dx^2.

6. At a minimum point explain why $\dfrac{dy}{dx} = 0$ and $\dfrac{d^2y}{dx^2} > 0$.

7. Bending down $\left(\dfrac{d^2y}{dx^2} < 0\right)$ changes to bending up $\left(\dfrac{d^2y}{dx^2} > 0\right)$ at a point

   of _____ : At this point $\dfrac{d^2y}{dx^2} = 0$ Does $y = \sin x$ have such a point ?

8. Suppose $x + X = 12$. What is the maximum of x times X ?

   This question asks for the maximum of $y = x(12 - x) = 12x - x^2$.

   Find where the slope $\dfrac{dy}{dx} = 12 - 2x$ is zero. What is x times X ?

The Big Picture of Integrals

Key problem   Recover the integral $y(x)$ from its derivative $\dfrac{dy}{dx}$

Find the total distance traveled from a record of the speed

Find Function (1) = total height knowing Function (2) = slope since the start

Simplest way   Recognize $\dfrac{dy}{dx}$ as derivative of a known $y(x)$

If $\dfrac{dy}{dx} = x^3$ then its **integral** $y(x)$ was $\dfrac{1}{4}x^4 + C = $ **Function (1)**

If $\dfrac{dy}{dx} = e^{2x}$ then $y = \dfrac{1}{2}e^{2x} + C$

Integral Calculus is the reverse of Differential Calculus

$y(x) = \displaystyle\int \dfrac{dy}{dx}\,dx$ adds up the whole history of slopes $\dfrac{dy}{dx}$ to find $y(x)$

Integral is like sum   Derivative is like difference

Derivative of the Sine and Cosine

This lecture shows that $\dfrac{d}{dx}(\sin x) = \cos x$ and $\dfrac{d}{dx}(\cos x) = -\sin x$

We have to measure the angle x in **radians** 2π radians $=$ full 360 degrees

All the way around the circle (2π radians)   **Length $= 2\pi$** when the radius is 1
Part way around the circle (x radians)   **Length $= x$** when the radius is 1

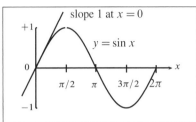

Slope $\cos x$

at $x = 0$	slope $1 = \cos 0$
at $x = \pi/2$	slope $0 = \cos \pi/2$
at $x = \pi$	slope $-1 = \cos \pi$

Slope $-\sin x$

at $x = 0$	slope $= 0 = -\sin 0$
at $x = \pi/2$	slope $-1 = -\sin \pi/2$
at $x = \pi$	slope $= 0 = -\sin \pi$

Problem: $\dfrac{\Delta y}{\Delta x} = \dfrac{\sin(x + \Delta x) - \sin x}{\Delta x}$ is not as simple as $\dfrac{(x + \Delta x)^2 - x^2}{\Delta x}$

Good idea to start at $x = 0$ Show $\dfrac{\Delta y}{\Delta x} = \dfrac{\sin \Delta x}{\Delta x}$ **approaches 1**

Draw a right triangle with angle Δx to see $\sin \Delta x \le \Delta x$

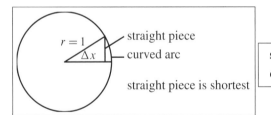

straight piece
curved arc

straight piece is shortest

straight length $= \sin \Delta x$
curved length $= \Delta x$

IDEA $\dfrac{\sin \Delta x}{\Delta x} < 1$ and $\dfrac{\sin \Delta x}{\Delta x} > \cos \Delta x$ will **squeeze** $\dfrac{\sin \Delta x}{\Delta x} \to 1$ as $\Delta x \to 0$

To prove $\dfrac{\sin \Delta x}{\Delta x} > \cos \Delta x$ which is $\tan \Delta x > \Delta x$ **Go to a bigger triangle**

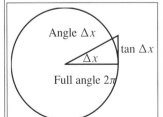

Angle Δx

$\tan \Delta x$

Δx

Full angle 2π

Triangle area $= \dfrac{1}{2}(1)(\tan \Delta x)$ greater than

Circular area $= \left(\dfrac{\Delta x}{2\pi}\right)$ (whole circle) $= \dfrac{1}{2}(\Delta x)$

The squeeze $\cos \Delta x < \dfrac{\sin \Delta x}{\Delta x} < 1$ tells us that $\dfrac{\sin \Delta x}{\Delta x}$ **approaches 1**

$\dfrac{(\sin \Delta x)^2}{(\Delta x)^2} < 1$ means $\dfrac{(1 - \cos \Delta x)}{\Delta x}(1 + \cos \Delta x) < \Delta x$

So $\dfrac{1 - \cos \Delta x}{\Delta x} \to 0$ **Cosine curve has slope $= 0$**

For the slope at other x remember a formula from trigonometry: $\sin(x + \Delta x) = \sin x \cos \Delta x + \cos x \sin \Delta x$

We want $\Delta y = \sin(x + \Delta x) - \sin x$ Divide that by Δx

$\dfrac{\Delta y}{\Delta x} = (\sin x)\left(\dfrac{\cos \Delta x - 1}{\Delta x}\right) + (\cos x)\left(\dfrac{\sin \Delta x}{\Delta x}\right)$ Now let $\Delta x \to 0$

In the limit $\dfrac{dy}{dx} = (\sin x)(0) + (\cos x)(1) = \cos x =$ Derivative of $\sin x$

For $y = \cos x$ the formula for $\cos(x + \Delta x)$ leads similarly to $\dfrac{dy}{dx} = -\sin x$

Practice Questions

1. What is the slope of $y = \sin x$ at $x = \pi$ and at $x = 2\pi$?

2. What is the slope of $y = \cos x$ at $x = \pi/2$ and $x = 3\pi/2$?

3. The slope of $(\sin x)^2$ is $2 \sin x \cos x$. The slope of $(\cos x)^2$ is $-2 \cos x \sin x$. Combined, the slope of $(\sin x)^2 + (\cos x)^2$ is **zero**. Why is this true ?

4. What is the **second derivative** of $y = \sin x$ (derivative of the derivative) ?

5. At what angle x does $y = \sin x + \cos x$ have zero slope ?

---

6. Here are amazing infinite series for $\sin x$ and $\cos x$. $e^{ix} = \cos x + i \sin x$

$$\sin x = \frac{x}{1} - \frac{x^3}{3 \cdot 2 \cdot 1} + \frac{x^5}{5 \cdot 4 \cdot 3 \cdot 2 \cdot 1} - \cdots \quad (odd\ powers\ of\ x)$$

$$\cos x = 1 - \frac{x^2}{2 \cdot 1} + \frac{x^4}{4 \cdot 3 \cdot 2 \cdot 1} - \cdots \quad (even\ powers\ of\ x)$$

7. Take the derivative of the sine series to see the cosine series.

8. Take the derivative of the cosine series to see **minus** the sine series.

9. Those series tell us that for small angles $\sin x \approx x$ and $\cos x \approx 1 - \frac{1}{2}x^2$. With these approximations check that $(\sin x)^2 + (\cos x)^2$ is close to 1.

Product and Quotient Rules

Goal To find the derivative of $y = f(x)g(x)$ from $\dfrac{df}{dx}$ and $\dfrac{dg}{dx}$

Idea Write $\Delta y = f(x+\Delta x)\,g(x+\Delta x) - f(x)g(x)$ by separating Δf and Δg

That same Δy is $f(x+\Delta x)[g(x+\Delta x) - g(x)] + g(x)[f(x+\Delta x) - f(x)]$

$$\frac{\Delta y}{\Delta x} = f(x+\Delta x)\frac{\Delta g}{\Delta x} + g(x)\frac{\Delta f}{\Delta x} \qquad \textbf{Product Rule} \quad \frac{dy}{dx} = f(x)\frac{dg}{dx} + g(x)\frac{df}{dx}$$

Example $y = x^2 \sin x$ Product Rule $\dfrac{dy}{dx} = x^2 \cos x + 2x \sin x$

A picture shows the two unshaded pieces of $\Delta y = f(x+\Delta x)\Delta g + g(x)\Delta f$

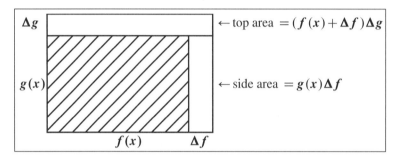

$\mathbf{\Delta g}$ $\leftarrow$ top area $= (f(x) + \Delta f)\Delta g$

$g(x)$ $\leftarrow$ side area $= g(x)\Delta f$

$f(x)$ Δf

Example $f(x) = x^n$ $g(x) = x$ $y = f(x)g(x) = x^{n+1}$

Product Rule $\dfrac{dy}{dx} = x^n \dfrac{dx}{dx} + x\dfrac{dx^n}{dx} = x^n + xnx^{n-1} = (n+1)x^n$

The correct derivative of x^n leads to the correct derivative of x^{n+1}

Quotient Rule If $y = \dfrac{f(x)}{g(x)}$ then $\dfrac{dy}{dx} = \left(g(x)\dfrac{df}{dx} - f(x)\dfrac{dg}{dx} \right)\bigg/ g^2$

EXAMPLE $\dfrac{d}{dx}\left(\dfrac{\sin x}{\cos x} \right) = (\cos x(\cos x) - \sin x(-\sin x)) \Big/ \cos^2 x$

This says that $\dfrac{d}{dx}\tan x = \dfrac{1}{\cos^2 x} = \sec^2 x$ (Notice $(\cos x)^2 + (\sin x)^2 = 1$)

EXAMPLE $\dfrac{d}{dx}\left(\dfrac{1}{x^4} \right) = \dfrac{x^4 \text{ times } 0 - 1 \text{ times } 4x^3}{x^8} = \dfrac{-4}{x^5}$ This is nx^{n-1}

Prove the Quotient Rule $\Delta y = \dfrac{f(x+\Delta x)}{g(x+\Delta x)} - \dfrac{f(x)}{g(x)} = \dfrac{f+\Delta f}{g+\Delta g} - \dfrac{f}{g}$

Write this Δy as $\dfrac{g(f+\Delta f) - f(g+\Delta g)}{g(g+\Delta g)} = \dfrac{g\Delta f - f\Delta g}{g(g+\Delta g)}$

Now divide that Δy by Δx As $\Delta x \to 0$ we have the Quotient Rule

Practice Questions

1. Product Rule: Find the derivative of $y = (x^3)(x^4)$. Simplify and explain.

2. Product Rule: Find the derivative of $y = (x^2)(x^{-2})$. Simplify and explain.

3. Quotient Rule: Find the derivative of $y = \dfrac{\cos x}{\sin x}$.

4. Quotient Rule: Show that $y = \dfrac{\sin x}{x}$ has a maximum (zero slope) at $x = 0$.

5. Product and Quotient! Find the derivative of $y = \dfrac{x \sin x}{\cos x}$.

6. $g(x)$ has a minimum when $\dfrac{dg}{dx} = 0$ and $\dfrac{d^2 g}{dx^2} > 0$ The graph is bending up

   $y = \dfrac{1}{g(x)}$ has a *maximum* at that point: Show that $\dfrac{dy}{dx} = 0$ and $\dfrac{d^2 y}{dx^2} < 0$

Chain Rule for the Slope of $f(g(x))$

$y = g(x) \quad z = f(y) \quad \longrightarrow \quad$ the chain is $\quad z = f(g(x))$

$y = x^5 \quad\;\; z = y^4 \quad \longrightarrow \quad$ the chain is $\quad z = (x^5)^4 = x^{20}$

Average slope $\quad \dfrac{\Delta z}{\Delta x} = \left(\dfrac{\Delta z}{\Delta y} \right) \left(\dfrac{\Delta y}{\Delta x} \right)$ Just cancel Δy

Instant slope $\quad \dfrac{dz}{dx} = \dfrac{dz}{dy} \dfrac{dy}{dx} = $ **CHAIN RULE** (like cancelling dy)

You MUST change y to $g(x)$ in the final answer

Example of chain $\quad z = y^4 = (x^5)^4 \quad \dfrac{dz}{dy} = 4y^3 \quad \dfrac{dy}{dx} = 5x^4$

Chain rule $\quad \dfrac{dz}{dx} = \left(\dfrac{dz}{dy} \right) \left(\dfrac{dy}{dx} \right) = (4y^3)(5x^4) = 20y^3 x^4$

Replace y by x^5 to get only x $\qquad \dfrac{dz}{dx} = 20(x^5)^3 x^4 = 20 x^{19}$

CHECK $\quad z = (x^5)^4 = x^{20}$ does have $\dfrac{dz}{dx} = 20 x^{19}$

1. Find $\dfrac{dz}{dx}$ for $z = \cos(4x)$ $\qquad$ Write $y = 4x$ and $z = \cos y$ so $\dfrac{dz}{dx} =$

2. Find $\dfrac{dz}{dx}$ for $z = (1 + 4x)^2$ $\qquad$ Write $y = 1 + 4x$ and $z = y^2$ so $\dfrac{dz}{dx} =$

CHECK $\quad (1 + 4x)^2 = 1 + 8x + 16x^2$ so $\dfrac{dz}{dx} =$

Practice Questions

3. Find $\dfrac{dh}{dx}$ for $h(x) = (\sin 3x)(\cos 3x)$

Product rule first   Then the Chain rule for each factor

$$\frac{dh}{dx} = (\sin 3x)\frac{d}{dx}(\cos 3x) + (\cos 3x)\frac{d}{dx}(\sin 3x)$$

$$= (\sin 3x)(\text{CHAIN}) + (\cos 3x)(\text{CHAIN}) = \quad ?$$

4. Tough challenge: Find the **second derivative** of $z(x) = f(g(x))$

FIRST DERIV $\dfrac{dz}{dx} = \left(\dfrac{dz}{dy}\right)\left(\dfrac{dy}{dx}\right)$ Function of $y(x)$ times function of x

PRODUCT RULE $\dfrac{d^2z}{dx^2} = \left(\dfrac{dz}{dy}\right)\dfrac{d}{dx}\left(\dfrac{dy}{dx}\right) + \left(\dfrac{dy}{dx}\right)\dfrac{d}{dx}\left(\dfrac{dz}{dy}\right)$

SECOND DERIV $\left(\dfrac{dz}{dy}\right)\left(\dfrac{d^2y}{dx^2}\right) + \left(\dfrac{dy}{dx}\right)\left(\dfrac{d^2z}{dy^2}\right)\left(\dfrac{dy}{dx}\right)$ $\dfrac{dy}{dx}$ *twice!*

Check $y = x^5$ $z = y^4 = x^{20}$ $\dfrac{dz}{dx} = 20x^{19}$ $\dfrac{d^2z}{dx^2} = 380x^{18}$

SECOND DERIV $(4y^3)(20x^3) + (5x^4)(12y^2)(5x^4)$ $80 + 300 = 380$ OK

Inverse Functions and Logarithms

A function assigns an **output** $y = f(x)$ to each **input** x

A one-to-one function has different outputs y for different inputs x

For the **inverse function** the input is y and the output is $x = f^{-1}(y)$

Example   If $y = f(x) = x^5$ then $x = f^{-1}(y) = y^{\frac{1}{5}}$

KEY   If $y = ax + b$ then solve for $x = \dfrac{y - b}{a}$ = inverse function

Notice that $x = f^{-1}(f(x))$ and $y = f(f^{-1}(y))$

The **chain rule** will connect the derivatives of f^{-1} and f

---

The great function of calculus is $y = e^x$

Its inverse function is the **"natural logarithm"** $x = \ln y$

Remember that x is the exponent in $y = e^x$

The rule $e^x e^X = e^{x+X}$ tells us that $\ln(yY) = \ln y + \ln Y$

Add logarithms because you add exponents: $\ln(e^2 e^3) = 5$

$(e^x)^n = e^{nx}$ (multiply exponent) tells us that $\ln(y^n) = n \ln y$

We can change from base e to base 10: New function $y = 10^x$

The inverse function is the logarithm to base 10   Call it log: $x = \log y$

Then $\log 100 = 2$ and $\log \dfrac{1}{100} = -2$ and $\log 1 = 0$

We will soon find the beautiful derivative of $\ln y$ $\dfrac{d}{dy}(\ln y) = \dfrac{1}{y}$

You can change letters to write that as $\dfrac{d}{dx}(\ln x) = \dfrac{1}{x}$

Practice Questions

1. What is $x = f^{-1}(y)$ if $y = 50x$?

2. What is $x = f^{-1}(y)$ if $y = x^4$? Why do we keep $x \ge 0$?

3. Draw a graph of an increasing function $y = f(x)$. This has different outputs y for different x. **Flip the graph (switch the axes) to see $x = f^{-1}(y)$**

4. This graph has the same y **from two x's. There is no $f^{-1}(y)$**

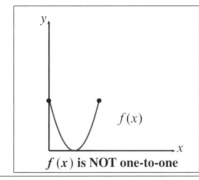

$f(x)$

$f(x)$ is NOT one-to-one

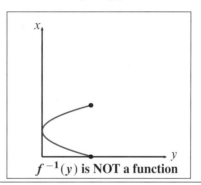

$f^{-1}(y)$ is NOT a function

---

5. The natural logarithm of $y = 1/e$ is $\ln(e^{-1}) = ?$ What is $\ln(\sqrt{e})$?

6. The natural logarithm of $y = 1$ is $\ln 1 = ?$ and also base 10 has $\log 1 = ?$

7. The natural logarithm of $(e^2)^{50}$ is ? The base 10 logarithm of $(10^2)^{50}$ is ?

8. I believe that $\ln y = (\ln 10)(\log y)$ because we can write y in two ways $y = e^{\ln y}$ and also $y = 10^{\log y} = e^{(\ln 10)(\log y)}$. Explain those last steps.

9. Change from base e and base 10 to **base 2**. Now $y = 2^x$ means $x = \log_2 y$. What are $\log_2 32$ and $\log_2 2$? Why is $\log_2(e) > 1$?

Growth Rates and Log Graphs

In order of fast growth as x gets large

$\log x$	x, x^2, x^3	$2^x, e^x, 10^x$	$x!, x^x$
logarithmic	polynomial	exponential	factorial

Choose $x = 1000 = 10^3$ so that $\log x = 3$　OK to use $x! \approx \dfrac{x^x}{e^x}$

　$\log 1000 = 3$　　$10^3, 10^6, 10^9$　　　　$10^{300}, 10^{434}, 10^{1000}$　　$10^{2566}, 10^{3000}$

Why is $1000^{1000} = 10^{3000}$?　　　　Logarithms are best for big numbers

Logarithms are exponents!　　$\log 10^9 = 9$　　$\log \log x$ is VERY slow

Logarithms　　**3, 6, 9**　　　　**300, 434, 1000**　　　　**2566, 3000**

Polynomial growth $\ll$ Exponential growth $\ll$ Factorial growth

Decay to zero for NEGATIVE powers and exponents

$\dfrac{1}{x^2} = x^{-2}$ decays much more slowly than the exponential $\dfrac{1}{e^x} = e^{-x}$

Logarithmic scale shows $x = 1, 10, 100$ equally spaced. NO ZERO!

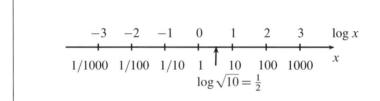

Question　If $x = 1, 2, 4, 8$ are plotted, what would you see ?
Answer　　THEY ARE EQUALLY SPACED TOO!

log-log graphs (log scale up and also across)

If $y = Ax^n$, how to see A and n on the graph ?

Plot $\log y$ versus $\log x$ to get a straight line

$\log y = \log A + n \log x$　　**Slope on a log-log graph is the exponent n**

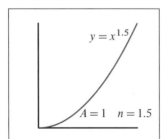

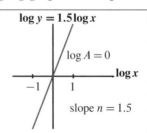

For $y = Ab^x$ use **semilog** (x versus $\log y$ is now a line)　　$\log y = \log A + x \log b$

New type of question   How quickly does $\dfrac{\Delta f}{\Delta x}$ approach $\dfrac{df}{dx}$ as $\Delta x \to 0$?

The error $E = \dfrac{\Delta f}{\Delta x} - \dfrac{df}{dx}$ will be $E \approx A(\Delta x)^n$ What is n ?

Usual one-sided $\dfrac{\Delta f}{\Delta x} = \dfrac{f(x + \Delta x) - f(x)}{\Delta x}$ only has $n = 1$

Centered difference $\dfrac{f(x + \Delta x) - f(x - \Delta x)}{2\Delta x}$ has $n = 2$

Centered is much better than one-sided $E \approx (\Delta x)^2$ vs $E \approx \Delta x$

$\begin{bmatrix} \text{IDEA FOR } f(x) = e^x \\ \text{PROJECT} \quad \text{at } x = 0 \end{bmatrix}$ One-sided E vs centered E

Graph $\log E$ vs $\log \Delta x$ Should see slope 1 or 2

Practice Questions

1. Does x^{100} grow faster or slower than e^x as x gets large ?

2. Does $100 \ln x$ grow faster or slower than x as x gets large ?

3. Put these in increasing order for large n:

   $$\dfrac{1}{n}, \quad n \log n, \quad n^{1.1}, \quad \dfrac{10^n}{n!}$$

4. Put these in increasing order for large x:

   $$2^{-x}, \quad e^{-x}, \quad \dfrac{1}{x^2}, \quad \dfrac{1}{x^{10}}$$

---

5. Describe the log-log graph of $y = 10x^5$ (graph $\log y$ vs $\log x$)

Why don't we see $y = 0$ at $x = 0$ on this graph ?

What is the slope of the straight line on the log-log graph ?

The line crosses the vertical axis when $x =$ _____ and $y =$ _____

Then $\log x = 0$ and $\log y =$ _____

The line crosses the horizontal axis when $x =$ _____ and $y = 1$

Then $\log x =$ _____ and $\log y = 0$

6. Draw the semilog graph (a line) of $y = 10e^x$ (graph $\log y$ versus x)

7. That line cross the $x = 0$ axis at which $\log y$? What is the slope ?

EXAMPLE 1 $\dfrac{d^2y}{dt^2}+6\dfrac{dy}{dt}+8y=0$ $m=1$ and $2r=6$ and $k=8$

$\lambda_1,\lambda_2=\dfrac{-r\pm\sqrt{r^2-km}}{m}$ is $-3\pm\sqrt{9-8}$ Then $\begin{aligned}\lambda_1&=-2\\\lambda_2&=-4\end{aligned}$

Solution $y=Ce^{-2t}+De^{-4t}$ Overdamping with no oscillation

EXAMPLE 2   Change to $k=10$ $\lambda=-3\pm\sqrt{9-10}$ has $\begin{aligned}\lambda_1&=-3+i\\\lambda_2&=-3-i\end{aligned}$

Oscillations from the imaginary part of λ **Decay** from the real part -3

Solution $y=Ce^{\lambda_1 t}+De^{\lambda_2 t}=Ce^{(-3+i)t}+De^{(-3-i)t}$
$e^{it}=\cos t+i\sin t$ leads to $y=(C+D)e^{-3t}\cos t+(C-D)e^{-3t}\sin t$

EXAMPLE 3   Change to $k=9$ Now $\lambda=-3,-3$ (repeated root)

Solution $y=Ce^{-3t}+Dte^{-3t}$ includes the factor t

Practice Questions

1. For $\dfrac{d^2y}{dt^2}=4y$ find two solutions $y=Ce^{at}+De^{bt}$. What are a and b ?

2. For $\dfrac{d^2y}{dt^2}=-4y$ find two solutions $y=C\cos\omega t+D\sin\omega t$. What is ω ?

3. For $\dfrac{d^2y}{dt^2}=0y$ find two solutions $y=Ce^{0t}$ and (???)

4. Put $y=e^{\lambda t}$ into $2\dfrac{d^2y}{dt^2}+3\dfrac{dy}{dt}+y=0$ to find λ_1 and λ_2 (**real** numbers)

5. Put $y=e^{\lambda t}$ into $2\dfrac{d^2y}{dt^2}+5\dfrac{dy}{dt}+3y=0$ to find λ_1 and λ_2 (**complex** numbers)

6. Put $y=e^{\lambda t}$ into $\dfrac{d^2y}{dt^2}+2\dfrac{dy}{dt}+y=0$ to find λ_1 and λ_2 (**equal** numbers)

   Now $y=Ce^{\lambda_1 t}+Dte^{\lambda_1 t}$. The factor t appears when $\lambda_1=\lambda_2$

Power Series and Euler's Formula

At $x = 0$, the nth derivative of x^n is the number $n!$.   Other derivatives are 0.

Multiply the nth derivatives of $f(x)$ by $x^n/n!$ to match function with series

TAYLOR SERIES
$$f(x) = f(0) + f'(0)\frac{x}{1} + f''(0)\frac{x^2}{2} + \cdots + f^{(n)}(0)\frac{x^n}{n!} + \cdots$$

EXAMPLE 1 $f(x) = e^x$ All derivatives $= 1$ at $x = 0$ Match with $x^n/n!$

Taylor Series
Exponential Series
$$= e^x = 1 + 1\frac{x}{1} + 1\frac{x^2}{2} + \cdots + 1\frac{x^n}{n!} + \cdots$$

EXAMPLE 2 $f = \sin x$ $f' = \cos x$ $f'' = -\sin x$ $f''' = -\cos x$

At $x = 0$ this is 0   1   0 -1 0   1   0 -1 REPEAT

$$\sin x = 1 \cdot \frac{x}{1} - 1\frac{x^3}{3!} + 1\frac{x^5}{5!} - \cdots$$ ODD POWERS $\sin(-x) = -\sin x$

EXAMPLE 3 $f = \cos x$ produces 1   0 -1 0   1   0 -1 0   REPEAT

$$\cos x = 1 - 1\frac{x^2}{2!} + 1\frac{x^4}{4!} - \cdots$$ EVEN POWERS $\frac{d}{dx}(\cos x) = -\sin x$

Imaginary $i^2 = -1$ and then $i^3 = -i$ **Find the exponential e^{ix}**

$$e^{ix} = 1 + ix + \frac{1}{2!}(ix)^2 + \frac{1}{3!}(ix)^3 + \cdots$$

Those are

$$= \left(1 - \frac{x^2}{2!} + \cdots\right) + i\left(x - \frac{x^3}{3!} + \cdots\right)$$ $\cos x + i \sin x$

EULER'S GREAT FORMULA $e^{ix} = \cos x + i \sin x$

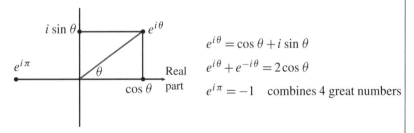

$e^{i\theta} = \cos\theta + i\sin\theta$

$e^{i\theta} + e^{-i\theta} = 2\cos\theta$

$e^{i\pi} = -1$ combines 4 great numbers

Two more examples of Power Series (Taylor Series for $f(x)$)

$$f(x) = \frac{1}{1-x} = 1 + x + x^2 + x^3 + \cdots$$ "Geometric series"

$$f(x) = -\ln(1-x) = \frac{x}{1} + \frac{x^2}{2} + \frac{x^3}{3} + \frac{x^4}{4} + \cdots$$ "Integral of geometric series"

Summary: Six Functions, Six Rules, Six Theorems

Integrals	Six Functions	Derivatives
$x^{n+1}/(n+1)$, $n \neq -1$	x^n	nx^{n-1}
$-\cos x$	$\sin x$	$\cos x$
$\sin x$	$\cos x$	$-\sin x$
e^{cx}/c	e^{cx}	ce^{cx}
$x \ln x - x$	$\ln x$	$1/x$
Ramp function	**Step function**	**Delta function**

Ramp function (graph with line rising from 0, labeled x)

Step function (graph with value 1)

Delta function (Infinite spike has area $= 1$)

Six Rules of Differential Calculus

1. The derivative of $af(x) + bg(x)$ is $a\dfrac{df}{dx} + b\dfrac{dg}{dx}$ **Sum**

2. The derivative of $f(x)g(x)$ is $f(x)\dfrac{dg}{dx} + g(x)\dfrac{df}{dx}$ **Product**

3. The derivative of $\dfrac{f(x)}{g(x)}$ is $\left(g\dfrac{df}{dx} - f\dfrac{dg}{dx}\right)\Big/ g^2$ **Quotient**

4. The derivative of $f(g(x))$ is $\dfrac{df}{dy}\dfrac{dy}{dx}$ where $y = g(x)$ **Chain**

5. The derivative of $x = f^{-1}(y)$ is $\dfrac{dx}{dy} = \dfrac{1}{dy/dx}$ **Inverse**

6. When $f(x) \to 0$ and $g(x) \to 0$ as $x \to a$, what about $f(x)/g(x)$? **l'Hôpital**

$\lim \dfrac{f(x)}{g(x)} = \lim \dfrac{df/dx}{dg/dx}$ if these limits exist. Normally this is $\dfrac{f'(a)}{g'(a)}$

Fundamental Theorem of Calculus

If $f(x) = \displaystyle\int_a^x s(t)dt$ then **derivative of integral** $= \dfrac{df}{dx} = s(x)$

If $\dfrac{df}{dx} = s(x)$ then **integral of derivative** $= \displaystyle\int_a^b s(x)dx = f(b) - f(a)$

Both parts assume that $s(x)$ is a continuous function.

All Values Theorem Suppose $f(x)$ is a continuous function for $a \leqslant x \leqslant b$. Then on that interval, $f(x)$ reaches its maximum value M and its minimum m. And $f(x)$ takes all values between m and M (there are no jumps).

Mean Value Theorem   If $f(x)$ has a derivative for $a \leqslant x \leqslant b$ then

$$\frac{f(b) - f(a)}{b - a} = \frac{df}{dx}(c) \text{ at some } c \text{ between } a \text{ and } b$$

"At some moment c, instant speed $=$ average speed"

Taylor Series   Match all the derivatives $f^{(n)} = d^n f / dx^n$ at the basepoint $x = a$

$$f(x) = f(a) + f'(a)(x - a) + \frac{1}{2} f''(a)(x - a)^2 + \cdots$$

$$= \sum_{n=0}^{\infty} \frac{1}{n!} f^{(n)}(a) (x - a)^n$$

Stopping at $(x - a)^n$ leaves the error $f^{n+1}(c)(x - a)^{n+1} / (n + 1)!$

[c is somewhere between a and x] [$n = 0$ is the Mean Value Theorem]

The Taylor series looks best around $a = 0$ $f(x) = \sum_{n=0}^{\infty} \frac{1}{n!} f^{(n)}(0) x^n$

Binomial Theorem shows Pascal's triangle

$$(1 + x) \qquad\qquad 1 + 1x$$
$$(1 + x)^2 \qquad\qquad 1 + 2x + 1x^2$$
$$(1 + x)^3 \qquad 1 + 3x + 3x^2 + 1x^3$$
$$(1 + x)^4 \quad 1 + 4x + 6x^2 + 4x^3 + 1x^4$$

Those are just the Taylor series for $f(x) = (1 + x)^p$ when $p = 1, 2, 3, 4$

$$f^{(n)}(x) = (1 + x)^p \quad p(1 + x)^{p-1} \quad p(p-1)(1 + x)^{p-2} \cdots$$
$$f^{(n)}(0) = \quad 1 \qquad\qquad p \qquad\qquad p(p-1) \qquad\qquad \cdots$$

Divide by $n!$ to find the Taylor coefficients $=$ **Binomial coefficients**

$$\frac{1}{n!} f^{(n)}(0) = \frac{p(p-1)\cdots(p-n+1)}{n(n-1)\cdots(1)} = \frac{p!}{(p-n)! \, n!} = \binom{p}{n}$$

The series stops at x^n when $p = n$ Infinite series for other p

Every $(1 + x)^p = 1 + px + \dfrac{p(p-1)}{(2)(1)} x^2 + \dfrac{p(p-1)(p-2)}{(3)(2)(1)} x^3 + \cdots$

Practice Questions

1. Check that the derivative of $y = x \ln x - x$ is $dy/dx = \ln x$.

2. The "sign function" is $S(x) = \begin{cases} 1 & \text{for } x \geq 0 \\ -1 & \text{for } x < 0 \end{cases}$

   What ramp function $F(x)$ has $\dfrac{dF}{dx} = S(x)$? F is the integral of S.

   Why is the derivative $\dfrac{dS}{dx} = 2\,\mathbf{delta}(x)$? (Infinite spike at $x = 0$ with area 2)

3. (l'Hôpital) What is the limit of $\dfrac{2x + 3x^2}{5x + 7x^2}$ as $x \to 0$? What about $x \to \infty$?

4. l'Hôpital's Rule says that $\displaystyle \lim_{x \to 0} \dfrac{f(x)}{x} = ??$ when $f(0) = 0$. Here $g(x) = x$.

---

5. **Derivative is like Difference   Integral is like Sum**

   Difference of sums   If $f_n = s_1 + s_2 + \cdots + s_n$, what is $f_n - f_{n-1}$?

   Sums of differences   What is $(f_1 - f_0) + (f_2 - f_1) + \cdots + (f_n - f_{n-1})$?

   Those are the **Fundamental Theorems** of **"Difference Calculus"**

6. Draw a non-continuous graph for $0 \leq x \leq 1$ where your function does NOT reach its maximum value.

7. For $f(x) = x^2$, which in-between point c gives $\dfrac{f(5) - f(1)}{5 - 1} = \dfrac{df}{dx}(c)$?

8. If your average speed on the Mass Pike is 75, then at some instant your speedometer will read _____ .

9. Find three Taylor coefficients A, B, C for $\sqrt{1 + x}$ (around $x = 0$).

$$(1 + x)^{\frac{1}{2}} = A + Bx + Cx^2 + \cdots$$

10. Find the Taylor (= Binomial) series for $f = \dfrac{1}{1 + x}$ around $x = 0$ $(p = -1)$.

0.5 Graphs and Graphing Calculators

This book started with the sentence *"Calculus is about functions."* When these functions are given by formulas like $y = x + x^2$, we now know a formula for the slope (and even the slope of the slope). When we only have a rough graph of the function, we can't expect more than a rough graph of the slope. But graphs are very valuable in applications of calculus!

From a graph of $y(x)$, this section extracts the basic information about the growth rate (the slope) and the minimum/maximum and the bending (and area too). A big part of that information is contained in *a plus or minus sign*. Is $y(x)$ increasing? Is its slope increasing? Is the area under its graph increasing? In each case some number is greater than zero. The three numbers are dy/dx and d^2y/dx^2 and $y(x)$ itself.

When one of those numbers is *exactly zero* we always have a special situation. It is a good thing that mathematics invented zero, we need it.

This section is organized by two themes:

(**1**) Graphs that are drawn without a formula for $y(x)$. From that graph you can draw other graphs—the slope dy/dx, the second derivative d^2y/dx^2, the area $A(x)$ under the graph.

You can also identify where those functions are positive or negative—and especially the points where dy/dx or d^2y/dx^2 or $y(x)$ is *zero*.

(**2**) Graphs that are drawn by a calculator or computer. Now there is a formula for $y(x)$. The display allows us to guess rules for derivatives:

Chain Rule   Inverse Rule   l'Hôpital's Rule

These rules come into later chapters of the book. They are also explained in *Highlights of Calculus*, the video lectures that are available to everyone. One specific goal is to see how the derivative of 2^x is proportional to 2^x.

This section was much improved by ideas that were offered by Benjamin Goldstein.

GRAPH WITHOUT FORMULAS

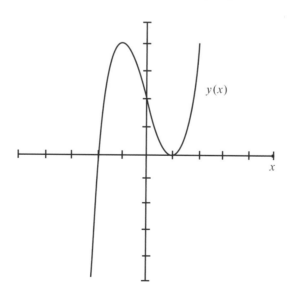

$y(x)$

x

1. Suppose this is the graph of some function $y(x)$

   a. At what value(s) of x does $y(x)$ have a local minimum?

   b. At what value(s) of x does $y(x)$ have a local maximum?

   c. At what value(s) of x does $y(x)$ have an inflection point? (Estimate.)

2. Let's change the problem. Suppose this is the graph of dy/dx, the derivative of $y(x)$. Answer the following questions about $y(x)$, the original function.

   a. At what value(s) of x does $y(x)$ have a local minimum?

   b. At what value(s) of x does $y(x)$ have a local maximum?

   c. At what value(s) of x does $y(x)$ have an inflection point?

3. One more variation. Suppose this is the graph of the second derivative d^2y/dx^2 (slope of the slope). If any of these questions can't be answered, explain why.

   a. At what value(s) of x does $y(x)$ have a local minimum?

   b. At what value(s) of x does $y(x)$ have a local maximum?

   c. At what value(s) of x does $y(x)$ have an inflection point?

4. Answer the same 9 questions for this second graph.

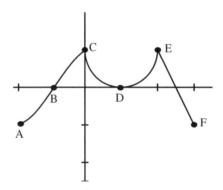

5. The following table shows the velocity of a car at selected times.

time	0	5	10	15	20	25	30	35
velocity	45	40	30	40	45	40	30	25

   a. Was there any time t when the car was moving with acceleration $d^2y/dt^2 = 0$? Justify your answer.

   b. If $y(t)$ represents the car's position as a function of time, was there ever a time when $d^3y/dt^3 = 0$? Justify your answer. The third derivative is sometimes referred to as '*jerk*' because it indicates the jerkiness of the motion. This is *important* to roller-coaster designers.

   c. What assumptions have you made about $y(t)$ and (more importantly) dy/dt in your answers to parts (a) and (b)? Are the assumptions reasonable?

THE CHAIN RULE ON A CALCULATOR

a. On your calculator, graph $Y_1 = \sin(X)$ and its slope $Y_2 = \text{nDeriv}(Y_1, X, X)$. Make sure you are in radian mode, and select the trigonometric viewing window.

1. What function does Y_2 appear to be?

2. Change Y_1 to $Y_1 = \sin(2X)$. Now what function does Y_2 appear to be? Check your guess by graphing the true derivative.

3. Finally, change Y_1 to $Y_1 = \sin(3X)$. What does Y_2 appear to be this time?

4. Conjecture: If k is some constant, then the derivative of $\sin(kx)$ is _____.

b. Those functions are *chains* (also called *compositions*). They can be written in the form $Y = f(g(x))$. For $\sin(kx)$ the outer function is $f(x) =$ _____ and the inner function is $g(x) =$ _____.

c. So far the inner function $g(x)$ has been linear, but it doesn't have to be. Let $Y = \sin(\sqrt{x})$.

Conjecture: $\dfrac{dY}{dx} =$ _____ when $g(x) = \sqrt{x}$.

Check your conjecture by graphing Y and comparing to the graph of the numerical derivative.

d. Now we generalize. Suppose $g(x)$ is any function. If $y = \sin(g(x))$, then $dy/dx =$ _____.

e. There is nothing magical about the sine function. Whenever we have a composition of an outer and an inner function, the chain rule applies. Predict the following derivatives and check by graphing the numerical derivative on your calculator.

1. $y = (2x + 4)^3; \, dy/dx =$
2. $y = \cos^2 x = (\cos x)^2; \, dy/dx =$
3. $y = \cos(x^2); \, dy/dx =$
4. $y = [\sin(x^2 + 1)]^3; \, dy/dx =$

COMPUTING IN CALCULUS

Software is available for calculus courses—a lot of it. The packages keep getting better. Which program to use (if any) depends on cost and convenience and purpose. *How* to use it is a much harder question. These pages identify some of the goals. Our aim is to support, with examples, the effort to use computing to help learning.

For calculus, ***the greatest advantage of the computer is to offer graphics***. You see the function, not just the formula. As you watch, $f(x)$ reaches a maximum or a minimum or zero. A separate graph shows its derivative. Those statements are not 100% true, as everybody learns right away—as soon as a few functions are typed in. But the power to *see this subject* is enormous, because it is adjustable. If we don't like the picture we change to a new viewing window.

This is computer-based graphics. It combines **numerical** computation with **graphical** computation. You get pictures as well as numbers—a powerful combination. The computer offers the experience of actually working with a function. The domain and range are not just abstract ideas. *You choose them.* May I give a few examples.

Can you find v if you know f, and vice versa, and how? If we know the velocity over the whole history of the car, we should be able to compute the total distance traveled. In other words, if the speedometer record is complete but the odometer is missing, its information could be recovered. One way to do it (without calculus) is to put in a new odometer and drive the car all over again at the right speeds. That seems like a hard way; calculus may be easier. But the point is that *the information is there.* If we know everything about v, there must be a method to find f.

What happens in the opposite direction, when f is known? If you have a complete record of distance, could you recover the complete velocity? In principle you could drive the car, repeat the history, and read off the speed. Again there must be a better way.

The whole subject of calculus is built on the relation between v and f. The question we are raising here is not some kind of joke, after which the book will get serious and the mathematics will get started. On the contrary, *I am serious now*—and the mathematics has already started. We need to know how to find the velocity from a record of the distance. (That is called **differentiation**, and it is the central idea of **differential calculus**.) We also want to compute the distance from a history of the velocity. (That is **integration**, and it is the goal of **integral calculus**.)

Differentiation goes from f to v; integration goes from v to f. We look first at examples in which these pairs can be computed and understood.

CONSTANT VELOCITY

Suppose the velocity is fixed at $v = 60$ (miles per hour). Then f increases at this constant rate. After two hours the distance is $f = 120$ (miles). After four hours $f = 240$ and after t hours $f = 60t$. We say that f increases **linearly** with time—its graph is a straight line.

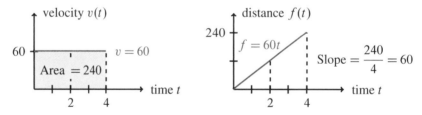

Fig. 1.2   Constant velocity $v = 60$ and linearly increasing distance $f = 60t$.

Notice that this example starts the car at full velocity. No time is spent picking up speed. (The velocity is a "step function.") Notice also that the distance starts at zero; the car is new. Those decisions make the graphs of v and f as neat as possible. One is the horizontal line $v = 60$. The other is the sloping line $f = 60t$. This v, f, t relation needs algebra but not calculus:

$$\text{If } v \text{ is constant and } f \text{ starts at zero then } f = vt.$$

The opposite is also true. When f increases linearly, v is constant. *The division by time gives the slope.* The distance is $f_1 = 120$ miles when the time is $t_1 = 2$ hours. Later $f_2 = 240$ miles at $t_2 = 4$ hours. At both points, the ratio f/t is 60 miles/hour. Geometrically, *the velocity is the slope of the distance graph*:

$$\text{slope} = \frac{\text{change in distance}}{\text{change in time}} = \frac{vt}{t} = v.$$

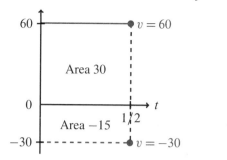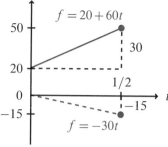

Fig. 1.3 Straight lines $f = 20 + 60t$ (slope 60) and $f = -30t$ (slope -30).

The slope of the f-graph gives the v-graph. Figure 1.3 shows two more possibilities:

1. The distance starts at 20 instead of 0. The distance formula changes from $60t$ to $20 + 60t$. The number 20 cancels when we compute *change* in distance—so the slope is still 60.

2. When v is *negative*, the graph of f goes *downward*. The car goes backward and the slope of $f = -30t$ is $v = -30$.

I don't think speedometers go below zero. But driving backwards, it's not that safe to watch. If you go fast enough, Toyota says they measure "absolute values"—the speedometer reads $+30$ when the velocity is -30. For the odometer, as far as I know it just stops. It should go backward.†

VELOCITY *vs.* DISTANCE: SLOPE *vs.* AREA

How do you compute f from v? The point of the question is to see $f = vt$ *on the graphs*. We want to start with the graph of v and discover the graph of f. Amazingly, the opposite of slope is *area*.

The distance f is the area under the v-graph. When v is constant, the region under the graph is a rectangle. Its height is v, its width is t, and its area is v times t. This is *integration*, to go from v to f by computing the area. We are glimpsing two of the central facts of calculus.

> **1A The slope of the f-graph gives the velocity v. The area under the v-graph gives the distance f.**

That is certainly not obvious, and I hesitated a long time before I wrote it down in this first section. The best way to understand it is to look first at more examples. The whole point of calculus is to deal with velocities that are *not* constant, and from now on v has several values.

EXAMPLE (*Forward and back*) There is a motion that you will understand right away. The car goes forward with velocity V, and comes back at the same speed. To say it more correctly, the *velocity in the second part is* $-V$. If the forward part lasts until $t = 3$, and the backward part continues to $t = 6$, **the car will come back where it started**. The total distance after both parts will be $f = 0$.

---

†This actually happened in *Ferris Bueller's Day Off*, when the hero borrowed his father's sports car and ran up the mileage. At home he raised the car and drove in reverse. I forget if it worked.

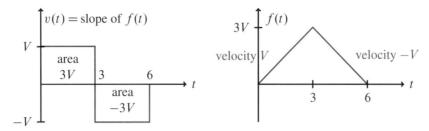

Fig. 1.4   Velocities $+V$ and $-V$ gives motion forward and back, ending at $f(6) = 0$.

The v-graph shows velocities $+V$ and $-V$. The distance starts up with slope $+V$ and reaches $f = 3V$. Then the car starts backward. The distance goes down with slope $-V$ and returns to $f = 0$ at $t = 6$.

Notice what that means. The total area "under" the v-graph is zero! A negative velocity makes the distance graph go *downward* (*negative slope*). The car is moving backward. ***Area below the axis in the v-graph is counted as negative.***

FUNCTIONS

This forward-back example gives practice with a crucially important idea—the concept of a "***function***." We seize this golden opportunity to explain functions:

The number $v(t)$ is the value of the function v at the time t.

The time t is the ***input*** to the function. The velocity $v(t)$ at that time is the ***output***. Most people say "*v of t*" when they read $v(t)$. The number "*v of 2*" is the velocity when $t = 2$. The forward-back example has $v(2) = +V$ and $v(4) = -V$. The function contains the whole history, like a memory bank that has a record of v at each t.

It is simple to convert forward-back motion into a formula. Here is $v(t)$:

$$v(t) = \begin{cases} +V & \text{if} \quad 0 < t < 3 \\ \ \ ? & \text{if} \quad t = 3 \\ -V & \text{if} \quad 3 < t < 6 \end{cases}$$

The right side contains the instructions for finding $v(t)$. The input t is converted into the output $+V$ or $-V$. The velocity $v(t)$ depends on t. In this case the function is "discontinuous," because the needle jumps at $t = 3$. *The velocity is not defined at that instant.* There is no $v(3)$. (You might argue that v is zero at the jump, but that leads to trouble.) The graph of f has a corner, and we can't give its slope.

The problem also involves a second function, namely the distance. The principle behind $f(t)$ is the same: $f(t)$ **is the distance at time** t. It is the net distance forward, and again the instructions change at $t = 3$. In the forward motion, $f(t)$ equals Vt as before. In the backward half, a calculation is built into the formula for $f(t)$:

$$f(t) = \begin{cases} Vt & \text{if} \quad 0 \leqslant t \leqslant 3 \\ V(6-t) & \text{if} \quad 3 \leqslant t \leqslant 6 \end{cases}$$

At the switching time the right side gives two instructions (one on each line). This would be bad except that they agree: $f(3) = 3V$.† The distance function is "*continuous*." There is no jump in f, even when there is a jump in v. After $t = 3$ the distance decreases because of $-Vt$. At $t = 6$ the second instruction correctly gives $f(6) = 0$.

---

† A function is only allowed *one value* $f(t)$ or $v(t)$ at each time t.

Notice something more. The functions were given by graphs before they were given by formulas. The graphs tell you f and v at every time t—sometimes more clearly than the formulas. The values $f(t)$ and $v(t)$ can also be given by tables or equations or a set of instructions. (In some way all functions are instructions—the function tells how to find f at time t.) Part of knowing f is knowing all its inputs and outputs—its *domain* and *range*:

The domain of a function is the set of inputs. The range is the set of outputs.

The domain of f consists of all times $0 \leqslant t \leqslant 6$. The range consists of all distances $0 \leqslant f(t) \leqslant 3V$. (The range of v contains only the two velocities $+V$ and $-V$.) We mention now, and repeat later, that every "linear" function has a formula $f(t) = vt + C$. Its graph is a line and v is the slope. The constant C moves the line up and down. It adjusts the line to go through any desired starting point.

SUMMARY: MORE ABOUT FUNCTIONS

May I collect together the ideas brought out by this example? We had two functions v and f. One was *velocity*, the other was *distance*. Each function had a *domain*, and a *range*, and most important a *graph*. For the f-graph we studied the slope (which agreed with v). For the v-graph we studied the area (which agreed with f). Calculus produces functions in pairs, and the best thing a book can do early is to show you more of them.

$$
\begin{array}{l}
\textit{\textbf{in}} \\
\textit{\textbf{the}} \\
\textit{\textbf{domain}}
\end{array}
\left\{
\begin{array}{lcccc}
\textit{input } t & \rightarrow & \textit{function } f & \rightarrow & \textit{output } f(t) \\
\textit{input } 2 & \rightarrow & \textit{function } v & \rightarrow & \textit{output } v(2) \\
\textit{input } 7 & \rightarrow & f(t) = 2t + 6 & \rightarrow & f(7) = 20
\end{array}
\right\}
\begin{array}{l}
\textit{\textbf{in}} \\
\textit{\textbf{the}} \\
\textit{\textbf{range}}
\end{array}
$$

Note about the definition of a function. The idea behind the symbol $f(t)$ is absolutely crucial to mathematics. Words don't do it justice! By definition, a function is a "rule" that assigns one member of the range to each member of the domain. Or, a function is a set of pairs $(t, f(t))$ with no t appearing twice. (These are "ordered pairs" because we write t before $f(t)$.) Both of those definitions are correct—but somehow they are too passive.

In practice what matters is the active part. The number $f(t)$ is produced from the number t. We read a graph, plug into a formula, solve an equation, run a computer program. The input t is "mapped" to the output $f(t)$, which changes as t changes. Calculus is about the *rate of change*. This rate is our other function v.

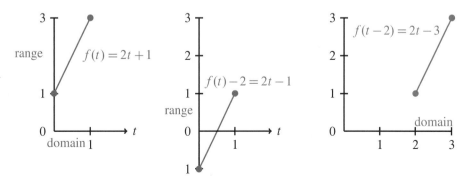

Fig. 1.5 Subtracting 2 from f affects the range. Subtracting 2 from t affects the domain.

It is quite hard at the beginning, and not automatic, to see the difference between $f(t)-2$ and $f(t-2)$. Those are both new functions, created out of the original $f(t)$. In $f(t)-2$, we subtract 2 from all the distances. That moves the whole graph *down*. In $f(t-2)$, we subtract 2 from the time. That moves the graph over *to the right*. Figure 1.5 shows both movements, starting from $f(t)=2t+1$. The formula to find $f(t-2)$ is $2(t-2)+1$, which is $2t-3$.

A graphing calculator also moves the graph, when you change the viewing window. You can pick any rectangle $A \le t \le B, C \le f(t) \le D$. The screen shows that part of the graph. But on the calculator, *the function $f(t)$ remains the same*. It is the axes that get renumbered. In our figures the axes stay the same and the function is changed.

There are two more basic ways to change a function. (We are always creating new functions—that is what mathematics is all about.) Instead of subtracting or adding, we can *multiply* the distance by 2. Figure 1.6 shows $2f(t)$. And instead of shifting the time, we can *speed it up*. The function becomes $f(2t)$. Everything happens twice as fast (and takes half as long). On the calculator those changes correspond to a "*zoom*" —on the f axis or the t axis. We soon come back to zooms.

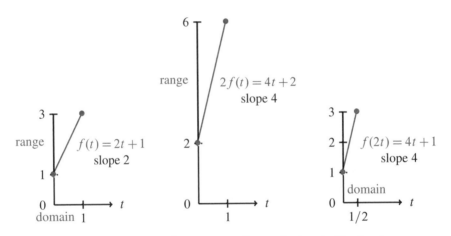

Fig. 1.6   Doubling the distance or speeding up the time doubles the slope.

1.1   EXERCISES

Each section of the book contains read-through questions. They allow you to outline the section yourself—more actively than reading a summary. This is probably the best way to remember the important ideas.

Starting from $f(0)=0$ at constant velocity v, the distance function is $f(t)=$ __a__. When $f(t)=55t$ the velocity is $v=$ __b__. When $f(t)=55t+1000$ the velocity is still __c__ and the starting value is $f(0)=$ __d__. In each case v is the __e__ of the graph of f. When __f__ is negative, the graph of __g__ goes downward. In that case area in the v-graph counts as __h__.

Forward motion from $f(0)=0$ to $f(2)=10$ has $v=$ __i__. Then backward motion to $f(4)=0$ has $v=$ __j__. The distance function is $f(t)=5t$ for $0 \le t \le 2$ and then $f(t)=$ __k__

(not $-5t$). The slopes are __l__ and __m__. The distance $f(3)=$ __n__. The area under the v-graph up to time 1.5 is __o__. The domain of f is the time interval __p__, and the range is the distance interval __q__. The range of $v(t)$ is only __r__.

The value of $f(t)=3t+1$ at $t=2$ is $f(2)=$ __s__. The value 19 equals $f($ __t__$)$. The difference $f(4)-f(1)=$ __u__. That is the change in distance, when $4-1$ is the change in __v__. The ratio of those changes equals __w__, which is the __x__ of the graph. The formula for $f(t)+2$ is $3t+3$ whereas $f(t+2)$ equals __y__. Those functions have the same __z__ as f: the graph of $f(t)+2$ is shifted __A__ and $f(t+2)$ is shifted __B__. The formula for $f(5t)$ is __C__. The formula for $5f(t)$ is __D__. The slope has jumped from 3 to __E__.

The set of inputs to a function is its __F__. The set of outputs is its __G__. The functions $f(t) = 7 + 3(t-2)$ and $f(t) = vt + C$ are __H__. Their graphs are __I__ with slopes equal to __J__ and __K__. They are the same function, if $v = $ __L__ and $C = $ __M__.

Draw the velocity graph that goes with each distance graph.

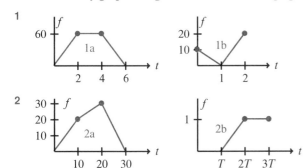

3 Write down three-part formulas for the velocities $v(t)$ in Problem 2, starting from $v(t) = 2$ for $0 < t < 10$.

4 The distance in 1b starts with $f(t) = 10 - 10t$ for $0 \leqslant t \leqslant 1$. Give a formula for the second part.

5 In the middle of graph 2a find $f(15)$ and $f(12)$ and $f(t)$.

6 In graph 2b find $f(1.4T)$. If $T = 3$ what is $f(4)$?

7 Find the *average speed* between $t = 0$ and $t = 5$ in graph 1a. What is the speed at $t = 5$?

8 What is the average speed between $t = 0$ and $t = 2$ in graph 1b? The average speed is zero between $t = \frac{1}{2}$ and $t = $ _____.

9 (recommended) A car goes at speed $v = 20$ into a brick wall at distance $f = 4$. Give two-part formulas for $v(t)$ and $f(t)$ (before and after), and draw the graphs.

10 Draw any reasonable graphs of $v(t)$ and $f(t)$ when

    (a) the driver backs up, stops to shift gear, then goes fast;

    (b) the driver slows to 55 for a police car;

    (c) in a rough gear change, the car accelerates in jumps;

    (d) the driver waits for a light that turns green.

11 Your bank account earns simple interest on the opening balance $f(0)$. What are the interest rates per year?

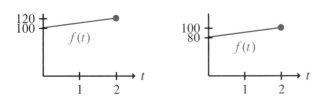

12 The earth's population is growing at $v = 100$ million a year, starting from $f = 5.2$ billion in 1990. Graph $f(t)$ and find $f(2000)$.

Draw the distance graph that goes with each velocity graph. Start from $f = 0$ at $t = 0$ and mark the distance.

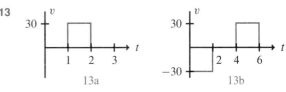

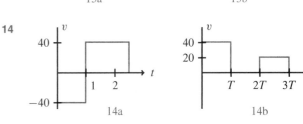

15 Write down formulas for $v(t)$ in Problem 14, starting with $v = -40$ for $0 < t < 1$. Find the average velocities to $t = 2.5$ and $t = 3T$.

16 Give 3-part formulas for the areas $f(t)$ under $v(t)$ in 13.

17 The distance in 14a starts with $f(t) = -40t$ for $0 \leqslant t \leqslant 1$. Find $f(t)$ in the other part, which passes through $f = 0$ at $t = 2$.

18 Draw the velocity and distance graphs if $v(t) = 8$ for $0 < t < 2$, $f(t) = 20 + t$ for $2 \leqslant t \leqslant 3$.

19 Draw rough graphs of $y = \sqrt{x}$ and $y = \sqrt{x-4}$ and $y = \sqrt{x} - 4$. They are "half-parabolas" with infinite slope at the start.

20 What is the break-even point if x yearbooks cost $\$1,200 + 30x$ to produce and the income is $40x$? The slope of the cost line is _____ (cost per additional book). If it goes above _____ you can't break even.

21 What are the domains and ranges of the distance functions in 14a and 14b—all values of t and $f(t)$ if $f(0) = 0$?

22 What is the range of $v(t)$ in 14b? Why is $t = 1$ not in the domain of $v(t)$ in 14a?

Problems 23–28 involve *linear functions* $f(t) = vt + C$. Find the constants v and C.

23 What linear function has $f(0) = 3$ and $f(2) = -11$?

24 Find *two* linear functions whose domain is $0 \leqslant t \leqslant 2$ and whose range is $1 \leqslant f(t) \leqslant 9$.

25 Find the linear function with $f(1) = 4$ and slope 6.

26 What functions have $f(t+1) = f(t) + 2$?

27 Find the linear function with $f(t+2) = f(t) + 6$ and $f(1) = 10$.

28 Find the only $f = vt$ that has $f(2t) = 4f(t)$. Show that every $f = \frac{1}{2}at^2$ has this property. To go _____ times as far in twice the time, you must accelerate.

Note The letter j is sometimes useful to tell which number in f we are looking at. For this example the zeroth number is $f_0 = 0$ and the jth number is $f_j = j^2$. This is a part of algebra, to give a formula for the f's instead of a list of numbers. We can also use j to tell which difference we are looking at. The first v is the first odd number $v_1 = 1$. The jth difference is the jth odd number $v_j = 2j - 1$. (Thus v_4 is $8 - 1 = 7$.) It is better to start the differences with $j = 1$, since there is no zeroth odd number v_0.

With this notation the jth *difference* is $v_j = f_j - f_{j-1}$. Sooner or later you will get comfortable with subscripts like j and $j - 1$, but it can be later. The important point is that the sum of the v's equals $f_{\text{last}} - f_{\text{first}}$. We now connect the v's to slopes and the f's to areas.

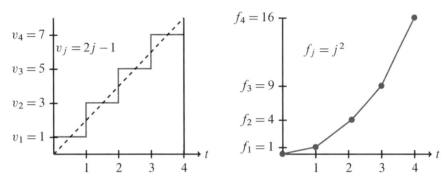

Fig. 1.7   Linear increase in $v = 1, 3, 5, 7$. Squares in the distances $f = 0, 1, 4, 9, 16$.

Figure 1.7 shows a natural way to graph Example 2, with the odd numbers in v and the squares in f. Notice an important difference between the v-graph and the f-graph. The graph of f is *"piecewise linear."* We plotted the numbers in f and connected them by straight lines. The graph of v is *"piecewise constant."* We plotted the differences as constant over each piece. This reminds us of the distance-velocity graphs, when the distance $f(t)$ is a straight line and the velocity $v(t)$ is a horizontal line.

Now make the connection to slopes:

$$\text{The slope of the } f\text{-graph is } \frac{\text{distance up}}{\text{distance across}} = \frac{\text{change in } f}{\text{change in } t} = v.$$

Over each piece, the change in t (across) is 1. The change in f (upward) is the difference that we are calling v. The ratio is the slope $v/1$ or just v. The slope makes a sudden change at the breakpoints $t = 1, 2, 3, \ldots$. At those special points the slope of the f-graph is not defined—we connected the v's by vertical lines but this is very debatable. **The main idea is that between the breakpoints, the slope of $f(t)$ is $v(t)$.**

Now make the connection to areas:

The total area under the v-graph is $f_{\text{last}} - f_{\text{first}}$.

This area, underneath the staircase in Figure 1.7, is composed of rectangles. The base of every rectangle is 1. The heights of the rectangles are the v's. So the areas also equal the v's, and the total area is the sum of the v's. This area is $f_{\text{last}} - f_{\text{first}}$.

Even more is true. We could start at any time and end at any later time— not necessarily at the special times $t = 0, 1, 2, 3, 4$. Suppose we stop at $t = 3.5$. Only half of the last rectangular area (under $v = 7$) will be counted. The total area is $1 + 3 + 5 + \frac{1}{2}(7) = 12.5$. This still agrees with $f_{\text{last}} - f_{\text{first}} = 12.5 - 0$. At this new ending time $t = 3.5$, we are only halfway up the last step in the f-graph. Halfway between 9 and 16 is 12.5.

The piecewise linear sine has slopes __p__ . Those form a piecewise __q__ cosine. Both functions have __r__ equal to 6, which means that $f(t+6) =$ __s__ for every t. The velocities $v = 1, 2, 4, 8, \ldots$ have $v_j =$ __t__ . In that case $f_0 = 1$ and $f_j =$ __u__ . The sum of $1, 2, 4, 8, 16$ is __v__ . The difference $2^j - 2^{j-1}$ equals __w__ . After a burst of speed V to time T, the distance is __x__ . If $f(T) = 1$ and V increases, the burst lasts only to $T =$ __y__ . When V approaches infinity, $f(t)$ approaches a __z__ function. The velocities approach a __A__ function, which is concentrated at $t = 0$ but has area __B__ under its graph. The slope of a step function is __C__ .

Problems 1–4 are about numbers f and differences v.

1 From the numbers $f = 0, 2, 7, 10$ find the differences v and the sum of the three v's. Write down another f that leads to the same v's. For $f = 0, 3, 12, 10$ the sum of the v's is still _____ .

2 Starting from $f = 1, 3, 2, 4$ draw the f-graph (linear pieces) and the v-graph. What are the areas "under" the v-graph that add to $4 - 1$? If the next number in f is 11, what is the area under the next v?

3 From $v = 1, 2, 1, 0, -1$ find the f's starting at $f_0 = 3$. Graph v and f. The maximum value of f occurs when $v =$ _____ . Where is the maximum f when $v = 1, 2, 1, -1$?

4 For $f = 1, b, c, 7$ find the differences v_1, v_2, v_3 and add them up. Do the same for $f = a, b, c, 7$. Do the same for $f = a, b, c, d$.

Problems 5–11 are about linear functions and constant slopes.

5 Write down the slopes of these linear functions:

(a) $f(t) = 1.1t$ (b) $f(t) = 1 - 2t$ (c) $f(t) = 4 + 5(t - 6)$.

Compute $f(6)$ and $f(7)$ for each function and confirm that $f(7) - f(6)$ equals the slope.

6 If $f(t) = 5 + 3(t-1)$ and $g(t) = 1.5 + 2.5(t-1)$ what is $h(t) = f(t) - g(t)$? Find the slopes of f, g and h.

7 Suppose $v(t) = 2$ for $t < 5$ and $v(t) = 3$ for $t > 5$.

(a) If $f(0) = 0$ find a two-part formula for $f(t)$.

(b) Check that $f(10)$ equals the area under the graph of $v(t)$ (two rectangles) up to $t = 10$.

8 Suppose $v(t) = 10$ for $t < 1/10, v(t) = 0$ for $t > 1/10$. Starting from $f(0) = 1$ find $f(t)$ in two pieces.

9 Suppose $g(t) = 2t + 1$ and $f(t) = 4t$. Find $g(3)$ and $f(g(3))$ and $f(g(t))$. How is the slope of $f(g(t))$ related to the slopes of f and g?

10 For the same functions, what are $f(3)$ and $g(f(3))$ and $g(f(t))$? When t is changed to $4t$, distance increases _____ times as fast and the velocity is multiplied by _____ .

11 Compute $f(6)$ and $f(8)$ for the functions in Problem 5. Confirm that the slopes v agree with

$$\text{slope} = \frac{f(8) - f(6)}{8 - 6} = \frac{\text{change in } f}{\text{change in } t}.$$

Problems 12–18 are based on Example 3 about income taxes.

12 What are the income taxes on $x = \$10,000$ and $x = \$30,000$ and $x = \$50,000$?

13 What is the equation for income tax $f(x)$ in the second bracket $\$20,350 \leqslant x \leqslant \$49,300$? How is the number $11,158.50$ connected with the other numbers in the tax instructions?

14 Write the tax function $F(x)$ for a married couple if the IRS treats them as two single taxpayers each with taxable income $x/2$. (This is not done.)

15 In the 15% bracket, with 5% state tax as a deduction, the combined rate is not 20% but _____ . Think about the tax on an extra \$100.

16 A piecewise linear function is **continuous** when $f(t)$ at the end of each interval equals $f(t)$ at the start of the following interval. If $f(t) = 5t$ up to $t = 1$ and $v(t) = 2$ for $t > 1$, define f beyond $t = 1$ so it is (a) continuous (b) discontinuous. (c) Define a tax function $f(x)$ with rates .15 and .28 so you would lose by earning an extra dollar beyond the breakpoint.

17 The difference between a tax *credit* and a *deduction* from income is the difference between $f(x) - c$ and $f(x - d)$. Which is more desirable, a credit of $c = \$1000$ or a deduction of $d = \$1000$, and why? Sketch the tax graphs when $f(x) = .15x$.

18 The average tax rate on the taxable income x is $a(x) = f(x)/x$. This is the slope between $(0, 0)$ and the point $(x, f(x))$. Draw a rough graph of $a(x)$. The average rate a is below the marginal rate v because _____ .

Problems 19–30 involve numbers $f_0, f_1, f_2, \ldots$ and their differences $v_j = f_j - f_{j-1}$. They give practice with subscripts $0, \ldots, j$.

19 Find the velocities v_1, v_2, v_3 and formulas for v_j and f_j:

(a) $f = 1, 3, 5, 7, \ldots$ (b) $f = 0, 1, 0, 1, \ldots$ (c) $f = 0, \frac{1}{2}, \frac{3}{4}, \frac{7}{8}, \ldots$

20 Find f_1, f_2, f_3 and a formula for f_j with $f_0 = 0$:

(a) $v = 1, 2, 4, 8, \ldots$ (b) $v = -1, 1, -1, 1, \ldots$

21 The areas of these nested squares are $1^2, 2^2, 3^2, \ldots$. What are the areas of the L-shaped bands (the differences between squares)? How does the figure show that $1 + 3 + 5 + 7 = 4^2$?

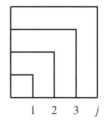
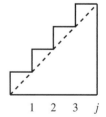

22 From the area under the staircase (by rectangles and then by triangles) show that the first j whole numbers 1 to j add up to $\frac{1}{2}j^2 + \frac{1}{2}j$. Find $1 + 2 + \cdots + 100$.

23 If $v = 1, 3, 5, \ldots$ then $f_j = j^2$. If $v = 1, 1, 1, \ldots$ then $f_j =$ _____ . Add those to find the sum of $2, 4, 6, \ldots, 2j$. Divide by 2 to find the sum of $1, 2, 3, \ldots, j$. (Compare Problem 22.)

24 *True* (with reason) *or false* (with example).

(a) When the f's are increasing so are the v's.

(b) When the v's are increasing so are the f's.

(c) When the f's are periodic so are the v's.

(d) When the v's are periodic so are the f's.

25 If $f(t) = t^2$, compute $f(99)$ and $f(101)$. Between those times, what is the increase in f divided by the increase in t ?

26 If $f(t) = t^2 + t$, compute $f(99)$ and $f(101)$. Between those times, what is the increase in f divided by the increase in t ?

27 If $f_j = j^2 + j + 1$ find a formula for v_j.

28 Suppose the v's increase by 4 at every step. Show by example and then by algebra that the "second difference" $f_{j+1} - 2f_j + f_{j-1}$ equals 4.

29 Suppose $f_0 = 0$ and the v's are $1, \frac{1}{2}, \frac{1}{2}, \frac{1}{4}, \frac{1}{4}, \frac{1}{4}, \frac{1}{4}, \ldots$. For which j does $f_j = 5$?

30 Show that $a_j = f_{j+1} - 2f_j + f_{j-1}$ always equals $v_{j+1} - v_j$. If v is velocity then a stands for _____ .

Problems 31–34 involve periodic f's and v's (like $\sin t$ and $\cos t$).

31 For the discrete sine $f = 0, 1, 1, 0, -1, -1, 0$ find the second differences $a_1 = f_2 - 2f_1 + f_0$ and $a_2 = f_3 - 2f_2 + f_1$ and a_3. Compare a_j with f_j.

32 If the sequence $v_1, v_2, \ldots$ has period 6 and $w_1, w_2, \ldots$ has period 10, what is the period of $v_1 + w_1, v_2 + w_2, \ldots$?

33 Draw the graph of $f(t)$ starting from $f_0 = 0$ when $v = 1, -1, -1, 1$. If v has period 4 find $f(12), f(13), f(100.1)$.

34 Graph $f(t)$ from $f_0 = 0$ to $f_4 = 4$ when $v = 1, 2, 1, 0$. If v has period 4, find $f(12)$ and $f(14)$ and $f(16)$. Why *doesn't f have period 4* ?

Problems 35–42 are about exponential v's and f's.

35 Find the v's for $f = 1, 3, 9, 27$. Predict v_4, and v_j. Algebra gives $3^j - 3^{j-1} = (3-1)3^{j-1}$.

36 Find $1 + 2 + 4 + \cdots + 32$ and also $1 + \frac{1}{2} + \frac{1}{4} + \cdots + \frac{1}{32}$.

37 Estimate the slope of $f(t) = 2^t$ at $t = 0$. Use a calculator to compute (increase in f)/(increase in t) when t is small:

$$\frac{f(t) - f(0)}{t} = \frac{2 - 1}{1} \text{ and } \frac{2^{\cdot 1} - 1}{.1} \text{ and } \frac{2^{\cdot 01} - 1}{.01} \text{ and } \frac{2^{\cdot 001} - 1}{.001}.$$

38 Suppose $f_0 = 1$ and $v_j = 2f_{j-1} = v_j$. Find f_4.

39 (a) From $f = 1, \frac{1}{2}, \frac{1}{4}, \frac{1}{8}$ find v_1, v_2, v_3, and predict v_j.

(b) Check $f_3 - f_0 = v_1 + v_2 + v_3$ and $f_j - f_{j-1} = v_j$.

40 Suppose $v_j = r^j$. Show that $f_j = (r^{j+1} - 1)/(r - 1)$ starts from $f_0 = 1$ and has $f_j - f_{j-1} = v_j$. (Then this is the correct $f_j = 1 + r + \cdots + r^j =$ sum of a geometric series.)

41 From $f_j = (-1)^j$ compute v_j. What is $v_1 + v_2 + \cdots + v_j$?

42 Estimate the slope of $f(t) = e^t$ at $t = 0$. Use a calculator that knows e (or else take $e = 2.78$) to compute

$$\frac{f(t) - f(0)}{t} = \frac{e - 1}{1} \text{ and } \frac{e^{\cdot 1} - 1}{.1} \text{ and } \frac{e^{\cdot 01} - 1}{.01}.$$

Problems 43–47 are about $U(t) =$ step from 0 to 1 at $t = 0$.

43 Graph the four functions $U(t - 1)$ and $U(t) - 2$ and $U(3t)$ and $4U(t)$. Then graph $f(t) = 4U(3t - 1) - 2$.

44 Graph the square wave $U(t) - U(t - 1)$. If this is the velocity $v(t)$, graph the distance $f(t)$. If this is the distance $f(t)$, graph the velocity.

45 Two bursts of speed lead to the same distance $f = 10$:

$$v = \underline{\quad} \text{ to } t = .001 \qquad v = V \text{ to } t = \underline{\quad}.$$

As $V \to \infty$ the limit of the $f(t)$'s is _____ .

46 Draw the staircase function $U(t) + U(t - 1) + U(t - 2)$. Its slope is a sum of three _____ functions.

47 Which capital letters like **L** are the graphs of functions when steps are allowed ? The slope of **L** is minus a delta function. Graph the slopes of the others.

48 Write a subroutine FINDV whose input is a sequence $f_0, f_1, \ldots, f_N$ and whose output is $v_1, v_2, \ldots, v_N$. Include graphical output if possible. Test on $f_j = 2j$ and j^2 and 2^j.

49 Write a subroutine FINDF whose input is $v_1, \ldots, v_N$. and f_0, and whose output is $f_0, f_1, \ldots, f_N$. The default value of f_0 is zero. Include graphical output if possible. Test $v_j = j$.

50 If FINDV is applied to the output of FINDF, what sequence is returned ? If FINDF is applied to the output of FINDV, what sequence is returned ? Watch f_0.

51 Arrange $2j$ and j^2 and 2^j and $\sqrt{j}$ in increasing order

(a) when j is large: $j = 9$　　(b) when j is small: $j = \frac{1}{9}$.

52 The average age of your family since 1970 is a piecewise linear function $A(t)$. Is it continuous or does it jump ? What is its slope ? Graph it the best you can.

1.3 The Velocity at an Instant

We have arrived at the central problems that calculus was invented to solve. There are two questions, in opposite directions, and I hope you could see them coming.

1. If the velocity is changing, *how can you compute the distance traveled?*
2. If the graph of $f(t)$ is not a straight line, *what is its slope?*

Find the distance from the velocity, find the velocity from the distance. Our goal is to do both—but not in one section. Calculus may be a good course, but it is not magic. The first step is to let the velocity change in the steadiest possible way.

Question 1 *Suppose the velocity at each time t is $v(t) = 2t$. Find $f(t)$.*

With $v = 2t$, a physicist would say that the acceleration is constant (it equals 2). The driver steps on the gas, the car accelerates, and the speedometer goes steadily up. The distance goes up too—faster and faster. If we measure t in seconds and v in feet per second, the distance f comes out in feet. After 10 seconds the speed is 20 feet per second. After 44 seconds the speed is 88 feet/second (which is 60 miles/hour). The acceleration is clear, **but how far has the car gone?**

Question 2 *The distance traveled by time t is $f(t) = t^2$. Find the velocity $v(t)$.*

The graph of $f(t) = t^2$ is on the right of Figure 1.12. It is a *parabola*. The curve starts at zero, when the car is new. At $t = 5$ the distance is $f = 25$. By $t = 10$, f reaches 100.

Velocity is distance divided by time, but what happens when the speed is changing? Dividing $f = 100$ by $t = 10$ gives $v = 10$—the *average velocity* over the first ten seconds. Dividing $f = 121$ by $t = 11$ gives the average speed over 11 seconds. But how do we find the **instantaneous velocity**—the reading on the speedometer at the exact instant when $t = 10$?

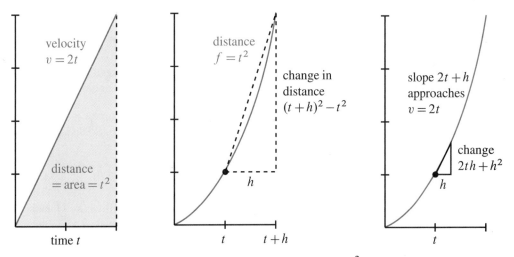

Fig. 1.12 The velocity $v = 2t$ is linear. The distance $f = t^2$ is quadratic.

I hope you see the problem. As the car goes faster, the graph of t^2 gets steeper—because more distance is covered in each second. The average velocity between $t = 10$ and $t = 11$ is a good approximation—but only an approximation—to the speed at the moment $t = 10$. Averages are easy to find:

distance at $t = 10$ is $f(10) = 10^2 = 100$ distance at $t = 11$ is $f(11) = 11^2 = 121$

$$\textit{average velocity is } \frac{f(11) - f(10)}{11 - 10} = \frac{121 - 100}{1} = 21.$$

The car covered 21 feet in that 1 second. Its average speed was 21 feet/second. Since it was gaining speed, the velocity at the beginning of that second was below 21.

Geometrically, what is the average? It is a slope, but not the slope of the curve. **The average velocity is the slope of a straight line**. The line goes between two points on the curve in Figure 1.12. When we compute an average, we pretend the velocity is constant—so we go back to the easiest case. It only requires a division of distance by time:

$$\text{average velocity} = \frac{\text{change in } f}{\text{change in } t}. \tag{1}$$

Calculus and the Law You enter a highway at $1:00$. If you exit 150 miles away at $3:00$, your average speed is 75 miles per hour. I'm not sure if the police can give you a ticket. You could say to the judge, "When was I doing 75?" The police would have to admit that they have no idea—but they would have a definite feeling that you must have been doing 75 sometime.†

We return to the central problem—computing $v(10)$ at the instant $t = 10$. The average velocity over the next second is 21. We can also find the average over the *half-second* between $t = 10.0$ and $t = 10.5$. Divide the change in distance by the change in time:

$$\frac{f(10.5) - f(10.0)}{10.5 - 10.0} = \frac{(10.5)^2 - (10.0)^2}{.5} = \frac{110.25 - 100}{.5} = 20.5.$$

That average of 20.5 is closer to the speed at $t = 10$. It is still not exact.

The way to find $v(10)$ is to *keep reducing the time interval*. This is the basis for Chapter 2, and the key to differential calculus. **Find the slope between points that are closer and closer on the curve**. The "limit" is the slope at a single point.

Algebra gives the average velocity between $t = 10$ and any later time $t = 10 + h$. The distance increases from 10^2 to $(10 + h)^2$. The change in time is h. So divide:

$$v_{\text{average}} = \frac{(10 + h)^2 - 10^2}{h} = \frac{100 + 20h + h^2 - 100}{h} = 20 + h. \tag{2}$$

This formula fits our previous calculations. The interval from $t = 10$ to $t = 11$ had $h = 1$, and the average was $20 + h = 21$. When the time step was $h = \frac{1}{2}$, the average was $20 + \frac{1}{2} = 20.5$. Over a millionth of a second the average will be 20 plus $1/1,000,000$—which is very near 20.

Conclusion: ***The velocity at*** $t = 10$ ***is*** $v = 20$. *That is the slope of the curve*. It agrees with the v-graph on the left side of Figure 1.12, which also has $v(10) = 20$.

---

† This is our first encounter with the much despised "Mean Value Theorem." If the judge can prove the theorem, you are dead. A few v-graphs and f-graphs will confuse the situation.

We now show that the two graphs match at all times. If $f(t) = t^2$ then $v(t) = 2t$. You are seeing the key computation of calculus, and we can put it into words before equations. Compute the distance at time $t + h$, *subtract* the distance at time t, and *divide* by h. That gives the average velocity:

$$v_{\text{ave}} = \frac{f(t+h) - f(t)}{h} = \frac{(t+h)^2 - t^2}{h} = \frac{t^2 + 2th + h^2 - t^2}{h} = 2t + h. \quad (3)$$

This fits the previous calculation, where t was 10. The average was $20 + h$. Now the average is $2t + h$. It depends on the time step h, because the velocity is changing. But we can see what happens *as h approaches zero*. The average is closer and closer to the speedometer reading of $2t$, at the exact moment when the clock shows time t:

1E As h approaches zero, the average velocity $2t + h$ approaches $v(t) = 2t$.

Note   The computation (3) shows how calculus needs algebra. If we want the whole v-graph, we have to let time be a "*variable*." It is represented by the letter t. Numbers are enough at the specific time $t = 10$ and the specific step $h = 1$—but algebra gets beyond that. The average between any t and any $t + h$ is $2t + h$. Please don't hesitate to put back numbers for the letters—that checks the algebra.

There is also a step beyond algebra! Calculus requires the *limit of the average*. As h shrinks to zero, the points on the graph come closer. "Average over an interval" becomes "velocity at an instant." The general theory of limits is not particularly simple, but here we don't need it. (It isn't particularly hard either.) In this example *the limiting value is easy to identify*. The average $2t + h$ approaches $2t$, as $h \to 0$.

What remains to do in this section? We answered Question 2—to find velocity from distance. We have not answered Question 1. If $v(t) = 2t$ increases linearly with time, what is the distance? This goes in the opposite direction (it is *integration*).

The Fundamental Theorem of Calculus says that no new work is necessary. *If the slope of $f(t)$ leads to $v(t)$, then the area under that v-graph leads back to the f-graph*. The odometer readings $f = t^2$ produced speedometer readings $v = 2t$. By the Fundamental Theorem, the area under $2t$ should be t^2. But we have certainly not proved any fundamental theorems, so it is better to be safe—by actually computing the area.

Fortunately, it is the area of a triangle. The base of the triangle is t and the height is $v = 2t$. The area agrees with $f(t)$:

$$\text{area} = \tfrac{1}{2}(\text{base})(\text{height}) = \tfrac{1}{2}(t)(2t) = t^2. \quad (4)$$

EXAMPLE 1   The graphs are *shifted in time*. The car doesn't start until $t = 1$. Therefore $v = 0$ and $f = 0$ up to that time. After the car starts we have $v = 2(t - 1)$ and $f = (t - 1)^2$. You see how the time delay of 1 enters the formulas. Figure 1.13 shows how it affects the graphs.

EXAMPLE 2   The acceleration changes from 2 to another constant a. The velocity changes from $v = 2t$ to $v = at$. *The acceleration is the slope of the velocity curve!* The distance is also proportional to a, but notice the factor $\tfrac{1}{2}$:

$$\text{acceleration } a \quad \Leftrightarrow \quad \text{velocity } v = at \quad \Leftrightarrow \quad \text{distance } f = \tfrac{1}{2}at^2.$$

If a equals 1, then $v = t$ and $f = \tfrac{1}{2}t^2$. That is one of the most famous pairs in calculus. If a equals the gravitational constant g, then $v = gt$ is the velocity of a

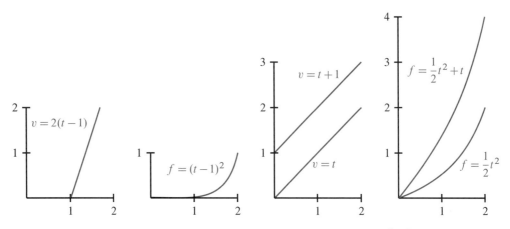

Fig. 1.13   Delayed velocity and distance. The pairs $v = at + b$ and $f = \frac{1}{2}at^2 + bt$.

falling body. The speed doesn't depend on the mass (tested by Galileo at the Leaning Tower of Pisa). Maybe he saw the distance $f = \frac{1}{2}gt^2$ more easily than the speed $v = gt$. Anyway, this is the most famous pair in physics.

EXAMPLE 3   Suppose $f(t) = 3t + t^2$. The average velocity from t to $t + h$ is

$$v_{\text{ave}} = \frac{f(t + h) - f(t)}{h} = \frac{3(t + h) + (t + h)^2 - 3t - t^2}{h}.$$

The change in distance has an extra $3h$ (coming from $3(t + h)$ minus $3t$). The velocity contains an additional 3 (coming from $3h$ divided by h). When $3t$ is added to the distance, 3 is added to the velocity. If Galileo had thrown a weight instead of dropping it, the starting velocity v_0 would have added $v_0 t$ to the distance.

FUNCTIONS ACROSS TIME

The idea of slope is not difficult—for one straight line. Divide the change in f by the change in t. In Chapter 2, divide the change in y by the change in x. Experience shows that the hard part is to see what happens to the slope as the line moves.

Figure 1.14a shows the line between points A and B on the curve. This is a "secant line." Its slope is an *average* velocity. What calculus does is to bring that point B *down the curve toward* A.

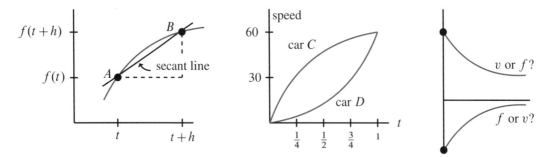

Fig. 1.14   Slope of line, slope of curve. Two velocity graphs. *Which is which?*

Question 1   What happens to the "change in f"—the height of B above A?
Answer   The change in f decreases to zero. So does the change in t.

Question 2   As B approaches A, does the slope of the line increase or decrease?
Answer   I am not going to answer that question. It is too important. Draw another secant line with B closer to A. Compare the slopes.

This question was created by Steve Monk at the University of Washington—where 57% of the class gave the right answer. Probably 97% would have found the right slope from a formula. Figure 1.14b shows the opposite problem. We know the velocity, not the distance. But calculus answers questions about both functions.

Question 3   Which car is going faster at time $t = 3/4$?
Answer   Car C has higher speed. Car D has greater acceleration.

Question 4   If the cars start together, is D catching up to C at the end? Between $t = \frac{1}{2}$ and $t = 1$, do the cars get closer or further apart?
Answer   This time more than half the class got it wrong. You won't but you can see why they did. You have to look at the speed graph and imagine the distance graph. When car C is going faster, the distance between them _____ .
To repeat: The cars start together, but they don't finish together. They reach the same speed at $t = 1$, not the same distance. Car C went faster. You really should draw their distance graphs, to see how they bend.
These problems help to emphasize one more point. Finding the speed (or slope) is entirely different from finding the distance (or area):

1. To find the *slope* of the f-graph at a particular time t, you *don't* have to know the whole history.
2. To find the *area* under the v-graph up to a particular time t, you *do* have to know the whole history.

A short record of distance is enough to recover $v(t)$. Point B moves toward point A. The problem of slope is *local*—the speed is completely decided by $f(t)$ near point A.
In contrast, a short record of speed is *not enough* to recover the total distance. We have to know what the mileage was earlier. Otherwise we can only know the *increase* in mileage, not the total.

1.3 EXERCISES

Read-through questions

Between the distances $f(2) = 100$ and $f(6) = 200$, the average velocity is __a__ . If $f(t) = \frac{1}{4}t^2$ then $f(6) = $ __b__ and $f(8) = $ __c__ . The average velocity in between is __d__ . The instantaneous velocities at $t = 6$ and $t = 8$ are __e__ and __f__ .

The average velocity is computed from $f(t)$ and $f(t+h)$ by $v_{ave} = $ __g__ . If $f(t) = t^2$ then $v_{ave} = $ __h__ . From $t = 1$ to $t = 1.1$ the average is __i__ . The instantaneous velocity is the __j__ of v_{ave}. If the distance is $f(t) = \frac{1}{2}at^2$ then the velocity is $v(t) = $ __k__ and the acceleration is __l__ .

On the graph of $f(t)$, the average velocity between A and B is the slope of __m__ . The velocity at A is found by __n__ . The velocity at B is found by __o__ . When the velocity is positive,

the distance is __p__ . When the velocity is increasing, the car is __q__ .

1 Compute the average velocity between $t = 5$ and $t = 8$:
  (a) $f(t) = 6t$ (b) $f(t) = 6t + 2$
  (c) $f(t) = \frac{1}{2}at^2$ (d) $f(t) = t - t^2$
  (e) $f(t) = 6$ (f) $v(t) = 2t$

2 For the same functions compute $[f(t+h) - f(t)]/h$. This depends on t and h. Find the limit as $h \to 0$.

3 If the odometer reads $f(t) = t^2 + t$ (f in miles or kilometers, t in hours), find the average speed between
  (a) $t = 1$ and $t = 2$
  (b) $t = 1$ and $t = 1.1$

(c) $t = 1$ and $t = 1 + h$

(d) $t = 1$ and $t = .9$ (note $h = -.1$)

4 For the same $f(t) = t^2 + t$, find the average speed between

(a) $t = 0$ and 1 (b) $t = 0$ and $\frac{1}{2}$ (c) $t = 0$ and h.

5 In the answer to 3(c), find the limit as $h \to 0$. What does that limit tell us ?

6 Set $h = 0$ in your answer to 4(c). Draw the graph of $f(t) = t^2 + t$ and show its slope at $t = 0$.

7 Draw the graph of $v(t) = 1 + 2t$. From geometry find the area under it from 0 to t. Find the slope of that area function $f(t)$.

8 Draw the graphs of $v(t) = 3 - 2t$ and the area $f(t)$.

9 *True or false*

(a) If the distance $f(t)$ is positive, so is $v(t)$.

(b) If the distance $f(t)$ is increasing, so is $v(t)$.

(c) If $f(t)$ is positive, $v(t)$ is increasing.

(d) If $v(t)$ is positive, $f(t)$ is increasing.

10 If $f(t) = 6t^2$ find the slope of the f-graph and also the v-graph. The slope of the v-graph is the _____ .

11 If $f(t) = t^2$ what is the average velocity between $t = .9$ and $t = 1.1$? What is the average between $t - h$ and $t + h$?

12 (a) Show that for $f(t) = \frac{1}{2}at^2$ the average velocity between $t - h$ and $t + h$ is exactly the velocity at t.

(b) The area under $v(t) = at$ from $t - h$ to $t + h$ is exactly the base $2h$ times _____ .

13 Find $f(t)$ from $v(t) = 20t$ if $f(0) = 12$. Also if $f(1) = 12$.

14 *True or false*, for any distance curves.

(a) The slope of the line from A to B is the average velocity between those points.

(b) Secant lines have smaller slopes than the curve.

(c) If $f(t)$ and $F(t)$ start together and finish together, the average velocities are equal.

(d) If $v(t)$ and $V(t)$ start together and finish together, the increases in distance are equal.

15 When you jump up and fall back your height is $y = 2t - t^2$ in the right units.

(a) Graph this parabola and its slope.

(b) Find the time in the air and maximum height.

(c) *Prove*: Half the time you are above $y = \frac{3}{4}$.

Basketball players "hang" in the air partly because of (c).

16 Graph $f(t) = t^2$ and $g(t) = f(t) - 2$ and $h(t) = f(2t)$, all from $t = 0$ to $t = 1$. Find the velocities.

17 (Recommended) An up and down velocity is $v(t) = 2t$ for $t \leqslant 3, v(t) = 12 - 2t$ for $t \geqslant 3$. Draw the piecewise parabola $f(t)$. Check that $f(6) =$ area under the graph of $v(t)$.

18 Suppose $v(t) = t$ for $t \leqslant 2$ and $v(t) = 2$ for $t \geqslant 2$. Draw the graph of $f(t)$ out to $t = 3$.

19 Draw $f(t)$ up to $t = 4$ when $v(t)$ increases linearly from

(a) 0 to 2   (b) -1 to 1   (c) -2 to 0.

20 (Recommended) Suppose $v(t)$ is the piecewise linear sine function of Section 1.2. (In Figure 1.8 it was the distance.) Find the area under $v(t)$ between $t = 0$ and $t = 1, 2, 3, 4, 5, 6$. Plot those points $f(1), \ldots, f(6)$ and draw the complete piecewise parabola $f(t)$.

21 Draw the graph of $f(t) = |1 - t^2|$ for $0 \leqslant t \leqslant 2$. Find a three-part formula for $v(t)$.

22 Draw the graphs of $f(t)$ for these velocities (to $t = 2$):

(a) $v(t) = 1 - t$

(b) $v(t) = |1 - t|$

(c) $v(t) = (1 - t) + |1 - t|$.

23 When does $f(t) = t^2 - 3t$ reach 10 ? Find the average velocity up to that time and the instantaneous velocity at that time.

24 If $f(t) = \frac{1}{2}at^2 + bt + c$, what is $v(t)$? What is the slope of $v(t)$? When does $f(t)$ equal 41, if $a = b = c = 1$?

25 If $f(t) = t^2$ then $v(t) = 2t$. Does the speeded-up function $f(4t)$ have velocity $v(4t)$ or $4v(t)$ or $4v(4t)$?

26 If $f(t) = t - t^2$ find $v(t)$ and $f(3t)$. Does the slope of $f(3t)$ equal $v(3t)$ or $3v(t)$ or $3v(3t)$?

27 For $f(t) = t^2$ find $v_{\text{ave}}(t)$ between 0 and t. Graph $v_{\text{ave}}(t)$ and $v(t)$.

28 If you know the average velocity $v_{\text{ave}}(t)$, how can you find the distance $f(t)$? Start from $f(0) = 0$.

1.4 Circular Motion

This section introduces completely new distances and velocities—*the sines and cosines from trigonometry*. As I write that last word, I ask myself how much trigonometry it is essential to know. There will be the basic picture of a right triangle, with sides $\cos t$ and $\sin t$ and 1. There will also be the crucial equation $(\cos t)^2 + (\sin t)^2 = 1$, which is Pythagoras' law $a^2 + b^2 = c^2$. The squares of two sides add to the square of the hypotenuse (and the 1 is really 1^2). Nothing else is needed immediately. If you don't know trigonometry, don't stop—an important part can be learned now.

You will recognize the wavy graphs of the sine and cosine. *We intend to find the slopes of those graphs.* That can be done without using the formulas for $\sin(x + y)$ and $\cos(x + y)$—which later give the same slopes in a more algebraic way. Here it is only basic things that are needed.† And anyway, how complicated can a triangle be ?

Remark You might think trigonometry is only for surveyors and navigators (people with triangles). Not at all! By far the biggest applications are to ***rotation*** and ***vibration*** and ***oscillation.*** It is fantastic that sines and cosines are so perfect for "repeating motion"—around a circle or up and down.

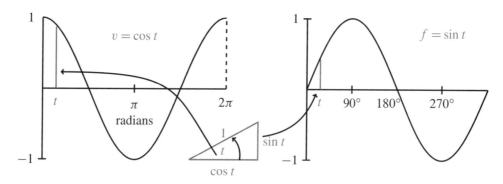

Fig. 1.15 As the angle t changes, the graphs show the sides of the right triangle.

Our underlying goal is to offer one more example in which the velocity can be computed by common sense. Calculus is mainly an extension of common sense, but here that extension is not needed. We will find the slope of the sine curve. The straight line $f = vt$ was easy and the parabola $f = \frac{1}{2}at^2$ was harder. The new example also involves realistic motion, seen every day. We start with ***circular motion***, in which the position is given and the velocity will be found.

A ball goes around a circle of radius one. The center is at $x = 0, y = 0$ (the origin). The x and y coordinates satisfy $x^2 + y^2 = 1^2$, to keep the ball on the circle. We specify its position in Figure 1.16a by giving its angle with the horizontal. And we make the ball travel with constant speed, by requiring that ***the angle is equal to the time*** t. The ball goes counterclockwise. At time 1 it reaches the point where the angle equals 1. The angle is measured in ***radians*** rather than degrees, so a full circle is completed at $t = 2\pi$ instead of $t = 360$.

The ball starts on the x axis, where the angle is zero. Now find it at time t:

The ball is at the point where $x = \cos t$ ***and*** $y = \sin t.$

---

† Sines and cosines are so important that I added a review of trigonometry in Section 1.5. But the concepts in this section can be more valuable than formulas.

This is where trigonometry is useful. The cosine oscillates between 1 and -1, as the ball goes from far right to far left and back again. The sine also oscillates between 1 and -1, starting from $\sin 0 = 0$. At time $\pi/2$ the sine (the height) increases to one. The cosine is zero and the ball reaches the top point $x = 0, y = 1$. At time π the cosine is -1 and the sine is back to zero—the coordinates are $(-1,0)$. At $t = 2\pi$ the circle is complete (the angle is also 2π), and $x = \cos 2\pi = 1$, $y = \sin 2\pi = 0$.

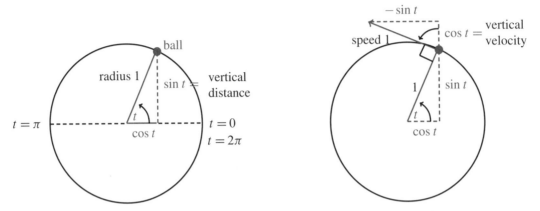

Fig. 1.16   Circular motion with speed 1, angle t, height $\sin t$, upward velocity $\cos t$.

Important point: The distance around the circle (its circumference) is $2\pi r = 2\pi$, because the radius is 1. The ball travels a distance 2π in a time 2π. **The speed equals 1**. It remains to find the velocity, which involves not only speed but *direction*.

Degrees vs. radians   A full circle is 360 degrees and 2π radians. Therefore

$$1 \text{ radian} = 360/2\pi \text{ degrees} \approx 57.3 \text{ degrees}$$
$$1 \text{ degree} = 2\pi/360 \text{ radians} \approx .01745 \text{ radians}$$

Radians were invented to avoid those numbers! The speed is exactly 1, reaching t radians at time t. The speed would be .01745, if the ball only reached t degrees. The ball would complete the circle at time $T = 360$. We cannot accept the division of the circle into 360 pieces (by whom ?), which produces these numbers.

To check degree mode vs. radian mode, verify that $\sin 1° \approx .017$ and $\sin 1 \approx .84$.

VELOCITY OF THE BALL

At time t, which direction is the ball going ? Calculus watches the motion between t and $t + h$. For a ball on a string, we don't need calculus—just let go. **The direction of motion is tangent to the circle.** With no force to keep it on the circle, *the ball goes off on a tangent.* If the ball is the moon, the force is gravity. If it is a hammer swinging around on a chain, the force is from the center. When the thrower lets go, the hammer takes off—and it is an art to pick the right moment. (I once saw a friend hit by a hammer at MIT. He survived, but the thrower quit track.) Calculus will find that same tangent direction, when the points at t and $t + h$ come close.

The "***velocity triangle***" is in Figure 1.16b. It is the same as the position triangle, but rotated through $90°$. The hypotenuse is tangent to the circle, in the direction the ball is moving. Its length equals 1 (the speed). The angle t still appears, but now it is the angle with the vertical. **The upward component of velocity is $\cos t$, when the upward component of position is** $\sin t$. That is our common sense calculation, based on a figure rather than a formula. The rest of this section depends on it—and we check $v = \cos t$ at special points.

At the starting time $t = 0$, the movement is all upward. The height is $\sin 0 = 0$ and the upward velocity is $\cos 0 = 1$. At time $\pi/2$, the ball reaches the top. The height is $\sin \pi/2 = 1$ and the upward velocity is $\cos \pi/2 = 0$. At that instant the ball is not moving up or down.

The horizontal velocity contains a minus sign. At first the ball travels to the *left*. The value of x is $\cos t$, but *the speed in the x direction is* $-\sin t$. Half of trigonometry is in that figure (the good half), and you see how $\sin^2 t + \cos^2 t = 1$ is so basic. That equation applies to position and velocity, at every time.

Application of plane geometry: The right triangles in Figure 1.16 are the same size and shape. They look congruent and they are—the angle t above the ball equals the angle t at the center. That is because the three angles at the ball add to $180°$.

OSCILLATION: UP AND DOWN MOTION

We now use circular motion to study *straight-line motion*. That line will be the y axis. Instead of a ball going around a circle, a mass will move up and down. It oscillates between $y = 1$ and $y = -1$. **The mass is the "shadow of the ball,"** as we explain in a moment.

There is a jumpy oscillation that we do not want, with $v = 1$ and $v = -1$. That "bang-bang" velocity is like a billiard ball, bouncing between two walls without slowing down. If the distance between the walls is 2, then at $t = 4$ the ball is back to the start. The distance graph is a zigzag (or sawtooth) from Section 1.2.

We prefer a smoother motion. Instead of velocities that jump between $+1$ and -1, a real oscillation *slows down to zero* and gradually builds up speed again. The mass is on a spring, which pulls it back. The velocity drops to zero as the spring is fully stretched. Then v is negative, as the mass goes the same distance in the opposite direction. **Simple harmonic motion** is the most important back and forth motion, while $f = vt$ and $f = \frac{1}{2}at^2$ are the most important one-way motions.

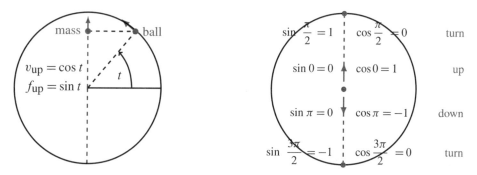

Fig. 1.17 Circular motion of the ball and harmonic motion of the mass (its shadow).

How do we describe this oscillation ? The best way is to match it with the ball on the circle. **The height of the ball will be the height of the mass**. The "shadow of the ball" goes up and down, level with the ball. As the ball passes the top of the circle, the mass stops at the top and starts down. As the ball goes around the bottom, the mass stops and turns back up the y axis. Halfway up (or down), the speed is 1.

Figure 1.17a shows the mass at a typical time t. The height is $y = f(t) = \sin t$, level with the ball. This height oscillates between $f = 1$ and $f = -1$. But the mass does not move with constant speed. ***The speed of the mass is changing although the speed of the ball is always*** 1. The time for a full cycle is still 2π, but within that cycle the mass speeds up and slows down. The problem is to find the changing velocity v. Since the distance is $f = \sin t$, the velocity will be the *slope of the sine curve*.

THE SLOPE OF THE SINE CURVE

At the top and bottom ($t = \pi/2$ and $t = 3\pi/2$) the ball changes direction and $v = 0$. *The slope at the top and bottom of the sine curve is zero.*† At time zero, when the ball is going straight up, the slope of the sine curve is $v = 1$. At $t = \pi$, when the ball and mass and f-graph are going down, the velocity is $v = -1$. The mass goes fastest at the center. The mass goes slowest (in fact it stops) when the height reaches a maximum or minimum. The velocity triangle yields v at every time t.

To find the upward velocity of the mass, look at the upward velocity of the ball. Those velocities are the same! The mass and ball stay level, and we know v from circular motion: ***The upward velocity is*** $v = \cos t$.

Figure 1.18 shows the result we want. On the right, $f = \sin t$ gives the height. On the left is the velocity $v = \cos t$. That velocity is the slope of the f-curve. The height and velocity (red lines) are oscillating together, but they are out of phase—just as the position triangle and velocity triangle were at right angles. This is absolutely fantastic, that in calculus the two most famous functions of trigonometry form a pair: ***The slope of the sine curve is given by the cosine curve***.

> ***When the distance is*** $f(t) = \sin t$, ***the velocity is*** $v(t) = \cos t$.

Admission of guilt: The slope of $\sin t$ was not computed in the standard way. Previously we compared $(t + h)^2$ with t^2, and divided that distance by h. This average velocity approached the slope $2t$ as h became small. ***For*** $\sin t$ ***we could have done the same***:

$$\text{average velocity} = \frac{\text{change in } \sin t}{\text{change in } t} = \frac{\sin(t + h) - \sin t}{h}. \tag{1}$$

This is where we need the formula for $\sin(t + h)$, coming soon. Somehow the ratio in (1) should approach $\cos t$ as $h \to 0$. (It does.) The sine and cosine fit the same pattern as t^2 and $2t$—our shortcut was to watch the shadow of motion around a circle.

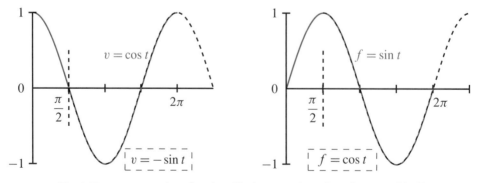

Fig. 1.18 $v = \cos t$ when $f = \sin t$ (blue); $v = -\sin t$ when $f = \cos t$ (black).

---

†That looks easy but you will see later that it is extremely important. ***At a maximum or minimum the slope is zero***. The curve levels off.

Question 1 *What if the ball goes twice as fast, to reach angle* $2t$ *at time* t *?*

Answer   The speed is now 2. The time for a full circle is only π. The ball's position is $x = \cos 2t$ and $y = \sin 2t$. The velocity is still tangent to the circle—but the tangent is at angle $2t$ where the ball is. Therefore $\cos 2t$ enters the upward velocity and $-\sin 2t$ enters the horizontal velocity. The difference is that *the velocity triangle is twice as big*. The upward velocity is not $\cos 2t$ but $2\cos 2t$. The horizontal velocity is $-2\sin 2t$. Notice these 2's!

Question 2 *What is the area under the cosine curve from* $t = 0$ *to* $t = \pi/2$?

You can answer that, if you accept the Fundamental Theorem of Calculus—*computing areas is the opposite of computing slopes*. The slope of $\sin t$ is $\cos t$, so the area under $\cos t$ is the increase in $\sin t$. No reason to believe that yet, but we use it anyway.

From $\sin 0 = 0$ to $\sin \pi/2 = 1$, the increase is 1. Please realize the power of calculus. No other method could compute the area under a cosine curve so fast.

THE SLOPE OF THE COSINE CURVE

I cannot resist uncovering another distance and velocity (another f-v pair) with no extra work. This time f is the cosine. The time clock starts *at the top of the circle*. The old time $t = \pi/2$ is now $t = 0$. The dotted lines in Figure 1.18 show the new start. But the shadow has exactly the same motion—the ball keeps going around the circle, and the mass follows it up and down. The f-graph and v-graph are still correct, both with a time shift of $\pi/2$.

The new f-graph is the cosine. The new v-graph is **minus the sine**. The slope of the cosine curve follows the *negative* of the sine curve. That is another famous pair, twins of the first:

> *When the distance is* $f(t) = \cos t$, *the velocity is* $v(t) = -\sin t$.

You could see that coming, by watching the ball go left and right (instead of up and down). Its distance across is $f = \cos t$. Its velocity across is $v = -\sin t$. That twin pair completes the calculus in Chapter 1 (trigonometry to come). We review the ideas:

> v is the **velocity**
>   the **slope** of the distance curve
>   the **limit** of average velocity over a short time
>   the **derivative** of f.
> f is the **distance**
>   the **area** under the velocity curve
>   the **limit** of total distance over many short times
>   the **integral** of v.

Differential calculus: *Compute v from f*.   *Integral calculus*: *Compute f from v*.

With constant velocity, f equals vt. With constant acceleration, $v = at$ and $f = \frac{1}{2}at^2$. In harmonic motion, $v = \cos t$ and $f = \sin t$. One part of our goal is to extend that list—for which we need the tools of calculus. Another and more important part is to put these ideas to use.

Before the chapter ends, may I add a note about the book and the course? The book is more personal than usual, and I hope readers will approve. What I write is very close to what I would say, if you were in this room. The sentences are spoken before they are written.† Calculus is alive and moving forward—it needs to be taught that way.

One new part of the subject has come with the computer. It works with a finite step h, not an "infinitesimal" limit. What it can do, it does quickly—even if it cannot find exact slopes or areas. The result is an overwhelming growth in the range of problems that can be solved. We landed on the moon because f and v were so accurate. (The moon's orbit has sines and cosines, the spacecraft starts with $v = at$ and $f = \frac{1}{2}at^2$. Only the computer can account for the atmosphere and the sun's gravity and the changing mass of the spacecraft.) **_Modern mathematics is a combination of exact formulas and approximate computations_**. Neither part can be ignored, and I hope you will see numerically what we derive algebraically. The exercises are to help you master both parts.

The course has made a quick start—not with an abstract discussion of sets or functions or limits, but with the concrete questions that led to those ideas. You have seen a distance function f and a limit v of average velocities. We will meet more functions and more limits (and their definitions!) but it is crucial to study important examples early. There is a lot to do, but the course has definitely begun.

1.4 EXERCISES

Read-through questions

A ball at angle t on the unit circle has coordinates $x =$ __a__ and $y =$ __b__. It completes a full circle at $t =$ __c__. Its speed is __d__. Its velocity points in the direction of the __e__, which is __f__ to the radius coming out from the center. The upward velocity is __g__ and the horizontal velocity is __h__.

A mass going up and down level with the ball has height $f(t) =$ __i__. This is called simple __j__ motion. The velocity is $v(t) =$ __k__. When $t = \pi/2$ the height is $f =$ __l__ and the velocity is $v =$ __m__. If a speeded-up mass reaches $f = \sin 2t$ at time t, its velocity is $v =$ __n__. A shadow traveling _under_ the ball has $f = \cos t$ and $v =$ __o__. When f is distance $=$ area $=$ integral, v is __p__ $=$ __q__ $=$ __r__.

1 For a ball going around a unit circle with speed 1,

    (a) how long does it take for 5 revolutions?

    (b) at time $t = 3\pi/2$ where is the ball?

    (c) at $t = 22$ where is the ball (approximately)?

2 For the same motion find the exact x and y coordinates at $t = 2\pi/3$. At what time would the ball hit the x axis, if it goes off on the tangent at $t = 2\pi/3$?

3 A ball goes around a circle of radius 4. At time t (when it reaches angle t) find

    (a) its x and y coordinates

    (b) the speed and the distance traveled

    (c) the vertical and horizontal velocity.

4 On a circle of radius R find the x and y coordinates at time t (and angle t). Draw the velocity triangle and find the x and y velocities.

5 A ball travels around a unit circle (radius 1) with speed 3, starting from angle zero. At time t,

    (a) what angle does it reach?

    (b) what are its x and y coordinates?

    (c) what are its x and y velocities? This part is harder.

6 If another ball stays $\pi/2$ radians ahead of the ball with speed 3, find its angle, its x and y coordinates, and its vertical velocity at time t.

7 A mass moves on the x axis under or over the original ball (on the unit circle with speed 1). What is the position $x = f(t)$? Find x and v at $t = \pi/4$. Plot x and v up to $t = \pi$.

---

†On television you know immediately when the words are live. The same with writing.

8 Does the new mass (under or over the ball) meet the old mass (level with the ball)? What is the distance between the masses at time t?

9 Draw graphs of $f(t) = \cos 3t$ and $\cos 2\pi t$ and $2\pi \cos t$, marking the time axes. How long until each f repeats?

10 Draw graphs of $f = \sin(t + \pi)$ and $v = \cos(t + \pi)$. This oscillation stays level with what ball?

11 Draw graphs of $f = \sin(\pi/2 - t)$ and $v = -\cos(\pi/2 - t)$. This oscillation stays level with a ball going which way starting where?

12 Draw a graph of $f(t) = \sin t + \cos t$. Estimate its greatest height (maximum f) and the time it reaches that height. By computing f^2 check your estimate.

13 How fast should you run across the circle to meet the ball again? It travels at speed 1.

14 A mass falls from the top of the unit circle when the ball of speed 1 passes by. What acceleration a is necessary to meet the ball at the bottom?

Find the area under $v = \cos t$ from the change in $f = \sin t$:

15 from $t = 0$ to $t = \pi$

16 from $t = 0$ to $t = \pi/6$

17 from $t = 0$ to $t = 2\pi$

18 from $t = \pi/2$ to $t = 3\pi/2$.

19 The distance curve $f = \sin 4t$ yields the velocity curve $v = 4\cos 4t$. Explain both 4's.

20 The distance curve $f = 2\cos 3t$ yields the velocity curve $v = -6\sin 3t$. Explain the -6.

21 The velocity curve $v = \cos 4t$ yields the distance curve $f = \frac{1}{4}\sin 4t$. Explain the $\frac{1}{4}$.

22 The velocity $v = 5\sin 5t$ yields what distance?

23 Find the slope of the sine curve at $t = \pi/3$ from $v = \cos t$. Then find an average slope by dividing $\sin \pi/2 - \sin \pi/3$ by the time difference $\pi/2 - \pi/3$.

24 The slope of $f = \sin t$ at $t = 0$ is $\cos 0 = 1$. Compute average slopes $(\sin t)/t$ for $t = 1, .1, .01, .001$.

The ball at $x = \cos t, y = \sin t$ circles (1) counterclockwise (2) with radius 1 (3) starting from $x = 1, y = 0$ (4) at speed 1. Find (1)(2)(3)(4) for the motions 25–30.

25 $x = \cos 3t, \ y = -\sin 3t$

26 $x = 3\cos 4t, \ y = 3\sin 4t$

27 $x = 5\sin 2t, \ y = 5\cos 2t$

28 $x = 1 + \cos t, \ y = \sin t$

29 $x = \cos(t + 1), \ y = \sin(t + 1)$

30 $x = \cos(-t), \ y = \sin(-t)$

The oscillation $x = 0, y = \sin t$ goes (1) up and down (2) between -1 and 1 (3) starting from $x = 0, y = 0$ (4) at velocity $v = \cos t$. Find (1)(2)(3)(4) for the oscillations 31–36.

31 $x = \cos t, \ y = 0$

32 $x = 0, \ y = \sin 5t$

33 $x = 0, \ y = 2\sin(t + \theta)$

34 $x = \cos t, \ y = \cos t$

35 $x = 0, \ y = -2\cos \frac{1}{2}t$

36 $x = \cos^2 t, \ y = \sin^2 t$

37 If the ball on the unit circle reaches t *degrees* at time t, find its position and speed and upward velocity.

38 Choose the number k so that $x = \cos kt, y = \sin kt$ completes a rotation at $t = 1$. Find the speed and upward velocity.

39 If a pitcher doesn't pause before starting to throw, a balk is called. The American League decided mathematically that there is always a stop between backward and forward motion, even if the time is too short to see it. (Therefore no balk.) Is that true?

1.5 A Review of Trigonometry

Trigonometry begins with a right triangle. The size of the triangle is not as important as the angles. We focus on one particular angle—call it θ—and on the *ratios* between the three sides x, y, r. The ratios don't change if the triangle is scaled to another size. Three sides give six ratios, which are the basic functions of trigonometry:

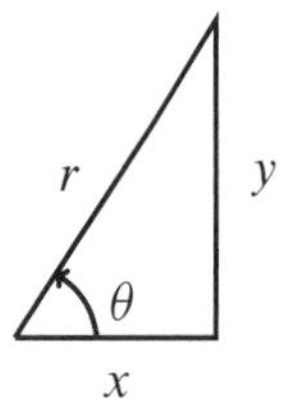

$$\cos \theta = \frac{x}{r} = \frac{\text{near side}}{\text{hypotenuse}} \qquad \sec \theta = \frac{r}{x} = \frac{1}{\cos \theta}$$

$$\sin \theta = \frac{y}{r} = \frac{\text{opposite side}}{\text{hypotenuse}} \qquad \csc \theta = \frac{r}{y} = \frac{1}{\sin \theta}$$

$$\tan \theta = \frac{y}{x} = \frac{\text{opposite side}}{\text{near side}} \qquad \cot \theta = \frac{x}{y} = \frac{1}{\tan \theta}$$

Fig. 1.19

Of course those six ratios are not independent. The three on the right come directly from the three on the left. And the tangent is the sine divided by the cosine:

$$\tan \theta = \frac{\sin \theta}{\cos \theta} = \frac{y/r}{x/r} = \frac{y}{x}.$$

Note that "tangent of an angle" and "tangent to a circle" and "tangent line to a graph" are different uses of the same word. As the cosine of θ goes to zero, the tangent of θ goes to infinity. The side x becomes zero, θ approaches $90°$, and the triangle is infinitely steep. The sine of $90°$ is $y/r = 1$.

Triangles have a serious limitation. They are excellent for angles up to $90°$, and they are OK up to $180°$, but after that they fail. We cannot put a $240°$ angle into a triangle. Therefore we change now to a circle.

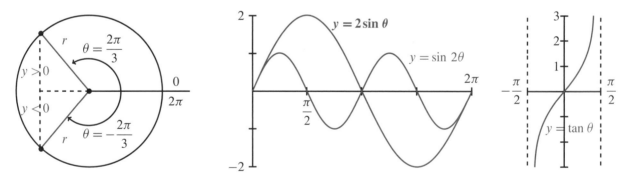

Fig. 1.20   Trigonometry on a circle. Compare $2 \sin \theta$ with $\sin 2\theta$ and $\tan \theta$ (periods $2\pi, \pi, \pi$).

Angles are measured from the positive x axis (counterclockwise). Thus $90°$ is straight up, $180°$ is to the left, and $360°$ is in the same direction as $0°$. (Then $450°$ is the same as $90°$.) Each angle yields a point on the circle of radius r. The coordinates x and y of that point can be negative (*but never r*). As the point goes around the circle, the six ratios $\cos \theta, \sin \theta, \tan \theta, \ldots$ trace out six graphs. The cosine waveform is the same as the sine waveform—just shifted by $90°$.

One more change comes with the move to a circle. Degrees are out. Radians are in. The distance around the whole circle is $2\pi r$. The distance around to other points is θr. **We measure the angle by that multiple θ.** For a half-circle the distance is πr, so the angle is π radians—which is 180°. A quarter-circle is $\pi/2$ radians or 90°. **The distance around to angle θ is r times θ.**

When $r = 1$ this is the ultimate in simplicity: *The distance is θ.* A 45° angle is $\frac{1}{8}$ of a circle and $2\pi/8$ radians—and the length of the circular arc is $2\pi/8$. Similarly for 1°:

$$360° = 2\pi \text{ radians} \qquad 1° = 2\pi/360 \text{ radians} \qquad 1 \text{ radian} = 360/2\pi \text{ degrees}.$$

An angle going clockwise is *negative.* The angle $-\pi/3$ is $-60°$ and takes us $\frac{1}{6}$ of the *wrong* way around the circle. What is the effect on the six functions?

Certainly the radius r is not changed when we go to $-\theta$. Also x is not changed (see Figure 1.20a). But y reverses sign, because $-\theta$ is below the axis when $+\theta$ is above. This change in y affects y/r and y/x but not x/r:

$$\cos(-\theta) = \cos\theta \qquad \sin(-\theta) = -\sin\theta \qquad \tan(-\theta) = -\tan\theta.$$

The cosine is *even* (no change). The sine and tangent are *odd* (change sign).

The same point is $\frac{5}{6}$ of the *right* way around. Therefore $\frac{5}{6}$ of 2π radians (or 300°) gives the same direction as $-\pi/3$ radians or $-60°$. **A difference of 2π makes no difference to x, y, r.** Thus $\sin\theta$ and $\cos\theta$ and the other four functions have period 2π. We can go five times or a hundred times around the circle, adding 10π or 200π to the angle, and the six functions repeat themselves.

EXAMPLE Evaluate the six trigonometric functions at $\theta = 2\pi/3$ (or $\theta = -4\pi/3$).

This angle is shown in Figure 1.20a (where $r = 1$). The ratios are

$$\cos\theta = x/r = -1/2 \qquad \sin\theta = y/r = \sqrt{3}/2 \qquad \tan\theta = y/x = -\sqrt{3}$$

$$\sec\theta = -2 \qquad \csc\theta = 2/\sqrt{3} \qquad \cot\theta = -1/\sqrt{3}$$

Those numbers illustrate basic facts about the sizes of four functions:

$$|\cos\theta| \leqslant 1 \qquad |\sin\theta| \leqslant 1 \qquad |\sec\theta| \geqslant 1 \qquad |\csc\theta| \geqslant 1.$$

The tangent and cotangent can fall anywhere, as long as $\cot\theta = 1/\tan\theta$.

The numbers reveal more. The tangent $-\sqrt{3}$ is the ratio of sine to cosine. The secant -2 is $1/\cos\theta$. Their squares are 3 and 4 (differing by 1). That may not seem remarkable, but it is. There are three relationships in the *squares* of those six numbers, and they are the key identities of trigonometry:

$$\cos^2\theta + \sin^2\theta = 1 \qquad 1 + \tan^2\theta = \sec^2\theta \qquad \cot^2\theta + 1 = \csc^2\theta$$

Everything flows from the Pythagoras formula $x^2 + y^2 = r^2$. Dividing by r^2 gives $(x/r)^2 + (y/r)^2 = 1$. That is $\cos^2\theta + \sin^2\theta = 1$. Dividing by x^2 gives the second identity, which is $1 + (y/x)^2 = (r/x)^2$. Dividing by y^2 gives the third. All three will be needed throughout the book—and the first one has to be unforgettable.

DISTANCES AND ADDITION FORMULAS

To compute the distance between points we stay with Pythagoras. The points are in Figure 1.21a. They are known by their x and y coordinates, and d is the distance between them. The third point completes a right triangle.

For the x distance along the bottom we don't need help. It is $x_2 - x_1$ (or $|x_2 - x_1|$ since distances can't be negative). The distance up the side is $|y_2 - y_1|$. Pythagoras immediately gives the distance d:

$$\textit{distance between points} = d = \sqrt{(x_2 - x_1)^2 + (y_2 - y_1)^2}. \qquad (1)$$

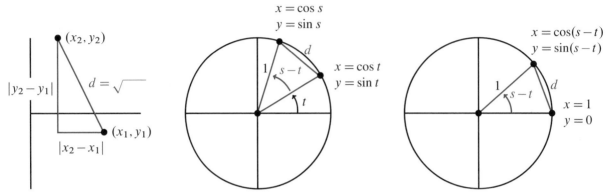

Fig. 1.21 Distance between points and equal distances in two circles.

By applying this distance formula in two identical circles, we discover the cosine of $s - t$. (Subtracting angles is important.) In Figure 1.21b, the distance squared is

$$d^2 = (\text{change in } x)^2 + (\text{change in } y)^2$$
$$= (\cos s - \cos t)^2 + (\sin s - \sin t)^2. \qquad (2)$$

Figure 1.21c shows the same circle and triangle (but rotated). The same distance squared is

$$d^2 = (\cos(s - t) - 1)^2 + (\sin(s - t))^2. \qquad (3)$$

Now multiply out the squares in equations (2) and (3). Whenever $(\text{cosine})^2 + (\sin e)^2$ appears, replace it by 1. The distances are the same, so $(2) = (3)$:

$$(2) = 1 + 1 - 2 \cos s \cos t - 2 \sin s \sin t$$
$$(3) = 1 + 1 - 2 \cos(s - t).$$

After canceling $1 + 1$ and then -2, we have the "**addition formula**" for $\cos(s - t)$:

$$\textbf{The cosine of } s - t \textbf{ equals } \cos s \; \cos t + \sin s \; \sin t. \qquad (4)$$
$$\textbf{The cosine of } s + t \textbf{ equals } \cos s \; \cos t - \sin s \; \sin t. \qquad (5)$$

The easiest is $t = 0$. Then $\cos t = 1$ and $\sin t = 0$. The equations reduce to $\cos s = \cos s$.

To go from (4) to (5) in all cases, replace t by $-t$. No change in $\cos t$, but a "minus" appears with the sine. In the special case $s = t$, we have $\cos(t + t) = (\cos t)(\cos t) - (\sin t)(\sin t)$. This is a much-used formula for $\cos 2t$:

$$\textit{Double angle}: \cos 2t = \cos^2 t - \sin^2 t = 2\cos^2 t - 1 = 1 - 2\sin^2 t. \qquad (6)$$

I am constantly using $\cos^2 t + \sin^2 t = 1$, to switch between sines and cosines.

We also need addition formulas and double-angle formulas for the *sine* of $s - t$ and $s + t$ and $2t$. For that we connect sine to cosine, rather than $(\sin e)^2$ to $(\text{cosine})^2$. The connection goes back to the ratio y/r in our original triangle. This is the sine of the angle θ and also the cosine of the *complementary angle* $\pi/2 - \theta$:

$$\sin\theta = \cos(\pi/2 - \theta) \qquad \text{and} \qquad \cos\theta = \sin(\pi/2 - \theta). \tag{7}$$

The complementary angle is $\pi/2 - \theta$ because the two angles add to $\pi/2$ (a right angle). By making this connection in Problem 19, formulas (4–5–6) move from cosines to sines:

$$\sin(s - t) = \sin s \cos t - \cos s \sin t \tag{8}$$

$$\sin(s + t) = \sin s \cos t + \cos s \sin t \tag{9}$$

$$\sin 2t = \sin(t + t) = 2 \sin t \cos t \tag{10}$$

I want to stop with these ten formulas, even if more are possible. Trigonometry is full of identities that connect its six functions—basically because all those functions come from a single right triangle. The x, y, r ratios and the equation $x^2 + y^2 = r^2$ can be rewritten in many ways. But you have now seen the formulas that are needed by calculus.† They give derivatives in Chapter 2 and integrals in Chapter 5. And it is typical of our subject to add something of its own—a limit in which an angle approaches zero. *The essence of calculus is in that limit.*

Review of the ten formulas Figure 1.22 shows $d^2 = (0 - \frac{1}{2})^2 + (1 - \sqrt{3}/2)^2$.

$$\cos\frac{\pi}{6} = \cos\frac{\pi}{2}\cos\frac{\pi}{3} + \sin\frac{\pi}{2}\sin\frac{\pi}{3} \quad (s-t) \qquad \sin\frac{\pi}{6} = \sin\frac{\pi}{2}\cos\frac{\pi}{3} - \cos\frac{\pi}{2}\sin\frac{\pi}{3}$$

$$\cos\frac{5\pi}{6} = \cos\frac{\pi}{2}\cos\frac{\pi}{3} - \sin\frac{\pi}{2}\sin\frac{\pi}{3} \quad (s+t) \qquad \sin\frac{5\pi}{6} = \sin\frac{\pi}{2}\cos\frac{\pi}{3} + \cos\frac{\pi}{2}\sin\frac{\pi}{3}$$

$$\cos 2\frac{\pi}{3} = \cos^2\frac{\pi}{3} - \sin^2\frac{\pi}{3} \qquad\qquad (2t) \qquad \sin 2\frac{\pi}{3} = 2\sin\frac{\pi}{3}\cos\frac{\pi}{3}$$

$$\cos\frac{\pi}{6} = \sin\frac{\pi}{3} = \sqrt{3}/2 \qquad\qquad \left(\frac{\pi}{2} - \theta\right) \qquad \sin\frac{\pi}{6} = \cos\frac{\pi}{3} = 1/2$$

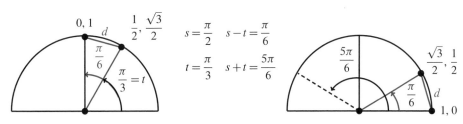

Fig. 1.22

Question 1 Draw graphs for equations $y = \sin 2x$, $y = 2\sin\pi x$, $y = \frac{1}{2}\cos 2\pi x$, $y = \sin x + \cos x$, and mark three points.

Question 2 Which of the six trigonometric functions are infinite at what angles?

Question 3 Draw rough graphs or computer graphs of $t \sin t$ and $\sin 4t \sin t$ from 0 to 2π.

---

† Calculus turns (6) around to $\cos^2 t = \frac{1}{2}(1 + \cos 2t)$ and $\sin^2 t = \frac{1}{2}(1 - \cos 2t)$.

Read-through questions

Starting with a __a__ triangle, the six basic functions are the __b__ of the sides. Two ratios (the cosine x/r and the __c__) are below 1. Two ratios (the secant r/x and the __d__) are above 1. Two ratios (the __e__ and the __f__) can take any value. The six functions are defined for all angles θ, by changing from a triangle to a __g__.

The angle θ is measured in __h__. A full circle is $\theta =$ __i__, when the distance around is $2\pi r$. The distance to angle θ is __j__. All six functions have period __k__. Going clockwise changes the sign of θ and __l__ and __m__. Since $\cos(-\theta) = \cos\theta$, the cosine is __n__.

Coming from $x^2 + y^2 = r^2$ are the three identities $\sin^2\theta + \cos^2\theta = 1$ and __o__ and __p__. (Divide by r^2 and __q__ and __r__.) The distance from $(2,5)$ to $(3,4)$ is $d =$ __s__. The distance from $(1,0)$ to $(\cos(s-t), \sin(s-t))$ leads to the addition formula $\cos(s-t) =$ __t__. Changing the sign of t gives $\cos(s+t) =$ __u__. Choosing $s = t$ gives $\cos 2t =$ __v__ or __w__. Therefore $\frac{1}{2}(1 + \cos 2t) =$ __x__, a formula needed in calculus.

1 In a $60 - 60 - 60$ triangle show why $\sin 30° = \frac{1}{2}$.

2 Convert $\pi, 3\pi, -\pi/4$ to degrees and $60°, 90°, 270°$ to radians. What angles between 0 and 2π correspond to $\theta = 480°$ and $\theta = -1°$?

3 Draw graphs of $\tan\theta$ and $\cot\theta$ from 0 to 2π. What is their (shortest) period?

4 Show that $\cos 2\theta$ and $\cos^2\theta$ have period π and draw them on the same graph.

5 At $\theta = 3\pi/2$ compute the six basic functions and check $\cos^2\theta + \sin^2\theta$, $\sec^2\theta - \tan^2\theta$, $\csc 2\theta - \cot^2\theta$.

6 Prepare a table showing the values of the six basic functions at $\theta = 0, \pi/4, \pi/3, \pi/2, \pi$.

7 The area of a circle is πr^2. What is the area of the sector that has angle θ? It is a fraction _____ of the whole area.

8 Find the distance from $(1,0)$ to $(0,1)$ along (a) a straight line (b) a quarter-circle (c) a semicircle centered at $\left(\frac{1}{2}, \frac{1}{2}\right)$.

9 Find the distance d from $(1,0)$ to $\left(\frac{1}{2}, \sqrt{3}/2\right)$ and show on a circle why $6d$ is less than 2π.

10 In Figure 1.22 compute d^2 and (with calculator) $12d$. Why is $12d$ close to and below 2π?

11 Decide whether these equations are true or false:

(a) $\dfrac{\sin\theta}{1 - \cos\theta} = \dfrac{1 + \cos\theta}{\sin\theta}$

(b) $\dfrac{\sec\theta + \csc\theta}{\tan\theta + \cot\theta} = \sin\theta + \cos\theta$

(c) $\cos\theta - \sec\theta = \sin\theta\tan\theta$

(d) $\sin(2\pi - \theta) = \sin\theta$

12 Simplify $\sin(\pi - \theta), \cos(\pi - \theta), \sin(\pi/2 + \theta), \cos(\pi/2 + \theta)$.

13 From the formula for $\cos(2t + t)$ find $\cos 3t$ in terms of $\cos t$.

14 From the formula for $\sin(2t + t)$ find $\sin 3t$ in terms of $\sin t$.

15 By averaging $\cos(s - t)$ and $\cos(s + t)$ in (4–5) find a formula for $\cos s \cos t$. Find a similar formula for $\sin s \sin t$.

16 Show that $(\cos t + i \sin t)^2 = \cos 2t + i \sin 2t$, if $i^2 = -1$.

17 Draw $\cos\theta$ and $\sec\theta$ on the same graph. Find all points where $\cos\theta = \sec\theta$.

18 Find all angles s and t between 0 and 2π where $\sin(s + t) = \sin s + \sin t$.

19 Complementary angles have $\sin\theta = \cos(\pi/2 - \theta)$. Write $\sin(s + t)$ as $\cos(\pi/2 - s - t)$ and apply formula (4) with $\pi/2 - s$ instead of s. In this way derive the addition formula (9).

20 If formula (9) is true, how do you prove (8)?

21 Check the addition formulas (4–5) and (8–9) for $s = t = \pi/4$.

22 Use (5) and (9) to find a formula for $\tan(s + t)$.

In 23–28 find *every* θ that satisfies the equation.

23 $\sin\theta = -1$ **24** $\sec\theta = -2$

25 $\sin\theta = \cos\theta$ **26** $\sin\theta = \theta$

27 $\sec^2\theta + \csc^2\theta = 1$ **28** $\tan\theta = 0$

29 Rewrite $\cos\theta + \sin\theta$ as $\sqrt{2}\sin(\theta + \phi)$ by choosing the correct "phase angle" ϕ. (Make the equation correct at $\theta = 0$. Square both sides to check.)

30 Match $a\sin x + b\cos x$ with $A\sin(x + \phi)$. From equation (9) show that $a = A\cos\phi$ and $b = A\sin\phi$. Square and add to find $A =$ _____. Divide to find $\tan\phi = b/a$.

31 Draw the base of a triangle from the origin $O = (0,0)$ to $P = (a,0)$. The third corner is at $Q = (b\cos\theta, b\sin\theta)$. What are the side lengths OP and OQ? From the distance formula (1) show that the side PQ has length

$$d^2 = a^2 + b^2 - 2ab\cos\theta \quad \text{(law of cosines)}.$$

32 Extend the same triangle to a parallelogram with its fourth corner at $R = (a + b\cos\theta, b\sin\theta)$. Find the length squared of the other diagonal OR.

1.6 A Thousand Points of Light

The figures drawn below show $y = \sin n$. This is very different from $y = \sin x$. The graph of $\sin x$ is one continuous curve. By the time it reaches $x = 10{,}000$, the curve has gone up and down $10{,}000/2\pi$ times. Those 1591 oscillations would be so crowded that you couldn't see anything. The graph of $\sin n$ has picked **10,000 points** from the curve—and for some reason those points seem to lie on more than 40 separate sine curves.

The second graph shows the first 1000 points. They *don't* seem to lie on sine curves. Most people see hexagons. *But they are the same thousand points!* It is hard to believe that the graphs are the same, but I have learned what to do. ***Tilt the second graph and look from the side at a narrow angle.*** Now the first graph appears. I believe you will see "diamonds." The narrow angle compresses the x axis—back to the scale of the first graph.

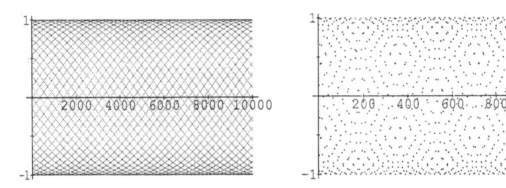

The effect of scale is something we don't think of. We understand it for maps. Computers can zoom in or zoom out—those are changes of scale. What our eyes see depends on what is "close." We think we see sine curves in the 10,000 point graph, and they raise several questions:

1. Which points are near $(0,0)$?
2. How many sine curves are there?
3. Where does the middle curve, going upward from $(0,0)$, come back to zero?

A point near $(0,0)$ really means that $\sin n$ is close to zero. That is certainly not true of $\sin 1$ (1 is one radian!). In fact $\sin 1$ is up the axis at .84, at the start of the seventh sine curve. Similarly $\sin 2$ is .91 and $\sin 3$ is .14. (The numbers 3 and .14 make us think of π. The sine of 3 equals the sine of $\pi - 3$. Then $\sin .14$ is near .14.) Similarly $\sin 4, \sin 5, \ldots, \sin 21$ are not especially close to zero.

The first point to come close is $\sin 22$. This is because $22/7$ is near π. Then 22 is close to 7π, whose sine is zero:

$$\sin 22 = \sin(7\pi - 22) \approx \sin(-.01) \approx -.01.$$

That is the first point to the right of $(0,0)$ and slightly below. You can see it on graph 1, and more clearly on graph 2. It begins a curve downward.

The next point to come close is $\sin 44$. This is because 44 is just past 14π.

$$44 \approx 14\pi + .02 \qquad \text{so} \qquad \sin 44 \approx \sin .02 \approx .02.$$

This point $(44, \sin 44)$ ***starts the middle sine curve***. Next is $(88, \sin 88)$.

Now we know something. ***There are* 44 *curves*.** They begin near the heights sin 0, sin 1, ..., sin 43. Of these 44 curves, 22 start upward and 22 start downward. I was confused at first, because I could only find 42 curves. The reason is that sin 11 equals −0.99999 and sin 33 equals .9999. Those are so close to the bottom and top that you can't see their curves. The sine of 11 is near −1 because sin 22 is near zero. It is almost impossible to follow a single curve past the top—coming back down it is not the curve you think it is.

The points on the middle curve are at $n = 0$ and 44 and 88 and every number $44N$. Where does that curve come back to zero? In other words, when does $44N$ come *very close* to a multiple of π? We know that 44 is $14\pi + .02$. More exactly 44 is $14\pi + .0177$. So we multiply .0177 until we reach π:

$$\text{if } N = \pi/.0177 \quad \text{then} \quad 44N = (14\pi + .0177)N = 14\pi N + \pi.$$

This gives $N = 177.5$. At that point $44N = 7810$. *This is half the period of the sine curve.* The sine of 7810 is very near zero.

If you follow the middle sine curve, you will see it come back to zero above 7810. The actual points on that curve have $n = 44 \cdot 177$ and $n = 44 \cdot 178$, with sines just above and below zero. Halfway between is $n = 7810$. ***The equation for the middle sine curve is*** $y = \sin(\pi x/7810)$. Its period is $15,620$—beyond our graph.

Question The fourth point on that middle curve looks the same as the fourth point coming down from sin 3. What is this "double point?"
Answer 4 times 44 is 176. On the curve going up, the point is $(176, \sin 176)$. On the curve coming down it is $(179, \sin 179)$. ***The sines of*** 176 ***and*** 179 ***differ only by*** .00003.

The second graph spreads out this double point. Look above 176 and 179, at the center of a hexagon. You can follow the sine curve all the way across graph 2.

Only a little question remains. Why does graph 2 have hexagons? *I don't know.* The problem is with your eyes. To understand the hexagons, Doug Hardin plotted points on straight lines as well as sine curves. Graph 3 shows $y = $ fractional part of $n/2\pi$. Then he made a second copy, turned it over, and placed it on top. That produced graph 4—with hexagons. Graphs 3 and 4 are on the next page.

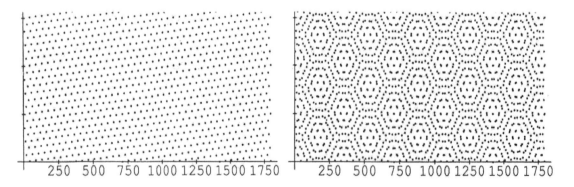

This is called a ***Moiré pattern***. If you can get a transparent copy of graph 3, and turn it slowly over the original, you will see fantastic hexagons. They come from interference between periodic patterns—in our case $44/7$ and $25/4$ and $19/3$ are near 2π. This interference is an enemy of printers, when color screens don't line up. It can cause vertical lines on a TV. Also in making cloth, operators get dizzy from seeing Moiré patterns move. There are good applications in engineering and optics—but we have to get back to calculus.

CHAPTER 2

Derivatives

This chapter begins with the definition of the derivative. Two examples were in Chapter 1. When the distance is t^2, the velocity is $2t$. When $f(t) = \sin t$ we found $v(t) = \cos t$. *The velocity is now called the **derivative** of $f(t)$.* As we move to a more formal definition and new examples, we use new symbols f' and df/dt for the derivative.

2A At time t, the *derivative* $f'(t)$ or df/dt or $v(t)$ is

$$f'(t) = \lim_{\Delta t \to 0} \frac{f(t + \Delta t) - f(t)}{\Delta t}. \tag{1}$$

The ratio on the right is the average velocity over a short time Δt. The derivative, on the left side, is its limit as the step Δt (*delta t*) approaches zero.

Go slowly and look at each piece. The distance at time $t + \Delta t$ is $f(t + \Delta t)$. The distance at time t is $f(t)$. Subtraction gives the ***change in distance***, between those times. We often write Δf for this difference: $\Delta f = f(t + \Delta t) - f(t)$. ***The average velocity is the ratio*** $\Delta f / \Delta t$—change in distance divided by change in time.

The limit of the average velocity is the derivative, if this limit exists:

$$\frac{df}{dt} = \lim_{\Delta t \to 0} \frac{\Delta f}{\Delta t}. \tag{2}$$

This is the neat notation that Leibniz invented: $\Delta f / \Delta t$ *approaches* df/dt. Behind the innocent word "*limit*" is a process that this course will help you understand.

Note that Δf is not Δ times f! ***It is the change in*** f. Similarly Δt is not Δ times t. It is the time step, positive or negative and eventually small. To have a one-letter symbol we replace Δt by h.

The right sides of (1) and (2) contain average speeds. On the graph of $f(t)$, the distance *up* is divided by the distance *across*. That gives the average slope $\Delta f / \Delta t$.

The left sides of (1) and (2) are ***instantaneous*** speeds df/dt. They give the slope at the instant t. This is the derivative df/dt (when Δt and Δf shrink to zero). Look again at the calculation for $f(t) = t^2$:

$$\frac{\Delta f}{\Delta t} = \frac{f(t + \Delta t) - f(t)}{\Delta t} = \frac{t^2 + 2t\,\Delta t + (\Delta t)^2 - t^2}{\Delta t} = 2t + \Delta t. \qquad (3)$$

Important point: Those steps are taken before Δt goes to zero. **If we set $\Delta t = 0$ too soon, we learn nothing.** The ratio $\Delta f / \Delta t$ becomes $0/0$ (which is meaningless). The numbers Δf and Δt must approach zero together, not separately. Here their ratio is $2t + \Delta t$, the average speed.

To repeat: Success came by writing out $(t + \Delta t)^2$ and subtracting t^2 and dividing by Δt. Then and only then can we approach $\Delta t = 0$. The limit is the derivative $2t$.

There are several new things in formulas (1) and (2). Some are easy but important, others are more profound. The idea of a function we will come back to, and the definition of a limit. But the notations can be discussed right away. They are used constantly and you also need to know how to read them aloud:

$$f(t) = \text{``}f \text{ of } t\text{''} = \text{the value of the function } f \text{ at time } t$$
$$\Delta t = \text{``delta } t\text{''} = \text{the time step forward or backward from } t$$
$$f(t + \Delta t) = \text{``}f \text{ of } t \text{ plus delta } t\text{''} = \text{the value of } f \text{ at time } t + \Delta t$$
$$\Delta f = \text{``delta } f\text{''} = \text{the change } f(t + \Delta t) - f(t)$$
$$\Delta f / \Delta t = \text{``delta } f \text{ over delta } t\text{''} = \text{the average velocity}$$
$$f'(t) = \text{``}f \text{ prime of } t\text{''} = \text{the value of the derivative at time } t$$
$$df/dt = \text{``d } f \text{ d } t\text{''} = \text{the same as } f' \text{ (the instantaneous velocity)}$$
$$\lim_{\Delta \to 0} = \text{``limit as delta } t \text{ goes to zero''} = \text{the process that starts with}$$
$$\text{numbers } \Delta f / \Delta t \text{ and produces the number } df/dt.$$

From those last words you see what lies behind the notation df/dt. The symbol Δt indicates a nonzero (usually short) length of time. The symbol dt indicates an infinitesimal (even shorter) length of time. Some mathematicians work separately with df and dt, and df/dt is their ratio. For us df/dt is a single notation (don't cancel d and don't cancel Δ). The derivative df/dt is the limit of $\Delta f / \Delta t$. *When that notation df/dt is awkward, use f' or v.*

Remark   The notation hides one thing we should mention. The time step can be *negative* just as easily as positive. We can compute the average $\Delta f / \Delta t$ over a time interval *before* the time t, instead of after. This ratio also approaches df/dt.

The notation also hides another thing: **The derivative might not exist.** The averages $\Delta f / \Delta t$ might not approach a limit (it has to be the same limit going forward and backward from time t). In that case $f'(t)$ is not defined. At that instant there is no clear reading on the speedometer. This will happen in Example 2.

EXAMPLE 1   (Constant velocity $V = 2$) The distance f is V times t. The distance at time $t + \Delta t$ is V times $t + \Delta t$. **The difference Δf is V times Δt:**

$$\frac{\Delta f}{\Delta t} = \frac{V \Delta t}{\Delta t} = V \quad \text{so the limit is} \quad \frac{df}{dt} = V.$$

The derivative of Vt is V. The derivative of $2t$ is 2. The averages $\Delta f / \Delta t$ are always $V = 2$, in this exceptional case of a constant velocity.

EXAMPLE 2 Constant velocity 2 up to time $t = 3$, *then stop*.

For small times we still have $f(t) = 2t$. But after the stopping time, the distance is fixed at $f(t) = 6$. The graph is flat beyond time 3. Then $f(t + \Delta t) = f(t)$ and $\Delta f = 0$ and *the derivative of a constant function is zero*:

$$t > 3: \quad f'(t) = \lim_{\Delta t \to 0} \frac{f(t + \Delta t) - f(t)}{\Delta t} = \lim_{\Delta t \to 0} \frac{0}{\Delta t} = 0. \qquad (4)$$

In this example *the derivative is not defined at the instant when* $t = 3$. The velocity falls suddenly from 2 to zero. The ratio $\Delta f / \Delta t$ depends, at that special moment, on whether Δt is positive or negative. The average velocity *after* time $t = 3$ is zero. The average velocity *before* that time is 2. When the graph of f has a corner, the graph of v has a *jump*. It is a *step function*.

One new part of that example is the notation (df/dt or f' instead of v). Please look also at the third figure. It shows how the function takes t (on the left) to $f(t)$. Especially it shows Δt and Δf. At the start, $\Delta f / \Delta t$ is 2. After the stop at $t = 3$, all t's go to the same $f(t) = 6$. So $\Delta f = 0$ and $df/dt = 0$.

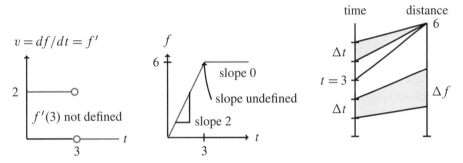

Fig. 2.1   The derivative is 2 then 0. It does not exist at $t = 3$.

THE DERIVATIVE OF $1/t$

Here is a completely different slope, for the "demand function" $f(t) = 1/t$. The demand is $1/t$ when the price is t. A high price t means a low demand $1/t$. Increasing the price reduces the demand. The calculus question is: *How quickly does* $1/t$ *change when* t *changes*? The "marginal demand" is the slope of the demand curve.

The big thing is to find the derivative of $1/t$ once and for all. It is $-1/t^2$.

EXAMPLE 3 $f(t) = \dfrac{1}{t}$ has $\Delta f = \dfrac{1}{t + \Delta t} - \dfrac{1}{t}$. This equals $\dfrac{t - (t + \Delta t)}{t(t + \Delta t)} = \dfrac{-\Delta t}{t(t + \Delta t)}$.

Divide by Δt and let $\Delta t \to 0$: $\dfrac{\Delta f}{\Delta t} = \dfrac{-1}{t(t + \Delta t)}$ approaches $\dfrac{df}{dt} = \dfrac{-1}{t^2}$.

Line 1 is algebra, line 2 is calculus. The first step in line 1 subtracts $f(t)$ from $f(t + \Delta t)$. The difference is $1/(t + \Delta t)$ minus $1/t$. The common denominator is t times $t + \Delta t$—this makes the algebra possible. We can't set $\Delta t = 0$ in line 2, until we have divided by Δt.

The average is $\Delta f / \Delta t = -1/t(t + \Delta t)$. Now set $\Delta t = 0$. The derivative is $-1/t^2$. Section 2.4 will discuss the first of many cases when substituting $\Delta t = 0$ is not possible, and the idea of a limit has to be made clearer.

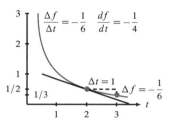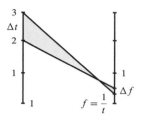

Fig. 2.2   Average slope is $-\frac{1}{6}$, true slope is $-\frac{1}{4}$. Increase in t produces decrease in f.

Check the algebra at $t = 2$ and $t + \Delta t = 3$. The demand $1/t$ drops from $1/2$ to $1/3$. The difference is $\Delta f = -1/6$, which agrees with $-1/(2)(3)$ in line 1. As the steps Δf and Δt get smaller, their ratio approaches $-1/(2)(2) = -1/4$.

This derivative is negative. The function $1/t$ is *decreasing*, and Δf is below zero. The graph is going *downward* in Figure 2.2, and its slope is negative:

*An increasing $f(t)$ **has positive slope**.   A decreasing $f(t)$ **has negative slope**.*

The slope $-1/t^2$ is very negative for small t. A price increase severely cuts demand.

The next figure makes a small but important point. There is nothing sacred about t. Other letters can be used—especially x. A quantity can depend on ***position instead of time***. The height changes as we go west. The area of a square changes as the side changes. Those are not affected by the passage of time, and there is no reason to use t. You will often see $y = f(x)$, with x across and y up—connected by a function f.

Similarly, f is not the only possibility. Not every function is named f! That letter is useful because it stands for the word function—but we are perfectly entitled to write $y(x)$ or $y(t)$ instead of $f(x)$ or $f(t)$. The distance up is a function of the distance across. This relationship "y of x" is all-important to mathematics.

The slope is also a function. Calculus is about two functions, $y(x)$ and dy/dx.

Question   If we add 1 to $y(x)$, what happens to the slope ? *Answer* Nothing.

Question   If we add 1 to the slope, what happens to the height ? *Answer* _____.

The symbols t and x represent ***independent variables***—they take any value they want to (in the domain). Once they are set, $f(t)$ and $y(x)$ are determined. Thus f and y represent ***dependent variables***—they *depend* on t and x. A change Δt produces a

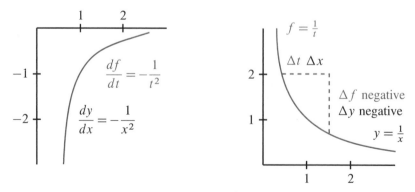

Fig. 2.3   The derivative of $1/t$ is $-1/t^2$. The slope of $1/x$ is $-1/x^2$.

change Δf. A change Δx produces Δy. The *in*dependent variable goes *in*side the parentheses in $f(t)$ and $y(x)$. It is not the letter that matters, it is the idea:

$$\text{independent variable } t \text{ or } x$$
$$\text{dependent variable } f \text{ or } g \text{ or } y \text{ or } z \text{ or } u$$
$$\text{derivative } df/dt \text{ or } df/dx \text{ or } dy/dx \text{ or } \cdots$$

The derivative dy/dx comes from [change in y] divided by [change in x]. The time step becomes a space step, forward or backward. The slope is the rate at which y changes with x. ***The derivative of a function is its "rate of change."***

I mention that physics books use $x(t)$ for distance. Darn it.

To emphasize the definition of a derivative, here it is again with y and x:

$$\frac{\Delta y}{\Delta x} = \frac{y(x + \Delta x) - y(x)}{\Delta x} = \frac{\text{distance up}}{\text{distance across}} \qquad \frac{dy}{dx} = \lim_{\Delta x \to 0} \frac{\Delta y}{\Delta x} = y'(x).$$

The notation $y'(x)$ pins down the point x where the slope is computed. In dy/dx that extra precision is omitted. This book will try for a reasonable compromise between logical perfection and ordinary simplicity. The notation $dy/dx(x)$ is not good; $y'(x)$ is better; when x is understood it need not be written in parentheses.

You are allowed to say that the function is $y = x^2$ and the derivative is $y' = 2x$— even if the strict notation requires $y(x) = x^2$ and $y'(x) = 2x$. You can even say that the function is x^2 and its derivative is $2x$ and its ***second derivative*** is 2—provided everybody knows what you mean.

Here is an example. It is a little early and optional but terrific. You get excellent practice with letters and symbols, and out come new derivatives.

EXAMPLE 4   If $u(x)$ has slope du/dx, what is the slope of $f(x) = (u(x))^2$?

From the derivative of x^2 this will give the derivative of x^4. In that case $u = x^2$ and $f = x^4$. First point: ***The derivative of u^2 is not*** $(du/dx)^2$. We do not square the derivative $2x$. To find the "square rule" we start as we have to—with $\Delta f = f(x + \Delta x) - f(x)$:

$$\Delta f = (u(x + \Delta x))^2 - (u(x))^2 = [u(x + \Delta x) + u(x)][u(x + \Delta x) - u(x)].$$

This algebra puts Δf in a convenient form. We factored $a^2 - b^2$ into $[a + b]$ times $[a - b]$. Notice that we don't have $(\Delta u)^2$. We have Δf, the change in u^2. Now divide by Δx and take the limit:

$$\frac{\Delta f}{\Delta x} = [u(x + \Delta x) + u(x)]\left[\frac{u(x + \Delta x) - u(x)}{\Delta x}\right] \text{ approaches } 2u(x)\frac{du}{dx}. \quad (5)$$

This is the *square rule*: ***The derivative of $(u(x))^2$ is $2u(x)$ times du/dx***. From the derivatives of x^2 and $1/x$ and $\sin x$ (all known) the examples give new derivatives.

EXAMPLE 5 $(u = x^2)$ The derivative of x^4 is $2u\, du/dx = 2(x^2)(2x) = 4x^3$.

EXAMPLE 6 $(u = 1/x)$ The derivative of $1/x^2$ is $2u\, du/dx = (2/x)(-1/x^2) = -2/x^3$.

EXAMPLE 7 $(u = \sin x,\ du/dx = \cos x)$ The derivative of $u^2 = \sin^2 x$ is $2 \sin x \cos x$.

Mathematics is really about ideas. The notation is created to express those ideas. Newton and Leibniz invented calculus independently, and Newton's friends spent a lot of time proving that he was first. He was, but it was Leibniz who thought of writing dy/dx—which caught on. It is the perfect way to suggest the limit of $\Delta y/\Delta x$. Newton was one of the great scientists of all time, and calculus was one of the great inventions of all time-but the notation must help. You now can write and speak about the derivative. What is needed is a longer list of functions and derivatives.

2.1 EXERCISES

Read-through questions

The derivative is the __a__ of $\Delta f/\Delta t$ as Δt approaches __b__. Here Δf equals __c__. The step Δt can be positive or __d__. The derivative is written v or __e__ or __f__. If $f(x) = 2x + 3$ and $\Delta x = 4$ then $\Delta f =$ __g__. If $\Delta x = -1$ then $\Delta f =$ __h__. If $\Delta x = 0$ then $\Delta f =$ __i__. The slope is not 0/0 but $df/dx =$ __j__.

The derivative does not exist where $f(t)$ has a __k__ and $v(t)$ has a __l__. For $f(t) = 1/t$ the derivative is __m__. The slope of $y = 4/x$ is $dy/dx =$ __n__. A decreasing function has a __o__ derivative. The __p__ variable is t or x and the __q__ variable is f or y. The slope of y^2 (is) (is not) $(dy/dx)^2$. The slope of $(u(x))^2$ is __r__ by the square rule. The slope of $(2x + 3)^2$ is __s__.

1 Which of the following numbers (*as is*) gives df/dt at time t? If in doubt test on $f(t) = t^2$.

(a) $\dfrac{f(t + \Delta t) - f(t)}{\Delta t}$

(b) $\lim\limits_{h \to 0} \dfrac{f(t + 2h) - f(t)}{2h}$

(c) $\lim\limits_{\Delta t \to 0} \dfrac{f(t - \Delta t) - f(t)}{-\Delta t}$

(d) $\lim\limits_{t \to 0} \dfrac{f(t + \Delta t) - f(t)}{\Delta t}$

2 Suppose $f(x) = x^2$. Compute each ratio and set $h = 0$:

(a) $\dfrac{f(x + h) - f(x)}{h}$

(b) $\dfrac{f(x + 5h) - f(x)}{5h}$

(c) $\dfrac{f(x + h) - f(x - h)}{2h}$

(d) $\dfrac{f(x + 1) - f(x)}{h}$

3 For $f(x) = 3x$ and $g(x) = 1 + 3x$, find $f(4 + h)$ and $g(4 + h)$ and $f'(4)$ and $g'(4)$. Sketch the graphs of f and g—why do they have the same slope?

4 Find three functions with the same slope as $f(x) = x^2$.

5 For $f(x) = 1/x$, sketch the graphs of $f(x) + 1$ and $f(x + 1)$. Which one has the derivative $-1/x^2$?

6 Choose c so that the line $y = x$ is tangent to the parabola $y = x^2 + c$. They have the same slope where they touch.

7 Sketch the curve $y(x) = 1 - x^2$ and compute its slope at $x = 3$.

8 If $f(t) = 1/t$, what is the average velocity between $t = \frac{1}{2}$ and $t = 2$? What is the average between $t = \frac{1}{2}$ and $t = 1$? What is the average (to one decimal place) between $t = \frac{1}{2}$ and $t = 101/200$?

9 Find $\Delta y/\Delta x$ for $y(x) = x + x^2$. Then find dy/dx.

10 Find $\Delta y/\Delta x$ and dy/dx for $y(x) = 1 + 2x + 3x^2$.

11 When $f(t) = 4/t$, simplify the difference $f(t + \Delta t) - f(t)$, divide by Δt, and set $\Delta t = 0$. The result is $f'(t)$.

12 Find the derivative of $1/t^2$ from $\Delta f(t) = 1/(t + \Delta t)^2 - 1/t^2$. Write Δf as a fraction with the denominator $t^2(t + \Delta t)^2$. Divide the numerator by Δt to find $\Delta f/\Delta t$. Set $\Delta t = 0$.

13 Suppose $f(t) = 7t$ to $t = 1$. Afterwards $f(t) = 7 + 9(t - 1)$.

(a) Find df/dt at $t = \frac{1}{2}$ and $t = \frac{3}{2}$.

(b) Why doesn't $f(t)$ have a derivative at $t = 1$?

14 Find the derivative of the derivative (the *second derivative*) of $y = 3x^2$. What is the third derivative?

15 Find numbers A and B so that the straight line $y = x$ fits smoothly with the curve $Y = A + Bx + x^2$ at $x = 1$. Smoothly means that $y = Y$ and $dy/dx = dY/dx$ at $x = 1$.

16 Find numbers A and B so that the horizontal line $y = 4$ fits smoothly with the curve $y = A + Bx + x^2$ at the point $x = 2$.

17 *True* (with reason) *or false* (with example):

(a) If $f(t) < 0$ then $df/dt < 0$.

(b) The derivative of $(f(t))^2$ is $2df/dt$.

(c) The derivative of $2f(t)$ is $2df/dt$.

(d) The derivative is the limit of Δf divided by the limit of Δt.

18 For $f(x) = 1/x$ the *centered difference* $f(x + h) - f(x - h)$ is $1/(x + h) - 1/(x - h)$. Subtract by using the common denominator $(x + h)(x - h)$. Then divide by $2h$ and set $h = 0$. Why divide by $2h$ to obtain the correct derivative?

19 Suppose $y = mx + b$ for negative x and $y = Mx + B$ for $x \geq 0$. The graphs meet if _____. The two slopes are _____. The slope at $x = 0$ is _____ (what is possible?).

20 The slope of $y = 1/x$ at $x = 1/4$ is $y' = -1/x^2 = -16$. At $h = 1/12$, which of these ratios is closest to -16?

$$\frac{y(x + h) - y(x)}{h} \qquad \frac{y(x) - y(x - h)}{h} \qquad \frac{y(x + h) - y(x - h)}{2h}$$

21 Find the average slope of $y = x^2$ between $x = x_1$ and $x = x_2$. What does this average approach as x_2 approaches x_1?

22 Redraw Figure 2.1 when $f(t) = 3 - 2t$ for $t \leqslant 2$ and $f(t) = -1$ for $t \geqslant 2$. Include df/dt.

23 Redraw Figure 2.3 for the function $y(x) = 1 - (1/x)$. Include dy/dx.

24 The limit of $0/\Delta t$ as $\Delta t \to 0$ is not $0/0$. Explain.

25 Guess the limits by an informal working rule. Set $\Delta t = 0.1$ and -0.1 and imagine Δt becoming smaller:

(a) $\dfrac{1 + \Delta t}{2 + \Delta t}$ (b) $\dfrac{|\Delta t|}{\Delta t}$

(c) $\dfrac{\Delta t + (\Delta t)^2}{\Delta t - (\Delta t)^2}$ (d) $\dfrac{t + \Delta t}{t - \Delta t}$

*26 Suppose $f(x)/x \to 7$ as $x \to 0$. Deduce that $f(0) = 0$ and $f'(0) = 7$. Give an example other than $f(x) = 7x$.

27 What is $\lim\limits_{x \to 0} \dfrac{f(3+x) - f(3)}{x}$ if it exists ? What if $x \to 1$?

Problems 28–31 use the square rule: $d(u^2)/dx = 2u(du/dx)$.

28 Take $u = x$ and find the derivative of x^2 (a new way).

29 Take $u = x^4$ and find the derivative of x^8 (using $du/dx = 4x^3$).

30 If $u = 1$ then $u^2 = 1$. Then $d1/dx$ is 2 times $d1/dx$. How is this possible ?

31 Take $u = \sqrt{x}$. The derivative of $u^2 = x$ is $1 = 2u(du/dx)$. So what is du/dx, the derivative of $\sqrt{x}$?

32 The left figure shows $f(t) = t^2$. Indicate distances $f(t + \Delta t)$ and Δt and Δf. Draw lines that have slope $\Delta f/\Delta t$ and $f'(t)$.

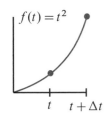
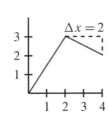

33 The right figure shows $f(x)$ and Δx. Find $\Delta f/\Delta x$ and $f'(2)$.

34 Draw $f(x)$ and Δx so that $\Delta f/\Delta x = 0$ but $f'(x) \neq 0$.

35 If $f = u^2$ then $df/dx = 2u$ du/dx. If $g = f^2$ then $dg/dx = 2f$ df/dx. Together those give $g = u^4$ and $dg/dx = \underline{\qquad}$.

36 *True or false*, assuming $f(0) = 0$:

(a) If $f(x) \leqslant x$ for all x, then $df/dx \leqslant 1$.
(b) If $df/dx \leqslant 1$ for all x, then $f(x) \leqslant x$.

37 The graphs show Δf and $\Delta f/h$ for $f(x) = x^2$. Why is $2x + h$ the equation for $\Delta f/h$? If h is cut in half, draw in the new graphs.

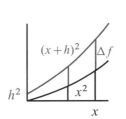
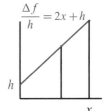

38 Draw the corresponding graphs for $f(x) = \frac{1}{2}x$.

39 Draw $1/x$ and $1/(x+h)$ and $\Delta f/h$—either by hand with $h = \frac{1}{2}$ or by computer to show $h \to 0$.

40 For $y = e^x$, show on computer graphs that $dy/dx = y$.

41 *Explain the derivative in your own words.*

2.2 Powers and Polynomials

This section has two main goals. One is to find the derivatives of $f(x) = x^3$ and x^4 and x^5 (and more generally $f(x) = x^n$). The *power* or *exponent* n is at first a positive integer. Later we allow x^π and $x^{2.2}$ and every x^n.

The other goal is different. While computing these derivatives, we look ahead to their applications. In using calculus, we meet *equations with derivatives in them*— "*differential equations*." It is too early to solve those equations. But it is not too early to see the purpose of what we are doing. Our examples come from economics and biology.

With $n = 2$, the derivative of x^2 is $2x$. With $n = -1$, the slope of x^{-1} is $-1x^{-2}$. Those are two pieces in a beautiful pattern, which it will be a pleasure to discover. We begin with x^3 and its derivative $3x^2$, before jumping to x^n.

EXAMPLE 1   If $f(x) = x^3$ then $\Delta f = (x+h)^3 - x^3 = (x^3 + 3x^2h + 3xh^2 + h^3) - x^3$.

Step 1: Cancel x^3.   **Step 2:** Divide by h.   **Step 3:** h goes to zero.

$$\frac{\Delta f}{h} = 3x^2 + 3xh + h^2 \quad \text{approaches} \quad \frac{df}{dx} = 3x^2.$$

That is straightforward, and you see the crucial step. The power $(x+h)^3$ yields four separate terms $x^3 + 3x^2h + 3xh^2 + h^3$. (Notice 1, 3, 3, 1.) After x^3 is subtracted, we can divide by h. At the limit ($h = 0$) we have $3x^2$.

For $f(x) = x^n$ the plan is the same. A step of size h leads to $f(x+h) = (x+h)^n$. One reason for algebra is to calculate powers like $(x+h)^n$, and if you have forgotten the binomial formula we can recapture its main point. Start with $n = 4$:

$$(x+h)(x+h)(x+h)(x+h) = x^4 + \ ??? \ + h^4. \tag{1}$$

Multiplying the four x's gives x^4. Multiplying the four h's gives h^4. These are the easy terms, but not the crucial ones. The subtraction $(x+h)^4 - x^4$ will remove x^4, and the limiting step $h \to 0$ will wipe out h^4 (even after division by h). ***The products that matter are those with exactly one h.*** In Example 1 with $(x+h)^3$, this key term was $3x^2h$. Division by h left $3x^2$.

With only one h, there are n places it can come from. Equation (1) has four h's in parentheses, and four ways to produce x^3h. Therefore the key term is $4x^3h$. (Division by h leaves $4x^3$.) In general there are n parentheses and n ways to produce $x^{n-1}h$, so the ***binomial formula*** contains $nx^{n-1}h$:

$$(x+h)^n = x^n + \underline{nx^{n-1}h} + \cdots + h^n. \tag{2}$$

> **2B** For $n = 1, 2, 3, 4, \ldots$, the derivative of x^n is nx^{n-1}.

Subtract x^n from (2). Divide by h. The key term is nx^{n-1}. The rest disappears as $h \to 0$:

$$\frac{\Delta f}{\Delta x} = \frac{(x+h)^n - x^n}{h} = \frac{nx^{n-1}h + \cdots + h^n}{h} \quad \text{so} \quad \frac{df}{dx} = nx^{n-1}.$$

The terms replaced by the dots involve h^2 and h^3 and higher powers. After dividing by h, they still have at least one factor h. All those terms vanish as h approaches zero.

EXAMPLE 2 $(x+h)^4 = x^4 + \underline{4x^3h} + 6x^2h^2 + 4xh^3 + h^4$. This is $n = 4$ in detail.

Subtract x^4, divide by h, let $h \to 0$. The derivative is $4x^3$. The coefficients $1, 4, 6, 4, 1$ are in Pascal's triangle below. For $(x+h)^5$ the next row is $1, 5, 10, \underline{\ ?\ }$.

Remark The missing terms in the binomial formula (replaced by the dots) contain all the products $x^{n-j}h^j$. An x or an h comes from each parenthesis. The binomial coefficient "n *choose* j" is **the number of ways to choose** j h's **out of** n **parentheses**. It involves n *factorial*, which is $n(n-1)\cdots(1)$. Thus $5! = 5 \cdot 4 \cdot 3 \cdot 2 \cdot 1 = 120$.

These are numbers that gamblers know and love:

$$\text{"}n \text{ choose } j\text{"} = \binom{n}{j} = \frac{n!}{j!(n-j)!}$$

1	*Pascal's*
1 1	*triangle*
1 2 1	
1 3 3 1	$n = 3$
1 4 6 4 1	$n = 4$

In the last row, the coefficient of x^3h is $4!/1!3! = 4 \cdot 3 \cdot 2 \cdot 1/1 \cdot 3 \cdot 2 \cdot 1 = 4$. For the x^2h^2 term, with $j = 2$, there are $4 \cdot 3 \cdot 2 \cdot 1/2 \cdot 1 \cdot 2 \cdot 1 = 6$ ways to choose two h's. Notice that $1 + 4 + 6 + 4 + 1$ equals 16, which is 2^4. Each row of Pascal's triangle adds to a power of 2.

Choosing 6 numbers out of 49 in a lottery, the odds are $49 \cdot 48 \cdot 47 \cdot 46 \cdot 45 \cdot 44/6!$ to 1. That number is $N = $ "49 choose 6" $= 13,983,816$. It is the coefficient of $x^{43}h^6$ in $(x+h)^{49}$. If λ times N tickets are bought, the expected number of winners is λ. The chance of no winner is $e^{-\lambda}$. The chance of *one* winner is $\lambda e^{-\lambda}$. See Section 8.4.

Florida's lottery in September 1990 (these rules) had six winners out of 109, 163, 978 tickets.

DERIVATIVES OF POLYNOMIALS

Now we have an infinite list of functions and their derivatives:

$$x \quad x^2 \quad x^3 \quad x^4 \quad x^5 \quad \cdots \qquad 1 \quad 2x \quad 3x^2 \quad 4x^3 \quad 5x^4 \quad \cdots$$

The derivative of x^n is n times the next lower power x^{n-1}. That rule extends beyond these integers $1, 2, 3, 4, 5$ to all powers:

$$f = 1/x \quad \text{has} \quad f' = -1/x^2: \qquad \text{Example 3 of section 2.1} \quad (n = -1)$$
$$f = 1/x^2 \quad \text{has} \quad f' = -2/x^3: \qquad \text{Example 6 of section 2.1} \quad (n = -2)$$
$$f = \sqrt{x} \quad \text{has} \quad f' = \tfrac{1}{2}x^{-1/2}: \qquad \text{true but not yet checked} \quad (n = \tfrac{1}{2})$$

Remember that x^{-2} means $1/x^2$ and $x^{-1/2}$ means $1/\sqrt{x}$. Negative powers lead to *decreasing* functions, approaching zero as x gets large. Their slopes have minus signs.

Question What are the derivatives of x^{10} and $x^{2.2}$ and $x^{-1/2}$?
Answer $10x^9$ and $2.2x^{1.2}$ and $-\frac{1}{2}x^{-3/2}$. Maybe $(x+h)^{2.2}$ is a little unusual. Pascal's triangle can't deal with this fractional power, but the formula stays firm: **After $x^{2.2}$ comes $2.2x^{1.2}h$.** The complete binomial formula is in Section 10.5.

That list is a good start, but plenty of functions are left. What comes next is really simple. A tremendous number of new functions are "linear combinations" like

$$f(x) = 6x^3 \quad \text{or} \quad 6x^3 + \frac{1}{2}x^2 \quad \text{or} \quad 6x^3 - \frac{1}{2}x^2.$$

What are their derivatives? The answers are known for x^3 and x^2, and we want to multiply by 6 or divide by 2 or add or subtract. *Do the same to the derivatives*:

$$f'(x) = 18x^2 \quad \text{or} \quad 18x^2 + x \quad \text{or} \quad 18x^2 - x.$$

2C The derivative of c times $f(x)$ is c times $f'(x)$.

2D The derivative of $f(x) + g(x)$ is $f'(x) + g'(x)$.

The number c can be any constant. We can add (or subtract) any functions. The rules allow any combination of f and g: ***The derivative of*** $9f(x) - 7g(x)$ ***is*** $9f'(x) - 7g'(x)$.

    The reasoning is direct. When $f(x)$ is multiplied by c, so is $f(x+h)$. The difference Δf is also multiplied by c. All averages $\Delta f / h$ contain c, so their limit is cf'. The only incomplete step is the last one (the limit). ***We still have to say what*** "*limit*" ***means***.

    Rule 2D is similar. Adding $f + g$ means adding $\Delta f + \Delta g$. Now divide by h. In the limit as $h \to 0$ we reach $f' + g'$—because a limit of sums is a sum of limits. Any example is easy and so is the proof—it is the definition of limit that needs care (Section 2.6).

    You can now find the derivative of every polynomial. A "polynomial" is a combination of $1, x, x^2, \ldots, x^n$—for example $9 + 2x - x^5$. That particular polynomial has slope $2 - 5x^4$. Note that the derivative of 9 is zero! A constant just raises or lowers the graph, without changing its slope. It alters the mileage before starting the car.

    The disappearance of constants is one of the nice things in differential calculus. The reappearance of those constants is one of the headaches in integral calculus. When you find v from f, the starting mileage doesn't matter. The constant in f has no effect on v. (Δf *is measured by a trip meter*; Δt *comes from a stopwatch*.) To find distance from velocity, you need to know the mileage at the start.

A LOOK AT DIFFERENTIAL EQUATIONS (FIND y FROM dy/dx)

We know that $y = x^3$ has the derivative $dy/dx = 3x^2$. Starting with the function, we found its slope. Now reverse that process. ***Start with the slope and find the function***. This is what science does all the time—and it seems only reasonable to say so.

    Begin with $dy/dx = 3x^2$. The slope is given, the function y is not given.

Question   Can you go backward to reach $y = x^3$?
Answer   Almost but not quite. You are only entitled to say that $y = x^3 + C$. The constant C is the starting value of y (when $x = 0$). Then the ***differential equation*** $dy/dx = 3x^2$ is solved.

    Every time you find a derivative, you can go backward to solve a differential equation. The function $y = x^2 + x$ has the slope $dy/dx = 2x + 1$. In reverse, the slope $2x + 1$ produces $x^2 + x$—and all the other functions $x^2 + x + C$, shifted up and down. After going from distance f to velocity v, we return to $f + C$. But there is a lot more to differential equations. Here are two crucial points:

    1. We reach dy/dx by way of $\Delta y / \Delta x$, but we have no system to go backward. With $dy/dx = (\sin x)/x$ we are lost. What function has this derivative?

2. Many equations have the same solution $y = x^3$. Economics has $dy/dx = 3y/x$. Geometry has $dy/dx = 3y^{2/3}$. These equations involve y as well as dy/dx. Function and slope are mixed together! This is typical of differential equations.

To summarize: Chapters 2-4 compute and use derivatives. Chapter 5 goes in reverse. Integral calculus discovers the function from its slope. Given dy/dx we find $y(x)$. Then Chapter 6 solves the differential equation $dy/dt = y$, function mixed with slope. Calculus moves from *derivatives* to *integrals* to *differential equations*.

This discussion of the purpose of calculus should mention a specific example. Differential equations are applied to an epidemic (like AIDS). In most epidemics the number of cases grows exponentially. The peak is quickly reached by e^t, and the epidemic dies down. Amazingly, exponential growth is not happening with AIDS—the best fit to the data through 1988 is a **cubic polynomial** (*Los Alamos Science*, 1989):

The number of cases fits a cubic within 2% : $y = 174.6(t - 1981.2)^3 + 340$.

This is dramatically different from other epidemics. Instead of $dy/dt = y$ we have $dy/dt = 3y/t$. Before this book is printed, we may know what has been preventing e^t (fortunately). Eventually the curve will turn away from a cubic—I hope that mathematical models will lead to knowledge that saves lives.

Added in proof: In 1989 the curve for the U.S. dropped from t^3 to t^2.

MARGINAL COST AND ELASTICITY IN ECONOMICS

First point about economics: The **marginal** cost and **marginal** income are crucially important. The average cost of making automobiles may be $10,000. But it is the $8,000 cost of the *next car* that decides whether Ford makes it. "*The average describes the past, the marginal predicts the future*." For bank deposits or work hours or wheat, which come in smaller units, the amounts are continuous variables. Then the word "marginal" says one thing: **Take the derivative**.†

The average pay over all the hours we ever worked may be low. We wouldn't work another hour for that! This average is rising, but the pay for each additional hour rises faster—possibly it jumps. When $10/hour increases to $15/hour after a 40-hour week, a 50-hour week pays $550. The average income is $11/hour. The marginal income is $15/hour—the overtime rate.

Concentrate next on cost. Let $y(x)$ be the cost of producing x tons of steel. The cost of $x + \Delta x$ tons is $y(x + \Delta x)$. The extra cost is the difference Δy. Divide by Δx, the number of extra tons. The ratio $\Delta y / \Delta x$ is **the average cost per extra ton**. When Δx is an ounce instead of a ton, we are near the marginal cost dy/dx.

Example: When the cost is x^2, the average cost is $x^2/x = x$. The marginal cost is $2x$. Figure 2.4 has increasing slope—an example of "diminishing returns to scale."

This raises another point about economics. The units are arbitrary. In yen per kilogram the numbers look different. The way to correct for arbitrary units is to work with *percentage change* or **relative change**. An increase of Δx tons is a relative increase of $\Delta x/x$. A cost increase Δy is a relative increase of $\Delta y/y$. Those are *dimensionless*, the same in tons/tons or dollars/dollars or yen/yen.

A third example is *the demand y at price x*. Now dy/dx is negative. But again the units are arbitrary. The demand is in liters or gallons, the price is in dollars or pesos.

---

†These paragraphs show how calculus applies to economics. You do *not* have to be an economist to understand them. Certainly the author is not, probably the instructor is not, possibly the student is not. We can all use dy/dx.

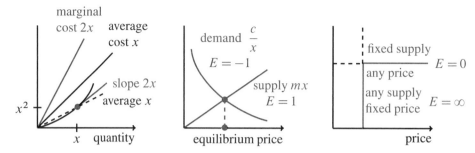

Fig. 2.4 Marginal exceeds average. Constant elasticity $E = \pm 1$. Perfectly elastic to perfectly inelastic (Γ curve).

Relative changes are better. When the price goes up by 10%, the demand may drop by 5%. If that ratio stays the same for small increases, *the elasticity of demand is* $\frac{1}{2}$.

Actually this number should be $-\frac{1}{2}$. The price rose, the demand dropped. In our definition, the elasticity *will* be $-\frac{1}{2}$. In conversation between economists the minus sign is left out (I hope not forgotten).

DEFINITION   The elasticity of the demand function $y(x)$ is

$$E(x) = \lim_{\Delta x \to 0} \frac{\Delta y/y}{\Delta x/x} = \frac{dy/dx}{y/x}. \tag{3}$$

Elasticity is "marginal" divided by "average." $E(x)$ is also relative change in y divided by relative change in x. Sometimes $E(x)$ is the same at all prices—this important case is discussed below.

EXAMPLE 4   Suppose the demand is $y = c/x$ when the price is x. The derivative $dy/dx = -c/x^2$ comes from calculus. The division $y/x = c/x^2$ is only algebra. *The ratio is $E = -1$:*

> *For the demand $y = c/x$, the elasticity is $(-c/x^2)/(c/x^2) = -1$.*

All demand curves are compared with this one. The demand is ***inelastic*** when $|E| < 1$. It is ***elastic*** when $|E| > 1$. The demand $20/\sqrt{x}$ is inelastic ($E = -\frac{1}{2}$), while x^{-3} is elastic ($E = -3$). ***The power $y = cx^n$, whose derivative we know, is the function with constant elasticity n:***

> if $y = cx^n$ then $dy/dx = cnx^{n-1}$ and $E = cnx^{n-1}/(cx^n/x) = n$.

It is because $y = cx^n$ sets the standard that we could come so early to economics.

In the special case when $y = c/x$, consumers spend the same at all prices. Price x times quantity y remains constant at $xy = c$.

EXAMPLE 5   The supply curve has $E > 0$—supply increases with price. Now the baseline case is $y = cx$. The slope is c and the average is $y/x = c$. *The elasticity is $E = c/c = 1$.*

Compare $E = 1$ with $E = 0$ and $E = \infty$. A constant supply is "perfectly inelastic." The power n is zero and the slope is zero: $y = c$. No more is available when the harvest is over. Whatever the price, the farmer cannot suddenly grow more wheat. Lack of elasticity makes farm economics difficult.

The other extreme $E = \infty$ is "perfectly elastic." The supply is unlimited at a fixed price x. Once this seemed true of water and timber. In reality the steep curve

$x = $ constant is leveling off to a flat curve $y = $ constant. Fixed price is changing to fixed supply, $E = \infty$ is becoming $E = 0$, and the supply of water follows a "gamma curve" shaped like Γ.

EXAMPLE 6 Demand is an increasing function of *income*—more income, more demand. The *income elasticity* is $E(I) = (dy/dI)/(y/I)$. A luxury has $E > 1$ (elastic). Doubling your income more than doubles the demand for caviar. A necessity has $E < 1$ (inelastic). The demand for bread does not double. Please recognize how the central ideas of calculus provide a language for the central ideas of economics.

Important note on supply $=$ demand This is the basic equation of microeconomics. Where the supply curve meets the demand curve, the economy finds the equilibrium price. *Supply $=$ demand assumes perfect competition.* With many suppliers, no one can raise the price. If someone tries, the customers go elsewhere.

The opposite case is a **monopoly**—no competition. Instead of many small producers of wheat, there is one producer of electricity. An airport is a monopolist (and maybe the National Football League). If the price is raised, some demand remains.

Price fixing occurs when several producers act like a monopoly-which antitrust laws try to prevent. The price is not set by supply $=$ demand. The calculus problem is different—*to maximize profit*. Section 3.2 locates the maximum where the marginal profit (the slope!) is zero.

Question on income elasticity   From an income of $\$10,000$ you save $\$500$. The income elasticity of savings is $E = 2$. Out of the next dollar what fraction do you save ?

Answer   The savings is $y = cx^2$ because $E = 2$. The number c must give $500 = c(10,000)^2$, so c is $5 \cdot 10^{-6}$. Then the slope dy/dx is $2cx = 10 \cdot 10 \cdot 10^{-6} \cdot 10^4 = \frac{1}{10}$. This is the marginal savings, ten cents on the dollar. *Average savings is* 5%, *marginal savings is* 10%, *and $E = 2$.*

2.2 EXERCISES

Read-through questions

The derivative of $f = x^4$ is $f' = $ __a__ . That comes from expanding $(x+h)^4$ into the five terms __b__ . Subtracting x^4 and dividing by h leaves the four terms __c__ . This is $\Delta f/h$, and its limit is __d__ .

The derivative of $f = x^n$ is $f' = $ __e__ . Now $(x+h)^n$ comes from the __f__ theorem. The terms to look for are $x^{n-1}h$, containing only one __g__ . There are __h__ of those terms, so $(x+h)^n = x^n + $ __i__ $+ \cdots$. After subtracting __j__ and dividing by h, the limit of $\Delta f/h$ is __k__ . The coefficient of $x^{n-j}h^j$, not needed here, is "n choose j" $= $ __l__ , where $n!$ means __m__ .

The derivative of x^{-2} is __n__ . The derivative of $x^{1/2}$ is __o__ . The derivative of $3x + (1/x)$ is __p__ , which uses the following rules: The derivative of $3f(x)$ is __q__ and the

derivative of $f(x) + g(x)$ is __r__ . Integral calculus recovers __s__ from dy/dx. If $dy/dx = x^4$ then $y(x) = $ __t__ .

1 Starting with $f = x^6$, write down f' and then f''. (This is "f double prime," the derivative of f'.) After _____ derivatives of x^6 you reach a constant. What constant ?

2 Find a function that has x^6 as its derivative.

Find the derivatives of the functions in 3–10. Even if n is negative or a fraction, the derivative of x^n is nx^{n-1}.

3 $x^2 + 7x + 5$

4 $1 + (7/x) + (5/x^2)$

5 $1 + x + x^2 + x^3 + x^4$

6 $(x^2+1)^2$

7 $x^n + x^{-n}$

8 $x^n/n!$

9 $1 + x + \frac{1}{2}x^2 + \frac{1}{6}x^3 + \frac{1}{24}x^4$

10 $\frac{2}{3}x^{3/2} + \frac{2}{5}x^{5/2}$

11 Name two functions with $df/dx = 1/x^2$.

12 *Find the mistake*: x^2 is $x + x + \cdots + x$ (with x terms). Its derivative is $1 + 1 + \cdots + 1$ (also x terms). So the derivative of x^2 seems to be x.

13 What are the derivatives of $3x^{1/3}$ and $-3x^{-1/3}$ and $(3x^{1/3})^{-1}$?

14 The slope of $x + (1/x)$ is zero when $x = $ _____ . What does the graph do at that point ?

15 Draw a graph of $y = x^3 - x$. Where is the slope zero ?

16 If df/dx is negative, is $f(x)$ always negative ? Is $f(x)$ negative for large x ? If you think otherwise, give examples.

17 A rock thrown upward with velocity 16 ft/sec reaches height $f = 16t - 16t^2$ at time t.

(a) Find its average speed $\Delta f/\Delta t$ from $t = 0$ to $t = \frac{1}{2}$.

(b) Find its average speed $\Delta f/\Delta t$ from $t = \frac{1}{2}$ to $t = 1$.

(c) What is df/dt at $t = \frac{1}{2}$?

18 When f is in feet and t is in seconds, what are the units of f' and its derivative f'' ? In $f = 16t - 16t^2$, the first 16 is ft/sec but the second 16 is _____ .

19 Graph $y = x^3 + x^2 - x$ from $x = -2$ to $x = 2$ and estimate where it is decreasing. Check the transition points by solving $dy/dx = 0$.

20 At a point where $dy/dx = 0$, what is special about the graph of $y(x)$? Test case: $y = x^2$.

21 Find the slope of $y = \sqrt{x}$ by algebra (then $h \to 0$):

$$\frac{\Delta y}{h} = \frac{\sqrt{x+h} - \sqrt{x}}{h} = \frac{\sqrt{x+h} - \sqrt{x}}{h} = \frac{\sqrt{x+h} + \sqrt{x}}{\sqrt{x+h} + \sqrt{x}}.$$

22 Imitate Problem 21 to find the slope of $y = 1/\sqrt{x}$.

23 Complete Pascal's triangle for $n = 5$ and $n = 6$. Why do the numbers across each row add to 2^n ?

24 Complete $(x + h)^5 = x^5 + $ _____ . What are the binomial coefficients $\binom{5}{1}$ and $\binom{5}{2}$ and $\binom{5}{3}$?

25 Compute $(x + h)^3 - (x - h)^3$, divide by $2h$, and set $h = 0$. Why divide by $2h$ to find this slope ?

26 Solve the differential equation $y'' = x$ to find $y(x)$.

27 For $f(x) = x^2 + x^3$, write out $f(x + \Delta x)$ and $\Delta f/\Delta x$. What is the limit at $\Delta x = 0$ and what rule about sums is confirmed ?

28 The derivative of $(u(x))^2$ is _____ from Section 2.1. Test this rule on $u = x^n$.

29 What are the derivatives of $x^7 + 1$ and $(x + 1)^7$? Shift the graph of x^7.

30 If df/dx is $v(x)$, what functions have these derivatives ?

(a) $4v(x)$

(b) $v(x) + 1$

(c) $v(x + 1)$

(d) $v(x) + v'(x)$.

31 What function $f(x)$ has fourth derivative equal to 1 ?

32 What function $f(x)$ has nth derivative equal to 1 ?

33 Suppose $df/dx = 1 + x + x^2 + x^3$. Find $f(x)$.

34 Suppose $df/dx = x^{-2} - x^{-3}$. Find $f(x)$.

35 $f(x)$ *can be its own derivative*. In the infinite polynomial $f = 1 + x + \frac{1}{2}x^2 + \frac{1}{6}x^3 + $ _____ , what numbers multiply x^4 and x^5 if df/dx equals f ?

36 Write down a differential equation $dy/dx = $ _____ that is solved by $y = x^2$. Make the right side involve y (not just $2x$).

37 *True or false*: (a) The derivative of x^π is πx^π.

(b) The derivative of ax^n/bx^n is a/b.

(c) If $df/dx = x^4$ and $dg/dx = x^4$ then $f(x) = g(x)$.

(d) $(f(x) - f(a))/(x - a)$ approaches $f'(a)$ as $x \to a$.

(e) The slope of $y = (x - 1)^3$ is $y' = 3(x - 1)^2$.

Problems 38–44 are about calculus in economics.

38 When the cost is $y = y_0 + cx$, find $E(x) = (dy/dx)/(y/x)$. It approaches _____ for large x.

39 From an income of $x = \$10,000$ you spend $y = \$1,200$ on your car. If $E = \frac{1}{2}$, what fraction of your next dollar will be spent on the car ? Compare dy/dx (marginal) with y/x (average).

40 Name a product whose price elasticity is

(a) high

(b) low

(c) negative (?)

41 The demand $y = c/x$ has $dy/dx = -y/x$. Show that $\Delta y/\Delta x$ is *not* $-y/x$. (Use numbers or algebra.) Finite steps miss the special feature of infinitesimal steps.

42 The demand $y = x^n$ has $E = $ _____ . The revenue xy (price times demand) has elasticity $E = $ _____ .

43 $y = 2x + 3$ grows with marginal cost 2 from the fixed cost 3. Draw the graph of $E(x)$.

44 From an income I we save $S(I)$. The *marginal* propensity to save is _____ . Elasticity is not needed because S and I have the same _____ . Applied to the whole economy this is (microeconomics) (macroeconomics).

45 2^t is doubled when t increases by _____ . t^3 is doubled when t increases to _____ t. The doubling time for AIDS is proportional to t.

46 Biology also leads to $dy/y = n\,dx/x$, for the relative growth of the head (dy/y) and the body (dx/x). Is $n > 1$ or $n < 1$ for a child ?

47 What functions have $df/dx = x^9$ and $df/dx = x^n$? Why does $n = -1$ give trouble ?

48 The slope of $y = x^3$ comes from this identity:

$$\frac{(x+h)^3 - x^3}{h} = (x+h)^2 + (x+h)x + x^2.$$

(a) Check the algebra. Find dy/dx as $h \to 0$.

(b) Write a similar identity for $y = x^4$.

49 (Computer graphing) Find all the points where $y = x^4 + 2x^3 - 7x^2 + 3 = 0$ and where $dy/dx = 0$.

50 The graphs of $y_1(x) = x^4 + x^3$ and $y_2(x) = 7x - 5$ touch at the point where $y_3(x) = \underline{\quad\quad} = 0$. Plot $y_3(x)$ to see what is special. What does the graph of $y(x)$ do at a point where $y = y' = 0$?

51 In the Massachusetts lottery you choose 6 numbers out of 36. What is your chance to win?

52 In what circumstances would it pay to buy a lottery ticket for every possible combination, so one of the tickets would win?

2.3 The Slope and the Tangent Line

Chapter 1 started with straight line graphs. The velocity was constant (at least piecewise). The distance function was linear. Now we are facing polynomials like $x^3 - 2$ or $x^4 - x^2 + 3$, with other functions to come soon. Their graphs are definitely curved. Most functions are not close to linear—except if you focus all your attention near a single point. That is what we will do.

Over a very short range a curve looks straight. Look through a microscope, or zoom in with a computer, and there is no doubt. The graph of distance versus time becomes nearly linear. Its slope is the velocity at that moment. We want to find the line that the graph stays closest to—the "*tangent line*"—before it curves away.

The tangent line is easy to describe. We are at a particular point on the graph of $y = f(x)$. At that point x equals a and y equals $f(a)$ and the slope equals $f'(a)$. *The tangent line goes through that point $x = a, y = f(a)$ with that slope $m = f'(a)$*. Figure 2.5 shows the line more clearly than any equation, but we have to turn the geometry into algebra. We need the equation of the line.

EXAMPLE 1   Suppose $y = x^4 - x^2 + 3$. At the point $x = a = 1$, the height is $y = f(a) = 3$. The slope is $dy/dx = 4x^3 - 2x$. At $x = 1$ the slope is $4 - 2 = 2$. That is $f'(a)$:

> *The numbers $x = 1, y = 3, dy/dx = 2$ determine the tangent line.*

The equation of the tangent line is $y - 3 = 2(x - 1)$, and this section explains why.

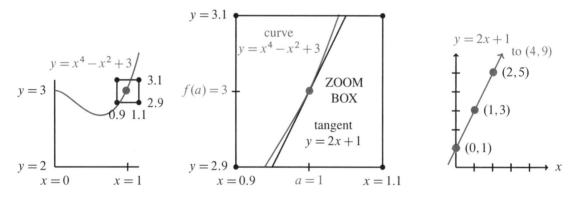

Fig. 2.5   The tangent line has the same slope 2 as the curve (especially after zoom).

THE EQUATION OF A LINE

A straight line is determined by two conditions. We know the line if we know two of its points. (We still have to write down the equation.) Also, if we know *one point and the slope*, the line is set. That is the situation for the tangent line, which has a known slope at a known point:

1. The equation of a line has the form $y = mx + b$
2. The number m is the slope of the line, because $dy/dx = m$
3. The number b adjusts the line to go through the required point.

I will take those one at a time—first $y = mx + b$, then m, then b.

1. The graph of $y = mx + b$ is not curved. How do we know? For the specific example $y = 2x + 1$, take two points whose coordinates x, y satisfy the equation:

$$x = 0, y = 1 \quad \text{and} \quad x = 4, y = 9 \quad \text{both satisfy} \quad y = 2x + 1.$$

Those points $(0, 1)$ and $(4, 9)$ lie on the graph. *The point halfway between has $x = 2$ and $y = 5$.* That point also satisfies $y = 2x + 1$. **The halfway point is on the graph.** If we subdivide again, the midpoint between $(0, 1)$ and $(2, 5)$ is $(1, 3)$. This also has $y = 2x + 1$. The graph contains all halfway points and must be straight.

2. What is the correct slope m for the tangent line? In our example it is $m = f'(a) = 2$.

   The curve and its tangent line have the same slope at the crucial point: $dy/dx = 2$.

   Allow me to say in another way why the line $y = mx + b$ has slope m. At $x = 0$ its height is $y = b$. At $x = 1$ its height is $y = m + b$. The graph has gone *one unit across* (0 to 1) *and m units up* (b to $m + b$). The whole idea is

$$\text{slope} = \frac{\text{distance up}}{\text{distance across}} = \frac{m}{1}. \tag{1}$$

Each unit across means m units up, to $2m + b$ or $3m + b$. A straight line keeps a constant slope, whereas the slope of $y = x^4 - x^2 + 3$ equals 2 only at $x = 1$.

3. Finally we decide on b. The tangent line $y = 2x + b$ must go through $x = 1$, $y = 3$. Therefore $b = 1$. With letters instead of numbers, $y = mx + b$ leads to $f(a) = ma + b$. So we know b:

2E The equation of the tangent line has $b = f(a) - ma$:

$$y = mx + f(a) - ma \quad \text{or} \quad y - f(a) = m(x - a). \tag{2}$$

That last form is the best. You see immediately what happens at $x = a$. The factor $x - a$ is zero. Therefore $y = f(a)$ as required. This is the **point-slope form** of the equation, and we use it constantly:

$$y - 3 = 2(x - 1) \quad \text{or} \quad \frac{y - 3}{x - 1} = \frac{\text{distance up}}{\text{distance across}} = \text{slope } 2.$$

EXAMPLE 2 The curve $y = x^3 - 2$ goes through $y = 6$ when $x = 2$. At that point $dy/dx = 3x^2 = 12$. The point-slope equation of the tangent line uses 2 and 6 and 12:

$$y - 6 = 12(x - 2) \quad \text{which is also} \quad y = 12x - 18.$$

There is another important line. It is *perpendicular* to the tangent line and *perpendicular* to the curve. This is the **normal line** in Figure 2.6. Its new feature is its slope. When the tangent line has slope m, the normal line has slope $-1/m$. (Rule: Slopes of perpendicular lines multiply to give -1.) Example 2 has $m = 12$, so the normal line has slope $-1/12$:

$$\text{tangent line: } y - 6 = 12(x - 2) \qquad \text{normal line: } y - 6 = -\tfrac{1}{12}(x - 2).$$

Light rays travel in the normal direction. So do brush fires—they move perpendicular to the fire line. Use the point-slope form! The tangent is $y = 12x - 18$, the normal is not $y = -\tfrac{1}{12}x - 18$.

EXAMPLE 3   You are on a roller-coaster whose track follows $y = x^2 + 4$. You see
a friend at $(0,0)$ and want to get there quickly. Where do you step off?

Solution Your path will be the tangent line (at high speed). The problem is *to choose*
$x = a$ *so the tangent line passes through* $x = 0, y = 0$. When you step off at $x = a$,

the height is $y = a^2 + 4$ and the slope is $2a$

the equation of the tangent line is $y - (a^2 + 4) = 2a(x - a)$

this line goes through $(0,0)$ if $-(a^2 + 4) = -2a^2$ or $a = \pm 2$.

The same problem is solved by spacecraft controllers and baseball pitchers. Releasing
a ball at the right time to hit a target 60 feet away is an amazing display of calculus.
Quarterbacks with a moving target should read Chapter 4 on related rates.

Here is a better example than a roller-coaster. Stopping at a red light wastes gas. It
is smarter to slow down early, and then accelerate. When a car is waiting in front of
you, the timing needs calculus:

EXAMPLE 4   How much must you slow down when a red light is 72 meters away?
In 4 seconds it will be green. The waiting car will accelerate at 3 meters/sec^2. You
cannot pass the car.

Strategy Slow down immediately to the speed V at which you will just catch that car.
(If you wait and brake later, your speed will have to go below V.) At the catchup time
T, the cars have the same speed and same distance. *Two conditions*, so the distance
functions in Figure 2.6d are tangent.

Solution   At time T, the other car's speed is $3(T - 4)$. That shows the delay of 4
seconds. Speeds are equal when $3(T - 4) = V$ or $T = \frac{1}{3}V + 4$. Now require equal
distances. Your distance is V times T. The other car's distance is $72 + \frac{1}{2}at^2$:

$$72 + \frac{1}{2} \cdot 3(T - 4)^2 = VT \quad \text{becomes} \quad 72 + \frac{1}{2} \cdot \frac{1}{3}V^2 = V\left(\frac{1}{3}V + 4\right).$$

The solution is $V = 12$ meters/second. This is 43 km/hr or 27 miles per hour.

Without the other car, you only slow down to $V = 72/4 = 18$ meters/second. As
the light turns green, you go through at 65 km/hr or 40 miles per hour. Try it.

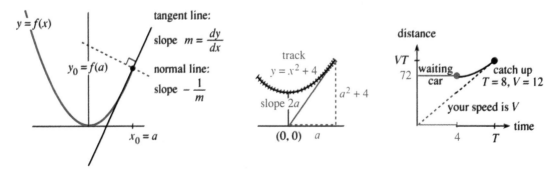

Fig. 2.6   Tangent line $y - y_0 = m(x - x_0)$. Normal line $y - y_0 = -\dfrac{1}{m}(x - x_0)$. Leaving a
roller-coaster and catching up to a car.

THE SECANT LINE CONNECTING TWO POINTS ON A CURVE

Instead of the tangent line through one point, consider the *secant line through two points*. For the tangent line the points came together. Now spread them apart. The point-slope form of a linear equation is replaced by the *two-point form*.

The equation of the curve is still $y = f(x)$. The first point remains at $x = a$, $y = f(a)$. The other point is at $x = c$, $y = f(c)$. The secant line goes between them, and we want its equation. This time we don't start with the slope—but m is easy to find.

EXAMPLE 5 The curve $y = x^3 - 2$ goes through $x = 2$, $y = 6$. It also goes through $x = 3$, $y = 25$. The slope between those points is

$$m = \frac{\text{change in } y}{\text{change in } x} = \frac{25 - 6}{3 - 2} = 19.$$

The point-slope form (at the first point) is $y - 6 = 19(x - 2)$. This line automatically goes through the second point $(3, 25)$. Check: $25 - 6$ equals $19(3 - 2)$. The secant has the right slope 19 to reach the second point. It is the *average slope* $\Delta y / \Delta x$.

A look ahead The second point is going to approach the first point. The secant slope $\Delta y / \Delta x$ will approach the tangent slope dy/dx. We discover the derivative (in the limit). That is the main point now—but not forever.

Soon you will be fast at derivatives. The exact dy/dx will be much easier than $\Delta y / \Delta x$. The situation is turned around as soon as you know that x^9 has slope $9x^8$. Near $x = 1$, the distance *up* is about 9 times the distance *across*. To find $\Delta y = 1.001^9 - 1^9$, just multiply $\Delta x = .001$ by 9. The quick approximation is .009, the calculator gives $\Delta y = .009036$. It is easier to follow the tangent line than the curve.

Come back to the secant line, and change numbers to letters. What line connects $x = a$, $y = f(a)$ to $x = c$, $y = f(c)$? A mathematician puts formulas ahead of numbers, and reasoning ahead of formulas, and ideas ahead of reasoning:

(1) The slope is $m = \dfrac{\text{distance up}}{\text{distance across}} = \dfrac{f(c) - f(a)}{c - a}$

(2) The height is $y = f(a)$ at $x = a$

(3) The height is $y = f(c)$ at $x = c$ (automatic with correct slope).

2F The *two-point form* uses the slope between the points:

$$\text{secant line}: \quad y - f(a) = \left(\frac{f(c) - f(a)}{c - a} \right)(x - a). \qquad (3)$$

At $x = a$ the right side is zero. So $y = f(a)$ on the left side. At $x = c$ the right side has two factors $c - a$. They cancel to leave $y = f(c)$. With equation (2) for the tangent line and equation (3) for the secant line, we are ready for the moment of truth.

THE SECANT LINE APPROACHES THE TANGENT LINE

What comes now is pretty basic. It matches what we did with velocities:

$$\text{average velocity} = \frac{\Delta \text{ distance}}{\Delta \text{ time}} = \frac{f(t + \Delta t) - f(t)}{\Delta t}.$$

The limit is df/dt. We now do exactly the same thing with slopes. **The secant line turns into the tangent line as c approaches a:**

$$\text{slope of secant line:} \quad \frac{\Delta f}{\Delta x} = \frac{f(c) - f(a)}{c - a}$$

$$\text{slope of tangent line:} \quad \frac{df}{dx} = \text{limit of } \frac{\Delta f}{\Delta x}.$$

There stands the fundamental idea of differential calculus! You have to imagine more secant lines than I can draw in Figure 2.7, as c comes close to a. Everybody recognizes $c - a$ as Δx. Do you recognize $f(c) - f(a)$ as $f(x + \Delta x) - f(x)$? It is Δf, the change in height. All lines go through $x = a, y = f(a)$. **Their limit is the tangent line.**

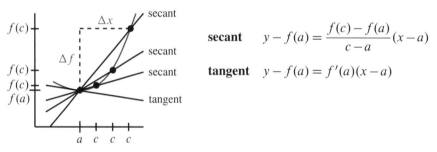

secant $\quad y - f(a) = \dfrac{f(c) - f(a)}{c - a}(x - a)$

tangent $\quad y - f(a) = f'(a)(x - a)$

Fig. 2.7 Secants approach tangent as their slopes $\Delta f/\Delta x$ approach df/dx.

Intuitively, the limit is pretty clear. The two points come together, and the tangent line touches the curve at *one* point. (It could touch again at faraway points.) Mathematically this limit can be tricky—it takes us from algebra to calculus. Algebra stays away from $0/0$, but calculus gets as close as it can.

The new limit for df/dx looks different, but it is the same as before:

$$f'(a) = \lim_{c \to a} \frac{f(c) - f(a)}{c - a}. \tag{4}$$

EXAMPLE 6 Find the secant lines and tangent line for $y = f(x) = \sin x$ at $x = 0$.

The starting point is $x = 0, y = \sin 0$. This is the origin $(0,0)$. The ratio of distance up to distance across is $(\sin c)/c$:

$$\text{secant equation} \quad y = \frac{\sin c}{c}x \qquad \text{tangent equation} \quad y = 1x.$$

As c approaches zero, the secant line becomes the tangent line. The limit of $(\sin c)/c$ is not $0/0$, which is meaningless, but 1, which is dy/dx.

EXAMPLE 7 The gold you own will be worth $\sqrt{t}$ million dollars in t years. When does the rate of increase drop to 10% of the current value, so you should sell the gold and buy a bond? At $t = 25$, how far does that put you ahead of $\sqrt{t} = 5$?

Solution The rate of increase is the derivative of $\sqrt{t}$, which is $1/2\sqrt{t}$. That is 10% of the current value $\sqrt{t}$ when $1/2\sqrt{t} = \sqrt{t}/10$. Therefore $2t = 10$ or $t = 5$. At that time you sell the gold, leave the curve, and go onto the tangent line:

$$y - \sqrt{5} = \frac{\sqrt{5}}{10}(t - 5) \quad \text{becomes} \quad y - \sqrt{5} = 2\sqrt{5} \quad \text{at} \quad t = 25.$$

With straight interest on the bond, not compounded, you have reached $y = 3\sqrt{5} = 6.7$ million dollars. The gold is worth a measly five million.

2.3 EXERCISES

Read-through questions

A straight line is determined by __a__ points, or one point and the __b__. The slope of the tangent line equals the slope of the __c__. The point-slope form of the tangent equation is $y - f(a) = $__d__.

The tangent line to $y = x^3 + x$ at $x = 1$ has slope __e__. Its equation is __f__. It crosses the y axis at __g__ and the x axis at __h__. The normal line at this point $(1, 2)$ has slope __i__. Its equation is $y - 2 = $__j__. The secant line from $(1, 2)$ to $(2,$__k__$)$ has slope __l__. Its equation is $y - 2 = $__m__.

The point $(c, f(c))$ is on the line $y - f(a) = m(x - a)$ provided $m = $__n__. As c approaches a, the slope m approaches __o__. The secant line approaches the __p__ line.

1 (a) Find the slope of $y = 12/x$. Find the slope of $y = 12/x$.
   (b) Find the equation of the tangent line at $(2, 6)$.
   (c) Find the equation of the normal line at $(2, 6)$.
   (d) Find the equation of the secant line to $(4, 3)$.

2 For $y = x^2 + x$ find equations for
   (a) the tangent line and normal line at $(1, 2)$;
   (b) the secant line to $x = 1 + h$, $y = (1 + h)^2 + (1 + h)$.

3 A line goes through $(1, -1)$ and $(4, 8)$. Write its equation in point-slope form. Then write it as $y = mx + b$.

4 The tangent line to $y = x^3 + 6x$ at the origin is $y = $_____. Does it cross the curve again?

5 The tangent line to $y = x^3 - 3x^2 + x$ at the origin is $y = $_____. It is also the secant line to the point _____.

6 Find the tangent line to $x = y^2$ at $x = 4$, $y = 2$.

7 For $y = x^2$ the secant line from (a, a^2) to (c, c^2) has the equation _____. Do the division by $c - a$ to find the tangent line as c approaches a.

8 Construct a function that has the same slope at $x = 1$ and $x = 2$. Then find two points where $y = x^4 - 2x^2$ has the same tangent line (draw the graph).

9 Find a curve that is tangent to $y = 2x - 3$ at $x = 5$. Find the normal line to that curve at $(5, 7)$.

10 For $y = 1/x$ the secant line from $(a, 1/a)$ to $(c, 1/c)$ has the equation _____. Simplify its slope and find the limit as c approaches a.

11 What are the equations of the tangent line and normal line to $y = \sin x$ at $x = \pi/2$?

12 If c and a both approach an in-between value $x = b$, then the secant slope $(f(c) - f(a))/(c - a)$ approaches _____.

13 At $x = a$ on the graph of $y = 1/x$, compute
   (a) the equation of the tangent line
   (b) the points where that line crosses the axes.

The triangle between the tangent line and the axes always has area _____.

14 Suppose $g(x) = f(x) + 7$. The tangent lines to f and g at $x = 4$ are _____. *True or false*: The distance between those lines is 7.

15 Choose c so that $y = 4x$ is tangent to $y = x^2 + c$. Match heights as well as slopes.

16 Choose c so that $y = 5x - 7$ is tangent to $y = x^2 + cx$.

17 For $y = x^3 + 4x^2 - 3x + 1$, find all points where the tangent is horizontal.

18 $y = 4x$ can't be tangent to $y = cx^2$. Try to match heights and slopes, or draw the curves.

19 Determine c so that the straight line joining $(0, 3)$ and $(5, -2)$ is tangent to the curve $y = c/(x + 1)$.

20 Choose b, c, d so that the two parabolas $y = x^2 + bx + c$ and $y = dx - x^2$ are tangent to each other at $x = 1$, $y = 0$.

21 The graph of $f(x) = x^3$ goes through $(1, 1)$.
   (a) Another point is $x = c = 1 + h$, $y = f(c) = $_____.
   (b) The change in f is $\Delta f = $_____.
   (c) The slope of the secant is $m = $_____.
   (d) As h goes to zero, m approaches _____.

22 Construct a function $y = f(x)$ whose tangent line at $x = 1$ is the same as the secant that meets the curve again at $x = 3$.

23 Draw two curves bending away from each other. Mark the points P and Q where the curves are closest. At those points, the tangent lines are _____ and the normal lines are _____ .

*24 If the parabolas $y = x^2 + 1$ and $y = x - x^2$ come closest at $(a, a^2 + 1)$ and $(c, c - c^2)$, set up two equations for a and c.

25 A light ray comes down the line $x = a$. It hits the parabolic reflector $y = x^2$ at $P = (a, a^2)$.

(a) Find the tangent line at P. Locate the point Q where that line crosses the y axis.

(b) Check that P and Q are the same distance from the focus at $F = (0, \frac{1}{4})$.

(c) Show from (b) that the figure has equal angles.

(d) What law of physics makes every ray reflect off the parabola to the focus at F ?

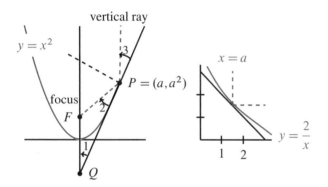

26 In a bad reflector $y = 2/x$, a ray down one special line $x = a$ is reflected horizontally. What is a ?

27 For the parabola $4py = x^2$, where is the slope equal to 1 ? At that point a vertical ray will reflect horizontally. So the focus is at $(0,$ _____ $)$.

28 Why are these statements wrong ? Make them right.

(a) If $y = 2x$ is the tangent line at $(1, 2)$ then $y = -\frac{1}{2}x$ is the normal line.

(b) As c approaches a, the secant slope $(f(c) - f(a))/(c - a)$ approaches $(f(a) - f(a))/(a - a)$.

(c) The line through $(2, 3)$ with slope 4 is $y - 2 = 4(x - 3)$.

29 A ball goes around a circle: $x = \cos t, y = \sin t$. At $t = 3\pi/4$ the ball flies off on the tangent line. Find the equation of that line and the point where the ball hits the ground ($y = 0$).

30 If the tangent line to $y = f(x)$ at $x = a$ is the same as the tangent line to $y = g(x)$ at $x = b$, find two equations that must be satisfied by a and b.

31 Draw a circle of radius 1 resting in the parabola $y = x^2$. At the touching point (a, a^2), the equation of the normal line is _____ . That line has $x = 0$ when $y =$ _____ . The distance to (a, a^2) equals the radius 1 when $a =$ _____ . This locates the touching point.

32 Follow Problem 31 for the flatter parabola $y = \frac{1}{2}x^2$ and explain where the circle rests.

33 You are applying for a $1000 scholarship and your time is worth $10 an hour. If the chance of success is $1 - (1/x)$ from x hours of writing, when should you stop ?

34 Suppose $|f(c) - f(a)| \leqslant |c - a|$ for every pair of points a and c. Prove that $|df/dx| \leqslant 1$.

35 From which point $x = a$ does the tangent line to $y = 1/x^2$ hit the x axis at $x = 3$?

36 If $u(x)/v(x) = 7$ find $u'(x)/v'(x)$. Also find $(u(x)/v(x))'$.

37 Find $f(c) = 1.001^{10}$ in two ways—by calculator and by $f(c) - f(a) \approx f'(a)(c - a)$. Choose $a = 1$ and $f(x) = x^{10}$.

38 At a distance Δx from $x = 1$, how far is the curve $y = 1/x$ above its tangent line ?

39 At a distance Δx from $x = 2$, how far is the curve $y = x^3$ above its tangent line ?

40 Based on Problem 38 or 39, the distance between curve and tangent line grows like what power $(\Delta x)^p$?

41 The tangent line to $f(x) = x^2 - 1$ at $x_0 = 2$ crosses the x axis at $x_1 =$ _____ . The tangent line at x_1, crosses the x axis at $x_2 =$ _____ . Draw the curve and the two lines, which are the beginning of *Newton's method* to solve $f(x) = 0$.

42 (Puzzle) The equation $y = mx + b$ requires *two* numbers, the point-slope form $y - f(a) = f'(a)(x - a)$ requires *three*, and the two-point form requires *four*: $a, f(a), c, f(c)$. How can this be ?

43 Find the time T at the tangent point in Example 4, when you catch the car in front.

44 If the waiting car only accelerates at 2 meters/sec^2, what speed V must you slow down to ?

45 A thief 40 meters away runs toward you at 8 meters per second. What is the smallest acceleration so that $v = at$ keeps you in front ?

46 With 8 meters to go in a relay race, you slow down badly ($f = -8 + 6t - \frac{1}{2}t^2$). How fast should the next runner start (choose v in $f = vt$) so you can just pass the baton ?

2.4 The Derivative of the Sine and Cosine

This section does two things. One is to compute the derivatives of $\sin x$ and $\cos x$. The other is to explain why these functions are so important. They describe *oscillation*, which will be expressed in words and equations. You will see a "*differential equation*." It involves the derivative of an unknown function $y(x)$.

The differential equation will say that the *second* derivative—*the derivative of the derivative*—is equal and opposite to y. In symbols this is $y'' = -y$. Distance in one direction leads to acceleration in the other direction. That makes y and y' and y'' all oscillate. The solutions to $y'' = -y$ are $\sin x$ and $\cos x$ and all their combinations.

We begin with the slope. The derivative of $y = \sin x$ is $y' = \cos x$. There is no reason for that to be a mystery, but I still find it beautiful. Chapter 1 followed a ball around a circle; the shadow went up and down. Its height was $\sin t$ and its velocity was $\cos t$. We now find that derivative by *the standard method of limits*, when $y(x) = \sin x$:

$$\frac{dy}{dx} = \text{limit of } \frac{\Delta y}{\Delta x} = \lim_{h \to 0} \frac{\sin(x+h) - \sin x}{h}. \tag{1}$$

The sine is harder to work with than x^2 or x^3. Where we had $(x+h)^2$ or $(x+h)^3$, we now have $\sin(x+h)$. This calls for one of the basic "addition formulas" from trigonometry, reviewed in Section 1.5:

$$\sin(x+h) = \sin x \cos h + \cos x \sin h \tag{2}$$
$$\cos(x+h) = \cos x \cos h - \sin x \sin h. \tag{3}$$

Equation (2) puts $\Delta y = \sin(x+h) - \sin x$ in a new form:

$$\frac{\Delta y}{\Delta x} = \frac{\sin x \cos h + \cos x \sin h - \sin x}{h} = \sin x \left(\frac{\cos h - 1}{h} \right) + \cos x \left(\frac{\sin h}{h} \right). \tag{4}$$

The ratio splits into two simpler pieces on the right. Algebra and trigonometry got us this far, and now comes the calculus problem. *What happens as $h \to 0$?* It is no longer easy to divide by h. (I will not even mention the unspeakable crime of writing $(\sin h)/h = \sin$.) There are two critically important limits—the first is zero and the second is one:

$$\lim_{h \to 0} \frac{\cos h - 1}{h} = 0 \quad \text{and} \quad \lim_{h \to 0} \frac{\sin h}{h} = 1. \tag{5}$$

The careful reader will object that limits have not been defined! You may further object to computing these limits separately, before combining them into equation (4). Nevertheless—following the principle of *ideas now, rigor later*—I would like to proceed. It is entirely true that the limit of (4) comes from the two limits in (5):

$$\frac{dy}{dx} = (\sin x)(\text{first limit}) + (\cos x)(\text{second limit}) = 0 + \cos x. \tag{6}$$

The secant slope $\Delta y / \Delta x$ has approached the tangent slope dy/dx.

2G The derivative of $y = \sin x$ is $dy/dx = \cos x$.

We cannot pass over the crucial step—the two limits in (5). They contain the real ideas. ***Both ratios become*** $0/0$ ***if we just substitute*** $h = 0$. Remember that the cosine of a zero angle is 1, and the sine of a zero angle is 0. Figure 2.8a shows a small angle h (as near to zero as we could reasonably draw). The edge of length $\sin h$ is close to zero, and the edge of length $\cos h$ is near 1. Figure 2.8b shows how the *ratio* of $\sin h$ to h (both headed for zero) gives the slope of the sine curve at the start.

When two functions approach zero, their ratio might do anything. We might have

$$\frac{h^2}{h} \to 0 \quad \text{or} \quad \frac{h}{h} \to 1 \quad \text{or} \quad \frac{\sqrt{h}}{h} \to \infty.$$

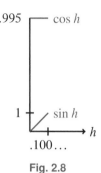

No clue comes from $0/0$. When matters is *whether the top or bottom goes to zero more quickly*. Roughly speaking, we want to show that $(\cos h - 1)/h$ is like h^2/h and $(\sin h)/h$ is like h/h.

Time out The graph of $\sin x$ is in Figure 2.9 (in black). The graph of $\sin(x + \Delta x)$ sits just beside it (in red). The height difference is Δf when the shift distance is Δx.

Fig. 2.8

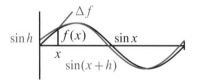

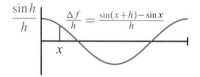

Fig. 2.9 $\sin(x + h)$ with $h = 10° = \pi/18$ radians. $\Delta f/\Delta x$ is close to $\cos x$.

Now divide by that small number Δx (or h). The second figure shows $\Delta f/\Delta x$. It is close to $\cos x$. (Look how it starts—it is not quite $\cos x$.) Mathematics will prove that the limit is $\cos x$ exactly, when $\Delta x \to 0$. Curiously, the reasoning concentrates on only one point ($x = 0$). The slope at that point is $\cos 0 = 1$.

We now prove this: $\sin \Delta x$ divided by Δx goes to 1. The sine curve starts with slope 1. By the addition formula for $\sin(x + h)$, this answer at one point will lead to the slope $\cos x$ at all points.

Question Why does the graph of $f(x + \Delta x)$ shift left from $f(x)$ when $\Delta x > 0$?
Answer   When $x = 0$, the shifted graph is already showing $f(\Delta x)$. In Figure 2.9a, the red graph is shifted *left* from the black graph. The red graph shows $\sin h$ when the black graph shows $\sin 0$.

THE LIMIT OF $(\sin h)/h$ IS 1

There are several ways to find this limit. The direct approach is to let a computer draw a graph. Figure 2.10a is very convincing. ***The function*** $(\sin h)/h$ ***approaches*** 1 ***at the key point*** $h = 0$. So does $(\tan h)/h$. In practice, the only danger is that you might get a message like "undefined function" and no graph. (The machine may refuse to divide by zero at $h = 0$. Probably you can get around that.) Because of the importance of this limit, I want to give a mathematical proof that it equals 1.

Figure 2.10b indicates, but still only graphically, that $\sin h$ *stays below* h. (The first graph shows that too; $(\sin h)/h$ is below 1.) We also see that $\tan h$ *stays above* h. Remember that the tangent is the ratio of sine to cosine. Dividing by the cosine is enough to push the tangent above h. The crucial inequalities (to be proved when h is small and positive) are

$$\sin h < h \qquad \text{and} \qquad \tan h > h. \tag{7}$$

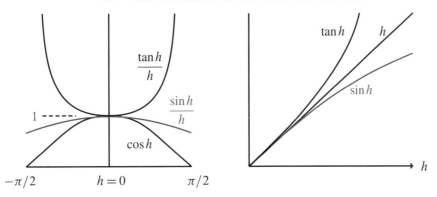

Fig. 2.10 $(\sin h)/h$ squeezed between $\cos x$ and 1; $(\tan h)/h$ decreases to 1.

Since $\tan h = (\sin h)/(\cos h)$, those are the same as

$$\frac{\sin h}{h} < 1 \qquad \text{and} \qquad \frac{\sin h}{h} > \cos h. \tag{8}$$

What happens as h goes to zero? **The ratio $(\sin h)/h$ is squeezed between $\cos h$ and 1.** But $\cos h$ is approaching 1! The squeeze as $h \to 0$ leaves only one possibility for $(\sin h)/h$, which is caught in between: *The ratio $(\sin h)/h$ approaches 1.*

Figure 2.10 shows that "squeeze play." **If two functions approach the same limit, so does any function caught in between.** This is proved at the end of Section 2.6.

For negative values of h, which are absolutely allowed, the result is the same. To the left of zero, h reverses sign and $\sin h$ reverses sign. The ratio $(\sin h)/h$ is unchanged. (The sine is an odd function: $\sin(-h) = -\sin h$.) The ratio is an *even* function, symmetric around zero and approaching 1 from both sides.

The proof depends on $\sin h < h < \tan h$, which is displayed by the graph but not explained. We go back to right triangles.

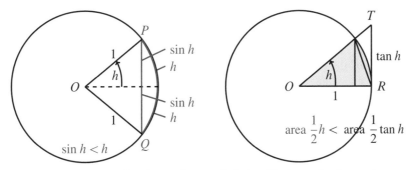

Fig. 2.11 Line shorter than arc: $2 \sin h < 2h$. Areas give $h < \tan h$.

Figure 2.11a shows why $\sin h < h$. The straight line PQ has length $2 \sin h$. The circular arc must be longer, because the shortest distance between two points is a straight line.† The arc PQ has length $2h$. (Important: *When the radius is 1, the arc length equals the angle.* The full circumference is 2π and the full angle is also 2π.) **The straight distance $2 \sin h$ is less than the circular distance $2h$, so $\sin h < h$.**

Figure 2.11b shows why $h < \tan h$. This time we look at **areas**. The triangular area is $\frac{1}{2}$(base)(height) $= \frac{1}{2}(1)(\tan h)$. Inside that triangle is the shaded sector of the circle.

---

† If we try to prove that, we will be here all night. Accept it as true.

Its area is $h/2\pi$ times the area of the whole circle (because the angle is that fraction of the whole angle). The circle has area $\pi r^2 = \pi$, so multiplication by $h/2\pi$ gives $\frac{1}{2}h$ for the area of the sector. Comparing with the triangle around it, $\frac{1}{2}\tan h > \frac{1}{2}h$.

The inequalities $\sin h < h < \tan h$ are now proved. The squeeze in equation (8) produces $(\sin h)/h \to 1$. Q.E.D. Problem 13 shows how to prove $\sin h < h$ from areas.

Note   All angles x and h are being measured in radians. *In degrees, $\cos x$ is not the derivative of $\sin x$.* A degree is much less than a radian, and dy/dx is reduced by the factor $2\pi/360$.

THE LIMIT OF $(\cos h - 1)/h$ IS 0

This second limit is different. We will show that $1 - \cos h$ shrinks to zero *more quickly* than h. Cosines are connected to sines by $(\sin h)^2 + (\cos h)^2 = 1$. We start from the known fact $\sin h < h$ and work it into a form involving cosines:

$$(1 - \cos h)(1 + \cos h) = 1 - (\cos h)^2 = (\sin h)^2 < h^2. \tag{9}$$

Note that everything is positive. Divide through by h and also by $1 + \cos h$:

$$0 < \frac{1 - \cos h}{h} < \frac{h}{1 + \cos h}. \tag{10}$$

Our ratio is caught in the middle. *The right side goes to zero because $h \to 0$.* This is another "*squeeze*"—there is no escape. Our ratio goes to zero.

For $\cos h - 1$ or for negative h, the signs change but minus zero is still zero. This confirms equation (6). The slope of $\sin x$ is $\cos x$.

Remark   Equation (10) also shows that $1 - \cos h$ is approximately $\frac{1}{2}h^2$. The 2 comes from $1 + \cos h$. This is a basic purpose of calculus—to find simple approximations like $\frac{1}{2}h^2$. A "tangent parabola" $1 - \frac{1}{2}h^2$ is close to the top of the cosine curve.

THE DERIVATIVE OF THE COSINE

This will be easy. The quick way to differentiate $\cos x$ is to shift the sine curve by $\pi/2$. That yields the cosine curve (solid line in Figure 2.12b). The derivative also shifts by $\pi/2$ (dotted line). ***The derivative of*** $\cos x$ ***is*** $-\sin x$.

Notice how the dotted line (the slope) goes below zero when the solid line turns downward. The slope equals zero when the solid line is level. ***Increasing functions have positive slopes. Decreasing functions have negative slopes.*** That is important, and we return to it.

There is more information in dy/dx than "function rising" or "function falling." The slope tells *how quickly* the function goes up or down. It gives the *rate of change*. The slope of $y = \cos x$ can be computed in the normal way, as the limit of $\Delta y/\Delta x$:

$$\frac{\Delta y}{\Delta x} = \frac{\cos(x + h) - \cos x}{h} = \cos x \left(\frac{\cos h - 1}{h}\right) - \sin x \left(\frac{\sin h}{h}\right)$$

$$\frac{dy}{dx} = (\cos x)(0) - (\sin x)(1) = -\sin x. \tag{11}$$

The first line came from formula (3) for $\cos(x + h)$. The second line took limits, reaching 0 and 1 as before. This confirms the graphical proof that the slope of $\cos x$ is $-\sin x$.

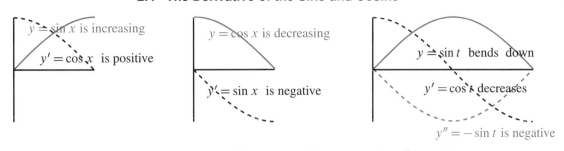

Fig. 2.12 $y(x)$ increases where y' is positive. $y(x)$ bends up where y'' is positive.

THE SECOND DERIVATIVES OF THE SINE AND COSINE

We now introduce ***the derivative of the derivative***. That is the ***second derivative*** of the original function. It tells how fast the slope is changing, not how fast y itself is changing. The second derivative is the "rate of change of the velocity." A straight line has constant slope (constant velocity), so its second derivative is zero:

$$f(t) = 5t \quad \text{has} \quad df/dt = 5 \quad \text{and} \quad d^2 f/dt^2 = 0.$$

The parabola $y = x^2$ has slope $2x$ (linear) which has slope 2 (constant). Similarly

$$f(t) = \tfrac{1}{2}at^2 \quad \text{has} \quad df/dt = at \quad \text{and} \quad d^2 f/dt^2 = a.$$

There stands the notation $d^2 f/dt^2$ (or $d^2 y/dx^2$) for the second derivative. A short form is f'' or y''. (This is pronounced f *double prime* or y *double prime*). Example: The second derivative of $y = x^3$ is $y'' = 6x$.

In the distance-velocity problem, f'' is *acceleration*. It tells how fast v is changing, while v tells how fast f is changing. Where df/dt was distance/time, the second derivative is distance/(time)². The acceleration due to gravity is about 32 ft/sec² or 9.8 m/sec², which means that v increases by 32 ft/sec in one second. It does not mean that the distance increases by 32 feet!

The graph of $y = \sin t$ increases at the start. Its derivative $\cos t$ is positive. However the second derivative is $-\sin t$. ***The curve is bending down while going up***. The arch is "***concave down***" because $y'' = -\sin t$ is negative.

At $t = \pi$ the curve reaches zero and goes negative. The second derivative becomes positive. *Now the curve bends upward*. The lower arch is "***concave up***."

$y'' > 0$ means that y' increases so y bends upward (concave up)

$y'' < 0$ means that y' decreases so y bends down (concave down).

Chapter 3 studies these things properly—here we get an advance look for $\sin t$.

The remarkable fact about the sine and cosine is that $y'' = -y$. That is unusual and special: *acceleration $= -$distance*. The greater the distance, the greater the force pulling back:

$$y = \sin t \quad \text{has} \quad dy/dt = +\cos t \quad \text{and} \quad d^2 y/dt^2 = -\sin t = -y.$$
$$y = \cos t \quad \text{has} \quad dy/dt = -\sin t \quad \text{and} \quad d^2 y/dt^2 = -\cos t = -y.$$

Question Does $d^2 y/dt^2 < 0$ mean that the distance $y(t)$ is decreasing ?
Answer No. Absolutely not! It means that dy/dt is decreasing, not necessarily y. At the start of the sine curve, y is still increasing but $y'' < 0$.

Sines and cosines give *simple harmonic motion*—up and down, forward and back, out and in, tension and compression. Stretch a spring, and the restoring force pulls it back. Push a swing up, and gravity brings it down. These motions are controlled by a **differential equation**:

$$\frac{d^2 y}{dt^2} = -y. \tag{12}$$

All solutions are combinations of the sine and cosine: $y = A \sin t + B \cos t$.

This is not a course on differential equations. But you have to see the purpose of calculus. It models events by equations. It models oscillation by equation (12). Your heart fills and empties. Balls bounce. Current alternates. The economy goes up and down:

high prices $\rightarrow$ high production $\rightarrow$ low prices $\rightarrow$ $\cdots$

We can't live without oscillations (or differential equations).

2.4 EXERCISES

Read-through questions

The derivative of $y = \sin x$ is $y' =$ __a__ . The second derivative (the __b__ of the derivative) is $y'' =$ __c__ . The fourth derivative is $y'''' =$ __d__ . Thus $y = \sin x$ satisfies the differential equations $y'' =$ __e__ and $y'''' =$ __f__ . So does $y = \cos x$, whose second derivative is __g__ .

All these derivatives come from one basic limit: $(\sin h)/h$ approaches __h__ . The sine of .01 radians is very close to __i__ . So is the __j__ of .01. The cosine of .01 is not .99, because $1 - \cos h$ is much __k__ than h. The ratio $(1 - \cos h)/h^2$ approaches __l__ . Therefore $\cos h$ is close to $1 - \frac{1}{2}h^2$ and $\cos .01 \approx$ __m__ . We can replace h by x.

The differential equation $y'' = -y$ leads to __n__ . When y is positive, y'' is __o__ . Therefore y' is __p__ . Eventually y goes below zero and y'' becomes __q__ . Then y' is __r__ . Examples of oscillation in real life are __s__ and __t__ .

1 Which of these ratios approach 1 as $h \rightarrow 0$?

(a) $\dfrac{h}{\sin h}$ (b) $\dfrac{\sin^2 h}{h^2}$ (c) $\dfrac{\sin h}{\sin 2h}$ (d) $\dfrac{\sin(-h)}{h}$

2 (Calculator) Find $(\sin h)/h$ at $h = 0.5$ and 0.1 and $.01$. Where does $(\sin h)/h$ go above $.99$?

3 Find the limits as $h \rightarrow 0$ of

(a) $\dfrac{\sin^2 h}{h}$ (b) $\dfrac{\sin 5h}{5h}$ (c) $\dfrac{\sin 5h}{h}$ (d) $\dfrac{\sin h}{5h}$

4 Where does $\tan h = 1.01h$? Where does $\tan h = h$?

5 $y = \sin x$ has period 2π, which means that $\sin x =$ ___ . The limit of $(\sin(2\pi + h) - \sin 2\pi)/h$ is 1 because ___ . This gives dy/dx at $x =$ ___ .

6 Draw $\cos(x + \Delta x)$ next to $\cos x$. Mark the height difference Δy. Then draw $\Delta y/\Delta x$ as in Figure 2.9.

7 The key to trigonometry is $\cos^2 \theta = 1 - \sin^2 \theta$. Set $\sin \theta \approx \theta$ to find $\cos^2 \theta \approx 1 - \theta^2$. *The square root is* $\cos \theta \approx 1 - \frac{1}{2}\theta^2$. Reason: Squaring gives $\cos^2 \theta \approx$ ___ and the correction term ___ is very small near $\theta = 0$.

8 (Calculator) Compare $\cos \theta$ with $1 - \frac{1}{2}\theta^2$ for

(a) $\theta = 0.1$ (b) $\theta = 0.5$ (c) $\theta = 30°$ (d) $\theta = 30°$.

9 Trigonometry gives $\cos \theta = 1 - 2\sin^2 \frac{1}{2}\theta$. The approximation $\sin \frac{1}{2}\theta \approx$ ___ leads directly to $\cos \theta \approx 1 - \frac{1}{2}\theta^2$.

10 Find the limits as $h \rightarrow 0$:

(a) $\dfrac{1 - \cos h}{h^2}$ (b) $\dfrac{1 - \cos^2 h}{h^2}$

(c) $\dfrac{1 - \cos^2 h}{\sin^2 h}$ (d) $\dfrac{1 - \cos 2h}{h}$

11 Find by calculator or calculus:

(a) $\lim\limits_{h \rightarrow 0} \dfrac{\sin 3h}{\sin 2h}$ (b) $\lim\limits_{h \rightarrow 0} \dfrac{1 - \cos 2h}{1 - \cos h}$.

12 Compute the slope at $x = 0$ directly from limits:

(a) $y = \tan x$ (b) $y = \sin(-x)$

13 The unmarked points in Figure 2.11 are P and S. Find the height PS and the area of triangle OPR. Prove by areas that $\sin h < h$.

14 The slopes of $\cos x$ and $1 - \frac{1}{2}x^2$ are $-\sin x$ and ___ . The slopes of $\sin x$ and ___ are $\cos x$ and $1 - \frac{1}{2}x^2$.

15 Chapter 10 gives an infinite series for $\sin x$:

$$\sin x = \frac{x}{1} - \frac{x^3}{3 \cdot 2 \cdot 1} + \frac{x^5}{5 \cdot 4 \cdot 3 \cdot 2 \cdot 1} - \cdots .$$

From the derivative find the series for $\cos x$. Then take *its* derivative to get back to $-\sin x$.

16 A *centered difference* for $f(x) = \sin x$ is

$$\frac{f(x+h) - f(x-h)}{2h} = \frac{\sin(x+h) - \sin(x-h)}{2h} = ?$$

Use the addition formula (2). Then let $h \to 0$.

17 Repeat Problem 16 to find the slope of $\cos x$. Use formula (3) to simplify $\cos(x+h) - \cos(x-h)$.

18 Find the tangent line to $y = \sin x$ at

(a) $x = 0$ (b) $x = \pi$ (c) $x = \pi/4$

19 Where does $y = \sin x + \cos x$ have zero slope ?

20 Find the derivative of $\sin(x+1)$ in two ways:

(a) Expand to $\sin x \cos 1 + \cos x \sin 1$. Compute dy/dx.
(b) Divide $\Delta y = \sin(x+1+\Delta x) - \sin(x+1)$ by Δx. Write X instead of $x+1$. Let Δx go to zero.

21 Show that $(\tan h)/h$ is squeezed between 1 and $1/\cos h$. As $h \to 0$ the limit is _____ .

22 For $y = \sin 2x$, the ratio $\Delta y/h$ is

$$\frac{\sin 2(x+h) - \sin 2x}{h} = \frac{\sin 2x(\cos 2h - 1) + \cos 2x \sin 2h}{h}.$$

Explain why the limit dy/dx is $2 \cos 2x$.

23 Draw the graph of $y = \sin \frac{1}{2}x$. State its slope at $x = 0, \pi/2, \pi$, and $2/\pi$. Does $\frac{1}{2}\sin x$ have the same slopes ?

24 Draw the graph of $y = \sin x + \sqrt{3} \cos x$. Its maximum value is $y =$ _____ at $x =$ _____ . The slope at that point is _____ .

25 By combining $\sin x$ and $\cos x$, find a combination that starts at $x = 0$ from $y = 2$ with slope 1. This combination also solves $y'' =$ _____ .

26 *True or false*, with reason:

(a) The derivative of $\sin^2 x$ is $\cos^2 x$
(b) The derivative of $\cos(-x)$ is $\sin x$
(c) A positive function has a negative second derivative.
(d) If y' is increasing then y'' is positive.

27 Find solutions to $dy/dx = \sin 3x$ and $dy/dx = \cos 3x$.

28 If $y = \sin 5x$ then $y' = 5\cos 5x$ and $y'' = -25 \sin 5x$. So this function satisfies the differential equation $y'' =$ _____ .

29 If h is measured in degrees, find $\lim_{h \to 0} (\sin h)/h$. You could set your calculator in degree mode.

30 Write down a ratio that approaches dy/dx at $x = \pi$. For $y = \sin x$ and $\Delta x = .01$ compute that ratio.

31 By the square rule, the derivative of $(u(x))^2$ is $2u\, du/dx$. Take the derivative of each term in $\sin^2 x + \cos^2 x = 1$.

32 Give an example of oscillation that does not come from physics. Is it simple harmonic motion (one frequency only) ?

33 Explain the second derivative in your own words.

2.5 The Product and Quotient and Power Rules

What are the derivatives of $x + \sin x$ and $x \sin x$ and $1/\sin x$ and $x/\sin x$ and $\sin^n x$? Those are made up from the familiar pieces x and $\sin x$, but we need new rules. Fortunately they are rules that apply to every function, so they can be established once and for all. If we know the separate derivatives of two functions u and v, then the derivatives of $u + v$ and uv and $1/v$ and u/v and u^n are immediately available.

This is a straightforward section, with those five rules to learn. It is also an important section, containing most of the working tools of differential calculus. But I am afraid that five rules and thirteen examples (which we need—the eyes glaze over with formulas alone) make a long list. At least the easiest rule comes first. *When we add functions, we add their derivatives*.

Sum Rule

The derivative of the sum $u(x) + v(x)$ is $\dfrac{d}{dx}(u+v) = \dfrac{du}{dx} + \dfrac{dv}{dx}$.   (1)

EXAMPLE 1   The derivative of $x + \sin x$ is $1 + \cos x$. That is tremendously simple, but it is fundamental. The interpretation for distances may be more confusing (and more interesting) than the rule itself:

> Suppose a train moves with velocity 1. The distance at time t is t. On the train a professor paces back and forth (in simple harmonic motion). His distance from his seat is $\sin t$. Then the total distance from his starting point is $t + \sin t$, and his velocity (train speed plus walking speed) is $1 + \cos t$.

If you add distances, you add velocities. Actually that example is ridiculous, because the professor's maximum speed equals the train speed ($= 1$). He is running like mad, not pacing. Occasionally he is standing still with respect to the ground.

The sum rule is a special case of a bigger rule called "*linearity*." It applies when we add or subtract functions and multiply them by constants—as in $3x - 4\sin x$. By linearity the derivative is $3 - 4\cos x$. The rule works for all functions $u(x)$ and $v(x)$. A *linear combination* is $y(x) = au(x) + bv(x)$, where a and b are any real numbers. Then $\Delta y/\Delta x$ is

$$\frac{au(x+\Delta x) + bv(x+\Delta x) - au(x) - bv(x)}{\Delta x} = a\frac{u(x+\Delta x) - u(x)}{\Delta x} + b\frac{v(x+\Delta x) - v(x)}{\Delta x}.$$

The limit on the left is dy/dx. The limit on the right is $a\,du/dx + b\,dv/dx$. We are allowed to take limits separately and add. The result is what we hope for:

Rule of Linearity

The derivative of $au(x) + bv(x)$ is $\dfrac{d}{dx}(au+bv) = a\dfrac{du}{dx} + b\dfrac{dv}{dx}$.   (2)

The **product rule** comes next. It can't be so simple—products are not linear. The sum rule is what you would have done anyway, but products give something new. **The derivative of u times v is not du/dx times dv/dx.** Example: The derivative of x^5 is $5x^4$. Don't multiply the derivatives of x^3 and x^2. ($3x^2$ times $2x$ is not $5x^4$.) *For a product of two functions, the derivative has two terms.*

Product Rule (the key to this section)

The derivative of $u(x)v(x)$ is $\dfrac{d}{dx}(uv) = u\dfrac{dv}{dx} + v\dfrac{du}{dx}$.   (3)

EXAMPLE 2 $u = x^3$ times $v = x^2$ is $uv = x^5$. The product rule leads to $5x^4$:

$$x^3 \frac{dv}{dx} + x^2 \frac{du}{dx} = x^3(2x) + x^2(3x^2) = 2x^4 + 3x^4 = 5x^4.$$

EXAMPLE 3 In the slope of $x \sin x$, I don't write $dx/dx = 1$ but it's there:

$$\frac{d}{dx}(x \sin x) = x \cos x + \sin x.$$

EXAMPLE 4 If $u = \sin x$ and $v = \sin x$ then $uv = \sin^2 x$. We get two equal terms:

$$\sin x \frac{d}{dx}(\sin x) + \sin x \frac{d}{dx}(\sin x) = 2 \sin x \cos x.$$

This confirms the "square rule" $2u \, du/dx$, when u is the same as v. Similarly the slope of $\cos^2 x$ is $-2 \cos x \sin x$ (minus sign from the slope of the cosine).

Question Those answers for $\sin^2 x$ and $\cos^2 x$ have opposite signs, so the derivative of $\sin^2 x + \cos^2 x$ is zero (sum rule). How do you see that more quickly?

EXAMPLE 5 The derivative of uvw is $uvw' + uv'w + u'vw$—one derivative at a time. The derivative of xxx is $xx + xx + xx$.

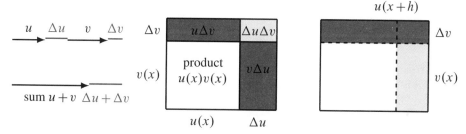

Fig. 2.13 Change in length $= \Delta u + \Delta v$. Change in area $= u \, \Delta v + v \, \Delta u + \Delta u \, \Delta v$.

After those examples we prove the product rule. Figure 2.13 explains it best. The area of the big rectangle is uv. *The important changes in area are the two strips $u \, \Delta v$ and $v \, \Delta u$.* The corner area $\Delta u \, \Delta v$ is much smaller. When we divide by Δx, the strips give $u \, \Delta v/\Delta x$ and $v \, \Delta u/\Delta x$. The corner gives $\Delta u \, \Delta v/\Delta x$, which approaches zero.

Notice how the sum rule is in one dimension and the product rule is in two dimensions. The rule for uvw would be in three dimensions.

The extra area comes from the whole top strip plus the side strip. By algebra,

$$u(x+h)v(x+h) - u(x)v(x) = u(x+h)[v(x+h) - v(x)] + v(x)[u(x+h) - u(x)]. \quad (4)$$

This increase is $u(x+h)\Delta v + v(x)\Delta u$—top plus side. *Now divide by h (or Δx) and let $h \to 0$*. The left side of equation (4) becomes the derivative of $u(x)v(x)$. The right side becomes $u(x)$ times dv/dx—*we can multiply the two limits*—plus $v(x)$ times du/dx. That proves the product rule—definitely useful.

We could go immediately to the quotient rule for $u(x)/v(x)$. But start with $u = 1$. The derivative of $1/x$ is $-1/x^2$ (known). What is the derivative of $1/v(x)$?

Reciprocal Rule

The derivative of $\dfrac{1}{v(x)}$ is $\dfrac{-dv/dx}{v^2}$. $\quad (5)$

The proof starts with $(v)(1/v) = 1$. The derivative of 1 is 0. Apply the product rule:

$$v \frac{d}{dx}\left(\frac{1}{v}\right) + \frac{1}{v}\frac{dv}{dx} = 0 \quad \text{so that} \quad \frac{d}{dx}\left(\frac{1}{v}\right) = \frac{-dv/dx}{v^2}. \tag{6}$$

It is worth checking the units—in the reciprocal rule and others. A test of dimensions is automatic in science and engineering, and a good idea in mathematics. The test ignores constants and plus or minus signs, but it prevents bad errors. If v is in dollars and x is in hours, dv/dx is in *dollars per hour*. Then dimensions agree:

$$\frac{d}{dx}\left(\frac{1}{v}\right) \approx \frac{(1/\text{dollars})}{\text{hour}} \quad \text{and also} \quad \frac{-dv/dx}{v^2} \approx \frac{\text{dollars/hour}}{(\text{dollars})^2}.$$

From this test, the derivative of $1/v$ cannot be $1/(dv/dx)$. A similar test shows that Einstein's formula $e = mc^2$ is dimensionally possible. The theory of relativity might be correct! Both sides have the dimension of $(\text{mass})(\text{distance})^2/(\text{time})^2$, when mass is converted to energy.†

EXAMPLE 6   The derivatives of x^{-1}, x^{-2}, x^{-n} are $-1x^{-2}, -2x^{-3}, -nx^{-n-1}$.

Those come from the reciprocal rule with $v = x$ and x^2 and any x^n:

$$\frac{d}{dx}(x^{-n}) = \frac{d}{dx}\left(\frac{1}{x^n}\right) = -\frac{nx^{n-1}}{(x^n)^2} = -nx^{-n-1}.$$

The beautiful thing is that this answer $-nx^{-n-1}$ fits into the same pattern as x^n. *Multiply by the exponent and reduce it by one.*

> *For negative and positive exponents the derivative of x^n is nx^{n-1}.*   (7)

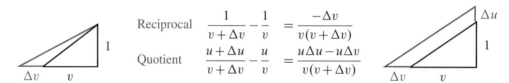

Reciprocal	$\dfrac{1}{v+\Delta v} - \dfrac{1}{v} = \dfrac{-\Delta v}{v(v+\Delta v)}$	
Quotient	$\dfrac{u+\Delta u}{v+\Delta v} - \dfrac{u}{v} = \dfrac{u\Delta u - u\Delta v}{v(v+\Delta v)}$	

Fig. 2.14   Reciprocal rule from $(-\Delta v)/v^2$. Quotient rule from $(v\Delta u - u\Delta v)/v^2$.

EXAMPLE 7   The derivatives of $\dfrac{1}{\cos x}$ and $\dfrac{1}{\sin x}$ are $\dfrac{+\sin x}{\cos^2 x}$ and $\dfrac{-\cos x}{\sin^2 x}$.

Those come directly from the reciprocal rule. In trigonometry, $1/\cos x$ is the ***secant*** of the angle x, and $1/\sin x$ is the ***cosecant*** of x. Now we have their derivatives:

$$\frac{d}{dx}(\sec x) = \frac{\sin x}{\cos^2 x} = \frac{1}{\cos x}\frac{\sin x}{\cos x} = \sec x \tan x. \tag{8}$$

$$\frac{d}{dx}(\csc x) = -\frac{\cos x}{\sin^2 x} = -\frac{1}{\sin x}\frac{\cos x}{\sin x} = -\csc x \cot x. \tag{9}$$

---

†But only Einstein knew that the constant is 1.

Those formulas are often seen in calculus. If you have a good memory they are worth storing. Like most mathematicians, I have to check them every time before using them (maybe once a year). It is really the rules that are basic, not the formulas.

The next rule applies to the quotient $u(x)/v(x)$. That is u times $1/v$. Combining the product rule and reciprocal rule gives something new and important:

Quotient Rule

The derivative of $\dfrac{u(x)}{v(x)}$ is $\dfrac{1}{v}\dfrac{du}{dx} - u\dfrac{dv/dx}{v^2} = \dfrac{v\,du/dx - u\,dv/dx}{v^2}$.

You *must* memorize that last formula. The v^2 is familiar. The rest is new, but not very new. If $v = 1$ the result is du/dx (of course). For $u = 1$ we have the reciprocal rule. Figure 2.14b shows the difference $(u + \Delta u)/(v + \Delta v) - (u/v)$. The denominator $v(v + \Delta v)$ is responsible for v^2.

EXAMPLE 8   (only practice) If $u/v = x^5/x^3$ (which is x^2) the quotient rule gives $2x$:

$$\frac{d}{dx}\left(\frac{x^5}{x^3}\right) = \frac{x^3(5x^4) - x^5(3x^2)}{x^6} = \frac{5x^7 - 3x^7}{x^6} = 2x.$$

EXAMPLE 9   (important) For $u = \sin x$ and $v = \cos x$, the quotient is $\sin x/\cos x = \tan x$. **The derivative of** $\tan x$ **is** $\sec^2 x$. Use the quotient rule and $\cos^2 x + \sin^2 x = 1$:

$$\frac{d}{dx}\left(\frac{\sin x}{\cos x}\right) = \frac{\cos x\,(\cos x) - \sin x\,(-\sin x)}{\cos^2 x} = \frac{1}{\cos^2 x} = \sec^2 x. \tag{11}$$

Again to memorize: $(\tan x)' = \sec^2 x$. At $x = 0$, this slope is 1. The graphs of $\sin x$ and x and $\tan x$ all start with this slope (then they separate). At $x = \pi/2$ the sine curve is flat ($\cos x = 0$) and the tangent curve is vertical ($\sec^2 x = \infty$).

The slope generally blows up faster than the function. We divide by $\cos x$, once for the tangent and twice for its slope. The slope of $1/x$ is $-1/x^2$. The slope is more sensitive than the function, because of the square in the denominator.

EXAMPLE 10 $\dfrac{d}{dx}\left(\dfrac{\sin x}{x}\right) = \dfrac{x\cos x - \sin x}{x^2}.$

That one I hesitate to touch at $x = 0$. Formally it becomes $0/0$. In reality it is more like $0^3/0^2$, and the true derivative is zero. Figure 2.10 showed graphically that $(\sin x)/x$ is flat at the center point. The function is *even* (symmetric across the y axis) so its derivative can only be zero.

This section is full of rules, and I hope you will allow one more. It goes beyond x^n to $(u(x))^n$. A power of x changes to a power of $u(x)$—as in $(\sin x)^6$ or $(\tan x)^7$ or $(x^2 + 1)^8$. The derivative contains nu^{n-1} (copying nx^{n-1}), but **there is an extra factor** du/dx. Watch that factor in $6(\sin x)^5\cos x$ and $7(\tan x)^6\sec^2 x$ and $8(x^2 + 1)^7(2x)$:

Power Rule

The derivative of $\Big[u(x)\Big]^n$ is $n\Big[u(x)\Big]^{n-1}\dfrac{du}{dx}.$ \hfill (12)

For $n = 1$ this reduces to $du/dx = du/dx$. For $n = 2$ we get the square rule $2u\,du/dx$. Next comes u^3. The best approach is to use **mathematical induction**, which goes from each n to the next power $n+1$ by the product rule:

$$\frac{d}{dx}(u^{n+1}) = \frac{d}{dx}(u^n u) = u^n \frac{du}{dx} + u\left(nu^{n-1}\frac{du}{dx}\right) = (n+1)u^n \frac{du}{dx}.$$

That is exactly equation (12) for the power $n+1$. We get all positive powers this way, going up from $n=1$—then the negative powers come from the reciprocal rule.

Figure 2.15 shows the power rule for $n=1,2,3$. The cube makes the point best. The three thin slabs are u by u by Δu. **The change in volume is essentially $3u^2\Delta u$.** From multiplying out $(u+\Delta u)^3$, the exact change in volume is $3u^2\Delta u + 3u(\Delta u)^2 + (\Delta u)^3$—which also accounts for three narrow boxes and a midget cube in the corner. This is the binomial formula in a picture.

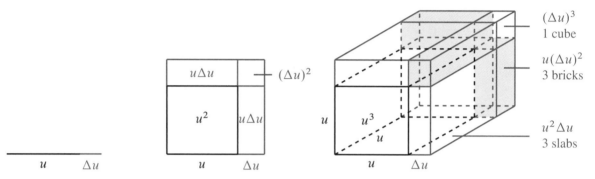

Fig. 2.15   Length change $= \Delta u$; area change $\approx 2u\Delta u$; volume change $\approx 3u^2\Delta u$.

EXAMPLE 11 $\dfrac{d}{dx}(\sin x)^n = n(\sin x)^{n-1}\cos x$. The extra factor $\cos x$ is du/dx.

Our last step finally escapes from a very undesirable restriction—that n must be a whole number. We want to allow fractional powers $n=p/q$, and keep the same formula. *The derivative of x^n is still nx^{n-1}.*

To deal with square roots I can write $(\sqrt{x})^2 = x$. Its derivative is $2\sqrt{x}(\sqrt{x})' = 1$. Therefore $(\sqrt{x})'$ is $1/2\sqrt{x}$, which fits the formula when $n=\frac{1}{2}$. Now try $n=p/q$:

Fractional powers   Write $u=x^{p/q}$ as $u^q = x^p$. Take derivatives, assuming they exist:

$$qu^{q-1}\frac{du}{dx} = px^{p-1} \qquad \text{(power rule on both sides)}$$

$$\frac{du}{dx} = \frac{px^{-1}}{qu^{-1}} \qquad \text{(cancel } x^p \text{ with } u^q)$$

$$\frac{du}{dx} = nx^{n-1} \qquad \text{(replace } p/q \text{ by } n \text{ and } u \text{ by } x^n)$$

EXAMPLE 12   The slope of $x^{1/3}$ is $\frac{1}{3}x^{-2/3}$. The slope is infinite at $x=0$ and zero at $x=\infty$. But the curve in Figure 2.16 keeps climbing. It doesn't stay below an "asymptote."

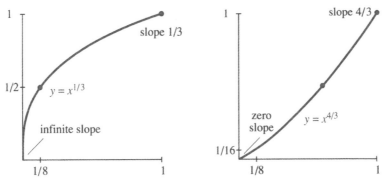

Fig. 2.16 Infinite slope of x^n versus zero slope: the difference between $0 < n < 1$ and $n > 1$.

EXAMPLE 13 The slope of $x^{4/3}$ is $\frac{4}{3}x^{1/3}$. The slope is zero at $x = 0$ and infinite at $x = \infty$. The graph climbs faster than a line and slower than a parabola ($\frac{4}{3}$ is between 1 and 2). Its slope follows the cube root curve (times $\frac{4}{3}$).

WE STOP NOW! I am sorry there were so many rules. A computer can memorize them all, but it doesn't know what they mean and you do. Together with the chain rule that dominates Chapter 4, they achieve virtually all the derivatives ever computed by mankind. We list them in one place for convenience.

Rule of Linearity	$(au + bv)' = au' + bv'$
Product Rule	$(uv)' = uv' + vu'$
Reciprocal Rule	$(1/v)' = -v'/v^2$
Quotient Rule	$(u/v)' = (vu' - uv')/v^2$
Power Rule	$(u^n)' = nu^{n-1}u'$

The power rule applies when n is *negative*, or a *fraction*, or **any real number**. The derivative of x^π is $\pi x^{\pi - 1}$, according to Chapter 6. The derivative of $(\sin x)^\pi$ is _____. And the derivatives of all six trigonometric functions are now established:

$$(\sin x)' = \cos x \qquad (\tan x)' = \sec^2 x \qquad (\sec x)' = \sec x \tan x$$
$$(\cos x)' = -\sin x \qquad (\cot x)' = -\csc^2 x \qquad (\csc x)' = -\csc x \cot x.$$

2.5 EXERCISES

Read-through questions

The derivatives of $\sin x \cos x$ and $1/\cos x$ and $\sin x/\cos x$ and $\tan^3 x$ come from the __a__ rule, __b__ rule, __c__ rule, and __d__ rule. The product of $\sin x$ times $\cos x$ has $(uv)' = uv' + $ __e__ $=$ __f__ . The derivative of $1/v$ is __g__ , so the slope of $\sec x$ is __h__ . The derivative of u/v is __i__ , so the slope of $\tan x$ is __j__ . The derivative of $\tan^3 x$ is __k__ . The slope of x^n is __l__ and the slope of $(u(x))^n$ is __m__ . With $n = -1$ the derivative of $(\cos x)^{-1}$ is __n__ , which agrees with the rule for $\sec x$.

Even simpler is the rule of __o__ , which applies to $au(x) + bv(x)$. The derivative is __p__ . The slope of $3 \sin x + 4 \cos x$ is __q__ . The derivative of $(3 \sin x + 4 \cos x)^2$ is __r__ . The derivative of __s__ is $4 \sin^3 x \cos x$.

Find the derivatives of the functions in 1–26.

1 $(x + 1)(x - 1)$

2 $(x^2 + 1)(x^2 - 1)$

3 $\dfrac{1}{1 + x} + \dfrac{1}{1 + \sin x}$

4 $\dfrac{1}{1 + x^2} + \dfrac{1}{1 - \sin x}$

2 Derivatives

5 $(x-1)(x-2)(x-3)$

6 $(x-1)^2(x-2)^2$

7 $x^2 \cos x + 2x \sin x$

8 $x^{1/2}(x + \sin x)$

9 $\dfrac{x^3+1}{x+1} + \dfrac{\cos x}{\sin x}$

10 $\dfrac{x^2+1}{x^2-1} + \dfrac{\sin x}{\cos x}$

11 $x^{1/2}\sin^2 x + (\sin x)^{1/2}$

12 $x^{3/2}\sin^3 x + (\sin x)^{3/2}$

13 $x^4 \cos x + x\cos^4 x$

14 $\sqrt{x}(\sqrt{x}+1)(\sqrt{x}+2)$

15 $\frac{1}{2}x^2 \sin x - x \cos x + \sin x$

16 $(x-6)^{10} + \sin^{10} x$

17 $\sec^2 x - \tan^2 x$

18 $\csc^2 x - \cot^2 x$

19 $\dfrac{4}{(x-5)^{2/3}} + \dfrac{4}{(5-x)^{2/3}}$

20 $\dfrac{\sin x - \cos x}{\sin x + \cos x}$

21 $(\sin x \cos x)^3 + \sin 2x$

22 $x \cos x \csc x$

23 $u(x)v(x)w(x)z(x)$

24 $[u(x)]^2 [v(x)]^2$

25 $\dfrac{1}{\tan x} - \dfrac{1}{\cot x}$

26 $x \sin x + \cos x$

27 A growing box has length t, width $1/(1+t)$, and height $\cos t$.

   (a) What is the rate of change of the volume ?

   (b) What is the rate of change of the surface area ?

28 With two applications of the product rule show that the derivative of uvw is $uvw' + uv'w + u'vw$. When a box with sides u, v, w grows by $\Delta u, \Delta v, \Delta w$, three slabs are added with volume $uv\,\Delta w$ and _____ and _____ .

29 Find the velocity if the distance is $f(t) = $

   $5t^2$ for $t \leqslant 10,$ $500 + 100\sqrt{t-10}$ for $t \geqslant 10$.

30 A cylinder has radius $r = \dfrac{t^{3/2}}{1+t^{3/2}}$ and height $h = \dfrac{1}{1+t}$.

   (a) What is the rate of change of its volume ?

   (b) What is the rate of change of its surface area (including top and base) ?

31 The height of a model rocket is $f(t) = t^3/(1+t)$.

   (a) What is the velocity $v(t)$?

   (b) What is the acceleration dv/dt ?

32 Apply the product rule to $u(x)u^2(x)$ to find the power rule for $u^3(x)$.

33 Find the *second* derivative of the product $u(x)v(x)$. Find the *third* derivative. Test your formulas on $u = v = x$.

34 Find functions $y(x)$ whose derivatives are

   (a) x^3 (b) $1/x^3$ (c) $(1-x)^{3/2}$ (d) $\cos^2 x \sin x$

35 Find the distances $f(t)$, starting from $f(0) = 0$, to match these velocities:

   (a) $v(t) = \cos t \sin t$ (b) $v(t) = \tan t \sec^2 t$

   (c) $v(t) = \sqrt{1+t}$

36 Apply the quotient rule to $(u(x))^3/(u(x))^2$ and $-v'/v^2$. The latter gives the second derivative of _____ .

37 Draw a figure like 2.13 to explain the *square rule*.

38 Give an example where $u(x)/v(x)$ is increasing but $du/dx = dv/dx = 1$.

39 **True or false**, with a good reason:

   (a) The derivative of x^{2n} is $2nx^{2n-1}$.

   (b) By linearity the derivative of $a(x)u(x) + b(x)v(x)$ is $a(x)du/dx + b(x)dv/dx$.

   (c) The derivative of $|x|^3$ is $3|x|^2$.

   (d) $\tan^2 x$ and $\sec^2 x$ have the same derivative.

   (e) $(uv)' = u'v'$ is true when $u(x) = 1$.

40 The cost of u shares of stock at v dollars per share is uv dollars. Check dimensions of $d(uv)/dt$ and $u\,dv/dt$ and $v\,du/dt$.

41 If $u(x)/v(x)$ is a ratio of polynomials of degree n, what are the degrees for its derivative ?

42 For $y = 5x + 3$, is $(dy/dx)^2$ the same as d^2y/dx^2 ?

43 If you change from $f(t) = t \cos t$ to its tangent line at $t = \pi/2$, find the two-part function df/dt.

44 Explain in your own words why the derivative of $u(x)v(x)$ has two terms.

45 A plane starts its descent from height $y = h$ at $x = -L$ to land at $(0,0)$. Choose a,b,c,d so its landing path $y = ax^3 + bx^2 + cx + d$ is **smooth**. With $dx/dt = V = $ constant, find dy/dt and d^2y/dt^2 at $x = 0$ and $x = -L$. (To keep d^2y/dt^2 small, a coast-to-coast plane starts down $L > 100$ miles from the airport.)

2.6 Limits

You have seen enough limits to be ready for a definition. It is true that we have survived this far without one, and we could continue. But this seems a reasonable time to define limits more carefully. The goal is to achieve rigor without rigor mortis.

First you should know that limits of $\Delta y/\Delta x$ are by no means the only limits in mathematics. Here are five completely different examples. They involve $n \to \infty$, not $\Delta x \to 0$:

1. $a_n = (n-3)/(n+3)$ (for large n, ignore the 3's and find $a_n \to 1$)
2. $a_n = \frac{1}{2}a_{n-1}+4$ (start with any a_1 and always $a \to 8$)
3. $a_n =$ probability of living to year n (unfortunately $a_n \to 0$)
4. $a_n =$ fraction of zeros among the first n digits of π ($a_n \to \frac{1}{10}$?)
5. $a_1 = .4, a_2 = .49, a_3 = .493, \ldots$ *No matter what the remaining decimals are, the a's converge to a limit.* Possibly $a_n \to .493000\ldots$, but not likely.

The problem is to say what the limit symbol $\to$ really means.

A good starting point is to ask about convergence to *zero*. When does a sequence of positive numbers approach zero? What does it mean to write $a_n \to 0$? The numbers $a_1, a_2, a_3, \ldots$, must become "small," but that is too vague. We will propose four definitions of **convergence to zero**, and I hope the right one will be clear.

1. *All the numbers a_n are below 10^{-10}.* That may be enough for practical purposes, but it certainly doesn't make the a_n approach zero.

2. *The sequence is getting closer to zero*—each a_{n+1} is smaller than the preceding a_n. This test is met by $1.1, 1.01, 1.001, \ldots$ which converges to 1 instead of 0.

3. *For any small number you think of, at least one of the a_n's is smaller.* That pushes something toward zero, but not necessarily the whole sequence. The condition would be satisfied by $1, \frac{1}{2}, 1, \frac{1}{3}, 1, \frac{1}{4}, \ldots$, which does not approach zero.

4. *For any small number you think of, the a_n's eventually go below that number and **stay below**.* This is the correct definition.

I want to repeat that. To test for convergence to zero, start with a small number—say 10^{-10}. The a_n's must go *below that number*. They may come back up and go below again—the first million terms make absolutely no difference. Neither do the next billion, but eventually all terms must go below 10^{-10}. After waiting longer (possibly a lot longer), all terms drop below 10^{-20}. The tail end of the sequence decides everything.

Question 1   Does the sequence $10^{-3}, 10^{-2}, 10^{-6}, 10^{-5}, 10^{-9}, 10^{-8}, \ldots$ approach 0?
Answer   Yes, These up and down numbers eventually stay below any ε.

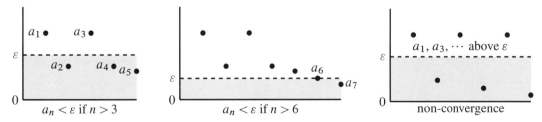

Fig. 2.17   Convergence means: Only a finite number of a's are outside any strip around L.

Question 2   Does $10^{-4}, 10^{-6}, 10^{-4}, 10^{-8}, 10^{-4}, 10^{-10}, \ldots$ approach zero ?
Answer   No. This sequence goes below 10^{-4} but does not stay below.

There is a recognized symbol for "an arbitrarily small positive number." By worldwide agreement, it is the Greek letter ε (***epsilon***). Convergence to zero means that ***the sequence eventually goes below ε and stays there***. The smaller the ε, the tougher the test and the longer we wait. Think of ε as the tolerance, and keep reducing it.

To emphasize that ε comes from outside, Socrates can choose it. Whatever ε he proposes, the a's must eventually be smaller. *After some a_N, all the a's are below the tolerance ε.* Here is the exact statement:

for any ε there is an N such that $a_n < \varepsilon$ if $n > N$.

Once you see that idea, the rest is easy. Figure 2.17 has $N = 3$ and then $N = 6$.

EXAMPLE 1   The sequence $\frac{1}{2}, \frac{4}{4}, \frac{9}{8}, \ldots$ starts upward but goes to zero. Notice that $1, 4, 9, \ldots, 100, \ldots$ are squares, and $2, 4, 8, \ldots, 1024, \ldots$ are powers of 2. Eventually 2^n grows faster than n^2, as in $a_{10} = 100/1024$. The ratio goes below any ε.

EXAMPLE 2 $1, 0, \frac{1}{2}, 0, \frac{1}{3}, 0, \ldots$ approaches zero. These a's do not decrease steadily (the mathematical word for steadily is "monotonically") but still their limit is zero. The choice $\varepsilon = 1/10$ produces the right response: *Beyond a_{2001} all terms are below* $1/1000$. So $N = 2001$ for that ε.

The sequence $1, \frac{1}{2}, \frac{1}{2}, \frac{1}{3}, \frac{1}{3}, \frac{1}{3}, \ldots$ is much slower—but it also converges to zero.

Next we allow the numbers a_n to be *negative* as well as positive. They can converge upward toward zero, or they can come in from both sides. The test still requires the a_n to go inside any strip near zero (and stay there). But now the strip starts at $-\varepsilon$.

The distance from zero is the absolute value $|a_n|$. Therefore $a_n \to 0$ means $|a_n| \to 0$. The previous test can be applied to $|a_n|$:

for any ε there is an N such that $|a_n| < \varepsilon$ if $n > N$.

EXAMPLE 3 $1, -\frac{1}{2}, \frac{1}{3}, -\frac{1}{4}, \ldots$ converges to zero because $1, \frac{1}{2}, \frac{1}{3}, \frac{1}{4}, \ldots$ converges to zero.

It is a short step to limits other than zero. ***The limit is L if the numbers $a_n - L$ converge to zero***. Our final test applies to the absolute value $|a_n - L|$:

for any ε there is an N such that $|a_n - L| < \varepsilon$ if $n > N$.

This is the definition of convergence! Only a finite number of a's are outside any strip around L (Figure 2.18). We write $a_n \to L$ or $\lim a_n = L$ or $\lim_{n \to \infty} a_n = L$.

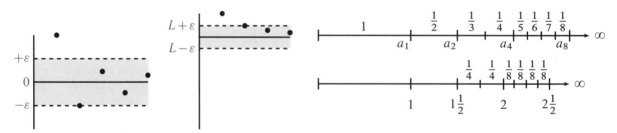

Fig. 2.18 $a_n \to 0$ in Example 3; $a_n \to 1$ in Example 4; $a_n \to \infty$ in Example 5 (but $a_{n+1} - a_n \to 0$).

EXAMPLE 4   The numbers $\frac{3}{2}, \frac{5}{4}, \frac{7}{6}, \ldots$ converge to $L = 1$. After subtracting 1 the differences $\frac{1}{2}, \frac{1}{4}, \frac{1}{6}, \ldots$ converge to zero. Those difference are $|a_n - L|$.

EXAMPLE 5   *The sequence* $1, 1 + \frac{1}{2}, 1 + \frac{1}{2} + \frac{1}{3}, 1 + \frac{1}{2} + \frac{1}{3} + \frac{1}{4}, \ldots$ *fails to converge.*

The distance between terms is getting smaller. But those numbers $a_1, a_2, a_3, a_4, \ldots$ go past any proposed limit L. The second term is $1\frac{1}{2}$. The fourth term adds on $\frac{1}{3} + \frac{1}{4}$, so a_4 goes past 2. The eighth term has four new fractions $\frac{1}{5} + \frac{1}{6} + \frac{1}{7} + \frac{1}{8}$, totaling more than $\frac{1}{8} + \frac{1}{8} + \frac{1}{8} + \frac{1}{8} = \frac{1}{2}$. Therefore a_8 exceeds $2\frac{1}{2}$. Eight more terms will add more than 8 times $\frac{1}{16}$, so a_{16} is beyond 3. The lines in Figure 2.18c are infinitely long, not stopping at any L.

In the language of Chapter 10, the harmonic series $1 + \frac{1}{2} + \frac{1}{3} + \ldots$ *does not converge.* The sum is infinite, because the "partial sums" a_n go beyond every limit L (a_{5000} is past $L = 9$). We will come back to infinite series, but this example makes a subtle point: The steps between the a_n can go to zero while still $a_n \to \infty$.

Thus the condition $a_{n+1} - a_n \to 0$ is **not sufficient** for convergence. However this condition is **necessary**. If we do have convergence, then $a_{n+1} - a_n \to 0$. That is a good exercise in the logic of convergence, emphasizing the difference between "sufficient" and "necessary." We discuss this logic below, after proving that [statement A] implies [statement B]:

$$\text{If } [a_n \text{ converges to } L] \text{ then } [a_{n+1} - a_n \text{ converges to zero}]. \tag{1}$$

Proof Because the a_n converge, there is a number N beyond which $|a_n - L| < \varepsilon$ and also $|a_{n+1} - L| < \varepsilon$. Since $a_{n+1} - a_n$ is the sum of $a_{n+1} - L$ and $L - a_n$, its absolute value cannot exceed $\varepsilon + \varepsilon = 2\varepsilon$. Therefore $a_{n+1} - a_n$ approaches zero.

Objection by Socrates: *We only got below* 2ε *and he asked for* ε. *Our reply*: If he particularly wants $|a_{n+1} - a_n| < 1/10$, we start with $\varepsilon = 1/20$. Then $2\varepsilon = 1/10$. But this juggling is not necessary. To stay below 2ε is just as convincing as to stay below ε.

THE LOGIC OF "IF" AND "ONLY IF"

The following page is inserted to help with the language of mathematics. In ordinary language we might say "I will come if you call." Or we might say "I will come only if you call." That is different! A mathematician might even say "I will come *if and only if* you call." Our goal is to think through the logic, because it is important and not so familiar.[†]

Statement A above implies statement B. Statement A is $a_n \to L$; statement B is $a_{n+1} - a_n \to 0$. Mathematics has at least five ways of writing down $A \Rightarrow B$, and I though you might like to see them together. It seems excessive to have so many expressions for the same idea, but authors get desperate for a little variety. Here are the five ways that come to mind:

$$A \Rightarrow B$$

A implies B

if A **then** B

A is a **sufficient** condition for B

B is true **if** A is true

[†] Logical thinking is much more important than ε and δ.

EXAMPLES *If* [positive numbers are decreasing] *then* [they converge to a limit].
   If [sequences a_n and b_n converge] *then* [the sequence $a_n + b_n$ converges].
   If [$f(x)$ is the integral of $v(x)$] *then* [$v(x)$ is the derivative of $f(x)$].

Those are all true, but not proved. A is the hypothesis, B is the conclusion.

   Now we go in the other direction. (It is called the "converse," not the inverse.) *We
exchange A and B.* Of course stating the converse does not make it true! B might
imply A, or it might not. In the first two examples the converse was false—the a_n
can converge without decreasing, and $a_n + b_n$ can converge when the separate
sequences do not. The converse of the third statement is true—and there are five
more ways to state it:

$$A \Leftarrow B$$

A is implied by B

if B then A

A is a *necessary* condition for B

B is true *only if* A is true

   Those words "necessary" and "sufficient" are not always easy to master. The same
is true of the deceptively short phrase "if and only if." The two statements $A \Rightarrow B$ and
$A \Leftarrow B$ are completely different and *they both require proof.* That means two separate proofs. But they can be stated together for convenience (when both are true):

$$A \Leftrightarrow B$$

A implies B and B implies A

A is *equivalent* to B

A is a *necessary and sufficient* condition for B

A is true *if and only if* B is true

EXAMPLES $[a_n \to L] \Leftrightarrow [2a_n \to 2L] \Leftrightarrow [a_n + 1 \to L + 1] \Leftrightarrow [a_n - L \to 0]$.

RULES FOR LIMITS

Calculus needs a *definition of limits*, to define dy/dx. That derivative contains two
limits: $\Delta x \to 0$ and $\Delta y / \Delta x \to dy/dx$. Calculus also needs *rules for limits*, to prove
the sum rule and product rule for derivatives. We started on the definition, and now
we start on the rules.

   Given *two convergent sequences,* $a_n \to L$ and $b_n \to M$, other sequences also
converge:

   Addition: $a_n + b_n \to L + M$ Subtraction: $a_n - b_n \to L - M$

Multiplication: $a_n b_n \to LM$ Division: $a_n / b_n \to L/M$ (provided $M \neq 0$)

We check the multiplication rule, which uses a convenient identity:

$$a_n b_n - LM = (a_n - L)(b_n - M) + M(a_n - L) + L(b_n - M). \qquad (2)$$

Suppose $|a_n - L| < \varepsilon$ beyond some point N, and $|b_n - M| < \varepsilon$ beyond some other
point N'. Then beyond the larger of N and N', the right side of (2) is small. It is less
than $\varepsilon \cdot \varepsilon + M \varepsilon + L \varepsilon$. This proves that (2) gives $a_n b_n \to LM$.

   An important special case is $ca_n \to cL$. (The sequence of b's is $c, c, c, c, \ldots$)
Thus a constant can be brought "outside" the limit, to give $\lim ca_n = c \lim a_n$.

THE LIMIT OF $f(x)$ AS $x \to a$

The final step is to replace sequences by functions. Instead of $a_1, a_2, \ldots$ there is a continuum of values $f(x)$. The limit is taken as x approaches a specified point a (instead of $n \to \infty$). Example: As x approaches $a = 0$, the function $f(x) = 4 - x^2$ approaches $L = 4$. As x approaches $a = 2$, the function $5x$ approaches $L = 10$. Those statements are fairly obvious, but we have to say what they mean. Somehow it must be this:

if x is close to a then $f(x)$ is close to L.

If $x - a$ is small, then $f(x) - L$ should be small. As before, the word *small* does not say everything. We really mean "arbitrarily small," or "below any ε." The difference $f(x) - L$ must become *as small as anyone wants*, when x gets near a. In that case $\lim_{x \to a} f(x) = L$. Or we write $f(x) \to L$ as $x \to a$.

The statement is awkward because it involves *two limits*. The limit $x \to a$ is forcing $f(x) \to L$. (Previously $n \to \infty$ forced $a \to L$.) But it is wrong to expect the same ε in both limits. We cannot require that $|x - a| < \varepsilon$ produces $|f(x) - L| < \varepsilon$. *It may be necessary to push x extremely close to a* (closer than ε). We must guarantee that if x is close enough to a, then $|f(x) - L| < \varepsilon$.

We have come to the "*epsilon-delta definition*" of limits. First, Socrates chooses ε. He has to be shown that $f(x)$ is within ε of L, for every x near a. Then somebody else (maybe Plato) replies with a number δ. That gives the meaning of "near a." Plato's goal is to get $f(x)$ within ε of L, by keeping x within δ of a:

$$\text{if} \quad 0 < |x - a| < \delta \quad \text{then} \quad |f(x) - L| < \varepsilon. \tag{3}$$

The input tolerance is δ (delta), the output tolerance is ε. When Plato can find a δ for every ε, Socrates concedes that the limit is L.

EXAMPLE Prove that $\lim_{x \to 2} 5x = 10$. In this case $a = 2$ and $L = 10$.

Socrates asks for $|5x - 10| < \varepsilon$. Plato responds by requiring $|x - 2| < \delta$. What δ should he choose? In this case $|5x - 10|$ is exactly 5 times $|x - 2|$. So Plato picks δ below $\varepsilon/5$ (a smaller δ is always OK). Whenever $|x - 2| < \varepsilon/5$, multiplication by 5 shows that $|5x - 10| < \varepsilon$.

Remark 1 In Figure 2.19, Socrates chooses the height of the box. It extends above and below L, by the small number ε. Second, Plato chooses the width. He must make the box narrow enough for the graph to go *out the sides*. Then $|f(x) - L| < \varepsilon$.

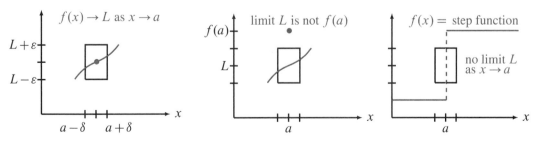

Fig. 2.19 S chooses height 2ε, then P chooses width 2δ. Graph must go out the sides.

When $f(x)$ has a jump, the box can't hold it. A step function has no limit as x approaches the jump, because the graph goes through the top or bottom—no matter how thin the box.

Remark 2   The second figure has $f(x) \to L$, because in taking limits **we ignore the final point** $x = a$. The value $f(a)$ can be anything, with no effect on L. The first figure has more: $f(a)$ equals L. Then a special name applies—f is **continuous**. The left figure shows a continuous function, the other figures do not.

We soon come back to continuous functions.

Remark 3   In the example with $f = 5x$ and $\delta = \varepsilon/5$, the number 5 was the *slope*. That choice barely kept the graph in the box—it goes out the corners. A little narrower, say $\delta = \varepsilon/10$, and the graph goes safely out the sides. *A reasonable choice is to divide ε by $2|f'(a)|$.* (We double the slope for safety.) I want to say why this δ works—even if the $\varepsilon - \delta$ test is seldom used in practice.

The ratio of $f(x) - L$ to $x - a$ is distance up over distance across. This is $\Delta f/\Delta x$, close to the slope $f'(a)$. When the distance across is δ, the distance up or down is near $\delta|f'(a)|$. That equals $\varepsilon/2$ for our "reasonable choice" of δ—so we are safely below ε. This choice solves most exercises. But Example 7 shows that a limit might exist even when the slope is infinite.

EXAMPLE 7 $\lim\limits_{x \to 1^+} \sqrt{x - 1} = 0$ (*a one-sided limit*).

Notice the plus sign in the symbol $x \to 1^+$. The number x approaches $a = 1$ only from above. An ordinary limit $x \to 1$ requires us to accept x on both sides of 1 (the exact value $x = 1$ is not considered). Since negative numbers are not allowed by the square root, we have a *one-sided limit*. It is $L = 0$.

Suppose ε is $1/10$. Then the response could be $\delta = 1/100$. A number below $1/100$ has a square root below $1/10$. In this case the box must be made extremely narrow, δ much smaller than ε, because the square root starts with infinite slope.

Those examples show the point of the $\varepsilon - \delta$ definition. (Given ε, look for δ. This came from Cauchy in France, not Socrates in Greece.) We also see its bad feature: The test is not convenient. Mathematicians do not go around proposing ε's and replying with δ's. We may live a strange life, but not that strange.

It is easier to establish once and for all that $5x$ approaches its obvious limit $5a$. The same is true for other familiar functions: $x^n \to a^n$ and $\sin x \to \sin a$ and $(1-x)^{-1} \to (1-a)^{-1}$—except at $a = 1$. **The correct limit L comes by substituting $x = a$ into the function**. This is exactly the property of a "**continuous function**." Before the section on continuous functions, we prove the Squeeze Theorem using ε and δ.

2H Squeeze Theorem   Suppose $f(x) \leqslant g(x) \leqslant h(x)$ for x near a. If $f(x) \to L$ and $h(x) \to L$ as $x \to a$, then the limit of $g(x)$ is also L.

Proof $g(x)$ is squeezed between $f(x)$ and $h(x)$. After subtracting L, $g(x) - L$ is between $f(x) - L$ and $h(x) - L$. Therefore

$$|g(x) - L| < \varepsilon \quad \text{if} \quad |f(x) - L| < \varepsilon \quad \text{and} \quad |h(x) - L| < \varepsilon.$$

For any ε, the last two inequalities hold in some region $0 < |x - a| < \delta$. So the first one also holds. This proves that $g(x) \to L$. Values at $x = a$ are not involved—until we get to continuous functions.

Read-through questions

The limit of $a_n = (\sin n)/n$ is __a__. The limit of $a_n = n^4/2^n$ is __b__. The limit of $a_n = (-1)^n$ is __c__. The meaning of $a_n \to 0$ is: Only __d__ of the numbers $|a_n|$ can be __e__. The meaning of $a_n \to L$ is: For every __f__ there is an __g__ such that __h__ if $n > $__i__. The sequence $1, 1+\frac{1}{2}, 1+\frac{1}{2}+\frac{1}{3}, \ldots$ is not __j__ because eventually those sums go past __k__.

The limit of $f(x) = \sin x$ as $x \to a$ is __l__. The limit of $f(x) = x/|x|$ as $x \to -2$ is __m__, but the limit as $x \to 0$ does not __n__. This function only has __o__-sided limits. The meaning of $\lim_{x \to a} f(x) = L$ is: For every ε there is a δ such that $|f(x) - L| < \varepsilon$ whenever __p__.

Two rules for limits, when $a_n \to L$ and $b_n \to M$, are $a_n + b_n \to$ __q__ and $a_n b_n \to$ __r__. The corresponding rules for functions, when $f(x) \to L$ and $g(x) \to M$ as $x \to a$, are __s__ and __t__. In all limits, $|a_n - L|$ or $|f(x) - L|$ must eventually go below and __u__ any positive __v__.

$A \Rightarrow B$ means that A is a __w__ condition for B. Then B is true __x__ A is true. $A \Leftrightarrow B$ means that A is a __y__ condition for B. Then B is true __z__ A is true.

1 What is a_4 and what is the limit L? After which N is $|a_n - L| < \frac{1}{10}$? (Calculator allowed)

(a) $-1, +\frac{1}{2}, -\frac{1}{3}, \ldots$ (b) $\frac{1}{2}, \frac{1}{2} + \frac{1}{4}, \frac{1}{2} + \frac{1}{4} + \frac{1}{6}, \ldots$

(c) $\frac{1}{2}, \frac{2}{4}, \frac{3}{8}, \ldots a_n = n/2^n$ (d) $1.1, 1.11, 1.111, \ldots$

(e) $a_n = \sqrt[n]{n}$ (f) $a_n = \sqrt{n^2 + n} - n$

(g) $1 + 1, (1 + \frac{1}{2})^2, (1 + \frac{1}{3})^3, \ldots$

2 Show by example that these statements are false:

(a) If $a_n \to L$ and $b_n \to L$ then $a_n/b_n \to 1$

(b) $a_n \to L$ if and only if $a_n^2 \to L^2$

(c) If $a_n < 0$ and $a_n \to L$ then $L < 0$

(d) If infinitely many a_n's are inside every strip around zero then $a_n \to 0$.

3 Which of these statements are equivalent to $B \Rightarrow A$?

(a) If A is true so is B

(b) A is true if and only if B is true

(c) B is a sufficient condition for A

(d) A is a necessary condition for B.

4 Decide whether $A \Rightarrow B$ or $B \Rightarrow A$ or neither or both:

(a) $A = [a_n \to 1]$ $B = [-a_n \to -1]$

(b) $A = [a_n \to 0]$ $B = [a_n - a_{n-1} \to 0]$

(c) $A = [a_n \le n]$ $B = [a_n = n]$

(d) $A = [a_n \to 0]$ $B = [\sin a_n \to 0]$

(e) $A = [a_n \to 0]$ $B = [1/a_n \text{ fails to converge}]$

(f) $A = [a_n < n]$ $B = [a_n/n \text{ converges}]$

*5 If the sequence $a_1, a_2, a_3, \ldots$ approaches zero, prove that we can put those numbers in any order and the new sequence still approaches zero.

*6 Suppose $f(x) \to L$ and $g(x) \to M$ as $x \to a$. Prove from the definitions that $f(x) + g(x) \to L + M$ as $x \to a$.

Find the limits 7–24 if they exist. An $\varepsilon - \delta$ test is not required.

7 $\lim_{t \to 2} \dfrac{t+3}{t^2 - 2}$

8 $\lim_{t \to 2} \dfrac{t^2 + 3}{t - 2}$

9 $\lim_{x \to 0} \dfrac{f(x+h) - f(x)}{h}$ (careful)

10 $\lim_{h \to 0} \dfrac{f(1+h) - f(1)}{h}$

11 $\lim_{h \to 0} \dfrac{\sin^2 h \cos^2 h}{h^2}$

12 $\lim_{x \to 0} \dfrac{2x \tan x}{\sin x}$

13 $\lim_{x \to 0^+} \dfrac{|x|}{x}$ (one-sided)

14 $\lim_{x \to 0^-} \dfrac{|x|}{x}$ (one-sided)

15 $\lim_{x \to 1} \dfrac{\sin x}{x}$

16 $\lim_{c \to a} \dfrac{f(c) - f(a)}{c - a}$

17 $\lim_{x \to 5} \dfrac{x^2 + 25}{x - 5}$

18 $\lim_{x \to 5} \dfrac{x^2 - 25}{x - 5}$

19 $\lim_{x \to 0} \dfrac{\sqrt{1+x} - 1}{x}$ (test $x = .01$)

20 $\lim_{x \to 2} \dfrac{\sqrt{4-x}}{\sqrt{6+x}}$

21 $\lim_{x \to a} [f(x) - f(a)]$ (?)

22 $\lim_{x \to \pi/2} (\sec x - \tan x)$

23 $\lim_{x \to 0} \dfrac{\sin x}{\sin x/2}$

24 $\lim_{x \to 1} \dfrac{\sin(x-1)}{x^2 - 1}$

25 Choose δ so that $|f(x)| < \frac{1}{100}$ if $0 < x < \delta$.

$f(x) = 10x$ $f(x) = \sqrt{x}$ $f(x) = \sin 2x$ $f(x) = x \sin x$

26 Which does the definition of a limit require?

(1) $|f(x) - L| < \varepsilon \Rightarrow 0 < |x - a| < \delta$.

(2) $|f(x) - L| < \varepsilon \Leftarrow 0 < |x - a| < \delta$.

(3) $|f(x) - L| < \varepsilon \Leftrightarrow 0 < |x - a| < \delta$.

27 The definition of "$f(x) \to L$ as $x \to \infty$" is this: For any ε there is an X such that _____ $< \varepsilon$ if $x > X$. Give an example in which $f(x) \to 4$ as $x \to \infty$.

28 Give a correct definition of "$f(x) \to 0$ as $x \to -\infty$."

29 The limit of $f(x) = (\sin x)/x$ as $x \to \infty$ is _____. For $\varepsilon = .01$ find a point X beyond which $|f(x)| < \varepsilon$.

30 The limit of $f(x) = 2x/(1+x)$ as $x \to \infty$ is $L = 2$. For $\varepsilon = .01$ find a point X beyond which $|f(x) - 2| < \varepsilon$.

31 The limit of $f(x) = \sin x$ as $x \to \infty$ does not exist. Explain why not.

32 (Calculator) Estimate the limit of $\left(1 + \dfrac{1}{x}\right)^x$ as $x \to \infty$.

33 For the polynomial $f(x) = 2x - 5x^2 + 7x^3$ find

(a) $\lim\limits_{x \to 1} f(x)$

(b) $\lim\limits_{x \to \infty} f(x)$

(c) $\lim\limits_{x \to \infty} \dfrac{f(x)}{x^3}$

(d) $\lim\limits_{x \to -\infty} \dfrac{f(x)}{x^3}$

34 For $f(x) = 6x^3 + 1000x$ find

(a) $\lim\limits_{x \to \infty} \dfrac{f(x)}{x}$

(b) $\lim\limits_{x \to \infty} \dfrac{f(x)}{x^2}$

(c) $\lim\limits_{x \to \infty} \dfrac{f(x)}{x^4}$

(d) $\lim\limits_{x \to \infty} \dfrac{f(x)}{x^3 + 1}$

Important rule As $x \to \infty$ the ratio of polynomials $f(x)/g(x)$ has the same limit as the ratio of their *leading terms*. $f(x) = x^3 - x + 2$ has leading term x^3 and $g(x) = 5x^6 + x + 1$ has leading term $5x^6$. Therefore $f(x)/g(x)$ behaves like $x^3/5x^6 \to 0$, $g(x)/f(x)$ behaves like $5x^6/x^3 \to \infty$, $(f(x))^2/g(x)$ behaves like $x^6/5x^6 \to 1/5$.

35 Find the limit as $x \to \infty$ if it exists:

$$\frac{3x^2 + 2x + 1}{3 + 2x + x^2} \qquad \frac{x^4}{x^3 + x^2} \qquad \frac{x^2 + 1000}{x^3 - 1000} \qquad x \sin \frac{1}{x}.$$

36 If a particular δ achieves $|f(x) - L| < \varepsilon$, why is it OK to choose a smaller δ ?

37 The sum of $1 + r + r^2 + \cdots + r^{n-1}$ is $a_n = (1 - r^n)/(1 - r)$. What is the limit of a_n as $n \to \infty$? For which r does the limit exist ?

38 If $a_n \to L$ prove that there is a number N with this property: If $n > N$ and $m > N$ then $|a_n - a_m| < 2\varepsilon$. This is Cauchy's test for convergence.

39 No matter what decimals come later, $a_1 = .4, a_2 = .49,$ $a_3 = .493, \ldots$ approaches a limit L. How do we know (when we can't know L) ? *Cauchy's test* is passed: the a's get closer to each other.

(a) From a_4 onwards we have $|a_n - a_m| < \underline{\hspace{1cm}}$.

(b) After which a_N is $|a_m - a_n| < 10^{-7}$?

40 Choose decimals in Problem 39 so the limit is $L = .494$. Choose decimals so that your professor can't find L.

41 If every decimal in $.abcde \cdots$ is picked at random from $0, 1, \ldots, 9$, what is the "average" limit L ?

42 If every decimal is 0 or 1 (at random), what is the average limit L ?

43 Suppose $a_n = \frac{1}{2} a_{n-1} + 4$ and start from $a_1 = 10$. Find a_2 and a_3 and a connection between $a_n - 8$ and $a_{n-1} - 8$. Deduce that $a_n \to 8$.

44 "For every δ there is an ε such that $|f(x)| < \varepsilon$ if $|x| < \delta$." That test is twisted around. Find ε when $f(x) = \cos x$, which does *not* converge to zero.

45 Prove the Squeeze Theorem for sequences, using ε: If $a_n \to L$ and $c_n \to L$ and $a_n \leqslant b_n \leqslant c_n$ for $n > N$, then $b_n \to L$.

46 Explain in 110 words the difference between "we will get there if you hurry" and "we will get there only if you hurry" and "we will get there if and only if you hurry."

2.7 Continuous Functions

This will be a brief section. It was originally included with limits, but the combination was too long. We are still concerned with the limit of $f(x)$ as $x \to a$, but a new number is involved. That number is $f(a)$, *the value of f at $x = a$*. For a "limit," x approached a but never reached it—so $f(a)$ was ignored. For a "continuous function," this final number $f(a)$ must be right.

May I summarize the usual (good) situation as x approaches a ?

1. The number $f(a)$ exists   (f is defined at a)

2. The limit of $f(x)$ exists   (it was called L)

3. The limit L equals $f(a)$ ($f(a)$ is the right value)

In such a case, $f(x)$ is ***continuous*** at $x = a$. These requirements are often written in a single line: $f(x) \to f(a)$ as $x \to a$. By way of contrast, start with four functions that are *not* continuous at $x = 0$.

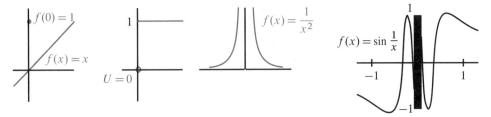

Fig. 2.20   Four types of discontinuity (others are possible) at $x = 0$.

In Figure 2.20, the first function would be continuous if it had $f(0) = 0$. But it has $f(0) = 1$. After changing $f(0)$ to the right value, the problem is gone. The discontinuity is *removable*. Examples $2, 3, 4$ are more important and more serious. There is no "correct" value for $f(0)$:

2. $f(x) = $ step function   (jump from 0 to 1 at $x = 0$)
3. $f(x) = 1/x^2$ (infinite limit as $x \to 0$)
4. $f(x) = \sin(1/x)$ (infinite oscillation as $x \to 0$).

The graphs show how the limit fails to exist. The step function has a ***jump discontinuity***. It has ***one-sided limits***, from the left and right. It does not have an ordinary (two-sided) limit. The limit from the left ($x \to 0^-$) is 0. The limit from the right ($x \to 0^+$) is 1. Another step function is $x/|x|$, which jumps from -1 to 1.

In the graph of $1/x^2$, the only reasonable limit is $L = +\infty$. I cannot go on record as saying that this limit exists. Officially, it doesn't—but we often write it anyway: $1/x^2 \to \infty$ as $x \to 0$. This means that $1/x^2$ goes (and stays) above every L as $x \to 0$.

In the same unofficial way we write one-sided limits for $f(x) = 1/x$:

$$\text{From the left, } \lim_{x \to 0^-} \frac{1}{x} = -\infty. \quad \text{From the right, } \lim_{x \to 0^+} \frac{1}{x} = +\infty. \qquad (1)$$

Remark $1/x$ has a "***pole***" at $x = 0$. So has $1/x^2$ (a double pole). The function $1/(x^2 - x)$ has poles at $x = 0$ and $x = 1$. In each case the denominator goes to zero and the function goes to $+\infty$ or $-\infty$. Similarly $1/\sin x$ has a pole at every multiple of π (where $\sin x$ is zero). Except for $1/x^2$ these poles are "*simple*"—the functions are completely smooth at $x = 0$ when we multiply them by x:

$$(x)\left(\frac{1}{x}\right) = 1 \text{ and } (x)\left(\frac{1}{x^2 - x}\right) = \frac{1}{x - 1} \text{ and } (x)\left(\frac{1}{\sin x}\right) \text{ are continuous at } x = 0.$$

$1/x^2$ has a double pole, since it needs multiplication by x^2 (not just x). A ratio of polynomials $P(x)/Q(x)$ has poles where $Q = 0$, provided any common factors like $(x + 1)/(x + 1)$ are removed first.

Jumps and poles are the most basic discontinuities, but others can occur. The fourth graph shows that $\sin(1/x)$ has no limit as $x \to 0$. This function does not blow up; the sine never exceeds 1. At $x = \frac{1}{3}$ and $\frac{1}{4}$ and $\frac{1}{1000}$ it equals $\sin 3$ and $\sin 4$ and $\sin 1000$. Those numbers are positive and negative and (?). As x gets small and $1/x$ gets large, the sine oscillates faster and faster. Its graph won't stay in a small box of height ε, no matter how narrow the box.

CONTINUOUS FUNCTIONS

DEFINITION f is "***continuous at $x = a$***" if $f(a)$ is defined and $f(x) \to f(a)$ as $x \to a$. If f is continuous at every point where it is defined, it is a ***continuous function***.

Objection   The definition makes $f(x) = 1/x$ a continuous function! It is not defined at $x = 0$, so its continuity can't fail. The logic requires us to accept this, but we don't have to like it. Certainly there is no $f(0)$ that would make $1/x$ continuous at $x = 0$.

It is amazing but true that the definition of "continuous function" is still debated (*Mathematics Teacher*, May 1989). You see the reason—we speak about a discontinuity of $1/x$, and at the same time call it a continuous function. The definition misses the difference between $1/x$ and $(\sin x)/x$. *The function $f(x) = (\sin x)/x$ can be made continuous at all x.* Just set $f(0) = 1$.

We call a function "***continuable***" if its definition can be extended *to all x* in a way that makes it continuous. Thus $(\sin x)/x$ and $\sqrt{x}$ are continuable. The functions $1/x$ and $\tan x$ are not continuable. This suggestion may not end the debate, but I hope it is helpful.

EXAMPLE 1 $\sin x$ and $\cos x$ and all polynomials $P(x)$ are continuous functions.

EXAMPLE 2   The absolute value $|x|$ is continuous. Its slope jumps (not continuable).

EXAMPLE 3   Any rational function $P(x)/Q(x)$ is continuous except where $Q = 0$.

EXAMPLE 4   The function that jumps between 1 at fractions and 0 at non-fractions is ***discontinuous everywhere***. There is a fraction between every pair of non-fractions and vice versa. (Somehow there are many more non-fractions.)

EXAMPLE 5   The function 0^{x^2} is zero for every x, except that 0^0 is not defined. So define it as zero and this function is continuous. But see the next paragraph where 0^0 has to be 1.

We could fill the book with proofs of continuity, but usually the situation is clear. "A function is continuous if you can draw its graph without lifting up your pen." At a jump, or an infinite limit, or an infinite oscillation, there is no way across the discontinuity except to start again on the other side. The function x'' is continuous for $n > 0$. It is not continuable for $n < 0$. The function x^0 equals 1 for every x, except that 0^0 is not defined. This time continuity requires $0^0 = 1$.

The interesting examples are the close ones—we have seen two of them:

EXAMPLE 6 $\dfrac{\sin x}{x}$ and $\dfrac{1 - \cos x}{x}$ are both continuable at $x = 0$.

Those were crucial for the slope of $\sin x$. The first approaches 1 and the second approaches 0. Strictly speaking we must give these functions the correct values (1 and 0) at the limiting point $x = 0$—which of course we do.

It is important to know what happens when the denominators change to x^2.

EXAMPLE 7 $\dfrac{\sin x}{x^2}$ blows up but $\dfrac{1 - \cos x}{x^2}$ has the limit $\dfrac{1}{2}$ at $x = 0$.

Since $(\sin x)/x$ approaches 1, dividing by another x gives a function like $1/x$. There is a simple pole. It is an example of $0/0$, in which the zero from x^2 is reached more quickly than the zero from $\sin x$. The "***race to zero***" produces almost all interesting problems about limits.

For $1 - \cos x$ and x^2 the race is almost even. Their ratio is 1 to 2:

$$\frac{1 - \cos x}{x^2} = \frac{1 - \cos^2 x}{x^2(1 + \cos x)} = \frac{\sin^2 x}{x^2} \cdot \frac{1}{1 + \cos x} \to \frac{1}{1 + 1} \quad \text{as} \quad x \to 0.$$

This answer $\frac{1}{2}$ will be found again (more easily) by "l'Hôpital's rule." Here I emphasize not the answer but the problem. A central question of differential calculus is *to know how fast the limit is approached*. **The speed of approach is exactly the information in the derivative**.

These three examples are all continuous at $x = 0$. The race is controlled by the slope—because $f(x) - f(0)$ is nearly $f'(0)$ times x:

derivative of $\sin x$ is 1 $\leftrightarrow$ $\sin x$ decreases like x

derivative of $\sin^2 x$ is 0 $\leftrightarrow$ $\sin^2 x$ decreases faster than x

derivative of $x^{1/3}$ is ∞ $\leftrightarrow$ $x^{1/3}$ decreases more slowly than x.

DIFFERENTIABLE FUNCTIONS

The absolute value $|x|$ is continuous at $x = 0$ but has no derivative. The same is true for $x^{1/3}$. **Asking for a derivative is more than asking for continuity**. The reason is fundamental, and carries us back to the key definitions:

Continuous at x: $f(x + \Delta x) - f(x) \to 0$ as $\Delta x \to 0$

Derivative at x: $\dfrac{f(x + \Delta x) - f(x)}{\Delta x} \to f'(x)$ as $\Delta x \to 0$.

In the first case, Δf goes to zero (maybe slowly). In the second case, Δf goes to zero *as fast as* Δx (because $\Delta f / \Delta x$ has a limit). That requirement is stronger:

2I At a point where $f(x)$ has a derivative, the function must be continuous. But $f(x)$ can be continuous with no derivative.

Proof The limit of $\Delta f = (\Delta x)(\Delta f / \Delta x)$ is $(0)(df/dx) = 0$. So $f(x + \Delta x) - f(x) \to 0$.

The continuous function $x^{1/3}$ has no derivative at $x = 0$, because $\frac{1}{3}x^{-2/3}$ blows up. The absolute value $|x|$ has no derivative because its slope jumps. The remarkable function $\frac{1}{2}\cos 3x + \frac{1}{4}\cos 9x + \cdots$ is continuous at *all points* and has a derivative at *no points*. You can draw its graph without lifting your pen (but not easily—it turns at every point). To most people, it belongs with space-filling curves and unmeasurable areas—in a box of curiosities. Fractals used to go into the same box! They are beautiful shapes, with boundaries that have no tangents. The theory of fractals is very alive, for good mathematical reasons, and we touch on it in Section 3.7.

I hope you have a clear idea of these basic definitions of calculus:

1 *Limit* $(n \to \infty$ or $x \to a)$ **2 *Continuity*** (at $x = a$)   **3 *Derivative*** (at $x = a$).

Those go back to ε and δ, but it is seldom necessary to follow them so far. In the same way that economics describes many transactions, or history describes many events, a function comes from many values $f(x)$. A few points may be special, like market crashes or wars or discontinuities. At other points df/dx is the best guide to the function.

This chapter ends with two essential facts about *a continuous function on a closed interval*. The interval is $a \leqslant x \leqslant b$, written simply as $[a, b]$. † At the endpoints a and b we require $f(x)$ to approach $f(a)$ and $f(b)$.

Extreme Value Property   A continuous function on the finite interval $[a, b]$ has a maximum value M and a minimum value m. There are points $x_{\max}$ and $x_{\min}$ in $[a, b]$ where it reaches those values:

$$f(x_{\max}) = M \geqslant f(x) \geqslant f(x_{\min}) = m \quad \text{for all } x \text{ in } [a, b].$$

Intermediate Value Property   If the number F is between $f(a)$ and $f(b)$, there is a point c between a and b where $f(c) = F$. Thus if F is between the minimum m and the maximum M, there is a point c between $x_{\min}$ and $x_{\max}$ where $f(c) = F$.

Examples show why we require closed intervals and continuous functions. For $0 < x \leqslant 1$ the function $f(x) = x$ never reaches its minimum (zero). If we close the interval by defining $f(0) = 3$ (discontinuous) the minimum is still not reached. Because of the jump, the intermediate value $F = 2$ is also not reached. The idea of continuity was inescapable, after Cauchy defined the idea of a limit.

---

† The interval $[a, b]$ is ***closed*** (endpoints included). The interval (a, b) is ***open*** (a and b left out). The infinite interval $[0, \infty)$ contains all $x \geqslant 0$.

2.7 EXERCISES

Read-through questions

Continuity requires the __a__ of $f(x)$ to exist as $x \to a$ and to agree with __b__. The reason that $x/|x|$ is not continuous at $x = 0$ is __c__. This function does have __d__ limits. The reason that $1/\cos x$ is discontinuous at __e__ is __f__. The reason that $\cos(1/x)$ is discontinuous at $x = 0$ is __g__. The function $f(x) = $ __h__ has a simple pole at $x = 3$, where f^2 has a __i__ pole.

The power x^n is continuous at all x provided n is __j__. It has no derivative at $x = 0$ when n is __k__. $f(x) = \sin(-x)/x$ approaches __l__ as $x \to 0$, so this is a __m__ function provided we define $f(0) = $ __n__. A "continuous function" must be continuous at all __o__. A "continuable function" can be extended to every point x so that __p__.

If f has a derivative at $x = a$ then f is necessarily __q__ at $x = a$. The derivative controls the speed at which $f(x)$ approaches __r__. On a closed interval $[a, b]$, a continuous f has the __s__ value property and the __t__ value property. It reaches its __u__ M and its __v__ m, and it takes on every value __w__.

In Problems 1–20, find the numbers c that make $f(x)$ into (A) a continuous function and (B) a differentiable function. In one case $f(x) \to f(a)$ at every point, in the other case $\Delta f / \Delta x$ has a limit at every point.

1 $f(x) = \begin{cases} \sin x & x < 1 \\ c & x \geq 1 \end{cases}$

2 $f(x) = \begin{cases} \cos^3 x & x \neq \pi \\ c & x = \pi \end{cases}$

3 $f(x) = \begin{cases} cx & x < 0 \\ 2cx & x \geq 0 \end{cases}$

4 $f(x) = \begin{cases} cx & x < 1 \\ 2cx & x \geq 1 \end{cases}$

5 $f(x) = \begin{cases} c + x & x < 0 \\ c^2 + x^2 & x \geq 0 \end{cases}$

6 $f(x) = \begin{cases} x^3 & x \neq c \\ -8 & x = c \end{cases}$

7 $f(x) = \begin{cases} 2x & x < c \\ x + 1 & x \geq c \end{cases}$

8 $f(x) = \begin{cases} x^c & x \neq 0 \\ 0 & x = 0 \end{cases}$

9 $f(x) = \begin{cases} (\sin x)/x^2 & x \neq 0 \\ c & x = 0 \end{cases}$

10 $f(x) = \begin{cases} x + c & x \leq c \\ 1 & x > c \end{cases}$

11 $f(x) = \begin{cases} c & x \neq 4 \\ 1/x^3 & x = 4 \end{cases}$

12 $f(x) = \begin{cases} c & x \leq 0 \\ \sec x & x \geq 0 \end{cases}$

13 $f(x) = \begin{cases} \dfrac{x^2 + c}{x - 1} & x \neq 1 \\ 2 & x = 1 \end{cases}$

14 $f(x) = \begin{cases} \dfrac{x^2 - 1}{x - c} & x \neq c \\ 2c & x = c \end{cases}$

15 $f(x) = \begin{cases} (\tan x)/x & x \neq 0 \\ c & x = 0 \end{cases}$

16 $f(x) = \begin{cases} x^2 & x \leq c \\ 2x & x > c \end{cases}$

17 $f(x) = \begin{cases} (c + \cos x)/x & x \neq 0 \\ 0 & x = 0 \end{cases}$

18 $f(x) = |x + c|$

19 $f(x) = \begin{cases} (\sin x - x)/x^c & x \neq 0 \\ 0 & x = 0 \end{cases}$

20 $f(x) = |x^2 + c^2|$

Construct your own $f(x)$ with these discontinuities at $x = 1$.

21 Removable discontinuity

22 Infinite oscillation

23 Limit for $x \to 1^+$, no limit for $x \to 1^-$

24 A double pole

25 $\displaystyle \lim_{x \to 1^-} f(x) = 4 + \lim_{x \to 1^+} f(x)$

26 $\displaystyle \lim_{x \to 1} f(x) = \infty$ but $\displaystyle \lim_{x \to 1} (x - 1)f(x) = 0$

27 $\displaystyle \lim_{x \to 1} (x - 1)f(x) = 5$

28 The statement "$3x \to 7$ as $x \to 1$" is false. Choose an ε for which no δ can be found. The statement "$3x \to 3$ as $x \to 1$" is true. For $\varepsilon = \frac{1}{2}$ choose a suitable δ.

29 How many derivatives $f', f'', \ldots$ are continuable functions?

(a) $f = x^{3/2}$ (b) $f = x^{3/2} \sin x$ (c) $f = (\sin x)^{5/2}$

30 Find one-sided limits at points where there is no two-sided limit. Give a 3-part formula for function (c).

(a) $\dfrac{|x|}{7x}$ (b) $\sin |x|$ (c) $\dfrac{d}{dx} |x^2 - 1|$

31 Let $f(1) = 1$ and $f(-1) = 1$ and $f(x) = (x^2 - x)/(x^2 - 1)$ otherwise. Decide whether f is continuous at

(a) $x = 1$ (b) $x = 0$ (c) $x = -1$

*32 Let $f(x) = x^2 \sin 1/x$ for $x \neq 0$ and $f(0) = 0$. If the limits exist, find

(a) $\displaystyle \lim_{x \to 0} f(x)$ (b) df/dx at $x = 0$ (c) $\displaystyle \lim_{x \to 0} f'(x)$

33 If $f(0) = 0$ and $f'(0) = 3$, rank these functions from smallest to largest as x decreases to zero:

$f(x), \quad x, \quad xf(x), \quad f(x) + 2x, \quad 2(f(x) - x), \quad (f(x))^2.$

34 Create a discontinuous function $f(x)$ for which $f^2(x)$ is continuous.

35 *True or false*, with an example to illustrate:

(a) If $f(x)$ is continuous at all x, it has a maximum value M.

(b) If $f(x) \leq 7$ for all x, then f reaches its maximum.

(c) If $f(1) = 1$ and $f(2) = -2$, then somewhere $f(x) = 0$.

(d) If $f(1) = 1$ and $f(2) = -2$ and f is continuous on $[1, 2]$, then somewhere on that interval $f(x) = 0$.

36 The functions $\cos x$ and $2x$ are continuous. Show from the _____ property that $\cos x = 2x$ at some point between 0 and 1.

37 Show by example that these statements are false:

(a) If a function reaches its maximum and minimum then the function is continuous.

(b) If $f(x)$ reaches its maximum and minimum and all values between $f(0)$ and $f(1)$, it is continuous at $x = 0$.

(c) (mostly for instructors) If $f(x)$ has the intermediate value property between all points a and b, it must be continuous.

38 Explain with words and a graph why $f(x) = x\sin(1/x)$ is continuous but has no derivative at $x = 0$. Set $f(0) = 0$.

39 Which of these functions are _continuable_, and why ?

$$f_1(x) = \begin{cases} \sin x & x < 0 \\ \cos x & x > 1 \end{cases} \qquad f_2(x) = \begin{cases} \sin 1/x & x < 0 \\ \cos 1/x & x > 1 \end{cases}$$

$$f_3(x) = \frac{x}{\sin x} \text{ when } \sin x \neq 0 \qquad f_4(x) = x^0 + 0^{x^2}$$

40 Explain the difference between a continuous function and a continuable function. Are continuous functions always continuable ?

*$*41$ $f(x)$ is any continuous function with $f(0) = f(1)$.

(a) Draw a typical $f(x)$. Mark where $f(x) = f(x + \frac{1}{2})$.

(b) Explain why $g(x) = f(x + \frac{1}{2}) - f(x)$ has $g(\frac{1}{2}) = -g(0)$.

(c) Deduce from (b) that (a) is always possible: There _must_ be a point where $g(x) = 0$ and $f(x) = f(x + \frac{1}{2})$.

42 Create an $f(x)$ that is continuous only at $x = 0$.

43 If $f(x)$ is continuous and $0 \leqslant f(x) \leqslant 1$ for all x, then there is a point where $f(x^*) = x^*$. Explain with a graph and prove with the intermediate value theorem.

44 In the ε–δ definition of a limit, change $0 < |x - a| < \delta$ to $|x - a| < \delta$. Why is $f(x)$ now _continuous_ at $x = a$?

45 A function has a _____ at $x = 0$ if and only if $(f(x) - f(0))/x$ is _____ at $x = 0$.

CHAPTER 3

Applications of the Derivative

Chapter 2 concentrated on computing derivatives. This chapter concentrates on *using* them. Our computations produced dy/dx for functions built from x^n and $\sin x$ and $\cos x$. Knowing the slope, and if necessary also the second derivative, we can answer the questions about $y = f(x)$ that this subject was created for:

1. How does y change when x changes?
2. What is the maximum value of y? Or the minimum?
3. How can you tell a maximum from a minimum, using derivatives?

The information in dy/dx is entirely *local*. It tells what is happening close to the point and nowhere else. In Chapter 2, Δx and Δy went to zero. Now we want to get them back. The local information explains the larger picture, *because Δy is approximately dy/dx times Δx*.

The problem is to connect the finite to the infinitesimal—the average slope to the instantaneous slope. Those slopes are close, and occasionally they are equal. Points of equality are assured by the Mean Value Theorem—which is the local-global connection at the center of differential calculus. But we cannot predict *where dy/dx* equals $\Delta y/\Delta x$. Therefore we now find other ways to recover a function from its derivatives—or to estimate distance from velocity and acceleration.

It may seem surprising that we learn about y from dy/dx. All our work has been going the other way! We struggled with y to squeeze out dy/dx. Now we use dy/dx to study y. That's life. Perhaps it really is life, to understand one generation from later generations.

3.1 Linear Approximation

The book started with a straight line $f = vt$. The distance is linear when the velocity is constant. As soon as v begins to change, $f = vt$ falls apart. Which velocity do we choose, when $v(t)$ is not constant? The solution is to take very short time intervals, in which v is nearly constant:

$$f = vt \qquad \textit{is completely false}$$
$$\Delta f = v \Delta t \qquad \textit{is nearly true}$$
$$df = v dt \qquad \textit{is exactly true}$$

For a brief moment the function $f(t)$ is linear—and stays near its tangent line.

In Section 2.3 we found the tangent line to $y = f(x)$. At $x = a$, the slope of the curve and the slope of the line are $f'(a)$. For points on the line, start at $y = f(a)$. Add the slope times the "increment" $x - a$:

$$Y = f(a) + f'(a)(x - a). \tag{1}$$

We write a capital Y for the line and a small y for the curve. The whole point of tangents is that they are close (*provided we don't move too far from a*):

$$y \approx Y \qquad \text{or} \qquad f(x) \approx f(a) + f'(a)(x - a). \tag{2}$$

That is the all-purpose **linear approximation**. Figure 3.1 shows the square root function $y = \sqrt{x}$ and its tangent line at $x = a = 100$. At the point $y = \sqrt{100} = 10$, the slope is $1/2\sqrt{x} = 1/20$. The table beside the figure compares $y(x)$ with $Y(x)$.

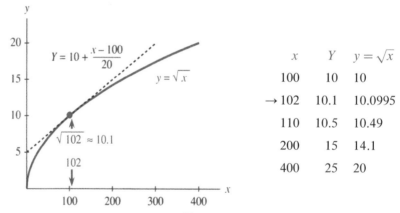

x	Y	$y = \sqrt{x}$
100	10	10
→ 102	10.1	10.0995
110	10.5	10.49
200	15	14.1
400	25	20

Fig. 3.1 $Y(x)$ is the linear approximation to $\sqrt{x}$ near $x = a = 100$.

The accuracy gets worse as x departs from 100. The tangent line leaves the curve. The arrow points to a good approximation at 102, and at 101 it would be even better. In this example Y is larger than y—the straight line is above the curve. The slope of the line stays constant, and the slope of the curve is decreasing. Such a curve will soon be called "concave downward," and its tangent lines are above it.

Look again at $x = 102$, where the approximation is good. In Chapter 2, when we were approaching dy/dx, we started with $\Delta y / \Delta x$:

$$\text{slope} \approx \frac{\sqrt{102} - \sqrt{100}}{102 - 100}. \tag{3}$$

Now that is turned around! The slope is $1/20$. **What we don't know is** $\sqrt{102}$:

$$\sqrt{102} \approx \sqrt{100} + (\text{slope})(102 - 100). \tag{4}$$

You work with what you have. Earlier we didn't know dy/dx, so we used (3). Now we are experts at dy/dx, and we use (4). After computing $y' = 1/20$ once and for all, the tangent line stays near $\sqrt{x}$ for every number near 100. When that nearby number is $100 + \Delta x$, notice the error as the approximation is squared:

$$\left(\sqrt{100} + \frac{1}{20}\Delta x\right)^2 = 100 + \Delta x + \frac{1}{400}(\Delta x)^2.$$

The desired answer is $100 + \Delta x$, and we are off by the last term involving $(\Delta x)^2$. The whole point of linear approximation is to ignore every term after Δx.

There is nothing magic about $x = 100$, except that it has a nice square root. Other points and other functions allow $y \approx Y$. I would like to express this same idea in different symbols. *Instead of starting from a and going to x, we start from x and go a distance Δx to $x + \Delta x$.* The letters are different but the mathematics is identical.

3A At any point x, and for any smooth function $y = f(x)$,

$$\text{slope at } x \approx \frac{f(x + \Delta x) - f(x)}{\Delta x}. \tag{5}$$

For the approximation to $f(x + \Delta x)$, multiply both sides by Δx and add $f(x)$:

$$f(x + \Delta x) \approx f(x) + (\text{slope at } x)(\Delta x). \tag{6}$$

EXAMPLE 1 *An important linear approximation*: $(1 + x)^n \approx 1 + nx$ for x near zero.

EXAMPLE 2 *A second important approximation*: $1/(1 + x)^n \approx 1 - nx$ for x near zero.

Discussion Those are really the same. By changing n to $-n$ in Example 1, it becomes Example 2. These are linear approximations using the slopes n and $-n$ at $x = 0$:

$$(1 + x)^n \approx 1 + (\text{slope at zero}) \text{ times } (x - 0) = 1 + nx.$$

Here is the same thing with $f(x) = x^n$. The basepoint in equation (6) is now 1 or x:

$$(1 + \Delta x)^n \approx 1 + n\Delta x \qquad (x + \Delta x)^n \approx x^n + nx^{n-1}\Delta x.$$

Better than that, here are numbers. For $n = 3$ and -1 and 100, take $\Delta x = .01$:

$$(1.01)^3 \approx 1.03 \qquad \frac{1}{1.01} \approx .99 \qquad \left(1 + \frac{1}{100}\right)^{100} \approx 2$$

Actually that last number is no good. The 100th power is too much. Linear approximation gives $1 + 100\Delta x = 2$, but a calculator gives $(1.01)^{100} = 2.7\ldots$. This is close to e, the all-important number in Chapter 6. The binomial formula shows why the approximation failed:

$$(1 + \Delta x)^{100} = 1 + 100\Delta x + \frac{(100)(99)}{(2)(1)}(\Delta x)^2 + \cdots.$$

Linear approximation forgets the $(\Delta x)^2$ term. For $\Delta x = 1/100$ that error is nearly $\frac{1}{2}$. It is too big to overlook. The exact error is $\frac{1}{2}(\Delta x)^2 f''(c)$, where the Mean Value Theorem in Section 3.8 places c between x and $x + \Delta x$. You already see the point:

$$y - Y \text{ is of order } (\Delta x)^2. \textbf{ Linear approximation, quadratic error.}$$

DIFFERENTIALS

There is one more notation for this linear approximation. It has to be presented, because it is often used. The notation is suggestive and confusing at the same time—it keeps the same symbols dx and dy that appear in the derivative. Earlier we took great pains to emphasize that dy/dx is not an ordinary fraction.† Until this paragraph, dx and dy have had no independent meaning. Now they become separate variables, like x and y but with their own names. These quantities dx and dy are called **differentials**.

The symbols dx and dy measure changes *along the tangent line*. They do for the approximation $Y(x)$ exactly what Δx and Δy did for $y(x)$. Thus dx and Δx both measure distance across.

Figure 3.2 has $\Delta x = dx$. But the change in y does not equal the change in Y. One is Δy (exact for the function). The other is dy (exact for the tangent line). **The differential dy is equal to ΔY, the change along the tangent line**. Where Δy is the true change, dy is its linear approximation $(dy/dx)dx$.

You often see dy written as $f'(x)dx$.

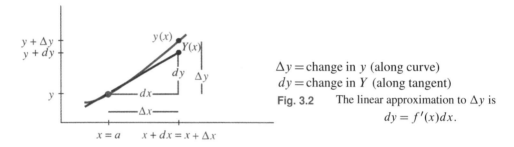

$\Delta y =$ change in y (along curve)
$dy =$ change in Y (along tangent)

Fig. 3.2   The linear approximation to Δy is
$$dy = f'(x)dx.$$

EXAMPLE 3 $y = x^2$ has $dy/dx = 2x$ so $dy = 2x\,dx$. The table has basepoint $x = 2$. The prediction dy differs from the true Δy by exactly $(\Delta x)^2 = .01$ and $.04$ and $.09$.

	dx	dy	Δx	Δy	
$y = x^2$	.1	0.4	.1	0.41	$\Delta y = (2+\Delta x)^2 - 2^2$
$dy = 4dx$	.2	0.8	.2	0.84	$\Delta y = 4\Delta x + (\Delta x)^2$
	.3	1.2	.3	1.29	

The differential $dy = f'(x)dx$ is consistent with the derivative $dy/dx = f'(x)$. We finally have $dy = (dy/dx)dx$, but this is not as obvious as it seems! It looks like cancellation—it is really a definition. Entirely new symbols could be used, but dx and dy have two advantages: They suggest small steps and they satisfy $dy = f'(x)dx$. Here are three examples and three rules:

$$d(x^n) = nx^{n-1}dx \qquad d(f+g) = df + dg$$
$$d(\sin x) = \cos x\,dx \qquad d(cf) = c\,df$$
$$d(\tan x) = \sec^2 x\,dx \qquad d(fg) = f\,dg + g\,df$$

Science and engineering and virtually all applications of mathematics depend on linear approximation. The true function is "**linearized**," using its slope v:

---

†Fraction or not, it is absolutely forbidden to cancel the d's

Increasing the time by Δt increases the distance by $\approx v\Delta t$

Increasing the force by Δf increases the deflection by $\approx v\Delta f$

Increasing the production by Δp increases its value by $\approx v\Delta p$.

The goal of dynamics or statics or economics is to predict this multiplier v—the derivative that equals the slope of the tangent line. The multiplier gives a *local prediction* of the change in the function. The exact law is nonlinear—but Ohm's law and Hooke's law and Newton's law are linear approximations.

ABSOLUTE CHANGE, RELATIVE CHANGE, PERCENTAGE CHANGE

The change Δy or Δf can be measured in three ways. So can Δx:

Absolute change	Δf	Δx
Relative change	$\dfrac{\Delta f}{f(x)}$	$\dfrac{\Delta x}{x}$
Percentage change	$\dfrac{\Delta f}{f(x)} \times 100$	$\dfrac{\Delta x}{x} \times 100$

Relative change is often more realistic than absolute change. If we know the distance to the moon within three miles, that is more impressive than knowing our own height within one inch. Absolutely, one inch is closer than three miles. Relatively, three miles is much closer:

$$\frac{3 \text{ miles}}{300,000 \text{ miles}} < \frac{1 \text{ inch}}{70 \text{ inches}} \quad \text{or} \quad .001\% < 1.4\%.$$

EXAMPLE 4 The radius of the Earth is within 80 miles of $r = 4,000$ miles.
(a) Find the variation dV in the volume $V = \frac{4}{3}\pi r^3$, using linear approximation.
(b) Compute the relative variations dr/r and dV/V and $\Delta V/V$.

Solution The job of calculus is to produce the derivative. After $dV/dr = 4\pi r^2$, its work is done. The variation in volume is $dV = 4\pi(4000)^2(80)$ cubic miles. A 2% relative variation in r gives a 6% relative variation in V:

$$\frac{dr}{r} = \frac{80}{4000} = 2\% \qquad \frac{dV}{V} = \frac{4\pi(4000)^2(80)}{4\pi(4000)^3/3} = 6\%.$$

Without calculus we need the exact volume at $r = 4000 + 80$ (also at $r = 3920$):

$$\frac{\Delta V}{V} = \frac{4\pi(4080)^3/3 - 4\pi(4000)^3/3}{4\pi(4000)^3/3} \approx 6.1\%$$

One comment on $dV = 4\pi r^2 dr$. This is (area of sphere) times (change in radius). It is the volume of a thin shell around the sphere. The shell is added when the radius grows by dr. The exact $\Delta V/V$ is $3917312/640000\%$, but calculus just calls it 6%.

Read-through questions

On the graph, a linear approximation is given by the __a__
line. At $x = a$, the equation for that line is $Y = f(a) +$ __b__.
Near $x = a = 10$, the linear approximation to $y = x^3$ is
$Y = 1000 +$ __c__. At $x = 11$ the exact value is $(11)^3 =$ __d__.
The approximation is $Y =$ __e__. In this case $\Delta y =$ __f__ and
$dy =$ __g__. If we know $\sin x$, then to estimate $\sin(x + \Delta x)$ we
add __h__.

In terms of x and Δx, linear approximation is
$f(x + \Delta x) \approx f(x) +$ __i__. The error is of order $(\Delta x)^p$ or
$(x - a)^p$ with $p =$ __j__. The differential dy equals __k__ times
the differential __l__. Those movements are along the __m__
line, where Δy is along the __n__.

Find the linear approximation Y to $y = f(x)$ near $x = a$:

1 $f(x) = x + x^4$, $a = 0$ 2 $f(x) = 1/x$, $a = 2$

3 $f(x) = \tan x$, $a = \pi/4$ 4 $f(x) = \sin x$, $a = \pi/2$

5 $f(x) = x \sin x$, $a = 2\pi$ 6 $f(x) = \sin^2 x$, $a = 0$

**Compute 7–12 within .01 by deciding on $f(x)$, choosing the
basepoint a, and evaluating $f(a) + f'(a)(x - a)$. A calculator
shows the error.**

7 $(2.001)^6$ 8 $\sin(.02)$

9 $\cos(.03)$ 10 $(15.99)^{1/4}$

11 $1/.98$ 12 $\sin(3.14)$

**Calculate the numerical error in these linear approximations
and compare with $\frac{1}{2}(\Delta x)^2 f''(x)$:**

13 $(1.01)^3 \approx 1 + 3(.01)$ 14 $\cos(.01) \approx 1 + 0(.01)$

15 $(\sin .01)^2 \approx 0 + 0(.01)$ 16 $(1.01)^{-3} \approx 1 - 3(.01)$

17 $\left(1 + \frac{1}{10}\right)^{10} \approx 2$ 18 $\sqrt{8.99} \approx 3 + \frac{1}{6}(-.01)$

Confirm the approximations 19–21 by computing $f'(0)$:

19 $\sqrt{1 - x} \approx 1 - \frac{1}{2}x$

20 $1/\sqrt{1 - x^2} \approx 1 + \frac{1}{2}x^2$ (use $f = 1/\sqrt{1 - u}$, then put $u = x^2$)

21 $\sqrt{c^2 + x^2} \approx c + \dfrac{1}{2}\dfrac{x^2}{c}$ (use $f(u) = \sqrt{c^2 + u}$, then put
$u = x^2$)

22 Write down the differentials df for $f(x) = \cos x$ and
$(x + 1)/(x - 1)$ and $(x^2 + 1)^2$.

**In 23–27 find the linear change dV in the volume or dA in the
surface area.**

23 dV if the sides of a cube change from 10 to 10.1.

24 dA if the sides of a cube change from x to $x + dx$.

25 dA if the radius of a sphere changes by dr.

26 dV if a circular cylinder with $r = 2$ changes height from 3
to 3.05 (recall $V = \pi r^2 h$).

27 dV if a cylinder of height 3 changes from $r = 2$ to $r = 1.9$.
Extra credit: *What is dV if r and h both change $(dr$ and $dh)$?*

28 In relativity the mass is $m_0/\sqrt{1 - (v/c)^2}$ at velocity v. By
Problem 20 this is near $m_0 +$ _____ for small v. Show that
the kinetic energy $\frac{1}{2}mv^2$ and the change in mass satisfy
Einstein's equation $e = (\Delta m)c^2$.

29 Enter 1.1 on your calculator. Press the square root key 5
times (slowly). What happens each time to the number after the
decimal point? This is because $\sqrt{1 + x} \approx$ _____.

30 In Problem 29 the numbers you see are less than
$1.05, 1.025, \ldots$. The second derivative of $\sqrt{1 + x}$ is _____ so the
linear approximation is higher than the curve.

31 Enter 0.9 on your calculator and press the square root
key 4 times. Predict what will appear the fifth time and press
again. You now have the _____ root of 0.9. How many decimals
agree with $1 - \frac{1}{32}(0.1)$?

3.2 Maximum and Minimum Problems

Our goal is to learn about $f(x)$ from df/dx. We begin with two quick questions. If df/dx is positive, what does that say about f? If the slope is negative, how is that reflected in the function? Then the third question is the critical one:

How do you identify a **_maximum_** or **_minimum_**?
Normal answer: **_The slope is zero._**

This may be the most important application of calculus, to reach $df/dx = 0$.

Take the easy questions first. Suppose df/dx is *positive* for every x between a and b. All tangent lines slope upward. *The function $f(x)$ is **increasing** as x goes from a to b.*

> **3B** If $df/dx > 0$ then $f(x)$ is *increasing*. If $df/dx < 0$ then $f(x)$ is *decreasing*.

To define increasing and decreasing, look at any two points $x < X$. "Increasing" requires $f(x) < f(X)$. "Decreasing" requires $f(x) > f(X)$. *A positive slope does not mean a positive function.* The function itself can be positive or negative.

EXAMPLE 1 $f(x) = x^2 - 2x$ has slope $2x - 2$. This slope is positive when $x > 1$ and negative when $x < 1$. The function increases after $x = 1$ and decreases before $x = 1$.

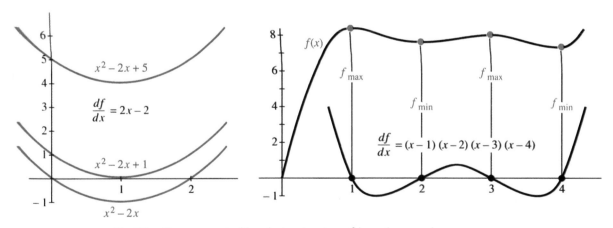

Fig. 3.3 Slopes are $- +$. Slope is $+ - + - +$ so f is up-down-up-down-up.

We say that without computing $f(x)$ at any point! The parabola in Figure 3.3 goes down to its minimum at $x = 1$ and up again.

EXAMPLE 2 $x^2 - 2x + 5$ has the same slope. Its graph is shifted up by 5, a number that disappears in df/dx. All functions with slope $2x - 2$ are parabolas $x^2 - 2x + C$, shifted up or down according to C. Some parabolas cross the x axis (those crossings are solutions to $f(x) = 0$). Other parabolas stay above the axis. The solutions to $x^2 - 2x + 5 = 0$ are complex numbers and we don't see them. The special parabola $x^2 - 2x + 1 = (x-1)^2$ grazes the axis at $x = 1$. It has a "double zero," where $f(x) = df/dx = 0$.

EXAMPLE 3   Suppose $df/dx = (x-1)(x-2)(x-3)(x-4)$. This slope is positive beyond $x=4$ and up to $x=1$ ($df/dx=24$ at $x=0$). And df/dx is positive again between 2 and 3. At $x=1,2,3,4$, this slope is zero and $f(x)$ changes direction.

Here $f(x)$ is a fifth-degree polynomial, because $f'(x)$ is fourth-degree. The graph of f goes up-down-up-down-up. It might cross the x axis five times. *It must cross at least once* (like this one). When complex numbers are allowed, every fifth-degree polynomial has five roots.

You may feel that "*positive slope implies increasing function*" is obvious—perhaps it is. But there is still something delicate. Starting from $df/dx > 0$ at every *single* point, we have to deduce $f(X) > f(x)$ at *pairs* of points. That is a "local to global" question, to be handled by the Mean Value Theorem. It could also wait for the Fundamental Theorem of Calculus: *The difference $f(X) - f(x)$ equals the area under the graph of df/dx.* That area is positive, so $f(X)$ exceeds $f(x)$.

MAXIMA AND MINIMA

Which x makes $f(x)$ as large as possible? Where is the smallest $f(x)$? Without calculus we are reduced to computing values of $f(x)$ and comparing. With calculus, the information is in df/dx.

Suppose the maximum or minimum is at a particular point x. It is possible that the graph has a corner—and no derivative. *But if df/dx exists, it must be zero.* The tangent line is level. The parabolas in Figure 3.3 change from decreasing to increasing. The slope changes from negative to positive. At this crucial point *the slope is zero.*

3C *Local Maximum or Minimum*   Suppose the maximum or minimum occurs at a point x inside an interval where $f(x)$ and df/dx are defined. Then $f'(x) = 0$.

The word "*local*" allows the possibility that in other intervals, $f(x)$ goes higher or lower. *We only look near x, and we use the definition of df/dx.*

Start with $f(x + \Delta x) - f(x)$. If $f(x)$ is the maximum, this difference is negative or zero. The step Δx can be forward or backward:

$$\text{if } \Delta x > 0: \quad \frac{f(x+\Delta x)-f(x)}{\Delta x} = \frac{\text{negative}}{\text{positive}} \leqslant 0 \quad \text{and in the limit} \quad \frac{df}{dx} \leqslant 0.$$

$$\text{if } \Delta x < 0: \quad \frac{f(x+\Delta x)-f(x)}{\Delta x} = \frac{\text{negative}}{\text{negative}} \geqslant 0 \quad \text{and in the limit} \quad \frac{df}{dx} \geqslant 0.$$

Both arguments apply. Both conclusions $df/dx \leqslant 0$ and $df/dx \geqslant 0$ are correct. Thus $df/dx = 0$.

Maybe Richard Feynman said it best. He showed his friends a plastic curve that was made in a special way—"*no matter how you turn it, the tangent at the lowest point is horizontal.*" They checked it out. It was true.

Surely You're Joking, Mr. Feynman! is a good book (but rough on mathematicians).

EXAMPLE 3 (continued)   Look back at Figure 3.3b. The points that stand out are not the "ups" or "downs" but the "turns." Those are *stationary points*, where $df/dx = 0$. We see two maxima and two minima. None of them are absolute maxima or minima, because $f(x)$ starts at $-\infty$ and ends at $+\infty$.

EXAMPLE 4 $f(x) = 4x^3 - 3x^4$ has slope $12x^2 - 12x^3$. That derivative is zero when x^2 equals x^3, at the two points $x = 0$ and $x = 1$. To decide between minimum and maximum (local or absolute), the first step is to evaluate $f(x)$ at these *stationary points*. We find $f(0) = 0$ and $f(1) = 1$.

Now look at large x. The function goes down to $-\infty$ in both directions. (*You can mentally substitute $x = 1000$ and $x = -1000$*). For large x, $-3x^4$ dominates $4x^3$.

Conclusion $f = 1$ is an absolute maximum. $f = 0$ is not a maximum or minimum (local or absolute). We have to recognize this exceptional possibility, that a curve (or a car) can pause for an instant ($f' = 0$) and continue in the same direction. The reason is the "double zero" in $12x^2 - 12x^3$, from its double factor x^2.

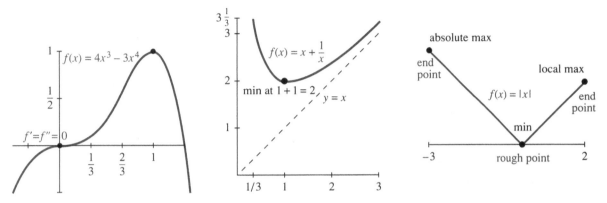

Fig. 3.4   The graphs of $4x^3 - 3x^4$ and $x + x^{-1}$. Check rough points and endpoints.

EXAMPLE 5   Define $f(x) = x + x^{-1}$ for $x > 0$. Its derivative $1 - 1/x^2$ is zero at $x = 1$. At that point $f(1) = 2$ is the minimum value. Every combination like $\frac{1}{3} + 3$ or $\frac{2}{3} + \frac{3}{2}$ is larger than $f_{min} = 2$. Figure 3.4 shows that *the maximum of $x + x^{-1}$ is $+\infty$.*†

Important   The maximum always occurs at a ***stationary point*** (where $df/dx = 0$) or a ***rough point*** (no derivative) or an ***endpoint*** of the domain. These are the three types of ***critical points***. All maxima and minima occur at critical points! At every other point $df/dx > 0$ or $df/dx < 0$. Here is the procedure:

1. Solve $df/dx = 0$ to find the stationary points $f(x)$.
2. Compute $f(x)$ at every critical point—*stationary point, rough point, endpoint*.
3. Take the maximum and minimum of those critical values of $f(x)$.

EXAMPLE 6   (***Absolute value*** $f(x) = |x|$) The minimum is zero at a rough point. The maximum is at an endpoint. There are no stationary points.

The derivative of $y = |x|$ is never zero. Figure 3.4 shows the maximum and minimum on the interval $[-3, 2]$. This is typical of piecewise linear functions.

Question   Could the minimum be zero when the function never reaches $f(x) = 0$?
Answer   *Yes*, $f(x) = 1/(1 + x)^2$ approaches but never reaches zero as $x \to \infty$.

---

†A good word is *approach* when $f(x) \to \infty$. Infinity is not reached. But I still say "the maximum is ∞."

Remark 1 $x \to \pm\infty$ and $f(x) \to \pm\infty$ are avoided when f is **continuous on a closed interval** $a \leqslant x \leqslant b$. Then $f(x)$ reaches its maximum and its minimum (*Extreme Value Theorem*). But $x \to \infty$ and $f(x) \to \infty$ are too important to rule out. You test $x \to \infty$ by considering large x. You recognize $f(x) \to \infty$ by going above every finite value.

Remark 2   Note the difference between critical *points* (specified by x) and critical *values* (specified by $f(x)$). The example $x + x^{-1}$ had the minimum *point* $x = 1$ and the minimum *value* $f(1) = 2$.

MAXIMUM AND MINIMUM IN APPLICATIONS

To find a maximum or minimum, solve $f'(x) = 0$. The slope is zero at the top and bottom of the graph. The idea is clear—and then check rough points and endpoints. But to be honest, that is not where the problem starts.

In a real application, the first step (often the hardest) is to choose the unknown and *find the function*. It is we ourselves who decide on x and $f(x)$. The equation $df/dx = 0$ comes in the middle of the problem, not at the beginning. I will start on a new example, with a question instead of a function.

EXAMPLE 7   Where should you get onto an expressway for minimum driving time, if the expressway speed is 60 mph and ordinary driving speed is 30 mph?

I know this problem well—it comes up every morning. The Mass Pike goes to MIT and I have to join it somewhere. There is an entrance near Route 128 and another entrance further in. I used to take the second one, now I take the first. Mathematics should decide which is faster—some mornings I think they are maxima.

Most models are simplified, to focus on the key idea. We will allow the expressway to be entered at *any point* x (Figure 3.5). Instead of two entrances (a discrete problem) we have a continuous choice (a calculus problem). The trip has two parts, at speeds 30 and 60:

a distance $\sqrt{a^2 + x^2}$ up to the expressway, in $\sqrt{a^2 + x^2}/30$ hours

a distance $b - x$ on the expressway, in $(b - x)/60$ hours

Problem Minimize $f(x) = $ total time $= \dfrac{1}{30}\sqrt{a^2 + x^2} + \dfrac{1}{60}(b - x)$.

We have the function $f(x)$. Now comes calculus. The first term uses the power rule: *The derivative of* $u^{1/2}$ *is* $\frac{1}{2}u^{-1/2}\,du/dx$. Here $u = a^2 + x^2$ has $du/dx = 2x$:

$$f'(x) = \frac{1}{30}\frac{1}{2}(a^2 + x^2)^{-1/2}(2x) - \frac{1}{60}. \tag{1}$$

To solve $f'(x) = 0$, multiply by 60 and square both sides:

$$(a^2 + x^2)^{-1/2}(2x) = 1 \quad \text{gives} \quad 2x = (a^2 + x^2)^{1/2} \quad \text{and} \quad 4x^2 = a^2 + x^2. \tag{2}$$

Thus $3x^2 = a^2$. This yields two candidates, $x = a/\sqrt{3}$ and $x = -a/\sqrt{3}$. But a negative x would mean useless driving on the expressway. In fact f' is *not zero* at $x = -a/\sqrt{3}$. That false root entered when we squared $2x$.

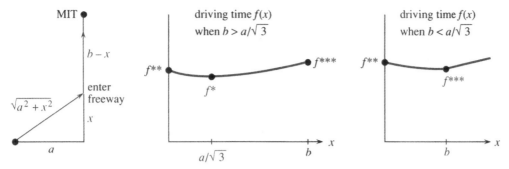

Fig. 3.5 Join the freeway at x—minimize the driving time $f(x)$.

I notice something surprising. The stationary point $x = a/\sqrt{3}$ does not depend on b. The total time includes the constant $b/60$, which disappeared in df/dx. Somehow b must enter the answer, and this is a warning to go carefully. The minimum might occur at a rough point or an endpoint. Those are the other critical points of f, and our drawing may not be realistic. Certainly we expect $x \leqslant b$, or we are entering the expressway beyond MIT.

Continue with calculus. Compute the driving time $f(x)$ for an entrance at $x^* = a/\sqrt{3}$:

$$f(x) = \frac{1}{30}\sqrt{a^2 + (a^2/3)} + \frac{1}{60}\left(b - \frac{a}{\sqrt{3}}\right) = \frac{\sqrt{3}\,a}{60} + \frac{b}{60} = f^*.$$

The square root of $4a^2/3$ is $2a/\sqrt{3}$. We combined $2/30 - 1/60 = 3/60$ and divided by $\sqrt{3}$. **Is this stationary value f^* a minimum?** You must look also at *endpoints*:

enter at $x = 0$: travel time is $a/30 + b/60 = f^{**}$

enter at $x = b$: travel time is $\sqrt{a^2 + b^2}/30 = f^{***}$.

The comparison $f^* < f^{**}$ should be automatic. Entering at $x = 0$ was a candidate and calculus didn't choose it. The derivative is not zero at $x = 0$. It is not smart to go perpendicular to the expressway.

The second comparison has $x = b$. We drive directly to MIT at speed 30. This option has to be taken seriously. In fact it is optimal when b is small or a is large.

This choice $x = b$ can arise mathematically in two ways. If all entrances are between 0 and b, then b is an **endpoint**. If we can enter beyond MIT, then b is a **rough point**. The graph in Figure 3.5c has a corner at $x = b$, where the derivative jumps. The reason is that distance on the expressway is the *absolute value* $|b - x|$— never negative.

Either way $x = b$ is a critical point. **The optimal x is the smaller of $a/\sqrt{3}$ and b.**

if $a/\sqrt{3} \leqslant b$: stationary point wins, enter at $x = a/\sqrt{3}$, total time f^*

if $a/\sqrt{3} \geqslant b$: no stationary point, drive directly to MIT, time f^{***}

The heart of this subject is in "word problems." All the calculus is in a few lines, computing f' and solving $f'(x) = 0$. The formulation took longer. Step 1 usually does:

1. Express the quantity to be minimized or maximized as a function $f(x)$. *The variable x has to be selected.*

2. Compute $f'(x)$, solve $f'(x) = 0$, check critical points for $f_{\min}$ and $f_{\max}$.

A picture of the problem (and the graph of $f(x)$) makes all the difference.

EXAMPLE 7 (continued)   Choose x as an *angle* instead of a distance. Figure 3.6 shows the triangle with angle x and side a. The driving distance to the expressway is $a \sec x$. The distance on the expressway is $b - a \tan x$. Dividing by the speeds 30 and 60, the driving time has a nice form:

$$f(x) = \text{total time} = \frac{a \sec x}{30} + \frac{b - a \tan x}{60}. \tag{3}$$

The derivatives of $\sec x$ and $\tan x$ go into df/dx:

$$\frac{df}{dx} = \frac{a}{30} \sec x \tan x - \frac{a}{60} \sec^2 x. \tag{4}$$

Now set $df/dx = 0$, divide by a, and multiply by $30\cos^2 x$:

$$\sin x = \tfrac{1}{2}. \tag{5}$$

This answer is beautiful. The angle x is $30°$! That optimal angle ($\pi/6$ radians) has $\sin x = \frac{1}{2}$. The triangle with side a and hypotenuse $a/\sqrt{3}$ is a 30–60–90 right triangle.

I don't know whether you prefer $\sqrt{a^2 + x^2}$ or trigonometry. The minimum is exactly as before—either at $30°$ or going directly to MIT.

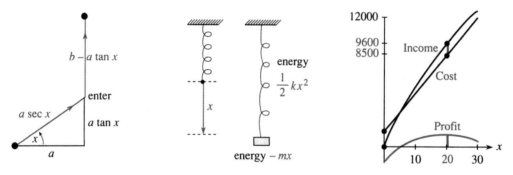

Fig. 3.6   (a) Driving at angle x. (b) Energies of spring and mass. (c) Profit = income − cost.

EXAMPLE 8   In mechanics, *nature chooses minimum energy*. A spring is pulled down by a mass, the energy is $f(x)$, and $df/dx = 0$ gives equilibrium. It is a philosophical question why so many laws of physics involve minimum energy or minimum time—which makes the mathematics easy.

The energy has two terms—for the spring and the mass. The spring energy is $\frac{1}{2}kx^2$—positive in stretching ($x > 0$ is downward) and also positive in compression ($x < 0$). The potential energy of the mass is taken as $-mx$—decreasing as the mass goes down. The balance is at the minimum of $f(x) = \frac{1}{2}kx^2 - mx$.

I apologize for giving you such a small problem, but it makes a crucial point. *When $f(x)$ is quadratic, the equilibrium equation $df/dx = 0$ is linear.*

$$df/dx = kx - m = 0.$$

Graphically, $x = m/k$ is at the bottom of the parabola. Physically, $kx = m$ is a balance of forces—the spring force against the weight. *Hooke's law* for the spring force is elastic constant k times displacement x.

EXAMPLE 9 *Derivative of cost = marginal cost* (our first management example).

The paper to print x copies of this book might cost $C = 1000 + 3x$ dollars. The derivative is $dC/dx = 3$. This is the *marginal cost* of paper for each additional book. If x increases by one book, the cost C increases by $3. The marginal cost is like the velocity and the total cost is like the distance.

    Marginal cost is in dollars per book. Total cost is in dollars. On the plus side, the income is $I(x)$ and the marginal income is dI/dx. To apply calculus, we overlook the restriction to whole numbers.

    Suppose the number of books increases by dx.† The cost goes up by $(dC/dx)dx$. The income goes up by $(dI/dx)dx$. If we skip all other costs, then *profit $P(x)$ = income $I(x)$ − cost $C(x)$.* In most cases P increases to a maximum and falls back.

    At the high point on the profit curve, *the marginal profit is zero*:

$$dP/dx = 0 \quad \text{or} \quad dI/dx = dC/dx. \tag{6}$$

Profit is maximized when marginal income I' equals marginal cost C'.

    This basic rule of economics comes directly from calculus, and we give an example:

$$C(x) = cost \ of \ x \ advertisements = 900 + 400x - x^2$$

setup cost 900, print cost $400x$, volume savings x^2

$$I(x) = income \ due \ to \ x \ advertisements = 600x - 6x^2$$

sales 600 per advertisement, subtract $6x^2$ for diminishing returns

$$optimal \ decision \ dC/dx = dI/dx \quad \text{or} \quad 400 - 2x = 600 - 12x \quad \text{or} \quad x = 20$$

$$profit = income - cost = 9600 - 8500 = 1100.$$

The next section shows how to verify that this profit is a maximum not a minimum.

    The first exercises ask you to solve $df/dx = 0$. Later exercises also look for $f(x)$.

3.2 EXERCISES

Read-through questions

If $df/dx > 0$ in an interval then $f(x)$ is __a__. If a maximum or minimum occurs at x then $f'(x) = $__b__. Points where $f'(x) = 0$ are called __c__ points. The function $f(x) = 3x^2 - x$ has a (minimum)(maximum) at $x = $__d__. A stationary point that is not a maximum or minimum occurs for $f(x) = $__e__.

    Extreme values can also occur where __f__ is not defined or at the __g__ of the domain. The minima of $|x|$ and $5x$ for $-2 \leqslant x \leqslant 2$ are at $x = $__h__ and $x = $__i__, even though df/dx is not zero. x^* is an absolute __j__ when $f(x^*) \geqslant f(x)$ for all x. A __k__ minimum occurs when $f(x^*) \leqslant f(x)$ for all x near x^*.

The minimum of $\frac{1}{2}ax^2 - bx$ is __l__ at $x = $__m__.

Find the stationary points and rough points and endpoints. Decide whether each point is a local or absolute minimum or maximum.

1   $f(x) = x^2 + 4x + 5, -\infty < x < \infty$

2   $f(x) = x^3 - 12x, -\infty < x < \infty$

3   $f(x) = x^2 + 3, -1 \leqslant x \leqslant 4$

4   $f(x) = x^2 + (2/x), 1 \leqslant x \leqslant 4$

5   $f(x) = (x - x^2)^2, -1 \leqslant x \leqslant 1$

---

†Maybe dx is a differential calculus book. I apologize for that.

6 $f(x) = 1/(x - x^2), 0 < x < 1$

7 $f(x) = 3x^4 + 8x^3 - 18x^2, -\infty < x < \infty$

8 $f(x) = \{x^2 - 4x \text{ for } 0 \leqslant x \leqslant 1, x^2 - 4 \text{ for } 1 \leqslant x \leqslant 2\}$

9 $f(x) = \sqrt{x-1} + \sqrt{9-x}, 1 \leqslant x \leqslant 9$

10 $f(x) = x + \sin x, 0 \leqslant x \leqslant 2\pi$

11 $f(x) = x^3(1-x)^6, -\infty < x < \infty$

12 $f(x) = x/(1+x), 0 \leqslant x \leqslant 100$

13 $f(x) = $ distance from $x \geqslant 0$ to nearest whole number

14 $f(x) = $ distance from $x \geqslant 0$ to nearest prime number

15 $f(x) = |x+1| + |x-1|, -3 \leqslant x \leqslant 2$

16 $f(x) = x\sqrt{1-x^2}, 0 \leqslant x \leqslant 1$

17 $f(x) = x^{1/2} - x^{3/2}, 0 \leqslant x \leqslant 4$

18 $f(x) = \sin x + \cos x, 0 \leqslant x \leqslant 2\pi$

19 $f(x) = x + \sin x, 0 \leqslant x \leqslant 2\pi$

20 $f(\theta) = \cos^2\theta \sin \theta, -\pi \leqslant \theta \leqslant \pi$

21 $f(\theta) = 4\sin \theta - 3\cos \theta, 0 \leqslant \theta \leqslant 2\pi$

22 $f(x) = \{x^2 + 1 \text{ for } x \leqslant 1, x^2 - 4x + 5 \text{ for } x \geqslant 1\}$.

In applied problems, choose metric units if you prefer.

23 The airlines accept a box if length + width + height = $l + w + h \leqslant 62''$ or 158 cm. If h is fixed show that the maximum volume $(62 - w - h)wh$ is $V = h(31 - \frac{1}{2}h)^2$. Choose h to maximize V. The box with greatest volume is a _____.

24 If a patient's pulse measures 70, then 80, then 120, what least squares value minimizes $(x - 70)^2 + (x - 80)^2 + (x - 120)^2$? If the patient got nervous, assign 120 a lower weight and minimize $(x - 70)^2 + (x - 80)^2 + \frac{1}{2}(x - 120)^2$.

25 At speed v, a truck uses $av + (b/v)$ gallons of fuel per mile. How many miles per gallon at speed v? Minimize the fuel consumption. Maximize the number of miles per gallon.

26 A limousine gets $(120 - 2v)/5$ miles per gallon. The chauffeur costs \$10/hour, the gas costs \$1/gallon.

    (a) Find the cost per mile at speed v.

    (b) Find the cheapest driving speed.

27 You should shoot a basketball at the angle θ requiring minimum speed. Avoid line drives and rainbows. Shooting from $(0,0)$ with the basket at (a,b), minimize $f(\theta) = 1/(a\sin\theta\cos\theta - b\cos^2\theta)$.

    (a) If $b = 0$ you are level with the basket. Show that $\theta = 45°$ is best (Jabbar sky hook).

    (b) Reduce $df/d\theta = 0$ to $\tan 2\theta = -a/b$. Solve when $a = b$.

    (c) Estimate the best angle for a free throw.

The same angle allows the largest margin of error (*Sports Science* by Peter Brancazio). Section 12.2 gives the flight path.

28 On the longest and shortest days, in June and December, why does the length of day change the least?

29 Find the shortest **Y** connecting P, Q, and B in the figure. Originally B was a birdfeeder. The length of **Y** is $L(x) = (b - x) + 2\sqrt{a^2 + x^2}$.

    (a) Choose x to minimize L(not allowing $x > b$).

    (b) Show that the center of the **Y** has 120° angles.

    (c) The best **Y** becomes a **V** when $a/b = $ _____ .

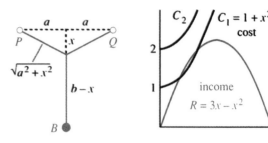

30 If the distance function is $f(t) = (1 + 3t)/(1 + 3t^2)$, when does the forward motion end? How far have you traveled? Extra credit: Graph $f(t)$ and df/dt.

In 31–34, we make and sell x pizzas. The income is $R(x) = ax + bx^2$ and the cost is $C(x) = c + dx + ex^2$.

31 The profit is $\Pi(x) = $ _____ . The average profit per pizza is $=$ _____ . The marginal profit per additional pizza is $d\Pi/dx = $ _____ . We should maximize the (profit)(average profit)(marginal profit).

32 We receive $R(x) = ax + bx^2$ when the price per pizza is $p(x) = $ _____ . In reverse: When the price is p we sell $x = $ _____ pizzas (a function of p). We expect $b < 0$ because _____ .

33 Find x to maximize the profit $\Pi(x)$. At that x the marginal profit is $d\Pi/dx = $ _____ .

34 Figure B shows $R(x) = 3x - x^2$ and $C_1(x) = 1 + x^2$ and $C_2(x) = 2 + x^2$. With cost C_1, which sales x makes a profit? Which x makes the most profit? With higher fixed cost in C_2, the best plan is _____ .

The cookie box and popcorn box were created by Kay Dundas from a $12'' \times 12''$ square. A box with no top is a calculus classic.

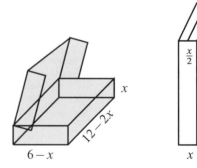

35 Choose x to find the maximum volume of the cookie box.

36 Choose x to maximize the volume of the popcorn box.

37 A high-class chocolate box adds a strip of width x down across the front of the cookie box. Find the new volume $V(x)$ and the x that maximizes it. Extra credit: Show that V_{max} is reduced by more than 20%.

38 For a box with no top, cut four squares of side x from the corners of the $12''$ square. Fold up the sides so the height is x. Maximize the volume.

Geometry provides many problems, more applied than they seem.

39 A wire four feet long is cut in two pieces. One piece forms a circle of radius r, the other forms a square of side x. Choose r to minimize the sum of their areas. Then choose r to maximize.

40 A fixed wall makes one side of a rectangle. We have 200 feet of fence for the other three sides. Maximize the area A in 4 steps:

    1 Draw a picture of the situation.
    2 Select one unknown quantity as x (but not A!).
    3 Find all other quantities in terms of x.
    4 Solve $dA/dx = 0$ and check endpoints.

41 With no fixed wall, the sides of the rectangle satisfy $2x + 2y = 200$. Maximize the area. Compare with the area of a circle using the same fencing.

42 Add 200 meters of fence to an existing straight 100–meter fence, to make a rectangle of maximum area (invented by Professor Klee).

43 How large a rectangle fits into the triangle with sides $x = 0, y = 0$, and $x/4 + y/6 = 1$? Find the point on this third side that maximizes the area xy.

44 The largest rectangle in Problem 43 may not sit straight up. Put one side along $x/4 + y/6 = 1$ and maximize the area.

45 The distance around the rectangle in Problem 43 is $P = 2x + 2y$. Substitute for y to find $P(x)$. Which rectangle has $P_{max} = 12$?

46 Find the right circular cylinder of largest volume that fits in a sphere of radius 1.

47 How large a cylinder fits in a cone that has base radius R and height H? For the cylinder, choose r and h on the sloping surface $r/R + h/H = 1$ to maximize the volume $V = \pi r^2 h$.

48 The cylinder in Problem 47 has side area $A = 2\pi rh$. Maximize A instead of V.

49 Including top and bottom, the cylinder has area

$$A = 2\pi rh + 2\pi r^2 = 2\pi rH(1 - (r/R)) + 2\pi r^2.$$

Maximize A when $H > R$. Maximize A when $R > H$.

*50 A wall 8 feet high is 1 foot from a house. Find the length L of the shortest ladder over the wall to the house. Draw a triangle with height y, base $1 + x$, and hypotenuse L.

51 Find the closed cylinder of volume $V = \pi r^2 h = 16\pi$ that has the least surface area.

52 Draw a kite that has a triangle with sides $1, 1, 2x$ next to a triangle with sides $2x, 2, 2$. Find the area A and the x that maximizes it. *Hint*: In dA/dx simplify $\sqrt{1 - x^2} - x^2/\sqrt{1 - x^2}$ to $(1 - 2x^2)/\sqrt{1 - x^2}$.

In 53–56, x and y are nonnegative numbers with $x + y = 10$. Maximize and minimize:

53 xy **54** $x^2 + y^2$ **55** $y - (1/x)$ **56** $\sin x \sin y$

57 Find the total distance $f(x)$ from A to X to C. Show that $df/dx = 0$ leads to $\sin a = \sin c$. Light reflects at an equal angle to minimize travel time.

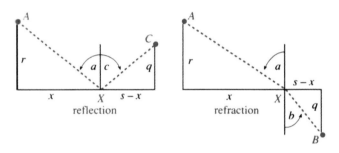

reflection   refraction

58 Fermat's principle says that light travels from A to B on the quickest path. Its velocity above the x axis is v and below the x axis is w.

    (a) Find the time $T(x)$ from A to X to B. On AX, time = distance/velocity = $\sqrt{r^2 + x^2}/v$.
    (b) Find the equation for the minimizing x.
    (c) Deduce *Snell's law* $(\sin a)/v = (\sin b)/w$.

"Closest point problems" are models for many applications.

59 Where is the parabola $y = x^2$ closest to $x = 0$, $y = 2$?

60 Where is the line $y = 5 - 2x$ closest to $(0,0)$?

61 What point on $y = -x^2$ is closest to what point on $y = 5 - 2x$? At the nearest points, the graphs have the same slope. Sketch the graphs.

62 Where is $y = x^2$ closest to $(0, \frac{1}{3})$? Minimizing $x^2 + (y - \frac{1}{3})^2 = y + (y - \frac{1}{3})^2$ gives $y < 0$. What went wrong?

63 Draw the line $y = mx$ passing near $(2, 3), (1, 1)$, and $(-1, 1)$. For a least squares fit, minimize

$$(3 - 2m)^2 + (1 - m)^2 + (1 + m)^2.$$

64 A triangle has corners $(-1,1), (x,x^2)$, and $(3,9)$ on the parabola $y = x^2$. Find its maximum area for x between -1 and 3. *Hint*: The distance from (X,Y) to the line $y = mx + b$ is $|Y - mX - b|/\sqrt{1+m^2}$.

65 Submarines are located at $(2,0)$ and $(1,1)$. Choose the slope m so the line $y = mx$ goes between the submarines but stays as far as possible from the nearest one.

Problems 66–72 go back to the theory.

66 To find where the graph of $y(x)$ has greatest slope, solve _____ . For $y = 1/(1+x^2)$ this point is _____ .

67 When the difference between $f(x)$ and $g(x)$ is smallest, their slopes are _____ . Show this point on the graphs of $f = 2 + x^2$ and $g = 2x - x^2$.

68 Suppose y is fixed. The minimum of $x^2 + xy - y^2$ (a function of x) is $m(y) =$ _____ . Find the maximum of $m(y)$.

Now x is fixed. The maximum of $x^2 + xy - y^2$ (a function of y) is $M(x) =$ _____ . Find the minimum of $M(x)$.

69 For each m the minimum value of $f(x) - mx$ occurs at $x = m$. What is $f(x)$?

70 $y = x + 2x^2 \sin(1/x)$ has slope 1 at $x = 0$. But show that y is not increasing on an *interval* around $x = 0$, by finding points where $dy/dx = 1 - 2\cos(1/x) + 4x\sin(1/x)$ is negative.

71 *True or false*, with a reason: Between two local minima of a smooth function $f(x)$ there is a local maximum.

72 Create a function $y(x)$ that has its maximum at a rough point and its minimum at an endpoint.

73 Draw a circular pool with a lifeguard on one side and a drowner on the opposite side. The lifeguard swims with velocity v and runs around the rest of the pool with velocity $w = 10v$. If the swim direction is at angle θ with the direct line, choose θ to minimize and maximize the arrival time.

3.3 Second Derivatives: Bending and Acceleration

When $f'(x)$ is positive, $f(x)$ is increasing. When dy/dx is negative, $y(x)$ is decreasing. That is clear, but what about the *second* derivative? From looking at the curve, can you decide the sign of $f''(x)$ or d^2y/dx^2? The answer is *yes* and the key is in the **bending**.

A straight line doesn't bend. The slope of $y = mx + b$ is m(a constant). The second derivative is zero. We have to go to curves, to see a changing slope. Changes in the derivative show up in $f''(x)$:

$f = x^2$ has $f' = 2x$ and $f'' = 2$ (this parabola bends *up*)

$y = \sin x$ has $dy/dx = \cos x$ and $d^2y/dx^2 = -\sin x$ (the sine bends *down*)

The slope $2x$ gets larger *even when the parabola is falling*. The sign of f or f' is not revealed by f''. The second derivative tells about **change in slope**.

A function with $f''(x) > 0$ is **concave up**. It bends upward as the slope increases. It is also called *convex*. A function with decreasing slope—this means $f''(x) < 0$—is **concave down**. Note how $\cos x$ and $1 + \cos x$ and even $1 + \frac{1}{2}x + \cos x$ change from concave down to concave up at $x = \pi/2$. At that point $f'' = -\cos x$ changes from negative to positive. The extra $1 + \frac{1}{2}x$ tilts the graph but the bending is the same.

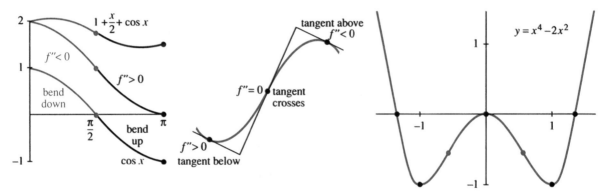

Fig. 3.7 Increasing slope = concave up ($f'' > 0$). Concave down is $f'' < 0$. Inflection point $f'' = 0$.

Here is another way to see the sign of f''. **Watch the tangent lines**.When the curve is concave up, the tangent stays below it. A linear approximation is too low. This section computes a *quadratic* approximation—which includes the term with $f'' > 0$. When the curve bends down ($f'' < 0$), the opposite happens—the tangent lines are above the curve. The linear approximation is too high, and f'' lowers it.

In physical motion, $f''(t)$ is the *acceleration*—in units of distance/(time)^2. Acceleration is rate of change of velocity. The oscillation $\sin 2t$ has $v = 2\cos 2t$ (maximum speed 2) and $a = -4\sin 2t$ (maximum acceleration 4).

An increasing population means $f' > 0$. **An increasing growth rate means** $f'' > 0$. Those are different. The rate can slow down while the growth continues.

MAXIMUM VS. MINIMUM

Remember that $f'(x) = 0$ locates a stationary point. That may be a *minimum* or a *maximum*. **The second derivative decides**! Instead of computing $f(x)$ at many points, we compute $f''(x)$ at one point—the stationary point. It is a minimum if $f''(x) > 0$.

> **3D** When $f'(x) = 0$ and $f''(x) > 0$, there is a *local minimum* at x.
> When $f'(x) = 0$ and $f''(x) < 0$, there is a *local maximum* at x.

To the left of a minimum, the curve is falling. After the minimum, the curve rises. The slope has changed from negative to positive. The graph bends upward and $f''(x) > 0$.

At a maximum the slope drops from positive to negative. In the exceptional case, when $f'(x) = 0$ and also $f''(x) = 0$, anything can happen. An example is x^3, which pauses at $x = 0$ and continues up (its slope is $3x^2 \geqslant 0$). However x^4 pauses and goes down (with a very flat graph).

We emphasize that the information from $f'(x)$ and $f''(x)$ is only "*local*." To be certain of an *absolute* minimum or maximum, we need information over the whole domain.

EXAMPLE 1 $f(x) = x^3 - x^2$ has $f'(x) = 3x^2 - 2x$ and $f''(x) = 6x - 2$.

To find the maximum and/or minimum, solve $3x^2 - 2x = 0$. The stationary points are $x = 0$ and $x = \frac{2}{3}$. At those points we need the second derivative. It is $f''(0) = -2$ (local maximum) and $f''(\frac{2}{3}) = +2$ (local minimum).

Between the maximum and minimum is the **inflection point**. *That is where* $f''(x) = 0$. The curve changes from concave down to concave up. This example has $f'' = 6x - 2$, so the inflection point is at $x = \frac{1}{3}$.

INFLECTION POINTS

In mathematics it is a special event when a function passes through zero. When the function is f, its graph crosses the axis. When the function is f', the tangent line is horizontal. When f'' goes through zero, we have an *inflection point*.

The direction of bending changes at an inflection point. Your eye picks that out in a graph. For an instant the graph is straight (straight lines have $f'' = 0$). It is easy to see crossing points and stationary points and inflection points. Very few people can recognize where $f''' = 0$ or $f'''' = 0$. I am not sure if those points have names.

There is a genuine maximum or minimum when $f'(x)$ changes sign. Similarly, there is a genuine inflection point when $f''(x)$ changes sign. ***The graph is concave down on one side of an inflection point and concave up on the other side.***[†] The tangents are above the curve on one side and below it on the other side. At an inflection point, *the tangent line crosses the curve* (Figure 3.7b).

Notice that a parabola $y = ax^2 + bx + c$ has no inflection points: y'' is constant. A cubic curve has one inflection point, because f'' is linear. A fourth-degree curve might or might not have inflection points—the quadratic $f''(x)$ might or might not cross the axis.

EXAMPLE 2 $x^4 - 2x^2$ is W-shaped, $4x^3 - 4x$ has two bumps, $12x^2 - 4$ is U-shaped. The table shows the signs at the important values of x:

x	$-\sqrt{2}$	-1	$-1/\sqrt{3}$	0	$1/\sqrt{3}$	1	$\sqrt{2}$
$f(x)$	0	$-$	$-$	$0,0$	$-$	$-$	0
$f'(x)$		0	$+$	0	$-$	0	
$f''(x)$			0	$-$	0		

[†] That rules out $f(x) = x^4$, which has $f'' = 12x^2 > 0$ on both sides of zero. Its tangent line is the x axis. The line stays below the graph—so no inflection point

Between zeros of $f(x)$ come zeros of $f'(x)$ (stationary points). Between zeros of $f'(x)$ come zeros of $f''(x)$ (inflection points). In this example $f(x)$ has a double zero at the origin, so a single zero of f' is caught there. It is a local maximum, since $f''(0) < 0$.

Inflection points are important—not just for mathematics. We know the world population will keep rising. We don't know if the *rate* of growth will slow down. Remember: ***The rate of growth stops growing at the inflection point***. Here is the 1990 report of the UN Population Fund.

> The next ten years will decide whether the world population trebles or merely doubles before it finally stops growing. This may decide the future of the earth as a habitation for humans. The population, now 5.3 billion, is increasing by a quarter of a million every day. Between 90 and 100 million people will be added every year during the 1990s; a billion people—a whole China—over the decade. The fastest growth will come in the poorest countries.
>
> A few years ago it seemed as if the rate of population growth was slowing† everywhere except in Africa and parts of South Asia. The world's population seemed set to stabilize around 10.2 billion towards the end of the next century.
>
> Today, the situation looks less promising. The world has overshot the marker points of the 1984 "most likely" medium projection. It is now on course for an eventual total that will be closer to 11 billion than to 10 billion.
>
> If fertility reductions continue to be slower than projected, the mark could be missed again. In that case the world could be headed towards a total of up to 14 billion people.

Starting with a census, the UN follows each age group in each country. They estimate the death rate and fertility rate—the medium estimates are published. This report is saying that we are not on track with the estimate.

Section 6.5 will come back to population, with an equation that predicts 10 billion. It assumes we are now at the inflection point. But China's second census just started on July 1, 1990. When it's finished we will know if the inflection point is still ahead.

You now understand the meaning of $f''(x)$. Its sign gives the direction of bending—the change in the slope. *The rest of this section computes **how much** the curve bends*—using the *size* of f'' and not just its sign. We find quadratic approximations based on $f''(x)$. In some courses they are optional—the main points are highlighted.

CENTERED DIFFERENCES AND SECOND DIFFERENCES

Calculus begins with average velocities, computed on either side of x:

$$\frac{f(x + \Delta x) - f(x)}{\Delta x} \quad \text{and} \quad \frac{f(x) - f(x - \Delta x)}{\Delta x} \quad \text{are close to } f'(x) \quad (1)$$

We never mentioned it, but a better approximation to $f'(x)$ comes from *averaging those two averages*. This produces a ***centered difference***, which is based on $x + \Delta x$ and $x - \Delta x$. It divides by $2\Delta x$:

$$f'(x) \approx \frac{1}{2}\left[\frac{f(x + \Delta x) - f(x)}{\Delta x} + \frac{f(x) - f(x - \Delta x)}{\Delta x} \right] = \frac{f(x + \Delta x) - f(x - \Delta x)}{2\Delta x}. \quad (2)$$

We claim this is better. The test is to try it on powers of x.

---

†The United Nations watches the second derivative!

For $f(x) = x$ these ratios all give $f' = 1$ (exactly). For $f(x) = x^2$, only the centered difference correctly gives $f' = 2x$. The one-sided ratio gave $2x + \Delta x$ (in Chapter 1 it was $2t + h$). It is only "first-order accurate." But centering leaves no error. We are averaging $2x + \Delta x$ with $2x - \Delta x$. Thus the centered difference is "second-order accurate."

I ask now: *What ratio converges to the second derivative?* One answer is to take differences of the first derivative. Certainly $\Delta f'/\Delta x$ approaches f''. But we want a ratio involving f itself. A natural idea is to take *differences of differences*, which brings us to "**second differences**":

$$\frac{\dfrac{f(x+\Delta x) - f(x)}{\Delta x} - \dfrac{f(x) - f(x - \Delta x)}{\Delta x}}{\Delta x} = \frac{f(x + \Delta x) - 2f(x) + f(x - \Delta x)}{(\Delta x)^2} \rightarrow \frac{d^2 f}{dx^2}. \quad (3)$$

On the top, the difference of the difference is $\Delta(\Delta f) = \Delta^2 f$. It corresponds to $d^2 f$. On the bottom, $(\Delta x)^2$ corresponds to dx^2. This explains the way we place the 2's in $d^2 f/dx^2$. To say it differently: dx is squared, df is not squared—as in distance/(time)^2.

Note that $(\Delta x)^2$ becomes much smaller than Δx. If we divide Δf by $(\Delta x)^2$, the ratio blows up. It is the extra cancellation in the second difference $\Delta^2 f$ that allows the limit to exist. That limit is $f''(x)$.

Application The great majority of equations can't be solved exactly. A typical case is $f''(x) = -\sin f(x)$ (the pendulum equation). To compute a solution, I would replace $f''(x)$ by the second difference in equation (3). Approximations at points spaced by Δx are a very large part of scientific computing.

To test the accuracy of these differences, here is an experiment on $f(x) = \sin x + \cos x$. The table shows the errors at $x = 0$ from formulas (1), (2), (3):

step length Δx	one-sided errors	centered errors	second difference errors
1/4	.1347	.0104	−.0052
1/8	.0650	.0026	−.0013
1/16	.0319	.0007	−.0003
1/32	.0158	.0002	−.0001

The one-sided errors are cut in half when Δx is cut in half. The other columns decrease like $(\Delta x)^2$. Each reduction divides those errors by 4. ***The errors from one-sided differences are $O(\Delta x)$ and the errors from centered differences are $O(\Delta x)^2$.***

The "big O" notation When the errors are of order Δx, we write $E = O(\Delta x)$. This means that $E \leqslant C\Delta x$ for some constant C. We don't compute C—in fact we don't want to deal with it. The statement "one-sided errors are Oh of delta x" captures what is important. The main point of the other columns is $E = O(\Delta x)^2$.

LINEAR APPROXIMATION VS. QUADRATIC APPROXIMATION

The second derivative gives a tremendous improvement over linear approximation $f(a) + f'(a)(x - a)$. A tangent line starts out close to the curve, but *the line has no way to bend.* After a while it overshoots or undershoots the true function (see Figure 3.8). That is especially clear for the model $f(x) = x^2$, when the tangent is the x axis and the parabola curves upward.

You can almost guess the term with bending. *It should involve f''*, and also $(\Delta x)^2$. It might be exactly $f''(x)$ times $(\Delta x)^2$ but it is not. The model function x^2 has $f'' = 2$. There must be a factor $\frac{1}{2}$ to cancel that 2:

> **3E** The **quadratic approximation** to a smooth function $f(x)$ near $x = a$ is
>
> $$f(x) \approx f(a) + f'(a)(x - a) + \tfrac{1}{2} f''(a)(x - a)^2. \qquad (4)$$

At the basepoint this is $f(a) = f(a)$. The derivatives also agree at $x = a$. Furthermore *the second derivatives agree*. On both sides of (4), the second derivative at $x = a$ is $f''(a)$.

The quadratic approximation bends with the function. It is not the absolutely final word, because there is a cubic term $\frac{1}{6} f'''(a)(x - a)^3$ and a fourth-degree term $\frac{1}{24} f''''(a)(x - a)^4$ and so on. The whole infinite sum is a "Taylor series." Equation (4) carries that series through the quadratic term—which for practical purposes gives a terrific approximation. You will see that in numerical experiments.

Two things to mention. First, equation (4) shows why $f'' > 0$ brings the curve above the tangent line. The linear part gives the line, while the quadratic part is positive and bends upward. Second, equation (4) comes from (2) and (3). Where one-sided differences give $f(x + \Delta x) \approx f(x) + f'(x)\Delta x$, centered differences give the quadratic:

$$\text{from}(2): \quad f(x + \Delta x) \approx f(x - \Delta x) + 2f'(x)\Delta x$$

$$\text{from}(3): \quad f(x + \Delta x) \approx 2f(x) - f(x - \Delta x) + f''(x)(\Delta x)^2.$$

Add and divide by 2. The result is $f(x + \Delta x) \approx f(x) + f'(x)\Delta x + \frac{1}{2} f''(x)(\Delta x)^2$. This is correct through $(\Delta x)^2$ and misses by $(\Delta x)^3$, as examples show:

EXAMPLE 3 $(x + \Delta x)^3 \approx (x^3) + (3x^2)(\Delta x) + \frac{1}{2}(6x)(\Delta x)^2 + \text{error}(\Delta x)^3.$

EXAMPLE 4 $(1 + x)^n \approx 1 + nx + \frac{1}{2} n(n - 1) x^2.$

The first derivative at $x = 0$ is n. The second derivative is $n(n - 1)$. The cubic term would be $\frac{1}{6} n(n - 1)(n - 2) x^3$. We are just producing the binomial expansion!

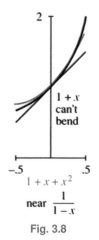

2

$1 + x$
can't
bend

$-.5$ $.5$
$1 + x + x^2$

near $\dfrac{1}{1 - x}$

Fig. 3.8

EXAMPLE 5 $\dfrac{1}{1 - x} \approx 1 + x + x^2 = $ start of a geometric series.

$1/(1 - x)$ has derivative $1/(1 - x)^2$. Its second derivative is $2/(1 - x)^3$. At $x = 0$ those equal $1, 1, 2$. The factor $\frac{1}{2}$ cancels the 2, which leaves $1, 1, 1$. This explains $1 + x + x^2$.

The next terms are x^3 and x^4. The whole series is $1/(1 - x) = 1 + x + x^2 + x^3 + \cdots$.

Numerical experiment $1/\sqrt{1 + x} \approx 1 - \frac{1}{2}x + \frac{3}{8}x^2$ is tested for accuracy. Dividing x by 2 almost divides the error by 8. If we only keep the linear part $1 - \frac{1}{2}x$, the error is only divided by 4. Here are the errors at $x = \frac{1}{4}, \frac{1}{8}$ and $\frac{1}{16}$:

$$\text{linear approximation}\left(\text{error} \approx \frac{3}{8}x^2\right): \quad .0194 \quad .0053 \quad .0014$$

$$\text{quadratic approximation}\left(\text{error} \approx \frac{-5}{16}x^3\right): -.00401 \quad -.00055 \quad -.00007$$

Read-through questions

The direction of bending is given by the sign of ___a___ . If the second derivative is ___b___ in an interval, the function is concave up (or convex). The graph bends ___c___ . The tangent lines are ___d___ the graph. If $f''(x) < 0$ then the graph is concave ___e___ , and the slope is ___f___ .

At a point where $f'(x) = 0$ and $f''(x) > 0$, the function has a ___g___ . At a point where ___h___ , the function has a maximum. A point where $f''(x) = 0$ is an ___i___ point, provided f'' changes sign. The tangent line ___j___ the graph.

The centered approximation to $f'(x)$ is $[\_\_k\_\_]/2\Delta x$. The 3-point approximation to $f''(x)$ is $[\_\_l\_\_]/(\Delta x)^2$. The second order approximation to $f(x + \Delta x)$ is $f(x) + f'(x)\Delta x + \_\_m\_\_$. Without that extra term this is just the ___n___ approximation. With that term the error is O(___o___).

1 A graph that is concave upward is inaccurately said to "hold water." Sketch a graph with $f''(x) > 0$ that would not hold water.

2 Find a function that is concave down for $x < 0$ and concave up for $0 < x < 1$ and concave down for $x > 1$.

3 Can a function be always concave down and never cross zero? Can it be always concave down and positive? Explain.

4 Find a function with $f''(2) = 0$ and no other inflection point.

True or false, when $f(x)$ is a 9th degree polynomial with $f'(1) = 0$ and $f'(3) = 0$. Give (or draw) a reason.

5 $f(x) = 0$ somewhere between $x = 1$ and $x = 3$.

6 $f''(x) = 0$ somewhere between $x = 1$ and $x = 3$.

7 There is no absolute maximum at $x = 3$.

8 There are seven points of inflection.

9 If $f(x)$ has nine zeros, it has seven inflection points.

10 If $f(x)$ has seven inflection points, it has nine zeros.

In 11–16 decide which stationary points are maxima or minima.

11 $f(x) = x^2 - 6x$ **12** $f(x) = x^3 - 6x^2$

13 $f(x) = x^4 - 6x^3$ **14** $f(x) = x^{11} - 6x^{10}$

15 $f(x) = \sin x - \cos x$ **16** $f(x) = x + \sin 2x$

Locate the inflection points and the regions where $f(x)$ is concave up or down.

17 $f(x) = x + x^2 - x^3$ **18** $f(x) = \sin x + \tan x$

19 $f(x) = (x-2)^2(x-4)^2$ **20** $f(x) = \sin x + (\sin x)^3$

21 If $f(x)$ is an even function, the centered difference $[f(\Delta x) - f(-\Delta x)]/2\Delta x$ exactly equals $f'(0) = 0$. Why?

22 If $f(x)$ is an odd function, the second difference $[f(\Delta x) - 2f(0) + f(-\Delta x)]/(\Delta x)^2$ exactly equals $f''(0) = 0$. Why?

Write down the quadratic $f(0) + f'(0)x + \frac{1}{2}f''(0)x^2$ in 23–26.

23 $f(x) = \cos x + \sin x$ **24** $f(x) = \tan x$

25 $f(x) = (\sin x)/x$ **26** $f(x) = 1 + x + x^2$

In 26, find $f(1) + f'(1)(x-1) + \frac{1}{2}f''(1)(x-1)^2$ around $a = 1$.

27 Find A and B in $\sqrt{1-x} \approx 1 + Ax + Bx^2$.

28 Find A and B in $1/(1-x)^2 \approx 1 + Ax + Bx^2$.

29 Substitute the quadratic approximation into $[f(x + \Delta x) - f(x)]/\Delta x$, to estimate the error in this one-sided approximation to $f'(x)$.

30 What is the quadratic approximation at $x = 0$ to $f(-\Delta x)$?

31 Substitute for $f(x + \Delta x)$ and $f(x - \Delta x)$ in the centered approximation $[f(x + \Delta x) - f(x - \Delta x)]/2\Delta x$, to get $f'(x) +$ error. Find the Δx and $(\Delta x)^2$ terms in this error. Test on $f(x) = x^3$ at $x = 0$.

32 Guess a third-order approximation $f(\Delta x) \approx f(0) + f'(0)\Delta x + \frac{1}{2}f''(0)(\Delta x)^2 + \_\_\_\_\_$. Test it on $f(x) = x^3$.

Construct a table as in the text, showing the actual errors at $x = 0$ in one-sided differences, centered differences, second differences, and quadratic approximations. By hand take two values of Δx, by calculator take three, by computer take four.

33 $f(x) = x^3 + x^4$ **34** $f(x) = 1/(1-x)$

35 $f(x) = x^2 + \sin x$

36 Example 5 was $1/(1-x) \approx 1 + x + x^2$. What is the error at $x = 0.1$? What is the error at $x = 2$?

37 Substitute $x = .01$ and $x = -0.1$ in the geometric series $1/(1-x) = 1 + x + x^2 + \cdots$ to find $1/.99$ and $1/1.1$—first to four decimals and then to all decimals.

38 Compute $\cos 1°$ by equation (4) with $a = 0$. OK to check on a calculator. Also compute $\cos 1$. Why so far off?

39 Why is $\sin x \approx x$ not only a linear approximation but also a quadratic approximation? $x = 0$ is an _____ point.

40 If $f(x)$ is an even function, find its quadratic approximation at $x = 0$. What is the equation of the tangent line?

41 For $f(x) = x + x^2 + x^3$, what is the centered difference $[f(3) - f(1)]/2$, and what is the true slope $f'(2)$?

42 For $f(x) = x + x^2 + x^3$, what is the second difference $[f(3) - 2f(2) + f(1)]/1^2$, and what is the exact $f''(2)$?

43 The error in $f(a) + f'(a)(x - a)$ is approximately $\frac{1}{2} f''(a)(x - a)^2$. This error is positive when the function is _____ . Then the tangent line is _____ the curve.

44 Draw a piecewise linear $y(x)$ that is concave up. Define "concave up" without using the test $d^2 y/dx^2 \geq 0$. If derivatives don't exist, a new definition is needed.

45 What do these sentences say about f or f' or f'' or f'''?

  1. The population is growing more slowly.

  2. The plane is landing smoothly.

  3. The economy is picking up speed.

  4. The tax rate is constant.

  5. A bike accelerates faster but a car goes faster.

  6. Stock prices have peaked.

  7. The rate of acceleration is slowing down.

  8. This course is going downhill.

46 (Recommended) Draw a curve that goes up-down-up. Below it draw its derivative. Then draw its second derivative. *Mark the same points on all curves*—the maximum, minimum, and inflection points of the first curve.

47 Repeat Problem 46 on a printout showing $y(x) = x^3 - 4x^2 + x + 2$ and dy/dx and $d^2 y/dx^2$ on the same graph.

3.4 Graphs

Reading a graph is like appreciating a painting. Everything is there, but you have to know what to look for. One way to learn is by sketching graphs yourself, and in the past that was almost the only way. Now it is obsolete to spend weeks drawing curves—a computer or graphing calculator does it faster and better. That doesn't remove the need to appreciate a graph (or a painting), since a curve displays a tremendous amount of information.

This section combines two approaches. One is to study actual machine-produced graphs (especially electrocardiograms). The other is to understand the mathematics of graphs—slope, concavity, asymptotes, shifts, and scaling. We introduce the ***centering transform*** and ***zoom transform***. These two approaches are like the rest of calculus, where special derivatives and integrals are done by hand and day-to-day applications are by computer. Both are essential—the machine can do experiments that we could never do. But without the mathematics our instructions miss the point. To create good graphs you have to know a few of them personally.

READING AN ELECTROCARDIOGRAM (ECG or EKG)

REFERENCE

The graphs of an ECG show the electrical potential during a heartbeat. There are twelve graphs—six from leads attached to the chest, and six from leads to the arms and left leg. (It doesn't hurt, but everybody is nervous. You have to lie still, because contraction of other muscles will mask the reading from the heart.) The graphs record electrical impulses, as the cells depolarize and the heart contracts.

What can I explain in two pages? The graph shows the fundamental pattern of the ECG. ***Note the*** P ***wave***, ***the*** QRS ***complex***, ***and the*** T ***wave***. Those patterns, seen differently in the twelve graphs, tell whether the heart is normal or out of rhythm—or suffering an infarction (a heart attack).

HEART RATE (3 CYCLES FROM REFERENCE ARROW—CHART PAPER SPEED: 25 mm./sec.)

500 —
400 —
300 —
200 —
175 —
150 —
140 —
130 —
120 —
110 —
100 —
95 —
90 —
85 —
80 —
75 —
70 —
65 —
60 —
55 —
50 —
45 —
40 —
35 —

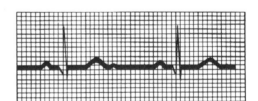

First of all the graphs show the ***heart rate***. The dark vertical lines are by convention $\frac{1}{5}$ second apart. The light lines are $\frac{1}{25}$ second apart. If the heart beats every $\frac{1}{5}$ second (one dark line) the rate is 5 beats per second or 300 per minute. That is extreme *tachycardia*—not compatible with life. The normal rate is between three dark lines per beat ($\frac{3}{5}$ second, or 100 beats per minute) and five dark lines (one second between beats, 60 per minute). A baby has a faster rate, over 100 per minute. In this figure the rate is _____ . A rate below 60 is *bradycardia*, not in itself dangerous. For a resting athlete that is normal.

Doctors memorize the six rates $300, 150, 100, 75, 60, 50$. Those correspond to $1, 2, 3, 4, 5, 6$ dark lines between heartbeats. The distance is easiest to measure between spikes (the peaks of the R wave). Many doctors put a printed scale next to the chart. One textbook emphasizes that "Where the next wave falls determines the rate. No mathematical computation is necessary." But you see where those numbers come from.

The next thing to look for is ***heart rhythm***. The regular rhythm is set by the pacemaker, which produces the P wave. A constant distance between waves is good—and then each beat is examined. When there is a block in the pathway, it shows as a delay in the graph. Sometimes the pacemaker fires irregularly. Figure 3.10 shows *sinus arrythmia* (fairly normal). The time between peaks is changing. In disease or emergency, there are potential pacemakers in all parts of the heart.

I should have pointed out the main parts. We have four chambers, an atrium ventricle pair on the left and right. The SA node should be the pacemaker. The stimulus spreads from the atria to the ventricles— from the small chambers that "prime the pump" to the powerful chambers that drive blood through the body. The P wave comes with contraction of the atria. There is a pause of $\frac{1}{10}$ second at the AV node. Then the big QRS wave starts contraction of the ventricles, and the T wave is when the ventricles relax. The cells switch back to negative charge and the heart cycle is complete.

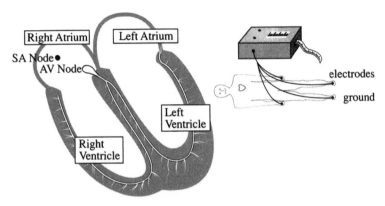

Fig. 3.9   Happy person with a heart and a normal electrocardiogram.

The ECG shows when the pacemaker goes wrong. Other pacemakers take over—the AV node will pace at 60/minute. An early firing in the ventricle can give a wide spike in the QRS complex, followed by a long pause. The impulses travel by a slow path. Also the pacemaker can suddenly speed up (paroxysmal tachycardia is $150 - 250$/minute). But the most critical danger is *fibrillation*.

Figure 3.10b shows a dying heart. The ECG indicates irregular contractions—no normal PQRST sequence at all. What kind of heart would generate such a rhythm? The muscles are quivering or "fibrillating" independently. The pumping action is nearly gone, which means emergency care. The patient needs immediate CPR—someone to do the pumping that the heart can't do. Cardio-pulmonary resuscitation is a combination of chest pressure and air pressure (hand and mouth) to restart the rhythm. CPR can be done on the street. A hospital applies a defibrillator, which shocks the heart back to life. It depolarizes *all* the heart cells, so the timing can be reset. Then the charge spreads normally from SA node to atria to AV node to ventricles.

This discussion has not used all twelve graphs to locate the problem. That needs *vectors*. Look ahead at Section 11.1 for the heart vector, and especially at Section 11.2 for its *twelve projections*. Those readings distinguish between atrium and ventricle, left and right, forward and back. This information is of vital importance in the event of a heart attack. A "heart attack" is a *myocardial infarction* (MI).

An MI occurs when part of an artery to the heart is blocked (a coronary occlusion).

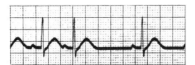 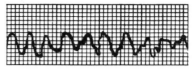

Fig. 3.10   Doubtful rhythm. Serious fibrillation. Signals of a heart attack.

An area is without blood supply—therefore without oxygen or glucose. Often the attack is in the thick left ventricle, which needs the most blood. The cells are first ischemic, then injured, and finally infarcted (dead). The classical ECG signals involve those three I's:

> **I***schemia*: Reduced blood supply, upside-down T wave in the chest leads.
> **I***njury*: An elevated segment between S and T means a recent attack.
> **I***nfarction*: The Q wave, normally a tiny dip or absent, is as wide as a small square ($\frac{1}{25}$ second). It may occupy a third of the entire QRS complex.

The Q wave gives the diagnosis. You can find all three I's in Figure 3.10c.

It is absolutely amazing how much a good graph can do.

THE MECHANICS OF GRAPHS

From the meaning of graphs we descend to the mechanics. A formula is now given for $f(x)$. The problem is *to create the graph*. It would be too old-fashioned to evaluate $f(x)$ by hand and draw a curve through a dozen points. A computer has a much better idea of a parabola than an artist (who tends to make it asymptotic to a straight line). There are some things a computer knows, and other things an artist knows, and still others that you and I know—because we understand derivatives.

Our job is to apply calculus. We extract information from f' and f'' as well as f. Small movements in the graph may go unnoticed, but the important properties come through. Here are the main tests:

1. The sign of $f(x)$ (above or below axis: $f = 0$ at *crossing point*)
2. The sign of $f'(x)$ (increasing or decreasing: $f' = 0$ at *stationary point*)
3. The sign of $f''(x)$ (concave up or down: $f'' = 0$ at *injection point*)
4. The behavior of $f(x)$ as $x \to \infty$ and $x \to -\infty$
5. The points at which $f(x) \to \infty$ or $f(x) \to -\infty$
6. Even or odd?   Periodic?   Jumps in f or f'?   Endpoints? $f(0)$?

EXAMPLE 1 $f(x) = \dfrac{x^2}{1 - x^2}$ $f'(x) = \dfrac{2x}{(1 - x^2)^2}$ $f''(x) = \dfrac{2 + 6x^2}{(1 - x^2)^3}$

The sign of $f(x)$ depends on $1 - x^2$. Thus $f(x) > 0$ in the inner interval where $x^2 < 1$. The graph bends upwards ($f''(x) > 0$) in that same interval. There are no inflection points, since f'' is never zero. The stationary point where f' vanishes is $x = 0$. We have a *local minimum* at $x = 0$.

The guidelines (or **asymptotes**) meet the graph at infinity. For large x the important terms are x^2 and $-x^2$. Their ratio is $+x^2/-x^2 = -1$—which is the limit as $x \to \infty$, and $x \to -\infty$. **The horizontal asymptote is the line $y = -1$.**

The other infinities, where f blows up, occur when $1 - x^2$ is zero. That happens at $x = 1$ and $x = -1$. **The vertical asymptotes are the lines $x = 1$ and $x = -1$.** The graph in Figure 3.11a approaches those lines.

if $f(x) \to b$ as $x \to \infty$ or $-\infty$, the line $y = b$ is a *horizontal asymptote*

if $f(x) \to +\infty$ or $-\infty$ as $x \to a$, the line $x = a$ is a *vertical asymptote*

if $f(x) - (mx + b) \to 0$ as $x \to +\infty$ or $\to -\infty$, the line $y = mx + b$ is a *sloping asymptote*.

Finally comes the vital fact that this function is **even**: $f(x) = f(-x)$ because squaring x obliterates the sign. The graph is symmetric across the y axis.

To summarize the effect of dividing by $1 - x^2$: No effect near $x = 0$. Blowup at 1 and -1 from zero in the denominator. The function approaches -1 as $|x| \to \infty$.

EXAMPLE 2 $\quad f(x) = \dfrac{x^2}{x-1} \qquad f' = \dfrac{x^2 - 2x}{(x-1)^2} \qquad f'' = \dfrac{2}{(x-1)^3}$

This example divides by $x - 1$. Therefore $x = 1$ is a vertical asymptote, where $f(x)$ becomes infinite. Vertical asymptotes come mostly from *zero denominators*.

Look beyond $x = 1$. Both $f(x)$ and $f''(x)$ are positive for $x > 1$. The slope is zero at $x = 2$. That must be a local minimum.

What happens as $x \to \infty$? Dividing x^2 by $x - 1$, the leading term is x. The function becomes large. It grows linearly—we expect a **sloping asymptote**. To find it, do the division properly:

$$\frac{x^2}{x-1} = x + 1 + \frac{1}{x-1}. \tag{1}$$

The last term goes to zero. The function approaches $y = x + 1$ as the asymptote.

This function is not odd or even. Its graph is in Figure 3.11b. With **zoom out** you see the asymptotes. **Zoom in** for $f = 0$ or $f' = 0$ or $f'' = 0$.

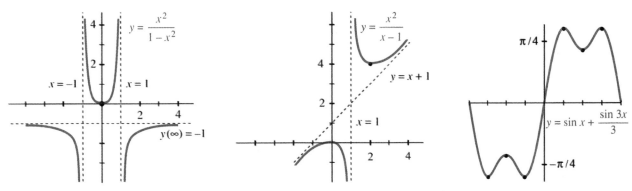

Fig. 3.11 The graphs of $x^2/(1-x^2)$ and $x^2/(x-1)$ and $\sin x + \frac{1}{3}\sin 3x$.

EXAMPLE 3 $\quad f(x) = \sin x + \frac{1}{3}\sin 3x \quad$ has the slope $\quad f'(x) = \cos x + \cos 3x$.

Above all these functions are **periodic**. If x increases by 2π, nothing changes. The graphs from 2π to 4π are repetitions of the graphs from 0 to 2π. Thus $f(x + 2\pi) = f(x)$ and the period is 2π. Any interval of length 2π will show a complete picture, and Figure 3.11c picks the interval from $-\pi$ to π.

The second outstanding property is that f is **odd**. The sine functions satisfy $f(-x) = -f(x)$. The graph is symmetric through the origin. By reflecting the right half through the origin, you get the left half. In contrast, the cosines in $f'(x)$ are even.

To find the zeros of $f(x)$ and $f'(x)$ and $f''(x)$, rewrite those functions as

$$f(x) = 2\sin x - \tfrac{4}{3}\sin^3 x \quad f'(x) = -2\cos x + 4\cos^3 x \quad f''(x) = -10\sin x + 12\sin^3 x.$$

We changed $\sin 3x$ to $3\sin x - 4\sin^3 x$. For the derivatives use $\sin^2 x = 1 - \cos^2 x$. Now find the zeros—the *crossing points*, *stationary points*, and *inflection points*:

$$f = 0 \quad 2\sin x = \tfrac{4}{3}\sin^3 x \Rightarrow \sin x = 0 \ \text{or} \ \sin^2 x = \tfrac{3}{2} \Rightarrow x = 0, \pm\pi$$

$$f' = 0 \quad 2\cos x = 4\cos^3 x \Rightarrow \cos x = 0 \ \text{or} \ \cos^2 x = \tfrac{1}{2} \Rightarrow x = \pm\pi/4, \pm\pi/2, \pm 3\pi/4$$

$$f'' = 0 \quad 5\sin x = 6\sin^3 x \Rightarrow \sin x = 0 \ \text{or} \ \sin^2 x = \tfrac{5}{6} \Rightarrow x = 0, \pm 66°, \pm 114°, \pm\pi$$

That is more than enough information to sketch the graph. The stationary points $\pi/4, \pi/2, 3\pi/4$ are evenly spaced. At those points $f(x)$ is $\sqrt{8}/3$ (maximum), $2/3$ (local minimum), $\sqrt{8}/3$ (maximum). Figure 3.11c shows the graph.

I would like to mention a beautiful continuation of this same pattern:

$$f(x) = \sin x + \tfrac{1}{3}\sin 3x + \tfrac{1}{5}\sin 5x + \cdots \qquad f'(x) = \cos x + \cos 3x + \cos 5x + \cdots$$

If we stop after ten terms, $f(x)$ is extremely close to a **step function**. If we don't stop, *the exact step function contains infinitely many sines*. It jumps from $-\pi/4$ to $+\pi/4$ as x goes past zero. More precisely it is a "**square wave**," because the graph jumps back down at π and repeats. The slope $\cos x + \cos 3x + \cdots$ also has period 2π. **Infinitely many cosines add up to a delta function**! (The slope at the jump is an infinite spike.) These sums of sines and cosines are **Fourier series**.

GRAPHS BY COMPUTERS AND CALCULATORS

We have come to a topic of prime importance. If you have **graphing software** for a computer, or if you have a **graphing calculator**, you can bring calculus to life. A graph presents $y(x)$ in a new way—different from the formula. Information that is buried in the formula is clear on the graph. *But don't throw away $y(x)$ and dy/dx.* The derivative is far from obsolete.

These pages discuss how calculus and graphs go together. We work on a crucial problem of applied mathematics—to find where $y(x)$ reaches its minimum. There is no need to tell you a hundred applications. Begin with the formula. How do you find the point x^* where $y(x)$ is smallest?

First, draw the graph. That shows the main features. We should see (roughly) where x^* lies. There may be several minima, or possibly none. But what we see depends on a decision that is ours to make—*the range of x and y in the viewing window.*

If nothing is known about $y(x)$, the range is hard to choose. We can accept a default range, and zoom in or out. We can use the autoscaling program in Section 1.7. Somehow x^* can be observed on the screen. Then the problem is to compute it.

I would like to work with a specific example. We solved it by calculus—to find the best point x^* to enter an expressway. The speeds in Section 3.2 were 30 and 60. The length of the fast road will be $b = 6$. *The range of reasonable values for the entering point is $0 \leq x \leq 6$.* The distance to the road in Figure 3.12 is $a = 3$. We drive a distance $\sqrt{3^2 + x^2}$ at speed 30 and the remaining distance $6 - x$ at speed 60:

$$driving\ time \quad y(x) = \frac{1}{30}\sqrt{3^2 + x^2} + \frac{1}{60}(6 - x). \tag{2}$$

This is the function to be minimized. Its graph is extremely flat.

It may seem unusual for the graph to be so level. On the contrary, it is common. *A flat graph is the whole point of $dy/dx = 0$.*

The graph near the minimum looks like $y = Cx^2$. It is a parabola sitting on a horizontal tangent. At a distance of $\Delta x = .01$, we only go up by $C(\Delta x)^2 = .0001C$. Unless C is a large number, this Δy can hardly be seen.

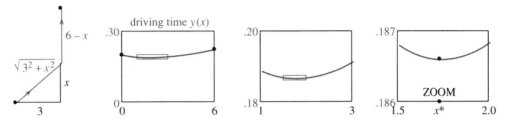

Fig. 3.12   Enter at x. The graph of driving time $y(x)$. Zoom boxes locate x^*.

*The solution is to change scale. **Zoom in on** x^*.* The tangent line stays flat, since dy/dx is still zero. But the bending from C is increased. Figure 3.12 shows the *zoom box* blown up into a new graph of $y(x)$.

A calculator has one or more ways to find x^*. With a TRACE mode, you direct a cursor along the graph. From the display of y values, read y_{max} and x^* to the nearest pixel. A zoom gives better accuracy, because it stretches the axes—each pixel represents a smaller Δx and Δy. The TI-81 stretches by 4 as default. Even better, let the whole process be graphical—*draw the actual* ZOOM BOX *on the screen*. Pick two opposite corners, press ENTER, and the box becomes the new viewing window (Figure 3.12).

The first zoom narrows the search for x^*. It lies between $x = 1$ and $x = 3$. We build a new ZOOM BOX and zoom in again. Now $1.5 \leqslant x^* \leqslant 2$. Reasonable accuracy comes quickly. High accuracy does not come quickly. It takes time to create the box and execute the zoom.

Question 1   What happens as we zoom in, if all boxes are square (equal scaling) ?

Answer   The picture gets flatter and flatter. We are zooming in to the tangent line. Changing x to $X/4$ and y to $Y/4$, the parabola $y = x^2$ flattens to $Y = X^2/4$. To see any bending, *we must use a long thin zoom box*.

I want to change to a totally different approach. Suppose we have a formula for dy/dx. *That derivative was produced by an infinite zoom!* The limit of $\Delta y/\Delta x$ came by brainpower alone:

$$\frac{dy}{dx} = \frac{x}{30\sqrt{3^2 + x^2}} = -\frac{1}{60}. \qquad \text{Call this } f(x).$$

This function is zero at x^*. The computing problem is completely changed: Solve $f(x) = 0$. *It is easier to find a root of $f(x)$ than a minimum of $y(x)$.* The graph of $f(x)$ crosses the x axis. The graph of $y(x)$ goes flat—this is harder to pinpoint.

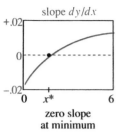

slope dy/dx

zero slope
at minimum

Fig. 3.13

Take the model function $y = x^2$ for $|x| < .01$. The slope $f = 2x$ changes from $-.02$ to $+.02$. The value of x^2 moves only by .0001 —its minimum point is hard to see.

To repeat: Minimization is easier with dy/dx. The screen shows an order of magnitude improvement, when we trace or zoom on $f(x) = 0$. In calculus, we have been taking the derivative for granted. It is natural to get blasé about $dy/dx = 0$. We forget how intelligent it is, to work with the slope instead of the function.

Question 2   How do you get another order of magnitude improvement?

Answer   Use the next derivative! With a formula for df/dx, which is d^2y/dx^2, the convergence is even faster. In two steps the error goes from .01 to .0001 to .00000001. Another infinite zoom went into the formula for df/dx, and *Newton's method* takes account of it. Sections 3.6 and 3.7 study $f(x) = 0$.

The expressway example allows perfect accuracy. We can solve $dy/dx = 0$ by algebra. The equation simplifies to $60x = 30\sqrt{3^2 + x^2}$. Dividing by 30 and squaring yields $4x^2 = 3^2 + x^2$. Then $3x^2 = 3^2$. The exact solution is $x^* = \sqrt{3} = 1.73205\ldots$

A model like this is a benchmark, to test competing methods. It also displays what we never appreciated—the extreme flatness of the graph. The difference in driving time between entering at $x^* = \sqrt{3}$ and $x = 2$ is *one second*.

THE CENTERING TRANSFORM AND ZOOM TRANSFORM

For a photograph we do two things—point the right way and stand at the right distance. Then take the picture. Those steps are the same for a graph. First we pick the new center point. The graph is *shifted*, to move that point from (a,b) to $(0,0)$. Then we decide how far the graph should reach. It fits in a rectangle, just like the photograph. *Rescaling* to x/c and y/d puts the desired section of the curve into the rectangle.

A good photographer does more (like an artist). The subjects are placed and the camera is focused. For good graphs those are necessary too. But an everyday calculator or computer or camera is built to operate without an artist—just aim and shoot. I want to explain how to aim at $y = f(x)$.

We are doing exactly what a calculator does, with one big difference. *It doesn't change coordinates. We do.* When $x = 1$, $y = -2$ moves to the center of the viewing window, the calculator still shows that point as $(1, -2)$. When the **centering transform** acts on $y + 2 = m(x - 1)$, those numbers disappear. This will be confusing unless x and y also change. *The new coordinates are $X = x - 1$ and $Y = y + 2$. Then the new equation is $Y = mX$.*

The main point (for humans) is to make the algebra simpler. The computer has no preference for $Y = mX$ over $y - y_0 = m(x - x_0)$. It accepts $2x^2 - 4x$ as easily as x^2. But we do prefer $Y = mX$ and $y = x^2$, partly because their graphs go through $(0, 0)$. Ever since zero was invented, mathematicians have liked that number best.

3F A **centering transform** shifts left by a and down by b:

$$X = x - a \text{ and } Y = y - b \text{ change } y = f(x) \text{ into } Y + b = f(X + a).$$

EXAMPLE 4   The parabola $y = 2x^2 - 4x$ has its minimum when $dy/dx = 4x - 4 = 0$. Thus $x = 1$ and $y = -2$. Move this bottom point to the center:

$y = 2x^2 - 4x$ is

$$Y + 2 = 2(X - 1)^2 - 4(X - 1) \quad \text{or} \quad Y = 2X^2.$$

The new parabola $Y = 2X^2$ has its bottom at $(0,0)$. It is the same curve, shifted across and up. The only simpler parabola is $y = x^2$. This final step is the job of the zoom.

Next comes scaling. We may want more detail (zoom in to see the tangent line). We may want a big picture (zoom out to check asymptotes). We might stretch one axis more than the other, if the picture looks like a pancake or a skyscraper.

3G A *zoom transform* scales the X and Y axes by c and d:

$$\mathbf{x} = cX \quad \text{and} \quad \mathbf{y} = dY \quad \text{change} \quad Y = F(X) \quad \text{to} \quad \mathbf{y} = dF(\mathbf{x}/c).$$

The new $\mathbf{x}$ and $\mathbf{y}$ are boldface letters, and the graph is rescaled. Often $c = d$.

EXAMPLE 5 Start with $Y = 2X^2$. Apply a square zoom with $c = d$. In the new $\mathbf{xy}$ coordinates, the equation is $\mathbf{y}/c = 2(\mathbf{x}/c)^2$. The number 2 disappears if $c = d = 2$. With the right centering and the right zoom, every parabola that opens upward is $\mathbf{y} = \mathbf{x}^2$.

Question 3 What happens to the derivatives (*slope and bending*) after a zoom?
Answer The slope (first derivative) is multiplied by d/c. Apply the chain rule to $\mathbf{y} = dF(\mathbf{x}/c)$. A square zoom has $d/c = 1$—*lines keep their slope*. The second derivative is multiplied by d/c^2, which changes the bending. A zoom out divides by small numbers $c = d$, so the big picture is more, curved.

Combining the centering and zoom transforms, as we do in practice, gives $\mathbf{y}$ in terms of $\mathbf{x}$:

$$y = f(x) \quad \text{becomes} \quad Y = f(X + a) - b \quad \text{and then} \quad \mathbf{y} = d\left[f\left(\frac{\mathbf{x}}{c} + a \right) - b \right].$$

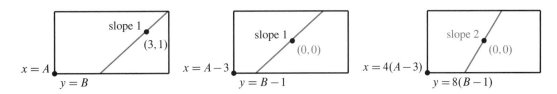

Fig. 3.14 Change of coordinates by centering and zoom. Calculators still show (x, y).

Question 4 Find x and y ranges after two transforms. Start between -1 and 1.
Answer The window after centering is $-1 \leqslant x - a \leqslant 1$ and $-1 \leqslant y - b \leqslant 1$. The window after zoom is $-1 \leqslant c(x - a) \leqslant 1$ and $-1 \leqslant d(y - b) \leqslant 1$. The point $(1, 1)$ was originally in the corner. The point $(c^{-1} + a, d^{-1} + b)$ is now in the corner.

The numbers a, b, c, d are chosen to produce a simpler function (like $\mathbf{y} = \mathbf{x}^2$). Or else—this is important in applied mathematics—they are chosen to make $\mathbf{x}$ and $\mathbf{y}$ "dimensionless." An example is $y = \frac{1}{2} \cos 8t$. The frequency 8 has dimension 1/time. The amplitude $\frac{1}{2}$ is a distance. With $d = 2$ cm and $c = 8$ sec, the units are removed and $\mathbf{y} = \cos \mathbf{t}$.

May I mention one transform that *does* change the slope? It is a ***rotation***. The whole plane is turned. A photographer might use it—but normally people are supposed to be upright. You use rotation when you turn a map or straighten a picture. In the next section, an unrecognizable hyperbola is turned into $Y = 1/X$.

3.4 EXERCISES

Read-through questions

The position, slope, and bending of $y = f(x)$ are decided by __a__, __b__ and __c__. If $|f(x)| \to \infty$ as $x \to a$, the line $x = a$ is a vertical __d__. If $f(x) \to b$ for large x, then $y = b$ is a __e__. If $f(x) - mx \to b$ for large x, then $y = mx + b$ is a __f__. The asymptotes of $y = x^2/(x^2 - 4)$ are __g__. This function is even because $y(-x) =$ __h__. The function $\sin kx$ has period __i__.

Near a point where $dy/dx = 0$, the graph is extremely __j__. For the model $y = Cx^2$, $x = .1$ gives $y =$ __k__. A box around the graph looks long and __l__. We __m__ in to that box for another digit of x^*. But solving $dy/dx = 0$ is more accurate, because its graph __n__ the x axis. The slope of dy/dx is __o__. Each derivative is like an __p__ zoom.

To move (a,b) to $(0,0)$, shift the variables to $X =$ __q__ and $Y =$ __r__. This __s__ transform changes $y = f(x)$ to $Y =$ __t__. The original slope at (a,b) equals the new slope at __u__. To stretch the axes by c and d, set $\mathbf{x} = cX$ and __v__. The __w__ transform changes $Y = F(X)$ to $\mathbf{y} =$ __x__. Slopes are multiplied by __y__. Second derivatives are multiplied by __z__.

1 Find the pulse rate when heartbeats are $\frac{1}{2}$ second or two dark lines or x seconds apart.

2 Another way to compute the heart rate uses marks for 6-second intervals. Doctors count the cycles in an interval.

(a) How many dark lines in 6 seconds?

(b) With 8 beats per interval, find the rate.

(c) Rule: Heart rate = cycles per interval times _____.

Which functions in 3–18 are even or odd or periodic? Find all asymptotes: $y = b$ or $x = a$ or $y = mx + b$. Draw roughly by hand or smoothly by computer.

3 $f(x) = x - (9/x)$

4 $f(x) = x^n$ (any integer n)

5 $f(x) = \dfrac{1}{1 - x^2}$

6 $f(x) = \dfrac{x^3}{4 - x^2}$

7 $f(x) = \dfrac{x^2 + 3}{x^2 + 1}$

8 $f(x) = \dfrac{x^2 + 3}{x + 1}$

9 $f(x) = (\sin x)(\sin 2x)$

10 $f(x) = \cos x + \cos 3x + \cos 5x$

11 $f(x) = \dfrac{x \sin x}{x^2 - 1}$

12 $f(x) = \dfrac{x}{\sin x}$

13 $f(x) = \dfrac{1}{x^3 + x^2}$

14 $f(x) = \dfrac{1}{x - 1} - 2x$

15 $f(x) = \dfrac{x^3 + 1}{x^3 - 1}$

16 $f(x) = \dfrac{\sin x + \cos x}{\sin x - \cos x}$

17 $f(x) = x - \sin x$

18 $f(x) = (1/x) - \sqrt{x}$

In 19–24 construct $f(x)$ with exactly these asymptotes.

19 $x = 1$ and $y = 2$

20 $x = 1$, $x = 2$, $y = 0$

21 $y = x$ and $x = 4$

22 $y = 2x + 3$ and $x = 0$

23 $y = x (x \to \infty)$, $y = -x (x \to -\infty)$

24 $x = 1, x = 3, y = x$

25 For $P(x)/Q(x)$ to have $y = 2$ as asymptote, the polynomials P and Q must be _____.

26 For $P(x)/Q(x)$ to have a sloping asymptote, the degrees of P and Q must be _____.

27 For $P(x)/Q(x)$ to have the asymptote $y = 0$, the degrees of P and Q must _____. The graph of $x^4/(1 + x^2)$ has what asymptotes?

28 Both $1/(x - 1)$ and $1/(x - 1)^2$ have $x = 1$ *and* $y = 0$ as asymptotes. The most obvious difference in the graphs is _____.

29 If $f'(x)$ has asymptotes $x = 1$ and $y = 3$ then $f(x)$ has asymptotes _____.

30 **True** (with reason) or **false** (with example).

(a) Every ratio of polynomials has asymptotes

(b) If $f(x)$ is even so is $f''(x)$

(c) If $f''(x)$ is even so is $f(x)$

(d) Between vertical asymptotes, $f'(x)$ touches zero.

31 Construct an $f(x)$ that is "even around $x = 3$."

32 Construct $g(x)$ to be "odd around $x = \pi$."

Create graphs of 33–38 on a computer or calculator.

33 $y(x) = (1 + 1/x)^x, -3 \leqslant x \leqslant 3$

34 $y(x) = x^{1/x}, 0.1 \leqslant x \leqslant 2$

35 $y(x) = \sin(x/3) + \sin(x/5)$

36 $y(x) = (2 - x)/(2 + x), -3 \leqslant x \leqslant 3$

37 $y(x) = 2x^3 + 3x^2 - 12x + 5$ on $[-3, 3]$ and $[2.9, 3.1]$

38 $100[\sin(x + .1) - 2\sin x + \sin(x - .1)]$

In 39–40 show the asymptotes on large-scale computer graphs.

39 (a) $y = \dfrac{x^3 + 8x - 15}{x^2 - 2}$ (b) $y = \dfrac{x^4 - 6x^3 + 1}{2x^4 + x^2}$

40 (a) $y = \dfrac{x^2 - 2}{x^3 + 8x - 15}$ (b) $y = \dfrac{x^2 - x + 2}{x^2 - 2x + 1}$

41 Rescale $y = \sin x$ so X is in degrees, not radians, and Y changes from meters to centimeters.

Problems 42–46 minimize the driving time $y(x)$ in the text. Some questions may not fit your software.

42 Trace along the graph of $y(x)$ to estimate x^*. Choose an xy range or use the default.

43 Zoomin by $c = d = 4$. How many zooms until you reach $x^* = 1.73205$ or 1.7320508 ?

44 Ask your program for the minimum of $y(x)$ and the solution of $dy/dx = 0$. Same answer ?

45 What are the scaling factors c and d for the two zooms in Figure 3.12 ? They give the stretching of the x and y axes.

46 Show that $dy/dx = -1/60$ and $d^2y/dx^2 = 1/90$ at $x = 0$. Linear approximation gives $dy/dx \approx -1/60 + x/90$. So the slope is zero near $x = $ _____ . This is Newton's method, using the next derivative.

Change the function to $y(x) = \sqrt{15 + x^2}/30 + (10 - x)/60$.

47 Find x^* using only the graph of $y(x)$.

48 Find x^* using also the graph of dy/dx.

49 What are the xy and XY and **xy** equations for the line in Figure 3.14 ?

50 Define $f_n(x) = \sin x + \frac{1}{3}\sin 3x + \frac{1}{5}\sin 5x + \cdots$ (n terms). Graph f_5 and f_{10} from $-\pi$ to π. Zoom in and describe the *Gibbs phenomenon* at $x = 0$.

On the graphs of 51–56, zoom in to all maxima and minima (3 significant digits). *Estimate inflection points.*

51 $y = 2x^5 - 16x^4 + 5x^3 - 37x^2 + 21x + 683$

52 $y = x^5 - x^4 - \sqrt{3x + 1} - 2$

53 $y = x(x - 1)(x - 2)(x - 4)$

54 $y = 7\sin 2x + 5\cos 3x$

55 $y = (x^3 - 2x + 1)/(x^4 - 3x^2 - 15), -3 \leqslant x \leqslant 5$

56 $y = x\sin(1/x), 0.1 \leqslant x \leqslant 1$

57 A 10-digit computer shows $y = 0$ and $dy/dx = .01$ at $x^* = 1$. This root should be correct to about (8 digits) (10 digits) (12 digits). *Hint*: Suppose $y = .01(x - 1 + \text{error})$. What errors don't show in 10 digits of y ?

58 Which is harder to compute accurately: Maximum point or inflection point ? First derivative or second derivative ?

3.5 Parabolas, Ellipses, and Hyperbolas

Here is a list of the most important curves in mathematics, so you can tell what is coming. It is not easy to rank the top four:

1. *straight lines*
2. *sines and cosines* (oscillation)
3. *exponentials* (growth and decay)
4. *parabolas*, *ellipses*, *and hyperbolas* (using $1, x, y, x^2, xy, y^2$).

The curves that I wrote last, the Greeks would have written first. It is so natural to go from linear equations to quadratic equations. Straight lines use $1, x, y$. Second degree curves include x^2, xy, y^2. If we go on to x^3 and y^3, the mathematics gets complicated. We now study equations of second degree, and the curves they produce.

It is quite important to see both the **equations** and the **curves**. This section connects two great parts of mathematics—*analysis* of the equation and *geometry* of the curve. Together they produce "**analytic geometry**." You already know about functions and graphs. Even more basic: Numbers correspond to points. We speak about "*the point* $(5, 2)$." Euclid might not have understood.

Where Euclid drew a $45°$ line through the origin, Descartes wrote down $y = x$. Analytic geometry has become central to mathematics—we now look at one part of it.

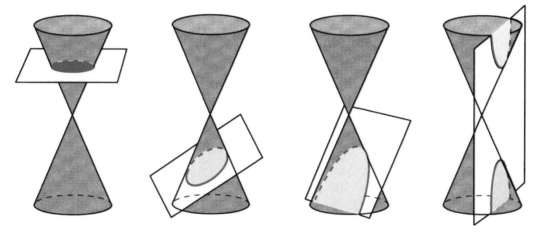

Fig. 3.15   The cutting plane gets steeper: circle to ellipse to parabola to hyperbola.

CONIC SECTIONS

The parabola and ellipse and hyperbola have absolutely remarkable properties. The Greeks discovered that all these curves come from **slicing a cone by a plane**. The curves are "conic sections." A level cut gives a **circle**, and a moderate angle produces an **ellipse**. A steep cut gives the two pieces of a **hyperbola** (Figure 3.15d). At the borderline, when the slicing angle matches the cone angle, the plane carves out a **parabola**. It has one branch like an ellipse, but it opens to infinity like a hyperbola.

Throughout mathematics, parabolas are on the border between ellipses and hyperbolas.

To repeat: We can slice through cones or we can look for equations. For a cone of light, we see an ellipse on the wall. (The wall cuts into the light cone.) For an equation $Ax^2 + Bxy + Cy^2 + Dx + Ey + F = 0$, we will work to make it simpler.

The graph will be centered and rescaled (and rotated if necessary), aiming for an equation like $y = x^2$. Eccentricity and polar coordinates are left for Chapter 9.

THE PARABOLA $y = ax^2 + bx + c$

You knew this function long before calculus. The graph crosses the x axis when $y = 0$. The quadratic formula solves $y = 3x^2 - 4x + 1 = 0$, and so does factoring into $(x - 1)(3x - 1)$. The crossing points $x = 1$ and $x = \frac{1}{3}$ come from algebra.

The other important point is found by calculus. It is the *minimum* point, where $dy/dx = 6x - 4 = 0$. The x coordinate is $\frac{4}{6} = \frac{2}{3}$, halfway between the crossing points. The height is $y_{min} = -\frac{1}{3}$. This is the **vertex** V in Figure 3.16a—at the bottom of the parabola.

A parabola has no asymptotes. The slope $6x - 4$ doesn't approach a constant.

To center the vertex Shift left by $\frac{2}{3}$ and up by $\frac{1}{3}$. So introduce the new variables $X = x - \frac{2}{3}$ and $Y = y + \frac{1}{3}$. Then $x = \frac{2}{3}$ and $y = -\frac{1}{3}$ correspond to $X = Y = 0$—which is the new vertex:

$$y = 3x^2 - 4x + 1 \quad \text{becomes} \quad Y = 3X^2. \tag{1}$$

Check the algebra. $Y = 3X^2$ is the same as $y + \frac{1}{3} = 3\left(x - \frac{2}{3}\right)^2$. That simplifies to the original equation $y = 3x^2 - 4x + 1$. The second graph shows the centered parabola $Y = 3X^2$, with the vertex moved to the origin.

To zoom in on the vertex Rescale X and Y by the zoom factor a:

$$Y = 3X^2 \quad \text{becomes} \quad \mathbf{y}/a = 3(\mathbf{x}/a)^2.$$

The final equation has $\mathbf{x}$ and $\mathbf{y}$ in boldface. With $a = 3$ we find $\mathbf{y} = \mathbf{x}^2$—the graph is magnified by 3. In two steps we have reached the model parabola opening upward.

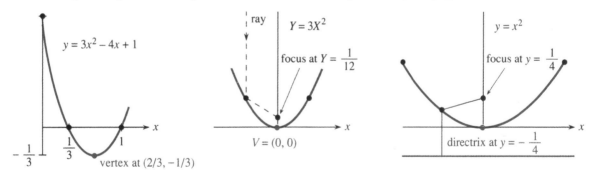

Fig. 3.16 Parabola with minimum at V. Rays reflect to focus. Centered in (b), rescaled in (c).

A parabola has another important point—the *focus*. Its distance from the vertex is called p. The special parabola $\mathbf{y} = \mathbf{x}^2$ has $p = 1/4$, and other parabolas $Y = aX^2$ have $p = 1/4a$. You magnify by a factor a to get $\mathbf{y} = \mathbf{x}^2$. The beautiful property of a parabola is that *every ray coming straight down is reflected to the focus*.

Problem 2.3.25 located the focus F—here we mention two applications. A solar collector and a TV dish are parabolic. They concentrate sun rays and TV signals onto a point—a heat cell or a receiver collects them at the focus. The 1982 *UMAP Journal* explains how radar and sonar use the same idea. Car headlights turn the idea around, and send the light outward.

Here is a classical fact about parabolas. **From each point on the curve, the distance to the focus equals the distance to the "directrix."** The directrix is the line $y = -p$ below the vertex (so the vertex is halfway between focus and directrix). With $p = \frac{1}{4}$, the distance down from any (x, y) is $y + \frac{1}{4}$. Match that with the distance to the focus at $\left(0, \frac{1}{4}\right)$— this is the square root below. Out comes the special parabola $y = x^2$:

$$y + \tfrac{1}{4} = \sqrt{x^2 + \left(y - \tfrac{1}{4}\right)^2} \quad \longrightarrow \quad \text{(square both sides)} \quad \longrightarrow \quad y = x^2. \quad (2)$$

The exercises give practice with all the steps we have taken—center the parabola to $Y = aX^2$, rescale it to $\mathbf{y} = \mathbf{x}^2$, locate the vertex and focus and directrix.

Summary for other parabolas $\quad y = ax^2 + bx + c$ has its vertex where dy/dx is zero. Thus $2ax + b = 0$ and $x = -b/2a$. Shifting across to that point is "completing the square":

$$ax^2 + bx + c \quad \text{equals} \quad a\left(x + \frac{b}{2a}\right)^2 + C. \quad (3)$$

Here $C = c - (b^2/4a)$ is the height of the vertex. The centering transform $X = x + (b/2a), Y = y - C$ produces $Y = aX^2$. It moves the vertex to $(0,0)$, where it belongs.

For the ellipse and hyperbola, our plan of attack is the same:

1. Center the curve to remove any linear terms Dx and Ey.
2. Locate each focus and discover the reflection property.
3. Rotate to remove Bxy if the equation contains it.

$$\textbf{ELLIPSES} \quad \frac{x^2}{a^2} + \frac{y^2}{b^2} = 1 \quad \textbf{(CIRCLES HAVE } a = b\textbf{)}$$

This equation makes the ellipse symmetric about $(0,0)$—the center. Changing x to $-x$ or y to $-y$ leaves the same equation. No extra centering or rotation is needed.

The equation also shows that x^2/a^2 and y^2/b^2 cannot exceed one. (They add to one and can't be negative.) Therefore $x^2 \leqslant a^2$, and x stays between $-a$ and a. Similarly y stays between b and $-b$. The ellipse is inside a rectangle.

By solving for y we get a function (or two functions!) of x:

$$\frac{y^2}{b^2} = 1 - \frac{x^2}{a^2} \quad \text{gives} \quad \frac{y}{b} = \pm\sqrt{1 - \frac{x^2}{a^2}} \quad \text{or} \quad y = \pm\frac{b}{a}\sqrt{a^2 - x^2}.$$

The graphs are the top half $(+)$ and bottom half $(-)$ of the ellipse. To draw the ellipse, plot them together. They meet when $y = 0$, at $x = a$ on the far right of Figure 3.17 and at $x = -a$ on the far left. The maximum $y = b$ and minimum $y = -b$ are at the top and bottom of the ellipse, where we bump into the enclosing rectangle.

A circle is a special case of an ellipse, when $a = b$. The circle equation $x^2 + y^2 = r^2$ is the ellipse equation with $a = b = r$. This circle is centered at $(0,0)$; other circles are centered at $x = h, y = k$. The circle is determined by its *radius r* and its *center (h,k)*:

$$\textit{Equation of circle}: \quad (x - h)^2 + (y - k)^2 = r^2. \quad (4)$$

In words, the distance from (x, y) on the circle to (h, k) at the center is r. The equation has linear terms $-2hx$ and $-2ky$—they disappear when the center is $(0,0)$.

EXAMPLE 1 Find the circle that has a diameter from $(1,7)$ to $(5,7)$.

Solution The center is halfway at $(3,7)$. So $r = 2$ and $(x-3)^2 + (y-7)^2 = 2^2$.

EXAMPLE 2 Find the center and radius of the circle $x^2 - 6x + y^2 - 14y = -54$.

Solution Complete $x^2 - 6x$ to the square $(x-3)^2$ by adding 9. Complete $y^2 - 14y$ to $(y-7)^2$ by adding 49. Adding 9 and 49 to both sides of the equation leaves $(x-3)^2 + (y-7)^2 = 4$—the same circle as in Example 1.

Quicker Solution Match the given equation with (4). Then $h = 3, k = 7$, and $r = 2$:
$x^2 - 6x + y^2 - 14y = -54$ must agree with $x^2 - 2hx + h^2 + y^2 - 2ky + k^2 = r^2$.

The change to $X = x - h$ and $Y = y - k$ moves the center of the circle from (h,k) to $(0,0)$. This is equally true for an ellipse:

$$\text{The ellipse } \frac{(x-h)^2}{a^2} + \frac{(y-k)^2}{b^2} = 1 \quad \text{becomes} \quad \frac{X^2}{a^2} + \frac{Y^2}{b^2} = 1$$

When we rescale by $x = X/a$ and $y = Y/b$, we get the unit circle $x^2 + y^2 = 1$.

The unit circle has area π. **The ellipse has area πab** (proved later in the book). The distance around the circle is 2π. The distance around an ellipse does not rescale—it has no simple formula.

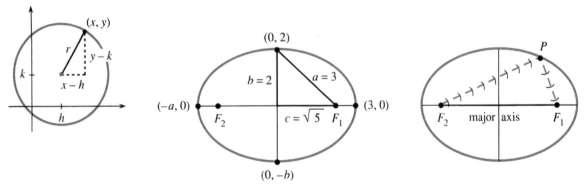

Fig. 3.17 Uncentered circle. Centered ellipse $x^2/3^2 + y^2/2^2 = 1$. *The distance from center to far right is also $a = 3$. All rays from F_2 reflect to F_1.*

Now we leave circles and concentrate on ellipses. They have **two foci** (pronounced *fo-sigh*). For a parabola, the second focus is at infinity. For a circle, both foci are at the center. The foci of an ellipse are on its longer axis (its *major* axis), one focus on each side of the center:

$$F_1 \text{ is at } x = c = \sqrt{a^2 - b^2} \quad \text{and} \quad F_2 \text{ is at } x = -c.$$

The right triangle in Figure 3.17 has sides a, b, c. From the top of the ellipse, the distance to each focus is a. From the endpoint at $x = a$, the distances to the foci are $a + c$ and $a - c$. Adding $(a+c) + (a-c)$ gives $2a$. *As you go around the ellipse, the distance to F_1 plus the distance to F_2 is constant* (always $2a$).

3H At all points on the ellipse, the sum of distances from the foci is $2a$. This is another equation for the ellipse:

$$\text{from } F_1 \text{ and } F_2 \text{ to } (x,y): \sqrt{(x-c)^2 + y^2} + \sqrt{(x+c)^2 + y^2} = 2a. \quad (5)$$

To draw an ellipse, tie a string of length $2a$ to the foci. Keep the string taut and your moving pencil will create the ellipse. This description uses a and c—the other form uses a and b (remember $b^2 + c^2 = a^2$). Problem 24 asks you to simplify equation (5) until you reach $x^2/a^2 + y^2/b^2 = 1$.

The "whispering gallery" of the United States Senate is an ellipse. If you stand at one focus and speak quietly, you can be heard at the other focus (and nowhere else). Your voice is reflected off the walls to the other focus—following the path of the string. For a parabola the rays come in to the focus from infinity—where the second focus is.

A hospital uses this reflection property to split up kidney stones. The patient sits inside an ellipse with the kidney stone at one focus. At the other focus a *lithotripter* sends out hundreds of small shocks. You get a spinal anesthetic (I mean the patient) and the stones break into tiny pieces.

The most important focus is the Sun. The ellipse is the orbit of the Earth. See Section 12.4 for a terrible printing mistake by the Royal Mint, on England's last pound note. They put the Sun at the center.

Question 1   Why do the whispers (and shock waves) arrive together at the second focus?

Answer   Whichever way they go, the distance is $2a$. Exception: straight path is $2c$.

Question 2   Locate the ellipse with equation $4x^2 + 9y^2 = 36$.

Answer   Divide by 36 to change the constant to 1. Now identify a and b:

$$\frac{x^2}{9} + \frac{y^2}{4} = 1 \text{ so } a = \sqrt{9} \text{ and } b = \sqrt{4}. \text{ Foci at } \pm\sqrt{9-4} = \pm\sqrt{5}.$$

Question 3   Shift the center of that ellipse across and down to $x = 1$, $y = -5$.

Answer   Change x to $x - 1$. Change y to $y + 5$. The equation becomes $(x-1)^2/9 + (y+5)^2/4 = 1$. In practice we start with this uncentered ellipse and go the other way to center it.

$$\textbf{HYPERBOLAS} \quad \frac{y^2}{a^2} - \frac{x^2}{b^2} = 1$$

Notice the minus sign for a hyperbola. That makes all the difference. Unlike an ellipse, x and y can both be large. The curve goes out to infinity. It is still symmetric, since x can change to $-x$ and y to $-y$.

The center is at $(0,0)$. Solving for y again yields two functions ($+$ and $-$):

$$\frac{y^2}{a^2} - \frac{x^2}{b^2} = 1 \quad \text{gives} \quad \frac{y}{a} = \pm\sqrt{1 + \frac{x^2}{b^2}} \quad \text{or} \quad y = \pm\frac{a}{b}\sqrt{b^2 + x^2}. \tag{6}$$

The hyperbola has two branches that never meet. The upper branch, with a plus sign, has $y \geq a$. The *vertex V_1* is at $x = 0$, $y = a$—the lowest point on the branch. Much further out, when x is large, the hyperbola climbs up beside its *sloping asymptotes*:

$$\text{if } \frac{x^2}{b^2} = 1000 \text{ then } \frac{y^2}{a^2} = 1001. \text{ So } \frac{y}{a} \text{ is close to } \frac{x}{b} \text{ or } -\frac{x}{b}.$$

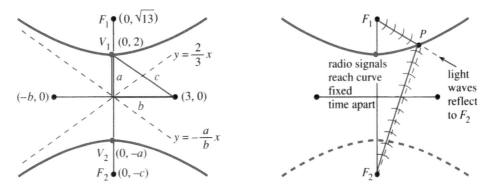

Fig. 3.18   The hyperbola $\frac{1}{4}y^2 - \frac{1}{9}x^2 = 1$ has $a = 2, b = 3, c = \sqrt{4+9}$. The distances to F_1 and F_2 differ by $2a = 4$.

The asymptotes are the lines $y/a = x/b$ and $y/a = -x/b$. Their slopes are a/b and $-a/b$. You can't miss them in Figure 3.18.

For a hyperbola, the foci are inside the two branches. Their distance from the center is still called c. But now $c = \sqrt{a^2 + b^2}$, which is larger than a and b. The vertex is a distance $c - a$ from one focus and $c + a$ from the other. The *difference* (not the sum) is $(c + a) - (c - a) = 2a$.

All points on the hyperbola have this property: ***The difference between distances to the foci is constantly*** $2a$. A ray coming in to one focus is reflected toward the other. The reflection is on the *outside* of the hyperbola, and the *inside* of the ellipse.

Here is an application to navigation. Radio signals leave two fixed transmitters at the same time. A ship receives the signals a millisecond apart. Where is the ship? *Answer*: It is on a hyperbola with foci at the transmitters. Radio signals travel 186 miles in a millisecond, so $186 = 2a$. This determines the curve. In Long Range Navigation (LORAN) a third transmitter gives another hyperbola. Then the ship is located exactly.

Question 4   How do hyperbolas differ from parabolas, far from the center?
Answer   Hyperbolas have asymptotes. Parabolas don't.

The hyperbola has a natural rescaling. The appearance of x/b is a signal to change to X. Similarly y/a becomes Y. Then $Y = 1$ at the vertex, and we have a standard hyperbola:

$$y^2/a^2 - x^2/b^2 = 1 \quad \text{becomes} \quad Y^2 - X^2 = 1.$$

A $90°$ turn gives $X^2 - Y^2 = 1$—the hyperbola opens to the sides. A $45°$ turn produces $2XY = 1$. We show below how to recognize $x^2 + xy + y^2 = 1$ as an ellipse and $x^2 + 3xy + y^2 = 1$ as a hyperbola. (They are not circles because of the xy term.) When the xy coefficient increases past 2, $x^2 + y^2$ no longer indicates an ellipse.

Question 5   Locate the hyperbola with equation $9y^2 - 4x^2 = 36$.
Answer   Divide by 36. Then $y^2/4 - x^2/9 = 1$. Recognize $a = \sqrt{4}$ and $b = \sqrt{9}$.

Question 6   Locate the uncentered hyperbola $9y^2 - 18y - 4x^2 - 4x = 28$.
Answer   Complete $9y^2 - 18y$ to $9(y-1)^2$ by adding 9. Complete $4x^2 + 4x$ to $4(x+\frac{1}{2})^2$ by adding $4\left(\frac{1}{2}\right)^2 = 1$. The equation is rewritten as $9(y-1)^2 - 4(x+\frac{1}{2})^2 = 28 + 9 - 1$. This is the hyperbola in Question 5—except its center is $(-\frac{1}{2}, 1)$.

To summarize: Find the center by completing squares. Then read off a and b.

THE GENERAL EQUATION $Ax^2 + Bxy + Cy^2 + Dx + Ey + F = 0$

This equation is of second degree, containing any and all of $1, x, y, x^2, xy, y^2$. A plane is cutting through a cone. *Is the curve a parabola or ellipse or hyperbola*? Start with the most important case $Ax^2 + Bxy + Cy^2 = 1$.

3I The equation $Ax^2 + Bxy + cy^2 = 1$ produces a hyperbola if $B^2 > 4AC$ and an ellipse if $B^2 < 4AC$. A parabola has $B^2 = 4AC$.

To recognize the curve, we remove Bxy by *rotating the plane*. This also changes A and C—but the combination $B^2 - 4AC$ is not changed (proof omitted). An example is $2xy = 1$, with $B^2 = 4$. It rotates to $y^2 - x^2 = 1$, with $-4AC = 4$. That positive number 4 signals a hyperbola—since $A = -1$ and $C = 1$ have opposite signs.

Another example is $x^2 + y^2 = 1$. It is a circle (a special ellipse). However we rotate, the equation stays the same. The combination $B^2 - 4AC = 0 - 4 \cdot 1 \cdot 1$ is negative, as predicted for ellipses.

To rotate by an angle α, change x and y to new variables x' and y':

$$\begin{array}{ll} x = x' \cos \alpha - y' \sin \alpha & \\ y = x' \sin \alpha + y' \cos \alpha & \text{and} \end{array} \quad \begin{array}{l} x' = x \cos \alpha + y \sin \alpha \\ y' = -y \sin \alpha + x \cos \alpha. \end{array} \tag{7}$$

Substituting for x and y changes $Ax^2 + Bxy + Cy^2 = 1$ to $A'x'^2 + B'x'y' + C'y'^2 = 1$. The formulas for A', B', C' are painful so I go to the key point:

B' *is zero if the rotation angle α has* $\tan 2\alpha = B/(A - C)$.

With $B' = 0$, the curve is easily recognized from $A'x'^2 + C'y'^2 = 1$. It is a hyperbola if A' and C' have opposite signs. Then $B'^2 - 4A'C'$ is positive. The original $B^2 - 4AC$ was also positive, because this special combination stays constant during rotation.

After the xy term is gone, we deal with x and y—by *centering*. To find the center, complete squares as in Questions 3 and 6. For total perfection, rescale to one of the model equations $\mathbf{y = x^2}$ or $\mathbf{x^2 + y^2 = 1}$ or $\mathbf{y^2 - x^2 = 1}$.

The remaining question is about $F = 0$. What is the graph of $Ax^2 + Bxy + Cy^2 = 0$? The ellipse-hyperbola-parabola have disappeared. But if the Greeks were right, the cone is still cut by a plane. The degenerate case $F = 0$ occurs when the plane cuts *right through the sharp point of the cone*.

A level cut hits only that one point $(0, 0)$. The equation shrinks to $x^2 + y^2 = 0$, a circle with radius zero. A steep cut gives two lines. The hyperbola becomes $y^2 - x^2 = 0$, leaving only its asymptotes $y = \pm x$. A cut at the exact angle of the cone gives only one line, as in $x^2 = 0$. A *single point, two lines*, and *one line* are very extreme cases of an ellipse, hyperbola, and parabola.

All these "conic sections" come from planes and cones. The beauty of the geometry, which Archimedes saw, is matched by the importance of the equations. Galileo discovered that projectiles go along parabolas (Chapter 12). Kepler discovered that the Earth travels on an ellipse (also Chapter 12). Finally Einstein discovered that light travels on hyperbolas. That is in four dimensions, and not in Chapter 12.

equation	*vertices*	*foci*
P $\quad y = ax^2 + bx + c$	$\left(-\dfrac{b}{2a}, c - \dfrac{b^2}{4a}\right)$	$\dfrac{1}{4a}$ above vertex, also infinity
E $\quad \dfrac{x^2}{a^2} + \dfrac{y^2}{b^2} = 1, a > b$	$(a,0)$ and $(-a,0)$	$(c,0)$ and $(-c,0): c = \sqrt{a^2 - b^2}$
H $\quad \dfrac{y^2}{a^2} - \dfrac{x^2}{b^2} = 1$	$(0,a)$ and $(0,-a)$	$(0,c)$ and $(0,-c): c = \sqrt{a^2 + b^2}$

3.5 EXERCISES

Read-through questions

The graph of $y = x^2 + 2x + 5$ is a __a__. Its lowest point (the vertex) is $(x, y) = ($__b__$)$. Centering by $X = x + 1$ and $Y =$ __c__ moves the vertex to $(0,0)$. The equation becomes $Y =$ __d__. The focus of this centered parabola is __e__. All rays coming straight down are __f__ to the focus.

The graph of $x^2 + 4y^2 = 16$ is an __g__. Dividing by __h__ leaves $x^2/a^2 + y^2/b^2 = 1$ with $a =$ __i__ and $b =$ __j__. The graph lies in the rectangle whose sides are __k__. The area is $\pi ab =$ __l__. The foci are at $x = \pm c =$ __m__. The sum of distances from the foci to a point on this ellipse is always __n__. If we rescale to $X = x/4$ and $Y = y/2$ the equation becomes __o__ and the graph becomes a __p__.

The graph of $y^2 - x^2 = 9$ is a __q__. Dividing by 9 leaves $y^2/a^2 - x^2/b^2 = 1$ with $a =$ __r__ and $b =$ __s__. On the upper branch $y \geqslant$ __t__. The asymptotes are the lines __u__. The foci are at $y = \pm c =$ __v__. The __w__ of distances from the foci to a point on this hyperbola is __x__.

All these curves are conic sections—the intersection of a __y__ and a __z__. A steep cutting angle yields a __A__. At the borderline angle we get a __B__. The general equation is $Ax^2 +$ __C__ $+ F = 0$. If $D = E = 0$ the center of the graph is at __D__. The equation $Ax^2 + Bxy + Cy^2 = 1$ gives an ellipse when __E__. The graph of $4x^2 + 5xy + 6y^2 = 1$ is a __F__.

1 The vertex of $y = ax^2 + bx + c$ is at $x = -b/2a$. What is special about this x? Show that it gives $y = c - (b^2/4a)$.

2 The parabola $y = 3x^2 - 12x$ has $x_{min} =$ _____. At this minimum, $3x^2$ is _____ as large as $12x$. Introducing $X = x - 2$ and $Y = y + 12$ centers the equation to _____.

Draw the curves 3–14 by hand or calculator or computer. Locate the vertices and foci.

3 $y = x^2 - 2x - 3$

4 $y = (x-1)^2$

5 $4y = -x^2$

6 $4x = y^2$

7 $(x-1)^2 + (y-1)^2 = 1$

8 $x^2 + 9y^2 = 9$

9 $9x^2 + y^2 = 9$

10 $x^2/4 - (y-1)^2 = 1$

11 $y^2 - 4x^2 = 1$

12 $(y-1)^2 - 4x^2 = 1$

13 $y^2 - x^2 = 0$

14 $xy = 0$

Problems 15–20 are about parabolas, 21–34 are about ellipses, 35–41 are about hyperbolas.

15 Find the parabola $y = ax^2 + bx + c$ that goes through $(0,0)$ and $(1,1)$ and $(2,12)$.

16 $y = x^2 - x$ has vertex at _____. To move the vertex to $(0,0)$ set $X =$ _____ and $Y =$ _____. Then $Y = X^2$.

17 (a) In equation (2) change $\frac{1}{4}$ to p. Square and simplify.

(b) Locate the focus and directrix of $Y = 3X^2$. Which points are a distance 1 from the directrix and focus?

18 The parabola $y = 9 - x^2$ opens _____ with vertex at _____. Centering by $Y = y - 9$ yields $Y = -x^2$.

19 Find equations for all parabolas which

(a) open to the right with vertex at $(0,0)$

(b) open upwards with focus at $(0,0)$

(c) open downwards and go through $(0,0)$ and $(1,0)$.

20 A projectile is at $x = t$, $y = t - t^2$ at time t. Find dx/dt and dy/dt at the start, the maximum height, and an xy equation for the path.

21 Find the equation of the ellipse with extreme points at $(\pm 2, 0)$ and $(0, \pm 1)$. Then shift the center to $(1,1)$ and find the new equation.

22 On the ellipse, $x^2/a^2 + y^2/b^2 = 1$, solve for y when $x = c = \sqrt{a^2 - b^2}$. This height above the focus will be valuable in proving Kepler's third law.

23 Find equations for the ellipses with these properties:

(a) through $(5,0)$ with foci at $(\pm 4, 0)$

(b) with sum of distances to $(1,1)$ and $(5,1)$ equal to 12

(c) with both foci at $(0,0)$ and sum of distances $= 2a = 10$.

24 Move a square root to the right side of equation (5) and square both sides. Then isolate the remaining square root and square again. Simplify to reach the equation of an ellipse.

25 Decide between circle-ellipse-parabola-hyperbola, based on the XY equation with $X = x - 1$ and $Y = y + 3$.

(a) $x^2 - 2x + y^2 + 6y = 6$
(b) $x^2 - 2x - y^2 - 6y = 6$
(c) $x^2 - 2x + 2y^2 + 12y = 6$
(d) $x^2 - 2x - y = 6$.

26 A tilted cylinder has equation $(x - 2y - 2z)^2 + (y - 2x - 2z)^2 = 1$. Show that the water surface at $z = 0$ is an ellipse. What is its equation and what is $B^2 - 4AC$?

27 $(4, 9/5)$ is above the focus on the ellipse $x^2/25 + y^2/9 = 1$. Find dy/dx at that point and the equation of the tangent line.

28 (a) Check that the line $xx_0 + yy_0 = r^2$ is tangent to the circle $x^2 + y^2 = r^2$ at (x_0, y_0).

(b) For the ellipse $x^2/a^2 + y^2/b^2 = 1$ show that the tangent equation is $xx_0/a^2 + yy_0/b^2 = 1$. (Check the slope.)

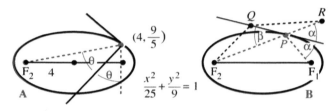

29 The slope of the normal line in Figure **A** is $s = -1/(\text{slope of tangent}) = $ _____ . The slope of the line from F_2 is $S = $ _____ . By the reflection property,

$$S = \cot 2\theta = \frac{1}{2}(\cot\theta - \tan\theta) = \frac{1}{2}\left(s - \frac{1}{s}\right).$$

Test your numbers s and S against this equation.

30 Figure **B** proves the reflecting property of an ellipse. R is the mirror image of F_1 in the tangent line; Q is any other point on the line. Deduce steps 2,3,4 from 1,2,3:

1. $PF_1 + PF_2 < QF_1 + QF_2$ (*left side* $= 2a$, Q *is outside*)
2. $PR + PF_2 < QR + QF_2$
3. P is on the straight line from F_2 to R
4. $\alpha = \beta$: the reflecting property is proved.

31 The ellipse $(x-3)^2/4 + (y-1)^2/4 = 1$ is really a _____ with center at _____ and radius _____ . Choose X and Y to produce $X^2 + Y^2 = 1$.

32 Compute the area of a square that just fits inside the ellipse $x^2/a^2 + y^2/b^2 = 1$.

33 Rotate the axes of $x^2 + xy + y^2 = 1$ by using equation (7) with $\sin\alpha = \cos\alpha = 1/\sqrt{2}$. The $x'y'$ equation should show an ellipse.

34 What are a, b, c for the Earth's orbit around the sun ?

35 Find an equation for the hyperbola with

(a) vertices $(0, \pm 1)$, foci $(0, \pm 2)$
(b) vertices $(0, \pm 3)$, asymptotes $y = \pm 2x$
(c) $(2, 3)$ on the curve, asymptotes $y = \pm x$

36 Find the slope of $y^2 - x^2 = 1$ at (x_0, y_0). Show that $yy_0 - xx_0 = 1$ goes through this point with the right slope (it has to be the tangent line).

37 If the distances from (x, y) to $(8, 0)$ and $(-8, 0)$ differ by 10, what hyperbola contains (x, y) ?

38 If a cannon was heard by Napoleon and one second later by the Duke of Wellington, the cannon was somewhere on a _____ with foci at _____ .

39 $y^2 - 4y$ is part of $(y - 2)^2 = $ _____ and $2x^2 + 12x$ is part of $2(x + 3)^2 = $ _____ . Therefore $y^2 - 4y - 2x^2 - 12x = 0$ gives the hyperbola $(y - 2)^2 - 2(x + 3)^2 = $ _____ . Its center is _____ and it opens to the _____ .

40 Following Problem 39 turn $y^2 + 2y = x^2 + 10x$ into $Y^2 = X^2 + C$ with X, Y, and C equal to _____ .

41 Draw the hyperbola $x^2 - 4y^2 = 1$ and find its foci and asymptotes.

Problems 42–46 are about second-degree curves (conics).

42 For which A, C, F does $Ax^2 + Cy^2 + F = 0$ have no solution (empty graph) ?

43 Show that $x^2 + 2xy + y^2 + 2x + 2y + 1 = 0$ is the equation (squared) of a single line.

44 Given any _____ points in the plane, a second-degree curve $Ax^2 + \cdots + F = 0$ goes through those points.

45 (a) When the plane $z = ax + by + c$ meets the cone $z^2 = x^2 + y^2$, eliminate z by squaring the plane equation. Rewrite in the form $Ax^2 + Bxy + Cy^2 + Dx + Ey + F = 0$.

(b) Compute $B^2 - 4AC$ in terms of a and b.

(c) Show that the plane meets the cone in an ellipse if $a^2 + b^2 < 1$ and a hyperbola if $a^2 + b^2 > 1$ (*steeper*).

46 The roots of $ax^2 + bx + c = 0$ also involve the special combination $b^2 - 4ac$. This quadratic equation has two real roots if _____ and no real roots if _____ . The roots come together when $b^2 = 4ac$, which is the borderline case like a parabola.

3.6 Iterations $x_{n+1} = F(x_n)$

Iteration means repeating the same function. Suppose the function is $F(x) = \cos x$. Choose any starting value, say $x_0 = 1$. Take its cosine: $x_1 = \cos x_0 = .54$. **Then take the cosine of x_1.** That produces $x_2 = \cos .54 = .86$. *The iteration is* $x_{n+1} = \cos x_n$. I am in radian mode on a calculator, pressing "cos" each time. The early numbers are not important, what is important is the output after 12 or 30 or 100 steps:

EXAMPLE 1 $x_{12} = .75$, $x_{13} = .73$, $x_{14} = .74$,   ..., $x_{29} = .7391$, $x_{30} = .7391$.

The goal is to explain why the x's approach $x^* = .739085\ldots$. Every starting value x_0 leads to this same number x^*. **What is special about** .7391 ?

Note on iterations   Do $x_1 = \cos x_0$, and $x_2 = \cos x_1$, mean that $x_2 = \cos^2 x_0$? Absolutely not! Iteration creates a new and different function $\cos(\cos x)$. It uses the cos button, not the squaring button. The third step creates $F(F(F(x)))$. As soon as you can, iterate with $x_{n+1} = \frac{1}{2}\cos x_n$. What limit do the x's approach ? Is it $\frac{1}{2}(.7931)$?

Let me slow down to understand these questions. **The central idea is expressed by the equation $x_{n+1} = F(x_n)$.** Substituting x_0 into F gives x_1. This output x_1 is the input that leads to x_2. In its turn, x_2 is the input and out comes $x_3 = F(x_2)$. This is *iteration*, and it produces the sequence $x_0, x_1, x_2, \ldots$.

The x's may approach a limit x^*, depending on the function F. Sometimes x^* also depends on the starting value x_0. Sometimes there is *no* limit. Look at a second example, which does not need a calculator.

EXAMPLE 2 $x_{n+1} = F(x_n) = \frac{1}{2}x_n + 4$. Starting from $x_0 = 0$ the sequence is

$$x_1 = \tfrac{1}{2}\cdot 0 + 4 = 4, \ \ x_2 = \tfrac{1}{2}\cdot 4 + 4 = 6, \ \ x_3 = \tfrac{1}{2}\cdot 6 + 4 = 7, \ \ x_4 = \tfrac{1}{2}\cdot 7 + 4 = 7\tfrac{1}{2}, \ \ \ldots.$$

Those numbers $0, 4, 6, 7, 7\frac{1}{2}, \ldots$ seem to be approaching $x^* = 8$. A computer would convince us. So will mathematics, when we see what is special about 8:

> When the x's approach x^*, the limit of $x_{n+1} = \frac{1}{2}x_n + 4$
>
> is $x^* = \frac{1}{2}x^* + 4$. This limiting equation yields $x^* = 8$.

8 is the "steady state" where *input equals output*: $8 = F(8)$. It is the **fixed point**.

If we start at $x_0 = 8$, the sequence is $8, 8, 8, \ldots$. When we start at $x_0 = 12$, the sequence goes back toward 8:

$$x_1 = \tfrac{1}{2}\cdot 12 + 4 = 10, \quad x_2 = \tfrac{1}{2}\cdot 10 + 4 = 9, \quad x_3 = \tfrac{1}{2}\cdot 9 + 4 = 8.5, \quad \ldots.$$

Equation for limit: **If the iterations $x_{n+1} = F(x_n)$ converge to x^*, then $x^* = F(x^*)$.**

To repeat: 8 is special because it equals $\frac{1}{2}\cdot 8 + 4$. The number $.7391\ldots$ is special because it equals $\cos .7391 \ldots$. **The graphs of $y = x$ and $y = F(x)$ intersect at** x^*. To explain *why* the x's converge (or why they don't) is the job of calculus.

EXAMPLE 3 $x_{n+1} = x_n^2$ has two fixed points: $0 = 0^2$ and $1 = 1^2$. Here $F(x) = x^2$.

Starting from $x_0 = \frac{1}{2}$ the sequence $\frac{1}{4}, \frac{1}{16}, \frac{1}{256}, \ldots$ goes quickly to $x^* = 0$. The only approaches to $x^* = 1$ are from $x_0 = 1$ (of course) and from $x_0 = -1$. Starting from $x_0 = 2$ we get $4, 16, 256, \ldots$ and *the sequence diverges to* $+\infty$.

Each limit x^* has a "***basin of attraction***." The basin contains all starting points x_0 that lead to x^*. For Examples 1 and 2, every x_0 led to .7391 and 8. The basins were the whole line (that is still to be proved). Example 3 had three basins—the interval $-1 < x_0 < 1$, the two points $x_0 = \pm 1$, and all the rest. The outer basin $|x_0| > 1$ led to $\pm \infty$. I challenge you to find the limits and the basins of attraction (by calculator) for $F(x) = x - \tan x$.

In Example 3, $x^* = 0$ is ***attracting***. Points near x^* move toward x^*. The fixed point $x^* = 1$ is ***repelling***. Points near 1 move away. We now find the rule that decides whether x^* is attracting or repelling. ***The key is the slope dF/dx at x^*.***

3J Start from any x_0 near a fixed point $x^* = F(x^*)$:

$$x^* \text{ is } \textbf{\textit{attracting}} \text{ if } |dF/dx| \text{ is below 1 at } x^*$$
$$x^* \text{ is } \textbf{\textit{repelling}} \text{ if } |dF/dx| \text{ is above 1 at } x^*.$$

First I will give a calculus proof. Then comes a picture of convergence, by "*cobwebs*." Both methods throw light on this crucial test for attraction: $|dF/dx| < 1$.

First proof: Subtract $x^* = F(x^*)$ from $x_{n+1} = F(x_n)$. The difference $x_{n+1} - x^*$ is the same as $F(x_n) - F(x^*)$. This is ΔF. ***The basic idea of calculus is that ΔF is close to $F' \Delta x$:***

$$x_{n+1} - x^* = F(x_n) - F(x^*) \approx F'(x^*)(x_n - x^*). \tag{1}$$

The "error" $x_n - x^*$ is multiplied by the slope dF/dx. The next error $x_{n+1} - x^*$ is smaller or larger, based on $|F'| < 1$ or $|F'| > 1$ at x^*. Every step multiplies approximately by $F'(x^*)$. ***Its size controls the speed of convergence.***

In Example 1, $F(x)$ is $\cos x$ and $F'(x)$ is $-\sin x$. There is attraction to .7391 because $|\sin x^*| < 1$. In Example 2, F is $\frac{1}{2}x + 4$ and F' is $\frac{1}{2}$. There is attraction to 8. In Example 3, F is x^2 and F' is $2x$. There is superattraction to $x^* = 0$ (where $F' = 0$). There is repulsion from $x^* = 1$ (where $F' = 2$).

I admit one major difficulty. The approximation in equation (1) only holds *near* x^*. If x_0 is far away, does the sequence still approach x^*? When there are several attracting points, which x^* do we reach? This section starts with good iterations, which solve the equation $x^* = F(x^*)$ or $f(x) = 0$. At the end we discover ***Newton's method***. The next section produces crazy but wonderful iterations, not converging and not blowing up. They lead to "*fractals*" and "*Cantor sets*" and "*chaos*."

The mathematics of iterations is not finished. It may never be finished, but we are converging on the answers. Please choose a function and join in.

THE GRAPH OF AN ITERATION: COBWEBS

The iteration $x_{n+1} = F(x_n)$ involves two graphs at the same time. One is the graph of $y = F(x)$. The other is the graph of $y = x$ (the 45° line). The iteration jumps back and forth between these graphs. It is a very convenient way to see the whole process.

Example 1 was $x_{n+1} = \cos x_n$. Figure 3.19 shows the graph of $\cos x$ and the "*cobweb*." Starting at (x_0, x_0) on the 45° line, the rule is based on $x_1 = F(x_0)$:

From (x_0, x_0) go up or down to (x_0, x_1) ***on the curve***.

From (x_0, x_1) go across to (x_1, x_1) ***on the*** 45° ***line***.

These steps are repeated forever. From x_1 go up to the curve at $F(x_1)$. That height is x_2. Now cross to the 45° line at (x_2, x_2). The iterations are aiming for $(x^*, x^*) = (.7391, .7391)$. This is the *crossing point* of the two graphs $y = F(x)$ and $y = x$.

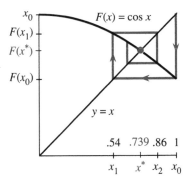
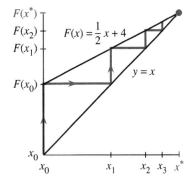

Fig. 3.19 Cobwebs go from (x_0, x_0) to (x_0, x_1) to (x_1, x_1)—line to curve to line.

Example 2 was $x_{n+1} = \frac{1}{2}x_n + 4$. Both graphs are straight lines. The cobweb is one-sided, from $(0,0)$ to $(0,4)$ to $(4,4)$ to $(4,6)$ to $(6,6)$. Notice how y changes (vertical line) and then x changes (horizontal line). The slope of $F(x)$ is $\frac{1}{2}$, so the distance to 8 is multiplied by $\frac{1}{2}$ at every step.

Example 3 was $x_{n+1} = x_n^2$. The graph of $y = x^2$ crosses the 45° line at two fixed points: $0^2 = 0$ and $1^2 = 1$. Figure 3.20a starts the iteration close to 1, but it quickly goes away. This fixed point is repelling because $F'(1) = 2$. Distance from $x^* = 1$ is doubled (at the start). One path moves down to $x^* = 0$—which is *superattractive* because $F' = 0$. The path from $x_0 > 1$ diverges to infinity.

EXAMPLE 4 $F(x)$ has two attracting points x^* (a repelling x^* is always between).

Figure 3.20b shows two crossings with slope zero. The iterations and cobwebs converge quickly. In between, the graph of $F(x)$ must cross the 45° line from below. That requires a slope greater than one. Cobwebs diverge from this unstable point, which separates the basins of attraction. The fixed point $x = \pi$ is in a basin by itself!

Note 1   To draw cobwebs on a calculator, graph $y = F(x)$ on top of $y = x$. On a Casio, one way is to plot (x_0, x_0) and give the command **LINE: PLOT X, Y** followed by **EXE**. Now move the cursor vertically to $y = F(x)$ and press **EXE**. Then move horizontally to $y = x$ and press **EXE**. Continue. Each step draws a line.

For the TI-81 (and also the Casio) a short program produces a cobweb. Store $F(x)$ in the $Y =$ function slot **Y₁**. Set the range (square window or autoscaling). Run the program and answer the prompt with x_0:

```
PrgmC:COBWEB  :Disp ''INITIAL XØ''  :Input X  :All-off

:Y₁-On  :''X''→Y₄  :Lbl 1  :X→S  :Y₁→T  :Line (S,S,S,T)

:Line(S,T,T,T)  :T→X  :Pause  :Goto 1
```

Note 2   The x's approach x^* from one side when $0 < dF/dx < 1$.

Note 3   A basin of attraction can include faraway x_0's (basins can come in infinitely many pieces). This makes the problem interesting. If no fixed points are attracting, see Section 3.7 for "cycles" and "chaos."

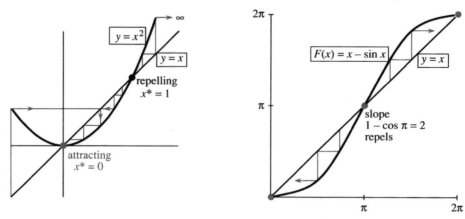

Fig. 3.20   Converging and diverging cobwebs: $F(x) = x^2$ and $F(x) = x - \sin x$.

THE ITERATION $x_{n+1} = X_n - cf(x_n)$

At this point we offer the reader a choice. One possibility is to jump ahead to the next section on "Newton's Method." That method is an iteration to solve $f(x) = 0$. The function $F(x)$ combines x_n and $f(x_n)$ and $f'(x_n)$ into an optimal formula for x_{n+1}. We will see how quickly Newton's method works (when it works). It is *the* outstanding algorithm to solve equations, and it is totally built on tangent approximations.

The other possibility is to understand (through calculus) a whole family of iterations. This family depends on a number c, which is at our disposal. ***The best choice of c produces Newton's method***. I emphasize that iteration is by no means a new and peculiar idea. *It is a fundamental technique in scientific computing.*

We start by recognizing that there are many ways to reach $f(x^*) = 0$. (I write x^* for the solution.) A good algorithm may switch to Newton as it gets close. The iterations use $f(x_n)$ to decide on the next point x_{n+1}:

$$x_{n+1} = F(x_n) = x_n - cf(x_n). \qquad (2)$$

Notice how $F(x)$ is constructed from $f(x)$—they are different! We move f to the right side and multiply by a "preconditioner" c. *The choice of c (or c_n, if it changes from step to step) is absolutely critical.* The starting guess x_0 is also important—but its accuracy is not always under our control.

Suppose the x_n converge to x^*. Then the limit of equation (2) is

$$x^* = x^* - cf(x^*). \qquad (3)$$

That gives $f(x^*) = 0$. If the x_n's have a limit, it solves the right equation. It is a fixed point of F (we can assume $c_n \to c \neq 0$ and $f(x_n) \to f(x^*)$). There are two key questions, and both of them are answered by the slope $F'(x^*)$:

1. How quickly does x_n approach x^* (or do the x_n diverge)?

2. What is a good choice of c (or c_n)?

EXAMPLE 5 $f(x) = ax - b$ is zero at $x^* = b/a$. The iteration $x_{n+1} = x_n - c(ax_n - b)$ intends to find b/a without actually dividing. (Early computers could not divide; they used iteration.) Subtracting x^* from both sides leaves an equation for the error:

$$x_{n+1} - x^* = x_n - x^* - c(ax_n - b).$$

Replace b by ax^*. The right side is $(1 - ca)(x_n - x^*)$. This "error equation" is

$$(\text{error})_{n+1} = (1 - ca)(\text{error})_n. \tag{4}$$

At every step the error is multiplied by $(1 - ca)$, which is F'. *The error goes to zero if* $|F'|$ *is less than* 1. The absolute value $|1 - ca|$ decides everything:

$$x_n \text{ converges to } x^* \text{ if and only if } -1 < 1 - ca < 1. \tag{5}$$

The perfect choice (if we knew it) is $c = 1/a$, which turns the multiplier $1 - ca$ into zero. Then one iteration gives the exact answer: $x_1 = x_0 - (1/a)(ax_0 - b) = b/a$. That is the horizontal line in Figure 3.21a, converging in one step. But look at the other lines.

This example did not need calculus. Linear equations never do. The key idea is that *close to x^* the nonlinear equation $f(x) = 0$ is nearly linear.* We apply the tangent approximation. You are seeing how calculus is used, in a problem that doesn't start by asking for a derivative.

THE BEST CHOICE OF c

The immediate goal is to study the errors $x_n - x^*$. They go quickly to zero, if the multiplier is small. To understand $x_{n+1} = x_n - cf(x_n)$, subtract the equation $x^* = x^* - cf(x^*)$:

$$x_{n+1} - x^* = x_n - x^* - c(f(x_n) - f(x^*)). \tag{6}$$

Now calculus enters. ***When you see a difference of f's think of df/dx.*** Replace $f(x_n) - f(x^*)$ by $A(x_n - x^*)$, where A stands for the slope df/dx at x^*:

$$x_{n+1} - x^* \approx (1 - cA)(x_n - x^*). \tag{7}$$

This is the ***error equation***. The new error at step $n + 1$ is approximately the old error multiplied by $m = 1 - cA$. This corresponds to $m = 1 - ca$ in the linear example. We keep returning to the basic test $|m| = |F'(x^*)| < 1$:

3K Starting near x^*, the errors $x_n - x^*$ go to zero if multiplier has $|m| < 1$. The perfect choice is $c = 1/A = 1/f'(x^*)$. Then $m = 1 - cA = 0$.

There is only one difficulty: *We don't know x^*.* Therefore we don't know the perfect c. It depends on the slope $A = f'(x^*)$ at the unknown solution. However we can come close, by using the slope at x_n:

Choose $c_n = 1/f'(x_n)$. Then $x_{n+1} = x_n - f(x_n)/f'(x_n) = F(x_n)$.

This is Newton's method. The multiplier $m = 1 - cA$ is as near to zero as we can make it. By building df/dx into $F(x)$, Newton speeded up the convergence of the iteration.

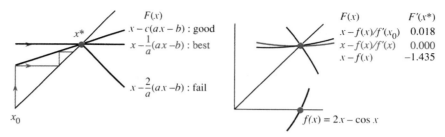

Fig. 3.21   The error multiplier is $m = 1 - cf'(x^*)$. Newton has $c = 1/f'(x_n)$ and $m \to 0$.

EXAMPLE 6   Solve $f(x) = 2x - \cos x = 0$ with different iterations (different c's).

The line $y = 2x$ crosses the cosine curve somewhere near $x = \frac{1}{2}$. The intersection point where $2x^* = \cos x^*$ has no simple formula. We start from $x_0 = \frac{1}{2}$ and iterate $x_{n+1} = x_n - c(2x_n - \cos x_n)$ with *three different choices* of c.

Take $c = 1$ or $c = 1/f'(x_0)$ or update c by Newton's rule $c_n = 1/f'(x_n)$:

$x_0 = .50$	$c = 1$	$c = 1/f'(x_0)$	$c_n = 1/f'(x_n)$
$x_1 =$	.38	.45063	.45062669
$x_2 =$	.55	.45019	.45018365
$x_3 =$	.30	.45018	.45018361...

The column with $c = 1$ is diverging (repelled from x^*). The second column shows convergence (attracted to x^*). The third column (Newton's method) approaches x^* so quickly that $.4501836$ *and seven more digits* are exact for x_3.

How does this convergence match the prediction? Note that $f'(x) = 2 + \sin x$ so $A = 2.435$. Look to see whether the actual errors $x_n - x^*$, going down each column, are multiplied by the predicted m below that column:

	$c = 1$	$c = 1/(2 + \sin \frac{1}{2})$	$c_n = 1/(2 + \sin x_n)$
$x_0 - x^* =$	0.05	$4.98 \cdot 10^{-2}$	$4.98 \cdot 10^{-2}$
$x_1 - x^* =$	-0.07	$4.43 \cdot 10^{-4}$	$4.43 \cdot 10^{-4}$
$x_2 - x^* =$	0.10	$7.88 \cdot 10^{-6}$	$3.63 \cdot 10^{-8}$
$x_3 - x^* =$	-0.15	$1.41 \cdot 10^{-7}$	$2.78 \cdot 10^{-16}$
multiplier	$m = -1.4$	$m = .018$	$m \to 0$ (Newton)

The first column shows a multiplier below -1. The errors grow at every step. Because m is negative the errors change sign—the cobweb goes outward.

The second column shows convergence with $m = .018$. It takes one genuine Newton step, then c is fixed. After n steps the error is closely proportional to $m^n = (.018)^n$— that is "*linear convergence*" with a good multiplier.

The third column shows the "*quadratic convergence*" of Newton's method. Multiplying the error by m is more attractive than ever, because $m \to 0$. In fact m itself is proportional to the error, so *at each step the error is squared*. Problem 3.8.31 will show that $(error)_{n+1} \leq M(error)_n^2$. This squaring carries us from 10^{-2} to 10^{-4} to 10^{-8} to "machine ε" in three steps. The number of correct digits is doubled at every step as Newton converges.

Note 1   The choice $c = 1$ produces $x_{n+1} = x_n - f(x_n)$. This is "successive substitution." The equation $f(x) = 0$ is rewritten as $x = x - f(x)$, and each x_n is substituted back to produce x_{n+1}. Iteration with $c = 1$ does not always fail!

Note 2   Newton's method is successive substitution for f/f', not f. Then $m \approx 0$.

Note 3   Edwards and Penney happened to choose the same example $2x = \cos x$. But they cleverly wrote it as $x_{n+1} = \frac{1}{2}\cos x_n$, which has $|F'| = |\frac{1}{2}\sin x| < 1$. This iteration fits into our family with $c = \frac{1}{2}$, and it succeeds. We asked earlier if its limit is $\frac{1}{2}(.7391)$. *No*, it is $x^* = .450\ldots$.

Note 4   The choice $c = 1/f'(x_0)$ is "**modified Newton**." After one step of Newton's method, c is fixed. The steps are quicker, because they don't require a new $f'(x_n)$. But we need more steps. Millions of dollars are spent on Newton's method, so speed is important. In all its forms, $f(x) = 0$ is the central problem of computing.

3.6 EXERCISES

Read-through questions

$x_{n+1} = x_n^3$ describes, an ___a___. After one step $x_1 =$ ___b___. After two steps $x_2 = F(x_1) =$ ___c___. If it happens that input = output, or $x^* =$ ___d___, then x^* is a ___e___ point. $F = x^3$ has ___f___ fixed points, at $x^* =$ ___g___. Starting near a fixed point, the x_n will converge to it if ___h___ < 1. That is because $x_{n+1} - x^* = F(x_n) - F(x^*) \approx$ ___i___. The point is called ___j___. The x_n are repelled if ___k___. For $F = x^3$ the fixed points have $F' =$ ___l___. The cobweb goes from (x_0, x_0) to $(\,,\,)$ to $(\,,\,)$ and converges to $(x^*, x^*) =$ ___m___. This is an intersection of $y = x^3$ and $y =$ ___n___, and it is superattracting because ___o___.

$f(x) = 0$ can be solved iteratively by $x_{n+1} = x_n - cf(x_n)$, in which case $F'(x^*) =$ ___p___. Subtracting $x^* = x^* - cf(x^*)$, the error equation is $x_{n+1} - x^* \approx m$ (___q___). The multiplier is $m =$ ___r___. The errors approach zero if ___s___. The choice $c_n =$ ___t___ produces Newton's method. The choice $c = 1$ is "successive ___u___" and $c =$ ___v___ is modified Newton. Convergence to x^* is ___w___ certain.

We have three ways to study iterations $x_{n+1} = F(x_n)$: (**1**) compute $x_1, x_2, \ldots$ from different x_0 (**2**) find the fixed points x^* and test $|dF/dx| < 1$ (**3**) draw cobwebs.

In Problems 1–8 start from $x_0 = .6$ and $x_0 = 2$. Compute $x_1, x_2, \ldots$ to test convergence:

1 $x_{n+1} = x_n^2 - \frac{1}{2}$

2 $x_{n+1} = 2x_n(1 - x_n)$

3 $x_{n+1} = \sqrt{x_n}$

4 $x_{n+1} = 1/\sqrt{x_n}$

5 $x_{n+1} = 3x_n(1 - x_n)$

6 $x_{n+1} = x_n^2 + x_n - 2$

7 $x_{n+1} = \frac{1}{2}x_n - 1$

8 $x_{n+1} = |x_n|$

9 Check dF/dx at all fixed points in Problems 1–6. Are they attracting or repelling?

10 From $x_0 = -1$ compute the sequence $x_{n+1} = -x_n^3$. Draw the cobweb with its "cycle." Two steps produce $x_{n+2} = x_n^9$, which has the fixed points _____.

11 Draw the cobwebs for $x_{n+1} = \frac{1}{2}x_n - 1$ and $x_{n+1} = 1 - \frac{1}{2}x_n$ starting from $x_0 = 2$. Rule: Cobwebs are two-sided when dF/dx is _____.

12 Draw the cobweb for $x_{n+1} = x_n^2 - 1$ starting from the periodic point $x_0 = 0$. Another periodic point is _____. Start nearby at $x_0 = .1$ to see if the iterations are attracted to $0, -1, 0, -1, \ldots$.

Solve equations 13–16 within 1% by iteration.

13 $x = \cos \frac{1}{2}x$

14 $x = \cos^2 x$

15 $x = \cos \sqrt{x}$

16 $x = 2x - 1(??)$

17 For which numbers a does $x_{n+1} = a(x_n - x_n^2)$ converge to $x^* = 0$?

18 For which numbers a does $x_{n+1} = a(x_n - x_n^2)$ converge to $x^* = (a-1)/a$?

19 Iterate $x_{n+1} = 4(x_n - x_n^2)$ to see chaos. Why don't the x_n approach $x^* = \frac{3}{4}$?

20 One fixed point of $F(x) = x^2 - \frac{1}{2}$ is attracting, the other is repelling. By experiment or cobwebs, find the basin of x_0's that go to the attractor.

21 (important) Find the fixed point for $F(x) = ax + s$. When is it attracting?

22 What happens in the linear case $x_{n+1} = ax_n + 4$ when $a = 1$ and when $a = -1$?

23 Starting with \$1,000, you spend half your money each year and a rich but foolish aunt gives you a new \$1,000. What is your steady state balance x^*? What is x^* if you start with a million dollars?

24 The US national debt was once \$1 trillion. Inflation reduces its real value by 5% each year (so multiply by $a = .95$), but overspending adds another \$100 billion. What is the steady state debt x^* ?

25 $x_{n+1} = b/x_n$ has the fixed point $x^* = \sqrt{b}$. Show that $|dF/dx| = 1$ at that point—what is the sequence starting from x_0 ?

26 Show that both fixed points of $x_{n+1} = x_n^2 + x_n - 3$ are repelling. What do the iterations do ?

27 A \$5 calculator takes square roots but not cube roots. Explain why $x_{n+1} = \sqrt{2/x_n}$ converges to $\sqrt[3]{2}$.

28 Start the cobwebs for $x_{n+1} = \sin x_n$ and $x_{n+1} = \tan x_n$. In both cases $dF/dx = 1$ at $x^* = 0$. (a) Do the iterations converge ? (b) Propose a theory based on F'' for cases when $F' = 1$.

Solve $f(x) = 0$ in 29–32 by the iteration $x_{n+1} = x_n - cf(x_n)$, to find a c that succeeds and a c that fails.

29 $f(x) = x^2 - 4$

30 $f(x) = x^2 - 4x + 3$

31 $f(x) = (x-2)^9 - 1$

32 $f(x) = (1-x)^{-1} - 3$

33 Newton's method computes a new $c = 1/f'(x_n)$ at each step. Write out the iteration formulas for $f(x) = x^3 - 2 = 0$ and $f(x) = \sin x - \frac{1}{2} = 0$.

34 Apply Problem 33 to find the first six decimals of $\sqrt[3]{2}$ and $\pi/6$.

35 By experiment find each x^* and its basin of attraction, when Newton's method is applied to $f(x) = x^2 - 5x + 4$.

36 Test Newton's method on $x^2 - 1 = 0$, starting far out at $x_0 = 10^6$. At first the error is reduced by about $m = \frac{1}{2}$. Near $x^* = 1$ the multiplier approaches $m = 0$.

37 Find the multiplier m at each fixed point of $x_{n+1} = x_n - c(x_n^2 - x_n)$. Predict the convergence for different c (to which x^* ?).

38 Make a table of iterations for $c = 1$ and $c = 1/f'(x_0)$ and $c = 1/f'(x_n)$, when $f(x) = x^2 - \frac{1}{2}$ and $x_0 = 1$.

39 In the iteration for $x^2 - 2 = 0$, find dF/dx at x^*:

$$x_{n+1} = \frac{1}{2}\left(x_n + \frac{2}{x_n}\right).$$

(b) Newton's iteration has $F(x) = x - f(x)/f'(x)$. Show that $F' = 0$ when $f(x) = 0$. *The multiplier for Newton is* $m = 0$.

40 What are the solutions of $f(x) = x^2 + 2 = 0$ and why is Newton's method sure to fail ? But carry out the iteration to see whether $x_n \to \infty$.

41 *Computer project* $F(x) = x - \tan x$ has fixed points where $\tan x^* = 0$. So x^* is any multiple of π. From $x_0 = 2.0$ and 1.8 and 1.9, which multiple do you reach ? Test points in $1.7 < x_0 < 1.9$ to find basins of attraction to $\pi, 2\pi, 3\pi, 4\pi$.

Between any two basins there are basins for *every* multiple of π. And more basins between these (*a fractal*). Mark them on the line from 0 to π. Magnify the picture around $x_0 = 1.9$ (in color ?).

42 Graph $\cos x$ and $\cos(\cos x)$ and $\cos(\cos(\cos x))$. Also $(\cos)^8 x$. What are these graphs approaching ?

43 Graph $\sin x$ and $\sin(\sin x)$ and $(\sin)^8 x$. What are these graphs approaching ? Why so slow ?

3.7 Newton's Method (and Chaos)

The equation to be solved is $f(x) = 0$. Its solution x^* is the point where the graph crosses the x axis. Figure 3.22 shows x^* and a starting guess x_0. Our goal is to come as close as possible to x^*, *based on the information $f(x_0)$ and $f'(x_0)$.*

Section 3.6 reached Newton's formula for x_1 (the next guess). We now do that directly.

What do we see at x_0? The graph has height $f(x_0)$ and slope $f'(x_0)$. We know where we are, and which direction the curve is going. We don't know if the curve bends (we don't have f''). The best plan is *to follow the tangent line*, which uses all the information we have.

Newton replaces $f(x)$ by its linear approximation (= tangent approximation):

$$f(x) \approx f(x_0) + f'(x_0)(x - x_0). \tag{1}$$

We want the left side to be zero. The best we can do is to make the right side zero! The tangent line crosses the axis at x_1, while the curve crosses at x^*. The new guess x_1 comes from $f(x_0) + f'(x_0)(x_1 - x_0) = 0$. Dividing by $f'(x_0)$ and solving for x_1, this is step 1 of Newton's method:

$$x_1 = x_0 - \frac{f(x_0)}{f'(x_0)}. \tag{2}$$

At this new point, compute $f(x_1)$ and $f'(x_1)$—the height and slope at x_1. They give a new tangent line, which crosses at x_2. *At every step we want $f(x_{n+1}) = 0$ and we settle for $f(x_n) + f'(x_n)(x_{n+1} - x_n) = 0$.* After dividing by $f'(x_n)$, the formula for x_{n+1} is Newton's method.

3L The tangent line from x_n crosses the axis at x_{n+1}:

$$\textbf{\textit{Newton's method}} \qquad x_{n+1} = x_n - \frac{f(x_n)}{f'(x_n)}. \tag{3}$$

Usually this iteration $x_{n+1} = F(x_n)$ converges quickly to x^*.

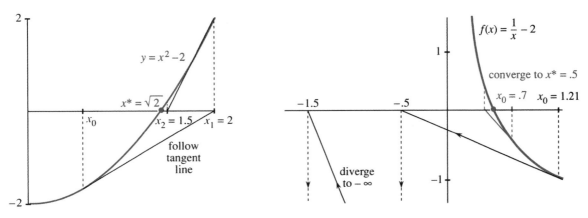

Fig. 3.22 Newton's method along tangent lines from x_0 to x_1 to x_2.

Linear approximation involves three numbers. They are Δx (across) and Δf (up) and the slope $f'(x)$. If we know two of those numbers, we can estimate the third. It is remarkable to realize that calculus has now used all three calculations—they are the key to this subject:

1. *Estimate the slope $f'(x)$ from $\Delta f / \Delta x$*   (Section 2.1)
2. *Estimate the change Δf from $f'(x)\Delta x$*   (Section 3.1)
3. *Estimate the change Δx from $\Delta f / f'(x)$*   (Newton's method)

The desired Δf is $-f(x_n)$. Formula (3) is exactly $\Delta x = -f(x_n)/f'(x_n)$.

EXAMPLE 1   (Square roots) $f(x) = x^2 - b$ is zero at $x^* = \sqrt{b}$ and also at $-\sqrt{b}$. Newton's method is a quick way to find square roots—probably built into your calculator. The slope is $f'(x_n) = 2x_n$, and formula (3) for the new guess becomes

$$x_{n+1} = x_n - \frac{x_n^2 - b}{2x_n} = x_n - \frac{1}{2}x_n + \frac{b}{2x_n}. \qquad (4)$$

This simplifies to $x_{n+1} = \frac{1}{2}(x_n + b/x_n)$. **Guess the square root, divide into b, and average the two numbers**. The ancient Babylonians had this same idea, without knowing functions or slopes. They iterated $x_{n+1} = F(x_n)$:

$$F(x) = \frac{1}{2}\left(x + \frac{b}{x}\right) \quad \text{and} \quad F'(x) = \frac{1}{2}\left(1 - \frac{b}{x^2}\right). \qquad (5)$$

The Babylonians did exactly the right thing. The slope F' is zero *at the solution*, when $x^2 = b$. That makes Newton's method converge at high speed. The convergence test is $|F'(x^*)| < 1$. Newton achieves $F'(x^*) = 0$—which is *superconvergence*.

   To find $\sqrt{4}$, start the iteration $x_{n+1} = \frac{1}{2}(x_n + 4/x_n)$ at $x_0 = 1$. Then $x_1 = \frac{1}{2}(1 + 4)$:

$$x_1 = 2.5 \quad x_2 = 2.05 \quad x_3 = 2.0006 \quad x_4 = 2.000000009.$$

The wrong decimal is twice as far out at each step. **The error is squared**. Subtracting $x^* = 2$ from both sides of $x_{n+1} = F(x_n)$ gives an *error equation* which displays that square:

$$x_{n+1} - 2 = \frac{1}{2}\left(x_n + \frac{4}{x_n}\right) - 2 = \frac{1}{2x_n}(x_n - 2)^2. \qquad (6)$$

This is $(error)_{n+1} \approx \frac{1}{4}(error)_n^2$. It explains the speed of Newton's method.

Remark 1   You can't start this iteration at $x_0 = 0$. The first step computes $4/0$ and blows up. Figure 3.22a shows why—the tangent line at zero is horizontal. It will never cross the axis.

Remark 2   Starting at $x_0 = -1$, Newton converges to $-\sqrt{2}$ instead of $+\sqrt{2}$. That is the other x^*. Often it is difficult to predict which x^* Newton's method will choose. Around every solution is a "basin of attraction," but other parts of the basin may be far away. Numerical experiments are needed, with many starts x_0. Finding basins of attraction was one of the problems that led to fractals.

EXAMPLE 2 *Solve* $\dfrac{1}{x} - a = 0$ *to find* $x^* = \dfrac{1}{a}$ *without dividing by a.*

Here $f(x) = (1/x) - a$. Newton uses $f'(x) = -1/x^2$. Surprisingly, we don't divide:

$$x_{n+1} = x_n - \frac{(1/x_n) - a}{-1/x_n^2} = x_n + x_n - ax_n^2. \tag{7}$$

Do these iterations converge? I will take $a = 2$ and aim for $x^* = \frac{1}{2}$. Subtracting $\frac{1}{2}$ from both sides of (7) changes the iteration into the error equation:

$$x_{n+1} = 2x_n - 2x_n^2 \quad \text{becomes} \quad x_{n+1} - \tfrac{1}{2} = -2\left(x_n - \tfrac{1}{2}\right)^2. \tag{8}$$

At each step the error is squared. This is terrific if (and only if) you are close to $x^* = \frac{1}{2}$. Otherwise squaring a large error and multiplying by -2 is not good:

$$x_0 = .70 \quad x_1 = .42 \quad x_2 = .487 \quad x_3 = .4997 \quad x_4 = .4999998$$
$$x_0 = 1.21 \quad x_1 = -.5 \quad x_2 = -1.5 \quad x_3 = -7.5 \quad x_4 = -127.5$$

The algebra in Problem 18 confirms those experiments. There is fast convergence if $0 < x_0 < 1$. There is divergence if x_0 is negative or $x_0 > 1$. The tangent line goes to a negative x_1. After that Figure 3.22 shows a long trip backwards.

In the previous section we drew $F(x)$. The iteration $x_{n+1} = F(x_n)$ converged to the 45° line, where $x^* = F(x^*)$. In this section we are drawing $f(x)$. Now x^* is the point on the axis where $f(x^*) = 0$.

To repeat: It is $f(x^*) = 0$ that we aim for. But it is the slope $F'(x^*)$ that decides whether we get there. Example 2 has $F(x) = 2x - 2x^2$. The fixed points are $x^* = \frac{1}{2}$ (our solution) and $x^* = 0$ (not attractive). The slopes $F'(x^*)$ are zero (typical Newton) and 2 (typical repeller). *The key to Newton's method is $F' = 0$ at the solution:*

The slope of $F(x) = x - \dfrac{f(x)}{f'(x)}$ *is* $\dfrac{f(x)f''(x)}{(f'(x))^2}$. *Then* $F'(x) = 0$ *when* $f(x) = 0$.

The examples $x^2 = b$ and $1/x = a$ show fast convergence or failure. In Chapter 13, and in reality, Newton's method solves much harder equations. Here I am going to choose a third example that came from pure curiosity about what might happen. The results are absolutely amazing. The equation is $x^2 = -1$.

EXAMPLE 3 *What happens to Newton's method if you ask it to solve* $f(x) = x^2 + 1 = 0$?

The only solutions are the imaginary numbers $x^* = i$ and $x^* = -i$. There is no real square root of -1. Newton's method might as well give up. But it has no way to know that! The tangent line still crosses the axis at a new point x_{n+1}, even if the curve $y = x^2 + 1$ never crosses. Equation (5) still gives the iteration for $b = -1$:

$$x_{n+1} = \frac{1}{2}\left(x_n - \frac{1}{x_n}\right) = F(x_n). \tag{9}$$

The x's cannot approach i or $-i$ (nothing is imaginary). So what do they do?

The starting guess $x_0 = 1$ is interesting. It is followed by $x_1 = 0$. Then x_2 divides by zero and blows up. I expected other sequences to go to infinity. But the experiments showed something different (and mystifying). When x_n is large, x_{n+1} is less than half as large. After $x_n = 10$ comes $x_{n+1} = \frac{1}{2}(10 - \frac{1}{10}) = 4.95$. After much indecision and a long wait, a number near zero eventually appears. Then the next guess divides by that small number and goes far out again. This reminded me of "chaos."

It is tempting to retreat to ordinary examples, where Newton's method is a big success. By trying exercises from the book or equations of your own, you will see that the fast convergence to $\sqrt{4}$ is very typical. The function can be much more complicated than $x^2 - 4$ (in practice it certainly is). The iteration for $2x = \cos x$ was in the previous section, and the error was squared at every step. If Newton's method starts close to x^*, its convergence is overwhelming. That has to be the main point of this section: ***Follow the tangent line***.

Instead of those good functions, may I stay with this strange example $x^2 + 1 = 0$? It is not so predictable, and maybe not so important, but somehow it is more interesting. There is no real solution x^*, and Newton's method $x_{n+1} = \frac{1}{2}(x_n - 1/x_n)$ bounces around. We will now discover x_n.

A FORMULA FOR x_n

The key is an exercise from trigonometry books. Most of those problems just give practice with sines and cosines, but this one exactly fits $\frac{1}{2}(x_n - 1/x_n)$:

$$\frac{1}{2}\left(\frac{\cos\theta}{\sin\theta} - \frac{\sin\theta}{\cos\theta}\right) = \frac{\cos 2\theta}{\sin 2\theta} \quad \text{or} \quad \frac{1}{2}\left(\cot\theta - \frac{1}{\cot\theta}\right) = \cot 2\theta$$

In the left equation, the common denominator is $2\sin\theta\cos\theta$ (which is $\sin 2\theta$). The numerator is $\cos^2\theta - \sin^2\theta$ (which is $\cos 2\theta$). Replace cosine/sine by cotangent, and the identity says this:

If $\quad x_0 = \cot\theta \quad$ then $\quad x_1 = \cot 2\theta$. Then $\quad x_2 = \cot 4\theta \quad$ Then $\quad x_n = \cot 2^n\theta$.

This is the formula. *Our points are on the cotangent curve.* Figure 3.23 starts from $x_0 = 2 = \cot\theta$, and every iteration doubles the angle.

Example A The sequence $x_0 = 1, x_1 = 0, x_2 = \infty$ matches the cotangents of $\pi/4, \pi/2$, and π. This sequence blows up because x_2 has a division by $x_1 = 0$.

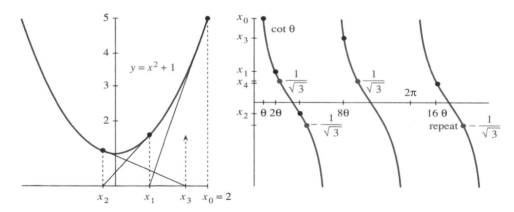

Fig. 3.23 Newton's method for $x^2 + 1 = 0$. Iteration gives $x_n = \cot 2^n\theta$.

Example B The sequence $1/\sqrt{3}, -1/\sqrt{3}, 1/\sqrt{3}$ matches the cotangents of $\pi/3, 2\pi/3$, and $4\pi/3$. This sequence *cycles forever* because $x_0 = x_2 = x_4 = \ldots$.

Example C Start with a large x_0 (a small θ). Then x_1 is about half as large (at 2θ). Eventually one of the angles $4\theta, 8\theta, \ldots$ hits on a large cotangent, and the x's go far out again. *This is typical.* Examples A and B were special, when θ/π was $\frac{1}{4}$ or $\frac{1}{3}$.

What we have here is **chaos**. The x's can't converge. They are strongly repelled by all points. They are also extremely sensitive to the value of θ. After ten steps θ is multiplied by $2^{10} = 1024$. The starting angles $60°$ and $61°$ look close, but now they are different by $1024°$. If that were a multiple of $180°$, the cotangents would still be close. In fact the x_{10}'s are 0.6 and 14.

This chaos in mathematics is also seen in nature. The most familiar example is the weather, which is much more delicate than you might think. The headline "Forecasting Pushed Too Far" appeared in *Science* (1989). The article said that the snowballing of small errors destroys the forecast after six days. We can't follow the weather equations for a month—the flight of a plane can change everything. This is a revolutionary idea, that a simple rule can lead to answers that are too sensitive to compute.

We are accustomed to complicated formulas (or no formulas). We are not accustomed to innocent-looking formulas like $\cot 2^n \theta$, which are absolutely hopeless after 100 steps.

CHAOS FROM A PARABOLA

Now I get to tell you about new mathematics. First I will change the iteration $x_{n+1} = \frac{1}{2}(x_n - 1/x_n)$ into one that is even simpler. By switching from x to $z = 1/(1 + x^2)$, each new z turns out to involve only the old z and z^2:

$$z_{n+1} = 4z_n - 4z_n^2. \tag{10}$$

This is the most famous quadratic iteration in the world. There are books about it, and Problem 28 shows where it comes from. Our formula for x_n leads to z_n:

$$z_n = \frac{1}{1 + x_n^2} = \frac{1}{1 + (\cot 2^n \theta)^2} = (\sin 2^n \theta)^2. \tag{11}$$

The sine is just as unpredictable as the cotangent, when $2^n \theta$ gets large. The new thing is to locate this quadratic as the last member (when $a = 4$) of the family

$$z_{n+1} = az_n - az_n^2, \quad 0 \leqslant a \leqslant 4. \tag{12}$$

Example 2 happened to be the middle member $a = 2$, converging to $\frac{1}{2}$. I would like to give a brief and very optional report on this iteration, for different a's.

The general principle is to start with a number z_0 between 0 and 1, and compute $z_1, z_2, z_3, \ldots$. It is fascinating to watch the behavior change as a increases. **You can see it on your own computer**. Here we describe some things to look for. All numbers stay between 0 and 1 and they may approach a limit. That happens when a is small:

for $0 \leqslant a \leqslant 1$ the z_n approach $z^* = 0$
for $1 \leqslant a \leqslant 3$ the z_n approach $z^* = (a-1)/a$

Those limit points are the solutions of $z = F(z)$. They are the fixed points where $z^* = az^* - a(z^*)^2$. But remember the test for approaching a limit: *The slope at z^* cannot be larger than one.* Here $F = az - az^2$ has $F' = a - 2az$. It is easy to check $|F'| \leqslant 1$ at the limits predicted above. The hard problem—sometimes impossible— is to predict what happens above $a = 3$. Our case is $a = 4$.

The z's cannot approach a limit when $|F'(z^*)| > 1$. Something has to happen, and there are at least three possibilities:

The z_n's can cycle or fill the whole interval $(0, 1)$ or approach a Cantor set.

I start with a random number z_0, take 100 steps, and write down steps 101 to 105:

	$a = 3.4$	$a = 3.5$	$a = 3.8$	$a = 4.0$
$z_{101} =$	.842	.875	.336	.169
$z_{102} =$	.452	.383	.848	.562
$z_{103} =$	.842	.827	.491	.985
$z_{104} =$	.452	.501	.950	.060
$z_{105} =$	.842	.875	.182	.225

The first column is converging to a "2-cycle." It alternates between $x = .842$ and $y = .452$. Those satisfy $y = F(x)$ and $x = F(y) = F(F(x))$. If we look at a *double step* when $a = 3.4, x$ and y are fixed points of the double iteration $z_{n+2} = F(F(z_n))$. When a increases past 3.45, this cycle becomes unstable.

At that point the period doubles from 2 to 4. With $a = 3.5$ you see a "4-cycle" in the table—it repeats after four steps. The sequence bounces from .875 to .383 to .827 to .501 *and back to* .875. This cycle must be attractive or we would not see it. But it also becomes unstable as a increases. Next comes an 8-cycle, which is stable in a little window (you could compute it) around $a = 3.55$. *The cycles are stable for shorter and shorter intervals of a*'s. Those stability windows are reduced by the *Feigenbaum shrinking factor* 4.6692.... Cycles of length 16 and 32 and 64 can be seen in physical experiments, but they are all unstable before $a = 3.57$. What happens then?

The new and unexpected behavior is between 3.57 and 4. Down each line of Figure 3.24, the computer has plotted the values of z_{1001} to z_{2000}—omitting the first thousand points to let a stable period (or chaos) become established. No points appeared in the big white wedge. I don't know why. In the window for period 3, you see only three z's. Period 3 is followed by $6, 12, 24, \dots$. There is *period doubling* at the end of every window (including all the windows that are too small to see). You can reproduce this figure by iterating $z_{n+1} = az_n - az_n^2$ from any z_0 and plotting the results.

CANTOR SETS AND FRACTALS

I can't tell what happens at $a = 3.8$. There may be a stable cycle of some long period. The z's may come close to every point between 0 and 1. A third possibility is to approach a very thin limit set, which looks like the famous **Cantor set**:

> To construct the Cantor set, divide $[0, 1]$ into three pieces and remove the open interval $\left(\frac{1}{3}, \frac{2}{3}\right)$. Then remove $\left(\frac{1}{9}, \frac{2}{9}\right)$ and $\left(\frac{7}{9}, \frac{8}{9}\right)$ from what remains. At each step *take out the middle thirds*. The points that are left form the Cantor set.

All the endpoints $\frac{1}{3}, \frac{2}{3}, \frac{1}{9}, \frac{2}{9}, \dots$ are in the set. So is $\frac{4}{3}$ (Problem 42). Nevertheless the lengths of the removed intervals add to 1 and the Cantor set has "measure zero." What is especially striking is its *self-similarity*: *Between 0 and $\frac{1}{3}$ you see the same Cantor set three times smaller.* From 0 to $\frac{1}{9}$ the Cantor set is there again, scaled down by 9. Every section, when blown up, copies the larger picture.

Fractals That self-similarity is typical of a *fractal*. There is an infinite sequence of scales. A mathematical snowflake starts with a triangle and adds a bump in the middle of each side. At every step the bumps lengthen the sides by $4/3$. The final boundary is self-similar, like an infinitely long coastline.

The period 2, 4, ... is the number of z's in a cycle.

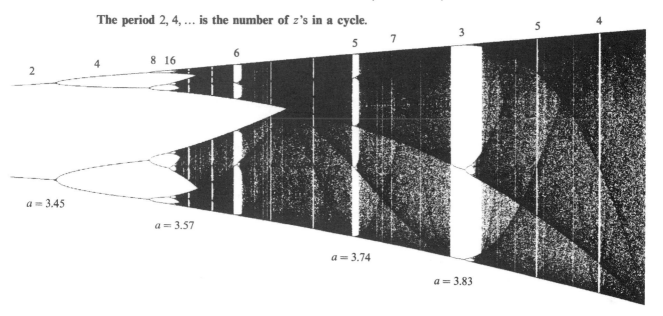

Fig. 3.24 Period doubling and chaos from iterating $F(z)$ (stolen by special permission from *Introduction to Applied Mathematics* by Gilbert Strang, Wellesley-Cambridge Press).

The word "fractal" comes from ***fractional dimension***. The snowflake boundary has dimension larger than 1 and smaller than 2. The Cantor set has dimension larger than 0 and smaller than 1. Covering an ordinary line segment with circles of radius r would take c/r circles. For fractals it takes c/r^D circles—and D is the dimension.

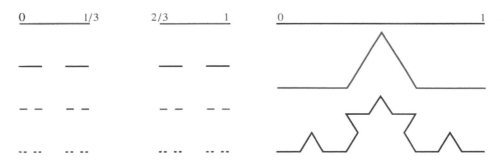

Fig. 3.25 Cantor set (middle thirds removed). Fractal snowflake (infinite boundary).

Our iteration $z_{n+1} = 4z_n - 4z_n^2$ has $a = 4$, at the end of Figure 3.24. The sequence $z_0, z_1, \ldots$ goes everywhere and nowhere. Its behavior is chaotic, and statistical tests find no pattern. For all practical purposes the numbers are random.

Think what this means in an experiment (or the stock market). If simple rules produce chaos, there is *absolutely no way* to predict the results. No measurement can ever be sufficiently accurate. The newspapers report that Pluto's orbit is chaotic— even though it obeys the law of gravity. The motion is totally unpredictable over long times. I don't know what that does for astronomy (or astrology).

The most readable book on this subject is Gleick's best-seller *Chaos: Making a New Science*. The most dazzling books are *The Beauty of Fractals* and *The Science of Fractal Images*, in which Peitgen and Richter and Saupe show photographs that have been in art museums around the world. The most original books are Mandelbrot's *Fractals* and *Fractal Geometry*. Our cover has a fractal from Figure 13.11.

We return to friendlier problems in which calculus is not helpless.

NEWTON'S METHOD VS. SECANT METHOD: CALCULATOR PROGRAMS

The hard part of Newton's method is to find df/dx. We need it for the slope of the tangent line. But calculus can approximate by $\Delta f/\Delta x$—using the values of $f(x)$ already computed at x_n and x_{n-1}.

The **secant method** follows the secant line instead of the tangent line:

$$\textbf{Secant}: \quad x_{n+1} = x_n - \frac{f(x_n)}{(\Delta f/\Delta x)_n} \quad \text{where} \quad \left(\frac{\Delta f}{\Delta x}\right)_n = \frac{f(x_n) - f(x_{n-1})}{x_n - x_{n-1}}. \quad (13)$$

The secant line connects the two latest points on the graph of $f(x)$. Its equation is $y - f(x_n) = (\Delta f/\Delta x)(x - x_n)$. Set $y = 0$ to find equation (13) for the new $x = x_{n+1}$, where the line crosses the axis.

Prediction: *Three* secant steps are about as good as *two* Newton steps. Both should give four times as many correct decimals: $(error) \to (error)^4$. Probably the secant method is also chaotic for $x^2 + 1 = 0$.

These Newton and secant programs are for the TI-81. Place the formula for $f(x)$ in slot Y_1 and the formula for $f'(x)$ in slot Y_2 on the $Y=$ function edit screen. Answer the prompt with the initial $x_0 = X\emptyset$. The programs pause to display each approximation x_n, the value $f(x_n)$, and the difference $x_n - x_{n-1}$. Press **ENTER** to continue or press **ON** and select item **2:Quit** to break. If $f(x_n) = 0$, the programs display **ROOT AT** and the root x_n.

```
PrgmN:NEWTON      :Disp"ENTER FOR MORE"      PrgmS:SECANT       :Y→T
:Disp"XØ="        :Disp"ON 2 TO BREAK"       :Disp"XØ="         :Y1→Y
:Input X          :Disp"  "                  :Input X           :Disp"ENTER FOR MORE"
:X→S              :Disp"XN FXN XN-XNM1"      :X→S               :Disp"XN FXN XN-XNM1"
:Y1→Y             :Disp X                    :Y1→T              :Disp X
:Lbl 1            :Disp Y                    :Disp "X1="        :Disp Y
:X-Y/Y2→X         :Disp D                    :Input X           :Disp D
:X-S→D            :Pause                     :Y1→Y              :Pause
:X→S              :If Y≠Ø                    :Lbl 1             :If Y≠Ø
:Y1→Y             :Goto 1                    :X-S→D             :Goto 1
                  :Disp "ROOT AT"            :X→S               :Disp "ROOT AT"
                  :Disp X                    :X-YD/(Y-T)→X      :Disp X
```

3.7 EXERCISES

Read-through questions

When $f(x) = 0$ is linearized to $f(x_n) + f'(x_n)(x - x_n) = 0$, the solution $x = $ __a__ is Newton's x_{n+1}. The __b__ to the curve crosses the axis at x_{n+1}, while the __c__ crosses at x^*. The errors at x_n and x_{n+1} are normally related by $(error)_{n+1} \approx M$__d__. This is __e__ convergence. The number of correct decimals __f__ at every step.

For $f(x) = x^2 - b$, Newton's iteration is $x_{n+1} = $ __g__. The x_n converge to __h__ if $x_0 > 0$ and to __i__ if $x_0 < 0$. For $f(x) = x^2 + 1$, the iteration becomes $x_{n+1} = $ __j__. This cannot converge to __k__. Instead it leads to chaos. Changing to $z = 1/(x^2 + 1)$ yields the parabolic iteration $z_{n+1} = $ __l__.

For $a \leqslant 3$, $z_{n+1} = az_n - az_n^2$ converges to a single __m__. After $a = 3$ the limit is a 2-cycle, which means __n__. Later the limit is a Cantor set, which is a one-dimensional example of a __o__. The cantor set is self-__p__.

1 To solve $f(x) = x^3 - b = 0$, what iteration comes from Newton's method?

2 For $f(x) = (x-1)/(x+1)$ Newton's formula is $x_{n+1} = F(x_n) = $ _____. Solve $x^* = F(x^*)$ and find $F'(x^*)$. What limit do the x_n's approach?

3 I believe that Newton only applied his method in public to one equation $x^3 - 2x - 5 = 0$. Raphson carried the idea forward but got partial credit at best. After two steps from $x_0 = 2$, how many decimals in $x^* = 2.09455148$ are correct?

4 Show that Newton's method for $f(x) = x^{1/3}$ gives the strange formula $x_{n+1} = -2x_n$. Draw a graph to show the iterations.

5 Find x_1 if (a) $f(x_0) = 0$; (b) $f'(x_0) = 0$.

6 Graph $f(x) = x^3 - 3x - 1$ and estimate its roots x^*. Run Newton's method starting from 0, 1, $-\frac{1}{2}$, and 1.1. Experiment to decide which x_0 converge to which root.

7 Solve $x^2 - 6x + 5 = 0$ by Newton's method with $x_0 = 2.5$ and 3. Draw a graph to show which x_0 lead to which root.

8 If $f(x)$ is increasing and concave up ($f' > 0$ and $f'' > 0$) show by a graph that Newton's method converges. From which side?

Solve 9–17 to four decimal places by Newton's method with a computer or calculator. Choose any x_0 except x^*.

9 $x^2 - 10 = 0$

10 $x^4 - 100 = 0$ (faster or slower than Problem 9?)

11 $x^2 - x = 0$ (which x_0 to which root?)

12 $x^3 - x = 0$ (which x_0 to which root?)

13 $x + 5\cos x = 0$ (this has three roots)

14 $x + \tan x = 0$ (find two roots) (are there more?)

15 $1/(1-x) = 2$

16 $1 + x + x^2 + x^3 + x^4 = 2$

17 $x^3 + (x+1)^3 = 10^3$

18 (a) Show that $x_{n+1} = 2x_n - 2x_n^2$ in Example 2 is the same as $(1 - 2x_{n+1}) = (1 - 2x_n)^2$.

(b) Prove divergence if $|1 - 2x_0| > 1$. Prove convergence if $|1 - 2x_0| < 1$ or $0 < x_0 < 1$.

19 With $a = 3$ in Example 2, experiment with the Newton iteration $x_{n+1} = 2x_n - 3x_n^2$ to decide which x_0 lead to $x^* = \frac{1}{3}$.

20 Rewrite $x_{n+1} = 2x_n - ax_n^2$ as $(1 - ax_{n+1}) = (1 - ax_n)^2$. For which x_0 does the sequence $1 - ax_n$ approach zero (so $x \to 1/a$)?

21 What is Newton's method to find the kth root of 7? Calculate $\sqrt[7]{7}$ to 7 places.

22 Find all solutions of $x^3 = 4x - 1$ (5 decimals).

Problems 23–29 are about $x^2 + 1 = 0$ and chaos.

23 For $\theta = \pi/16$ when does $x_n = \cot 2^n \theta$ blow up? For $\theta = \pi/7$ when does $\cot 2^n \theta = \cot \theta$? (The angles $2^n \theta$ and θ differ by a multiple of π.)

24 For $\theta = \pi/9$ follow the sequence until $x_n = x_0$.

25 For $\theta = 1$, x_n never returns to $x_0 = \cot 1$. The angles 2^n and 1 never differ by a multiple of π because _____.

26 If z_0 equals $\sin^2\theta$, show that $z_1 = 4z_0 - 4z_0^2$ equals $\sin^2 2\theta$.

27 If $y = x^2 + 1$, each new y is

$$y_{n+1} = x_{n+1}^2 + 1 = \frac{1}{4}\left(x_n - \frac{1}{x_n}\right)^2 + 1.$$

Show that this equals $y_n^2/4(y_n - 1)$.

28 Turn Problem 27 upside down, $1/y_{n+1} = 4(y_n - 1)/y_n^2$, to find the quadratic iteration (10) for $z_n = 1/y_n = 1/(1 + x_n^2)$.

29 If $F(z) = 4z - 4z^2$ what is $F(F(z))$? How many solutions to $z = F(F(z))$? How many are not solutions to $z = F(z)$?

30 Apply Newton's method to $x^3 - .64x - .36 = 0$ to find the basin of attraction for $x^* = 1$. Also find a pair of points for which $y = F(z)$ and $z = F(y)$. In this example Newton does not always find a root.

31 Newton's method solves $x/(1-x) = 0$ by $x_{n+1} = $ _____. From which x_0 does it converge? The distance to $x^* = 0$ is exactly squared.

Problems 33–41 are about competitors of Newton.

32 At a double root, Newton only converges linearly. What is the iteration to solve $x^2 = 0$?

33 To speed up Newton's method, find the step Δx from $f(x_n) + \Delta x f'(x_n) + \frac{1}{2}(\Delta x)^2 f''(x_n) = 0$. Test on $f(x) = x^2 - 1$ from $x_0 = 0$ and explain.

34 Halley's method uses $f_n + \Delta x f_n' + \frac{1}{2}\Delta x(-f_n/f_n')f_n'' = 0$. For $f(x) = x^2 - 1$ and $x_0 = 1 + \varepsilon$, show that $x_1 = 1 + O(\varepsilon^3)$— which is *cubic* convergence.

35 Apply the secant method to $f(x) = x^2 - 4 = 0$, starting from $x_0 = 1$ and $x_1 = 2.5$. Find $\Delta f/\Delta x$ and the next point x_2 by hand. Newton uses $f'(x_1) = 5$ to reach $x_2 = 2.05$. Which is closer to $x^* = 2$?

36 Draw a graph of $f(x) = x^2 - 4$ to show the secant line in Problem 35 and the point x_2 where it crosses the axis.

Bisection method **If** $f(x)$ **changes sign between** x_0 **and** x_1, **find its sign at the midpoint** $x_2 = \frac{1}{2}(x_0 + x_1)$. **Decide whether** $f(x)$ **changes sign between** x_0 **and** x_2 **or** x_2 **and** x_1. **Repeat on that half-length (bisected) interval.** Continue. **Switch to a faster method when the interval is small enough.**

37 $f(x) = x^2 - 4$ is negative at $x = 1$, positive at $x = 2.5$, and negative at the midpoint $x = 1.75$. So x^* lies in what interval? Take a second step to cut the interval in half again.

38 Write a code for the bisection method. At each step print out an interval that contains x^*. The inputs are x_0 and x_1; lbr the code calls $f(x)$. Stop if $f(x_0)$ and $f(x_1)$ have the same sign.

39 Three bisection steps reduce the interval by what factor? Starting from $x_0 = 0$ and $x_1 = 8$, take three steps for $f(x) = x^2 - 10$.

40 A direct method is to *zoom in* where the graph crosses the axis. Solve $10x^3 - 8.3x^2 + 2.295x - .21141 = 0$ by several zooms.

41 If the zoom factor is 10, then the number of correct decimals _____ for every zoom. Compare with Newton.

42 The number $\frac{3}{4}$ equals $\frac{2}{3}(1 + \frac{1}{9} + \frac{1}{81} + \cdots)$. Show that it is in the Cantor set. It survives when middle thirds are removed.

43 The solution to $f(x) = (x - 1.9)/(x - 2.0) = 0$ is $x^* = 1.9$. Try Newton's method from $x_0 = 1.5, 2.1$, and 1.95. Extra credit: Which x_0's give convergence?

44 Apply the secant method to solve $\cos x = 0$ from $x_0 = .308$.

45 Try Newton's method on $\cos x = 0$ from $x_0 = .308$. If $\cot x_0$ is exactly π, show that $x_1 = x_0 + \pi$ (and $x_2 = x_1 + \pi$). From $x_0 = .308169071$ does Newton's method ever stop?

46 Use the Newton and secant programs to solve $x^3 - 10x^2 + 22x + 6 = 0$ from $x_0 = 2$ and 1.39.

47 Newton's method for $\sin x = 0$ is $x_{n+1} = x_n - \tan x_n$. Graph $\sin x$ and three iterations from $x_0 = 2$ and $x_0 = 1.8$. Predict the result for $x_0 = 1.9$ and test. This leads to the *computer project* in Problem 3.6.41, which finds fractals.

48 Graph $Y_1(x) = 3.4(x - x^2)$ and $Y_2(x) = Y_1(Y_1(x))$ in the square window $(0, 0) \leqslant (x, y) \leqslant (1, 1)$. Then graph $Y_3(x) = Y_2(Y_1(x))$ and $Y_4, \ldots, Y_9$. The cycle is from .842 to .452.

49 Repeat Problem 48 with 3.4 changed to 2 or 3.5 or 4.

3.8 The Mean Value Theorem and l'Hôpital's Rule

Now comes one of the cornerstones of calculus: the *Mean Value Theorem*. It connects the local picture (slope at a point) to the global picture (average slope across an interval). In other words it relates df/dx to $\Delta f/\Delta x$. Calculus depends on this connection, which we saw first for velocities. If the average velocity is 75, is there a moment when the instantaneous velocity is 75 ?

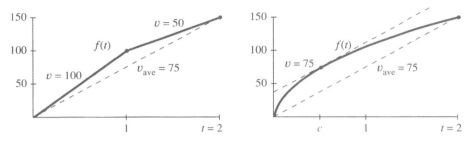

Fig. 3.26   (a) v jumps over v_{average}. (b) v equals v_{average}.

Without more information, the answer to that question is *no*. The velocity could be 100 and then 50—averaging 75 but never equal to 75. If we allow a jump in velocity, it can jump right over its average. At that moment the velocity does not exist. (The distance function in Figure 3.26a has no derivative at $x = 1$.) We will take away this cheap escape by requiring a derivative at all points inside the interval.

In Figure 3.26b the distance increases by 150 when t increases by 2. There is a derivative df/dt at all interior points (but an infinite slope at $t = 0$). The average velocity is

$$\frac{\Delta f}{\Delta t} = \frac{f(2) - f(0)}{2 - 0} = \frac{150}{2} = 75.$$

The conclusion of the theorem is that $df/dt = 75$ at some point inside the interval. There is at least one point where $f'(c) = 75$.

This is not a constructive theorem. The value of c is not known. We don't find c, we just claim (with proof) that such a point exists.

3M *Mean Value Theorem* Suppose $f(x)$ is continuous in the closed interval $a \leqslant x \leqslant b$ and has a derivative everywhere in the open interval $a < x < b$. Then

$$\frac{f(b) - f(a)}{b - a} = f'(c) \text{ at some point } a < c < b. \tag{1}$$

The left side is the average slope $\Delta f/\Delta x$. It equals df/dx at c. The notation for a closed interval [with endpoints] is $[a, b]$. For an open interval (without endpoints) we write (a, b). Thus f' is defined in (a, b), and f remains continuous at a and b. A derivative is allowed at those endpoints too—but the theorem doesn't require it.

The proof is based on a special case—when $f(a) = 0$ and $f(b) = 0$. *Suppose the function starts at zero and returns to zero*. The average slope or velocity is zero. We have to prove that $f'(c) = 0$ at a point in between. This special case (keeping the assumptions on $f(x)$) is called *Rolle's theorem*.

Geometrically, if f goes away from zero and comes back, then $f' = 0$ *at the turn*.

> **3N Rolle's theorem**   Suppose $f(a) = f(b) = 0$ (zero at the ends). Then $f'(c) = 0$ at some point with $a < c < b$.

Proof At a point inside the interval where $f(x)$ reaches its maximum or minimum, df/dx must be zero. That is an acceptable point c. Figure 3.27a shows the difference between $f = 0$ (assumed at a and b) and $f' = 0$ (proved at c).

Small problem: The maximum could be reached at the ends a and b, if $f(x) < 0$ in between. At those endpoints df/dx might not be zero. But in that case the *minimum* is reached at an interior point c, which is equally acceptable. The key to our proof is that *a continuous function on $[a, b]$ reaches its maximum and minimum*. This is the *Extreme Value Theorem*.†

It is ironic that Rolle himself did not believe the logic behind calculus. He may not have believed his own theorem! Probably he didn't know what it meant—the language of "evanescent quantities" (Newton) and "infinitesimals" (Leibniz) was exciting but frustrating. Limits were close but never reached. Curves had infinitely many flat sides. Rolle didn't accept that reasoning, and what was really serious, he didn't accept the conclusions. The Académie des Sciences had to stop his battles (he fought against ordinary mathematicians, not Newton and Leibniz). So he went back to number theory, but his special case when $f(a) = f(b) = 0$ leads directly to the big one.

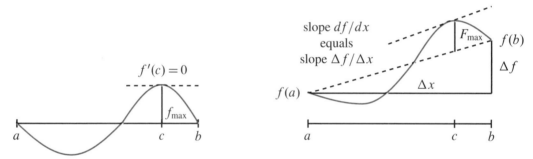

Fig. 3.27   Rolle's theorem is when $f(a) = f(b) = 0$ in the Mean Value Theorem.

Proof of the Mean Value Theorem We are looking for a point where df/dx equals $\Delta f/\Delta x$. The idea is *to tilt the graph back to Rolle's special case* (when Δf was zero). In Figure 3.27b, the distance $F(x)$ between the curve and the dotted secant line comes from subtraction:

$$F(x) = f(x) - \left[f(a) + \frac{\Delta f}{\Delta x}(x - a) \right]. \tag{2}$$

At a and b, this distance is $F(a) = F(b) = 0$. *Rolle's theorem applies to $F(x)$*. There is an interior point where $F'(c) = 0$. At that point take the derivative of equation (2): $0 = f'(c) - (\Delta f/\Delta x)$. The desired point c is found, proving the theorem.

EXAMPLE 1   The function $f(x) = \sqrt{x}$ goes from zero at $x = 0$ to ten at $x = 100$. Its average slope is $\Delta f/\Delta x = 10/100$. The derivative $f'(x) = 1/2\sqrt{x}$ exists in the open interval $(0, 100)$, even though it blows up at the end $x = 0$. By the Mean Value Theorem there must be a point where $10/100 = f'(c) = 1/2\sqrt{c}$. That point is $c = 25$.

---

†If $f(x)$ doesn't reach its maximum M, then $1/(M - f(x))$ would be continuous but also approach infinity. Essential fact: *A continuous function on $[a, b]$ cannot approach infinity*.

The truth is that nobody cares about the exact value of c. Its existence is what matters. Notice how it affects the linear approximation $f(x) \approx f(a) + f'(a)(x - a)$, which was basic to this chapter. Close becomes exact ($\approx$ becomes $=$) when f' is computed at c instead of a:

3O The derivative at c gives an exact prediction of $f(x)$:
$$f(x) = f(a) + f'(c)(x - a). \tag{3}$$
The Mean Value Theorem is rewritten here as $\Delta f = f'(c) \Delta x$. Now $a < c < x$.

EXAMPLE 2 The function $f(x) = \sin x$ starts from $f(0) = 0$. The linear prediction (tangent line) uses the slope $\cos 0 = 1$. The exact prediction uses the slope $\cos c$ at an unknown point between 0 and x:

$$(approximate) \ \sin x \approx x \qquad (exact) \ \sin x = (\cos c) x. \tag{4}$$

The approximation is useful, because everything is computed at $x = a = 0$. The exact formula is interesting, because $\cos c \leqslant 1$ proves again that $\sin x \leqslant x$. The slope is below 1, so the sine graph stays below the $45°$ line.

EXAMPLE 3 *If $f'(c) = 0$ at all points in an interval then $f(x)$ is constant.*

Proof When f' is everywhere zero, the theorem gives $\Delta f = 0$. Every pair of points has $f(b) = f(a)$. The graph is a horizontal line. That deceptively simple case is a key to the Fundamental Theorem of Calculus.

Most applications of $\Delta f = f'(c) \Delta x$ do not end up with a number. They end up with another theorem (like this one). The goal is to connect derivatives (local) to differences (global). But the next application—*l'Hôpital's Rule*—manages to produce a number out of $0/0$.

L'HÔPITAL'S RULE

When $f(x)$ and $g(x)$ both approach zero, what happens to their ratio $f(x)/g(x)$?

$$\frac{f(x)}{g(x)} = \frac{x^2}{x} \quad \text{or} \quad \frac{\sin x}{x} \quad \text{or} \quad \frac{x - \sin x}{1 - \cos x} \quad \text{all become} \quad \frac{0}{0} \quad \text{at} \quad x = 0.$$

Since $0/0$ is meaningless, we cannot work separately with $f(x)$ and $g(x)$. This is a "*race toward zero,*" in which two functions become small while their ratio might do anything. The problem is to find the limit of $f(x)/g(x)$.

One such limit is already studied. *It is the derivative!* $\Delta f / \Delta x$ automatically builds in a race toward zero, whose limit is df/dx:

$$\frac{f(x) - f(a)}{x - a} \to \frac{0}{0} \quad \text{but} \quad \lim_{x \to a} \frac{f(x) - f(a)}{x - a} = f'(a). \tag{5}$$

The idea of l'Hôpital is to use f'/g' to handle f/g. The derivative is the special case $g(x) = x - a$, with $g' = 1$. The Rule is followed by examples and proofs.

3P *l'Hôpital's Rule* Suppose $f(x)$ and $g(x)$ both approach zero as $x \to a$. Then $f(x)/g(x)$ approaches the same limit as $f'(x)/g'(x)$, if that second limit exists:

$$\lim_{x \to a} \frac{f(x)}{g(x)} = \lim_{x \to a} \frac{f'(x)}{g'(x)}. \qquad \text{Normally this limit is} \qquad \frac{f'(a)}{g'(a)}. \tag{6}$$

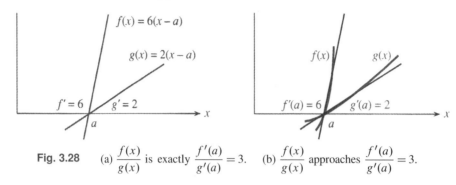

Fig. 3.28   (a) $\dfrac{f(x)}{g(x)}$ is exactly $\dfrac{f'(a)}{g'(a)} = 3.$ (b) $\dfrac{f(x)}{g(x)}$ approaches $\dfrac{f'(a)}{g'(a)} = 3.$

This is not the quotient rule! The derivatives of $f(x)$ and $g(x)$ are taken separately. Geometrically, l'Hôpital is saying that *when functions go to zero their slopes control their size*. An easy case is $f = 6(x - a)$ and $g = 2(x - a)$. The ratio f/g is exactly $6/2$, the ratio of their slopes. Figure 3.28 shows these straight lines dropping to zero, controlled by 6 and 2.

The next figure shows the same limit $6/2$, when the curves are ***tangent*** to the lines. That picture is the key to l'Hôpital's rule.

Generally the limit of f/g can be a finite number L or $+\infty$ or $-\infty$. (Also the limit point $x = a$ can represent a finite number or $+\infty$ or $-\infty$. We keep it finite.) The one absolute requirement is that $f(x)$ and $g(x)$ must separately approach zero—we insist on $0/0$. Otherwise there is no reason why equation (6) should be true. With $f(x) = x$ and $g(x) = x - 1$, ***don't*** use l'Hôpital:

$$\frac{f(x)}{g(x)} \to \frac{a}{a - 1} \qquad \text{but} \qquad \frac{f'(x)}{g'(x)} = \frac{1}{1}.$$

Ordinary ratios approach $\lim f(x)$ divided by $\lim g(x)$. l'Hôpital enters only for $0/0$.

EXAMPLE 4   (an old friend) $\displaystyle\lim_{x \to 0} \frac{1 - \cos x}{x}$ equals $\displaystyle\lim_{x \to 0} \frac{\sin x}{1}$. This equals zero.

EXAMPLE 5 $\dfrac{f}{g} = \dfrac{\tan x}{\sin x}$ leads to $\dfrac{f'}{g'} = \dfrac{\sec^2 x}{\cos x}$. At $x = 0$ the limit is $\dfrac{1}{1}$.

EXAMPLE 6 $\dfrac{f}{g} = \dfrac{x - \sin x}{1 - \cos x}$ leads to $\dfrac{f'}{g'} = \dfrac{1 - \cos x}{\sin x}$. At $x = 0$ this is still $\dfrac{0}{0}$.

Solution   *Apply the Rule to f'/g'*. It has the same limit as f''/g'':

$$\text{if } \frac{f}{g} \to \frac{0}{0} \text{ and } \frac{f'}{g'} \to \frac{0}{0} \text{ then compute } \frac{f''(x)}{g''(x)} = \frac{\sin x}{\cos x} \to \frac{0}{1} = 0.$$

The reason behind l'Hôpital's Rule is that the following fractions are the same:

$$\frac{f(x)}{g(x)} = \frac{f(x) - f(a)}{x - a} \bigg/ \frac{g(x) - g(a)}{x - a}. \tag{7}$$

That is just algebra; the limit hasn't happened yet. The factors $x - a$ cancel, and the numbers $f(a)$ and $g(a)$ are zero by assumption. Now take the limit on the right side of (7) as x approaches a.

What normally happens is that one part approaches f' at $x = a$. The other part approaches $g'(a)$. We hope $g'(a)$ is not zero. In this case we can divide one limit by the other limit. That gives the "normal" answer

$$\lim_{x \to a} \frac{f(x)}{g(x)} = \text{limit of } (7) = \frac{f'(a)}{g'(a)}. \tag{8}$$

This is also l'Hôpital's answer. When $f'(x) \to f'(a)$ and separately $g'(x) \to g'(a)$, his overall limit is $f'(a)/g'(a)$. He published this rule in the first textbook ever written on differential calculus. (That was in 1696—the limit was actually discovered by his teacher Bernoulli.) Three hundred years later we apply his name to other cases permitted in (6), when f'/g' might approach a limit even if the separate parts do not.

To prove this more general form of l'Hôpital's Rule, we need a more general Mean Value Theorem. *I regard the discussion below as optional in a calculus course* (but required in a calculus book). The important idea already came in equation (8).

Remark **The basic "indeterminate" is** $\infty - \infty$. If $f(x)$ and $g(x)$ approach infinity, anything is possible for $f(x) - g(x)$. We could have $x^2 - x$ or $x - x^2$ or $(x + 2) - x$. Their limits are ∞ and $-\infty$ and 2.

At the next level are $0/0$ and ∞/∞ and $0 \cdot \infty$. To find the limit in these cases, try l'Hôpital's Rule. See Problem 24 when $f(x)/g(x)$ approaches ∞/∞. When $f(x) \to 0$ and $g(x) \to \infty$, apply the $0/0$ rule to $f(x)/(1/g(x))$.

The next level has 0^0 and 1^∞ and ∞^0. Those come from limits of $f(x)^{g(x)}$. If $f(x)$ approaches $0, 1$, or ∞ while $g(x)$ approaches $0, \infty$, or 0, we need more information. A really curious example is $x^{1/\ln x}$, which shows all three possibilities 0^0 and 1^∞ and ∞^0. This function is actually a constant! It equals e.

To go back down a level, take logarithms. Then $g(x) \ln f(x)$ returns to $0/0$ and $0 \cdot \infty$ and l'Hôpital's Rule. But logarithms and e have to wait for Chapter 6.

THE GENERALIZED MEAN VALUE THEOREM

The MVT can be extended to *two functions*. The extension is due to Cauchy, who cleared up the whole idea of limits. You will recognize the special case $g = x$ as the ordinary Mean Value Theorem.

3Q **Generalized MVT** If $f(x)$ and $g(x)$ are continuous on $[a, b]$ and differentiable on (a, b), there is a point $a < c < b$ where

$$[f(b) - f(a)]g'(c) = [g(b) - g(a)]f'(c). \tag{9}$$

The proof comes by constructing a new function that has $F(a) = F(b)$:

$$F(x) = [f(b) - f(a)]g(x) - [g(b) - g(a)]f(x).$$

The ordinary Mean Value Theorem leads to $F'(c) = 0$—which is equation (9).

Application 1 (Proof of l'Hôpital's Rule) The rule deals with $f(a)/g(a) = 0/0$. Inserting those zeros into equation (9) leaves $f(b)g'(c) = g(b)f'(c)$. Therefore

$$\frac{f(b)}{g(b)} = \frac{f'(c)}{g'(c)}. \tag{10}$$

As b approaches a, so does c. The point c is squeezed between a and b. The limit of equation (10) as $b \to a$ and $c \to a$ is l'Hôpital's Rule.

Application 2 (Error in linear approximation)   Section 3.2 stated that the distance between a curve and its tangent line grows like $(x-a)^2$. Now we can prove this, and find out more. Linear approximation is

$$f(x) = f(a) + f'(a)(x-a) + error\ e(x). \tag{11}$$

The pattern suggests an error involving $f''(x)$ and $(x-a)^2$. The key example $f = x^2$ shows the need for a factor $\frac{1}{2}$ (to cancel $f'' = 2$). **The error in linear approximation is**

$$e(x) = \tfrac{1}{2} f''(c)(x-a)^2 \quad \text{with} \quad a < c < x. \tag{12}$$

Key idea   Compare the error $e(x)$ to $(x-a)^2$. Both are zero at $x = a$:

$$e = f(x) - f(a) - f'(a)(x-a) \qquad e' = f'(x) - f'(a) \qquad e'' = f''(x)$$
$$g = (x-a)^2 \qquad\qquad\qquad g' = 2(x-a) \qquad\quad g'' = 2$$

The Generalized Mean Value Theorem finds a point C between a and x where $e(x)/g(x) = e'(C)/g'(C)$. This is equation (10) with different letters. After checking $e'(a) = g'(a) = 0$, apply the same theorem to $e'(x)$ and $g'(x)$. It produces a point c between a and C—certainly between a and x—where

$$\frac{e'(C)}{g'(C)} = \frac{e''(c)}{g''(c)} \quad \text{and therefore} \quad \frac{e(x)}{g(x)} = \frac{e''(c)}{g''(c)}.$$

With $g = (x-a)^2$ and $g'' = 2$ and $e'' = f''$, the equation on the right is $e(x) = \frac{1}{2} f''(c)(x-a)^2$. The error formula is proved. A very good approximation is $\frac{1}{2} f''(c)(x-a)^2$.

EXAMPLE 7 $f(x) = \sqrt{x}$ near $a = 100$: $\sqrt{102} \approx 10 + \left(\dfrac{1}{20}\right)2 + \dfrac{1}{2}\left(\dfrac{-1}{4000}\right)2^2.$

That last term predicts $e = -.0005$. The actual error is $\sqrt{102} - 10.1 = -.000496$.

3.8 EXERCISES

Read-through questions

The Mean Value Theorem equates the average slope $\Delta f/\Delta x$ over an __a__ $[a,b]$ to the slope df/dx at an unknown __b__. The statement is __c__. It requires $f(x)$ to be __d__ on the __e__ interval $[a,b]$, with a __f__ on the open interval (a,b). Rolle's theorem is the special case when $f(a) = f(b) = 0$, and the point c satisfies __g__. The proof chooses c as the point where f reaches its __h__.

Consequences of the Mean Value Theorem include: If $f'(x) = 0$ everywhere in an interval then $f(x) = $__i__. The prediction $f(x) = f(a) + $ __j__ $(x-a)$ is exact for some c between a and x. The quadratic prediction $f(x) = f(a) + f'(a)(x-a) + $ __k__ $(x-a)^2$ is exact for another c. The error in $f(a) + f'(a)(x-a)$ is less than $\frac{1}{2} M(x-a)^2$ where M is the maximum of __l__.

A chief consequence is l'Hôpital's Rule, which applies when $f(x)$ and $g(x) \to$ __m__ as $x \to a$. In that case the limit of $f(x)/g(x)$ equals the limit of __n__, provided this limit exists. Normally this limit is $f'(a)/g'(a)$. If this is also 0/0, go on to the limit of __o__.

Find all points $0 < c < 2$ where $f(2) - f(0) = f'(c)(2-0)$.

1 $f(x) = x^3$ 2 $f(x) = \sin \pi x$

3 $f(x) = \tan 2\pi x$ 4 $f(x) = 1 + x + x^2$

5 $f(x) = (x-1)^{10}$ 6 $f(x) = (x-1)^9$

In 7–10 show that no point c yields $f(1) - f(-1) = f'(c)(2)$. Explain why the Mean Value Theorem fails to apply.

7 $f(x) = |x - \frac{1}{2}|$ 8 $f(x) = $ unit step function

9 $f(x) = |x|^{1/2}$ 10 $f(x) = 1/x^2$

11 Show that $\sec^2 x$ and $\tan^2 x$ have the same derivative, and draw a conclusion about $f(x) = \sec^2 x - \tan^2 x$.

12 Show that $\csc^2 x$ and $\cot^2 x$ have the same derivative and find $f(x) = \csc^2 x - \cot^2 x$.

Evaluate the limits in 13–22 by l'Hôpital's Rule.

13 $\lim\limits_{x \to 3} \dfrac{x^2 - 9}{x - 3}$

14 $\lim\limits_{x \to 3} \dfrac{x^2 - 9}{x + 3}$

15 $\lim\limits_{x \to 0} \dfrac{(1+x)^{-2} - 1}{x}$

16 $\lim\limits_{x \to 0} \dfrac{\sqrt{1 - \cos x}}{x}$

17 $\lim\limits_{x \to \pi} \dfrac{x - \pi}{\sin x}$

18 $\lim\limits_{x \to 1} \dfrac{x - 1}{\sin x}$

19 $\lim\limits_{x \to 0} \dfrac{(1+x)^n - 1}{x}$

20 $\lim\limits_{x \to 0} \dfrac{(1+x)^n - 1 - nx}{x^2}$

21 $\lim\limits_{x \to 0} \dfrac{\sin x - \tan x}{x^3}$

22 $\lim\limits_{x \to 0} \dfrac{\sqrt{1+x} - \sqrt{1-x}}{x}$

23 For $f = x^2 - 4$ and $g = x + 2$, the ratio f'/g' approaches 4 as $x \to 2$. What is the limit of $f(x)/g(x)$? What goes wrong in l'Hôpital's Rule?

24 *l'Hôpital's Rule still holds for* $f(x)/g(x) \to \infty/\infty$: L **is**

$$\lim \frac{f(x)}{g(x)} = \lim \frac{1/g(x)}{1/f(x)} = \lim \frac{g'(x)/g^2(x)}{f'(x)/f^2(x)} = L^2 \lim \frac{g'(x)}{f'(x)}.$$

Then L equals $\lim[f'(x)/g'(x)]$ if this limit exists. Where did we use the rule for $0/0$? What other limit rule was used?

25 Compute $\lim\limits_{x \to 0} \dfrac{1 + (1/x)}{1 - (1/x)}$.

26 Compute $\lim\limits_{x \to \infty} \dfrac{x^2 + x}{2x^2}$.

27 Compute $\lim\limits_{x \to \infty} \dfrac{x + \cos x}{x + \sin x}$ by common sense. Show that l'Hôpital gives no answer.

28 Compute $\lim\limits_{x \to \infty} \dfrac{\csc x}{\cot x}$ by common sense or trickery.

29 The Mean Value Theorem applied to $f(x) = x^3$ guarantees that some number c between 1 and 4 has a certain property. Say what the property is and find c.

30 If $|df/dx| \leq 1$ at all points, prove this fact:

$$|f(x) - f(y)| \leq 1 \text{ at all } x \text{ and } y.$$

31 The error in Newton's method is squared at each step: $|x_{n+1} - x^*| \leq M|x_n - x^*|^2$. The proof starts from $0 = f(x^*) = f(x_n) + f'(x_n)(x^* - x_n) + \frac{1}{2}f''(c)(x^* - x_n)^2$. Divide by $f'(x_n)$, recognize x_{n+1}, and estimate M.

32 (Rolle's theorem backward) Suppose $f'(c) = 0$. Are there necessarily two points around c where $f(a) = f(b)$?

33 Suppose $f(0) = 0$. If $f(x)/x$ has a limit as $x \to 0$, that limit is better known to us as _____. L'Hôpital's Rule looks instead at the limit of _____.

Conclusion from l'Hôpital: The limit of $f'(x)$, if it exists, agrees with $f'(0)$. Thus $f'(x)$ cannot have a "removable _____."

34 It is possible that $f'(x)/g'(x)$ has *no* limit but $f(x)/g(x) \to L$. This is why l'Hôpital included an "if."

(a) Find L as $x \to 0$ when $f(x) = x^2 \cos(1/x)$ and $g(x) = x$. Remember that cosines are below 1.

(b) From the formula $f'(x) = \sin(1/x) + 2x \cos(1/x)$ show that f'/g' has no limit as $x \to 0$.

35 Stein's calculus book asks for the limiting ratio of $f(x) =$ triangular area ABC to $g(x) =$ curved area ABC.
(a) Guess the limit of f/g as the angle x goes to zero.
(b) Explain why $f(x)$ is $\frac{1}{2}(\sin x - \sin x \cos x)$ and $g(x)$ is $\frac{1}{2}(x - \sin x \cos x)$. (c) Compute the true limit of $f(x)/g(x)$.

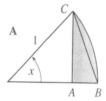 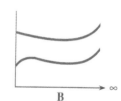

36 If you drive $3{,}000$ miles from New York to L.A. in 100 hours (sleeping and eating and going backwards are allowed) then at some moment your speed is _____.

37 As $x \to \infty$ l'Hôpital's Rule still applies. The limit of $f(x)/g(x)$ equals the limit of $f'(x)/g'(x)$, if that limit exists. What is the limit as the graphs become parallel in Figure B?

38 Prove that $f(x)$ is increasing when its slope is positive: *If* $f'(c) > 0$ *at all points* c, *then* $f(b) > f(a)$ *at all pairs of points* $b > a$.

CHAPTER 4

Derivatives by the Chain Rule

You remember that the derivative of $f(x)g(x)$ is not $(df/dx)(dg/dx)$. The derivative of $\sin x$ times x^2 is not $\cos x$ times $2x$. The product rule gave two terms, not one term. But there is another way of combining the sine function f and the squaring function g into a single function. The derivative of that new function does involve the cosine times $2x$ (but with a certain twist). We will first explain the new function, and then find the "***chain rule***" for its derivative.

May I say here that the chain rule is important. It is easy to learn, and you will use it often. I see it as the third basic way to find derivatives of new functions from derivatives of old functions. (So far the old functions are x^n, $\sin x$, and $\cos x$. Still ahead are e^x and $\log x$.) When f and g are added and multiplied, derivatives come from the *sum rule* and *product rule*. This section combines f and g in a third way.

The new function is $\sin(x^2)$—***the sine of*** x^2. It is created out of the two original functions: if $x = 3$ then $x^2 = 9$ and $\sin(x^2) = \sin 9$. There is a "chain" of functions, combining $\sin x$ and x^2 into the composite function $\sin(x^2)$. You start with x, *then find* $g(x)$, ***then find*** $f(g(x))$:

> The squaring function gives $y = x^2$. This is $g(x)$.
> The sine function produces $z = \sin y = \sin(x^2)$. This is $f(g(x))$.

The "***inside function***" $g(x)$ gives y. *This is the input to the* "***outside function***" $f(y)$. That is called ***composition***. It starts with x and ends with z. The composite function is sometimes written $f \circ g$ (the circle shows the difference from an ordinary product fg). More often you will see $f(g(x))$:

$$z(x) = f \circ g(x) = f(g(x)). \tag{1}$$

Other examples are $\cos 2x$ and $(2x)^3$, with $g = 2x$. *On a calculator you input* x, *then push the* "g" *button, then push the* "f" *button*:

> ***From*** x ***compute*** $y = g(x)$ ***From*** y ***compute*** $z = f(y)$.

There is not a button for every function! But the squaring function and sine function are on most calculators, and they are used ***in that order***. Figure 4.1a shows how squaring will stretch and squeeze the sine function.

That graph of $\sin x^2$ is a crazy FM signal (the Frequency is Modulated). The wave goes up and down like $\sin x$, but not at the same places. Changing to $\sin g(x)$ moves the peaks left and right. Compare with a product $g(x)\sin x$, which is an AM signal (the Amplitude is Modulated).

Remark $f(g(x))$ is usually different from $g(f(x))$. **The order of f and g is usually important**. For $f(x)=\sin x$ and $g(x)=x^2$, the chain in the opposite order $g(f(x))$ gives something different:

First apply the sine function: $y=\sin x$
Then apply the squaring function: $z=(\sin x)^2$.

That result is often written $\sin^2 x$, to save on parentheses. It is never written $\sin x^2$, which is totally different. Compare them in Figure 4.1.

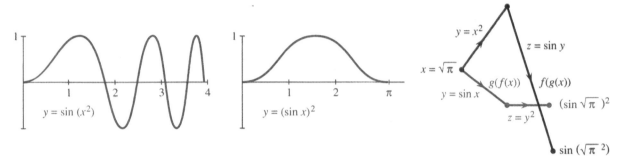

Fig. 4.1 $f(g(x))$ is different from $g(f(x))$. Apply g then f, or f then g.

EXAMPLE 1   The composite function $f \circ g$ can be deceptive. If $g(x)=x^3$ and $f(y)=y^4$, how does $f(g(x))$ differ from the ordinary product $f(x)g(x)$? The ordinary product is x^7. The chain starts with $y=x^3$, and then $z=y^4=x^{12}$. The composition of x^3 and y^4 gives $f(g(x))=x^{12}$.

EXAMPLE 2   In Newton's method, $F(x)$ is composed with itself. This is *iteration*. Every output x_n is fed back as input, to find $x_{n+1}=F(x_n)$. The example $F(x)=\frac{1}{2}x+4$ has $F(F(x))=\frac{1}{2}(\frac{1}{2}x+4)+4$. That produces $z=\frac{1}{4}x+6$.

The derivative of $F(x)$ is $\frac{1}{2}$. The derivative of $z=F(F(x))$ is $\frac{1}{4}$, which is $\frac{1}{2}$ times $\frac{1}{2}$. **We multiply derivatives**. This is a special case of the chain rule.

An extremely special case is $f(x)=x$ and $g(x)=x$. The ordinary product is x^2. The chain $f(g(x))$ produces only x! The output from the "*identity function*" is $g(x)=x$.† When the second identity function operates on x it produces x again. The derivative is 1 times 1. I can give more composite functions in a table:

$y=g(x)$	$z=f(y)$	$z=f(g(x))$
x^2-1	$\sqrt{y}$	$\sqrt{x^2-1}$
$\cos x$	y^3	$(\cos x)^3$
2^x	2^y	2^{2^x}
$x+5$	$y-5$	x

† A calculator has no button for the identity function. It wouldn't do anything.

The last one adds 5 to get y. Then it subtracts 5 to reach z. So $z = x$. Here output equals input: $f(g(x)) = x$. These "***inverse functions***" are in Section 4.3. The other examples create new functions $z(x)$ and we want their derivatives.

THE DERIVATIVE OF $f(g(x))$

What is the derivative of $z = \sin x^2$? It is the limit of $\Delta z / \Delta x$. Therefore we look at a nearby point $x + \Delta x$. That change in x produces a change in $y = x^2$—which moves to $y + \Delta y = (x + \Delta x)^2$. From this change in y, there is a change in $z = f(y)$. It is a "domino effect," in which each changed input yields a changed output: Δx ***produces*** Δy ***produces*** Δz. We have to connect the final Δz to the original Δx.

   The key is to write $\Delta z / \Delta x$ ***as*** $\Delta z / \Delta y$ ***times*** $\Delta y / \Delta x$. Then let Δx approach zero. In the limit, dz/dx is given by the "chain rule":

$$\frac{\Delta z}{\Delta x} = \frac{\Delta z}{\Delta y} \frac{\Delta y}{\Delta x} \textit{ becomes the chain rule } \frac{dz}{dx} = \frac{dz}{dy} \frac{dy}{dx}. \tag{2}$$

As Δx goes to zero, the ratio $\Delta y / \Delta x$ approaches dy/dx. Therefore Δy must be going to zero, and $\Delta z / \Delta y$ approaches dz/dy. The limit of a product is the product of the separate limits (end of quick proof). *We multiply derivatives*:

4A *Chain Rule* Suppose $g(x)$ has a derivative at x and $f(y)$ has a derivative at $y = g(x)$. Then the derivative of $z = f(g(x))$ is

$$\frac{dz}{dx} = \frac{dz}{dy} \frac{dy}{dx} = f'(g(x)) g'(x). \tag{3}$$

The slope at x is df/dy (at y) times dg/dx (at x).

Caution The chain rule does *not* say that the derivative of $\sin x^2$ is $(\cos x)(2x)$. True, $\cos y$ is the derivative of $\sin y$. The point is that $\cos y$ *must be evaluated at* y (not at x). We do not want df/dx at x, we want df/dy at $y = x^2$:

$$\textit{The derivative of } \sin x^2 \textit{ is } (\cos x^2) \textit{ times } (2x). \tag{4}$$

EXAMPLE 3   If $z = (\sin x)^2$ then $dz/dx = (2 \sin x)(\cos x)$. Here $y = \sin x$ is *inside*.

In this order, $z = y^2$ leads to $dz/dy = 2y$. *It does not lead to* $2x$. The inside function $\sin x$ produces $dy/dx = \cos x$. The answer is $2y \cos x$. We have not yet found the function whose derivative is $2x \cos x$.

EXAMPLE 4   The derivative of $z = \sin 3x$ is $\dfrac{dz}{dx} = \dfrac{dz}{dy} \dfrac{dy}{dx} = 3 \cos 3x$.

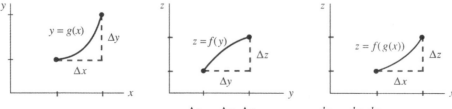

Fig. 4.2   The chain rule: $\dfrac{\Delta z}{\Delta x} = \dfrac{\Delta z}{\Delta y} \dfrac{\Delta y}{\Delta x}$ approaches $\dfrac{dz}{dx} = \dfrac{dz}{dy} \dfrac{dy}{dx}$.

The outside function is $z = \sin y$. The inside function is $y = 3x$. Then $dz/dy = \cos y$—this is $\cos 3x$, not $\cos x$. Remember the other factor $dy/dx = 3$.

I can explain that factor 3, especially if x is switched to t. The distance is $z = \sin 3t$. That oscillates like $\sin t$ except *three times as fast*. The speeded-up function $\sin 3t$ completes a wave at time $2\pi/3$ (instead of 2π). Naturally the velocity contains the extra factor 3 from the chain rule.

EXAMPLE 5 Let $z = f(y) = y^n$. Find the derivative of $f(g(x)) = [g(x)]^n$.

In this case dz/dy is ny^{n-1}. The chain rule multiplies by dy/dx:

$$\frac{dz}{dx} = ny^{n-1}\frac{dy}{dx} \qquad \text{or} \qquad \frac{d}{dx}[g(x)]^n = n[g(x)]^{n-1}\frac{dg}{dx}. \tag{5}$$

This is the ***power rule***! It was already discovered in Section 2.5. Square roots (when $n = 1/2$) are frequent and important. Suppose $y = x^2 - 1$:

$$\frac{d}{dx}\sqrt{x^2 - 1} = \frac{1}{2}(x^2 - 1)^{-1/2}(2x) = \frac{x}{\sqrt{x^2 - 1}}. \tag{6}$$

Question A Buick uses $1/20$ of a gallon of gas per mile. You drive at 60 miles per hour. How many gallons per hour?

Answer (***Gallons/hour***) = (***gallons/mile***)(***miles/hour***). The chain rule is $(dy/dt) = (dy/dx)(dx/dt)$. The answer is $(1/20)(60) = 3$ gallons/hour.

Proof of the chain rule The discussion above was correctly based on

$$\frac{\Delta z}{\Delta x} = \frac{\Delta z}{\Delta y}\frac{\Delta y}{\Delta x} \qquad \text{and} \qquad \frac{dz}{dx} = \frac{dz}{dy}\frac{dy}{dx}. \tag{7}$$

It was here, over the chain rule, that the "battle of notation" was won by Leibniz. His notation practically tells you what to do: Take the limit of each term. (I have to mention that when Δx is approaching zero, it is theoretically possible that Δy might *hit* zero. If that happens, $\Delta z/\Delta y$ becomes $0/0$. We have to assign it the correct meaning, which is dz/dy.) As $\Delta x \to 0$,

$$\frac{\Delta y}{\Delta x} \to g'(x) \qquad \text{and} \qquad \frac{\Delta z}{\Delta y} \to f'(y) = f'(g(x)).$$

Then $\Delta z/\Delta x$ approaches $f'(y)$ times $g'(x)$, which is the chain rule $(dz/dy)(dy/dx)$. In the table below, the derivative of $(\sin x)^3$ is $3(\sin x)^2\cos x$. That extra factor $\cos x$ is easy to forget. It is even easier to forget the -1 in the last example.

$$z = (x^3 + 1)^5 \quad dz/dx = 5(x^3 + 1)^4 \quad \text{times } 3x^2$$
$$z = (\sin x)^3 \quad dz/dx = 3\sin^2 x \quad \text{times } \cos x$$
$$z = (1 - x)^2 \quad dz/dx = 2(1 - x) \quad \text{times } -1$$

Important All kinds of letters are used for the chain rule. We named the output z. Very often it is called y, and the inside function is called u:

$$\textbf{\textit{The derivative of }} y = \sin u(x) \textbf{\textit{ is }} \frac{dy}{dx} = \cos u\frac{du}{dx}.$$

Examples with du/dx are extremely common. I have to ask you to accept whatever letters may come. What never changes is the key idea—*derivative of outside function times derivative of inside function*.

EXAMPLE 6　The chain rule is barely needed for $\sin(x-1)$. Strictly speaking the inside function is $u = x - 1$. Then du/dx is just 1 (not -1). **If** $y = \sin(x-1)$ **then** $dy/dx = \cos(x-1)$. The graph is shifted and the slope shifts too.
　　Notice especially: The cosine is computed at $x - 1$ and not at the unshifted x.

RECOGNIZING $f(y)$ AND $g(x)$

A big part of the chain rule is *recognizing the chain*. The table started with $(x^3 + 1)^5$. You look at it for a second. Then you see it as u^5. The inside function is $u = x^3 + 1$. With practice this decomposition (the opposite of composition) gets easy:

$$\cos(2x+1) \text{ is } \cos u \qquad \sqrt{1 + \sin t} \text{ is } \sqrt{u} \qquad x \sin x \text{ is } \ldots \text{ (product rule!)}$$

In calculations, the careful way is to write down all the functions:

$$z = \cos u \quad u = 2x + 1 \quad dz/dx = (-\sin u)(2) = -2\sin(2x+1).$$

The quick way is to keep in your mind "the derivative of what's inside." The slope of $\cos(2x+1)$ is $-\sin(2x+1)$, *times 2 from the chain rule*. The derivative of $2x + 1$ is remembered—without z or u or f or g.

EXAMPLE 7　$\sin\sqrt{1-x}$ is a chain of $z = \sin y$, $y = \sqrt{u}, u = 1 - x$ (*three functions*).

With that triple chain you will have the hang of the chain rule:

$$\text{The derivative of } \sin\sqrt{1-x} \text{ is } (\cos\sqrt{1-x})\left(\frac{1}{2\sqrt{1-x}}\right)(-1).$$

This is $(dz/dy)(dy/du)(du/dx)$. Evaluate them at the right places y, u, x.
　　Finally there is the question of *second derivatives*. The chain rule gives dz/dx as a product, so d^2z/dx^2 needs the product rule:

$$\frac{dz}{dx} = \frac{dz}{dy}\frac{dy}{dx} \qquad \text{leads to} \qquad \frac{d^2z}{dx^2} = \frac{dz}{dy}\frac{d^2y}{dx^2} + \frac{d}{dx}\left(\frac{dz}{dy}\right)\frac{dy}{dx}. \qquad (8)$$

$$ u \quad v u \quad v' + \quad u' \quad v$$

That last term needs the chain rule again. It becomes d^2z/dy^2 times $(dy/dx)^2$.

EXAMPLE 8　The derivative of $\sin x^2$ is $2x \cos x^2$. Then the product rule gives $d^2z/dx^2 = 2\cos x^2 - 4x^2\sin x^2$. In this case $y'' = 2$ and $(y')^2 = 4x^2$.

4.1　EXERCISES

Read-through questions

$z = f(g(x))$ comes from $z = f(y)$ and $y = $ __a__. At $x = 2$, the chain $(x^2 - 1)^3$ equals __b__. Its inside function is $y = $__c__, its outside function is $z = $__d__. Then dz/dx equals __e__. The first factor is evaluated at $y = $__f__ (not at $y = x$). For $z = \sin(x^4 - 1)$ the derivative is __g__. The triple chain $z = \cos(x + 1)^2$ has a shift and a __h__ and a cosine. Then $dz/dx = $__i__.

　　The proof of the chain rule begins with $\Delta z / \Delta x = ($__j__$)($__k__$)$ and ends with __l__. Changing letters, $y = \cos u(x)$ has $dy/dx = $ __m__. The power rule for $y = [u(x)]^n$ is the chain rule $dy/dx = $__n__. The slope of $5g(x)$ is __o__ and the slope of $g(5x)$ is __p__. When $f = $ cosine and $g = $ sine and $x = 0$, the numbers $f(g(x))$ and $g(f(x))$ and $f(x)g(x)$ are __q__.

In 1–10 identify $f(y)$ and $g(x)$. **From their derivatives find $\dfrac{dz}{dx}$.**

1　$z = (x^2 - 3)^3$

2　$z = (x^3 - 3)^2$

3　$z = \cos(x^3)$

4　$z = \tan 2x$

5 $z = \sqrt{\sin x}$ **6** $z = \sin \sqrt{x}$

7 $z = \tan(1/x) + 1/\tan x$ **8** $z = \sin(\cos x)$

9 $z = \cos(x^2 + x + 1)$ **10** $z = \sqrt{x^2}$

In 11–16 write down dz/dx. **Don't write down** f **and** g.

11 $z = \sin(17x)$ **12** $z = \tan(x + 1)$

13 $z = \cos(\cos x)$ **14** $z = (x^2)^{3/2}$

15 $z = x^2 \sin x$ **16** $z = (9x + 4)^{3/2}$

Problems 17–22 involve three functions $z(y)$, $y(u)$, **and** $u(x)$. **Find** dz/dx **from** $(dz/dy)(dy/du)(du/dx)$.

17 $z = \sin \sqrt{x + 1}$ **18** $z = \sqrt{\sin(x + 1)}$

19 $z = \sqrt{1 + \sin x}$ **20** $z = \sin(\sqrt{x} + 1)$

21 $z = \sin(1/\sin x)$ **22** $z = (\sin x^2)^2$

In 23–26 find dz/dx **by the chain rule and also by rewriting** z.

23 $z = ((x^2)^2)^2$ **24** $z = (3x)^3$

25 $z = (x + 1)^2 + \sin(x + \pi)$ **26** $z = \sqrt{1 - \cos^2 x}$

27 If $f(x) = x^2 + 1$ what is $f(f(x))$? If $U(x)$ is the unit step function (from 0 to 1 at $x = 0$) draw the graphs of $\sin U(x)$ and $U(\sin x)$. If $R(x)$ is the *ramp function* $\frac{1}{2}(x + |x|)$, draw the graphs of $R(x)$ and $R(\sin x)$.

28 (Recommended) If $g(x) = x^3$ find $f(y)$ so that $f(g(x)) = x^3 + 1$. Then find $h(y)$ so that $h(g(x)) = x$. Then find $k(y)$ so that $k(g(x)) = 1$.

29 If $f(y) = y - 2$ find $g(x)$ so that $f(g(x)) = x$. Then find $h(x)$ so that $f(h(x)) = x^2$. Then find $k(x)$ so that $f(k(x)) = 1$.

30 Find two different pairs $f(y), g(x)$ so that $f(g(x)) = \sqrt{1 - x^2}$.

31 The derivative of $f(f(x))$ is _____. Is it $(df/dx)^2$? Test your formula on $f(x) = 1/x$.

32 If $f(3) = 3$ and $g(3) = 5$ and $f'(3) = 2$ and $g'(3) = 4$, find the derivative at $x = 3$ if possible for

    (a) $f(x)g(x)$ (b) $f(g(x))$ (c) $g(f(x))$ (d) $f(f(x))$

33 For $F(x) = \frac{1}{2}x + 8$, show how iteration gives $F(F(x)) = \frac{1}{4}x + 12$. Find $F(F(F(x)))$—also called $F^{(3)}(x)$. The derivative of $F^{(4)}(x)$ is _____.

34 In Problem 33 the limit of $F^{(n)}(x)$ is a constant $C =$ _____. From any start (try $x = 0$) the iterations $x_{n+1} = F(x_n)$ converge to C.

35 Suppose $g(x) = 3x + 1$ and $f(y) = \frac{1}{3}(y - 1)$. Then $f(g(x)) =$ _____ and $g(f(y)) =$ _____. These are *inverse functions*.

36 Suppose $g(x)$ is continuous at $x = 4$, say $g(4) = 7$. Suppose $f(y)$ is continuous at $y = 7$, say $f(7) = 9$. Then $f(g(x))$ is continuous at $x = 4$ and $f(g(4)) = 9$.

Proof ε is given. Because _____ is continuous, there is a δ such that $|f(g(x)) - 9| < \varepsilon$ whenever $|g(x) - 7| < \delta$. Then because _____ is continuous, there is a θ such that $|g(x) - 7| < \delta$ whenever $|x - 4| < \theta$. Conclusion: If $|x - 4| < \theta$ then _____. This shows that $f(g(x))$ approaches $f(g(4))$.

37 Only six functions can be constructed by compositions (in any sequence) of $g(x) = 1 - x$ and $f(x) = 1/x$. Starting with g and f, find the other four.

38 If $g(x) = 1 - x$ then $g(g(x)) = 1 - (1 - x) = x$. If $g(x) = 1/x$ then $g(g(x)) = 1/(1/x) = x$. Draw graphs of those g's and explain from the graphs why $g(g(x)) = x$. Find two more g's with this special property.

39 Construct functions so that $f(g(x))$ is always zero, but $f(y)$ is not always zero.

40 **True or false**

    (a) If $f(x) = f(-x)$ then $f'(x) = f'(-x)$.

    (b) The derivative of the identity function is zero.

    (c) The derivative of $f(1/x)$ is $-1/(f(x))^2$.

    (d) The derivative of $f(1 + x)$ is $f'(1 + x)$.

    (e) The second derivative of $f(g(x))$ is $f''(g(x))g''(x)$.

41 On the same graph draw the parabola $y = x^2$ and the curve $z = \sin y$ (keep y upwards, with x and z across). Starting at $x = 3$ find your way to $z = \sin 9$.

42 On the same graph draw $y = \sin x$ and $z = y^2$ (y upwards for both). Starting at $x = \pi/4$ find $z = (\sin x)^2$ on the graph.

43 Find the second derivative of

    (a) $\sin(x^2 + 1)$ (b) $\sqrt{x^2 - 1}$ (c) $\cos \sqrt{x}$

44 Explain why $\dfrac{d}{dx}\left(\dfrac{dz}{dy}\right) = \left(\dfrac{d^2 z}{dy^2}\right)\left(\dfrac{dy}{dx}\right)$ in equation (8). Check this when $z = y^2$, $y = x^3$.

Final practice with the chain rule and other rules (and other letters!). Find the x **or** t **derivative of** z **or** y.

45 $z = f(u(t))$ **46** $z = u^3, u = x^3$

47 $y = \sin u(x) \cos u(x)$ **48** $y = \sqrt{u(t)}$

49 $y = x^2 u(x)$ **50** $y = f(x^2) = (f(x))^2$

51 $z = \sqrt{1 - u}, u = \sqrt{1 - x}$ **52** $z = 1/u^n(t)$

53 $z = f(u), u = v^2, v = \sqrt{t}$ **54** $y = u, u = x, x = 1/t$

55 If $f = x^4$ and $g = x^3$ then $f' = 4x^3$ and $g' = 3x^2$. The chain rule multiplies derivatives to get $12x^5$. But $f(g(x)) = x^{12}$ and its derivative is not $12x^5$. Where is the flaw?

56 The derivative of $y = \sin(\sin x)$ is $dy/dx =$

$\cos(\cos x)$ $\sin(\cos x)\cos x$ $\cos(\sin x)\cos x$ $\cos(\cos x)\cos x$.

57 (a) A book has 400 words per page. There are 9 pages per section. So there are _____ words per section.

(b) You read 200 words per minute. So you read _____ pages per minute. How many minutes per section?

58 (a) You walk in a train at 3 miles per hour. The train moves at 50 miles per hour. Your ground speed is _____ miles per hour.

(b) You walk in a train at 3 miles per hour. The train is shown on TV (1 mile train = 20 inches on TV screen). Your speed across the screen is _____ inches per hour.

59 Coke costs 1/3 dollar per bottle. The buyer gets _____ bottles per dollar. If $dy/dx = 1/3$ then $dx/dy =$ _____ .

60 (Computer) Graph $F(x) = \sin x$ and $G(x) = \sin(\sin x)$—not much difference. Do the same for $F'(x)$ and $G'(x)$. Then plot $F''(x)$ and $G''(x)$ to see where the difference shows up.

4.2 Implicit Differentiation and Related Rates

We start with the equations $xy = 2$ and $y^5 + xy = 3$. As x changes, these y's will change—to keep (x, y) on the curve. **We want to know dy/dx at a typical point.** For $xy = 2$ that is no trouble, but the slope of $y^5 + xy = 3$ requires a new idea.

In the first case, solve for $y = 2/x$ and take its derivative: $dy/dx = -2/x^2$. The curve is a hyperbola. At $x = 2$ the slope is $-2/4 = -1/2$.

The problem with $y^5 + xy = 3$ is that it can't be solved for y. Galois proved that there is no solution formula for fifth-degree equations.† **The function $y(x)$ cannot be given explicitly.** All we have is the *implicit* definition of y, as a solution to $y^5 + xy = 3$. The point $x = 2$, $y = 1$ satisfies the equation and lies on the curve, but how to find dy/dx?

This section answers that question. It is a situation that often occurs. Equations like $\sin y + \sin x = 1$ or $y \sin y = x$ (maybe even $\sin y = x$) are difficult or impossible to solve directly for y. Nevertheless we can find dy/dx at any point.

The way out is **implicit differentiation**. Work with the equation as it stands. **Find the x derivative of every term in $y^5 + xy = 3$.** That includes the constant term 3, whose derivative is zero.

EXAMPLE 1   The power rule for y^5 and the product rule for xy yield

$$5y^4 \frac{dy}{dx} + x\frac{dy}{dx} + y = 0. \tag{1}$$

Now substitute the typical point $x = 2$ and $y = 1$, and solve for dy/dx:

$$5\frac{dy}{dx} + 2\frac{dy}{dx} + 1 = 0 \quad \text{produces} \quad \frac{dy}{dx} = -\frac{1}{7}. \tag{2}$$

This is implicit differentiation (**ID**), and you see the idea: Include dy/dx from the chain rule, even if y is not known explicitly as a function of x.

EXAMPLE 2 $\sin y + \sin x = 1$ leads to $\cos y\dfrac{dy}{dx} + \cos x = 0$

EXAMPLE 3 $y \sin y = x$ leads to $y \cos y\dfrac{dy}{dx} = \sin y\dfrac{dy}{dx} = 1$

Knowing the slope makes it easier to draw the curve. We still need points (x, y) that satisfy the equation. Sometimes we can solve for x. Dividing $y^5 + xy = 3$ by y gives $x = 3/y - y^4$. Now the derivative (the x derivative!) is

$$1 = \left(-\frac{3}{y^2} - 4y^3\right)\frac{dy}{dx} = -7\frac{dy}{dx} \text{ at } y = 1. \tag{3}$$

Again $dy/dx = -1/7$. All these examples confirm the main point of the section:

4B (**Implicit differentiation**) An equation $F(x, y) = 0$ can be differentiated directly by the chain rule, without solving for y in terms of x.

The example $xy = 2$, done implicitly, gives $x\,dy/dx + y = 0$. The slope dy/dx is $-y/x$. That agrees with the explicit slope $-2/x^2$.

ID is explained better by examples than theory (maybe everything is). The essential theory can be boiled down to one idea: "**Go ahead and differentiate.**"

---

† That was before he went to the famous duel, and met his end. Fourth-degree equations do have a solution formula, but it is practically never used.

EXAMPLE 4   Find the tangent direction to the circle $x^2 + y^2 = 25$.

We can solve for $y = \pm\sqrt{25 - x^2}$, or operate directly on $x^2 + y^2 = 25$:

$$2x + 2y\frac{dy}{dx} = 0 \qquad \text{or} \qquad \frac{dy}{dx} = -\frac{x}{y}. \tag{4}$$

Compare with the radius, which has slope y/x. The radius goes across x and up y. The tangent goes across $-y$ and up x. The slopes multiply to give $(-x/y)(y/x) = -1$.

To emphasize implicit differentiation, go on to the *second derivative*. The top of the circle is concave down, so d^2y/dx^2 is negative. Use the quotient rule on $-x/y$:

$$\frac{dy}{dx} = -\frac{x}{y} \quad \text{so} \quad \frac{d^2y}{dx^2} = -\frac{y\,dx/dx - x\,dy/dx}{y^2} = -\frac{y + (x^2/y)}{y^2} = -\frac{y^2 + x^2}{y^3}. \tag{5}$$

RELATED RATES

There is a group of problems that has never found a perfect place in calculus. They seem to fit here—as applications of the chain rule. The problem is to compute df/dt, but the odd thing is that *we are given another derivative dg/dt*. To find df/dt, we need a relation between f and g.

The chain rule is $df/dt = (df/dg)(dg/dt)$. Here the variable is t because that is typical in applications. From the rate of change of g we find *the rate of change of f*. This is the problem of **related rates**, and examples will make the point.

EXAMPLE 5   The radius of a circle is growing by $dr/dt = 7$. How fast is the circumference growing? Remember that $C = 2\pi r$ (this relates C to r).

Solution
$$\frac{dC}{dt} = \frac{dC}{dr}\frac{dr}{dt} = (2\pi)(7) = 14\pi.$$

That is pretty basic, but its implications are amazing. Suppose you want to put a rope around the earth that any 7-footer can walk under. If the distance is 24,000 miles, what is the additional length of the rope? Answer: Only 14π feet.

More realistically, if two lanes on a circular track are separated by 5 feet, how much head start should the outside runner get? Only 10π feet. If your speed around a turn is 55 and the car in the next lane goes 56, who wins? See Problem 14.

Examples 6–8 are from the 1988 *Advanced Placement Exams* (copyright 1989 by the College Entrance Examination Board). Their questions are carefully prepared.

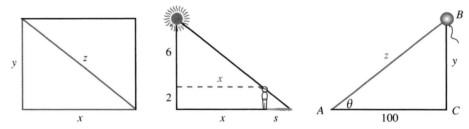

Fig. 4.3   Rectangle for Example 6, shadow for Example 7, balloon for Example 8.

EXAMPLE 6   The sides of the rectangle increase in such a way that $dz/dt = 1$ and $dx/dt = 3dy/dt$. At the instant when $x = 4$ and $y = 3$, what is the value of dx/dt?

Solution   The key relation is $x^2 + y^2 = z^2$. Take its derivative (*implicitly*):

$$2x\frac{dx}{dt} + 2y\frac{dy}{dt} = 2z\frac{dz}{dt} \quad \text{produces} \quad 8\frac{dx}{dt} + 6\frac{dy}{dt} = 10.$$

We used all information, including $z = 5$, except for $dx/dt = 3dy/dt$. The term $6dy/dt$ equals $2dx/dt$, so we have $10dx/dt = 10$. Answer: $dx/dt = 1$.

EXAMPLE 7   A person 2 meters tall walks directly away from a streetlight that is 8 meters above the ground. If the person's shadow is lengthening at the rate of $4/9$ meters per second, at what rate in meters per second is the person walking?

Solution   Draw a figure! You must relate the shadow length s to the distance x from the streetlight. The problem gives $ds/dt = 4/9$ and asks for dx/dt:

$$\text{By similar triangles} \quad \frac{x}{6} = \frac{s}{2} \quad \text{so} \quad \frac{dx}{dt} = \frac{6}{2}\frac{ds}{dt} = (3)\left(\frac{4}{9}\right) = \frac{4}{3}.$$

Note   This problem was hard. I drew three figures before catching on to x and s. It is interesting that *we never knew x or s or the angle*.

EXAMPLE 8   An observer at point A is watching balloon B as it rises from point C. (*The figure is given.*) The balloon is rising at a constant rate of 3 meters per second (*this means $dy/dt = 3$*) and the observer is 100 meters from point C.

(a) Find the rate of change in z at the instant when $y = 50$. (*They want dz/dt.*)

$$z^2 = y^2 + 100^2 \Rightarrow 2z\frac{dz}{dt} = 2y\frac{dy}{dt}$$

$$z = \sqrt{50^2 + 100^2} = 50\sqrt{5} \Rightarrow \frac{dz}{dt} = \frac{2 \cdot 50 \cdot 3}{2 \cdot 50\sqrt{5}} = \frac{3\sqrt{5}}{5}.$$

(b) Find the rate of change in the area of right triangle BCA when $y = 50$.

$$A = \frac{1}{2}(100)(y) = 50y \qquad \frac{dA}{dt} = 50\frac{dy}{dt} = 50 \cdot 3 = 150.$$

(c) Find the rate of change in θ when $y = 50$. (*They want $d\theta/dt$.*)

$$y = 50 \Rightarrow \cos\theta = \frac{100}{50\sqrt{5}} = \frac{2}{\sqrt{5}}$$

$$\tan\theta = \frac{y}{100} \Rightarrow \sec^2\theta\frac{d\theta}{dt} = \frac{1}{100}\frac{dy}{dt} \Rightarrow \frac{d\theta}{dt} = \left(\frac{2}{\sqrt{5}}\right)^2\frac{3}{100} = \frac{3}{125}$$

In all problems I first wrote down a relation from the figure. Then I took its derivative. Then I substituted known information. (The substitution is *after* taking the derivative of $\tan\theta = y/100$. If we substitute $y = 50$ too soon, the derivative of $50/100$ is useless.)

   "Candidates are advised to show their work in order to minimize the risk of not receiving credit for it." 50% solved Example 6 and 21% solved Example 7. From $12,000$ candidates, the average on Example 8 (free response) was 6.1 out of 9.

EXAMPLE 9 A is a lighthouse and BC is the shoreline (same figure as the balloon). The light at A turns once a second ($d\theta/dt = 2\pi$ radians/second). How quickly does the receiving point B move up the shoreline?

Solution   The figure shows $y = 100 \tan \theta$. The speed dy/dt is $100 \sec^2\theta \, d\theta/dt$. This is $200\pi \sec^2\theta$, so B speeds up as $\sec \theta$ increases.

Paradox   When θ approaches a right angle, $\sec \theta$ approaches infinity. So does dy/dt. B **moves faster than light**! This contradicts Einstein's theory of relativity. The paradox is resolved (I hope) in Problem 18.

   If you walk around a light at A, your shadow at B seems to go faster than light. Same problem. This speed is impossible—something has been forgotten.

Smaller paradox (not destroying the theory of relativity). The figure shows $y = z \sin \theta$. Apparently $dy/dt = (dz/dt)\sin \theta$. **This is totally wrong**. Not only is it wrong, the exact opposite is true: $dz/dt = (dy/dt)\sin \theta$. If you can explain that (Problem 15), then **ID** and related rates hold no terrors.

4.2   EXERCISES

Read-through questions

For $x^3 + y^3 = 2$ the derivative dy/dx comes from __a__ differentiation. We don't have to solve for __b__. Term by term the derivative is $3x^2 + $ __c__ $= 0$. Solving for dy/dx gives __d__. At $x = y = 1$ this slope __e__. The equation of the tangent line is $y - 1 = $ __f__.

A second example is $y^2 = x$. The x derivative of this equation is __g__. Therefore $dy/dx = $ __h__. Replacing y by $\sqrt{x}$, this is $dy/dx = $ __i__.

In related rates, we are given dg/dt and we want df/dt. We need a relation between f and __j__. If $f = g^2$, then $(df/dt) = $ __k__ (dg/dt). If $f^2 + g^2 = 1$, then $df/dt = $ __l__. If the sides of a cube grow by $ds/dt = 2$, then its volume grows by $dV/dt = $ __m__. To find a number (8 is wrong), you also need to know __n__.

By implicit differentiation find dy/dx in $1 - 10$.

1   $y^n + x^n = 1$

2   $x^2y + y^2x = 1$

3   $(x - y)^2 = 4$

4   $\sqrt{x} + \sqrt{y} = 3$ at $x = 4$

5   $x = F(y)$

6   $f(x) + F(y) = xy$

7   $x^2y = y^2x$

8   $x = \sin y$

9   $x = \tan y$

10   $y^n = x$ at $x = 1$

11   Show that the hyperbolas $xy = C$ are perpendicular to the hyperbolas $x^2 - y^2 = D$. (Perpendicular means that the product of slopes is -1.)

12   Show that the circles $(x - 2)^2 + y^2 = 2$ and $x^2 + (y - 2)^2 = 2$ are tangent at the point $(1, 1)$.

13   At 25 meters/second, does your car turn faster or slower than a car traveling 5 meters further out at 26 meters/second? Your radius is (a) 50 meters (b) 100 meters.

14   Equation (4) is $2x + 2y \, dy/dx = 0$ (on a circle). Directly by **ID** reach d^2y/dx^2 in equation (5).

Problems 15–18 resolve the speed of light paradox in Example 9.

15   (Small paradox first) The right triangle has $z^2 = y^2 + 100^2$. Take the t derivative to show that $z' = y' \sin \theta$.

16   (Even smaller paradox) As B moves up the line, why is dy/dt larger than dz/dt? Certainly z is larger than y, But as θ increases they become _____.

17   (Faster than light) The derivative of $y = 100 \tan \theta$ in Example 9 is $y' = 100\sec^2\theta \, \theta' = 200\pi \sec^2\theta$. Therefore y' passes c (the speed of light) when $\sec^2\theta$ passes _____. Such a speed is impossible—we forget that light takes time to reach B.

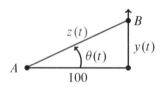

θ increases by 2π in 1 second

t is arrival time of light

θ is different from $2\pi t$

18   (Explanation by **ID**) Light travels from A to B in time z/c, distance over speed. Its arrival time is $t = \theta/2\pi + z/c$ so $\theta'/2\pi = 1 - z'/c$. Then $z' = y' \sin \theta$ and $y' = 100\sec^2\theta \, \theta'$ (all these are **ID**) lead to

$$y' = 200\pi c / (c\cos^2\theta + 200\pi \sin \theta)$$

As θ approaches $\pi/2$, this speed approaches _____ .

Note: y' still exceeds c for some negative angle. That is for Einstein to explain. See the 1985 *College Math Journal*, page 186, and the 1960 *Scientific American*, "Things that go faster than light."

19 If a plane follows the curve $y = f(x)$, and its ground speed is $dx/dt = 500$ mph, how fast is the plane going up? How fast is the plane going?

20 Why can't we differentiate $x = 7$ and reach $1 = 0$?

Problems 21–29 are applications of related rates.

21 (Calculus classic) The bottom of a 10-foot ladder is going away from the wall at $dx/dt = 2$ feet per second. How fast is the top going down the wall? Draw the right triangle to find dy/dt when the height y is (a) 6 feet (b) 5 feet (c) zero.

22 The top of the 10-foot ladder can go faster than light. At what height y does $dy/dt = -c$?

23 How fast does the level of a Coke go down if you drink a cubic inch a second? The cup is a cylinder of radius 2 inches—first write down the volume.

24 A jet flies at 8 miles up and 560 miles per hour. How fast is it approaching you when (a) it is 16 miles from you; (b) its shadow is 8 miles from you (the sun is overhead); (c) the plane is 8 miles from you (exactly above)?

25 Starting from a $3-4-5$ right triangle, the short sides increase by 2 meters/second but the angle between them decreases by 1 radian/second. How fast does the area increase or decrease?

26 A pass receiver is at $x = 4$, $y = 8t$. The ball thrown at $t = 3$ is at $x = c(t-3)$, $y = 10c(t-3)$.

(a) Choose c so the ball meets the receiver.

*(b) At that instant the distance D between them is changing at what rate?

27 A thief is 10 meters away (8 meters ahead of you, across a street 6 meters wide). The thief runs on that side at 7 meters/second, you run at 9 meters/second. How fast are you approaching if (a) you follow on your side; (b) you run toward the thief; (c) you run away on your side?

28 A spherical raindrop evaporates at a rate equal to twice its surface area. Find dr/dt.

29 Starting from $P = V = 5$ and maintaining $PV = T$, find dV/dt if $dP/dt = 2$ and $dT/dt = 3$.

30 (a) The crankshaft AB turns twice a second so $d\theta/dt = $ _____ .

(b) Differentiate the cosine law $6^2 = 3^2 + x^2 - 2(3x \cos \theta)$ to find the piston speed dx/dt when $\theta = \pi/2$ and $\theta = \pi$.

31 A camera turns at C to follow a rocket at R.

(a) Relate dz/dt to dy/dt when $y = 10$.

(b) Relate $d\theta/dt$ to dy/dt based on $y = 10 \tan \theta$.

(c) Relate $d^2\theta/dt^2$ to d^2y/dt^2 and dy/dt.

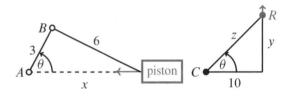

4.3 Inverse Functions and Their Derivatives

There is a remarkable special case of the chain rule. It occurs when $f(y)$ and $g(x)$ are "*inverse functions.*" That idea is expressed by a very short and powerful equation: $f(g(x)) = x$. Here is what that means.

Inverse functions: Start with any input, say $x = 5$. *Compute* $y = g(x)$, say $y = 3$. Then compute $f(y)$, and *the answer must be* 5. What one function does, the inverse function undoes. If $g(5) = 3$ then $f(3) = 5$. *The inverse function* f *takes the output* y *back to the input* x.

EXAMPLE 1 $g(x) = x - 2$ and $f(y) = y + 2$ are inverse functions. Starting with $x = 5$, the function g subtracts 2. That produces $y = 3$. Then the function f adds 2. *That brings back $x = 5$. To say it directly: **The inverse of $y = x - 2$ is $x = y + 2$.***

EXAMPLE 2 $y = g(x) = \frac{5}{9}(x - 32)$ and $x = f(y) = \frac{9}{5}y + 32$ are inverse functions (for temperature). Here x is degrees Fahrenheit and y is degrees Celsius. From $x = 32$ (freezing in Fahrenheit) you find $y = 0$ (freezing in Celsius). The inverse function takes $y = 0$ back to $x = 32$. Figure 4.4 shows how $x = 50°F$ matches $y = 10°C$.

Notice that $\frac{5}{9}(x - 32)$ subtracts 32 *first*. The inverse $\frac{9}{5}y + 32$ adds 32 *last*. In the same way g multiplies last by $\frac{5}{9}$ while f multiplies first by $\frac{9}{5}$.

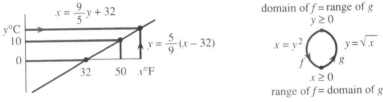

Fig. 4.4   °F to °C to °F. Always $g^{-1}(g(x)) = x$ and $g(g^{-1} = (y)) = y$. If $f = g^{-1}$ then $g = f^{-1}$.

*The inverse function is written $f = g^{-1}$ and pronounced "g inverse." **It is not** $1/g(x)$.*

If the demand y is a function of the price x, then the price is a function of the demand. Those are inverse functions. ***Their derivatives obey a fundamental rule***: dy/dx *times dx/dy equals* 1. In Example 2, dy/dx is $5/9$ and dx/dy is $9/5$.

There is another important point. When f and g are applied in the *opposite order*, they still come back to the start. First f adds 2, then g subtracts 2. The chain $g(f(y)) = (y + 2) - 2$ brings back y. **If f is the inverse of g then g is the inverse of f.** The relation is completely symmetric, and so is the definition:

Inverse function: **If $y = g(x)$ then $x = g^{-1}(y)$. If $x = g^{-1}(y)$ then $y = g(x)$.**

The loop in the figure goes from x to y to x. The composition $g^{-1}(g(x))$ is the "identity function." Instead of a new point z it returns to the original x. This will make the chain rule particularly easy—leading to $(dy/dx)(dx/dy) = 1$.

EXAMPLE 3 $y = g(x) = \sqrt{x}$ and $x = f(y) = y^2$ are inverse functions.

Starting from $x = 9$ we find $y = 3$. The inverse gives $3^2 = 9$. The square of $\sqrt{x}$ is $f(g(x)) = x$. In the opposite direction, the square root of y^2 is $g(f(y)) = y$.

Caution That example does not allow x to be negative. The domain of g—the set of numbers with square roots—is restricted to $x \geqslant 0$. This matches the range of g^{-1}. The outputs y^2 are nonnegative. With *domain of g = range of g^{-1}*, the equation $x = (\sqrt{x})^2$ is possible and true. The nonnegative x goes into g and comes out of g^{-1}.

In this example y is also nonnegative. You might think we could square anything, but y must come back as the square root of y^2. So $y \geqslant 0$.

To summarize: ***The domain of a function matches the range of its inverse***. The inputs to g^{-1} are the outputs from g. The inputs to g are the outputs from g^{-1}.

If $g(x) = y$ then solving that equation for x gives $x = g^{-1}(y)$:

$$\text{if } y = 3x - 6 \quad \text{then } x = \tfrac{1}{3}(y + 6) \quad (\text{this is } g^{-1}(y))$$
$$\text{if } y = x^3 + 1 \quad \text{then } x = \sqrt[3]{y - 1} \quad (\text{this is } g^{-1}(y))$$

In practice that is how g^{-1} is computed: *Solve $g(x) = y$*. This is the reason inverses are important. Every time we solve an equation we are computing a value of g^{-1}.

Not all equations have one solution. ***Not all functions have inverses***. For each y, the equation $g(x) = y$ is only allowed to produce one x. That solution is $x = g^{-1}(y)$. If there is a second solution, then g^{-1} will not be a function—because a function cannot produce two x's from the same y.

EXAMPLE 4 There is more than one solution to $\sin x = \tfrac{1}{2}$. Many angles have the same sine. On the interval $0 \leqslant x \leqslant \pi$, the inverse of $y = \sin x$ is not a function. Figure 4.5 shows how two x's give the same y.

Prevent x from passing $\pi/2$ and the sine has an inverse. Write $x = \sin^{-1} y$.

The function g has no inverse if two points x_1 and x_2 give $g(x_1) = g(x_2)$. Its inverse would have to bring the same y back to x_1 and x_2. No function can do that; $g^{-1}(y)$ cannot equal both x_1 and x_2. There must be only one x for each y.

To be invertible over an interval, g must be steadily increasing or steadily decreasing.

Fig. 4.5　Inverse exists (one x for each y). No inverse function (two x's for one y).

THE DERIVATIVE OF g^{-1}

It is time for calculus. Forgive me for this very humble example.

EXAMPLE 5　(ordinary multiplication) The inverse of $y = g(x) = 3x$ is $x = f(y) = \tfrac{1}{3}y$.

This shows with special clarity the rule for derivatives: ***The slopes $dy/dx = 3$ and $dx/dy = \tfrac{1}{3}$ multiply to give*** 1. This rule holds for all inverse functions, even if their slopes are not constant. It is a crucial application of the chain rule to the derivative of $f(g(x)) = x$.

4C (***Derivative of inverse function***)   From $f(g(x)) = x$ the chain rule gives $f'(g(x))g'(x) = 1$. Writing $y = g(x)$ and $x = f(y)$, this rule looks better:

$$\frac{dx}{dy}\frac{dy}{dx} = 1 \qquad or \qquad \frac{dx}{dy} = \frac{1}{dy/dx}. \qquad (1)$$

The slope of $x = g^{-1}(y)$ times the slope of $y = g(x)$ equals one.

This is the chain rule with a special feature. Since $f(g(x)) = x$, *the derivative of both sides is* 1. If we know g' we now know f'. That rule will be tested on a familiar example. In the next section it leads to totally new derivatives.

EXAMPLE 6   The inverse of $y = x^3$ is $x = y^{1/3}$. We can find dx/dy two ways:

$$\text{directly}: \frac{dx}{dy} = \frac{1}{3}y^{-2/3} \qquad \text{indirectly}: \frac{dx}{dy} = \frac{1}{dy/dx} = \frac{1}{3x^2} = \frac{1}{3y^{2/3}}.$$

The equation $(dx/dy)(dy/dx) = 1$ is not ordinary algebra, but it is true. Those derivatives are limits of fractions. The fractions are $(\Delta x/\Delta y)(\Delta y/\Delta x) = 1$ and we let $\Delta x \to 0$.

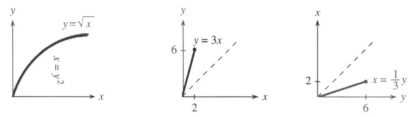

Fig. 4.6   Graphs of inverse functions: $x = \frac{1}{3}y$ is the mirror image of $y = 3x$.

Before going to new functions, I want to draw graphs. Figure 4.6 shows $y = \sqrt{x}$ and $y = 3x$. What is special is that *the same graphs also show the inverse functions*. The inverse of $y = \sqrt{x}$ is $x = y^2$. The pair $x = 4, y = 2$ is the same for both. That is the whole point of inverse functions—if $2 = g(4)$ then $4 = g^{-1}(2)$. Notice that the graphs go steadily up.

The only problem is, the graph of $x = g^{-1}(y)$ is on its side. To change the slope from 3 to $\frac{1}{3}$, you would have to turn the figure. After that turn there is another problem—the axes don't point to the right and up. You also have to look in a mirror! (The typesetter refused to print the letters backward. He thinks it's crazy but it's not.) To keep the book in position, and the typesetter in position, we need a better idea.

The graph of $x = \frac{1}{3}y$ comes from ***turning the picture across the*** 45° ***line***. The y axis becomes horizontal and x goes upward. The point $(2, 6)$ on the line $y = 3x$ goes into the point $(6, 2)$ on the line $x = \frac{1}{3}y$. The eyes see a reflection across the 45° line (Figure 4.6c). The mathematics sees the same pairs x and y. The special properties of g and g^{-1} allow us to know two functions—and draw two graphs—at the same time.† ***The graph of*** $x = g^{-1}(y)$ ***is the mirror image of the graph of*** $y = g(x)$.

---

†I have seen graphs with $y = g(x)$ and also $y = g^{-1}(x)$. For me that is wrong: it has to be $x = g^{-1}(y)$. If $y = \sin x$ then $x = \sin^{-1} y$.

EXPONENTIALS AND LOGARITHMS

I would like to add two more examples of inverse functions, because they are so important. Both examples involve the *exponential* and the *logarithm*. One is made up of linear pieces that imitate 2^x; it appeared in Chapter 1. The other is the true function 2^x, which is not yet defined—and it is not going to be defined here. The functions b^x and $\log_b y$ are so overwhelmingly important that they deserve and will get a whole chapter of the book (at least). But you have to see the graphs.

The slopes in the linear model are powers of 2. So are the heights y at the start of each piece. *The slopes* $1, 2, 4, \ldots$ *equal the heights* $1, 2, 4, \ldots$ *at those special points*.

The inverse is a discrete model for the logarithm (to base 2). The logarithm of 1 is 0, because $2^0 = 1$. The logarithm of 2 is 1, because $2^1 = 2$. The logarithm of 2^j is the exponent j. Thus the model gives the correct $x = \log_2 y$ at the breakpoints $y = 1, 2, 4, 8, \ldots$. The slopes are $1, \frac{1}{2}, \frac{1}{4}, \frac{1}{8}, \ldots$ because $dx/dy = 1/(dy/dx)$.

The model is good, but the real thing is better. The figure on the right shows the true exponential $y = 2^x$. At $x = 0, 1, 2, \ldots$ the heights y are the same as before. But now the height at $x = \frac{1}{2}$ is the number $2^{1/2}$, which is $\sqrt{2}$. The height at $x = .10$ is the tenth root $2^{1/10} = 1.07\ldots$. *The slope at $x = 0$ is no longer 1—it is closer to* $\Delta y/\Delta x = .07/.10$. The exact slope is a number c (near .7) that we are not yet prepared to reveal.

The special property of $y = 2^x$ is that the slope at all points is cy. *The slope is proportional to the function*. The exponential solves $dy/dx = cy$.

Now look at the inverse function—*the logarithm*. Its graph is the mirror image:

If $y = 2^x$ then $x = \log_2 y$. If $2^{1/10} \approx 1.07$ then $\log_2 1.07 \approx 1/10$.

What the exponential does, the logarithm undoes—and vice versa. *The logarithm of 2^x is the exponent x*. Since the exponential starts with slope c, the logarithm must start with slope $1/c$. Check that numerically. The logarithm of 1.07 is near 1/10. The slope is near $.10/.07$. The beautiful property is that $dx/dy = 1/cy$.

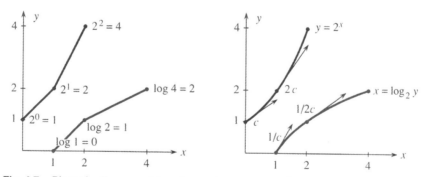

Fig. 4.7 Piecewise linear models and smooth curves: $y = 2^x$ and $x = \log_2 y$. Base $b = 2$.

I have to mention that calculus avoids logarithms to base 2. The reason lies in that mysterious number c. It is the "natural logarithm" of 2, which is $.693147\ldots$ —and who wants that? Also $1/.693147\ldots$ enters the slope of $\log_2 y$. Then $(dx/dy)(dy/dx) = 1$. The right choice is to use "natural logarithms" throughout. In place of 2, they are based on the special number e:

$$y = e^x \text{ is the inverse of } x = \ln y. \tag{2}$$

The derivatives of those functions are sensational—they are saved for Chapter 6. Together with x^n and $\sin x$ and $\cos x$, they are the backbone of calculus.

Note   It is almost possible to go directly to Chapter 6. The inverse functions $x = \sin^{-1} y$ and $x = \tan^{-1} y$ can be done quickly. The reason for including integrals first (Chapter 5) is that they solve differential equations with no guesswork:

$$\frac{dy}{dx} = y \quad \text{or} \quad \frac{dx}{dy} = \frac{1}{y} \quad \text{leads to} \quad \int dx = \int \frac{dy}{y} \quad \text{or} \quad x = \ln y + C.$$

Integrals have applications of all kinds, spread through the rest of the book. But do not lose sight of 2^x and e^x. They solve $dy/dx = cy$—the key to applied calculus.

THE INVERSE OF A CHAIN $h(g(x))$

The functions $g(x) = x - 2$ and $h(y) = 3y$ were easy to invert. For g^{-1} we added 2, and for h^{-1} we divided by 3. Now the question is: If we create the composite function $z = h(g(x))$, or $z = 3(x-2)$, what is its inverse?

Virtually all known functions are created in this way, from chains of simpler functions. *The problem is to invert a chain using the inverse of each piece.* The answer is one of the fundamental rules of mathematics:

4D The inverse of $z = h(g(x))$ is a chain of inverses *in the opposite order*:

$$x = g^{-1}(h^{-1}(z)). \tag{3}$$

h^{-1} is applied first because h was applied last: $g^{-1}(h^{-1}(h(g(x)))) = x$.

That last equation looks like a mess, but it holds the key. In the middle you see h^{-1} and h. That part of the chain does nothing! The inverse functions cancel, to leave $g^{-1}(g(x))$. *But that is x.* The whole chain collapses, when g^{-1} and h^{-1} are in the correct order—which is opposite to the order of $h(g(x))$.

EXAMPLE 7 $z = h(g(x)) = 3(x - 2)$ and $x = g^{-1}(h^{-1}(z)) = \frac{1}{3}z + 2$.

First h^{-1} divides by 3. Then g^{-1} adds 2. The inverse of $h \circ g$ is $g^{-1} \circ h^{-1}$. *It can be found directly by solving $z = 3(x - 2)$.* A chain of inverses is like writing in prose—we do it without knowing it.

EXAMPLE 8   Invert $z = \sqrt{x - 2}$ by writing $z^2 = x - 2$ and then $x = z^2 + 2$.

The inverse adds 2 and takes the square—*but not in that order.* That would give $(z + 2)^2$, which is wrong. The correct order is $z^2 + 2$.

The domains and ranges are explained by Figure 4.8. We start with $x \geqslant 2$. Subtracting 2 gives $y \geqslant 0$. Taking the square root gives $z \geqslant 0$. Taking the square brings back $y \geqslant 0$. Adding 2 brings back $x \geqslant 2$—which is in the original domain of g.

Fig. 4.8   The chain $g^{-1}(h^{-1}(h(g(x)))) = x$ is one-to-one at every step.

EXAMPLE 9 Inverse matrices $(AB)^{-1} = B^{-1}A^{-1}$ (this linear algebra is optional).

Suppose a vector x is multiplied by a square matrix B: $y = g(x) = Bx$. The inverse function multiplies by the **inverse matrix**: $x = g^{-1}(y) = B^{-1}y$. It is like multiplication by $B = 3$ and $B^{-1} = 1/3$, except that x and y are vectors.

Now suppose a second function multiplies by another matrix A: $z = h(g(x)) = ABx$. The problem is to recover x from z. The first step is to invert A, because that came last: $Bx = A^{-1}z$. Then the second step multiplies by B^{-1} and brings back $x = B^{-1}A^{-1}z$. **The product $B^{-1}A^{-1}$ inverts the product AB**. The rule for matrix inverses is like the rule for function inverses—in fact it is a special case.

I had better not wander too far from calculus. The next section introduces the inverses of the sine and cosine and tangent, and finds their derivatives. Remember that the ultimate source is the chain rule.

4.3 EXERCISES

Read-through questions

The functions $g(x) = x - 4$ and $f(y) = y + 4$ are ___a___ functions, because $f(g(x)) = $___b___. Also $g(f(y)) = $___c___. The notation is $f = g^{-1}$ and $g = $___d___. The composition ___e___ is the identity function. By definition $x = g^{-1}(y)$ if and only if $y = $___f___. When y is in the range of g, it is in the ___g___ of g^{-1}. Similarly x is in the ___h___ of g when it is in the ___i___ of g^{-1}. If g has an inverse then $g(x_1)$ ___j___ $g(x_2)$ at any two points. The function g must be steadily ___k___ or steadily ___l___.

The chain rule applied to $f(g(x)) = x$ gives $(df/dy)($___m___$)$ $= $___n___. The slope of g^{-1} times the slope of g equals ___o___. More directly $dx/dy = 1/$___p___. For $y = 2x + 1$ and $x = \frac{1}{2}(y - 1)$, the slopes are $dy/dx = $___q___ and $dx/dy = $___r___. For $y = x^2$ and $x = $___s___, the slopes are $dy/dx = $___t___ and $dx/dy = $___u___. Substituting x^2 for y gives $dx/dy = $___v___. Then $(dx/dy)(dy/dx) = $___w___.

The graph of $y = g(x)$ is also the graph of $x = $___x___, but with x across and y up. For an ordinary graph of g^{-1}, take the reflection in the line ___y___. If $(3,8)$ is on the graph of g, then its mirror image (___z___) is on the graph of g^{-1}. Those particular points satisfy $8 = 2^3$ and $3 = $___A___.

The inverse of the chain $z = h(g(x))$ is the chain $x = $___B___. If $g(x) = 3x$ and $h(y) = y^3$ then $z = $___C___. Its inverse is $x = $___D___, which is the composition of ___E___ and ___F___.

Solve equations 1–10 for x, **to find the inverse function** $x = g^{-1}(y)$. **When more than one** x **gives the same** y, **write "no inverse."**

1 $y = 3x - 6$

2 $y = Ax + B$

3 $y = x^2 - 1$

4 $y = x/(x - 1)$ [solve $xy - y = x$]

5 $y = 1 + x^{-1}$

6 $y = |x|$

7 $y = x^3 - 1$

8 $y = 2x + |x|$

9 $y = \sin x$

10 $y = x^{1/5}$ [draw graph]

11 Solving $y = \dfrac{1}{x - a}$ gives $xy - ay = 1$ or $x = \dfrac{1 + ay}{y}$. Now solve that equation for y.

12 Solving $y = \dfrac{x + 1}{x - 1}$ gives $xy - y = x + 1$ or $x = \dfrac{y + 1}{y - 1}$. Draw the graph to see why f and f^{-1} are the same. Compute dy/dx and dx/dy.

13 Suppose f is increasing and $f(2) = 3$ and $f(3) = 5$. What can you say about $f^{-1}(4)$?

14 Suppose $f(2) = 3$ and $f(3) = 5$ and $f(5) = 5$. What can you say about f^{-1}?

15 Suppose $f(2) = 3$ and $f(3) = 5$ and $f(5) = 0$. How do you know that there is no function f^{-1}?

16 **Vertical line test**: If no vertical line touches its graph twice then $f(x)$ is a **function** (one y for each x). **Horizontal line test**: If no horizontal line touches its graph twice then $f(x)$ is **invertible** because _____.

17 If $f(x)$ and $g(x)$ are increasing, which two of these might not be increasing?

$f(x) + g(x)$ $\quad$ $f(x)g(x)$ $\quad$ $f(g(x))$ $\quad$ $f^{-1}(x)$ $\quad$ $1/f(x)$

18 If $y = 1/x$ then $x = 1/y$. If $y = 1 - x$ then $x = 1 - y$. The graphs are their own mirror images in the 45° line. Construct two more functions with this property $f = f^{-1}$ or $f(f(x)) = x$.

19 For which numbers m are these functions invertible?

(a) $y = mx + b$ $\quad$ (b) $y = mx + x^3$ $\quad$ (c) $y = mx + \sin x$

20 From its graph show that $y = |x| + cx$ is invertible if $c > 1$ and also if $c < -1$. The inverse of a piecewise linear function is piecewise _____.

In 21–26 find dy/dx in terms of x and dx/dy in terms of y.

21 $y = x^5$

22 $y = 1/(x-1)$

23 $y = x^3 - 1$

24 $y = 1/x^3$

25 $y = \dfrac{x}{x-1}$

26 $y = \dfrac{ax+b}{cx+d}$

27 If $dy/dx = 1/y$ then $dx/dy =$ _____ and $x =$ _____ .

28 If $dx/dy = 1/y$ then $dy/dx =$ _____ (these functions are $y = e^x$ and $x = \ln y$, soon to be honored properly).

29 The slopes of $f(x) = \frac{1}{3}x^3$ and $g(x) = -1/x$ are x^2 and $1/x^2$. Why isn't $f = g^{-1}$? What is g^{-1}? Show that $g'(g^{-1})' = 1$.

30 At the points x_1, x_2, x_3 a piecewise constant function jumps to y_1, y_2, y_3. Draw its graph starting from $y(0) = 0$. The mirror image is piecewise constant with jumps at the points _____ to the heights _____. Why isn't this the inverse function?

In 31–38 draw the graph of $y = g(x)$. Separately draw its mirror image $x = g^{-1}(y)$.

31 $y = 5x - 10$

32 $y = \cos x, 0 \leqslant x \leqslant \pi$

33 $y = 1/(x+1)$

34 $y = |x| - 2x$

35 $y = 10^x$

36 $y = \sqrt{1-x^2}, 0 \leqslant x \leqslant 1$

37 $y = 2^{-x}$

38 $y = 1/\sqrt{1-x^2}, 0 \leqslant x < 1$

In 39–42 find dx/dy at the given point.

39 $y = \sin x$ at $x = \pi/6$

40 $y = \tan x$ at $x = \pi/4$

41 $y = \sin x^2$ at $x = 3$

42 $y = x - \sin x$ at $x = 0$

43 If y is a decreasing function of x, then x is a _____ function of y. Prove by graphs and by the chain rule.

44 If $f(x) > x$ for all x, show that $f^{-1}(y) < y$.

45 *True or false*, with example:

(a) If $f(x)$ is invertible so is $h(x) = (f(x))^2$.

(b) If $f(x)$ is invertible so is $h(x) = f(f(x))$.

(c) $f^{-1}(y)$ has a derivative at every y.

In the ehains 46–51 write down $g(x)$ and $f(y)$ and their inverses. Then find $x = g^{-1}(f^{-1}(z))$.

46 $z = 5(x-4)$

47 $z = (x^m)^n$

48 $z = (6+x)^3$

49 $z = 6 + x^3$

50 $z = \frac{1}{2}(\frac{1}{2}x + 4) + 4$

51 $z = \log(10^x)$

52 Solving $f(x) = 0$ is a large part of applied mathematics. Express the solution x^* in terms of f^{-1}: $x^* =$ _____ .

53 (a) Show by example that d^2x/dy^2 is not $1/(d^2y/dx^2)$.

(b) If y is in meters and x is in seconds, then d^2y/dx^2 is in _____ and d^2x/dy^2 is in _____ .

54 Newton's method solves $f(x^*) = 0$ by applying a linear approximation to f^{-1}:

$$f^{-1}(0) \approx f^{-1}(y) + (df^{-1}/dy)(0 - y).$$

For $y = f(x)$ this is Newton's equation $x^* \approx x +$ _____ .

55 If the demand is $1/(p+1)^2$ when the price is p, then the demand is y when the price is _____ . If the range of prices is $p \geqslant 0$, what is the range of demands?

56 If $dF/dx = f(x)$ show that the derivative of $G(y) = yf^{-1}(y) - F(f^{-1}(y))$ is $f^{-1}(y)$.

57 For each number y find the maximum value of $yx - 2x^4$. This maximum is a function $G(y)$. Verify that the derivatives of $G(y)$ and $2x^4$ are inverse functions.

58 (for professors only) If $G(y)$ is the maximum value of $yx - F(x)$, prove that $F(x)$ is the maximum value of $xy - G(y)$. Assume that $f(x) = dF/dx$ is increasing, like $8x^3$ in Problem 57.

59 Suppose the richest x percent of people in the world have $10\sqrt{x}$ percent of the wealth. Then y percent of the wealth is held by _____ percent of the people.

4.4 Inverses of Trigonometric Functions

Mathematics is built on basic functions like the sine, and on basic ideas like the inverse. Therefore *it is totally natural to invert the sine function*. The graph of $x = \sin^{-1} y$ is a mirror image of $y = \sin x$. This is a case where we pay close attention to the domains, since the sine goes up and down infinitely often. We only want *one piece* of that curve, in Figure 4.9.

For the bold line the domain is restricted. *The angle x lies between $-\pi/2$ and $+\pi/2$*. On that interval the sine is increasing, so *each y comes from exactly one angle x*. If the whole sine curve is allowed, infinitely many angles would have $\sin x = 0$. The sine function could not have an inverse. By restricting to an interval where $\sin x$ is increasing, we make the function invertible.

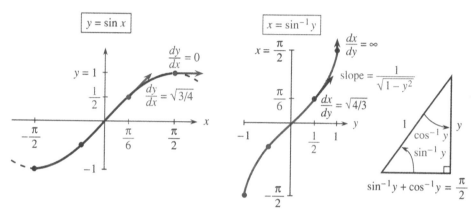

Fig. 4.9 Graphs of $\sin x$ and $\sin^{-1} y$. Their slopes are $\cos x$ and $1/\sqrt{1-y^2}$.

The inverse function brings y back to x. It is $x = \sin^{-1} y$ (the *inverse sine*):

$$x = \sin^{-1} y \text{ when } y = \sin x \text{ and } |x| \leqslant \pi/2. \tag{1}$$

The inverse starts with a number y between -1 and 1. It produces an angle $x = \sin^{-1} y$—*the angle whose sine is y*. The angle x is between $-\pi/2$ and $\pi/2$, with the required sine. Historically x was called the "arc sine" of y, and *arcsin* is used in computing. The mathematical notation is $\sin^{-1}$. *This has nothing to do with $1/\sin x$.*

The figure shows the 30° angle $x = \pi/6$. Its sine is $y = \frac{1}{2}$. *The inverse sine of $\frac{1}{2}$ is $\pi/6$*. Again: The symbol $\sin^{-1}(1)$ stands for the angle whose sine is 1 (this angle is $x = \pi/2$). We are seeing $g^{-1}(g(x)) = x$:

$$\sin^{-1}(\sin x) = x \text{ for } -\frac{\pi}{2} \leqslant x \leqslant \frac{\pi}{2} \qquad \sin(\sin^{-1} y) = y \text{ for } -1 \leqslant y \leqslant 1.$$

EXAMPLE 1   (important) If $\sin x = y$ find a formula for $\cos x$.

Solution   We are given the sine, we want the cosine. The key to this problem must be $\cos^2 x = 1 - \sin^2 x$. When the sine is y, the cosine is the square root of $1 - y^2$:

$$\cos x = \cos(\sin^{-1} y) = \sqrt{1 - y^2}. \tag{2}$$

This formula is crucial for computing derivatives. We use it immediately.

THE DERIVATIVE OF THE INVERSE SINE

The calculus problem is to find the slope of the inverse function $f(y) = \sin^{-1} y$. The chain rule gives (*slope of inverse function*) $= 1/$(*slope of original function*). Certainly the slope of $\sin x$ is $\cos x$. To switch from x to y, use equation (2):

$$y = \sin x \text{ gives } \frac{dy}{dx} = \cos x \text{ so that } \frac{dx}{dy} = \frac{1}{\cos x} = \frac{1}{\sqrt{1 - y^2}}. \qquad (3)$$

This derivative $1/\sqrt{1 - y^2}$ gives a new v–f pair that is extremely valuable in calculus:

$$\text{velocity} \quad v(t) = 1/\sqrt{1 - t^2} \qquad \text{distance} \quad f(t) = \sin^{-1} t.$$

Inverse functions will soon produce two more pairs, from the derivatives of $\tan^{-1} y$ and $\sec^{-1} y$. *The table at the end lists all the essential facts*.

EXAMPLE 2   The slope of $\sin^{-1} y$ at $y = 1$ is *infinite*: $1/\sqrt{1 - y^2} = 1/0$. Explain.

At $y = 1$ the graph of $y = \sin x$ is horizontal. The slope is zero. So its mirror image is vertical. The slope $1/0$ is an extreme case of the chain rule.

Question What is $d/dx(\sin^{-1} x)$?   Answer $1/\sqrt{1 - x^2}$. I just changed letters.

THE INVERSE COSINE AND ITS DERIVATIVE

Whatever is done for the sine can be done for the cosine. But the domain and range have to be watched. The graph cannot be allowed to go up and down. Each y from -1 to 1 should be the cosine of *only one angle x*. That puts x between 0 and π. Then the cosine is steadily decreasing and $y = \cos x$ has an inverse:

$$\cos^{-1}(\cos x) = x \text{ and } \cos(\cos^{-1} y) = y. \qquad (4)$$

The cosine of the angle $x = 0$ is the number $y = 1$. The inverse cosine of $y = 1$ is the angle $x = 0$. Those both express the same fact, that $\cos 0 = 1$.

For the slope of $\cos^{-1} y$, we could copy the calculation that succeeded for $\sin^{-1} y$. The chain rule could be applied as in (3). But there is a faster way, because of a special relation between $\cos^{-1} y$ and $\sin^{-1} y$. *Those angles always add to a right angle*:

$$\cos^{-1} y + \sin^{-1} y = \pi/2. \qquad (5)$$

Figure 4.9c shows the angles and Figure 4.10c shows the graphs. The sum is $\pi/2$ (the dotted line), and its derivative is zero. So the derivatives of $\cos^{-1} y$ and $\sin^{-1} y$ must add to zero. *Those derivatives have opposite sign*. There is a **minus** for the inverse cosine, and its graph goes downward:

$$\textit{The derivative of } x = \cos^{-1} y \textit{ is } dx/dy = -1/\sqrt{1 - y^2}. \qquad (6)$$

Question How can two functions $x = \sin^{-1} y$ and $x = -\cos^{-1} y$ have the *same derivative*?

Answer $\sin^{-1} y$ must be the same as $-\cos^{-1} y + C$. Equation (5) gives $C = \pi/2$.

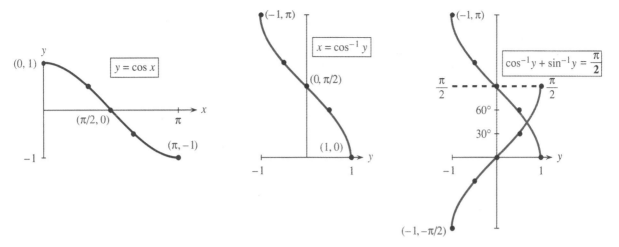

Fig. 4.10 The graphs of $y = \cos x$ and $x = \cos^{-1} y$. Notice the domain $0 \leqslant x \leqslant \pi$.

THE INVERSE TANGENT AND ITS DERIVATIVE

The tangent is $\sin x / \cos x$. The inverse tangent is *not* $\sin^{-1} y / \cos^{-1} y$. The inverse function produces ***the angle whose tangent is*** y. Figure 4.11 shows that angle, which is between $-\pi/2$ and $\pi/2$. The tangent can be any number, but the inverse tangent is in the *open interval* $-\pi/2 < x < \pi/2$. (The interval is "open" because its endpoints are not included.) The tangents of $\pi/2$ and $-\pi/2$ are not defined.

The slope of $y = \tan x$ is $dy/dx = \sec^2 x$. What is the slope of $x = \tan^{-1} y$?

$$\text{By the chain rule } \frac{dx}{dy} = \frac{1}{\sec^2 x} = \frac{1}{1 + \tan^2 x} = \frac{1}{1 + y^2}. \qquad (7)$$

4E *The derivative of* $f(y) = \tan^{-1} y$ *is* $\dfrac{df}{dy} = \dfrac{1}{1 + y^2}.$ $\qquad (8)$

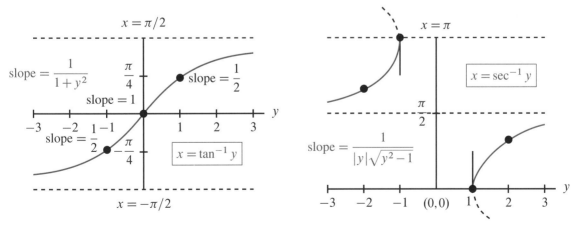

Fig. 4.11 $x = \tan^{-1} y$ has slope $1/(1 + y^2)$. $x = \sec^{-1} y$ has slope $1/|y|\sqrt{y^2 - 1}$.

EXAMPLE 3 The tangent of $x = \pi/4$ is $y = 1$. We check slopes. On the inverse tangent curve, $dx/dy = 1/(1 + y^2) = \frac{1}{2}$. On the tangent curve, $dy/dx = \sec^2 x$. At

$\pi/4$ the secant squared equals 2. The slopes $dx/dy = \frac{1}{2}$ and $dy/dx = 2$ multiply to give 1.

Important   Soon will come the following question. *What function has the derivative* $1/(1+x^2)$? One reason for reading this section is to learn the answer. The function is in equation (8)—*if we change letters. It is* $f(x) = \tan^{-1}x$ *that has slope* $1/(1+x^2)$.

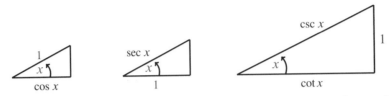

Fig. 4.12 $\cos^2 x + \sin^2 x = 1$ and $1 + \tan^2 x = \sec^2 x$ and $1 + \cot^2 x = \csc^2 x$.

INVERSE COTANGENT, INVERSE SECANT, INVERSE COSECANT

There is no way we can avoid completing this miserable list! But it can be painless. The idea is to use $1/(dy/dx)$ for $y = \cot x$ and $y = \sec x$ and $y = \csc x$:

$$\frac{dx}{dy} = \frac{-1}{\csc^2 x} \quad \text{and} \quad \frac{dx}{dy} = \frac{1}{\sec x \tan x} \quad \text{and} \quad \frac{dx}{dy} = \frac{-1}{\csc x \cot x}. \qquad (9)$$

In the middle equation, replace $\sec x$ by y and $\tan x$ by $\pm\sqrt{y^2 - 1}$. Choose the sign for positive slope (compare Figure 4.11). That gives the middle equation in (10):

The derivatives of $\cot^{-1}y$ *and* $\sec^{-1}y$ *and* $\csc^{-1}y$ *are*

$$\frac{d}{dy}(\cot^{-1}y) = \frac{-1}{1+y^2} \quad \frac{d}{dy}(\sec^{-1}y) = \frac{1}{|y|\sqrt{y^2-1}} \quad \frac{d}{dy}(\csc^{-1}y) = \frac{-1}{|y|\sqrt{y^2-1}}. \qquad (10)$$

Note about the inverse secant   When y is negative there is a choice for $x = \sec^{-1}y$. We selected the angle in the second quadrant (between $\pi/2$ and π). Its cosine is negative, so its secant is negative. This choice makes $\sec^{-1}y = \cos^{-1}(1/y)$, which matches $\sec x = 1/\cos x$. It also makes $\sec^{-1}y$ an increasing function, where $\cos^{-1}y$ is a decreasing function. So we needed the absolute value $|y|$ in the derivative.

Some mathematical tables make a different choice. The angle x could be in the third quadrant (between $-\pi$ and $-\pi/2$). Then the slope omits the absolute value.

Summary   For the six inverse functions it is only necessary to learn three derivatives. The other three just have minus signs, as we saw for $\sin^{-1}y$ and $\cos^{-1}y$. Each inverse function and its "cofunction" add to $\pi/2$, so their derivatives add to zero. Here are the six functions for quick reference, with the three new derivatives.

function $f(y)$	*inputs* y	*outputs* x	*slope* dx/dy
$\sin^{-1}y, \cos^{-1}y$	$|y| \leqslant 1$	$\left[-\dfrac{\pi}{2}, \dfrac{\pi}{2}\right], [0, \pi]$	$\pm\dfrac{1}{\sqrt{1-y^2}}$
$\tan^{-1}y, \cot^{-1}y$	all y	$\left(-\dfrac{\pi}{2}, \dfrac{\pi}{2}\right), (0, \pi)$	$\pm\dfrac{1}{1+y^2}$
$\sec^{-1}y, \csc^{-1}y$	$|y| \geqslant 1$	$[0, \pi]^*, \left[-\dfrac{\pi}{2}, \dfrac{\pi}{2}\right]^*$	$\pm\dfrac{1}{|y|\sqrt{y^2-1}}$

If $y = \cos x$ or $y = \sin x$ then $|y| \leqslant 1$. For $y = \sec x$ and $y = \csc x$ the opposite is true; we must have $|y| \geqslant 1$. The graph of $\sec^{-1} y$ misses all the points $-1 < y < 1$.

Also, that graph misses $x = \pi/2$—where the cosine is zero. The secant of $\pi/2$ would be $1/0$ (impossible). Similarly $\csc^{-1} y$ misses $x = 0$, because $y = \csc 0$ cannot be $1/\sin 0$. The asterisks in the table are to remove those points $x = \pi/2$ and $x = 0$.

The column of derivatives is what we need and use in calculus.

4.4 EXERCISES

Read-through questions

The relation $x - \sin^{-1} y$ means that __a__ is the sine of __b__. Thus x is the angle whose sine is __c__. The number y lies between __d__ and __e__. The angle x lies between __f__ and __g__. (If we want the inverse to exist, there cannot be two angles with the same sine.) The cosine of the angle $\sin^{-1} y$ is $\sqrt{\underline{\quad h \quad}}$. The derivative of $x = \sin^{-1} y$ is $dx/dy = \underline{\quad i \quad}$

The relation $x = \cos^{-1} y$ means that y equals __j__. Again the number y lies between __k__ and __l__. This time the angle x lies between __m__ and __n__ (so that each y comes from only one angle x). The sum $\sin^{-1} y + \cos^{-1} y = \underline{\quad o \quad}$. (The angles are called __p__, and they add to a __q__ angle.) Therefore the derivative of $x = \cos^{-1} y$ is $dx/dy = \underline{\quad r \quad}$, the same as for $\sin^{-1} y$ except for a __s__ sign.

The relation $x = \tan^{-1} y$ means that $y = \underline{\quad t \quad}$. The number y lies between __u__ and __v__. The angle x lies between __w__ and __x__. The derivative is $dx/dy = \underline{\quad y \quad}$. Since $\tan^{-1} y + \cot^{-1} y = \underline{\quad z \quad}$, the derivative of $\cot^{-1} y$ is the same except for a __A__ sign.

The relation $x = \sec^{-1} y$ means that __B__. The number y *never* lies between __C__ and __D__. The angle x lies between __E__ and __F__, but never at $x = \underline{\quad G \quad}$. The derivative of $x = \sec^{-1} y$ is $dx/dy = \underline{\quad H \quad}$.

In 1–4, find the angles $\sin^{-1} y$ **and** $\cos^{-1} y$ **and** $\tan^{-1} y$ **in radians.**

1 $y = 0$ 2 $y = -1$ 3 $y = 1$ 4 $y = \sqrt{3}$

5 We know that $\sin \pi = 0$. Why isn't $\pi = \sin^{-1} 0$?

6 Suppose $\sin x = y$. Under what restriction is $x = \sin^{-1} y$?

7 Sketch the graph of $x = \sin^{-1} y$ and locate the points with slope $dx/dy = 2$.

8 Find dx/dy if $x = \sin^{-1} \frac{1}{2} y$. Draw the graph.

9 If $y = \cos x$ find a formula for $\sin x$. First draw a right triangle with angle x and near side y—what are the other two sides?

10 If $y = \sin x$ find a formula for $\tan x$. First draw a right triangle with angle x and far side y—what are the other sides?

11 Take the x derivative of $\sin^{-1}(\sin x) = x$ by the chain rule. Check that $d(\sin^{-1} y)/dy = -1/\sqrt{1 - y^2}$ gives a correct result.

12 Take the y derivative of $\cos(\cos^{-1} y) = y$ by the chain rule. Check that $d(\cos^{-1} y)/dy = -1/\sqrt{1 - y^2}$ gives a correct result.

13 At $y = 0$ and $y = 1$, find the slope dx/dy of $x = \sin^{-1} y$ and $x = \cos^{-1} y$ and $x = \tan^{-1} y$.

14 At $x = 0$ and $x = 1$, find the slope dx/dy of $x = \sin^{-1} y$ and $x = \cos^{-1} y$ and $x = \tan^{-1} y$.

15 **True or false**, with reason:
   (a) $(\sin^{-1} y)^2 + (\cos^{-1} y)^2 = 1$
   (b) $\sin^{-1} y = \cos^{-1} y$ has no solution
   (c) $\sin^{-1} y$ is an increasing function
   (d) $\sin^{-1} y$ is an odd function
   (e) $\sin^{-1} y$ and $-\cos^{-1} y$ have the same slope—so they are the same.
   (f) $\sin(\cos x) = \cos(\sin x)$

16 Find $\tan(\cos^{-1}(\sin x))$ by drawing a triangle with sides $\sin x, \cos x, 1$.

Compute the derivatives in 17–28 (using the letters as given).

17 $u = \sin^{-1} x$

18 $u = \tan^{-1} 2x$

19 $z = \sin^{-1}(\sin 3x)$

20 $z = \sin^{-1}(\cos x)$

21 $z = (\sin^{-1} x)^2$

22 $z = (\sin^{-1} x)^{-1}$

23 $z = \sqrt{1 - y^2} \sin^{-1} y$

24 $z = (1 + x^2) \tan^{-1} x$

25 $x = \sec^{-1}(y + 1)$

26 $u = \sec^{-1}(\sec x^2)$

27 $u = \sin^{-1} y / \cos^{-1} \sqrt{1 - y^2}$

28 $u = \sin^{-1} y + \cos^{-1} y + \tan^{-1} y$

29 Draw a right triangle to show why $\tan^{-1} y + \cot^{-1} y = \pi/2$.

30 Draw a right triangle to show why $\tan^{-1} y = \cot^{-1}(1/y)$.

31 If $y = \tan x$ find $\sec x$ in terms of y.

32 Draw the graphs of $y = \cot x$ and $x = \cot^{-1} y$.

33 Find the slope dx/dy of $x = \tan^{-1} y$ at
   (a) $y = -3$ (b) $x = 0$ (c) $x = -\pi/4$

34 Find a function $u(t)$ whose slope satisfies $u' + t^2 u' = 1$.

35 What is the second derivative d^2x/dy^2 of $x = \sin^{-1} y$?

36 What is d^2u/dy^2 for $u = \tan^{-1} y$?

Find the derivatives in 37–44.

37 $y = \sec \frac{1}{2} x$

38 $x = \sec^{-1} 2y$

39 $u = \sec^{-1}(x^n)$

40 $u = \sec^{-1}(\tan x)$

41 $\tan y = (x-1)/(x+1)$

42 $z = (\sin x)(\sin^{-1} x)$

43 $y = \sec^{-1}\sqrt{x^2 + 1}$

44 $z = \sin(\cos^{-1} x) - \cos(\sin^{-1} x)$

45 Differentiate $\cos^{-1}(1/y)$ to find the slope of $\sec^{-1} y$ in a new way.

46 The domain and range of $x = \csc^{-1} y$ are _____.

47 Find a function $u(y)$ such that $du/dy = 4/\sqrt{1-y^2}$.

48 Solve the differential equation $du/dx = 1/(1+4x^2)$.

49 If $du/dx = 2/\sqrt{1-x^2}$ find $u(1) - u(0)$.

50 (recommended) With $u(x) = (x-1)/(x+1)$, find the derivative of $\tan^{-1} u(x)$. This is also the derivative of _____. So the difference between the two functions is a _____.

51 Find $u(x)$ and $\tan^{-1} u(x)$ and $\tan^{-1} x$ at $x = 0$ and $x = \infty$. Conclusion based on Problem 50: $\tan^{-1} u(x) - \tan^{-1} x$ equals the number

52 Find $u(x)$ and $\tan^{-1} u(x)$ and $\tan^{-1} x$ as $x \to -\infty$. Now $\tan^{-1} u(x) - \tan^{-1} x$ equals _____. Something has happened to $\tan^{-1} u(x)$. At what x do $u(x)$ and $\tan^{-1} u(x)$ change instantly?

CHAPTER 5

Integrals

This chapter is about the *idea* of integration, and also about the *technique* of integration. We explain how it is done *in principle*, and then how it is done *in practice*. Integration is a problem of adding up infinitely many things, each of which is infinitesimally small. Doing the addition is not recommended. The whole point of calculus is to offer a better way.

The problem of integration is to find a limit of sums. The key is to work backward from a limit of differences (which is the derivative). **We can integrate $v(x)$ *if it turns up as the derivative of another function* $f(x)$.** The integral of $v = \cos x$ is $f = \sin x$. The integral of $v = x$ is $f = \frac{1}{2}x^2$. Basically, $f(x)$ is an "*antiderivative*". The list of f's will grow much longer (Section 5.4 is crucial). A selection is inside the cover of this book. If we don't find a suitable $f(x)$, numerical integration can still give an excellent answer.

I could go directly to the formulas for integrals, which allow you to compute areas under the most amazing curves. (Area is the clearest example of adding up infinitely many infinitely thin rectangles, so it always comes first. It is certainly not the only problem that integral calculus can solve.) But I am really unwilling just to write down formulas, and skip over all the ideas. Newton and Leibniz had an absolutely brilliant intuition, and there is no reason why we can't share it.

They started with something simple. We will do the same.

SUMS AND DIFFERENCES

Integrals and derivatives can be mostly explained by working (very briefly) with sums and differences. Instead of functions, we have n ordinary numbers. The key idea is nothing more than a basic fact of algebra. In the limit as $n \to \infty$, it becomes the basic fact of calculus. The step of "going to the limit" is the essential difference between algebra and calculus! It has to be taken, in order to add up infinitely many infinitesimals—but we start out this side of it.

To see what happens before the limiting step, we need *two sets of n numbers*. The first set will be $v_1, v_2, \ldots, v_n$, where v suggests velocity. The second set of numbers will be $f_1, f_2, \ldots, f_n$, where f recalls the idea of distance. You might think d would be a better symbol for distance, but that is needed for the dx and dy of calculus.

229

A first example has $n = 4$:

$$v_1, v_2, v_3, v_4 = 1, 2, 3, 4 \qquad f_1, f_2, f_3, f_4 = 1, 3, 6, 10.$$

The relation between the v's and f's is seen in that example. When you are given $1, 3, 6, 10$, how do you produce $1, 2, 3, 4$? **By taking differences**. The difference between 10 and 6 is 4. Subtracting $6 - 3$ is 3. The difference $f_2 - f_1 = 3 - 1$ is $v_2 = 2$. Each v is the difference between two f's:

$$v_j \text{ is the difference } f_j - f_{j-1}.$$

This is the discrete form of the derivative. I admit to a small difficulty at $j = 1$, from the fact that there is no f_0. The first v should be $f_1 - f_0$, and the natural idea is to agree that f_0 is zero. This need for a starting point will come back to haunt us (or help us) in calculus.

Now look again at those same numbers—but start with v. From $v = 1, 2, 3, 4$ how do you produce $f = 1, 3, 6, 10$? **By taking sums**. The first two v's add to 3, which is f_2. The first three v's add to $f_3 = 6$. The sum of all four v's is $1 + 2 + 3 + 4 = 10$. **Taking sums is the opposite of taking differences**.

That idea from algebra is the key to calculus. The sum f_j involves all the numbers $v_1 + v_2 + \cdots + v_j$. The difference v_j involves only the *two* numbers $f_j - f_{j-1}$. The fact that one reverses the other is the "Fundamental Theorem." Calculus will change sums to integrals and differences to derivatives—but why not let the key idea come through now?

5A *Fundamental Theorem of Calculus* (before limits):

$$\textbf{\textit{If each }} v_j = f_j - f_{j-1}, \textbf{\textit{then }} v_1 + v_2 + \cdots + v_n = f_n - f_0.$$

The differences of the f's add up to $f_n - f_0$. All f's in between are canceled, leaving only the last f_n and the starting f_0. *The sum "telescopes"*:

$$v_1 + v_2 + v_3 + \cdots + v_n = (f_1 - f_0) + (f_2 - f_1) + (f_3 - f_2) + \cdots + (f_n - f_{n-1}).$$

The number f_1 is canceled by $-f_1$. Similarly $-f_2$ cancels f_2 and $-f_3$ cancels f_3. Eventually f_n and $-f_0$ are left. When f_0 is zero, the sum is the final f_n.

That completes the algebra. **We add the v's by finding the f's**.

Question How do you add the odd numbers $1 + 3 + 5 + \cdots + 99$ (the v's)?
Answer They are the differences between $0, 1, 4, 9, \ldots$. These f's are squares. By the Fundamental Theorem, the sum of 50 odd numbers is $(50)^2$.

The tricky part is to discover the right f's! Their differences must produce the v's. In calculus, the tricky part is to find the right $f(x)$. Its derivative must produce $v(x)$. It is remarkable how often f can be found—more often for integrals than for sums. Our next step is to understand how **the integral is a limit of sums**.

SUMS APPROACH INTEGRALS

Suppose you start a successful company. The rate of income is increasing. After x years, the income per year is $\sqrt{x}$ million dollars. In the first four years you reach $\sqrt{1}, \sqrt{2}, \sqrt{3}$, and $\sqrt{4}$ million dollars. Those numbers are displayed in a bar graph (Figure 5.1a, for investors). I realize that most start-up companies make losses, but your company is an exception. If the example is too good to be true, please keep reading.

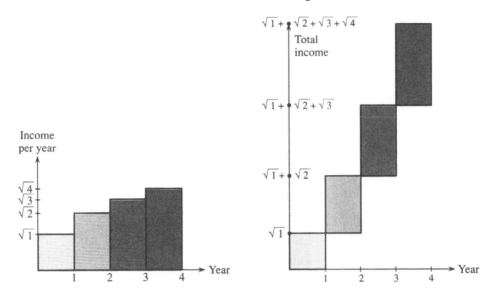

Fig. 5.1   Total income = total area of rectangles = 6.15.

The graph shows four rectangles, of heights $\sqrt{1}, \sqrt{2}, \sqrt{3}, \sqrt{4}$. Since the base of each rectangle is one year, those numbers are also the **areas** of the rectangles. One investor, possibly weak in arithmetic, asks a simple question: *What is the total income for all four years?* There are two ways to answer, and I will give both.

The first answer is $\sqrt{1} + \sqrt{2} + \sqrt{3} + \sqrt{4}$. Addition gives 6.15 million dollars. Figure 5.1b shows this total—which is reached at year 4. This is exactly like velocities and distances, but now v is the **income per year** and f is the **total income**. Algebraically, f_j is still $v_1 + \cdots + v_j$.

The second answer comes from geometry. *The **total income** is the **total area** of the rectangles*. We are emphasizing the correspondence between **addition** and **area**. That point may seem obvious, but it becomes important when a second investor (smarter than the first) asks a harder question.

Here is the problem. ***The incomes as stated are false***. The company did not make a million dollars the first year. After three months, when x was $1/4$, the rate of income was only $\sqrt{x} = 1/2$. The bar graph showed $\sqrt{1} = 1$ for the whole year, but that was an overstatement. The income in three months was not more than $1/2$ times $1/4$, the rate multiplied by the time.

All other quarters and years were also overstated. Figure 5.2a is closer to reality, with 4 years divided into 16 quarters. It gives a new estimate for total income.

Again there are two ways to find the total. We add $\sqrt{1/4} + \sqrt{2/4} + \cdots + \sqrt{16/4}$, remembering to multiply them all by $1/4$ (because each rate applies to $1/4$ year). This is also the area of the 16 rectangles. The area approach is better because the $1/4$ is automatic. Each rectangle has base $1/4$, so that factor enters each area. The total area is now 5.56 million dollars, closer to the truth.

You see what is coming. The next step divides time into weeks. After one week the rate $\sqrt{x}$ only $\sqrt{1/52}$. That is the height of the first rectangle—its base is $\Delta x = 1/52$. There is a rectangle for every week. Then a hard-working investor divides time into days, and the base of each rectangle is $\Delta x = 1/365$. At that point there are $4 \times 365 = 1460$ rectangles, or 1461 because of leap year, with a total area below $5\frac{1}{2}$

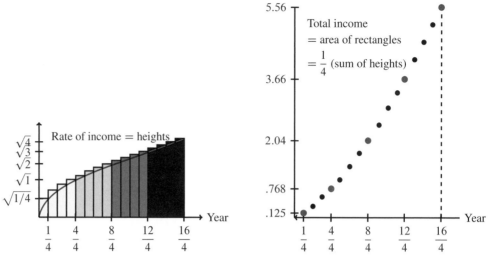

Fig. 5.2 Income = sum of areas (not heights)

$$= \frac{1}{4}\left(\sqrt{\frac{1}{4}}+\sqrt{\frac{2}{4}}+\cdots+\sqrt{\frac{16}{4}}\right)$$

million dollars. The calculation is elementary but depressing—adding up thousands of square roots, each multiplied by Δx from the base. There has to be a better way.

The better way, in fact the best way, is calculus. The whole idea is to allow for ***continuous change***. ***The geometry problem is to find the area under the square root curve***. That question cannot be answered by arithmetic, because it involves a *limit*. The rectangles have base Δx and heights $\sqrt{\Delta x}, \sqrt{\Delta 2x}, \ldots, \sqrt{4}$. There are $4/\Delta x$ rectangles—more and more terms from thinner and thinner rectangles. ***The area is the limit of the sum as $\Delta x \to 0$***.

This limiting area is the "integral." We are looking for a number below $5\frac{1}{2}$.

Algebra (*area of n rectangles*): Compute $v_1 + \cdots + v_n$ by finding f's.
Key idea: If $v_j = f_j - f_{j-1}$, then the sum is $f_n - f_0$.
Calculus (*area under curve*): Compute the limit of $\Delta x[v(\Delta x) + v(2\Delta x) + \cdots]$.
Key idea: ***If $v(x) = df/dx$ then area = integral to be explained next***.

5.1 EXERCISES

Read-through questions

The problem of summation is to add $v_1 + \cdots + v_n$. It is solved if we find f's such that $v_j =$ __a__ . Then $v_1 + \cdots + v_n$ equals __b__ . The cancellation in $(f_1 - f_0) + (f_2 - f_1) + \cdots + (f_n - f_{n-1})$ leaves only __c__ . Taking sums is the __d__ of taking differences.

The differences between $0, 1, 4, 9$ are $v_1, v_2, v_3 =$ __e__ . For $f_j = j^2$ the difference between f_{10}, and f_9 is $v_{10} =$ __f__ . From this pattern $1 + 3 + 5 + \cdots + 19$ equals __g__ .

For functions, finding the integral is the reverse of __h__ . If the derivative of $f(x)$ is $v(x)$, then the __i__ of $v(x)$ is $f(x)$. If $v(x) = 10x$ then $f(x) =$ __j__ . This is the __k__ of a triangle with base x and height $10x$.

Integrals begin with sums. The triangle under $v = 10x$ out to $x = 4$ has area __l__ . It is approximated by four rectangles of heights $10, 20, 30, 40$ and area __m__ . It is better approximated by eight rectangles of heights __n__ and area __o__ . For n rectangles covering the triangle the area is the sum of __p__ . As $n \to \infty$

this sum should approach the number __q__ . *That is the integral of $v = 10x$ from 0 to 4.*

Problems 1–6 are about sums f_j and differences v_j.

1 With $v = 1, 2, 4, 8$, the formula for v_j _____ is (not 2^j). Find f_1, f_2, f_3, f_4, starting from $f_0 = 0$. What is f_7?

2 The same $v = 1, 2, 4, 8, \ldots$ are the differences between $f = 1, 2, 4, 8, 16, \ldots$. Now $f_0 = 1$ and $f_j = 2^j$. (a) Check that $2^5 - 2^4$ equal v_5. (b) What is $1 + 2 + 4 + 8 + 16$?

3 The differences between $f = 1, 1/2, 1/4, 1/8$ are $v = -1/2, -1/4, -1/8$. These negative v's do not add up to these positive f's. Verify that $v_1 + v_2 + v_3 + v_4 = f_4 - f_0$ is still true.

4 Any constant C can be added to the antiderivative $f(x)$ because the _____ of a constant is zero. Any C can be added to $f_0, f_1, \ldots$ because the _____ between the f's is not changed.

5 Show that $f_j = r^j / (r - 1)$ has $f_j - f_{j-1} = r^{j-1}$. Therefore the geometric series $1 + r + \cdots + r^{j-1}$ adds up to _____ (remember to subtract f_0).

6 The sums $f_j = (r^j - 1)/(r - 1)$ also have $f_j - f_{j-1} = r^{j-1}$. Now $f_0 = $ _____ . Therefore $1 + r + \cdots + r^{j-1}$ adds up to f_j. The sum $1 + r + \cdots + r^n$ equals _____ .

7 Suppose $v(x) = 3$ for $x < 1$ and $v(x) = 7$ for $x > 1$. Find the area $f(x)$ from 0 to x, under the graph of $v(x)$. (Two pieces.)

8 If $v = 1, -2, 3, -4, \ldots$, write down the f's starting from $f_0 = 0$. Find formulas for v_j and f_j when j is odd and j is even.

Problems 9–16 are about the company earning $\sqrt{x}$ per year.

9 When time is divided into weeks there are $4 \times 52 = 208$ rectangles. Write down the first area, the 208th area, and the jth area.

10 How do you know that the sum over 208 weeks is smaller than the sum over 16 quarters?

11 A pessimist would use $\sqrt{x}$ at the *beginning* of each time period as the income rate for that period. Redraw Figure 5.1 (both parts) using heights $\sqrt{0}, \sqrt{1}, \sqrt{2}, \sqrt{3}$. How much lower is the estimate of total income?

12 The same pessimist would redraw Figure 5.2 with heights $0, \sqrt{1/4}, \ldots$. What is the height of the last rectangle? How much does this change reduce the total rectangular area 5.56?

13 At every step from years to weeks to days to hours, the pessimist's area goes _____ and the optimist's area goes _____ . The difference between them is the area of the last _____ .

14 The optimist and pessimist arrive at the same limit as years are divided into weeks, days, hours, seconds. Draw the $\sqrt{x}$ curve between the rectangles to show why the pessimist is always too low and the optimist is too high.

15 (Important) Let $f(x)$ be the area under the $\sqrt{x}$ curve, above the interval from 0 to x. The area to $x + \Delta x$ is $f(x + \Delta x)$. The extra area is $\Delta f = $ _____ . This is almost a rectangle with base _____ and height $\sqrt{x}$. So $\Delta f / \Delta x$ is close to _____ . As $\Delta x \to 0$ we suspect that $df/dx = $ _____ .

16 Draw the $\sqrt{x}$ curve from $x = 0$ to 4 and put triangles below to prove that the area under it is more than 5. Look left and right from the point where $\sqrt{1} = 1$.

Problems 17–22 are about a company whose expense rate $v(x) = 6 - x$ is decreasing.

17 The expenses drop to zero at $x = $ _____ . The total expense during those years equals _____ . This is the area of _____ .

18 The rectangles of heights $6, 5, 4, 3, 2, 1$ give a total estimated expense of _____ . Draw them enclosing the triangle to show why this total is too high.

19 How many rectangles (enclosing the triangle) would you need before their areas are within 1 of the correct triangular area?

20 The accountant uses 2-year intervals and computes $v = 5, 3, 1$ at the midpoints (the odd-numbered years). What is her estimate, how accurate is it, and why?

21 What is the area $f(x)$ under the line $v(x) = 6 - x$ above the interval from 2 to x? What is the derivative of this $f(x)$?

22 What is the area $f(x)$ under the line $v(x) = 6 - x$ above the interval from x to 6? What is the derivative of this $f(x)$?

23 With $\Delta x = 1/3$, find the area of the three rectangles that enclose the graph of $v(x) = x^2$.

24 Draw graphs of $v = \sqrt{x}$ and $v = x^2$ from 0 to 1. Which areas add to 1? The same is true for $v = x^3$ and $v = $ _____ .

25 From x to $x + \Delta x$, the area under $v = x^2$ is Δf. This is almost a rectangle with base Δx and height _____ . So $\Delta f / \Delta x$ is close to _____ . In the limit we find $df/dx = x^2$ and $f(x) = $ _____ .

26 Compute the area of 208 rectangles under $v(x) = \sqrt{x}$ from $x = 0$ to $x = 4$.

5.2 Antiderivatives

The symbol $\int$ was invented by Leibniz to represent the integral. It is a stretched-out **S**, from the Latin word for sum. This symbol is a powerful reminder of the whole construction: *Sum approaches integral*, **S** *approaches* $\int$, *and rectangular area approaches curved area*:

$$curved\ area = \int v(x)\,dx = \int \sqrt{x}\,dx. \tag{1}$$

The rectangles of base Δx lead to this limit—the integral of $\sqrt{x}$. The "dx" indicates that Δx approaches zero. The heights v_j of the rectangles are the heights $v(x)$ of the curve. The sum of v_j times Δx approaches "*the integral of v of x dx.*" You can imagine an infinitely thin rectangle above every point, instead of ordinary rectangles above special points.

We now find the area under the square root curve. The "*limits of integration*" are 0 and 4. The lower limit is $x = 0$, where the area begins. (*The start could be any point* $x = a$.) The upper limit is $x = 4$, since we stop after four years. (*The finish could be any point* $x = b$.) The area of the rectangles is a sum of base Δx times heights $\sqrt{x}$. The curved area is the limit of this sum. *That limit is the integral of* $\sqrt{x}$ *from* 0 *to* 4:

$$\lim_{\Delta x \to 0}\left[(\sqrt{\Delta x})(\Delta x) + (\sqrt{2\Delta x})(\Delta x) + \cdots + (\sqrt{4})(\Delta x)\right] = \int_{x=0}^{x=4}\sqrt{x}\,dx. \tag{2}$$

The outstanding problem of integral calculus is still to be solved. *What is this limiting area?* We have a symbol for the answer, involving $\int$ and $\sqrt{x}$ and dx—but we don't have a number.

THE ANTIDERIVATIVE

I wish I knew who discovered the area under the graph of $\sqrt{x}$. It may have been Newton. The answer was available earlier, but the key idea was shared by Newton and Leibniz. They understood the parallels between sums and integrals, and between differences and derivatives. I can give the answer, by following that analogy. I can't give the proof (yet)—it is the Fundamental Theorem of Calculus.

In algebra the difference $f_j - f_{j-1}$ is v_j. When we add, the sum of the v's is $f_n - f_0$. *In calculus the derivative of* $f(x)$ *is* $v(x)$. When we integrate, *the area under the* $v(x)$ *curve is* $f(x)$ *minus* $f(0)$. Our problem asks for the area out to $x = 4$:

> **5B** (Discrete vs. continuous, rectangles vs. curved areas, addition vs. integration)
> *The integral of* $v(x)$ *is the difference in* $f(x)$:
>
> $$\text{If }\ df/dx = \sqrt{x}\ \text{ then area} = \int_{x=0}^{x=4}\sqrt{x}\,dx = f(4) - f(0). \tag{3}$$

What is $f(x)$? Instead of the derivative of $\sqrt{x}$, we need its "*antiderivative*." We have to find a function $f(x)$ whose derivative is $\sqrt{x}$. It is the opposite of Chapters $2-4$, and requires us to *work backwards*. The derivative of x^n is nx^{n-1}—now we need the antiderivative. The quick formula is $f(x) = x^{n+1}/(n+1)$—we aim to understand it.

Solution   Since the derivative lowers the exponent, the antiderivative *raises* it. We go from $x^{1/2}$ to $x^{3/2}$. But then the derivative is $(3/2)x^{1/2}$. It contains an unwanted factor $3/2$. *To cancel that factor, put* $2/3$ *into the antiderivative*:

$f(x) = \frac{2}{3}x^{3/2}$ **has the required derivative** $v(x) = x^{1/2} = \sqrt{x}$.

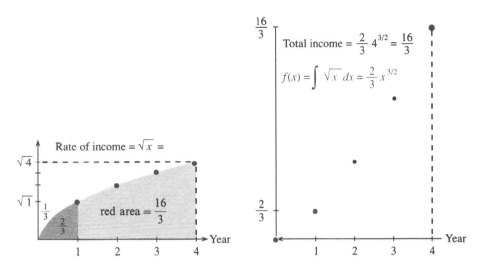

Fig. 5.3   The integral of $v(x) = \sqrt{x}$ is the exact area $16/3$ under the curve.

There you see the key to integrals: Work backward from derivatives (and adjust).

Now comes a number—the exact area. At $x = 4$ we find $x^{3/2} = 8$. Multiply by $2/3$ to get $16/3$. Then subtract $f(0) = 0$:

$$\int_{x=0}^{x=4} \sqrt{x}\, dx = \frac{2}{3}(4)^{3/2} - \frac{2}{3}(0)^{3/2} = \frac{2}{3}(8) = \frac{16}{3} \tag{4}$$

The total income over four years is $16/3 = 5\frac{1}{3}$ **million dollars.** This is $f(4) - f(0)$. The sum from thousands of rectangles was slowly approaching this exact area $5\frac{1}{3}$.

Other areas The income in the first year, at $x = 1$, is $\frac{2}{3}(1)^{3/2} = \frac{2}{3}$ million dollars. (The false income was 1 million dollars.) The total income after x years is $\frac{2}{3}(x)^{3/2}$, which is the antiderivative $f(x)$. *The square root curve covers* $2/3$ *of the overall rectangle it sits in.* The rectangle goes out to x and up to $\sqrt{x}$, with area $x^{3/2}$, and $2/3$ of that rectangle is below the curve. ($1/3$ is above.)

Other antiderivatives The derivative of x^5 is $5x^4$. Therefore the antiderivative of x^4 is $x^5/5$. Divide by 5 (or $n+1$) to cancel the 5 (or $n+1$) from the derivative. And don't allow $n + 1 = 0$:

The derivative $v(x) = x^n$ **has the antiderivative** $f(x) = x^{n+1}/(n+1)$.

EXAMPLE 1   The antiderivative of x^2 is $\frac{1}{3}x^3$. This is the area under the parabola $v(x) = x^2$. The area out to $x = 1$ is $\frac{1}{3}(1)^3 - \frac{1}{3}(0)^3$, or $1/3$.

Remark on $\sqrt{x}$ *and* x^2 The $2/3$ from $\sqrt{x}$ and the $1/3$ from x^2 add to 1. Those are the areas below and above the $\sqrt{x}$ curve, in the corner of Figure 5.3. If you turn the curve by $90°$, it becomes the parabola. The functions $y = \sqrt{x}$ and $x = y^2$ are inverses! The areas for these inverse functions add to a square of area 1.

AREA UNDER A STRAIGHT LINE

You already know the area of a triangle. The region is below the diagonal line $v = x$ in Figure 5.4. The base is 4, the height is 4, and the area is $\frac{1}{2}(4)(4) = 8$. Integration is

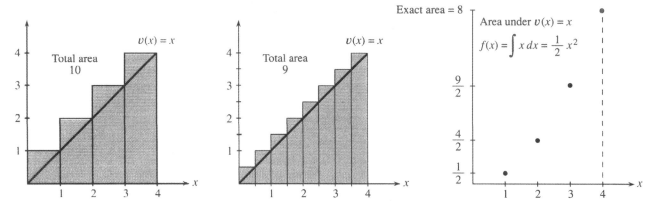

Fig. 5.4   Triangular area 8 as the limit of rectangular areas $10, 9, 8\frac{1}{2}, \ldots$.

not required! But if you allow calculus to repeat that answer, and build up the integral $f(x) = \frac{1}{2}x^2$ as the limiting area of many rectangles, you will have the beginning of something important.

The four rectangles have area $1 + 2 + 3 + 4 = 10$. That is greater than 8, because the triangle is inside. 10 is a first approximation to the triangular area 8, and to improve it we need more rectangles.

The next rectangles will be thinner, of width $\Delta x = 1/2$ instead of the original $\Delta x = 1$. There will be eight rectangles instead of four. They extend above the line, so the answer is still too high. The new heights are $1/2, 1, 3/2, 2, 5/2, 3, 7/2, 4$. The total area in Figure 5.4b is the sum of the base $\Delta x = 1/2$ times those heights:

$$\text{area} = \tfrac{1}{2}\left(\tfrac{1}{2} + 1 + \tfrac{3}{2} + 2 + \cdots + 4\right) = 9 \text{ (which is closer to 8).}$$

Question What is the area of 16 rectangles? Their heights are $\frac{1}{4}, \frac{1}{2}, \ldots, 4$.
Answer   With base $\Delta x = \frac{1}{4}$ the area is $\frac{1}{4}\left(\frac{1}{4} + \frac{1}{2} + \cdots + 4\right) = 8\frac{1}{2}$.

The effort of doing the addition is increasing. A formula for the sums is needed, and will be established soon. (The next answer would be $8\frac{1}{4}$.) But more important than the formula is the idea. *We are carrying out a limiting process, one step at a time.* The area of the rectangles is approaching the area of the triangle, as Δx decreases. The same limiting process will apply to other areas, in which the region is much more complicated. Therefore we pause to comment on what is important.

Area Under a Curve

What requirements are imposed on those thinner and thinner rectangles? It is not essential that they all have the same width. And it is not required that they cover the triangle completely. The rectangles could lie *below* the curve. The limiting answer will still be 8, even if the widths Δx are unequal and the rectangles fit inside the triangle or across it. We only impose two rules:

1. The largest width $\Delta x_{\max}$ must approach zero.
2. The top of each rectangle must touch or cross the curve.

The area under the graph is defined to be the limit of these rectangular areas, if that limit exists. For the straight line, the limit does exist and equals 8. That limit is independent of the particular widths and heights—as we absolutely insist it should be.

Section 5.5 allows any continuous $v(x)$. The question will be the same—***Does the limit exist***? The answer will be the same—***Yes***. That limit will be the *integral of* $v(x)$, and it will be the area under the curve. It will be $f(x)$.

EXAMPLE 2   The triangular area from 0 to x is $\frac{1}{2}(\text{base})(\text{height}) = \frac{1}{2}(x)(x)$. That is $f(x) = \frac{1}{2}x^2$. Its derivative is $v(x) = x$. But notice that $\frac{1}{2}x^2 + 1$ has the *same derivative*. So does $f = \frac{1}{2}x^2 + C$, for any constant C. There is a "***constant of integration***" in $f(x)$, which is wiped out in its derivative $v(x)$.

EXAMPLE 3   Suppose the velocity is decreasing: $v(x) = 4 - x$. If we sample v at $x = 1, 2, 3, 4$, the rectangles lie *under* the graph. Because v is decreasing, the right end of each interval gives v_{min}. Then the rectangular area $3 + 2 + 1 + 0 = 6$ is less than the exact area 8. The rectangles are *inside* the triangle, and eight rectangles with base $\frac{1}{2}$ come closer:

$$\text{rectangular area } = \tfrac{1}{2}(3\tfrac{1}{2} + 3 + \cdots + \tfrac{1}{2} + 0) = 7.$$

Sixteen rectangles would have area $7\frac{1}{2}$. We repeat that the rectangles need not have the same widths Δx, but it makes these calculations easier.

What is the area out to an arbitrary point (like $x = 3$ or $x = 1$)? We could insert rectangles, but the Fundamental Theorem offers a faster way. Any antiderivative of $4 - x$ will give the area. ***We look for a function whose derivative is*** $4 - x$. The derivative of $4x$ is 4, the derivative of $\frac{1}{2}x^2$ is x, so work backward:

$$\text{to achieve } df/dx = 4 - x \text{ choose } f(x) = 4x - \tfrac{1}{2}x^2.$$

Calculus skips past the rectangles and computes $f(3) = 7\frac{1}{2}$. ***The area between $x = 1$ and $x = 3$ is the difference*** $7\frac{1}{2} - 3\frac{1}{2} = 4$. In Figure 5.5, this is the area of the trapezoid.

The f-curve flattens out when the v-curve touches zero. No new area is being added.

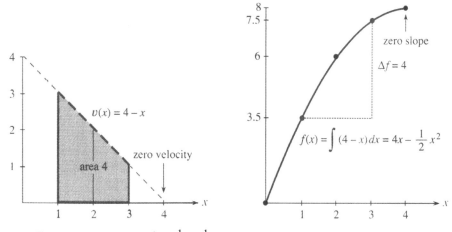

Fig. 5.5   The area is $\Delta f = 7\frac{1}{2} - 3\frac{1}{2} = 4$. Since $v(x)$ decreases, $f(x)$ bends down.

INDEFINITE INTEGRALS AND DEFINITE INTEGRALS

We have to distinguish two different kinds of integrals. They both use the antiderivative $f(x)$. The definite one involves the limits 0 and 4, the indefinite one doesn't:

The **indefinite integral** is a **function** $f(x) = 4x - \frac{1}{2}x^2$.

The **definite integral** from $x = 0$ to $x = 4$ is the **number** $f(4) - f(0)$.

The definite integral is definitely 8. But the indefinite integral is not necessarily $4x - \frac{1}{2}x^2$. **We can change $f(x)$ by a constant without changing its derivative** (since the derivative of a constant is zero). The following functions are also antiderivatives:

$$f(x) = 4x - \tfrac{1}{2}x^2 + 1, \qquad f(x) = 4x - \tfrac{1}{2}x^2 - 9, \qquad f(x) = 4x - \tfrac{1}{2}x^2 + C.$$

The first two are particular examples. The last is the general case. The constant C can be anything (including zero), to give all functions with the required derivative. The theory of calculus will show that there are no others. The indefinite integral is the most general antiderivative (with no limits):

$$\textit{indefinite integral} \quad f(x) = \int v(x)\,dx = 4x - \tfrac{1}{2}x^2 + C. \tag{5}$$

By contrast, the definite integral is a number. It contains no arbitrary constant C. More that that, it contains no variable x. The definite integral is determined by the function $v(x)$ and the limits of integration (also known as the **endpoints**). It is the area under the graph between those endpoints.

To see the relation of indefinite to definite, answer this question: *What is the definite integral between $x = 1$ and $x = 3$?* The indefinite integral gives $f(3) = 7\frac{1}{2} + C$ and $f(1) = 3\frac{1}{2} + C$. To find the area between the limits, **subtract f at one limit from f at the other limit**:

$$\int_{x=1}^{3} v(x)\,dx = f(3) - f(1) = (7\tfrac{1}{2} + C) - (3\tfrac{1}{2} + C) = 4. \tag{6}$$

The constant cancels itself! The definite integral is the *difference* between the values of the indefinite integral. C disappears in the subtraction.

The difference $f(3) - f(1)$ is like $f_n - f_0$. The sum of v_j from 1 to n has become "**the integral of $v(x)$ from 1 to 3.**" Section 5.3 computes other areas from sums, and 5.4 computes many more from antiderivatives. Then we come back to the definite integral and the Fundamental Theorem:

$$\int_{a}^{b} v(x)\,dx = \int_{a}^{b} \frac{df}{dx}\,dx = f(b) - f(a). \tag{7}$$

5.2 EXERCISES

Read-through questions

Integration yields the __a__ under a curve $y = v(x)$. It starts from rectangles with base __b__ and heights $v(x)$ and areas __c__. As $\Delta x \to 0$ the area $v_1 \Delta x + \cdots + v_n \Delta x$ becomes the __d__ of $v(x)$. The symbol for the indefinite integral of $v(x)$ is __e__.

The problem of integration is solved if we find $f(x)$ such that __f__. Then f is the __g__ of v, and $\int_2^6 v(x)\,dx$ equals __h__ minus __i__. The limits of integration are __j__. This is a __k__ integral, which is a __l__ and not a function $f(x)$.

The example $v(x) = x$ has $f(x) = $ __m__. It also has $f(x) = $ __n__. The area under $v(x)$ from 2 to 6 is __o__. The constant is canceled in computing the difference __p__ minus __q__. If $v(x) = x^8$ then $f(x) = $ __r__.

The sum $v_1 + \cdots + v_n = f_n - f_0$ leads to the Fundamental Theorem $\int_a^b v(x)\,dx = \underline{\quad s \quad}$. The $\underline{\quad t \quad}$ integral is $f(x)$ and the $\underline{\quad u \quad}$ integral is $f(b) - f(a)$. Finding the $\underline{\quad v \quad}$ under the v-graph is the opposite of finding the $\underline{\quad w \quad}$ of the f-graph.

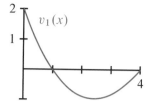

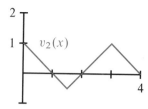

Find an antiderivative $f(x)$ for $v(x)$ in 1–14. Then compute the definite integral $\int_0^1 v(x)\,dx = f(1) - f(0)$.

1 $5x^4 + 4x^5$

2 $x + 12x^2$

3 $1/\sqrt{x}$ (or $x^{-1/2}$)

4 $(\sqrt{x})^3$ (or $x^{3/2}$)

5 $x^{1/3} + (2x)^{1/3}$

6 $x^{1/3}/x^{2/3}$

7 $2\sin x + \sin 2x$

8 $\sec^2 x + 1$

9 $x \cos x$ (by experiment)

10 $x \sin x$ (by experiment)

11 $\sin x \cos x$

12 $\sin^2 x \cos x$

13 0 (find all f)

14 -1 (find all f)

15 If $df/dx = v(x)$ then the definite integral of $v(x)$ from a to b is _____. If $f_j - f_{j-1} = v_j$ then the definite sum of $v_3 + \cdots + v_7$ is _____.

16 The areas include a factor Δx, the base of each rectangle. So the sum of v's is multiplied by _____ to approach the integral. The difference of f's is divided by _____ to approach the derivative.

17 The areas of 4, 8, and 16 rectangles were 10, 9, and $8\frac{1}{2}$, containing the triangle out to $x = 4$. Find a formula for the area A_N of N rectangles and test it for $N = 3$ and $N = 6$.

18 Draw four rectangles with base 1 *below* the $y = x$ line, and find the total area. What is the area with N rectangles?

19 Draw $y = \sin x$ from 0 to π. Three rectangles (base $\pi/3$) and six rectangles (base $\pi/6$) contain an arch of the sine function. Find the areas and guess the limit.

20 Draw an example where three lower rectangles under a curve (heights m_1, m_2, m_3) have less area than two rectangles.

21 Draw $y = 1/x^2$ for $0 < x < 1$ with two rectangles under it (base $1/2$). What is their area, and what is the area for four rectangles? Guess the limit.

22 Repeat Problem 21 for $y = 1/x$.

23 (with calculator) For $v(x) = 1/\sqrt{x}$ take enough rectangles over $0 \leqslant x \leqslant 1$ to convince any reasonable professor that the area is 2. Find $f(x)$ and verify that $f(1) - f(0) = 2$.

24 Find the area under the parabola $v = x^2$ from $x = 0$ to $x = 4$. Relate it to the area 16/3 below $\sqrt{x}$.

25 For v_1 and v_2 in the figure estimate the areas $f(2)$ and $f(4)$. Start with $f(0) = 0$.

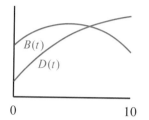

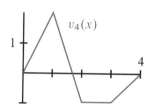

26 Draw $y = v(x)$ so that the area $f(x)$ increases until $x = 1$, stays constant to $x = 2$, and decreases to $f(3) = 1$.

27 Describe the indefinite integrals of v_1 and v_2. Do the areas increase? Increase then decrease? ...

28 For $v_4(x)$ find the area $f(4) - f(1)$. Draw $f_4(x)$.

29 The graph of $B(t)$ shows the birth rate: births per unit time at time t. $D(t)$ is the death rate. In what way do these numbers appear on the graph?

1. The change in population from $t = 0$ to $t = 10$.
2. The time T when the population was largest.
3. The time t^* when the population increased fastest.

30 Draw the graph of a function $y_4(x)$ whose area function is $v_4(x)$.

31 If $v_2(x)$ is an antiderivative of $y_2(x)$, draw $y_2(x)$.

32 Suppose $v(x)$ increases from $v(0) = 0$ to $v(3) = 4$. The area under $y = v(x)$ plus the area on the left side of $x = v^{-1}(y)$ equals _____.

33 *True or false*, when $f(x)$ is an antiderivative of $v(x)$.

(a) $2f(x)$ is an antiderivative of $2v(x)$ (*try examples*)
(b) $f(2x)$ is an antiderivative of $v(2x)$
(c) $f(x) + 1$ is an antiderivative of $v(x) + 1$
(d) $f(x + 1)$ is an antiderivative of $v(x + 1)$.
(e) $(f(x))^2$ is an antiderivative of $(v(x))^2$.

5.3 Summation versus Integration

This section does integration the hard way. We find explicit formulas for $f_n = v_1 + \cdots + v_n$. From areas of rectangles, the limits produce the area $f(x)$ under a curve. According to the Fundamental Theorem, df/dx should return us to $v(x)$—and we verify in each case that it does.

May I recall that there is sometimes an easier way? If we can find an $f(x)$ whose derivative is $v(x)$, then the integral of v is f. Sums and limits are not required, when f is spotted directly. The next section, which explains how to look for $f(x)$, will displace this one. (If we can't find an antiderivative we fall back on summation.) Given a successful f, adding any constant produces another f—since the derivative of the constant is zero. The right constant achieves $f(0) = 0$, with no extra effort.

This section constructs $f(x)$ from sums. The next section searches for antiderivatives.

THE SIGMA NOTATION

In a section about sums, there has to be a decent way to express them. Consider $1^2 + 2^2 + 3^2 + 4^2$. The individual terms are $v_j = j^2$. Their sum can be written in **summation notation**, using the capital Greek letter Σ (pronounced sigma):

$$1^2 + 2^2 + 3^2 + 4^2 \text{ is written } \sum_{j=1}^{4} j^2.$$

Spoken aloud, that becomes "**the sum of j^2 from $j = 1$ to 4.**" It equals 30. The limits on j (written below and above Σ) indicate where to start and stop:

$$v_1 + \cdots + v_n = \sum_{j=1}^{n} v_j \quad \text{and} \quad v_3 + \cdots + v_9 = \sum_{k=3}^{9} v_k. \tag{1}$$

The k at the end of (1) makes an additional point. There is nothing special about the letter j. That is a "*dummy variable*," no better and no worse than k (or i). Dummy variables are only on one side (the side with Σ), and they have no effect on the sum. *The upper limit n is on both sides.* Here are six sums:

$$\sum_{k=1}^{n} k = 1 + 2 + 3 + \cdots + n \qquad \sum_{j=1}^{4} (-1)^j = -1 + 1 - 1 + 1 = 0$$

$$\sum_{j=1}^{5} (2j - 1) = 1 + 3 + 5 + 7 + 9 = 5^2 \qquad \sum_{i=0}^{0} v_i = v_0 \left[\text{only one term}\right]$$

$$\sum_{i=1}^{4} j^2 = \left[\text{meaningless?}\right] \qquad \sum_{k=0}^{\infty} \frac{1}{2^k} = 1 + \frac{1}{2} + \frac{1}{4} + \cdots = 2 \left[\text{infinite series}\right]$$

The numbers 1 and n or 1 and 4 (or 0 and ∞) are the **lower limit** and **upper limit**. The dummy variable i or j or k is the **index** of summation. I hope it seems reasonable that the infinite series $1 + \frac{1}{2} + \frac{1}{4} + \cdots$ adds to 2. We will come back to it in Chapter 10.[†]

A sum like $\Sigma_{j=1}^{n} 6$ looks meaningless, but it is actually $6 + 6 + \cdots + 6 = 6n$. It follows the rules. In fact $\Sigma_{i=1}^{4} j^2$ is not meaningless either. Every term is j^2 and by

---

[†] Zeno the Greek believed it was impossible to get anywhere, since he would only go halfway and then half again and half again. Infinite series would have changed his whole life.

the same rules, that sum is $4j^2$. However the i was probably intended to be j. Then the sum is $1+4+9+16=30$.

Question What happens to these sums when the upper limits are changed to n?

Answer *The sum depends on the stopping point n.* A formula is required (when possible). Integrals stop at x, sums stop at n, and we now look for special cases when $f(x)$ or f_n can be found.

A SPECIAL SUMMATION FORMULA

How do you add the first 100 whole numbers? The problem is to compute

$$\sum_{j=1}^{100} j = 1+2+3+\cdots+98+99+100 =?$$

If you were Gauss, you would see the answer at once. (He solved this problem at a ridiculous age, which gave his friends the idea of getting him into another class.) His solution was to combine $1+100$, and $2+99$, and $3+98$, *always adding to* 101. There are fifty of those combinations. Thus the sum is $(50)(101)=5050$.

The sum from 1 to n uses the same idea. The first and last terms add to $n+1$. The next terms $n-1$ and 2 also add to $n+1$. If n is even (as 100 was) then there are $\frac{1}{2}n$ parts. Therefore the sum is $\frac{1}{2}n$ times $n+1$:

$$\sum_{j=1}^{n} j = 1+2+\cdots+(n-1)+n = \frac{1}{2}n(n+1). \tag{2}$$

The important term is $\frac{1}{2}n^2$, but the exact sum is $\frac{1}{2}n^2 + \frac{1}{2}n$.

What happens if n is an odd number (like $n=99$)? Formula (2) remains true. The combinations $1+99$ and $2+98$ still add to $n+1=100$. There are $\frac{1}{2}(99)=49\frac{1}{2}$ such pairs, because the middle term (which is 50) has nothing to combine with. Thus $1+2+\cdots+99$ equals $49\frac{1}{2}$ times 100, or 4950.

Remark That sum had to be 4950, because it is 5050 minus 100. The sum up to 99 equals the sum up to 100 with the last term removed. Our key formula $f_n - f_{n-1} = v_n$ has turned up again!

EXAMPLE Find the sum $101+102+\cdots+200$ of the *second* hundred numbers.

First solution This is the sum from 1 to 200 minus the sum from 1 to 100:

$$\sum_{101}^{200} j = \sum_{1}^{200} j - \sum_{1}^{100} j. \tag{3}$$

The middle sum is $\frac{1}{2}(200)(201)$ and the last is $\frac{1}{2}(100)(101)$. Their difference is 15050.

Note! I left out "$j=$" in the limits. It is there, but not written.

Second solution The answer 15050 is exactly the sum of the first hundred numbers (which was 5050) plus an additional 10000. Believing that a number like 10000 can never turn up by accident, we look for a reason. It is found through ***changing the limits of summation***:

$$\sum_{j=101}^{200} j \text{ is the same sum as } \sum_{k=1}^{100} (k + 100). \qquad (4)$$

This is important, to be able to shift limits around. Often the lower limit is moved to zero or one, for convenience. Both sums have 100 terms (that doesn't change). The dummy variable j is replaced by another dummy variable k. They are related by $j = k + 100$ or equivalently by $k = j - 100$.

   The variable must change everywhere—in the lower limit and the upper limit as well as inside the sum. If j starts at 101, then $k = j - 100$ starts at 1. If j ends at 200, k ends at 100. If j appears in the sum, it is replaced by $k + 100$ (and if j^2 appeared it would become $(k + 100)^2$).

   From equation (4) you see why the answer is 15050. The sum $1 + 2 + \cdots + 100$ is 5050 as before. 100 ***is added to each of those*** 100 ***terms***. That gives 10000.

EXAMPLES OF CHANGING THE VARIABLE (and the limits)

$\displaystyle\sum_{i=0}^{3} 2^i$ equals $\displaystyle\sum_{j=1}^{4} 2^{j-1}$ (here $i = j - 1$). Both sums are $1 + 2 + 4 + 8$

$\displaystyle\sum_{i=3}^{n} v_i$ equals $\displaystyle\sum_{j=0}^{n-3} v_{j+3}$ (here $i = j + 3$ and $j = i - 3$). Both sums are $v_3 + \cdots + v_n$.

Why change n to $n - 3$? Because the upper limit is $i = n$. So $j + 3 = n$ and $j = n - 3$.

   A final step is possible, and you will often see it. ***The new variable j can be changed back to i.*** Dummy variables have no meaning of their own, but at first the result looks surprising:

$$\sum_{i=0}^{5} 2^i \text{ equals } \sum_{j=1}^{6} 2^{j-1} \text{ equals } \sum_{i=1}^{6} 2^{i-1}.$$

With practice you might do that in one step, skipping the temporary letter j. Every i on the left becomes $i - 1$ on the right. Then $i = 0, \ldots, 5$ changes to $i = 1, \ldots, 6$. (At first two steps are safer.) This may seem a minor point, but soon we will be changing the limits on *integrals* instead of sums. Integration is parallel to summation, and it is better to see a "change of variable" here first.

Note about $1 + 2 + \cdots + n$. The good thing is that Gauss found the sum $\frac{1}{2}n(n+1)$. The bad thing is that his method looked too much like a trick. I would like to show how this fits the fundamental rule connecting sums and differences:

$$\textit{if } v_1 + v_2 + \cdots + v_n = f_n \textbf{ then } v_n = f_n - f_{n-1}. \qquad (5)$$

Gauss says that f_n is $\frac{1}{2}n(n+1)$. Reducing n by 1, his formula for f_{n-1} is $\frac{1}{2}(n-1)n$. ***The difference $f_n - f_{n-1}$ should be the last term n in the sum***:

$$f_n - f_{n-1} = \tfrac{1}{2}n(n+1) - \tfrac{1}{2}(n-1)n = \tfrac{1}{2}(n^2 + n - n^2 + n) = n. \qquad (6)$$

This is the one term $v_n = n$ that is included in f_n but not in f_{n-1}.

   There is a deeper point here. For any sum f_n, there are two things to check. The f's must *begin* correctly and they must *change* correctly. The underlying idea is ***mathematical induction: Assume the statement is true below n. Prove it for n.***

Goal: *To prove that* $1 + 2 + \cdots + n = \frac{1}{2}n(n+1)$. *This is the guess* f_n.

Proof by induction: *Check* f_1 (*it equals* 1). *Check* $f_n - f_{n-1}$(*it equals* n).

For $n = 1$ the answer $\frac{1}{2}n(n+1) = \frac{1}{2} \cdot 1 \cdot 2$ is correct. For $n = 2$ this formula $\frac{1}{2} \cdot 2 \cdot 3$ agrees with $1 + 2$. But that separate test is not necessary! *If f_1 is right, and if the change $f_n - f_{n-1}$ is right for every n, then f_n must be right*. Equation (6) was the key test, to show that the change in f's agrees with v.

That is the logic behind mathematical induction, but I am not happy with most of the exercises that use it. There is absolutely no excitement. The answer is given by some higher power (like Gauss), and it is proved correct by some lower power (like us). It is much better when we lower powers find the answer for ourselves.† Therefore I will try to do that for the second problem, which is the *sum of squares*.

THE SUM OF j^2 AND THE INTEGRAL OF x^2

An important calculation comes next. It is the area in Figure 5.6. One region is made up of rectangles, so its area is a sum of n pieces. The other region lies under the parabola $v = x^2$. It cannot be divided into rectangles, and calculus is needed.

The first problem is to find $f_n = 1^2 + 2^2 + 3^2 + \cdots + n^2$. This is a sum of squares, with $f_1 = 1$ and $f_2 = 5$ and $f_3 = 14$. The goal is to find the pattern in that sequence. By trying to guess f_n we are copying what will soon be done for integrals.

Calculus looks for an $f(x)$ whose derivative is $v(x)$. There f is an *antiderivative*

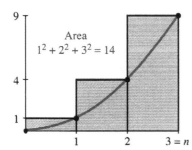
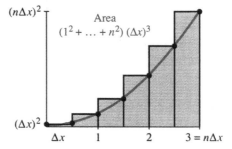
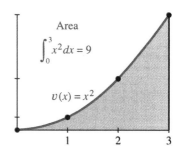

Fig. 5.6   Rectangles enclosing $v = x^2$ have area $\left(\frac{1}{3}n^3 + \frac{1}{2}n^2 + \frac{1}{6}n\right)(\Delta x)^3 \approx \frac{1}{3}(n\Delta x)^3 = \frac{1}{3}x^3$.

(or an integral). Algebra looks for f_n's whose differences produce v_n. Here f_n could be called an *antidifference* (better to call it a sum).

The best start is a good guess. Copying directly from integrals, we might try $f_n = \frac{1}{3}n^3$. To test if it is right, check whether $f_n - f_{n-1}$ produces on $v_n = n^2$:

$$\tfrac{1}{3}n^3 - \tfrac{1}{3}(n-1)^3 = \tfrac{1}{3}n^3 - \tfrac{1}{3}(n^3 - 3n^2 + 3n - 1) = n^2 - n + \tfrac{1}{3}.$$

We see n^2, but also $-n + \frac{1}{3}$. The guess $\frac{1}{3}n^3$ needs *correction terms*. To cancel $\frac{1}{3}$ in the difference, I subtract $\frac{1}{3}n$ from the sum. To put back n in the difference, I add $1 + 2 + \cdots + n = \frac{1}{2}n(n+1)$ to the sum. The new guess (which should be right) is

$$f_n = \tfrac{1}{3}n^3 + \tfrac{1}{2}n(n+1) - \tfrac{1}{3}n = \tfrac{1}{3}n^3 + \tfrac{1}{2}n^2 + \tfrac{1}{6}n. \tag{7}$$

To check this answer, verify first that $f_1 = 1$. Also $f_2 = 5$ and $f_3 = 14$. To be certain, verify that $f_n - f_{n-1} = n^2$. For calculus the important term is $\frac{1}{3}n^3$:

† The goal of real teaching is for the *student* to find the answer. And also the problem.

The sum $\sum_{j=1}^{n} j^2$ of the first n squares is $\frac{1}{3}n^3$ plus corrections $\frac{1}{2}n^2$ and $\frac{1}{6}n$.

In practice $\frac{1}{3}n^3$ is an excellent estimate. The sum of the first 100 squares is approximately $\frac{1}{3}(100)^3$, or a third of a million. If we need the exact answer, equation (7) is available: the sum is $338,350$. Many applications (example: the number of steps to solve 100 linear equations) can settle for $\frac{1}{3}n^3$.

What is fascinating is the contrast with calculus. **Calculus *has no correction terms*!** They get washed away in the limit of thin rectangles. When the sum is replaced by the integral (the area), we get an absolutely clean answer:

The integral of $v = x^2$ from $x = 0$ to $x = n$ is exactly $\frac{1}{3}n^3$.

The area under the parabola, out to the point $x = 100$, is precisely a third of a million. We have to explain why, with many rectangles.

The idea is to approach an infinite number of infinitely thin rectangles. A hundred rectangles gave an area of $338,350$. Now take a thousand rectangles. Their heights are $\left(\frac{1}{10}\right)^2, \left(\frac{2}{10}\right)^2, \ldots$ because the curve is $v = x^2$. The base of every rectangle is $\Delta x = \frac{1}{10}$, and we add heights times base:

$$\text{area of rectangles} = \left(\frac{1}{10}\right)^2 \left(\frac{1}{10}\right) + \left(\frac{2}{10}\right)^2 \left(\frac{1}{10}\right) + \cdots + \left(\frac{1000}{10}\right)^2 \left(\frac{1}{10}\right).$$

Factor out $\left(\frac{1}{10}\right)^3$. What you have left is $1^2 + 2^2 + \cdots + 1000^2$, which fits the sum of squares formula. The exact area of the thousand rectangles is $333,833.5$. I could try to guess ten thousand rectangles but I won't.

Main point: The area is approaching $333,333.333 \ldots$. But the calculations are getting worse. It is time for algebra—which means that we keep "Δx" and avoid numbers.

The interval of length 100 is divided into n pieces of length Δx. (Thus $n = 100/\Delta x$.) The jth rectangle meets the curve $v = x^2$, so its height is $(j\Delta x)^2$. Its base is Δx, and we add areas:

$$\boldsymbol{area} = (\Delta x)^2(\Delta x) + (2\Delta x)^2(\Delta x) + \cdots + (n\Delta x)^2(\Delta x) = \sum_{j=1}^{n}(j\Delta x)^2(\Delta x).$$

$$\tag{8}$$

Factor out $(\Delta x)^3$, leaving a sum of n squares. The area is $(\Delta x)^3$ times f_n, and $n = \dfrac{100}{\Delta x}$:

$$(\Delta x)^3 \left[\frac{1}{3}\left(\frac{100}{\Delta x}\right)^3 + \frac{1}{2}\left(\frac{100}{\Delta x}\right)^2 + \frac{1}{6}\left(\frac{100}{\Delta x}\right) \right] = \frac{1}{3}100^3 + \frac{1}{2}100^2(\Delta x) + \frac{1}{6}100(\Delta x)^2.$$

$$\tag{9}$$

This equation shows what is happening. The leading term is a third of a million, as predicted. The other terms are approaching zero! They contain Δx, and as the rectangles get thinner they disappear. They only account for the small corners of rectangles that lie above the curve. The vanishing of those corners will eventually be proved for any continuous functions—*the area from the correction terms goes to zero*—but here in equation (9) you see it explicitly.

The area under the curve came from the central idea of integration: $100/\Delta x$ rectangles of width Δx approach the limiting area$= \frac{1}{3}(100)^3$. ***The rectangular area is $\Sigma v_j \Delta x$. The exact area is $\int v(x)dx$. In the limit Σ becomes $\int$ and v_j becomes $v(x)$ and Δx becomes dx.***

That completes the calculation for a parabola. It used the formula for a sum of squares, which was special. But the underlying idea is much more general. The limit of the sums agrees with the antiderivative: **The antiderivative of $v(x) = x^2$ is $f(x) = \frac{1}{3}x^3$.** According to the Fundamental Theorem, the area under $v(x)$ is $f(x)$:

$$\int_0^{100} v(x)\, dx = f(100) - f(0) = \tfrac{1}{3}(100)^3.$$

That Fundamental Theorem is not yet proved! I mean it is not proved by us. Whether Leibniz or Newton managed to prove it, I am not quite sure. But it can be done. Starting from sums of differences, the difficulty is that we have too many limits at once. The sums of $v_j \Delta x$ are approaching the integral. The differences $\Delta f / \Delta x$ approach the derivative. A real proof has to separate those steps, and Section 5.7 will do it.

Proved or not, you are seeing the main point. What was true for the numbers f_j and v_j is true in the limit for $v(x)$ and $f(x)$. Now $v(x)$ can vary continuously, but it is still the slope of $f(x)$. **The reverse of slope is area**.

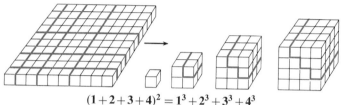

$$(1+2+3+4)^2 = 1^3 + 2^3 + 3^3 + 4^3$$

Proof without words by Roger Nelsen (*Mathematics Magazine* 1990).

Finally we review the area under $v = x$. The sum of $1 + 2 + \cdots + n$ is $\frac{1}{2}n^2 + \frac{1}{2}n$. This gives the area of $n = 4/\Delta x$ rectangles, going out to $x = 4$. The heights are $j\Delta x$, the bases are Δx, and we add areas:

$$\sum_{j=1}^{4/\Delta x} (j\Delta x)(\Delta x) = (\Delta x)^2 \left[\frac{1}{2}\left(\frac{4}{\Delta x}\right)^2 + \frac{1}{2}\left(\frac{4}{\Delta x}\right) \right] = 8 + 2\Delta x. \qquad (10)$$

With $\Delta x = 1$ the area is $1 + 2 + 3 + 4 = 10$. With eight rectangles and $\Delta x = \frac{1}{2}$, the area was $8 + 2\Delta x = 9$. Sixteen rectangles of width $\frac{1}{4}$ brought the correction $2\Delta x$ down to $\frac{1}{2}$. The exact area is 8. **The error is proportional to Δx.**

Important note There you see a question in applied mathematics. If there is an error, what size is it? How does it behave as $\Delta x \to 0$? The Δx term disappears in the limit, and $(\Delta x)^2$ disappears faster. But to get an error of 10^{-6} we need **eight million rectangles**:

$$2\Delta x = 2 \cdot 4/8,000,000 = 10^{-6}.$$

That is horrifying! The numbers $10, 9, 8\frac{1}{2}, 8\frac{1}{4}, \ldots$ seem to approach the area 8 in a satisfactory way, but the convergence is **much too slow**. It takes twice as much work to get one more binary digit in the answer—which is absolutely unacceptable. Somehow the Δx term must be removed. If the correction is $(\Delta x)^2$ instead of Δx, then a thousand rectangles will reach an accuracy of 10^{-6}.

The problem is that the rectangles are unbalanced. Their right sides touch the graph of v, but their left sides are much too high. The best is to cross the graph in the *middle* of the interval—this is the **midpoint rule**. Then the rectangle sits halfway across the line $v = x$, and the error is zero. Section 5.8 comes back to this rule—and to Simpson's rule that fits parabolas and removes the $(\Delta x)^2$ term and is built into many calculators.

Finally we try the quick way. The area under $v = x$ is $f = \frac{1}{2}x^2$, because df/dx is v. The area out to $x = 4$ is $\frac{1}{2}(4)^2 = 8$. Done.

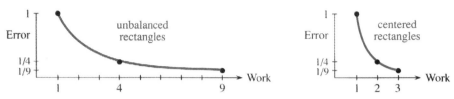

Fig. 5.7 Endpoint rules: error $\sim 1/(\text{work}) \sim 1/n$. Midpoint rule is better: error $\sim 1/(\text{work})^2$.

Optional: pth ***powers*** Our sums are following a pattern. First, $1 + \cdots + n$ is $\frac{1}{2}n^2$ plus $\frac{1}{2}n$. The sum of squares is $\frac{1}{3}n^3$ plus correction terms. *The sum of pth powers is*

$$1^p + 2^p + \cdots + n^p = \frac{1}{p+1}n^{p+1} \text{ plus correction terms.} \tag{11}$$

192 The correction involves lower powers of n, and you know what is coming. ***Those corrections disappear in calculus.*** The area under $v = x^p$ from 0 to n is

$$\int_{x=0}^{n} x^p \, dx = \lim_{\Delta x \to 0} \sum_{j=1}^{n/\Delta x} (j\Delta x)^p (\Delta x) = \frac{1}{p+1}n^{p+1}. \tag{12}$$

Calculus doesn't care if the upper limit n is an integer, and it doesn't care if the power p is an integer. We only need $p + 1 > 0$ to be sure n^{p+1} is genuinely the leading term. ***The antiderivative of $v = x^p$ is*** $f = x^{p+1}/(p+1)$.

We are close to interesting experiments. The correction terms disappear and the sum approaches the integral. Here are actual numbers for $p = 1$, when the sum and integral are easy: $S_n = 1 + \cdots + n$ and $I_n = \int x \, dx = \frac{1}{2}n^2$. The difference is $D_n = \frac{1}{2}n$. The thing to watch is the ***relative error*** $E_n = D_n/I_n$:

n	S_n	I_n	$D_n = S_n - I_n$	$E_n = D_n/I_n$
100	5050	5000	50	.010
200	20100	20000	100	.005

The number $20,100$ is $\frac{1}{2}(200)(201)$. Please write down the next line $n = 400$, ***and please find a formula for*** E_n. You can guess E_n from the table, or you can derive it from knowing S_n and I_n. The formula should show that E_n goes to zero. More important, it should show how quick (or slow) that convergence will be.

One more number—a third of a million—was mentioned earlier. It came from integrating x^2 from 0 to 100, which compares to the sum S_{100} of 100 squares:

n	p	S_n	$I_n = \frac{1}{3}n^3$	$D = S - I$	$E = D/I$
100	2	338350	$333333\frac{1}{3}$	$5016\frac{2}{3}$	.01505
200	2	2686700	$2666666\frac{2}{3}$	$20033\frac{1}{3}$	.0075125

These numbers suggest a new idea, ***to keep n fixed and change p***. The computer can find sums without a formula! With its help we go to fourth powers and square roots:

n	p	$S = 1^p + \cdots + n^p$	$I = n^{p+1}/(p+1)$	$D = S - I$	$E_{n,p} = D/I$
100	4	2050333330	$\frac{1}{5}(100)^5$	50333330	0.0252
100	$\frac{1}{2}$	671.4629	$\frac{2}{3}(100)^{3/2}$	4.7963	0.0072

In this and future tables we don't expect exact values. The last entries are rounded off, and the goal is to see the pattern. The errors $E_{n,p}$ are sure to obey a systematic rule—they are proportional to $1/n$ and to an unknown number $C(p)$ that depends on p. I hope you can push the experiments far enough to discover $C(p)$. This is not an exercise with an answer in the back of the book—it is mathematics.

5.3 EXERCISES

Read-through questions

The Greek letter __a__ indicates summation. In $\Sigma_1^n v_j$ the dummy variable is __b__. The limits are __c__, so the first term is __d__ and the last term is __e__. When $v_j = j$ this sum equals __f__. For $n = 100$ the leading term is __g__. The correction term is __h__. The leading term equals the integral of $v = x$ from 0 to 100, which is written __i__. The sum is the total __j__ of 100 rectangles. The correction term is the area between the __k__ and the __l__.

The sum $\Sigma_{i=3}^6 i^2$ is the same as $\Sigma_{j=1}^4$ __m__ and equals __n__. The sum $\Sigma_{i=4}^5 v_i$ is the same as __o__ v_{i+4} and equals __p__. For $f_n = \Sigma_{j=1}^n v_j$ the difference $f_n - f_{n-1}$ equals __q__.

The formula for $1^2 + 2^2 + \cdots + n^2$ is $f_n =$ __r__. To prove it by mathematical induction, check $f_1 =$ __s__ and check $f_n - f_{n-1} =$ __t__. The area under the parabola $v = x^2$ from $x = 0$ to $x = 9$ is __u__. This is close to the area of __v__ rectangles of base Δx. The correction terms approach zero very __w__.

1 Compute the numbers $\sum_{n=1}^4 1/n$ and $\sum_{i=2}^5 (2i-3)$.

2 Compute $\sum_{j=0}^3 (j^2 - j)$ and $\sum_{j=1}^n 1/2^j$.

3 Evaluate the sum $\sum_{i=0}^6 2^i$ and $\sum_{i=0}^n 2^i$.

4 Evaluate $\sum_{i=1}^6 (-1)^i i$ and $\sum_{j=1}^n (-1)^j j$.

5 Write these sums in sigma notation and compute them:

$2+4+6+\cdots+100 \quad 1+3+5+\cdots+199 \quad 1-\frac{1}{2}+\frac{1}{3}-\frac{1}{4}$

6 Express these sums in sigma notation:

$v_1 - v_2 + v_3 - v_4 \quad v_1 w_1 + v_2 w_2 + \cdots + v_n w_n \quad v_1 + v_3 + v_5$

7 Convert these sums to sigma notation:

$a_0 + a_1 x + \cdots + a_n x^n \quad \sin\frac{2\pi}{n} + \sin\frac{4\pi}{n} + \cdots + \sin 2\pi$

8 The binomial formula uses coefficients $\binom{n}{j} = \frac{n!}{j!(n-j)!}$:

$(a+b)^n = \binom{n}{0}a^n + \binom{n}{1}a^{n-1}b + \cdots + \binom{n}{n}b^n = \sum_{j=0}^n \underline{\quad} b^j.$

9 With electronic help compute $\sum_1^{100} 1/j$ and $\sum_1^{1000} 1/j$.

10 On a computer find $\sum_0^{10}(-1)^j/j!$ times $\sum_0^{10} 1/j!$.

11 Simplify $\sum_{i=1}^n (a_i + b_i)^2 + \sum_{i=1}^n (a_i - b_i)^2$ to $\sum_{i=1}^n \underline{\quad}$.

12 Show that $\left(\sum_{i=1}^n a_i\right)^2 \neq \sum_{i=1}^n a_i^2$ and $\sum_{i=1}^n a_i b_i \neq \sum_{j=1}^n a_j \sum_{k=1}^n b_k$.

13 "*Telescope*" the sums $\sum_{k=1}^n (2^k - 2^{k-1})$ and $\sum_{j=1}^{10} \left(\frac{1}{j+1} - \frac{1}{j}\right)$. All but two terms cancel.

14 Simplify the sums $\sum_{j=1}^n (f_j - f_{j-1})$ and $\sum_{j=3}^{12} (f_{j+1} - f_j)$.

15 *True or false*: (a) $\sum_{j=4}^8 v_j = \sum_{i=2}^6 v_{i-2}$ (b) $\sum_{i=1}^9 v_i = \sum_{i=3}^{11} v_{i-2}$

16 $\sum_{i=1}^n v_i = \sum_{j=0}^{n-1} \underline{\quad}$ and $\sum_{i=0}^6 i^2 = \sum_{i=2}^8 \underline{\quad}$.

17 The antiderivative of $d^2 f/dx^2$ is df/dx. What is the sum $(f_2 - 2f_1 + f_0) + (f_3 - 2f_2 + f_1) + \cdots + (f_9 - 2f_8 + f_7)$?

18 *Induction*: Verify that $1^2 + 2^2 + \cdots + n^2$ is $f_n = n(n+1)(2n+1)/6$ by checking that f_1 is correct and $f_n - f_{n-1} = n^2$.

19 Prove by induction: $1 + 3 + \cdots + (2n-1) = n^2$.

20 Verify that $1^3 + 2^3 + \cdots + n^3$ is $f_n = \frac{1}{4}n^2(n+1)^2$ by checking f_1 and $f_n - f_{n-1}$. The text has a *proof without words*.

21 Suppose f_n has the form $an + bn^2 + cn^3$. If you know $f_1 = 1$, $f_2 = 5$, $f_3 = 14$, turn those into three equations for a, b, c. The solutions $a = \frac{1}{6}, b = \frac{1}{2}, c = \frac{1}{3}$ give what formula?

22 Find q in the formula $1^8 + \cdots + n^8 = qn^9 +$ correction.

23 Add $n = 400$ to the table for $S_n = 1 + \cdots + n$ and find the relative error E_n. Guess and prove a formula for E_n.

24 Add $n = 50$ to the table for $S_n = 1^2 + \cdots + n^2$ and compute E_{50}. Find an approximate formula for E_n.

25 Add $p = \frac{1}{3}$ and $p = 3$ to the table for $S_{100,p} = 1^p + \cdots + 100^p$. Guess an approximate formula for $E_{100,p}$.

26 Guess $C(p)$ in the formula $E_{n,p} \approx C(p)/n$.

27 Show that $|1-5| < |1| + |-5|$. Always $|v_1 + v_2| < |v_1| + |v_2|$ unless _____ .

28 Let S be the sum $1 + x + x^2 + \cdots$ of the (infinite) geometric series. Then $xS = x + x^2 + x^3 + \cdots$ is the same as S minus _____ . Therefore $S =$ _____ . None of this makes sense if $x = 2$ because _____ .

29 The **double sum** $\sum\limits_{i=1}^{2} \left[\sum\limits_{j=1}^{3} (i+j) \right]$ is $v_1 = \sum\limits_{i=1}^{3} (1+j)$ plus $v_2 = \sum\limits_{j=1}^{3} (2+j)$. Compute v_1 and v_2 and the double sum.

30 The double sum $\sum\limits_{i=1}^{2} \left(\sum\limits_{j=1}^{3} w_{i,j} \right)$ is $(w_{1,1} + w_{1,2} + w_{1,3}) +$ _____ . The double sum $\sum\limits_{j=1}^{3} \left(\sum\limits_{i=1}^{2} w_{i,j} \right)$ is $(w_{1,1} + w_{2,1}) + (w_{1,2} + w_{2,2}) +$ _____ . Compare.

31 Find the flaw in the proof that $2^n = 1$ for every $n = 0, 1, 2, \ldots$. For $n = 0$ we have $2^0 = 1$. If $2^n = 1$ for every $n < N$, then $2^N = 2^{N-1} \cdot 2^{N-1}/2^{N-2} = 1 \cdot 1/1 = 1$.

32 Write out all terms to see why the following are true:

$$\sum_1^3 4v_j = 4\sum_1^3 v_j \qquad \sum_{i=1}^{2}\left(\sum_{j=1}^{3} u_i v_j\right) = \left(\sum_1^2 u_i\right)\left(\sum_1^3 v_j\right)$$

33 The average of $6, 11, 4$ is $\bar{v} = \frac{1}{3}(6 + 11 + 4)$. Then $(6 - \bar{v}) + (11 - \bar{v}) + (4 - \bar{v}) =$ _____ . The average of $v_1, \ldots, v_n$ is $\bar{v} =$ _____ . Prove that $\Sigma(v_i - \bar{v}) = 0$.

34 The *Schwarz inequality* is $\left(\sum_1^n a_i b_i\right)^2 \leqslant \left(\sum_1^n a_i^2\right)\left(\sum_1^n b_i^2\right)$. Compute both sides if $a_1 = 2$, $a_2 = 3$, $b_1 = 1$, $b_2 = 4$. Then compute both sides for any a_1, a_2, b_1, b_2. The proof in Section 11.1 uses vectors.

35 Suppose n rectangles with base Δx touch the graph of $v(x)$ at the points $x = \Delta x, 2\Delta x, \ldots, n\Delta x$. Express the total rectangular area in sigma notation.

36 If $1/\Delta x$ rectangles with base Δx touch the graph of $v(x)$ at the *left* end of each interval (thus at $x = 0, \Delta x, 2\Delta x, \ldots$) express the total area in sigma notation.

37 The sum $\Delta x \sum\limits_{j=1}^{1/\Delta x} \dfrac{f(j\Delta x) - f((j-1)\Delta x)}{\Delta x}$ equals _____ . In the limit this becomes $\int_0^1$ _____ $dx =$ _____ .

5.4 Indefinite Integrals and Substitutions

This section integrates the easy way, by looking for antiderivatives. We leave aside sums of rectangular areas, and their limits as $\Delta x \to 0$. Instead we search for an $f(x)$ with the required derivative $v(x)$. In practice, this approach is more or less independent of the approach through sums—but it gives the same answer. And also, *the search for an antiderivative may not succeed.* We may not find f. In that case we go back to rectangles, or on to something better in Section 5.8.

A computer is ready to integrate v, but not by discovering f. It integrates between specified limits, to obtain a **number** (the definite integral). Here we hope to find a **function** (the indefinite integral). That requires a symbolic integration code like MACSYMA or *Mathematica* or MAPLE, or a reasonably nice $v(x)$, or both. An expression for $f(x)$ can have tremendous advantages over a list of numbers.

Thus our goal is to find antiderivatives and use them. The techniques will be further developed in Chapter 7—this section is short but good. First we write down what we know. *On each line, $f(x)$ is an antiderivative of $v(x)$ because $df/dx = v(x)$.*

Known pairs	**Function** $v(x)$	**Antiderivative** $f(x)$
Powers of x	x^n	$x^{n+1}/(n+1) + C$

$n = -1$ is not included, because $n+1$ would be zero. $v = x^{-1}$ will lead us to $f = \ln x$.

Trigonometric functions		
	$\cos x$	$\sin x + C$
	$\sin x$	$-\cos x + C$
	$\sec^2 x$	$\tan x + C$
	$\csc^2 x$	$-\cot x + C$
	$\sec x \tan x$	$\sec x + C$
	$\csc x \cot x$	$-\csc x + C$
Inverse functions	$1/\sqrt{1-x^2}$	$\sin^{-1} x + C$
	$1/(1+x^2)$	$\tan^{-1} x + C$
	$1/|x|\sqrt{x^2-1}$	$\sec^{-1} x + C$

You recognize that each integration formula came directly from a differentiation formula. The integral of the cosine is the sine, because the derivative of the sine is the cosine. For emphasis we list three derivatives above three integrals:

$$\frac{d}{dx}(\text{constant}) = 0 \qquad \frac{d}{dx}(x) = 1 \qquad \frac{d}{dx}\left(\frac{x^{n+1}}{n+1}\right) = x^n$$

$$\int 0\,dx = C \qquad \int 1\,dx = x + C \qquad \int x^n dx = \frac{x^{n+1}}{n+1} + C$$

There are two ways to make this list longer. One is to find the derivative of a new $f(x)$. Then f goes in one column and $v = df/dx$ goes in the other column.† The other possibility is to use rules for derivatives to find rules for integrals. That is the way to extend the list, enormously and easily.

---

† We will soon meet e^x, which goes in *both columns*. It is $f(x)$ and also $v(x)$.

RULES FOR INTEGRALS

Among the rules for derivatives, three were of supreme importance. They were
linearity, the *product rule*, and the *chain rule*. Everything flowed from those three.
In the reverse direction (from v to f) this is still true. The three basic methods of
differential calculus also dominate integral calculus:

$$\textit{linearity of derivatives} \rightarrow \textit{linearity of integrals}$$

$$\textit{product rule for derivatives} \rightarrow \textit{integration by parts}$$

$$\textit{chain rule for derivatives} \rightarrow \textit{integrals by substitution}$$

The easiest is linearity, which comes first. Integration by parts will be left for
Section 7.1. This section starts on substitutions, reversing the chain rule to make an
integral simpler.

LINEARITY OF INTEGRALS

What is the integral of $v(x) + w(x)$? Add the two separate integrals. The graph of
$v + w$ has two regions below it, the area under v and the area from v to $v + w$.
Adding areas gives the sum rule. Suppose f and g are antiderivatives of v and w:

sum rule: $f + g$ is an antiderivative of $v + w$

constant rule: cf is an antiderivative of cv

linearity: $af + bg$ is an antiderivative of $av + bw$

This is a case of overkill. The first two rules are special cases of the third, so logically
the last rule is enough. However it is so important to deal quickly with constants— just
"*factor them out*"—that the rule $cv \leftrightarrow cf$ is stated separately. The proofs come from
the linearity of derivatives: $(af + bg)'$ equals $af' + bg'$ which equals $av + bw$.
The rules can be restated with integral signs:

sum rule: $\int \left[v(x) + w(x) \right] dx = \int v(x)\, dx + \int w(x)\, dx$

constant rule: $\int cv(x)\, dx = c \int v(x)\, dx$

linearity: $\int \left[av(x) + bw(x) \right] dx = a \int v(x)\, dx + b \int w(x)\, dx$

Note about the constant in $f(x) + C$. All antiderivatives allow the addition of a
constant. For a combination like $av(x) + bw(x)$, the antiderivative is
$af(x) + bg(x) + C$. *The constants for each part combine into a single constant.*
To give all possible antiderivatives of a function, just remember to write "$+C$" after
one of them. The real problem is to find that one antiderivative.

EXAMPLE 1   The antiderivative of $v = x^2 + x^{-2}$ is $f = x^3/3 + (x^{-1})/(-1) + C$.

EXAMPLE 2   The antiderivative of $6 \cos t + 7 \sin t$ is $6 \sin t - 7 \cos t + C$.

EXAMPLE 3   Rewrite $\dfrac{1}{1 - \sin x}$ as $\dfrac{1 - \sin x}{1 - \sin^2 x} = \dfrac{1 - \sin x}{\cos^2 x} = \sec^2 x - \sec x \tan x$.

The antiderivative is $\tan x - \sec x + C$. That rewriting is done by a symbolic algebra
code (or by you). Differentiation is often simple, so most people check that
$df/dx = v(x)$.

Question   How to integrate $\tan^2 x$?
Method   Write it as $\sec^2 x - 1$.   Answer $\tan x - x + C$.

INTEGRALS BY SUBSTITUTION

We now present the most valuable technique in this section—**substitution**. To see the idea, you have to remember the chain rule:

$$f(g(x)) \quad \text{has derivative} \quad f'(g(x))(dg/dx)$$
$$\sin x^2 \quad \text{has derivative} \quad (\cos x^2)(2x)$$
$$(x^3+1)^5 \quad \text{has derivative} \quad 5(x^3+1)^4(3x^2)$$

If the function on the right is given, the function on the left is its antiderivative! There are two points to emphasize right away:

1. **Constants are no problem—they can always be fixed.** Divide by 2 or 15:

$$\int x \cos(x)^2 dx = \frac{1}{2}\sin(x^2)+C \qquad \int x^2(x^3+1)^4 dx = \frac{1}{15}(x^3+1)^5+C$$

Notice the 2 from x^2, the 5 from the fifth power, and the 3 from x^3.

2. **Choosing the inside function g (or u) commits us to its derivative:**

the integral of $2x \cos x^2$ is $\sin x^2 + C$ $(g = x^2, dg/dx = 2x)$

the integral of $\cos x^2$ is (*failure*)   (no dg/dx)

the integral of $x^2 \cos x^2$ is (*failure*)   (wrong dg/dx)

To substitute g for x^2, we need its derivative. The trick is to spot an inside function whose derivative is present. We can fix constants like 2 or 15, but otherwise dg/dx has to be there. *Very often the inside function g is written u.* We use that letter to state the **substitution rule**, when f is the integral of v:

$$\int v(u(x))\frac{du}{dx}\,dx = f(u(x))+C. \tag{1}$$

EXAMPLE 4 $\int \sin x \cos x \, dx = \frac{1}{2}(\sin x)^2 + C$ $u = \sin x$ (compare Example 6)

EXAMPLE 5 $\int \sin^2 x \cos x \, dx = \frac{1}{3}(\sin x)^3 + C$ $u = \sin x$

EXAMPLE 6 $\int \cos x \sin x \, dx = -\frac{1}{2}(\cos x)^2 + C$ $u = \cos x$ (compare Example 4)

EXAMPLE 7 $\int \tan^4 x \sec^2 x \, dx = \frac{1}{5}(\tan x)^5 + C$ $u = \tan x$

The next example has $u = x^2 - 1$ and $du/dx = 2x$. The key step is choosing u:

EXAMPLE 8 $\int x \, dx/\sqrt{x^2-1} = \sqrt{x^2-1} + C$ $\int x\sqrt{x^2-1}\, dx = \frac{1}{3}(x^2-1)^{3/2} + C$

A **shift** of x (to $x+2$) or a **multiple** of x (rescaling to $2x$) is particularly easy:

EXAMPLES 9–10 $\int (x+2)^3 dx = \frac{1}{4}(x+2)^4 + C$ $\int \cos 2x \, dx = \frac{1}{2}\sin 2x + C$

You will soon be able to do those in your sleep. Officially the derivative of $(x+2)^4$ uses the chain rule. But the inside function $u = x+2$ has $du/dx = 1$. The "1" is there automatically, and the graph shifts over—as in Figure 5.8b.

For Example 10 the inside function is $u = 2x$. Its derivative is $du/dx = 2$. This

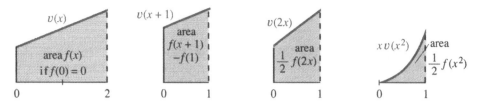

Fig. 5.8   Substituting $u = x + 1$ and $u = 2x$ and $u = x^2$. The last graph has half of $du/dx = 2x$.

required factor 2 is missing in $\int \cos 2x \, dx$, but we put it there by multiplying and dividing by 2. Check the derivative of $\frac{1}{2} \sin 2x$: the 2 from the chain rule cancels the $\frac{1}{2}$. The rule for any nonzero constant is similar:

$$\int v(x+c) \, dx = f(x+c) \quad \text{and} \quad \int v(cx) \, dx = \frac{1}{c} f(cx). \tag{2}$$

Squeezing the graph by c divides the area by c. Now $3x + 7$ rescales *and* shifts:

EXAMPLE 11 $\int \cos(3x+7) \, dx = \frac{1}{3} \sin(3x+7) + C \quad \int (3x+7)^2 dx = \frac{1}{3} \cdot \frac{1}{3}(3x+7)^3 + C$

Remark on writing down the steps   When the substitution is complicated, it is a good idea to get du/dx where you need it. Here $3x^2 + 1$ needs $6x$:

$$\int 7x(3x^2+1)^4 dx = \frac{7}{6} \int (3x^2+1)^4 6x \, dx = \frac{7}{6} \int u^4 \frac{du}{dx} \, dx$$

Now integrate: $\dfrac{7}{6}\dfrac{u^5}{5} + C = \dfrac{7}{6}\dfrac{(3x^2+1)^5}{5} + C.$ (3)

Check the derivative at the end. The exponent 5 cancels 5 in the denominator, $6x$ from the chain rule cancels 6, and $7x$ is what we started with.

Remark on differentials   In place of $(du/dx)dx$, many people just write du:

$$\int (3x^2+1)^4 6x \, dx = \int u^4 du = \tfrac{1}{5}u^5 + C. \tag{4}$$

This really shows how substitution works. *We switch from x to u,* **and we also switch from** dx **to** du. The most common mistake is to confuse dx with du. The factor du/dx from the chain rule is absolutely needed, to reach du. The change of variables (dummy variables anyway!) leaves an easy integral, and then u turns back into $3x^2 + 1$. Here are the four steps to substitute u for x:

1. Choose $u(x)$ and compute du/dx
2. Locate $v(u)$ times du/dx times dx, or $v(u)$ times du
3. Integrate $\int v(u) \, du$ to find $f(u) + C$
4. Substitute $u(x)$ back into this antiderivative f.

EXAMPLE 12 $\int (\cos \sqrt{x}) \, dx/2\sqrt{x} = \int \cos u \, du = \sin u + C = \sin \sqrt{x} + C$
                                    (put in u)   *(integrate)*   *(put back x)*

The choice of u must be right, to change everything from x to u. With ingenuity, some remarkable integrals are possible. But most will remain impossible forever. The functions $\cos x^2$ and $1/\sqrt{4 - \sin^2 x}$ have no "elementary" antiderivative. Those integrals are well defined and they come up in applications—the latter gives the distance around an ellipse. That can be computed to tremendous accuracy, but not to perfect accuracy.

The exercises concentrate on substitutions, which need and deserve practice. We give a **nonexample**—$\int (x^2+1)^2dx$ does not equal $\frac{1}{3}(x^2+1)^3$—to emphasize the need for du/dx. Since $2x$ is missing, $u=x^2+1$ does not work. But we can fix up π:

$$\int \sin \pi x \, dx = \int \sin u \frac{du}{\pi} = -\frac{1}{\pi}\cos u + C = -\frac{1}{\pi}\cos \pi x + C.$$

5.4 EXERCISES

Read-through questions

Finding integrals by substitution is the reverse of the __a__ rule. The derivative of $(\sin x)^3$ is __b__. Therefore the antiderivative of __c__ is __d__. To compute $\int(1+\sin x)^2 \cos x \, dx$, substitute $u=$__e__. Then $du/dx=$__f__ so substitute $du=$__g__. In terms of u the integral is $\int$__h__ $=$__i__. Returning to x gives the final answer.

The best substitutions for $\int \tan(x+3)\sec^2(x+3)\,dx$ and $\int(x^2+1)^{10}x\,dx$ are $u=$__j__ and $u=$__k__. Then $du=$__l__ and __m__. The answers are __n__ and __o__. The antiderivative of $v\,dv/dx$ is __p__. $\int 2x\,dx/(1+x^2)$ leads to $\int$__q__, which we don't yet know. The integral $\int dx/(1+x^2)$ is known immediately as __r__.

Find the indefinite integrals in 1–20.

1 $\int \sqrt{2+x}\,dx$ (add $+C$)
2 $\int \sqrt{3-x}\,dx$ (always $+C$)
3 $\int (x+1)^n\,dx$
4 $\int (x+1)^{-n}\,dx$
5 $\int (x^2+1)^5 x\,dx$
6 $\int \sqrt{1-3x}\,dx$
7 $\int \cos^3 x \sin x\,dx$
8 $\int \cos x\,dx/\sin^3 x$
9 $\int \cos^3 2x \sin 2x\,dx$
10 $\int \cos^3 x \sin 2x\,dx$
11 $\int dt/\sqrt{1-t^2}$
12 $\int t\sqrt{1-t^2}\,dt$
13 $\int t^3 dt/\sqrt{1+t^2}$
14 $\int t^3\sqrt{1-t^2}\,dt$
15 $\int (1+\sqrt{x})\,dx/\sqrt{x}$
16 $\int (1+x^{3/2})\sqrt{x}\,dx$
17 $\int \sec x \tan x\,dx$
18 $\int \sec^2 x \tan^2 x\,dx$
19 $\int \cos x \tan x\,dx$
20 $\int \sin^3 x\,dx$

In 21–32 find a function $y(x)$ that solves the differential equation.

21 $dy/dx = x^2+\sqrt{x}$
22 $dy/dx = y^2$ (try $y=cx^n$)
23 $dy/dx = \sqrt{1-2x}$
24 $dy/dx = 1/\sqrt{1-2x}$
25 $dy/dx = 1/y$
26 $dy/dx = x/y$
27 $d^2y/dx^2 = 1$
28 $d^5y/dx^5 = 1$
29 $d^2y/dx^2 = -y$
30 $dy/dx = \sqrt{xy}$
31 $d^2y/dx^2 = \sqrt{x}$
32 $(dy/dx)^2 = \sqrt{x}$

33 **True or false**, when f is an antiderivative of v:
(a) $\int v(u(x))\,dx = f(u(x))+C$
(b) $\int v^2(x)\,dx = \frac{1}{3}f^3(x)+C$
(c) $\int v(x)(du/dx)\,dx = f(u(x))+C$
(d) $\int v(x)(dv/dx)\,dx = \frac{1}{2}f^2(x)+C$

34 **True or false**, when f is an antiderivative of v:
(a) $\int f(x)(dv/dx)\,dx = \frac{1}{2}f^2(x)+C$
(b) $\int v(v(x))(dv/dx)\,dx = f(v(x))+C$
(c) Integral is inverse to derivative so $f(v(x))=x$
(d) Integral is inverse to derivative so $\int(df/dx)\,dx = f(x)$

35 If $df/dx = v(x)$ then $\int v(x-1)\,dx = $____ and $\int v(x/2)\,dx = $____.

36 If $df/dx = v(x)$ then $\int v(2x-1)\,dx = $____ and $\int v(x^2)x\,dx = $____.

37 $\dfrac{x^2}{1+x^2} = 1 - \dfrac{1}{1+x^2}$ so $\int \dfrac{x^2dx}{1+x^2} = $____.

38 $\int (x^2+1)^2 dx$ is not $\frac{1}{3}(x^2+1)^3$ but ____.

39 $\int 2x\,dx/(x^2+1)$ is $\int$____ du which will soon be $\ln u$.

40 Show that $\int 2x^3 dx/(1+x^2)^3 = \int (u-1)du/u^3 = $____.

41 The acceleration $d^2 f/dt^2 = 9.8$ gives $f(t) = $____ (two integration constants).

42 The solution to $d^4 y/dx^4 = 0$ is ____ (four constants).

43 If $f(t)$ is an antiderivative of $v(t)$, find antiderivatives of
(a) $v(t+3)$ (b) $v(t)+3$ (c) $3v(t)$ (d) $v(3t)$.

5.5 The Definite Integral

The integral of $v(x)$ is an antiderivative $f(x)$ plus a constant C. This section takes two steps. First, we choose C. Second, we construct $f(x)$. The object is *to define the integral*—in the most frequent case when a suitable $f(x)$ is not directly known.

The indefinite integral contains "$+C$." The constant is not settled because $f(x) + C$ has the same slope for every C. When we care only about the derivative, C makes no difference. When the goal is a number—a *definite integral*—C can be assigned a definite value at the starting point.

For mileage traveled, *we subtract the reading at the start*. This section does the same for area. Distance is $f(t)$ and area is $f(x)$—while the definite integral is $f(b) - f(a)$. Don't pay attention to t or x, pay attention to the great formula of integral calculus:

$$\int_a^b v(t)\,dt = \int_a^b v(x)\,dx = f(b) - f(a). \tag{1}$$

> *Viewpoint* 1 : When f is known, the equation gives the area from a to b.
> *Viewpoint* 2 : When f is *not* known, the equation defines f from the area.

For a typical $v(x)$, we can't find $f(x)$ by guessing or substitution. But still $v(x)$ has an "area" under its graph—and this yields the desired integral $f(x)$.

Most of this section is theoretical, leading to the definition of the integral. You may think we should have defined integrals before computing them—which is logically true. But the idea of area (and the use of rectangles) was already pretty clear in our first examples. Now we go much further. *Every continuous function $v(x)$ has an integral* (also some discontinuous functions). Then the Fundamental Theorem completes the circle: The integral leads back to $df/dx = v(x)$. The area up to x is the antiderivative that we couldn't otherwise discover.

THE CONSTANT OF INTEGRATION

Our goal is to turn $f(x) + C$ into a definite integral— the area between a and b. The first requirement is to have *area = zero* at the start:

$$f(a) + C = \text{starting area} = 0 \quad \text{so} \quad C = -f(a). \tag{2}$$

For the area up to x (moving endpoint, indefinite integral), use t as the dummy variable:

> *the area from a to x is* $\int_a^x v(t)\,dt = f(x) - f(a)$ (*indefinite integral*)
>
> *the area from a to b is* $\int_a^b v(x)\,dx = f(b) - f(a)$ (*definite integral*)

EXAMPLE 1   The area under the graph of $5(x+1)^4$ from a to b has $f(x) = (x+1)^5$:

$$\int_a^b 5(x+1)^4 dx = (x+1)^5 \Big]_a^b = (b+1)^5 - (a+1)^5.$$

The calculation has two separate steps—first find $f(x)$, then substitute b and a. After the first step, check that df/dx is v. The upper limit in the second step gives *plus*

$f(b)$, the lower limit gives *minus* $f(a)$. Notice the brackets (or the vertical bar):

$$f(x)\Big]_a^b = f(b) - f(a) \quad x^3\big|_1^2 = 8 - 1 \quad \Big[\cos x\Big]_0^{2t} = \cos 2t - 1.$$

Changing the example to $f(x) = (x+1)^5 - 1$ gives an equally good antiderivative—and now $f(0) = 0$. But $f(b) - f(a)$ stays the same, because the -1 disappears:

$$\Big[(x+1)^5 - 1\Big]_a^b = ((b+1)^5 - 1) - ((a+1)^5 - 1) = (b+1)^5 - (a+1)^5.$$

EXAMPLE 2 When $v = 2x \sin x^2$ we recognize $f = -\cos x^2$. The area from 0 to 3 is

$$\int_0^3 2x \sin x^2 \, dx = -\cos x^2 \Big]_0^3 = -\cos 9 + \cos 0.$$

The upper limit copies the minus sign. The lower limit gives $-(-\cos 0)$, which is $+\cos 0$. *That example shows the right form for solving exercises on definite integrals.*

Example 2 jumped directly to $f(x) = -\cos x^2$. But most problems involving the chain rule go more slowly—by *substitution*. Set $u = x^2$, with $du/dx = 2x$:

$$\int_0^3 2x \sin x^2 \, dx = \int_0^3 \sin u \frac{du}{dx} \, dx = \int_?^? \sin u \, du. \qquad (3)$$

We need new limits when u replaces x^2. Those limits on u are a^2 and b^2. (In this case $a^2 = 0^2$ and $b^2 = 3^2 = 9$.) *If x goes from a to b, then u goes from $u(a)$ to $u(b)$.*

$$\int_a^b v(u(x)) \frac{du}{dx} \, dx = \int_{u(a)}^{u(b)} v(u) \, du = f(u(b)) - f(u(a)). \qquad (4)$$

EXAMPLE 3 $$\int_{x=0}^1 (x^2+5)^3 x \, dx = \int_{u=5}^6 u^3 \frac{du}{2} = \frac{u^4}{8}\Big]_5^6 = \frac{6^4}{8} - \frac{5^4}{8}.$$

In this case $u = x^2 + 5$. Therefore $du/dx = 2x$ (or $du = 2x \, dx$ for differentials). We have to account for the missing 2. The integral is $\frac{1}{8}u^4$. The limits on $u = x^2 + 5$ are $u(0) = 0^2 + 5$ and $u(1) = 1^2 + 5$. That is why the u-integral goes from 5 to 6. The alternative is to find $f(x) = \frac{1}{8}(x^2+5)^4$ in one jump (*and check it*).

EXAMPLE 4 $\int_0^1 \sin x^2 \, dx = ??$ (*no elementary function gives this integral*).

If we try $\cos x^2$, the chain rule produces an extra $2x$—no adjustment will work. Does $\sin x^2$ still have an antiderivative? *Yes!* Every continuous $v(x)$ has an $f(x)$. Whether $f(x)$ has an algebraic formula or not, we can write it as $\int v(x)dx$. To define that integral, we now take the limit of rectangular areas.

INTEGRALS AS LIMITS OF "RIEMANN SUMS"

We have come to the *definition of the integral*. The chapter started with the integrals of x and x^2, from formulas for $1 + \cdots + n$ and $1^2 + \cdots + n^2$. We will not go back to those formulas. But for other functions, too irregular to find exact sums, the rectangular areas also approach a limit.

That limit is the integral. This definition is a major step in the theory of calculus. It can be studied in detail, or understood in principle. The truth is that the definition is not so painful—you virtually know it already.

Problem Integrate the continuous function $v(x)$ over the interval $[a,b]$.

Step 1 Split $[a,b]$ into n subintervals $[a,x_1],[x_1,x_2],\ldots,[x_{n-1},b]$.

The "meshpoints" $x_1,x_2,\ldots$ divide up the interval from a to b. The endpoints are $x_0=a$ and $x_n=b$. The length of subinterval k is $\Delta x_k=x_k-x_{k-1}$. In that smaller interval, the minimum of $v(x)$ is m_k. The maximum is M_k.

Now construct rectangles. The "*lower rectangle*" over interval k has height m_k. The "*upper rectangle*" reaches to M_k. Since v is continuous, there are points $x_{\min}$ and $x_{\max}$ where $v=m_k$ and $v=M_k$ (extreme value theorem). ***The graph of $v(x)$ is in between***.

Important: The area under $v(x)$ contains the area "s" of the lower rectangles:

$$\int_a^b v(x)\,dx \geqslant m_1\,\Delta x_1 + m_2\,\Delta x_2 + \cdots + m_n\,\Delta x_n = s. \qquad (5)$$

The area under $v(x)$ is contained in the area "S" of the upper rectangles:

$$\int_a^b v(x)\,dx \leqslant M_1\,\Delta x_1 + M_2\,\Delta x_2 + \cdots + M_n\,\Delta x_n = S. \qquad (6)$$

The ***lower sum s*** and the ***upper sum S*** were computed earlier in special cases—when v was x or x^2 and the spacings Δx were equal. Figure 5.9a shows why $s\leqslant area\leqslant S$.

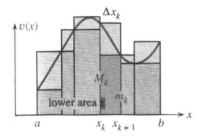
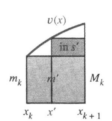
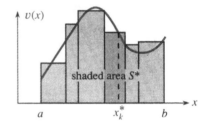

Fig. 5.9 Area of lower rectangles $=s$. Upper sum S includes top pieces. Riemann sum S^* is in between.

Notice an important fact. When a new dividing point x' is added, *the lower sum increases*. The minimum in one piece can be greater (see second figure) than the original m_k. Similarly *the upper sum decreases*. The maximum in one piece can be below the overall maximum. ***As new points are added, s goes up and S comes down***. So the sums come closer together:

$$s \leqslant s' \leqslant \qquad \leqslant S' \leqslant S. \qquad (7)$$

I have left space in between for the curved area—the integral of $v(x)$.

Now add more and more meshpoints in such a way that $\Delta x_{\max}\to 0$. The lower sums increase and the upper sums decrease. They never pass each other. ***If $v(x)$ is continuous, those sums close in on a single number A***. That number is the definite integral—the area under the graph.

DEFINITION The area A is the common limit of the lower and upper sums:

$$s \to A \text{ and } S \to A \text{ as } \Delta x_{\max} \to 0. \qquad (8)$$

This limit A exists for all continuous $v(x)$, and also for some discontinuous functions. When it exists, A is the "***Riemann integral***" of $v(x)$ from a to b.

REMARKS ON THE INTEGRAL

As for derivatives, so for integrals: The definition involves a limit. Calculus is built on limits, and we always add "if the limit exists." That is the delicate point. I hope the next five remarks (increasingly technical) will help to distinguish functions that are *Riemann integrable* from functions that are not.

Remark 1   The sums s and S may fail to approach the same limit. A standard example has $V(x) = 1$ at all fractions $x = p/q$, and $V(x) = 0$ at all other points. Every interval contains rational points (fractions) and irrational points (nonrepeating decimals). *Therefore $m_k = 0$ and $M_k = 1$.* The lower sum is always $s = 0$. The upper sum is always $S = b - a$ (the sum of 1's times Δx's). **The gap in equation** (7) **stays open**. This function $V(x)$ is not Riemann integrable. The area under its graph is not defined (at least by Riemann—see Remark 5).

Remark 2   The *step function* $U(x)$ is discontinuous but still integrable. On every interval the minimum m_k equals the maximum M_k—except on the interval containing the jump. That jump interval has $m_k = 0$ and $M_k = 1$. But when we multiply by Δx_k, and require $\Delta x_{\max} \to 0$, the difference between s and S goes to zero. The area under a step function is clear—the rectangles fit exactly.

Remark 3   With patience another key step could be proved: **If $s \to A$ and $S \to A$ for one sequence of meshpoints, then this limit A is approached by every choice of meshpoints with $\Delta x_{\max} \to 0$.** The integral is the lower bound of all upper sums S, and it is the upper bound of all possible s—provided those bounds are equal. The gap must close, to define the integral.

The same limit A is approached by "in-between rectangles." The height $v(x_k^*)$ can be computed at any point x_k^* in subinterval k. See Figures 5.9c and 5.10. Then the total rectangular area is a "***Riemann sum***" between s and S:

$$S^* = v(x_1^*)\,\Delta x_1 + v(x_2^*)\,\Delta x_2 + \cdots + v(x_n^*)\,\Delta x_n. \tag{9}$$

We cannot tell whether the true area is above or below S^*. Very often A is closer to S^* than to s or S. The *midpoint rule* takes x^* in the middle of its interval (Figure 5.10), and Section 5.8 will establish its extra accuracy. The extreme sums s and S are used in the definition while S^* is used in computation.

$v(x)$

   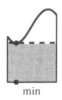

left   right   mid   min   max   any x_k^*

Fig. 5.10   Various positions for x_k^* in the base. The rectangles have height $v(x_k^*)$.

Remark 4   *Every continuous function is Riemann integrable*. The proof is optional (in my class), but it belongs here for reference. It starts with continuity at x^*: "*For any ε there is a δ ....*" When the rectangles sit between $x^* - \delta$ and $x^* + \delta$, the bounds M_k and m_k differ by less than 2ε. Multiplying by the base Δx_k, the areas differ by less than $2\varepsilon(\Delta x_k)$. Combining all rectangles, the upper and lower sums differ by less than $2\varepsilon(\Delta x_1 + \Delta x_2 + \cdots + \Delta x_n) = 2\varepsilon(b - a)$.

As $\varepsilon \to 0$ we conclude that S comes arbitrarily close to s. They squeeze in on a single number A. The Riemann sums approach the Riemann integral, *if v is continuous*.

Two problems are hidden by that reasoning. One is at the end, where S and s come together. We have to know that the line of real numbers has no "holes," so there is a number A to which these sequences converge. That is true.

Any increasing sequence, if it is bounded above, approaches a limit.

The decreasing sequence S, bounded below, converges to the same limit. So A exists.

The other problem is about continuity. We assumed without saying so that the width 2δ is the same around every point x^*. We did not allow for the possibility that δ might approach zero where $v(x)$ is rapidly changing—in which case an infinite number of rectangles could be needed. Our reasoning requires that

$v(x)$ is **uniformly continuous**: δ *depends on* ε **but not on the position of** x^*.

For each ε there is a δ that works at all points in the interval. *A continuous function on a closed interval is **uniformly** continuous.* This fact (proof omitted) makes the reasoning correct, and $v(x)$ is integrable.

On an infinite interval, even $v = x^2$ is not uniformly continuous. It changes across a subinterval by $(x^* + \delta)^2 - (x^* - \delta)^2 = 4x^*\delta$. As x^* gets larger, δ must get smaller— to keep $4x^*\delta$ below ε. No single δ succeeds at all x^*. But on a finite interval $[0, b]$, the choice $\delta = \varepsilon/4b$ works everywhere—so $v = x^2$ is uniformly continuous.

Remark 5   If those four remarks were fairly optional, this one is totally at your discretion. Modern mathematics needs to integrate the zero-one function $V(x)$ in the first remark. Somehow V has more 0's than 1's. The fractions (where $V(x) = 1$) can be put in a list, but the irrational numbers (where $V(x) = 0$) are "uncountable." The integral ought to be zero, but Riemann's upper sums all involve $M_k = 1$.

Lebesgue discovered a major improvement. He allowed *infinitely many subintervals* (smaller and smaller). Then all fractions can be covered with intervals of total width ε. (Amazing, when the fractions are packed so densely.) The idea is to cover $1/q, 2/q, \ldots, q/q$ by narrow intervals of total width $\varepsilon/2^q$. Combining all $q = 1, 2, 3, \ldots,$ the total width to cover all fractions is no more than $\varepsilon\left(\frac{1}{2} + \frac{1}{4} + \frac{1}{8} + \cdots\right) = \varepsilon$. Since $V(x) = 0$ everywhere else, the upper sum S is only ε. And since ε was arbitrary, the "*Lebesgue integral*" is zero as desired.

That completes a fair amount of theory, possibly more than you want or need— but it is satisfying to get things straight. The definition of the integral is still being studied by experts (and so is the derivative, again to allow more functions). By contrast, the *properties* of the integral are used by everybody. Therefore the next section turns from definition to properties, collecting the rules that are needed in applications. They are very straightforward.

Read-through questions

In $\int_a^x v(t)\,dt = f(x) + C$, the constant C equals __a__. Then at $x = a$ the integral is __b__. At $x = b$ the integral becomes __c__. The notation $f(x)]_a^b$ means __d__. Thus $\cos x]_0^\pi$ equals __e__. Also $[\cos x + 3]_0^\pi$ equals __f__, which shows why the antiderivative includes an arbitrary __g__. Substituting $u = 2x - 1$ changes $\int_1^3 \sqrt{2x-1}\,dx$ into __h__ (with limits on u).

The integral $\int_a^b v(x)\,dx$ can be defined for any __i__ function $v(x)$, even if we can't find a simple __j__. First the meshpoints $x_1, x_2, \ldots$ divide $[a,b]$ into subintervals of length $\Delta x_k = $ __k__. The upper rectangle with base Δx_k has height $M_k = $ __l__. The upper sum S is equal to __m__. The lower sum s is __n__. The __o__ is between s and S. As more meshpoints are added, S __p__ and s __q__. If S and s approach the same __r__, that defines the integral. The intermediate sums S^*, named after __s__, use rectangles of height $v(x_k^*)$. Here x_k^* is any point between __t__, and $S^* = $ __u__ approaches the area.

If $v(x) = df/dx$, what constants C make 1–10 true?

1 $\int_2^b v(x)\,dx = f(b) + C$

2 $\int_1^4 v(x)\,dx = f(4) + C$

3 $\int_x^3 v(t)\,dt = -f(x) + C$

4 $\int_{\pi/2}^b v(\sin x)\cos x\,dx = f(\sin b) + C$

5 $\int_1^x v(t)\,dt = f(t) + C$ (careful)

6 $df/dx = v(x) + C$

7 $\int_0^1 (x^2 - 1)^3 2x\,dx = \int_{-1}^C u^3\,du.$

8 $\int_0^{x^2} v(t)\,dt = f(x^2) + C$

9 $\int_a^b v(-x)\,dx = C$ (change $-x$ to t; also dx and limits)

10 $\int_0^2 v(x)\,dx = C \int_0^1 v(2t)\,dt.$

Choose $u(x)$ in 11–18 and *change limits*. Compute the integral in 11-16.

11 $\int_0^1 (x^2 + 1)^{10} x\,dx$ 12 $\int_0^{\pi/2} \sin^8 x \cos x\,dx$

13 $\int_0^{\pi/4} \tan x \sec^2 x\,dx$ 14 $\int_0^2 x^{2n+1}\,dx$ (take $u = x^2$)

15 $\int_0^{\pi/4} \sec^2 x \tan x\,dx$ 16 $\int_0^1 x\,dx/\sqrt{1-x^2}$

17 $\int_1^2 dx/x$ (take $u = 1/x$) 18 $\int_0^1 x^3(1-x)^3\,dx$ $(u = 1-x)$

With $\Delta x = \frac{1}{2}$ in 19–22, find the maximum M_k and minimum m_k and upper and lower sums S and s.

19 $\int_0^1 (x^2 + 1)^4\,dx$ 20 $\int_0^1 \sin 2\pi x\,dx$

21 $\int_0^2 x^3\,dx$ 22 $\int_{-1}^1 x\,dx.$

23 Repeat 19 and 20 with $\Delta x = \frac{1}{4}$ and compare with the correct answer.

24 The difference $S - s$ in 21 is the area $2^3 \Delta x$ of the far right rectangle. Find Δx so that $S < 4.001$.

25 If $v(x)$ is *increasing* for $a \leqslant x \leqslant b$, the difference $S - s$ is the area of the _____ rectangle minus the area of the _____ rectangle. Those areas approach zero. *So every increasing function on $[a,b]$ is Riemann integrable.*

26 Find the Riemann sum S^* for $V(x)$ in Remark 1, when $\Delta x = 1/n$ and each x_k^* is the midpoint. This S^* is well-behaved but still $V(x)$ is not Riemann integrable.

27 $W(x)$ equals 1 at $x = \frac{1}{2}, \frac{1}{4}, \frac{1}{8}, \ldots$, and elsewhere $W(x) = 0$. For $\Delta x = .01$ find the upper sum S. Is $W(x)$ integrable?

28 Suppose $M(x)$ is a *multistep* function with jumps of $\frac{1}{2}, \frac{1}{4}, \frac{1}{8}, \ldots$, at the points $x = \frac{1}{2}, \frac{1}{4}, \frac{1}{8}, \ldots$. Draw a rough graph with $M(0) = 0$ and $M(1) = 1$. With $\Delta x = \frac{1}{3}$ find S and s.

29 For $M(x)$ in Problem 28 find the difference $S - s$ (which approaches zero as $\Delta x \to 0$). What is the area under the graph?

30 If $df/dx = -v(x)$ and $f(1) = 0$, explain $f(x) = \int_x^1 v(t)\,dt$.

31 (a) If $df/dx = +v(x)$ and $f(0) = 3$, find $f(x)$.

(b) If $df/dx = +v(x)$ and $f(3) = 0$, find $f(x)$.

32 In your own words define the integral of $v(x)$ from a to b.

33 *True or false*, with reason or example.

(a) Every continuous $v(x)$ has an antiderivative $f(x)$.

(b) If $v(x)$ is not continuous, S and s approach different limits.

(c) If S and s approach A as $\Delta x \to 0$, then all Riemann sums S^* in equation (9) also approach A.

(d) If $v_1(x) + v_2(x) = v_3(x)$, their upper sums satisfy $S_1 + S_2 = S_3$.

(e) If $v_1(x) + v_2(x) = v_3(x)$, their Riemann sums at the midpoints x_k^* satisfy $S_1^* + S_2^* = S_3^*$.

(f) The midpoint sum is the average of S and s.

(g) One x_k^* in Figure 5.10 gives the exact area.

5.6 Properties of the Integral and Average Value

The previous section reached the definition of $\int_a^b v(x)\,dx$. But the subject cannot stop there. The integral was defined in order to be used. Its properties are important, and its applications are even more important. The definition was chosen so that the integral has properties that make the applications possible.

One direct application is to the **average value** of $v(x)$. The average of n numbers is clear, and the integral extends that idea—it produces the average of a whole continuum of numbers $v(x)$. This develops from the last rule in the following list (Property 7). We now collect together **seven basic properties of definite integrals**.

The addition rule for $\int [v(x) + w(x)]\,dx$ will not be repeated—even though this property of linearity is the most fundamental. We start instead with a different kind of addition. There is only one function $v(x)$, but now there are two intervals.

The integral from a to b is added to its neighbor from b to c. Their sum is the integral from a to c. That is the first (not surprising) property in the list.

Property **1**   Areas over neighboring intervals add to the area over the combined interval:

$$\int_a^b v(x)\,dx + \int_b^c v(x)\,dx = \int_a^c v(x)\,dx. \tag{1}$$

This sum of areas is graphically obvious (Figure 5.11a). It also comes from the formal definition of the integral. Rectangular areas obey (1)—with a meshpoint at $x = b$ to make sure. When $\Delta x_{\max}$ approaches zero, their limits also obey (1). *All the normal rules for rectangular areas are obeyed in the limit by integrals.*

Property **1** is worth pursuing. It indicates how to define the integral when $a = b$. The integral "*from b to b*" is the area over a point, which we expect to be zero. It is.

Property **2** $\int_b^b v(x)\,dx = 0.$

That comes from Property **1** when $c = b$. Equation (1) has two identical integrals, so the one from b to b must be zero. Next we see what happens if $c = a$—which makes the second integral go from b to a.

What happens when **an integral goes backward**? The "lower limit" is now the larger number b. The "upper limit" a is smaller. Going backward reverses the sign:

Property **3** $\int_b^a v(x)\,dx = -\int_a^b v(x)\,dx = f(a) - f(b).$

Proof When $c = a$ the right side of (1) is zero. Then the integrals on the left side must cancel, which is Property **3**. *In going from b to a the steps Δx are negative.* That justifies a minus sign on the rectangular areas, and a minus sign on the integral (Figure 5.11b). *Conclusion*: Property **1** holds for any ordering of a, b, c.

EXAMPLES $\displaystyle\int_x^0 t^2\,dt = -\frac{x^3}{3}$ $\displaystyle\int_1^0 dt = -1$ $\displaystyle\int_2^2 \frac{dt}{t} = 0$

Property **4**   For odd functions $\int_{-a}^a v(x)\,dx = 0$. "**Odd**" means that $v(-x) = -v(x)$. For even functions $\int_{-a}^a v(x)\,dx = 2\int_0^a v(x)\,dx$. "**Even**" means that $v(-x) = +v(x)$.

The functions $x, x^3, x^5, \ldots$ are odd. If x changes sign, these powers change sign. The functions $\sin x$ and $\tan x$ are also odd, together with their inverses. This is an important family of functions, and ***the integral of an odd function from $-a$ to a equals zero***. Areas cancel:

$$\int_{-a}^{a} 6x^5 \, dx = x^6]_{-a}^{a} = a^6 - (-a)^6 = 0.$$

If $v(x)$ is odd then $f(x)$ is even! All powers $1, x^2, x^4, \ldots$ are even functions. ***Curious fact***: Odd function times even function is *odd*, but odd number times even number is *even*.

For even functions, areas add: $\int_{-a}^{a} \cos x \, dx = \sin a - \sin(-a) = 2 \sin a$.

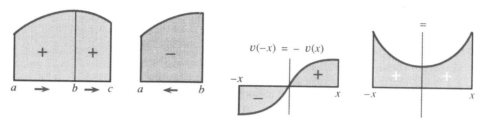

Fig. 5.11 Properties **1–4**: Add areas, change sign to go backward, odd cancels, even adds.

The next properties involve inequalities. If $v(x)$ is positive, the area under its graph is positive (not surprising). Now we have a proof: The lower sums s are positive and they increase toward the area integral. So the integral is positive:

Property 5   If $v(x) > 0$ for $a < x < b$ then $\int_a^b v(x) \, dx > 0$.

A positive velocity means a positive distance. A positive v lies above a positive area. A more general statement is true. Suppose $v(x)$ stays between a lower function $l(x)$ and an upper function $u(x)$. Then the rectangles for v stay between the rectangles for l and u. In the limit, the area under v (Figure 5.12) is between the areas under l and u:

Property 6   If $l(x) \leqslant v(x) \leqslant u(x)$ for $a \leqslant x \leqslant b$ then

$$\int_a^b l(x) \, dx \leqslant \int_a^b v(x) \, dx \leqslant \int_a^b u(x) \, dx. \tag{2}$$

EXAMPLE 1 $\cos t \leqslant 1 \quad \Rightarrow \quad \int_0^x \cos t \, dt \leqslant \int_0^x 1 \, dt \quad \Rightarrow \quad \sin x \leqslant x$

EXAMPLE 2 $1 \leqslant \sec^2 t \quad \Rightarrow \quad \int_0^x 1 \, dt \leqslant \int_0^x \sec^2 t \, dt \quad \Rightarrow \quad x \leqslant \tan x$

EXAMPLE 3   Integrating $\dfrac{1}{1+x^2} \leqslant 1$ leads to $\tan^{-1} x \leqslant x$.

All those examples are for $x > 0$. You may remember that Section 2.4 used geometry to prove $\sin h < h < \tan h$. Examples 1–2 seem to give new and shorter proofs. But I think the reasoning is doubtful. The inequalities were needed to compute the derivatives (therefore the integrals) in the first place.

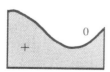 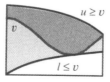 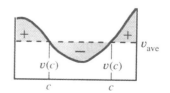

Fig. 5.12 Properties **5–7**: v above zero, v between l and u, average value (+ balances −).

Property 7   (*Mean Value Theorem for Integrals*)   If $v(x)$ is continuous, there is a point c between a and b where $v(c)$ equals the average value of $v(x)$:

$$v(c) = \frac{1}{b-a} \int_a^b v(x)\,dx = \text{“\textit{average value of} } v(x)\text{.”}} \qquad (3)$$

This is the same as the ordinary Mean Value Theorem (for the derivative of $f(x)$):

$$f'(c) = \frac{f(b) - f(c)}{b-a} = \text{“average slope of } f\text{.”} \qquad (4)$$

With $f' = v$, (3) and (4) are the same equation. But honesty makes me admit to a flaw in the logic. We need the Fundamental Theorem of Calculus to guarantee that $f(x) = \int_a^x v(t)\,dt$ really gives $f' = v$.

A direct proof of (3) places one rectangle across the interval from a to b. Now raise the top of that rectangle, starting at v_{min} (the bottom of the curve) and moving up to v_{max} (the top of the curve). At some height the area will be just right—equal to the area under the curve. Then the rectangular area, which is $(b-a)$ times $v(c)$, equals the curved area $\int_a^b v(x)\,dx$. This is equation (3).

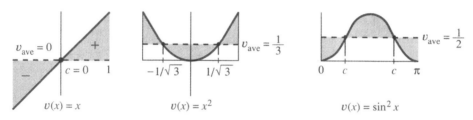

Fig. 5.13   Mean Value Theorem for integrals: area$/(b-a)$ = average height $= v(c)$ at some c.

That direct proof uses the *intermediate value theorem*: A continuous function $v(x)$ takes on every height between v_{min} and v_{max}. At some point (at two points in Figure 5.12c) the function $v(x)$ equals its own average value.

Figure 5.13 shows equal areas above and below the average height $v(c) =$ v_{ave}.

EXAMPLE 4   The average value of an odd function is *zero* (between -1 and 1):

$$\frac{1}{2} \int_{-1}^1 x\,dx = \left. \frac{x^2}{4} \right]_{-1}^1 = \frac{1}{4} - \frac{1}{4} = 0 \qquad \left(\text{note } \frac{1}{b-a} = \frac{1}{2} \right)$$

For once we know c. It is the center point $x = 0$, where $v(c) = v_{ave} = 0$.

EXAMPLE 5   The average value of x^2 is $\frac{1}{3}$ (between 1 and -1):

$$\frac{1}{2} \int_{-1}^1 x^2\,dx = \left. \frac{x^3}{6} \right]_{-1}^1 = \frac{1}{6} - \left(-\frac{1}{6} \right) = \frac{1}{3} \qquad \left(\text{note } \frac{1}{b-a} = \frac{1}{2} \right)$$

Where does this function x^2 equal its average value $\frac{1}{3}$? That happens when $c^2 = \frac{1}{3}$, so c can be either of the points $1/\sqrt{3}$ and $-1/\sqrt{3}$ in Figure 5.13b. Those are the *Gauss points*, which are terrific for numerical integration as Section 5.8 will show.

EXAMPLE 6 The average value of $\sin^2 x$ over a period (zero to π) is $\frac{1}{2}$:

$$\frac{1}{\pi} \int_0^\pi \sin x^2 \, dx = \frac{x - \sin x \cos x}{2\pi} \Big]_0^\pi = \frac{1}{2} \qquad \left(\text{note } \frac{1}{b-a} = \frac{1}{\pi} \right)$$

The point c is $\pi/4$ or $3\pi/4$, where $\sin^2 c = \frac{1}{2}$. **The graph of $\sin^2 x$ oscillates around its average value $\frac{1}{2}$.** See the figure or the formula:

$$\sin^2 x = \frac{1}{2} - \frac{1}{2} \cos 2x. \tag{5}$$

The steady term is $\frac{1}{2}$, the oscillation is $-\frac{1}{2} \cos 2x$. The integral is $f(x) = \frac{1}{2}x - \frac{1}{4} \sin 2x$, which is the same as $\frac{1}{2}x - \frac{1}{2} \sin x \cos x$. *This integral of $\sin^2 x$ will be seen again.* Please verify that $df/dx = \sin^2 x$.

THE AVERAGE VALUE AND EXPECTED VALUE

The "average value" from a to b is the integral divided by the length $b - a$. This was computed for x and x^2 and $\sin^2 x$, but not explained. It is a major application of the integral, and it is guided by the ordinary average of n numbers:

$$v_{\text{ave}} = \frac{1}{b-a} \int_a^b v(x) \, dx \qquad \text{comes from} \qquad v_{\text{ave}} = \frac{1}{n}(v_1 + v_2 + \cdots + v_n).$$

Integration is parallel to summation! Sums approach integrals. Discrete averages approach continuous averages. The average of $\frac{1}{3}, \frac{2}{3}, \frac{3}{3}$ is $\frac{2}{3}$. The average of $\frac{1}{5}, \frac{2}{5}, \frac{3}{5}, \frac{4}{5}, \frac{5}{5}$ is $\frac{3}{5}$. The average of n numbers from $1/n$ to n/n is

$$v_{\text{ave}} = \frac{1}{n} \left(\frac{1}{n} + \frac{2}{n} + \cdots + \frac{n}{n} \right) = \frac{n+1}{2n}. \tag{7}$$

The middle term gives the average, when n is odd. Or we can do the addition. As $n \to \infty$ the sum approaches an integral (do you see the rectangles?). The ordinary average of numbers becomes the continuous average of $v(x) = x$:

$$\frac{n+1}{2n} \to \frac{1}{2} \quad \text{and} \quad \int_0^1 x \, dx = \frac{1}{2} \quad \left(\text{note } \frac{1}{b-a} = 1 \right)$$

In ordinary language: "The average value of the numbers between 0 and 1 is $\frac{1}{2}$." Since a whole continuum of numbers lies between 0 and 1, that statement is meaningless until we have integration.

The average value of the *squares* of those numbers is $(x^2)_{\text{ave}} = \int x^2 \, dx/(b-a) = \frac{1}{3}$. *If you pick a number randomly between 0 and 1, its expected value is $\frac{1}{2}$ and its expected square is $\frac{1}{3}$.*

To me that sentence is a puzzle. First, we don't expect the number to be exactly $\frac{1}{2}$—so we need to define "expected value." Second, if the expected value is $\frac{1}{2}$, why is the expected square equal to $\frac{1}{3}$ instead of $\frac{1}{4}$? The ideas come from probability theory, and calculus is leading us to **continuous probability**. We introduce it briefly here, and come back to it in Chapter 8.

PREDICTABLE AVERAGES FROM RANDOM EVENTS

Suppose you throw a pair of dice. The outcome is not predictable. Otherwise why throw them? *But the average over more and more throws is totally predictable.* We don't know what will happen, but we know its probability.

For dice, we are adding two numbers between 1 and 6. The outcome is between 2 and 12. The probability of 2 is the chance of two ones: $(1/6)(1/6) = 1/36$. Beside each outcome we can write its probability:

$$2\left(\frac{1}{36}\right) 3\left(\frac{2}{36}\right) 4\left(\frac{3}{36}\right) 5\left(\frac{4}{36}\right) 6\left(\frac{5}{36}\right) 7\left(\frac{6}{36}\right) 8\left(\frac{5}{36}\right) 9\left(\frac{4}{36}\right) 10\left(\frac{3}{36}\right) 11\left(\frac{2}{36}\right) 12\left(\frac{1}{36}\right)$$

To repeat, one roll is unpredictable. Only the probabilities are known, and they add to 1. (Those fractions add to 36/36; all possibilities are covered.) The total from a million rolls is even more unpredictable—it can be anywhere between $2,000,000$ and $12,000,000$. Nevertheless the *average* of those million outcomes is almost completely predictable. This *expected value* is found by adding the products in that line above:

Expected value: *multiply* (*outcome*) *times* (*probability of outcome*) *and add*:

$$\frac{2}{36} + \frac{6}{36} + \frac{12}{36} + \frac{20}{36} + \frac{30}{36} + \frac{42}{36} + \frac{40}{36} + \frac{36}{36} + \frac{30}{36} + \frac{22}{36} + \frac{12}{36} = 7.$$

If you throw the dice 1000 times, and the average is not between 6.9 and 7.1, you get an A. Use the random number generator on a computer and round off to integers.

Now comes **continuous probability**. Suppose all numbers between 2 and 12 are equally probable. This means all numbers—not just integers. What is the probability of hitting the particular number $x = \pi$? *It is zero!* By any reasonable measure, π has *no chance* to occur. In the continuous case, every x has probability zero. But an *interval of x's* has a nonzero probability:

the probability of an outcome between 2 and 3 is $1/10$
the probability of an outcome between x and $x + \Delta x$ is $\Delta x/10$

To find the average, add up each outcome times the probability of that outcome. First divide 2 to 12 into intervals of length $\Delta x = 1$ and probability $p = 1/10$. If we round off x, the average is $6\frac{1}{2}$:

$$2\left(\frac{1}{10}\right) + 3\left(\frac{1}{10}\right) + \cdots + 11\left(\frac{1}{10}\right) = 6.5.$$

Here all outcomes are integers (as with dice). It is more accurate to use 20 intervals of length $1/2$ and probability $1/20$. The average is $6\frac{3}{4}$, and you see what is coming. These are rectangular areas (Riemann sums). As $\Delta x \to 0$ they approach an integral. The probability of an outcome between x and $x + dx$ is $p(x)\, dx$, and this problem has $p(x) = 1/10$. **The average outcome in the continuous case is not a sum but an integral**:

$$\text{\emph{expected value} } E(x) = \int_2^{12} x p(x)\, dx = \int_2^{12} x \frac{dx}{10} = \frac{x^2}{20}\Big]_2^{12} = 7.$$

That is a big jump. From the point of view of integration, it is a limit of sums. From the point of view of probability, the chance of each outcome is zero but the **probability**

density at x is $p(x) = 1/10$. The integral of $p(x)$ is 1, because some outcome must happen. The integral of $xp(x)$ is $x_{ave} = 7$, the expected value. Each choice of x is random, but the average is predictable.

This completes a first step in probability theory. The second step comes after more calculus. Decaying probabilities use e^{-x} and e^{-x^2}—then the chance of a large x is very small. Here we end with the expected values of x^n and $1/\sqrt{x}$ and $1/x$, for a random choice between 0 and 1 (so $p(x) = 1$):

$$E(x^n) = \int_0^1 x^n \, dx = \frac{1}{n+1} \qquad E\left(\frac{1}{\sqrt{x}}\right) = \int_0^1 \frac{dx}{\sqrt{x}} = 2 \qquad E\left(\frac{1}{x}\right) = \int_0^1 \frac{dx}{x} = \infty(!)$$

A CONFUSION ABOUT "EXPECTED" CLASS SIZE

A college can advertise an average class size of 29, while most students are in large classes most of the time. I will show quickly how that happens.

Suppose there are 95 classes of 20 students and 5 classes of 200 students. The total enrollment in 100 classes is $1900 + 1000 = 2900$. A random professor has expected class size 29. But a random student sees it differently. The probability is $1900/2900$ of being in a small class and $1000/2900$ of being in a large class. Adding class size times probability gives the expected class size *for the student*:

$$(20)\left(\frac{1900}{2900}\right) + (200)\left(\frac{1000}{2900}\right) = 82 \text{ students in the class.}$$

Similarly, the average waiting time at a restaurant seems like 40 minutes (to the customer). To the hostess, who averages over the whole day, it is 10 minutes. If you came at a random time it *would* be 10, but if you are a random customer it is 40.

Traffic problems could be eliminated by raising the average number of people per car to 2.5, or even 2. But that is virtually impossible. Part of the problem is the difference between (a) the percentage of cars with one person and (b) the percentage of people alone in a car. Percentage (b) is smaller. In practice, most people would be in crowded cars. See Problems $37 - 38$.

5.6 EXERCISES

Read-through questions

The integrals $\int_0^b v(x) \, dx$ and $\int_b^5 v(x) \, dx$ add to __a__. The integral $\int_1^3 v(x) \, dx$ equals __b__. The reason is __c__. If $v(x) \leqslant x$ then $\int_0^1 v(x) \, dx \leqslant$ __d__. The average value of $v(x)$ on the interval $1 \leqslant x \leqslant 9$ is defined by __e__. It is equal to $v(c)$ at a point $x = c$ which is __f__. The rectangle across this interval with height $v(c)$ has the same area as __g__. The average value of $v(x) = x + 1$ on the interval $1 \leqslant x \leqslant 9$ is __h__.

If x is chosen from $1, 3, 5, 7$ with equal probabilities $\frac{1}{4}$, its expected value (average) is __i__. The expected value of x^2 is __j__. If x is chosen from $1, 2, \ldots, 8$ with probabilities $\frac{1}{8}$, its expected value is __k__. If x is chosen from $1 \leqslant x \leqslant 9$, the chance of hitting an integer is __l__. The chance of falling between x and $x + dx$ is $p(x) \, dx =$ __m__. The expected value $E(x)$ is the integral __n__. It equals __o__.

In 1–6 find the average value of $v(x)$ between a and b, and find all points c where $v_{ave} = v(c)$.

1   $v = x^4, a = -1, b = 1$ 2 $v = x^5, a = -1, b = 1$

3   $v = \cos^2 x, a = 0, b = \pi$ 4 $v = \sqrt{x}, a = 0, b = 4$

5   $v = 1/x^2, a = 1, b = 2$ 6 $v = (\sin x)^9, a = -\pi, b = \pi$.

7   At $x = 8$, $F(x) = \int_3^x v(t) \, dt + \int_x^5 v(t) \, dt$ is _____.

8   $\int_1^3 x \, dx + \int_3^5 x \, dx - \int_5^1 x \, dx =$ _____.

Are 9–16 true or false? Give a reason or an example.

9 The minimum of $\int_4^x v(t)\,dt$ is at $x = 4$.

10 The value of $\int_x^{x+3} v(t)\,dt$ does not depend on x.

11 The average value from $x = 0$ to $x = 3$ equals

$$\tfrac{1}{3}(v_{\text{ave}} \text{ on } 0 \leqslant x \leqslant 1) + \tfrac{2}{3}(v_{\text{ave}} \text{ on } 1 \leqslant x \leqslant 3).$$

12 The ratio $(f(b) - f(a))/(b - a)$ is the average value of $f(x)$ on $a \leqslant x \leqslant b$.

13 On the symmetric interval $-1 \leqslant x \leqslant 1$, $v(x) - v_{\text{ave}}$ is an odd function.

14 If $l(x) \leqslant v(x) \leqslant u(x)$ then $dl/dx \leqslant dv/dx \leqslant du/dx$.

15 The average of $v(x)$ from 0 to 2 plus the average from 2 to 4 equals the average from 0 to 4.

16 (a) Antiderivatives of even functions are odd functions.

   (b) Squares of odd functions are odd functions.

17 What number $\bar{v}$ gives $\int_a^b (v(x) - \bar{v})\,dx = 0$?

18 If $f(2) = 6$ and $f(6) = 2$ then the average of df/dx from $x = 2$ to $x = 6$ is _____ .

19 (a) The averages of $\cos x$ and $|\cos x|$ from 0 to π are _____ .

   (b) The average of the numbers $v_1, \ldots, v_n$ is _____ than the average of $|v_1|, \ldots, |v_n|$.

20 (a) Which property of integrals proves
$\int_0^1 v(x)\,dx \leqslant \int_0^1 |v(x)|\,dx$?

   (b) Which property proves $-\int_0^1 v(x)\,dx \leqslant \int_0^1 |v(x)|\,dx$?

Together these are **Property 8**: $|\int_0^1 v(x)\,dx| \leqslant \int_0^1 |v(x)|\,dx$.

21 What function has v_{ave} (from 0 to x) equal to $\tfrac{1}{3}v(x)$ at all x? What functions have $v_{\text{ave}} = v(x)$ at all x?

22 (a) If $v(x)$ is increasing, explain from Property **6** why $\int_0^x v(t)\,dt \leqslant xv(x)$ for $x > 0$.

   (b) Take derivatives of both sides for a second proof.

23 The average of $v(x) = 1/(1 + x^2)$ on the interval $[0, b]$ approaches _____ as $b \to \infty$. The average of $V(x) = x^2/(1 + x^2)$ approaches _____ .

24 If the positive numbers v_n approach zero as $n \to \infty$ prove that their average $(v_1 + \cdots + v_n)/n$ also approaches zero.

25 Find the average distance from $x = a$ to points in the interval $0 \leqslant x \leqslant 2$. Is the formula different if $a < 2$?

26 (Computer experiment) Choose random numbers x between 0 and 1 until the average value of x^2 is between .333 and .334. How many values of x^2 are above and below? If possible repeat ten times.

27 A point P is chosen randomly along a semicircle (see figure: equal probability for equal arcs). What is the average distance y from the x axis? The radius is 1.

28 A point Q is chosen randomly between -1 and 1.

   (a) What is the average distance Y up to the semicircle?

(b) Why is this different from Problem 27?

Buffon needle

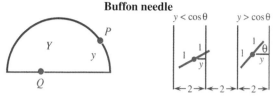

29 (A classic way to compute π) A $2''$ needle is tossed onto a floor with boards $2''$ wide. Find the probability of falling across a crack. (This happens when $\cos\theta > y =$ distance from midpoint of needle to nearest crack. In the rectangle $0 \leqslant \theta \leqslant \pi/2, 0 \leqslant y \leqslant 1$, shade the part where $\cos\theta > y$ and find the fraction of area that is shaded.)

30 If Buffon's needle has length $2x$ instead of 2, find the probability $P(x)$ of falling across the same cracks.

31 If you roll *three* dice at once, what are the probabilities of each outcome between 3 and 18? What is the expected value?

32 If you choose a random point in the square $0 \leqslant x \leqslant 1$, $0 \leqslant y \leqslant 1$, what is the chance that its coordinates have $y^2 \leqslant x$?

33 The voltage $V(t) = 220 \cos 2\pi t/60$ has frequency 60 hertz and amplitude 220 volts. Find V_{ave} from 0 to t.

34 (a) Show that $v_{\text{even}}(x) = \tfrac{1}{2}(v(x) + v(-x))$ is always even.

   (b) Show that $v_{\text{odd}}(x) = \tfrac{1}{2}(v(x) - v(-x))$ is always odd.

35 By Problem 34 or otherwise, write $(x + 1)^3$ and $1/(x + 1)$ as an even function plus an odd function.

36 Prove from the definition of df/dx that it is an odd function if $f(x)$ is even.

37 Suppose four classes have 6, 8, 10, and 40 students, averaging _____ . The chance of being in the first class is _____ . The expected class size (for the student) is

$$E(x) = 6\left(\frac{6}{64}\right) + 8\left(\frac{8}{64}\right) + 10\left(\frac{10}{64}\right) + 40\left(\frac{40}{64}\right) = \underline{\quad\quad}.$$

38 With groups of sizes $x_1, \ldots, x_n$ adding to G, the average size is _____ . The chance of an individual belonging to group 1 is _____ . The expected size of his or her group is $E(x) = x_1(x_1/G) + \cdots + x_n(x_n/G)$. * Prove $\Sigma_1^n x_i^2/G \geqslant G/n$.

39 **True or false**, 15 seconds each:

   (a) If $f(x) \leqslant g(x)$ then $df/dx \leqslant dg/dx$.

   (b) If $df/dx \leqslant dg/dx$ then $f(x) \leqslant g(x)$.

   (c) $xv(x)$ is odd if $v(x)$ is even.

   (d) If $v_{\text{ave}} \leqslant w_{\text{ave}}$ on all intervals then $v(x) \leqslant w(x)$ at all points.

40 If $v(x) = \begin{cases} 2x & \text{for } x < 3 \\ -2x & \text{for } x > 3 \end{cases}$ then $f(x) = \begin{cases} x^2 & \text{for } x < 3 \\ -x^2 & \text{for } x > 3 \end{cases}$.

Thus $\int_0^4 v(x)\,dx = f(4) - f(0) = -16$. Correct the mistake.

41 If $v(x) = |x - 2|$ find $f(x)$. Compute $\int_0^5 v(x)\,dx$.

42 Why are there equal areas above and below v_{ave}?

5.7 The Fundamental Theorem and Its Applications

When the endpoints are fixed at a and b, we have a *definite* integral. When the upper limit is a variable point x, we have an *indefinite* integral. More generally: When the endpoints depend in any way on x, **the integral is a function of** x. Therefore we can find its derivative. This requires the Fundamental Theorem of Calculus.

The essence of the Theorem is: **Derivative of integral of** v **equals** v. We also compute the derivative when the integral goes from $a(x)$ to $b(x)$—both limits variable.

Part **2** of the Fundamental Theorem reverses the order: **Integral of derivative of** f **equals** $f + C$. That will follow quickly from Part **1**, with help from the Mean Value Theorem. It is Part **2** that we use most, since integrals are harder than derivatives.

After the proofs we go to new applications, beyond the standard problem of area under a curve. Integrals can add up rings and triangles and shells—not just rectangles. The answer can be a volume or a probability—not just an area.

THE FUNDAMENTAL THEOREM, PART 1

Start with a continuous function v. Integrate it from a fixed point a to a variable point x. For each x, this integral $f(x)$ is a number. We do not require or expect a formula for $f(x)$—it is the area out to the point x. It is a function of x! The Fundamental Theorem says that this area function has a derivative (another limiting process). **The derivative** df/dx **equals the original** $v(x)$.

5C (*Fundamental Theorem, Part* **1**) Suppose $v(x)$ is a continuous function:

$$\text{If} \quad f(x) = \int_a^x v(t)\, dt \quad \text{then} \quad df/dx = v(x).$$

The dummy variable is written as t, so we can concentrate on the limits. The value of the integral depends on the limits a and x, not on t.

To find df/dx, start with $\Delta f = f(x + \Delta x) - f(x) = $ *difference of areas*:

$$\Delta f = \int_a^{x+\Delta x} v(t)\, dt - \int_a^x v(t)\, dt = \int_x^{x+\Delta x} v(t)\, dt. \tag{1}$$

Officially, this is Property **1**. The area out to $x + \Delta x$ minus the area out to x equals the small part from x to $x + \Delta x$. Now divide by Δx:

$$\frac{\Delta f}{\Delta x} = \frac{1}{\Delta x} \int_x^{x+\Delta x} v(t)\, dt = \textbf{average value} = v(c). \tag{2}$$

This is Property **7**, the Mean Value Theorem for integrals. The average value on this short interval equals $v(c)$. This point c is somewhere between x and $x + \Delta x$ (exact position not known), and we let Δx approach zero. That squeezes c toward x, so $v(c)$ approaches $u(x)$—remember that v is continuous. The limit of equation (2) is the Fundamental Theorem:

$$\frac{\Delta f}{\Delta x} \to \frac{df}{dx} \quad \text{and} \quad v(c) \to v(x) \quad \text{so} \quad \frac{df}{dx} = v(x). \tag{3}$$

If Δx is negative the reasoning still holds. Why assume that $v(x)$ is continuous? Because if v is a step function, then $f(x)$ has a corner where df/dx is not $v(x)$.

We could skip the Mean Value Theorem and simply bound v above and below:

for t between x and $x + \Delta x$: $v_{\min} \leqslant v(t) \leqslant v_{\max}$

integrate over that interval: $v_{\min} \Delta x \leqslant \Delta f \leqslant v_{\max} \Delta x$ (4)

divide by Δx: $v_{\min} \leqslant \Delta f / \Delta x \leqslant v_{\max}$

As $\Delta x \to 0$, $v_{\min}$ and $v_{\max}$ approach $v(x)$. In the limit df/dx again equals $v(x)$.

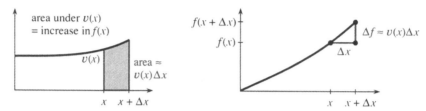

Fig. 5.14   Fundamental Theorem, Part **1**: (thin area Δf)/(base length Δx)$\to$ height $v(x)$.

Graphical meaning   The f-graph gives the area under the v-graph. The thin strip in Figure 5.14 has area Δf. ***That area is approximately $v(x)$ times Δx.*** Dividing by its base, $\Delta f / \Delta x$ is close to the height $v(x)$. When $\Delta x \to 0$ and the strip becomes infinitely thin, the expression "close to" converges to "equals." Then df/dx is the height at $v(x)$.

DERIVATIVES WITH VARIABLE ENDPOINTS

When the upper limit is x, the derivative is $v(x)$. Suppose the *lower limit* is x. The integral goes from x to b, instead of a to x. When x moves, the lower limit moves. The change in area is on the left side of Figure 5.15. *As x goes forward, **area is removed***. So there is a minus sign in the derivative of area:

The derivative of $g(x) = \displaystyle\int_{x}^{b} v(t)\, dt$ ***is*** $\dfrac{dg}{dx} = -v(x).$ (5)

The quickest proof is to reverse b and x, which reverses the sign (Property **3**):

$$g(x) = -\int_{b}^{x} v(t)\, dt \quad \text{so by Part } \mathbf{1} \quad \frac{dg}{dx} = -v(x).$$

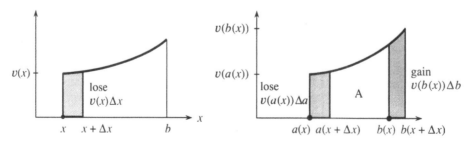

Fig. 5.15   Area from x to b has $dg/dx = -v(x)$. Area $v(b)db$ is added, area $v(a)da$ is lost

The general case is messier but not much harder (it is quite useful). Suppose *both limits* are changing. The upper limit $b(x)$ is not necessarily x, but it depends on x.

The lower limit $a(x)$ can also depend on x (Figure 5.15b). The area A between those limits changes as x changes, and we want dA/dx:

$$\textbf{If} \quad A = \int_{a(x)}^{b(x)} v(t)\,dt \quad \textbf{then} \quad \frac{dA}{dx} = v(b(x))\frac{db}{dx} - v(a(x))\frac{da}{dx}. \quad (6)$$

The figure shows two thin strips, one added to the area and one subtracted.

First check the two cases we know. When $a = 0$ and $b = x$, we have $da/dx = 0$ and $db/dx = 1$. The derivative according to (6) is $v(x)$ times 1—the Fundamental Theorem. The other case has $a = x$ and $b = $ constant. Then the lower limit in (6) produces $-v(x)$. When the integral goes from $a = 2x$ to $b = x^3$, its derivative is new:

EXAMPLE 1 $\qquad A = \int_{2x}^{x^3} \cos t\,dt = \sin x^3 - \sin 2x$

$$dA/dx = (\cos x^3)(3x^2) - (\cos 2x)(2).$$

That fits with (6), because db/dx is $3x^2$ and da/dx is 2 (with minus sign). It also looks like the chain rule—which it is! To prove (6) we use the letters v and f:

$$A = \int_{a(x)}^{b(x)} v(t)\,dt = f(b(x)) - f(a(x)) \qquad \text{(by Part \textbf{2} below)}$$

$$\frac{dA}{dx} = f'(b(x))\frac{db}{dx} - f'(a(x))\frac{da}{dx} \qquad \text{(by the chain rule)}$$

Since $f' = v$, equation (6) is proved. In the next example the area turns out to be constant, although it seems to depend on x. Note that $v(t) = 1/t$ so $v(3x) = 1/3x$.

EXAMPLE 2 $\quad A = \int_{2x}^{3x} \frac{1}{t}\,dt$ has $\dfrac{dA}{dx} = \left(\dfrac{1}{3x}\right)(3) - \left(\dfrac{1}{2x}\right)(2) = 0.$

Question $\quad A = \int_{-x}^{x} v(t)\,dt$ has $\dfrac{dA}{dx} = v(x) + v(-x).$ *Why does $v(-x)$ have a plus sign?*

THE FUNDAMENTAL THEOREM, PART 2

We have used a hundred times the Theorem that is now to be proved. It is the key to integration. "*The integral of df/dx is $f(x) + C$.*" The application starts with $v(x)$. We search for an $f(x)$ with this derivative. If $df/dx = v(x)$, the Theorem says that

$$\int v(x)\,dx = \int \frac{df}{dx}\,dx = f(x) + C.$$

We can't rely on knowing formulas for v and f—only the definitions of $\int$ and d/dx.

The proof rests on one extremely special case: df/dx is the **zero function**. We easily find $f(x) = $ *constant*. The problem is to prove that there are no other possibilities: f *must* be constant. When the slope is zero, the graph *must* be flat. Everybody knows this is true, but intuition is not the same as proof.

Assume that $df/dx = 0$ in an interval. If $f(x)$ is not constant, there are points where $f(a) \neq f(b)$. By the Mean Value Theorem, there is a point c where

$$f'(c) = \frac{f(b) - f(a)}{b - a} \qquad \text{(this is not zero because } f(a) \neq f(b)).$$

But $f'(c) \neq 0$ directly contradicts $df/dx = 0$. Therefore $f(x)$ must be constant.

Note the crucial role of the Mean Value Theorem. A *local* hypothesis ($df/dx = 0$ at each point) yields a *global* conclusion ($f = $ constant in the whole interval). The derivative narrows the field of view, the integral widens it. The Mean Value Theorem connects instantaneous to average, local to global, points to intervals. This special case (the zero function) applies when $A(x)$ and $f(x)$ have the same derivative:

$$\text{\textit{If} } dA/dx = df/dx \text{ \textit{on an interval, then} } A(x) = f(x) + C. \tag{7}$$

Reason: The derivative of $A(x) - f(x)$ is zero. So $A(x) - f(x)$ must be constant.

Now comes the big theorem. It assumes that $v(x)$ is continuous, and integrates using $f(x)$:

5D (***Fundamental Theorem, Part 2***) If $v(x) = \dfrac{df}{dx}$ then $\displaystyle\int_a^b v(x)\,dx = f(b) - f(a)$.

Proof The antiderivative is $f(x)$. But Part **1** gave another antiderivative for the same $v(x)$. It was the integral—constructed from rectangles and now called $A(x)$:

$$A(x) = \int_a^x v(t)\,dt \quad \textit{also has} \quad \frac{dA}{dx} = v(x).$$

Since $A' = v$ and $f' = v$, the special case in equation (7) states that $A(x) = f(x) + C$. That is the essential point: ***The integral from rectangles equals*** $f(x) + C$.

At the lower limit the area integral is $A = 0$. So $f(a) + C = 0$. At the upper limit $f(b) + C = A(b)$. Subtract to find $A(b)$, the definite integral:

$$A(b) = \int_a^b v(x)\,dx = f(b) - f(a).$$

Calculus is beautiful—its Fundamental Theorem is also its most useful theorem.

Another proof of Part **2** starts with $f' = v$ and looks at subintervals:

$$f(x_1) - f(a) = v(x_1^*)(x_1 - a) \qquad \text{(by the Mean Value Theorem)}$$
$$f(x_2) - f(x_1) = v(x_2^*)(x_2 - x_1) \qquad \text{(by the Mean Value Theorem)}$$
$$\ldots = \ldots$$
$$f(b) - f(x_{n-1}) = v(x_n^*)(b - x_{n-1}) \qquad \text{(by the Mean Value Theorem).}$$

The left sides add to $f(b) - f(a)$. The sum on the right, as $\Delta x \to 0$, is $\int_a^b v(x)\,dx$.

APPLICATIONS OF INTEGRATION

Up to now the integral has been the area under a curve. There are many other applications, quite different from areas. ***Whenever addition becomes*** "*continuous*," ***we have integrals instead of sums***. Chapter 8 has space to develop more applications, but four examples can be given immediately—which will make the point.

We stay with geometric problems, rather than launching into physics or engineering or biology or economics. All those will come. The goal here is to take a first step away from rectangles.

EXAMPLE 3 (for circles) *The area A and circumference C are related by $dA/dr = C$.*

The question is why. The area is πr^2. Its derivative $2\pi r$ is the circumference. By the Fundamental Theorem, the integral of C is A. What is missing is the geometrical reason. Certainly πr^2 is the integral of $2\pi r$, but what is the *real* explanation for $A = \int C(r)dr$?

My point is that *the pieces are not rectangles.* We could squeeze rectangles under a circular curve, but their heights would have nothing to do with C. Our intuition has to take a completely different direction, and add up the *thin rings* in Figure 5.16.

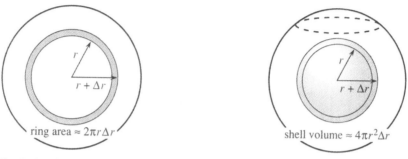

Fig. 5.16 Area of circle = integral over rings. Volume of sphere = integral over shells.

Suppose the ring thickness is Δr. Then the ring area is close to C times Δr. This is precisely the kind of approximation we need, because its error is of higher order $(\Delta r)^2$. *The integral adds ring areas* just as it added rectangular areas:

$$A = \int_0^r C \, dr = \int_0^r 2\pi r \, dr = \pi r^2.$$

That is our first step toward freedom, away from rectangles to rings.

The ring area ΔA can be checked exactly—it is the difference of circles:

$$\Delta A = \pi(r + \Delta r)^2 - \pi r^2 = 2\pi r \, \Delta r + \pi(\Delta r)^2.$$

This is $C\Delta r$ plus a correction. Dividing both sides by $\Delta r \to 0$ leaves $dA/dr = C$.

Finally there is a geometrical reason. The ring unwinds into a thin strip. Its width is Δr and its length is close to C. The inside and outside circles have different perimeters, so this is not a true rectangle—but the area is near $C\Delta r$.

EXAMPLE 4 For a sphere, surface area and volume satisfy $A = dV/dr$.

What worked for circles will work for spheres. The thin rings become *thin shells*. A shell goes from radius r to radius $r + \Delta r$, so its thickness is Δr. We want the volume of the shell, but we don't need it exactly. The surface area is $4\pi r^2$, so the volume is about $4\pi r^2 \, \Delta r$. That is close enough!

Again we are correct except for $(\Delta r)^2$. Infinitesimally speaking $dV = A \, dr$:

$$V = \int_0^r A \, dr = \int_0^r 4\pi r^2 \, dr = \frac{4}{3}\pi r^3.$$

This is the volume of a sphere. The derivative of V is A, and the shells explain why. Main point: *Integration is not restricted to rectangles.*

EXAMPLE 5 The distance around a square is $4s$. Why does the area have $dA/ds = 2s$?

The side is s and the area is s^2. Its derivative $2s$ goes only **half way around the square**. I tried to understand that by drawing a figure. Normally this works, but in the figure dA/ds looks like $4s$. Something is wrong. The bell is ringing so I leave this as an exercise.

EXAMPLE 6 Find the area under $v(x) = \cos^{-1}x$ from $x = 0$ to $x = 1$.

That is a conventional problem, but we have no antiderivative for $\cos^{-1}x$. We could look harder, and find one. However there is another solution—unconventional but correct. **The region can be filled with horizontal rectangles** (not vertical rectangles). Figure 5.17b shows a typical strip of length $x = \cos v$ (the curve has $v = \cos^{-1}x$). As the thickness Δv approaches zero, the total area becomes $\int x\, dv$. We are integrating upward, so *the limits are on v not on x*:

$$\text{area} = \int_0^{\pi/2} \cos v\, dv = \sin v \Big]_0^{\pi/2} = 1.$$

The exercises ask you to set up other integrals—not always with rectangles. Archimedes used triangles instead of rings to find the area of a circle.

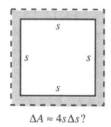

$\Delta A \approx 4s\Delta s?$

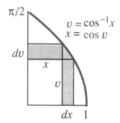

Fig. 5.17   Trouble with a square. Success with horizontal strips and triangles.

5.7 EXERCISES

Read-through questions

The area $f(x) = \int_a^x v(t)\, dt$ is a function of __a__. By Part **1** of the Fundamental Theorem, its derivative is __b__. In the proof, a small change Δx produces the area of a thin __c__. This area Δf is approximately __d__ times __e__. So the derivative of $\int_a^x t^2\, dt$ is __f__.

The integral $\int_x^b t^2\, dt$ has derivative __g__. The minus sign is because __h__. When both limits $a(x)$ and $b(x)$ depend on x, the formula for df/dx becomes __i__ minus __j__. In the example $\int_2^{3x} t\, dt$, the derivative is __k__.

By Part **2** of the Fundamental Theorem, the integral of df/dx is __l__. In the special case when $df/dx = 0$, this says that __m__. From this special case we conclude: If $dA/dx = dB/dx$ then $A(x) =$ __n__. If an antiderivative of $1/x$ is $\ln x$ (whatever that is), then automatically $\int_a^b dx/x =$ __o__.

The square $0 \le x \le s$, $0 \le y \le s$ has area $A =$ __p__. If s is increased by Δs, the extra area has the shape of __q__. That area ΔA is approximately __r__. So $dA/ds =$ __s__.

Find the derivatives of the following functions $F(x)$.

1 $\int_1^x \cos^2 t\, dt$ **2** $\int_x^1 \cos 3t\, dt$

3 $\int_0^2 t^n\, dt$ **4** $\int_0^2 x^n\, dt$

5 $\int_1^{x^2} u^3\, du$ **6** $\int_{-x}^{x/2} v(u)\, du$

7 $\int_x^{x+1} v(t)\, dt$ (a "*running average*" of v)

8 $\dfrac{1}{x}\displaystyle\int_0^x v(t)\, dt$ (the average of v; use product rule)

9 $\dfrac{1}{x}\displaystyle\int_0^x \sin^2 t\, dt$ **10** $\dfrac{1}{2}\displaystyle\int_x^{x+2} t^3\, dt$

11 $\int_0^x \left[\int_0^t v(u)\, du\right] dt$

12 $\int_0^x (df/dt)^2\, dt$

13 $\int_0^x v(t)\, dt + \int_x^1 v(t)\, dt$

14 $\int_0^x v(-t)\, dt$

15 $\int_{-x}^x \sin t^2\, dt$

16 $\int_{-x}^x \sin t\, dt$

17 $\int_0^x u(t)v(t)\, dt$

18 $\int_{a(x)}^{b(x)} 5\, dt$

19 $\int_0^{\sin x} \sin^{-1} t\, dt$

20 $\int_0^{f(x)} \dfrac{df}{dt}\, dt$

21 True or false

(a) If $df/dx = dg/dx$ then $f(x) = g(x)$.

(b) If $d^2 f/dx^2 = d^2 g/dx^2$ then $f(x) = g(x) + C$.

(c) If $3 > x$ then the derivative of $\int_3^x v(t)\, dt$ is $-v(x)$.

(d) The derivative of $\int_1^3 v(x)\, dx$ is zero.

22 For $F(x) = \int_x^{2x} \sin t\, dt$, locate $F(\pi + \Delta x) - F(\pi)$ on a sine graph. Where is $F(\Delta x) - F(0)$?

23 Find the function $v(x)$ whose average value between 0 and x is $\cos x$. Start from $\int_0^x v(t)\, dt = x \cos x$.

24 Suppose $df/dx = 2x$. We know that $d(x^2)/dx = 2x$. How do we prove that $f(x) = x^2 + C$?

25 If $\int_{-x}^0 v(t)\, dt = \int_0^x v(t)\, dt$ (equal areas left and right of zero), then $v(x)$ is an _____ function. Take derivatives to prove it.

26 Example 2 said that $\int_{2x}^{3x} dt/t$ does not really depend on x (or t!). Substitute xu for t and watch the limits on u.

27 True or false, with reason:

(a) All continuous functions have derivatives.

(b) All continuous functions have antiderivatives.

(c) All antiderivatives have derivatives.

(d) $A(x) = \int_{2x}^{3x} dt/t^2$ has $dA/dx = 0$.

Find $\int_1^x v(t)\, dt$ from the facts in 28–29.

28 $\dfrac{d(x^n)}{dx} = v(x)$

29 $\int_0^x v(t)\, dt = \dfrac{x}{x+2}$.

30 What is wrong with Figure 5.17? It seems to show that $dA = 4s\, ds$, which would mean $A = \int 4s\, ds = 2s^2$.

31 The cube $0 \leqslant x, y, z \leqslant s$ has volume $V = $ _____. The three square faces with $x = s$ or $y = s$ or $z = s$ have total area $A = $ _____. If s is increased by Δs, the extra volume has the shape of _____. That volume ΔV is approximately _____. So $dV/ds = $ _____.

32 The four-dimensional cube $0 \leqslant x, y, z, t \leqslant s$ has hypervolume $H = $ _____. The face with $x = s$ is really a _____. Its volume is $V = $ _____. The total volume of the four faces with $x = s$, $y = s$, $z = s$, or $t = s$ is _____. When s is increased by Δs, the extra hypervolume is $\Delta H \approx$ _____. So $dH/ds = $ _____.

33 The hypervolume of a four-dimensional sphere is $H = \frac{1}{2}\pi^2 r^4$. Therefore the area (volume?) of its three-dimensional surface $x^2 + y^2 + z^2 + t^2 = r^2$ is _____ .

34 The area above the parabola $y = x^2$ from $x = 0$ to $x = 1$ is $\frac{2}{3}$. Draw a figure with horizontal strips and integrate.

35 The wedge in Figure (a) has area $\frac{1}{2}r^2 d\theta$. One reason: It is a fraction $d\theta/2\pi$ of the total area πr^2. Another reason: It is close to a triangle with small base $r\, d\theta$ and height _____ . Integrating $\frac{1}{2}r^2 d\theta$ from $\theta = 0$ to $\theta = $ _____ gives the area _____ of a quarter-circle.

36 $A = \int_0^r \sqrt{r^2 - x^2}\, dx$ is also the area of a quarter-circle. Show why, with a graph and thin rectangles. Calculate this integral by substituting $x = r \sin\theta$ and $dx = r \cos\theta\, d\theta$.

(a) (b) (c)

37 The distance r in Figure (b) is related to θ by $r = $ _____ . Therefore the area of the thin triangle is $\frac{1}{2}r^2 d\theta = $ _____ . Integration to $\theta = $ _____ gives the total area $\frac{1}{2}$.

38 The x and y coordinates in Figure (c) add to $r \cos\theta + r \sin\theta = $ _____ . Without integrating explain why

$$\int_0^{\pi/2} \dfrac{d\theta}{(\cos\theta + \sin\theta)^2} = 1.$$

39 The horizontal strip at height y in Figure (d) has width dy and length $x = $ _____ . So the area up to $y = 2$ is _____ . What length are the vertical strips that give the same area?

40 Use thin rings to find the area between the circles $r = 2$ and $r = 3$. Draw a picture to show why thin rectangles would be extra difficult.

 (d)

(e)

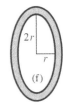 (f)

41 The length of the strip in Figure (e) is approximately _____ . The width is _____ . Therefore the triangle has area $\int_0^1$ _____ da (do you get $\frac{1}{2}$?).

42 The area of the ellipse in Figure (f) is $2\pi r^2$. Its derivative is $4\pi r$. But this is not the correct perimeter. Where does the usual reasoning go wrong?

43 The derivative of the integral of $v(x)$ is $v(x)$. What is the corresponding statement for sums and differences of the numbers v_j? Prove that statement.

44 The integral of the derivative of $f(x)$ is $f(x)+C$. What is the corresponding statement for sums of differences of f_j? Prove that statement.

45 Does $d^2f/dx^2 = a(x)$ lead to $\int_0^1 (\int_0^x a(t)\,dt)dx = f(1)-f(0)$?

46 The mountain $y = -x^2+t$ has an area $A(t)$ above the x axis. As t increases so does the area. Draw an xy graph of the mountain at $t=1$. What line gives dA/dt? Show with words or derivatives that $d^2A/dt^2 > 0$.

5.8 Numerical Integration

This section concentrates on definite integrals. The inputs are $y(x)$ and two endpoints a and b. The output is the integral I. Our goal is to find that number $\int_a^b y(x)\,dx = I$, accurately and in a short time. Normally this goal is achievable—as soon as we have a good method for computing integrals.

Our two approaches so far have weaknesses. The search for an antiderivative succeeds in important cases, and Chapter 7 extends that range—but generally $f(x)$ is not available. The other approach (by rectangles) is in the right direction but too crude. The height is set by $y(x)$ at the right and left end of each small interval. The *right and left rectangle rules* add the areas (Δx times y):

$$R_n = (\Delta x)(y_1 + y_2 + \cdots + y_n) \quad \text{and} \quad L_n = (\Delta x)(y_0 + y_1 + \cdots + y_{n-1}).$$

The value of $y(x)$ at the end of interval j is y_j. The extreme left value $y_0 = y(a)$ enters L_n. With n equal intervals of length $\Delta x = (b-a)/n$, the extreme right value is $y_n = y(b)$. It enters R_n. Otherwise the sums are the same—simple to compute, easy to visualize, but very inaccurate.

This section goes from slow methods (*rectangles*) to better methods (*trapezoidal and midpoint*) to good methods (*Simpson and Gauss*). Each improvement cuts down the error. You could discover the formulas without the book, by integrating x and x^2 and x^4. The rule R_n would come out on one side of the answer, and L_n would be on the other side. You would figure out what to do next, to come closer to the exact integral. The book can emphasize one key point:

> *The quality of a formula depends on how many integrals*
> $\int 1\,dx, \int x\,dx, \int x^2\,dx, \ldots,$ *it computes exactly. If* $\int x^p\,dx$
> *is the first to be wrong, the order of accuracy* is p.

By testing the integrals of $1, x, x^2, \ldots$, we decide how accurate the formulas are.

Figure 5.18 shows the rectangle rules R_n and L_n. **They are already wrong when** $y = x$. These methods are *first-order*: $p = 1$. The errors involve the first power of Δx—where we would much prefer a higher power. A larger p in $(\Delta x)^p$ means a smaller error.

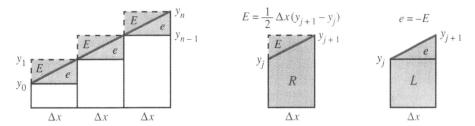

Fig. 5.18   Errors E and e in R_n and L_n are the areas of triangles.

When the graph of $y(x)$ is a straight line, the integral I is known. The error triangles E and e have base Δx. Their heights are the differences $y_{j+1} - y_j$. The areas are $\frac{1}{2}$(base)(height), and the only difference is a minus sign. (L is too low, so the error $L - I$ is negative.) The total error in R_n is the sum of the E's:

$$R_n - I = \tfrac{1}{2}\Delta x(y_1 - y_0) + \cdots + \tfrac{1}{2}\Delta x(y_n - y_{n-1}) = \tfrac{1}{2}\Delta x(y_n - y_0). \quad (1)$$

All y's between y_0 and y_n cancel. Similarly for the sum of the e's:

$$L_n - I = -\tfrac{1}{2}\Delta x(y_n - y_0) = -\tfrac{1}{2}\Delta x[y(b) - y(a)]. \quad (2)$$

The greater the slope of $y(x)$, the greater the error—since rectangles have zero slope.

Formulas (1) and (2) are nice—*but those errors are large.* To integrate $y = x$ from $a = 0$ to $b = 1$, the error is $\frac{1}{2}\Delta x(1 - 0)$. It takes $500,000$ rectangles to reduce this error to $1/1,000,000$. This accuracy is reasonable, but that many rectangles is unacceptable.

The beauty of the error formulas is that they are "*asymptotically correct*" for all functions. When the graph is curved, the errors don't fit exactly into triangles. But the ratio of predicted error to actual error approaches 1. As $\Delta x \to 0$, the graph is almost straight in each interval—this is linear approximation.

The error prediction $\frac{1}{2}\Delta x[y(b) - y(a)]$ is so simple that we test it on $y(x) = \sqrt{x}$:

$I = \int_0^1 \sqrt{x}\, dx = \frac{2}{3}$	$n =$	1	10	100	1000
	error $R_n - I =$	.33	.044	.0048	.00049
	error $L_n - I =$	$-.67$	$-.056$	$-.0052$	$-.00051$

The error decreases along each row. So does $\Delta x = .1, .01, .001, .0001$. Multiplying n by 10 divides Δx by 10. The error is also divided by 10 (almost). **The error is nearly proportional to Δx**—typical of first-order methods.

The predicted error is $\frac{1}{2}\Delta x$, since here $y(1) = 1$ and $y(0) = 0$. The computed errors in the table come closer and closer to $\frac{1}{2}\Delta x = .5, .05, .005, .0005$. The prediction is the "leading term" in the actual error.

The table also shows a curious fact. Subtracting the last row from the row above gives exact numbers $1, .1, .01$, and $.001$. This is $(R_n - I) - (L_n - I)$, which is $R_n - L_n$. It comes from an extra rectangle at the right, included in R_n but not L_n. Its height is 1 and its area is $1, .1, .01, .001$.

The errors in R_n and L_n almost cancel. The average $T_n = \frac{1}{2}(R_n + L_n)$ has less error—it is the "trapezoidal rule." First we give the rectangle prediction two final tests:

		$n = 1$	$n = 10$	$n = 100$	$n = 1000$
$\int (x^2 - x)\, dx:$	errors	$1.7 \cdot 10^{-1}$	$1.7 \cdot 10^{-3}$	$1.7 \cdot 10^{-5}$	$1.7 \cdot 10^{-7}$
$\int dx/(10 + \cos 2\pi x):$	errors	$-1 \cdot 10^{-3}$	$2 \cdot 10^{-14}$	"0"	"0"

Those errors are falling *faster* than Δx. For $y = x^2 - x$ the prediction explains why: $y(0)$ equals $y(1)$. The leading term, with $y(b)$ minus $y(a)$, is *zero*. The exact errors are $\frac{1}{6}(\Delta x)^2$, dropping from 10^{-1} to 10^{-3} to 10^{-5} to 10^{-7}. In these examples L_n is identical to R_n (and also to T_n), because the end rectangles are the same. We will see these $\frac{1}{6}(\Delta x)^2$ errors in the trapezoidal rule.

The last row in the table is more unusual. It shows practically no error. Why do the rectangle rules suddenly achieve such an outstanding success?

The reason is that $y(x) = 1/(10 + \cos 2\pi x)$ is *periodic*. The leading term in the error is zero, because $y(0) = y(1)$. Also the next term will be zero, because $y'(0) = y'(1)$. Every power of Δx is multiplied by zero, when we integrate over a complete period. So the errors go to zero exponentially fast.

Personal note I tried to integrate $1/(10 + \cos 2\pi x)$ by hand and failed. Then I was embarrassed to discover the answer in my book on applied mathematics. The method was a special trick using complex numbers, which applies over an exact period. Finally I found the antiderivative (quite complicated) in a handbook of integrals, and verified the area $1/\sqrt{99}$.

THE TRAPEZOIDAL AND MIDPOINT RULES

We move to integration formulas that are exact when $y = x$. They have **second-order accuracy**. The Δx error term disappears. The formulas give the correct area under straight lines. The predicted error is a multiple of $(\Delta x)^2$. That multiple is found by testing $y = x^2$—for which the answers are not exact.

The first formula combines R_n and L_n. To cancel as much error as possible, take the average $\frac{1}{2}(R_n + L_n)$. This yields the **trapezoidal rule**, which approximates $\int y(x)\, dx$ by T_n:

$$T_n = \tfrac{1}{2}R_n + \tfrac{1}{2}L_n = \Delta x\left(\tfrac{1}{2}y_0 + y_1 + \cdots + y_{n-1} + \tfrac{1}{2}y_n\right). \tag{3}$$

Another way to find T_n is from the area of the "trapezoid" below $y = x$ in Figure 5.19a.

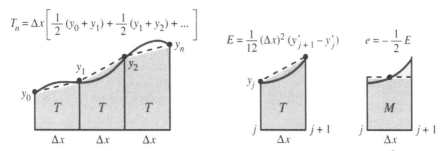

Fig. 5.19 Second-order accuracy: The error prediction is based on $v = x^2$.

The base is Δx and the sides have heights y_{j-1} and y_j. Adding those areas gives $\frac{1}{2}(L_n + R_n)$ in formula (3)—the coefficients of y_j combine into $\frac{1}{2} + \frac{1}{2} = 1$. Only the first and last intervals are missing a neighbor, so the rule has $\frac{1}{2}y_0$ and $\frac{1}{2}y_n$. Because trapezoids (unlike rectangles) fit under a sloping line, T_n is exact when $y = x$.

What is the difference from rectangles? The trapezoidal rule gives weight $\frac{1}{2}\Delta x$ to y_0 and y_n. The rectangle rule R_n gives full weight Δx to y_n (and no weight to y_0). $R_n - T_n$ is exactly the leading error $\frac{1}{2}y_n - \frac{1}{2}y_0$. The change to T_n knocks out that error.

Another important formula is exact for $y(x) = x$. A rectangle has the same area as a trapezoid, if the height of the rectangle is halfway between y_{j-1} and y_j. On a straight line graph that is achieved at the *midpoint* of the interval. By evaluating $y(x)$ at the halfway points $\frac{1}{2}\Delta x, \frac{3}{2}\Delta x, \frac{5}{2}\Delta x, \ldots$, we get much better rectangles. This leads to the **midpoint rule** M_n:

$$M_n = \Delta x\left(y_{1/2} + y_{3/2} + \cdots + y_{n-1/2}\right). \tag{4}$$

For $\int_0^4 x\, dx$, trapezoids give $\frac{1}{2}(0) + 1 + 2 + 3 + \frac{1}{2}(4) = 8$. The midpoint rule gives $\frac{1}{2} + \frac{3}{2} + \frac{5}{2} + \frac{7}{2} = 8$, again correct. The rules become different when $y = x^2$, because $y_{1/2}$ is no longer the average of y_0 and y_1. Try both second-order rules on x^2:

$I = \int_0^1 x^2\, dx$	$n =$	1	10	100
error $T_n - I =$		$1/6$	$1/600$	$1/60000$
error $M_n - I =$		$-1/12$	$-1/1200$	$-1/120000$

The errors fall by 100 when n is multiplied by 10. The midpoint rule is twice as good ($-1/12$ vs. $1/6$). Since all smooth functions are close to parabolas (quadratic approximation in each interval), the leading errors come from Figure 5.19. The trapezoidal error is exactly $\frac{1}{6}(\Delta x)^2$ when $y(x)$ is x^2 (the 12 in the formula divides the 2 in y'):

$$T_n - I \approx \frac{1}{12}(\Delta x)^2 \left[(y_1' - y_0') + \cdots + (y_n' - y_{n-1}')\right] = \frac{1}{12}(\Delta x)^2 \left[y_n' - y_0'\right] \quad (5)$$

$$M_n - I \approx -\frac{1}{24}(\Delta x)^2 \left[y_n' - y_0'\right] = -\frac{1}{24}(\Delta x)^2 \left[y'(b) - y'(a)\right] \quad (6)$$

For exact error formulas, change $y'(b) - y'(a)$ to $(b-a)y''(c)$. The location of c is unknown (as in the Mean Value Theorem). In practice these formulas are not much used—they involve the pth derivative at an unknown location c. The main point about the error is the factor $(\Delta x)^p$.

One crucial fact is easy to overlook in our tests. *Each value of $y(x)$ can be extremely hard to compute*. Every time a formula asks for y_j, a computer calls a subroutine. The goal of numerical integration is to get below the error tolerance, while calling for *a minimum number of values of y*. Second-order rules need about a thousand values for a typical tolerance of 10^{-6}. The next methods are better.

FOURTH-ORDER RULE: SIMPSON

The trapezoidal error is nearly twice the midpoint error ($1/6$ vs. $-1/12$). So a good combination will have twice as much of M_n as T_n. That is **Simpson's rule**:

$$S_n = \frac{1}{3}T_n + \frac{2}{3}M_n = \frac{1}{6}\Delta x \left[y_0 + 4y_{1/2} + 2y_1 + 4y_{3/2} + 2y_2 + \cdots + 4y_{n-1/2} + y_n\right]. \quad (7)$$

Multiply the midpoint values by $2/3 = 4/6$. The endpoint values are multiplied by $2/6$, except at the far ends a and b (with heights y_0 and y_n). This $1 - 4 - 2 - 4 - 2 - 4 - 1$ pattern has become famous.

Simpson's rule goes deeper than a combination of T and M. It comes from a *parabolic* approximation to $y(x)$ in each interval. When a parabola goes through y_0, $y_{1/2}, y_1$, the area under it is $\frac{1}{6}\Delta x(y_0 + 4y_{1/2} + y_1)$. This is S over the first interval. **All our rules are constructed this way**: **Integrate correctly as many powers $1, x, x^2, \ldots$ as possible**. Parabolas are better than straight lines, which are better than flat pieces. S beats M, which beats R. Check Simpson's rule on powers of x, with $\Delta x = 1/n$:

	$n = 1$	$n = 10$	$n = 100$
error if $y = x^2$	0	0	0
error if $y = x^3$	0	0	0
error if $y = x^4$	$8.33 \cdot 10^{-3}$	$8.33 \cdot 10^{-7}$	$8.33 \cdot 10^{-11}$

Exact answers for x^2 are no surprise. S_n was selected to get parabolas right. But the zero errors for x^3 were not expected. The accuracy has jumped to **fourth order**, with errors proportional to $(\Delta x)^4$. That explains the popularity of Simpson's rule.

To understand why x^3 is integrated exactly, look at the interval $[-1, 1]$. The odd function x^3 has zero integral, and Simpson agrees by symmetry:

$$\int_{-1}^{1} x^3 \, dx = \frac{1}{4}x^4 \Big]_{-1}^{1} = 0 \quad \text{and} \quad \frac{2}{6}\left[(-1)^3 + 4(0)^3 + 1^3\right] = 0. \quad (8)$$

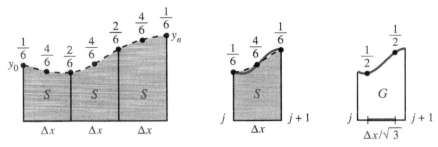

Fig. 5.20 Simpson versus Gauss: $E = c(\Delta x)^4(y'''_{j+1} - y'''_j)$ with $c_S = 1/2880$ and $c_G = -1/4320$.

THE GAUSS RULE (OPTIONAL)

We need a competitor for Simpson, and Gauss can compete with anybody. He calculated integrals in astronomy, and discovered that *two points are enough for a fourth-order method*. From -1 to 1 (a single interval) his rule is

$$\int_{-1}^{1} y(x)\, dx \approx y(-1/\sqrt{3}) + y(1/\sqrt{3}). \tag{9}$$

Those "**Gauss points**" $x = -1/\sqrt{3}$ and $x = 1/\sqrt{3}$ can be found directly. By placing them symmetrically, all odd powers $x, x^3, \ldots$ are correctly integrated. The key is in $y = x^2$, whose integral is $2/3$. The Gauss points $-x_G$ and $+x_G$ get this integral right:

$$\frac{2}{3} = (-x_G)^2 + (x_G)^2, \text{ so } x_G^2 = \frac{1}{3} \text{ and } x_G = \pm\frac{1}{\sqrt{3}}.$$

Figure 5.20c shifts to the interval from 0 to Δx. The Gauss points are $(1 \pm 1/\sqrt{3})\Delta x/2$. They are not as convenient as Simpson's (which hand calculators prefer). Gauss is good for thousands of integrations over one interval. Simpson is good when intervals go back to back—then Simpson also uses two y's per interval. For $y = x^4$, you see both errors drop by 10^{-4} in comparing $n = 1$ to $n = 10$:

$I = \int_0^1 x^4\, dx$	Simpson error	$8.33 \cdot 10^{-3}$	$8.33 \cdot 10^{-7}$
	Gauss error	$-5.56 \cdot 10^{-3}$	$-5.56 \cdot 10^{-7}$

DEFINITE INTEGRALS ON A CALCULATOR

It is fascinating to know how numerical integration is actually done. The points are not equally spaced! For an integral from 0 to 1, Hewlett-Packard machines might internally replace x by $3u^2 - 2u^3$ (the limits on u are also 0 and 1). The machine remembers to change dx. For example,

$$\int_0^1 \frac{dx}{\sqrt{x}} \text{ becomes } \int_0^1 \frac{6(u - u^2)du}{\sqrt{3u^2 - 2u^3}} = \int_0^1 \frac{6(1 - u)du}{\sqrt{3 - 2u}}.$$

Algebraically that looks worse—but the infinite value of $1/\sqrt{x}$ at $x = 0$ disappears at $u = 0$. The differential $6(u - u^2)du$ was chosen to vanish at $u = 0$ and $u = 1$. We don't need $y(x)$ at the endpoints—where infinity is most common. In the u variable the integration points are equally spaced—therefore in x they are not.

When a difficult point is *inside* $[a, b]$, break the interval in two pieces. And chop off integrals that go out to infinity. The integral of e^{-x^2} should be stopped by $x = 10$, since the tail is so thin. (It is bad to go too far.) Rapid oscillations are among the toughest—the answer depends on cancellation of highs and lows, and the calculator requires many integration points.

The change from x to u affects periodic functions. I thought equal spacing was good, since $1/(10+\cos 2\pi x)$ was integrated above to enormous accuracy. But there is a danger called **aliasing**. If $\sin 8\pi x$ is sampled with $\Delta x = 1/8$, it is always zero. A high frequency 8 is confused with a low frequency 0 (its "alias" which agrees at the sample points). With unequal spacing the problem disappears. *Notice how any integration method can be deceived*:

> Ask for the integral of $y = 0$ and specify the accuracy. The calculator samples y at $x_1, \ldots, x_k$. (With a PAUSE key, the x's may be displayed.) Then integrate $Y(x) = (x - x_1)^2 \cdots (x - x_k)^2$. That also returns the answer zero (now wrong), because the calculator follows the same steps.

On the calculator you enter the function, the endpoints, and the accuracy. The variable x can be named or not (see the margin). The outputs 4.67077 and 4.7E-5 are the requested integral $\int_1^2 e^x \, dx$ and the estimated error bound. Your input accuracy .00001 guarantees

$$\text{relative error in } y = \left| \frac{\text{true } y - \text{computed } y}{\text{computed } y} \right| \leqslant .00001.$$

The machine estimates accuracy based on its experience in sampling $y(x)$. If you guarantee e^x within .00000000001, it thinks you want high accuracy and takes longer.

In consulting for HP, William Kahan chose formulas using $1, 3, 7, 15, \ldots$ sample points. Each new formula uses the samples in the previous formula. The calculator stops when answers are close.

Read-through questions

To integrate $y(x)$, divide $[a, b]$ into n pieces of length $\Delta x = $___a___. R_n and L_n place a ___b___ over each piece, using the height at the right or ___c___ endpoint: $R_n = \Delta x(y_1 + \cdots + y_n)$ and $L_n = $___d___. These are ___e___ order methods, because they are incorrect for $y = $___f___. The total error on $[0, 1]$ is approximately ___g___. For $y = \cos \pi x$ this leading term is ___h___. For $y = \cos 2\pi x$ the error is very small because $[0, 1]$ is a complete ___i___.

A much better method is $T_n = \frac{1}{2} R_n + $___j___ $= \Delta x[\frac{1}{2} y_0 + $___k___ $y_1 + \cdots + $___l___ $y_n]$. This ___m___ rule is ___n___-order because the error for $y = x$ is ___o___. The error for $y = x^2$ from a to b is ___p___. The ___q___ rule is twice as accurate, using $M_n = \Delta x[$___r___$]$.

Simpson's method is $S_n = \frac{2}{3} M_n + $___s___. It is ___t___-order, because the powers ___u___ are integrated correctly. The coefficients of $y_0, y_{1/2}, y_1$ are ___v___ times Δx. Over three intervals the weights are $\Delta x/6$ times $1-4-$___w___. Gauss uses ___x___ points in each interval, separated by $\Delta x/\sqrt{3}$. For a method of order p the error is nearly proportional to ___y___.

1 What is the difference $L_n - T_n$? Compare with the leading error term in (2).

2 If you cut Δx in half, by what factor is the trapezoidal error reduced (approximately)? By what factor is the error in Simpson's rule reduced?

3 Compute R_n and L_n for $\int_0^1 x^3 dx$ and $n = 1, 2, 10$. Either verify (with computer) or use (without computer) the formula $1^3 + 2^3 + \cdots + n^3 = \frac{1}{4} n^2 (n+1)^2$.

4 One way to compute T_n is by averaging $\frac{1}{2}(L_n + R_n)$. Another way is to add $\frac{1}{2} y_0 + y_1 + \cdots + \frac{1}{2} y_n$. Which is more efficient? Compare the number of operations.

5 Test three different rules on $I = \int_0^1 x^4 dx$ for $n = 2, 4, 8$.

6 Compute π to six places as $4\int_0^1 dx/(1+x^2)$, using any rule.

7 Change Simpson's rule to $\Delta x(\frac{1}{4} y_0 + \frac{1}{2} y_{1/2} + \frac{1}{4} y_1)$ in each interval and find the order of accuracy p.

8 Demonstrate superdecay of the error when $1/(3+\sin x)$ is integrated from 0 to 2π.

9 Check that $(\Delta x)^2 (y'_{j+1} - y'_j)/12$ is the correct error for $y = 1$ and $y = x$ and $y = x^2$ from the first trapezoid ($j = 0$). Then it is correct for every parabola over every interval.

10 Repeat Problem 9 for the midpoint error $-(\Delta x)^2 (y'_{j+1} - y'_j)/24$. Draw a figure to show why the rectangle M has the same area as any trapezoid through the midpoint (including the trapezoid tangent to $y(x)$).

11 In principle $\int_{-\infty}^{\infty} \sin^2 x \, dx/x^2 = \pi$. With a symbolic algebra code or an HP-28S, how many decimal places do you get? Cut off the integral to $\int_{-A}^{A}$ and test large and small A.

12 These four integrals all equal π:

$$\int_0^1 \frac{dx}{\sqrt{x(1-x)}} \quad \int_{-\infty}^{\infty} \frac{\sin x}{x} dx \quad \frac{8}{3} \int_0^{\pi} \sin^4 x \, dx \quad \int_0^{\infty} \frac{x^{-1/2} dx}{1+x}$$

(a) Apply the midpoint rule to two of them until $\pi \approx 3.1416$.

(b) Optional: Pick the other two and find $\pi \approx 3$.

13 To compute $\ln 2 = \int_1^2 dx/x = .69315$ with error less than .001, how many intervals should T_n need? Its leading error is $(\Delta x)^2 [y'(b) - y'(a)]/12$. Test the actual error with $y = 1/x$.

14 Compare T_n with M_n for $\int_0^1 \sqrt{x} \, dx$ and $n = 1, 10, 100$. The error prediction breaks down because $y'(0) = \infty$.

15 Take $f(x) = \int_0^x y(x) \, dx$ in error formula **3R** to prove that $\int_0^{\Delta x} y(x) dx - y(0) \Delta x$ is exactly $\frac{1}{2}(\Delta x)^2 y'(c)$ some point c.

16 For the periodic function $y(x) = 1/(2 + \cos 6\pi x)$ from -1 to 1, compare T and S and G for $n = 2$.

17 For $I = \int_0^1 \sqrt{1 - x^2} \, dx$, the leading error in the trapezoidal rule is _____. Try $n = 2, 4, 8$ to defy the prediction.

18 Change to $x = \sin \theta, \sqrt{1 - x^2} = \cos \theta, dx = \cos \theta \, d\theta$, and repeat T_4 on $\int_0^{\pi/2} \cos^2 \theta \, d\theta$. What is the predicted error after the change to θ?

19 Write down the three equations $Ay(0) + By(\frac{1}{2}) + Cy(1) = I$ for the three integrals $I = \int_0^1 1 \, dx, \int_0^1 x \, dx, \int_0^1 x^2 \, dx$. Solve for A, B, C and name the rule.

20 Can you invent a rule using $Ay_0 + By_{1/4} + Cy_{1/2} + Dy_{3/4} + Ey_1$ to reach higher accuracy than Simpson's?

21 Show that T_n is the only combination of L_n and R_n that has second-order accuracy.

22 Calculate $\int e^{-x^2} dx$ with ten intervals from 0 to 5 and 0 to 20 and 0 to 400. The integral from 0 to ∞ is $\frac{1}{2}\sqrt{\pi}$. What is the best point to chop off the infinite integral?

23 The graph of $y(x) = 1/(x^2 + 10^{-10})$ has a sharp spike and a long tail. Estimate $\int_0^1 y \, dx$ from T_{10} and T_{100} (don't expect much). Then substitute $x = 10^{-5} \tan \theta, dx = 10^{-5} \sec^2 \theta \, d\theta$ and integrate 10^5 from 0 to $\pi/4$.

24 Compute $\int_0^4 |x - \pi| \, dx$ from T_4 and compare with the divide and conquer method of separating $\int_0^{\pi} |x - \pi| \, dx$ from $\int_{\pi}^4 |x - \pi| \, dx$.

25 Find a, b, c so that $y = ax^2 + bx + c$ equals $1, 3, 7$ at $x = 0, \frac{1}{2}, 1$ (three equations). Check that $\frac{1}{6} \cdot 1 + \frac{4}{6} \cdot 3 + \frac{1}{6} \cdot 7$ equals $\int_0^1 y \, dx$.

26 Find c in $S - I = c(\Delta x)^4 [y'''(1) - y'''(0)]$ by taking $y = x^4$ and $\Delta x = 1$.

27 Find c in $G - I = c(\Delta x)^4 [y'''(1) - y'''(-1)]$ by taking $y = x^4$, $\Delta x = 2$, and $G = (-1/\sqrt{3})^4 + (1/\sqrt{3})^4$.

28 What condition on $y(x)$ makes $L_n = R_n = T_n$ for the integral $\int_a^b y(x) \, dx$?

29 Suppose $y(x)$ is *concave up*. Show from a picture that the trapezoidal answer is too high and the midpoint answer is too low. How does $y'' > 0$ make equation (5) positive and (6) negative?

CHAPTER 6

Exponentials and Logarithms

This chapter is devoted to exponentials like 2^x and 10^x and above all e^x. The goal is to understand them, differentiate them, integrate them, solve equations with them, and invert them (to reach the logarithm). The overwhelming importance of e^x makes this a crucial chapter in pure and applied mathematics.

In the traditional order of calculus books, e^x waits until other applications of the integral are complete. I would like to explain why it is placed earlier here. I believe that the equation $dy/dx = y$ has to be emphasized above techniques of integration. The laws of nature are expressed by *differential equations*, and at the center is e^x. Its applications are to life sciences and physical sciences and economics and engineering (and more—wherever change is influenced by the present state). The model produces a differential equation and I want to show what calculus can do.

The key is always $b^{m+n} = (b^m)(b^n)$. Section 6.1 applies that rule in three ways:

1. to understand the *logarithm* as the *exponent*;
2. to draw *graphs* on ordinary and semilog and log-log paper;
3. to find *derivatives*. The slope of b^x will use $b^{x+\Delta x} = (b^x)(b^{\Delta x})$.

6.1 An Overview

There is a good chance you have met logarithms. They turn multiplication into addition, which is a lot simpler. They are the basis for slide rules (not so important) and for graphs on log paper (very important). Logarithms are mirror images of exponentials—and those I know you have met.

Start with exponentials. The numbers 10 and 10^2 and 10^3 are basic to the decimal system. For completeness I also include 10^0, which is "ten to the zeroth power" or 1. **The logarithms of those numbers are the exponents**. The logarithms of 1 and 10 and 100 and 1000 are 0 and 1 and 2 and 3. These are logarithms "*to base* 10," because the powers are powers of 10.

Question When the base changes from 10 to b, what is the logarithm of 1 ?
Answer Since $b^0 = 1$, $\log_b 1$ is always **zero**. To base b, **the logarithm of b^n is n**. Negative powers are also needed. The number 10^x is positive, but its exponent x can be negative. The first examples are $1/10$ and $1/100$, which are the same as 10^{-1} and 10^{-2}. **The logarithms are the exponents -1 and -2:**

$$1000 = 10^3 \qquad \text{and} \qquad \log 1000 = 3$$
$$1/1000 = 10^{-3} \qquad \text{and} \qquad \log 1/1000 = -3.$$

Multiplying 1000 times $1/1000$ gives $1 = 10^0$. Adding logarithms gives $3 + (-3) = 0$. Always 10^m times 10^n equals 10^{m+n}. In particular 10^3 times 10^2 produces five tens:

$$(10)(10)(10) \ \text{times} \ (10)(10) \ \text{equals} \ (10)(10)(10)(10)(10) = 10^5.$$

The law for b^m times b^n extends to all exponents, as in $10^{4.6}$ times 10^π. Furthermore the law applies to all bases (we restrict the base to $b > 0$ and $b \neq 1$). In every case **multiplication of numbers is addition of exponents**.

> **6A** b^m times b^n equals b^{m+n}, so logarithms (exponents) add
> b^m divided by b^n equals b^{m-n}, so logarithms (exponents) subtract
>
> $$\log_b(yz) = \log_b y + \log_b z \qquad \text{and} \qquad \log_b(y/z) = \log_b y - \log_b z. \quad (1)$$

Historical note In the days of slide rules, 1.2 and 1.3 were multiplied by sliding one edge across to 1.2 and reading the answer under 1.3. A slide rule made in Germany would give the third digit in 1.56. Its photograph shows the numbers on a log scale. The distance from 1 to 2 equals the distance from 2 to 4 and from 4 to 8. By sliding the edges, you add distances and multiply numbers.

Division goes the other way. *Notice how $1000/10 = 100$ matches $3 - 1 = 2$.* To divide 1.56 by 1.3, look back along line D for the answer 1.2.

The second figure, though smaller, is the important one. **When x increases by 1, 2^x is multiplied by 2**. *Adding to x multiplies y.* This rule easily gives $y = 1, 2, 4, 8$, but look ahead to calculus—which doesn't stay with whole numbers.

Calculus will add Δx. Then y is multiplied by $2^{\Delta x}$. This number is near 1. If $\Delta x = \frac{1}{10}$ then $2^{\Delta x} \approx 1.07$—the tenth root of 2. **To find the slope, we have to consider** $(2^{\Delta x} - 1)/\Delta x$. The limit is near $(1.07 - 1)/\frac{1}{10} = .7$, but the exact number will take time.

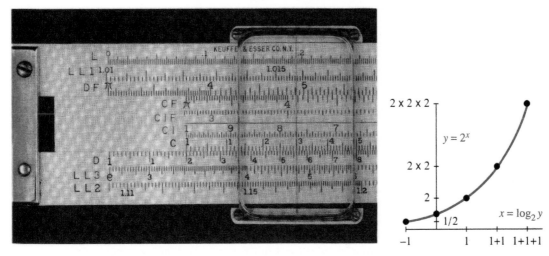

Fig. 6.1 An ancient relic (the slide rule). When exponents x add, powers 2^x multiply.

Base Change Bases other than 10 and exponents other than $1, 2, 3, \ldots$ are needed for applications. The population of the world x years from now is predicted to grow by a factor close to 1.02^x. Certainly x does not need to be a whole number of years. And certainly the base 1.02 should not be 10 (or we are in real trouble). This prediction will be refined as we study the differential equations for growth. It can be rewritten to base 10 if that is preferred (*but look at the exponent*):

$$1.02^x \quad \textit{is the same as} \quad 10^{(\log 1.02)x}.$$

When the base changes from 1.02 to 10, the exponent is multiplied—as we now see.

For practice, start with base b and change to base a. The logarithm to base a will be written "log." Everything comes from the rule that logarithm = exponent:

$$\textit{base change for numbers}: \quad b = a^{\log b}.$$

Now raise both sides to the power x. You see the change in the exponent:

$$\textit{base change for exponentials}: \quad b^x = a^{(\log b)x}.$$

Finally set $y = b^x$. Its logarithm to base b is x. Its logarithm to base a is the exponent on the right hand side: $\log_a y = (\log_a b)x$. Now replace x by $\log_b y$:

$$\textit{base change for logarithms}: \quad \log_a y = (\log_a b)(\log_b y).$$

We absolutely need this ability to change the base. An example with $a = 2$ is

$$b = 8 = 2^3 \qquad 8^2 = (2^3)^2 = 2^6 \qquad \log_2 64 = 3 \cdot 2 = (\log_2 8)(\log_8 64).$$

The rule behind base changes is $(a^m)^x = a^{mx}$. When the mth power is raised to the xth power, *the exponents multiply*. The square of the cube is the sixth power:

$$(a)(a)(a) \ \text{times} \ (a)(a)(a) \ \text{equals} \ (a)(a)(a)(a)(a)(a): \quad (a^3)^2 = a^6.$$

Another base will soon be more important than 10—here are the rules for base changes:

> **6B** To change *numbers*, *powers*, and *logarithms* from base b to base a, use
>
> $$b = a^{\log_a b} \qquad b^x = a^{(\log_a b)x} \qquad \log_a y = (\log_a b)(\log_b y) \qquad (2)$$

The first is the definition. The second is the xth power of the first. The third is the logarithm of the second (remember y is b^x). An important case is $y = a$:

$$\log_a a = (\log_a b)(\log_b a) = 1 \quad \text{so} \quad \log_a b = 1/\log_b a. \qquad (3)$$

EXAMPLE $8 = 2^3$ means $8^{1/3} = 2$. Then $(\log_2 8)(\log_8 2) = (3)(1/3) = 1$.

This completes the algebra of logarithms. The addition rules **6A** came from $(b^m)(b^n) = b^{m+n}$. The multiplication rule **6B** came from $(a^m)^x = a^{mx}$. *We still need to define b^x and a^x for all real numbers x. When x is a fraction, the defi-nition is easy. The square root of a^8 is a^4 ($m = 8$ times $x = 1/2$). When x is not a fraction, as in 2^π, the graph suggests one way to fill in the hole.*

We could define 2^π as the limit of $2^3, 2^{31/10}, 2^{314/100}, \ldots$. As the fractions r approach π, the powers 2^r approach 2^π. This makes $y = 2^x$ into a continuous function, with the desired properties $(2^m)(2^n) = 2^{m+n}$ and $(2^m)^x = 2^{mx}$—whether m and n and x are integers or not. But the ε's and δ's of continuity are not attractive, and we eventually choose (in Section 6.4) a smoother approach based on integrals.

GRAPHS OF b^x AND $\log_b y$

It is time to draw graphs. In principle one graph should do the job for both functions, because $y = b^x$ means the same as $x = \log_b y$. *These are inverse functions*. What one function does, its inverse undoes. The logarithm of $g(x) = b^x$ is x:

$$g^{-1}(g(x)) = \log_b(b^x) = x. \qquad (4)$$

In the opposite direction, the exponential of the logarithm of y is y:

$$g(g^{-1}(y)) = b^{(\log_b y)} = y. \qquad (5)$$

This holds for every base b, and it is valuable to see $b = 2$ and $b = 4$ on the same graph. Figure 6.2a shows $y = 2^x$ and $y = 4^x$. Their mirror images in the 45° line give the logarithms to base 2 and base 4, which are in the right graph.

When x is negative, $y = b^x$ is still positive. If the first graph is extended to the left, it stays above the x axis. **Sketch it in with your pencil**. Also extend the second graph down, to be the mirror image. Don't cross the vertical axis.

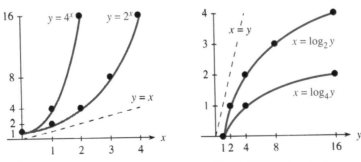

Fig. 6.2 Exponentials and mirror images (logarithms). Different scales for x and y.

There are interesting relations within the left figure. All exponentials start at 1, because b^0 is always 1. At the height $y = 16$, one graph is above $x = 2$ (because $4^2 = 16$). The other graph is above $x = 4$ (because $2^4 = 16$). **Why does 4^x in one graph equal 2^{2x} in the other**? This is the base change for powers, since $4 = 2^2$.

The figure on the right shows the mirror image—the logarithm. All logarithms start from zero at $y = 1$. The graphs go down to $-\infty$ at $y = 0$. (Roughly speaking $2^{-\infty}$ is zero.) Again x in one graph corresponds to $2x$ in the other (base change for logarithms). Both logarithms climb slowly, since the exponentials climb so fast.

The number $\log_2 10$ is between 3 and 4, because 10 is between 2^3 and 2^4. The slope of 2^x is proportional to 2^x—which never happened for x^n. But there are two practical difficulties with those graphs:

1. 2^x and 4^x increase too fast. The curves turn virtually straight up.
2. The most important fact about Ab^x is the value of b—and the base doesn't stand out in the graph.

There is also another point. In many problems we don't know the function $y = f(x)$. We are looking for it! All we have are measured values of y (with errors mixed in). When the values are plotted on a graph, we want to discover $f(x)$.

Fortunately there is a solution. **Scale the y axis differently**. On ordinary graphs, each unit upward adds a fixed amount to y. **On a log scale each unit multiplies y by a fixed amount**. The step from $y = 1$ to $y = 2$ is the same length as the step from 3 to 6 or 10 to 20.

On a log scale, $y = 11$ is not halfway between 10 and 12. And $y = 0$ is not there at all. Each step down divides by a fixed amount—*we never reach zero*. This is completely satisfactory for Ab^x, which also never reaches zero.

Figure 6.3 is on **semilog paper** (also known as *log-linear*), with an ordinary x axis. **The graph of $y = Ab^x$ is a straight line**. To see why, take logarithms of that equation:

$$\log y = \log A + x \log b. \tag{6}$$

The relation between x and $\log y$ is linear. It is really $\log y$ that is plotted, so the graph is straight. The markings on the y axis allow you to enter y without looking up its logarithm—you get an ordinary graph of $\log y$ against x.

Figure 6.3 shows two examples. One graph is an exact plot of $y = 2 \cdot 10^x$. It goes upward with slope 1, because a unit across has the same length as multiplication by 10 going up. 10^x **has slope** 1 **and** $10^{(\log b)x}$ (which is b^x) **will have slope** $\log b$. The crucial number $\log b$ can be measured directly as the slope.

The second graph in Figure 6.3 is more typical of actual practice, in which we start with measurements and look for $f(x)$. Here are the data points:

$$x = 0.0 \quad 0.2 \quad 0.4 \quad 0.6 \quad 0.8 \quad 1.0$$
$$y = 4.0 \quad 3.2 \quad 2.4 \quad 2.0 \quad 1.6 \quad 1.3$$

We don't know in advance whether these values fit the model $y = Ab^x$. The graph is strong evidence that they do. The points lie close to a line with negative slope—indicating $\log b < 0$ and $b < 1$. The slope down is half of the earlier slope up, so the model is consistent with

$$y = A \cdot 10^{-x/2} \quad \text{or} \quad \log y = \log A - \tfrac{1}{2}x. \tag{7}$$

When x reaches 2, y drops by a factor of 10. At $x = 0$ we see $A \approx 4$.

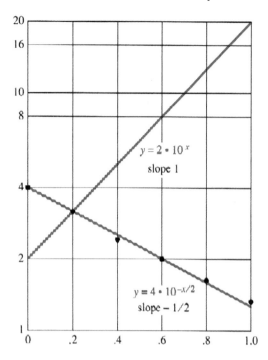

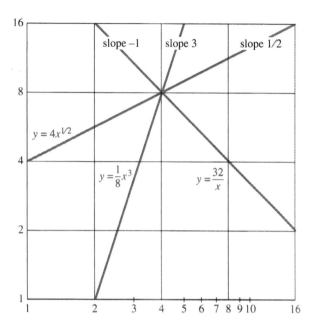

Fig. 6.3 $2 \cdot 10^x$ and $4 \cdot 10^{-x/2}$ on semilog paper.   **Fig. 6.4** Graphs of Ax^k on log-log paper.

Another model—a ***power*** $y = Ax^k$ instead of an exponential—also stands out with logarithmic scaling. This time we use ***log-log paper***, with both axes scaled. The logarithm of $y = Ax^k$ gives a linear relation between $\log y$ and $\log x$:

$$\log y = \log A + k \log x. \tag{8}$$

The exponent k becomes the slope on log-log paper. The base b makes no difference. We just measure the slope, and a straight line is a lot more attractive than a power curve.

The graphs in Figure 6.4 have slopes 3 and $\frac{1}{2}$ and -1. They represent Ax^3 and $A\sqrt{x}$ and A/x. To find the A's, look at one point on the line. At $x = 4$ the height is 8, so adjust the A's to make this happen: The functions are $x^3/8$ and $4\sqrt{x}$ and $32/x$. On semilog paper those graphs would not be straight!

You can buy log paper or create it with computer graphics.

THE DERIVATIVES OF $y = b^x$ AND $x = \log_b y$

This is a calculus book. *We have to ask about slopes*. The algebra of exponents is done, the rules are set, and on log paper the graphs are straight. Now come limits.

The central question is the derivative. ***What is dy/dx when $y = b^x$? What is dx/dy when x is the logarithm*** $\log_b y$? Those questions are closely related, because b^x and $\log_b y$ are inverse functions. If one slope can be found, the other is known from $dx/dy = 1/(dy/dx)$. The problem is to find one of them, and the exponential comes first.

You will now see that those questions have quick (and beautiful) answers, *except for a mysterious constant*. There is a multiplying factor c which needs more time. I think it is worth separating out the part that can be done immediately, leaving c

in dy/dx and $1/c$ in dx/dy. Then Section 6.2 discovers c by studying the special number called e (but $c \neq e$).

6C The derivative of b^x is a multiple cb^x. The number c depends on the base b.

The product and power and chain rules do not yield this derivative. We are pushed all the way back to the original definition, the limit of $\Delta y/\Delta x$:

$$\frac{dy}{dx} = \lim_{h \to 0} \frac{y(x+h) - y(x)}{h} = \lim_{h \to 0} \frac{b^{x+h} - b^x}{h}. \tag{9}$$

Key idea: Split b^{x+h} into b^x times b^h. Then the crucial quantity b^x factors out. More than that, b^x comes *outside the limit* because it does not depend on h. The remaining limit, inside the brackets, is the number c that we don't yet know:

$$\frac{dy}{dx} = \lim_{h \to 0} \frac{b^x b^h - b^x}{h} = b^x \left[\lim_{h \to 0} \frac{b^h - 1}{h} \right] = cb^x. \tag{10}$$

This equation is central to the whole chapter: dy/dx **equals** cb^x **which equals** cy. *The rate of change of y is proportional to y.* The slope increases in the same way that b^x increases (except for the factor c). A typical example is money in a bank, where interest is proportional to the principal. The rich get richer, and the poor get slightly richer. We will come back to compound interest, and identify b and c.

The inverse function is $x = \log_b y$. Now the unknown factor is $1/c$:

6D The slope of $\log_b y$ is $1/cy$ with the same c (depending on b).

Proof   If $dy/dx = cb^x$ then $dx/dy = 1/cb^x = 1/cy$. $\tag{11}$

That proof was like a Russian toast, powerful but too quick! We go more carefully:

$$\begin{aligned}
f(b^x) &= x & &\text{(logarithm of exponential)} \\
f'(b^x)(cb^x) &= 1 & &\text{(x derivative by chain rule)} \\
f'(b^x) &= 1/cb^x & &\text{(divide by cb^x)} \\
f'(y) &= 1/cy & &\text{(identify b^x as y)}
\end{aligned}$$

The logarithm gives another way to find c. From its slope we can discover $1/c$. *This is the way that finally works* (next section).

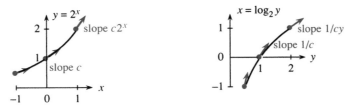

Fig. 6.5   The slope of 2^x is about $.7 \cdot 2^x$. The slope of $\log_2 y$ is about $1/.7y$.

Final remark It is extremely satisfying to meet an $f(y)$ whose derivative is $1/cy$. At last the "-1 power" has an antiderivative. Remember that $\int x^n dx = x^{n+1}/(n+$

1) is a failure when $n = -1$. The derivative of x^0 (a constant) does not produce x^{-1}. **We had no integral for x^{-1}, and the logarithm fills that gap.** If y is replaced by x or t (all dummy variables) then

$$\frac{d}{dx} \log_b x = \frac{1}{cx} \quad \text{and} \quad \frac{d}{dt} \log_b t = \frac{1}{ct}. \tag{12}$$

The base b can be chosen so that $c = 1$. Then the derivative is $1/x$. This final touch comes from the magic choice $b = e$—the highlight of Section 6.2.

6.1　EXERCISES

Read-through questions

In $10^4 = 10,000$, the exponent 4 is the __a__ of $10,000$. The base is $b =$ __b__. The logarithm of 10^m times 10^n is __c__. The logarithm of $10^m/10^n$ is __d__. The logarithm of $10,000^x$ is __e__. If $y = b^x$ then $x =$ __f__. Here x is any number, and y is always __g__.

A base change gives $b = a^{\underline{}h}$ and $b^x = a^{\underline{}i}$. Then 8^5 is 2^{15}. In other words $\log_2 y$ is __j__ times $\log_8 y$. When $y = 2$ it follows that $\log_2 8$ times $\log_8 2$ equals __k__.

On ordinary paper the graph of $y =$ __l__ is a straight line. Its slope is __m__. On semilog paper the graph of $y =$ __n__ is a straight line. Its slope is __o__. On log-log paper the graph of $y =$ __p__ is a straight line. Its slope is __q__.

The slope of $y = b^x$ is $dy/dx =$ __r__, where c depends on b. The number c is the limit as $h \to 0$ of __s__. Since $x = \log_b y$ is the inverse, $(dx/dy)(dy/dx) =$ __t__. Knowing $dy/dx = cb^x$ yields $dx/dy =$ __u__. Substituting b^x for y, the slope of $\log_b y$ is __v__. With a change of letters, the slope of $\log_b x$ is __w__.

Problems 1–10 use the rules for logarithms.

1 Find these logarithms (or exponents):

(a) $\log_2 32$　(b) $\log_2(1/32)$　(c) $\log_{32}(1/32)$

(d) $\log_{32} 2$　(e) $\log_{10}(10\sqrt{10})$　(f) $\log_2(\log_2 16)$

2 Without a calculator find the values of

(a) $3^{\log_3 5}$　　(b) $3^{2\log_3 5}$

(c) $\log_{10} 5 + \log_{10} 2$　(d) $(\log_3 b)(\log_b 9)$

(e) $10^5 10^{-4} 10^3$　(f) $\log_2 56 - \log_2 7$

3 Sketch $y = 2^{-x}$ and $y = \frac{1}{2}(4^x)$ from -1 to 1 on the same graph. Put their mirror images $x = -\log_2 y$ and $x = \log_4 2y$ on a second graph.

4 Following Figure 6.2 sketch the graphs of $y = \left(\frac{1}{2}\right)^x$ and $x = \log_{1/2} y$. What are $\log_{1/2} 2$ and $\log_{1/2} 4$?

5 Compute without a computer:

(a) $\log_2 3 + \log_2 \frac{2}{3}$　　(b) $\log_2 \left(\frac{1}{2}\right)^{10}$

(c) $\log_{10} 100^{40}$　　(d) $(\log_{10} e)(\log_e 10)$

(e) $2^{2^3}/(2^2)^3$　　(f) $\log_e(1/e)$

6 Solve the following equations for x:

(a) $\log_{10}(10^x) = 7$　　(b) $\log 4x - \log 4 = \log 3$

(c) $\log_x 10 = 2$　　(d) $\log_2(1/x) = 2$

(e) $\log x + \log x = \log 8$　(f) $\log_x(x^x) = 5$

7 The logarithm of $y = x^n$ is $\log_b y =$ _____.

*8 Prove that $(\log_b a)(\log_d c) = (\log_d a)(\log_b c)$.

9 2^{10} is close to 10^3 (1024 versus 1000). If they were equal then $\log_2 10$ would be _____. Also $\log_{10} 2$ would be _____ instead of 0.301.

10 The number 2^{1000} has approximately how many (decimal) digits?

Questions 11–19 are about the graphs of $y = b^x$ and $x = \log_b y$.

11 By hand draw the axes for semilog paper and the graphs of $y = 1.1^x$ and $y = 10(1.1)^x$.

12 Display a set of axes on which the graph of $y = \log_{10} x$ is a straight line. What other equations give straight lines on those axes?

13 When noise is measured in *decibels*, amplifying by a factor A increases the decibel level by $10 \log A$. If a whisper is 20db and a shout is 70db then $10 \log A = 50$ and $A =$ _____.

14 Draw semilog graphs of $y = 10^{1-x}$ and $y = \frac{1}{2}(\sqrt{10})^x$.

15 The Richter scale measures earthquakes by $\log_{10}(I/I_0) = R$. What is R for the standard earthquake of intensity I_0? If the 1989 San Francisco earthquake measured $R = 7$, how did its intensity I compare to I_0? The 1906 San Francisco quake had $R = 8.3$. The record quake was four times as intense with $R =$ _____.

16 The frequency of A above middle C is 440/second. The frequency of the next higher A is _____. Since $2^{7/12} \approx 1.5$, the note with frequency 660/sec is _____.

17 Draw your own semilog paper and plot the data

$$y = 7, 11, 16, 28, 44 \quad \text{at} \quad x = 0, 1/2, 1, 3/2, 2.$$

Estimate A and b in $y = Ab^x$.

18 Sketch log-log graphs of $y = x^2$ and $y = \sqrt{x}$.

19 On log-log paper, printed or homemade, plot $y = 4, 11,$ $21, 32, 45$ at $x = 1, 2, 3, 4, 5$. Estimate A and k in $y = Ax^k$.

Questions 20–29 are about the derivative $dy/dx = cb^x$.

20 $g(x) = b^x$ has slope $g' = cg$. Apply the chain rule to $g(f(y)) = y$ to prove that $df/dy = 1/cy$.

21 If the slope of $\log x$ is $1/cx$, find the slopes of $\log(2x)$ and $\log(x^2)$ and $\log(2^x)$.

22 What is the equation (including c) for the tangent line to $y = 10^x$ at $x = 0$? Find also the equation at $x = 1$.

23 What is the equation for the tangent line to $x = \log_{10} y$ at $y = 1$? Find also the equation at $y = 10$.

24 With $b = 10$, the slope of 10^x is $c10^x$. Use a calculator for small h to estimate $c = \lim (10^h - 1)/h$.

25 The unknown constant in the slope of $y = (.1)^x$ is $L = \lim (.1^h - 1)/h$. (a) Estimate L by choosing a small h. (b) Change h to $-h$ to show that $L = -c$ from Problem 24.

26 Find a base b for which $(b^h - 1)/h \approx 1$. Use $h = 1/4$ by hand or $h = 1/10$ and $1/100$ by calculator.

27 Find the second derivative of $y = b^x$ and also of $x = \log_b y$.

28 Show that $C = \lim (100^h - 1)/h$ is twice as large as $c = \lim (10^h - 1)/h$. (Replace the last h's by $2h$.)

29 In 28, the limit for $b = 100$ is twice as large as for $b = 10$. So c probably involves the _____ of b.

6.2 The Exponential e^x

The last section discussed b^x and $\log_b y$. The base b was arbitrary—it could be 2 or 6 or 9.3 or any positive number except 1. But in practice, only a few bases are used. I have never met a logarithm to base 6 or 9.3. Realistically there are two leading candidates for b, and 10 is one of them. This section is about the other one, which is an extremely remarkable number. This number is not seen in arithmetic or algebra or geometry, where it looks totally clumsy and out of place. In calculus it comes into its own.

The number is e. That symbol was chosen by Euler (initially in a fit of selfishness, but he was a wonderful mathematician). It is the base of the **natural logarithm**. It also controls the exponential e^x, which is much more important than $\ln x$. Euler also chose π to stand for perimeter—anyway, our first goal is to find e.

Remember that the derivatives of b^x and $\log_b y$ include a constant c that depends on b. Equations (10) and (11) in the previous section were

$$\frac{d}{dx} b^x = cb^x \qquad \text{and} \qquad \frac{d}{dy} \log_b y = \frac{1}{cy}. \tag{1}$$

At $x = 0$, the graph of b^x starts from $b^0 = 1$. The slope is c. At $y = 1$, the graph of $\log_b y$ starts from $\log_b 1 = 0$. The logarithm has slope $1/c$. **With the right choice of the base b those slopes will equal** 1 (because c will equal 1).

For $y = 2^x$ the slope c is near .7. We already tried $\Delta x = .1$ and found $\Delta y \approx .07$. The base has to be larger than 2, for a starting slope of $c = 1$.

We begin with a direct computation of the slope of $\log_b y$ at $y = 1$:

$$\frac{1}{c} = \text{slope at } 1 = \lim_{h \to 0} \frac{1}{h} \left[\log_b(1+h) - \log_b 1 \right] = \lim_{h \to 0} \log_b \left[(1+h)^{1/h} \right]. \tag{2}$$

Always $\log_b 1 = 0$. The fraction in the middle is $\log_b(1+h)$ times the number $1/h$. This number can go up into the exponent, and it did.

The quantity $(1+h)^{1/h}$ is unusual, to put it mildly. As $h \to 0$, the number $1 + h$ is approaching 1. At the same time, $1/h$ is approaching infinity. **In the limit we have** 1^∞. But that expression is meaningless (like $0/0$). Everything depends on the balance between "nearly 1" and "nearly ∞." This balance produces the extraordinary number e:

DEFINITION   *The number e is equal to* $\lim\limits_{h \to 0} (1+h)^{1/h}$. *Equivalently*

$$e = \lim_{n \to \infty} \left(1 + \frac{1}{n} \right)^n.$$

Before computing e, look again at the slope $1/c$. At the end of equation (2) is the logarithm of e:

$$1/c = \log_b e. \tag{3}$$

When the base is $b = e$, the slope is $\log_e e = 1$. That base e has $c = 1$ as desired:

> **The derivative of e^x is $1 \cdot e^x$ and the derivative of $\log_e y$ is** $\dfrac{1}{1 \cdot y}$. $\tag{4}$

This is why the base e is all-important in calculus. It makes $c = 1$.

To compute the actual number e from $(1+h)^{1/h}$, choose $h = 1, 1/10, 1/100, \ldots$. Then the exponents $1/h$ are $n = 1, 10, 100, \ldots$. (All limits and derivatives will become official in Section 6.4.) The table shows $(1+h)^{1/h}$ approaching e as $h \to 0$ and $n \to \infty$:

n	$h = \dfrac{1}{n}$	$1 + h = 1 + \dfrac{1}{n}$	$(1+h)^{1/h} = \left(1 + \dfrac{1}{n}\right)^n$
1	1.0	2.0	2.0
2	0.5	1.5	2.25
10	0.1	1.1	2.593742
100	0.01	1.01	2.704814
1000	0.001	1.001	2.716924
10000	0.0001	1.0001	2.718146

The last column is converging to e (not quickly). There is an infinite series that converges much faster. We know $125,000$ digits of e (and a billion digits of π). There are no definite patterns, although you might think so from the first sixteen digits:

$$e = 2.7 \ 1828 \ 1828 \ 45 \ 90 \ 45 \cdots \quad (\text{and } 1/e \approx .37).$$

The powers of e produce $y = e^x$. At $x = 2.3$ and 5, we are close to $y = 10$ and 150.

The logarithm is the inverse function. The logarithms of 150 and 10, to the base e, are close to $x = 5$ and $x = 2.3$. There is a special name for this logarithm—the *natural logarithm*. There is also a special notation "ln" to show that the base is e:

$\ln y$ *means the same as* $\log_e y$. *The natural logarithm is the exponent in* $e^x = y$.

The notation $\ln y$ (or $\ln x$—it is the function that matters, not the variable) is standard in calculus courses. After calculus, the base is generally assumed to be e. In most of science and engineering, the natural logarithm is the automatic choice. The symbol "$\exp(x)$" means e^x, and the truth is that the symbol "$\log x$" generally means $\ln x$. Base e is understood even without the letters $\ln$. But in any case of doubt—on a calculator key for example—the symbol "$\ln x$" emphasizes that the base is e.

THE DERIVATIVES OF e^x AND $\ln x$

Come back to derivatives and slopes. The derivative of b^x is cb^x, and the derivative of $\log_b y$ is $1/cy$. If $b = e$ *then* $c = 1$. For all bases, equation (3) is $1/c = \log_b e$. This gives c—the slope of b^x at $x = 0$:

> **6E** *The number c is* $1/\log_b e = \log_e b$. ***Thus c equals*** $\ln b$. (5)

$c = \ln b$ **is the mysterious constant that was not available earlier**. The slope of 2^x is $\ln 2$ times 2^x. The slope of e^x is $\ln e$ times e^x (but $\ln e = 1$). We have the derivatives on which this chapter depends:

> **6F** The derivatives of e^x and $\ln y$ are e^x and $1/y$. For other bases
>
> $$\frac{d}{dx} b^x = (\ln b) b^x \quad \text{and} \quad \frac{d}{dy} \log_b y = \frac{1}{(\ln b) y}. \tag{6}$$

To make clear that those derivatives come from the functions (and not at all from the dummy variables), we rewrite them using t and x:

$$\frac{d}{dt} e^t = e^t \quad \text{and} \quad \frac{d}{dx} \ln x = \frac{1}{x}. \tag{7}$$

Remark on slopes at $x = 0$: It would be satisfying to see directly that the slope of 2^x is below 1, and the slope of 4^x is above 1. Quick proof: e is between 2 and 4. But the idea is to see the slopes graphically. This is a small puzzle, which is fun to solve but can be skipped.

2^x rises from 1 at $x = 0$ to 2 at $x = 1$. On that interval its average slope is 1. Its slope at the beginning is *smaller* than average, so it must be less than 1—as desired. On the other hand 4^x rises from $\frac{1}{2}$ at $x = -\frac{1}{2}$ to 1 at $x = 0$. Again the average slope is $\frac{1}{2}/\frac{1}{2} = 1$. Since $x = 0$ comes at the *end* of this new interval, the slope of 4^x at that point exceeds 1. Somewhere between 2^x and 4^x is e^x, which starts out with slope 1.

This is the graphical approach to e. There is also the infinite series, and a fifth definition through integrals which is written here for the record:

1. e is the number such that e^x has slope 1 at $x = 0$
2. e is the base for which $\ln y = \log_e y$ has slope 1 at $y = 1$
3. e is the limit of $\left(1 + \dfrac{1}{n}\right)^n$ as $n \to \infty$
4. $e = \dfrac{1}{0!} + \dfrac{1}{1!} + \dfrac{1}{2!} + \dfrac{1}{3!} + \cdots = 1 + 1 + \dfrac{1}{2} + \dfrac{1}{6} + \cdots$
5. the area $\int_1^e x^{-1} dx$ equals 1.

The connections between **1**, **2**, and **3** have been made. The slopes are 1 when e is the limit of $(1 + 1/n)^n$. Multiplying this out wlll lead to **4**, the infinite series in Section 6.6. The official definition of $\ln x$ comes from $\int dx/x$, and then **5** says that $\ln e = 1$. This approach to e (Section 6.4) seems less intuitive than the others.

Figure 6.6b shows the graph of e^{-x}. It is the mirror image of e^x across the vertical axis. Their product is $e^x e^{-x} = 1$. Where e^x grows exponentially, e^{-x} decays exponentially—or it grows as x approaches $-\infty$. ***Their growth and decay are faster than any power of*** x. Exponential growth is more rapid than polynomial growth, so that e^x/x^n goes to infinity (Problem 59). It is the fact that e^x has slope e^x which keeps the function climbing so fast.

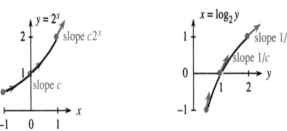

Fig. 6.6 e^x grows between 2^x and 4^x. Decay of e^{-x}, faster decay of $e^{-x^2/2}$.

The other curve is $y = e^{-x^2/2}$. This is the famous "*bell-shaped curve*" of probability theory. After dividing by $\sqrt{2\pi}$, it gives the ***normal distribution***, which applies to so many averages and so many experiments. The Gallup Poll will be an example in Section 8.4. The curve is symmetric around its mean value $x = 0$, since changing x to $-x$ has no effect on x^2.

About two thirds of the area under this curve is between $x = -1$ and $x = 1$. If you pick points at random below the graph, $2/3$ of all samples are expected in that interval. The points $x = -2$ and $x = 2$ are "two standard deviations" from the center, enclosing 95% of the area. There is only a 5% chance of landing beyond. The decay is even faster than an ordinary exponential, because $\frac{1}{2}x^2$ has replaced x.

THE DERIVATIVES OF e^{cx} AND $e^{u(x)}$

The slope of e^x is e^x. This opens up a whole world of functions that calculus can deal with. The chain rule gives the slope of e^{3x} and $e^{\sin x}$ and every $e^{u(x)}$:

6G The derivative of $e^{u(x)}$ is $e^{u(x)}$ times du/dx. (8)

Special case $u = cx$: The derivative of e^{cx} is ce^{cx}. (9)

EXAMPLE 1 The derivative of e^{3x} is $3e^{3x}$ (here $c = 3$). The derivative of $e^{\sin x}$ is $e^{\sin x} \cos x$ (here $u = \sin x$). The derivative of $f(u(x))$ is df/du times du/dx. Here $f = e^u$ so $df/du = e^u$. **The chain rule demands that second factor du/dx.**

EXAMPLE 2 $e^{(\ln 2)x}$ is the same as 2^x. Its derivative is $\ln 2$ times 2^x. The chain rule rediscovers our constant $c = \ln 2$. In the slope of b^x it rediscovers the factor $c = \ln b$.

Generally e^{cx} is preferred to the original b^x. The derivative just brings down the constant c. **It is better to agree on e as the base**, and put all complications (like $c = \ln b$) up in the exponent. The second derivative of e^{cx} is $c^2 e^{cx}$.

EXAMPLE 3 The derivative of $e^{-x^2/2}$ is $-xe^{-x^2/2}$ (here $u = -x^2/2$ so $du/dx = -x$).

EXAMPLE 4 The second derivative of $f = e^{-x^2/2}$, by the chain rule and product rule, is

$$f'' = (-1) \cdot e^{-x^2/2} + (-x)^2 e^{-x^2/2} = (x^2 - 1)e^{-x^2/2}. \tag{10}$$

Notice how **the exponential survives**. With every derivative it is multiplied by more factors, but it is still there to dominate growth or decay. The *points of inflection*, where the bell-shaped curve has $f'' = 0$ in equation (10), are $x = 1$ and $x = -1$.

EXAMPLE 5 ($u = n \ln x$). Since $e^{n \ln x}$ is x^n in disguise, its slope must be nx^{n-1}:

$$\text{slope} = e^{n \ln x} \frac{d}{dx}(n \ln x) = x^n \left(\frac{n}{x}\right) = nx^{n-1}. \tag{11}$$

This slope is correct for all n, integer or not. Chapter 2 produced $3x^2$ and $4x^3$ from the binomial theorem. Now nx^{n-1} comes from ln and exp and the chain rule.

EXAMPLE 6 An extreme case is $x^x = (e^{\ln x})^x$. Here $u = x \ln x$ and we need du/dx:

$$\frac{d}{dx}(x^x) = e^{x \ln x}\left(\ln x + x \cdot \frac{1}{x}\right) = x^x(\ln x + 1).$$

INTEGRALS OF e^{cx} AND $e^u \, du/dx$

The integral of e^x is e^x. *The integral of e^{cx} is not e^{cx}.* The derivative multiplies by c so the integral divides by c. **The integral of e^{cx} is e^{cx}/c (plus a constant).**

EXAMPLES $\displaystyle\int e^{2x}\,dx = \frac{1}{2}e^{2x} + C \qquad \int b^x\,dx = \frac{b^x}{\ln b} + C$

$$\int e^{3(x+1)}\,dx = \frac{1}{3}e^{3(x+1)} + C \qquad \int e^{-x^2/2}\,dx \to \text{failure}$$

The first one has $c = 2$. The second has $c = \ln b$—remember again that $b^x = e^{(\ln b)x}$. The integral divides by $\ln b$. In the third one, $e^{3(x+1)}$ is e^{3x} times the number e^3

and that number is carried along. Or more likely we see $e^{3(x+1)}$ as e^u. The missing $du/dx = 3$ is fixed by dividing by 3. The last example fails because du/dx is not there. **We cannot integrate without du/dx:**

6H *The indefinite integral* $\displaystyle\int e^u \frac{du}{dx}\, dx$ *equals* $e^{u(x)} + C.$

Here are three examples with du/dx and one without it:

$$\int e^{\sin x} \cos x\, dx = e^{\sin x} + C \qquad \int x e^{x^2/2}\, dx = e^{x^2/2} + C$$

$$\int \frac{e^{\sqrt{x}}\, dx}{\sqrt{x}} = 2e^{\sqrt{x}} + C \qquad \int \frac{e^x\, dx}{(1+e^x)^2} = \frac{-1}{1+e^x} + C$$

The first is a pure $e^u\, du$. So is the second. The third has $u = \sqrt{x}$ and $du/dx = 1/2\sqrt{x}$,

so only the factor 2 had to be fixed. The fourth example does not belong with the others. It is the integral of du/u^2, not the integral of $e^u\, du$. I don't know any way to tell you which substitution is best—except that *the complicated part is $1+e^x$ and it is natural to substitute u*. If it works, good.

Without an extra e^x for du/dx, the integral $\int dx/(1+e^x)^2$ looks bad. But $u = 1+e^x$ is still worth trying. It has $du = e^x dx = (u-1)dx$:

$$\int \frac{dx}{(1+e^x)^2} = \int \frac{du}{(u-1)u^2} = \int du \left(\frac{1}{u-1} - \frac{1}{u} - \frac{1}{u^2} \right). \qquad (12)$$

That last step is "*partial fractions*." The integral splits into simpler pieces (explained in Section 7.4) and we integrate each piece. Here are three other integrals:

$$\int e^{1/x} dx \qquad \int e^x (4+e^x) dx \qquad \int e^{-x}(4+e^x) dx$$

The first can change to $-\int e^u du/u^2$, which is not much better. (It is just as impossible.) The second is actually $\int u\, du$, but I prefer a split: $\int 4e^x$ and $\int e^{2x}$ are safer to do separately. The third is $\int (4e^{-x} + 1)dx$, which also separates. The exercises offer practice in reaching $e^u\, du/dx$—ready to be integrated.

Warning about definite integrals When the lower limit is $x = 0$, there is a natural tendency to expect $f(0) = 0$—in which case the lower limit contributes nothing. For a power $f = x^3$ that is true. For an exponential $f = e^{3x}$ it is definitely *not true*, because $f(0) = 1$:

$$\int_0^1 e^{3x} dx = \frac{1}{3} e^{3x} \Big]_0^1 = \frac{1}{3}(e^3 - 1) \qquad \int_0^1 x e^{x^2} dx = \frac{1}{2} e^{x^2} \Big]_0^1 = \frac{1}{2}(e - 1).$$

Read-through questions

The number e is approximately __a__. It is the limit of $(1+h)$ to the power __b__. This gives 1.01^{100} when $h =$ __c__. An equivalent form is $e = \lim(\_\_d\_\_)^n$.

When the base is $b = e$, the constant c in Section 6.1 is __e__. Therefore the derivative of $y = e^x$ is $dy/dx =$ __f__. The derivative of $x = \log_e y$ is $dx/dy =$ __g__. The slopes at $x = 0$ and $y = 1$ are both __h__. The notation for $\log_e y$ is __i__, which is the __j__ logarithm of y.

The constant c in the slope of b^x is $c =$ __k__. The function b^x can be rewritten as __l__. Its derivative is __m__. The derivative of $e^{u(x)}$ is __n__. The derivative of $e^{\sin x}$ is __o__. The derivative of e^{cx} brings down a factor __p__.

The integral of e^x is __q__. The integral of e^{cx} is __r__. The integral of $e^{u(x)}du/dx$ is __s__. In general the integral of $e^{u(x)}$ by itself is __t__ to find.

Find the derivatives of the functions in 1–18.

1 $7e^{7x}$

2 $-7e^{-7x}$

3 $(e^x)^8$

4 $(x^{-x})^{-8}$

5 3^x

6 $e^{x\ln 3}$

7 $(2/3)^x$

8 4^{4x}

9 $1/(1+e^x)$

10 $e^{1/(1+x)}$

11 $e^{\ln x} + x^{\ln e}$

12 $xe^{1/x}$

13 $xe^x - e^x$

14 $x^2 e^x - 2xe^x + 2e^x$

15 $\dfrac{e^x - e^{-x}}{e^x + e^{-x}}$

16 $e^{\ln(x^2)} + \ln(e^{x^2})$

17 $e^{\sin x} + \sin e^x$

18 $x^{-1/x}$ (which is e—)

19 The difference between e and $(1+1/n)^n$ is approximately Ce/n. Subtract the calculated values for $n = 10, 100, 1000$ from 2.7183 to discover the number C.

20 By algebra or a calculator find the limits of $(1+1/n)^{2n}$ and $(1+1/n)^{\sqrt{n}}$.

21 The limit of $(11/10)^{10}, (101/100)^{100}, \dots$ is e. So the limit of $(10/11)^{10}, (100/101)^{100}, \dots$ is _____. So the limit of $(10/11)^{11}, (100/101)^{101}, \dots$ is _____. The last sequence is $(1-1/n)^n$.

22 Compare the number of correct decimals of e for $(1.001)^{1000}$ and $(1.0001)^{10000}$ and if possible $(1.00001)^{100000}$. Which power n would give all the decimals in 2.71828?

23 The function $y = e^x$ solves $dy/dx = y$. Approximate this equation by $\Delta Y/\Delta x = Y$, which is $Y(x+h) - Y(x) = hY(x)$. With $h = \frac{1}{10}$ find $Y(h)$ after one step starting from $Y(0) = 1$. What is $Y(1)$ after ten steps?

24 The function that solves $dy/dx = -y$ starting from $y = 1$ at $x = 0$ is _____. Approximate by $Y(x+h) - Y(x) = -hY(x)$. If $h = \frac{1}{4}$ what is $Y(h)$ after one step and what is $Y(1)$ after four steps?

25 Invent three functions f, g, h such that for $x > 10$
$$(1+1/x)^x < f(x) < e^x < g(x) < e^{2x} < h(x) < x^x.$$

26 Graph e^x and $\sqrt{e^x}$ at $x = -2, -1, 0, 1, 2$. Another form of $\sqrt{e^x}$ is _____.

Find antiderivatives for the functions in 27–36.

27 $e^{3x} + e^{7x}$

28 $(e^{3x})(e^{7x})$

29 $1^x + 2^x + 3^x$

30 2^{-x}

31 $(2e)^x + 2e^x$

32 $(1/e^x) + (1/x^e)$

33 $xe^{x^2} + xe^{-x^2}$

34 $(\sin x)e^{\cos x} + (\cos x)e^{\sin x}$

35 $\sqrt{e^x} + (e^x)^2$

36 xe^x (trial and error)

37 Compare e^{-x} with e^{-x^2}. Which one decreases faster near $x = 0$? Where do the graphs meet again? When is the ratio of e^{-x^2} to e^{-x} less than $1/100$?

38 Compare e^x with x^x: Where do the graphs meet? What are their slopes at that point? Divide x^x by e^x and show that the ratio approaches infinity.

39 Find the tangent line to $y = e^x$ at $x = a$. From which point on the graph does the tangent line pass through the origin?

40 By comparing slopes, prove that if $x > 0$ then

(a) $e^x > 1 + x$

(b) $e^{-x} > 1 - x$.

41 Find the minimum value of $y = x^x$ for $x > 0$. Show from d^2y/dx^2 that the curve is concave upward.

42 Find the slope of $y = x^{1/x}$ and the point where $dy/dx = 0$. Check d^2y/dx^2 to show that the maximum of $x^{1/x}$ is _____.

43 If $dy/dx = y$ find the derivative of $e^{-x}y$ by the product rule. Deduce that $y(x) = Ce^x$ for some constant C.

44 Prove that $x^e = e^x$ has only one positive solution.

Evaluate the integrals in 45–54. With infinite limits, 49–50 are "improper."

45 $\int_0^1 e^{2x}\,dx$

46 $\int_0^\pi \sin x\, e^{\cos x}\,dx$

47 $\int_{-1}^1 2^x\,dx$

48 $\int_{-1}^1 2^{-x}\,dx$

49 $\int_0^\infty e^{-x}\,dx$

50 $\int_0^\infty x e^{-x^2}\,dx$

51 $\int_0^1 e^{1+x}\,dx$

52 $\int_0^1 e^{1+x^2} x\,dx$

53 $\int_0^\pi 2^{\sin x}\cos x\,dx$

54 $\int_0^1 (1-e^x)^{10} e^x\,dx$

55 Integrate the integrals that can be integrated:

$$\int \frac{e^u}{du/dx}\,dx \qquad \int \frac{du/dx}{e^u}\,dx$$

$$\int e^u \left(\frac{du}{dx}\right)^2 dx \qquad \int (e^u)^2 \frac{du}{dx}\,dx$$

56 Find a function that solves $y'(x)=5y(x)$ with $y(0)=2$.

57 Find a function that solves $y'(x)=1/y(x)$ with $y(0)=2$.

58 With electronic help graph the function $(1+1/x)^x$. What are its asymptotes? Why?

59 This exercise shows that $F(x)=x^n/e^x \to 0$ as $x\to\infty$.

(a) Find dF/dx. Notice that $F(x)$ decreases for $x>n>0$. The maximum of x^n/e^x, at $x=n$, is n^n/e^n.

(b) $F(2x)=(2x)^n/e^{2x}=2^n x^n/e^x \cdot e^x \leq 2^n n^n/e^n \cdot e^x$. Deduce that $F(2x)\to 0$ as $x\to\infty$. Thus $F(x)\to 0$.

60 With $n=6$, graph $F(x)=x^6/e^x$ on a calculator or computer. Estimate its maximum. Estimate x when you reach $F(x)=1$. Estimate x when you reach $F(x)=\frac12$.

61 Stirling's formula says that $n!\approx\sqrt{2\pi n}\,n^n/e^n$. Use it to estimate $6^6/e^6$ to the nearest whole number. Is it correct? How many decimal digits in 10!?

62 $x^6/e^x\to 0$ is also proved by l'Hôpital's rule (at $x=\infty$):
$$\lim x^6/e^x = \lim 6x^5/e^x = \underline{\text{fill this in}} = 0.$$

6.3 Growth and Decay in Science and Economics

The derivative of $y = e^{cx}$ has taken time and effort. **The result was $y' = ce^{cx}$, which means that $y' = cy$.** That computation brought others with it, virtually for free—the derivatives of b^x and x^x and $e^{u(x)}$. But I want to stay with $y' = cy$—which is the most important differential equation in applied mathematics.

Compare $y' = x$ with $y' = y$. The first only asks for an antiderivative of x. We quickly find $y = \frac{1}{2}x^2 + C$. The second has dy/dx equal to y itself—which we rewrite as $dy/y = dx$. The integral is in $y = x + C$. Then y itself is $e^x e^c$. Notice that the first solution is $\frac{1}{2}x^2$ *plus* a constant, and the second solution is e^x times a constant.

There is a way to graph slope x versus slope y. Figure 6.7 shows "*tangent arrows*," which give the slope at each x and y. For parabolas, the arrows grow steeper as x grows—because $y' = $ slope $= x$. For exponentials, the arrows grow steeper as y grows—the equation is $y' = $ slope $= y$. Now the arrows are connected by $y = Ae^x$. **A differential equation gives a field of arrows** (slopes). *Its solution is a curve that stays tangent to the arrows*—then the curve has the right slope.

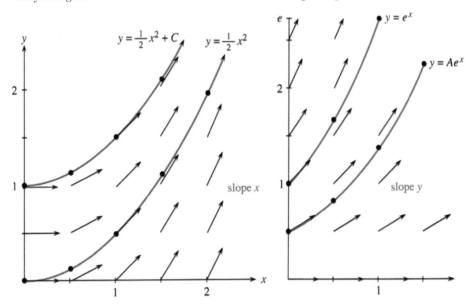

Fig. 6.7   The slopes are $y' = x$ and $y' = y$. The solution curves fit those slopes.

A field of arrows can show many solutions at once (this comes in a differential equations course). Usually a single y_0 is not sacred. To understand the equation we start from many y_0—on the left the parabolas stay parallel, on the right the heights stay proportional. For $y' = -y$ all solution curves go to zero.

From $y' = y$ it is a short step to $y' = cy$. To make c appear in the derivative, *put c into the exponent*. The derivative of $y = e^{cx}$ is ce^{cx}, which is c times y. We have reached the key equation, which comes with an *initial condition*—a starting value y_0:

$$dy/dt = cy \text{ with } y = y_0 \text{ at } t = 0. \tag{1}$$

A small change: x has switched to t. In most applications *time* is the natural variable, rather than space. The factor c becomes the "growth rate" or "decay rate"—and e^{cx} converts to e^{ct}.

The last step is to match the initial condition. The problem requires $y = y_0$ at $t = 0$. Our e^{ct} starts from $e^{c0} = 1$. *The constant of integration is needed now—the solutions are* $y = Ae^{ct}$. By choosing $A = y_0$, we match the initial condition and solve equation (1). *The formula to remember is* y_0e^{ct}.

> **6I** The **exponential law** $y = y_0e^{ct}$ solves $y' = cy$ starting from y_0.

The rate of growth or decay is c. May I call your attention to a basic fact? *The formula* y_0e^{ct} *contains three quantities* y_0, c, t. If two of them are given, plus one additional piece of information, the third is determined. Many applications have one of these three forms: **find** t, **find** c, **find** y_0.

1. Find the doubling time T if $c = 1/10$. At that time y_0e^{cT} equals $2y_0$:

$$e^{cT} = 2 \text{ yields } cT = \ln 2 \text{ so that } T = \frac{\ln 2}{c} \approx \frac{.7}{.1}. \tag{2}$$

The question asks for an exponent T. The answer involves logarithms. If a cell grows at a continuous rate of $c = 10\%$ per day, it takes about $.7/.1 = 7$ days to double in size. (Note that $.7$ is close to $\ln 2$.) If a savings account earns 10% continuous interest, it doubles in 7 years.

In this problem we knew c. In the next problem we know T.

2. Find the decay constant c for carbon-14 if $y = \frac{1}{2}y_0$ in $T = 5568$ years.

$$e^{cT} = \tfrac{1}{2} \text{ yields } cT = \ln \tfrac{1}{2} \text{ so that } c \approx \left(\ln \tfrac{1}{2}\right)/5568. \tag{3}$$

After the half-life $T = 5568$, the factor e^{cT} equals $\frac{1}{2}$. Now c is negative ($\ln \frac{1}{2} = -\ln 2$).

Question **1** was about growth. Question **2** was about decay. Both answers found e^{cT} as the ratio $y(T)/y(0)$. Then cT is its logarithm. Note how c sticks to T. T has the units of time, c has the units of "1/time."

Main point: The doubling time is $(\ln 2)/c$, because $cT = \ln 2$. The time to multiply by e is $1/c$. The time to multiply by 10 is $(\ln 10)/c$. The time to divide by e is $-1/c$, when a negative c brings decay.

3. Find the initial value y_0 if $c = 2$ and $y(1) = 5$:

$$y(t) = y_0e^{ct} \text{ yields } y_0 = y(t)e^{-ct} = 5e^{-2}.$$

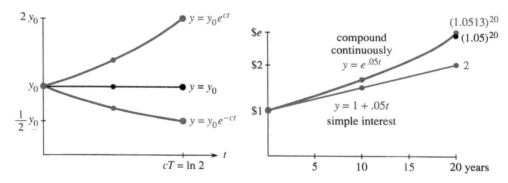

Fig. 6.8   Growth ($c > 0$) and decay ($c < 0$). Doubling time $T = (\ln 2)/c$. Future value at 5%.

Solution by Method 2 (slow way) The input y_0 produces the output $y_0 e^{ct}$. After t years any deposit is multiplied by e^{ct}. *That also applies to deposits made after the account is opened.* If the deposit enters at time T, the growing time is only $t - T$. Therefore the multiplying factor is only $e^{c(t-T)}$. This growth factor applies to the small deposit (amount $s\,dT$) made between time T and $T + dT$.

Now add up all outputs at time t. The output from y_0 is $y_0 e^{ct}$. The small deposit $s\,dT$ near time T grows to $e^{c(t-T)} s\,dT$. The total is an integral:

$$y(t) = y_0 e^{ct} + \int_{T=0}^{t} e^{c(t-T)} s\,dT. \tag{9}$$

This principle of Duhamel would still apply when the source s varies with time. Here s is constant, and the integral divides by c:

$$s \int_{T=0}^{t} e^{c(t-T)} dT = \frac{s e^{c(t-T)}}{-c} \Bigg]_0^t = -\frac{s}{c} + \frac{s}{c} e^{ct}. \tag{10}$$

That agrees with the source term from Method 1, at the end of equation (8). There we looked for "exponential plus constant," here we added up outputs.

Method 1 was easier. It succeeded because we knew the form $Ae^{ct} + B$—with "undetermined coefficients." Method 2 is more complete. The form for y is part of the output, not the input. The source s is a continuous supply of new deposits, all growing separately. Section 6.5 starts from scratch, by directly integrating $y' = cy + s$.

Remark Method 2 is often described in terms of an *integrating factor*. First write the equation as $y' - cy = s$. Then multiply by a magic factor that makes integration possible:

$$(y' - cy)e^{-ct} = se^{-ct} \qquad\qquad \text{multiply by the factor } e^{-ct}$$

$$ye^{-ct} \Bigg]_0^t = -\frac{s}{c} e^{-ct} \Bigg]_0^t \qquad\qquad \text{integrate both sides}$$

$$ye^{-ct} - y_0 = -\frac{s}{c}(e^{-ct} - 1) \qquad\qquad \text{substitute } 0 \text{ and } t$$

$$y = e^{ct} y_0 + \frac{s}{c}(e^{ct} - 1) \quad \text{isolate } y \text{ to reach formula (8)}$$

The integrating factor produced a perfect derivative in line 1. I prefer Duhamel's idea, that all inputs y_0 and s grow the same way. Either method gives formula (8) for y.

THE MATHEMATICS OF FINANCE (AT A CONTINUOUS RATE)

The question from finance is this: *What inputs give what outputs?* The inputs can come at the start by y_0, or continuously by s. The output can be paid at the end or continuously. There are six basic questions, two of which are already answered.

The future value is $y_0 e^{ct}$ from a deposit of y_0. To produce y in the future, deposit the present value $y e^{-ct}$. *Questions 3–6 involve the source term s.* We fix the continuous rate at 5% per year ($c = .05$), and start the account from $y_0 = 0$. The answers come fast from equation (8).

Question 3   With deposits of $s = \$1000/\text{year}$, how large is y after 20 years?

$$y = \frac{s}{c}(e^{ct} - 1) = \frac{1000}{.05}(e^{(.05)(20)} - 1) = 20,000(e - 1) \approx \$34,400.$$

One big deposit yields $20,000e \approx \$54,000$. The same $20,000$ via s yields $\$34,400$.

All we do is run the process backward. Start from 5 and go back to y_0. With time reversed, e^{ct} becomes e^{-ct}. The product of e^2 and e^{-2} is 1—growth forward and decay backward.

Equally important is $T + t$. **Go forward to time T and go on to $T + t$:**

$$y(T + t) \text{ is } y_0 e^{c(T+t)} \text{ which is } (y_0 e^{cT}) e^{ct}. \tag{4}$$

Ever step t, at the start or later, multiplies by the same e^{ct}. This uses the fundamental property of exponentials, that $e^{T+t} = e^T e^t$.

EXAMPLE 1 *Population growth from birth rate b and death rate d* (both constant):
$$dy/dt = by - dy = cy \quad \text{(the net rate is } c = b - d).$$

The population in this model is $y_0 e^{ct} = y_0 e^{bt} e^{-dt}$. It grows when $b > d$ (which makes $c > 0$). One estimate of the growth rate is $c = 0.02/\text{year}$:

> **The earth's population doubles in about $T = \dfrac{\ln 2}{c} \approx \dfrac{.7}{.02} = 35$ years.**

First comment: We predict the future based on c. We count the past population to find c. Changes in c are a serious problem for this model.

Second comment: $y_0 e^{ct}$ is not a whole number. You may prefer to think of bacteria instead of people. (***This section begins a major application of mathematics to economics and the life sciences***.) Malthus based his theory of human population on this equation $y' = cy$—and with large numbers a fraction of a person doesn't matter so much. To use calculus we go from discrete to continuous. The theory must fail when t is very large, since populations cannot grow exponentially forever. Section 6.5 introduces the *logistic* equation $y' = cy - by^2$, with a competition term $-by^2$ to slow the growth.

Third comment: The dimensions of b, c, d are "1/time." The dictionary gives birth rate = number of births per person in a unit of time. It is a *relative* rate—people divided by people and time. The product ct is dimensionless and e^{ct} makes sense (also dimensionless). Some texts replace c by λ (lambda). Then $1/\lambda$ is the growth time or decay time or drug elimination time or diffusion time.

EXAMPLE 2 *Radioactive dating* A gram of charcoal from the cave paintings in France gives 0.97 disintegrations per minute. A gram of living wood gives 6.68 disintegrations per minute. Find the age of those Lascaux paintings.

The charcoal stopped adding radiocarbon when it was burned (at $t = 0$). The amount has decayed to $y_0 e^{ct}$. In living wood this amount is still y_0, because cosmic rays maintain the balance. *Their ratio is $e^{ct} = 0.97/6.68$.* Knowing the decay rate c from Question **2** above, we know the present time t:

$$ct = \ln\left(\frac{0.97}{6.68}\right) \quad \text{yields} \quad t = \frac{5568}{-.7}\ln\left(\frac{0.97}{6.68}\right) = 14,400 \text{ years}.$$

Here is a related problem—*the age of uranium*. Right now there is 140 times as much U-238 as U-235. Nearly equal amounts were created, with half-lives of $(4.5)10^9$ and $(0.7)10^9$ years. **Question**: How long since uranium was created? *Answer*: Find t by substituting $c = (\ln\frac{1}{2})/(4.5)10^9$ and $C = (\ln\frac{1}{2})/(0.7)10^9$:

$$e^{ct}/e^{Ct} = 140 \Rightarrow ct - Ct = \ln 140 \Rightarrow t = \frac{\ln 140}{c - C} = 6(10^9) \text{ years}.$$

EXAMPLE 3   *Calculus in Economics*: *price inflation and the value of money*

We begin with two inflation rates—a *continuous rate* and an *annual rate*. For the price change Δy over a year, use the annual rate:

$$\Delta y = (annual\ rate)\ \text{times}\ (y)\ \text{times}\ (\Delta t). \tag{5}$$

Calculus applies the continuous rate to each instant dt. The price change is dy:

$$dy = (continuous\ rate)\ \text{times}\ (y)\ \text{times}\ (dt). \tag{6}$$

Dividing by dt, this is a differential equation for the price:

$$dy/dt = (\boldsymbol{continuous\ rate})\ \text{times}\ (y) = .05y.$$

The solution is $y_0 e^{.05t}$. Set $t = 1$. Then $e^{.05} \approx 1.0513$ and the annual rate is 5.13%.

When you ask a bank what interest they pay, they give both rates: 8% and 8.33%. The higher one they call the "effective rate." It comes from compounding (and depends how often they do it). If the compounding is continuous, every dt brings an increase of dy—and $e^{.08}$ is near 1.0833.

Section 6.6 returns to compound interest. The interval drops from a month to a day to a second. That leads to $(1 + 1/n)^n$, and in the limit to e. Here we compute the effect of 5% continuous interest:

Future value   A dollar now has the same value as $e^{.05T}$ dollars in T years.

Present value   A dollar in T years has the same value as $e^{-.05T}$ dollars now.

Doubling time   Prices double ($e^{.05T} = 2$) in $T = \ln 2/.05 \approx 14$ years.

With no compounding, the doubling time is 20 years. Simple interest adds on 20 times 5% = 100%. With continuous compounding the time is reduced by the factor $\ln 2 \approx .7$, regardless of the interest rate.

EXAMPLE 4   In 1626 the Indians sold Manhattan for \$24. Our calculations indicate that they knew what they were doing. Assuming 8% compound interest, the original \$24 is multiplied by $e^{.08t}$. After $t = 365$ years the multiplier is $e^{29.2}$ and the \$24 has grown to 115 trillion dollars. With that much money they could buy back the land and pay off the national debt.

This seems farfetched. Possibly there is a big flaw in the model. It is absolutely true that Ben Franklin left money to Boston and Philadelphia, to be invested for 200 years. In 1990 it yielded millions (not trillions, that takes longer). Our next step is a new model.

Question How can you estimate $e^{29.2}$ with a \$24 calculator (log but not ln) ?
Answer   Multiply 29.2 by $\log_{10} e = .434$ to get 12.7. This is the exponent to base 10. After that base change, we have $10^{12.7}$ or more than a trillion.

GROWTH OR DECAY WITH A SOURCE TERM

The equation $y' = y$ will be given a new term. Up to now, all growth or decay has started from y_0. No deposit or withdrawal was made later. The investment grew by itself—a pure exponential. ***The new term s allows you to add or subtract from the*** *account*. It is a "source"—or a "sink" if s is negative. The source $s = 5$ adds $5dt$, proportional to dt but not to y:

> **Constant source:** $dy/dt = y + 5$ starting from $y = y_0$.

Notice y on both sides! My first guess $y = e^{t+5}$ failed completely. Its derivative is e^{t+5} again, which is not $y + 5$. The class suggested $y = e^t + 5t$. But its derivative $e^t + 5$ is still not $y + 5$. We tried other ways to produce 5 in dy/dt. This idea is doomed to failure. *Finally we thought of $y = Ae^t - 5$. That has $y' = Ae^t = y + 5$* as required.

Important: *A is not y_0*. Set $t = 0$ to find $y_0 = A - 5$. The source contributes $5e^t - 5$:

> **The solution is $(y_0 + 5)e^t - 5$. *That is the same as $y_0 e^t + 5(e^t - 1)$.***

$s = 5$ multiplies the growth term $e^t - 1$ that starts at zero. $y_0 e^t$ grows as before.

EXAMPLE 5 $dy/dt = -y + 5$ has $y = (y_0 - 5)e^{-t} + 5$. This is $y_0 e^{-t} + 5(1 - e^{-t})$.

That final term from the source is still positive. The other term $y_0 e^{-t}$ decays to zero. *The limit as $t \to \infty$ is $y_\infty = 5$.* A negative c leads to a steady state y_∞.

Based on these examples with $c = 1$ and $c = -1$, we can find y for any c and s.

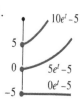

EQUATION WITH SOURCE $\dfrac{dy}{dt} = cy + s$ starts from $y = y_0$ at $t = 0$. (7)

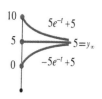

Fig. 6.9

The source could be a deposit of $s = \$1000/\text{year}$, after an initial investment of $y_0 = \$8000$. Or we can withdraw funds at $s = -\$200/\text{year}$. The units are "dollars per year" to match dy/dt. The equation feeds in \$1000 or removes \$200 *continuously*—not all at once.

Note again that $y = e^{(c+s)t}$ is not a solution. Its derivative is $(c + s)y$. The combination $y = e^{ct} + s$ is also not a solution (but closer). *The analysis of $y' = cy + s$ will be our main achievement for differential equations* (in this section). The equation is not restricted to finance—far from it—but that produces excellent examples.

I propose to find y in four ways. You may feel that one way is enough.† The first way is the fastest—only three lines—but please give the others a chance. There is no point in preparing for real problems if we don't solve them.

Solution by Method 1 (fast way) Substitute the combination $y = Ae^{ct} + B$. *The solution has this form—exponential plus constant*. From two facts we find A and B:

> the equation $y' = cy + s$ gives $cAe^{ct} = c(Ae^{ct} + B) + s$
>
> the initial value at $t = 0$ gives $A + B = y_0$.

The first line has cAe^{ct} on both sides. Subtraction leaves $cB + s = 0$, or $B = -s/c$. Then the second line becomes $A = y_0 - B = y_0 + (s/c)$:

KEY FORMULA $y = \left(y_0 + \dfrac{s}{c}\right)e^{ct} - \dfrac{s}{c}$ **or** $y = y_0 e^{ct} + \dfrac{s}{c}(e^{ct} - 1)$. (8)

With $s = 0$ this is the old solution $y_0 e^{ct}$ (no source). The example with $c = 1$ and $s = 5$ produced $(y_0 + 5)e^t - 5$. Separating the source term gives $y_0 e^t + 5(e^t - 1)$.

---

†My class says one way is *more* than enough. They just want the answer. Sometimes I cave in and write down the formula: y is $y_0 e^{ct}$ plus $s(e^{ct} - 1)/c$ from the source term.

Solution by Method 2 (slow way) The input y_0 produces the output y_0e^{ct}. After t years any deposit is multiplied by e^{ct}. *That also applies to deposits made after the account is opened.* If the deposit enters at time T, the growing time is only $t-T$. Therefore the multiplying factor is only $e^{c(t-T)}$. This growth factor applies to the small deposit (amount $s\,dT$) made between time T and $T+dT$.

Now add up all outputs at time t. The output from y_0 is y_0e^{ct}. The small deposit $s\,dT$ near time T grows to $e^{c(t-T)}s\,dT$. The total is an integral:

$$y(t) = y_0e^{ct} + \int_{T=0}^{t} e^{c(t-T)}s\,dT. \tag{9}$$

This principle of Duhamel would still apply when the source s varies with time. Here s is constant, and the integral divides by c:

$$s\int_{T=0}^{t} e^{c(t-T)}dT = \left.\frac{se^{c(t-T)}}{-c}\right]_0^t = -\frac{s}{c}+\frac{s}{c}e^{ct}. \tag{10}$$

That agrees with the source term from Method 1, at the end of equation (8). There we looked for "exponential plus constant," here we added up outputs.

Method 1 was easier. It succeeded because we knew the form $Ae^{ct}+B$—with "undetermined coefficients." Method 2 is more complete. The form for y is part of the output, not the input. The source s is a continuous supply of new deposits, all growing separately. Section 6.5 starts from scratch, by directly integrating $y'=cy+s$.

Remark Method 2 is often described in terms of an *integrating factor*. First write the equation as $y'-cy=s$. Then multiply by a magic factor that makes integration possible:

$$(y'-cy)e^{-ct}=se^{-ct} \qquad\qquad \text{multiply by the factor } e^{-ct}$$

$$ye^{-ct}\Big]_0^t = -\frac{s}{c}e^{-ct}\Big]_0^t \qquad\qquad \text{integrate both sides}$$

$$ye^{-ct}-y_0 = -\frac{s}{c}(e^{-ct}-1) \qquad\qquad \text{substitute } 0 \text{ and } t$$

$$y = e^{ct}y_0+\frac{s}{c}(e^{ct}-1) \quad \text{isolate } y \text{ to reach formula (8)}$$

The integrating factor produced a perfect derivative in line 1. I prefer Duhamel's idea, that all inputs y_0 and s grow the same way. Either method gives formula (8) for y.

THE MATHEMATICS OF FINANCE (AT A CONTINUOUS RATE)

The question from finance is this: *What inputs give what outputs?* The inputs can come at the start by y_0, or continuously by s. The output can be paid at the end or continuously. There are six basic questions, two of which are already answered.

The future value is y_0e^{ct} from a deposit of y_0. To produce y in the future, deposit the present value ye^{-ct}. *Questions 3–6 involve the source term s.* We fix the continuous rate at 5% per year ($c=.05$), and start the account from $y_0=0$. The answers come fast from equation (8).

Question 3 With deposits of $s=\$1000/\text{year}$, how large is y after 20 years?

$$y = \frac{s}{c}(e^{ct}-1) = \frac{1000}{.05}(e^{(.05)(20)}-1) = 20{,}000(e-1) \approx \$34{,}400.$$

One big deposit yields $20{,}000e \approx \$54{,}000$. The same $20{,}000$ via s yields $\$34{,}400$.

Notice a small by-product (for mathematicians). When the interest rate is $c = 0$, our formula $s(e^{ct} - 1)/c$ turns into $0/0$. We are absolutely sure that depositing $1000/year with no interest produces $20,000 after 20 years. But this is not obvious from $0/0$. By l'Hôpital's rule we take c-derivatives in the fraction:

$$\lim_{c \to 0} \frac{s(e^{ct} - 1)}{c} = \lim_{c \to 0} \frac{ste^{ct}}{1} = st. \text{ This is } (1000)(20) = 20,000. \quad (11)$$

Question 4 What continuous deposit of s per year yields $20,000 after 20 years?

$$20,000 = \frac{s}{.05}(e^{(.05)(20)} - 1) \text{ requires } s = \frac{1000}{e - 1} \approx 582.$$

Deposits of $582 over 20 years total $11,640. A single deposit of $y_0 = 20,000/e = \$7,360$ produces the same $20,000 at the end. Better to be rich at $t = 0$.

Questions **1** and **2** had $s = 0$ (no source). Questions **3** and **4** had $y_0 = 0$ (no initial deposit). Now we come to $y = 0$. In **5**, everything is paid out by an *annuity*. In **6**, everything is paid up on a *loan*.

Question 5 What deposit y_0 provides $1000/year for 20 years? End with $y = 0$.

$$y = y_0 e^{ct} + \frac{s}{c}(e^{ct} - 1) = 0 \text{ requires } y_0 = \frac{-s}{c}(1 - e^{-ct}).$$

Substituting $s = -1000$, $c = .05$, $t = 20$ gives $y_0 \approx 12,640$. If you win $20,000 in a lottery, and it is paid over 20 years, the lottery only has to put in $12,640. Even less if the interest rate is above 5%.

Question 6 What payments s will clear a loan of $y_0 = \$20,000$ in 20 years?

Unfortunately, s exceeds $1000 per year. The bank gives up more than the $20,000 to buy your car (and pay tuition). *It also gives up the interest on that money.* You pay that back too, but you don't have to stay even at every moment. Instead you repay at a *constant rate* for 20 years. Your payments mostly cover interest at the start and principal at the end. After $t = 20$ years you are even and your debt is $y = 0$.

This is like Question **5** (also $y = 0$), but now we know y_0 and we want s:

$$y = y_0 e^{ct} + \frac{s}{c}(e^{ct} - 1) = 0 \text{ requires } s = -cy_0 e^{ct}/(e^{ct} - 1).$$

The loan is $y_0 = \$20,000$, the rate is $c = .05/\text{year}$, the time is $t = 20$ years. Substituting in the formula for s, your payments are $1582 per year.

Puzzle How is $s = \$1582$ for loan payments related to $s = \$582$ for deposits?

$$0 \to \$582 \text{ per year} \to \$20,000 \quad \text{and} \quad \$20,000 \to -\$1582 \text{ per year} \to 0.$$

That difference of exactly 1000 cannot be an accident. 1582 and 582 came from

$$1000\,\frac{e}{e - 1} \text{ and } 1000\frac{1}{e - 1} \text{ with difference } 1000\frac{e - 1}{e - 1} = 1000.$$

Why? Here is the real reason. Instead of repaying 1582 we can pay only 1000 (to keep even with the interest on 20,000). The other 582 goes into a separate account. After 20 years the continuous 582 has built up to 20,000 (including interest as in Question **4**). From that account we pay back the loan.

Section 6.6 deals with daily compounding—which differs from continuous compounding by only a few cents. Yearly compounding differs by a few dollars.

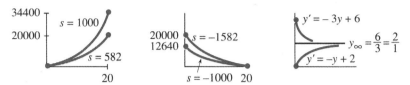

Fig. 6.10   Questions **3**–**4** deposit s. Questions **5**–**6** repay loan or annuity. Steady state $-s/c$.

TRANSIENTS VS. STEADY STATE

Suppose there is decay instead of growth. The constant c is negative and $y_0 e^{ct}$ dies out. That is the "*transient*" term, which disappears as $t \to \infty$. What is left is the "*steady state*." We denote that limit by y_∞.

Without a source, y_∞ is zero (total decay). When s is present, $y_\infty = -s/c$:

> **6J** The solution $y = \left(y_0 + \dfrac{s}{c}\right)e^{ct} - \dfrac{s}{c}$ approaches $y_\infty = -\dfrac{s}{c}$ when $e^{ct} \to 0$.

At this steady state, the source s exactly balances the decay cy. In other words $cy + s = 0$. From the left side of the differential equation, this means $dy/dt = 0$. *There is no change.* That is why y_∞ is steady.

Notice that y_∞ depends on the source and on c—*but not on y_0.*

EXAMPLE 6   Suppose Bermuda has a birth rate $b = .02$ and death rate $d = .03$. The net decay rate is $c = -.01$. There is also immigration from outside, of $s = 1200$/year. The initial population might be $y_0 = 5$ thousand or $y_0 = 5$ million, but that number has no effect on y_∞. *The steady state is independent of y_0.*

In this case $y_\infty = -s/c = 1200/.01 = 120,000$. The population grows to $120,000$ if y_0 is smaller. It decays to $120,000$ if y_0 is larger.

EXAMPLE 7   *Newton's Law of Cooling*: $dy/dt = c(y - y_\infty)$.   (12)

This is back to physics. The temperature of a body is y. The temperature around it is y_∞. Then y starts at y_0 and approaches y_∞, following Newton's rule: *The rate is proportional to $y - y_\infty$.* The bigger the difference, the faster heat flows.

The equation has $-cy_\infty$ where before we had s. That fits with $y_\infty = -s/c$. For the solution, replace s by $-cy_\infty$ in formula (8). Or use this new method:

Solution by Method 3   ***The new idea is to look at the difference*** $y - y_\infty$. Its derivative is dy/dt, since y_∞ is constant. But dy/dt is $c(y - y_\infty)$—this is our equation. The difference starts from $y_0 - y_\infty$, and grows or decays as a pure exponential:

$$\frac{d}{dt}(y - y_\infty) = c(y - y_\infty) \quad \textit{has the solution} \quad (y - y_\infty) = (y_0 - y_\infty)e^{ct}. \quad (13)$$

This solves the law of cooling. We repeat Method 3 using the letters s and c:

$$\frac{d}{dt}\left(y + \frac{s}{c}\right) = c\left(y + \frac{s}{c}\right) \quad \textit{has the solution} \quad \left(y + \frac{s}{c}\right) = \left(y_0 + \frac{s}{c}\right)e^{ct}. \quad (14)$$

Moving s/c to the right side recovers formula (8). There is a *constant term* and an *exponential term*. In a differential equations course, those are the "*particular solution*" and the "*homogeneous solution*." In a calculus course, it's time to stop.

EXAMPLE 8 In a $70°$ room, Newton's corpse is found with a temperature of $90°$. A day later the body registers $80°$. When did he stop integrating (at $98.6°$)?

Solution Here $y_\infty = 70$ and $y_0 = 90$. Newton's equation (13) is $y = 20e^{ct} + 70$. Then $y = 80$ at $t = 1$ gives $20e^c = 10$. The rate of cooling is $c = \ln \frac{1}{2}$. Death occurred when $20e^{ct} + 70 = 98.6$ or $e^{ct} = 1.43$. The time was $t = \ln 1.43 / \ln \frac{1}{2} =$ half a day earlier.

6.3 EXERCISES

Read-through exercises

If $y' = cy$ then $y(t) = $ __a__. If $dy/dt = 7y$ and $y_0 = 4$ then $y(t) = $ __b__. This solution reaches 8 at $t = $ __c__. If the doubling time is T then $c = $ __d__. If $y' = 3y$ and $y(1) = 9$ then y_0 was __e__. When c is negative, the solution approaches __f__ as $t \to \infty$.

The constant solution to $dy/dt = y + 6$ is $y = $ __g__. The general solution is $y = Ae^t - 6$. If $y_0 = 4$ then $A = $ __h__. The solution of $dy/dt = cy + s$ starting from y_0 is $y = Ae^{ct} + B = $ __i__. The output from the source s is __j__. An input at time T grows by the factor __k__ at time t.

At $c = 10\%$, the interest in time dt is $dy = $ __l__. This equation yields $y(t) = $ __m__. With a source term instead of y_0, a continuous deposit of $s = 4000$/year yields $y = $ __n__ after 10 years. The deposit required to produce $10,000$ in 10 years is $s = $ __o__ (exactly or approximately). An income of 4000/year forever (!) comes from $y_0 = $ __p__. The deposit to give 4000/year for 20 years is $y_0 = $ __q__. The payment rate s to clear a loan of $10,000$ in 10 years is __r__.

The solution to $y' = -3y + s$ approaches $y_\infty = $ __s__.

Solve 1–4 starting from $y_0 = 1$ and from $y_0 = -1$. Draw both solutions on the same graph.

1 $\dfrac{dy}{dt} = 2t$ 2 $\dfrac{dy}{dt} = -t$ 3 $\dfrac{dy}{dt} = 2y$ 4 $\dfrac{dy}{dt} = -y$

Solve 5–8 starting from $y_0 = 10$. At what time does y increase to 100 or drop to 1?

5 $\dfrac{dy}{dt} = 4y$ 6 $\dfrac{dy}{dt} = 4t$ 7 $\dfrac{dy}{dt} = e^{4t}$ 8 $\dfrac{dy}{dt} = e^{-4t}$

9 Draw a field of "tangent arrows" for $y' = -y$, with the solution curves $y = e^{-x}$ and $y = -e^{-x}$.

10 Draw a direction field of arrows for $y' = y - 1$, with solution curves $y = e^x + 1$ and $y = 1$.

Problems 11–27 involve $y_0 e^{ct}$. They ask for c or t or y_0.

11 If a culture of bacteria doubles in two hours, how many hours to multiply by 10? First find c.

12 If bacteria increase by factor of ten in ten hours, how many hours to increase by 100? What is c?

13 How old is a skull that contains $\frac{1}{5}$ as much radiocarbon as a modern skull?

14 If a relic contains 90% as much radiocarbon as new material, could it come from the time of Christ?

15 The population of Cairo grew from 5 million to 10 million in 20 years. From $y' = cy$ find c. When was $y = 8$ million?

16 The populations of New York and Los Angeles are growing at 1% and 1.4% a year. Starting from 8 million (NY) and 6 million (LA), when will they be equal?

17 Suppose the value of \$1 in Japanese yen decreases at 2% per year. Starting from \$1 = Y240, when will 1 dollar equal 1 yen?

18 The effect of advertising decays exponentially. If 40% remember a new product after three days, find c. How long will 20% remember it?

19 If $y = 1000$ at $t = 3$ and $y = 3000$ at $t = 4$ (exponential growth), what was y_0 at $t = 0$?

20 If $y = 100$ at $t = 4$ and $y = 10$ at $t = 8$ (exponential decay) when will $y = 1$? What was y_0?

21 Atmospheric pressure decreases with height according to $dp/dh = cp$. The pressures at $h = 0$ (sea level) and $h = 20$ km are 1013 and 50 millibars. Find c. Explain why $p = \sqrt{1013 \cdot 50}$ halfway up at $h = 10$.

22 For exponential decay show that $y(t)$ is the square root of $y(0)$ times $y(2t)$. How could you find $y(3t)$ from $y(t)$ and $y(2t)$?

23 Most drugs in the bloodstream decay by $y' = cy$ (*first-order kinetics*). (a) The half-life of morphine is 3 hours. Find its decay constant c (with units). (b) The half-life of nicotine is 2 hours. After a six-hour flight what fraction remains?

24 How often should a drug be taken if its dose is 3 mg, it is cleared at $c = .01$/hour, and 1 mg is required in the bloodstream at all times? (The doctor decides this level based on body size.)

25 The antiseizure drug dilantin has constant clearance rate $y' = -a$ until $y = y_1$. Then $y' = -ay/y_1$. Solve for $y(t)$ in two pieces from y_0. When does y reach y_1?

26 The actual elimination of nicotine is multiexponential: $y = Ae^{ct} + Be^{Ct}$. The first-order equation $(d/dt - c)y = 0$ changes to the second-order equation $(d/dt - c)(d/dt - C)y = 0$. Write out this equation starting with y'', and show that it is satisfied by the given y.

27 *True or false*. **If false, say what's true.**

(a) The time for $y = e^{ct}$ to double is $(\ln 2)/(\ln c)$.

(b) If $y' = cy$ and $z' = cz$ then $(y + z)' = 2c(y + z)$.

(c) If $y' = cy$ and $z' = cz$ then $(y/z)' = 0$.

(d) If $y' = cy$ and $z' = Cz$ then $(yz)' = (c + C)yz$.

28 A rocket has velocity v. Burnt fuel of mass Δm leaves at velocity $v - 7$. Total momentum is constant:

$$mv = (m - \Delta m)(v + \Delta v) + \Delta m(v - 7).$$

What differential equation connects m to v? Solve for $v(m)$ not $v(t)$, starting from $v_0 = 20$ and $m_0 = 4$.

Problems 29–36 are about solutions of $y' = cy + s$.

29 Solve $y' = 3y + 1$ with $y_0 = 0$ by assuming $y = Ae^{3t} + B$ and determining A and B.

30 Solve $y' = 8 - y$ starting from y_0 and $y = Ae^{-t} + B$.

Solve 31–34 with $y_0 = 0$ and graph the solution.

31 $\dfrac{dy}{dt} = y + 1$ **32** $\dfrac{dy}{dt} = y - 1$

33 $\dfrac{dy}{dt} = -y + 1$ **34** $\dfrac{dy}{dt} = -y - 1$

35 (a) What value $y = $ constant solves $dy/dt = -2y + 12$?

(b) Find the solution with an arbitrary constant A.

(c) What solutions start from $y_0 = 0$ and $y_0 = 10$?

(d) What is the steady state y_∞?

36 Choose $\pm$ signs in $dy/dt = \pm 3y \pm 6$ to achieve the following results starting from $y_0 = 1$. Draw graphs.

(a) y increases to ∞ (b) y increases to 2

(c) y decreases to -2 (d) y decreases to $-\infty$

37 What value $y = $ constant solves $dy/dt = 4 - y$? Show that $y(t) = Ae^{-t} + 4$ is also a solution. Find $y(1)$ and y_∞ if $y_0 = 3$.

38 Solve $y' = y + e^t$ from $y_0 = 0$ by Method 2, where the deposit e^T at time T is multiplied by e^{t-T}. The total output at time t is $y(t) = \int_0^t e^T e^{t-T} dT = $ _____ . Substitute back to check $y' = y + e^t$.

39 Rewrite $y' = y + e^t$ as $y' - y = e^t$. Multiplying by e^{-t}, the left side is the derivative of _____ . Integrate both sides from $y_0 = 0$ to find $y(t)$.

40 Solve $y' = -y + 1$ from $y_0 = 0$ by rewriting as $y' + y = 1$, multiplying by e^t, and integrating both sides.

41 Solve $y' = y + t$ from $y_0 = 0$ by assuming $y = Ae^t + Bt + C$.

Problems 42–57 are about the mathematics of finance.

42 Dollar bills decrease in value at $c = -.04$ per year because of inflation. If you hold $1000, what is the decrease in dt years? At what rate s should you print money to keep even?

43 If a bank offers annual interest of $7\frac{1}{2}\%$ or continuous interest of $7\frac{1}{4}\%$, which is better?

44 What continuous interest rate is equivalent to an annual rate of 9%? Extra credit: Telephone a bank for both rates and check their calculation.

45 At 100% interest ($c = 1$) how much is a continuous deposit of s per year worth after one year? What initial deposit y_0 would have produced the same output?

46 To have $50,000 for college tuition in 20 years, what gift y_0 should a grandparent make now? Assume $c = 10\%$. What continuous deposit should a parent make during 20 years? If the parent saves $s = \$1000$ per year, when does he or she reach $50,000 arid retire?

47 Income per person grows 3%, the population grows 2%, the total income grows _____ . Answer if these are (a) annual rates (b) continuous rates.

48 When $dy/dt = cy + 4$, how much is the deposit of $4dT$ at time T worth at the later time t? What is the value at $t = 2$ of deposits $4dT$ from $T = 0$ to $T = 1$?

49 Depositing $s = \$1000$ per year leads to $34,400 after 20 years (Question 3). To reach the same result, when should you deposit $20,000 all at once?

50 For how long can you withdraw $s = \$500$/year after depositing $y_0 = \$5000$ at 8%, before you run dry?

51 What continuous payment s clears a $1000 loan in 60 days, if a loan shark charges 1% per day continuously?

52 You are the loan shark. What is $1 worth after a year of continuous compounding at 1% per day?

53 You can afford payments of $s = \$100$ per month for 48 months. If the dealer charges $c = 6\%$, how much can you borrow?

54 Your income is $I_0 e^{2ct}$ per year. Your expenses are $E_0 e^{ct}$ per year. (a) At what future time are they equal? (b) If you borrow the difference until then, how much money have you borrowed?

55 If a student loan in your freshman year is repaid plus 20% four years later, what was the effective interest rate?

56 Is a variable rate mortgage with $c = .09 + .001t$ for 20 years better or worse than a fixed rate of 10%?

57 At 10% instead of 8%, the $24 paid for Manhattan is worth _____ after 365 years.

Problems 58–65 approach a steady state y_∞ as $t \to \infty$.

58 If $dy/dt = -y + 7$ what is y_∞? What is the derivative of $y - y_\infty$? Then $y - y_\infty$ equals $y_0 - y_\infty$ times _____.

59 Graph $y(t)$ when $y' = 3y - 12$ and y_0 is
 (a) below 4   (b) equal to 4   (c) above 4

60 The solutions to $dy/dt = c(y - 12)$ converge to $y_\infty =$ _____ provided c is _____.

61 Suppose the time unit in $dy/dt = cy$ changes from minutes to hours. How does the equation change? How does $dy/dt = -y + 5$ change? How does y_∞ change?

62 *True or false*, when y_1 and y_2 both satisfy $y' = cy + s$.
 (a) The sum $y = y_1 + y_2$ also satisfies this equation.
 (b) The average $y = \frac{1}{2}(y_1 + y_2)$ satisfies the same equation.
 (c) The derivative $y = y_1'$ satisfies the same equation.

63 If Newton's coffee cools from 80° to 60° in 12 minutes (room temperature 20°), find c. When was the coffee at 100°?

64 If $y_0 = 100$ and $y(1) = 90$ and $y(2) = 84$, what is y_∞?

65 If $y_0 = 100$ and $y(1) = 90$ and $y(2) = 81$, what is y_∞?

66 To cool down coffee, should you add milk now or later? The coffee is at 70°C, the milk is at 10°, the room is at 20°.
 (a) Adding 1 part milk to 5 parts coffee makes it 60°. With $y_\infty = 20°$, the white coffee cools to $y(t) =$ _____.
 (b) The black coffee cools to $y_c(t) =$ _____. The milk warms to $y_m(t) =$ _____. Mixing at time t gives $(5y_c + y_m)/6 =$ _____.

6.4 Logarithms

We have given first place to e^x and a lower place to $\ln x$. In applications that is absolutely correct. But logarithms have one important theoretical advantage (plus many applications of their own). The advantage is that the derivative of $\ln x$ is $1/x$, whereas the derivative of e^x is e^x. We can't define e^x as its own integral, without circular reasoning. But we can and do define $\ln x$ (the **natural logarithm**) as the integral of the "-1 power" which is $1/x$:

$$\ln x = \int_1^x \frac{1}{x}\,dx \qquad \text{or} \qquad \ln y = \int_1^y \frac{1}{u}\,du. \qquad (1)$$

Note the dummy variables, first x then u. Note also the live variables, first x then y. Especially note the lower limit of integration, which is 1 and not 0. **The logarithm is the area measured from** 1. Therefore $\ln 1 = 0$ at that starting point—as required.

Earlier chapters integrated all powers except this "-1 power." The logarithm is that missing integral. The curve in Figure 6.11 has height $y = 1/x$—it is a hyperbola. At $x = 0$ the height goes to infinity and the area becomes infinite: $\log 0 = -\infty$. The minus sign is because the integral goes backward from 1 to 0. The integral does not extend past zero to negative x. We are defining $\ln x$ only for $x > 0$.†

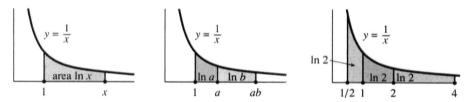

Fig. 6.11   *Logarithm as area*. Neighbors $\ln a + \ln b = \ln ab$.
Equal areas: $-\ln \frac{1}{2} = \ln 2 = \frac{1}{2}\ln 4$.

With this new approach, $\ln x$ has a direct definition. *It is an integral* (or an area). Its two key properties must follow from this definition. That step is a beautiful application of the theory behind integrals.

Property 1: $\ln ab = \ln a + \ln b$. The areas from 1 to a and from a to ab combine into a single area (1 to ab in the middle figure):

$$\text{Neighboring areas}: \int_1^a \frac{1}{x}\,dx + \int_a^{ab} \frac{1}{x}\,dx = \int_1^{ab} \frac{1}{x}\,dx. \qquad (2)$$

The right side is $\ln ab$, from definition (1). The first term on the left is $\ln a$. The problem is to show that the second integral (a to ab) is $\ln b$:

$$\int_a^{ab} \frac{1}{x}\,dx \overset{(?)}{=} \int_1^b \frac{1}{u}\,du = \ln b. \qquad (3)$$

We need $u = 1$ when $x = a$ (the lower limit) and $u = b$ when $x = ab$ (the upper limit). The choice $u = x/a$ satisfies these requirements. Substituting $x = au$ and $dx = a\,du$ yields $dx/x = du/u$. Equation (3) gives $\ln b$, and equation (2) is $\ln a + \ln b = \ln ab$.

---

† The logarithm of -1 is πi (an imaginary number). That is because $e^{\pi i} = -1$. The logarithm of i is also imaginary—it is $\frac{1}{2}\pi i$. In general, logarithms are complex numbers.

Property 2: $\ln b^n = n \ln b$. These are the left and right sides of

$$\int_1^{b^n} \frac{1}{x} \, dx \overset{(?)}{=} n \int_1^b \frac{1}{u} \, du. \tag{4}$$

This comes from the substitution $x = u^n$. The lower limit $x = 1$ corresponds to $u = 1$, and $x = b^n$ corresponds to $u = b$. The differential dx is $nu^{n-1} du$. Dividing by $x = u^n$ leaves $dx/x = n \, du/u$. Then equation (4) becomes $\ln b^n = n \ln b$.

Everything comes logically from the definition as an area. Also definite integrals:

EXAMPLE 1   Compute $\displaystyle\int_x^{3x} \frac{1}{t} \, dt$.   Solution: $\ln 3x - \ln x = \ln \dfrac{3x}{x} = \ln 3$.

EXAMPLE 2   Compute $\displaystyle\int_{.1}^1 \frac{1}{x} \, dx$.   Solution: $\ln 1 - \ln .1 = \ln 10$. (Why?)

EXAMPLE 3   Compute $\displaystyle\int_1^{e^2} \frac{1}{u} \, du$.   Solution: $\ln e^2 = 2$. The area from 1 to e^2 is 2.

Remark   While working on the theory this is a chance to straighten out old debts. The book has discussed and computed (and even differentiated) the functions e^x and b^x and x^n, without defining them properly. When the exponent is an irrational number like π, ***how do we multiply*** e ***by itself*** π ***times***? One approach (not taken) is to come closer and closer to π by rational exponents like $22/7$. Another approach (taken now) is to determine the number $e^{\pi} = 23.1 \ldots$ by its logarithm.† Start with e itself:

e is (by definition) the number whose logarithm is 1

e^{π} is (by definition) the number whose logarithm is π.

When the area in Figure 6.12 ***reaches*** 1, ***the basepoint is*** e. When the area reaches π, the basepoint is e^{π}. We are constructing the inverse function (which is e^x). But how do we know that the area reaches π or 1000 or -1000 at exactly one point? (The area is 1000 far out at e^{1000}. The area is -1000 very near zero at e^{-1000}.) To define e we have to know that somewhere the area equals 1!

For a proof in two steps, go back to Figure 6.11c. The area from 1 to 2 is *more than* $\frac{1}{2}$ (because $1/x$ is more than $\frac{1}{2}$ on that interval of length one). The combined area from 1 to 4 is more than 1. ***We come to*** area $= 1$ ***before reaching*** 4. (Actually at $e = 2.718\ldots$.) Since $1/x$ is positive, the area is increasing and never comes back to 1.

To double the area we have to square the distance. The logarithm creeps upwards:

$$\ln x \to \infty \quad \text{but} \quad \frac{\ln x}{x} \to 0. \tag{5}$$

The logarithm grows slowly because e^x ***grows so fast*** (and vice versa—they are inverses). Remember that e^x goes past every power x^n. Therefore $\ln x$ is passed by every root $x^{1/n}$. Problems 60 and 61 give two proofs that $(\ln x)/x^{1/n}$ approaches zero.

We might compare $\ln x$ with $\sqrt{x}$. At $x = 10$ they are close (2.3 versus 3.2). But out at $x = e^{10}$ the comparison is 10 against e^5, and $\ln x$ loses to $\sqrt{x}$.

---

†Chapter 9 goes on to *imaginary exponents*, and proves the remarkable formula $e^{\pi i} = -1$.

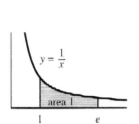

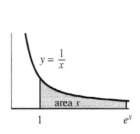

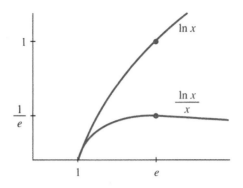

Fig. 6.12   Area is logarithm of basepoint.

Fig. 6.13   ln x grows more slowly than x.

APPROXIMATION OF LOGARITHMS

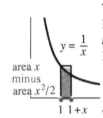

The limiting cases $\ln 0 = -\infty$ and $\ln \infty = +\infty$ are important. More important are logarithms near the starting point $\ln 1 = 0$. Our question is: **What is $\ln(1+x)$ for x near zero?** The exact answer is an area. The approximate answer is much simpler. If x (positive or negative) is small, then

$$\ln(1+x) \approx x \qquad \text{and} \qquad e^x \approx 1+x. \tag{6}$$

The calculator gives $\ln 1.01 = .0099503$. This is close to $x = .01$. Between 1 and $1+x$ the area under the graph of $1/x$ is nearly a rectangle. Its base is x and its height is 1. So the curved area $\ln(1+x)$ is close to the rectangular area x. Figure 6.14 shows how a small triangle is chopped off at the top.

The difference between .0099503 (actual) and .01 (linear approximation) is $-.0000497$. That is predicted almost exactly by the second derivative: $\frac{1}{2}$ times $(\Delta x)^2$ times $(\ln x)''$ is $\frac{1}{2}(.01)^2(-1) = -.00005$. **This is the area of the small triangle!**

$$\ln(1+x) \approx \textit{rectangular area minus triangular area} = x - \tfrac{1}{2}x^2.$$

The remaining mistake of .0000003 is close to $\frac{1}{3}x^3$ (Problem 65).

May I switch to e^x? Its slope starts at $e^0 = 1$, so its linear approximation is $1 + x$. Then $\ln(e^x) \approx \ln(1+x) \approx x$. *Two wrongs do make a right*: $\ln(e^x) = x$ exactly.

The calculator gives $e^{.01}$ as 1.0100502 (actual) instead of 1.01 (approximation). The second-order correction is again a small triangle: $\frac{1}{2}x^2 = .00005$. The complete series for $\ln(1+x)$ and e^x are in Sections 10.1 and 6.6:

$$\ln(1+x) = x - x^2/2 + x^3/3 - \dots \qquad e^x = 1 + x + x^2/2 + x^3/6 + \dots.$$

DERIVATIVES BASED ON LOGARITHMS

Logarithms turn up as antiderivatives very often. To build up a collection of integrals, we now differentiate $\ln u(x)$ by the chain rule.

6K The derivative of $\ln x$ is $\dfrac{1}{x}$.   ***The derivative of*** $\ln u(x)$ ***is*** $\dfrac{1}{u}\dfrac{du}{dx}$.

Fig. 6.14

The slope of $\ln x$ was hard work in Section 6.2. With its new definition (the integral of $1/x$) the work is gone. By the Fundamental Theorem, the slope must be $1/x$.

For $\ln u(x)$ the derivative comes from the chain rule. The inside function is u, the outside function is $\ln$. (Keep $u > 0$ to define $\ln u$.) The chain rule gives

$$\frac{d}{dx}\ln cx = \frac{1}{cx}c = \frac{1}{x}(!) \qquad \frac{d}{dx}\ln x^3 = 3x^2/x^3 = \frac{3}{x}$$

$$\frac{d}{dx}\ln(x^2+1) = 2x/(x^2+1) \qquad \frac{d}{dx}\ln\cos x = \frac{-\sin x}{\cos x} = -\tan x$$

$$\frac{d}{dx}\ln e^x = e^x/e^x = 1 \qquad \frac{d}{dx}\ln(\ln x) = \frac{1}{\ln x}\frac{1}{x}.$$

Those are worth another look, especially the first. Any reasonable person would expect the slope of $\ln 3x$ to be $3/x$. *Not so*. The 3 cancels, and $\ln 3x$ has the same slope as $\ln x$. (The real reason is that $\ln 3x = \ln 3 + \ln x$.) The antiderivative of $3/x$ is not $\ln 3x$ but $3\ln x$, which is $\ln x^3$.

Before moving to integrals, here is a new method for derivatives: *logarithmic differentiation* or **LD**. It applies to *products* and *powers*. The product and power rules are always available, but sometimes there is an easier way.

Main idea: The logarithm of a product $p(x)$ is a *sum of logarithms*. Switching to $\ln p$, the sum rule just adds up the derivatives. But there is a catch at the end, as you see in the example.

EXAMPLE 4 Find dp/dx if $p(x) = x^x\sqrt{x-1}$. Here $\ln p(x) = x\ln x + \frac{1}{2}\ln(x-1)$.

Take the derivative of $\ln p$: $\dfrac{1}{p}\dfrac{dp}{dx} = x\cdot\dfrac{1}{x} + \ln x + \dfrac{1}{2(x-1)}.$

Now multiply by $p(x)$: $\dfrac{dp}{dx} = p\left(1 + \ln x + \dfrac{1}{2(x-1)}\right).$

The catch is that last step. Multiplying by p complicates the answer. This can't be helped—logarithmic differentiation contains no magic. The derivative of $p = fg$ is the same as from the product rule: $\ln p = \ln f + \ln g$ gives

$$\frac{p'}{p} = \frac{f'}{f} + \frac{g'}{g} \quad\text{and}\quad p' = p\left(\frac{f'}{f} + \frac{g'}{g}\right) = f'g + fg'. \tag{7}$$

For $p = xe^x\sin x$, with three factors, the sum has three terms:

$$\ln p = \ln x + x + \ln\sin x \text{ and } p' = p\left[\frac{1}{x} + 1 + \frac{\cos x}{\sin x}\right].$$

We multiply p times p'/p (the derivative of $\ln p$). ***Do the same for powers***:

EXAMPLE 5 $p = x^{1/x} \Rightarrow \ln p = \dfrac{1}{x}\ln x \Rightarrow \dfrac{dp}{dx} = p\left[\dfrac{1}{x^2} - \dfrac{\ln x}{x^2}\right].$

EXAMPLE 6 $p = x^{\ln x} \Rightarrow \ln p = (\ln x)^2 \Rightarrow \dfrac{dp}{dx} = p\left[\dfrac{2\ln x}{x}\right].$

EXAMPLE 7 $p = x^{1/\ln x} \Rightarrow \ln p = \dfrac{1}{\ln x}\ln x = 1 \Rightarrow \dfrac{dp}{dx} = 0 \ (!)$

INTEGRALS BASED ON LOGARITHMS

Now comes an important step. Many integrals produce logarithms. The foremost example is $1/x$, whose integral is $\ln x$. In a certain way that is the only example, but its range is enormously extended by the chain rule. The derivative of $\ln u(x)$ is u'/u, so the integral goes from u'/u back to $\ln u$:

$$\int \frac{du/dx}{u(x)} \, dx = \ln u(x) \quad \text{or equivalently} \quad \int \frac{du}{u} = \ln u.$$

Try to choose $u(x)$ so that the integral contains du/dx divided by u.

EXAMPLES $\displaystyle\int \frac{dx}{x+7} = \ln|x+7|$ $\displaystyle\int \frac{dx}{cx+7} = \frac{1}{c}\ln|cx+7|$

Final remark When u is negative, $\ln u$ cannot be the integral of $1/u$. The logarithm is not defined when $u < 0$. But the integral can go forward by switching to $-u$:

$$\int \frac{du/dx}{u} \, dx = \int \frac{-du/dx}{-u} \, dx = \ln(-u). \tag{8}$$

Thus $\ln(-u)$ succeeds when $\ln u$ fails.† ***The forbidden case is $u = 0$.*** The integrals $\ln u$ and $\ln(-u)$, on the plus and minus sides of zero, can be combined as $\ln|u|$. Every integral that gives a logarithm allows $u < 0$ by changing to the absolute value $|u|$:

$$\int_{-e}^{-1} \frac{dx}{x} = \Big[\ln|x|\Big]_{-e}^{-1} = \ln 1 - \ln e \qquad \int_{2}^{4} \frac{dx}{x-5} = \Big[\ln|x-5|\Big]_{2}^{4} = \ln 1 - \ln 3.$$

The areas are -1 and $-\ln 3$. The graphs of $1/x$ and $1/(x-5)$ are below the x axis. We do *not* have logarithms of negative numbers, and we will not integrate $1/(x-5)$ from 2 to 6. That crosses the forbidden point $x = 5$, with infinite area on both sides.

The ratio du/u leads to important integrals. When $u = \cos x$ or $u = \sin x$, we are integrating the ***tangent*** and ***cotangent***. When there is a possibility that $u < 0$, write the integral as $\ln|u|$.

$$\int \tan x \, dx = \int \frac{\sin x}{\cos x} \, dx = -\ln|\cos x| \qquad \int \frac{x \, dx}{x^2+7} = \frac{1}{2}\ln(x^2+7)$$

$$\int \cot x \, dx = \int \frac{\cos x}{\sin x} \, dx = \ln|\sin x| \qquad \int \frac{dx}{x \ln x} = \ln|\ln x|$$

Now we report on the ***secant*** and ***cosecant***. The integrals of $1/\cos x$ and $1/\sin x$ also surrender to an attack by logarithms—based on a crazy trick:

$$\int \sec x \, dx = \int \sec x \left(\frac{\sec x + \tan x}{\sec x + \tan x}\right) dx = \ln|\sec x + \tan x|. \tag{9}$$

$$\int \csc x \, dx = \int \csc x \left(\frac{\csc x - \cot x}{\csc x - \cot x}\right) dx = \ln|\csc x + \cot x|. \tag{10}$$

Here $u = \sec x + \tan x$ is in the denominator; $du/dx = \sec x \tan x + \sec^2 x$ is above it. The integral is $\ln|u|$. Similarly (10) contains du/dx over $u = \csc x - \cot x$.

---

† The integral of $1/x$ (odd function) is $\ln|x|$ (even function). Stay clear of $x = 0$.

In closing we integrate $\ln x$ itself. The derivative of $x \ln x$ is $\ln x + 1$. To remove the extra 1, subtract x from the integral: $\int \ln x \, dx = x \ln x - x$.

In contrast, the area under $1/(\ln x)$ has no elementary formula. Nevertheless it is the key to the greatest approximation in mathematics—the **prime number theorem**. *The area $\int_a^b dx / \ln x$ is approximately the number of primes between a and b.* Near e^{1000}, about $1/1000$ of the integers are prime.

6.4 EXERCISES

Read-through questions

The natural logarithm of x is $\int_1^x$ __a__ . This definition leads to $\ln xy =$ __b__ and $\ln x^n =$ __c__ . Then e is the number whose logarithm (area under $1/x$ curve) is __d__ . Similarly e^x is now defined as the number whose natural logarithm is __e__ . As $x \to \infty$, $\ln x$ approaches __f__ . But the ratio $(\ln x)/\sqrt{x}$ approaches __g__ . The domain and range of $\ln x$ are __h__ .

The derivative of $\ln x$ is __i__ . The derivative of $\ln(1+x)$ is __j__ . The tangent approximation to $\ln(1+x)$ at $x = 0$ is __k__ . The quadratic approximation is __l__ . The quadratic approximation to e^x is __m__ .

The derivative of $\ln u(x)$ by the chain rule is __n__ . Thus $(\ln \cos x)' =$ __o__ . An antiderivative of $\tan x$ is __p__ . The product $p = x e^{5x}$ has $\ln p =$ __q__ . The derivative of this equation is __r__ . Multiplying by p gives $p' =$ __s__ , which is **LD** or logarithmic differentiation.

The integral of $u'(x)/u(x)$ is __t__ . The integral of $2x/(x^2 + 4)$ is __u__ . The integral of $1/cx$ is __v__ . The integral of $1/(ct + s)$ is __w__ . The integral of $1/\cos x$, after a trick, is __x__ . We should write $\ln |x|$ for the antiderivative of $1/x$, since this allows __y__ . Similarly $\int du/u$ should be written __z__ .

Find the derivative dy/dx in 1–10.

1 $y = \ln(2x)$

2 $y = \ln(2x + 1)$

3 $y = (\ln x)^{-1}$

4 $y = (\ln x)/x$

5 $y = x \ln x - x$

6 $y = \log_{10} x$

7 $y = \ln(\sin x)$

8 $y = \ln(\ln x)$

9 $y = 7 \ln 4x$

10 $y = \ln((4x)^7)$

Find the indefinite (or definite) integral in 11–24.

11 $\displaystyle \int \frac{dt}{3t}$

12 $\displaystyle \int \frac{dx}{1+x}$

13 $\displaystyle \int_0^1 \frac{dx}{3+x}$

14 $\displaystyle \int_0^1 \frac{dt}{3+2t}$

15 $\displaystyle \int_0^2 \frac{x \, dx}{x^2+1}$

16 $\displaystyle \int_0^2 \frac{x^3 \, dx}{x^2+1}$

17 $\displaystyle \int_2^e \frac{dx}{x(\ln x)}$

18 $\displaystyle \int_2^e \frac{dx}{x(\ln x)^2}$

19 $\displaystyle \int \frac{\cos x \, dx}{\sin x}$

20 $\displaystyle \int_0^{\pi/4} \tan x \, dx$

21 $\displaystyle \int \tan 3x \, dx$

22 $\displaystyle \int \cot 3x \, dx$

23 $\displaystyle \int \frac{(\ln x)^2 dx}{x}$

24 $\displaystyle \int \frac{dx}{x(\ln x)(\ln \ln x)}$

25 Graph $y = \ln(1+x)$

26 Graph $y = \ln(\sin x)$

Compute dy/dx by differentiating $\ln y$. This is LD:

27 $y = \sqrt{x^2 + 1}$

28 $y = \sqrt{x^2+1}\sqrt{x^2-1}$

29 $y = e^{\sin x}$

30 $y = x^{-1/x}$

31 $y = e^{(e^x)}$

32 $y = x^e$

33 $y = x^{(e^x)}$

34 $y = (\sqrt{x})(\sqrt[3]{x})(\sqrt[6]{x})$

35 $y = x^{-1/\ln x}$

36 $y = e^{-\ln x}$

Evaluate 37–42 by any method.

37 $\displaystyle \int_5^{10} \frac{dt}{t} - \int_{5x}^{10x} \frac{dt}{t}$

38 $\displaystyle \int_1^{e^\pi} \frac{dx}{x} + \int_{-2}^{-1} \frac{dx}{x}$

39 $\displaystyle \frac{d}{dx} \int_x^1 \frac{dt}{t}$

40 $\displaystyle \frac{d}{dx} \int_x^{x^2} \frac{dt}{t}$

41 $\displaystyle \frac{d}{dx} \ln(\sec x + \tan x)$

42 $\displaystyle \int \frac{\sec^2 x + \sec x \tan x}{\sec x + \tan x} dx$

Verify the derivatives 43–46, which give useful antiderivatives:

43 $\displaystyle \frac{d}{dx} \ln(x + \sqrt{x^2+1}) = \frac{1}{\sqrt{1+x^2}}$

44 $\displaystyle \frac{d}{dx} \ln\left(\frac{x-a}{x+a}\right) = \frac{2a}{(x^2-a^2)}$

45 $\displaystyle \frac{d}{dx} \ln\left(\frac{1+\sqrt{1-x^2}}{x}\right) = \frac{-1}{x\sqrt{1-x^2}}$

46 $\displaystyle \frac{d}{dx} \ln(x + \sqrt{x^2-a^2}) = \frac{1}{\sqrt{x^2-a^2}}$

Estimate 47–50 to linear accuracy, then quadratic accuracy, by $e^x \approx 1 + x + \frac{1}{2}x^2$. **Then use a calculator**.

47 $\ln(1.1)$ **48** $e^{.1}$ **49** $\ln(.99)$ **50** e^2

51 Compute $\lim\limits_{x \to 0} \dfrac{\ln(1+x)}{x}$ **52** Compute $\lim\limits_{x \to 0} \dfrac{e^x - 1}{x}$

53 Compute $\lim\limits_{x \to 0} \dfrac{\log_b(1+x)}{x}$ **54** Compute $\lim\limits_{x \to 0} \dfrac{b^x - 1}{x}$

55 Find the area of the "hyperbolic quarter-circle" enclosed by $x = 2$ and $y = 2$ above $y = 1/x$.

56 Estimate the area under $y = 1/x$ from 4 to 8 by four upper rectangles and four lower rectangles. Then average the answers (trapezoidal rule). What is the exact area?

57 Why is $\dfrac{1}{2} + \dfrac{1}{3} + \cdots + \dfrac{1}{n}$ near $\ln n$? Is it above or below?

58 Prove that $\ln x \leqslant 2(\sqrt{x} - 1)$ for $x > 1$. Compare the integrals of $1/t$ and $1/\sqrt{t}$, from 1 to x.

59 Dividing by x in Problem 58 gives $(\ln x)/x \leqslant 2(\sqrt{x} - 1)/x$. Deduce that $(\ln x)/x \to 0$ as $x \to \infty$. Where is the maximum of $(\ln x)/x$?

60 Prove that $(\ln x)/x^{1/n}$ also approaches zero. (Start with $(\ln x^{1/n})/x^{1/n} \to 0$.) Where is its maximum?

61 For any power n, Problem 6.2.59 proved $e^x > x^n$ for large x. Then by logarithms, $x > n \ln x$. Since $(\ln x)/x$ goes below $1/n$ and stays below, it converges to _____.

62 Prove that $y \ln y$ approaches zero as $y \to 0$, by changing y to $1/x$. Find the limit of y^y (take its logarithm as $y \to 0$). What is $.1^{.1}$ on your calculator?

63 Find the limit of $\ln x / \log_{10} x$ as $x \to \infty$.

64 We know the integral $\int_1^x t^{h-1} dt = [t^h/h]_1^x = (x^h - 1)/h$. Its limit as $h \to 0$ is _____.

65 Find linear approximations near $x = 0$ for e^{-x} and 2^x.

66 The x^3 correction to $\ln(1+x)$ yields $x - \frac{1}{2}x^2 + \frac{1}{3}x^3$. Check that $\ln 1.01 \approx .0099503$ and find $\ln 1.02$.

67 An ant crawls at 1 foot/second along a rubber band whose original length is 2 feet. The band is being stretched at 1 foot/second by pulling the other end. At what time T, *if ever*, does the ant reach the other end?

One approach: The band's length at time t is $t + 2$. Let $y(t)$ be the *fraction* of that length which the ant has covered, and explain

(a) $y' = 1/(t+2)$ (b) $y = \ln(t+2) - \ln 2$ (c) $T = 2e - 2$.

68 If the rubber band is stretched at 8 feet/second, when if ever does the same ant reach the other end?

69 A weaker ant slows down to $2/(t+2)$ feet/second, so $y' = 2/(t+2)^2$. Show that the other end is never reached.

70 The slope of $p = x^x$ comes two ways from $\ln p = x \ln x$:

1 Logarithmic differentiation (**LD**): Compute $(\ln p)'$ and multiply by p.

2 Exponential differentiation (**ED**): Write x^x as $e^{x \ln x}$, take its derivative, and put back x^x.

71 If $p = 2^x$ then $\ln p =$ _____. **LD** gives $p' = (p)(\ln p)' =$ _____. **ED** gives $p = e$ and then $p' =$ _____.

72 Compute $\ln 2$ by the trapezoidal rule and/or Simpson's rule, to get five correct decimals.

73 Compute $\ln 10$ by either rule with $\Delta x = 1$, and compare with the value on your calculator.

74 Estimate $1/\ln 90,000$, the fraction of numbers near $90,000$ that are prime. (879 of the next $10,000$ numbers are actually prime.)

75 Find a pair of positive integers for which $x^y = y^x$. Show how to change this equation to $(\ln x)/x = (\ln y)/y$. So look for two points at the same height in Figure 6.13. Prove that you have discovered all the integer solutions.

****76** Show that $(\ln x)/x = (\ln y)/y$ is satisfied by

$$x = \left(\frac{t+1}{t}\right)^t \text{ and } y = \left(\frac{t+1}{1}\right)^{t+1}$$

with $t \neq 0$. Graph those points to show the curve $x^y = y^x$. It crosses the line $y = x$ at $x =$ _____, where $t \to \infty$.

6.5 Separable Equations Including the Logistic Equation

This section begins with the integrals that solve two basic differential equations:

$$\frac{dy}{dt} = cy \qquad \text{and} \qquad \frac{dy}{dt} = cy + s. \tag{1}$$

We already know the solutions. What we don't know is how to discover those solutions, when a suggestion "*try e^{ct}*" has not been made. Many important equations, including these, separate into a y-integral and a t-integral. The answer comes directly from the two separate integrations. When a differential equation is reduced that far—to integrals that we know or can look up—it is solved.

One particular equation will be emphasized. The **logistic equation** describes the speedup and slowdown of growth. Its solution is an **S-curve**, which starts slowly, rises quickly, and levels off. (The 1990's are near the middle of the **S**, if the prediction is correct for the world population.) **S**-curves are solutions to **nonlinear** equations, and we will be solving our first nonlinear model. It is highly important in biology and all life sciences.

SEPARABLE EQUATIONS

The equations $dy/dt = cy$ and $dy/dt = cy + s$ (with constant source s) can be solved by a direct method. **The idea is to separate y from t:**

$$\frac{dy}{y} = c\,dt \qquad \text{and} \qquad \frac{dy}{y + (s/c)} = c\,dt. \tag{2}$$

All y's are on the left side. All t's are on the right side (and c can be on either side). This separation would not be possible for $dy/dt = y + t$.

Equation (2) contains differentials. They suggest integrals. The t-integrals give ct and the y-integrals give logarithms:

$$\ln y = ct + \text{constant} \qquad \text{and} \qquad \ln\left(y + \frac{s}{c}\right) = ct + \text{constant}. \tag{3}$$

The constant is determined by the initial condition. At $t = 0$ we require $y = y_0$, and the right constant will make that happen:

$$\ln y = ct + \ln y_0 \qquad \text{and} \qquad \ln\left(y + \frac{s}{c}\right) = ct + \ln\left(y_0 + \frac{s}{c}\right). \tag{4}$$

Then the final step isolates y. The goal is a formula for y itself, not its logarithm, so take the exponential of both sides ($e^{\ln y}$ is y):

$$y = y_0 e^{ct} \qquad \text{and} \qquad y + \frac{s}{c} = \left(y_0 + \frac{s}{c}\right)e^{ct}. \tag{5}$$

It is wise to substitute y back into the differential equation, as a check.

This is our fourth method for $y' = cy + s$. Method 1 assumed from the start that $y = Ae^{ct} + B$. Method 2 multiplied all inputs by their growth factors $e^{c(t-T)}$ and added up outputs. Method 3 solved for $y - y_\infty$. Method 4 is **separation of variables** (and all methods give the same answer). This separation method is so useful that we repeat its main idea, and then explain it by using it.

To solve $dy/dt = u(y)v(t)$, *separate* $dy/u(y)$ *from* $v(t)dt$ *and integrate both sides:*

$$\int dy/u(y) = \int v(t)dt + C. \qquad (6)$$

Then substitute the initial condition to determine C, *and solve for* $y(t)$.

EXAMPLE 1 $dy/dt = y^2$ separates into $dy/y^2 = dt$. Integrate to reach $-1/y = t + C$. Substitute $t = 0$ and $y = y_0$ to find $C = -1/y_0$. Now solve for y:

$$-\frac{1}{y} = t - \frac{1}{y_0} \qquad \text{and} \qquad y = \frac{y_0}{1 - ty_0}.$$

This solution blows up (Figure 6.15a) when t reaches $1/y_0$. If the bank pays interest on your deposit *squared* ($y' = y^2$), you soon have all the money in the world.

EXAMPLE 2 $dy/dt = ty$ separates into $dy/y = t\, dt$. Then by integration $\ln y = \frac{1}{2}t^2 + C$. Substitute $t = 0$ and $y = y_0$ to find $C = \ln y_0$. The exponential of $\frac{1}{2}t^2 + \ln y_0$ gives $y = y_0 e^{t^2/2}$. When the interest rate is $c = t$, the exponent is $t^2/2$.

EXAMPLE 3 $dy/dt = y + t$ is *not separable*. Method 1 survives by assuming $y = Ae^t + B + Dt$—with an extra coefficient D in Problem 23. Method 2 also succeeds—but not the separation method.

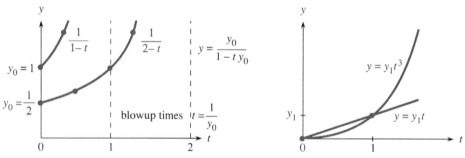

Fig. 6.15 The solutions to separable equations $\dfrac{dy}{dt} = y^2$ and $\dfrac{dy}{dt} = n\dfrac{y}{t}$ or $\dfrac{dy}{y} = n\dfrac{dt}{t}$.

EXAMPLE 4 Separate $dy/dt = ny/t$ into $dy/y = n\, dt/t$. By integration $\ln y = n \ln t + C$. Substituting $t = 0$ produces $\ln 0$ and disaster. This equation cannot start from time zero (it divides by t). However y can start from y_1 at $t = 1$, which gives $C = \ln y_1$. *The solution is a power function* $y = y_1 t^n$.

This was the first differential equation in the book (Section 2.2). The ratio of dy/y to dt/t is the "*elasticity*" in economics. These relative changes have units like dollars/dollars—they are dimensionless, and $y = t^n$ has constant elasticity n.

On log–log paper the graph of $\ln y = n \ln t + C$ is a *straight line with slope* n.

THE LOGISTIC EQUATION

The simplest model of population growth is $dy/dt = cy$. The growth rate c is the birth rate minus the death rate. If c is constant the growth goes on forever—beyond the point where the model is reasonable. A population can't grow all the way to infinity! Eventually there is competition for food and space, and $y = e^{ct}$ must slow down.

The true rate c depends on the population size y. It is a function $c(y)$ not a constant. The choice of the model is at least half the problem:

> ***Problem in biology or ecology***:   Discover $c(y)$.
>
> ***Problem in mathematics***:   Solve $dy/dt = c(y)y$.

Every model looks linear over a small range of y's—but not forever. When the rate drops off, two models are of the greatest importance. The *Michaelis-Menten* equation has $c(y) = c/(y + K)$. The *logistic* equation has $c(y) = c - by$. It comes first.

The nonlinear effect is from "interaction." For two populations of size y and z, the number of interactions is proportional to y times z. ***The Law of Mass Action produces a quadratic term byz.*** It is the basic model for interactions and competition. Here we have one population competing within itself, so z is the same as y. This competition slows down the growth, because $-by^2$ goes into the equation.

The basic model of *growth versus competition* is known as the ***logistic equation***:

$$dy/dt = cy - by^2. \tag{7}$$

Normally b is very small compared to c. The growth begins as usual (close to e^{ct}). The competition term by^2 is much smaller than cy, *until y itself gets large*. Then by^2 (with its minus sign) slows the growth down. The solution follows an **S**-curve that we can compute exactly.

What are the numbers b and c for human population? Ecologists estimate the natural growth rate as $c = .029/\text{year}$. That is not the actual rate, because of b. About 1930, the world population was 3 billion. The cy term predicts a yearly increase of $(.029)(3 \text{ billion}) = 87$ million. The actual growth was more like $dy/dt = 60$ million/year. That difference of 27 million/year was by^2:

$$27 \text{ million/year} = b(3 \text{ billion})^2 \text{ leads to } b = 3 \cdot 10^{-12}/\text{year}.$$

Certainly b is a small number (three trillionths) but its effect is not small. It reduces 87 to 60. What is fascinating is to calculate the ***steady state***, when the new term by^2 equals the old term cy. When these terms cancel each other, $dy/dt = cy - by^2$ is zero. The loss from competition balances the gain from new growth: $cy = by^2$ and $y = c/b$. The growth stops at this equilibrium point—the top of the **S**-curve:

$$y_\infty = \frac{c}{b} = \frac{.029}{3} 10^{12} \approx 10 \text{ billion people}.$$

According to Verhulst's logistic equation, *the world population is converging to* 10 *billion*. That is from the model. From present indications we are growing much faster. We will very probably go beyond 10 billion. The United Nations report in Section 3.3 predicts 11 billion to 14 billion.

Notice a special point halfway to $y_\infty = c/b$. (In the model this point is at 5 billion.) It is the *inflection point* where the **S**-curve begins to bend down. The second derivative d^2y/dt^2 is zero. The slope dy/dt is a maximum. It is easier to find this point from the differential equation (which gives dy/dt) than from y. Take one more derivative:

$$y'' = (cy - by^2)' = cy' - 2byy' = (c - 2by)y'. \tag{8}$$

The factor $c - 2by$ is zero at the inflection point $y = c/2b$, halfway up the **S**-curve.

THE S-CURVE

The logistic equation is solved by separating variables y and t:

$$dy/dt = cy - by^2 \text{ becomes } \int dy/(cy - by^2) = \int dt. \qquad (9)$$

The first question is whether we recognize this y-integral. *No.* The second question is whether it is listed in the cover of the book. *No.* The nearest is $\int dx/(a^2 - x^2)$, which can be reached with considerable manipulation (Problem 21). The third question is whether a general method is available. *Yes.* "Partial fractions" is perfectly suited to $1/(cy - by^2)$, and Section 7.4 gives the following integral of equation (9):

$$\ln \frac{y}{c-by} = ct + C \qquad \text{and then} \qquad \ln \frac{y_0}{c-by_0} = C. \qquad (10)$$

That constant C makes the solution correct at $t = 0$. The logistic equation is integrated, but the solution can be improved. Take exponentials of both sides to remove the logarithms:

$$\frac{y}{c-by} = e^{ct} \frac{y_0}{c-by_0}. \qquad (11)$$

This contains the same growth factor e^{ct} as in linear equations. But the logistic equation is not linear—it is not y that increases so fast. According to (11), it is $y/(c - by)$ that grows to infinity. This happens when $c - by$ approaches zero.

The growth stops at $y = c/b$. That is the final population of the world (10 billion?).

We still need a formula for y. *The perfect S-curve is the graph of $y = 1/(1 + e^{-t})$.* It equals 1 when $t = \infty$, it equals $\frac{1}{2}$ when $t = 0$, it equals 0 when $t = -\infty$. It satisfies $y' = y - y^2$, with $c = b = 1$. The general formula cannot be so beautiful, because it allows any c, b, and y_0. To find the **S**-curve, multiply equation (11) by $c - by$ and solve for y:

$$y = \frac{c}{b + e^{-ct}(c - by_0)/y_0} \qquad \text{or} \qquad y = \frac{c}{b + de^{-ct}}. \qquad (12)$$

When t approaches infinity, e^{-ct} approaches zero. The complicated part of the formula disappears. Then y approaches its steady state c/b, the asymptote in Figure 6.16. The **S**-shape comes from the inflection point halfway up.

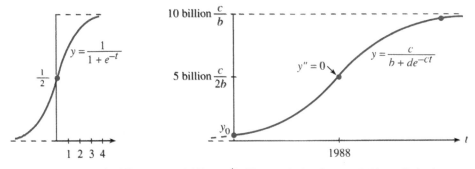

Fig. 6.16   The standard **S**-curve $y = 1/(1 + e^{-t})$. The population **S**-curve (with prediction).

Surprising observation: $z = 1/y$ *satisfies a linear equation. By calculus* $z' = -y'/y^2$. So

$$z' = \frac{-cy + by^2}{y^2} = -\frac{c}{y} + b = -cz + b. \tag{13}$$

This equation $z' = -cz + b$ is solved by an exponential e^{-ct} plus a constant:

$$z = Ae^{-ct} + \frac{a}{b} = \left(\frac{1}{y_0} - \frac{b}{c}\right)e^{-ct} + \frac{b}{c}. \tag{14}$$

Turned upside down, $y = 1/z$ is the **S**-curve (12). As z approaches b/c, the **S**-curve approaches c/b. Notice that z starts at $1/y_0$.

EXAMPLE 1 (United States population) The table shows the actual population and the model. Pearl and Reed used census figures for $1790, 1850$, and 1910 to compute c and b. In between, the fit is good but not fantastic. One reason is war—another is depression. Probably more important is immigration.† In fact the Pearl-Reed steady state c/b is below 200 million, which the US has already passed. Certainly their model can be and has been improved. **The** 1990 **census predicted a stop before** 300 **million**. For constant immigration s we could still solve $y' = cy - by^2 + s$ by partial fractions—but in practice the computer has taken over. The table comes from Braun's book *Differential Equations* (Springer 1975).

Year	US Population		Model
1790	3.9	=	3.9
1800	5.3		5.3
1810	7.2		7.2
1820	9.6		9.8
1830	12.9		13.1
1840	17.1		17.5
1850	23.2	=	23.2
1860	31.4		30.4
1870	38.6		39.4
1880	50.2		50.2
1890	62.9		62.8
1900	76.0		76.9
1910	92.0	=	92.0
1920	105.7		107.6
1930	122.8		123.1
1940	131.7	≠	136.7
1950	150.7		149.1

Remark For good science the y^2 term should be explained and justified. It gave a nonlinear model that could be completely solved, but simplicity is not necessarily truth. The basic justification is this: In a population of size y, *the number of encounters is proportional to* y^2. If those encounters are fights, the term is $-by^2$. If those encounters *increase* the population, as some like to think, the sign is changed. There is a cooperation term $+by^2$, and the population increases very fast.

EXAMPLE 5 $y' = cy + by^2$: *y goes to infinity in a finite time.*

EXAMPLE 6 $y' = -dy + by^2$: *y dies to zero if* $y_0 < d/b$.

In Example 6 death wins. A small population dies out before the cooperation by^2 can save it. A population below d/b is an endangered species.

The logistic equation can't predict oscillations—those go beyond $dy/dt = f(y)$.

The y line Here is a way to understand every nonlinear equation $y' = f(y)$. Draw a "y line." Add arrows to show the sign of $f(y)$. When $y' = f(y)$ is positive, y is increasing (***it follows the arrow to the right***). When f is negative, y goes to the left. When f is zero, the equation is $y' = 0$ and y is stationary:

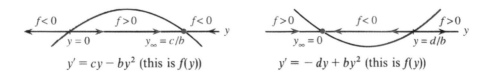

The arrows take you left or right, to the steady state or to infinity. Arrows go *toward* stable steady states. The arrows go *away*, when the stationary point is unstable. The ***y line*** shows which way y moves and where it stops.

---

†Immigration does not enter for the world population model (at least not yet).

The terminal velocity of a falling body is $v_\infty = \sqrt{g}$ in Problem 6.7.54. For $f(y) = \sin y$ there are several steady states:

falling body: $dv/dt = g - v^2$ $dy/dt = \sin y$

EXAMPLE 7   Kinetics of a chemical reaction $mA + nB \to pC$.

The reaction combines m molecules of A with n molecules of B to produce p molecules of C. The numbers m, n, p are $1, 1, 2$ for hydrogen chloride: $H_2 + Cl_2 = 2\,HCl$. The *Law of Mass Action* says that the reaction rate is proportional to the product of the concentrations $[A]$ and $[B]$. Then $[A]$ decays as $[C]$ grows:

$$d[A]/dt = -r[A][B] \quad \text{and} \quad d[C]/dt = +k[A][B]. \tag{15}$$

Chemistry measures r and k. Mathematics solves for $[A]$ and $[C]$. Write y for the concentration $[C]$, the number of molecules in a unit volume. Forming those y molecules drops the concentration $[A]$ from a_0 to $a_0 - (m/p)y$. Similarly $[B]$ drops from b_0 to $b_0 - (n/p)y$. The mass action law (15) contains y^2:

$$\frac{dy}{dt} = k\left(a_0 - \frac{m}{p}y\right)\left(b_0 - \frac{n}{p}y\right). \tag{16}$$

This fits our nonlinear model (Problem $33 - 34$). We now find this same mass action in biology. You recognize it whenever there is a product of two concentrations.

THE MM EQUATION $dy/dt = -cy/(y + K)$

Biochemical reactions are the keys to life. They take place continually in every living organism. Their mathematical description is not easy! Engineering and physics go far with linear models, while biology is quickly nonlinear. It is true that $y' = cy$ is extremely effective in first-order kinetics (Section 6.3), but nature builds in a nonlinear regulator.

It is *enzymes* that speed up a reaction. Without them, your life would be in slow motion. Blood would take years to clot. Steaks would take decades to digest. Calculus would take centuries to learn. The whole system is awesomely beautiful—DNA tells amino acids how to combine into useful proteins, and we get enzymes and elephants and Isaac Newton.

Briefly, the enzyme enters the reaction and comes out again. It is the *catalyst*. Its combination with the substrate is an unstable intermediate, which breaks up into a new product and the enzyme (which is ready to start over).

Here are examples of catalysts, some good and some bad.

1. The platinum in a catalytic converter reacts with pollutants from the car engine. (But platinum also reacts with lead—ten gallons of leaded gasoline and you can forget the platinum.)
2. Spray propellants (CFC's) catalyze the change from ozone (O_3) into ordinary oxygen (O_2). This wipes out the ozone layer—our shield in the atmosphere.
3. Milk becomes yoghurt and grape juice becomes wine.

4. Blood clotting needs a whole cascade of enzymes, amplifying the reaction at every step. In hemophilia—the "Czar's disease"—the enzyme called Factor VIII is missing. A small accident is disaster; the bleeding won't stop.

5. Adolph's Meat Tenderizer is a protein from papayas. It predigests the steak. The same enzyme (chymopapain) is injected to soften herniated disks.

6. Yeast makes bread rise. Enzymes put the sour in sourdough.

Of course, it takes enzymes to make enzymes. The maternal egg contains the material for a cell, and also half of the DNA. The fertilized egg contains the full instructions.

We now look at the Michaelis–Menten (MM) equation, to describe these reactions. It is based on the **Law of Mass Action**. An enzyme in concentration z converts a substrate in concentration y by $dy/dt = -byz$. The rate constant is b, and you see the product of "enzyme times substrate." A similar law governs the other reactions (some go backwards). The equations are nonlinear, with no exact solution. It is typical of applied mathematics (and nature) that a pattern can still be found.

What happens is that the enzyme concentration $z(t)$ quickly drops to $z_0 K/(y + K)$. The *Michaelis constant K* depends on the rates (like b) in the mass action laws. Later the enzyme reappears ($z_\infty = z_0$). But by then the first reaction is over. Its law of mass action is effectively

$$\frac{dy}{dt} = -byz = -\frac{cy}{y + K} \tag{17}$$

with $c = bz_0 K$. This is the **Michaelis–Menten equation**—basic to biochemistry.

The rate dy/dt is all-important in biology. Look at the function $cy/(y + K)$:

when y is large, $dy/dt \approx -c$ when y is small, $dy/dt \approx -cy/K$.

The start and the finish operate at different rates, depending whether y dominates K or K dominates y. The fastest rate is c.

A biochemist solves the MM equation by separating variables:

$$\int \frac{y + K}{y} \, dy = -\int c \, dt \quad \text{gives} \quad y + K \ln y = -ct + C. \tag{18}$$

Set $t = 0$ as usual. Then $C = y_0 + K \ln y_0$. The exponentials of the two sides are

$$e^y y^K = e^{-ct} e^{y_0} y_0^K. \tag{19}$$

We don't have a simple formula for y. We are lucky to get this close. A computer can quickly graph $y(t)$—and we see the dynamics of enzymes.

Problems $27 - 32$ follow up the Michaelis–Menten theory. In science, concentrations and rate constants come with units. In mathematics, variables can be made dimensionless and constants become 1. We solve $dY/dT = Y/(Y + 1)$ and then switch back to y, t, c, K. This idea applies to other equations too.

Essential point: *Most applications of calculus come through differential equations*. That is the language of mathematics—with populations and chemicals and epidemics obeying the same equation. Running parallel to $dy/dt = cy$ are the difference equations that come next.

6.5 EXERCISES

Read-through questions

The equations $dy/dt = cy$ and $dy/dt = cy + s$ and $dy/dt = u(y)v(t)$ are called __a__ because we can separate y from t. Integration of $\int dy/y = \int c\,dt$ gives __b__. Integration of $\int dy/(y + s/c) = \int c\,dt$ gives __c__. The equation $dy/dx = -x/y$ leads to __d__. Then $y^2 + x^2 =$ __e__ and the solution stays on a circle.

The logistic equation is $dy/dt =$ __f__. The new term $-by^2$ represents __g__ when cy represents growth. Separation gives $\int dy/(cy - by^2) = \int dt$, and the y-integral is $1/c$ times ln __h__. Substituting y_0 at $t = 0$ and taking exponentials produces $y/(c - by) = e^{ct}($ __i__ $)$. As $t \to \infty$, y approaches __j__. That is the steady state where $cy - by^2 =$ __k__. The graph of y looks like an __l__, because it has an inflection point at $y =$ __m__.

In biology and chemistry, concentrations y and z react at a rate proportional to y times __n__. This is the Law of __o__. In a model equation $dy/dt = c(y)y$, the rate c depends on __p__. The MM equation is $dy/dt =$ __q__. Separating variables yields $\int$ __r__ $dy =$ __s__ $= -ct + C$.

Separate, integrate, and solve equations 1–8.

1 $dy/dt = y + 5$, $y_0 = 2$

2 $dy/dt = 1/y$, $y_0 = 1$

3 $dy/dx = x/y^2$, $y_0 = 1$

4 $dy/dx = y^2 + 1$, $y_0 = 0$

5 $dy/dx = (y + 1)/(x + 1)$, $y_0 = 0$

6 $dy/dx = \tan y \cos x$, $y_0 = 1$

7 $dy/dt = y \sin t$, $y_0 = 1$

8 $dy/dt = e^{t-y}$, $y_0 = e$

9 Suppose the rate of growth is proportional to $\sqrt{y}$ instead of y. Solve $dy/dt = c\sqrt{y}$ starting from y_0.

10 The equation $dy/dx = ny/x$ for constant elasticity is the same as $d(\ln y)/d(\ln x) =$ _____ . The solution is $\ln y =$ _____ .

11 When $c = 0$ in the logistic equation, the only term is $y' = -by^2$. What is the steady state y_∞? How long until y drops from y_0 to $\frac{1}{2}y_0$?

12 Reversing signs in Problem 11, suppose $y' = +by^2$. At what time does the population explode to $y = \infty$, starting from $y_0 = 2$ (Adam+Eve)?

Problems 13–26 deal with logistic equations $y' = cy - by^2$.

13 Show that $y = 1/(1 + e^{-t})$ solves the equation $y' = y - y^2$. Draw the graph of y from starting values $\frac{1}{2}$ and $\frac{1}{3}$.

14 (a) What logistic equation is solved by $y = 2/(1 + e^{-t})$?

(b) Find c and b in the equation solved by $y = 1/(1 + e^{-3t})$.

15 Solve $z' = -z + 1$ with $z_0 = 2$. Turned upside down as in (13), what is $y = 1/z$?

16 By algebra find the S-curve (12) from $y = 1/z$ in (14).

17 How many years to grow from $y_0 = \frac{1}{2}c/b$ to $y = \frac{3}{4}c/b$? Use equation (10) for the time t since the inflection point in 1988. When does y reach 9 billion $= .9c/b$?

18 Show by differentiating $u = y/(c - by)$ that if $y' = cy - by^2$ then $u' = cu$. This explains the logistic solution (11)—it is $u = u_0 e^{ct}$.

19 Suppose Pittsburgh grows from $y_0 = 1$ million people in 1900 to $y = 3$ million in the year 2000. If the growth rate is $y' = 12,000/\text{year}$ in 1900 and $y' = 30,000/\text{year}$ in 2000, substitute in the logistic equation to find c and b. What is the steady state? Extra credit: When does $y = y_\infty/2 = c/2b$?

20 Suppose $c = 1$ but $b = -1$, giving cooperation $y' = y + y^2$. Solve for $y(t)$ if $y_0 = 1$. When does y become infinite?

21 Draw an S-curve through $(0,0)$ with horizontal asymptotes $y = -1$ and $y = 1$. Show that $y = (e^t - e^{-t})/(e^t + e^{-t})$ has those three properties. The graph of y^2 is shaped like _____ .

22 To solve $y' = cy - by^3$ change to $u = 1/y^2$. Substitute for y' in $u' = -2y'/y^3$ to find a linear equation for u. Solve it as in (14) but with $u_0 = 1/y_0^2$. Then $y = 1/\sqrt{u}$.

23 With $y = rY$ and $t = sT$, the equation $dy/dt = cy - by^2$ changes to $dY/dT = Y - Y^2$. Find r and s.

24 In a change to $y = rY$ and $t = sT$, how are the initial values y_0 and y_0' related to Y_0 and Y_0'?

25 A rumor spreads according to $y' = y(N - y)$. If y people know, then $N - y$ don't know. The product $y(N - y)$ measures the number of meetings (to pass on the rumor).

(a) Solve $dy/dt = y(N - y)$ starting from $y_0 = 1$.

(b) At what time T have $N/2$ people heard the rumor?

(c) This model is terrible because T goes to _____ as $N \to \infty$. A better model is $y' = by(N - y)$.

26 Suppose b and c are both multiplied by 10. Does the middle of the S-curve get steeper or flatter?

Problems 27–34 deal with mass action and the MM equation $y' = -cy/(y + K)$.

27 Most drugs are eliminated acording to $y' = -cy$ but aspirin follows the MM equation. With $c = K = y_0 = 1$, does aspirin decay faster?

28 If you take aspirin at a constant rate d (the maintenance dose), find the steady state level where $d = cy/(y + K)$. Then $y' = 0$.

29 Show that the rate $R = cy/(y + K)$ in the MM equation increases as y increases, and find the maximum as $y \to \infty$.

30 Graph the rate R as a function of y for $K = 1$ and $K = 10$. (Take $c = 1$.) As the Michaelis constant increases, the rate _____. *At what value of y is $R = \frac{1}{2}c$?*

31 With $y = KY$ and $ct = KT$, find the "nondimensional" MM equation for dY/dT. From the solution $e^Y Y = e^{-T}e^{Y_0}Y_0$ recover the y, t solution (19).

32 Graph $y(t)$ in (19) for different c and K (by computer).

33 The Law of Mass Action for $A + B \to C$ is $y' = k(a_0 - y)(b_0 - y)$. Suppose $y_0 = 0$, $a_0 = b_0 = 3$, $k = 1$. Solve for y and find the time when $y = 2$.

34 In addition to the equation for $d[C]/dt$, the mass action law gives $d[A]/dt =$ _____ .

35 Solve $y' = y + t$ from $y_0 = 0$ by assuming $y = Ae^t + B + Dt$. Find A, B, D.

36 Rewrite $cy - by^2$ as $a^2 - x^2$, with $x = \sqrt{b}y - c/2\sqrt{b}$ and $a =$ _____ . Substitute for a and x in the integral taken from tables, to obtain the y-integral in the text:

$$\int \frac{dx}{a^2 - x^2} = \frac{1}{2a}\ln\frac{a+x}{a-x} \qquad \int \frac{dy}{cy - by^2} = \frac{1}{c}\ln\frac{y}{c - by}$$

37 (Important) Draw the y-lines (with arrows as in the text) for $y' = y/(1-y)$ and $y' = y - y^3$. Which steady states are approached from which initial values y_0 ?

38 Explain in your own words how the y-line works.

39 (a) Solve $y' = \tan y$ starting from $y_0 = \pi/6$ to find $\sin y = \frac{1}{2}e^t$.

(b) Explain why $t = 1$ is never reached.

(c) Draw arrows on the y-line to show that y approaches $\pi/2$—when does it get there ?

40 Write the logistic equation as $y' = cy(1 - y/K)$. As y' approaches zero, y approaches _____ . Find y, y', y'' at the inflection point.

6.6 Powers Instead of Exponentials

You may remember our first look at e. It is the special base for which e^x has slope 1 at $x = 0$. That led to the great equation of exponential growth: *The derivative of e^x equals e^x.* But our look at the actual number $e = 2.71828...$ was very short. It appeared as the limit of $(1 + 1/n)^n$. This seems an unnatural way to write down such an important number.

I want to show how $(1 + 1/n)^n$ and $(1 + x/n)^n$ arise naturally. They give ***discrete growth in finite steps***—with applications to compound interest. Loans and life insurance and money market funds use the discrete form of $y' = cy + s$. (We include extra information about bank rates, hoping this may be useful some day.) The applications in science and engineering are equally important. Scientific computing, like accounting, has *difference equations* in parallel with differential equations.

Knowing that this section will be full of formulas, I would like to jump ahead and tell you the best one. It is an infinite series for e^x. What makes the series beautiful is that ***its derivative is itself***.

Start with $y = 1 + x$. This has $y = 1$ and $y' = 1$ at $x = 0$. But y'' is zero, not one. Such a simple function doesn't stand a chance! No polynomial can be its own derivative, because the highest power x^n drops down to nx^{n-1}. The only way is *to have no highest power*. We are forced to consider infinitely many terms—*a power series*—to achieve "derivative equals function."

To produce the derivative $1 + x$, we need $1 + x + \frac{1}{2}x^2$. Then $\frac{1}{2}x^2$ is the derivative of $\frac{1}{6}x^3$, which is the derivative of $\frac{1}{24}x^4$. The best way is to write the whole series at once:

$$\textbf{\textit{Infinite series}} \quad e^x = 1 + x + \tfrac{1}{2}x^2 + \tfrac{1}{6}x^3 + \tfrac{1}{24}x^4 + \cdots. \tag{1}$$

This must be the greatest power series ever discovered. Its derivative is itself:

$$de^x/dx = 0 + 1 + x + \tfrac{1}{2}x^2 + \tfrac{1}{6}x^3 + \cdots = e^x. \tag{2}$$

The derivative of each term is the term before it. The integral of each term is the one after it (so $\int e^x dx = e^x + C$). The approximation $e^x \approx 1 + x$ appears in the first two terms. Other properties like $(e^x)(e^x) = e^{2x}$ are not so obvious. (Multiplying series is hard but interesting.) *It is not even clear why the sum is $2.718...$ when $x = 1$.* Somehow $1 + 1 + \frac{1}{2} + \frac{1}{6} + \cdots$ equals e. That is where $(1 + 1/n)^n$ will come in.

Notice that x^n is divided by the product $1 \cdot 2 \cdot 3 \cdots \cdots n$. This is "*n factorial*." Thus x^4 *is divided by* $1 \cdot 2 \cdot 3 \cdot 4 = 4! = 24$, and x^5 is divided by $5! = 120$. The derivative of $x^5/120$ is $x^4/24$, because 5 from the derivative cancels 5 from the factorial. In general $x^n/n!$ has derivative $x^{n-1}/(n-1)!$ Surprisingly 0! is 1.

Chapter 10 emphasizes that $x^n/n!$ becomes extremely small as n increases. The infinite series adds up to a finite number—which is e^x. We turn now to discrete growth, which produces the same series in the limit.

This headline was on page one of the New York Times for May 27, 1990.

213 Years After Loan, Uncle Sam is Dunned

San Antonio, May 26—More than 200 years ago, a wealthy Pennsylvania merchant named Jacob DeHaven lent $450,000 to the Continental Congress to rescue the troops at Valley Forge. That loan was apparently never repaid.

So Mr. DeHaven's descendants are taking the United States Government to court to collect what they believe they are owed. The total: $141 billion if the interest is compounded daily at 6 percent, the going rate at the time. If compounded yearly, the bill is only $98 billion.

The thousands of family members scattered around the country say they are not being greedy. "It's not the money—it's the principle of the thing," said Carolyn Cokerham, a DeHaven on her father's side who lives in San Antonio.
"You have to wonder whether there would even be a United States if this man had not made the sacrifice that he did. He gave everything he had."

The descendants say that they are willing to be flexible about the amount of settlement. But they also note that interest is accumulating at $190 a second.

"None of these people have any intention of bankrupting the Government," said Jo Beth Kloecker, a lawyer from Stafford, Texas. Fresh out of law school, Ms. Kloecker accepted the case for less than the customary 30 percent contingency.

It is unclear how many descendants there are. Ms. Kloecker estimates that based on 10 generations with four children in each generation, there could be as many as half a million.

The initial suit was dismissed on the ground that the statute of limitations is six years for a suit against the Federal Government. The family's appeal asserts that this violates Article 6 of the Constitution, which declares as valid all debts owed by the Government before the Constitution was adopted.

Mr. DeHaven died penniless in 1812. He had no children.

COMPOUND INTEREST

The idea of compound interest can be applied right away. Suppose you invest $1000 at a rate of 100% (hard to do). If this is the *annual rate*, the interest after a year is another $1000. You receive $2000 in all. But if the interest is *compounded* you receive more:

> after six months: Interest of $500 is reinvested to give $1500
>
> end of year: New interest of $750 (50% of 1500) gives $2250 total.

The bank multiplied twice by 1.5 (1000 to 1500 to 2250). Compounding *quarterly* multiplies *four times* by 1.25 (1 for principal, .25 for interest):

> after one quarter the total is $1000 + (.25)(1000) = 1250$
>
> after two quarters the total is $1250 + (.25)(1250) = 1562.50$
>
> after nine months the total is $1562.50 + (.25)(1562.50) = 1953.12$
>
> after a full year the total is $1953.12 + (.25)(1953.12) = 2441.41$

Each step multiplies by $1 + (1/n)$, to add one nth of a year's interest—still at 100%:

> quarterly conversion: $(1 + 1/4)^4 \times 1000 = 2441.41$
>
> monthly conversion: $(1 + 1/12)^{12} \times 1000 = 2613.04$
>
> daily conversion: $(1 + 1/365)^{365} \times 1000 = 2714.57.$

Many banks use 360 days in a year, although computers have made that obsolete. Very few banks use minutes (525,600 per year). Nobody compounds every second

($n = 31,536,000$). But some banks offer **continuous compounding**. This is the limiting case ($n \to \infty$) that produces e:

$$\left(1 + \frac{1}{n}\right)^n \times 1000 \ \textit{approaches} \ e \times 1000 = 2718.28.$$

1. *Quick method for* $(1 + 1/n)^n$: **Take its logarithm. Use** $\ln(1+x) \approx x$ **with** $x = \frac{1}{n}$:

$$\ln\left(1 + \frac{1}{n}\right)^n = n \ln\left(1 + \frac{1}{n}\right) \approx n\left(\frac{1}{n}\right) = 1. \tag{3}$$

As $1/n$ gets smaller, this approximation gets better. The limit is 1. Conclusion: $(1 + 1/n)^n$ approaches the number whose logarithm is 1. Sections 6.2 and 6.4 define the same number (which is e).

2. *Slow method for* $(1 + 1/n)^n$: **Multiply out all the terms. Then let** $n \to \infty$.

This is a brutal use of the binomial theorem. It involves nothing smart like logarithms, but the result is a fantastic new formula for e.

$$\text{Practice for } n = 3: \quad \left(1 + \frac{1}{3}\right)^3 = 1 + 3\left(\frac{1}{3}\right) + \frac{3 \cdot 2}{1 \cdot 2}\left(\frac{1}{3}\right)^2 + \frac{3 \cdot 2 \cdot 1}{1 \cdot 2 \cdot 3}\left(\frac{1}{3}\right)^3.$$

Binomial theorem for any positive integer n:

$$\left(1 + \frac{1}{n}\right)^n = 1 + n\left(\frac{1}{n}\right) + \frac{n(n-1)}{1 \cdot 2}\left(\frac{1}{n}\right)^2 + \frac{n(n-1)(n-2)}{1 \cdot 2 \cdot 3}\left(\frac{1}{n}\right)^3 + \cdots + \left(\frac{1}{n}\right)^n. \tag{4}$$

Each term in equation (4) approaches a limit as $n \to \infty$. Typical terms are

$$\frac{n(n-1)}{1 \cdot 2}\left(\frac{1}{n}\right)^2 \to \frac{1}{1 \cdot 2} \quad \text{and} \quad \frac{n(n-1)(n-2)}{1 \cdot 2 \cdot 3}\left(\frac{1}{n}\right)^3 \to \frac{1}{1 \cdot 2 \cdot 3}.$$

Next comes $1/1 \cdot 2 \cdot 3 \cdot 4$. The sum of all those limits in (4) is our new formula for e:

$$\lim\left(1 + \frac{1}{n}\right)^n = 1 + 1 + \frac{1}{1 \cdot 2} + \frac{1}{1 \cdot 2 \cdot 3} + \frac{1}{1 \cdot 2 \cdot 3 \cdot 4} + \cdots = e. \tag{5}$$

In summation notation this is $\Sigma_{k=0}^{\infty} 1/k! = e$. The factorials give fast convergence:

$$1 + 1 + .5 + .16667 + .04167 + .00833 + .00139 + .00020 + .00002 = 2.71828.$$

Those nine terms give an accuracy that was not reached by $n = 365$ compoundings. A limit is still involved (to add up the whole series). **You never see** e **without a limit**! It can be defined by derivatives or integrals or powers $(1 + 1/n)^n$ or by an infinite series. Something goes to zero or infinity, and care is required.

All terms in equation (4) are below (or equal to) the corresponding terms in (5). **The power** $(1 + 1/n)^n$ **approaches** e **from below**. There is a steady increase with n. Faster compounding yields more interest. Continuous compounding at 100% yields e, as each term in (4) moves up to its limit in (5).

Remark Change $(1+1/n)^n$ to $(1+x/n)^n$. Now the binomial theorem produces e^x:

$$\left(1+\frac{x}{n}\right)^n = 1 + n\left(\frac{x}{n}\right) + \frac{n(n-1)}{1\cdot 2}\left(\frac{x}{n}\right)^2 + \cdots \text{ approaches } 1 + x + \frac{x^2}{1\cdot 2} + \cdots. \quad (6)$$

Please recognize e^x on the right side! It is the infinite power series in equation (1). The next term is $x^3/6$ (x can be positive or negative). This is a final formula for e^x:

6L The limit of $(1+x/n)^n$ is e^x. At $x = 1$ we find e.

The logarithm of that power is $n\,\ln(1+x/n) \approx n(x/n) = x$. The power approaches e^x.

To summarize: The quick method proves $(1+1/n)^n \to e$ by logarithms. The slow method (multiplying out every term) led to the infinite series. Together they show the agreement of all our definitions of e.

DIFFERENCE EQUATIONS VS. DIFFERENTIAL EQUATIONS

We have the chance to see an important part of applied mathematics. This is not a course on differential equations, and it cannot become a course on difference equations. But it is a course with a purpose—*we aim to use what we know.* Our main application of e was to solve $y' = cy$ and $y' = cy + s$. Now we solve the corresponding difference equations.

Above all, the goal is to see the connections. *The purpose of mathematics is to understand and explain patterns.* The path from "discrete to continuous" is beautifully illustrated by these equations. Not every class will pursue them to the end, but I cannot fail to show the pattern in a ***difference equation***:

$$y(t+1) = ay(t). \quad (7)$$

Each step multiplies by the same number a. The starting value y_0 is followed by ay_0, $a^2 y_0$, and $a^3 y_0$. The solution at discrete times $t = 0, 1, 2, \ldots$ is $y(t) = a^t y_0$.
This formula $a^t y_0$ replaces the continuous solution $e^{ct} y_0$ of the differential equation.

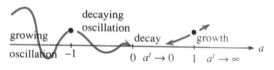

Fig. 6.17 Growth for $|a| > 1$, decay for $|a| < 1$. Growth factor a compares to e^c.

A source or sink (birth or death, deposit or withdrawal) is like $y' = cy + s$:

$$y(t+1) = ay(t) + s. \quad (8)$$

Each step multiplies by a *and adds s*. The first outputs are

$$y(1) = ay_0 + s, \quad y(2) = a^2 y_0 + as + s, \quad y(3) = a^3 y_0 + a^2 s + as + s.$$

We saw this pattern for differential equations—*every input s becomes a new starting point*. It is multiplied by powers of a. Since s enters later than y_0, the powers stop at $t-1$. Algebra turns the sum into a clean formula by adding the geometric series:

$$y(t) = a^t y_0 + s\left[a^{t-1} + a^{t-2} + \cdots + a + 1\right] = a^t y_0 + s(a^t - 1)/(a - 1). \quad (9)$$

EXAMPLE 1   Interest at 8% from annual IRA deposits of $s = \$2000$ (here $y_0 = 0$).

The first deposit is at year $t = 1$. In a year it is multiplied by $a = 1.08$, because 8% is added. At the same time a new $s = 2000$ goes in. At $t = 3$ the first deposit has been multiplied by $(1.08)^2$, the second by 1.08, and there is another $s = 2000$. After year t,

$$y(t) = 2000(1.08^t - 1)/(1.08 - 1). \quad (10)$$

With $t = 1$ this is 2000. With $t = 2$ it is 2000 $(1.08 + 1)$—two deposits. Notice how $a - 1$ (the interest rate .08) appears in the denominator.

EXAMPLE 2   Approach to steady state when $|a| < 1$. Compare with $c < 0$.

With $a > 1$, everything has been increasing. That corresponds to $c > 0$ in the differential equation (which is growth). But things die, and money is spent, so a can be smaller than one. In that case $a^t y_0$ approaches zero—the starting balance disappears. What happens if there is also a source ? *Every year half of the balance* $y(t)$ *is spent and a new* $\$2000$ *is deposited*. Now $a = \frac{1}{2}$:

$$y(t+1) = \tfrac{1}{2}y(t) + 2000 \quad \text{yields} \quad y(t) = \left(\tfrac{1}{2}\right)^t y_0 + 2000\left[\left(\left(\tfrac{1}{2}\right)^t - 1\right)/\left(\tfrac{1}{2} - 1\right)\right].$$

The limit as $t \to \infty$ is an equilibrium point. As $\left(\tfrac{1}{2}\right)^t$ goes to zero, $y(t)$ stabilizes to

$$y_\infty = 2000(0 - 1)/\left(\tfrac{1}{2} - 1\right) = 4000 = \textbf{\textit{steady state}}. \quad (11)$$

Why is 4000 steady ? Because half is lost and the new 2000 makes it up again. **The iteration is** $y_{n+1} = \frac{1}{2}y_n + 2000$. **Its fixed point is where** $y_\infty = \frac{1}{2}y_\infty + 2000$.

In general the steady equation is $y_\infty = a y_\infty + s$. Solving for y_∞ gives $s/(1-a)$. Compare with the steady differential equation $y' = cy + s = 0$:

$$y_\infty = -\frac{s}{c}\text{(differential equation)} \quad vs. \quad y_\infty = \frac{s}{1-a} \text{ (difference equation)}. \quad (12)$$

EXAMPLE 3   Demand equals supply when the price is right.

Difference equations are basic to economics. Decisions are made every year (by a farmer) or every day (by a bank) or every minute (by the stock market). There are three assumptions:

1. Supply next time depends on price this time: $S(t+1) = cP(t)$.
2. Demand next time depends on price next time: $D(t+1) = -dP(t+1) + b$.
3. Demand next time equals supply next time: $D(t+1) = S(t+1)$.

Comment on **3**: the price sets itself to make **demand = supply**. The demand slope $-d$ is negative. The supply slope c is positive. Those lines intersect at the competitive price, where supply equals demand. To find the difference equation, substitute **1** and **2** into **3**:

Difference equation : $-dP(t+1) + b = cP(t)$

Steady state price : $-dP_\infty + b = cP_\infty$. Thus $P_\infty = b/(c+d)$.

If the price starts above P_∞, the difference equation brings it down. If below, the price goes up. When the price is P_∞, it stays there. This is not news—economic theory depends on approach to a steady state. But convergence only occurs if $c < d$. *If supply is less sensitive than demand, the economy is stable.*

Blow-up example: $c = 2, b = d = 1$. The difference equation is $-P(t+1)+1 = 2P(t)$. From $P(0) = 1$ the price oscillates as it grows: $P = -1, 3, -5, 11, \ldots$.

Stable example: $c = 1/2$, $b = d = 1$. The price moves from $P(0) = 1$ to $P(\infty) = 2/3$:

$$-P(t+1)+1 = \frac{1}{2}P(t) \text{ yields } P = 1, \frac{1}{2}, \frac{3}{4}, \frac{5}{8}, \ldots, \text{ approaching } \frac{2}{3}.$$

Increasing d gives greater stability. That is the effect of price supports. For $d = 0$ (fixed demand regardless of price) the economy is out of control.

THE MATHEMATICS OF FINANCE

It would be a pleasure to make this supply-demand model more realistic—with curves, not straight lines. Stability depends on the slope—calculus enters. But we also have to be realistic about class time. I believe the most practical application is to solve *the fundamental problems of finance*. Section 6.3 answered six questions about continuous interest. We now answer the same six questions when the annual rate is $x = .05 = 5\%$ and *interest is compounded n times a year*.

First we compute *effective rates*, higher than .05 because of compounding:

$$\text{compounded } quarterly \left(1 + \frac{.05}{4}\right)^4 = 1.0509 \quad \left[\textit{effective rate } .0509 = 5.09\%\right]$$

$$\text{compounded } continuously \quad e^{.05} = 1.0513 \quad \left[\textit{effective rate } 5.13\%\right]$$

Now come the six questions. Next to the new answer (discrete) we write the old answer (continuous). One is algebra, the other is calculus. The time period is 20 years, so simple interest on y_0 would produce $(.05)(20)(y_0)$. That equals y_0—money doubles in 20 years at 5% simple interest.

Questions **1** and **2** ask for the *future value* y and *present value* y_0 with compound interest n times a year:

1. y growing from y_0: $\qquad y = \left(1 + \frac{.05}{n}\right)^{20n} y_0 \qquad y = e^{(.05)(20)} y_0$

2. deposit y_0 to reach y: $\qquad y_0 = \left(1 + \frac{.05}{n}\right)^{-20n} y \qquad y_0 = e^{-(.05)(20)} y$

Each step multiplies by $a = (1 + .05/n)$. There are $20n$ steps in 20 years. Time goes backward in Question **2**. We divide by the growth factor instead of multiplying. The future value is greater than the present value (unless the interest rate is negative!). As $n \to \infty$ the discrete y on the left approaches the continuous y on the right.

Questions **3** and **4** connect y to s (with $y_0 = 0$ at the start). As soon as each s is deposited, it starts growing. Then $y = s + as + a^2s + \cdots$.

3. y growing from deposits s: $\qquad y = s\left[\frac{(1+.05/n)^{20n} - 1}{.05/n}\right] \qquad y = s\left[\frac{e^{(.05)(20)} - 1}{.05}\right]$

4. deposits s to reach y: $\qquad s = y\left[\frac{.05/n}{(1+.05/n)^{20n} - 1}\right] \qquad s = y\left[\frac{.05}{e^{(.05)(20)} - 1}\right]$

Questions **5** and **6** connect y_0 to s. This time y is zero—*there is nothing left at the end*. Everything is paid. The deposit y_0 is just enough to allow payments of s. This is an **annuity**, where the bank earns interest on your y_0 while it pays you s (n times a year for 20 years). So your deposit in Question **5** is less than $20ns$.

Question **6** is the opposite—a **loan**. At the start you borrow y_0 (instead of giving the bank y_0). You can earn interest on it as you pay it back. Therefore your payments have to total more than y_0. This is the calculation for car loans and mortgages.

5. **Annuity**: Deposit y_0 to receive $20n$ payments of s:

$$y_0 = s\left[\frac{1-(1+.05/n)^{-20n}}{.05/n}\right] \qquad y_0 = s\left[\frac{1-e^{-(.05)(20)}}{.05}\right]$$

6. **Loan**: Repay y_0 with $20n$ payments of s:

$$s = y_0\left[\frac{.05/n}{1-(1+.05/n)^{-20n}}\right] \qquad s = y_0\left[\frac{.05}{1-e^{-(.05)(20)}}\right]$$

Questions **2,4,6** are the inverses of **1,3,5**. Notice the pattern: There are three numbers y, y_0, and s. *One of them is zero each time*. If all three are present, go back to equation (9).

The algebra for these lines is in the exercises. *It is not calculus because Δt is not dt*. All factors in brackets [ ] are listed in tables, and the banks keep copies. It might also be helpful to know their symbols. If a bank has interest rate i per period over N periods, then in our notation $a = 1 + i = 1 + .05/n$ and $t = N = 20n$:

future value of $y_0 = \$1$ (line **1**) : $y(N) = (1+i)^N$

present value of $y = \$1$ (line **2**) : $y_0 = (1+i)^{-N}$

future value of $s = \$1$ (line **3**) : $y(N) = s_{N\rceil i} = \left[(1+i)^N - 1\right]/i$

present value of $s = \$1$ (line **5**) : $y_0 = a_{N\rceil i} = \left[1-(1+i)^{-N}\right]/i$

To tell the truth, I never knew the last two formulas until writing this book. The mortgage on my home has $N = (12)(25)$ monthly payments with interest rate $i = .07/12$. In 1972 the present value was $\$42,000 =$ amount borrowed. I am now going to see if the bank is honest.†

Remark   In many loans, the bank computes interest on the amount paid back instead of the amount received. This is called **discounting**. A loan of $\$1000$ at 5% for one year costs $\$50$ interest. Normally you receive $\$1000$ and pay back $\$1050$. *With discounting you receive $\$950$ (called the proceeds) and you pay back $\$1000$*. The true interest rate is higher than 5%—because the $\$50$ interest is paid on the smaller amount $\$950$. In this case the "discount rate" is $50/950 = 5.26\%$.

SCIENTIFIC COMPUTING: DIFFERENTIAL EQUATIONS BY DIFFERENCE EQUATIONS

In biology and business, most events are discrete. In engineering and physics, time and space are continuous. Maybe at some quantum level it's all the same, but the

---

† It's not. s is too big. I knew it.

equations of physics (starting with Newton's law $F = ma$) are differential equations. The great contribution of calculus is to model the rates of change we see in nature. But to *solve that model with a computer*, it needs to be made digital and discrete.

These paragraphs work with $dy/dt = cy$. It is the test equation that all analysts use, as soon as a new computing method is proposed. Its solution is $y = e^{ct}$, starting from $y_0 = 1$. Here we test Euler's method (nearly ancient, and not well thought of). He replaced dy/dt by $\Delta y/\Delta t$:

$$\textbf{Euler's Method} \qquad \frac{y(t + \Delta t) - y(t)}{\Delta t} = cy(t). \qquad (13)$$

The left side is dy/dt, in the limit $\Delta t \to 0$. We stop earlier, when $\Delta t > 0$.

The problem is to solve (13). Multiplying by Δt, the equation is

$$y(t + \Delta t) = (1 + c\Delta t)y(t) \qquad \text{(with } y(0) = 1).$$

Each step multiplies by $a = 1 + c\Delta t$, so n steps multiply by a^n:

$$y = a^n = (1 + c\Delta t)^n \text{ at time } n\Delta t. \qquad (14)$$

This is growth or decay, *depending on a*. The correct e^{ct} is growth or decay, *depending on c*. **The question is whether a^n and e^{ct} stay close.** Can one of them grow while the other decays? We expect the difference equation to copy $y' = cy$, but we might be wrong.

A good example is $y' = -y$. Then $c = -1$ and $y = e^{-t}$—the true solution decays. The calculator gives the following answers a^n for $n = 2, 10, 20$:

Δt	$a = 1 + c\Delta t$	a^2	a^{10}	a^{20}
3	-2	4	1024	1048576
1	0	0	0	0
1/10	.90	.81	.35	.12
1/20	.95	.90	.60	.36

The big step $\Delta t = 3$ shows total instability (top row). The numbers blow up when they should decay. The row with $\Delta t = 1$ is equally useless (all zeros). In practice the magnitude of $c\Delta t$ must come down to .10 or .05. For accurate calculations it would have to be even smaller, unless we change to a better difference equation. That is the right thing to do.

Notice the two reasonable numbers. They are .35 and .36, approaching $e^{-1} = .37$. They come from $n = 10$ (with $\Delta t = 1/10$) and $n = 20$ (with $\Delta t = 1/20$). Those have the same clock time $n\Delta t = 1$:

$$\left(1 - \frac{1}{10}\right)^{10} = .35 \qquad \left(1 - \frac{1}{20}\right)^{20} = .36 \qquad \left(1 - \frac{1}{n}\right)^n \to e^{-1} = .37.$$

The main diagonal of the table is executing $(1 + x/n)^n \to e^x$ in the case $x = -1$.

Final question: **How quickly are .35 and .36 converging to** $e^{-1} = .37$? With $\Delta t = .10$ the error is .02. With $\Delta t = .05$ the error is .01. Cutting the time step in half cuts the error in half. We are not keeping enough digits to be sure, but *the* *error seems close to* $\frac{1}{5}\Delta t$. To test that, apply the "quick method" and estimate $a^n = (1 - \Delta t)^n$ from its logarithm:

$$\ln(1 - \Delta t)^n = n \ln(1 - \Delta t) \approx n\left[-\Delta t - \frac{1}{2}(\Delta t)^2\right] = -1 - \frac{1}{2}\Delta t. \qquad (15)$$

The clock time is $n\Delta t = 1$. Now take exponentials of the far left and right:

$$a^n = (1 - \Delta t)^n \approx e^{-1}e^{-\Delta t/2} \approx e^{-1}\left(1 - \frac{1}{2}\Delta t\right). \qquad (16)$$

The difference between a^n and e^{-1} is the last term $\frac{1}{2}\Delta t e^{-1}$. Everything comes down to one question: Is that error the same as $\frac{1}{5}\Delta t$? *The answer is yes*, because $e^{-1}/2$ is $1/5$. If we keep only one digit, the prediction is perfect!

That took an hour to work out, and I hope it takes longer than Δt to read. I wanted you to see *in use* the properties of $\ln x$ and e^x. The exact property $\ln a^n = n \ln a$ came first. In the middle of (15) was the key approximation $\ln(1 + x) \approx x - \frac{1}{2}x^2$, with $x = -\Delta t$. That x^2 term uses the second derivative (Section 6.4). At the very end came $e^x \approx 1 + x$.

A linear approximation shows convergence: $(1 + x/n)^n \to e^x$. A quadratic shows the error: proportional to $\Delta t = 1/n$. It is like using rectangles for areas, with error proportional to Δx. This minimal accuracy was enough to define the integral, and here it is enough to define e. It is completely unacceptable for scientific computing.

The trapezoidal rule, for integrals or for $y' = cy$, has errors of order $(\Delta x)^2$ and $(\Delta t)^2$. All good software goes further than that. Euler's first-order method could not predict the weather before it happens.

$$\textbf{Euler's Method for } \frac{dy}{dt} = F(y,t): \quad \frac{y(t + \Delta t) - y(t)}{\Delta t} = F(y(t), t).$$

6.6 EXERCISES

Read-through questions

The infinite series for e^x is ___a___ . Its derivative is ___b___ . The denominator $n!$ is called "___c___" and it equals ___d___. At $x = 1$ the series for e is ___e___ .

To match the original definition of e, multiply out $(1 + 1/n)^n =$ ___f___ (first three terms). As $n \to \infty$ those terms approach ___g___ in agreement with e. The first three terms of $(1 + x/n)^n$ are ___h___ . As $n \to \infty$ they approach ___i___ in agreement with e^x. Thus $(1 + x/n)^n$ approaches ___j___ . A quicker method computes $\ln(1 + x/n)^n \approx$ ___k___ (first term only) and takes the exponential.

Compound interest (n times in one year at annual rate x) multiplies by $(\_\_l\_\_)^n$. As $n \to \infty$, continuous compounding multiplies by ___m___ . At $x = 10\%$ with continuous compounding, $1 grows to ___n___ in a year.

The difference equation $y(t + 1) = ay(t)$ yields $y(t) =$ ___o___ times y_0. The equation $y(t + 1) = ay(t) + s$ is solved by $y = a^t y_0 + s[1 + a + \cdots + a^{t-1}]$. The sum in brackets is ___p___ .

When $a = 1.08$ and $y_0 = 0$, annual deposits of $s = 1$ produce $y =$ ___q___ after t years. If $a = \frac{1}{2}$ and $y_0 = 0$, annual deposits of $s = 6$ leave ___r___ after t years, approaching $y_\infty =$ ___s___ . The steady equation $y_\infty = ay_\infty + s$ gives $y_\infty =$ ___t___ .

When $i =$ interest rate per period, the value of $y_0 = \$1$ after N periods is $y(N) =$ ___u___ . The deposit to produce $y(N) = 1$ is $y_0 =$ ___v___ . The value of $s = \$1$ deposited after each period grows to $y(N) =$ ___w___ . The deposit to reach $y(N) = 1$ is $s =$ ___x___ .

Euler's method replaces $y' = cy$ by $\Delta y = cy\Delta t$. Each step multiplies y by ___y___ . Therefore y at $t = 1$ is $(1 + c\Delta t)^{1/t}y_0$, which converges to ___z___ as $\Delta t \to 0$. The error is proportional to ___A___ , which is too ___B___ for scientific computing.

1 Write down a power series $y = 1 - x + \cdots$ whose derivative is $-y$.

2 Write down a power series $y = 1 + 2x + \cdots$ whose derivative is $2y$.

3 Find *two* series that are equal to their second derivatives.

4 By comparing $e = 1 + 1 + \frac{1}{2} + \frac{1}{6} + \frac{1}{24} + \cdots$ with a larger series (whose sum is easier) show that $e < 3$.

5 At 5% interest compute the output from $1000 in a year with 6-month and 3-month and weekly compounding.

6 With the quick method $\ln(1+x) \approx x$, estimate $\ln(1-1/n)^n$ and $\ln(1+2/n)^n$. Then take exponentials to find the two limits.

7 With the slow method multiply out the three terms of $(1 - \frac{1}{2})^2$ and the five terms of $(1 - \frac{1}{4})^4$. What are the first three terms of $(1 - 1/n)^n$, and what are their limits as $n \to \infty$?

8 The slow method leads to $1 - 1 + 1/2! - 1/3! + \cdots$ for the limit of $(1-1/n)^n$. What is the sum of this infinite series—the exact sum and the sum after five terms ?

9 Knowing that $(1+1/n)^n \to e$, explain $(1+1/n)^{2n} \to e^2$ and $(1+2/N)^N \to e^2$.

10 What are the limits of $(1+1/n^2)^n$ and $(1+1/n)^{n^2}$? OK to use a calculator to guess these limits.

11 (a) The power $(1+1/n)^n$ (decreases) (increases) with n, as we compound more often. (b) The derivative of $f(x) = x \ln(1+1/x)$, which is _____, should be $(<0)(>0)$. This is confirmed by Problem 12.

12 Show that $\ln(1+1/x) > 1/(x+1)$ by drawing the graph of $1/t$. The area from $t = 1$ to $1+1/x$ is _____. The rectangle inside it has area _____.

13 Take three steps of $y(t+1) = 2y(t)$ from $y_0 = 1$.

14 Take three steps of $y(t+1) = 2y(t) + 1$ from $y_0 = 0$.

Solve the difference equations 15–22.

15 $y(t+1) = 3y(t), y_0 = 4$ **16** $y(t+1) = \frac{1}{2}y(t), y_0 = 1$

17 $y(t+1) = y(t) + 1, y_0 = 0$ **18** $y(t+1) = y(t) - 1, y_0 = 0$

19 $y(t+1) = 3y(t) + 1, y_0 = 0$ **20** $y(t+1) = 3y(t) + s, y_0 = 1$

21 $y(t+1) = ay(t) + s, y_0 = 0$ **22** $y(t+1) = ay(t) + s, y_0 = 5$

In 23–26, which initial value produces $y_1 = y_0$ (steady state) ?

23 $y(t+1) = 2y(t) - 6$ **24** $y(t+1) = \frac{1}{2}y(t) - 6$

25 $y(t+1) = -y(t) + 6$ **26** $y(t+1) = -\frac{1}{2}y(t) + 6$

27 In Problems 23 and 24, start from $y_0 = 2$ and take three steps to reach y_3. Is this approaching a steady state ?

28 For which numbers a does $(1-a^t)/(1-a)$ approach a limit as $t \to \infty$ and what is the limit ?

29 The price P is determined by supply = demand or $-dP(t+1) + b = cP(t)$. Which price P is not changed from one year to the next ?

30 Find $P(t)$ from the supply-demand equation with $c = 1, d = 2, b = 8, P(0) = 0$. What is the steady state as $t \to \infty$?

Assume 10% **interest (so $a = 1 + i = 1.1$) in Problems** $31 - 38$.

31 At 10% interest compounded quarterly, what is the effective rate ?

32 At 10% interest compounded daily, what is the effective rate ?

33 Find the future value in 20 years of $100 deposited now.

34 Find the present value of $1000 promised in twenty years.

35 For a mortgage of $100,000 over 20 years, what is the monthly payment ?

36 For a car loan of $10,000 over 6 years, what is the monthly payment ?

37 With annual compounding of deposits $s = \$1000$, what is the balance in 20 years ?

38 If you repay $s = \$1000$ annually on a loan of $8000, when are you paid up ? (Remember interest.)

39 Every year two thirds of the available houses are sold, and 1000 new houses are built. What is the steady state of the housing market—how many are available ?

40 If a loan shark charges 5% interest a month on the $1000 you need for blackmail, and you pay $60 a month, how much do you still owe after one month (and after a year) ?

41 Euler charges $c = 100\%$ interest on his $1 fee for discovering e. What do you owe (including the $1) after a year with (a) no compounding; (b) compounding every week; (c) continuous compounding ?

42 Approximate $(1+1/n)^n$ as in (15) and (16) to show that you owe Euler about $e - e/2n$. Compare Problem 6.2.5.

43 My Visa statement says monthly rate = 1.42% and yearly rate = 17%. What is the true yearly rate, since Visa compounds the interest ? Give a formula or a number.

44 You borrow $y_0 = \$80,000$ at 9% to buy a house.

(a)   What are your monthly payments s over 30 years ?

(b)   How much do you pay altogether ?

6.7 Hyperbolic Functions

This section combines e^x with e^{-x}. Up to now those functions have gone separate ways—one increasing, the other decreasing. But two particular combinations have earned names of their own ($\cosh x$ and $\sinh x$):

hyperbolic cosine $\cosh x = \dfrac{e^x + e^{-x}}{2}$ *hyperbolic sine* $\sinh x = \dfrac{e^x - e^{-x}}{2}$

The first name rhymes with "gosh". The second is usually pronounced "cinch".

The graphs in Figure 6.18 show that $\cosh x > \sinh x$. For large x both hyperbolic functions come extremely close to $\frac{1}{2}e^x$. When x is large and *negative*, it is e^{-x} that dominates. Cosh x still goes up to $+\infty$ while $\sinh x$ goes down to $-\infty$ (because $\sinh x$ has a minus sign in front of e^{-x}).

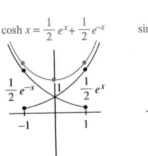

$\cosh x = \dfrac{1}{2}e^x + \dfrac{1}{2}e^{-x}$ $\sinh x = \dfrac{1}{2}e^x - \dfrac{1}{2}e^{-x}$

Fig. 6.18   Cosh x and sinh x. The hyperbolic functions combine $\frac{1}{2}e^x$ and $\frac{1}{2}e^{-x}$.

Fig. 6.19   Gateway Arch courtesy of the St. Louis Visitors Commission.

The following facts come directly from $\frac{1}{2}\left(e^x + e^{-x}\right)$ and $\frac{1}{2}\left(e^x - e^{-x}\right)$:

$\cosh(-x) = \cosh x$ and $\cosh 0 = 1$ ($\cosh$ is *even* like the cosine)

$\sinh(-x) = -\sinh x$ and $\sinh 0 = 0$ ($\sinh$ is *odd* like the sine)

The graph of $\cosh x$ corresponds to a *hanging cable* (hanging under its weight). Turned upside down, it has the shape of the Gateway Arch in St. Louis. That must be the largest upside-down cosh function ever built. A cable is easier to construct than an arch, because gravity does the work. With the right axes in Problem 55, the height of the cable is a stretched-out cosh function called a *catenary*:

$$y = a \cosh (x/a) \qquad \text{(cable tension/cable density} = a).$$

Busch Stadium in St. Louis has 96 catenary curves, to match the Arch.

The properties of the hyperbolic functions come directly from the definitions. There are too many properties to memorize—and no reason to do it! One rule is the most important. **Every fact about sines and cosines is reflected in a corresponding fact about** $\sinh x$ **and** $\cosh x$. Often the only difference is a minus sign. Here are four properties:

1. $(\cosh x)^2 - (\sinh x)^2 = 1$ $\left[\text{instead of } (\cos x)^2 + (\sin x)^2 = 1\right]$

Check: $\left[\dfrac{e^x + e^{-x}}{2}\right]^2 - \left[\dfrac{e^x - e^{-x}}{2}\right]^2 = \dfrac{e^{2x} + 2 + e^{-2x} - e^{2x} + 2 - e^{-2x}}{4} = 1$

2. $\dfrac{d}{dx}(\cosh x) = \sinh x$ $\left[\text{instead of } \dfrac{d}{dx}(\cos x) = -\sin x\right]$

3. $\dfrac{d}{dx}(\sinh x) = \cosh x$ $\left[\text{like } \dfrac{d}{dx}\sin x = \cos x\right]$

4. $\displaystyle\int \sinh x\, dx = \cosh x + C$ and $\displaystyle\int \cosh x\, dx = \sinh x + C$

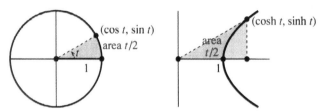

Fig. 6.20 The unit circle $\cos^2 t + \sin^2 t = 1$ and the unit hyperbola $\cosh^2 t - \sinh^2 t = 1$.

Property 1 is the connection to hyperbolas. It is responsible for the "h" in cosh and sinh. Remember that $(\cos x)^2 + (\sin x)^2 = 1$ puts the point $(\cos x, \sin x)$ onto a *unit circle*. As x varies, the point goes around the circle. The ordinary sine and cosine are "circular functions." Now look at $(\cosh x, \sinh x)$. Property 1 is $(\cosh x)^2 - (\sinh x)^2 = 1$, so this point travels on the *unit hyperbola* in Figure 6.20.

You will guess the definitions of the other four hyperbolic functions:

$$\tanh x = \frac{\sinh x}{\cosh x} = \frac{e^x - e^{-x}}{e^x + e^{-x}} \qquad \coth x = \frac{\cosh x}{\sinh x} = \frac{e^x + e^{-x}}{e^x - e^{-x}}$$

$$\text{sech } x = \frac{1}{\cosh x} = \frac{2}{e^x + e^{-x}} \qquad \text{csch } x = \frac{1}{\sinh x} = \frac{2}{e^x - e^{-x}}$$

I think "tanh" is pronounceable, and "sech" is easy. The others are harder. Their properties come directly from $\cosh^2 x - \sinh^2 x = 1$. Divide by $\cosh^2 x$ and $\sinh^2 x$:

$$1 - \tanh^2 x = \text{sech}^2 x \qquad \text{and} \quad \coth^2 x - 1 = \text{csch}^2 x$$

$$(\tanh x)' = \text{sech}^2 x \qquad \text{and} \quad (\text{sech } x)' = -\text{sech } x \tanh x$$

$$\int \tanh x\, dx = \int \frac{\sinh x}{\cosh x}\, dx = \ln(\cosh x) + C.$$

INVERSE HYPERBOLIC FUNCTIONS

You remember the angles $\sin^{-1}x$ and $\tan^{-1}x$ and $\sec^{-1}x$. In Section 4.4 we differentiated those inverse functions by the chain rule. The main application was to *integrals*. If we happen to meet $\int dx/(1 + x^2)$, it is $\tan^{-1}x + C$. The situation for

$\sinh^{-1} x$ and $\tanh^{-1} x$ and $\mathrm{sech}^{-1} x$ is the same except for sign changes—which are expected for hyperbolic functions. We write down the ***three new derivatives***:

$$y = \sinh^{-1} x \ (\text{meaning } x = \sinh y) \text{ has } \frac{dy}{dx} = \frac{1}{\sqrt{x^2 + 1}} \qquad (1)$$

$$y = \tanh^{-1} x \ (\text{meaning } x = \tanh y) \text{ has } \frac{dy}{dx} = \frac{1}{1 - x^2} \qquad (2)$$

$$y = \mathrm{sech}^{-1} x \ (\text{meaning } x = \mathrm{sech}\, y) \text{ has } \frac{dy}{dx} = \frac{-1}{x\sqrt{1 - x^2}} \qquad (3)$$

Problems $44 - 46$ compute dy/dx from $1/(dx/dy)$. The alternative is to use logarithms. Since $\ln x$ is the inverse of e^x, we can express $\sinh^{-1} x$ and $\tanh^{-1} x$ and $\mathrm{sech}^{-1} x$ as logarithms. Here is $y = \tanh^{-1} x$:

$$y = \frac{1}{2} \ln\left[\frac{1+x}{1-x}\right] \text{ has slope } \frac{dy}{dx} = \frac{1}{2}\frac{1}{1+x} - \frac{1}{2}\frac{1}{1-x} = \frac{1}{1-x^2}. \qquad (4)$$

The last step is an ordinary derivative of $\frac{1}{2}\ln(1+x) - \frac{1}{2}\ln(1-x)$. Nothing is new except the answer. But where did the logarithms come from? In the middle of the following identity, multiply above and below by $\cosh y$:

$$\frac{1+x}{1-x} = \frac{1+\tanh y}{1-\tanh y} = \frac{\cosh y + \sinh y}{\cosh y - \sinh y} = \frac{e^y}{e^{-y}} = e^{2y}.$$

Then $2y$ is the logarithm of the left side. This is the first equation in (4), and it is the third formula in the following list:

$$\sinh^{-1} x = \ln\left[x + \sqrt{x^2 + 1}\right] \qquad \cosh^{-1} x = \ln\left[x + \sqrt{x^2 - 1}\right]$$

$$\tanh^{-1} x = \frac{1}{2}\ln\left[\frac{1+x}{1-x}\right] \qquad \mathrm{sech}^{-1} x = \ln\left[\frac{1 + \sqrt{1 - x^2}}{x}\right]$$

Remark 1   Those are listed *only for reference*. If possible do not memorize them. The derivatives in equations (1), (2), (3) offer a choice of antiderivatives—either inverse functions or logarithms (most tables prefer logarithms). The inside cover of the book has

$$\int \frac{dx}{1-x^2} = \frac{1}{2}\ln\left[\frac{1+x}{1-x}\right] + C \quad (\text{in place of } \tanh^{-1} x + C).$$

Remark 2   Logarithms were not seen for $\sin^{-1} x$ and $\tan^{-1} x$ and $\sec^{-1} x$. You might wonder why. How does it happen that $\tanh^{-1} x$ is expressed by logarithms, when the parallel formula for $\tan^{-1} x$ was missing? Answer: *There must be a parallel formula*. To display it I have to reveal a secret that has been hidden throughout this section.

The secret is one of the great equations of mathematics. **What formulas for $\cos x$ *and* $\sin x$ *correspond to* $\frac{1}{2}(e^x + e^{-x})$ *and* $\frac{1}{2}(e^x - e^{-x})$?** With so many analogies (circular vs. hyperbolic) you would expect to find something. The formulas do exist, but ***they involve imaginary numbers***. Fortunately they are very simple and there is no reason to withhold the truth any longer:

$$\cos x = \frac{1}{2}\left(e^{ix} + e^{-ix}\right) \quad \text{and} \quad \sin x = \frac{1}{2i}\left(e^{ix} - e^{-ix}\right). \tag{5}$$

It is the imaginary exponents that kept those identities hidden. Multiplying $\sin x$ by i and adding to $\cos x$ gives Euler's unbelievably beautiful equation

$$\cos x + i \sin x = e^{ix}. \tag{6}$$

That is parallel to the non-beautiful hyperbolic equation $\cosh x + \sinh x = e^x$.

I have to say that (6) is infinitely more important than anything hyperbolic will ever be. The sine and cosine are far more useful than the sinh and cosh. So we end our record of the main properties, with exercises to bring out their applications.

6.7 EXERCISES

Read-through questions

Cosh $x = \underline{\quad a \quad}$ and $\sinh x = \underline{\quad b \quad}$ and $\cosh^2 x - \sinh^2 x = \underline{\quad c \quad}$. Their derivatives are $\underline{\quad d \quad}$ and $\underline{\quad e \quad}$ and $\underline{\quad f \quad}$. The point $(x, y) = (\cosh t, \sinh t)$ travels on the hyperbola $\underline{\quad g \quad}$. A cable hangs in the shape of a catenary $y = \underline{\quad h \quad}$.

The inverse functions $\sinh^{-1} x$ and $\tanh^{-1} x$ are equal to $\ln[x + \sqrt{x^2 + 1}]$ and $\frac{1}{2}\ln\underline{\quad i \quad}$. Their derivatives are $\underline{\quad j \quad}$ and $\underline{\quad k \quad}$. So we have two ways to write the anti $\underline{\quad l \quad}$. The parallel to $\cosh x + \sinh x = e^x$ is Euler's formula $\underline{\quad m \quad}$. The formula $\cos x = \frac{1}{2}\left(e^{ix} + e^{-ix}\right)$ involves $\underline{\quad n \quad}$ exponents. The parallel formula for $\sin x$ is $\underline{\quad o \quad}$.

1 Find $\cosh x + \sinh x$, $\cosh x - \sinh x$, and $\cosh x \sinh x$.

2 From the definitions of $\cosh x$ and $\sinh x$, find their derivatives.

3 Show that both functions satisfy $y'' = y$.

4 By the quotient rule, verify $(\tanh x)' = \text{sech}^2 x$.

5 Derive $\cosh^2 x + \sinh^2 x = \cosh 2x$, from the definitions.

6 From the derivative of Problem 5 find $\sinh 2x$.

7 The parallel to $(\cos x + i \sin x)^n = \cos nx + i \sin nx$ is a hyperbolic formula $(\cosh x + \sinh x)^n = \cosh nx + \underline{\quad\quad}$.

8 Prove $\sinh(x + y) = \sinh x \cosh y + \cosh x \sinh y$ by changing to exponentials. Then the x-derivative gives $\cosh(x + y) = \underline{\quad\quad}$.

Find the derivatives of the functions 9–18:

9 $\cosh(3x + 1)$

10 $\sinh x^2$

11 $1/\cosh x$

12 $\sinh(\ln x)$

13 $\cosh^2 x + \sinh^2 x$

14 $\cosh^2 x - \sinh^2 x$

15 $\tanh \sqrt{x^2 + 1}$

16 $(1 + \tanh x)/(1 - \tanh x)$

17 $\sinh^6 x$

18 $\ln(\text{sech } x + \tanh x)$

19 Find the minimum value of $\cosh(\ln x)$ for $x > 0$.

20 From $\tanh x = \frac{3}{5}$ find $\text{sech } x$, $\cosh x$, $\sinh x$, $\coth x$, $\text{csch } x$.

21 Do the same if $\tanh x = -12/13$.

22 Find the other five values if $\sinh x = 2$.

23 Find the other five values if $\cosh x = 1$.

24 Compute $\sinh(\ln 5)$ and $\tanh(2 \ln 4)$.

Find antiderivatives for the functions in 25–32:

25 $\cosh(2x + 1)$

26 $x \cosh(x^2)$

27 $\cosh^2 x \sinh x$

28 $\tanh^2 x \, \text{sech}^2 x$

29 $\dfrac{\sinh x}{1 + \cosh x}$

30 $\coth x = \dfrac{e^x + e^{-x}}{e^x - e^{-x}}$

31 $\sinh x + \cosh x$

32 $(\sinh x + \cosh x)^n$

33 The triangle in Figure 6.20 has area $\frac{1}{2}\cosh t \sinh t$.

    (a) Integrate to find the shaded area below the hyperbola

    (b) For the area A in red verify that $dA/dt = \frac{1}{2}$

    (c) Conclude that $A = \frac{1}{2}t + C$ and show $C = 0$.

Sketch graphs of the functions in 34–40.

34 $y = \tanh x$ (with inflection point)

35 $y = \coth x$ (in the limit as $x \to \infty$)

36 $y = \text{sech } x$

37 $y = \sinh^{-1} x$

38 $y = \cosh^{-1} x$ for $x \geqslant 1$

39 $y = \operatorname{sech}^{-1} x$ for $0 < x \leqslant 1$

40 $y = \tanh^{-1} x = \frac{1}{2} \ln\left(\dfrac{1+x}{1-x}\right)$ for $|x| < 1$

41 (a) Multiplying $x = \sinh y = \frac{1}{2}(e^y - e^{-y})$ by $2e^y$ gives $(e^y)^2 - 2x(e^y) - 1 = 0$. Solve as a quadratic equation for e^y.

(b) Take logarithms to find $y = \sinh^{-1} x$ and compare with the text.

42 (a) Multiplying $x = \cosh y = \frac{1}{2}(e^y + e^{-y})$ by $2e^y$ gives $(e^y)^2 - 2x(e^y) + 1 = 0$. Solve for e^y.

(b) Take logarithms to find $y = \cosh^{-1} x$ and compare with the text.

43 Turn (4) upside down to prove $y' = -1/(1 - x^2)$, if $y = \coth^{-1} x$.

44 Compute $dy/dx = 1/\sqrt{x^2 + 1}$ by differentiating $x = \sinh y$ and using $\cosh^2 y - \sinh^2 y = 1$.

45 Compute $dy/dx = 1/(1 - x^2)$ if $y = \tanh^{-1} x$ by differentiating $x = \tanh y$ and using $\operatorname{sech}^2 y + \tanh^2 y = 1$.

46 Compute $dy/dx = -1/x\sqrt{1 - x^2}$ for $y = \operatorname{sech}^{-1} x$, by differentiating $x = \operatorname{sech} y$.

From formulas (1), (2), (3) or otherwise, find antiderivatives in 47–52:

47 $\displaystyle\int dx/(4 - x^2)$ **48** $\displaystyle\int dx/(a^2 - x^2)$

49 $\displaystyle\int dx/\sqrt{x^2 + 1}$ **50** $\displaystyle\int x\,dx/\sqrt{x^2 + 1}$

51 $\displaystyle\int dx/x\sqrt{1 - x^2}$ **52** $\displaystyle\int dx/\sqrt{1 - x^2}$

53 Compute $\displaystyle\int_0^{1/2} \frac{dx}{1 - x^2}$ and $\displaystyle\int_0^1 \frac{dx}{1 - x^2}$.

54 A falling body with friction equal to velocity squared obeys $dv/dt = g - v^2$.

(a) Show that $v(t) = \sqrt{g}\tanh\sqrt{g}t$ satisfies the equation.

(b) Derive this v yourself, by integrating $dv/(g - v^2) = dt$.

(c) Integrate $v(t)$ to find the distance $f(t)$.

55 A cable hanging under its own weight has slope $S = dy/dx$ that satisfies $dS/dx = c\sqrt{1 + S^2}$. The constant c is the ratio of cable density to tension.

(a) Show that $S = \sinh cx$ satisfies the equation.

(b) Integrate $dy/dx = \sinh cx$ to find the cable height $y(x)$, if $y(0) = 1/c$.

(c) Sketch the cable hanging between $x = -L$ and $x = L$ and find how far it sags down at $x = 0$.

56 The simplest nonlinear wave equation (Burgers' equation) yields a waveform $W(x)$ that satisfies $W'' = WW' - W'$. One integration gives $W' = \frac{1}{2}W^2 - W$.

(a) Separate variables and integrate: $dx = dW/(\frac{1}{2}W^2 - W) = -dW/(2 - W) - dW/W$.

(b) Check $W' = \frac{1}{2}W^2 - W$.

57 A solitary water wave has a shape satisfying the KdV equation $y'' = y' - 6yy'$.

(a) Integrate once to find y''. Multiply the answer by y'.

(b) Integrate again to find y' (all constants of integration are zero).

(c) Show that $y = \frac{1}{2}\operatorname{sech}^2(x/2)$ gives the shape of the "soliton."

58 Derive $\cos ix = \cosh x$ from equation (5). What is the cosine of the imaginary angle $i = \sqrt{-1}$?

59 Derive $\sin ix = i\sinh x$ from (5). What is $\sin i$?

60 The derivative of $e^{ix} = \cos x + i\sin x$ is _____ .

CHAPTER 7

Techniques of Integration

Chapter 5 introduced the integral as a limit of sums. The calculation of areas was started—by hand or computer. Chapter 6 opened a different door. Its new functions e^x and $\ln x$ led to differential equations. You might say that all along we have been solving the special differential equation $df/dx = v(x)$. *The solution is $f = \int v(x)\,dx$.* But the step to $dy/dx = cy$ was a breakthrough.

The truth is that we are able to do remarkable things. ***Mathematics has a language, and you are learning to speak it***. A short time ago the symbols dy/dx and $\int v(x)\,dx$ were a mystery. (My own class was not too sure about $v(x)$ itself—the symbol for a function.) It is easy to forget how far we have come, in looking ahead to what is next.

I do want to look ahead. For integrals there are two steps to take—more functions and more applications. ***By using mathematics we make it live***. The applications are most complete when we know the integral. This short chapter will widen (very much) the range of functions we can integrate. A computer with symbolic algebra widens it more.

Up to now, integration depended on recognizing derivatives. If $v(x) = \sec^2 x$ then $f(x) = \tan x$. To integrate $\tan x$ we use a substitution:

$$\int \frac{\sin x}{\cos x}\,dx = -\int \frac{du}{u} = -\ln u = -\ln \cos x.$$

What we need now are ***techniques for other integrals***, to change them around until we can attack them. Two examples are $\int x \cos x\,dx$ and $\int \sqrt{1 - x^2}\,dx$, which are not immediately recognizable. With integration by parts, and a new substitution, they become simple.

Those examples indicate where this chapter starts and stops. With reasonable effort (and the help of tables, which is fair) you can integrate important functions. With intense effort you could integrate even more functions. In older books that extra exertion was made—it tended to dominate the course. They had integrals like $\int (x+1)\,dx/\sqrt{2x^2 - 6x + 4}$, which we could work on if we had to. ***Our time is too valuable for that***! Like long division, the ideas are for us and their intricate elaboration is for the computer.

Integration by parts comes first. Then we do new substitutions. Partial fractions is a useful idea (already applied to the logistic equation $y' = cy - by^2$). In the last section x goes to infinity or $y(x)$ goes to infinity—but the area stays finite. These improper integrals are quite common. Chapter 8 brings the applications.

7.1 Integration by Parts

There are two major ways to manipulate integrals (with the hope of making them easier). Substitutions are based on the chain rule, and more are ahead. Here we present the other method, based on the **product rule**. The reverse of the product rule, to find integrals not derivatives, is **integration by parts**.

We have mentioned $\int \cos^2 x \, dx$ and $\int \ln x \, dx$. Now is the right time to compute them (plus more examples). You will see how $\int \ln x \, dx$ is exchanged for $\int 1 \, dx$—a definite improvement. Also $\int xe^x \, dx$ is exchanged for $\int e^x \, dx$. The difference between the harder integral and the easier integral is a known term—that is the point.

One note before starting: Integration by parts is *not just a trick* with no meaning. On the contrary, it expresses basic physical laws of equilibrium and force balance. It is a foundation for the theory of differential equations (and even delta functions). The final paragraphs, which are completely optional, illustrate those points too.

We begin with the product rule for the derivative of $u(x)$ times $v(x)$:

$$u(x)\frac{dv}{dx} + v(x)\frac{du}{dx} = \frac{d}{dx}(u(x)v(x)). \tag{1}$$

Integrate both sides. On the right, integration brings back $u(x)v(x)$. On the left are two integrals, and one of them moves to the other side (with a minus sign):

$$\int u(x)\frac{dv}{dx}\, dx = u(x)v(x) - \int v(x)\frac{du}{dx}\, dx. \tag{2}$$

That is the key to this section—not too impressive at first, but very powerful. It is **integration by parts** (u and v are the parts). In practice we write it without x's:

> **7A**   The integration by parts formula is $\int u \, dv = uv - \int v \, du.$ (3)

The problem of integrating $u \, dv/dx$ is changed into the problem of integrating $v \, du/dx$. There is a minus sign to remember, and there is the "integrated term" $u(x)v(x)$. In the *definite integral*, that product $u(x)v(x)$ is evaluated at the endpoints a and b:

$$\int_a^b u\frac{dv}{dx}\, dx = u(b)v(b) - u(a)v(a) - \int_a^b v\frac{du}{dx}\, dx. \tag{4}$$

The key is in choosing u and v. The goal of that choice is to make $\int v \, du$ easier than $\int u \, dv$. This is best seen by examples.

EXAMPLE 1   For $\int \ln x \, dx$ choose $u = \ln x$ and $dv = dx$ (so $v = x$):

$$\int \ln x \, dx = uv - \int v \, du = x \ln x - \int x\frac{1}{x}\, dx.$$

I used the basic formula (3). Instead of working with $\ln x$ (searching for an antiderivative), we now work with the right hand side. There x times $1/x$ is 1. The integral of 1 is x. Including the minus sign and the integrated term $uv = x \ln x$ and the constant C, the answer is

$$\int \ln x \, dx = x \ln x - x + C. \tag{5}$$

For safety, **take the derivative**. The product rule gives $\ln x + x(1/x) - 1$, which is $\ln x$. The area under $y = \ln x$ from 2 to 3 is $3 \ln 3 - 3 - 2 \ln 2 + 2$.

To repeat: We exchanged the integral of ln x for the integral of 1.

EXAMPLE 2 For $\int x \cos x \, dx$ choose $u = x$ and $dv = \cos x \, dx$ (so $v(x) = \sin x$):

$$\int x \cos x \, dx = uv - \int v \, du = x \sin x - \int \sin x \, dx. \tag{6}$$

Again the right side has a simple integral, which completes the solution:

$$\int x \cos x \, dx = x \sin x + \cos x + C. \tag{7}$$

Note   The new integral is not always simpler. We could have chosen $u = \cos x$ and $dv = x \, dx$. Then $v = \frac{1}{2}x^2$. Integration using those parts give the true but useless result

$$\int x \cos x \, dx = uv - \int v \, du = \frac{1}{2}x^2 \cos x + \int \frac{1}{2}x^2 \sin x \, dx.$$

The last integral is harder instead of easier (x^2 is worse than x). In the forward direction this is no help. But in the opposite direction it simplifies $\int \frac{1}{2}x^2 \sin x \, dx$. The idea in choosing u and v is this: ***Try to give u a nice derivative and dv a nice integral.***

EXAMPLE 3 For $\int (\cos x)^2 \, dx$ choose $u = \cos x$ and $dv = \cos x \, dx$ (so $v = \sin x$):

$$\int (\cos x)^2 \, dx = uv - \int v \, du = \cos x \sin x + \int (\sin x)^2 \, dx.$$

The integral of $(\sin x)^2$ is no better and no worse than the integral of $(\cos x)^2$. But we never see $(\sin x)^2$ without thinking of $1 - (\cos x)^2$. So substitute for $(\sin x)^2$:

$$\int (\cos x)^2 \, dx = \cos x \sin x + \int 1 \, dx - \int (\cos x)^2 \, dx.$$

The last integral on the right joins its twin on the left, and $\int 1 \, dx = x$:

$$2 \int (\cos x)^2 \, dx = \cos x \sin x + x.$$

Dividing by 2 gives the answer, which is definitely not $\frac{1}{3}(\cos x)^3$. Add any C:

$$\int (\cos x)^2 \, dx = \frac{1}{2}(\cos x \sin x + x) + C. \tag{8}$$

Question Integrate $(\cos x)^2$ from 0 to 2π. Why should the area be π ?

Answer   The definite integral is $\frac{1}{2}(\cos x \sin x + x)\big]_0^{2\pi}$. This does give π. That area can also be found by common sense, starting from $(\cos x)^2 + (\sin x)^2 = 1$. The area under 1 is 2π. The areas under $(\cos x)^2$ and $(\sin x)^2$ are the same. So each one is π.

EXAMPLE 4   Evaluate $\int \tan^{-1} x \, dx$ by choosing $u = \tan^{-1} x$ and $v = x$:

$$\int \tan^{-1} x \, dx = uv - \int v \, du = x \tan^{-1} x - \int x \frac{dx}{1 + x^2}. \tag{9}$$

The last integral has $w = 1 + x^2$ below and almost has $dw = 2x \, dx$ above:

$$\int \frac{x \, dx}{1 + x^2} = \frac{1}{2} \int \frac{dw}{w} = \frac{1}{2} \ln w = \frac{1}{2} \ln(1 + x^2).$$

Substituting back into (9) gives $\int \tan^{-1} x \, dx$ as $x \tan^{-1} x - \frac{1}{2} \ln(1 + x^2)$. All the familiar inverse functions can be integrated by parts (take $v = x$, and add "$+C$" at the end).

Our final example shows how ***two integrations by parts*** may be needed, when the first one only simplifies the problem half way.

EXAMPLE 5   For $\int x^2 e^x\,dx$ choose $u = x^2$ and $dv = e^x dx$ (so $v = e^x$):

$$\int x^2 e^x\,dx = uv - \int v\,du = x^2 e^x - \int e^x(2x\,dx). \tag{10}$$

The last integral involves xe^x. This is better than $x^2 e^x$, but it still needs work:

$$\int xe^x dx = uv - \int v\,du = xe^x - \int e^x\,dx \quad \text{(now } u = x). \tag{11}$$

Finally e^x is alone. After *two* integrations by parts, we reach $\int e^x dx$. In equation (11), **the integral of xe^x is $xe^x - e^x$.** Substituting back into (10),

$$\int x^2 e^x\,dx = x^2 e^x - 2[xe^x - e^x] + C. \tag{12}$$

These five examples are in the list of prime candidates for integration by parts:

$$x^n e^x,\ \ x^n \sin x,\ \ x^n \cos x,\ \ x^n \ln x,\ \ e^x \sin x,\ \ e^x \cos x,\ \ \sin^{-1} x,\ \ \tan^{-1} x,\ \ldots.$$

This concludes the presentation of the method—brief and straightforward. Figure 7.1a shows how the areas $\int u\,dv$ and $\int v\,du$ fill out the difference between the big area $u(b)v(b)$ and the smaller area $u(a)v(a)$.

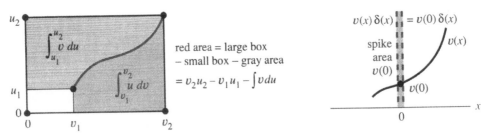

Fig. 7.1   The geometry of integration by parts. Delta function (area 1) multiplies $v(x)$ at $x=0$.

In the movie *Stand and Deliver*, the Los Angeles teacher Jaime Escalante computed $\int x^2 \sin x\,dx$ with two integrations by parts. His success was through exercises—plus insight in choosing u and v. (Notice the difference from $\int x \sin x^2\,dx$. That falls the other way—to a substitution.) The class did extremely well on the Advanced Placement Exam. If you saw the movie, you remember that the examiner didn't believe it was possible. I spoke to him long after, and he confirms that practice was the key.

THE DELTA FUNCTION

From the most familiar functions we move to the least familiar. ***The delta function is the derivative of a step function***. The step function $U(x)$ jumps from 0 to 1 at $x = 0$. We write $\delta(x) = dU/dx$, recognizing as we do it that there is no genuine derivative at the jump. The delta function is the limit of higher and higher spikes—from the "burst of speed" in Section 1.2. They approach an infinite spike concentrated at a single point (where U jumps). This "non-function" may be unconventional—*it is certainly optional*—but it is important enough to come back to.

The slope dU/dx is zero except at $x = 0$, where the step function jumps. Thus $\delta(x) = 0$ except at that one point, where the delta function has a "spike." We cannot give a value for δ at $x = 0$, but *we know its integral across the jump*. On every interval from $-A$ to A, the integral of dU/dx brings back U:

$$\int_{-A}^{A} \delta(x)\,dx = \int_{-A}^{A} \frac{dU}{dx}\,dx = U(x)\Big]_{-A}^{A} = 1. \tag{13}$$

"The area under the infinitely tall and infinitely thin spike $\delta(x)$ equals 1."

So far so good. The integral of $\delta(x)$ is $U(x)$. We now integrate by parts for a crucial purpose—*to find the area under* $v(x)\delta(x)$. This is an ordinary function times the delta function. In some sense $v(x)$ times $\delta(x)$ equals $v(0)$ times $\delta(x)$—because away from $x = 0$ the product is always zero. Thus $e^x\delta(x)$ equals $\delta(x)$, and $\sin x\,\delta(x) = 0$.

The area under $v(x)\delta(x)$ is $v(0)$—which integration by parts will prove:

> **7B** The integral of $v(x)$ times $\delta(x)$ is $\int_{-A}^{A} v(x)\delta(x)dx = v(0)$.

The area is $v(0)$ because the spike is multiplied by $v(0)$—the value of the smooth function $v(x)$ *at the spike*. But multiplying infinity is dangerous, to say the least. (Two times infinity is infinity). We cannot deal directly with the delta function. *It is only known by its integrals!* As long as the applications produce integrals (as they do), we can avoid the fact that δ is not a true function.

The integral of $v(x)\delta(x) = v(x)dU/dx$ is computed "by parts:"

$$\int_{-A}^{A} v(x)\delta(x)\,dx = v(x)U(x)\Big]_{-A}^{A} - \int_{-A}^{A} U(x)\frac{dv}{dx}\,dx. \tag{14}$$

Remember that $U = 0$ or $U = 1$. The right side of (14) is our area $v(0)$:

$$v(A)\cdot 1 - \int_{0}^{A} 1\frac{dv}{dx}\,dx = v(A) - (v(A) - v(0)) = v(0). \tag{15}$$

When $v(x) = 1$, this answer matches $\int \delta\,dx = 1$. We give three examples:

$$\int_{-2}^{2} \cos x\,\delta(x)\,dx = 1 \qquad \int_{-5}^{6}(U(x) + \delta(x))dx = 7 \qquad \int_{-1}^{1}(\delta(x))^2 dx = \infty.$$

A nightmare question occurs to me. *What is the derivative of the delta function?*

INTEGRATION BY PARTS IN ENGINEERING

Physics and engineering and economics frequently involve *products*. Work is force times distance. Power is voltage times current. Income is price times quantity. When there are several forces or currents or sales, we add the products. When there are infinitely many, we integrate (probably by parts).

I start with differential equations for the displacement u at point x in a bar:

$$-\frac{dv}{dx} = f(x) \text{ with } v(x) = k\frac{du}{dx}. \tag{16}$$

This describes a hanging bar pulled down by a force $f(x)$. Each point x moves through a distance $u(x)$. The top of the bar is fixed, so $u(0) = 0$. The stretching in the bar is du/dx. The internal force created by stretching is $v = k\,du/dx$. (This is Hooke's law.) Equation (16) is a *balance of forces* on the small piece of the bar in Figure 7.2.

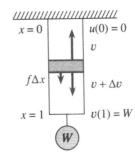

Fig. 7.2 Difference in internal force balances external force

$$-\Delta v = f\Delta x \text{ or } -dv/dx = f(x)$$

$v = W$ at $x = 1$ balances hanging weight

EXAMPLE 6 Suppose $f(x) = F$, a constant force per unit length. We can solve (16):

$$v(x) = -Fx + C \quad \text{and} \quad ku(x) = -\tfrac{1}{2}Fx^2 + Cx + D. \tag{17}$$

The constants C and D are settled at the endpoints (as usual for integrals). At $x = 0$ we are given $u = 0$ so $D = 0$. At $x = 1$ we are given $v = W$ so $C = W + F$. Then $v(x)$ and $u(x)$ give force and displacement in the bar.

To see integration by parts, multiply $-dv/dx = f(x)$ by $u(x)$ and integrate:

$$\int_0^1 f(x)u(x)\,dx = -\int_0^1 \frac{dv}{dx}u(x)\,dx = -u(x)v(x)\Big]_0^1 + \int_0^1 v(x)\frac{du}{dx}\,dx. \tag{18}$$

The left side is force times displacement, or *external work*. The last term is internal force times stretching—or *internal work*. The integrated term has $u(0) = 0$—the fixed support does no work. It also has $-u(1)W$, the work by the hanging weight. The balance of forces has been replaced by a **balance of work**.

This is a touch of engineering mathematics, and here is the main point. Integration by parts makes physical sense! When $-dv/dx = f$ is multiplied by other functions—called *test functions* or virtual displacements—then equation (18) becomes **the principle of virtual work**. It is absolutely basic to mechanics.

7.1 EXERCISES

Read-through questions

Integration by parts is the reverse of the __a__ rule. It changes $\int u\,dv$ into __b__ minus __c__. In case $u = x$ and $dv = e^{2x}\,dx$, it changes $\int xe^{2x}\,dx$ to __d__ minus __e__. The definite integral $\int_0^2 xe^{2x}\,dx$ becomes __f__ minus __g__.

In choosing u and dv, the __h__ of u and the __i__ of dv/dx should be as simple as possible. Normally $\ln x$ goes into __j__ and e^x goes into __k__. Prime candidates are $u = x$ or x^2 and $v = \sin x$ or __l__ or __m__. When $u = x^2$ we need __n__ integrations by parts. For $\int \sin^{-1} x\,dx$, the choice $dv = dx$ leads to __o__ minus __p__.

If U is the unit step function, $dU/dx = \delta$ is the unit __q__ function. The integral from $-A$ to A is $U(A) - U(-A) = $__r__. The integral of $v(x)\delta(x)$ equals __s__. The integral $\int_{-1}^1 \cos x\,\delta(x)\,dx$ equals __t__. In engineering, the balance of forces $-dv/dx = f$ is multiplied by a displacement $u(x)$ and integrated to give a balance of __u__.

Integrate 1–16, usually by parts (sometimes twice).

1　$\int x \sin x\,dx$　　　　　　2　$\int xe^{4x}\,dx$

3　$\int xe^{-x}\,dx$　　　　　　4　$\int x\cos 3x\,dx$

5　$\int x^2 \cos x\,dx$ (use Problem 1)

6　$\int x \ln x\,dx$　　　　　　7　$\int \ln(2x+1)\,dx$

8　$\int x^2 e^{4x}\,dx$ (use Problem 2)

9　$\int e^x \sin x\,dx$　　　　　10　$\int e^x \cos x\,dx$

[9 and 10 need two integrations. I think e^x can be u or v.]

11 $\int e^{ax} \sin bx \, dx$ 12 $\int xe^{-x^2} dx$

13 $\int \sin(\ln x) \, dx$ 14 $\int \cos(\ln x) \, dx$

15 $\int (\ln x)^2 dx$ 16 $\int x^2 \ln x \, dx$

17 $\int \sin^{-1} x \, dx$ 18 $\int \cos^{-1}(2x) \, dx$

19 $\int x \tan^{-1} x \, dx$

20 $\int x^2 \sin x \, dx$ (from the movie)

21 $\int x^3 \cos x \, dx$ 22 $\int x^3 \sin x \, dx$

23 $\int x^3 e^x \, dx$ 24 $\int x \sec^{-1} x \, dx$

25 $\int x \sec^2 x \, dx$ 26 $\int x \cosh x \, dx$

Compute the definite integrals 27–34.

27 $\int_0^1 \ln x \, dx$ 28 $\int_0^1 e^{\sqrt{x}} dx$ (let $u = \sqrt{x}$)

29 $\int_0^1 x e^{-2x} dx$ 30 $\int_1^e \ln(x^2) dx$

31 $\int_0^\pi x \cos x \, dx$ 32 $\int_{-\pi}^\pi x \sin x \, dx$

33 $\int_0^3 \ln(x^2 + 1) dx$ 34 $\int_0^{\pi/2} x^2 \sin x \, dx$

In 35–40 derive "reduction formulas" from higher to lower powers.

35 $\int x^n e^x \, dx = x^n e^x - n \int x^{n-1} e^x dx$

36 $\int x^n e^{ax} \, dx =$ _____

37 $\int x^n \cos x \, dx = x^n \sin x - n \int x^{n-1} \sin x \, dx$

38 $\int x^n \sin x \, dx =$ _____

39 $\int (\ln x)^n dx = x(\ln x)^n - n \int (\ln x)^{n-1} dx$

40 $\int x(\ln x)^n \, dx =$ _____

41 How would you compute $\int x \sin x \, e^x dx$ using Problem 9? Not necessary to do it.

42 How would you compute $\int x e^x \tan^{-1} x \, dx$? Don't do it.

43 (a) Integrate $\int x^3 \sin x^2 dx$ by substitution and parts.

    (b) The integral $\int x^n \sin x^2 dx$ is possible if n is _____ .

44–54 are about optional topics at the end of the section.

44 For the delta function $\delta(x)$ find these integrals:

    (a) $\int_{-1}^1 e^{2x} \delta(x) dx$ (b) $\int_{-1}^3 v(x) \delta(x) dx$ (c) $\int_2^4 \cos x \, \delta(x) dx$.

45 Solve $dy/dx = 3\delta(x)$ and $dy/dx = 3\delta(x) + y(x)$.

46 Strange fact: $\delta(2x)$ *is different from* $\delta(x)$. Integrate them both from -1 to 1.

47 The integral of $\delta(x)$ is the unit step $U(x)$. Graph the next integrals $R(x) = \int U(x) dx$ and $Q(x) = \int R(x) dx$. The ramp R and quadratic spline Q are zero at $x = 0$.

48 In $\delta(x - \frac{1}{2})$, the spike shifts to $x = \frac{1}{2}$. It is the derivative of the shifted step $U(x - \frac{1}{2})$. The integral of $v(x)\delta(x - \frac{1}{2})$ equals the value of v at $x = \frac{1}{2}$. Compute

    (a) $\int_0^1 \delta(x - \frac{1}{2}) dx$;   (b) $\int_0^1 e^x \delta(x - \frac{1}{2}) dx$;

    (c) $\int_{-1}^1 \delta(x)\delta(x - \frac{1}{2}) dx$.

49 The derivative of $\delta(x)$ is extremely singular. It is a "dipole" known by its integrals. Integrate by parts in (b) and (c):

(a) $\int_{-1}^1 \frac{d\delta}{dx} dx$ (b) $\int_{-1}^1 x \frac{d\delta}{dx} dx$ (c) $\int_{-1}^1 v(x) \frac{d\delta}{dx} dx = -v'(0)$.

50 Why is $\int_{-1}^1 U(x)\delta(x) dx$ equal to $\frac{1}{2}$? (By parts.)

51 Choose limits of integration in $v(x) = \int f(x) dx$ so that $dv/dx = -f(x)$ and $v = 0$ at $x = 1$.

52 Draw the graph of $v(x)$ if $v(1) = 0$ and $-dv/dx = f(x)$:

    (a) $f = x$;   (b) $f = U(x - \frac{1}{2})$;   (c) $f = \delta(x - \frac{1}{2})$.

53 What integral $u(x)$ solves $k \, du/dx = v(x)$ with end condition $u(0) = 0$? Find $u(x)$ for the three v's (not f's) in Problem 52, and graph the three u's.

54 Draw the graph of $\Delta U/\Delta x = [U(x + \Delta x) - U(x)]/\Delta x$. What is the area under this graph?

Problems 55–62 need more than one integration.

55 Two integrations by parts lead to $V =$ integral of v:

$$\int uv' dx = uv - Vu' + \int Vu'' dx.$$

Test this rule on $\int x^2 \sin x \, dx$.

56 After n integrations by parts, $\int u(dv/dx) dx$ becomes

$$uv - u^{(1)}v_{(1)} + u^{(2)}v_{(2)} - \cdots + (-1)^n \int u^{(n)} v_{(n-1)} dx.$$

$u^{(n)}$ is the nth derivative of u, and $v_{(n)}$ is the nth integral of v. Integrate the last term by parts to stretch this formula to $n + 1$ integrations.

57 Use Problem 56 to find $\int x^3 e^x dx$.

58 From $f(x) - f(0) = \int_0^x f'(t) dt$, integrate by parts (notice dt not dx) to reach $f(x) = f(0) + f'(0)x + \int_0^x f''(t)(x - t) dt$. Continuing as in Problem 56 produces *Taylor's formula*:

$$f(x) = f(0) + f'(0)x + \frac{1}{2!} f''(0)x^2 + \cdots + \int_0^x f^{(n+1)}(t) \frac{(x-t)^n}{n!} dt.$$

59 What is the difference between $\int_0^1 uw'' dx$ and $\int_0^1 u''w \, dx$?

60 Compute the areas $A = \int_1^e \ln x \, dx$ and $B = \int_0^1 e^y dy$. Mark them on the rectangle with corners $(0,0), (e,0), (e,1), (0,1)$.

61 *Find the mistake.* I don't believe $e^x \cosh x = e^x \sinh x$:

$$\int e^x \sinh x \, dx = e^x \cosh x - \int e^x \cosh x \, dx$$
$$= e^x \cosh x - e^x \sinh x + \int e^x \sinh x \, dx.$$

62 Choose C and D to make the derivative of $C e^{ax} \cos bx + D e^{ax} \sin bx$ equal to $e^{ax} \cos bx$. Is this easier than integrating $e^{ax} \cos bx$ twice by parts?

7.2 Trigonometric Integrals

The next section will put old integrals into new forms. For example $\int x^2\sqrt{1-x^2}\,dx$ will become $\int \sin^2\theta\cos^2\theta\,d\theta$. That looks simpler because the square root is gone. But still $\sin^2\theta\cos^2\theta$ has to be integrated. **This brief section integrates any product of sines and cosines and secants and tangents.**

There are two methods to choose from. One uses integration by parts, the other is based on trigonometric identities. Both methods try to make the integral easy (but that may take time). We follow convention by changing the letter θ back to x.

Notice that $\sin^4 x \cos x\,dx$ *is easy to integrate.* It is $u^4\,du$. This is the goal in Example 1—to separate out $\cos x\,dx$. It becomes du, and $\sin x$ is u.

EXAMPLE 1 $\int \sin^2 x\cos^3 x\,dx$ (the exponent 3 is *odd*)

Solution   Keep $\cos x\,dx$ as du. Convert the other $\cos^2 x$ to $1-\sin^2 x$:

$$\int \sin^2 x\cos^3 x\,dx = \int \sin^2 x(1-\sin^2 x)\cos x\,dx = \frac{\sin^3 x}{3} - \frac{\sin^5 x}{5} + C.$$

EXAMPLE 2 $\int \sin^5 x\,dx$ (the exponent 5 is *odd*)

Solution   Keep $\sin x\,dx$ and convert everything else to cosines. **The conversion is always based on** $\sin^2 x + \cos^2 x = 1$:

$$\int (1-\cos^2 x)^2 \sin x\,dx = \int (1-2\cos^2 x+\cos^4 x)\sin x\,dx.$$

Now $\cos x$ is u and $-\sin x\,dx$ is du. We have $\int(-1+2u^2-u^4)du$.

General method for $\int \sin^m x \cos^n x\,dx$*, when m or n is odd*

If n is odd, separate out a single $\cos x\,dx$. That leaves an even number of cosines.
   Convert them to sines. Then $\cos x\,dx$ is du and the sines are u's.
If m is odd, separate out a single $\sin x\,dx$ as du. Convert the rest to cosines.
If m and n are both odd, use either method.
If m and n are both even, a new method is needed. Here are two examples.

EXAMPLE 3 $\int \cos^2 x\,dx$ $(m=0, n=2$, both even$)$

There are two good ways to integrate $\cos^2 x$, but substitution is not one of them. If u equals $\cos x$, then du is not here. The successful methods are integration by parts and double-angle formulas. Both answers are in equation (2) below—I don't see either one as the obvious winner.

Integrating $\cos^2 x$ by parts was Example 3 of Section 7.1. The other approach, by double angles, is based on these formulas from trigonometry:

$$\cos^2 x = \tfrac{1}{2}(1+\cos 2x) \qquad \sin^2 x = \tfrac{1}{2}(1-\cos 2x) \tag{1}$$

The integral of $\cos 2x$ is $\frac{1}{2}\sin 2x$. So these formulas can be integrated directly. They give the only integrals you should memorize—either the integration by parts form, or the result from these double angles:

$$\int \cos^2 x\,dx \ \textbf{equals} \quad \tfrac{1}{2}(x+\sin x\cos x) \quad \textit{or} \quad \tfrac{1}{2}x+\tfrac{1}{4}\sin 2x \quad (\text{plus } C). \tag{2}$$
$$\int \sin^2 x\,dx \ \textbf{equals} \quad \tfrac{1}{2}(x-\sin x\cos x) \quad \textit{or} \quad \tfrac{1}{2}x-\tfrac{1}{4}\sin 2x \quad (\text{plus } C). \tag{3}$$

EXAMPLE 4 $\int \cos^4 x \, dx$ $(m = 0, n = 4,$ both are even$)$

Changing $\cos^2 x$ to $1 - \sin^2 x$ gets us nowhere. All exponents stay even. Substituting $u = \sin x$ won't simplify $\sin^4 x \, dx$, without du. Integrate by parts or switch to $2x$.

First solution Integrate by parts.   Take $u = \cos^3 x$ and $dv = \cos x \, dx$:

$$\int (\cos^3 x)(\cos x \, dx) = uv - \int v \, du = \cos^3 x \, \sin x - \int (\sin x)(-3\cos^2 x \sin x \, dx).$$

The last integral has even powers $\sin^2 x$ and $\cos^2 x$. This looks like no progress. Replacing $\sin^2 x$ by $1 - \cos^2 x$ produces $\cos^4 x$ on the right-hand side also:

$$\int \cos^4 x \, dx = \cos^3 x \, \sin x + 3 \int \cos^2 x (1 - \cos^2 x) dx.$$

Always even powers in the integrals. But now move $3 \int \cos^4 x \, dx$ to the left side:

Reduction $4 \int \cos^4 x \, dx = \cos^3 x \, \sin x + 3 \int \cos^2 x \, dx.$ (4)

Partial success—the problem is *reduced* from $\cos^4 x$ to $\cos^2 x$. Still an even power, but a lower power. The integral of $\cos^2 x$ is already known. Use it in equation (4):

$$\int \cos^4 x \, dx = \tfrac{1}{4}\cos^3 x \, \sin x + \tfrac{3}{4} \cdot \tfrac{1}{2}\left(x + \sin x \, \cos x\right) + C.$$ (5)

Second solution Substitute the double-angle formula $\cos^2 x = \tfrac{1}{2} + \tfrac{1}{2} \cos 2x$:

$$\int \cos^4 x \, dx = \int (\tfrac{1}{2} + \tfrac{1}{2}\cos 2x)^2 dx = \tfrac{1}{4} \int (1 + 2\cos 2x + \cos^2 2x) dx.$$

Certainly $\int dx = x$. Also $2 \int \cos 2x \, dx = \sin 2x$. That leaves the cosine squared:

$$\int \cos^2 2x = \int \tfrac{1}{2}(1 + \cos 4x) dx = \tfrac{1}{2}x + \tfrac{1}{8}\sin 4x + C.$$

The integral of $\cos^4 x$ using double angles is

$$\tfrac{1}{4}\left[x + \sin 2x + \tfrac{1}{2}x + \tfrac{1}{8}\sin 4x\right] + C.$$ (6)

That solution looks different from equation (5), but it can't be. There all angles were x, here we have $2x$ and $4x$. We went from $\cos^4 x$ to $\cos^2 2x$ to $\cos 4x$, which was integrated immediately. The powers were cut in half as the angle was doubled.

Double-angle method for $\int \sin^m x \, \cos^n x \, dx$, **when** m **and** n **are even.**

Replace $\sin^2 x$ by $\tfrac{1}{2}(1 - \cos 2x)$ and $\cos^2 x$ by $\tfrac{1}{2}(1 + \cos 2x)$. The exponents drop to $m/2$ and $n/2$. If those are even, repeat the idea ($2x$ goes to $4x$). If $m/2$ or $n/2$ is odd, switch to the "general method" using substitution. With an odd power, we have du.

EXAMPLE 5   (Double angle) $\int \sin^2 x \, \cos^2 x \, dx = \int \tfrac{1}{4}(1 - \cos 2x)(1 + \cos 2x) dx.$

This leaves $1 - \cos^2 2x$ in the last integral. That is familiar but not necessarily easy. We can look it up (safest) or remember it (quickest) or use double angles again:

$$\frac{1}{4}\int (1 - \cos^2 2x) dx = \frac{1}{4}\int \left(1 - \frac{1}{2} - \frac{1}{2}\cos 4x\right) dx = \frac{x}{8} - \frac{\sin 4x}{32} + C.$$

Conclusion *Every* $\sin^m x \, \cos^n x$ *can be integrated*. This includes negative m and n— see tangents and secants below. Symbolic codes like MACSYMA or *Mathematica* give the answer directly. Do they use double angles or integration by parts?

You may prefer the answer from integration by parts (I usually do). It avoids $2x$ and $4x$. But it makes no sense to go through every step every time. Either a computer does the algebra, or we use a "reduction formula" from n to $n-2$:

Reduction $n \int \cos^n x \, dx = \cos^{n-1} x \sin x + (n-1) \int \cos^{n-2} x \, dx.$ (7)

For $n=2$ this is $\int \cos^2 x \, dx$—the integral to learn. For $n=4$ the reduction produces $\cos^2 x$. The integral of $\cos^6 x$ goes to $\cos^4 x$. There are similar reduction formulas for $\sin^m x$ and also for $\sin^m x \cos^n x$. I don't see a good reason to memorize them.

INTEGRALS WITH ANGLES px AND qx

Instead of $\sin^8 x$ times $\cos^6 x$, suppose you have $\sin 8x$ times $\cos 6x$. How do you integrate? Separately a sine and cosine are easy. The new question is *the integral of the product*:

EXAMPLE 6   *Find* $\int_0^{2\pi} \sin 8x \cos 6x \, dx$.   *More generally find* $\int_0^{2\pi} \sin px \cos qx \, dx$.

This is not for the sake of making up new problems. I believe these are the most important definite integrals in this chapter (p and q are $0, 1, 2, \ldots$). They may be the most important in all of mathematics, especially because the question has such a beautiful answer. ***The integrals are zero***. On that fact rests the success of Fourier series, and the whole industry of signal processing.

One approach (the slow way) is to replace $\sin 8x$ and $\cos 6x$ by powers of cosines. That involves $\cos^{14} x$. The integration is not fun. A better approach, which applies to all angles px and qx, is to use the identity

$$\sin px \cos qx = \tfrac{1}{2} \sin(p+q)x + \tfrac{1}{2} \sin(p-q)x. \qquad (8)$$

Thus $\sin 8x \cos 6x = \tfrac{1}{2} \sin 14x + \tfrac{1}{2} \sin 2x$. Separated like that, sines are easy to integrate:

$$\int_0^{2\pi} \sin 8x \cos 6x \, dx = \left[-\frac{1}{2} \frac{\cos 14x}{14} - \frac{1}{2} \frac{\cos 2x}{2} \right]_0^{2\pi} = 0.$$

Since $\cos 14x$ is periodic, it has the same value at 0 and 2π. Subtraction gives zero. The same is true for $\cos 2x$. The integral of sine times cosine is always zero over a complete period (like 0 to 2π).

What about $\sin px \sin qx$ and $\cos px \cos qx$? Their integrals are also zero, *provided p is different from q*. When $p = q$ we have a perfect square. There is no negative area to cancel the positive area. The integral of $\cos^2 px$ or $\sin^2 px$ is π.

EXAMPLE 7 $\int_0^{2\pi} \sin 8x \sin 7x \, dx = 0$ and $\int_0^{2\pi} \sin^2 8x \, dx = \pi$.

With two sines or two cosines (instead of sine times cosine), we go back to the addition formulas of Section 1.5. Problem 24 derives these formulas:

$$\sin px \sin qx = -\tfrac{1}{2} \cos(p+q)x + \tfrac{1}{2} \cos(p-q)x \qquad (9)$$

$$\cos px \cos qx = \tfrac{1}{2} \cos(p+q)x + \tfrac{1}{2} \cos(p-q)x. \qquad (10)$$

With $p = 8$ and $q = 7$, we get $\cos 15x$ and $\cos x$. Their definite integrals are zero. With $p = 8$ and $q = 8$, we get $\cos 16x$ and $\cos 0x$ (which is 1). Formulas (9) and (10) also give a factor $\frac{1}{2}$. The integral of $\frac{1}{2}$ is π:

$$\int_0^{2\pi} \sin 8x \, \sin 7x \, dx = -\frac{1}{2}\int_0^{2\pi} \cos 15x \, dx + \frac{1}{2}\int_0^{2\pi} \cos \ x \, dx = 0 + 0$$

$$\int_0^{2\pi} \sin 8x \, \sin 8x \, dx = -\frac{1}{2}\int_0^{2\pi} \cos 16x \, dx + \frac{1}{2}\int_0^{2\pi} \cos 0x \, dx = 0 + \pi$$

The answer zero is memorable. The answer π appears constantly in Fourier series. No ordinary numbers are seen in these integrals. The case $p = q = 1$ brings back $\int \cos^2 x \, dx = \frac{1}{2} + \frac{1}{4}\sin 2x$.

SECANTS AND TANGENTS

When we allow *negative* powers m and n, the main fact remains true. All integrals $\int \sin^m x \cos^n x \, dx$ can be expressed by known functions. The novelty for negative powers is that *logarithms appear*. That happens right at the start, for $\sin x / \cos x$ and for $1/\cos x$ (tangent and secant):

$$\int \tan x \, dx = -\int du/u = -\ln|\cos x| \qquad \text{(here } u = \cos x)$$

$$\int \sec x \, dx = \ \ \int du/u = \ \ \ln|\sec x + \tan x| \quad \text{(here } u = \sec x + \tan x).$$

For higher powers there is one key identity: $1 + \tan^2 x = \sec^2 x$. That is the old identity $\cos^2 x + \sin^2 x = 1$ in disguise (just divide by $\cos^2 x$). We switch tangents to secants just as we switched sines to cosines. Since $(\tan x)' = \sec^2 x$ and $(\sec x)' = \sec x \tan x$, nothing else comes in.

EXAMPLE 8 $\int \tan^2 x \, dx = \int (\sec^2 x - 1) dx = \tan x - x + C.$

EXAMPLE 9 $\int \tan^3 x \, dx = \int \tan x (\sec^2 x - 1) dx.$

The first integral on the right is $\int u \, du = \frac{1}{2}u^2$, with $u = \tan x$. The last integral is $-\int \tan x \, dx$. The complete answer is $\frac{1}{2}(\tan x)^2 + \ln|\cos x| + C$. By taking absolute values, a negative cosine is also allowed. Avoid $\cos x = 0$.

EXAMPLE 10 *Reduction* $\displaystyle\int (\tan x)^m dx = \frac{(\tan x)^{m-1}}{m-1} - \int (\tan x)^{m-2} dx$

Same idea—separate off $(\tan x)^2$ as $\sec^2 x - 1$. Then integrate $(\tan x)^{m-2}\sec^2 x \, dx$, which is $u^{m-2} du$. This leaves the integral on the right, with the exponent lowered by 2. Every power $(\tan x)^m$ is eventually reduced to Example 8 or 9.

EXAMPLE 11 $\int \sec^3 x \, dx = uv - \int v \, du = \sec x \tan x - \int \tan^2 x \, \sec x \, dx$

This was integration by parts, with $u = \sec x$ and $v = \tan x$. In the integral on the right, replace $\tan^2 x$ by $\sec^2 x - 1$ (this identity is basic):

$$\int \sec^3 x \, dx = \sec x \tan x - \int \sec^3 x \, dx + \int \sec x \, dx.$$

Bring $\int \sec^3 x \, dx$ to the left side. That reduces the problem from $\sec^3 x$ to $\sec x$.

I believe those examples make the point—*trigonometric integrals are computable*. Every product $\tan^m x \sec^n x$ can be reduced to one of these examples. If n is even we substitute $u = \tan x$. If m is odd we set $u = \sec x$. If m is even and n is odd, use a reduction formula (and always use $\tan^2 x = \sec^2 x - 1$.)

I mention very briefly a completely different substitution $u = \tan \frac{1}{2}x$. This seems to all students and instructors (quite correctly) to come out of the blue:

$$\sin x = \frac{2u}{1+u^2} \quad \text{and} \quad \cos x = \frac{1-u^2}{1+u^2} \quad \text{and} \quad dx = \frac{2du}{1+u^2}. \quad (11)$$

The x-integral can involve sums as well as products—not only $\sin^m x \cos^n x$ but also $1/(5 + \sin x - \tan x)$. (No square roots.) The u-integral is a *ratio of ordinary polynomials*. It is done by *partial fractions*.

Application of $\int \sec x\, dx$ to distance on a map (Mercator projection)

The strange integral $\ln(\sec x + \tan x)$ has an everyday application. It measures the distance from the equator to latitude x, on a Mercator map of the world.

All mapmakers face the impossibility of putting part of a sphere onto a flat page. You can't preserve distances, when an orange peel is flattened. But angles can be preserved, and Mercator found a way to do it. His map came before Newton and Leibniz. Amazingly, and accidentally, somebody matched distances on the map with a table of logarithms—and discovered $\int \sec x\, dx$ *before calculus*. You would not be surprised to meet $\sin x$, but who would recognize $\ln(\sec x + \tan x)$?

The map starts with strips at all latitudes x. The heights are dx, the lengths are proportional to $\cos x$. **We stretch the strips by** $1/\cos x$—then Figure 7.3c lines up evenly on the page. When dx is also divided by $\cos x$, angles are preserved—a small

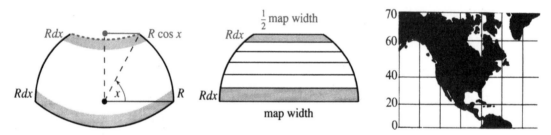

Fig. 7.3   Strips at latitude x are scaled by $\sec x$, making Greenland too large.

square becomes a bigger square. The distance north adds up the strip heights $dx/\cos x$. This gives $\int \sec x\, dx$.

The distance to the North Pole is infinite! Close to the Pole, maps are stretched totally out of shape. When sailors wanted to go from A to B at a constant angle with the North Star, they looked on Mercator's map to find the angle.

7.2 EXERCISES

Read-through questions

To integrate $\sin^4 x \cos^3 x$, replace $\cos^2 x$ by __a__. Then $(\sin^4 x - \sin^6 x)\cos x\, dx$ is __b__ du. In terms of $u = \sin x$ the integral is __c__. This idea works for $\sin^m x \cos^n x$ if either m or n is __d__.

If both m and n are __e__, one method is integration by __f__. For $\int \sin^4 x\, dx$, split off $dv = \sin x\, dx$. Then $-\int v\, du$

is __g__. Replacing $\cos^2 x$ by __h__ creates a new $\sin^4 x\, dx$ that combines with the original one. The result is a *reduction* to $\int \sin^2 x\, dx$, which is known to equal __i__.

The second method uses the double-angle formula $\sin^2 x = $ __j__. Then $\sin^4 x$ involves $\cos^2$ __k__. Another doubling comes from $\cos^2 2x = $ __l__. The integral contains the sine of __m__.

To integrate $\sin 6x \cos 4x$, rewrite it as $\frac{1}{2}\sin 10x +$ __n__. The indefinite integral is __o__. The definite integral from 0 to 2π is __p__. The product $\cos px \cos qx$ is written as $\frac{1}{2}\cos(p+q)x +$ __q__. Its integral is also zero, except if __r__ when the answer is __s__.

With $u = \tan x$, the integral of $\tan^9 x \sec^2 x$ is __t__. Similarly $\int \sec^9 x\,(\sec x \tan x\,dx) =$ __u__. For the combination $\tan^m x \sec^n x$ we apply the identity $\tan^2 x =$ __v__. After reduction we may need $\int \tan x\,dx =$ __w__ and $\int \sec x\,dx =$ __x__.

Compute 1–8 by the "general method," when m or n is odd.

1 $\int \sin^3 x\,dx$

2 $\int \cos^3 x\,dx$

3 $\int \sin x \cos x\,dx$

4 $\int \cos^5 x\,dx$

5 $\int \sin^5 x \cos^2 x\,dx$

6 $\int \sin^3 x \cos^3 x\,dx$

7 $\int \sqrt{\sin x}\,\cos x\,dx$

8 $\int \sqrt{\sin x}\,\cos^3 x\,dx$

9 Repeat Problem 6 starting with $\sin x \cos x = \frac{1}{2}\sin 2x$.

10 Find $\int \sin^2 ax \cos ax\,dx$ and $\int \sin ax \cos ax\,dx$.

In 11–16 use the double-angle formulas (m, n even).

11 $\int_0^\pi \sin^2 x\,dx$

12 $\int_0^\pi \sin^4 x\,dx$

13 $\int \cos^2 3x\,dx$

14 $\int \sin^2 x \cos^2 x\,dx$

15 $\int \sin^2 x\,dx + \int \cos^2 x\,dx$

16 $\int \sin^2 x \cos^2 2x\,dx$

17 Use the reduction formula (7) to integrate $\cos^6 x$.

18 For $n > 1$ use formula (7) to prove

$$\int_0^{\pi/2} \cos^n x\,dx = \frac{n-1}{n}\int_0^{\pi/2} \cos^{n-2}x\,dx.$$

19 For $n = 2, 4, 6, \ldots$ deduce from Problem 18 that

$$\int_0^{\pi/2} \cos^n x\,dx = \frac{\pi}{2}\frac{(1)(3)\cdots(n-1)}{(2)(4)\cdots(n)}.$$

20 For $n = 3, 5, 7, \ldots$ deduce from Problem 18 that

$$\int_0^{\pi/2} \cos^n x\,dx = \frac{(2)(4)\cdots(n-1)}{(3)(5)\cdots(n)}.$$

21 (a) Separate $dv = \sin x\,dx$ from $u = \sin^{n-1}x$ and integrate $\int \sin^n x\,dx$ by parts.

(b) Substitute $1 - \sin^2 x$ for $\cos^2 x$ to find a reduction formula like equation (7).

22 For which n does symmetry give $\int_0^\pi \cos^n x\,dx = 0$?

23 Are the integrals (a)–(f) positive, negative, or zero?

(a) $\int_0^\pi \cos 3x \sin 3x\,dx$

(b) $\int_0^\pi \cos x \sin 2x\,dx$

(c) $\int_{-2\pi}^0 \cos x \sin x\,dx$

(d) $\int_0^\pi (\cos^2 x - \sin^2 x)\,dx$

(e) $\int_\pi^{3\pi} \cos px \sin qx\,dx$

(f) $\int_\pi^0 \cos^4 x\,dx$

24 Write down equation (9) for $p = q = 1$, and (10) for $p = 2$, $q = 1$. Derive (9) from the addition formulas for $\cos(s+t)$ and $\cos(s-t)$ in Section 1.5.

In 25–32 compute the indefinite integrals first, then the definite integrals.

25 $\int_0^{2\pi} \cos x \sin 2x\,dx$

26 $\int_0^\pi \sin 3x \sin 5x\,dx$

27 $\int_0^\pi \cos 99x \sin 101x\,dx$

28 $\int_{-\pi}^\pi \cos^2 3x\,dx$

29 $\int_0^\pi \cos 99x \cos 101x\,dx$

30 $\int_0^{2\pi} \sin x \sin 2x \sin 3x\,dx$

31 $\int_0^{4\pi} \cos x/2 \sin x/2\,dx$

32 $\int_0^\pi x \cos x\,dx$ (by parts)

33 Suppose a *Fourier sine series* $A\sin x + B\sin 2x + C\sin 3x + \cdots$ adds up to x on the interval from 0 to π. Find A by multiplying all those functions (including x) by $\sin x$ and integrating from 0 to π. (B and C will disappear.)

34 Suppose a Fourier sine series $A\sin x + B\sin 2x + C\sin 3x + \cdots$ adds up to 1 on the interval from 0 to π. Find C by multiplying all functions (including 1) by $\sin 3x$ and integrating from 0 to π. (A and B will disappear.)

35 In 33, the series also equals x from $-\pi$ to 0, because all functions are odd. Sketch the "sawtooth function," which equals x from $-\pi$ to π and then has period 2π. What is the sum of the sine series at $x = \pi$?

36 In 34, the series equals -1 from $-\pi$ to 0, because sines are odd functions. Sketch the "square wave," which is alternately -1 and $+1$, and find A and B.

37 The area under $y = \sin x$ from 0 to π is positive. Which frequencies p have $\int_0^\pi \sin px\,dx = 0$?

38 Which frequencies q have $\int_0^\pi \cos qx\,dx = 0$?

39 For which p, q is $\int_0^\pi \sin px \cos qx\,dx = 0$?

40 Show that $\int_0^\pi \sin px \sin qx\,dx$ is always zero.

Compute the indefinite integrals 41–52.

41 $\int \sec x \tan x\,dx$

42 $\int \tan 5x\,dx$

43 $\int \tan^2 x \sec^2 x\,dx$

44 $\int \tan^2 x \sec x\,dx$

45 $\int \tan x \sec^3 x\,dx$

46 $\int \sec^4 x\,dx$

47 $\int \tan^4 x\,dx$

48 $\int \tan^5 x\,dx$

49 $\int \cot x\,dx$

50 $\int \csc x\,dx$

51 $\int \dfrac{\sin x}{\cos^3 x}\,dx$

52 $\int \dfrac{\sin^6 x}{\cos^3 x}\,dx$

53 Choose A so that $\cos x - \sin x = A\cos(x + \pi/4)$. Then integrate $1/(\cos x - \sin x)$.

54 Choose A so that $\cos x - \sqrt{3}\sin x = A\cos(x + \pi/3)$. Then integrate $1/(\cos x - \sqrt{3}\sin x)^2$.

55 Evaluate $\int_0^{2\pi} |\cos x - \sin x|\,dx$.

56 Show that $a \cos x + b \sin x = \sqrt{a^2 + b^2} \cos(x - \alpha)$ and find the correct *phase angle* α.

57 If a square Mercator map shows 1000 miles at latitude $30°$, how many miles does it show at latitude $60°$?

58 When lengths are scaled by $\sec x$, area is scaled by _____ . Why is the area from the equator to latitude x proportional to $\tan x$?

59 Use substitution (11) to find $\int dx/(1 + \cos x)$.

60 Explain from areas why $\int_0^\pi \sin^2 x\, dx = \int_0^\pi \cos^2 x\, dx$. These integrals add to $\int_0^\pi 1\, dx$, so they both equal _____ .

61 What product $\sin px \sin qx$ is graphed below ? Check that $(p \cos px \sin qx - q \sin px \cos qx)/(q^2 - p^2)$ has this derivative.

62 Finish $\int \sec^3 x\, dx$ in Example 11. This is needed for the length of a parabola and a spiral (Problem 7.3.8 and Sections 8.2 and 9.3).

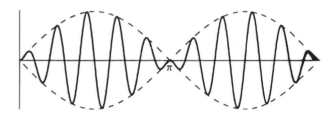

7.3 Trigonometric Substitutions

The most powerful tool we have, for integrating with pencil and paper and brain, is the *method of substitution*. To make it work, we have to think of good substitutions—which make the integral simpler. This section concentrates on the single most valuable collection of substitutions. They are the only ones you should memorize, and two examples are given immediately.

To integrate $\sqrt{1-x^2}$, substitute $x = \sin\theta$. Do not set $u = 1 - x^2$ $\left(\dfrac{du}{dx} \text{ is missing}\right)$

$$\int \sqrt{1-x^2}\,dx \rightarrow \int (\cos\theta)(\cos\theta\,d\theta) \qquad \int \frac{dx}{\sqrt{1-x^2}} \rightarrow \int \frac{\cos\theta\,d\theta}{\cos\theta}$$

The expression $\sqrt{1-x^2}$ is awkward as a function of x. It becomes graceful as a function of θ. We are practically invited to use the equation $1 - (\sin\theta)^2 = (\cos\theta)^2$. Then the square root is simply $\cos\theta$—provided this cosine is positive.

Notice the change in dx. When x is $\sin\theta$, dx is $\cos\theta\,d\theta$. Figure 7.4a shows the original area with new letters. Figure 7.4b shows an equal area, after rewriting $\int (\cos\theta)(\cos\theta\,d\theta)$ as $\int (\cos^2\theta)\,d\theta$. Changing from x to θ gives a new height and a new base. There is no change in area—that is the point of substitution.

To put it bluntly: If we go from $\sqrt{1-x^2}$ to $\cos\theta$, and forget the difference between dx and $d\theta$, and just compute $\int \cos\theta\,d\theta$, the answer is totally wrong.

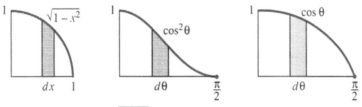

Fig. 7.4   Same area for $\sqrt{1-x^2}\,dx$ and $\cos^2\theta\,d\theta$. Third area is wrong: $dx \neq d\theta$

We still need the integral of $\cos^2\theta$. This was Example 3 of integration by parts, and also equation 7.2.6. It is worth memorizing. The example shows this θ integral, and returns to x:

EXAMPLE 1 $\int \cos^2\theta\,d\theta = \frac{1}{2}\sin\theta\cos\theta + \frac{1}{2}\theta$ is after substitution

$\int \sqrt{1-x^2}\,dx = \frac{1}{2}x\sqrt{1-x^2} + \frac{1}{2}\sin^{-1}x$ is the original problem.

We changed $\sin\theta$ back to x and $\cos\theta$ to $\sqrt{1-x^2}$. *Notice that θ is $\sin^{-1}x$.* The answer is trickier than you might expect for the area under a circular arc. Figure 7.5 shows how the two pieces of the integral are the areas of a pie-shaped wedge and a triangle.

EXAMPLE 2 $\int \dfrac{dx}{\sqrt{1-x^2}} = \int \dfrac{\cos\theta\,d\theta}{\cos\theta} = \theta + C = \sin^{-1}x + C.$

Remember: We already know $\sin^{-1}x$. Its derivative $1/\sqrt{1-x^2}$ was computed in Section 4.4. That solves the example. But instead of matching this special problem

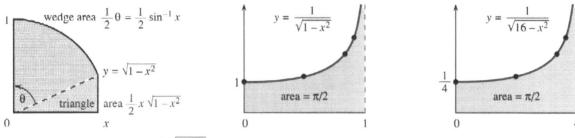

Fig. 7.5 $\int \sqrt{1-x^2}\,dx$ is a sum of simpler areas. Infinite graph but finite area.

with a memory from Chapter 4, the substitution $x = \sin\theta$ makes the solution automatic. From $\int d\theta = \theta$ we go back to $\sin^{-1} x$.

The rest of this section is about other substitutions. They are more complicated than $x = \sin\theta$ (but closely related). A table will display the three main choices— $\sin\theta, \tan\theta, \sec\theta$—and their uses.

TRIGONOMETRIC SUBSTITUTIONS

After working with $\sqrt{1-x^2}$, the next step is $\sqrt{4-x^2}$. The change $x = \sin\theta$ simplified the first, but it does nothing for the second: $4 - \sin^2\theta$ is not familiar. Nevertheless a factor of 2 makes everything work. Instead of $x = \sin\theta$, *the idea is to substitute $x = 2\sin\theta$*:

$$\sqrt{4-x^2} = \sqrt{4 - 4\sin^2\theta} = 2\cos\theta \quad \text{and} \quad dx = 2\cos\theta\,d\theta.$$

Notice both 2's. The integral is $4\int \cos^2\theta\,d\theta = 2\sin\theta\cos\theta + 2\theta$. But watch closely. This is not 4 times the previous $\int \cos^2\theta\,d\theta$! *Since x is $2\sin\theta$, θ is now $\sin^{-1}(x/2)$.*

EXAMPLE 3 $\int \sqrt{4-x^2}\,dx = 4\int \cos^2\theta\,d\theta = x\sqrt{1 - (x/2)^2} + 2\sin^{-1}(x/2).$

Based on $\sqrt{1-x^2}$ and $\sqrt{4-x^2}$, here is the general rule for $\sqrt{a^2 - x^2}$. Substitute $x = a\sin\theta$. Then the a's separate out:

$$\sqrt{a^2 - x^2} = \sqrt{a^2 - a^2\sin^2\theta} = a\cos\theta \quad \text{and} \quad dx = a\cos\theta\,d\theta.$$

That is the automatic substitution to try, whenever the square root appears.

EXAMPLE 4 $\displaystyle\int_{x=0}^{4} \frac{dx}{\sqrt{16-x^2}} = \int_{\theta=0}^{\pi/2} \frac{4\cos\theta\,d\theta}{\sqrt{4^2 - 4^2(\sin\theta)^2}} = \int_{\theta=0}^{\pi/2} d\theta = \frac{\pi}{2}.$

Here $a^2 = 16$. Then $a = 4$ and $x = 4\sin\theta$. The integral has $4\cos\theta$ above and below, so it is $\int d\theta$. The antiderivative is just θ. For the definite integral notice that $x = 4$ means $\sin\theta = 1$, and this means $\theta = \pi/2$.

A table of integrals would hide that substitution. The table only gives $\sin^{-1}(x/4)$. There is no mention of $\int d\theta = \theta$. *But what if $16 - x^2$ changes to $x^2 - 16$?*

EXAMPLE 5 $\displaystyle\int_{x=4}^{8} \frac{dx}{\sqrt{x^2 - 16}} = \,?$

Notice the two changes—the sign in the square root and the limits on x. Example 4 stayed *inside* the interval $|x| \leqslant 4$, where $16 - x^2$ has a square root. Example 5 stays *outside*, where $x^2 - 16$ has a square root. The new problem cannot use $x = 4 \sin \theta$, because we don't want the square root of $-\cos^2\theta$.

The new substitution is $x = 4 \sec \theta$. This turns the square root into $4 \tan \theta$:

$$x = 4 \sec \theta \text{ gives } dx = 4 \sec \theta \tan \theta \, d\theta \text{ and } x^2 - 16 = 16 \sec^2\theta - 16 = 16 \tan^2\theta.$$

This substitution solves the example, when the limits are changed to θ:

$$\int_0^{\pi/3} \frac{4 \sec \theta \tan d\theta}{4 \tan \theta} = \int_0^{\pi/3} \sec \theta \, d\theta = \ln(\sec \theta + \tan \theta) \Big]_0^{\pi/3} = \ln(2 + \sqrt{3}).$$

I want to emphasize the three steps. First came the substitution $x = 4 \sec \theta$. An unrecognizable integral became $\int \sec \theta \, d\theta$. Second came the new limits ($\theta = 0$ when $x = 4$, $\theta = \pi/3$ when $x = 8$). Then I integrated $\sec \theta$.

Example 6 has the same $x^2 - 16$. So the substitution is again $x = 4 \sec \theta$:

EXAMPLE 6
$$\int_{x=8}^{\infty} \frac{16 \, dx}{(x^2 - 16)^{3/2}} = \int_{\theta=\pi/3}^{\pi/2} \frac{64 \sec \theta \tan \theta \, d\theta}{(4 \tan \theta)^3} = \int_{\pi/3}^{\pi/2} \frac{\cos \theta \, d\theta}{\sin^2\theta}.$$

Step one substitutes $x = 4 \sec \theta$. Step two changes the limits to θ. The upper limit $x = \infty$ becomes $\theta = \pi/2$, where the secant is infinite. The limit $x = 8$ again means $\theta = \pi/3$. To get a grip on the integral, I also changed to sines and cosines.

The integral of $\cos \theta/\sin^2\theta$ needs another substitution! (Or else recognize $\cot \theta \csc \theta$.) With $u = \sin \theta$ we have $\int du/u^2 = -1/u = -1/\sin \theta$:

Solution
$$\int_{\pi/3}^{\pi/2} \frac{\cos \theta \, d\theta}{\sin^2\theta} = \frac{-1}{\sin \theta} \Big]_{\pi/3}^{\pi/2} = -1 + \frac{2}{\sqrt{3}}.$$

Warning With lower limit $\theta = 0$ (or $x = 4$) this integral would be a disaster. It divides by $\sin 0$, which is zero. *This area is infinite.*

(Warning)^2 Example 5 also blew up at $x = 4$, but the area was *not* infinite. To make the point directly, compare $x^{-1/2}$ to $x^{-3/2}$. Both blow up at $x = 0$, but the first one has finite area:

$$\int_0^1 \frac{1}{\sqrt{x}} \, dx = 2\sqrt{x} \Big]_0^1 = 2 \qquad \int_0^1 \frac{1}{x^{3/2}} \, dx = \frac{-2}{\sqrt{x}} \Big]_0^1 = \infty.$$

Section 7.5 separates finite areas (slow growth of $1/\sqrt{x}$) from infinite areas (fast growth of $x^{-3/2}$).

Last substitution Together with $16 - x^2$ and $x^2 - 16$ comes the possibility $16 + x^2$. (You might ask about $-16 - x^2$, but for obvious reasons we don't take its square root.) This third form $16 + x^2$ requires a third substitution $x = 4 \tan \theta$. Then $16 + x^2 = 16 + 16 \tan^2\theta = 16 \sec^2\theta$. Here is an example:

EXAMPLE 7
$$\int_{x=0}^{\infty} \frac{dx}{16 + x^2} = \int_{\theta=0}^{\pi/2} \frac{4 \sec^2\theta \, d\theta}{16 \sec^2\theta} = \frac{1}{4}\theta \Big]_0^{\pi/2} = \frac{\pi}{8}.$$

Table of substitutions for $a^2 - x^2$, $a^2 + x^2$, $x^2 - a^2$

$x = a \sin \theta$ replaces $a^2 - x^2$ by $a^2\cos^2\theta$ and dx by $a \cos \theta \, d\theta$

$x = a \tan \theta$ replaces $a^2 + x^2$ by $a^2\sec^2\theta$ and dx by $a \sec^2\theta \, d\theta$

$x = a \sec \theta$ replaces $x^2 - a^2$ by $a^2\tan^2\theta$ and dx by $a \sec \theta \tan \theta \, d\theta$

Note There is a subtle difference between changing x to $\sin\theta$ and changing $\sin\theta$ to u:

in Example 1, dx was replaced by $\cos\theta\,d\theta$(new method)

in Example 6, $\cos\theta\,d\theta$ was already there and became du (old method).

The combination $\cos\theta\,d\theta$ was put into the first and pulled out of the second.

My point is that Chapter 5 needed du/dx inside the integral. Then $(du/dx)dx$ became du. ***Now it is not necessary to see so far ahead***. We can try any substitution. If it works, we win. In this section, $x=\sin\theta$ or $\sec\theta$ or $\tan\theta$ is bound to succeed.

$$\textbf{NEW }\int\frac{dx}{1+x^2}=\int d\theta \text{ by trying } x=\tan\theta \qquad \textbf{OLD }\int\frac{x\,dx}{1+x^2}=\int\frac{du}{2u}\text{ by seeing }du$$

We mention the ***hyperbolic substitutions*** $\tanh\theta, \sinh\theta,$ and $\cosh\theta$. The table below shows their use. They give new forms for the same integrals. If you are familiar with hyperbolic functions the new form might look simpler—as it does in Example 8.

$x=a\tanh\theta$	replaces	a^2-x^2	by	$a^2\operatorname{sech}^2\theta$	and	dx	by	$a\operatorname{sech}^2\theta\,d\theta$
$x=a\sinh\theta$	replaces	a^2+x^2	by	$a^2\cosh^2\theta$	and	dx	by	$a\cosh\theta\,d\theta$
$x=a\cosh\theta$	replaces	x^2-a^2	by	$a^2\sinh^2\theta$	and	dx	by	$a\sinh\theta\,d\theta$

EXAMPLE 8 $$\int\frac{dx}{\sqrt{x^2-1}}=\int\frac{\sinh\theta\,d\theta}{\sinh\theta}=\theta+C=\cosh^{-1}x+C.$$

$\int d\theta$ is simple. The bad part is $\cosh^{-1}x$ at the end. Compare with $x=\sec\theta$:

$$\int\frac{dx}{\sqrt{x^2-1}}=\int\frac{\sec\theta\tan\theta\,d\theta}{\tan\theta}=\ln(\sec\theta+\tan\theta)+C=\ln(x+\sqrt{x^2-1})+C.$$

This way looks harder, but most tables prefer that final logarithm. It is clearer than $\cosh^{-1}x$, even if it takes more space. All answers agree if Problem 35 is correct.

COMPLETING THE SQUARE

We have not said what to do for $\sqrt{x^2-2x+2}$ or $\sqrt{-x^2+2x}$. Those square roots contain a *linear term*—a multiple of x. The device for removing linear terms is worth knowing. It is called ***completing the square***, and two examples will begin to explain it:

$$x^2-2x+2=(x-1)^2+1=u^2+1$$
$$-x^2+2x=-(x-1)^2+1=1-u^2.$$

The idea has three steps. First, get the x^2 and x terms into one square. Here that square was $(x-1)^2=x^2-2x+1$. Second, fix up the constant term. Here we recover the original functions by adding 1. Third, set $u=x-1$ to leave no linear term. Then the integral goes forward based on the substitutions of this section:

$$\int\frac{dx}{\sqrt{x^2-2x+1}}=\int\frac{du}{\sqrt{u^2+1}} \qquad \int\frac{dx}{\sqrt{2x-x^2}}=\int\frac{du}{\sqrt{1-u^2}}$$

The same idea applies to any quadratic that contains a linear term $2bx$:

rewrite $x^2 + 2bx + c$ as $(x+b)^2 + C$, with $C = c - b^2$

rewrite $-x^2 + 2bx + c$ as $-(x-b)^2 + C$, with $C = c + b^2$

To match the quadratic with the square, we fix up the constant:

$$x^2 + 10x + 16 = (x+5)^2 + C \text{ leads to } C = 16 - 25 = -9$$
$$-x^2 + 10x + 16 = -(x-5)^2 + C \text{ leads to } C = 16 + 25 = 41.$$

EXAMPLE 9 $$\int \frac{dx}{x^2 + 10x + 16} = \int \frac{dx}{(x+5)^2 - 9} = \int \frac{du}{u^2 - 9}.$$

Here $u = x + 5$ and $du = dx$. Now comes a choice—struggle on with $u = 3 \sec \theta$ or look for $\int du/(u^2 - a^2)$ inside the front cover. Then set $a = 3$:

$$\int \frac{du}{u^2 - 9} = \frac{1}{6} \ln \left| \frac{u-3}{u+3} \right| = \frac{1}{6} \ln \left| \frac{x+2}{x+8} \right|.$$

Note   If the quadratic starts with $5x^2$ or $-5x^2$, factor out the 5 first:

$$5x^2 - 10x + 25 = 5(x^2 - 2x + 5) = (\text{complete the square}) = 5[(x-1)^2 + 4].$$

Now $u = x - 1$ produces $5[u^2 + 4]$. This is ready for table lookup or $u = 2 \tan \theta$:

EXAMPLE 10 $$\int \frac{dx}{5x^2 - 10x + 25} = \int \frac{du}{5[u^2 + 4]} = \int \frac{2 \sec^2 \theta \, d\theta}{5[4 \sec^2 \theta]} = \frac{1}{10} \int d\theta.$$

This answer is $\theta/10 + C$. Now go backwards: $\theta/10 = (\tan^{-1} \frac{1}{2} u)/10 = (\tan^{-1} \frac{1}{2}(x - 1))/10$. Nobody could see that from the start. A double substitution takes practice, from x to u to θ. Then go backwards from θ to u to x.

Final remark   For $u^2 + a^2$ we substitute $u = a \tan \theta$. For $u^2 - a^2$ we substitute $u = a \sec \theta$. This big dividing line depends on whether the constant C (after completing the square) is positive or negative. We either have $C = a^2$ or $C = -a^2$. The same dividing line in the original $x^2 + 2bx + c$ is between $c > b^2$ and $c < b^2$. In between, $c = b^2$ yields the perfect square $(x+b)^2$— and no trigonometric substitution at all.

7.3 EXERCISES

Read-through questions

The function $\sqrt{1-x^2}$ suggests the substitution $x = \underline{\quad a \quad}$. The square root becomes $\underline{\quad b \quad}$ and dx changes to $\underline{\quad c \quad}$. The integral $\int (1-x^2)^{3/2} dx$ becomes $\int \underline{\quad d \quad} d\theta$. The interval $\frac{1}{2} \leqslant x \leqslant 1$ changes to $\underline{\quad e \quad} \leqslant \theta \leqslant \underline{\quad f \quad}$.

For $\sqrt{a^2 - x^2}$ the substitution is $x = \underline{\quad g \quad}$ with $dx = \underline{\quad h \quad}$. For $x^2 - a^2$ we use $x = \underline{\quad i \quad}$ with $dx = \underline{\quad j \quad}$. Then $\int dx/(1 + x^2)$ becomes $\int d\theta$, because $1 + \tan^2 \theta = \underline{\quad k \quad}$. The answer is $\theta = \tan^{-1} x$. We already knew that $\underline{\quad l \quad}$ is the derivative of $\tan^{-1} x$.

The quadratic $x^2 + 2bx + c$ contains a $\underline{\quad m \quad}$ term $2bx$. To remove it we $\underline{\quad n \quad}$ the square. This gives $(x+b)^2 + C$ with $C = \underline{\quad o \quad}$. The example $x^2 + 4x + 9$ becomes $\underline{\quad p \quad}$. Then $u = x + 2$. In case x^2 enters with a minus sign, $-x^2 + 4x + 9$ becomes $(\underline{\quad q \quad})^2 + \underline{\quad r \quad}$. When the quadratic contains $4x^2$, start by factoring out $\underline{\quad s \quad}$.

Integrate 1–20 by substitution. Change θ back to x.

1   $\displaystyle \int \frac{dx}{\sqrt{4 - x^2}}$ 2 $\displaystyle \int \frac{dx}{\sqrt{x^2 - a^2}}$

3 $\displaystyle\int \sqrt{4-x^2}\,dx$

4 $\displaystyle\int \frac{dx}{\sqrt{1+9x^2}}$

5 $\displaystyle\int \frac{x^2\,dx}{\sqrt{1-x^2}}$

6 $\displaystyle\int \frac{dx}{x^2\sqrt{1-x^2}}$

7 $\displaystyle\int \frac{dx}{(1+x^2)^2}$

8 $\displaystyle\int \sqrt{x^2+a^2}\,dx$ (see 7.2.62)

9 $\displaystyle\int \frac{\sqrt{x^2-25}}{x}\,dx$

10 $\displaystyle\int \frac{x^3\,dx}{\sqrt{9-x^2}}$

11 $\displaystyle\int \frac{dx}{\sqrt{x^6-x^4}}$

12 $\displaystyle\int \sqrt{x^6-x^8}\,dx$

13 $\displaystyle\int \frac{dx}{(1+x^2)^{3/2}}$

14 $\displaystyle\int \frac{dx}{(1-x^2)^{3/2}}$

15 $\displaystyle\int \frac{dx}{(x^2-9)^{3/2}}$

*16 $\displaystyle\int \frac{\sqrt{1+x^2}\,dx}{x}$

*17 $\displaystyle\int \frac{x^2\,dx}{\sqrt{x^2-1}}$

18 $\displaystyle\int \frac{x^2\,dx}{x^2+4}$

19 $\displaystyle\int \frac{dx}{x^2\sqrt{x^2+1}}$

*20 $\displaystyle\int \frac{x^2\,dx}{\sqrt{1+x^2}}$

21 (Important) This section started with $x=\sin\theta$ and

$$\int dx/\sqrt{1-x^2}=\int d\theta=\theta=\sin^{-1}x.$$

(a) Use $x=\cos\theta$ to get a different answer.

(b) How can the same integral give two answers?

22 Compute $\int dx/x\sqrt{x^2-1}$ with $x=\sec\theta$. Recompute with $x=\csc\theta$. How can both answers be correct?

23 Integrate $x/(x^2+1)$ with $x=\tan\theta$, and also directly as a logarithm. Show that the results agree.

24 Show that $\int dx/x\sqrt{x^4-1}=\frac{1}{2}\sec^{-1}(x^2)$.

Calculate the definite integrals 25–32.

25 $\displaystyle\int_{-a}^{a} \sqrt{a^2-x^2}\,dx=$ area of _____

26 $\displaystyle\int_{-1}^{1} (1-x^2)^{3/2}\,dx$

27 $\displaystyle\int_{.5}^{1} \frac{dx}{\sqrt{1-x^2}}$

28 $\displaystyle\int_{1}^{4} \frac{dx}{\sqrt{x^2-1}}$

29 $\displaystyle\int_{2}^{\infty} \frac{dx}{(x^2-1)^{3/2}}$

30 $\displaystyle\int_{-1}^{1} \frac{x\,dx}{x^2+1}$

31 $\displaystyle\int_{-\infty}^{\infty} \frac{dx}{x^2+9}$

32 $\displaystyle\int_{1/2}^{1} \sqrt{1-x^2}\,dx=$ area of _____.

33 Combine the integrals to prove the reduction formula ($n\neq 0$):

$$\int \frac{x^{n+1}}{x^2+1}\,dx=\frac{x^n}{n}-\int \frac{x^{n-1}}{x^2+1}\,dx.$$

34 Integrate $1/\cos x$ and $1/(1+\cos x)$ and $\sqrt{1+\cos x}$.

35 (a) $x=$ _____ gives $\int dx/\sqrt{x^2-1}=\ln(\sec\theta+\tan\theta)$.
(b) From the triangle, this answer is $f=\ln(x+\sqrt{x^2-1})$. Check that $df/dx=1/\sqrt{x^2-1}$.
(c) Verify that $\cosh f=\frac{1}{2}(e^f+e^{-f})=x$. Then $f=\cosh^{-1}x$, the answer in Example 8.

36 (a) $x=$ _____ gives $\int dx/\sqrt{x^2+1}=\ln(\sec\theta+\tan\theta)$.
(b) The second triangle converts this answer to $g=\ln(x+\sqrt{x^2+1})$. Check that $dg/dx=1/\sqrt{x^2+1}$.
(c) Verify that $\sinh g=\frac{1}{2}(e^g-e^{-g})=x$ so $g=\sinh^{-1}x$.
(d) Substitute $x=\sinh g$ directly into $\int dx/\sqrt{x^2+1}$ and integrate.

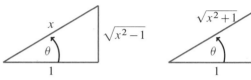

In 37–42 substitute $x=\sinh\theta$, $\cosh\theta$, **or** $\tanh\theta$. **After integration change back to** x.

37 $\displaystyle\int \frac{dx}{\sqrt{x^2-1}}$

38 $\displaystyle\int \frac{dx}{x\sqrt{1-x^2}}$

39 $\displaystyle\int \sqrt{x^2-1}\,dx$

40 $\displaystyle\int \frac{\sqrt{x^2-1}}{x^2}\,dx$

41 $\displaystyle\int \frac{dx}{1-x^2}$

42 $\displaystyle\int \frac{\sqrt{1+x^2}}{x^2}\,dx$

Rewrite 43–48 as $(x+b)^2+C$ **or** $-(x-b)^2+C$ **by completing the square.**

43 x^2-4x+8

44 $-x^2+2x+8$

45 x^2-6x

46 $-x^2+10$

47 x^2+2x+1

48 $x^2+4x-12$

49 For the three functions $f(x)$ in Problems 43, 45, 47 integrate $1/f(x)$.

50 For the three functions $g(x)$ in Problems 44, 46, 48 integrate $1/\sqrt{g(x)}$.

51 For $\int dx/(x^2+2bx+c)$ why does the answer have different forms for $b^2>c$ and $b^2<c$? What is the answer if $b^2=c$?

52 What substitution $u=x+b$ or $u=x-b$ will remove the linear term?

(a) $\displaystyle\int \frac{dx}{x^2-4x+c}$

(b) $\displaystyle\int \frac{dx}{3x^2+6x}$

(c) $\displaystyle\int \frac{dx}{-x^2+10x+c}$

(d) $\displaystyle\int \frac{dx}{2x^2-x}$

53 *Find the mistake.* With $x=\sin\theta$ and $\sqrt{1-x^2}=\cos\theta$, substituting $dx=\cos\theta\,d\theta$ changes

$$\int_0^{2\pi} \cos^2\theta\,d\theta \quad \text{into} \quad \int_0^0 \sqrt{1-x^2}\,dx.$$

54 (a) If $x = \tan\theta$ then $\int \sqrt{1+x^2}\,dx = \int$ _____ $d\theta$.

(b) Convert $\frac{1}{2}[\sec\theta\,\tan\theta + \ln(\sec\theta + \tan\theta)]$ back to x.

(c) If $x = \sinh\theta$ then $\int \sqrt{1+x^2}\,dx = \int$ _____ $d\theta$.

(d) Convert $\frac{1}{2}[\sinh\theta\,\cosh\theta + \theta]$ back to x.

These answers agree. In Section 8.2 they will give the length of a parabola. Compare with Problem 7.2.62.

55 Rescale x and y in Figure 7.5b to produce the equal area $\int y\,dx$ in Figure 7.5c. What happens to y and what happens to dx?

56 Draw $y = 1/\sqrt{1-x^2}$ and $y = 1/\sqrt{16-x^2}$ to the same scale (1″ across and up; 4″ across and $\frac{1}{4}$″ up).

57 What is wrong, if anything, with

$$\int_0^2 \frac{dx}{\sqrt{1-x^2}} = \sin^{-1}2\,?$$

7.4 Partial Fractions

This section is about rational functions $P(x)/Q(x)$. Sometimes their integrals are also rational functions (ratios of polynomials). More often they are not. It is very common for the integral of P/Q to involve logarithms. We meet logarithms immediately in the simple case $1/(x-2)$, whose integral is $\ln|x-2|+C$. We meet them again in a sum of simple cases:

$$\int \left[\frac{1}{x-2}+\frac{3}{x+2}-\frac{4}{x}\right] dx = \ln|x-2|+3\ln|x+2|-4\ln|x|+C.$$

Our plan is to split P/Q into a sum like this—and integrate each piece.

Which rational function produced that particular sum ? It was

$$\frac{1}{x-2}+\frac{3}{x+2}-\frac{4}{x} = \frac{(x+2)(x)+3(x-2)(x)-4(x-2)(x+2)}{(x-2)(x+2)(x)} = \frac{-4x+16}{(x-2)(x+2)(x)}.$$

This is P/Q. It is a ratio of polynomials, degree 1 over degree 3. The pieces of P are collected into $-4x+16$. The common denominator $(x-2)(x+2)(x)=x^3-4x$ is Q. But I kept these factors separate, for the following reason. When we start with P/Q, and break it into a sum of pieces, *the first things we need are the factors of Q*.

In the standard problem P/Q is given. To integrate it, we break it up. The goal of partial fractions is to find the pieces—*to prepare for integration*. That is the technique to learn in this section, and we start right away with examples.

EXAMPLE 1   Suppose P/Q has the same Q but a different numerator P:

$$\frac{P}{Q} = \frac{3x^2+8x-4}{(x-2)(x+2)(x)} = \frac{A}{x-2}+\frac{B}{x+2}+\frac{C}{x}. \tag{1}$$

Notice the form of those pieces! They are the "***partial fractions***" that add to P/Q. Each one is a constant divided by a factor of Q. We know the factors $x-2$ and $x+2$ and x. We don't know the constants A,B,C. In the previous case they were $1,3,-4$. In this and other examples, there are two ways to find them.

Method **1** (slow) Put the right side of (1) over the common denominator Q:

$$\frac{3x^2+8x-4}{Q} = \frac{A(x+2)(x)+B(x-2)(x)+C(x-2)(x+2)}{(x-2)(x+2)(x)} \tag{2}$$

Why is A multiplied by $(x+2)(x)$? Because canceling those factors will leave $A/(x-2)$ as in equation (1). Similarly we have $B/(x+2)$ and C/x. **Choose the numbers A,B,C so that the numerators match**. As soon as they agree, the splitting is correct.

Method **2** (quicker) Multiply equation (1) by $x-2$. That leaves a space:

$$\frac{3x^2+8x-4}{(x+2)(x)} = A+\frac{B(x-2)}{x+2}+\frac{C(x-2)}{x}. \tag{3}$$

Now set $x=2$ and immediately you have A. The last two terms of (3) are zero, because $x-2$ is zero when $x=2$. On the left side, $x=2$ gives

$$\frac{3(2)^2+8(2)-4}{(2+2)(2)} = \frac{24}{8} = 3. \quad \text{(which is A)}.$$

Notice how multiplying by $x - 2$ produced a hole on the left side. Method **2** is the "***cover-up method***." *Cover up $x - 2$ and then substitute $x = 2$.* The result is $3 = A + 0 + 0$, just what we wanted.

In Method **1**, the numerators of equation (2) must agree. The factors that multiply B and C are again zero at $x = 2$. That leads to the same A—but the cover-up method avoids the unnecessary step of writing down equation (2).

Calculation of B Multiply equation (1) by $x + 2$, which covers up the $(x + 2)$:

$$\frac{3x^2 + 8x - 4}{(x - 2)\quad(x)} = \frac{A(x + 2)}{x - 2} + B + \frac{C(x + 2)}{x}. \tag{4}$$

Now set $x = -2$, so A and C are multiplied by zero:

$$\frac{3(-2)^2 + 8(-2) - 4}{(-2 - 2)\quad(-2)} = \frac{-8}{8} = -1 = B.$$

This is almost full speed, but (4) was not needed. *Just cover up in Q and give x the right value* (which makes the covered factor zero).

*Calculation of C (**quickest**)* In equation (1), cover up the factor (x) and set $x = 0$:

$$\frac{3(0)^2 + 8(0) - 4}{(0 - 2)(0 + 2)} = \frac{-4}{-4} = 1 = C. \tag{5}$$

To repeat: The same result $A = 3$, $B = -1$, $C = 1$ comes from Method **1**.

EXAMPLE 2 $\qquad \dfrac{x + 2}{(x - 1)(x + 3)} = \dfrac{A}{x - 1} + \dfrac{B}{x + 3}.$

First cover up $(x - 1)$ on the left and set $x = 1$. Next cover up $(x + 3)$ and set $x = -3$:

$$\frac{1 + 2}{(\quad)(1 + 3)} = \frac{3}{4} = A \qquad \frac{-3 + 2}{(-3 - 1)(\quad)} = \frac{-1}{-4} = B.$$

The integral is $\frac{3}{4} \ln|x - 1| + \frac{1}{4} \ln|x + 3| + C$.

EXAMPLE 3 This was needed for the logistic equation in Section 6.5:

$$\frac{1}{y(c - by)} = \frac{A}{y} + \frac{B}{c - by}. \tag{6}$$

First multiply by y. That covers up y in the first two terms and changes B to By. Then set $y = 0$. The equation becomes $1/c = A$.

To find B, multiply by $c - by$. That covers up $c - by$ in the outside terms. In the middle, A times $c - by$ will be zero at $y = c/b$. That leaves B on the right equal to $1/y = b/c$ on the left. Then $A = 1/c$ and $B = b/c$ give the integral announced in Equation 6.5.9:

$$\int \frac{dy}{cy - by^2} = \int \frac{dy}{cy} + \int \frac{b\,dy}{c(c - by)} = \frac{\ln y}{c} - \frac{\ln(c - by)}{c}. \tag{7}$$

It is time to admit that the general method of partial fractions can be very awkward. First of all, it requires the factors of the denominator Q. When Q is a quadratic $ax^2 + bx + c$, we can find its roots and its factors. In theory a cubic or a quartic can also be factored, but in practice only a few are possible—for example $x^4 - 1$ is $(x^2 - 1)(x^2 + 1)$. Even for this good example, *two of the roots are imaginary*. We can split $x^2 - 1$ into $(x + 1)(x - 1)$. We cannot split $x^2 + 1$ without introducing i.

The method of partial fractions can work directly with $x^2 + 1$, as we now see.

EXAMPLE 4 $\displaystyle\int \frac{3x^2 + 2x + 7}{x^2 + 1}\, dx$ (a quadratic over a quadratic).

This has another difficulty. The degree of P equals the degree of $Q(= 2)$. *Partial fractions cannot start until P has lower degree*. Therefore I divide the leading term x^2 into the leading term $3x^2$. That gives 3, which is separated off by itself:

$$\frac{3x^2 + 2x + 7}{x^2 + 1} = 3 + \frac{2x + 4}{x^2 + 1}. \tag{8}$$

Note how 3 really used $3x^2 + 3$ from the original numerator. That left $2x + 4$. *Partial fractions will accept a linear factor $2x + 4$ (or $Ax + B$, not just A) above a quadratic*.

This example contains $2x/(x^2 + 1)$, which integrates to $\ln(x^2 + 1)$. The final $4/(x^2 + 1)$ integrates to $4\tan^{-1} x$. When the denominator is $x^2 + x + 1$ we complete the square before integrating. The point of Sections 7.2 and 7.3 was to make that integration possible. This section gets the fraction ready—in parts.

The essential point is that we never have to go higher than quadratics. *Every denominator Q can be split into linear factors and quadratic factors*. There is no magic way to find those factors, and most examples begin by giving them. They go into their own fractions, and they have their own numerators—which are the A and B and $2x + 4$ we have been computing.

The one remaining question is *what to do if a factor is repeated*. This happens in Example 5.

EXAMPLE 5 $\displaystyle\frac{2x + 3}{(x - 1)^2} = \frac{A}{(x - 1)} + \frac{B}{(x - 1)^2}.$

The key is the new term $B/(x - 1)^2$. That is the right form to expect. With $(x - 1)$ $(x - 2)$ this term would have been $B/(x - 2)$. But when $(x - 1)$ is repeated, something new is needed. To find B, *multiply through by $(x - 1)^2$ and set $x = 1$*:

$$2x + 3 = A(x - 1) + B \quad \text{becomes} \quad 5 = B \quad \text{when} \quad x = 1.$$

This cover-up method gives B. Then $A = 2$ is easy, and the integral is $2\ln|x - 1| - 5/(x - 1)$. The fraction $5/(x - 1)^2$ has an integral without logarithms.

EXAMPLE 6 $\displaystyle\frac{2x^3 + 9x^2 + 4}{x^2(x^2 + 4)(x - 1)} = \frac{A}{x} + \frac{B}{x^2} + \frac{Cx + D}{x^2 + 4} + \frac{E}{x - 1}.$

This final example has almost everything! It is more of a game than a calculus problem. In fact calculus doesn't enter until we integrate (and nothing is new there). Before computing A, B, C, D, E, we write down the overall rules for partial fractions:

1. The degree of P must be less than the degree of Q. Otherwise divide their leading terms as in equation (8) to lower the degree of P. Here $3 < 5$.
2. Expect the fractions illustrated by Example 6. The linear factors x and $x + 1$ (*and the repeated x^2*) are underneath constants. The quadratic $x^2 + 4$ is under a linear term. A repeated $(x^2 + 4)^2$ would be under a new $Fx + G$.

3. Find the numbers $A, B, C, \ldots$ by any means, including cover-up.

4. Integrate each term separately and add.

We could prove that this method always works. It makes better sense to show that it works once, in Example 6.

To find E, cover up $(x-1)$ on the left and substitute $x=1$. Then $E=3$. To find B, cover up x^2 on the left and set $x=0$. Then $B=4/(0+4)(0-1)=-1$. The cover-up method has done its job, and there are several ways to find A, C, D. Compare the numerators, after multiplying through by the common denominator Q:

$$2x^3+9x^2+4 = Ax(x^2+4)(x-1)-(x^2+4)(x-1)+(Cx+D)(x^2)(x-1)+3x^2(x^2+4).$$

The known terms on the right, from $B=-1$ and $E=3$, can move to the left:

$$-3x^4+3x^3-4x^2+4x = Ax(x^2+4)(x-1)+(Cx+D)x^2(x-1).$$

We can divide through by x and $x-1$, which checks that B and E were correct:

$$-3x^2-4 = A(x^2+4)+(Cx+D)x.$$

Finally $x=0$ yields $A=-1$. This leaves $-2x^2=(Cx+D)x$. Then $C=-2$ and $D=0$.

You should never have to do such a problem! I never intend to do another one. It completely depends on expecting the right form and matching the numerators. They could also be matched by comparing coefficients of $x^4, x^3, x^2, x, 1$—to give five equations for A, B, C, D, E. That is an invitation to human error. Cover-up is the way to start, and usually the way to finish. With repeated factors and quadratic factors, match numerators at the end.

7.4 EXERCISES

Read-through questions

The idea of __a__ fractions is to express $P(x)/Q(x)$ as a __b__ of simpler terms, each one easy to integrate. To begin, the degree of P should be __c__ the degree of Q. Then Q is split into __d__ factors like $x-5$ (possibly repeated) and quadratic factors like x^2+x+1 (possibly repeated). The quadratic factors have two __e__ roots, and do not allow real linear factors.

A factor like $x-5$ contributes a fraction $A/$__f__. Its integral is __g__. To compute A, cover up __h__ in the denominator of P/Q. Then set $x=$__i__, and the rest of P/Q becomes A. An equivalent method puts all fractions over a common denominator (which is __j__). Then match the __k__. At the same point $x=$__l__ this matching gives A.

A repeated linear factor $(x-5)^2$ contributes not only $A/(x-5)$ but also $B/$__m__. A quadratic factor like x^2+x+1 contributes a fraction __n__$/(x^2+x+1)$ involving C and D. A repeated quadratic factor or a triple linear factor would bring in $(Ex+F)/(x^2+x+1)^2$ or $G/(x-5)^3$. The conclusion is that any P/Q can be split into partial __o__, which can always be integrated.

1 Find the numbers A and B to split $1/(x^2-x)$:

$$\frac{1}{x(x-1)} = \frac{A}{x}+\frac{B}{x-1}.$$

Cover up x and set $x=0$ to find A. Cover up $x-1$ and set $x=1$ to find B. Then integrate.

2 Find the numbers A and B to split $1/(x^2-1)$:

$$\frac{1}{x^2-1} = \frac{A}{x-1}+\frac{B}{x+1}.$$

Multiply by $x-1$ and set $x=1$. Multiply by $x+1$ and set $x=-1$. *Integrate*. Then find A and B again by method 1—with numerator $A(x+1)+B(x-1)$ equal to 1.

Express the rational functions 3–16 as partial fractions:

3 $\dfrac{1}{(x-3)(x-2)}$

4 $\dfrac{x}{(x-3)(x-2)}$

5 $\dfrac{x^2+1}{(x)(x+1)(x+2)}$

6 $\dfrac{1}{x^3-x}$

7 $\dfrac{3x+1}{x^2}$

8 $\dfrac{3x+1}{(x-1)^2}$

9 $\dfrac{3x^2}{x^2+1}$ (divide first)

10 $\dfrac{1}{(x-1)(x^2+1)}$

11 $\dfrac{1}{x^2(x-1)}$

12 $\dfrac{x}{x^2-4}$

13 $\dfrac{1}{x(x-1)(x-2)(x-3)}$

14 $\dfrac{x^2+1}{x+1}$ (divide first)

15 $\dfrac{1}{x^4-1} = \dfrac{1}{(x+1)(x-1)(x^2+1)}$

16 $\dfrac{1}{x^2(x-1)}$ $\left(\text{remember the } \dfrac{A}{x} \text{ term}\right)$

17 Apply Method **1** (matching numerators) to Example 3:

$$\frac{1}{cy-by^2} = \frac{A}{y} + \frac{B}{c-by} = \frac{A(c-by)+By}{y(c-by)}.$$

Match the numerators on the far left and far right. Why does $Ac=1$? *Why does* $-bA+B=0$? *What are* A *and* B?

18 What goes wrong if we look for A and B so that

$$\frac{x^2}{(x-3)(x+3)} = \frac{A}{x-3} + \frac{B}{x+3}?$$

Over a common denominator, try to match the numerators. What to do first?

19 Split $\dfrac{3x^2}{x^3-1} = \dfrac{3x^2}{(x-1)(x^2+x+1)}$ into $\dfrac{A}{x-1} + \dfrac{Bx+C}{x^2+x+1}$.

(a)   Cover up $x-1$ and set $x=1$ to find A.

(b)   Subtract $A/(x-1)$ from the left side. Find $Bx+C$.

(c)   Integrate all terms. Why do we already know

$$\ln(x^3-1) = \ln(x-1) + \ln(x^2+x+1)?$$

20 Solve $dy/dt = 1-y^2$ by separating $\int dy/1-y^2 = \int dt$. Then

$$\frac{1}{1-y^2} = \frac{1}{(1-y)(1+y)} = \frac{1/2}{1-y} + \frac{1/2}{1+y}.$$

Integration gives $\frac{1}{2}\ln$ ____ $= t + C$. With $y_0 = 0$ the constant is $C =$ ____. Taking exponentials gives ____. The solution is $y =$ ____. *This is the S-curve*.

By substitution change 21–28 to integrals of rational functions. Problem 23 integrates $1/\sin\theta$ **with no special trick.**

21 $\displaystyle\int \frac{e^x\,dx}{e^{2x}-e^x}$

22 $\displaystyle\int \frac{1-\sqrt{x}}{1+\sqrt{x}}\,dx$

23 $\displaystyle\int \frac{\sin\theta\,d\theta}{1-\cos^2\theta}$

24 $\displaystyle\int \frac{dt}{(e^t-e^{-t})^2}$

25 $\displaystyle\int \frac{1+e^x}{1-e^x}\,dx$

26 $\displaystyle\int \frac{\sqrt[3]{x-8}}{x}\,dx$

27 $\displaystyle\int \frac{dx}{1+\sqrt{x+1}}$

28 $\displaystyle\int \frac{dx}{\sqrt{x}+\sqrt[4]{x}}$

29 Multiply this partial fraction by $x-a$. Then let $x \to a$:

$$\frac{1}{Q(x)} = \frac{A}{x-a} + \cdots.$$

Show that $A = 1/Q'(a)$. When $x=a$ is a double root this fails because $Q'(a) =$ ____.

30 Find A in $\dfrac{1}{x^8-1} = \dfrac{A}{x-1} + \cdots$. Use Problem 29.

31 (for instructors only) Which rational functions P/Q are the derivatives of other rational functions (no logarithms)?

7.5 Improper Integrals

"**Improper**" means that some part of $\int_a^b y(x)\,dx$ becomes **infinite**. It might be b or a or the function y. The region under the graph reaches infinitely far—to the right or left or up or down. (Those come from $b = \infty$ and $a = -\infty$ and $y \to \infty$ and $y \to -\infty$.) Nevertheless the integral may "*converge*." Just because the **region** is infinite, it is not automatic that the **area** is infinite. That is the point of this section—to decide when improper integrals have proper answers.

The first examples show finite area when $b = \infty$, then $a = -\infty$, then $y = 1/\sqrt{x}$ at $x = 0$. The areas in Figure 7.6 are $1, 1, 2$:

$$\int_1^\infty \frac{dx}{x^2} = -\frac{1}{x}\Big]_1^\infty = 1 \qquad \int_{-\infty}^0 e^x dx = e^x\Big]_{-\infty}^0 = 1 \qquad \int_0^1 \frac{dx}{\sqrt{x}} = 2\sqrt{x}\Big]_0^1 = 2.$$

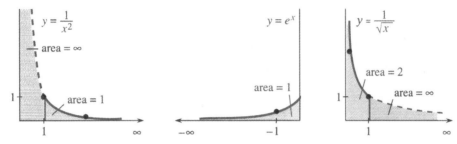

Fig. 7.6 The shaded areas are finite but the regions go to infinity.

In practice we substitute the dangerous limits and watch what happens. When the integral is $-1/x$, substituting $b = \infty$ gives "$-1/\infty = 0$." When the integral is e^x, substituting $a = -\infty$ gives "$e^{-\infty} = 0$." I think that is fair, and I know it is successful. But it is not completely precise.

The strict rules involve a limit. Calculus sneaks up on $1/\infty$ and $e^{-\infty}$ just as it sneaks up on $0/0$. Instead of swallowing an infinite region all at once, the formal definitions push out to the limit:

DEFINITION $\displaystyle\int_a^\infty y(x)dx = \lim_{b \to \infty} \int_a^b y(x)dx \qquad \int_{-\infty}^b y(x)dx = \lim_{a \to -\infty} \int_a^b y(x)dx.$

The conclusion is the same. The first examples converged to $1, 1, 2$. Now come two more examples going out to $b = \infty$:

The area under $1/x$ is *infinite*: $\displaystyle\int_1^\infty \frac{dx}{x} = \ln x\Big]_1^\infty = \infty$ \hfill (1)

The area under $1/x^p$ is *finite* if $p > 1$: $\displaystyle\int_1^\infty \frac{dx}{x^p} = \frac{x^{1-p}}{1-p}\Big]_1^\infty = \frac{1}{p-1}.$ \hfill (2)

The area under $1/x$ is like $1 + \frac{1}{2} + \frac{1}{3} + \frac{1}{4} + \cdots$, which is also infinite. In fact the sum approximates the integral—the curved area is close to the rectangular area. They go together (slowly to infinity).

A larger p brings the graph more quickly to zero. Figure 7.7a shows a finite area $1/(p-1) = 100$. The region is still infinite, but we can cover it with strips cut out of a square! The borderline for finite area is $p = 1$. I call it the borderline, but $p = 1$ is strictly on the side of divergence.

The borderline is also $p = 1$ *when the function climbs the y axis*. At $x = 0$, the graph of $y = 1/x^p$ goes to infinity. For $p = 1$, the area under $1/x$ is again infinite. But at $x = 0$ it is a *small p* (meaning $p < 1$) that produces finite area:

$$\int_0^1 \frac{dx}{x} = \ln x \Big]_0^1 = \infty \qquad \int_0^1 \frac{dx}{x^p} = \frac{x^{1-p}}{1-p}\Big]_0^1 = \frac{1}{1-p} \quad \text{if } p < 1. \qquad (3)$$

Loosely speaking "$-\ln 0 = \infty$." Strictly speaking we integrate from the point $x = a$ *near* zero, to get $\int_a^1 dx/x = -\ln a$. As a approaches zero, the area shows itself as infinite. For $y = 1/x^2$, which blows up faster, the area $-1/x]_0^1$ is again infinite.

For $y = 1/\sqrt{x}$, the area from 0 to 1 is 2. In that case $p = \frac{1}{2}$. For $p = 99/100$ the area is $1/(1-p) = 100$. Approaching $p = 1$ the borderline in Figure 7.7 seems clear. *But that cutoff is not as sharp as it looks.*

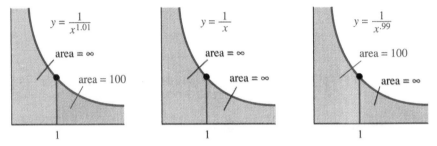

Fig. 7.7   Graphs of $1/x^p$ on both sides of $p = 1$. I drew the same curves!

Narrower borderline Under the graph of $1/x$, the area is infinite. When we divide by $\ln x$ or $(\ln x)^2$, the borderline is somewhere in between. One has infinite area (going out to $x = \infty$), the other area is finite:

$$\int_e^\infty \frac{dx}{x(\ln x)} = \ln(\ln x)\Big]_e^\infty = \infty \qquad \int_e^\infty \frac{dx}{x(\ln x)^2} = -\frac{1}{\ln x}\Big]_e^\infty = 1. \qquad (4)$$

The first is $\int du/u$ with $u = \ln x$. The logarithm of $\ln x$ does eventually make it to infinity. At $x = 10^{10}$, the logarithm is near 23 and $\ln(\ln x)$ is near 3. That is slow! Even slower is $\ln(\ln(\ln x))$ in Problem 11. No function is *exactly* on the borderline.

The second integral in equation (4) is convergent (to 1). It is $\int du/u^2$ with $u = \ln x$. At first I wrote it with x going from zero to infinity. That gave an answer I couldn't believe:

$$\int_0^\infty \frac{dx}{x(\ln x)^2} = -\frac{1}{\ln x}\Big]_0^\infty = 0 \text{ (??)}$$

There must be a mistake, because we are integrating a positive function. The area can't be zero. It is true that $1/\ln b$ goes to zero as $b \to \infty$. It is also true that $1/\ln a$ goes to zero as $a \to 0$. *But there is another infinity in this integral.* The trouble is at $x = 1$, where $\ln x$ is zero and the area is infinite.

EXAMPLE 1   The factor e^{-x} overrides any power x^p (but only as $x \to \infty$).

$$\int_0^\infty x^{50} e^{-x}\, dx = 50! \quad \text{but} \quad \int_0^\infty x^{-1} e^{-x}\, dx = \infty.$$

The first integral is $(50)(49)(48)\cdots(1)$. It comes from fifty integrations by parts (not recommended). Changing 50 to $\frac{1}{2}$, the integral defines "$\frac{1}{2}$ *factorial*." The product

$\frac{1}{2}\left(-\frac{1}{2}\right)\left(-\frac{3}{2}\right)\cdots$ has no way to stop, but somehow $\frac{1}{2}!$ is $\frac{1}{2}\sqrt{\pi}$. See Problem 28. The integral $\int_0^\infty x^0 e^{-x}\,dx = 1$ is the reason behind "*zero factorial*" $= 1$. That seems the most surprising of all.

The area under e^{-x}/x is $(-1)! = \infty$. The factor e^{-x} is absolutely no help at $x = 0$. That is an example (the first of many) in which we do not know an antiderivative—but still we get a decision. To integrate e^{-x}/x we need a computer. But to decide that an improper integral is infinite (in this case) or finite (in other cases), we rely on the following *comparison test*:

7C (**Comparison test**) Suppose that $0 \leqslant u(x) \leqslant v(x)$. Then the area under $u(x)$ is smaller than the area under $v(x)$:

$$\int u(x)\,dx < \infty \text{ if } \int v(x)\,dx < \infty \qquad \text{if } \int u(x)\,dx = \infty \text{ then } \int v(x)\,dx = \infty.$$

Comparison can decide if the area is finite. We don't get the exact area, but we learn about one function from the other. The trick is to construct a simple function (like $1/x^p$) which is on one side of the given function—and stays close to it:

EXAMPLE 2 $\qquad \displaystyle\int_1^\infty \frac{dx}{x^2 + 4x}$ converges by comparison with $\displaystyle\int_1^\infty \frac{dx}{x^2} = 1$.

EXAMPLE 3 $\qquad \displaystyle\int_1^\infty \frac{dx}{\sqrt{x}+1}$ diverges by comparison with $\displaystyle\int_1^\infty \frac{dx}{2\sqrt{x}} = \infty$.

EXAMPLE 4 $\qquad \displaystyle\int_0^1 \frac{dx}{x^2 + 4x}$ diverges by comparison with $\displaystyle\int_0^1 \frac{dx}{5x} = \infty$.

EXAMPLE 5 $\qquad \displaystyle\int_0^1 \frac{dx}{\sqrt{x}+1}$ converges by comparison with $\displaystyle\int_0^1 \frac{dx}{1} = 1$.

In Examples 2 and 5, the integral on the right is larger than the integral on the left. Removing $4x$ and $\sqrt{x}$ increased the area. Therefore the integrals on the left are somewhere between 0 and 1.

In Examples 3 and 4, we *increased* the denominators. The integrals on the right are smaller, but still they diverge. So the integrals on the left diverge. The idea of *comparing functions* is seen in the next examples and Figure 7.8.

EXAMPLE 6 $\qquad \displaystyle\int_0^\infty e^{-x^2}\,dx$ is below $\displaystyle\int_0^1 1\,dx + \int_1^\infty e^{-x}\,dx = 1 + 1$.

EXAMPLE 7 $\qquad \displaystyle\int_1^e \frac{dx}{\ln x}$ is above $\displaystyle\int_1^e \frac{dx}{x \ln x} = \infty$.

EXAMPLE 8 $\qquad \displaystyle\int_0^1 \frac{dx}{\sqrt{x - x^2}}$ is below $\displaystyle\int_0^1 \frac{dx}{\sqrt{x}} + \int_0^1 \frac{dx}{\sqrt{1-x}} = 2 + 2$.

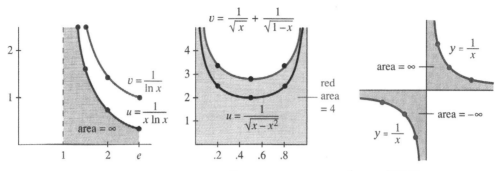

Fig. 7.8 Comparing $u(x)$ to $v(x)$: $\int_1^e dx/\ln x = \infty$ and $\int_0^1 dx/\sqrt{x-x^2} < 4$. But $\infty - \infty \neq 0$.

There are two situations not yet mentioned, and both are quite common. The first is an integral all the way from $a = -\infty$ to $b = +\infty$. That is split into two parts, and **each part must converge**. By definition, the limits at $-\infty$ and $+\infty$ are kept *separate*:

$$\int_{-\infty}^{\infty} y(x)\, dx = \int_{-\infty}^{0} y(x)\, dx + \int_{0}^{\infty} y(x)\, dx = \lim_{a \to -\infty} \int_a^0 y(x)\, dx + \lim_{b \to \infty} \int_0^b y(x)\, dx.$$

The bell-shaped curve $y = e^{-x^2}$ covers a finite area (exactly $\sqrt{\pi}$). The region extends to infinity in both directions, and the separate areas are $\frac{1}{2}\sqrt{\pi}$. But notice:

$$\int_{-\infty}^{\infty} x\, dx \text{ is not defined even though } \int_{-b}^{b} x\, dx = 0 \text{ for every } b.$$

The area under $y = x$ is $+\infty$ on one side of zero. The area is $-\infty$ on the other side. **We cannot accept** $\infty - \infty = 0$. The two areas must be separately finite, and in this case they are not.

EXAMPLE 9 $1/x$ has balancing regions left and right of $x = 0$. Compute $\int_{-1}^{1} dx/x$.

This integral does not exist. There is no answer, even for the region in Figure 7.8c. (They are mirror images because $1/x$ is an odd function.) You may feel that the combined integral from -1 to 1 should be zero. Cauchy agreed with that—his "principal value integral" is zero. But the rules say no: $\infty - \infty$ is not zero.

7.5 EXERCISES

Read-through questions

An improper integral $\int_a^b y(x)\, dx$ has lower limit $a = \underline{\quad a \quad}$ or upper limit $b = \underline{\quad b \quad}$ or y becomes $\underline{\quad c \quad}$ in the interval $a \leqslant x \leqslant b$. The example $\int_1^{\infty} dx/x^3$ is improper because $\underline{\quad d \quad}$. We should study the limit of $\int_1^b dx/x^3$ as $\underline{\quad e \quad}$. In practice we work directly with $-\frac{1}{2}x^{-2}]_1^{\infty} = \underline{\quad f \quad}$. For $p > 1$ the improper integral $\underline{\quad g \quad}$ is finite. For $p < 1$ the improper integral $\underline{\quad h \quad}$ is finite. For $y = e^{-x}$ the integral from 0 to ∞ is $\underline{\quad i \quad}$.

Suppose $0 \leqslant u(x) \leqslant v(x)$ for all x. The convergence of $\underline{\quad j \quad}$ implies the convergence of $\underline{\quad k \quad}$. The divergence of $\int u(x)\, dx \underline{\quad l \quad}$ the divergence of $\int v(x)\, dx$. From $-\infty$ to ∞, the integral of $1/(e^x + e^{-x})$ converges by comparison with $\underline{\quad m \quad}$. Strictly speaking we split $(-\infty, \infty)$ into $(\underline{\quad n \quad}, 0)$ and $(0, \underline{\quad o \quad})$. Changing to $1/(e^x - e^{-x})$ gives divergence, because $\underline{\quad p \quad}$. Also $\int_{-\pi}^{\pi} dx/\sin x$ diverges by comparison with $\underline{\quad q \quad}$.

The regions left and right of zero don't cancel because $\infty - \infty$ is __r__ .

Decide convergence or divergence in 1–16. Compute the integrals that converge.

1 $\int_1^\infty \dfrac{dx}{x^e}$

2 $\int_0^1 \dfrac{dx}{x^\pi}$

3 $\int_0^1 \dfrac{dx}{\sqrt{1-x}}$

4 $\int_0^1 \dfrac{dx}{1-x}$

5 $\int_{-\infty}^0 \dfrac{dx}{x^2+1}$

6 $\int_{-1}^1 \dfrac{dx}{\sqrt{1-x^2}}$

7 $\int_0^1 \dfrac{\ln x}{x}\,dx$

8 $\int_{-\infty}^\infty \sin x\,dx$

9 $\int_0^e \ln x\,dx$ (by parts)

10 $\int_0^\infty x\,e^{-x}\,dx$ (by parts)

11 $\int_{100}^\infty \dfrac{dx}{x(\ln x)(\ln \ln x)}$

12 $\int_{-\infty}^\infty \dfrac{x\,dx}{(x^2-1)^2}$

13 $\int_0^\infty \cos^2 x\,dx$

14 $\int_0^{\pi/2} \tan x\,dx$

15 $\int_0^\infty \dfrac{dx}{x^p}$

16 $\int_0^\infty \dfrac{e^x\,dx}{(e^x-1)^p}$

In 17–26, find a larger integral that converges or a smaller integral that diverges.

17 $\int_1^\infty \dfrac{dx}{x^6+1}$

18 $\int_0^1 \dfrac{dx}{x^6+1}$

19 $\int_0^\infty \dfrac{\sqrt{x}\,dx}{x^2+1}$

20 $\int_0^1 \dfrac{e^{-x}\,dx}{1-x}$

21 $\int_1^\infty e^{-x}\sin x\,dx$

22 $\int_1^\infty x^{-x}\,dx$

23 $\int_0^\infty e^{2x}e^{-x^2}\,dx$

24 $\int_0^1 \sqrt{-\ln x}\,dx$

25 $\int_0^\infty \dfrac{\sin^2 x}{x^2}\,dx$

26 $\int_1^\infty \left(\dfrac{1}{x}-\dfrac{1}{1+x}\right)dx$

27 If $p>0$, integrate by parts to show that

$$\int_0^\infty x^p e^{-x}\,dx = p \int_0^\infty x^{p-1}e^{-x}\,dx.$$

The first integral is the definition of $p!$ So the equation is $p! =$ ____ . In particular $0! =$ ____ . Another notation for $p!$ is $\Gamma(p+1)$—using the *gamma function* emphasizes that p need not be an integer.

28 Compute $\left(-\frac{1}{2}\right)!$ by substituting $x = u^2$:

$$\int_0^\infty x^{-1/2}e^{-x}\,dx = \underline{\quad\quad} = \sqrt{\pi}\text{ (known)}.$$

Then apply Problem 27 to find $\left(\frac{1}{2}\right)!$

29 Integrate $\int_0^\infty x^2 e^{-x^2}\,dx$ by parts.

30 The *beta function* $B(m,n) = \int_0^1 x^{m-1}(1-x)^{n-1}\,dx$ is finite when m and n are greater than ____ .

31 A *perpetual annuity* pays s dollars a year forever. With continuous interest rate c, its present value is $y_0 = \int_0^\infty s e^{-ct}\,dt$. To receive $1000/year at $c = 10\%$, you deposit $y_0 =$ ____ .

32 In a perpetual annuity that pays once a year, the present value is $y_0 = s/a + s/a^2 + \cdots =$ ____ . To receive $1000/year at 10% (now $a = 1.1$) you again deposit $y_0 =$ ____ . Infinite sums are like improper integrals.

33 The work to move a satellite (mass m) infinitely far from the Earth (radius R, mass M) is $W = \int_R^\infty GMm\,dx/x^2$. Evaluate W. What *escape velocity* at liftoff gives an energy $\frac{1}{2}mv_0^2$ that equals W?

34 The escape velocity for a black hole exceeds the speed of light: $v_0 > 3 \cdot 10^8$ m/sec. The Earth has $GM = 4 \cdot 10^{14}$m^3/sec^2. If it were compressed to radius $R =$ ____ , the Earth would be a black hole.

35 Show how the area under $y = 1/2^x$ can be covered (draw a graph) by rectangles of area $1 + \frac{1}{2} + \frac{1}{4} + \cdots = 2$. What is the exact area from $x = 0$ to $x = \infty$?

36 Explain this paradox:

$$\int_{-b}^b \dfrac{x\,dx}{1+x^2} = 0 \text{ for every } b \text{ but } \int_{-\infty}^\infty \dfrac{x\,dx}{1+x^2} \text{ diverges.}$$

37 Compute the area between $y = \sec x$ and $y = \tan x$ for $0 \leqslant x \leqslant \pi/2$. What is improper?

*38 Compute any of these integrals found by geniuses:

$$\int \dfrac{x^{-1/2}\,dx}{1+x} = \pi \qquad \int_0^\infty \dfrac{e^{-x}-e^{-2x}}{x}\,dx = \ln 2$$

$$\int_0^\infty x\,e^{-x}\cos x\,dx = 0 \qquad \int_0^\infty \cos x^2\,dx = \sqrt{\pi/8}.$$

39 For which p is $\displaystyle\int_0^\infty \dfrac{dx}{x^p + x^{-p}} = \infty$?

40 Explain from Figure 7.6c why the red area is 2, when Figure 7.6a has red area 1.

CHAPTER 8

Applications of the Integral

We are experts in one application of the integral—to find the area under a curve. The curve is the graph of $y = v(x)$, extending from $x = a$ at the left to $x = b$ at the right. The area between the curve and the x axis is the definite integral.

I think of that integral in the following way. The region is made up of ***thin strips***. Their width is dx and their height is $v(x)$. The area of a strip is $v(x)$ times dx. The area of all the strips is $\int_a^b v(x)\, dx$. Strictly speaking, the area of one strip is meaningless—genuine rectangles have width Δx. My point is that the picture of thin strips gives the correct approach.

We know what function to integrate (from the picture). We also know how (from this course or a calculator). The new applications to volume and length and surface area cut up the region in new ways. Again the small pieces tell the story. In this chapter, *what* to integrate is more important than *how*.

8.1 Areas and Volumes by Slices

This section starts with areas between curves. Then it moves to *volumes*, where the strips become *slices*. We are weighing a loaf of bread by adding the weights of the slices. The discussion is dominated by examples and figures—the theory is minimal. The real problem is to set up the right integral. At the end we look at a different way of cutting up volumes, into thin shells. *All formulas are collected into final table.*

Figure 8.1 shows *the area between two curves*. The upper curve is the graph of $y = v(x)$. The lower curve is the graph of $y = w(x)$. The strip height is $v(x) - w(x)$, from one curve down to the other. The width is dx (speaking informally again). The total area is the integral of "top minus bottom":

$$\textbf{\textit{area between two curves}} = \int_a^b \left[v(x) - w(x) \right] dx. \qquad (1)$$

EXAMPLE 1   The upper curve is $y = 6x$ (straight line). The lower curve is $y = 3x^2$ (parabola). The area lies between the points where those curves intersect.

To find the intersection points, solve $v(x) = w(x)$ or $6x = 3x^2$.

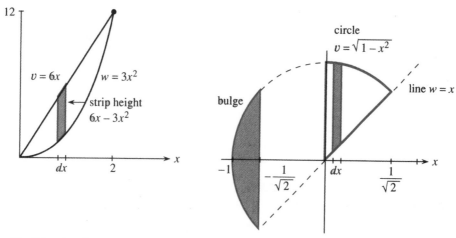

Fig. 8.1   Area between curves $=$ integral of $v - w$. Area in Example 2 starts with $x \geqslant 0$.

One crossing is at $x = 0$, the other is at $x = 2$. The area is an integral from 0 to 2:

$$\text{area} = \int_a^b (v - w) dx = \int_0^2 (6x - 3x^2) dx = 3x^2 - x^3 \Big]_0^2 = 4.$$

EXAMPLE 2   Find the area between the circle $v = \sqrt{1 - x^2}$ and the 45° line $w = x$.

First question: Which area and what limits? Start with the pie-shaped wedge in Figure 8.1b. The area begins at the y axis and ends where the circle meets the line. At the intersection point we have $v(x) = w(x)$:

$$\text{from } \sqrt{1 - x^2} = x \text{ squaring gives } 1 - x^2 = x^2 \text{ and then } 2x^2 = 1.$$

Thus $x^2 = \frac{1}{2}$. The endpoint is at $x = 1/\sqrt{2}$. Now integrate the strip height $v - w$:

$$\int_0^{1/\sqrt{2}} (\sqrt{1-x^2} - x)\, dx = \frac{1}{2}\sin^{-1}x + \frac{1}{2}x\sqrt{1-x^2} - \frac{1}{2}x^2 \Big]_0^{1/\sqrt{2}}$$

$$= \frac{1}{2}\sin^{-1}\left(\frac{1}{\sqrt{2}}\right) + \frac{1}{4} - \frac{1}{4} = \frac{1}{2}\left(\frac{\pi}{4}\right).$$

The area is $\pi/8$ (one eighth of the circle). To integrate $\sqrt{1-x^2}\, dx$ we apply the techniques of Chapter 7: Set $x = \sin\theta$, convert to $\int \cos^2\theta\, d\theta = \frac{1}{2}(\theta + \sin\theta\cos\theta)$, convert back using $\theta = \sin^{-1}x$. It is harder than expected, for a familiar shape.

Remark　Suppose the problem is to find the *whole area* between the circle and the line. The figure shows $v = w$ at two points, which are $x = 1/\sqrt{2}$ (already used) and also $x = -1/\sqrt{2}$. Instead of starting at $x = 0$, which gave $\frac{1}{8}$ of a circle, we now include the area to the left.

　Main point: *Integrating from $x = -1/\sqrt{2}$ to $x = 1/\sqrt{2}$ will give the wrong answer*. It misses the part of the circle that bulges out over itself, at the far left. In that part, the strips have height $2v$ instead of $v - w$. The figure is essential, to get the correct area of this half-circle.

HORIZONTAL STRIPS INSTEAD OF VERTICAL STRIPS

There is more than one way to slice a region. *Vertical slices give x integrals.* *Horizontal slices give y integrals.* We have a free choice, and sometimes the y integral is better.

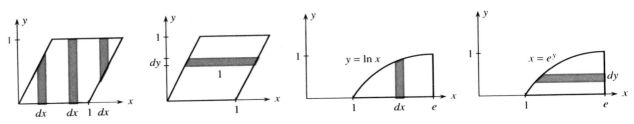

Fig. 8.2　Vertical slices (x integrals) vs. horizontal slices (y integrals).

Figure 8.2 shows a unit parallelogram, with base 1 and height 1. To find its area from vertical slices, three separate integrals are necessary. You should see why! With horizontal slices of length 1 and thickness dy, the area is just $\int_0^1 dy = 1$.

EXAMPLE 3　Find the area under $y = \ln x$ (or beyond $x = e^y$) out to $x = e$.

The x integral from vertical slices is in Figure 8.2c. The y integral is in Figure 8.2d. The area is a choice between two equal integrals (I personally would choose y):

$$\int_{x=1}^{e} \ln x\, dx = \Big[x\ln x - x\Big]_1^e = 1 \quad \text{or} \quad \int_{y=0}^1 (e - e^y)\, dy = \Big[ey - e^y\Big]_0^1 = 1.$$

VOLUMES BY SLICES

For the first time in this book, we now look at *volumes*. **The regions are three-dimensional solids**. There are three coordinates x, y, z—and many ways to cut up a solid.

Figure 8.3 shows one basic way—using **slices**. The slices have thickness dx, like strips in the plane. Instead of the height y of a strip, we now have **the area A of a cross-section**. This area is different for different slices: A depends on x. The volume of the slice is its area times its thickness: $dV = A(x)dx$. *The volume of the whole solid is the integral*:

$$\textbf{\textit{volume = integral of area times thickness}} = \int A(x)\,dx. \qquad (2)$$

Note An actual slice does not have the same area on both sides! Its thickness is Δx (not dx). Its volume is approximately $A(x)\Delta x$ (but not exactly). In the limit, the thickness approaches zero and the sum of volumes approaches the integral.

For a cylinder all slices are the same. Figure 8.3b shows a cylinder—*not circular*. The area is a fixed number A, so integration is trivial. *The volume is A times h*. The

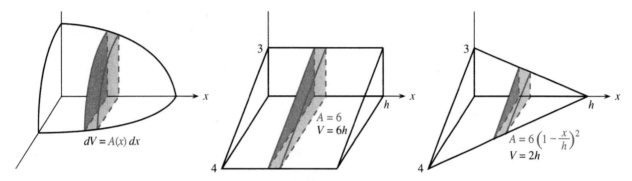

Fig. 8.3 Cross-sections have area $A(x)$. Volumes are $\int A(x)\,dx$.

letter h, which stands for *height*, reminds us that the cylinder often stands on its end. Then the slices are horizontal and the y integral or z integral goes from 0 to h.

When the cross-section is a circle, the cylinder has volume $\pi r^2 h$.

EXAMPLE 4 The *triangular wedge* in Figure 8.3b has constant cross-sections with area $A = \frac{1}{2}(3)(4) = 6$. The volume is $6h$.

EXAMPLE 5 For the *triangular pyramid* in Figure 8.3c, the area $A(x)$ drops from 6 to 0. It is a general rule for pyramids or cones that their volume has an extra factor $\frac{1}{3}$ (compared to cylinders). The volume is now $2h$ instead of $6h$. For a cone with base area πr^2, the volume is $\frac{1}{3}\pi r^2 h$. *Tapering the area to zero leaves only $\frac{1}{3}$ of the volume*.

Why the $\frac{1}{3}$? Triangles sliced from the pyramid have shorter sides. Starting from 3 and 4, the side lengths $3(1 - x/h)$ and $4(1 - x/h)$ drop to zero at $x = h$. *The area is $A = 6(1 - x/h)^2$*. Notice: The side lengths go down linearly, the area drops quadratically. The factor $\frac{1}{3}$ really comes from integrating x^2 to get $\frac{1}{3}x^3$:

$$\int_0^h A(x)\,dx = \int_0^h 6\left(1 - \frac{x}{h}\right)^2 dx = -2h\left(1 - \frac{x}{h}\right)^3\Bigg]_0^h = 2h.$$

EXAMPLE 6   A half-sphere of radius R has known volume $\frac{1}{2}(\frac{4}{3}\pi R^3)$. Its cross-sections are *semicircles*. The key relation is $x^2 + r^2 = R^2$, for the right triangle in Figure 8.4a. The area of the semicircle is $A = \frac{1}{2}\pi r^2 = \frac{1}{2}\pi(R^2 - x^2)$. So we integrate $A(x)$:

$$\text{volume} = \int_{-R}^{R} A(x)\,dx = \frac{1}{2}\pi(R^2 x - \frac{1}{3}x^3)\Big]_{-R}^{R} = \frac{2}{3}\pi R^3.$$

EXAMPLE 7   Find the volume of the same half-sphere using horizontal slices (Figure 8.4b). The sphere still has radius R. The new right triangle gives $y^2 + r^2 = R^2$. Since we have full circles the area is $\pi r^2 = \pi(R^2 - y^2)$. Notice that this is $A(y)$ not $A(x)$. But the y integral starts at zero:

$$\text{volume} = \int_0^R A(y)\,dy = \pi(R^2 y - \frac{1}{3}y^3)\Big]_0^R = \frac{2}{3}\pi R^3 \text{ (as before)}.$$

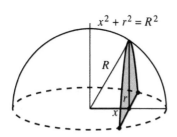 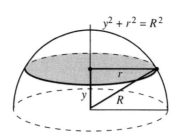 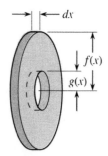

Fig. 8.4   A half-sphere sliced vertically or horizontally. Washer area $\pi f^2 - \pi g^2$.

SOLIDS OF REVOLUTION

Cones and spheres and circular cylinders are "solids of revolution." Rotating a horizontal line around the x axis gives a cylinder. Rotating a sloping line gives a cone. Rotating a semicircle gives a sphere. If a circle is moved away from the axis, rotation produces a torus (a doughnut). The rotation of any curve $y = f(x)$ produces a ***solid of revolution***.

The volume of that solid is made easier because ***every cross-section is a circle***. All slices are pancakes (or pizzas). Rotating the curve $y = f(x)$ around the x axis gives disks of radius y, so the area is $A = \pi y^2 = \pi[f(x)]^2$. We add the slices:

$$\boldsymbol{\text{volume of solid of revolution}} = \int_a^b \pi y^2\,dx = \int_a^b \pi[f(x)]^2\,dx.$$

EXAMPLE 8   Rotating $y = \sqrt{x}$ with $A = \pi(\sqrt{x}^2)^2$ produces a "headlight" (Figure 8.5a):

$$\text{volume of headlight} = \int_0^2 A\,dx = \int_0^2 \pi x\,dx = \frac{1}{2}\pi x^2\Big]_0^2 = 2\pi.$$

If the same curve is rotated around the y axis, it makes a champagne glass. *The slices are horizontal.* The area of a slice is πx^2 not πy^2. When $y = \sqrt{x}$ this area is πy^4. Integrating from $y = 0$ to $\sqrt{2}$ gives the champagne volume $\pi(\sqrt{2})^5/5$.

$$\boldsymbol{\text{revolution around the } y \text{ axis:}} \quad \text{volume} = \int \pi x^2\,dy.$$

EXAMPLE 9 The headlight has a hole down the center (Figure 8.5b). Volume $= ?$

The hole has radius 1. *All* of the $\sqrt{x}$ solid is removed, up to the point where $\sqrt{x}$ reaches 1. After that, from $x = 1$ to $x = 2$, each cross-section is a disk with a hole. The disk has radius $f = \sqrt{x}$ and the hole has radius $g = 1$. *The slice is a flat ring or a "**washer**."* Its area is the full disk minus the area of the hole:

$$\textit{area of washer} = \pi f^2 - \pi g^2 = \pi(\sqrt{x})^2 - \pi(1)^2 = \pi x - \pi.$$

This is the area $A(x)$ in the *method of washers*. Its integral is the volume:

$$\int_1^2 A\, dx = \int_1^2 (\pi x - \pi)\, dx = \left[\tfrac{1}{2}\pi x^2 - \pi x\right]_1^2 = \tfrac{1}{2}\pi.$$

Please notice: *The washer area is not $\pi(f - g)^2$. It is $A = \pi f^2 - \pi g^2$.*

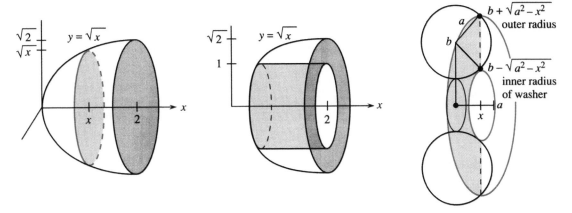

Fig. 8.5 $y = \sqrt{x}$ revolved; $y = 1$ revolved inside it; circle revolved to give torus.

EXAMPLE 10 (Doughnut sliced into washers) Rotate a circle of radius a around the x axis. The center of the circle stays out at a distance $b > a$. Show that the volume of the doughnut (or torus) is $2\pi^2 a^2 b$.

The outside half of the circle rotates to give the outside of the doughnut. The inside half gives the hole. The biggest slice (through the center plane) has outer radius $b + a$ and inner radius $b - a$.

Shifting over by x, the outer radius is $f = b + \sqrt{a^2 - x^2}$ and the inner radius is $g = b - \sqrt{a^2 - x^2}$. Figure 8.5c shows a slice (a washer) with area $\pi f^2 - \pi g^2$.

$$\text{area } A = \pi(b + \sqrt{a^2 - x^2})^2 - \pi(b - \sqrt{a^2 - x^2})^2 = 4\pi b\sqrt{a^2 - x^2}.$$

Now integrate over the washers to find the volume of the doughnut:

$$\int_{-a}^{a} A(x)\, dx = 4\pi b \int_{-a}^{a} \sqrt{a^2 - x^2}\, dx = (4\pi b)(\tfrac{1}{2}\pi a^2) = 2\pi^2 a^2 b.$$

That integral $\tfrac{1}{2}\pi a^2$ is the area of a semicircle. When we set $x = a\sin\theta$ the area is $\int a^2 \cos^2\theta\, d\theta$. Not for the last time do we meet $\cos^2\theta$.

The hardest part is visualizing the washers, because a doughnut usually breaks the other way. A better description is a **bagel**, sliced the long way to be buttered.

VOLUMES BY CYLINDRICAL SHELLS

Finally we look at a different way of cutting up a solid of revolution. So far it was cut into slices. The slices were perpendicular to the axis of revolution. Now the cuts are *parallel* to the axis, and each piece is a ***thin cylindrical shell***. The new formula gives the same volume, but the integral to be computed might be easier.

Figure 8.6a shows a solid cone. A shell is inside it. The inner radius is x and the outer radius is $x + dx$. *The shell is an outer cylinder minus an inner cylinder:*

$$\text{shell volume } \pi(x+dx)^2 h - \pi x^2 h = \pi x^2 h + 2\pi x (dx) h + \pi (dx)^2 h - \pi x^2 h. \tag{3}$$

The term that matters is $2\pi x (dx) h$. **The shell volume is essentially $2\pi x$ (the distance around) *times* dx (the thickness) *times* h (the height).** The volume of the solid comes from putting together the thin shells:

$$\textit{solid volume} = \textit{integral of shell volumes} = \int 2\pi xh \, dx. \tag{4}$$

This is the central formula of the shell method. The rest is examples.

Remark on this volume formula It is completely typical of integration that $(dx)^2$ and $(\Delta x)^2$ disappear. The reason is this. The number of shells grows like $1/\Delta x$. Terms of order $(\Delta x)^2$ add up to a volume of order Δx (approaching zero). ***The linear term involving Δx or dx is the one to get right***. Its limit gives the integral $\int 2\pi xh \, dx$. The key is to build the solid out of shells—and to find the area or volume of each piece.

EXAMPLE 11   Find the volume of a cone (base area πr^2, height b) cut into shells.

A tall shell at the center has h near b. A short shell at the outside has h near zero. In between the shell height h decreases linearly, reaching zero at $x = r$. The height in Figure 8.6a is $h = b - bx/r$. Integrating over all shells gives the volume of the cone (with the expected $\frac{1}{3}$):

$$\int_0^r 2\pi x \left(b - b\frac{x}{r}\right) dx = \left[\pi x^2 b - \frac{2\pi x^3 b}{3r}\right]_0^r = \frac{1}{3}\pi r^2 b.$$

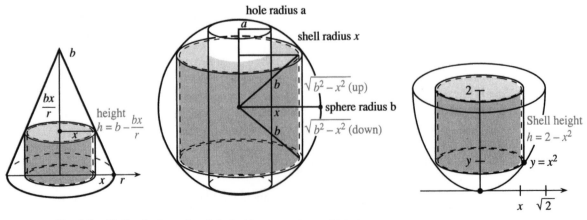

Fig. 8.6   Shells of volume $2\pi xh \, dx$ inside cone, sphere with hole, and paraboloid.

EXAMPLE 12 Bore a hole of radius a through a sphere of radius $b > a$.

The hole removes all points out to $x = a$, where the shells begin. The height of the shell is $h = 2\sqrt{b^2 - x^2}$. (The key is the right triangle in Figure 8.6b. The height upward is $\sqrt{b^2 - x^2}$—this is half the height of the shell.) Therefore the sphere-with-hole has

$$\text{volume} = \int_a^b 2\pi x h \, dx = \int_a^b 4\pi x \sqrt{b^2 - x^2} \, dx.$$

With $u = b^2 - x^2$ we almost see du. Multiplying $du = -2x \, dx$ is an extra factor -2π:

$$\text{volume} = -2\pi \int \sqrt{u} \, du = -2\pi \left(\tfrac{2}{3} u^{3/2}\right).$$

We can find limits on u, or we can put back $u = b^2 - x^2$:

$$\text{volume} = -\frac{4\pi}{3}(b^2 - x^2)^{3/2} \Big]_a^b = \frac{4\pi}{3}(b^2 - a^2)^{3/2}.$$

If $a = b$ (the hole is as big as the sphere) this volume is zero. If $a = 0$ (no hole) we have $4\pi b^3/3$ for the complete sphere.

Question What if the sphere-with-hole is cut into slices instead of shells ?
Answer Horizontal slices are washers (Problem 66). Vertical slices are not good.

EXAMPLE 13 Rotate the parabola $y = x^2$ around the y axis to form a bowl.

We go out to $x = \sqrt{2}$ (and up to $y = 2$). The shells in Figure 8.6c have height $h = 2 - x^2$. The bowl (or paraboloid) is the same as the headlight in Example 8, but we have shells not slices:

$$\int_0^{\sqrt{2}} 2\pi x (2 - x^2) \, dx = 2\pi x^2 - \frac{2\pi x^4}{4} \Big]_0^{\sqrt{2}} = 2\pi.$$

TABLE OF AREAS AND VOLUMES	***area between curves***: $A = \int (v(x) - w(x)) \, dx$ ***solid volume cut into slices***: $V = \int A(x) \, dx$ or $\int A(y) \, dy$ ***solid of revolution***: cross-section $A = \pi y^2$ or πx^2 ***solid with hole***: washer area $A = \pi f^2 - \pi g^2$ ***solid of revolution cut into shells***: $V = \int 2\pi x h \, dx$.

Which to use, slices or shells? Start with a vertical line going up to $y = \cos x$. Rotating the line around the x axis produces a *slice* (a circular disk). The radius is $\cos x$. Rotating the line around the y axis produces a *shell* (the outside of a cylinder). The height is $\cos x$. See Figure 8.7 for the slice and the shell. For volumes we just integrate $\pi \cos^2 x \, dx$ (the slice volume) or $2\pi x \cos x \, dx$ (the shell volume).

   This is the normal choice—slices through the x axis and shells around the y axis. Then $y = f(x)$ gives the disk radius and the shell height. The slice is a washer instead of a disk if there is also an inner radius $g(x)$. No problem—just integrate small volumes.

   What if you use slices for rotation around the y axis ? The disks are in Figure 8.7b, and *their radius is x*. This is $x = \cos^{-1} y$ in the example. It is $x = f^{-1}(y)$ in general. You have to solve $y = f(x)$ to find x in terms of y. Similarly for shells around the x axis: *The length of the shell is $x = f^{-1}(y)$.* Integrating may be difficult or impossible.

When $y = \cos x$ is rotated around the x axis, here are the choices for volume:

$$(\textbf{\textit{good by slices}}) \int \pi \cos^2 x \, dx \qquad (\textbf{\textit{bad by shells}}) \int 2\pi y \cos^{-1} y \, dy.$$

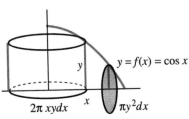
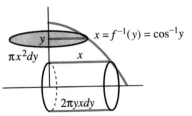

Fig. 8.7   Slices through x axis and shells around y axis (*good*). The opposite way needs $f^{-1}(y)$.

8.1 EXERCISES

Read-through questions

The area between $y = x^3$ and $y = x^4$ equals the integral of __a__. If the region ends where the curves intersect, we find the limits on x by solving __b__. Then the area equals __c__. When the area between $y = \sqrt{x}$ and the y axis is sliced horizontally, the integral to compute is __d__.

In three dimensions the volume of a slice is its thickness dx times its __e__. If the cross-sections are squares of side $1-x$, the volume comes from $\int$ __f__. From $x = 0$ to $x = 1$, this gives the volume __g__ of a square __h__. If the cross-sections are circles of radius $1-x$, the volume comes from $\int$ __i__. This gives the volume __j__ of a circular __k__.

For a solid of revolution, the cross-sections are __l__. Rotating the graph of $y = f(x)$ around the x axis gives a solid volume $\int$ __m__. Rotating around the y axis leads to $\int$ __n__. Rotating the area between $y = f(x)$ and $y = g(x)$ around the x axis, the slices look like __o__. Their areas are __p__ so the volume is $\int$ __q__.

Another method is to cut the solid into thin cylindrical __r__. Revolving the area under $y = f(x)$ around the y axis, a shell has height __s__ and thickness dx and volume __t__. The total volume is $\int$ __u__.

Find where the curves in 1–12 intersect, draw rough graphs, and compute the area between them.

1   $y = x^2 - 3$ and $y = 1$

2   $y = x^2 - 2$ and $y = 0$

3   $y^2 = x$ and $x = 9$

4   $y^2 = x$ and $x = y + 2$

5   $y = x^4 - 2x^2$ and $y = 2x^2$

6   $x = y^5$ and $y = x^4$

7   $y = x^2$ and $y = -x^2 + 18x$

8   $y = 1/x$ and $y = 1/x^2$ and $x = 3$

9   $y = \cos x$ and $y = \cos^2 x$

10   $y = \sin \pi x$ and $y = 2x$ and $x = 0$

11   $y = e^x$ and $y = e^{2x-1}$ and $x = 0$

12   $y = e$ and $y = e^x$ and $y = e^{-x}$

13   Find the area inside the three lines $y = 4 - x, y = 3x$, and $y = x$.

14   Find the area bounded by $y = 12 - x, y = \sqrt{x}$, and $y = 1$.

15   Does the parabola $y = 1 - x^2$ out to $x = 1$ sit inside or outside the unit circle $x^2 + y^2 = 1$? Find the area of the "skin" between them.

16   Find the area of the largest triangle with base on the x axis that fits (a) inside the unit circle (b) inside that parabola.

17   Rotate the ellipse $x^2/a^2 + y^2/b^2 = 1$ around the x axis to find the volume of a football. What is the volume around the y axis? If $a = 2$ and $b = 1$, locate a point (x, y, z) that is in one football but not the other.

18   What is the volume of the loaf of bread which comes from rotating $y = \sin x \, (0 \leqslant x \leqslant \pi)$ around the x axis?

19   What is the volume of the flying saucer that comes from rotating $y = \sin x \, (0 \leqslant x \leqslant \pi)$ around the y axis?

20   What is the volume of the galaxy that comes from rotating $y = \sin x \, (0 \leqslant x \leqslant \pi)$ around the x axis and then rotating the whole thing around the y axis?

Draw the region bounded by the curves in 21–28. Find the volume when the region is rotated (a) around the x axis (b) around the y axis.

21   $x + y = 8, x = 0, y = 0$

22   $y - e^x = 1, x = 1, y = 0, x = 0$

23   $y = x^4, y = 1, x = 0$

24 $y = \sin x, y = \cos x, x = 0$

25 $xy = 1, x = 2, y = 3$

26 $x^2 - y^2 = 9, x + y = 9$ (rotate the region where $y \geqslant 0$)

27 $x^2 = y^3, x^3 = y^2$

28 $(x-2)^2 + (y-1)^2 = 1$

In 29–34 find the volume and draw a typical slice.

29 A cap of height h is cut off the top of a sphere of radius R. Slice the sphere horizontally starting at $y = R - h$.

30 A pyramid P has height 6 and square base of side 2. Its volume is $\frac{1}{3}(6)(2)^2 = 8$.

    (a) Find the volume up to height 3 by horizontal slices. What is the length of a side at height y?

    (b) Recompute by removing a smaller pyramid from P.

31 The base is a disk of radius a. Slices perpendicular to the base are squares.

32 The base is the region under the parabola $y = 1 - x^2$. Slices perpendicular to the x axis are squares.

33 The base is the region under the parabola $y = 1 - x^2$. Slices perpendicular to the y axis are squares.

34 The base is the triangle with corners $(0,0), (1,0), (0,1)$. Slices perpendicular to the x axis are semicircles.

35 Cavalieri's principle for areas: If two regions have strips of equal length, then the regions have the same area. Draw a parallelogram and a curved region, both with the same strips as the unit square. Why are the areas equal?

36 Cavalieri's principle for volumes: If two solids have slices of equal area, the solids have the same volume. Find the volume of the tilted cylinder in the figure.

37 Draw another region with the same slice areas as the tilted cylinder. When all areas $A(x)$ are the same, the volumes $\int$ _____ are the same.

38 Find the volume common to two circular cylinders of radius a. One eighth of the region is shown (axes are perpendicular and horizontal slices are squares).

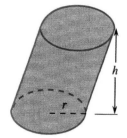

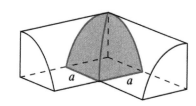

39 A wedge is cut out of a cylindrical tree (see figure). One cut is along the ground to the x axis. The second cut is at angle θ, also stopping at the x axis.

    (a) The curve C is part of a (circle) (ellipse) (parabola).

    (b) The height of point P in terms of x is _____.

    (c) The area $A(x)$ of the triangular slice is _____.

    (d) The volume of the wedge is _____.

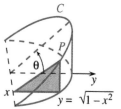

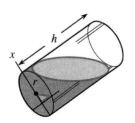

40 The same wedge is sliced perpendicular to the y axis.

    (a) The slices are now (triangles) (rectangles) (curved).

    (b) The slice area is _____ (slice height $y \tan \theta$).

    (c) The volume of the wedge is the integral _____.

    (d) Change the radius from 1 to r. The volume is multiplied by _____.

41 A cylinder of radius r and height h is half full of water. Tilt it so the water just covers the base.

    (a) Find the volume of water by common sense.

    (b) Slices perpendicular to the x axis are (rectangles) (trapezoids) (curved). I had to tilt an actual glass.

*42 Find the area of a slice in Problem 41. (The tilt angle has $\tan \theta = 2h/r$.) Integrate to find the volume of water.

The slices in 43–46 are washers. Find the slice area and volume.

43 The rectangle with sides $x = 1, x = 3, y = 2, y = 5$ is rotated around the x axis.

44 The same rectangle is rotated around the y axis.

45 The same rectangle is rotated around the line $y = 1$.

46 Draw the triangle with corners $(1,0), (1,1), (0,1)$. After rotation around the x axis, describe the solid and find its volume.

47 Bore a hole of radius a down the axis of a cone and through the base of radius b. If it is a $45°$ cone (height also b), what volume is left? Check $a = 0$ and $a = b$.

48 Find the volume common to two spheres of radius r if their centers are $2(r-h)$ apart. Use Problem 29 on spherical caps.

49 (Shells vs. disks) Rotate $y = 3 - x$ around the x axis from $x = 0$ to $x = 2$. Write down the volume integral by disks and then by shells.

50 (Shells vs. disks) Rotate $y = x^3$ around the y axis from $y = 0$ to $y = 8$. Write down the volume integral by shells and disks and compute both ways.

51 Yogurt comes in a solid of revolution. Rotate the line $y = mx$ around the y axis to find the volume between $y = a$ and $y = b$.

52 Suppose $y = f(x)$ decreases from $f(0) = b$ to $f(1) = 0$. The curve is rotated around the y axis. *Compare shells to disks*:

$$\int_0^1 2\pi x f(x)dx = \int_0^b \pi(f^{-1}(Y))^2 dy.$$

Substitute $y = f(x)$ in the second. Also substitute $dy = f'(x)dx$. Integrate by parts to reach the first.

53 If a roll of paper with inner radius 2 cm and outer radius 10 cm has about 10 thicknesses per centimeter, approximately how long is the paper when unrolled ?

54 Find the approximate volume of your brain. OK to include everything above your eyes (skull too).

Use shells to find the volumes in 55–63. The rotated regions lie between the curve and x axis.

55 $y = 1 - x^2, 0 \leqslant x \leqslant 1$ (around the y axis)

56 $y = 1/x, 1 \leqslant x \leqslant 100$ (around the y axis)

57 $y = \sqrt{1 - x^2}, 0 \leqslant x \leqslant 1$ (around either axis)

58 $y = 1/(1 + x^2), 0 \leqslant x \leqslant 3$ (around the y axis)

59 $y = \sin(x^2), 0 \leqslant x \leqslant \sqrt{\pi}$ (around the y axis)

60 $y = 1/\sqrt{1 - x^2}, 0 \leqslant x \leqslant 1$ (around the y axis)

61 $y = x^2, 0 \leqslant x \leqslant 2$ (around the x axis)

62 $y = e^x, 0 \leqslant x \leqslant 1$ (around the x axis)

63 $y = \ln x, 1 \leqslant x \leqslant e$ (around the x axis)

64 The region between $y = x^2$ and $y = x$ is revolved around the y axis. (a) Find the volume by cutting into shells. (b) Find the volume by slicing into washers.

65 The region between $y = f(x)$ and $y = 1 + f(x)$ is rotated around the y axis. The shells have height _____ . The volume out to $x = a$ is _____ . It equals the volume of a _____ because the shells are the same.

66 A horizontal slice of the sphere-with-hole in Figure 8.6b is a washer. Its area is $\pi x^2 - \pi a^2 = \pi(b^2 - y^2 - a^2)$.

(a) Find the upper limit on y (the top of the hole).

(b) Integrate the area to verify the volume in Example 12.

67 If the hole in the sphere has length 2, show that the volume is $4\pi/3$ regardless of the radii a and b.

*68 An upright cylinder of radius r is sliced by two parallel planes at angle α. One is a height h above the other.

(a) Draw a picture to show that the volume between the planes is $\pi r^2 h$.

(b) Tilt the picture by α, so the base and top are flat. What is the shape of the base ? What is its area A ? What is the height H of the tilted cylinder ?

69 *True or false*, with a reason.

(a) A cube can only be sliced into squares.

(b) A cube cannot be cut into cylindrical shells.

(c) The washer with radii r and R has area $\pi(R - r)^2$.

(d) The plane $w = \frac{1}{2}$ slices a 3-dimensional sphere out of a 4-dimensional sphere $x^2 + y^2 + z^2 + w^2 = 1$.

8.2 Length of a Plane Curve

The graph of $y = x^{3/2}$ is a curve in the x-y plane. **How long is that curve?** A definite integral needs endpoints, and we specify $x = 0$ and $x = 4$. The first problem is to know what "length function" to integrate.

The distance along a curve is the *arc length*. To set up an integral, we break the problem into small pieces. Roughly speaking, *small pieces of a smooth curve are nearly straight*. We know the exact length Δs of a straight piece, and Figure 8.8 shows how it comes close to a curved piece.

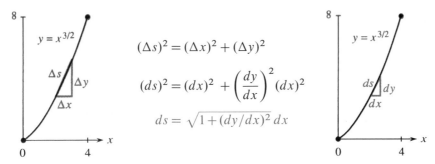

$$(\Delta s)^2 = (\Delta x)^2 + (\Delta y)^2$$

$$(ds)^2 = (dx)^2 + \left(\frac{dy}{dx}\right)^2 (dx)^2$$

$$ds = \sqrt{1 + (dy/dx)^2}\, dx$$

Fig. 8.8 Length Δs of short straight segment. Length ds of very short curved segment.

Here is the unofficial reasoning that gives the length of the curve. A straight piece has $(\Delta s)^2 = (\Delta x)^2 + (\Delta y)^2$. Within that right triangle, the height Δy is the slope $(\Delta y/\Delta x)$ times Δx. This secant slope is close to the slope of the curve. Thus Δy is approximately $(dy/dx)\, \Delta x$.

$$\Delta s \approx \sqrt{(\Delta x)^2 + (dy/dx)^2 (\Delta x)^2} = \sqrt{1 + (dy/dx)^2}\, \Delta x. \tag{1}$$

Now add these pieces and make them smaller. The infinitesimal triangle has $(ds)^2 = (dx)^2 + (dy)^2$. Think of ds as $\sqrt{1 + (dy/dx)^2}\, dx$ and integrate:

$$\textit{length of curve} = \int ds = \int \sqrt{1 + (dy/dx)^2}\, dx. \tag{2}$$

EXAMPLE 1 Keep $y = x^{3/2}$ and $dy/dx = \frac{3}{2} x^{1/2}$. Watch out for $\frac{3}{2}$ and $\frac{9}{4}$:

$$\text{length} = \int_0^4 \sqrt{1 + \tfrac{9}{4}x}\, dx = \left(\tfrac{2}{3}\right)\left(\tfrac{4}{9}\right)\left(1 + \tfrac{9}{4}x\right)^{3/2}\Big]_0^4 = \tfrac{8}{27}(10^{3/2} - 1^{3/2}). \tag{3}$$

This answer is just above 9. A straight line from $(0,0)$ to $(4,8)$ has exact length $\sqrt{80}$. Note $4^2 + 8^2 = 80$. Since $\sqrt{80}$ is just below 9, the curve is surprisingly straight.

You may not approve of those numbers (or the reasoning behind them). We can fix the reasoning, but nothing can be done about the numbers. This example $y = x^{3/2}$ had to be chosen carefully to make the integration possible at all. The length integral is difficult because of the square root. In most cases we integrate numerically.

EXAMPLE 2 The straight line $y = 2x$ from $x = 0$ to $x = 4$ has $dy/dx = 2$:

$$\text{length} = \int_0^4 \sqrt{1 + 4}\, dx = 4\sqrt{5} = \sqrt{80} \text{ as be-fore} \qquad \text{(just checking).}$$

We return briefly to the reasoning. The curve is the graph of $y = f(x)$. Each piece contains at least one point where secant slope equals tangent slope: $\Delta y/\Delta x = f'(c)$.

The Mean Value Theorem applies when the slope is continuous—this is required for a smooth curve. The straight length Δs is exactly $\sqrt{(\Delta x)^2 + (f'(c)\Delta x)^2}$. Adding the n pieces gives the length of the broken line (close to the curve):

$$\sum_1^n \Delta s_i = \sum_1^n \sqrt{1 + [f'(c_i)]^2}\, \Delta x_i.$$

As $n \to \infty$ and $\Delta x_{max} \to 0$ this approaches the integral that gives arc length.

8A The length of the curve $y = f(x)$ from $x = a$ to $x = 6$ is

$$s = \int ds = \int_a^b \sqrt{1 + [f'(x)]^2}\, dx = \int_a^b \sqrt{1 + (dy/dx)^2}\, dx. \qquad (4)$$

EXAMPLE 3   Find the length of the first quarter of the circle $y = \sqrt{1 - x^2}$.

Here $dy/dx = -x/\sqrt{1 - x^2}$. From Figure 8.9a, the integral goes from $x = 0$ to $x = 1$:

$$\text{length} = \int_0^1 \sqrt{1 + (dy/dx)^2}\, dx = \int_0^1 \sqrt{1 + \frac{x^2}{1 - x^2}}\, dx = \int_0^1 \frac{dx}{\sqrt{1 - x^2}}.$$

The antiderivative is $\sin^{-1} x$. It equals $\pi/2$ at $x = 1$. This length $\pi/2$ is a quarter of the full circumference 2π.

EXAMPLE 4   Compute the distance around a quarter of the *ellipse* $y^2 + 2x^2 = 2$.

The equation is $y = \sqrt{2 - 2x^2}$ and the slope is $dy/dx = -2x/\sqrt{2 - 2x^2}$. So $\int ds$ is

$$\int_0^1 \sqrt{1 + \frac{4x^2}{2 - 2x^2}}\, dx = \int_0^1 \sqrt{\frac{2 + 2x^2}{2 - 2x^2}}\, dx = \int_0^1 \sqrt{\frac{1 + x^2}{1 - x^2}}\, dx. \qquad (5)$$

That integral can't be done in closed form. *The length of an ellipse can only be computed numerically.* The denominator is zero at $x = 1$, so a blind application of the trapezoidal rule or Simpson's rule would give length $= \infty$. The midpoint rule gives length $= 1.91$ with thousands of intervals.

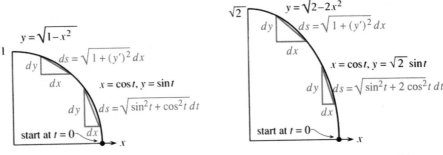

Fig. 8.9   Circle and ellipse, directly by $y = f(x)$ or parametrically by $x(t)$ and $y(t)$.

LENGTH OF A CURVE FROM PARAMETRIC EQUATIONS: $x(f)$ AND $y(f)$

We have met the unit circle in two forms. One is $x^2 + y^2 = 1$. The other is $x = \cos t$, $y = \sin t$. Since $\cos^2 t + \sin^2 t = 1$, this point goes around the correct circle. One advantage of the "***parameter***" t is to give extra information—it tells *where* the point is and also *when*. In Chapter 1, the parameter was the time and also the angle—because we moved around the circle with speed 1.

Using t is a natural way to give the position of a particle or a spacecraft. We can recover the velocity if we know x and y at every time t. An equation $y = f(x)$ tells the shape of the path, not the speed along it.

Chapter 12 deals with parametric equations for curves. Here we concentrate on the *path length*—which allows you to see the idea of a parameter t without too much detail. We give x as a function of t and y as a function of t. The curve is still approximated by straight pieces, and each piece has $(\Delta s)^2 = (\Delta x)^2 + (\Delta y)^2$. But instead of using $\Delta y \approx (dy/dx)\,\Delta x$, we approximate Δx and Δy separately:

$$\Delta x \approx (dx/dt)\Delta t, \quad \Delta y \approx (dy/dt)\Delta t, \quad \Delta s \approx \sqrt{(dx/dt)^2 + (dy/dt)^2}\,\Delta t.$$

8B The length of a parametric curve is an integral with respect to t:

$$\int ds = \int (ds/dt)\,dt = \int \sqrt{(dx/dt)^2 + (dy/dt)^2}\,dt. \qquad (6)$$

EXAMPLE 5   Find the length of the quarter-circle using $x = \cos t$ and $y = \sin t$:

$$\int_0^{\pi/2} \sqrt{(dx/dt)^2 + (dy/dt)^2}\,dt = \int_0^{\pi/2} \sqrt{\sin^2 t + \cos^2 t}\,dt = \int_0^{\pi/2} dt = \frac{\pi}{2}.$$

The integral is simpler than $1/\sqrt{1 - x^2}$, and there is one new advantage. *We can integrate around a whole circle with no trouble.* Parametric equations allow a path to close up or even cross itself. The time t keeps going and the point $(x(t), y(t))$ keeps moving. In contrast, curves $y = f(x)$ are limited to one y for each x.

EXAMPLE 6   Find the length of the quarter-ellipse: $x = \cos t$ and $y = \sqrt{2}\sin t$:

On this path $y^2 + 2x^2$ is $2\sin^2 t + 2\cos^2 t = 2$ (same ellipse). The ***non***-parametric equation $y = \sqrt{2 - 2x^2}$ comes from eliminating t. We keep t:

$$\text{length} = \int_0^{\pi/2} \sqrt{(dx/dt)^2 + (dy/dt)^2}\,dt = \int_0^{\pi/2} \sqrt{\sin^2 t + 2\cos^2 t}\,dt. \qquad (7)$$

This integral (7) must equal (5). If one cannot be done, neither can the other. They are related by $x = \cos t$, but (7) does not blow up at the endpoints. The trapezoidal rule gives 1.9101 with less than 100 intervals. Section 5.8 mentioned that calculators *automatically* do a substitution that makes (5) more like (7).

EXAMPLE 7   The path $x = t^2$, $y = t^3$ goes from $(0,0)$ to $(4,8)$. Stop at $t = 2$.

To find this path without the parameter t, first solve for $t = x^{1/2}$. Then substitute into the equation for y: $y = t^3 = x^{3/2}$. ***The non-parametric form*** (with t eliminated) ***is the same curve*** $y = x^{3/2}$ ***as in Example 1***.

The length from the t-integral equals the length from the x-integral. This is Problem 22.

EXAMPLE 8 *Special choice of parameter*: t is x. The curve becomes $x = t$, $y = t^{3/2}$.

If $x = t$ then $dx/dt = 1$. The square root in (6) is the same as the square root in (4). Thus the non-parametric form $y = f(x)$ is a special case of the parametric form—just take $t = x$.
  Compare $x = t$, $y = t^{3/2}$ with $x = t^2$, $y = t^3$. *Same curve, same length, different speed.*

EXAMPLE 9 Define "*speed*" by $\dfrac{\text{short distance}}{\text{short time}} = \dfrac{ds}{dt}$. It is $\sqrt{\left(\dfrac{dx}{dt}\right)^2 + \left(\dfrac{dy}{dt}\right)^2}$.

When a ball is thrown straight upward, dx/dt is zero. But the speed is not dy/dt. It is $|dy/dt|$. The speed is positive downward as well as upward.

8.2 EXERCISES

Read-through questions

The length of a straight segment (Δx across, Δy up) is $\Delta s = $ __a__ . Between two points of the graph of $y(x)$, Δy is approximately dy/dx times __b__ . The length of that piece is approximately $\sqrt{(\Delta x)^2 + }$ __c__ . An infinitesimal piece of the curve has length $ds = $ __d__ . Then the arc length integral is $\int$ __e__ .

For $y = 4 - x$ from $x = 0$ to $x = 3$ the arc length is $\int$ __f__ $ = $ __g__ . For $y = x^3$ the arc length integral is __h__ .

The curve $x = \cos t$, $y = \sin t$ is the same as __i__ . The length of a curve given by $x(t)$, $y(t)$ is $\int \sqrt{\ \text{j}\ }\, dt$. For example $x = \cos t$, $y = \sin t$ from $t = \pi/3$ to $t = \pi/2$ has length __k__ . The speed is $ds/dt = $ __l__ . For the special case $x = t$, $y = f(t)$ the length formula goes back to $\int \sqrt{\text{m}}\, dx$.

Find the lengths of the curves in Problems 1–8.

1 $y = x^{3/2}$ from $(0,0)$ to $(1,1)$

2 $y = x^{2/3}$ from $(0,0)$ to $(1,1)$ (compare with Problem 1 or put $u = \frac{4}{9} + x^{2/3}$ in the length integral)

3 $y = \frac{1}{3}(x^2 + 2)^{3/2}$ from $x = 0$ to $x = 1$

4 $y = \frac{1}{3}(x^2 - 2)^{3/2}$ from $x = 2$ to $x = 4$

5 $y = \dfrac{x^3}{3} + \dfrac{1}{4x}$ from $x = 1$ to $x = 3$

6 $y = \dfrac{x^4}{4} + \dfrac{1}{8x^2}$ from $x = 1$ to $x = 2$

7 $y = \frac{2}{3}x^{3/2} - \frac{1}{2}x^{1/2}$ from $x = 1$ to $x = 4$

8 $y = x^2$ from $(0,0)$ to $(1,1)$

9 The curve given by $x = \cos^3 t$, $y = \sin^3 t$ is an *astroid* (a hypocycloid). Its non-parametric form is $x^{2/3} + y^{2/3} = 1$. Sketch the curve from $t = 0$ to $t = \pi/2$ and find its length.

10 Find the length from $t = 0$ to $t = \pi$ of the curve given by $x = \cos t + \sin t$, $y = \cos t - \sin t$. Show that the curve is a circle (of what radius?).

11 Find the length from $t = 0$ to $t = \pi/2$ of the curve given by $x = \cos t$, $y = t - \sin t$.

12 What integral gives the length of Archimedes' spiral $x = t \cos t$, $y = t \sin t$?

13 Find the distance traveled in the first second (to $t = 1$) if $x = \frac{1}{2}t^2$, $y = \frac{1}{3}(2t + 1)^{3/2}$.

14 $x = (1 - \frac{1}{2}\cos 2t)\cos t$ and $y = (1 + \frac{1}{2}\cos 2t)\sin t$ lead to $4(1 - x^2 - y^2)^3 = 27(x^2 - y^2)^2$. Find the arc length from $t = 0$ to $\pi/4$.

Find the arc lengths in 15–18 by numerical integration.

15 One arch of $y = \sin x$, from $x = 0$ to $x = \pi$.

16 $y = e^x$ from $x = 0$ to $x = 1$.

17 $y = \ln x$ from $x = 1$ to $x = e$.

18 $x = \cos t$, $y = 3 \sin t$, $0 \leqslant x \leqslant 2\pi$.

19 Draw a rough picture of $y = x^{10}$. Without computing the length of $y = x^n$ from $(0,0)$ to $(1,1)$, find the limit as $n \to \infty$.

20 Which is longer between $(1,1)$ and $(2, \frac{1}{2})$, the hyperbola $y = 1/x$ or the graph of $x + 2y = 3$?

21 Find the speed ds/dt on the circle $x = 2 \cos 3t$, $y = 2 \sin 3t$.

22 Examples 1 and 7 were $y = x^{3/2}$ and $x = t^2$, $y = t^3$:

$$\text{length} = \int_0^4 \sqrt{1 + \tfrac{9}{4}x}\, dx, \quad \text{length} = \int_0^2 \sqrt{4t^2 + 9t^4}\, dt.$$

Show by substituting $x = $ _____ that these integrals agree.

23 Instead of $y = f(x)$ a curve can be given as $x = g(y)$. Then

$$ds = \sqrt{(dx)^2 + (dy)^2} = \sqrt{(dx/dy)^2 + 1}\, dy.$$

Draw $x = 5y$ from $y = 0$ to $y = 1$ and find its length.

24 The length of $x = y^{3/2}$ from $(0,0)$ to $(1,1)$ is $\int ds = \int \sqrt{\frac{9}{4}y + 1}\, dy$. Compare with Problem 1: Same length? Same curve?

25 Find the length of $x = \frac{1}{2}(e^y + e^{-y})$ from $y = -1$ to $y = 1$ and draw the curve.

26 The length of $x = g(y)$ is a special case of equation (6) with $y = t$ and $x = g(t)$. The length integral becomes _____.

27 Plot the point $x = 3\cos t$, $y = 4\sin t$ at the five times $t = 0$, $\pi/2$, π, $3\pi/2$, 2π. The equation of the curve is $(x/3)^2 + (y/4)^2 = 1$, not a circle but an _____. This curve cannot be written as $y = f(x)$ because _____.

28 (a) Find the length of $x = \cos^2 t$, $y = \sin^2 t$, $0 \leqslant y \leqslant \pi$.

(b) Why does this path stay on the line $x + y = 1$?

(c) Why isn't the path length equal to $\sqrt{2}$?

29 (important) The line $y = x$ is close to a staircase of pieces that go *straight across or straight up*. With 100 pieces of length $\Delta x = 1/100$ or $\Delta y = 1/100$, find the length of carpet on the staircase. (The length of the $45°$ line is $\sqrt{2}$. The staircase can be close when its length is not close.)

30 The area of an ellipse is πab. The area of a strip around it (width Δ) is $\pi(a+\Delta)(b+\Delta) - \pi ab \approx \pi(a+b)\Delta$. The distance around the ellipse seems to be $\pi(a+b)$. But this distance is impossible to find—what is wrong?

31 The point $x = \cos t$, $y = \sin t$, $z = t$ moves on a *space curve*.

(a) In three-dimensional space $(ds)^2$ equals $(dx)^2 + $ _____. In equation (6), ds is now _____ dt.

(b) This particular curve has $ds = $ _____. Find its length from $t = 0$ to $t = 2\pi$.

(c) Describe the curve and its shadow in the xy plane.

32 Explain in 50 words the difference between a non-parametric equation $y = f(x)$ and two parametric equations $x = x(t)$, $y = y(t)$.

33 Write down the integral for the length L of $y = x^2$ from $(0,0)$ to $(1,1)$. Show that $y = \frac{1}{2}x^2$ from $(0,0)$ to $(2,2)$ is exactly twice as long. If possible give a reason using the graphs.

34 (for professors) Compare the lengths of the parabola $y = x^2$ and the line $y = bx$ from $(0,0)$ to (b,b^2). Does the difference approach a limit as $b \to \infty$?

8.3 Area of a Surface of Revolution

This section starts by constructing surfaces. *A curve $y = f(x)$ is revolved around an axis*. That produces a "*surface of revolution*," which is symmetric around the axis. If we revolve a sloping line, the result is a cone. When the line is parallel to the axis we get a cylinder (a pipe). By revolving a curve we might get a lamp or a lamp shade (or even the light bulb).

Section 8.1 computed the volume inside that surface. **This section computes the surface area.** Previously we cut the solid into slices or shells. Now we need a good way to cut up the surface.

The key idea is *to revolve short straight line segments*. Their slope is $\Delta y / \Delta x$. They can be the same pieces of length Δs that were used to find length—now we compute area. When revolved, a straight piece produces a "*thin band*" (Figure 8.10). The curved surface, from revolving $y = f(x)$, is close to the bands. The first step is to compute *the surface area of a band*.

A small comment: Curved surfaces can also be cut into tiny patches. Each patch is nearly flat, like a little square. The sum of those patches leads to a double integral (with $dx\,dy$). Here the integral stays one-dimensional (dx or dy or dt). Surfaces of revolution are special—we approximate them by bands that go all the way around. A band is just a belt with a slope, and its slope has an effect on its area.

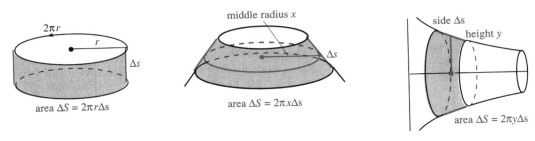

Fig. 8.10   Revolving a straight piece and a curve around the y axis and x axis.

Revolve a small straight piece (*length Δs not Δx*). The center of the piece goes around a circle of radius r. The band is *a slice of a cone*. When we flatten it out (Problems $11 - 13$) we discover its area. The area is the *side length Δs* times the *middle circumference $2\pi r$*:

> **The surface area of a band is $2\pi r \Delta s = 2\pi r \sqrt{1 + (\Delta y / \Delta x)^2}\, \Delta x$.**

For revolution around the y axis, the radius is $r = x$. For revolution around the x axis, the radius is the height: $r = y = f(x)$. Figure 8.10 shows both bands—the problem tells us which to use. The sum of band areas $2\pi r\,\Delta s$ is close to the area S of the curved surface. In the limit we integrate $2\pi r\,ds$:

> **8C** The surface area generated by revolving the curve $y = f(x)$ between $x = a$ and $x = b$ is
>
> $$S = \int_a^b 2\pi y \sqrt{1 + (dy/dx)^2}\, dx \quad \text{around the } x \text{ axis} \quad (r = y) \quad (1)$$
>
> $$S = \int_a^b 2\pi x \sqrt{1 + (dy/dx)^2}\, dx \quad \text{around the } y \text{ axis} \quad (r = x). \quad (2)$$

EXAMPLE 1 Revolve a complete semicircle $y = \sqrt{R^2 - x^2}$ around the x axis.

The surface of revolution is a **sphere**. Its area (known!) is $4\pi R^2$. The limits on x are $-R$ and R. The slope of $y = \sqrt{R^2 - x^2}$ is $dy/dx = -x/\sqrt{R^2 - x^2}$:

$$\text{area } S = \int_{-R}^{R} 2\pi \sqrt{R^2 - x^2} \sqrt{1 + \frac{x^2}{R^2 - x^2}} \, dx = \int_{-R}^{R} 2\pi R \, dx = 4\pi R^2.$$

EXAMPLE 2 Revolve a piece of the straight line $y = 2x$ around the x axis.

The surface is a **cone** with $(dy/dx)^2 = 4$. The band from $x = 0$ to $x = 1$ has area $2\pi\sqrt{5}$:

$$S = \int 2\pi y \, ds = \int_0^1 2\pi(2x)\sqrt{1+4} \, dx = 2\pi\sqrt{5}.$$

This answer must agree with the formula $2\pi r \, \Delta s$ (which it came from). The line from $(0,0)$ to $(1,2)$ has length $\Delta s = \sqrt{5}$. Its mid-point is $(\frac{1}{2}, 1)$. Around the x axis, the middle radius is $r = 1$ and the area is $2\pi\sqrt{5}$.

EXAMPLE 3 Revolve the same straight line segment around the y axis. Now the radius is x instead of $y = 2x$. The area in Example 2 is cut in half:

$$S = \int 2\pi x \, ds = \int_0^1 2\pi x\sqrt{1+4} \, dx = \pi\sqrt{5}.$$

For surfaces as for arc length, only a few examples have convenient answers. Watermelons and basketballs and light bulbs are in the exercises. Rather than stretching out this section, we give a final area formula and show how to use it.

The formula applies when there is a **parameter** t. Instead of $(x, f(x))$ the points on the curve are $(x(t), y(t))$. As t varies, we move along the curve. The length formula $(ds)^2 = (dx)^2 + (dy)^2$ is expressed *in terms of* t.

For the surface of revolution around the x axis, the area becomes a t-integral:

8D The surface area is $\int 2\pi y \, ds = \int 2\pi y(t)\sqrt{(dx/dt)^2 + (dy/dt)^2} \, dt.$ (3)

EXAMPLE 4 The point $x = \cos t$, $y = 5 + \sin t$ travels on a circle with center at $(0,5)$. Revolving that circle around the x axis produces a doughnut. Find its surface area.

Solution $(dx/dt)^2 + (dy/dt)^2 = \sin^2 t + \cos^2 t = 1$. The circle is complete at $t = 2\pi$:

$$\int 2\pi y \, ds = \int_0^{2\pi} 2\pi(5 + \sin t) dt = \left[2\pi(5t - \cos t) \right]_0^{2\pi} = 20\pi^2.$$

Read-through questions

A surface of revolution comes from revolving a __a__ around __b__. This section computes the __c__. When the curve is a short straight piece (length Δs), the surface is a __d__. Its area is $\Delta S =$ __e__. In that formula (Problem 13) r is the radius of __f__. The line from $(0,0)$ to $(1,1)$ has length __g__, and revolving it produces area __h__.

When the curve $y = f(x)$ revolves around the x axis, the surface area is the integral __i__. For $y = x^2$ the integral to compute is __j__. When $y = x^2$ is revolved around the y axis, the area is $S =$ __k__. For the curve given by $x = 2t, y = t^2$, change ds to __l__ dt.

Find the surface area when curves 1–6 revolve around the x axis.

1 $y = \sqrt{x}$, $2 \leqslant x \leqslant 6$

2 $y = x^3$, $0 \leqslant x \leqslant 1$

3 $y = 7x$, $-1 \leqslant x \leqslant 1$ (watch sign)

4 $y = \sqrt{4 - x^2}$, $0 \leqslant x \leqslant 2$

5 $y = \sqrt{4 - x^2}$, $-1 \leqslant x \leqslant 1$

6 $y = \cosh x$, $0 \leqslant x \leqslant 1$.

In 7–10 find the area of the surface of revolution around the y axis.

7 $y = x^2$, $0 \leqslant x \leqslant 2$ 8 $y = \frac{1}{2}x^2 + \frac{1}{2}$, $0 \leqslant x \leqslant 1$

9 $y = x + 1$, $0 \leqslant x \leqslant 3$ 10 $y = x^{1/3}$, $0 \leqslant x \leqslant 1$

11 A cone with base radius R and slant height s is laid out flat. Explain why the angle (in radians) is $\theta = 2\pi R/s$. Then the surface area is a fraction of a circle:

$$\text{area} = \left(\frac{\theta}{2\pi}\right)\pi s^2 = \left(\frac{R}{s}\right)\pi s^2 = \pi R s.$$

12 A band with slant height $\Delta s = s - s'$ and radii R and R' is laid out flat. Explain in one line why its surface area is $\pi R s - \pi R' s'$.

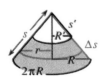

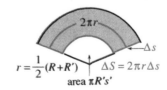

$r = \frac{1}{2}(R + R')$ $\Delta S = 2\pi r \Delta s$
area $\pi R's'$

13 By similar triangles $R/s = R'/s'$ or $Rs' = R's$. The middle radius r is $\frac{1}{2}(R + R')$. Substitute for r and Δs in the proposed *area formula* $2\pi r \Delta s$, to show that this gives the correct area $\pi R s - \pi R' s'$.

14 Slices of a basketball all have the same area of cover, if they have the same thickness.

(a) Rotate $y = \sqrt{1 - x^2}$ around the x axis. Show that $dS = 2\pi\, dx$.

(b) The area between $x = a$ and $x = a + h$ is _____.

(c) $\frac{1}{4}$ of the Earth's area is above latitude _____.

15 Change the circle in Example 4 to $x = a \cos t$ and $y = b + a \sin t$. Its radius is _____ and its center is _____. Find the surface area of a torus by revolving this circle around the x axis.

16 What part of the circle $x = R \cos t, y = R \sin t$ should rotate around the y axis to produce the top half of a sphere? Choose limits on t and verify the area.

17 The base of a lamp is constructed by revolving the quarter-circle $y = \sqrt{2x - x^2}$ ($x = 1$ to $x = 2$) around the y axis. Draw the quarter-circle, find the area integral, and compute the area.

18 The light bulb is a sphere of radius $1/2$ with its bottom sliced off to fit onto a cylinder of radius $1/4$ and length $1/3$. Draw the light bulb and find its surface area (ends of the cylinder not included).

19 The lamp shade is constructed by rotating $y = 1/x$ around the y axis, and keeping the part from $y = 1$ to $y = 2$. Set up the definite integral that gives its surface area.

20 Compute the area of that lamp shade.

21 Explain why the surface area is infinite when $y = 1/x$ is rotated around the x axis ($1 \leqslant x < \infty$). But the volume of "Gabriel's horn" is _____. It can't enough paint to paint its surface.

22 A disk of radius $1''$ can be covered by four strips of tape (width $\frac{1}{2}''$). If the strips are not parallel, prove that they can't cover the disk. **Hint**: Change to a unit sphere sliced by planes $\frac{1}{2}''$ apart. Problem 14 gives surface area π for each slice.

23 A watermelon (maybe a football) is the result of rotating half of the ellipse $x = \sqrt{2} \cos t, y = \sin t$ (which means $x^2 + 2y^2 = 2$). Find the surface area, parametrically or not.

24 Estimate the surface area of an egg.

8.4 Probability and Calculus

Discrete probability usually involves careful counting. Not many samples are taken and not many experiments are made. There is a list of possible outcomes, and a known probability for each outcome. But probabilities go far beyond red cards and black cards. The real questions are much more practical:

1. How often will too many passengers arrive for a flight ?

2. How many random errors do you make on a quiz ?

3. What is the chance of exactly one winner in a big lottery ?

Those are important questions and we will set up models to answer them.

There is another point. Discrete models do not involve calculus. The number of errors or bumped passengers or lottery winners is a small whole number. *Calculus enters for continuous probability*. Instead of results that exactly equal 1 or 2 or 3, calculus deals with results that fall in a range of numbers. Continuous probability comes up in at least two ways:

(A) An experiment is repeated many times and we take *averages*.

(B) The outcome lies anywhere in an *interval* of numbers.

In the continuous case, the probability p_n of hitting a particular value $x = n$ becomes zero. Instead we have a ***probability density*** $p(x)$—which is a key idea. *The chance that a random X falls between a and b is found by integrating the density p(x)*:

$$\text{Prob}\{a \leqslant X \leqslant b\} = \int_a^b p(x)\,dx. \tag{1}$$

Roughly speaking, $p(x)\,dx$ is the chance of falling between x and $x + dx$. Certainly $p(x) \geqslant 0$. If a and b are the extreme limits $-\infty$ and ∞, including all possible outcomes, the probability is necessarily one:

$$\text{Prob}\{-\infty < X < +\infty\} = \int_{-\infty}^{\infty} p(x)\,dx = 1. \tag{2}$$

This is a case where infinite limits of integration are natural and unavoidable. In studying probability they create no difficulty—areas out to infinity are often easier.

Here are typical questions involving continuous probability and calculus:

4. How conclusive is a $53\% - 47\%$ poll of 2500 voters ?

5. Are 16 random football players safe on an elevator with capacity 3600 pounds ?

6. How long before your car is in an accident ?

It is not so traditional for a calculus course to study these questions. They need extra thought, beyond computing integrals (so this section is harder than average). But probability is more important than some traditional topics, and also more interesting. Drug testing and gene identification and market research are major applications. Comparing Questions **1–3** with **4–6** brings out the relation of **discrete** to **continuous**— the differences between them, and the parallels.

It would be impossible to give here a full treatment of probability theory. I believe you will see the point (and the use of calculus) from our examples. Frank Morgan's lectures have been a valuable guide.

8 Applications of the Integral

DISCRETE RANDOM VARIABLES

A **discrete** random variable X has a list of possible values. For two dice the outcomes are $X = 2, 3, \ldots, 12$. For coin tosses (see below), the list is infinite: $X = 1, 2, 3, \ldots$.

A **continuous** variable lies in an interval $a \leqslant X \leqslant b$.

EXAMPLE 1   Toss a fair coin until heads come up. The outcome X is the **number of tosses**. The value of X is 1 or 2 or 3 or $\ldots$, and the probability is $\frac{1}{2}$ that $X = 1$ (heads on the first toss). The probability of tails then heads is $p_2 = \frac{1}{4}$. The probability that $X = n$ is $p_n = (\frac{1}{2})^n$—this is the chance of $n - 1$ tails followed by heads. **The sum of all probabilities is necessarily** 1:

$$p_1 + p_2 + p_3 + \cdots = \tfrac{1}{2} + \tfrac{1}{4} + \tfrac{1}{8} + \cdots = 1.$$

EXAMPLE 2   Suppose a student (not you) makes an average of 2 unforced errors per hour exam. The number of actual errors on the next exam is $X = 0$ or 1 or 2 or $\ldots$. A reasonable model for the probability of n errors—when they are random and independent—is the **Poisson model** (pronounced Pwason):

$$p_n = \textit{probability of } n \textit{ errors} = \frac{2^n}{n!} e^{-2}.$$

The probabilities of no errors, one error, and two errors are p_0, p_1, and p_2:

$$p_0 = \frac{2^0}{0!} e^{-2} = \frac{1}{1} e^{-2} \approx .135 \qquad p_1 = \frac{2^1}{1!} e^{-2} \approx .27 \qquad p_2 = \frac{2^2}{2!} e^{-2} \approx .27.$$

The probability of more than two errors is $1 - .135 - .27 - .27 = .325$.

This Poisson model can be derived theoretically or tested experimentally. The total probability is again 1, from the infinite series (Section 6.6) for e^2:

$$p_0 + p_1 + p_2 + \cdots = \left(\frac{2^0}{0!} + \frac{2^1}{1!} + \frac{2^2}{2!} + \cdots \right) e^{-2} = e^2 e^{-2} = 1. \qquad (3)$$

EXAMPLE 3   Suppose on average 3 out of 100 passengers with reservations don't show up for a flight. If the plane holds 98 passengers, **what is the probability that someone will be bumped**?

If the passengers come independently to the airport, use the Poisson model with 2 changed to 3. X is the number of no-shows, and $X = n$ happens with probability p_n:

$$p_n = \frac{3^n}{n!} e^{-3} \qquad p_0 = \frac{3^0}{0!} e^{-3} \qquad p_1 = \frac{3^1}{1!} e^{-3} = 3e^{-3}.$$

There are 98 seats and 100 reservations. Someone is bumped if $X = 0$ or $X = 1$:

$$\text{chance of bumping} = p_0 + p_1 = e^{-3} + 3e^{-3} \approx 4/20.$$

We will soon define the **average** or **expected value** or **mean** of X—this model has $\mu = 3$.

CONTINUOUS RANDOM VARIABLES

If X is the lifetime of a VCR, all numbers $X \geqslant 0$ are possible. If X is a score on the SAT, then $200 \leqslant X \leqslant 800$. If X is the fraction of computer owners in a poll of 600 people, X is between 0 and 1. You may object that the SAT score is a whole number and the fraction of computer owners must be 0 or $1/600$ or $2/600$ or .... But it is completely impractical to work with 601 discrete possibilities. Instead we take X to be a ***continuous random variable***, falling *anywhere* in the range $X \geqslant 0$ or $[200, 800]$ or $0 \leqslant X \leqslant 1$. Of course the various values of X are not equally probable.

EXAMPLE 4 The average lifetime of a VCR is 4 years. A reasonable model for breakdown time is an ***exponential random variable***. Its probability density is

$$p(x) = \tfrac{1}{4} e^{-x/4} \quad \text{for} \quad 0 \leqslant x < \infty.$$

The probability that the VCR will eventually break is 1:

$$\int_0^\infty \tfrac{1}{4} e^{-x/4} \, dx = \left[-e^{-x/4}\right]_0^\infty = 0 - (-1) = 1. \tag{4}$$

The probability of breakdown within 12 years (X from 0 to 12) is .95:

$$\int_0^{12} \tfrac{1}{4} e^{-x/4} \, dx = \left[-e^{-x/4}\right]_0^{12} = -e^{-3} + 1 \approx .95. \tag{5}$$

An exponential distribution has $p(x) = a e^{-ax}$. Its integral from 0 to x is $F(x) = 1 - e^{-ax}$. Figure 8.11 is the graph for $a = 1$. It shows the area up to $x = 1$.

To repeat: ***The probability that $a \leqslant X \leqslant b$ is the integral of $p(x)$ from a to b.***

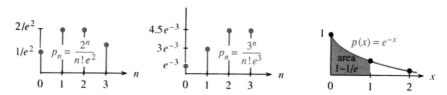

Fig. 8.11 Probabilities add to $\Sigma \, p_n = 1$. Continuous density integrates to $\int p(x) \, dx = 1$.

EXAMPLE 5 We now define the most important density function. Suppose the average SAT score is 500, and the *standard deviation* (defined below—it measures the spread around the average) is 200. Then the ***normal distribution*** of grades has

$$p(x) = \frac{1}{200\sqrt{2\pi}} e^{-(x-500)^2/2(200)^2} \quad \text{for} \quad -\infty < x < \infty.$$

This is the normal (or Gaussian) distribution with mean 500 and standard deviation 200. The graph of $p(x)$ is the famous ***bell-shaped curve*** in Figure 8.12.

A new objection is possible. The actual scores are between 200 and 800, while the density $p(x)$ extends all the way from $-\infty$ to ∞. I think the Educational Testing Service counts all scores over 800 as 800. The fraction of such scores is pretty small—in fact the normal distribution gives

$$\text{Prob}\{X \geqslant 800\} = \int_{800}^\infty \frac{1}{200\sqrt{2\pi}} e^{-(x-500)^2/2(200)^2} \, dx \approx .0013. \tag{6}$$

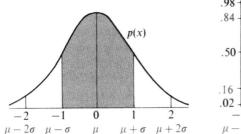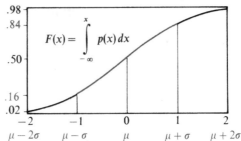

Fig. 8.12   The normal distribution (bell-shaped curve) and its cumulative density $F(x)$.

Regrettably, e^{-x^2} has no elementary antiderivative. We need numerical integration. But there is nothing the matter with that! The integral is called the *"error function,"* and special tables give its value to great accuracy. The integral of $e^{-x^2/2}$ from $-\infty$ to ∞ is exactly $\sqrt{2\pi}$. Then division by $\sqrt{2\pi}$ keeps $\int p(x)\,dx = 1$.

Notice that the normal distribution involves *two parameters*. They are the mean value (in this case $\mu = 500$) and the standard deviation (in this case $\sigma = 200$). Those numbers *mu* and *sigma* are often given the "normalized" values $\mu = 0$ and $\sigma = 1$:

$$p(x) = \frac{1}{\sigma\sqrt{2\pi}}\, e^{-(x-\mu)^2/2\sigma^2} \quad \text{becomes} \quad p(x) = \frac{1}{\sqrt{2\pi}}\, e^{-e^2/2}.$$

The bell-shaped graph of p is symmetric around the middle point $x = \mu$. The width of the graph is governed by the second parameter σ—which stretches the x axis and shrinks the y axis (leaving total area equal to 1). The axes are labeled to show the standard case $\mu = 0, \sigma = 1$ and also the graph for any other μ and σ.

We now give a name to the integral of $p(x)$. The limits will be $-\infty$ and x, so the integral $F(x)$ measures the ***probability that a random sample is below*** x:

$$\text{Prob}\{X \leqslant x\} = \int_{-\infty}^{x} p(x)\,dx = \textit{\textbf{cumulative density function}}\ F(x). \quad (7)$$

$F(x)$ accumulates the probabilities given by $p(x)$, so $dF/dx = p(x)$. The total probability is $F(\infty) = 1$. This integral from $-\infty$ to ∞ covers all outcomes.

Figure 8.12b shows the integral of the bell-shaped normal distribution. The middle point $x = \mu$ has $F = \frac{1}{2}$. By symmetry there is a $50 - 50$ chance of an outcome below the mean. The cumulative density $F(x)$ is near *.16* at $\mu - \sigma$ and near *.84* at $\mu + \sigma$. The chance of falling in between is $.84 - .16 = .68$. Thus 68% of the outcomes are less than one deviation σ away from the center μ.

Moving out to $\mu - 2\sigma$ and $\mu + 2\sigma$, 95% of the area is in between. **With** 95% **confidence** X **is less than two deviations from the mean**. Only one sample in 20 is further out (less than one in 40 on each side).

Note that $\sigma = 200$ is not the precise value for the SAT!

MEAN, VARIANCE, AND STANDARD DEVIATION

In Example 1, X was the number of coin tosses until the appearance of heads. The probabilities were $p_1 = \frac{1}{2}, p_2 = \frac{1}{4}, p_3 = \frac{1}{8}, \ldots$. *What is the **average** number of tosses?* We now find the "mean" μ of any distribution $p(x)$—not only the normal distribution, where symmetry guarantees that the built-in number μ is the mean.

To find μ, ***multiply outcomes by probabilities and add***:

$$\mu = \textit{\textbf{mean}} = \sum np_n = 1(p_1) + 2(p_2) + 3(p_3) + \cdots. \quad (8)$$

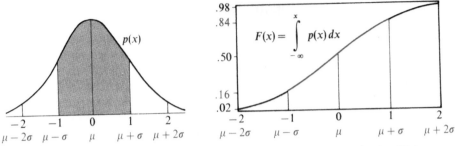

The average number of tosses is $1(\frac{1}{2}) + 2(\frac{1}{4}) + 3(\frac{1}{8}) + \cdots$. This series adds up (in Section 10.1) to $\mu = 2$. Please do the experiment 10 times. I am almost certain that the average will be near 2.

When the average is $\lambda = 2$ quiz errors or $\lambda = 3$ no-shows, the Poisson probabilities are $p_n = \lambda^n e^{-\lambda}/n!$ Check that the formula $\mu = \Sigma n p_n$ does give λ as the mean:

$$\left[1\frac{\lambda}{1!} + 2\frac{\lambda^2}{2!} + 3\frac{\lambda^3}{3!} + \cdots\right]e^{-\lambda} = \lambda\left[1 + \frac{\lambda}{1!} + \frac{\lambda^2}{2!} + \cdots\right]e^{-\lambda} = \lambda e^{\lambda}e^{-\lambda} = \lambda.$$

For continuous probability, the sum $\mu = \Sigma n p_n$ changes to $\mu = \int x p(x)\,dx$. We multiply outcome x by probability $p(x)$ and integrate. In the VCR model, integration by parts gives a mean breakdown time of $\mu = 4$ years:

$$\int x\, p(x)\,dx = \int_0^\infty x(\tfrac{1}{4}e^{-x/4})\,dx = \left[-xe^{-x/4} - 4e^{-x/4}\right]_0^\infty = 4. \qquad (9)$$

Together with the mean we introduce the **variance**. It is always written σ^2, and in the normal distribution that measured the "width" of the curve. When σ^2 was 200^2, SAT scores spread out pretty far. If the testing service changed to $\sigma^2 = 1^2$, the scores would be a disaster. 95% of them would be within ± 2 of the mean. When a teacher announces an average grade of 72, the variance should also be announced—if it is big then those with 60 can relax. At least they have company.

8E The mean μ is the expected value of X. The variance σ^2 is the expected value of $(X - \text{mean})^2 = (X - \mu)^2$. Multiply outcome times probability and add:

$$\mu = \sum n p_n \qquad\quad \sigma^2 = \sum (n - \mu)^2 p_n \qquad \text{(discrete)}$$

$$\mu = \int_{-\infty}^{\infty} x p(x)\,dx \quad \sigma^2 = \int_{-\infty}^{\infty}(x - \mu)^2\, p(x)\,dx \quad \text{(continuous)}$$

The **standard deviation** (written σ) is the square root of σ^2.

EXAMPLE 6 (Yes-no poll, one person asked) The probabilities are p and $1 - p$.

A fraction $p = \frac{1}{3}$ of the population thinks *yes*, the remaining fraction $1 - p = \frac{2}{3}$ thinks *no*. Suppose we only ask one person. If $X = 1$ for yes and $X = 0$ for no, the expected value of X is $\mu = p = \frac{1}{3}$. The variance is $\sigma^2 = p(1 - p) = \frac{2}{9}$:

$$\mu = 0\left(\frac{2}{3}\right) + 1\left(\frac{1}{3}\right) = \frac{1}{3} \quad \text{and} \quad \sigma^2 = \left(0 - \frac{1}{3}\right)^2\left(\frac{2}{3}\right) + \left(1 - \frac{1}{3}\right)^2\left(\frac{1}{3}\right) = \frac{2}{9}.$$

The standard deviation is $\sigma = \sqrt{2/9}$. When the fraction p is near one or near zero, the spread is smaller—and one person is more likely to give the right answer for everybody. The maximum of $\sigma^2 = p(1 - p)$ is at $p = \frac{1}{2}$, where $\sigma = \frac{1}{2}$.

The table shows μ and σ^2 for important probability distributions.

Model	*Mean*	*Variance*	*Application*
$p_1 = p, p_0 = 1 - p$	p	$p(1 - p)$	yes-no
Poisson $p_n = \lambda^n e^{-\lambda}/n!$	λ	λ	random occurrence
Exponential $p(x) = ae^{-ax}$	$1/a$	$1/a^2$	waiting time
Normal $p(x) = \dfrac{1}{\sqrt{2\pi}\sigma} e^{-(x-\mu)^2/2\sigma^2}$	μ	σ^2	distribution around mean

THE LAW OF AVERAGESAND THE CENTRAL LIMIT THEOREM

We come to the center of probability theory (without intending to give proofs). The key idea is to repeat an experiment many times—poll many voters, or toss many dice, or play considerable poker. Each independent experiment produces an outcome X, and the average from N experiments is $\bar{X}$. It is called "X bar":

$$\bar{X} = \frac{X_1 + X_2 + \cdots + X_N}{N} = \text{average outcome.}$$

All we know about $p(x)$ is its mean μ and variance σ^2. It is amazing how much information that gives about the average $\bar{X}$:

8F *Law of Averages*: $\bar{X}$ is almost sure to approach μ as $N \to \infty$.
Central Limit Theorem: The probability density $p_N(x)$ for $\bar{X}$ approaches a normal distribution with the same mean μ and variance σ^2/N.

No matter what the probabilities for X, the probabilities for $\bar{X}$ move toward the normal bell-shaped curve. The standard deviation is close to $\sigma/\sqrt{N}$ when the experiment is repeated N times. In the Law of Averages, "almost sure" means that the chance of $\bar{X}$ *not* approaching μ is zero. It can happen, but it won't.

Remark 1   The Boston Globe doesn't understand the Law of Averages. I quote from September 1988: "What would happen if a giant Red Sox slump arrived? What would happen if the fabled Law of Averages came into play, reversing all those can't miss decisions during the winning streak?" They think the Law of Averages evens everything up, favoring heads after a series of tails. See Problem 20.

EXAMPLE 7   *Yes-no poll of $N = 2500$ voters.* **Is a $53\% - 47\%$ outcome conclusive?**

The fraction p of "yes" voters in the whole population is *not known*. That is the reason for the poll. The deviation $\sigma = \sqrt{p(1-p)}$ is also not known, but for one voter this is never more than $\frac{1}{2}$ (when $p = \frac{1}{2}$). Therefore $\sigma/\sqrt{N}$ for 2500 voters is no larger than $\frac{1}{2}/\sqrt{2500}$, which is 1%.

The result of the poll was $\bar{X} = 53\%$. With 95% confidence, this sample is within two standard deviations (here 2%) of its mean. Therefore with 95% confidence, **the unknown mean $\mu = p$ of the whole population is between** 51% **and** 55%. This poll is conclusive.

If the true mean had been $p = 50\%$, the poll would have had only a .0013 chance of reaching 53%. The error margin on each side of a poll is amazingly simple; it is always $1/\sqrt{N}$.

Remark 2   The New York Times has better mathematicians than the Globe. Two days after Bush defeated Dukakis, their poll of $N = 11,645$ voters was printed with the following explanation. "In theory, in 19 cases out of 20 [*there is* 95%] the results should differ by no more than one percentage point [*there is* $1/\sqrt{N}$] from what would have been obtained by seeking out all voters in the United States."

EXAMPLE 8   Football players at Caltech (if any) have average weight $\mu = 210$ pounds and standard deviation $\sigma = 30$ pounds. Are $N = 16$ players safe on an elevator with capacity 3600 pounds? 16 times 210 is 3360.

The average weight $\bar{X}$ is approximately a normal random variable with $\bar{\mu} = 210$ and $\bar{\sigma} = 30/\sqrt{N} = 30/4$. There is only a 2% chance that $\bar{X}$ is above $\bar{\mu} + 2\bar{\sigma} = 225$ (see Figure 8.12b—weights below the mean are no problem on an elevator). Since 16 times 225 is 3600, a statistician would have 98% confidence that the elevator is safe. This is an example where 98% is not good enough—I wouldn't get on.

EXAMPLE 9 (The famous Weldon Dice) Weldon threw 12 dice 26,306 times and counted the 5's and 6's. They came up in 33.77% of the 315,672 separate rolls. Thus $\bar{X} = .3377$ instead of the expected fraction $p = \frac{1}{3}$ of 5's and 6's. Were the dice fair?

The variance in each roll is $\sigma^2 = p(1 - p) = 2/9$. The standard deviation of $\bar{X}$ is $\bar{\sigma} = \sigma/\sqrt{N} = \sqrt{2/9}/\sqrt{315672} \approx .00084$. For fair dice, there is a 95% chance that $\bar{X}$ will differ from $\frac{1}{3}$ by less than $2\bar{\sigma}$. (For Poisson probabilities that is false. Here $\bar{X}$ is *normal*.) But .3377 differs from .3333 by more than $5\bar{\sigma}$. The chance of falling 5 standard deviations away from the mean is only about 1 in 10,000.†

So the dice were unfair. The faces with 5 or 6 indentations were *lighter* than the others, and a little more likely to come up. Modern dice are made to compensate for that, but Weldon never tried again.

8.4 EXERCISES

Read-through questions

Discrete probability uses counting, __a__ probability uses calculus. The function $p(x)$ is the probability __b__. The chance that a random variable falls between a and b is __c__. The total probability is $\int_{-\infty}^{\infty} p(x)\,dx =$ __d__. In the discrete case $\Sigma\, p_n =$ __e__. The mean (or expected value) is $\mu = \int$ __f__ in the continuous case and $\mu = \Sigma\, np_n$ in the __g__.

The Poisson distribution with mean λ has $p_n =$ __h__. The sum $\Sigma\, p_n = 1$ comes from the __i__ series. The exponential distribution has $p(x) = e^{-x}$ or $2e^{-2x}$ or __j__. The standard Gaussian (or __k__) distribution has $\sqrt{2\pi}\, p(x) = e^{-x^2/2}$. Its graph is the well-known __l__ curve. The chance that the variable falls below x is $F(x) =$ __m__. F is the __n__ density function. The difference $F(x + dx) - F(x)$ is about __o__, which is the chance that X is between x and $x + dx$.

The *variance*, which measures the spread around μ, is $\sigma^2 = \int$ __p__ in the continuous case and $\sigma^2 = \Sigma$ __q__ in the discrete case. Its square root σ is the __r__. The normal distribution has $p(x) =$ __s__. If $\bar{X}$ is the __t__ of N samples from any population with mean μ and variance σ^2, the Law of Averages says that $\bar{X}$ will approach __u__. The Central Limit Theorem says that the distribution for $\bar{X}$ approaches __v__. Its mean is __w__ and its variance is __x__.

In a yes-no poll when the voters are 50-50, the mean for one voter is $\mu = 0(\frac{1}{2}) + 1(\frac{1}{2}) =$ __y__. The variance is $(0 - \mu)^2 p_0 + (1 - \mu)^2 p_1 =$ __z__. For a poll with $N = 100$, $\bar{\sigma}$ is __A__. There is a 95% chance that $\bar{X}$ (the fraction saying yes) will be between __B__ and __C__.

1 If $p_1 = \frac{1}{2}, p_2 = \frac{1}{4}, P_3 = \frac{1}{8}, \ldots$, what is the probability of an outcome $X < 4$? What are the probabilities of $X = 4$ and $X > 4$?

2 With the same $p_n = (\frac{1}{2})^n$, what is the probability that X is odd? Why is $p_n = (\frac{1}{3})^n$ an impossible set of probabilities? What multiple $c(\frac{1}{3})^n$ is possible?

3 Why is $p(x) = e^{-2x}$ not an acceptable probability density for $x \geq 0$? Why is $p(x) = 4e^{-2x} - e^{-x}$ not acceptable?

*4 If $p_n = (\frac{1}{2})^n$, show that the probability P that X is a prime number satisfies $6/16 \leq P \leq 7/16$.

5 If $p(x) = e^{-x}$ for $x \geq 0$, find the probability that $X \geq 2$ and the approximate probability that $1 \leq X \leq 1.01$.

6 If $p(x) = C/x^3$ is a probability density for $x \geq 1$, find the constant C and the probability that $X \leq 2$.

7 If you choose x completely at random between 0 and π, what is the density $p(x)$ and the cumulative density $F(x)$?

In 8–13 find the mean value $\mu = \Sigma\, np_n$ **or** $\mu = \int xp(x)\,dx$.

8 $p_0 = 1/2, p_1 = 1/4, p_2 = 1/4$

9 $p_1 = 1/7, p_2 = 1/7, \ldots, p_7 = 1/7$

10 $p_n = 1/n!e$ $(p_0 = 1/e, p_1 = 1/e, p_2 = 1/2e, \ldots)$

---

† Joe Di-Maggio's 56-game hitting streak was much more improbable—I think it is statistically the most exceptional record in major sports.

11 $p(x) = 2/\pi(1+x^2)$, $x \geqslant 0$

12 $p(x) = e^{-x}$ (integrate by parts)

13 $p(x) = ae^{-ax}$ (integrate by parts)

14 Show by substitution that

$$\int_{-\infty}^{\infty} e^{-x^2/2\sigma^2} dx = \sqrt{2}\,\sigma \int_{-\infty}^{\infty} e^{-u^2} du = \sqrt{2\pi}\,\sigma.$$

15 Find the cumulative probability F (the integral of p) in Problems 11, 12, 13. In terms of F, what is the chance that a random sample lies between a and b?

16 Can-Do Airlines books 100 passengers when their plane only holds 98. If the average number of no-shows is 2, what is the Poisson probability that someone will be bumped?

17 The waiting time for a bus has probability density $(1/10)e^{-x/10}$, with $\mu = 10$ minutes. What is the probability of waiting longer than 10 minutes?

18 You make a 3-minute telephone call. If the waiting time for the next incoming call has $p(x) = e^{-x}$, what is the probability that your phone will be busy?

19 Supernovas are expected about every 100 years. What is the probability that you will be alive for the next one? Use a Poisson model with $\lambda = .01$ and estimate your lifetime. (Supernovas actually occurred in 1054 (Crab Nebula), 1572, 1604, and 1987. But the future distribution doesn't depend on the date of the last one.)

20 (a) A fair coin comes up heads 10 times in a row. Will heads or tails be more likely on the next toss?

(b) The fraction of heads after N tosses is α. The expected fraction after $2N$ tosses is _____.

21 Show that the area between μ and $\mu + \sigma$ under the bell-shaped curve is a fixed number (near $1/3$), by substituting $y = $ _____ :

$$\int_{\mu}^{\mu+\sigma} \frac{1}{\sigma\sqrt{2\pi}} e^{-(x-\mu)^2/2\sigma^2} dx = \int_0^1 \frac{1}{\sqrt{2\pi}} e^{-y^2/2} dy.$$

What is the area between $\mu - \sigma$ and μ? The area outside $(\mu - \sigma, \mu + \sigma)$?

22 For a *yes-no* poll of two voters, explain why

$$p_0 = (1-p)^2, p_1 = 2p - 2p^2, p_2 = p^2.$$

Find μ and σ^2. N voters give the "*binomial distribution.*"

23 Explain the last step in this reorganization of the formula for σ^2:

$$\sigma^2 = \int (x-\mu)^2 p(x)\,dx = \int (x^2 - 2x\mu + \mu^2)p(x)\,dx$$
$$= \int x^2 p(x)\,dx - 2\mu \int xp(x)\,dx + \mu^2 \int p(x)\,dx$$
$$= \int x^2 p(x)\,dx - \mu^2.$$

24 Use $\int (x-\mu)^2 p(x)\,dx$ and also $\int x^2 p(x)\,dx - \mu^2$ to find σ^2 for the *uniform distribution*: $p(x) = 1$ for $0 \leqslant x \leqslant 1$.

25 Find σ^2 if $p_0 = 1/3, p_1 = 1/3, p_2 = 1/3$. Use $\Sigma(n-\mu)^2 p_n$ and also $\Sigma n^2 p_n - \mu^2$.

26 Use Problem 23 and integration by parts (equation 7.1.10) to find σ^2 for the *exponential distribution* $p(x) = 2e^{-2x}$ for $x \geqslant 0$, which has mean $\frac{1}{2}$.

27 The waiting time to your next car accident has probability density $p(x) = \frac{1}{2}e^{-x/2}$. What is μ? What is the probability of no accident in the next four years?

28 With $p = \frac{1}{2}, \frac{1}{4}, \frac{1}{8}, \dots$, find the average number μ of coin tosses by writing $p_1 + 2p_2 + 3p_3 + \cdots$ as $(p_1 + p_2 + p_3 + \cdots) + (p_2 + p_3 + p_4 + \cdots) + (p_3 + p_4 + p_5 + \cdots) + \cdots$.

29 In a poll of 900 Americans, 30 are in favor of war. What range can you give with 95% confidence for the percentage of peaceful Americans?

30 Sketch rough graphs of $p(x)$ for the fraction x of heads in 4 tosses of a fair coin, and in 16 tosses. The mean value is $\frac{1}{2}$.

31 A judge tosses a coin 2500 times. How many heads does it take to prove with 95% confidence that the coin is unfair?

32 Long-life bulbs shine an average of 2000 hours with standard deviation 150 hours. You can have 95% confidence that your bulb will fail between _____ and _____ hours.

33 Grades have a normal distribution with mean 70 and standard deviation 10. If 300 students take the test and passing is 55, how many are expected to fail? (Estimate from Figure 8.12b.) What passing grade will fail $1/10$ of the class?

34 The average weight of luggage is $\mu = 30$ pounds with deviation $\sigma = 8$ pounds. What is the probability that the luggage for 64 passengers exceeds 2000 pounds? How does the answer change for 256 passengers and 8000 pounds?

35 A thousand people try independently to guess a number between 1 and 1000. This is like a lottery.

(a) What is the chance that the first person fails?

(b) What is the chance P_0 that they all fail?

(c) Explain why P_0 is approximately $1/e$.

36 (a) In Problem 35, what is the chance that the first person is right and all others are wrong?

(b) Show that the probability P_1 of exactly one winner is also close to $1/e$.

(c) Guess the probability P_n of n winners (fishy question).

8.5 Masses and Moments

This chapter concludes with two sections related to engineering and physics. Each application starts with a finite number of masses or forces. Their sum is the total mass or total force. Then comes the "continuous case," in which the mass is spread out instead of lumped. Its distribution is given by a **density function** ρ (Greek rho), and the sum changes to an *integral*.

The first step (hardest step?) is to get the physical quantities straight. The second step is to move from sums to integrals (discrete to continuous, lumped to distributed). By now we hardly stop to think about it—although this is the key idea of integral calculus. The third step is to evaluate the integrals. For that we can use substitution or integration by parts or tables or a computer.

Figure 8.13 shows the one-dimensional case: *masses along the x axis*. The total mass is the sum of the masses. The new idea is that of *moments*—when the mass or force is multiplied by a *distance*:

moment of mass around the y axis $= mx =$ (mass) times (*distance to axis*).

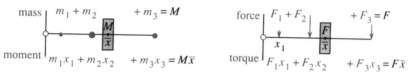

Fig. 8.13   The center of mass is at $\bar{x} =$ (total moment)/(total mass)= average distance.

The figure has masses $1, 3, 2$. The total mass is 6. The "lever arms" or "moment arms" are the distances $x = 1, 3, 7$. The masses have moments 1 and 9 and 14 (since mx is 2 times 7). The total moment is $1 + 9 + 14 = 24$. Then the balance point is at $\bar{x} = M_x/M = 24/6 = 4$.

The total mass is the sum of the m's. The total moment is the sum of m_n times x_n (negative on the other side of $x = 0$). If the masses are children on a seesaw, the balance point is the center of gravity $\bar{x}$—also called the **center of mass**:

DEFINITION
$$\bar{x} = \frac{\sum m_n x_n}{\sum m_n} = \frac{\text{total moment}}{\text{total mass}}. \tag{1}$$

If all masses are moved to $\bar{x}$, the total moment (6 times 4) is still 24. The moment equals the mass $\sum m_n$ times $\bar{x}$. **The masses act like a single mass at $\bar{x}$.**

Also: If we move the axis to $\bar{x}$, and leave the children where they are, the seesaw balances. The masses on the left of $\bar{x} = 4$ will offset the mass on the right. *Reason*: The distances to the new axis are $x_n - \bar{x}$. The moments add to zero by equation (1):

$$\text{moment around new axis} = \sum m_n(x_n - \bar{x}) = \sum m_n x_n - \sum m_n \bar{x} = 0.$$

Turn now to the *continuous case*, when mass is spread out along the line. Each piece of length Δx has an average density $\rho_n =$ (mass of piece)/(length of piece) $= \Delta m/\Delta x$. As the pieces get shorter, this approaches dm/dx—the density at the point. *The limit of (small mass)/(small length) is the density $\rho(x)$.*

Integrating that derivative $\rho = dm/dx$, we recover the total mass: $\sum \rho_n \Delta x$ becomes

$$\textit{total mass } M = \int \rho(x)\,dx. \tag{2}$$

When the mass is spread evenly, ρ is constant. Then $M = \rho L = $ *density times length*.

The moment formula is similar. For each piece, the moment is mass $\rho_n \Delta x$ multiplied by distance x—and we add. In the continuous limit, $\rho(x)dx$ is multiplied by x and we integrate:

$$\textit{total moment around y axis} = M_y = \int x\rho(x)\,dx. \tag{3}$$

Moment is mass times distance. Dividing by the total mass M gives "average distance":

$$\textit{center of mass } \bar{x} = \frac{\text{moment}}{\text{mass}} = \frac{M_y}{M} = \frac{\int x\rho(x)\,dx}{\int \rho(x)\,dx}. \tag{4}$$

Remark   If you studied Section 8.4 on probability, you will notice how the formulas match up. The mass $\int \rho(x)\,dx$ is like the total probability $\int p(x)\,dx$. The moment $\int x\rho(x)\,dx$ is like the mean $\int xp(x)\,dx$. The moment of inertia $\int (x-\bar{x})^2\rho(x)\,dx$ is the variance. Mathematics keeps hammering away at the same basic ideas! The only difference is that the total probability is always 1. The mean really corresponds to the *center* of mass $\bar{x}$, but in probability we didn't notice the division by $\int p(x)dx = 1$.

EXAMPLE 1   With constant density ρ from 0 to L, the mass is $M = \rho L$. The moment is

$$M_y = \int_0^L x\rho\,dx = \tfrac{1}{2}\rho x^2\big]_0^L = \tfrac{1}{2}\rho L^2.$$

The center of mass is $\bar{x} = M_y/M = L/2$. It is halfway along.

EXAMPLE 2   With density e^{-x} the mass is 1, the moment is 1, and $\bar{x}$ is 1:

$$\int_0^\infty e^{-x}\,dx = \big[-e^{-x}\big]_0^\infty = 1 \quad \text{and} \quad \int_0^\infty xe^{-x}\,dx = \big[-xe^{-x} - e^{-x}\big]_0^\infty = 1.$$

MASSES AND MOMENTS IN TWO DIMENSIONS

Instead of placing masses along the x axis, suppose m_1 is at the point (x_1, y_1) in the plane. Similarly m_n is at (x_n, y_n). Now there are *two moments* to consider. Around the y axis $M_y = \Sigma m_n x_n$ and around the x axis $M_x = \Sigma m_n y_n$. **Please notice that the x's go into the moment M_y**—because the x coordinate gives the distance from the y axis!

Around the x axis, the distance is y and the moment is M_x. The **center of mass** is the point $(\bar{x}, \bar{y})$ at which everything balances:

$$\bar{x} = \frac{M_y}{M} = \frac{\sum m_n x_n}{\sum m_n} \quad \text{and} \quad \bar{y} = \frac{M_x}{M} = \frac{\sum m_n y_n}{\sum m_n}. \tag{5}$$

In the continuous case these sums become two-dimensional integrals. The total mass is $\iint \rho(x, y)dx\,dy$, when the density is $\rho =$ mass per unit area. These "double integrals" are for the future (Section 14.1). Here we consider the most important case: $\rho = constant$. Think of a thin plate, made of material with constant density (say $\rho = 1$). To compute its mass and moments, the plate is cut into strips Figure 8.14:

$$\textit{mass } M = \text{area of plate} \tag{6}$$

$$\textit{moment } M_y = \int (\text{distance } x)(\text{length of vertical strip})\,dx \tag{7}$$

$$\textit{moment } M_x = \int (\text{height } y)(\text{length of horizontal strip})\,dy. \tag{8}$$

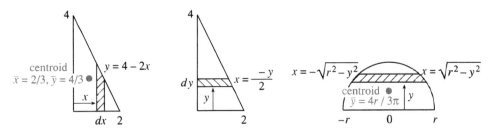

Fig. 8.14   Plates cut into strips to compute masses and moments and centroids.

The mass equals the area because $\rho = 1$. For moments, all points in a vertical strip are the same distance from the y axis. *That distance is x.* The moment is x times area, or x times length times dx—and the integral accounts for all strips.

Similarly the x-moment of a *horizontal* strip is y times strip length times dy.

EXAMPLE 3   A plate has sides $x = 0$ and $y = 0$ and $y = 4 - 2x$. Find M, M_y, M_x.

$$\text{mass } M = \text{area} = \int_0^2 y\, dx = \int_0^2 (4 - 2x)\, dx = \left[4x - x^2\right]_0^2 = 4.$$

The vertical strips go up to $y = 4 - 2x$, and the horizontal strips go out to $x = \frac{1}{2}(4 - y)$:

$$\text{moment } M_y = \int_0^2 x(4 - 2x)\, dx = \left[2x^2 - \frac{2}{3}x^3\right]_0^2 = \frac{8}{3}$$

$$\text{moment } M_x = \int_0^4 y\frac{1}{2}(4 - y)\, dy = \left[y^2 - \frac{1}{6}y^3\right]_0^4 = \frac{16}{3}.$$

The "center of mass" has $\bar{x} = M_y/M = 2/3$ and $\bar{y} = M_x/M = 4/3$. This is the **centroid** of the triangle (and also the "center of gravity"). With $\rho = 1$ these terms all refer to the same balance point $(\bar{x}, \bar{y})$. The plate will not tip over, if it rests on that point.

EXAMPLE 4   Find M_y and M_x for the half-circle below $x^2 + y^2 = r^2$.

$M_y = 0$ because the region is symmetric—Figure 8.14 balances on the y axis. In the x-moment we integrate y times the length of a horizontal strip (notice the factor 2):

$$M_x = \int_0^r y \cdot 2\sqrt{x^2 - y^2}\, dy = -\frac{2}{3}(r^2 - y^2)^{3/2}\Big]_0^r = \frac{2}{3}r^3.$$

Divide by the mass (the area $\frac{1}{2}\pi r^2$) to find the height of the centroid: $\bar{y} = M_x/M = 4r/3\pi$. This is less than $\frac{1}{2}r$ because the bottom of the semicircle is wider than the top.

MOMENT OF INERTIA

The **moment of inertia** comes from multiplying each mass by the **square** of its distance from the axis. Around the y axis, the distance is x. Around the origin, it is r:

$$I_y = \Sigma\, x_n^2 m_n \quad \text{and} \quad I_x = \Sigma\, y_n^2 m_n \quad \text{and} \quad I_0 = \Sigma\, r_n^2 m_n.$$

Notice that $I_x + I_y = I_0$ because $x_n^2 + y_n^2 = r_n^2$. In the continuous case we integrate.

The moment of inertia around the y axis is $I_y = \iint x^2 \rho(x, y)\, dx\, dy$. With a constant density $\rho = 1$, we again keep together the points on a strip. On a vertical strip they share the same x. On a horizontal strip they share y:

$$I_y = \int (x^2)\ (\text{vertical strip length})\, dx \quad \text{and} \quad I_x = \int (y^2)\ (\text{horizontal strip length})\, dy.$$

In engineering and physics, it is **rotation** that leads to the moment of inertia. Look at the energy of a mass m going around a circle of radius r. It has $I_0 = mr^2$.

$$\text{kinetic energy} = \tfrac{1}{2}mv^2 = \tfrac{1}{2}m(r\omega)^2 = \tfrac{1}{2}I_0\omega^2. \tag{9}$$

The angular velocity is ω (radians per second). The speed is $v = r\omega$ (meters per second).

An ice skater reduces I_0 by putting her arms up instead of out. She stays close to the axis of rotation (r is small). Since her rotational energy $\tfrac{1}{2}I_0\omega^2$ does not change, ω increases as I_0 decreases. Then she spins faster.

Another example: It takes force to turn a revolving door. More correctly, it takes **torque**. The force is multiplied by distance from the turning axis: $T = Fx$, so a push further out is more effective.

To see the physics, replace Newton's law $F = ma = m\, dv/dt$ by its rotational form: $T = I\, d\omega/dt$. Where F makes the mass move, the torque T makes it turn. Where m measures unwillingness to change speed, I measures unwillingness to change rotation.

EXAMPLE 5   Find the moment of inertia of a rod about (a) its end and (b) its center.

The distance x from the end of the rod goes from 0 to L. The distance from the center goes from $-L/2$ to $L/2$. Around the center, turning is easier because I is smaller:

$$I_{\text{end}} = \int_0^L x^2\, dx = \tfrac{1}{3}L^3 \qquad I_{\text{center}} = \int_{-L/2}^{L/2} x^2\, dx = \tfrac{1}{12}L^3. \tag{10}$$

  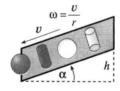

Fig. 8.15   Moment of inertia for rod and propeller. Rolling balls beat cylinders.

MOMENT OF INERTIA EXPERIMENT

Experiment: Roll a solid cylinder (a coin), a hollow cylinder (a ring), a solid ball (a marble), and a hollow ball (*not* a pingpong ball) down a slope. Galileo dropped things from the Leaning Tower—this experiment requires a Leaning Table. Objects that fall together from the tower don't roll together down the table.

Question 1   What is the order of finish ? *Record your prediction first!*

Question 2   Does size make a difference if shape and density are the same ?

Question 3   Does density make a difference if size and shape are the same ?

Question 4 Find formulas for the velocity v and the finish time T.

To compute v, the key is that potential energy plus kinetic energy is practically constant. Energy loss from rolling friction is very small. If the mass is m and the vertical drop is h, the energy at the top (all potential) is mgh. The energy at the bottom (all kinetic) has two parts: $\frac{1}{2}mv^2$ from movement along the plane plus $\frac{1}{2}I\omega^2$ from turning. *Important fact:* $v=\omega r$ for a rolling cylinder or ball of radius r.

Equate energies and set $\omega = v/r$:

$$mgh = \frac{1}{2}mv^2 + \frac{1}{2}I\omega^2 = \frac{1}{2}mv^2\left(1+\frac{I}{mr^2}\right). \tag{11}$$

The ratio I/mr^2 is critical. Call it J and solve (11) for v^2:

$$v^2 = \frac{2gh}{1+J} \quad (smaller\, J\; means\; larger\; velocity). \tag{12}$$

The order of J's, for different shapes and sizes, should decide the race. Apparently the density doesn't matter, because it is a factor in both I and m—so it cancels in $J = I/mr^2$. A hollow cylinder has $J = 1$, which is the largest possible—all its mass is at the full distance r from the axis. So the hollow cylinder should theoretically come in last. This experiment was developed by Daniel Drucker.

Problems $35 - 37$ find the other three J's. Problem 40 finds the time T by integration. Your experiment will show how close this comes to the measured time.

8.5 EXERCISES

Read-through questions

If masses m_n are at distances x_n, the total mass is $M =$ __a__. The total moment around $x = 0$ is $M_y =$ __b__. The center of mass is at $\bar{x} =$ __c__. In the continuous case, the mass distribution is given by the __d__ $\rho(x)$. The total mass is $M =$ __e__ and the center of mass is at $\bar{x} =$ __f__. With $\rho = x$, the integrals from 0 to L give $M =$ __g__ and $\int x\rho(x)\,dx =$ __h__ and $\bar{x} =$ __i__. The total moment is the same if the whole mass M is placed at __j__.

In a plane, with masses m_n at the points (x_n, y_n), the moment around the y axis is __k__. The center of mass has $\bar{x} =$ __l__ and $\bar{y} =$ __m__. For a plate with density $\rho = 1$, the mass M equals the __n__. If the plate is divided into vertical strips of height $y(x)$, then $M = \int y(x)\,dx$ and $M_y = \int$ __o__ dx. For a square plate $0 \leqslant x, y \leqslant L$, the mass is $M =$ __p__ and the moment around the y axis is $M_y =$ __q__. The center of mass is at $(\bar{x}, \bar{y}) =$ __r__. This point is the __s__, where the plate balances.

A mass m at a distance x from the axis has moment of inertia $I =$ __t__. A rod with $\rho = 1$ from $x = a$ to $x = b$ has $I_y =$ __u__. For a plate with $\rho = 1$ and strips of height $y(x)$, this becomes $I_y = \int$ __v__. The torque T is __w__ times __x__.

Compute the mass M along the x axis, the moment M_y around $x = 0$, and the center of mass $\bar{x} = M_y/M$.

1 $m_1 = 2$ at $x_1 = 1, m_2 = 4$ at $x_2 = 2$

2 $m = 3$ at $x = 0, 1, 2, 6$

3 $\rho = 1$ for $-1 \leqslant x \leqslant 3$

4 $\rho = x^2$ for $0 \leqslant x \leqslant L$

5 $\rho = 1$ for $0 \leqslant x < 1, \rho = 2$ for $1 \leqslant x \leqslant 2$

6 $\rho = \sin x$ for $0 \leqslant x \leqslant \pi$

Find the mass M, the moments M_y and M_x, and the center of mass $(\bar{x}, \bar{y})$.

7 Unit masses at $(x, y) = (1, 0), (0, 1)$, and $(1, 1)$

8 $m_1 = 1$ at $(1, 0), m_2 = 4$ at $(0, 1)$

9 $\rho = 7$ in the square $0 \leqslant x \leqslant 1, 0 \leqslant y \leqslant 1$.

10 $\rho = 3$ in the triangle with vertices $(0, 0), (a, 0)$, and $(0, b)$.

Find the area M and the centroid $(\bar{x}, \bar{y})$ inside curves 11–16.

11 $y = \sqrt{1-x^2}, y = 0, x = 0$ (quarter-circle)

12 $y = x$, $y = 2 - x$, $y = 0$ (triangle)

13 $y = e^{-2x}$, $y = 0$, $x = 0$ (infinite dagger)

14 $y = x^2$, $y = x$ (lens)

15 $x^2 + y^2 = 1$, $x^2 + y^2 = 4$ (ring)

16 $x^2 + y^2 = 1$, $x^2 + y^2 = 4$, $y = 0$ (half-ring).

Verify these engineering formulas for I_y with $\rho = 1$:

17 Rectangle bounded by $x = 0$, $x = a$, $y = 0$, $y = b$: $I_y = a^3 b/3$.

18 Square bounded by $x = -\frac{1}{2}a$, $x = \frac{1}{2}a$, $y = -\frac{1}{2}a$, $y = \frac{1}{2}a$: $I_y = a^4/12$.

19 Triangle bounded by $x = 0$, $y = 0$, $x + y = a$: $I_y = a^4/12$.

20 Disk of radius a centered at $x = y = 0$: $I_y = \pi a^4/4$.

21 The moment of inertia around the point $x = t$ of a rod with density $\rho(x)$ is $I = \int (x - t)^2 \rho(x)\, dx$. Expand $(x - t)^2$ and I into three terms. Show that $dI/dt = 0$ when $t = \bar{x}$. The moment of inertia is smallest around the center of mass.

22 A region has $\bar{x} = 0$ if $M_y = \int x(\text{height of strip})\, dx = 0$. The moment of inertia about any other axis $x = c$ is $I = \int (x - c)^2 (\text{height of strip})\, dx$. Show that $I = I_y + (\text{area})(c^2)$. This is the *parallel axis theorem*: I is smallest around the balancing axis $c = 0$.

23 (With thanks to Trivial Pursuit) In what state is the center of gravity of the United States—the "geographical center" or centroid?

24 Pappus (an ancient Greek) noticed that the volume is

$$V = \int 2\pi y(\text{strip width})\, dy = 2\pi M_x = 2\pi \bar{y} M$$

when a region of area M is revolved around the x axis. In the first step the solid was cut into _____.

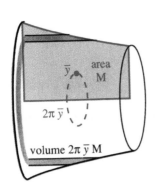

25 Use this theorem of Pappus to find the volume of a torus. Revolve a disk of radius a whose center is at height $\bar{y} = b > a$.

26 Rotate the triangle of Example 3 around the x axis and find the volume of the resulting cone—first from $V = 2\pi \bar{y} M$, second from $\frac{1}{3}\pi r^2 h$.

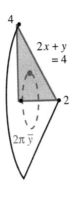

27 Find M_x and M_y for a thin wire along the semicircle $y = \sqrt{1 - x^2}$. Take $\rho = 1$ so $M = \text{length} = \pi$.

28 A second theorem of Pappus gives $A = 2\pi \bar{y} L$ as the surface area when a wire of length L is rotated around the x axis. Verify his formula for a horizontal wire along $y = 3$ ($x = 0$ to $x = L$) and a vertical wire ($y = 1$ to $y = L + 1$).

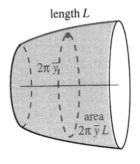
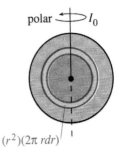

29 The surface area of a sphere is $A = 4\pi$ when $r = 1$. So $A = 2\pi \bar{y} L$ leads to $\bar{y} = $ _____ for the semicircular wire in Problem 27.

30 Rotating $y = mx$ around the x axis between $x = 0$ and $x = 1$ produces the surface area $A = $ _____ .

31 Put a mass m at the point $(x, 0)$. Around the origin the torque from gravity is the force mg times the distance x. This equals g times the _____ mx.

32 If ten equal forces F are alternately down and up at $x = 1, 2, \ldots, 10$, what is their torque?

33 The solar system has nine masses m_n at distances r_n with angular velocities ω_n. What is the moment of inertia around the sun? What is the rotational energy? What is the torque provided by the sun?

34 The disk $x^2 + y^2 \leqslant a^2$ has $I_0 = \int_0^a r^2 2\pi r\, dr = \frac{1}{2}\pi a^4$. Why is this different from I_y in Problem 20? Find the *radius of gyration* $\bar{r} = \sqrt{I_0/M}$. (The rotational energy $\frac{1}{2}I_0 \omega^2$ equals $\frac{1}{2}M\bar{r}^2 \omega^2$—when the whole mass is turning at radius $\bar{r}$.)

Questions 35–42 come from the moment of inertia experiment.

35 A solid cylinder of radius r is assembled from hollow cylinders of length l, radius x, and volume $(2\pi x)(l)(dx)$. The solid cylinder has

$$\text{mass } M = \int_0^r 2\pi x l \rho\, dx \quad \text{and} \quad I = \int_0^r x^2 2\pi x l \rho\, dx.$$

With $\rho = 7$ find M and I and $J = I/Mr^2$.

36 Problem 14.4.40 finds $J = 2/5$ for a solid ball. It is less than J for a solid cylinder because the mass of the ball is more concentrated near _____ .

37 Problem 14.4.39 finds $J = \frac{1}{2}\int_0^\pi \sin^3 \phi\, d\phi = $ _____ for a hollow ball. The four rolling objects finish in the order _____ .

38 By varying the density of the ball how could you make it roll faster than any of these shapes?

39 Answer Question 2 about the experiment.

40 For a vertical drop of y, equation (12) gives the velocity along the plane: $v^2 = 2gy/(1+J)$. Thus $v = cy^{1/2}$ for $c =$ _____ . The vertical velocity is $dy/dt = v \sin \alpha$:

$$dy/dt = cy^{1/2} \sin \alpha \quad \text{and} \quad \int y^{-1/2} dy = \int c \sin \alpha \, dt.$$

Integrate to find $y(t)$. Show that the bottom is reached ($y = h$) at time $T = 2\sqrt{h}/c \sin \alpha$.

41 What is the theoretical ratio of the four finishing times ?

42 True or false:

(a) Basketballs roll downhill faster than baseballs.

(b) The center of mass is always at the centroid.

(c) By putting your arms up you reduce I_x and I_y.

(d) The center of mass of a high jumper goes over the bar (on successful jumps).

8.6 Force, Work, and Energy

Chapter 1 introduced derivatives df/dt and df/dx. The independent variable could be t or x. For velocity it was natural to use the letter t. This section is about two important physical quantities—*force* and *work*—for which x is the right choice.

The basic formula is $W = Fx$. **Work equals force times distance moved** (distance in the direction of F). With a force of 100 pounds on a car that moves 20 feet, the work is 2000 foot-pounds. If the car is rolling forward and you are pushing backward, the work is -2000 foot-pounds. If your force is only 80 pounds and the car doesn't move, the work is zero. In these examples the force is constant.

$W = Fx$ is completely parallel to $f = vt$. When v is constant, we only need multiplication. It is a *changing velocity* that requires calculus. The integral $\int v(t)\,dt$ adds up small multiplications over short times. For a changing force, we add up small pieces of work $F\,dx$ over short distances:

$$W = Fx \quad \textbf{\textit{(constant force)}} \qquad W = \int F(x)\,dx \quad \textbf{\textit{(changing force)}}.$$

In the first case we lift a suitcase weighing $F = 30$ pounds up $x = 20$ feet of stairs. The work is $W = 600$ foot-pounds. The suitcase doesn't get heavier as we go up—it only seems that way. Actually it gets lighter (we study gravity below).

In the second case we stretch a spring, which needs more force as x increases. *Hooke's law says that* $F(x) = kx$. The force is proportional to the stretching distance x. Starting from $x = 0$, the work increases with the *square* of x:

$$F = kx \qquad \text{and} \qquad W = \int_0^x kx\,dx = \tfrac{1}{2}kx^2. \tag{1}$$

In metric units the force is measured in Newtons and the distance in meters. The unit of work is a Newton-meter (a joule). The 600 foot-pounds for an American suitcase would have been about 800 joules in France.

EXAMPLE 1 Suppose a force of $F = 20$ pounds stretches a spring 1 foot.

(a) *Find k.* The elastic constant is $k = F/x = 20$ pounds per foot.

(b) *Find W.* The work is $\tfrac{1}{2}kx^2 = \tfrac{1}{2} \cdot 20 \cdot 1^2 = 10$ foot-pounds.

(c) *Find x when $F = -10$* pounds. This is compression not stretching: $x = -\tfrac{1}{2}$ foot.

Compressing the same spring through the same distance requires the same work. For compression x and F are negative. But the work $W = \tfrac{1}{2}kx^2$ is still positive. Please note that W does not equal kx times x! That is the whole point of variable force (change Fx to $\int F(x)\,dx$).

May I add another important quantity from physics ? It comes from looking at the situation from the viewpoint of the spring. In its natural position, the spring rests comfortably. It feels no strain and has no energy. *Tension or compression gives it potential energy.* More stretching or more compression means more energy. **The change in energy equals the work**. The potential energy of the suitcase increases by 600 foot-pounds, when it is lifted 20 feet.

Write $V(x)$ for the potential energy. Here x is the height of the suitcase or the extension of the spring. In moving from $x = a$ to $x = b$, **work = increase in potential**:

$$W = \int_a^b F(x)\,dx = V(b) - V(a). \tag{2}$$

This is absolutely beautiful. The work W is the **definite integral**. The potential V is the **indefinite integral**. If we carry the suitcase up the stairs and back down, our total work is *zero*. We may feel tired, but the trip down should have given back our energy. (It was in the suitcase.) Starting with a spring that is compressed one foot, and ending with the spring extended one foot, again we have done no work. $V = \frac{1}{2}kx^2$ is the same for $x = -1$ and $x = 1$. But an extension from $x = 1$ to $x = 3$ requires work:

$$W = \text{change in } V = \frac{1}{2}k(3)^2 - \frac{1}{2}k(1)^2.$$

Indefinite integrals like V come with a property that we know well. *They include an arbitrary constant C*. The correct potential is not simply $\frac{1}{2}kx^2$, it is $\frac{1}{2}kx^2 + C$. To compute a *change* in potential, we don't need C. The constant cancels. But to determine V itself, we have to choose C. By fixing $V = 0$ at one point, the potential is determined at all other points. A common choice is $V = 0$ at $x = 0$. Sometimes $V = 0$ at $x = \infty$ (for gravity). Electric fields can be "grounded" at any point.

There is another connection between the potential V and the force F. According to (2), V is the indefinite integral of F. Therefore $F(x)$ **is the derivative of** $V(x)$. The fundamental theorem of calculus is also fundamental to physics:

$$\text{force exerted } on \text{ spring} : F = \quad dV/dx \tag{3a}$$

$$\text{force exerted } by \text{ spring} : F = -dV/dx \tag{3b}$$

Those lines say the same thing. One is our force pulling on the spring, the other is the "restoring force" pulling back. (3a) and (3b) are a warning that the sign of F depends on the point of view. Electrical engineers and physicists use the minus sign. In mechanics the plus sign is more common. It is one of the ironies of fate that $F = V'$, while distance and velocity have those letters reversed: $v = f'$. Note the change to capital letters and the change to x.

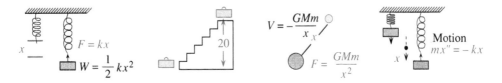

Fig. 8.16   Stretched spring; suitcase 20 feet up; moon of mass m; oscillating spring.

EXAMPLE 2   *Newton's law of gravitation* (inverse square law):

force to overcome gravity $= GMm/x^2$ force exerted by gravity $= -GMm/x^2$

An engine pushes a rocket forward. Gravity pulls it back. The gravitational constant is G and the Earth's mass is M. The mass of the rocket or satellite or suitcase is m, and the potential is the indefinite integral:

$$V(x) = \int F(x)\, dx = -GMm/x + C. \tag{4}$$

Usually $C = 0$, which makes the potential zero at $x = \infty$.

Remark   When carrying the suitcase upstairs, x changed by 20 feet. The weight was regarded as constant—which it nearly is. But an exact calculation of work uses the integral of $F(x)$, not just the multiplication 30 times 20. The serious difference comes when the suitcase is carried to $x = \infty$. With constant force that requires infinite work. With the correct (decreasing) force, the work equals V at infinity (which is zero) minus V at the pickup point x_0. The change in V is $W = GMm/x_0$.

KINETIC ENERGY

This optional paragraph carries the physics one step further. Suppose you release the spring or drop the suitcase. The external force changes to $F = 0$. But the internal force still acts on the spring, and gravity still acts on the suitcase. They both start moving. The potential energy of the suitcase is converted to *kinetic energy*, until it hits the bottom of the stairs.

Time enters the problem, either through Newton's law or Einstein's:

$$(\textbf{Newton})\ \ F = ma = m\frac{dv}{dt} \qquad (\textbf{Einstein})\ \ F = \frac{d}{dt}(mv). \qquad (5)$$

Here we stay with Newton, and pretend the mass is constant. Exercise 21 follows Einstein; the mass increases with velocity. There $m = m_0/\sqrt{1 - v^2/c^2}$ goes to infinity as v approaches c, the speed of light. That correction comes from the theory of relativity, and is not needed for suit-cases.

What happens as the suitcase falls? From $x = a$ at the top of the stairs to $x = b$ at the bottom, potential energy is lost. But kinetic energy $\frac{1}{2}mv^2$ is gained, as we see from integrating Newton's law:

$$\text{force } F = m\frac{dv}{dt} = m\frac{dv}{dx}\frac{dx}{dt} = mv\frac{dv}{dx}$$

$$\text{work } \int_a^b F\, dx = \int_a^b mv\frac{dv}{dx}dx = \frac{1}{2}mv^2(b) - \frac{1}{2}mv^2(a). \qquad (6)$$

This same force F is given by $-dV/dx$. So the work is also the change in V:

$$\text{work } \int_a^b F\, dx = \int_a^b \left(-\frac{dV}{dx}\right)dx = -V(b) + V(a). \qquad (7)$$

Since (6) = (7), the total energy $\frac{1}{2}mv^2 + V$ (*kinetic plus potential*) is constant:

$$\tfrac{1}{2}mv^2(b) + V(b) = \tfrac{1}{2}mv^2(a) + V(a). \qquad (8)$$

This is the law of **conservation of energy**. The total energy is conserved.

EXAMPLE 3 Attach a mass m to the end of a stretched spring and let go. The spring's
energy $V = \frac{1}{2}kx^2$ is gradually converted to kinetic energy of the mass. At $x = 0$ the change to kinetic energy is complete: the original $\frac{1}{2}kx^2$ has become $\frac{1}{2}mv^2$. Beyond $x = 0$ the potential energy increases, the force reverses sign and pulls back, and kinetic energy is lost. Eventually all energy is potential—when the mass reaches the other extreme. It is *simple harmonic motion*, exactly as in Chapter 1 (where the mass was the shadow of a circling ball). The equation of motion is the statement that **the rate of change of energy is zero** (and we cancel $v = dx/dt$):

$$\frac{d}{dt}\left(\frac{1}{2}mv^2 + \frac{1}{2}kx^2\right) = mv\frac{dv}{dt} + kx\frac{dx}{dt} = 0 \quad \text{or} \quad m\frac{d^2x}{dt^2} + kx = 0. \qquad (9)$$

That is $F = ma$ in disguise. For a spring, the solution $x = \cos\sqrt{k/m}\,t$ will be found in this book. For more complicated structures, engineers spend a billion dollars a year computing the solution.

PRESSURE AND HYDROSTATIC FORCE

Our forces have been concentrated at a single points. That is not the case for **pressure**. A fluid exerts a force all over the base and sides of its container. Suppose a water tank or swimming pool has constant depth h (in meters or feet). The water has weight-density $w \approx 9800 \text{ N/m}^3 \approx 62 \text{ lb/ft}^3$. On the base, the pressure is w times h. The force is wh times the base area A:

$$F = whA \text{ (pounds or Newtons)} \qquad p = F/A = wh \text{ (lb/ft}^2 \text{ or N/m}^2). \qquad (10)$$

Thus *pressure is force per unit area*. Here p and F are computed by multiplication, because the depth h is constant. Pressure is proportional to depth (as divers know). Down the side wall, h varies and we need calculus.

The pressure on the side is still wh—*the same in all directions*. We divide the side into horizontal strips of thickness Δh. Geometry gives the length $l(h)$ at depth h (Figure 8.17). The area of a strip is $l(h)\Delta h$. The pressure wh is nearly constant on the strip—the depth only changes by Δh. **The force on the strip is $\Delta F = whl\,\Delta h$.** Adding those forces, and narrowing the strips so that $\Delta h \to 0$, the total force approaches an integral:

$$\text{total force } F = \int whl(h)\,dh \qquad (11)$$

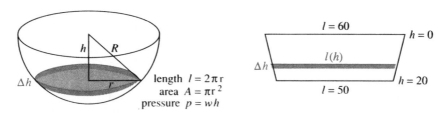

Fig. 8.17   Water tank and dam: length of side strip $= l$, area of layer $= A$.

EXAMPLE 4   Find the total force on the trapezoidal dam in Figure 8.17.

The side length is $l = 60$ when $h = 0$. The depth h increases from 0 to 20. The main problem is to find l at an in-between depth h. With straight sides the relation is linear: $l = 60 + ch$. We choose c to give $l = 50$ when $h = 20$. Then $50 = 60 + c(20)$ yields $c = -\frac{1}{2}$.

The total force is the integral of whl. So substitute $l = 60 - \frac{1}{2}h$:

$$F = \int_0^{20} wh(60 - \tfrac{1}{2}h)\,dh = \left[30wh^2 - \tfrac{1}{6}wh^3\right]_0^{20} = 12000w - \tfrac{1}{6}(8000w).$$

With distance in feet and $w = 62 \text{ lb/ft}^3$, F is in pounds. With distance in meters and $w = 9800 \text{ N/m}^3$, the force is in Newtons.

Note that (weight-density w) = (mass-density ρ) times $(g) = (1000)(9.8)$. These SI units were chosen to make the density of water at $0°C$ exactly $\rho = 1000 \text{ kg/m}^3$.

EXAMPLE 5   *Find the work to pump water out of a tank. The area at depth h is $A(h)$.*

Imagine lifting out *one layer of water at a time*. The layer weighs $wA(h)\Delta h$. The work to lift it to the top is its weight times the distance h, or $whA(h)\Delta h$. The work to empty the whole tank is the integral:

$$W = \int whA(h)\, dh. \tag{12}$$

Suppose the tank is the bottom half of a sphere of radius R. The cross-sectional area at depth h is $A = \pi(R^2 - h^2)$. Then the work is the integral (12) from 0 to R. It equals $W = \pi w R^4/4$.

Units: $w = \text{force}/(\text{distance})^3$ times $R^4 = (\text{distance})^4$ gives work $W = (\text{force})(\text{distance})$.

8.6 EXERCISES

Read-through questions

Work equals __a__ times __b__. For a spring the force is $F = $ __c__, proportional to the extension x (this is __d__ law). With this variable force, the work in stretching from 0 to x is $W = \int$ __e__ $=$ __f__. This equals the increase in the __g__ energy V. Thus W is a __h__ integral and V is the corresponding __i__ integral, which includes an arbitrary __j__. The derivative dV/dx equals __k__. The force of gravity is $F = $ __l__ and the potential is $V = $ __m__.

In falling, V is converted to __n__ energy $K = $ __o__. The total energy $K + V$ is __p__ (this is the law of __q__ when there is no external force).

Pressure is force per unit __r__. Water of density w in a pool of depth h and area A exerts a downward force $F = $ __s__ on the base. The pressure is $p = $ __t__. On the sides the __u__ is still wh at depth h, so the total force is $\int whl\, dh$, where l is __v__. In a cubic pool of side s, the force on the base is $F = $ __w__, the length around the sides is $l = $ __x__, and the total force on the four sides is $F = $ __y__. The work to pump the water out of the pool is $W = \int whA\, dh = $ __z__.

1 (a) Find the work W when a constant force $F = 12$ pounds moves an object from $x = .9$ feet to $x = 1.1$ feet.
(b) Compute W by integration when the force $F = 12/x^2$ varies with x.

2 A 12–inch spring is stretched to 15 inches by a force of 75 pounds.
(a) What is the spring constant k in *pounds per foot*?
(b) Find the work done in stretching the spring.
(c) Find the work to stretch it 3 more inches.

3 A shock-absorber is compressed 1 inch by a weight of 1 ton. Find its spring constant k in pounds per foot. What potential energy is stored in the shock-absorber?

4 A force $F = 20x - x^3$ stretches a nonlinear spring by x.
(a) What work is required to stretch it from $x = 0$ to $x = 2$?
(b) What is its potential energy V at $x = 2$, if $V(0) = 5$?
(c) What is $k = dF/dx$ for a small additional stretch at $x = 2$?

5 (a) A 120-lb person makes a scale go down x inches. How much work is done?
(b) If the same person goes x inches down the stairs, how much potential energy is lost?

6 A rocket burns its 100 kg of fuel at a steady rate to reach a height of 25 km.
(a) Find the weight of fuel left at height h.
(b) How much work is done lifting fuel?

7 Integrate to find the work in winding up a hanging cable of length 100 feet and weight density 5 lb/ft. How much additional work is caused by a 200-pound weight hanging at the end of the cable?

8 The great pyramid (height 500′—you can see it from Cairo) has a square base 800′ by 800′. Find the area A at height h. If the rock weighs $w = 100$ lb/ft^3, approximately how much work did it take to lift all the rock?

9 The force of gravity on a mass m is $F = -GMm/x^2$. With $G = 6 \cdot 10^{-17}$ and Earth mass $M = 6 \cdot 10^{24}$ and rocket mass $m = 1000$, compute the work to lift the rocket from $x = 6400$ to $x = 6500$. (The units are kgs and kms and Newtons, giving work in Newton-kms.)

10 The approximate work to lift a 30-pound suitcase 20 feet is 600 foot-pounds. The exact work is the change in the potential $V = -GmM/x$. Show that ΔV is 600 times a correction factor $R^2/(R^2 - 10^2)$, when x changes from $R - 10$ to $R + 10$. (This factor is practically 1, when $R = $ radius of the Earth.)

11 Find the work to lift the rocket in Problem 9 from $x = 6400$ out to $x = \infty$. If this work equals the original kinetic energy $\frac{1}{2}mv^2$, what was the original v (*the escape velocity*)?

12 The kinetic energy $\frac{1}{2}mv^2$ of a rocket is converted into potential energy $-GMm/x$. Starting from the Earth's radius $x = R$, what x does the rocket reach? If it reaches $x = \infty$ show that $v = \sqrt{2GM/R}$. This escape velocity is 25,000 miles per hour.

13 It takes 20 foot-pounds of work to stretch a spring 2 feet. How much work to stretch it one more foot?

14 A barrel full of beer is 4 feet high with a 1 foot radius and an opening at the bottom. How much potential energy is lost by the beer as it comes out of the barrel ?

15 A rectangular dam is 40 feet high and 60 feet wide. Compute the total side force F on the dam when   (a) the water is at the top (b) the water level is halfway up.

16 A triangular dam has an 80-meter base at a depth of 30 meters. If water covers the triangle, find
   (a)   the pressure at depth h
   (b)   the length l of the dam at depth h
   (c)   the total force on the dam.

17 A cylinder of depth H and cross-sectional area A stands full of water (density w). (a) Compute the work $W = \int wAh\,dh$ to lift all the water to the top. (b) Check the units of W. (c) What is the work W if the cylinder is only half full ?

18 In Problem 17, compute W in both cases if $H = 20$ feet, $w = 62$ lb/ft^3, and the base is a circle of radius $r = 5$ feet.

19 How much work is required to pump out a swimming pool, if the area of the base is 800 square feet, the water is 4 feet deep, and the top is one foot above the water level ?

20 For a cone-shaped tank the cross-sectional area increases with depth: $A = \pi r^2 h^2/H^2$. Show that the work to empty it is half the work for a cylinder with the same height and base. What is the ratio of volumes of water ?

21 In relativity the mass is $m = m_0/\sqrt{1 - v^2/c^2}$. Find the correction factor in Newton's equation $F = m_0 a$ to give Einstein's equation $F = d(mv)/dt = (d(mv)/dv)(dv/dt) = $ _____ $m_0 a$.

22 Estimate the depth of the *Titanic*, the pressure at that depth, and the force on a cabin door. Why doesn't every door collapse at the bottom of the Atlantic Ocean ?

23 A swimming pool is 4 meters wide, 10 meters long, and 2 meters deep. Find the weight of the water and the total force on the bottom.

24 If the pool in Problem 23 has a shallow end only one meter deep, what fraction of the water is saved ? Draw a cross-section (a trapezoid) and show the direction of force on the sides and the sloping bottom.

25 In what ways is work like a definite integral and energy like an indefinite integral ? Their derivative is the _____ .

CHAPTER 9

Polar Coordinates and Complex Numbers

9.1 Polar Coordinates

Up to now, points have been located by their x and y coordinates. But if you were a flight controller, and a plane appeared on the screen, you would not give its position that way. Instead of x and y, you would read off the **direction** of the plane and its **distance**. The direction is given by an angle θ. The distance is given by a positive number r. Those are the **polar coordinates** of the point, where x and y are the *rectangular coordinates*.

The angle θ is measured from the horizontal. Suppose the distance is 2 and the direction is $30°$ or $\pi/6$ (degrees preferred by flight controllers, radians by mathematicians). A pilot looking along the x axis would give the plane's direction as "11 o'clock." This totally destroys our system of units, by measuring direction in hours. But the angle and the distance locate the plane.

How far to a landing strip at $r = 1$ and $\theta = -\pi/2$? For that question polar coordinates are not good. They are perfect for distance from the origin (which equals r), but for most other distances I would switch to x and y. It is extremely simple to determine x and y from r and θ, and we will do it constantly. The most used formulas in this chapter come from Figure 9.1—where the right triangle has angle θ and hypotenuse r. *The sides of that triangle are x and y:*

$$x = r\cos\theta \quad \textbf{\textit{and}} \quad y = r\sin\theta. \tag{1}$$

The point at $r = 2, \theta = \pi/6$ has $x = 2\cos(\pi/6)$ and $y = 2\sin(\pi/6)$. The cosine of $\pi/6$ is $\sqrt{3}/2$ and the sine is $\frac{1}{2}$. So $x = \sqrt{3}$ and $y = 1$. Polar coordinates convert easily to xy coordinates—now we go the other way.

Always $x^2 + y^2 = r^2$. In this example $(\sqrt{3})^2 + (1)^2 = (2)^2$. Pythagoras produces r from x and y. The direction θ is also available, but the formula is not so beautiful:

$$r = \sqrt{x^2 + y^2} \quad \textbf{\textit{and}} \quad \tan\theta = \frac{y}{x} \quad \textbf{\textit{and}} \ (\textbf{\textit{almost}}) \quad \theta = \tan^{-1}\frac{y}{x}. \tag{2}$$

Our point has $y/x = 1/\sqrt{3}$. One angle with this tangent is $\theta = \tan^{-1}(1/\sqrt{3}) = \pi/6$.

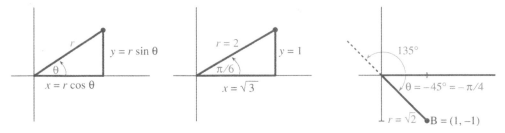

Fig. 9.1 Polar coordinates r, θ and rectangular coordinates $x = r\cos\theta, y = r\sin\theta$.

EXAMPLE 1 Point B in Figure 9.1c is at a *negative angle* $\theta = -\pi/4$. The x coordinate $r\cos(-\pi/4)$ is the same as $r\cos\pi/4$ (the cosine is even). But the y coordinate $r\sin(-\pi/4)$ is negative. Computing r and θ from $x = 1$ and $y = 1$, the distance is $r = \sqrt{1+1}$ and $\tan\theta$ is $-1/1$.

Warning To any angle θ we can add or subtract 2π—which goes a full $360°$ circle and keeps the same direction. Thus $-\pi/4$ or $-45°$ is the same angle as $7\pi/4$ or $315°$. So is $15\pi/4$ or $675°$.

If we add or subtract $180°$, the tangent doesn't change. The point $(1, -1)$ is on the $-45°$ line at $r = \sqrt{2}$. The point $(-1, 1)$ is on the $135°$ line also with $r = \sqrt{2}$. Both have $\tan\theta = -1$. We had to write "almost" in equation (2), because a point has many θ's and two points have the same r and $\tan\theta$.

Even worse, we could say that $B = (1, -1)$ is on the $135°$ line but at a *negative distance* $r = -\sqrt{2}$. A negative r carries the point *backward* along the $135°$ line, which is forward to B. In giving the position of B, I would always keep $r > 0$. But in drawing the graph of a polar equation, $r < 0$ is allowed. We move now to those graphs.

THE CIRCLE $r = \cos\theta$

The basis for Chapters 1–8 was $y = f(x)$. The key to this chapter is $r = F(\theta)$. That is a relation between the polar coordinates, and the points satisfying an equation like $r = \cos\theta$ produce a *polar graph*.

It is not obvious why $r = \cos\theta$ gives a circle. The equations $r = \cos 2\theta$ and $r = \cos^2\theta$ and $r = 1 + \cos\theta$ produce entirely different graphs—not circles. The direct approach is to take $\theta = 0°, 30°, 60°, \ldots$ and go out the distance $r = \cos\theta$ on each ray. The points are marked in Figure 9.2a, and connected into a curve. It seems to be a circle of radius $\frac{1}{2}$, with its center at the point $(\frac{1}{2}, 0)$. We have to be able to show mathematically that $r = \cos\theta$ represents a *shifted circle*.

One point must be mentioned. *The angles from 0 to π give the whole circle.* The number $r = \cos\theta$ becomes negative after $\pi/2$, and we go backwards along each ray.

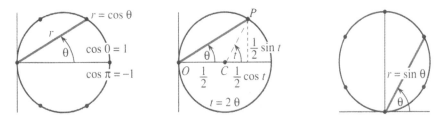

Fig. 9.2 The circle $r = \cos\theta$ and the switch to x and y. The circle $r = \sin\theta$.

At $\theta = \pi$ (to the *left* of the origin) the cosine is -1. Going backwards brings us to the same point as $\theta = 0$ and $r = +1$—which completes the circle.

When θ continues from π to 2π we go around again. The polar equation gives the circle *twice*. (Or more times, when θ continues past 2π.) If you don't like negative r's and multiple circles, restrict θ to the range from $-\pi/2$ to $\pi/2$. We still have to see why the graph of $r = \cos\theta$ is a circle.

***Method* 1** Multiply by r and convert to rectangular coordinates x and y:

$$r = \cos\theta \quad\Rightarrow\quad r^2 = r\cos\theta \quad\Rightarrow\quad x^2 + y^2 = x. \tag{3}$$

This is a circle because of $x^2 + y^2$. From rewriting as $(x - \frac{1}{2})^2 + y^2 = (\frac{1}{2})^2$ we recognize its center and radius. Center at $x = \frac{1}{2}$ and $y = 0$; radius $\frac{1}{2}$. Done.

***Method* 2** Write x and y *separately* as functions of θ. Then θ is a "*parameter*":

$$x = r\cos\theta = \cos^2\theta \quad\text{and}\quad y = r\sin\theta = \sin\theta\cos\theta. \tag{4}$$

These are not *polar* equations but *parametric* equations. The parameter θ is the angle, but it could be the time—the curve would be the same. Chapter 12 studies parametric equations in detail—here we stay with the circle.

To find the circle, square x and y and add. This produces $x^2 + y^2 = x$ in Problem 26. But here we do something new: **Start with the circle and find equation** (4). In case you don't reach Chapter 12, the idea is this. Add the vectors OC to the center and CP out the radius:

The point P in Figure 9.2 has $(x, y) = OC + CP = (\frac{1}{2}, 0) + (\frac{1}{2}\cos t, \frac{1}{2}\sin t)$.

The parameter t is the angle at the center of the circle. The equations are $x = \frac{1}{2} + \frac{1}{2}\cos t$ and $y = \frac{1}{2}\sin t$. A trigonometric person sees a double angle and sets $t = 2\theta$. The result is equation (4) for the circle:

$$x = \frac{1}{2} + \frac{1}{2}\cos 2\theta = \cos^2\theta \quad\text{and}\quad y = \frac{1}{2}\sin 2\theta = \sin\theta\cos\theta. \tag{5}$$

This step rediscovers a basic theorem of geometry: *The angle t at the center is twice the angle θ at the circumference.* End of quick introduction to parameters.

A second circle is $r = \sin\theta$, drawn in Figure 9.2c. A third circle is $r = \cos\theta + \sin\theta$, not drawn. Problem 27 asks you to find its xy equation and its radius. All calculations go back to $x = r\cos\theta$ and $y = r\sin\theta$—the basic facts of polar coordinates! The last exercise shows a parametric equation with beautiful graphs, because it may be possible to draw them now. Then the next section concentrates on $r = F(\theta)$—and goes far beyond circles.

9.1 EXERCISES

Read-through questions

Polar coordinates r and θ correspond to $x =$ __a__ and $y =$ __b__. The points with $r > 0$ and $\theta = \pi$ are located __c__. The points with $r = 1$ and $0 \leqslant \theta \leqslant \pi$ are located __d__. Reversing the sign of θ moves the point (x, y) to __e__.

Given x and y, the polar distance is $r =$ __f__. The tangent of θ is __g__. The point $(6, 8)$ has $r =$ __h__ and $\theta =$ __i__. Another point with the same θ is __j__. Another point with the same r is __k__. Another point with the same r and $\tan\theta$ is __l__.

The polar equation $r = \cos\theta$ produces a shifted __m__. The top point is at $\theta =$ __n__, which gives $r =$ __o__. When θ goes from 0 to 2π, we go __p__ times around the graph. Rewriting as $r^2 = r\cos\theta$ leads to the xy equation __q__. Substituting $r = \cos\theta$ into $x = r\cos\theta$ yields $x =$ __r__ and similarly $y =$ __s__. In this form x and y are functions of __t__ θ.

Find the polar coordinates $r \geqslant 0$ and $0 \leqslant \theta < 2\pi$ of these points.

1 $(x, y) = (0, 1)$

2 $(x, y) = (-4, 0)$

3 $(x, y) = (\sqrt{2}, \sqrt{2})$

4 $(x, y) = (-1, \sqrt{3})$

5 $(x, y) = (-1, -1)$

6 $(x, y) = (3, 4)$

Find rectangular coordinates (x, y) from polar coordinates.

7 $(r, \theta) = (2, \pi/2)$

8 $(r, \theta) = (1, 3\pi/2)$

9 $(r, \theta) = (\sqrt{20}, \pi/4)$

10 $(r, \theta) = (3\pi, 3\pi)$

11 $(r, \theta) = (2, -\pi/6)$

12 $(r, \theta) = (2, 5\pi/6)$

13 What is the distance from $(x, y) = (\sqrt{3}, 1)$ to $(1, -\sqrt{3})$?

14 How far is the point $r = 3, \theta = \pi/2$ from $r = 4, \theta = \pi$?

15 How far is $(x, y) = (r \cos \theta, r \sin \theta)$ from $(X, Y) = (R \cos \phi, R \sin \phi)$? Simplify $(x - X)^2 + (y - Y)^2$ by using $\cos(\theta - \phi) = \cos \theta \cos \phi + \sin \theta \sin \phi$.

16 Find a second set of polar coordinates (a different r or θ) for the points

$$(r, \theta) = (-1, \pi/2), \quad (-1, 3\pi/4), \quad (1, -\pi/2), \quad (0, 0).$$

17 Using polar coordinates describe (a) the half-plane $x > 0$; (b) the half-plane $y \leqslant 0$; (c) the ring with $x^2 + y^2$ between 4 and 5; (d) the wedge $x \geqslant |y|$.

18 True or false, with a reason or an example:

(a) Changing to $-r$ and $-\theta$ produces the same point.

(b) Each point has only one r and θ, when $r < 0$ is not allowed.

(c) The graph of $r = 1/\sin \theta$ is a straight line.

19 From x and θ find y and r.

20 Which other point has the same r and $\tan \theta$ as $x = \sqrt{3}, y = 1$ in Figure 9.1b ?

21 Convert from rectangular to polar equations:

(a) $y = x$ (b) $x + y = 1$ (c) $x^2 + y^2 = x + y$

22 Show that the triangle with vertices at $(0, 0), (r_1, \theta_1)$, and (r_2, θ_2) has area $A = \frac{1}{2} r_1 r_2 \sin(\theta_2 - \theta_1)$. Find the base and height assuming $0 \leqslant \theta_1 \leqslant \theta_2 \leqslant \pi$.

Problems 23–28 are about polar equations that give circles.

23 Convert $r = \sin \theta$ into an xy equation. Multiply first by r.

24 Graph $r = \sin \theta$ at $\theta = 0°, 30°, 60°, \ldots, 360°$. These thirteen values of θ give _____ different points on the graph. What range of θ's goes once around the circle ?

25 Substitute $r = \sin \theta$ into $x = r \cos \theta$ and $y = r \sin \theta$ to find x and y in terms of the parameter θ. Then compute $x^2 + y^2$ to reach the xy equation.

26 From the parametric equations $x = \cos^2 \theta$ and $y = \sin \theta \cos \theta$ in (4), recover the xy equation. Square, add, eliminate θ.

27 (a) Multiply $r = \cos \theta + \sin \theta$ by r to convert into an xy equation. (b) Rewrite the equation as $(x - \frac{1}{2})^2 + (y - \frac{1}{2})^2 = R^2$ to find the radius R. (c) Draw the graph.

28 Find the radius of $r = a \cos \theta + b \sin \theta$. (Multiply by r.)

29 Convert $x + y = 1$ into an $r\theta$ equation and solve for r. Then substitute this r into $x = r \cos \theta$ and $y = r \sin \theta$ to find parametric equations for the line.

30 The equations $x = \cos^2 \theta$ and $y = \sin^2 \theta$ also lead to $x + y = 1$—but they are different from the answer to Problem 29. Explanation: θ is no longer the polar angle and we should have written t. Find a point $x = \cos^2 \theta, y = \sin^2 \theta$ that is *not* at the angle θ.

31 Convert $r = \cos^2 \theta$ into an xy equation (of sixth degree!)

32 If you have a graphics package for parametric curves, graph some *hypocycloids*. The equations are $x = (1 - b) \cos t + b \cos(1 - b) t / b, y = (1 - b) \sin t - b \sin(1 - b) t / b$. The figure shows $b = \frac{3}{10}$ and part of $b = .31831$.

9.2 Polar Equations and Graphs

The most important equation in polar coordinates, by far, is $r = 1$. The angle θ does not even appear. The equation looks too easy, but that is the point! The graph is a circle around the origin (the unit circle). Compare with the line $x = 1$. More important, compare the simplicity of $r = 1$ with the complexity of $y = \pm\sqrt{1 - x^2}$. Circles are so common in applications that they created the need for polar coordinates.

This section studies polar curves $r = F(\theta)$. The cardioid is a sentimental favorite—maybe parabolas are more practical. The cardioid is $r = 1 + \cos\theta$, the parabola is $r = 1/(1 + \cos\theta)$. Section 12.2 adds cycloids and astroids. A graphics package can draw them and so can we.

Together with the circles $r = \text{constant}$ go the straight lines $\theta = \text{constant}$. The equation $\theta = \pi/4$ is a ray out from the origin, at that fixed angle. If we allow $r < 0$, as we do in drawing graphs, the one-directional ray changes to a full line. Important: *The circles are perpendicular to the rays*. We have "orthogonal coordinates"—more interesting than the $x - y$ grid of perpendicular lines. In principle x could be mixed with θ (non-orthogonal), but in practice that never happens.

Other curves are attractive in polar coordinates—we look first at five examples. Sometimes we switch back to $x = r\cos\theta$ and $y = r\sin\theta$, to recognize the graph.

EXAMPLE 1	The graph of $r = 1/\cos\theta$ is the ***straight line*** $x = 1$ (because $r\cos\theta = 1$).

EXAMPLE 2	The graph of $r = \cos 2\theta$ is the ***four-petal flower*** in Figure 9.3.

The points at $\theta = 30°$ and $-30°$ and $150°$ and $-150°$ are marked on the flower. They all have $r = \cos 2\theta = \frac{1}{2}$. ***There are three important symmetries—across the x axis, across the y axis, and through the origin***. This four-petal curve has them all. So does the vertical flower $r = \sin 2\theta$—but surprisingly, the tests it passes are different.

(*Across the x axis: y to $-y$*)	There are two ways to cross. First, change θ to $-\theta$. The equation $r = \cos 2\theta$ stays the same. Second, change θ to $\pi - \theta$ and also r to $-r$. The equation $r = \sin 2\theta$ stays the same. Both flowers have x axis symmetry.

(*Across the y axis: x to $-x$*)	There are two ways to cross. First, change θ to $\pi - \theta$. The equation $r = \cos 2\theta$ stays the same. Second, change θ to $-\theta$ and r to $-r$. Now $r = \sin 2\theta$ stays the same (the sine is odd). Both curves have y axis symmetry.

(*Through the origin*)	Now we change r to $-r$ ***or*** θ to $\theta + \pi$. The flower equations pass the second test only: $\cos 2(\theta + \pi) = \cos 2\theta$ and $\sin 2(\theta + \pi) = \sin 2\theta$. Every equation $r^2 = F(\theta)$ passes the first test, since $(-r)^2 = r^2$.

The circle $r = \cos\theta$ has x axis symmetry, but not y or r. The spiral $r = \theta^3$ has y axis symmetry, because $-r = (-\theta)^3$ is the same equation.

Question	What happens if you change r to $-r$ and also change θ to $\theta + \pi$?
Answer	*Nothing*—because (r, θ) and $(-r, \theta + \pi)$ are always the same point.

EXAMPLE 3	The graph of $r = \theta$ is a ***spiral of Archimedes***—or maybe two spirals.

The spiral adds new points as θ increases past 2π. Our other examples are "periodic"—$\theta = 2\pi$ gives the same point as $\theta = 0$. A periodic curve repeats itself. The spiral moves out by 2π each time it comes around. If we allow negative angles and negative $r = \theta$, a second spiral appears.

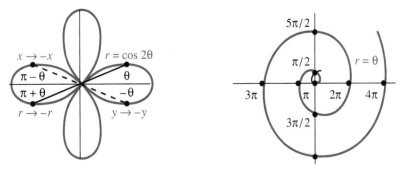

Fig. 9.3 The four-petal flower $r = \cos 2\theta$ and the spiral $r = \theta$ ($r > 0$ in red).

EXAMPLE 4 The graph of $r = 1 + \cos \theta$ is a *cardioid*. It is drawn in Figure 9.4c.

The cardioid has no simple xy equation. Still the curve is very attractive. It has a cusp at the origin and it is heart-shaped (hence its name). To draw it, plot $r = 1 + \cos \theta$ at $30°$ intervals and connect the points. For this curve r is never negative, since $\cos \theta$ never goes below -1.

It is a curious fact that the electrical vector in your heart almost traces out a cardioid. See Section 11.1 about electrocardiograms. If it is a perfect cardioid you are in a little trouble.

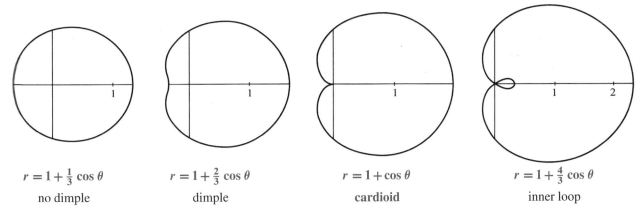

$r = 1 + \frac{1}{3}\cos \theta$	$r = 1 + \frac{2}{3}\cos \theta$	$r = 1 + \cos \theta$	$r = 1 + \frac{4}{3}\cos \theta$
no dimple	dimple	**cardioid**	inner loop

Fig. 9.4 Limaçons $r = 1 + b \cos \theta$, including a cardioid and Mars seen from Earth.

EXAMPLE 5 The graph of $r = 1 + b \cos \theta$ is a *limaçon* (a cardioid when $b = 1$).

Limaçon (soft c) is a French word for snail—not so well known as escargot but just as inedible. (*I am only referring to the shell. Excusez-moi!*) Figure 9.4 shows how a dimple appears as b increases. Then an inner loop appears beyond $b = 1$ (the cardioid at $b = 1$ is giving birth to a loop). For large b the curve looks more like two circles. The limiting case is a double circle, when the inner loop is the same as the outer loop. Remember that $r = \cos \theta$ goes around the circle twice.

We could magnify the limaçon by a factor c, changing to $r = c(1 + b \cos \theta)$. We could rotate $180°$ to $r = 1 - b \cos \theta$. But the real interest is whether these figures arise in applications, and Donald Saari showed me a nice example.

Mars seen from Earth The Earth goes around the Sun and so does Mars. Roughly speaking Mars is $1\frac{1}{2}$ times as far out, and completes its orbit in two Earth years.

We take the orbits as circles: $r = 2$ for Earth and $r = 3$ for Mars. Those equations tell *where* but not *when*. With time as a parameter, the coordinates of Earth and Mars are given at every instant t:

$$x_E = 2 \cos 2\pi t, \; y_E = 2 \sin 2\pi t \qquad \text{and} \qquad x_M = 3 \cos \pi t, \; y_M = 3 \sin \pi t.$$

At $t = 1$ year, the Earth completes a circle (angle $= 2\pi$) and Mars is halfway.

Now the key step. Subtract to find the position of Mars *relative to Earth*:

$$x_{M-E} = 3 \cos \pi t - 2 \cos 2\pi t \qquad \text{and} \qquad y_{M-E} = 3 \sin \pi t - 2 \sin 2\pi t.$$

Replacing $\cos 2\pi t$ by $2\cos^2 \pi t - 1$ and $\sin 2\pi t$ by $2 \sin \pi t \cos \pi t$, this is

$$x_{M-E} = (3 - 4 \cos \pi t) \cos \pi t + 2 \qquad \text{and} \qquad y_{M-E} = (3 - 4 \cos \pi t) \sin \pi t.$$

Seen from the Earth, Mars does a loop in the sky! There are two t's for which $3 - 4\cos \pi t = 0$ (or $\cos \pi t = \frac{3}{4}$). At both times, Mars is two units from Earth ($x_{M-E} = 2$ and $y_{M-E} = 0$). When we move the origin to that point, the 2 is subtracted away—the $M - E$ coordinates become $x = r \cos \pi t$ and $y = r \sin \pi t$ with $r = 3 - 4\cos \pi t$. That is a limaçon with a loop, like Figure 9.4d.

Note added in proof I didn't realize that a 3-to-2 ratio is also responsible for heating up two spots on opposite sides of Mercury. From the newspaper of June 13, 1990:

> "Astronomers today reported the first observations showing that Mercury has two extremely hot spots. That is because Mercury, the planet closest to the Sun, turns on its axis once every 59.6 days, which is a day on Mercury. It goes around the sun every 88 days, a Mercurian year. With this 3-to-2 ratio between spin and revolution, **the Sun appears to stop in the sky and move backward, describing a loop** over each of the hot spots."

CONIC SECTIONS IN POLAR COORDINATES

The exercises include other polar curves, like lemniscates and 200-petal flowers. But get serious. The most important curves are the ***ellipse*** and ***parabola*** and ***hyperbola***. In Section 3.5 their equations involved $1, x, y, x^2, xy, y^2$. With one focus at the origin, their polar equations are even better.

9A The graph of $r = A/(1 + e \cos \theta)$ is a conic section with "eccentricity" e:

circle if $e = 0$ ellipse if $0 < e < 1$ parabola if $e = 1$ hyperbola if $e > 1$.

EXAMPLE 6   ($e = 1$) The graph of $r = 1/(1 + \cos \theta)$ is a parabola. This equation is $r + r \cos \theta = 1$ or $r = 1 - x$. Squaring both sides gives $x^2 + y^2 = 1 - 2x + x^2$. Canceling x^2 leaves $y^2 = 1 - 2x$, the parabola in Figure 9.5b.

The amplifying factor A blows up all curves, with no change in shape.

EXAMPLE 7   ($e = 2$) The same steps lead from $r(1 + 2\cos \theta) = 1$ to $r = 1 - 2x$. Squaring gives $x^2 + y^2 = 1 - 4x + 4x^2$ and the x^2 terms do not cancel. Instead we have $y^2 - 3x^2 = 1 - 4x$. This is the hyperbola in Figure 9.5c, with a focus at $(0,0)$.

The hyperbola $y^2 - 3x^2 = 1$ (without the $-4x$) has its *center* at $(0,0)$.

EXAMPLE 8 $(e = \frac{1}{2})$ The same steps lead from $r(1 + \frac{1}{2}\cos\theta) = 1$ to $r = 1 - \frac{1}{2}x$. Squaring gives the ellipse $x^2 + y^2 = 1 - x + \frac{1}{4}x^2$. Polar equations look at conics in a new way, which happens to match the sun and planets perfectly. *The sun at $(0,0)$ is not the center of the system, but a focus.*

Finally $e = 0$ gives the circle $r = 1$. Center of circle = both foci = $(0,0)$.

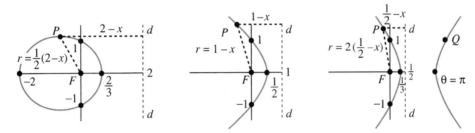

Fig. 9.5 $r = 1/(1 + e\cos\theta)$ is an ellipse for $e = \frac{1}{2}$, a parabola for $e = 1$, a hyperbola for $e = 2$.

The directrix The figure shows the line d (the "directrix") for each curve. All points P on the curve satisfy $r = |PF| = e|Pd|$. **The distance to the focus is e times the distance to the directrix.** (e is still the eccentricity, nothing to do with exponentials.) A geometer would start from this property $r = e|Pd|$ and construct the curve. We derive the property from the equation:

$$r = \frac{A}{1 + e\cos\theta} \quad \Rightarrow \quad r + ex = A \quad \Rightarrow \quad r = e\left(\frac{A}{e} - x\right). \qquad (1)$$

The directrix is the line at $x = A/e$. That last equation is exactly $|PF| = e|Pd|$.

Notice how two numbers determine these curves. Here the numbers are A and e. In Section 3.5 they were a and b. (The ellipse was $x^2/a^2 + y^2/b^2 = 1$.) Using A and e we go smoothly from ellipses through parabolas (at $e = 1$) and on to hyperbolas. With three more numbers we can move the focus to any point and rotate the curve through any angle. **Conics are determined by five numbers.**

9.2 EXERCISES

Read-through questions

The circle of radius 3 around the origin has polar equation __a__. The 45° line has polar equation __b__. Those graphs meet at an angle of __c__. Multiplying $r = 4\cos\theta$ by r yields the xy equation __d__. Its graph is a __e__ with center at __f__. The graph of $r = 4/\cos\theta$ is the line $x =$ __g__. The equation $r^2 = \cos 2\theta$ is not changed when $\theta \to -\theta$ (symmetric across __h__) and when $\theta \to \pi + \theta$ (or $r \to$ __i__). The graph of $r = 1 + \cos\theta$ is a __j__.

The graph of $r = A/($__k__$)$ is a conic section with one focus at __l__. It is an ellipse if __m__ and a hyperbola if __n__. The equation $r = 1/(1 + \cos\theta)$ leads to $r + x = 1$ which gives a __o__. Then $r =$ distance from origin equals $1 - x =$ distance from __p__. The equations $r = 3(1 - x)$ and

$r = \frac{1}{3}(1 - x)$ represents a __q__ and an __r__. Including a shift and rotation, conics are determined by __s__ numbers.

Convert to xy coordinates to draw and identify these curves.

1 $r\sin\theta = 1$ 2 $r(\cos\theta - \sin\theta) = 2$

3 $r = 2\cos\theta$ 4 $r = -2\sin\theta$

5 $r = 1/(2 + \cos\theta)$ 6 $r = 1/(1 + 2\cos\theta)$

In 7–14 sketch the curve and check for $x, y,$ and r symmetry.

7 $r^2 = 4\cos 2\theta$ (lemniscate)

8 $r^2 = 4\sin 2\theta$ (lemniscate)

9 $r = \cos 3\theta$ (three petals)

10 $r^2 = 10 + 6 \cos 4\theta$

11 $r = e^\theta$ (logarithmic spiral)

12 $r = 1/\theta$ (hyperbolic spiral)

13 $r = \tan\theta$

14 $r = 1 - 2\sin 3\theta$ (rose inside rose)

15 Convert $r = 6\sin\theta + 8\cos\theta$ to the xy equation of a circle (what radius, what center ?).

*16 Squaring and adding in the Mars-Earth equation gives $x^2_{M-E} + y^2_{M-E} = 13 - 12\cos\pi t$. The graph of $r^2 = 13 - 12\cos\theta$ is not at all like Figure 9.4d. What went wrong ?

In 17–23 find the points where the two curves meet.

17 $r = 2\cos\theta$ and $r = 1 + \cos\theta$
Warning: You might set $2\cos\theta = 1 + \cos\theta$ to find $\cos\theta = 1$. But the graphs have another meeting point—they reach it at different θ's. Draw graphs to find all meeting points.

18 $r^2 = \sin 2\theta$ and $r^2 = \cos 2\theta$

19 $r = 1 + \cos\theta$ and $r = 1 - \sin\theta$

20 $r = 1 + \cos\theta$ and $r = 1 - \cos\theta$

21 $r = 2$ and $r = 4\sin 2\theta$

22 $r^2 = 4\cos\theta$ and $r = 1 - \cos\theta$

23 $r\sin\theta = 1$ and $r\cos(\theta - \pi/4) = \sqrt{2}$ (straight lines)

24 When is there a dimple in $r = 1 + b\cos\theta$? From $x = (1 + b\cos\theta)\cos\theta$ find $dx/d\theta$ and $d^2x/d\theta^2$ at $\theta = \pi$. When that second derivative is negative the limaçon has a dimple.

25 How many petals for $r = \cos 5\theta$? For $r = \cos\theta$ there was one, for $r = \cos 2\theta$ there were four.

26 Explain why $r = \cos 100\,\theta$ has 200 petals but $r = \cos 101\,\theta$ only has 101. The other 101 petals are _____. What about $r = \cos\frac{1}{2}\theta$?

27 Find an xy equation for the cardioid $r = 1 + \cos\theta$.

28 (a) The flower $r = \cos 2\theta$ is symmetric across the x and y axes. Does that make it symmetric about the origin ? (Do two symmetries imply the third, so $-r = \cos 2\theta$ produces the same curve ?)
(b) How can $r = 1$, $\theta = \pi/2$ lie on the curve but fail to satisfy the equation ?

29 Find an xy equation for the flower $r = \cos 2\theta$.

30 Find equations for curves with these properties:

(a) Symmetric about the origin but not the x axis

(b) Symmetric across the $45°$ line but not symmetric in x or y or r

(c) Symmetric in x and y and r (like the flower) but changed when $x \leftrightarrow y$ (not symmetric across the $45°$ line).

Problems 31–37 are about conic sections—especially ellipses.

31 Find the top point of the ellipse in Figure 9.5a, by maximizing $y = r\sin\theta = \sin\theta/(1 + \frac{1}{2}\cos\theta)$.

32 (a) Show that all conics $r = 1/(1 + e\cos\theta)$ go through $x = 0, y = 1$.

(b) Find the second focus of the ellipse and hyperbola. For the parabola ($e = 1$) where is the second focus ?

33 The point Q in Figure 9.5c has $y = 1$. By symmetry find x and then r (negative!). Check that $x^2 + y^2 = r^2$ and $|QF| = 2|Qd|$.

34 The equations $r = A/(1 + e\cos\theta)$ and $r = 1/(C + D\cos\theta)$ are the same if $C = $ _____ and $D = $ _____. For the mirror image across the y axis replace θ by _____. This gives $r = 1/(C - D\cos\theta)$ as in Figure 12.10 for a planet around the sun.

35 The ellipse $r = A/(1 + e\cos\theta)$ has length $2a$ on the x axis. Add r at $\theta = 0$ to r at $\theta = \pi$ to prove that $a = A/(1 - e^2)$. The Earth's orbit has $a = 92,600,000$ miles $=$ one astronomical unit (AU).

36 The maximum height b occurs when $y = r\sin\theta = A\sin\theta/(1 + e\cos\theta)$ has $dy/d\theta = 0$. Show that $b = y_{max} = A/\sqrt{1 - e^2}$.

37 Combine a and b from Problems $35 - 36$ to find $c = \sqrt{a^2 - b^2} = Ae/(1 - e^2)$. *Then the eccentricity e is c/a.* Halley's comet is an ellipse with $a = 18.1$ AU and $b = 4.6$ AU so $e = $ _____.

Comets have large eccentricity, planets have much smaller e: **Mercury** .21, **Venus** .01, **Earth** .02, **Mars** .09, **Jupiter** .05, **Saturn** .05, **Uranus** .05, **Neptune** .01, **Pluto** .25, **Kohoutek** .9999.

38 If you have a computer with software to do polar graphs, start with these:

1. Flowers $r = A + \cos n\theta$ for $n = \frac{1}{2}, 3, 7, 8$; $A = 0, 1, 2$

2. Petals $r = (\cos m\theta + 4\cos n\theta)/\cos\theta$, $(m, n) = (5, 3)$, $(3, 5)$, $(9, 1)$, $(2, 3)$

3. Logarithmic spiral $r = e^{\theta/2\pi}$

4. Nephroid $r = 1 + 2\sin\frac{1}{2}\theta$ from the bottom of a teacup

5. Dr. Fay's butterfly $r = e^{\cos\theta} - 2\cos 4\theta + \sin^5(\theta/12)$

Then create and name your own curve.

9.3 Slope, Length, and Area for Polar Curves

The previous sections introduced polar coordinates and polar equations and polar graphs. There was no calculus! We now tackle the problems of *area* (integral calculus) and *slope* (differential calculus), when the equation is $r = F(\theta)$. The use of F instead of f is a reminder that the slope is *not* $dF/d\theta$ and the area is *not* $\int F(\theta)d\theta$.

Start with area. The region is always divided into small pieces—what is their shape? Look between the angles θ and $\theta + \Delta\theta$ in Figure 9.6a. Inside the curve is a narrow wedge—almost a triangle, with $\Delta\theta$ as its small angle. If the radius is constant

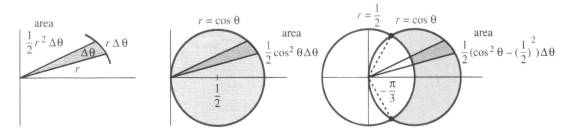

Fig. 9.6 Area of a wedge and a circle and an intersection of circles.

the wedge is a sector of a circle. It is a piece of pie cut at the extremely narrow angle $\Delta\theta$. The area of that piece is a fraction (the angle $\Delta\theta$ divided by the whole angle 2π) of the whole area πr^2 of the circle:

$$\textbf{\textit{area of wedge}} = \frac{\Delta\theta}{2\pi}\pi r^2 = \frac{1}{2}r^2\Delta\theta = \frac{1}{2}[F(\theta)]^2\,\Delta\theta. \tag{1}$$

We admit that the exact shape is not circular. The true radius $F(\theta)$ varies with θ—but in a narrow angle that variation is small. When we add up the wedges and let $\Delta\theta$ approach zero, the area becomes an integral.

9B The area inside the polar curve $r = F(\theta)$ is the limit of $\sum\frac{1}{2}r^2\Delta\theta = \sum\frac{1}{2}F^2\Delta\theta$:

$$area = \int\frac{1}{2}r^2\,d\theta = \int\frac{1}{2}[F(\theta)]^2\,d\theta. \tag{2}$$

EXAMPLE 1   Find the area inside the circle $r = \cos\theta$ of radius $\frac{1}{2}$ (Figure 9.6).

$$area = \int_0^{2\pi}\frac{1}{2}\cos^2\theta\,d\theta = \frac{\cos\theta\sin\theta+\theta}{4}\Big]_0^{2\pi} = \frac{2\pi}{4}.$$

That is wrong! The correct area of a circle of radius $\frac{1}{2}$ is $\pi/4$. The mistake is that we went *twice* around the circle as θ increased to 2π. Integrating from θ to π gives $\pi/4$.

EXAMPLE 2   Find the area between the circles $r = \cos\theta$ and $r = \frac{1}{2}$.

The circles cross at the points where $r = \cos\theta$ agrees with $r = \frac{1}{2}$. Figure 9.6 shows these points at $\pm 60°$, or $\theta = \pm\pi/3$. Those are the limits of integration, where $\cos\theta = \frac{1}{2}$.

The integral adds up the difference between two wedges, one out to $r = \cos\theta$ and a smaller one with $r = \frac{1}{2}$:

$$area = \int_{-\pi/3}^{\pi/3} \frac{1}{2}\left[(\cos\theta)^2 - \left(\frac{1}{2}\right)^2\right]d\theta. \qquad (3)$$

Note that chopped wedges have area $\frac{1}{2}(F_1^2 - F_2^2)\Delta\theta$ and not $\frac{1}{2}(F_1 - F_2)^2\Delta\theta$.

EXAMPLE 3 Find the area between the cardioid $r = 1 + \cos\theta$ and the circle $r = 1$.

$$area = \int_{-\pi/2}^{\pi/2} \frac{1}{2}\left[(1 + \cos\theta)^2 - 1^2\right]d\theta \quad \left(\text{limits } \theta = \pm\frac{\pi}{2} \text{ where } 1 + \cos\theta = 1\right)$$

SLOPE OF A POLAR CURVE

Where is the highest point on the cardioid $r = 1 + \cos\theta$? What is the slope at $\theta = \pi/4$? Those are not the most important questions in calculus, but still we should know how to answer them. I will describe the method quickly, by switching to rectangular coordinates:

$$x = r\cos\theta = (1 + \cos\theta)\cos\theta \qquad \text{and} \qquad y = r\sin\theta = (1 + \cos\theta)\sin\theta.$$

For the highest point, maximize y by setting its derivative to zero:

$$dy/d\theta = (1 + \cos\theta)(\cos\theta) + (-\sin\theta)(\sin\theta) = 0. \qquad (3)$$

Thus $\cos\theta + \cos 2\theta = 0$, which happens at $60°$. The height is $y = (1 + \frac{1}{2})(\sqrt{3}/2)$.

For the slope, use the chain rule $dy/d\theta = (dy/dx)(dx/d\theta)$:

$$\frac{dy}{dx} = \frac{dy/d\theta}{dx/d\theta} = \frac{(1 + \cos\theta)(\cos\theta) + (-\sin\theta)(\sin\theta)}{(1 + \cos\theta)(-\sin\theta) + (-\sin\theta)\cos\theta}. \qquad (4)$$

Equations (3) and (4) avoid the awkward (or impossible) step of eliminating θ. Instead of trying to find y as a function of x, *we keep x and y as functions of θ.* At $\theta = \pi/4$, the ratio in equation (4) yields $dy/dx = -1/(1 + \sqrt{2})$.

Problem 18 finds a general formula for the slope, using $dy/dx = (dy/d\theta)/(dx/d\theta)$. Problem 20 finds a more elegant formula, by looking at the question differently.

LENGTH OF A POLAR CURVE

The length integral always starts with $ds = \sqrt{(dx)^2 + (dy)^2}$. A polar curve has $x = r\cos\theta = F(\theta)\cos\theta$ and $y = F(\theta)\sin\theta$. Now take derivatives by the product rule:

$$dx = (F'(\theta)\cos\theta - F(\theta)\sin\theta)d\theta \qquad \text{and} \qquad dy = (F'(\theta)\sin\theta + F(\theta)\cos\theta)d\theta.$$

Squaring and adding (note $\cos^2\theta + \sin^2\theta$) gives the element of length ds:

$$ds = \sqrt{[F'(\theta)]^2 + [F(\theta)]^2}\, d\theta. \tag{5}$$

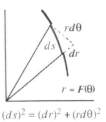

The figure shows $(ds)^2 = (dr)^2 + (rd\theta)^2$, the same formula with different letters. The total arc length is $\int ds$.

The area of a surface of revolution is $\int 2\pi y\, ds$ (around the x axis) or $\int 2\pi x\, ds$ (around the y axis). **Write x, y, and ds in terms of θ and $d\theta$.** Then integrate.

$(ds)^2 = (dr)^2 + (rd\theta)^2$

Fig. 9.7

EXAMPLE 4 The circle $r = \cos\theta$ has $ds = \sqrt{1}\, d\theta$. So its length is π (not 2π!!—don't go around twice). Revolved around the y axis the circle yields a doughnut with no hole. Since $x = r\cos\theta = \cos^2\theta$, the surface area of the doughnut is

$$\int 2\pi x\, ds = \int_0^\pi 2\pi \cos^2\theta\, d\theta = \pi^2.$$

EXAMPLE 5 The length of $r = 1 + \cos\theta$ is, by symmetry, double the integral from 0 to π:

$$\text{length of cardioid} = 2\int_0^\pi \sqrt{(-\sin\theta)^2 + (1 + \cos\theta)^2}\, d\theta$$

$$= 2\int_0^\pi \sqrt{2 + 2\cos\theta}\, d\theta = 4\int_0^\pi \cos\frac{\theta}{2}\, d\theta = 8.$$

We substituted $4\cos^2\frac{1}{2}\theta$ for $2 + 2\cos\theta$ in the square root. It is possible to skip symmetry and integrate from 0 to 2π—but that needs the absolute value $|\cos\frac{1}{2}\theta|$ to maintain a positive square root.

EXAMPLE 6 The logarithmic spiral $r = e^{-\theta}$ has $ds = \sqrt{e^{-2\theta} + e^{-2\theta}}\, d\theta$. It spirals to zero as θ goes to infinity, and the total length is finite:

$$\int ds = \int_0^\infty \sqrt{2}\, e^{-\theta}\, d\theta = -\sqrt{2}\, e^{-\theta}\Big]_0^\infty = \sqrt{2}.$$

Revolve this spiral for a mathematical seashell with area $\int_0^\infty (2\pi e^{-\theta}\cos\theta)\sqrt{2}\, e^{-\theta}\, d\theta$.

9.3 EXERCISES

Read-through questions

A circular wedge with angle $\Delta\theta$ is a fraction __a__ of a whole circle. If the radius is r, the wedge area is __b__. Then the area inside $r = F(\theta)$ is $\int$ __c__ . The area inside $r = \theta^2$ from 0 to π is __d__ . That spiral meets the circle $r = 1$ at $\theta =$ __e__ . The area inside the circle and outside the spiral is __f__ . A chopped wedge of angle $\Delta\theta$ between r_1 and r_2 has area __g__ .

The curve $r = F(\theta)$ has $x = r\cos\theta =$ __h__ and $y =$ __i__ . The slope dy/dx is $dy/d\theta$ divided by __j__ . For length $(ds)^2 = (dx)^2 + (dy)^2 =$ __k__ . The length of the spiral $r = \theta$ to $\theta = \pi$ is $\int$ __l__ (not to compute integrals). The surface area when $r = \theta$ is revolved around the x axis is $\int 2\pi y\, ds = \int$ __m__ . The volume of that solid is $\int \pi y^2\, dx = \int$ __n__ .

In 1–6 draw the curve and find the area inside.

1 $r = 1 + \cos\theta$

2 $r = \sin\theta + \cos\theta$ from 0 to π

3 $r = 2 + \cos\theta$

4 $r = 1 + 2\cos\theta$ (inner loop only)

5 $r = \cos 2\theta$ (one petal only)

6 $r = \cos 3\theta$ (one petal only)

Find the area between the curves in 7–12 after locating their intersections (draw them first).

7 circle $r = \cos\theta$ and circle $r = \sin\theta$

8 spiral $r = \theta$ and y axis (first arch)

9 outside cardioid $r = 1 + \cos\theta$ inside circle $r = 3\cos\theta$

10 lemniscate $r^2 = 4\cos 2\theta$ outside $r = \sqrt{2}$

11 circle $r = 8\cos\theta$ beyond line $r\cos\theta = 4$

12 circle $r = 10$ beyond line $r\cos\theta = 6$

13 Locate the mistake and find the correct area of the lemniscate
$r^2 = \cos 2\theta$:   area $= \int_0^\pi \frac{1}{2} r^2\, d\theta = \int_0^\pi \frac{1}{2}\cos 2\theta\, d\theta = 0$.

14 Find the area between the two circles in Example 2.

15 Compute the area between the cardioid and circle in Example 3.

16 Find the complete area (carefully) between the spiral $r = e^{-\theta}$ ($\theta \geqslant 0$) and the origin.

17 At what θ's does the cardioid $r = 1 + \cos\theta$ have infinite slope? Which points are furthest to the left (minimum x)?

18 Apply the chain rule $dy/dx = (dy/d\theta)/(dx/d\theta)$ to $x = F(\theta)\cos\theta$, $y = F(\theta)\sin\theta$. Simplify to reach

$$\frac{dy}{dx} = \frac{F + \tan\theta\, dF/d\theta}{-F\tan\theta + dF/d\theta}.$$

19 The groove in a record is nearly a spiral $r = c\theta$:

$$\text{length} = \int \sqrt{r^2 + (dr/d\theta)^2}\, d\theta = \int_6^{14} \sqrt{r^2 + c^2}\, dr/c.$$

Take $c = .002$ to give 636 turns between the outer radius 14 cm and the inner radius 6 cm ($14 - 6$ equals $.002(636)2\pi$).

(a) Omit c^2 and just integrate $r\, dr/c$.

(b) Compute the length integral. Tables and calculators allowed. You will never trust integrals again.

20 Show that the angle ψ between the ray from the origin and the tangent line has $\tan\psi = F/(dF/d\theta)$.
Hint: If the tangent line is at an angle ϕ with the horizontal, then $\tan\phi$ is the slope dy/dx in Problem 18. Therefore

$$\tan\psi = \tan(\phi - \theta) = \frac{\tan\phi - \tan\theta}{1 + \tan\phi\tan\theta}.$$

Substitute for $\tan\phi$ and simplify like mad.

21 The circle $r = F(\theta) = 4\sin\theta$ has $\psi = \theta$. Draw a figure including θ, ϕ, ψ and check $\tan\psi$.

22 Draw the cardioid $r = 1 - \cos\theta$, noticing the minus sign. Include the angles θ, ϕ, ψ and show that $\psi = \theta/2$.

23 The first limaçon in Figure 9.4 looks like a circle centered at $(\frac{1}{3}, 0)$. Prove that it isn't.

24 Find the equation of the tangent line to the circle $r = \cos\theta$ at $\theta = \pi/6$.

In 25–28 compute the length of the curve.

25 $r = \theta$ (θ from 0 to 2π)

26 $r = \sec\theta$ (θ from 0 to $\pi/4$)

27 $r = \sin^3(\theta/3)$ (θ from 0 to 3π)

28 $r = \theta^2$ (θ from 0 to π)

29 The narrow wedge in Figure 9.6 is almost a triangle. It was treated as a circular sector but triangles are more familiar. Why is the area approximately $\frac{1}{2}r^2\Delta\theta$?

30 In Example 4 revolve the circle around the x axis and find the surface area. *We really only revolve a semicircle.*

31 Compute the seashell area $2\pi\sqrt{2}\int_0^\infty e^{-2\theta}\cos\theta\, d\theta$ using two integrations by parts.

32 Find the surface area when the cardioid $r = 1 + \cos\theta$ is revolved around the x axis.

33 Find the surface area when the lemniscate $r^2 = \cos 2\theta$ is revolved around the x axis. What is θ after one petal?

34 When $y = f(x)$ is revolved around the x axis, the volume is $\int \pi y^2 dx$. When the circle $r = \cos\theta$ is revolved, switch to a θ-integral from 0 to $\pi/2$ and check the volume of a sphere.

35 Find the volume when the cardioid $r = 1 + \cos\theta$ is rotated around the x axis.

36 Find the surface area and volume when the graph of $r = 1/\cos\theta$ is rotated around the y axis ($0 \leqslant \theta \leqslant \pi/4$).

37 Show that the spirals $r = \theta$ and $r = 1/\theta$ are perpendicular when they meet at $\theta = 1$.

38 Draw three circles of radius 1 that touch each other and find the area of the curved triangle between them.

39 Draw the unit square $0 \leqslant x \leqslant 1, 0 \leqslant y \leqslant 1$. In polar coordinates its right side is $r = \underline{\hspace{1cm}}$. Find the area from $\int \frac{1}{2}r^2 d\theta$.

40 (Unravel the paradox) The area of the ellipse $x = 4\cos\theta$, $y = 3\sin\theta$ is $\pi \cdot 4 \cdot 3 = 12\pi$. But the integral of $\frac{1}{2}r^2 d\theta$ is

$$\int_0^{2\pi} \frac{1}{2}(16\cos^2\theta + 9\sin^2\theta)\, d\theta = 12\frac{1}{2}\pi.$$

9.4 Complex Numbers

Real numbers are sufficient for most of calculus. Starting from $x^2 + 4$, its integral $\frac{1}{3}x^3 + 4x + C$ is also real. If we are given $x^3 - 1$, its derivative $3x^2$ is real. *But the roots (or zeros) of those polynomials are complex numbers*:

$$x^2 + 4 = 0 \quad \textit{and} \quad x^3 - 1 = 0 \quad \textit{have complex solutions.}$$

We expect two square roots of -4. There are three cube roots of 1. Complex numbers are unavoidable, in order to find n roots for each polynomial of degree n.

This section explains how to work with complex numbers. You will see their relation to polar coordinates. At the end, we use them to solve differential equations.

Start with the imaginary number i. Everybody knows that $x^2 = -1$ has no real solution. When you square a real number, the result is never negative. So the world has agreed on a solution called i. (Except that electrical engineers call it j.) Imaginary numbers follow the normal rules of addition, subtraction, multiplication, and division, with one difference: *Whenever i^2 appears it is replaced by -1*. In particular $-i$ times $-i$ gives $+i^2 = -1$. In other words, $-i$ is also a square root of -1. There are two solutions (i and $-i$) to the equation $x^2 + 1 = 0$.

Finding cube roots of 1 will stretch us further. We need complex numbers—real plus imaginary.

9B A *complex number* (say $1 + 3i$) is the sum of a real number (1) and a pure imaginary number ($3i$). Addition keeps those parts separate; multiplication uses $i^2 = -1$:

$$\text{Addition:} \quad (1 + 3i) + (1 + 3i) = 1 + 1 + i(3 + 3) = 2 + 6i$$

$$\text{Multiplication:} \quad (1 + 3i)(1 + 3i) = 1 + 3i + 3i + 9i^2 = -8 + 6i.$$

Adding $1 + 3i$ to $5 - i$ is easy ($6 + 2i$). Multiplying is longer, but you see the rules:

$$(1 + 3i)(5 - i) = 5 + 15i - i - 3i^2 = 8 + 14i.$$

The point is this: We don't have to imagine any more new numbers. After accepting i, the rest is straightforward. A real number is just a complex number with no imaginary part! When $1 + 3i$ combines with its "*complex conjugate*" $1 - 3i$—adding or multiplying—the answer is real:

$$(1 + 3i) + (1 - 3i) = 2 \quad \textit{(real)}$$
$$(1 + 3i)(1 - 3i) = 1 - 3i + 3i - 9i^2 = 10. \quad \textit{(real)} \tag{1}$$

The complex conjugate offers a way to do division, by making the denominator real:

$$\frac{1}{1 + 3i} = \frac{1}{1 + 3i}\frac{1 - 3i}{1 - 3i} = \frac{1 - 3i}{10} \quad \text{and} \quad \frac{1}{x + iy} = \frac{1}{x + iy}\frac{x - iy}{x - iy} = \frac{x - iy}{x^2 + y^2}.$$

9C The complex number $x + iy$ has real part x and imaginary part y. Its complex conjugate is $x - iy$. The product $(x + iy)(x - iy)$ equals $x^2 + y^2 = r^2$. The *absolute value* (or modulus) is $r = |x + iy| = \sqrt{x^2 + y^2}$.

THE COMPLEX PLANE

Complex numbers correspond to points in a plane. The number $1 + 3i$ corresponds to the point $(1, 3)$. Similarly $x + iy$ is paired with (x, y)—which is x units along the "real axis" and y units up the "imaginary axis." The ordinary plane turns into the ***complex plane***. The absolute value r is the same as the polar coordinate r (Figure 9.8a).

The figure shows two more copies of the complex plane. The one in the middle is for addition and subtraction. It uses rectangular coordinates. The one on the right is for multiplication and division and squaring. It uses polar coordinates. In squaring a complex number, r is squared and θ is doubled—as the right figure and equation (3) both show.

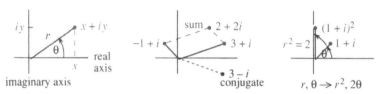

Fig. 9.8   The complex plane shows x, y, r, θ. Add with x and y, multiply with r and θ.

Adding complex numbers is like adding vectors (Chapter 11). The real parts give $3 - 1$ and the imaginary parts give $1 + 1$. The vector sum $(2, 2)$ corresponds to the complex sum $2 + 2i$. The complex conjugate $3 - i$ is the mirror image across the real axis (i reversed to $-i$). The connection to r and θ is the same as before (you see it in the triangle):

$$x = r \cos \theta \quad \textbf{\textit{and}} \quad y = r \sin \theta \quad \textbf{\textit{so that}} \quad x + iy = r(\cos \theta + i \sin \theta). \quad (2)$$

In the third figure, $1 + i$ has $r = \sqrt{2}$ and $\theta = \pi/4$. The polar form is $\sqrt{2} \cos \pi/4 + \sqrt{2} i \sin \pi/4$. When this number is squared, its $45°$ angle becomes $90°$. The square is $(1 + i)^2 = 1 + 2i - 1 = 2i$. Its polar form is $2 \cos \pi/2 + 2i \sin \pi/2$.

> **9D** Multiplication adds angles, division subtracts angles, and squaring doubles angles. The absolute values are multiplied, divided, and squared:
>
> $$(r \cos \theta + ir \sin \theta)^2 = r^2 \cos 2\theta + ir^2 \sin 2\theta. \quad (3)$$

For nth powers we reach r^n and $n\theta$. For square roots, r goes to $\sqrt{r}$ and θ goes to $\frac{1}{2}\theta$. The number -1 is at $180°$, so its square root i is at $90°$.

To see why θ is doubled in equation (3), factor out r^2 and multiply as usual:

$$(\cos \theta + i \sin \theta)(\cos \theta + i \sin \theta) = \cos^2 \theta - \sin^2 \theta + 2i \sin \theta \cos \theta.$$

The right side is $\cos 2\theta + i \sin 2\theta$. The double-angle formulas from trigonometry match the squaring of complex numbers. The cube would be $\cos 3\theta + i \sin 3\theta$ (because 2θ and θ add to 3θ, and r is still 1). The nth power is in ***de Moivre's formula***:

$$(\cos \theta + i \sin \theta)^n = \cos n\theta + i \sin n\theta. \quad (4)$$

With $n = -1$ we get $\cos(-\theta) + i \sin(-\theta)$—which is $\cos \theta - i \sin \theta$, the complex conjugate:

$$\frac{1}{\cos \theta + i \sin \theta} = \frac{1}{\cos \theta + i \sin \theta} \frac{\cos \theta - i \sin \theta}{\cos \theta - i \sin \theta} = \frac{\cos \theta - i \sin \theta}{1}. \tag{5}$$

We are almost touching **Euler's formula**, the key to all numbers on the unit circle:

$$\textbf{\textit{Euler's formula}} : \qquad \cos \theta + i \sin \theta = e^{i\theta}. \tag{6}$$

Squaring both sides gives $(e^{i\theta})(e^{i\theta}) = e^{2i\theta}$. That is equation (3). The -1 power is $1/e^{i\theta} = e^{-i\theta}$. That is equation (5). Multiplying any $e^{i\theta}$ by $e^{i\phi}$ produces $e^{i(\theta+\phi)}$. The special case $\phi = \theta$ gives the square, and the special case $\phi = -\theta$ gives $e^{i\theta}e^{-i\theta} = 1$.

Euler's formula appeared in Section 6.7, by changing x to $i\theta$ in the series for e^x:

$$e^x = 1 + x + \frac{x^2}{2} + \frac{x^3}{6} + \cdots \quad \text{becomes} \quad e^{i\theta} = 1 + i\theta - \frac{\theta^2}{2} - i\frac{\theta^3}{6} + \cdots.$$

A highlight of Chapter 10 is to recognize two new series on the right. The real terms $1 - \frac{1}{2}\theta^2 + \cdots$ add up to $\cos \theta$. The imaginary part $\theta - \frac{1}{6}\theta^3 + \cdots$ adds up to $\sin \theta$. Therefore $e^{i\theta}$ equals $\cos \theta + i \sin \theta$. It is fantastic that the most important periodic functions in all of mathematics come together in this graceful way.

We learn from Euler (pronounced *oiler*) that $e^{2\pi i} = 1$. The cosine of 2π is 1, the sine is zero. If you substitute $x = 2\pi i$ into the infinite series, somehow everything cancels except the 1—this is almost a miracle. From the viewpoint of angles, $\theta = 2\pi$ carries us around a full circle and back to $e^{2\pi i} = 1$.

Multiplying Euler's formula by r, we have a third way to write a complex number:

$$\textbf{\textit{Every complex number is}} \qquad x + iy = r \cos \theta + ir \sin \theta = re^{i\theta}. \tag{7}$$

EXAMPLE 1 $2e^{i\theta}$ times $3e^{i\theta}$ equals $6e^{2i\theta}$. For $\theta = \pi/2$, $2i$ times $3i$ is -6.

EXAMPLE 2 Find w^2 and w^4 and w^8 and w^{25} when $w = e^{i\pi/4}$.

Solution $e^{i\pi/4}$ is $1/\sqrt{2} + i/\sqrt{2}$. Note that $r^2 = \frac{1}{2} + \frac{1}{2} = 1$. Now watch angles:

$$w^2 = e^{i\pi/2} = i \quad w^4 = e^{i\pi} = -1 \quad w^8 = 1 \quad w^{25} = w^8 w^8 w^8 w = w.$$

Figure 9.9 shows the eight powers of w. **They are the eighth roots of** 1.

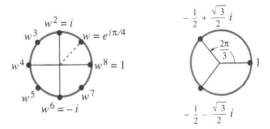

Fig. 9.9 The eight powers of w and the cube roots of 1.

EXAMPLE 3 $(x^2 + 4 = 0)$ The square roots of -4 are $2i$ and $-2i$. Instead of $(i)(i) = -1$ we have $(2i)(2i) = -4$. If Euler insists, we write $2i$ and $-2i$ as $2e^{i\pi/2}$ and $2e^{i3\pi/2}$.

EXAMPLE 4 (The cube roots of 1) In rectangular coordinates we have to solve $(x+iy)^3 = 1$, which is not easy. In polar coordinates this same equation is $r^3 e^{3i\theta} = 1$. Immediately $r = 1$. The angle θ can be $2\pi/3$ or $4\pi/3$ or $6\pi/3$—*the cube roots in the figure are evenly spaced*:

$$(e^{2\pi i/3})^3 = e^{2\pi i} = 1 \quad (e^{4\pi i/3})^3 = e^{4\pi i} = 1 \quad (e^{6\pi i/3})^3 = e^{6\pi i} = 1.$$

You see why the angle $8\pi/3$ gives nothing new. It completes a full circle back to $2\pi/3$.

> *The nth roots of 1 are $e^{2\pi i/n}$, $e^{4\pi i/n}$, ..., 1. There are n of them.*
> *They lie at angles $2\pi/n$, $4\pi/n$, ..., 2π around the unit circle.*

SOLUTION OF DIFFERENTIAL EQUATIONS

The algebra of complex numbers is now applied to the calculus of complex functions. The complex number is c, the complex function is e^{ct}. It will solve the equations $y'' = -4y$ and $y''' = y$, by connecting them to $c^2 = -4$ and $c^3 = 1$. Chapter 16 does the same for all linear differential equations with constant coefficients—this is an optional preview.

Please memorize the one key idea: **Substitute $y = e^{ct}$ into the differential equation and solve for c.** Each derivative brings a factor c, so $y' = ce^{ct}$ and $y'' = c^2 e^{ct}$:

$$d^2 y/dt^2 = -4y \text{ leads to } c^2 e^{ct} = -4e^{ct}, \text{ which gives } c^2 = -4. \tag{8}$$

For this differential equation, c must be a square root of -4. We know the candidates ($c = 2i$ and $c = -2i$). The equation has two "pure exponential solutions" e^{ct}:

$$y = e^{2it} \quad \text{and} \quad y = e^{-2it}. \tag{9}$$

Their combinations $y = Ae^{2it} + Be^{-2it}$ give all solutions. In Chapter 16 we will choose the two numbers A and B to match two initial conditions at $t = 0$.

The solution $y = e^{2it} = \cos 2t + i \sin 2t$ is complex. The differential equation is real. For real y's, **take the real and imaginary parts of the complex solutions**:

$$y_{\text{real}} = \cos 2t \quad \text{and} \quad y_{\text{imaginary}} = \sin 2t. \tag{10}$$

These are the "pure oscillatory solutions." When $y = e^{2it}$ travels around the unit circle, its imaginary part $\sin 2t$ moves up and down. (It is like the ball and its shadow in Section 1.4, but twice as fast because of $2t$.) The real part $\cos 2t$ goes backward and forward. By the chain rule, *the second derivative of* $\cos 2t$ *is* $-4\cos 2t$. Thus $d^2 y/dt^2 = -4y$ and we have real solutions.

EXAMPLE 5   Find three solutions and then three *real* solutions to $d^3 y/dt^3 = y$.

Key step: **Substitute $y = e^{ct}$.** The result is $c^3 e^{ct} = e^{ct}$. Thus $c^3 = 1$ and c is a cube root of 1. The candidate $c = 1$ gives $y = e^t$ (our first solution). The next c is complex:

$$c = e^{2\pi i/3} = -\frac{1}{2} + i\frac{\sqrt{3}}{2} \quad \textbf{\textit{yields}} \quad y = e^{ct} = e^{-t/2} e^{i\sqrt{3}t/2}. \tag{11}$$

The real part of the exponent leads to the absolute value $|y| = e^{-t/2}$. It decreases as t gets larger, so y moves toward zero. At the same time, the factor $e^{i\sqrt{3}t/2}$ goes around the unit circle. Therefore y spirals in to zero (Figure 9.10). So does its complex conjugate, which is the third exponential. Changing i to $-i$ in (11) gives the third cube root of 1 and the third solution $e^{-t/2}e^{-i\sqrt{3}t/2}$.

The first real solution is $y = e^t$. The others are the two parts of the spiral:

$$y_{\text{real}} = e^{-t/2}\cos\sqrt{3}t/2 \quad\text{and}\quad y_{\text{imaginary}} = e^{-t/2}\sin\sqrt{3}t/2. \quad (12)$$

That is $r\cos\theta$ and $r\sin\theta$. It is the ultimate use (until Chapter 16) of polar coordinates and complex numbers. We might have discovered $\cos 2t$ and $\sin 2t$ without help, for $y'' = -4y$. I don't think these solutions to $y''' = y$ would have been found.

EXAMPLE 6 Find four solutions to $d^4y/dt^4 = y$ by substituting $y = e^{ct}$.

Four derivatives lead to $c^4 = 1$. Therefore c is i or -1 or $-i$ or 1. The solutions are $y = e^{it}$, e^{-t}, e^{-it}, and e^t. If we want real solutions, e^{it} and e^{-it} combine into $\cos t$ and $\sin t$. In all cases $y'''' = y$.

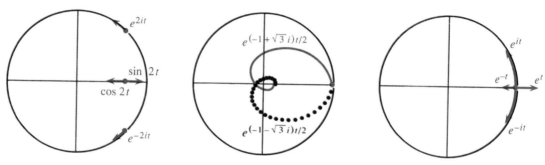

Fig. 9.10 Solutions move in the complex plane: $y'' = -4y$ and $y''' = y$ and $y'''' = y$.

9.4 EXERCISES

Read-through questions

The complex number $3 + 4i$ has real part __a__ and imaginary part __b__. Its absolute value is $r =$ __c__ and its complex conjugate is __d__. Its position in the complex plane is at (__e__). Its polar form is $r\cos\theta + ir\sin\theta =$ __f__ $e^{i\theta}$. Its square is __g__ $+ i$ __h__. Its nth power is __i__ $e^{in\theta}$.

The sum of $1 + i$ and $1 - i$ is __j__. The product of $1 + i$ and $1 - i$ is __k__. In polar form this is $\sqrt{2}e^{i\pi/4}$ times __l__. The quotient $(1+i)/(1-i)$ equals the imaginary number __m__. The number $(1+i)^8$ equals __n__. An eighth root of 1 is $w =$ __o__. The other eighth roots are __p__.

To solve $d^8y/dt^8 = y$, look for a solution of the form $y =$ __q__. Substituting and canceling e^{ct} leads to the equation __r__. There are __s__ choices for c, one of which is $(-1+i)/\sqrt{2}$. With that choice $|e^{ct}| =$ __t__. The real solutions are $\text{Re } e^{ct} =$ __u__ and $\text{Im } e^{ct} =$ __v__.

In 1–6 plot each number in the complex plane.

1 $2 + i$ and its complex conjugate $2 - i$ and their sum and product

2 $1 + i$ and its square $(1+i)^2$ and its reciprocal $1/(1+i)$

3 $2e^{i\pi/6}$ and its reciprocal $\frac{1}{2}e^{-i\pi/6}$ and their squares

4 The sixth roots of 1 (six of them)

5 $\cos 3\pi/4 + i\sin 3\pi/4$ and its square and cube

6 $4e^{i\pi/3}$ and its square roots

7 For complex numbers $c = x + iy = re^{i\theta}$ and their conjugates $\bar{c} = x - iy = re^{-i\theta}$, find all possible locations in the complex plane of (1) $c + \bar{c}$ (2) $c - \bar{c}$ (3) $c\bar{c}$ (4) $c/\bar{c}$.

8 Find x and y for the complex numbers $x + iy$ at angles $\theta = 45°, 90°, 135°$ on the unit circle. Verify directly that the square of the first is the second and the cube of the first is the third.

9 If $c = 2 + i$ and $d = 4 + 3i$ find cd and c/d. Verify that the absolute value $|cd|$ equals $|c|$ times $|d|$, and $|c/d|$ equals $|c|$ divided by $|d|$.

10 Find a solution x to $e^{ix} = i$ and a solution to $e^{ix} = 1/e$. Then find a second solution.

Find the sum and product of the numbers in 11–14.

11 $e^{i\theta}$ and $e^{-i\theta}$, also $e^{2\pi i/3}$ and $e^{4\pi i/3}$

12 $e^{i\theta}$ and $e^{i\phi}$, also $e^{\pi i/4}$ and $e^{-\pi i/4}$

13 The sixth roots of 1 (add and multiply all six)

14 The two roots of $c^2 - 4c + 5 = 0$

15 If $c = re^{i\theta}$ is not zero, what are c^4 and c^{-1} and c^{-4}?

16 Multiply out $(\cos\theta + i\sin\theta)^3 = e^{i3\theta}$, to find the real part $\cos 3\theta$ and the imaginary part $\sin 3\theta$ in terms of $\cos\theta$ and $\sin\theta$.

17 Plot the three cube roots of a typical number $re^{i\theta}$. Show why they add to zero. One cube root is $r^{1/3}e^{i\theta/3}$.

18 Prove that the four fourth roots of $re^{i\theta}$ multiply to give $-re^{i\theta}$.

In 19–22, find all solutions of the form $y = e^{ct}$.

19 $y'' + y = 0$ **20** $y''' + y = 0$

21 $y''' - y' = 0$ **22** $y'' + 6y' + 5y = 0$

Construct two real solutions from the real and imaginary parts of e^{ct} (first find c):

23 $y'' + 49y = 0$ **24** $y'' - 2y' + 2y = 0$

Sketch the path of $y = e^{ct}$ as t increases from zero, and mark $y = e^c$:

25 $c = 1 - i$ **26** $c = -1 + i$ **27** $c = \pi i/4$

28 What is the solution of $dy/dt = iy$ starting from $y_0 = 1$? For this solution, matching real parts and imaginary parts of $dy/dt = iy$ gives _____ and _____.

29 In Figure 9.10b, at what time t does the spiral cross the real axis at the far left? What does y equal at that time?

30 Show that $\cos\theta = \frac{1}{2}(e^{i\theta} + e^{-i\theta})$ and find a similar formula for $\sin\theta$.

31 *True or false*, with an example to show why:

(a) If $c_1 + c_2$ is real, the c's are complex conjugates.

(b) If $|c_1| = 2$ and $|c_2| = 4$ then $c_1 c_2$ has absolute value 8.

(c) If $|c_1| = 1$ and $|c_2| = 1$ then $|c_1 + c_2|$ is (at least 1) (at most 2) (equal to 2).

(d) If e^{ct} approaches zero as $t \to \infty$, then (c is negative) (the real part of c is negative) ($|c|$ is less than 1).

32 The polar form of $re^{i\theta}$ times $Re^{i\phi}$ is _____. The rectangular form is _____. Circle the terms that give $rR\cos(\theta + \phi)$.

33 The complex number $1/(re^{i\theta})$ has polar form _____ and rectangular form _____ and square roots _____.

34 Show that $\cos ix = \cosh x$ and $\sin ix = i\sinh x$. What is the cosine of i?

CHAPTER 10

Infinite Series

Infinite series can be a pleasure (sometimes). They throw a beautiful light on $\sin x$ and $\cos x$. They give famous numbers like π and e. Usually they produce totally unknown functions—which might be good. But on the painful side is the fact that an infinite series has infinitely many terms.

It is not easy to know the sum of those terms. More than that, it is not certain that there *is* a sum. We need tests, to decide if the series converges. We also need ideas, to discover what the series converges *to*. Here are examples of **convergence, divergence**, and **oscillation**:

$$1 + \tfrac{1}{2} + \tfrac{1}{4} + \cdots = 2 \quad 1 + 1 + 1 + \cdots = \infty \quad 1 - 1 + 1 - 1 \cdots = ?$$

The first series converges. Its next term is $1/8$, after that is $1/16$—and every step brings us halfway to 2. The second series (the sum of 1's) obviously diverges to infinity. The oscillating example (with 1's and -1's) also fails to converge.

All those and more are special cases of one infinite series which is absolutely the most important of all:

> *The geometric series is* $1 + x + x^2 + x^3 + \cdots = \dfrac{1}{1-x}$.

This is a series of *functions*. It is a "power series." When we substitute numbers for x, the series on the left may converge to the sum on the right. We need to know when it doesn't. Choose $x = \tfrac{1}{2}$ and $x = 1$ and $x = -1$:

$$1 + \tfrac{1}{2} + \left(\tfrac{1}{2}\right)^2 + \cdots \text{ is the convergent series. Its sum is } \frac{1}{1 - \tfrac{1}{2}} = 2$$

$$1 + 1 + 1 + \cdots \text{ is divergent. Its sum is } \frac{1}{1-1} = \frac{1}{0} = \infty$$

$$1 + (-1) + (-1)^2 + \cdots \text{ is the oscillating series. Its sum should be } \frac{1}{1 - (-1)} = \frac{1}{2}.$$

The last sum bounces between one and zero, so at least its average is $\tfrac{1}{2}$. At $x = 2$ there is no way that $1 + 2 + 4 + 8 + \cdots$ agrees with $1/(1-2)$.

This behavior is typical of a power series—to converge in an interval of x's and to diverge when x is large. The geometric series is safe for x between -1 and 1. Outside that range it diverges.

431

The next example shows a ***repeating decimal*** $1.111\ldots$:

$$\text{Set } x = \frac{1}{10}. \text{ The geometric series is } 1 + \frac{1}{10} + \left(\frac{1}{10}\right)^2 + \left(\frac{1}{10}\right)^3 + \cdots$$

The decimal $1.111\ldots$ is also the fraction $1/\left(1 - \frac{1}{10}\right)$, which is $10/9$. ***Every fraction leads to a repeating decimal. Every repeating decimal adds up*** (through the geometric series) ***to a fraction***.

To get $3.333\ldots$, just multiply by 3. This is $10/3$. To get $1.0101\ldots$, set $x = 1/100$. This is the fraction $1/\left(1 - \frac{1}{100}\right)$, which is $100/99$.

Here is an unusual decimal (which eventually repeats). I don't really understand it:

$$\frac{1}{243} = .004\ 115\ 226\ 337\ 448\ldots$$

Most numbers are not fractions (or repeating decimals). A good example is π:

$$\pi = 3 + \frac{1}{10} + \frac{4}{100} + \frac{1}{1000} + \frac{5}{10000} + \cdots.$$

This is $3.1415\ldots$, a series that certainly converges. We happen to know the first billion terms (the billionth is given below). Nobody knows the 2 billionth term. Compare that series with this one, which also equals π:

$$\pi = 4 - \frac{4}{3} + \frac{4}{5} - \frac{4}{7} + \cdots$$

That *alternating series* is really remarkable. It is typical of this chapter, because its pattern is clear. We know the 2 billionth term (it has a minus sign). This is not a geometric series, but in Section 10.1 it comes from a geometric series.

Question Does this series actually converge? What if all signs are $+$?
Answer   The alternating series converges to π (Section 10.3). The positive series diverges to infinity (Section 10.2). The terms go to zero, but their sum is infinite.

This example begins to show what the chapter is about. Part of the subject deals with special series, adding to $10/9$ or π or e^x. The other part is about series in general, adding to numbers or functions that nobody has heard of. The situation was the same for integrals—they give famous answers like $\ln x$ or unknown answers like $\int x^x\, dx$. The sum of $1 + 1/8 + 1/27 + \cdots$ is also unknown—although a lot of mathematicians have tried.

The chapter is not long, but it is full. The last half studies *power series*. We begin with a linear approximation like $1 + x$. Next is a quadratic approximation like $1 + x + x^2$. In the end we match *all* the derivatives of $f(x)$. This is the "***Taylor series,***" a new way to create functions—not by formulas or integrals but by infinite series.

No example can be better than $1/(1 - x)$, which dominates Section 10.1. Then we define convergence and test for it. (Most tests are really comparisons with a geometric series.) The second most important series in mathematics is the *exponential series* $e^x = 1 + x + \frac{1}{2}x^2 + \frac{1}{6}x^3 + \cdots$. It includes the series for $\sin x$ and $\cos x$, because of the formula $e^{ix} = \cos x + i \sin x$. Finally a whole range of new and old functions will come from Taylor series.

In the end, all the key functions of calculus appear as "*infinite polynomials*" (except the step function). This is the ultimate voyage from the linear function $y = mx + b$.

10.1 The Geometric Series

We begin by looking at both sides of the geometric series:

$$1 + x + x^2 + x^3 + \cdots = \frac{1}{1-x}. \tag{1}$$

How does the series on the left produce the function on the right? How does $1/(1-x)$ produce the series? Add up two terms of the series, then three terms, then n terms:

$$1 + x = \frac{1-x^2}{1-x} \quad 1 + x + x^2 = \frac{1-x^3}{1-x} \quad 1 + \cdots + x^{n-1} = \frac{1-x^n}{1-x}. \tag{2}$$

For the first, $1 + x$ times $1 - x$ equals $1 - x^2$ by ordinary algebra. The second begins to make the point: $1 + x + x^2$ times $1 - x$ gives $1 - x + x - x^2 + x^2 - x^3$. Between 1 at the start and $-x^3$ at the end, everything cancels. The same happens in all cases: $1 + \cdots + x^{n-1}$ times $1 - x$ leaves 1 at the start and $-x^n$ at the end. This proves equation (2)—the sum of n terms of the series.

For the whole series we will push n towards infinity. On a graph you can see what is happening. Figure 10.1 shows $n = 1$ and $n = 2$ and $n = 3$ and $n = \infty$.

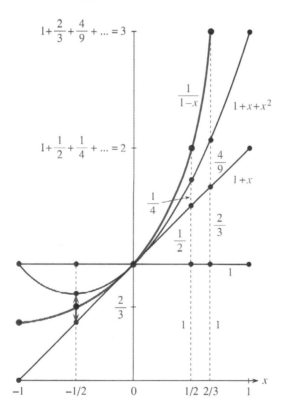

Fig. 10.1 Two terms, then three terms, then full series:

$$1 + x + x^2 + \cdots = \frac{1}{1-x}.$$

$$1 - x \overline{)\,1\phantom{\begin{array}{c}1+x+x^2+\cdots\end{array}}}$$

$$\begin{array}{r} 1 + x + x^2 + \cdots \\ \hline 1 \\ 1 - x \\ \hline x \\ x - x^2 \\ \hline x^2 \\ x^2 - x^3 \\ \hline \cdots \end{array}$$

The infinite sum gives a finite answer, provided x is between -1 and 1. Then x^n goes to zero:

$$\frac{1-x^n}{1-x} \to \frac{1}{1-x}.$$

Now start with the function $1/(1-x)$. How does it produce the series? One way is elementary but brutal, to do "long division" of $1 - x$ into 1 (next to the figure). Another way is to look up the binomial formula for $(1-x)^{-1}$. That is cheating— we want to discover the series, not just memorize it. The successful approach uses calculus. *Compute the derivatives of* $f(x) = 1/(1-x)$:

$$f' = (1-x)^{-2} \quad f'' = 2(1-x)^{-3} \quad f''' = 6(1-x)^{-4} \cdots \tag{3}$$

At $x = 0$ these derivatives are 1, 2, 6, 24, ... Notice how -1 from the chain rule keeps them positive. *The nth derivative at $x = 0$ is n factorial*:

$$f(0) = 1 \quad f'(0) = 1 \quad f''(0) = 2 \quad f'''(0) = 6 \quad \cdots \quad f^{(n)}(0) = n!.$$

Now comes the idea. ***To match the series with $1/(1-x)$, match all those derivatives at $x = 0$.*** Each power x^n gets one derivative right. Its derivatives at $x = 0$ are zero, except the nth derivative, which is $n!$ By adding all powers we get every derivative right—so the geometric series matches the function:

$$1 + x + x^2 + x^3 + \cdots \text{ *has the same derivatives at* } x = 0 \text{ *as* } 1/(1-x).$$

The linear approximation is $1 + x$. Then comes $\frac{1}{2} f''(0) x^2 = x^2$. The third derivative is supposed to be 6, and x^3 is just what we need. *Through its derivatives, the function produces the series.*

With that example, you have seen a part of this subject. The geometric series diverges if $|x| \geq 1$. Otherwise it adds up to the function it comes from (when $-1 < x < 1$). To get familiar with other series, we now apply algebra or calculus—to reach the square of $1/(1-x)$ or its derivative or its integral. The point is that these operations are applied *to the series*.

The best I know is to show you eight operations that produce something useful. At the end we discover series for $\ln 2$ and π.

1. *Multiply the geometric series by a or ax*:

$$a + ax + ax^2 + \cdots = \frac{a}{1-x} \quad ax + ax^2 + ax^3 + \cdots = \frac{ax}{1-x}. \tag{4}$$

The first series fits the decimal 3.333.... In that case $a = 3$. The geometric series for $x = \frac{1}{10}$ gave $1.111\ldots = 10/9$, and this series is just three times larger. Its sum is $10/3$.

The second series fits other decimals that are fractions in disguise. To get $12/99$, choose $a = 12$ and $x = 1/100$:

$$.121212\ldots = \frac{12}{100} + \frac{12}{100^2} + \frac{12}{100^3} + \cdots = \frac{12/100}{1 - 1/100} = \frac{12}{99}.$$

Problem 13 asks about $.8787\ldots$ and $.123123\ldots$. It is usual in precalculus to write $a + ar + ar^2 + \cdots = a/(1-r)$. But we use x instead of r to emphasize that *this is a function*—which we can now differentiate.

2. *The derivative of the geometric series $1 + x + x^2 + \cdots$ is $1/(1-x)^2$*:

$$1 + 2x + 3x^2 + 4x^3 + \cdots = \frac{d}{dx}\left(\frac{1}{1-x}\right) = \frac{1}{(1-x)^2}. \tag{5}$$

At $x = \frac{1}{10}$ the left side starts with 1.23456789. The right side is $1/(1 - \frac{1}{10})^2 = 1/(9/10)^2$, which is $100/81$. If you have a calculator, divide 100 by 81.

The answer should also be near $(1.11111111)^2$, which is 1.2345678987654321.

3. *Subtract $1 + x + x^2 + \cdots$ from $1 + 2x + 3x^2 + \cdots$ as you subtract functions*:

$$x + 2x^2 + 3x^3 + \cdots = \frac{1}{(1-x)^2} = \frac{1}{(1-x)} = \frac{x}{(1-x)^2}. \tag{6}$$

Curiously, the same series comes from multiplying (5) by x. It answers a question left open in Section 8.4—the average number of coin tosses until the result is heads. This is the sum $1(p_1) + 2(p_2) + \cdots$ from probability, with $x = \frac{1}{2}$:

$$1\left(\tfrac{1}{2}\right)+2\left(\tfrac{1}{2}\right)^2+3\left(\tfrac{1}{2}\right)^3+\cdots=\frac{\tfrac{1}{2}}{(1-\tfrac{1}{2})^2}=2. \tag{7}$$

The probability of waiting until the nth toss is $p_n=\left(\tfrac{1}{2}\right)^n$. The expected value is *two tosses*. I suggested experiments, but now this mean value is exact.

4. *Multiply series*: *the geometric series times itself is* $1/(1-x)$ *squared*:

$$(1+x+x^2+\cdots)(1+x+x^2+\cdots)=1+2x+3x^2+\cdots. \tag{8}$$

The series on the right is not new! In equation (5) it was the **derivative** of $y=1/(1-x)$. Now it is the **square** of the same y. The geometric series satisfies $dy/dx=y^2$, so the function does too. We have stumbled onto a differential equation.

Notice how the series was squared. A typical term in equation (8) is $3x^2$, coming from 1 times x^2 and x times x and x^2 times 1 on the left side. It is a lot quicker to square $1/(1-x)$—but other series can be multiplied when we don't know what functions they add up to.

5. *Solve* $dy/dx=y^2$ *from any starting value—a new application of series*:

Suppose the starting value is $y=1$ at $x=0$. The equation $y'=y^2$ gives 1^2 for the derivative. Now a key step: **The derivative of the equation gives** $y''=2yy'$. At $x=0$ that is $2\cdot1\cdot1$. Continuing upwards, the derivative of $2yy'$ is $2yy''+2(y')^2$. At $x=0$ that is $y'''=4+2=6$.

All derivatives are factorials: $1,2,6,24,\ldots$. We are matching the derivatives of the geometric series $1+x+x^2+x^3+\ldots$. Term by term, we rediscover the solution to $y'=y^2$. The solution starting from $y(0)=1$ is $y=1/(1-x)$.

A different starting value is -1. Then $y'=(-1)^2=1$ as before. The chain rule gives $y''=2yy'=-2$ and then $y'''=6$. With alternating signs to match these derivatives, the solution starting from -1 is

$$y=-1+x-x^2+x^3+\cdots=-1/(1+x). \tag{9}$$

It is a small challenge to recognize the function on the right from the series on the left. The series has $-x$ in place of x; then multiply by -1. The sum $y=-1/(1+x)$ also satisfies $y'=y^2$. *We can solve differential equations from all starting values by infinite series*. Essentially we substitute an unknown series into the equation, and calculate one term at a time.

6. *The integrals of* $1+x+x^2+\cdots$ *and* $1-x+x^2-\cdots$ *are logarithms*:

$$x+\frac{1}{2}x^2+\frac{1}{3}x^3+\cdots=\int_0^x\frac{dx}{1-x}=-\ln\,(1-x) \tag{10a}$$

$$x-\frac{1}{2}x^2+\frac{1}{3}x^3-\cdots=\int_0^x\frac{dx}{1+x}=+\ln\,(1+x) \tag{10b}$$

The derivative of (10a) brings back the geometric series. For logarithms we find $1/n$ not $1/n!$ The first term x and second term $\tfrac{1}{2}x^2$ give linear and quadratic approximations. Now we have the whole series. I cannot fail to substitute 1 and $\tfrac{1}{2}$, to find $\ln(1-1)$ and $\ln(1+1)$ and $\ln(1-\tfrac{1}{2})$:

$$x=1:\quad 1+\tfrac{1}{2}+\tfrac{1}{3}+\tfrac{1}{4}+\cdots=-\ln 0=+\infty \tag{11a}$$

$$x=1:\quad 1-\tfrac{1}{2}+\tfrac{1}{3}-\tfrac{1}{4}+\cdots=\quad \ln 2=.693 \tag{11b}$$

$$x=\tfrac{1}{2}:\quad \tfrac{1}{2}+\tfrac{1}{8}+\tfrac{1}{24}+\tfrac{1}{64}+\cdots=-\ln\tfrac{1}{2}=\ln 2. \tag{12}$$

The first series diverges to infinity. This **harmonic series** $1 + \frac{1}{2} + \frac{1}{3} + \cdots$ came into the earliest discussion of limits (Section 2.6). The second series has alternating signs and converges to ln 2. The third has plus signs and also converges to ln 2. These will be examples for a major topic in infinite series— tests for convergence.

For the first time in this book we are able to compute a logarithm! Something remarkable is involved. *The sums of numbers in* (11) *and* (12) *were discovered from the sums of functions in* (10). You might think it would be easier to deal only with numbers, to compute ln 2. But then we would never have integrated the series for $1/(1-x)$ and detected (10). It is better to work with x, and substitute special values like $\frac{1}{2}$ at the end.

There are two practical problems with these series. For ln 2 they converge slowly. For ln e they blow up. The correct answer is ln $e = 1$, but the series can't find it. Both problems are solved by adding (10a) to (10b), which cancels the even powers:

$$2\left(x + \frac{x^3}{3} + \frac{x^5}{5} + \cdots\right) = \ln(1+x) - \ln(1-x) = \ln\frac{1+x}{1-x}. \tag{13}$$

At $x = \frac{1}{3}$, the right side is $\ln\frac{4}{3} - \ln\frac{2}{3} = \ln 2$. Powers of $\frac{1}{3}$ are much smaller than powers of 1 or $\frac{1}{2}$, so ln 2 is quickly computed. All logarithms can be found from the improved series (13).

7. Change variables in the geometric series (replace x **by** x^2 **or** $-x^2$**):**

$$1 + x^2 + x^4 + x^6 + \cdots = 1/(1-x^2) \tag{14}$$

$$1 - x^2 + x^4 - x^6 + \cdots = 1/(1+x^2). \tag{15}$$

This produces new functions (always our goal). They involve even powers of x. The second series will soon be used to calculate π. Other changes are valuable:

$$\frac{x}{2} \text{ in place of } x: \quad 1 + \frac{x}{2} + \left(\frac{x}{2}\right)^2 + \cdots = \frac{1}{1-(x/2)} = \frac{2}{2-x} \tag{16}$$

$$\frac{1}{x} \text{ in place of } x: \quad 1 + \frac{1}{x} + \frac{1}{x^2} + \cdots = \frac{1}{1-(1/x)} = \frac{x}{x-1}. \tag{17}$$

Equation (17) is a series of *negative powers* x^{-n}. It converges when $|x|$ is *greater* than 1. Convergence in (17) is for large x. Convergence in (16) is for $|x| < 2$.

8. The integral of $1 - x^2 + x^4 - x^6 + \cdots$ **yields the inverse tangent of** x**:**

$$x - \frac{1}{3}x^3 + \frac{1}{5}x^5 - \frac{1}{7}x^7 + \cdots = \int \frac{dx}{1+x^2} = \tan^{-1} x. \tag{18}$$

We integrated (15) and got odd powers. The magical formula for π (discovered by Leibniz) comes when $x = 1$. The angle with tangent 1 is $\pi/4$:

$$1 - \frac{1}{3} + \frac{1}{5} - \frac{1}{7} + \cdots = \frac{\pi}{4}. \tag{19}$$

The first three terms give $\pi \approx 3.47$ (not very close). The 5000th term is still of size .0001, so the fourth decimal is still not settled. By changing to $x = 1/\sqrt{3}$, the astronomer Halley and his assistant found 71 correct digits of $\pi/6$ (while waiting for the comet). That is one step in the long and amazing story of calculating π. The

Chudnovsky brothers recently took the latest step with a supercomputer—*they have found more than* **one billion** *decimal places of* π (see *Science*, June 1989). The digits look completely random, as everyone expected. But so far we have no proof that all ten digits occur $\frac{1}{10}$ of the time.

Historical note Archimedes located π above 3.14 and below $3\frac{1}{7}$. Variations of his method (polygons in circles) reached as far as 34 digits—but not for 1800 years. Then Halley found 71 digits of $\pi/6$ with equation (18). For faster convergence that series was replaced by other inverse tangents, using smaller values of x:

$$\frac{\pi}{4} = \tan^{-1}\frac{1}{2} + \tan^{-1}\frac{1}{3} = 4\tan^{-1}\frac{1}{5} - \tan^{-1}\frac{1}{239}. \tag{20}$$

A prodigy named Dase, who could multiply 100-digit numbers in his head in 8 hours, finally passed 200 digits of π. The climax of hand calculation came when Shanks published 607 digits. I am sorry to say that only 527 were correct. (With years of calculation he went on to 707 digits, but still only 527 were correct.) The mistake was not noticed until 1945! Then Ferguson reached 808 digits with a desk calculator.

Now comes the computer. Three days on an ENIAC (1949) gave 2000 digits. A hundred minutes on an IBM 704 (1958) gave 10,000 digits. Shanks (no relation) reached 100,000 digits. Finally a million digits were found in a day in 1973, with a CDC 7600. All these calculations used variations of equation (20).

The record after that went between Cray and Hitachi and now IBM. But the method changed. The calculations rely on an incredibly accurate algorithm, based on the "arithmetic-geometric mean iteration" of Gauss. It is also incredibly simple, all things considered:

$$a_{n+1} = \frac{a_n + b_n}{2} \quad b_{n+1} = \sqrt{a_n b_n} \quad \pi_n = 2a_{n+1}^2 \Bigg/ \left(1 - \sum_{k=0}^{n} 2^k (a_k^2 - b_k^2)\right).$$

The number of correct digits more than doubles at every step. By $n = 9$ we are far beyond Shanks (the hand calculator). No end is in sight. Almost anyone can go past a billion digits, since with the Chudnovsky method we don't have to start over again.

It is time to stop. You may think (or hope) that nothing more could possibly be done with geometric series. We have gone a long way from $1/(1-x)$, but some functions can never be reached. One is e^x (and its relatives $\sin x$, $\cos x$, $\sinh x$, $\cosh x$). Another is $\sqrt{1-x}$ (and its relatives $1/\sqrt{1-x^2}$, $\sin^{-1} x$, $\sec^{-1} x$, ...). The exponentials are in 10.4, with series that converge for all x. The square-roots are in 10.5, closer to geometric series and converging for $|x| < 1$. Before that we have to say what convergence means.

The series came fast, but I hope you see what can be done (subtract, multiply, differentiate, integrate). Addition is easy, division is harder, all are legal. Some unexpected numbers are the sums of infinite series.

Added in proof By e-mail I just learned that the record for π is back in Japan: 2^{30} digits which is more than 1.07 billion. The elapsed time was 100 hours (75 hours of CPU time on an NEC machine). The billionth digit after the decimal point is 9.

Read-through questions

The geometric series $1 + x + x^2 + \cdots$ adds to __a__. It converges provided $|x| <$ __b__. The sum of n terms is __c__. The derivatives of the series match the derivatives of $1/(1-x)$ at the point $x =$ __d__, where the nth derivative is __e__. The decimal $1.111\ldots$ is the geometric series at $x =$ __f__ and equals the fraction __g__. The decimal $.666\ldots$ multiplies this by __h__. The decimal $.999\ldots$ is the same as __i__.

The derivative of the geometric series is __j__ $=$ __k__. This also comes from squaring the __l__ series. By choosing $x = .01$, the decimal 1.02030405 is close to __m__. The differential equation $dy/dx = y^2$ is solved by the geometric series, going term by term starting from $y(0) =$ __n__.

The integral of the geometric series is __o__ $=$ __p__. At $x = 1$ this becomes the __q__ series, which diverges. At $x =$ __r__ we find $\ln 2 =$ __s__. The change from x to $-x$ produces the series $1/(1+x) =$ __t__ and $\ln(1+x) =$ __u__.

In the geometric series, changing to x^2 or $-x^2$ gives $1/(1-x^2) =$ __v__ and $1/(1+x^2) =$ __w__. Integrating the last one yields $x - \frac{1}{3}x^3 + \frac{1}{5}x^5 \cdots =$ __x__. The angle whose tangent is $x = 1$ is $\tan^{-1} 1 =$ __y__. Then substituting $x = 1$ gives the series $\pi =$ __z__.

1 The geometric series is $1 + x + x^2 + \cdots = G$. Another way to discover G is to multiply by x. Then $x + x^2 + x^3 + \cdots = xG$, and this can be subtracted from the original series. What does that leave, and what is G?

2 A basketball is dropped 10 feet and bounces back 6 feet. After every fall it recovers $\frac{3}{5}$ of its height. What total distance does the ball travel, bouncing forever?

3 Find the sums of $\frac{1}{3} + \frac{1}{9} + \frac{1}{27} + \cdots$ and $1 - \frac{1}{4} + \frac{1}{16} - \cdots$ and $10 - 1 + .1 - .01\ldots$ and $3.040404\ldots$.

4 Replace x by $1 - x$ in the geometric series to find a series for $1/x$. Integrate to find a series for $\ln x$. These are power series "around the point $x = 1$." What is their sum at $x = 0$?

5 What is the *second derivative* of the geometric series, and what is its sum at $x = \frac{1}{2}$?

6 Multiply the series $(1 + x + x^2 + \cdots)(1 - x + x^2 - \cdots)$ and find the product by comparing with equation (14).

7 Start with the fraction $\frac{1}{7}$. Divide 7 into $1.000\ldots$ (by long division or calculator) until the numbers start repeating. Which is the first number to repeat? How do you know that the next _____ digits will be the same as the first?

Note about the fractions $1/q$, $10/q$, $100/q$, $\ldots$ All remainders are less than q so eventually two remainders are the same. By subtraction, q goes evenly into a power 10^N minus a smaller

power 10^{N-n}. Thus $qc = 10^N - 10^{N-n}$ for some c and $1/q$ has a repeating decimal:

$$\frac{1}{q} = \frac{c}{10^N - 10^{N-n}} = \frac{c}{10^N} \frac{1}{1 - 10^{-n}}$$
$$= \frac{c}{10^N} \left(1 + \frac{1}{10^n} + \frac{1}{10^{2n}} + \cdots \right).$$

Conclusion: Every fraction equals a repeating decimal.

8 Find the repeating decimal for $\frac{1}{13}$ and read off c. What is the number n of digits before it repeats?

9 From the fact that every q goes evenly into a power 10^N minus a smaller power, show that all primes except 2 or 5 go evenly into 9 or 99 or 999 or $\cdots$.

10 Explain why $.010010001\ldots$ cannot be a fraction (the number of zeros increases).

11 Show that $.123456789101112\ldots$ is not a fraction.

12 From the geometric series, the repeating decimal $1.065065\ldots$ equals what fraction? Explain why every repeating decimal equals a fraction.

13 Write $.878787\ldots$ and $.123123\ldots$ as fractions and as geometric series.

14 Find the square of $1.111\ldots$ as an infinite series.

Find the functions which equal the sums 15–24.

15 $x + x^3 + x^5 + \cdots$　　　　**16** $1 - 2x + 4x^2 - \cdots$

17 $x^3 + x^6 + x^9 + \cdots$　　　**18** $\frac{1}{2}x - \frac{1}{4}x^2 + \frac{1}{8}x^3 - \cdots$

19 $\ln x + (\ln x)^2 + (\ln x)^3 + \cdots$　　**20** $x - 2x^2 + 3x^3 - \cdots$

21 $\frac{1}{x} + \frac{1}{x^2} + \frac{1}{x^3} + \cdots$　　**22** $x + \frac{x}{1+x} + \frac{x}{(1+x)^2} + \cdots$

23 $\tan x - \frac{1}{3}\tan^3 x + \frac{1}{5}\tan^5 x - \cdots$　**24** $e^x + e^{2x} + e^{3x} + \cdots$

25 Multiply the series for $1/(1-x)$ and $1/(1+x)$ to find the coefficients of x, x^2, x^3 and x^n.

26 Compare the integral of $1 + x^2 + x^4 + \cdots$ to equation (13) and find $\int dx/(1 - x^2)$.

27 What fractions are close to $.2468$ and $.987654321$?

28 Find the first three terms in the series for $1/(1-x)^3$.

Add up the series 29–34. Problem 34 comes from (18).

29 $\frac{2}{3} + \frac{2}{3^2} + \frac{2}{3^3} + \cdots$　　　**30** $.1 + .02 + .003 + \cdots$

31 $.1 + \frac{1}{2}(.01) + \frac{1}{3}(.001) + \cdots$　　**32** $.1 - \frac{1}{2}(.01) + \frac{1}{3}(.001) - \cdots$

33 $.1 + \frac{1}{3}(.001) + \frac{1}{5}(.00001) + \cdots$　**34** $1 - \frac{1}{3 \cdot 3} + \frac{1}{5 \cdot 3^2} - \cdots$

35 Compute the nth derivative of $1+2x+3x^2+\cdots$ at $x=0$. Compute also the nth derivative of $(1-x)^{-2}$.

36 The differential equation $dy/dx=y^2$ starts from $y(0)=b$. From the equation and its derivatives find y', y'', y''' at $x=0$, and construct the start of a series that matches those derivatives. Can you recognize $y(x)$?

37 The equation $dy/dx=y^2$ has the differential form $dy/y^2=dx$. Integrate both sides and choose the integration constant so that $y=b$ at $x=0$. Solve for $y(x)$ and compare with Problem 36.

38 In a bridge game, what is the average number μ of deals until you get the best hand? The probability on the first deal is $p_1=\frac{1}{4}$. Then $p_2=\left(\frac{3}{4}\right)\left(\frac{1}{4}\right)=$ (probability of missing on the first) times (probability of winning on the second). Generally $p_n=\left(\frac{3}{4}\right)^{n-1}\left(\frac{1}{4}\right)$. The mean value μ is $p_1+2p_2+3p_3+\cdots=$ _____.

39 Show that $(\Sigma a_n)(\Sigma b_n)=\Sigma a_n b_n$ is ridiculous.

40 Find a series for $\ln\frac{1}{3}$ by choosing x in (10b). Find a series for $\ln 3$ by choosing x in (13). How is $\ln\frac{1}{3}$ related to $\ln 3$, and which series converges faster?

41 Compute $\ln 3$ to its second decimal place without a calculator (OK to check).

42 To four decimal places, find the angle whose tangent is $x=\frac{1}{10}$.

43 Two tennis players move to the net as they volley the ball. Starting together they each go forward 39 feet at 13 feet per second. The ball travels back and forth at 26 feet per second. How far does it travel before the collision at the net? (Look for an easy way and also an infinite series.)

44 How many terms of the series $1-\frac{1}{2}+\frac{1}{3}-\frac{1}{4}+\cdots$ are needed before the first decimal place doesn't change? Which power of $\frac{1}{4}$ equals the 100th power of $\frac{1}{2}$? Which power $1/a^n$ equals $1/2^{100}$?

45 If $\tan y=\frac{1}{2}$ and $\tan z=\frac{1}{3}$, then the tangent of $y+z$ is $(\tan y+\tan z)/(1-\tan y\tan z)=1$. If $\tan y=\frac{1}{5}$ and $\tan z=$ _____, again $\tan(y+z)=1$. Why is this not as good as equation (20), to find $\pi/4$?

46 Find one decimal of π beyond 3.14 from the series for $4\tan^{-1}\frac{1}{2}$ and $4\tan^{-1}\frac{1}{3}$. How many terms are needed in each series?

47 (Calculator) In the same way find one decimal of π beyond 3.14159. How many terms did you take?

48 From equation (10a) what is $\Sigma e^{in}/n$?

49 Zeno's Paradox is that if you go half way, and then half way, and then half way ..., you will never get there. In your opinion, does $\frac{1}{2}+\frac{1}{4}+\frac{1}{8}+\cdots$ add to 1 or not?

10.2 Convergence Tests: Positive Series

This is the third time we have stopped the calculations to deal with the definitions. Chapter 2 said what a derivative is. Chapter 5 said what an integral is. Now we say what the sum of a series is—*if it exists*. In all three cases *a limit is involved*. That is the formal, careful, cautious part of mathematics, which decides if the active and progressive parts make sense.

The series $\frac{1}{2} + \frac{1}{4} + \frac{1}{8} + \cdots$ converges to 1. The series $1 + \frac{1}{2} + \frac{1}{3} + \cdots$ diverges to infinity. The series $1 - \frac{1}{2} + \frac{1}{3} - \cdots$ converges to ln 2. When we speak about convergence or divergence of a series, we are really speaking about convergence or divergence of its "*partial sums*."

DEFINITION 1   The *partial sum* s_n of the series $a_1 + a_2 + a_3 + \cdots$ stops at a_n:

$$s_n = \textbf{\textit{sum of the first n terms}} = a_1 + a_2 + \cdots + a_n.$$

Thus s_n is part of the total sum. The example $\frac{1}{2} + \frac{1}{4} + \frac{1}{8} + \cdots$ has partial sums

$$s_1 = \frac{1}{2} \qquad s_2 = \frac{3}{4} \qquad s_3 = \frac{7}{8} \qquad s_n = 1 - \frac{1}{2^n}.$$

Those add up larger and larger parts of the series—what is the sum of the whole series? The answer is: *The series $\frac{1}{2} + \frac{1}{4} + \dots$ converges to 1 because its partial sums s_n converge to* 1. The series $a_1 + a_2 + a_3 + \dots$ converges to s when its partial sums—going further and further out—approach this limit s. **Add the a's, not the s's.**

DEFINITION 2   *The sum of a series is the limit of its partial sums s_n.*

We repeat: *if the limit exists*. The numbers s_n may have no limit. When the partial sums jump around, the whole series *has no sum*. Then the series does not converge. When the partial sums approach s, the distant terms a_n are approaching zero. More than that, the *sum* of distant terms is approaching zero.

The new idea ($\Sigma\, a_n = s$) has been converted to the old idea ($s_n \to s$).

EXAMPLE 1   The geometric series $\frac{1}{10} + \frac{1}{100} + \frac{1}{1000} + \cdots$ converges to $s = \frac{1}{9}$.

The partial sums s_1, s_2, s_3, s_4 are .1, .11, .111, .1111. They are approaching $s = \frac{1}{9}$. Note again the difference between the series of a's and the sequence of s's. The series $1 + 1 + 1 + \cdots$ diverges because the sequence of s's is 1, 2, 3, ... A sharper example is the harmonic series: $1 + \frac{1}{2} + \frac{1}{3} + \cdots$ diverges because its partial sums $1, 1\frac{1}{2}, \dots$ eventually go past every number s. We saw that in 2.6 and will see it again here.

Do not confuse $a_n \to 0$ with $s_n \to s$. You cannot be sure that a series converges, just on the basis that $a_n \to 0$. The harmonic series is the best example: $a_n = 1/n \to 0$ but still $s_n \to \infty$. This makes infinite series into a delicate game, which mathematicians enjoy. The line between divergence and convergence is hard to find and easy to cross. A slight push will speed up $a_n \to 0$ and make the s_n converge. Even though $a_n \to 0$ does not by itself guarantee convergence, it is the first requirement:

> **10A** If a series converges ($s_n \to s$) then its terms must approach zero ($a_n \to 0$).

Proof Suppose s_n approaches s (as required by convergence). Then also s_{n-1} approaches s, and the difference $s_n - s_{n-1}$ approaches zero. That difference is a_n. So $a_n \to 0$.

EXAMPLE 1 (continued) For the geometric series $1 + x + x^2 + \cdots$, the test $a_n \to 0$ is the same as $x^n \to 0$. The test is failed if $|x| \geqslant 1$, because the powers of x don't go to zero. Automatically the series diverges. The test is passed if $-1 < x < 1$. But to prove convergence, *we cannot rely on $a_n \to 0$*. It is the partial sums that must converge:

$$s_n = 1 + x + \cdots + x^{n-1} = \frac{1 - x^n}{1 - x} \quad \text{and} \quad s_n \to \frac{1}{1 - x}. \quad \text{This is } s.$$

For other series, first check that $a_n \to 0$ (otherwise there is no chance of convergence). The a_n will not have the special form x^n—so we need sharper tests.

The geometric series stays in our mind for this reason. *Many convergence tests are comparisons with that series.* The right comparison gives enough information:

If $|a_1| < \frac{1}{2}$ and $|a_2| < \frac{1}{4}$ and $\ldots$, then $a_1 + a_2 + \ldots$ converges faster than $\frac{1}{2} + \frac{1}{4} + \ldots$.

More generally, the terms in $a_1 + a_2 + a_3 + \ldots$ may be smaller than $ax + ax^2 + ax^3 + \ldots$. Provided $x < 1$, the second series converges. Then $\sum a_n$ also converges. We move now to *convergence by comparison* or *divergence by comparison*.

Throughout the rest of this section, all numbers a_n are assumed positive.

COMPARISON TEST AND INTEGRAL TEST

In practice it is rare to compute the partial sums $s_n = a_1 + \cdots + a_n$. Usually a simple formula can't be found. We may never know the exact limit s. But it is still possible to decide convergence—whether there *is* a sum—by comparison with another series that is known to converge.

10B (*Comparison test*) Suppose that $0 \leqslant a_n \leqslant b_n$ and $\sum b_n$ converges. Then $\sum a_n$ converges.

The smaller terms a_n add to a smaller sum: $\sum a_n$ is below $\sum b_n$ and must converge. On the other hand suppose $a_n \geqslant c_n$ and $\sum c_n = \infty$. This comparison forces $\sum a_n = \infty$. *A series diverges if it is above another divergent series.*

Note that a series of positive terms can only diverge "*to infinity*." It cannot oscillate, because each term moves it forward. Either the s_n creep up on s, passing every number below it, or they pass all numbers and diverge. *If an increasing sequence s_n is bounded above, it must converge.* The line of real numbers is complete, and has no holes.

The harmonic series $1 + \frac{1}{2} + \frac{1}{3} + \frac{1}{4} + \ldots$ diverges to infinity.

A comparison series is $1 + \frac{1}{2} + \frac{1}{4} + \frac{1}{4} + \frac{1}{8} + \frac{1}{8} + \frac{1}{8} + \frac{1}{8} + \ldots$. *The harmonic series is larger*. But this comparison series is really $1 + \frac{1}{2} + \frac{1}{2} + \frac{1}{2} + \ldots$, because $\frac{1}{2} = \frac{2}{4} = \frac{4}{8}$.

The comparison series diverges. The harmonic series, *above it*, must also diverge.

To apply the comparison test, we need something to compare with. In Example 2, we thought of another series. It was convenient because of those $\frac{1}{2}$'s. But a different series will need a different comparison, and where will it come from? There is an automatic way to think of a *comparison series*. It comes from the *integral test*.

Allow me to apply the integral test to the same example. *To understand the integral test, look at the areas in Figure 10.2*. The test compares rectangles with curved areas.

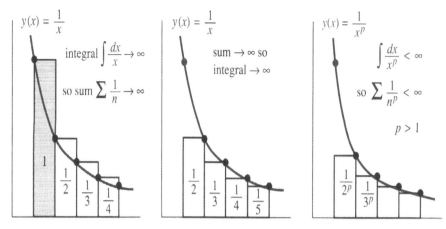

Fig. 10.2 *Integral test*: Sums and integrals both diverge ($p = 1$) and both converge ($p > 1$).

EXAMPLE 2 (again) *Compare* $1 + \frac{1}{2} + \frac{1}{3} + \dots$ *with the area under the curve* $y = 1/x$.

Every term $a_n = 1/n$ is the area of a rectangle. We are comparing it with a curved area c_n. Both areas are between $x = n$ and $x = n + 1$, and *the rectangle is above the curve*. So $a_n > c_n$:

$$\text{rectangular area } a_n = \frac{1}{n} \quad \text{exceeds curved area } c_n = \int_n^{n+1} \frac{dx}{x}.$$

Here is the point. Those c_n's look complicated, but *we can add them up*. The sum $c_1 + \dots + c_n$ is the whole area, from 1 to $n + 1$. It equals $\ln(n + 1)$—we know the integral of $1/x$. We also know that the logarithm goes to infinity.

The rectangular area $1 + 1/2 + \dots + 1/n$ is above the curved area. By comparison of areas, the harmonic series diverges to infinity—a little faster than $\ln(n + 1)$.

Remark   The integral of $1/x$ has another advantage over the series with $\frac{1}{2}$'s. First, the integral test was automatic. From $1/n$ in the series, we went to $1/x$ in the integral. Second, the comparison is *closer*. Instead of adding only $\frac{1}{2}$ when the number of terms is doubled, the true partial sums grow like $\ln n$. To prove that, put rectangles **under** the curve.

Rectangles *below* the curve give an area *below* the integral. Figure 10.2b omits the first rectangle, to get under the curve. Then we have the opposite to the first comparison—the sum is now smaller than the integral:

$$\frac{1}{2} + \frac{1}{3} + \dots + \frac{1}{n} < \int_1^n \frac{dx}{x} = \ln n.$$

Adding 1 to both sides, s_n *is below* $1 + \ln n$. From the previous test, s_n *is above* $\ln(n + 1)$. That is a narrow space—we have an excellent estimate of s_n. The sum of $1/n$ and the integral of $1/x$ diverge together. Problem 43 will show that the difference between s_n and $\ln n$ approaches "Euler's constant," which is $\gamma = .577\dots$.

Main point: Rectangular area is s_n. Curved area is close. We are using integrals to help with sums (it used to be the opposite).

Question   If a computer adds a million terms every second for a million years, how large is the partial sum of the harmonic series ?

Answer   The number of terms is $n = 60^2 \cdot 24 \cdot 365 \cdot 10^{12} < 3.2 \cdot 10^{19}$. Therefore $\ln n$ is less than $\ln 3.2 + 19 \ln 10 < 45$. By the integral test $s_n < 1 + \ln n$, the partial sum after a million years has not reached 46.

For other series, $1/x$ changes to a different function $y(x)$. At $x = n$ this function must equal a_n. Also $y(x)$ must be decreasing. Then a rectangle of height a_n is above the graph to the right of $x = n$, and below the graph to the left of $x = n$. **The series and the integral box each other in**: *left sum $\geqslant$ integral $\geqslant$ right sum.* The reasoning is the same as it was for $a_n = 1/n$ and $y(x) = 1/x$: There is finite area in the rectangles when there is finite area under the curve.

When we can't add the a's, we integrate $y(x)$ and compare areas:

10C (*Integral test*) If $y(x)$ is decreasing and $y(n)$ agrees with a_n, then

$$a_1 + a_2 + a_3 + \cdots \quad \text{and} \quad \int_1^\infty y(x)\,dx \quad \text{both converge or both diverge.}$$

EXAMPLE 3 The "p-series" $\dfrac{1}{2^p} + \dfrac{1}{3^p} + \dfrac{1}{4^p} + \cdots$ converges if $p > 1$. Integrate $y = \dfrac{1}{x^p}$:

$$\frac{1}{n^p} < \int_{n-1}^n \frac{dx}{x^p} \quad \text{so by addition} \quad \sum_{n=2}^\infty \frac{1}{n^p} < \int_1^\infty \frac{dx}{x^p}.$$

In Figure 10.2c, the area is finite if $p > 1$. The integral equals $\left[x^{1-p}/(1-p)\right]_1^\infty$, which is $1/(p-1)$. **Finite area means convergent series**. If $1/1^p$ is the first term, add 1 to the curved area:

$$\frac{1}{1^p} + \frac{1}{2^p} + \frac{1}{3^p} + \cdots < 1 + \frac{1}{p-1} = \frac{p}{p-1}.$$

The borderline case $p = 1$ is the harmonic series (divergent). By the comparison test, every $p < 1$ also produces divergence. Thus $\Sigma 1/\sqrt{n}$ diverges by comparison with $\int dx/\sqrt{x}$ (and also by comparison with $\Sigma 1/n$). Section 7.5 on improper integrals runs parallel to this section on "improper sums" (infinite series).

Notice the special cases $p = 2$ and $p = 3$. The series $1 + \frac{1}{4} + \frac{1}{9} + \cdots$ converges. Euler found $\pi^2/6$ as its sum. The series $1 + \frac{1}{8} + \frac{1}{27} + \cdots$ also converges. That is proved by comparing $\Sigma 1/n^3$ with $\Sigma 1/n^2$ or with $\int dx/x^3$. But the sum for $p = 3$ is unknown.

Extra credit problem The sum of the p-series leads to the most important problem in pure mathematics. The "zeta function" is $Z(p) = \Sigma 1/n^p$, so $Z(2) = \pi^2/6$ and $Z(3)$ is unknown. Riemann studied the complex numbers p where $Z(p) = 0$ (there are infinitely many). He conjectured that *the real part of those p is always $\frac{1}{2}$.* That has been tested for the first billion zeros, but never proved.

COMPARISON WITH THE GEOMETRIC SERIES

We can compare any new series $a_1 + a_2 + \cdots$ with $1 + x + \cdots$. Remember that the first million terms have nothing to do with convergence. It is further out, as $n \to \infty$, that the comparison stands or falls. We still assume that $a_n > 0$.

10D (*Ratio test*) If a_{n+1}/a_n approaches a limit $L < 1$, the series converges.

10E (*Root test*) If the nth root $(a_n)^{1/n}$ approaches $L < 1$, the series converges.

Roughly speaking, these tests make a_n comparable with L^n—therefore convergent. The tests also establish divergence if $L > 1$. They give no decision when $L = 1$. Unfortunately $L = 1$ is the most important and the hardest case.

On the other hand, you will now see that the ratio test is fairly easy.

EXAMPLE 4   The geometric series $x + x^2 + \cdots$ has ratio exactly x. The nth root is also exactly x. So $L = x$. There is convergence if $x < 1$ (known) and divergence if $x > 1$ (also known). The divergence of $1 + 1 + \cdots$ is too delicate (!) for the ratio test and root test, because $L = 1$.

EXAMPLE 5   The p-series has $a_n = 1/n^p$ and $a_{n+1}/a_n = n^p/(n+1)^p$. The limit as $n \to \infty$ is $L = 1$, for every p. The ratio test does not feel the difference between $p = 2$ (convergence) and $p = 1$ (divergence) or even $p = -1$ (extreme divergence). Neither does the root test. So the integral test is sharper.

EXAMPLE 6   A combination of p-series and geometric series can now be decided:

$$\frac{x}{1^p} + \frac{x^2}{2^p} + \cdots + \frac{x^n}{n^p} + \cdots \text{ has ratio } \frac{a_{n+1}}{a_n} = \frac{x^{n+1}}{(n+1)^p}\frac{n^p}{x^n} \text{ approaching } L = x.$$

It is $|x| < 1$ that decides convergence, not p. **The powers x^n are stronger than any** n^p. The factorials $n!$ will now prove stronger than any x^n.

EXAMPLE 7   *The exponential series $e^x = 1 + x + \frac{1}{2}x^2 + \frac{1}{6}x^3 + \cdots$ converges for all x.*

The terms of this series are $x^n/n!$ The ratio between neighboring terms is

$$\frac{x^{n+1}/(n+1)!}{x^n/n!} = \frac{x}{n+1}, \text{ which approaches } L = 0 \text{ as } n \to \infty.$$

With $x = 1$, this ratio test gives convergence of $\sum 1/n!$ *The sum is e.* With $x = 4$, the larger series $\sum 4^n/n!$ also converges. We know this sum too—it is e^4. Also the sum of $x^n n^p/n!$ converges for any x and p. Again $L = 0$—the ratio test is not even close. **The factorials take over, and give convergence.**

Here is the proof of convergence when the ratios approach $L < 1$. Choose x halfway from L to 1. Then $x < 1$. Eventually the ratios go below x and stay below:

$$a_{N+1}/a_N < x \qquad a_{N+2}/a_{N+1} < x \qquad a_{N+3}/a_{N+2} < x \qquad \cdots$$

Multiply the first two inequalities. Then multiply all three:

$$a_{N+1}/a_N < x \qquad a_{N+2}/a_N < x^2 \qquad a_{N+3}/a_N < x^3 \qquad \cdots$$

Therefore $a_{N+1} + a_{N+2} + a_{N+3} + \cdots$ is less than $a_N(x + x^2 + x^3 + \cdots)$. Since $x < 1$, comparison with the geometric series gives convergence.

EXAMPLE 8   The series $\sum 1/n^n$ is ideal for the root test. The nth root is $1/n$. Its limit is $L = 0$. Convergence is even faster than for $e = \sum 1/n!$ The root test is easily explained, since $(a_n)^{1/n} < x$ yields $a_n < x^n$ and x is close to $L < 1$. So we compare with the geometric series.

SUMMARY FOR POSITIVE SERIES

The convergence of geometric series and p-series and exponential series is settled. I will put these a_n's in a line, going from most divergent to most convergent.

The crossover to convergence is after $1/n$:

$$1 + 1 + \cdots \qquad (p < 1)\, \frac{1}{n^p} \quad \frac{1}{n} \quad \frac{1}{n^p}\,(p > 1) \qquad \frac{n}{2^n} \quad \frac{1}{2^n} \quad \frac{4^n}{n!} \quad \frac{1}{n!} \quad \frac{1}{n^n}$$

10A   **10B** and **10C**   **10D** and **10E**

$(a_n \not\to 0)$ (comparison and integral)   (ratio and root)

You should know that this crossover is not as sharp as it looks. On the convergent side, $1/n(\ln n)^2$ comes before all those p-series. On the divergent side, $1/n(\ln n)$ and $1/n(\ln n)(\ln \ln n)$ belong after $1/n$. For any divergent (or convergent) series, there is another that diverges (or converges) more slowly.

Thus there is no hope of an ultimate all-purpose comparison test. But comparison is the best method available. Every series in that line can be compared with its neighbors, and other series can be placed in between. It is a topic that is understood best by examples.

EXAMPLE 9 $\sum \dfrac{1}{\ln n}$ diverges because $\sum \dfrac{1}{n}$ diverges. The comparison uses $\ln n < n$.

EXAMPLE 10 $\displaystyle\sum \frac{1}{n(\ln n)^2} \approx \int \frac{dx}{x(\ln x)^2} < \infty \qquad \sum \frac{1}{n(\ln n)} \approx \int \frac{dx}{x(\ln x)} = \infty.$

The indefinite integrals are $-1/\ln x$ and $\ln(\ln x)$. The first goes to zero as $x \to \infty$; the integral and series both converge. The second integral $\ln(\ln x)$ goes to infinity—very slowly but it gets there. So the second series diverges. These examples squeeze new series into the line, closer to the crossover.

EXAMPLE 11 $\dfrac{1}{n^2 + 1} < \dfrac{1}{n^2}$ so $\dfrac{1}{2} + \dfrac{1}{5} + \dfrac{1}{10} + \cdots < \dfrac{1}{1} + \dfrac{1}{4} + \dfrac{1}{9} + \cdots$ (*convergence*).

The constant 1 in this denominator has no effect—and again in the next example.

EXAMPLE 12 $\dfrac{1}{2n - 1} > \dfrac{1}{2n}$ so $\dfrac{1}{1} + \dfrac{1}{3} + \dfrac{1}{5} + \cdots > \dfrac{1}{2} + \dfrac{1}{4} + \dfrac{1}{6} + \cdots.$

$\sum 1/2n$ is $1/2$ times $\sum 1/n$, so both series diverge. **Two series behave in the same way if the ratios a_n/b_n approach $L > 0$.** Examples $11 - 12$ have $n^2/(n^2 + 1) \to 1$ and $2n/(2n - 1) \to 1$. That leads to our final test:

10F (*Limit comparison test*) If the ratio a_n/b_n approaches a positive limit L, then $\sum a_n$ and $\sum b_n$ either both diverge or both converge.

Reason: a_n is smaller than $2Lb_n$, and larger than $\frac{1}{2}Lb_n$, at least when n is large. So the two series behave in the same way. For example $\sum \sin(7/n^p)$ converges for $p > 1$, not for $p \leqslant 1$. It behaves like $\sum 1/n^p$ (here $L = 7$). The tail end of a series (large n) controls convergence. The front end (small n) controls most of the sum.

There are many more series to be investigated by comparison.

Read-through questions

The convergence of $a_1 + a_2 + \cdots$ is decided by the partial sums $s_n = \underline{\quad a \quad}$. If the s_n approach s, then $\Sigma a_n = \underline{\quad b \quad}$. For the __c__ series $1 + x + \cdots$ the partial sums are $s_n = \underline{\quad d \quad}$. In that case $s_n \to 1/(1-x)$ if and only if __e__. In all cases the limit $s_n \to s$ requires that $a_n \to \underline{\quad f \quad}$. But the harmonic series $a_n = 1/n$ shows that we can have $a_n \to \underline{\quad g \quad}$ and still the series __h__.

The comparison test says that if $0 \leqslant a_n \leqslant b_n$ then __i__. In case a decreasing $y(x)$ agrees with a_n at $x = n$, we can apply the __j__ test. The sum Σa_n converges if and only if __k__. By this test the p-series $\Sigma 1/n^p$ converges if and only if p is __l__. For the harmonic series $(p = 1)$, $s_n = 1 + \cdots + 1/n$ is close to the integral $f(n) = \underline{\quad m \quad}$.

The __n__ test applies when $a_{n+1}/a_n \to L$. There is convergence if __o__, divergence if __p__, and no decision if __q__. The same is true for the __r__ test, when $(a_n)^{1/n} \to L$. For a geometric-p-series combination $a_n = x^n/n^p$, the ratio a_{n+1}/a_n equals __s__. Its limit is $L = \underline{\quad t \quad}$ so there is convergence if __u__. For the exponential $e^x = \Sigma x^n/n!$ the limiting ratio a_{n+1}/a_n is $L = \underline{\quad v \quad}$. This series always __w__ because $n!$ grows faster than any x^n or n^p.

There is no sharp line between __x__ and __y__. But if Σb_n converges and $a_n/b_n \to L$, it follows from the __z__ test that Σa_n also converges.

1 Here is a quick proof that a finite sum $1 + \frac{1}{2} + \frac{1}{3} + \cdots = s$ is impossible. Division by 2 would give $\frac{1}{2} + \frac{1}{4} + \frac{1}{6} + \cdots = \frac{1}{2}s$. Subtraction would leave $1 + \frac{1}{3} + \frac{1}{5} + \cdots = \frac{1}{2}s$. Those last two series cannot both add to $\frac{1}{2}s$ because _____.

2 Behind every decimal $s = .abc\ldots$ is a convergent series $a/10 + b/100 + \underline{\quad\quad} + \cdots$. By a comparison test prove convergence.

3 From these partial sums s_n, find a_n and also $s = \Sigma_1^\infty a_n$:

(a) $s_n = 1 - \dfrac{1}{n}$ (b) $s_n = 4n$ (c) $s_n = \ln \dfrac{2n}{n+1}$.

4 Find the partial sums $s_n = a_1 + a_2 + \cdots + a_n$:

(a) $a_n = 1/3^{n-1}$ (b) $a_n = \ln \dfrac{n}{n+1}$ (c) $a_n = n$

5 Suppose $0 < a_n < b_n$ and Σa_n converges. What can be deduced about Σb_n? Give examples.

6 (a) Suppose $b_n + c_n < a_n$ (all positive) and Σa_n converges. What can you say about Σb_n and Σc_n?

(b) Suppose $a_n < b_n + c_n$ (all positive) and Σa_n diverges. What can you say about Σb_n and Σc_n?

Decide convergence or divergence in 7–10 (and give a reason).

7 $\frac{1}{100} + \frac{1}{200} + \frac{1}{300} + \cdots$ **8** $\frac{1}{100} + \frac{1}{105} + \frac{1}{110} + \cdots$

9 $\frac{1}{101} + \frac{1}{104} + \frac{1}{109} + \cdots$ **10** $\frac{1}{101} + \frac{2}{108} + \frac{3}{127} + \cdots$

Establish convergence or divergence in 11–20 by a comparison test.

11 $\displaystyle\sum \frac{1}{n^2 + 10}$ **12** $\displaystyle\sum \frac{1}{\sqrt{n^2 + 10}}$

13 $\displaystyle\sum \frac{1}{n + \sqrt{n}}$ **14** $\displaystyle\sum \frac{\sqrt{n}}{n^2 + 4}$

15 $\displaystyle\sum \frac{n^3}{n^2 + n^4}$ **16** $\displaystyle\sum \frac{1}{n^2} \cos\left(\frac{1}{n}\right)$

17 $\displaystyle\sum \frac{1}{2^n - 1}$ **18** $\displaystyle\sum \sin^2\left(\frac{1}{n}\right)$

19 $\displaystyle\sum \frac{1}{3^n - 2^n}$ **20** $\displaystyle\sum \frac{1}{e^n - n^e}$

For 21–28 find the limit L in the ratio test or root test.

21 $\displaystyle\sum \frac{3^n}{n!}$ **22** $\displaystyle\sum \frac{1}{n^2}$

23 $\displaystyle\sum \frac{n^2 2^n}{n!}$ **24** $\displaystyle\sum \left(\frac{n-1}{n}\right)^n$

25 $\displaystyle\sum \frac{n}{2^n}$ **26** $\displaystyle\sum \frac{n!}{e^{n^2}}$

27 $\displaystyle\sum \left(\frac{n-1}{n}\right)^{n^2}$ **28** $\displaystyle\sum \frac{n!}{n^n}$

29 $(\frac{1}{1} - \frac{1}{2}) + (\frac{1}{2} - \frac{1}{3}) + (\frac{1}{3} - \frac{1}{4})$ is "telescoping" because $\frac{1}{2}$ and $\frac{1}{3}$ cancel $-\frac{1}{2}$ and $-\frac{1}{3}$. Add the infinite telescoping series

$$s = \sum_1^\infty \left(\frac{1}{n} - \frac{1}{n+1}\right) = \sum_1^\infty \left(\frac{1}{n(n+1)}\right).$$

30 Compute the sum s for other "telescoping series":

(a) $\left(\dfrac{1}{1} - \dfrac{1}{3}\right) + \left(\dfrac{1}{2} - \dfrac{1}{4}\right) + \left(\dfrac{1}{3} - \dfrac{1}{5}\right) \cdots$

(b) $\ln \frac{1}{2} + \ln \frac{2}{3} + \ln \frac{3}{4} + \cdots$

31 In the integral test, what sum is larger than $\int_1^n y(x)\,dx$ and what sum is smaller? Draw a figure to illustrate.

32 Comparing sums with integrals, find numbers larger and smaller than

$$s_n = 1 + \frac{1}{3} + \cdots + \frac{1}{2n - 1} \text{ and } s_n = 1 + \frac{1}{8} + \cdots + \frac{1}{n^3}.$$

33 Which integral test shows that $\displaystyle\sum_1^\infty 1/e^n$ converges? What is the sum?

34 Which integral test shows that $\displaystyle\sum_1^\infty n/e^n$ converges? What is the sum?

Decide for or against convergence in 35–42, based on $\int y(x)\,dx$.

35 $\displaystyle\sum \frac{1}{n^2+1}$

36 $\displaystyle\sum \frac{1}{3n+5}$

37 $\displaystyle\sum \frac{n}{n^2+1}$

38 $\displaystyle\sum \frac{\ln n}{n}$ $\left(\text{is } \dfrac{\ln x}{x} \text{ decreasing ?}\right)$

39 $\displaystyle\sum n^e/n^\pi$

40 $\displaystyle\sum_{2}^{\infty} \frac{1}{n(\ln n)(\ln\ln n)}$

41 $\displaystyle\sum e^n/\pi^n$

42 $\displaystyle\sum n/e^{n^2}$

43 (a) Explain why $D_n = \left(1+\dfrac{1}{2}+\cdots+\dfrac{1}{n}\right) - \ln n$ is positive by using rectangles as in Figure 10.2.

(b) Show that D_{n+1} is less than D_n by proving that

$$\frac{1}{n+1} < \int_{n}^{n+1} \frac{dx}{x}.$$

(c) (Calculator) The decreasing D_n's must approach a limit. Compute them until they go below .6 and below .58 (when ?). The limit of the D_n is *Euler's constant* $\gamma = .577\ldots$.

44 In the harmonic series, use $s_n \approx .577 + \ln n$ to show that $s_n = 1+\dfrac{1}{2}+\cdots+\dfrac{1}{n}$ needs more than 600 terms to reach $s_n > 7$. How many terms for $s_n > 10$?

45 (a) Show that $1 - \dfrac{1}{2} + \dfrac{1}{3} - \dfrac{1}{4} \cdots - \dfrac{1}{2n} = \dfrac{1}{n+1} + \cdots + \dfrac{1}{2n}$ by adding $2\left(\dfrac{1}{2}+\dfrac{1}{4}+\cdots+\dfrac{1}{2n}\right)$ to both sides.

(b) Why is the right side close to $\ln 2n - \ln n$? Deduce that $1 - \dfrac{1}{2} + \dfrac{1}{3} - \dfrac{1}{4} + \cdots$ approaches $\ln 2$.

46 Every second a computer adds a million terms of $\sum 1/(n\ln n)$. By comparison with $\int dx/(x\ln x)$, estimate the partial sum after a million years (see Question in text).

47 Estimate $\displaystyle\sum_{100}^{1000} \frac{1}{n^2}$ by comparison with an integral.

48 If Σa_n converges (all $a_n > 0$) show that Σa_n^2 converges.

49 If Σa_n converges (all $a_n > 0$) show that $\Sigma \sin a_n$ converges. How could $\Sigma \sin a_n$ converge when Σa_n diverges ?

50 The nth prime number p_n satisfies $p_n/n\ln n \to 1$. Prove that

$$\sum \frac{1}{p_n} = \frac{1}{2} + \frac{1}{3} + \frac{1}{5} + \frac{1}{7} + \frac{1}{11} + \cdots \text{ diverges.}$$

Construct a series Σa_n that converges faster than Σb_n but slower than Σc_n (meaning $a_n/b_n \to 0$, $a_n/c_n \to \infty$).

51 $b_n = 1/n^2,\ c_n = 1/n^3$

52 $b_n = n(\tfrac{1}{2})^n,\ c_n = (\tfrac{1}{2})^n$

53 $b_n = 1/n!,\ c_n = 1/n^n$

54 $b_n = 1/n^e,\ c_n = 1/e^n$

In Problem 53 use Stirling's formula $\sqrt{2\pi n}\,n^n/e^n n! \to 1$.

55 For the series $\dfrac{1}{2} + \dfrac{1}{2} + \dfrac{1}{4} + \dfrac{1}{4} + \dfrac{1}{8} + \dfrac{1}{8} + \cdots$ show that the ratio test fails. The roots $(a_n)^{1/n}$ do approach a limit L. Find L from the even terms $a_{2k} = 1/2^k$. Does the series converge ?

56 (For instructors) If the ratios a_{n+1}/a_n approach a positive limit L show that the roots $(a_n)^{1/n}$ also approach L.

Decide convergence in 57–66 and name your test.

57 $\displaystyle\sum \frac{1}{(\ln n)^n}$

58 $\displaystyle\sum \frac{1}{n^{\ln n}}$

59 $\displaystyle\sum \frac{1}{10^n}$

60 $\displaystyle\sum \frac{1}{\ln(10^n)}$

61 $\displaystyle\sum \ln\frac{n+2}{n+1}$

62 $\displaystyle\sum n^{-1/n}$

63 $\displaystyle\sum \frac{1}{(\ln n)^p}$ (test all p)

64 $\displaystyle\sum \frac{\ln n}{n^p}$ (test all p)

65 $\displaystyle\sum \frac{3^n}{4^n - 2^n}$

66 $\displaystyle\sum \frac{n^p}{(n!)^q}$ (test all p, q)

67 Suppose $a_n/b_n \to 0$ in the limit comparison test. Prove that Σa_n converges if Σb_n converges.

68 Can you invent a series whose convergence you and your instructor cannot decide ?

10.3 Convergence Tests: All Series

This section finally allows the numbers a_n to be negative. The geometric series $1 - \frac{1}{2} + \frac{1}{4} - \frac{1}{8} + \cdots = \frac{1}{3}$ is certainly allowed. So is the series $\pi = 4 - \frac{4}{3} + \frac{4}{5} - \frac{4}{7} + \cdots$. If we change all signs to $+$, the geometric series would still converge (to the larger sum 2). This is the first test, to bring back a positive series by taking the **absolute value** $|a_n|$ of every term.

DEFINITION   The series Σa_n is "**absolutely convergent**" if $\Sigma |a_n|$ is convergent.

Changing a negative number from a_n to $|a_n|$ increases the sum. Main point: The smaller series Σa_n is sure to converge if $\Sigma |a_n|$ converges.

10G   If $\Sigma |a_n|$ converges then Σa_n converges (absolutely). But Σa_n might converge, as in the series for π, even if $\Sigma |a_n|$ diverges to infinity.

EXAMPLE 1   Start with the positive series $\frac{1}{2} + \frac{1}{4} + \frac{1}{8} + \cdots$. *Change any signs to minus.* Then the new series converges (absolutely). The right choice of signs will make it converge to any number between -1 and 1.

EXAMPLE 2   Start with the alternating series $1 - \frac{1}{2} + \frac{1}{3} - \frac{1}{4} + \cdots$ which converges to $\ln 2$. Change to plus signs. The new series $1 + \frac{1}{2} + \frac{1}{3} + \cdots$ diverges to infinity. The original alternating series was not absolutely convergent. It was only "**conditionally convergent**." A series can converge (conditionally) by a careful choice of signs—even if $\Sigma |a_n| = \infty$.

    If $\Sigma |a_n|$ converges then Σa_n converges. Here is a quick proof. The numbers $a_n + |a_n|$ are either zero (if a_n is negative) or $2|a_n|$. By comparison with $\Sigma 2|a_n|$, which converges, $\Sigma (a_n + |a_n|)$ must converge. Now subtract the convergent series $\Sigma |a_n|$. The difference Σa_n also converges, completing the proof. All tests for positive series (integral, ratio, comparison, ...) apply immediately to absolute convergence, because we switch to $|a_n|$.

EXAMPLE 3   Start with the geometric series $\frac{1}{3} + \frac{1}{9} + \frac{1}{27} + \cdots$ which converges to $\frac{1}{2}$. *Change any of those signs to minus.* Then the new series must converge (absolutely). But the sign changes cannot achieve all sums between $-\frac{1}{2}$ and $\frac{1}{2}$. This time the sums belong to the famous (and very thin) *Cantor set* of Section 3.7.

EXAMPLE 4   (looking ahead) Suppose $\Sigma a_n x^n$ converges for a particular number x. Then for every x nearer to zero, it converges absolutely. This will be proved and used in Section 10.6 on power series, where it is the most important step in the theory.

EXAMPLE 5   Since $\Sigma 1/n^2$ converges, so does $\Sigma (\cos n)/n^2$. That second series has irregular signs, but it converges absolutely by comparison with the first series (since $|\cos n| < 1$). Probably $\Sigma (\tan n)/n^2$ does not converge, because the tangent does not stay bounded like the cosine.

ALTERNATING SERIES

The series $1 - \frac{1}{2} + \frac{1}{3} - \frac{1}{4} + \cdots$ converges to $\ln 2$. That was stated without proof. This is an example of an **alternating series**, in which the signs alternate between plus and minus. There is the additional property that the absolute values $1, \frac{1}{2}, \frac{1}{3}, \frac{1}{4}, \ldots$ decrease to zero. Those two facts—decrease to zero with alternating signs—guarantee convergence.

> **10H**   An alternating series $a_1 - a_2 + a_3 - a_4 \cdots$ converges (at least conditionally, maybe not absolutely) if every $a_{n+1} \leqslant a_n$ and $a_n \to 0$.

The best proof is in Figure 10.3. Look at $a_1 - a_2 + a_3$. It is below a_1, because a_3 (with plus sign) is smaller than a_2 (with minus sign). The sum of five terms is less than the

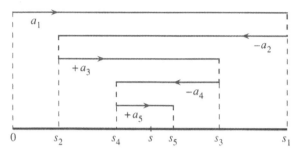

Fig. 10.3   An alternating series converges when the absolute values decrease to zero.

sum of three terms, because a_5 is smaller than a_4. These partial sums $s_1, s_3, s_5, \ldots$ with an odd number of terms are *decreasing*.

Now look at two terms $a_1 - a_2$, then four terms, then six terms. Adding on $a_3 - a_4$ increases the sum (because $a_3 \geqslant a_4$). Similarly s_6 is greater than s_4 (because it includes $a_5 - a_6$ which is positive). So the sums $s_2, s_4, s_6, \ldots$ are *increasing*.

The difference between s_{n-1} and s_n is the single number $\pm a_n$. It is required by 10H to approach zero. Therefore the decreasing sequence $s_1, s_3, \ldots$ approaches the *same* limit s as the increasing sequence $s_2, s_4, \ldots$. The series converges to s, which always lies between s_{n-1} and s_n.

This plus-minus pattern is special but important. The positive series Σa_n may not converge. *The alternating series is* $\Sigma (-1)^{n+1} a_n$.

EXAMPLE 6   The alternating series $4 - \frac{4}{3} + \frac{4}{5} - \frac{4}{7} \cdots$ is conditionally convergent (to π). The absolute values decrease to zero. Is this series absolutely convergent? *No.* With plus signs, $4(1 + \frac{1}{3} + \frac{1}{5} + \cdots)$ diverges like the harmonic series.

EXAMPLE 7   The alternating series $1 - 1 + 1 - 1 + \cdots$ is not convergent at all. *Which requirement in* 10H *is not met*? The partial sums $s_1, s_3, s_5, \ldots$ all equal 1 and $s_2, s_4, s_6, \ldots$ all equal 0—but they don't approach the same limit s.

MULTIPLYING AND REARRANGING SERIES

In Section 10.1 we added and subtracted and multiplied series. Certainly addition and subtraction are safe. If one series has partial sums $s_n \to s$ and the other has partial sums $t_n \to t$, then addition gives partial sums $s_n + t_n \to s + t$. But multiplication is more dangerous, because the *order* of the multiplication can make a difference. More exactly, **the order of terms is important when the series are conditionally convergent**. For absolutely convergent series, the order makes no difference. We can rearrange their terms and multiply them in any order, and the sum and product comes out right:

10I Suppose Σa_n converges absolutely. If $A_1, A_2, \ldots$ is any reordering of the a's, then $\Sigma A_n = \Sigma a_n$. In the new order ΣA_n also converges absolutely.

10J Suppose $\Sigma a_n = s$ and $\Sigma b_n = t$ converges absolutely. Then the infinitely many terms $a_i b_j$ in their product add (in any order) to st.

Rather than proving 10I and 10J, we show what happens when there is only conditional convergence. Our favorite is $1 - \frac{1}{2} + \frac{1}{3} - \frac{1}{4} + \cdots$, converging conditionally to ln 2. By rearranging, it will converge conditionally to *anything*! Suppose the desired sum is 1000. Take positive terms $1 + \frac{1}{3} + \cdots$ until they pass 1000. Then add negative terms $-\frac{1}{2} - \frac{1}{4} - \cdots$ until the subtotal drops below 1000. Then new positive terms bring it above 1000, and so on. All terms are eventually used, since at least one new term is needed at each step. The limit is $s = 1000$.

We also get strange products, when series fail to converge absolutely:

$$\left(1 - \frac{1}{\sqrt{2}} + \frac{1}{\sqrt{3}} \cdots\right)\left(1 - \frac{1}{\sqrt{2}} + \frac{1}{\sqrt{3}} \cdots\right) = 1 - \left(\frac{1}{\sqrt{2}} + \frac{1}{\sqrt{2}}\right) + \left(\frac{1}{\sqrt{3}} + \frac{1}{\sqrt{4}} + \frac{1}{\sqrt{3}}\right) \cdots .$$

On the left the series converge (conditionally). The alternating terms go to zero. On the right the series diverges. Its terms in parentheses don't even approach zero, and the product is completely wrong.

I close by emphasizing that it is absolute convergence that matters. ***The most important series are power series*** $\Sigma a_n x^n$. Like the geometric series (with all $a_n = 1$) there is absolute convergence over an interval of x's. They give *functions* of x, which is what calculus needs and wants.

We go next to the series for e^x, which is absolutely convergent everywhere. From the viewpoint of convergence tests it is too easy—the danger is gone. But from the viewpoint of calculus and its applications, e^x is unconditionally the best.

10.3 EXERCISES

Read-through questions

The series Σa_n is absolutely convergent if the series __a__ is convergent. Then the original series Σa_n is also __b__. But the series Σa_n can converge without converging absolutely. That is called __c__ convergence, and the series __d__ is an example.

For alternating series, the sign of each a_{n+1} is __e__ to the sign of a_n. With the extra conditions that __f__ and __g__, the series converges (at least conditionally). The partial sums $s_1, s_3, \ldots$ are __h__ and the partial sums $s_2, s_4, \ldots$ are __i__. The difference between s_n and s_{n-1} is __j__. Therefore the two series converge to the same number s. An alternating series that converges absolutely [conditionally] (not at all) is __k__ [__l__] (__m__). With absolute [conditional] convergence a reordering (can or cannot?) change the sum.

Do the series 1–12 converge *absolutely* or *conditionally*?

1 $\displaystyle\sum (-1)^{n+1}\frac{n}{n+3}$

2 $\displaystyle\sum (-1)^{n-1}/\sqrt{n+3}$

3 $\displaystyle\sum (-1)^{n+1}\frac{1}{n!}$

4 $\displaystyle\sum (-1)^{n+1}\frac{3^n}{n!}$

5 $\displaystyle\sum (-1)^{n+1}3\sqrt{n}/(n+1)$

6 $\displaystyle\sum (-1)^{n+1}\sin^2 n$

7 $\displaystyle\sum (-1)^{n+1}\ln\left(\frac{1}{n}\right)$

8 $\displaystyle\sum (-1)^{n+1}\frac{\sin^2 n}{n}$

9 $\displaystyle\sum (-1)^{n+1}n^2/(1+n^4)$

10 $\displaystyle\sum (-1)^{n+1}2^{1/n}$

11 $\displaystyle\sum (-1)^{n+1}n^{1/n}$

12 $\displaystyle\sum (-1)^{n+1}(1-n^{1/n})$

13 Suppose Σa_n converges absolutely. Explain why keeping the positive a's gives another convergent series.

14 Can Σa_n converge absolutely if all a_n are negative?

15 Show that the alternating series $1 - \frac{1}{2} + \frac{1}{2} - \frac{1}{4} + \frac{1}{3} - \frac{1}{6} + \cdots$ does not converge, by computing the partial sums $s_2, s_4, \ldots$. Which requirement of 10H is not met?

16 Show that $\frac{2}{3} - \frac{3}{5} + \frac{4}{7} - \frac{5}{9} + \cdots$ does not converge. Which requirement of 10H is not met?

17 (a) For an alternating series with terms decreasing to zero, why does the sum s always lie between s_{n-1} and s_n?

(b) Is $s - s_n$ positive or negative if s_n stops at a positive a_n?

18 Use Problem 17 to give a bound on the difference between $s_5 = 1 - \frac{1}{2} + \frac{1}{3} - \frac{1}{4} + \frac{1}{5}$ and the sum $s = \ln 2$ of the infinite series.

19 Find the sum $1 - \frac{1}{2!} + \frac{1}{3!} - \frac{1}{4!} + \cdots = s$. The partial sum s_4 is (above s)(below s) by less than _____.

20 Give a bound on the difference between $s_{100} = \frac{1}{1^2} - \frac{1}{2^2} + \frac{1}{3^2} \cdots - \frac{1}{100^2}$ and $s = \Sigma(-1)^{n+1}/n^2$.

21 Starting from $\frac{1}{1^2} + \frac{1}{2^2} + \frac{1}{3^2} + \cdots = \frac{\pi^2}{6}$, with plus signs, show that the alternating series in Problem 20 has $s = \pi^2/12$.

22 Does the alternating series in 20 or the positive series in 21 give π^2 more quickly? Compare $1/101^2 - 1/102^2 + \cdots$ with $1/101^2 + 1/102^2 + \cdots$.

23 If Σa_n does not converge show that $\Sigma |a_n|$ does not converge.

24 Find conditions which guarantee that $a_1 + a_2 - a_3 + a_4 + a_5 - a_6 + \cdots$ will converge (negative term follows two positive terms).

25 If the terms of $\ln 2 = 1 - \frac{1}{2} + \frac{1}{3} - \frac{1}{4} + \cdots$ are rearranged into $1 - \frac{1}{2} - \frac{1}{4} + \frac{1}{3} - \frac{1}{6} - \frac{1}{8} + \cdots$, show that this series now adds to $\frac{1}{2} \ln 2$. (Combine each positive term with the following negative term.)

26 Show that the series $1 + \frac{1}{3} - \frac{1}{2} + \frac{1}{5} + \frac{1}{7} - \frac{1}{4} + \cdots$ converges to $\frac{3}{2} \ln 2$.

27 What is the sum of $1 + \frac{1}{3} - \frac{1}{2} + \frac{1}{5} - \frac{1}{4} + \frac{1}{7} - \frac{1}{6} + \cdots$?

28 Combine $1 + \cdots + \frac{1}{n} - \ln n \to \gamma$ and $1 - \frac{1}{2} + \frac{1}{3} - \cdots \to \ln 2$ to prove $1 + \frac{1}{3} + \frac{1}{5} - \frac{1}{2} - \frac{1}{4} - \frac{1}{6} + \cdots = \ln 2$.

29 (a) Prove that this alternating series converges:

$$1 - \int_1^2 \frac{dx}{x} + \frac{1}{2} - \int_2^3 \frac{dx}{x} + \frac{1}{3} - \int_3^4 \frac{dx}{x} + \cdots$$

(b) Show that its sum is Euler's constant γ.

30 Prove that this series converges. Its sum is $\pi/2$.

$$\int_0^\pi \frac{\sin x}{x} dx + \int_\pi^{2\pi} \frac{\sin x}{x} dx + \cdots = \int_0^\infty \frac{\sin x}{x} dx.$$

31 The cosine of $\theta = 1$ radian is $1 - \frac{1}{2!} + \frac{1}{4!} - \cdots$. Compute $\cos 1$ to five correct decimals (how many terms?).

32 The sine of $\theta = \pi$ radians is $\pi - \frac{\pi^3}{3!} + \frac{\pi^5}{5!} - \cdots$. Compute $\sin \pi$ to eight correct decimals (how many terms?).

33 If Σa_n^2 and Σb_n^2 are convergent show that $\Sigma a_n b_n$ is absolutely convergent.
Hint: $(a \pm b)^2 \geqslant 0$ yields $2|ab| \leqslant a^2 + b^2$.

34 Verify the *Schwarz inequality* $(\Sigma a_n b_n)^2 \leqslant (\Sigma a_n^2)(\Sigma b_n^2)$ if $a_n = \left(\frac{1}{2}\right)^n$ and $b_n = \left(\frac{1}{3}\right)^n$.

35 Under what condition does $\sum_0^\infty (a_{n+1} - a_n)$ converge and what is its sum?

36 For a conditionally convergent series, explain how the terms could be rearranged so that the sum is $+\infty$. All terms must eventually be included, even negative terms.

37 Describe the terms in the product $(1 + \frac{1}{2} + \frac{1}{4} + \cdots)(1 + \frac{1}{3} + \frac{1}{9} + \cdots)$ and find their sum.

38 *True or false*:

(a) Every alternating series converges.

(b) Σa_n converges conditionally if $\Sigma |a_n|$ diverges.

(c) A convergent series with positive terms is absolutely convergent.

(d) If Σa_n and Σb_n both converge, so does $\Sigma(a_n + b_n)$.

39 Every number x between 0 and 2 equals $1 + \frac{1}{2} + \frac{1}{4} + \cdots$ with suitable terms deleted. Why?

40 Every number s between -1 and 1 equals $\pm \frac{1}{2} \pm \frac{1}{4} \pm \frac{1}{8} \pm \cdots$ with a suitable choice of signs. (Add $1 = \frac{1}{2} + \frac{1}{4} + \frac{1}{8} + \cdots$ to get Problem 39.) Which signs give $s = -1$ and $s = 0$ and $s = \frac{1}{3}$?

41 Show that no choice of signs will make $\pm \frac{1}{3} \pm \frac{1}{9} \pm \frac{1}{27} \pm \cdots$ equal to zero.

42 The sums in Problem 41 form a *Cantor set* centered at zero. What is the smallest positive number in the set? Choose signs to show that $\frac{1}{4}$ is in the set.

*43 Show that the tangent of $\theta = \frac{1}{2}(\pi - 1)$ is $\sin 1/(1 - \cos 1)$. This is the imaginary part of $s = -\ln(1 - e^i)$. From $s = \Sigma e^{in}/n$ deduce the remarkable sum $\Sigma(\sin n)/n = \frac{1}{2}(\pi - 1)$.

44 Suppose Σa_n converges and $|x| < 1$. Show that $\Sigma a_n x^n$ converges absolutely.

10.4 The Taylor Series for e^x, $\sin x$, and $\cos x$

This section goes back from numbers to functions. Instead of $\Sigma a_n = s$ it deals with $\Sigma a_n x^n = f(x)$. **The sum is a function of x.** The geometric series has all $a_n = 1$ (including a_0, the constant term) and its sum is $f(x) = 1/(1-x)$. The derivatives of $1 + x + x^2 + \cdots$ match the derivatives of f. Now we choose the a_n differently, to match a different function.

The new function is e^x. All its derivatives are e^x. At $x = 0$, this function and its derivatives equal 1. To match these 1's, we move factorials into the denominators. Term by term the series is

$$e^x = 1 + \frac{x}{1!} + \frac{x^2}{2!} + \frac{x^3}{3!} + \cdots. \tag{1}$$

$x^n/n!$ has the correct nth derivative $(=1)$. **From the derivatives at $x = 0$, we have built back the function!** At $x = 1$ the right side is $1 + 1 + \frac{1}{2} + \frac{1}{6} + \cdots$ and the left side is $e = 2.71828\ldots$. At $x = -1$ the series gives $1 - 1 + \frac{1}{2} - \frac{1}{6} + \cdots$, which is e^{-1}.

The same term-by-term idea works for differential equations, as follows.

EXAMPLE 1   Solve $dy/dx = -y$ starting from $y = 1$ at $x = 0$.

Solution   *The zeroth derivative at $x = 0$ is the function itself:* $y = 1$. Then the equation $y' = -y$ gives $y' = -1$ and $y'' = -y' = +1$. The alternating derivatives $1, -1, 1, -1, \ldots$ are matched by the alternating series for e^{-x}:

$$y = 1 - x + \tfrac{1}{2}x^2 - \tfrac{1}{6}x^3 + \cdots = e^{-x} \text{ (the correct solution to } y' = -y\text{).}$$

EXAMPLE 2   Solve $d^2y/dx^2 = -y$ starting from $y = 1$ and $y' = 0$ (the answer is $\cos x$).

Solution   The equation gives $y'' = -1$ (again at $x = 0$). The derivative of the equation gives $y''' = -y' = 0$. Then $y'''' = -y'' = +1$. The even derivatives are alternately $+1$ and -1, the odd derivatives are zero. This is matched by a series of even powers, which constructs $\cos x$:

$$y = 1 - \frac{1}{2!}x^2 + \frac{1}{4!}x^4 - \frac{1}{6!}x^6 + \cdots = \cos x.$$

The first terms $1 - \frac{1}{2}x^2$ came earlier in the book. Now we have the whole alternating series. It converges absolutely for all x, by comparison with the series for e^x (odd powers are dropped). The partial sums in Figure 10.4 reach further and further before they lose touch with $\cos x$.

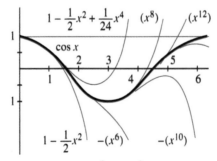

Fig. 10.4   The partial sums $1 - x^2/2 + x^4/24 - \cdots$ of the cosine series.

If we wanted plus signs instead of plus-minus, we could average e^x and e^{-x}. The differential equation for $\cosh x$ is $d^2 y / dx^2 = +y$, to give plus signs:

$$\frac{1}{2}(e^x + e^{-x}) = 1 + \frac{1}{2!}x^2 + \frac{1}{4!}x^4 + \frac{1}{6!}x^6 + \cdots \text{ (which is } \cosh x \text{)}.$$

TAYLOR SERIES

The idea of ***matching derivatives by powers*** is becoming central to this chapter. The derivatives are given at a basepoint (say $x = 0$). They are numbers $f(0), f'(0), \ldots$. The derivative $f^{(n)}(0)$ will be the nth derivative of $a_n x^n$, if we choose a_n to be $f^{(n)}(0)/n!$ Then the series $\Sigma a_n x^n$ has the same derivatives at the basepoint as $f(x)$:

10K The ***Taylor series*** that matches $f(x)$ and all its derivatives at $x = 0$ is

$$f(0) + f'(0)x + \frac{1}{2}f''(0)x^2 + \frac{1}{6}f'''(0)x^3 + \cdots = \sum_{n=0}^{\infty} \frac{f^{(n)}(0)}{n!}x^n.$$

The first terms give the linear and quadratic approximations that we know well. The x^3 term was mentioned earlier (but not used). Now we have *all* the terms—an "infinite approximation" that is intended to equal $f(x)$.

Two things are needed. First, the series must converge. Second, the function must do what the series predicts, away from $x = 0$. Those are true for e^x and $\cos x$ and $\sin x$; the series equals the function. We proceed on that basis.

The Taylor series with special basepoint $x = 0$ is also called the "*Maclaurin series*."

EXAMPLE 3 Find the Taylor series for $f(x) = \sin x$ around $x = 0$.

Solution   The numbers $f^{(n)}(0)$ are the values of $f = \sin x$, $f' = \cos x$, $f'' = -\sin x, \ldots$ at $x = 0$. Those values are $0, 1, 0, -1, 0, 1, \ldots$. All even derivatives are zero. To find the coefficients in the Taylor series, divide by the factorials:

$$\sin x = x - \tfrac{1}{6}x^3 + \tfrac{1}{120}x^5 - \cdots. \tag{2}$$

EXAMPLE 4 Find the Taylor series for $f(x) = (1 + x)^5$ around $x = 0$.

Solution   This function starts at $f(0) = 1$. Its derivative is $5(1+x)^4$, so $f'(0) = 5$. The second derivative is $5 \cdot 4 \cdot (1+x)^3$, so $f''(0) = 5 \cdot 4$. The next three derivatives are $5 \cdot 4 \cdot 3, 5 \cdot 4 \cdot 3 \cdot 2, 5 \cdot 4 \cdot 3 \cdot 2 \cdot 1$. *After that all derivatives are zero.* Therefore the Taylor series ***stops*** after the x^5 term:

$$1 + 5x + \frac{5 \cdot 4}{2!}x^2 + \frac{5 \cdot 4 \cdot 3}{3!}x^3 + \frac{5 \cdot 4 \cdot 3 \cdot 2}{4!}x^4 + \frac{5 \cdot 4 \cdot 3 \cdot 2 \cdot 1}{5!}x^5. \tag{3}$$

You may recognize $1, 5, 10, 10, 5, 1$. They are the ***binomial coefficients***, which appear in Pascal's triangle (Section 2.2). By matching derivatives, we see why $0!, 1!, 2!, \ldots$ are needed in the denominators.

There is no doubt that $x = 0$ is the nicest basepoint. But Taylor series can be constructed around other points $x = a$. The principle is the same—*match derivatives by powers*—but now the powers to use are $(x - a)^n$. The derivatives $f^{(n)}(a)$ are computed at the new basepoint $x = a$.

The Taylor series begins with $f(a) + f'(a)(x - a)$. This is the tangent approximation at $x = a$. The whole "infinite approximation" is centered at a— at that point it has the same derivatives as $f(x)$.

10L The *Taylor series* for $f(x)$ around the basepoint $x = a$ is

$$f(x) = f(a) + f'(a)(x - a) + \frac{1}{2}f''(a)(x - a)^2 + \cdots = \sum_{n=0}^{\infty} \frac{f^{(n)}(a)}{n!}(x - a)^n. \quad (4)$$

EXAMPLE 5 Find the Taylor series for $f(x) = (1 + x)^5$ around $x = a = 1$.

Solution   At $x = 1$, the function is $(1 + 1)^5 = 32$. Its first derivative $5(1 + x)^4$ is $5 \cdot 16 = 80$. We compute the nth derivative, divide by $n!$, and multiply by $(x - 1)^n$:

$$32 + 80(x - 1) + 80(x - 1)^2 + 40(x - 1)^3 + 10(x - 1)^4 + (x - 1)^5. \quad (5)$$

That Taylor series (which stops at $n = 5$) should agree with $(1 + x)^5$. It does. We could rewrite $1 + x$ as $2 + (x - 1)$, and take its fifth power directly. Then $32, 16, 8, 4, 2, 1$ will multiply the usual coefficients $1, 5, 10, 10, 5, 1$ to give our Taylor coefficients $32, 80, 80, 40, 10, 1$. The series stops as it will stop for any polynomial—because the high derivatives are zero.

EXAMPLE 6 Find the Taylor series for $f(x) = e^x$ around the basepoint $x = 1$.

Solution   At $x = 1$ the function and all its derivatives equal e. Therefore the Taylor series has that constant factor (note the powers of $x - 1$, not x):

$$e^x = e + e(x - 1) + \frac{e}{2!}(x - 1)^2 + \frac{e}{3!}(x - 1)^3 + \cdots. \quad (6)$$

DEFINING THE FUNCTION BY ITS SERIES

Usually, we define $\sin x$ and $\cos x$ from the sides of a triangle. But we could start instead with the series. Define $\sin x$ by equation (2). The logic goes backward, but it is still correct:

> First, prove that the series converges.
> Second, prove properties like $(\sin x)' = \cos x$.
> Third, connect the definitions by series to the sides of a triangle.

We don't plan to do all this. The usual definition was good enough. But note first: There is no problem with convergence. The series for $\sin x$ and $\cos x$ and e^x all have terms $\pm x^n/n!$. The factorials make the series converge for all x. The general rule for e^x times e^y can be based on the series. Equation (6) is typical: e is multiplied by powers of $(x - 1)$. Those powers add to e^{x-1}. So the series proves that $e^x = ee^{x-1}$. That is just one example of the multiplication $(e^x)(e^y) = e^{x+y}$:

$$\left(1 + x + \frac{x^2}{2} + \ldots\right)\left(1 + y + \frac{y^2}{2} + \ldots\right) = 1 + x + y + \frac{x^2}{2} + xy + \frac{y^2}{2} + \ldots. \quad (7)$$

Term by term, multiplication gives the series for e^{x+y}. Term by term, differentiating the series for e^x gives e^x. Term by term, the derivative of $\sin x$ is $\cos x$:

$$\frac{d}{dx}\left(x - \frac{x^3}{3!} + \frac{x^5}{5!} - \ldots\right) = 1 - \frac{x^2}{2!} + \frac{x^4}{4!} - \ldots. \quad (8)$$

We don't need the famous limit $(\sin x)/x \to 1$, by which geometry gave us the derivative. The identities of trigonometry become identities of infinite series. We could even define π as the first positive x at which $x - \frac{1}{6}x^3 + \cdots$ equals zero. But it is certainly not obvious that this sine series returns to zero—much less that the point of return is near 3.14.

*The function that **will** be defined by infinite series is $e^{i\theta}$.* This is the exponential of the ***imaginary number*** $i\theta$ (a multiple of $i = \sqrt{-1}$). The result $e^{i\theta}$ is a ***complex number***, and our goal is to identify it. (We will be confirming Section 9.4.) The technique is to treat $i\theta$ like all other numbers, real or complex, and simply put it into the series:

DEFINITION $e^{i\theta}$ is the sum of $1 + (i\theta) + \dfrac{1}{2!}(i\theta)^2 + \dfrac{1}{3!}(i\theta)^3 + \cdots$.   (9)

Now use $i^2 = -1$. The even powers are $i^4 = +1$, $i^6 = -1$, $i^8 = +1$, ... We are just multiplying -1 by -1 to get 1. The odd powers are $i^3 = -i$, $i^5 = +i$, ... Therefore $e^{i\theta}$ splits into a *real part* (with no i's) and an *imaginary part* (multiplying i):

$$e^{i\theta} = \left(1 - \frac{1}{2!}\theta^2 + \frac{1}{4!}\theta^4 - \cdots\right) + i\left(\theta - \frac{1}{3!}\theta^3 + \frac{1}{5!}\theta^5 - \cdots\right). \qquad (10)$$

You recognize those series. They are $\cos\theta$ and $\sin\theta$. Therefore:

Euler's formula is $e^{i\theta} = \cos\theta + i\sin\theta$. Note that $e^{2\pi i} = 1$.

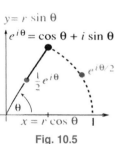

The real part is $x = \cos\theta$ and the imaginary part is $y = \sin\theta$. Those coordinates pick out the point $e^{i\theta}$ in the "complex plane." Its distance from the origin $(0,0)$ is $r = 1$, because $(\cos\theta)^2 + (\sin\theta)^2 = 1$. Its angle is θ, as shown in Figure 10.5. The number -1 is $e^{i\pi}$, at the distance $r = 1$ and the angle π. It is on the real axis to the left of zero. If $e^{i\theta}$ is multiplied by $r = 2$ or $r = \frac{1}{2}$ or any $r \geq 0$, the result is a complex number at a distance r from the origin:

Complex numbers: $re^{i\theta} = r(\cos\theta + i\sin\theta) = r\cos\theta + ir\sin\theta = x + iy$.

Fig. 10.5

With $e^{i\theta}$, a negative number has a logarithm. *The logarithm of -1 is imaginary* (it is $i\pi$, since $e^{i\pi} = -1$). A negative number also has fractional powers. The fourth root of -1 is $(-1)^{1/4} = e^{i\pi/4}$. More important for calculus: *The derivative of $x^{5/4}$ is $\frac{5}{4}x^{1/4}$.* That sounds old and familiar, but at $x = -1$ it was never allowed.

Complex numbers tie up the loose ends left by the limitations of real numbers.

The formula $e^{i\theta} = \cos\theta + i\sin\theta$ has been called "one of the greatest mysteries of undergraduate mathematics." Writers have used desperate methods to avoid infinite series. That proof in (10) may be the clearest (I remember sending it to a prisoner studying calculus) but here is a way to start from $d/dx(e^{ix}) = ie^{ix}$.

A different proof of Euler's formula   Any complex number is $e^{ix} = r(\cos\theta + i\sin\theta)$ for *some* r and θ. Take the x derivative of both sides, and substitute for ie^{ix}:

$$(\cos\theta + i\sin\theta)dr/dx + r(-\sin\theta + i\cos\theta)d\theta/dx = ir(\cos\theta + i\sin\theta).$$

Comparing the real parts and also the imaginary parts, we need $dr/dx = 0$ and $d\theta/dx = 1$. The starting values $r = 1$ and $\theta = 0$ are known from $e^{i0} = 1$. Therefore r is always 1 and θ is x. Substituting into the first sentence of the proof, we have Euler's formula $e^{i\theta} = 1(\cos\theta + i\sin\theta)$.

Read-through questions

The __a__ series is chosen to match $f(x)$ and all its __b__ at the basepoint. Around $x=0$ the series begins with $f(0)+$ __c__ $x+$ __d__ x^2. The coefficient of x^n is __e__. For $f(x)=e^x$ this series is __f__. For $f(x)=\cos x$ the series is __g__. For $f(x)=\sin x$ the series is __h__. If the signs were all positive in those series, the functions would be $\cosh x$ and __i__. Addition gives $\cosh x+\sinh x=$ __j__.

In the Taylor series for $f(x)$ around $x=a$, the coefficient of $(x-a)^n$ is $b_n=$ __k__. Then $b_n(x-a)^n$ has the same __l__ as f at the basepoint. In the example $f(x)=x^2$, the Taylor coefficients are $b_0=$ __m__, $b_1=$ __n__, $b_2=$ __o__. The series $b_0+b_1(x-a)+b_2(x-a)^2$ agrees with the original __p__. The series for e^x around $x=a$ has $b_n=$ __q__. Then the Taylor series reproduces the identity $e^x=($ __r__ $)($ __s__ $)$.

We define e^x, $\sin x$, $\cos x$, and also $e^{i\theta}$ by their series. The derivative $d/dx(1+x+\frac{1}{2}x^2+\cdots)=1+x+\cdots$ translates to __t__. The derivative of $1-\frac{1}{2}x^2+\cdots$ is __u__. Using $i^2=-1$ the series $1+i\theta+\frac{1}{2}(i\theta)^2+\cdots$ splits into $e^{i\theta}=$ __v__. Its square gives $e^{2i\theta}=$ __w__. Its reciprocal is $e^{-i\theta}=$ __x__. Multiplying by r gives $re^{i\theta}=$ __y__ $+i$ __z__, which connects the polar and rectangular forms of a __A__ number. The logarithm of $e^{i\theta}$ is __B__.

1 Write down the series for e^{2x} and compute all derivatives at $x=0$. Give a series of numbers that adds to e^2.

2 Write down the series for $\sin 2x$ and check the third derivative at $x=0$. Give a series of numbers that adds to $\sin 2\pi=0$.

In 3–8 find the derivatives of $f(x)$ at $x=0$ and the Taylor series (powers of x) with those derivatives.

3 $f(x)=e^{ix}$

4 $f(x)=1/(1+x)$

5 $f(x)=1/(1-2x)$

6 $f(x)=\cosh x$

7 $f(x)=\ln(1-x)$

8 $f(x)=\ln(1+x)$

Problems 9–14 solve differential equations by series.

9 From the equation $dy/dx=y-2$ find all the derivatives of y at $x=0$ starting from $y(0)=1$. Construct the infinite series for y, identify it as a known function, and verify that the function satisfies $y'=y-2$.

10 Differentiate the equation $y'=cy+s$ (c and s constant) to find all derivatives of y at $x=0$. If the starting value is $y_0=0$, construct the Taylor series for y and identify it with the solution of $y'=cy+s$ in Section 6.3.

11 Find the infinite series that solves $y''=-y$ starting from $y=0$ and $y'=1$ at $x=0$.

12 Find the infinite series that solves $y'=y$ starting from $y=1$ at $x=3$ (use powers of $x-3$). Identify y as a known function.

13 Find the infinite series (powers of x) that solves $y''=2y'-y$ starting from $y=0$ and $y'=1$ at $x=0$.

14 Solve $y''=y$ by a series with $y=1$ and $y'=0$ at $x=0$ and identify y as a known function.

15 Find the Taylor series for $f(x)=(1+x)^2$ around $x=a=0$ and around $x=a=1$ (powers of $x-1$). Check that both series add to $(1+x)^2$.

16 Find all derivatives of $f(x)=x^3$ at $x=a$ and write out the Taylor series around that point. Verify that it adds to x^3.

17 What is the series for $(1-x)^5$ with basepoint $a=1$?

18 Write down the Taylor series for $f=\cos x$ around $x=2\pi$ and also for $f=\cos(x-2\pi)$ around $x=0$.

In 19–24 compute the derivatives of f and its Taylor series around $x=1$.

19 $f(x)=1/x$

20 $f(x)=1/(2-x)$

21 $f(x)=\ln x$

22 $f(x)=x^4$

23 $f(x)=e^{-x}$

24 $f(x)=e^{2x}$

In 25–33 write down the first three nonzero terms of the Taylor series around $x=0$, from the series for e^x, $\cos x$, and $\sin x$.

25 xe^{2x}

26 $\cos\sqrt{x}$

27 $(1-\cos x)/x^2$

28 $\dfrac{\sin x}{x}$

29 $\displaystyle\int_0^x \dfrac{\sin x}{x}\,dx$

30 $\sin x^2$

31 e^{x^2}

32 $b^x=e^{x\ln b}$

33 $e^x\cos x$

*34 For $x<0$ the derivative of x^n is still nx^{n-1}:
$$\frac{d}{dx}(x^n)=\frac{d}{dx}(|x|^n e^{in\pi})=n|x|^{n-1}e^{in\pi}\frac{d|x|}{dx}.$$
What is $d|x|/dx$? Rewrite this answer as nx^{n-1}.

35 Why doesn't $f(x)=\sqrt{x}$ have a Taylor series around $x=0$? Find the first two terms around $x=1$.

36 Find the Taylor series for 2^x around $x=0$.

In 37–44 find the first three terms of the Taylor series around $x=0$.

37 $f(x)=\tan^{-1}x$

38 $f(x)=\sin^{-1}x$

39 $f(x)=\tan x$

40 $f(x)=\ln(\cos x)$

41 $f(x)=e^{\sin x}$

42 $f(x)=\tanh^{-1}x$

43 $f(x)=\cos^2 x$

44 $f(x)=\sec^2 x$

45 From $e^{i\theta} = \cos\theta + i\sin\theta$ and $e^{-i\theta} = \cos\theta - i\sin\theta$, add and subtract to find $\cos\theta$ and $\sin\theta$.

46 Does $(e^{i\theta})^2$ equal $\cos^2\theta + i\sin^2\theta$ or $\cos\theta^2 + i\sin\theta^2$?

47 Find the real and imaginary parts and the 99th power of $e^{i\pi}$, $e^{i\pi/2}$, $e^{i\pi/4}$ and $e^{-i\pi/6}$.

48 The three cube roots of 1 are 1, $e^{2\pi i/3}$, $e^{4\pi i/3}$.

(a) Find the real and imaginary parts of $e^{2\pi i/3}$.

(b) Explain why $(e^{2\pi i/3})^3 = 1$.

(c) Check this statement in rectangular coordinates.

49 The cube roots of $-1 = e^{i\pi}$ are $e^{i\pi/3}$ and _____ and _____ . Find their sum and their product.

50 Find the squares of $2e^{i\pi/3} = 1 + \sqrt{3}i$ and $4e^{i\pi/4} = 2\sqrt{2} + i2\sqrt{2}$ in both polar and rectangular coordinates.

51 Multiply $e^{is} = \cos s + i\sin s$ times $e^{it} = \cos t + i\sin t$ to find formulas for $\cos(s+t)$ and $\sin(s+t)$.

52 Multiply e^{is} times e^{-it} to find formulas for $\cos(s-t)$ and $\sin(s-t)$.

53 Find the logarithm of i. Then find another logarithm of i. (What can you add to the exponent of $e^{\ln i}$ without changing the result?)

54 (Proof that e is irrational) If $e = p/q$ then

$$N = p!\left[\frac{1}{e} - \left(1 - \frac{1}{1!} + \frac{1}{2!} - \cdots \pm \frac{1}{p!}\right)\right]$$

would be an integer. (Why?) The number in brackets—the distance from the alternating series to its sum $1/e$—is less than the last term which is $1/p!$. Deduce that $|N| < 1$ and reach a contradiction, which proves that e cannot equal p/q.

55 Solve $dy/dx = y$ by infinite series starting from $y = 2$ at $x = 0$.

10.5 Power Series

This section studies the properties of a power series. When the basepoint is zero, the powers are x^n. The series is $\Sigma a_n x^n$. When the basepoint is $x = a$, the powers are $(x - a)^n$. We want to know when and where (and how quickly) the series converges to the underlying function. For e^x and $\cos x$ and $\sin x$ there is convergence for all x—but that is certainly not true for $1/(1 - x)$. The convergence is best when the function is smooth.

First I emphasize that power series are not the only series. For many applications they are not the best choice. An alternative is a sum of sines, $f(x) = \Sigma b_n \sin nx$. That is a "*Fourier sine series*", which treats all x's equally instead of picking on a basepoint. A Fourier series allows jumps and corners in the graph—it takes the rough with the smooth. By contrast a power series is terrific near its basepoint, and gets worse as you move away. The Taylor coefficients a_n are totally determined *at the basepoint*—where all derivatives are computed. Remember the rule for Taylor series:

$$a_n = (n\text{th derivative at the basepoint})/n! = f^{(n)}(a)/n! \qquad (1)$$

A remarkable fact is the convergence in a *symmetric interval around $x = a$*.

10M A power series $\Sigma a_n x^n$ either converges for all x, or it converges only at the basepoint $x = 0$, or else it has a *radius of convergence* r:

$$\Sigma a_n x^n \text{ converges absolutely if } |x| < r \text{ and diverges if } |x| > r.$$

The series $\Sigma x^n/n!$ converges for all x (the sum is e^x). The series $\Sigma n! x^n$ converges for no x (except $x = 0$). The geometric series Σx^n converges absolutely for $|x| < 1$ and diverges for $|x| > 1$. *Its radius of convergence is $r = 1$*. Note that its sum $1/(1 - x)$ is perfectly good for $|x| > 1$—the function is all right but the series has given up. If something goes wrong at the distance r, a power series can't get past that point.

When the basepoint is $x = a$, the interval of convergence shifts over to $|x - a| < r$. The series converges for x between $a - r$ and $a + r$ (symmetric around a). We cannot say in advance whether the endpoints $a \pm r$ give divergence or convergence (absolute or conditional). *Inside* the interval, an easy comparison test will now prove convergence.

PROOF OF 10M Suppose $\Sigma a_n X^n$ converges at a particular point X. The proof will show that $\Sigma a_n x^n$ converges when $|x|$ is less than the number $|X|$. Thus convergence at X gives convergence at all closer points x (I mean closer to the basepoint 0). Proof: Since $\Sigma a_n X^n$ converges, its terms approach zero. Eventually $|a_n X^n| < 1$ and then

$$|a_n x^n| = |a_n X^n||x/X|^n < |x/X|^n.$$

Our series $\Sigma a_n x^n$ is absolutely convergent by comparison with the geometric series for $|x/X|$, which converges since $|x/X| < 1$.

EXAMPLE 1   The series $\Sigma n x^n/4^n$ has radius of convergence $r = 4$.

The ratio test and root test are best for power series. The ratios between terms approach $x/4$ (and so does the nth root of $n x^n/4^n$):

$$\frac{(n+1)x^{n+1}}{4^{n+1}} \bigg/ \frac{n x^n}{4^n} = \frac{x}{4}\frac{n+1}{n} \text{ approaches } L = \frac{x}{4}.$$

The ratio test gives convergence if $L < 1$, which means $|x| < 4$.

EXAMPLE 2 The sine series $x - \dfrac{x^3}{3!} + \dfrac{x^5}{5!} - \cdots$ has $r = \infty$ (it converges everywhere).

The ratio of $x^{n+2}/(n+2)!$ to $x^n/n!$ is $x^2/(n+2)(n+1)$. This approaches $L = 0$.

EXAMPLE 3 The series $\Sigma (x-5)^n/n^2$ has radius $r = 1$ *around its basepoint $a = 5$*.

The ratios between terms approach $L = x - 5$. (The fractions $n^2/(n+1)^2$ go toward 1.) There is absolute convergence if $|x - 5| < 1$. This is the interval $4 < x < 6$, symmetric around the basepoint. This series happens to converge at the endpoints 4 and 6, because of the factor $1/n^2$. That factor decides the delicate question—convergence at the endpoints—but all powers of n give the same *interval of convergence* $4 < x < 6$.

CONVERGENCE TO THE FUNCTION: REMAINDER TERM AND RADIUS r

Remember that a Taylor series starts with a function $f(x)$. The derivatives at the basepoint produce the series. Suppose the series converges: *Does it converge to the function*? This is a question about the *remainder* $R_n(x) = f(x) - s_n(x)$, which is the difference between f and the partial sum $s_n = a_0 + \cdots + a_n(x-a)^n$. *The remainder R_n is the error if we stop the series*, ending with the nth derivative term $a_n(x-a)^n$.

> **10N** Suppose f has an $(n+1)$st derivative from the basepoint a out to x. Then for some point c in between (position not known) the remainder at x equals
> $$R_n(x) = f(x) - s_n(x) = f^{(n+1)}(c)(x-a)^{n+1}/(n+1)! \qquad (2)$$

The error in stopping at the nth derivative is controlled by the $(n+1)$st derivative.

You will guess, correctly, that the unknown point c comes from the Mean Value Theorem. For $n = 1$ the proof is at the end of Section 3.8. That was the error $e(x)$ in *linear* approximation:

$$R_1(x) = f(x) - f(a) - f'(a)(x-a) = \tfrac{1}{2}f''(c)(x-a)^2.$$

For every n, the proof compares R_n with $(x-a)^{n+1}$. Their $(n+1)$st derivatives are $f^{(n+1)}$ and $(n+1)!$ The generalized Mean Value Theorem says that the ratio of R_n to $(x-a)^{n+1}$ equals the ratio of those derivatives, at the right point c. That is equation (2). The details can stay in Section 3.8 and Problem 23, because the main point is what we want. *The error is exactly like the next term $a_{n+1}(x-a)^{n+1}$*, except that the $(n+1)$st derivative is at c instead of the basepoint a.

EXAMPLE 4 When f is e^x, the $(n+1)$st derivative is e^x. Therefore the error is

$$R_n = e^x - \left(1 + x + \cdots + \frac{x^n}{n!}\right) = e^c \frac{x^{n+1}}{(n+1)!}. \qquad (3)$$

At $x = 1$ and $n = 2$, the error is $e - (1 + 1 + \tfrac{1}{2}) \approx .218$. The right side is $e^c/6$. The unknown point is $c = \ln(.218 \cdot 6) = .27$. Thus c lies between the basepoint $a = 0$

and the error point $x = 1$, as required. The series converges to the function, because $R_n \to 0$.

In practice, n is the number of derivatives to be calculated. We may aim for an error $|R_n|$ below 10^{-6}. Unfortunately, the high derivative in formula (2) is awkward to estimate (except for e^x). And high derivatives in formula (1) are difficult to compute. Most real calculations use only a *few terms* of a Taylor series. For more accuracy we move the basepoint closer, or switch to another series.

There is a direct connection between the function and the convergence radius r. A hint came for $f(x) = 1/(1-x)$. The function blows up at $x = 1$—which also ends the convergence interval for the series. Another hint comes for $f = 1/x$, if we expand around $x = a = 1$:

$$\frac{1}{x} = \frac{1}{1-(1-x)} = 1 + (1-x) + (1-x)^2 + \cdots. \qquad (4)$$

This geometric series converges for $|1-x| < 1$. Convergence stops at the end point $x = 0$—exactly where $1/x$ blows up. ***The failure of the function stops the convergence of the series***. But note that $1/(1+x^2)$, which never seems to fail, also has convergence radius $r = 1$:

$$1/(1+x^2) = 1 - x^2 + x^4 - x^6 + \cdots \text{ converges only for } |x| < 1.$$

When you see the reason, you will know why r is a "radius." There is a circle, and the function fails at the edge of the circle. The circle contains complex numbers as well as real numbers. The imaginary points i and $-i$ are at the edge of the circle. ***The function fails at those points because*** $1/(1+i^2) = \infty$.

Complex numbers are pulling the strings, out of sight. The circle of convergence reaches out to the nearest "singularity" of $f(x)$, real or imaginary or complex. For $1/(1+x^2)$, the singularities at i and $-i$ make $r = 1$. If we expand around $a = 3$, the distance to i and $-i$ is $r = \sqrt{10}$. If we change to $\ln(1+x)$, which blows up at $x = -1$, the radius of convergence of $x - \frac{1}{2}x^2 + \frac{1}{3}x^3 - \cdots$ is $r = 1$.

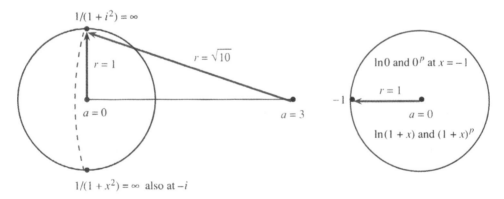

Fig. 10.6 Convergence radius r is distance from basepoint a to nearest singularity.

THE BINOMIAL SERIES

We close this chapter with one more series. It is the Taylor series for $(1+x)^p$, around the basepoint $x = 0$. A typical power is $p = \frac{1}{2}$, where we want the terms in

$$\sqrt{1+x} = 1 + \tfrac{1}{2}x + a_2 x^2 + \cdots.$$

The slow way is to square both sides, which gives $1 + x + (2a_2 + \frac{1}{4})x^2$ on the right. Since $1 + x$ is on the left, $a_2 = -\frac{1}{8}$ is needed to remove the x^2 term. Eventually a_3 can be found. The fast way is to match the derivatives of $f = (1+x)^{1/2}$:

$$f' = \tfrac{1}{2}(1+x)^{-1/2} \qquad f'' = \left(\tfrac{1}{2}\right)\left(-\tfrac{1}{2}\right)(1+x)^{-3/2} \qquad f''' = \left(\tfrac{1}{2}\right)\left(-\tfrac{1}{2}\right)\left(-\tfrac{3}{2}\right)(1+x)^{-5/2}.$$

At $x = 0$ those derivatives are $\frac{1}{2}, -\frac{1}{4}, \frac{3}{8}$. Dividing by $1!, 2!, 3!$ gives

$$a_1 = \frac{1}{2} \quad a_2 = -\frac{1}{8} \quad a_3 = \frac{1}{16} \qquad a_n = \frac{1}{n!}\left(\frac{1}{2}\right)\left(\frac{1}{2} - 1\right)\cdots\left(\frac{1}{2} - n + 1\right).$$

These are the **binomial coefficients** when the power is $p = \frac{1}{2}$.

Notice the difference from the binomials in Chapter 2. For those, the power p was a positive integer. The series $(1+x)^2 = 1 + 2x + x^2$ stopped at x^2. The coefficients for $p = 2$ were $1, 2, 1, 0, 0, 0, \ldots$. For fractional p or negative p those later coefficients are *not* zero, and we find them from the derivatives of $(1+x)^p$:

$$(1+x)^p \quad p(1+x)^{p-1} \quad p(p-1)(1+x)^{p-2} \quad f^{(n)} = p(p-1)\cdots(p-n+1)(1+x)^{p-n}.$$

Dividing by $0!, 1!, 2!, \ldots, n!$ at $x = 0$, the binomial coefficients are

$$1 \quad p \quad \frac{p(p-1)}{2} \quad \cdots \quad \frac{f^{(n)}(0)}{n!} = \frac{p(p-1)\cdots(p-n+1)}{n!}. \qquad (5)$$

For $p = n$ that last binomial coefficient is $n!/n! = 1$. It gives the final x^n at the end of $(1+x)^n$. For other values of p, the binomial series never stops. **It converges for $|x| < 1$:**

$$(1+x)^p = 1 + px + \frac{p(p-1)}{2}x^2 + \cdots = \sum_{n=0}^{\infty} \frac{p(p-1)\cdots(p-n+1)}{n!}x^n. \qquad (6)$$

When $p = 1, 2, 3, \ldots$ **the binomial coefficient $p!/n!(n-p)!$ counts the number of ways to select a group of n friends out of a group of p friends**. If you have 20 friends, you can choose 2 of them in $(20)(19)/2 = 190$ ways.

Suppose p is not a positive integer. What goes wrong with $(1+x)^p$, to stop the convergence at $|x| = 1$? The failure is at $x = -1$. If p is negative, $(1+x)^p$ blow up. If p is positive, as in $\sqrt{1+x}$, the higher derivatives blow up. Only for a positive integer $p = n$ does the convergence radius move out to $r = \infty$. In that case the series for $(1+x)^n$ stops at x^n, and f never fails.

A power series is a function in a new form. It is not a simple form, but sometimes it is the only form. To compute f we have to sum the series. To square f we have to multiply series. But the operations of calculus—derivative and integral—are easier. That explains why power series help to solve differential equations, which are a rich source of new functions. (Numerically the series are not always so good.) I should have said that the derivative and integral are easy *for each separate term $a_n x^n$*—and fortunately the convergence radius of the whole series is not changed.

If $f(x) = \Sigma a_n x^n$ has convergence radius r, so do its derivative and its integral:

$$df/dx = \Sigma n a_n x^{n-1} \quad \text{and} \quad \int f(x)dx = \Sigma a_n x^{n+1}/(n+1) \text{ also converge for } |x| < r.$$

EXAMPLE 5 The series for $1/(1-x)$ and its derivative $1/(1-x)^2$ and its integral $-\ln(1-x)$ all have $r = 1$ (because they all have trouble at $x = 1$). The series are Σx^n and $\Sigma n x^{n-1}$ and $\Sigma x^{n+1}/(n+1)$.

EXAMPLE 6　We can integrate e^{x^2} (previously impossible) by integrating every term in its series:

$$\int e^{x^2}\,dx = \int\left(1+x^2+\frac{1}{2!}x^4+\cdots\right)dx = x+\frac{x^3}{3}+\frac{1}{2!}\left(\frac{x^5}{5}\right)+\frac{1}{3!}\left(\frac{x^7}{7}\right)+\cdots.$$

This always converges ($r=\infty$). The derivative of e^{x^2} was never a problem.

10.5 EXERCISES

Read-through questions

If $|x|<|X|$ and $\Sigma a_n X^n$ converges, then the series $\Sigma a_n x^n$ also __a__. There is convergence in a __b__ interval around the __c__. For $\Sigma(2x)^n$ the convergence radius is $r=$__d__. For $\Sigma x^n/n!$ the radius is $r=$__e__. For $\Sigma(x-3)^n$ there is convergence for $|x-3|<$__f__. Then x is between __g__ and __h__.

Starting with $f(x)$, its Taylor series $\Sigma a_n x^n$ has $a_n=$__i__. With basepoint a, the coefficient of $(x-a)^n$ is __j__. The error after the x^n term is called the __k__ $R_n(x)$. It is equal to __l__ where the unknown point c is between __m__. Thus the error is controlled by the __n__ derivative.

The circle of convergence reaches out to the first point where $f(x)$ fails. For $f=4/(2-x)$, that point is $x=$__o__. Around the basepoint $a=5$, the convergence radius would be $r=$__p__. For $\sin x$ and $\cos x$ the radius is $r=$__q__.

The series for $\sqrt{1+x}$ is the __r__ series with $p=\frac{1}{2}$. Its coefficients are $a_n=$__s__. Its convergence radius is __t__. Its square is the very short series $1+x$.

In 1–6 find the Taylor series for $f(x)$ around $x=0$ and its radius of convergence r. At what point does $f(x)$ blow up?

1　$f(x)=1/(1-4x)$

2　$f(x)=1/(1-4x^2)$

3　$f(x)=e^{1-x}$

4　$f(x)=\tan x$ (through x^3)

5　$f(x)=\ln(e+x)$

6　$f(x)=1/(1+4x^2)$

Find the interval of convergence and the function in 7–10.

7　$f(x)=\displaystyle\sum_{0}^{\infty}\left(\frac{x-1}{2}\right)^n$

8　$f(x)=\displaystyle\sum_{0}^{\infty}n(x-a)^{n-1}$

9　$f(x)=\displaystyle\sum_{0}^{\infty}\frac{1}{n+1}(x-a)^{n+1}$

10　$f(x)=(x-2\pi)-\dfrac{(x-2\pi)^3}{3!}+\cdots$

11　Write down the Taylor series for $(e^x-1)/x$, based on the series for e^x. At $x=0$ the function is $0/0$. Evaluate the series at $x=0$. Check by l'Hôpital's Rule on $(e^x-1)/x$.

12　Write down the Taylor series for xe^x around $x=0$. Integrate and substitute $x=1$ to find the sum of $1/n!(n+2)$.

13　If $f(x)$ is an even function, so $f(-x)=f(x)$, what can you say about its Taylor coefficients in $f=\Sigma a_n x^n$?

14　Puzzle out the sums of the following series:

(a)　$x+x^2-x^3+x^4+x^5-x^6+\cdots$

(b)　$1+\dfrac{x^4}{4!}+\dfrac{x^8}{8!}+\cdots$

(c)　$(x-1)-\frac{1}{2}(x-1)^2+\frac{1}{3}(x-1)^3-\cdots$

15　From the series for $(1-\cos x)/x^2$ find the limit as $x\to0$ faster than l'Hôpital's rule.

16　Construct a power series that converges for $0<x<2\pi$.

17–24 are about remainders and 25–36 are about binomials.

17　If the cosine series stops before $x^8/8!$ show from (2) that the remainder R_7 is less than $x^8/8!$. Does this also follow because the series is alternating ?

18　If the sine series around $x=2\pi$ stops after the terms in problem 10, estimate the remainder from equation (2).

19　Estimate by (2) the remainder $R_n=x^{n+1}+x^{n+2}+\cdots$ in the geometric series. Then compute R_n exactly and find the unknown point c for $n=2$ and $x=\frac{1}{2}$.

20　For $-\ln(1-x)=x+\frac{1}{2}x^2+\frac{1}{3}x^3+R_3$, use equation (2) to show that $R_3\leqslant\frac{1}{8}$ at $x=\frac{1}{2}$.

21　Find R_n in Problem 20 and show that the series converges to the function at $x=\frac{1}{2}$ (prove that $R_n\to0$).

22　By estimating R_n prove that the Taylor series for e^x around $x=1$ converges to e^x as $n\to\infty$.

23　(Proof of the remainder formula when $n=2$)

(a)　At $x=a$ find R_2,R_2',R_2'',R_2'''.

(b)　At $x=a$ evaluate $g(x)=(x-a)^3$ and g',g'',g'''.

(c)　What rule gives $\dfrac{R_2(x)-R_2(a)}{g(x)-g(a)}=\dfrac{R_2'(c_1)}{g'(c_1)}$?

(d) In $\dfrac{R_2'(c_1) - R_2'(a)}{g'(c_1) - g'(a)} = \dfrac{R_2''(c_2)}{g''(c_2)}$ and

$\dfrac{R_2''(c_2) - R_2''(a)}{g''(c_2) - g''(a)} = \dfrac{R_2'''(c)}{g'''(c)}$ where are c_1 and c_2 and c ?

(e) Combine (a-b-c-d) into the remainder formula (2).

24 All derivatives of $f(x) = e^{-1/x^2}$ are zero at $x = 0$, including $f(0) = 0$. What is $f(.1)$? What is the Taylor series around $x = 0$? What is the radius of convergence? Where does the series converge to $f(x)$? For $x = 1$ and $n = 1$ what is the remainder estimate in (2)?

25 (a) Find the first three terms in the binomial series for $1/\sqrt{1 - x^2}$.

(b) Integrate to find the first three terms in the Taylor series for $\sin^{-1} x$.

26 Show that the binomial coefficients in $1/\sqrt{1 - x} = \sum a_n x^n$ are $a_n = 1 \cdot 3 \cdot 5 \cdots (2n - 1)/2^n n!$

27 For $p = -1$ and $p = -2$ find nice formulas for the binomial coefficients.

28 Change the dummy variable and add lower limits to make $\sum^\infty n x^{n-1} = \sum^\infty (n + 1) x^n$.

29 In $(1 - x)^{-1} = \sum x^n$ the coefficient of x^n is the number of groups of n friends that can be formed from 1 friend (not binomial—repetition is allowed!). The coefficient is 1 and there is only one group—the same friend n times.

(a) Describe all groups of n friends that can be formed from 2 friends. (There are $n + 1$ groups.)

(b) How many groups of 5 friends can be formed from 3 friends?

30 (a) What is the coefficient of x^n when $1 + x + x^2 + \cdots$ multiplies $1 + x + x^2 + \cdots$? Write the first three terms.

(b) What is the coefficient of x^5 in $(\sum x^k)^3$?

31 Show that the binomial series for $\sqrt{1 + 4x}$ has integer coefficients. (Note that x^n changes to $(4x)^n$. These coefficients are important in counting trees, paths, parentheses...)

32 In the series for $1/\sqrt{1 + 4x}$, show that the coefficient of x^n is $(2n)!$ divided by $(n!)^2$.

Use the binomial series to compute 33–36 with error less than 1/1000.

33 $(15)^{1/4}$

34 $(1001)^{1/3}$

35 $(1.1)^{1.1}$

36 $e^{1/1000}$

37 From $\sec x = 1/[1 - (1 - \cos x)]$ find the Taylor series of $\sec x$ up to x^6. What is the radius of convergence r (distance to blowup point)?

38 From $\sec^2 x = 1/[1 - \sin^2 x]$ find the Taylor series up to x^2. Check by squaring the secant series in Problem 37. Check by differentiating the tangent series in Problem 39.

39 (Division of series) Find $\tan x$ by long division of $\sin x / \cos x$:

$$\left(x - \frac{x^3}{6} + \frac{x^5}{120} \cdots \right) \bigg/ \left(1 - \frac{x^2}{2} + \frac{x^4}{24} \cdots \right) = x + \frac{x^3}{3} + \frac{2x^5}{15} + \cdots .$$

40 (Composition of series) If $f = a_0 + a_1 x + a_2 x^2 + \cdots$ and $g = b_1 x + b_2 x^2 + \cdots$ find the $1, x, x^2$ coefficients of $f(g(x))$. Test on $f = 1/(1 + x), g = x/(1 - x)$, with $f(g(x)) = 1 - x$.

41 (Multiplication of series) From the series for $\sin x$ and $1/(1 - x)$ find the first four terms for $f = \sin x/(1 - x)$.

42 (Inversion of series) If $f = a_1 x + a_2 x^2 + \cdots$ find coefficients b_1, b_2 in $g = b_1 x + b_2 x^2 + \cdots$ so that $f(g(x)) = x$. Compute b_1, b_2 for $f = e^x - 1, g = f^{-1} = \ln(1 + x)$.

43 From the multiplication $(\sin x)(\sin x)$ or the derivatives of $f(x) = \sin^2 x$ find the first three terms of the series. Find the first four terms for $\cos^2 x$ by an easy trick.

44 Somehow find the first six nonzero terms for $f = (1 - x)/(1 - x^3)$.

45 Find four terms of the series for $1/\sqrt{1 - x}$. Then square the series to reach a geometric series.

46 Compute $\int_0^1 e^{-x^2}\, dx$ to 3 decimals by integrating the power series.

47 Compute $\int_0^1 \sin^2 t\, dt$ to 4 decimals by power series.

48 Show that $\sum x^n / n$ converges at $x = -1$, even though its derivative $\sum x^{n-1}$ diverges. How can they have the same convergence radius?

49 Compute $\lim_{x \to 0} (\sin x - \tan x)/x^3$ from the series.

50 If the nth root of a_n approaches $L > 0$, explain why $\sum a_n x^n$ has convergence radius $r = 1/L$.

51 Find the convergence radius r around basepoints $a = 0$ and $a = 1$ from the blowup points of $(1 + \tan x)/(1 + x^2)$.

CHAPTER 11

Vectors and Matrices

This chapter opens up a new part of calculus. It is *multidimensional calculus*, because the subject moves into more dimensions. In the first ten chapters, all functions depended on time t or position x—but not both. We had $f(t)$ or $y(x)$. The graphs were curves in a plane. There was one independent variable (x or t) and one dependent variable (y or f). Now we meet functions $f(x,t)$ that depend on both x and t. Their graphs are *surfaces* instead of curves. This brings us to the *calculus of several variables*.

Start with the surface that represents the function $f(x,t)$ or $f(x,y)$ or $f(x,y,t)$. I emphasize functions, because that is what calculus is about.

EXAMPLE 1 $f(x,t) = \cos(x-t)$ is a traveling wave (cosine curve in motion).

At $t = 0$ the curve is $f = \cos x$. At a later time, the curve moves to the right (Figure 11.1). At each t we get a cross-section of the whole x-t surface. For a wave traveling along a string, the height depends on position as well as time.

A similar function gives a wave going around a stadium. Each person stands up and sits down. Somehow the wave travels.

EXAMPLE 2 $f(x,y) = 3x + y + 1$ is a sloping roof (fixed in time).

The surface is two-dimensional—you can walk around on it. It is flat because $3x + y + 1$ is a linear function. In the y direction the surface goes up at $45°$. If y increases by 1, so does f. That slope is 1. In the x direction the roof is steeper (slope 3). There is a direction in between where the roof is steepest (slope $\sqrt{10}$).

EXAMPLE 3 $f(x,y,t) = \cos(x-y-t)$ is an ocean surface with traveling waves.

This surface moves. At each time t we have a new x-y surface. There are three variables, x and y for position and t for time. I can't draw the function, it needs four dimensions! The base coordinates are x, y, t and the height is f. The alternative is a movie that shows the x-y surface changing with t.

At time $t = 0$ the ocean surface is given by $\cos(x-y)$. The waves are in straight lines. The line $x - y = 0$ follows a crest because $\cos 0 = 1$. The top of the next wave is on the parallel line $x - y = 2\pi$, because $\cos 2\pi = 1$. Figure 11.1 shows the ocean surface at a fixed time.

The line $x - y = t$ gives the crest at time t. The water goes up and down (like people in a stadium). ***The wave goes to shore, but the water stays in the ocean.***

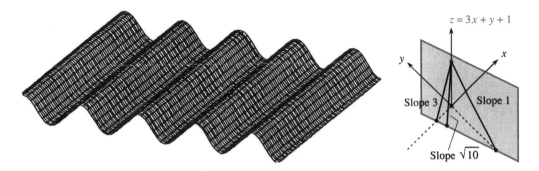

Fig. 11.1 Moving cosine with a small optical illusion—the darker bands seem to go from top to bottom as you turn.

Fig. 11.2 Linear functions give planes.

Of course multidimensional calculus is not only for waves. In business, demand is a function of price and date. In engineering, the velocity and temperature depend on position x and time t. Biology deals with many variables at once (and statistics is always looking for linear relations like $z = x + 2y$). A serious job lies ahead, to carry derivatives and integrals into more dimensions.

11.1 Vectors and Dot Products

In a plane, every point is described by two numbers. We measure across by x and up by y. Starting from the origin we reach the point with coordinates (x, y). I want to describe this movement by a ***vector***—the straight line that starts at $(0, 0)$ and ends at (x, y). This vector **v** has a ***direction***, which goes from $(0, 0)$ to (x, y) and not the other way.

In a picture, the vector is shown by an arrow. In algebra, **v** is given by its two components. For a *column vector*, write x above y:

$$\mathbf{v} = \begin{bmatrix} x \\ y \end{bmatrix} \quad (x \text{ and } y \text{ are the components of } \mathbf{v}). \tag{1}$$

Note that **v** is printed in boldface; its components x and y are in lightface.† The vector $-\mathbf{v}$ in the opposite direction changes signs. Adding **v** to $-\mathbf{v}$ gives the ***zero vector*** (different from the zero number and also in boldface):

$$-\mathbf{v} = \begin{bmatrix} -x \\ -y \end{bmatrix} \quad \text{and} \quad \mathbf{v} - \mathbf{v} = \begin{bmatrix} x - x \\ y - y \end{bmatrix} = \begin{bmatrix} 0 \\ 0 \end{bmatrix} = \mathbf{0}. \tag{2}$$

Notice how vector addition or subtraction is done separately on the x's and y's:

$$\mathbf{v} + \mathbf{w} = \begin{bmatrix} 3 \\ 1 \end{bmatrix} + \begin{bmatrix} -1 \\ 2 \end{bmatrix} = \begin{bmatrix} 2 \\ 3 \end{bmatrix}. \tag{3}$$

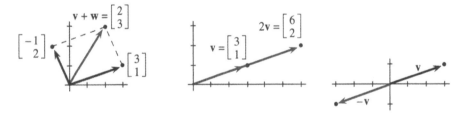

Fig. 11.3   Parallelogram for $\mathbf{v} + \mathbf{w}$, stretching for $2\mathbf{v}$, signs reversed for $-\mathbf{v}$.

The vector **v** has components $v_1 = 3$ and $v_2 = 1$. (I write v_1 for the first component and v_2 for the second component. I also write x and y, which is fine for two components.) The vector **w** has $w_1 = -1$ and $w_2 = 2$. To add the vectors, add the components. ***To draw this addition, place the start of* w *at the end of* v.** Figure 11.3 shows how **w** starts where **v** ends.

VECTORS WITHOUT COORDINATES

In that *head-to-tail addition* of $\mathbf{v} + \mathbf{w}$, we did something new. The vector **w** was moved away from the origin. Its length and direction were not changed! The new arrow is parallel to the old arrow—only the starting point is different. *The vector is the same as before.*

A vector can be defined without an origin and without x and y axes. The purpose of axes is to give the components—the separate distances x and y. Those numbers

---

† Another way to indicate a vector is $\vec{v}$. You will recognize vectors without needing arrows.

are necessary for calculations. But x and y coordinates are not necessary for head-to-tail addition $\mathbf{v}+\mathbf{w}$, or for stretching to $2\mathbf{v}$, or for linear combinations $2\mathbf{v}+3\mathbf{w}$. Some applications depend on coordinates, others don't.

Generally speaking, physics works without axes—it is "coordinate-free." A *velocity* has direction and magnitude, but it is not tied to a point. A *force* also has direction and magnitude, but it can act anywhere—not only at the origin. In contrast, a vector that gives the prices of five stocks is not floating in space. Each component has a meaning—there are five axes, and we know when prices are zero. After examples from geometry and physics (no axes), we return to vectors *with* coordinates.

EXAMPLE 1   (Geometry) Take any four-sided figure in space. Connect the midpoints of the four straight sides. ***Remarkable fact***: ***Those four midpoints lie in the same plane***. More than that, they form a *parallelogram*.

Frankly, this is amazing. Figure 11.4a cannot do justice to the problem, because it is printed on a flat page. Imagine the vectors $\mathbf{A}$ and $\mathbf{D}$ coming upward. $\mathbf{B}$ and $\mathbf{C}$ go down at different angles. Notice how easily we indicate the four sides as vectors, not caring about axes or origin.

I will prove that $\mathbf{V}=\mathbf{W}$. That shows that the midpoints form a parallelogram.

What is $\mathbf{V}$? It starts halfway along $\mathbf{A}$ and ends halfway along $\mathbf{B}$. The small triangle at the bottom shows $\mathbf{V}=\frac{1}{2}\mathbf{A}+\frac{1}{2}\mathbf{B}$. This is vector addition—the tail of $\frac{1}{2}\mathbf{B}$ is at the head of $\frac{1}{2}\mathbf{A}$. Together they equal the shortcut $\mathbf{V}$. For the same reason $\mathbf{W}=\frac{1}{2}\mathbf{C}+\frac{1}{2}\mathbf{D}$. The heart of the proof is to see these relationships.

One step is left. Why is $\frac{1}{2}\mathbf{A}+\frac{1}{2}\mathbf{B}$ equal to $\frac{1}{2}\mathbf{C}+\frac{1}{2}\mathbf{D}$? In other words, why is $\mathbf{A}+\mathbf{B}$ equal to $\mathbf{C}+\mathbf{D}$? (I multiplied by 2.) When the right question is asked, the answer jumps out. A head-to-tail addition $\mathbf{A}+\mathbf{B}$ brings us to the point R. Also $\mathbf{C}+\mathbf{D}$ ***brings us to*** R. The proof comes down to one line:

$$\mathbf{A}+\mathbf{B}=PR=\mathbf{C}+\mathbf{D}. \text{ Then } \mathbf{V}=\tfrac{1}{2}\mathbf{A}+\tfrac{1}{2}\mathbf{B} \text{ equals } \mathbf{W}=\tfrac{1}{2}\mathbf{C}+\tfrac{1}{2}\mathbf{D}.$$

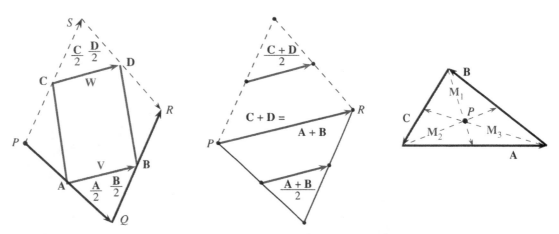

Fig. 11.4   Four midpoints form a parallelogram ($\mathbf{V}=\mathbf{W}$). Three medians meet at P.

EXAMPLE 2   (Also geometry) In any triangle, draw lines from the corners to the midpoints of the opposite sides. To prove by vectors: ***Those three lines meet at a point***. Problem 38 finds the meeting point in Figure 11.4c. Problem 37 says that ***the three vectors add to zero***.

EXAMPLE 3   (Medicine) An electrocardiogram shows the sum of many small vectors, the voltages in the wall of the heart. What happens to this sum—the *heart vector* **V**—in two cases that a cardiologist is watching for?

> *Case* 1. Part of the heart is dead (*infarction*).
> *Case* 2. Part of the heart is abnormally thick (*hypertrophy*).

A heart attack kills part of the muscle. A defective valve, or hypertension, overworks it. In **case 1** the cells die from the cutoff of blood (loss of oxygen). In **case 2** the heart wall can triple in size, from excess pressure. The causes can be chemical or mechanical. The effect we see is electrical.

    The machine is adding small vectors and "projecting" them in twelve directions. The leads on the arms, left leg, and chest give twelve directions in the body. Each graph shows the component of **V** in one of those directions. Three of the projections— two in the vertical plane, plus lead 2 for front-back—produce the "mean QRS vector" in Figure 11.5. That is the sum **V** when the ventricles start to contract. The left ventricle is larger, so the heart vector normally points down and to the left.

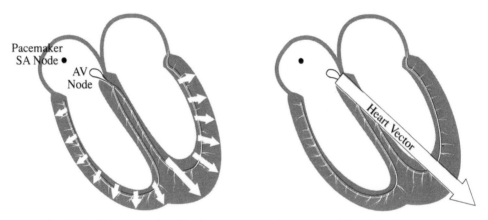

Fig. 11.5   **V** is a sum of small voltage vectors, at the moment of depolarization.

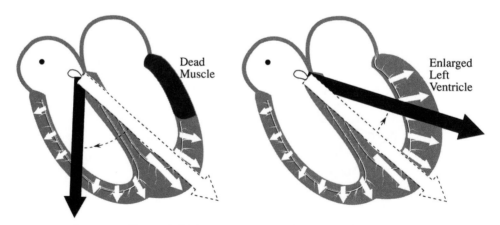

Fig. 11.6   Changes in **V** show dead muscle and overworked muscle.

We come soon to projections, but here the question is about **V** itself. How does the ECG identify the problem?

Case 1: *Heart attack* The dead cells make no contribution to the electrical potential. Some small vectors are missing. Therefore the sum **V** turns *away* from the infarcted part.

Case 2: *Hypertrophy* The overwork increases the contribution to the potential. Some vectors are larger than normal. Therefore **V** turns *toward* the thickened part.

When **V** points in an abnormal direction, the ECG graphs locate the problem. The P, Q, R, S, T waves on separate graphs can all indicate hypertrophy, in different regions of the heart. Infarctions generally occur in the left ventricle, which needs the greatest blood supply. When the supply of oxygen is cut back, that ventricle feels it first. The result can be a heart attack (= myocardial infarction = coronary occlusion). Section 11.2 shows how the projections on the ECG point to the location.

First come the basic facts about vectors—components, lengths, and dot products.

COORDINATE VECTORS AND LENGTH

To compute with vectors we need axes and coordinates. The picture of the heart is "coordinate-free," but calculations require numbers. A vector is known by its components. *The unit vectors along the axes are* **i** *and* **j** *in the plane and* **i**, **j**, **k** *in space*:

$$\textbf{in 2D:} \quad \mathbf{i} = \begin{bmatrix} 1 \\ 0 \end{bmatrix}, \mathbf{j} = \begin{bmatrix} 0 \\ 1 \end{bmatrix} \qquad \textbf{in 3D:} \quad \mathbf{i} = \begin{bmatrix} 1 \\ 0 \\ 0 \end{bmatrix}, \mathbf{j} = \begin{bmatrix} 0 \\ 1 \\ 0 \end{bmatrix}, \mathbf{k} = \begin{bmatrix} 0 \\ 1 \\ 0 \end{bmatrix}.$$

Notice how easily we moved into three dimensions! The only change is that vectors have three components. The combinations of **i** and **j** (or **i**, **j**, **k**) produce all vectors **v** in the plane (and all vectors **V** in space):

$$\mathbf{v} = 3\mathbf{i} + \mathbf{j} = \begin{bmatrix} 3 \\ 1 \end{bmatrix} \qquad \mathbf{V} = \mathbf{i} + 2\mathbf{j} - 2\mathbf{k} = \begin{bmatrix} 1 \\ 2 \\ -2 \end{bmatrix}.$$

Those vectors are also written $\mathbf{v} = (3, 1)$ *and* $\mathbf{V} = (1, 2, -2)$. The components of the vector are also the coordinates of a point. (The vector goes from the origin to the point.) This relation between point and vector is so close that we allow them the same notation: $P = (x, y, z)$ and $\mathbf{v} = (x, y, z) = x\mathbf{i} + y\mathbf{j} + z\mathbf{k}$.

The sum $\mathbf{v} + \mathbf{V}$ is totally meaningless. Those vectors live in different dimensions.

From the components we find the *length*. The length of $(3, 1)$ is $\sqrt{3^2 + 1^2} = \sqrt{10}$. This comes directly from a right triangle. In three dimensions, **V** has a third component to be squared and added. The length of $\mathbf{V} = (x, y, z)$ is $|\mathbf{V}| = \sqrt{x^2 + y^2 + z^2}$.

Vertical bars indicate length, which takes the place of absolute value. The length of $\mathbf{v} = 3\mathbf{i} + \mathbf{j}$ is the distance from the point $(0, 0)$ to the point $(3, 1)$:

$$|\mathbf{v}| = \sqrt{v_1^2 + v_2^2} = \sqrt{10} \qquad |\mathbf{V}| = \sqrt{1^2 + 2^2 + (-2)^2} = 3.$$

A unit vector is a vector of length one. Dividing **v** and **V** by their lengths produces unit vectors in the same directions:

$$\frac{\mathbf{v}}{|\mathbf{v}|} = \begin{bmatrix} 3/\sqrt{10} \\ 1/\sqrt{10} \end{bmatrix} \quad \text{and} \quad \frac{\mathbf{V}}{|\mathbf{V}|} = \begin{bmatrix} 1/3 \\ 2/3 \\ -2/3 \end{bmatrix} \quad \text{are unit vectors.}$$

11A Each nonzero vector has a positive length $|\mathbf{v}|$. The direction of $\mathbf{v}$ is given by a unit vector $\mathbf{u} = \mathbf{v}/|\mathbf{v}|$. The length times direction equals $\mathbf{v}$.

A unit vector in the plane is determined by its angle θ with the x axis:

$$\mathbf{u} = \begin{bmatrix} \cos\theta \\ \sin\theta \end{bmatrix} = (\cos\theta)\mathbf{i} + (\sin\theta)\mathbf{j} \text{ is a unit vector}: |\mathbf{u}|^2 = \cos^2\theta + \sin^2\theta = 1.$$

In 3-space the components of a unit vector are its "direction cosines":

$$\mathbf{U} = (\cos\alpha)\mathbf{i} + (\cos\beta)\mathbf{j} + (\cos\gamma)\mathbf{k}: \quad \alpha, \beta, \gamma = \text{angles with } x, y, z \text{ axes.}$$

Then $\cos^2\alpha + \cos^2\beta + \cos^2\gamma = 1$. We are doing algebra with numbers while we are doing geometry with vectors. It was the great contribution of Descartes to see how to study algebra and geometry at the same time.

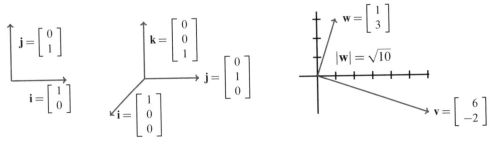

Fig. 11.7   Coordinate vectors $\mathbf{i}, \mathbf{j}, \mathbf{k}$. Perpendicular vectors $\mathbf{v}\cdot\mathbf{w} = (6)(1) + (-2)(3) = 0$.

THE DOT PRODUCT OF TWO VECTORS

There are two basic operations on vectors. First, vectors are added $(\mathbf{v} + \mathbf{w})$. Second, a vector is multiplied by a scalar ($7\mathbf{v}$ or $-2\mathbf{w}$). That leaves a natural question—how do you multiply two vectors ? The main part of the answer is—you don't. But there is an extremely important operation that begins with two vectors and produces a number. It is usually indicated by a dot between the vectors, as in $\mathbf{v}\cdot\mathbf{w}$, so it is called the **dot product**.

DEFINITION 1   *The dot product multiplies the lengths $|\mathbf{v}|$ times $|\mathbf{w}|$ times a cosine*:

$$\mathbf{v}\cdot\mathbf{w} = |\mathbf{v}||\mathbf{w}|\cos\theta, \qquad \theta = \text{angle between } \mathbf{v} \text{ and } \mathbf{w}.$$

EXAMPLE $\begin{bmatrix} 3 \\ 0 \end{bmatrix}$ has length 3, $\begin{bmatrix} 2 \\ 2 \end{bmatrix}$ has length $\sqrt{8}$, the angle is $45°$.

The dot product is $|\mathbf{v}||\mathbf{w}|\cos\theta = (3)(\sqrt{8})(1/\sqrt{2})$, which simplifies to 6. The square roots in the lengths are "canceled" by square roots in the cosine. For computing $\mathbf{v}\cdot\mathbf{w}$, a second and much simpler way involves no square roots in the first place.

DEFINITION 2   *The dot product* **v** · **w** *multiplies component by component and adds*:

$$\mathbf{v} \cdot \mathbf{w} = v_1 w_1 + v_2 w_2 \qquad \begin{bmatrix} 3 \\ 0 \end{bmatrix} \cdot \begin{bmatrix} 2 \\ 2 \end{bmatrix} = (3)(2) + (0)(2) = 6.$$

The first form $|\mathbf{v}||\mathbf{w}|\cos\theta$ is coordinate-free. The second form $v_1 w_1 + v_2 w_2$ computes with coordinates. Remark 4 explains why these two forms are equal.

11B The **dot product** or **scalar product** or **inner product** of three-dimensional vectors is

$$\mathbf{V} \cdot \mathbf{W} = |\mathbf{V}||\mathbf{W}|\cos\theta = V_1 W_1 + V_2 W_2 + V_3 W_3. \tag{4}$$

If the vectors are perpendicular then $\theta = 90°$ and $\cos\theta = 0$ and $\mathbf{V} \cdot \mathbf{W} = 0$.

$$\begin{bmatrix} 1 \\ 2 \\ 3 \end{bmatrix} \cdot \begin{bmatrix} 4 \\ 5 \\ 6 \end{bmatrix} = 32 \, (not \ perpendicular) \qquad \begin{bmatrix} 2 \\ 2 \\ -1 \end{bmatrix} \cdot \begin{bmatrix} -1 \\ 2 \\ 2 \end{bmatrix} = 0 \, (perpendicular).$$

These dot products 32 and 0 equal $|\mathbf{V}||\mathbf{W}|\cos\theta$. In the second one, $\cos\theta$ must be zero. The angle is $\pi/2$ or $-\pi/2$—in either case a right angle. Fortunately the cosine is the same for θ and $-\theta$, so we need not decide the sign of θ.

Remark 1   When $\mathbf{V} = \mathbf{W}$ the angle is zero but not the cosine! In this case $\cos\theta = 1$ and $\mathbf{V} \cdot \mathbf{V} = |\mathbf{V}|^2$. **The dot product of** $\mathbf{V}$ **with itself is the length squared**:

$$\mathbf{V} \cdot \mathbf{V} = (V_1, V_2, V_3) \cdot (V_1, V_2, V_3) = V_1^2 + V_2^2 + V_3^2 = |\mathbf{V}|^2. \tag{5}$$

Remark 2   The dot product of $\mathbf{i} = (1,0,0)$ with $\mathbf{j} = (0,1,0)$ is $\mathbf{i} \cdot \mathbf{j} = 0$. The axes are perpendicular. Similarly $\mathbf{i} \cdot \mathbf{k} = 0$ and $\mathbf{j} \cdot \mathbf{k} = 0$. Those are unit vectors: $\mathbf{i} \cdot \mathbf{i} = \mathbf{j} \cdot \mathbf{j} = \mathbf{k} \cdot \mathbf{k} = 1$.

Remark 3   The dot product has three properties that keep the algebra simple:

1. $\mathbf{V} \cdot \mathbf{W} = \mathbf{W} \cdot \mathbf{V}$ 2. $(c\mathbf{V}) \cdot \mathbf{W} = c(\mathbf{V} \cdot \mathbf{W})$ 3. $(\mathbf{U} + \mathbf{V}) \cdot \mathbf{W} = \mathbf{U} \cdot \mathbf{W} + \mathbf{V} \cdot \mathbf{W}$

When $\mathbf{V}$ is doubled ($c = 2$) the dot product is doubled. When $\mathbf{V}$ is split into $\mathbf{i}, \mathbf{j}, \mathbf{k}$ components, the dot product splits in three pieces. The same applies to $\mathbf{W}$, since $\mathbf{V} \cdot \mathbf{W} = \mathbf{W} \cdot \mathbf{V}$. The nine dot products of $\mathbf{i}, \mathbf{j}, \mathbf{k}$ are zeros and ones, and a giant splitting of both $\mathbf{V}$ and $\mathbf{W}$ gives back the correct $\mathbf{V} \cdot \mathbf{W}$:

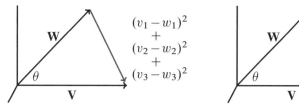

Fig. 11.8   Length squared $= (\mathbf{V} - \mathbf{W}) \cdot (\mathbf{V} - \mathbf{W})$, from coordinates and the cosine law.

$$\mathbf{V} \cdot \mathbf{W} = V_1 \mathbf{i} \cdot W_1 \mathbf{i} + V_2 \mathbf{j} \cdot W_2 \mathbf{j} + V_3 \mathbf{k} \cdot W_3 \mathbf{k} + \text{six zeroes} = V_1 W_1 + V_2 W_2 + V_3 W_3.$$

Remark 4   *The two forms of the dot product are equal*. This comes from computing $|\mathbf{V} - \mathbf{W}|^2$ by coordinates and also by the "law of cosines":

$$\text{with coordinates}: |\mathbf{V} - \mathbf{W}|^2 = (V_1 - W_1)^2 + (V_2 - W_2)^2 + (V_3 - W_3)^2$$
$$\text{from cosine law}: |\mathbf{V} - \mathbf{W}|^2 = |\mathbf{V}|^2 + |\mathbf{W}|^2 - 2|\mathbf{V}||\mathbf{W}| \cos \theta.$$

Compare those two lines. Line 1 contains V_1^2 and V_2^2 and V_3^2. Their sum matches $|\mathbf{V}|^2$ in the cosine law. Also $W_1^2 + W_2^2 + W_3^2$ matches $|\mathbf{W}|^2$. Therefore the terms containing -2 are the same (you can mentally cancel the -2). *The definitions agree*:

$$-2(V_1 W_1 + V_2 W_2 + V_3 W_3) \ \ equals \ -2|\mathbf{V}||\mathbf{W}| \cos \theta \ \ equals \ -2\mathbf{V} \cdot \mathbf{W}.$$

The cosine law is coordinate-free. It applies to all triangles (even in n dimensions). Its vector form in Figure 11.8 is $|\mathbf{V} - \mathbf{W}|^2 = |\mathbf{V}|^2 - 2\mathbf{V} \cdot \mathbf{W} + |\mathbf{W}|^2$. This application to $\mathbf{V} \cdot \mathbf{W}$ is its brief moment of glory.

Remark 5   The dot product is the best way to compute the cosine of θ:

$$\cos \theta = \frac{\mathbf{V} \cdot \mathbf{W}}{|\mathbf{V}||\mathbf{W}|}. \tag{6}$$

Here are examples of $\mathbf{V}$ and $\mathbf{W}$ with a range of angles from 0 to π:

i and 3**i** have the same direction	$\cos \theta = 1$	$\theta = 0$
$\mathbf{i} \cdot (\mathbf{i} + \mathbf{j}) = 1$ is positive	$\cos \theta = 1/\sqrt{2}$	$\theta = \pi/4$
i and **j** are perpendicular: $\mathbf{i} \cdot \mathbf{j} = 0$	$\cos \theta = 0$	$\theta = \pi/2$
$\mathbf{i} \cdot (-\mathbf{i} + \mathbf{j}) = -1$ is negative	$\cos \theta = -1/\sqrt{2}$	$\theta = 3\pi/4$
i and $-3\mathbf{i}$ have opposite directions	$\cos \theta = -1$	$\theta = \pi$

Remark 6   *The Cauchy-Schwarz inequality* $|\mathbf{V} \cdot \mathbf{W}| \leqslant |\mathbf{V}||\mathbf{W}|$ *comes from* $|\cos \theta| \leqslant 1$.

The left side is $|\mathbf{V}||\mathbf{W}|| \cos \theta|$. It never exceeds the right side $|\mathbf{V}||\mathbf{W}|$. This is a key inequality in mathematics, from which so many others follow:

Geometric mean $\sqrt{xy} \leqslant$ *arithmetic mean* $\frac{1}{2}(x + y)$ (true for any $x \geqslant 0$ and $y \geqslant 0$).

Triangle inequality $|\mathbf{V} + \mathbf{W}| \leqslant |\mathbf{V}| + |\mathbf{W}|$ ($|\mathbf{V}|, |\mathbf{W}|, |\mathbf{V} + \mathbf{W}|$ are lengths of sides).

These and other examples are in Problems 39 to 44. The Schwarz inequality $|\mathbf{V} \cdot \mathbf{W}| \leqslant |\mathbf{V}||\mathbf{W}|$ becomes an equality when $|\cos \theta| = 1$ and the vectors are _____ .

Read-through questions

A vector has length and ___a___ . If **v** has components 6 and -8, its length is $|\mathbf{v}| = $ ___b___ and its direction vector is $\mathbf{u} = $ ___c___ . The product of $|\mathbf{v}|$ with **u** is ___d___ . This vector goes from $(0,0)$ to the point $x = $ ___e___ , $y = $ ___f___ . A combination of the coordinate vectors $\mathbf{i} = $ ___g___ and $\mathbf{j} = $ ___h___ produces $\mathbf{v} = $ ___i___ $\mathbf{i} + $ ___j___ $\mathbf{j}$.

To add vectors we add their ___k___ . The sum of $(6,-8)$ and $(1,0)$ is ___l___ . To see $\mathbf{v}+\mathbf{i}$ geometrically, put the ___m___ of **i** at the ___n___ of **v**. The vectors form a ___o___ with diagonal $\mathbf{v}+\mathbf{i}$. (The other diagonal is ___p___ .) The vectors $2\mathbf{v}$ and $-\mathbf{v}$ are ___q___ and ___r___ . Their lengths are ___s___ and ___t___ .

In a space without axes and coordinates, the tail of **V** can be placed ___u___ . Two vectors with the same ___v___ are the same. If a triangle starts with **V** and continues with **W**, the third side is ___w___ . The vector connecting the midpoint of **V** to the midpoint of **W** is ___x___ . That vector is ___y___ the third side. In this coordinate-free form the dot product is $\mathbf{V}\cdot\mathbf{W} = $ ___z___ .

Using components, $\mathbf{V}\cdot\mathbf{W} = $ ___A___ and $(1,2,1)\cdot(2,-3,7) = $ ___B___ . The vectors are perpendicular if ___C___ . The vectors are parallel if ___D___ . $\mathbf{V}\cdot\mathbf{V}$ is the same as ___E___ . The dot product of $\mathbf{U}+\mathbf{V}$ with **W** equals ___F___ . The angle between **V** and **W** has $\cos\theta = $ ___G___ . When $\mathbf{V}\cdot\mathbf{W}$ is negative then θ is ___H___ . The angle between $\mathbf{i}+\mathbf{j}$ and $\mathbf{i}+\mathbf{k}$ is ___I___ . The Cauchy-Schwarz inequality is ___J___ , and for $\mathbf{V}=\mathbf{i}+\mathbf{j}$ and $\mathbf{W}=\mathbf{i}+\mathbf{k}$ it becomes $1 \leqslant$ ___K___ .

In 1−4 compute $\mathbf{V}+\mathbf{W}$ and $2\mathbf{V}-3\mathbf{W}$ and $|\mathbf{V}|^2$ and $\mathbf{V}\cdot\mathbf{W}$ and $\cos\theta$.

1 $\mathbf{V}=(1,1,1)$, $\mathbf{W}=(-1,-1,-1)$

2 $\mathbf{V}=\mathbf{i}+\mathbf{j}$, $\mathbf{W}=\mathbf{j}-\mathbf{k}$

3 $\mathbf{V}=\mathbf{i}-2\mathbf{j}+\mathbf{k}$, $\mathbf{W}=\mathbf{i}+\mathbf{j}-2\mathbf{k}$

4 $\mathbf{V}=(1,1,1,1)$, $\mathbf{W}=(1,2,3,4)$

5 (a) Find a vector that is perpendicular to (v_1, v_2).

(b) Find two vectors that are perpendicular to (v_1, v_2, v_3).

6 Find two vectors that are perpendicular to $(1,1,0)$ and to each other.

7 What vector is perpendicular to all 2-dimensional vectors? What vector is parallel to all 3-dimensional vectors?

8 In Problems $1-4$ construct unit vectors in the same direction as **V**.

9 If **v** and **w** are unit vectors, what is the geometrical meaning of $\mathbf{v}\cdot\mathbf{w}$? What is the geometrical meaning of $(\mathbf{v}\cdot\mathbf{w})\mathbf{v}$? Draw a figure with $\mathbf{v}=\mathbf{i}$ and $\mathbf{w}=(3/5)\mathbf{i}+(4/5)\mathbf{j}$.

10 Write down all unit vectors that make an angle θ with the vector $(1,0)$. Write down *all* vectors at that angle.

11 *True or false* in three dimensions:

1. If both **U** and **V** make a $30°$ angle with **W**, so does $\mathbf{U}+\mathbf{V}$.
2. If they make a $90°$ angle with **W**, so does $\mathbf{U}+\mathbf{V}$.
3. If they make a $90°$ angle with **W** they are perpendicular: $\mathbf{U}\cdot\mathbf{V}=0$.

12 From $\mathbf{W}=(1,2,3)$ subtract a multiple of $\mathbf{V}=(1,1,1)$ so that $\mathbf{W}-c\mathbf{V}$ is perpendicular to **V**. Draw **V** and **W** and $\mathbf{W}-c\mathbf{V}$.

13 (a) What is the sum **V** of the twelve vectors from the center of a clock to the hours?

(b) If the 4 o'clock vector is removed, find **V** for the other eleven vectors.

(c) If the vectors to $1,2,3$ are cut in half, find **V** for the twelve vectors.

14 (a) By removing one or more of the twelve clock vectors, make the length $|\mathbf{V}|$ as large as possible.

(b) Suppose the vectors start from the top instead of the center (the origin is moved to 12 o'clock, so $\mathbf{v}_{12}=\mathbf{0}$). What is the new sum $\mathbf{V}*$?

15 Find the angle POQ by vector methods if $P=(1,1,0)$, $O=(0,0,0)$, $Q=(1,2,-2)$.

16 (a) Draw the unit vectors $\mathbf{u}_1=(\cos\theta,\sin\theta)$ and $\mathbf{u}_2=(\cos\phi,\sin\phi)$. By dot products find the formula for $\cos(\theta-\phi)$.

(b) Draw the unit vector $\mathbf{u}_3$ from a $90°$ rotation of $\mathbf{u}_2$. By dot products find the formula for $\sin(\theta+\phi)$.

17 Describe all points (x,y) such that $\mathbf{v}=x\mathbf{i}+y\mathbf{j}$ satisfies

(a) $|\mathbf{v}|=2$ (b) $|\mathbf{v}-\mathbf{i}|=2$
(c) $\mathbf{v}\cdot\mathbf{i}=2$ (d) $\mathbf{v}\cdot\mathbf{i}=|\mathbf{v}|$

18 (Important) If **A** and **B** are non-parallel vectors from the origin, describe

(a) the endpoints of $t\mathbf{B}$ for all numbers t
(b) the endpoints of $\mathbf{A}+t\mathbf{B}$ for all t
(c) the endpoints of $s\mathbf{A}+t\mathbf{B}$ for all s and t
(d) the vectors **v** that satisfy $\mathbf{v}\cdot\mathbf{A}=\mathbf{v}\cdot\mathbf{B}$

19 (a) If $\mathbf{v}+2\mathbf{w}=\mathbf{i}$ and $2\mathbf{v}+3\mathbf{w}=\mathbf{j}$ find **v** and **w**.
(b) If $\mathbf{v}=\mathbf{i}+\mathbf{j}$ and $\mathbf{w}=3\mathbf{i}+4\mathbf{j}$ then $\mathbf{i}=$ _____ $\mathbf{v}+$ _____ $\mathbf{w}$.

20 If $P=(0,0)$ and $R=(0,1)$ choose Q so the angle PQR is $90°$. All possible Q's lie in a _____ .

21 (a) Choose d so that $\mathbf{A}=2\mathbf{i}+3\mathbf{j}$ is perpendicular to $\mathbf{B}=9\mathbf{i}+d\mathbf{j}$.

(b) Find a vector **C** perpendicular to $\mathbf{A}=\mathbf{i}+\mathbf{j}+\mathbf{k}$ and $\mathbf{B}=\mathbf{i}-\mathbf{k}$.

22 If a boat has velocity **V** with respect to the water and the water has velocity **W** with respect to the land, then _____ . The speed of the boat is not $|\mathbf{V}|+|\mathbf{W}|$ but _____ .

23 Find the angle between the diagonal of cube and (a) an edge (b) the diagonal of a face (c) another diagonal of the cube. Choose lines that meet.

24 Draw the triangle PQR in Example 1 (the four-sided figure in space). By geometry not vectors, show that PR is twice as long as $\mathbf{V}$. Similarly $|PR| = 2|\mathbf{W}|$. Also $\mathbf{V}$ is parallel to $\mathbf{W}$ because both are parallel to _____ . So $\mathbf{V} = \mathbf{W}$ as before.

25 (a) If $\mathbf{A}$ and $\mathbf{B}$ are unit vectors, show that they make equal angles with $\mathbf{A} + \mathbf{B}$.

(b) If $\mathbf{A}, \mathbf{B}, \mathbf{C}$ are unit vectors with $\mathbf{A} + \mathbf{B} + \mathbf{C} = 0$, they form a _____ triangle and the angle between any two is _____ .

26 (a) Find perpendicular unit vectors $\mathbf{I}$ and $\mathbf{J}$ in the plane that are different from $\mathbf{i}$ and $\mathbf{j}$.

(b) Find perpendicular unit vectors $\mathbf{I}, \mathbf{J}, \mathbf{K}$ different from $\mathbf{i}, \mathbf{j}, \mathbf{k}$.

27 If $\mathbf{I}$ and $\mathbf{J}$ are perpendicular, take their dot products with $\mathbf{A} = a\mathbf{I} + b\mathbf{J}$ to find a and b.

28 Suppose $\mathbf{I} = (\mathbf{i} + \mathbf{j})/\sqrt{2}$ and $\mathbf{J} = (\mathbf{i} - \mathbf{j})/\sqrt{2}$. Check $\mathbf{I} \cdot \mathbf{J} = 0$ and write $\mathbf{A} = 2\mathbf{i} + 3\mathbf{j}$ as a combination $a\mathbf{I} + b\mathbf{J}$. (Best method: use a and b from Problem 27. Alternative: Find $\mathbf{i}$ and $\mathbf{j}$ from $\mathbf{I}$ and $\mathbf{J}$ and substitute into $\mathbf{A}$.)

29 (a) Find the position vector OP and the velocity vector PQ when the point P moves around the unit circle (see figure) with speed 1. (b) Change to speed 2.

30 The sum $(\mathbf{A} \cdot \mathbf{i})^2 + (\mathbf{A} \cdot \mathbf{j})^2 + (\mathbf{A} \cdot \mathbf{k}^2)$ equals _____ .

31 In the semicircle find $\mathbf{C}$ and $\mathbf{D}$ in terms of $\mathbf{A}$ and $\mathbf{B}$. Prove that $\mathbf{C} \cdot \mathbf{D} = 0$ (they meet at right angles).

35 The vector from the earth's center to Seattle is $a\mathbf{i} + b\mathbf{j} + c\mathbf{k}$.

(a) Along the circle at the latitude of Seattle, what two functions of a, b, c stay constant ? $\mathbf{k}$ goes to the North Pole.

(b) On the circle at the longitude of Seattle—the meridian—what two functions of a, b, c stay constant ?

(c) Extra credit: Estimate a, b, c in your present position. The $0°$ meridian through Greenwich has $b = 0$.

36 If $|\mathbf{A} + \mathbf{B}|^2 = |\mathbf{A}|^2 + |\mathbf{B}|^2$, prove that $\mathbf{A}$ is perpendicular to $\mathbf{B}$.

37 In Figure 11.4, the medians go from the corners to the midpoints of the opposite sides. Express $\mathbf{M}_1, \mathbf{M}_2, \mathbf{M}_3$ in terms of $\mathbf{A}, \mathbf{B}, \mathbf{C}$. Prove that $\mathbf{M}_1 + \mathbf{M}_2 + \mathbf{M}_3 = 0$. What relation holds between $\mathbf{A}, \mathbf{B}, \mathbf{C}$?

38 The point $\frac{2}{3}$ of the way along is the same for all three medians. This means that $\mathbf{A} + \frac{2}{3}\mathbf{M}_3 = \frac{2}{3}\mathbf{M}_2 = $ _____ . Prove that those three vectors are equal.

39 (a) Verify the Schwarz inequality $|\mathbf{V} \cdot \mathbf{W}| \leqslant |\mathbf{V}||\mathbf{W}|$ for $\mathbf{V} = \mathbf{i} + 2\mathbf{j} + 2\mathbf{k}$ and $\mathbf{W} = 2\mathbf{i} + 2\mathbf{j} + \mathbf{k}$.

(b) What does the inequality become when $\mathbf{V} = (\sqrt{x}, \sqrt{y})$ and $\mathbf{W} = (\sqrt{y}, \sqrt{x})$?

40 By choosing the right vector $\mathbf{W}$ in the Schwarz inequality, show that $(V_1 + V_2 + V_3)^2 \leqslant 3(V_1^2 + V_2^2 + V_3^2)$. What is $\mathbf{W}$?

41 The Schwarz inequality for $a\mathbf{i} + b\mathbf{j}$ and $c\mathbf{i} + d\mathbf{j}$ says that $(a^2 + b^2)(c^2 + d^2) \geqslant (ac + bd)^2$. Multiply out to show that the difference is $\geqslant 0$.

42 The vectors $\mathbf{A}, \mathbf{B}, \mathbf{C}$ form a triangle if $\mathbf{A} + \mathbf{B} + \mathbf{C} = \mathbf{0}$. The **triangle inequality** $|\mathbf{A} + \mathbf{B}| \leqslant |\mathbf{A}| + |\mathbf{B}|$ says that any one side length is less than _____ . The proof comes from Schwarz:

$$|\mathbf{A} + \mathbf{B}|^2 = \mathbf{A} \cdot \mathbf{A} + 2\mathbf{A} \cdot \mathbf{B} + \mathbf{B} \cdot \mathbf{B}$$
$$\leqslant |\mathbf{A}|^2 + \text{_____} + |\mathbf{B}|^2 = (|\mathbf{A}| + |\mathbf{B}|^2).$$

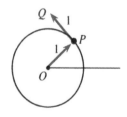
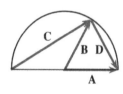
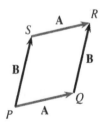

32 The diagonal PR has $|PR|^2 = (\mathbf{A} + \mathbf{B}) \cdot (\mathbf{A} + \mathbf{B}) = \mathbf{A} \cdot \mathbf{A} + \mathbf{A} \cdot \mathbf{B} + \mathbf{B} \cdot \mathbf{A} + \mathbf{B} \cdot \mathbf{B}$. Add $|QS|^2$ from the other diagonal to prove the parallelogram law: $|PR|^2 + |QS|^2 = $ sum of squares of the four side lengths.

33 If $(1,2,3), (3,4,7)$, and $(2,1,2)$ are corners of a parallelogram, find all possible fourth corners.

34 The diagonals of the parallelogram are $\mathbf{A} + \mathbf{B}$ and _____ . If they have the same length, prove that $\mathbf{A} \cdot \mathbf{B} = 0$ and the region is a _____ .

43 **True or false,** with reason or example:

(a) $|\mathbf{V} + \mathbf{W}|^2$ is never larger than $|\mathbf{V}|^2 + |\mathbf{W}|^2$

(b) In a real triangle $|\mathbf{V} + \mathbf{W}|$ never equals $|\mathbf{V}| + |\mathbf{W}|$

(c) $\mathbf{V} \cdot \mathbf{W}$ equals $\mathbf{W} \cdot \mathbf{V}$

(d) The vectors perpendicular to $\mathbf{i} + \mathbf{j} + \mathbf{k}$ lie along a line.

44 If $\mathbf{V} = \mathbf{i} + 2\mathbf{k}$ choose $\mathbf{W}$ so that $\mathbf{V} \cdot \mathbf{W} = |\mathbf{V}||\mathbf{W}|$ and $|\mathbf{V} + \mathbf{W}| = |\mathbf{V}| + |\mathbf{W}|$.

45 A methane molecule has a carbon atom at $(0,0,0)$ and hydrogen atoms at $(1,1,-1)$, $(1,-1,1)$, $(-1,1,1)$, and $(-1,-1,-1)$. Find

(a) the distance between hydrogen atoms

(b) the angle between vectors going out from the carbon atom to the hydrogen atoms.

46 (a) Find a vector $\mathbf{V}$ at a $45°$ angle with $\mathbf{i}$ and $\mathbf{j}$.

(b) Find $\mathbf{W}$ that makes a $60°$ angle with $\mathbf{i}$ and $\mathbf{j}$.

(c) Explain why no vector makes a $30°$ angle with $\mathbf{i}$ and $\mathbf{j}$.

11.2 Planes and Projections

The most important "curves" are straight lines. The most important functions are linear. Those sentences take us back to the beginning of the book—the graph of $mx + b$ is a line. The goal now is to move into three dimensions, where *graphs are surfaces.* Eventually the surfaces will be curved. But calculus starts with the flat surfaces that correspond to straight lines:

What are the most important surfaces ? *Planes*.
What are the most important functions ? *Still linear*.

The geometrical idea of a plane is turned into algebra, by finding *the equation of a plane*. Not just a general formula, but the particular equation of a particular plane.

A line is determined by one point (x_0, y_0) and the slope m. The point-slope equation is $y - y_0 = m(x - x_0)$. That is a linear equation, it is satisfied when $y = y_0$ and $x = x_0$, and dy/dx is m. For a plane, we start again with a particular point—which is now (x_0, y_0, z_0). But the slope of a plane is not so simple. Many planes climb at a $45°$ angle—with "slope 1"—and more information is needed.

The direction of a plane is described by a vector **N**. The vector is not *in* the plane, but *perpendicular* to the plane. In the plane, there are many directions. Perpendicular to the plane, there is only one direction. A vector in that perpendicular direction is a *normal vector*.

The normal vector **N** can point "up" or "down". The length of **N** is not crucial (we often make it a unit vector and call it **n**). Knowing **N** and the point $P_0 = (x_0, y_0, z_0)$, we know the plane (Figure 11.9). For its equation we switch to algebra and use the dot product—which is the key to perpendicularity.

N is described by its components (a, b, c). In other words **N** is $a\mathbf{i} + b\mathbf{j} + c\mathbf{k}$. *This vector is perpendicular to every direction in the plane*. A typical direction goes from

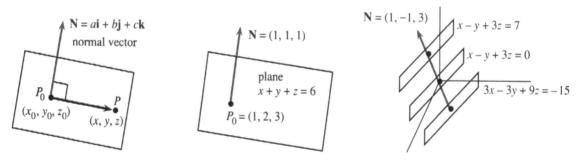

Fig. 11.9   The normal vector to a plane. Parallel planes have the same N.

P_0 to another point $P = (x, y, z)$ in the plane. The vector from P_0 to P has components $(x - x_0, y - y_0, z - z_0)$. This vector lies in the plane, so *its dot product with* **N** *is zero*:

11C   The plane through P_0 perpendicular to $\mathbf{N} = (a, b, c)$ has the equation

$$(a, b, c) \cdot (x - x_0, y - y_0, z - z_0) = 0 \quad \text{or}$$

$$a(x - x_0) + b(y - y_0) + c(z - z_0) = 0. \tag{1}$$

The point P lies on the plane when its coordinates x, y, z satisfy this equation.

EXAMPLE 1   The plane through $P_0 = (1, 2, 3)$ perpendicular to $\mathbf{N} = (1, 1, 1)$ has the equation $(x - 1) + (y - 2) + (z - 3) = 0$. That can be rewritten as $x + y + z = 6$.

Notice three things. First, P_0 lies on the plane because $1 + 2 + 3 = 6$. Second, $\mathbf{N} = (1, 1, 1)$ can be recognized from the x, y, z coefficients in $x + y + z = 6$. Third, we could change $\mathbf{N}$ to $(2, 2, 2)$ and we could change P_0 to $(8, 2, -4)$—because $\mathbf{N}$ is still perpendicular and P_0 is still in the plane: $8 + 2 - 4 = 6$.

The new normal vector $\mathbf{N} = (2, 2, 2)$ produces $2(x - 1) + 2(y - 2) + 2(z - 3) = 0$. That can be rewritten as $2x + 2y + 2z = 12$. Same normal direction, same plane.

The new point $P_0 = (8, 2, -4)$ produces $(x - 8) + (y - 2) + (z + 4) = 0$. That is another form of $x + y + z = 6$. All we require is a perpendicular $\mathbf{N}$ and a point P_0 in the plane.

EXAMPLE 2   The plane through $(1, 2, 4)$ with the same $\mathbf{N} = (1, 1, 1)$ has a different equation: $(x - 1) + (y - 2) + (z - 4) = 0$. This is $x + y + z = 7$ (instead of 6). ***These planes with 7 and 6 are parallel***.

Starting from $a(x - x_0) + b(y - y_0) + c(z - z_0) = 0$, we often move $ax_0 + by_0 + cz_0$ to the right hand side—and call this constant d:

11D   With the P_0 terms on the right side, the equation of the plane is $\mathbf{N} \cdot \mathbf{P} = d$:

$$ax + by + cz = ax_0 + by_0 + cz_0 = d. \tag{2}$$

A different d gives a ***parallel plane***; $d = 0$ gives a ***plane through the origin***.

EXAMPLE 3   The plane $x - y + 3z = 0$ goes through the origin $(0, 0, 0)$. The normal vector is read directly from the equation: $\mathbf{N} = (1, -1, 3)$. The equation is satisfied by $P_0 = (1, 1, 0)$ and $P = (1, 4, 1)$. Subtraction gives a vector $\mathbf{V} = (0, 3, 1)$ that is in the plane, and $\mathbf{N} \cdot \mathbf{V} = 0$.

The parallel planes $x - y + 3z = d$ have the same $\mathbf{N}$ but different d's. These planes miss the origin because d is not zero ($x = 0, y = 0, z = 0$ on the left side needs $d = 0$ on the right side). Note that $3x - 3y + 9z = -15$ is parallel to both planes. $\mathbf{N}$ is changed to $3\mathbf{N}$ in Figure 11.9, but its direction is not changed.

EXAMPLE 4   *The angle between two planes is the angle between their normal vectors.*

The planes $x - y + 3z = 0$ and $3y + z = 0$ are perpendicular, because $(1, -1, 3) \cdot (0, 3, 1) = 0$. The planes $z = 0$ and $y = 0$ are also perpendicular, because $(0, 0, 1) \cdot (0, 1, 0) = 0$. (Those are the xy plane and the xz plane.) The planes $x + y = 0$ and $x + z = 0$ make a $60°$ angle, because $\cos 60° = (1, 1, 0) \cdot (1, 0, 1) / \sqrt{2}\sqrt{2} = \frac{1}{2}$.

The cosine of the angle between two planes is $|\mathbf{N}_1 \cdot \mathbf{N}_2| / |\mathbf{N}_1| |\mathbf{N}_2|$. See Figure 11.10.

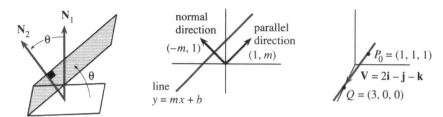

Fig. 11.10   Angle between planes = angle between normals. Parallel and perpendicular to a line. *A line in space through P_0 and Q.*

Remark 1   We gave the "point-slope" equation of a line (using m), and the "point-normal" equation of a plane (using $\mathbf{N}$). What is the normal vector $\mathbf{N}$ to a line?

The vector $\mathbf{V} = (1, m)$ is parallel to the line $y = mx + b$. The line goes across by 1 and up by m. ***The perpendicular vector is*** $\mathbf{N} = (-m, 1)$. The dot product $\mathbf{N} \cdot \mathbf{V}$ is $-m + m = 0$. Then the point-normal equation matches the point-slope equation:

$$-m(x - x_0) + 1(y - y_0) = 0 \text{ is the same as } y - y_0 = m(x - x_0). \tag{3}$$

Remark 2   What is the point-slope equation for a plane? The difficulty is that a plane has different slopes in the x and y directions. The function $f(x, y) = m(x - x_0) + M(y - y_0)$ has ***two derivatives*** m and M.

This remark has to stop. In Chapter 13, "slopes" become "***partial derivatives***."

A LINE IN SPACE

In three dimensions, a line is not as simple as a plane. ***A line in space needs two equations***. Each equation gives a plane, and the line is the ***intersection of two planes***.

> ***The equations $x + y + z = 3$ and $2x + 3y + z = 6$ determine a line.***

Two points on that line are $P_0 = (1, 1, 1)$ and $Q = (3, 0, 0)$. They satisfy both equations so they lie on both planes. Therefore they are on the line of intersection. The direction of that line, subtracting coordinates of P_0 from Q, is along the vector $\mathbf{V} = 2\mathbf{i} - \mathbf{j} - \mathbf{k}$.

> ***The line goes through $P_0 = (1, 1, 1)$ in the direction of $\mathbf{V} = 2\mathbf{i} - \mathbf{j} - \mathbf{k}$.***

Starting from $(x_0, y_0, z_0) = (1, 1, 1)$, ***add on any multiple*** $t\mathbf{V}$. Then $x = 1 + 2t$ and $y = 1 - t$ and $z = 1 - t$. Those are the components of the vector equation $\mathbf{P} = \mathbf{P}_0 + t\mathbf{V}$—which produces the line.

Here is the problem. The line needs two equations—or a vector equation with a *parameter t*. Neither form is as simple as $ax + by + cz = d$. Some books push ahead anyway, to give full details about both forms. After trying this approach, I believe that those details should wait. Equations with parameters are the subject of Chapter 12, and a line in space is the first example. Vectors and planes give plenty to do here—especially when a vector is projected onto another vector or a plane.

PROJECTION OF A VECTOR

What is the projection of a vector $\mathbf{B}$ onto another vector $\mathbf{A}$? One part of $\mathbf{B}$ goes ***along*** $\mathbf{A}$—that is the projection. The other part of $\mathbf{B}$ is ***perpendicular*** to $\mathbf{A}$. We now compute these two parts, which are $\mathbf{P}$ and $\mathbf{B} - \mathbf{P}$.

In geometry, projections involve $\cos\theta$. In algebra, we use the dot product (which is closely tied to $\cos\theta$). In applications, the vector **B** might be a *velocity* **V** or a *force* **F**:

> An airplane flies northeast, and a 100-mile per hour wind blows due east. What is the projection of $\mathbf{V} = (100, 0)$ in the flight direction **A**?

> Gravity makes a ball roll down the surface $2x + 2y + z = 0$. What are the projections of $\mathbf{F} = (0, 0, -mg)$ in the plane and perpendicular to the plane?

The component of **V** along **A** is the push from the wind (tail wind). The other component of **V** pushes sideways (crosswind). Similarly the force parallel to the surface makes the ball move. Adding the two components brings back **V** or **F**.

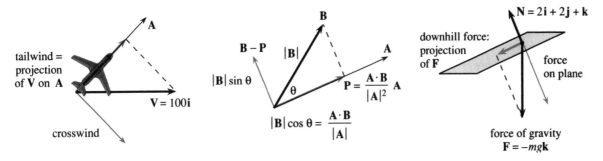

Fig. 11.11   Projections along **A** of wind velocity **V** and force **F** and vector **B**.

*We now compute the projection of **B** onto **A**.* Call this projection **P**. Since its direction is known—**P** is along **A**—we can describe **P** in two ways:

1) *Give the* **length** *of* **P** *along* **A**
2) *Give the* **vector** **P** *as a multiple of* **A**.

Figure 11.11b shows the projection **P** and its length. The hypotenuse is $|\mathbf{B}|$. The length is $|\mathbf{P}| = |\mathbf{B}|\cos\theta$. The perpendicular component $\mathbf{B} - \mathbf{P}$ has length $|\mathbf{B}|\sin\theta$. The cosine is positive for angles less than $90°$. The cosine (and **P**!) are zero when **A** and **B** are perpendicular. $|\mathbf{B}|\cos\theta$ is negative for angles greater than $90°$, and the projection points along $-\mathbf{A}$ (the length is $|\mathbf{B}||\cos\theta|$). Unless the angle is $0°$ or $30°$ or $45°$ or $60°$ or $90°$, we don't want to compute cosines—and we don't have to. The dot product does it automatically:

$$|\mathbf{A}||\mathbf{B}|\cos\theta = \mathbf{A}\cdot\mathbf{B} \ \textit{so the length of } \mathbf{P} \textit{ along } \mathbf{A} \textit{ is } |\mathbf{B}|\cos\theta = \frac{\mathbf{A}\cdot\mathbf{B}}{|\mathbf{A}|}. \qquad (4)$$

Notice that the length of **A** cancels out at the end of (4). If **A** is doubled, **P** is unchanged. But if **B** is doubled, the projection is doubled.

What is the vector **P**? Its length along **A** is $\mathbf{A}\cdot\mathbf{B}/|\mathbf{A}|$. If **A** is a unit vector, then $|\mathbf{A}| = 1$ and the projection is $\mathbf{A}\cdot\mathbf{B}$ times **A**. Generally **A** is not a unit vector, until we divide by $|\mathbf{A}|$. *Here is the projection* **P** *of* **B** *along* **A**:

$$\mathbf{P} = (\textit{length of } \mathbf{P})(\textit{unit vector}) = \left(\frac{\mathbf{A}\cdot\mathbf{B}}{|\mathbf{A}|}\right)\left(\frac{\mathbf{A}}{|\mathbf{A}|}\right) = \frac{\mathbf{A}\cdot\mathbf{B}}{|\mathbf{A}|^2}\mathbf{A}. \qquad (5)$$

EXAMPLE 5   For the wind velocity $\mathbf{V} = (100,0)$ and flying direction $\mathbf{A} = (1,1)$, find $\mathbf{P}$. Here $\mathbf{V}$ points east, $\mathbf{A}$ points northeast. The projection of $\mathbf{V}$ onto $\mathbf{A}$ is $\mathbf{P}$:

$$\text{length } |\mathbf{P}| = \frac{\mathbf{A}\cdot\mathbf{V}}{|\mathbf{A}|} = \frac{100}{\sqrt{2}} \quad \text{vector } \mathbf{P} = \frac{\mathbf{A}\cdot\mathbf{V}}{|\mathbf{A}|^2}\mathbf{A} = \frac{100}{2}(1,1) = (50,50).$$

EXAMPLE 6   Project $\mathbf{F} = (0,0,-mg)$ onto the plane with normal $\mathbf{N} = (2,2,1)$.

The projection of $\mathbf{F}$ along $\mathbf{N}$ is *not* the answer. But compute that first:

$$\frac{\mathbf{F}\cdot\mathbf{N}}{|\mathbf{N}|} = -\frac{mg}{3} \quad \mathbf{P} = \frac{\mathbf{F}\cdot\mathbf{N}}{|\mathbf{N}|^2}\mathbf{N} = -\frac{mg}{9}(2,2,1).$$

$\mathbf{P}$ is the component of $\mathbf{F}$ *perpendicular* to the plane. It does *not* move the ball. The in-plane component is the difference $\mathbf{F}-\mathbf{P}$. Any vector $\mathbf{B}$ has two projections, along $\mathbf{A}$ and perpendicular:

The projection $\mathbf{P} = \dfrac{\mathbf{A}\cdot\mathbf{B}}{|\mathbf{A}|^2}\mathbf{A}$ *is perpendicular to the remaining component* $\mathbf{B}-\mathbf{P}$.

EXAMPLE 7   Express $\mathbf{B} = \mathbf{i}-\mathbf{j}$ as the sum of a vector $\mathbf{P}$ parallel to $\mathbf{A} = 3\mathbf{i}+\mathbf{j}$ and a vector $\mathbf{B}-\mathbf{P}$ perpendicular to $\mathbf{A}$. Note $\mathbf{A}\cdot\mathbf{B} = 2$.

Solution $\mathbf{P} = \dfrac{\mathbf{A}\cdot\mathbf{B}}{|\mathbf{A}|^2}\mathbf{A} = \dfrac{2}{10}\mathbf{A} = \dfrac{6}{10}\mathbf{i} + \dfrac{2}{10}\mathbf{j}.$ Then $\mathbf{B}-\mathbf{P} = \dfrac{4}{10}\mathbf{i} - \dfrac{12}{10}\mathbf{j}.$

Check: $\mathbf{P}\cdot(\mathbf{B}-\mathbf{P}) = \left(\frac{6}{10}\right)\left(\frac{4}{10}\right) - \left(\frac{2}{10}\right)\left(\frac{12}{10}\right) = 0$. These projections of $\mathbf{B}$ are perpendicular.

Pythagoras: $|\mathbf{P}|^2 + |\mathbf{B}-\mathbf{P}|^2$ equals $|\mathbf{B}|^2$. Check that too: $0.4 + 1.6 = 2.0$.

Question   When is $\mathbf{P} = \mathbf{0}$?   Answer   When $\mathbf{A}$ and $\mathbf{B}$ are perpendicular.

EXAMPLE 8   *Find the nearest point to the origin on the plane* $x + 2y + 2z = 5$.

The shortest distance from the origin is along the normal vector $\mathbf{N}$. The vector $\mathbf{P}$ to the nearest point (Figure 11.12) is t times $\mathbf{N}$, for some unknown number t. We find t by requiring $\mathbf{P} = t\mathbf{N}$ to lie on the plane.

The plane $x + 2y + 2z = 5$ has normal vector $\mathbf{N} = (1,2,2)$. Therefore $\mathbf{P} = t\mathbf{N} = (t,2t,2t)$. To lie on the plane, this must satisfy $x + 2y + 2z = 5$:

$$t + 2(2t) + 2(2t) = 5 \quad \text{or} \quad 9t = 5 \quad \text{or} \quad t = \tfrac{5}{9}. \tag{6}$$

Then $\mathbf{P} = \frac{5}{9}\mathbf{N} = (\frac{5}{9}, \frac{10}{9}, \frac{10}{9})$. That locates the nearest point. The distance is $\frac{5}{9}|\mathbf{N}| = \frac{5}{3}$. This example is important enough to memorize, with letters not numbers:

11E On the plane $ax + by + cz = d$, the nearest point to $(0,0,0)$ is

$$P = \frac{(da,db,dc)}{a^2+b^2+c^2}. \quad \text{The distance is} \quad \frac{|d|}{\sqrt{a^2+b^2+c^2}}. \tag{7}$$

The steps are the same. $\mathbf{N}$ has components a,b,c. The nearest point on the plane is a multiple (ta,tb,tc). It lies on the plane if $a(ta) + b(tb) + c(tc) = d$.

Thus $t = d/(a^2 + b^2 + c^2)$. The point $(ta, tb, tc) = t\mathbf{N}$ is in equation (7). The distance to the plane is $|t\mathbf{N}| = |d|/|\mathbf{N}|$.

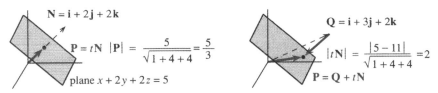

Fig. 11.12   Vector to the nearest point P is a multiple $t\mathbf{N}$. The distance is in (7) and (9).

Question   How far is the plane from an arbitrary point $Q = (x_1, y_1, z_1)$?

Answer   *The vector from Q to P is our multiple $t\mathbf{N}$. In vector form $\mathbf{P} = \mathbf{Q} + t\mathbf{N}$.* This reaches the plane if $\mathbf{P} \cdot \mathbf{N} = d$, and again we find t:

$$(\mathbf{Q} + t\mathbf{N}) \cdot \mathbf{N} = d \quad \text{yields} \quad t = (d - \mathbf{Q} \cdot \mathbf{N})/|\mathbf{N}|^2. \tag{8}$$

This new term $\mathbf{Q} \cdot \mathbf{N}$ enters the distance from Q to the plane:

$$\boldsymbol{distance} = |t\mathbf{N}| = |d - \mathbf{Q} \cdot \mathbf{N}|/|\mathbf{N}| = |d - ax_1 - by_1 - cz_1|/\sqrt{a^2 + b^2 + c^2}. \tag{9}$$

When the point is on the plane, that distance is zero—because $ax_1 + by_1 + cz_1 = d$. When $\mathbf{Q}$ is $\mathbf{i} + 3\mathbf{j} + 2\mathbf{k}$, the figure shows $\mathbf{Q} \cdot \mathbf{N} = 11$ and distance $= 2$.

PROJECTIONS OF THE HEART VECTOR

An electrocardiogram has leads to your right arm–left arm–left leg. *You produce the voltage.* The machine amplifies and records the readings. There are also six chest leads, to add a front-back dimension that is monitored across the heart. We will concentrate on the big "Einthoven triangle," named after the inventor of the ECG.

The graphs show voltage variations plotted against time. The first graph plots the voltage difference between the arms. Lead II connects the left leg to the right arm. Lead III completes the triangle, which has roughly equal sides (especially if you are a little lopsided). So the projections are based on 60° and 120° angles.

The heart vector $\mathbf{V}$ is the sum of many small vectors—all moved to the same origin. $\mathbf{V}$ is the net effect of action potentials from the cells—small dipoles adding to a single dipole. The pacemaker (S–A node) starts the impulse. The atria depolarize to give the P wave on the graphs. This is actually a P *loop* of the heart vector— the graphs only show its projections. The impulse reaches the AV node, pauses, and moves quickly through the ventricles. This produces the QRS complex—the large sharp movement on the graph.

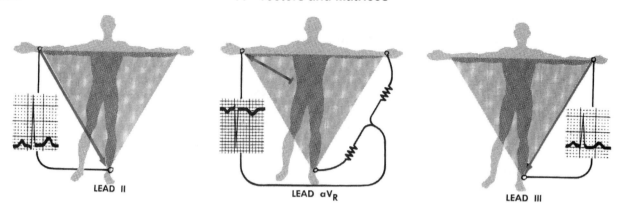

Fig. A   The graphs show the component of the moving heart vector along each lead. These figures art reproduced with permission from the CIBA Collection of Medical Illustrations by Frank H. Netter, M.D. Copyright 1978 CIBA-GEIGY, all rights reserved.

The total QRS *interval should not exceed* 1/10 *second* ($2\frac{1}{2}$ spaces on the printout). **V** points first toward the right shoulder. This direction is opposite to the leads, so the tracings go slightly down. That is the Q wave, small and negative. Then the heart vector sweeps toward the left leg. In positions 3 and 4, its projection on lead I (between the arms) is strongly positive. The R wave is this first upward deflection in each lead. Closing the loop, the S wave is negative (best seen in leads I and aVR).

Question 1   How many graphs from the arms and leg are really independent?

Answer   Only two! In a plane, the heart vector **V** has two components. If we know two projections, we can compute the others. (The ECG does that for us.) Different vectors show better in different projections. A mathematician would use 90° angles, with an electrode at your throat.

Question 2   How are the voltages related? What is the aVR lead?

Answer   Project the heart vector **V** onto the sides of the triangle:

> The lead vectors have $\mathbf{L_I} - \mathbf{L_{II}} + \mathbf{L_{III}} = \mathbf{0}$— they form a triangle.

> The projections have $\mathbf{V_I} - \mathbf{V_{II}} + \mathbf{V_{III}} = \mathbf{V} \cdot \mathbf{L_I} - \mathbf{V} \cdot \mathbf{L_{II}} + \mathbf{V} \cdot \mathbf{L_{III}} = 0$.

The aVR lead is $-\frac{1}{2}\mathbf{L_I} - \frac{1}{2}\mathbf{L_{II}}$. It is pure algebra (no wire). By vector addition it points toward the electrode on the right arm. Its length is $\sqrt{3}$ if the other lengths are 2.

Including aVL and aVF to the left arm and foot, there are *six leads intersecting at equal angles*. Visualize them going out from a single point (the origin in the chest).

Question 3   If the heart vector is $\mathbf{V} = 2\mathbf{i} - \mathbf{j}$, what voltage differences are recorded?

Answer   The leads around the triangle have length 2. The machine projects **V**:

> Lead I is the horizontal vector $2\mathbf{i}$. So $\mathbf{V} \cdot \mathbf{L_I} = 4$.
> Lead II is the $-60°$ vector $\mathbf{i} - \sqrt{3}\mathbf{j}$. So $\mathbf{V} \cdot \mathbf{L_{II}} = 2 + \sqrt{3}$.
> Lead III is the $-120°$ vector $-\mathbf{i} - \sqrt{3}\mathbf{j}$. So $\mathbf{V} \cdot \mathbf{L_{III}} = -2 + \sqrt{3}$.

The first and third add to the second. The largest R waves are in leads I and II. In aVR the projection of **V** will be negative (Problem 46), and will be labeled an S wave.

Question 4   What about the *potential* (not just its differences). Is it zero at the center?

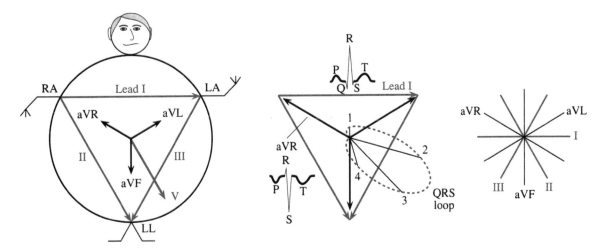

Fig. B Heart vector goes around the QRS loop. Projections are spikes on the ECG.

Answer *It is zero if we say so*. The potential contains an arbitrary constant C. (It is like an indefinite integral. Its differences are like definite integrals.) Cardiologists define a "central terminal" where the potential is zero.

The average of **V** over a loop is the *mean heart vector* **H**. This average requires $\int \mathbf{V}\,dt$, by Chapter 5. With no time to integrate, the doctor looks for a lead where the area under the QRS complex is zero. Then the direction of **H** (the *axis*) is perpendicular to that lead. There is so much to say about calculus in medicine.

11.2 EXERCISES

Read-through questions

A plane in space is determined by a point $P_0 = (x_0, y_0, z_0)$ and a __a__ vector **N** with components (a,b,c). The point $P = (x,y,z)$ is on the plane if the dot product of **N** with __b__ is zero. (*That answer was not* P!) The equation of this plane is $a(\underline{\ c\ }) + b(\underline{\ d\ }) + c(\underline{\ e\ }) = 0$. The equation is also written as $ax + by + cz = d$, where d equals __f__. A parallel plane has the same __g__ and a different __h__. A plane through the origin has $d = \underline{\ i\ }$.

The equation of the plane through $P_0 = (2,1,0)$ perpendicular to **N** $= (3,4,5)$ is __j__. A second point in the plane is $P = (0,0,\underline{\ k\ })$. The vector from P_0 to P is __l__, and it is __m__ to **N**. (Check by dot product.) The plane through $P_0 = (2,1,0)$ perpendicular to the z axis has **N** $= $ __n__ and equation __o__.

The component of **B** in the direction of **A** is __p__, where θ is the angle between the vectors. This is **A** · **B** divided by __q__. The projection vector **P** is $|\mathbf{B}|\cos\theta$ times a __r__ vector in the direction of **A**. Then **P** $= (|\mathbf{B}|\cos\theta)(\mathbf{A}/|\mathbf{A}|)$ simplifies to __s__. When **B** is doubled, **P** is __t__. When **A** is doubled, **P** is __u__. If **B** reverses direction then **P** __v__. If **A** reverses direction then **P** __w__.

When **B** is a velocity vector, **P** represents the __x__. When **B** is a force vector, **P** is __y__. The component of **B** perpendicular to **A** equals __z__. The shortest distance from $(0,0,0)$ to the plane $ax + by + cz = d$ is along the __A__ vector. The distance is __B__ and the closest point on the plane is $P = $ __C__. The distance from $\mathbf{Q} = (x_1, y_1, z_1)$ to the plane is __D__.

Find two points P **and** P_0 **on the planes 1–6 and a normal vector N. Verify that** $\mathbf{N} \cdot (P - P_0) = 0$.

1 $x + 2y + 3z = 0$ **2** $x + 2y + 3z = 6$ **3** the yz plane

4 the plane through $(0,0,0)$ perpendicular to $\mathbf{i}+\mathbf{j}-\mathbf{k}$

5 the plane through $(1,1,1)$ perpendicular to $\mathbf{i}+\mathbf{j}-\mathbf{k}$

6 the plane through $(0,0,0)$ and $(1,0,0)$ and $(0,1,1)$.

Find an $x - y - z$ **equation for planes 7–10.**

7 The plane through $P_0 = (1,2,-1)$ perpendicular to $\mathbf{N}=\mathbf{i}+\mathbf{j}$

8 The plane through $P_0 = (1,2,-1)$ perpendicular to $\mathbf{N}=\mathbf{i}+2\mathbf{j}-\mathbf{k}$

9 The plane through $(1,0,1)$ parallel to $x + 2y + z = 0$

10 The plane through (x_0, y_0, z_0) parallel to $x + y + z = 1$.

11 When is a plane with normal vector $\mathbf{N}$ parallel to the vector $\mathbf{V}$? When is it perpendicular to $\mathbf{V}$?

12 (a) If two planes are perpendicular (front wall and side wall), is every line in one plane perpendicular to every line in the other ?

(b) If a third plane is perpendicular to the first, it might be (parallel) (perpendicular) (at a $45°$ angle) to the second.

13 Explain why a plane cannot

(a) contain $(1,2,3)$ and $(2,3,4)$ and be perpendicular to $\mathbf{N} = \mathbf{i} + \mathbf{j}$

(b) be perpendicular to $\mathbf{N} = \mathbf{i} + \mathbf{j}$ and parallel to $\mathbf{V} = \mathbf{i} + \mathbf{k}$

(c) contain $(1,0,0), (0,1,0), (0,0,1)$, and $(1,1,1)$

(d) contain $(1,1,-1)$ if it has $\mathbf{N} = \mathbf{i} + \mathbf{j} - \mathbf{k}$ (*maybe it can*)

(e) go through the origin and have the equation $ax + by + cz = 1$.

14 The equation $3x + 4y + 7z - t = 0$ yields a hyperplane in four dimensions. Find its normal vector $\mathbf{N}$ and two points P, Q on the hyperplane. Check $(P - Q) \cdot \mathbf{N} = 0$.

15 The plane through (x, y, z) perpendicular to $a\mathbf{i} + b\mathbf{j} + c\mathbf{k}$ goes through $(0,0,0)$ if _____. The plane goes through (x_0, y_0, z_0) if _____.

16 A curve in three dimensions is the intersection of _____ surfaces. A line in four dimensions is the intersection of _____ hyperplanes.

17 (angle between planes) Find the cosine of the angle between $x + 2y + 2z = 0$ and (a) $x + 2z = 0$ (b) $x + 2z = 5$ (c) $x = 0$.

18 $\mathbf{N}$ is perpendicular to a plane and $\mathbf{V}$ is along a line. Draw the angle θ between the plane and the line, and explain why $\mathbf{V} \cdot \mathbf{N} / |\mathbf{V}||\mathbf{N}|$ is $\sin\theta$ not $\cos\theta$. Find the angle between the xy plane and $\mathbf{V} = \mathbf{i} + \mathbf{j} + \sqrt{2}\mathbf{k}$.

In 19–26 find the projection P of B along A. Also find $|P|$.

19 $\mathbf{A} = (4,2,4), \mathbf{B} = (1,-1,0)$

20 $\mathbf{A} = (1,-1,0), \mathbf{B} = (4,2,4)$

21 $\mathbf{B} = $ unit vector at $60°$ angle with $\mathbf{A}$

22 $\mathbf{B} = $ vector of length 2 at $60°$ angle with $\mathbf{A}$

23 $\mathbf{B} = -\mathbf{A}$ **24** $\mathbf{A} = \mathbf{i} + \mathbf{j}, \mathbf{B} = \mathbf{i} + \mathbf{k}$

25 $\mathbf{A}$ is perpendicular to $x - y + z = 0, \mathbf{B} = \mathbf{i} + \mathbf{j}$.

26 $\mathbf{A}$ is perpendicular to $x - y + z = 5, \mathbf{B} = \mathbf{i} + \mathbf{j} + 5\mathbf{k}$.

27 The force $\mathbf{F} = 3\mathbf{i} - 4\mathbf{k}$ acts at the point $(1,2,2)$. How much force pulls toward the origin ? How much force pulls vertically down ? Which direction does a mass move under the force $\mathbf{F}$?

28 The projection of $\mathbf{B}$ along $\mathbf{A}$ is $\mathbf{P} = $ _____ .The projection of $\mathbf{B}$ perpendicular to $\mathbf{A}$ is _____ . Check the dot product of the two projections.

29 $\mathbf{P} = (x, y, z)$ is on the plane $ax + by + cz = 5$ if $\mathbf{P} \cdot \mathbf{N} = |\mathbf{P}||\mathbf{N}| \cos\theta = 5$. Since the largest value of $\cos\theta$ is 1, the smallest value of $|\mathbf{P}|$ is _____. This is the distance between _____ .

30 If the air speed of a jet is 500 and the wind speed is 50, what information do you need to compute the jet's speed over land ? What is that speed ?

31 How far is the plane $x + y - z = 1$ from $(0,0,0)$ and also from $(1,1,-1)$? Find the nearest points.

32 Describe all points at a distance 1 from the plane $x + 2y + 2z = 3$.

33 The shortest distance from $Q = (2,1,1)$ to the plane $x + y + z = 0$ is along the vector _____. The point $\mathbf{P} = Q + t\mathbf{N} = (2+t, 1+t, 1+t)$ lies on the plane if $t = $ _____. Then $\mathbf{P} = $ _____ and the shortest distance is _____. (This distance is not $|\mathbf{P}|$.)

34 The plane through $(1,1,1)$ perpendicular to $\mathbf{N} = \mathbf{i} + 2\mathbf{j} + 2\mathbf{k}$ is a distance _____ from $(0,0,0)$.

35 (*Distance between planes*) $2x - 2y + z = 1$ is parallel to $2x - 2y + z = 3$ because _____. Choose a vector $\mathbf{Q}$ on the first plane and find t so that $\mathbf{Q} + t\mathbf{N}$ lies on the second plane. The distance is $|t\mathbf{N}| = $ _____ .

36 The distance between the planes $x + y + 5z = 7$ and $3x + 2y + z = 1$ is zero because _____ .

In Problems 37–41 all points and vectors are in the xy plane.

37 The line $3x + 4y = 10$ is perpendicular to the vector $\mathbf{N} = $ _____ . On the line, the closest point to the origin is $P = t\mathbf{N}$. Find t and P and $|P|$.

38 Draw the line $x + 2y = 4$ and the vector $\mathbf{N} = \mathbf{i} + 2\mathbf{j}$. The closest point to $Q = (3,3)$ is $P = Q + t\mathbf{N}$. Find t. Find P.

39 A new way to find P in Problem 37: minimize $x^2 + y^2 = x^2 + \left(\frac{10}{4} - \frac{3}{4}x\right)^2$. By calculus find the best x and y.

40 To catch a drug runner going from $(0,0)$ to $(4,0)$ at 8 meters per second, you must travel from $(0,3)$ to $(4,0)$ at _____ meters per second. The projection of your velocity vector onto his velocity vector will have length _____ .

41 Show by vectors that the distance from (x_1, y_1) to the line $ax + by = d$ is $|d - ax_1 - by_1|/\sqrt{a^2 + b^2}$.

42 It takes three points to determine a plane. So why does $ax + by + cz = d$ contain four numbers a, b, c, d ? When does $ex + fy + gz = 1$ represent the same plane ?

43 (projections by calculus) The dot product of $\mathbf{B} - t\mathbf{A}$ with itself is $|\mathbf{B}|^2 - 2t\mathbf{A} \cdot \mathbf{B} + t^2|\mathbf{A}|^2$. (a) This has a minimum at $t = $ _____ . (b) Then $t\mathbf{A}$ is the projection of _____ . A figure showing $\mathbf{B}, t\mathbf{A}$, and $\mathbf{B} - t\mathbf{A}$ is worth 1000 words.

44 From their equations, how can you tell if two planes are

(a) parallel (b) perpendicular (c) at a $45°$ angle ?

Problems 45–48 are about the ECG and heart vector.

45 The aVR lead is $-\frac{1}{2}\mathbf{L_I} - \frac{1}{2}\mathbf{L_{II}}$. Find the aVL and aVF leads toward the left arm and foot. Show that aVR+aVL+aVF $= \mathbf{0}$. They go out from the center at $120°$ angles.

46 Find the projection on the aVR lead of $\mathbf{V} = 2\mathbf{i} - \mathbf{j}$ in Question 3.

47 If the potentials are $\varphi_{RA} = 1$ (right arm) and $\varphi_{LA} = 2$ and $\varphi_{LL} = -3$, find the heart vector $\mathbf{V}$. The *differences* in potential are the projections of $\mathbf{V}$.

48 If $\mathbf{V}$ is perpendicular to a lead $\mathbf{L}$, the reading on that lead is _____ . If $\int \mathbf{V}(t)dt$ is perpendicular to lead $\mathbf{L}$, why is the *area* under the reading zero ?

11.3 Cross Products and Determinants

After saying that vectors are not multiplied, we offered the dot product. Now we contradict ourselves further, by defining the cross product. Where $\mathbf{A} \cdot \mathbf{B}$ was a number, *the cross product $\mathbf{A} \times \mathbf{B}$ is a vector*. It has length and direction:

The length is $|\mathbf{A}||\mathbf{B}||\sin \theta|$. The direction is perpendicular to $\mathbf{A}$ and $\mathbf{B}$.

The cross product (also called vector product) is defined in three dimensions only. $\mathbf{A}$ and $\mathbf{B}$ lie on a plane through the origin. $\mathbf{A} \times \mathbf{B}$ is along the normal vector $\mathbf{N}$, perpendicular to that plane. We still have to say whether it points "up" or "down" along $\mathbf{N}$.

The length of $\mathbf{A} \times \mathbf{B}$ depends on $\sin \theta$, where $\mathbf{A} \cdot \mathbf{B}$ involved $\cos \theta$. The dot product rewards vectors for being parallel ($\cos 0 = 1$). The cross product is largest when $\mathbf{A}$ is perpendicular to $\mathbf{B}$ ($\sin \pi/2 = 1$). At every angle

$$|\mathbf{A} \cdot \mathbf{B}|^2 + |\mathbf{A} \times \mathbf{B}|^2 = |\mathbf{A}|^2|\mathbf{B}|^2 \cos^2\theta + |\mathbf{A}|^2|\mathbf{B}|^2 \sin^2\theta = |\mathbf{A}|^2|\mathbf{B}|^2. \qquad (1)$$

That will be a bridge from geometry to algebra. *This section goes from definition to formula to volume to determinant*. Equations (6) and (14) are the key formulas for $\mathbf{A} \times \mathbf{B}$.

Notice that $\mathbf{A} \times \mathbf{A} = \mathbf{0}$. (This is the zero vector, not the zero number.) When $\mathbf{B}$ is parallel to $\mathbf{A}$, the angle is zero and the sine is zero. Parallel vectors have $\mathbf{A} \times \mathbf{B} = \mathbf{0}$. Perpendicular vectors have $\sin \theta = 1$ and $|\mathbf{A} \times \mathbf{B}| = |\mathbf{A}||\mathbf{B}| =$ area of rectangle with sides $\mathbf{A}$ and $\mathbf{B}$.

Here are four examples that lead to the cross product $\mathbf{A} \times \mathbf{B}$.

EXAMPLE 1   (From geometry) Find the area of a parallelogram and a triangle.

Vectors $\mathbf{A}$ and $\mathbf{B}$, going out from the origin, form two sides of a triangle. They produce the parallelogram in Figure 11.13, which is twice as large as the triangle.

The area of a parallelogram is base times height (perpendicular height not sloping height). The base is $|\mathbf{A}|$. The height is $|\mathbf{B}||\sin \theta|$. We take absolute values because height and area are not negative. Then the area is the length of the cross product:

$$\textit{area of parallelogram} = |\mathbf{A}||\mathbf{B}||\sin \theta| = |\mathbf{A} \times \mathbf{B}|. \qquad (2)$$

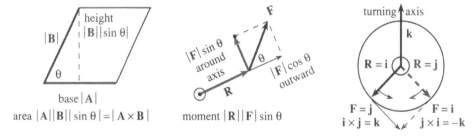

Fig. 11.13   Area $|\mathbf{A} \times \mathbf{B}|$ and moment $|\mathbf{R} \times \mathbf{F}|$. Cross products are perpendicular to the page.

EXAMPLE 2   (From physics) The torque vector $\mathbf{T} = \mathbf{R} \times \mathbf{F}$ produces rotation.

The force $\mathbf{F}$ acts at the point (x, y, z). When $\mathbf{F}$ is parallel to the position vector $\mathbf{R} = x\mathbf{i} + y\mathbf{j} + z\mathbf{k}$, the force pushes outward (*no turning*). When $\mathbf{F}$ is perpendicular to $\mathbf{R}$, the force creates *rotation*. For in-between angles there is an outward force $|\mathbf{F}| \cos \theta$

and a turning force $|\mathbf{F}|\sin\theta$. The turning force times the distance $|\mathbf{R}|$ is the *moment* $|\mathbf{R}||\mathbf{F}|\sin\theta$.

The moment gives the magnitude and sign of the *torque vector* $\mathbf{T} = \mathbf{R} \times \mathbf{F}$. The direction of $\mathbf{T}$ is along the axis of rotation, at right angles to $\mathbf{R}$ and $\mathbf{F}$.

EXAMPLE 3   Does the cross product go up or down ? *Use the right-hand rule.*

Forces and torques are probably just fine for physicists. Those who are not natural physicists want to see something turn.† We can visualize a record or compact disc rotating around its axis—which comes up through the center.

At a point on the disc, you give a push. When the push is outward (hard to do), nothing turns. Rotation comes from force "around" the axis. The disc can turn either way—depending on the angle between force and position. A sign convention is necessary, and it is the *right-hand rule*:

$\mathbf{A} \times \mathbf{B}$ *points along your right thumb when the fingers curl from* $\mathbf{A}$ *toward* $\mathbf{B}$.

This rule is simplest for the vectors $\mathbf{i}, \mathbf{j}, \mathbf{k}$ in Figure 11.14—which is all we need.

Suppose the fingers curl from $\mathbf{i}$ to $\mathbf{j}$. The thumb points along $\mathbf{k}$. *The x-y-z axes form a* "**right-handed triple**." Since $|\mathbf{i}| = 1$ and $|\mathbf{j}| = 1$ and $\sin \pi/2 = 1$, the length of $\mathbf{i} \times \mathbf{j}$ is 1. *The cross product is* $\mathbf{i} \times \mathbf{j} = \mathbf{k}$. The disc turns counterclockwise—its angular velocity is up—when the force acts at $\mathbf{i}$ in the direction $\mathbf{j}$.

Figure 11.14b reverses $\mathbf{i}$ and $\mathbf{j}$. The force acts at $\mathbf{j}$ and its direction is $\mathbf{i}$. The disc turns clockwise (the way records and compact discs actually turn). When the fingers curl from $\mathbf{j}$ to $\mathbf{i}$, the thumb points *down*. Thus $\mathbf{j} \times \mathbf{i} = -\mathbf{k}$. This is a special case of an amazing rule:

The cross product is anticommutative: $\mathbf{B} \times \mathbf{A} = -(\mathbf{A} \times \mathbf{B})$. $\qquad$ (3)

That is quite remarkable. Its discovery by Hamilton produced an intellectual revolution in 19th century algebra, which had been totally accustomed to $AB = BA$. This commutative law is old and boring for numbers (it is new and boring for dot products). Here we see its *opposite* for vector products $\mathbf{A} \times \mathbf{B}$. Neither law holds for matrix products.

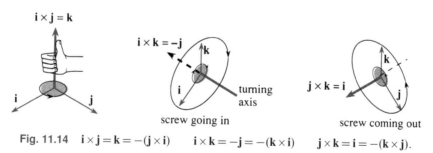

Fig. 11.14 $\mathbf{i} \times \mathbf{j} = \mathbf{k} = -(\mathbf{j} \times \mathbf{i})$ $\qquad$ $\mathbf{i} \times \mathbf{k} = -\mathbf{j} = -(\mathbf{k} \times \mathbf{i})$ $\qquad$ $\mathbf{j} \times \mathbf{k} = \mathbf{i} = -(\mathbf{k} \times \mathbf{j})$.

EXAMPLE 4   A screw goes into a wall or out, following the right-hand rule.

The disc was in the xy plane. So was the force. (We are not breaking records here.) The axis was up and down. To see the cross product more completely we need to turn a screw into a wall.

Figure 11.14b shows the xz plane as the wall. The screw is in the y direction. By turning from x toward z we drive the screw *into* the wall—which is the *negative y*

---

†Everybody is a natural mathematician. That is the axiom behind this book.

direction. In other words $\mathbf{i} \times \mathbf{k}$ equals *minus* $\mathbf{j}$. We turn the screw clockwise to make it go in. To take out the screw, twist from $\mathbf{k}$ toward $\mathbf{i}$. Then $\mathbf{k} \times \mathbf{i}$ equals *plus* $\mathbf{j}$.

To summarize: $\mathbf{k} \times \mathbf{i} = \mathbf{j}$ and $\mathbf{j} \times \mathbf{k} = \mathbf{i}$ have plus signs because $\mathbf{kij}$ and $\mathbf{jki}$ are in the same "**cyclic order**" as $\mathbf{ijk}$. (*Anticyclic is minus.*) The z-x-y and y-z-x axes form righthanded triples like x-y-z.

THE FORMULA FOR THE CROSS PRODUCT

We begin the algebra of $\mathbf{A} \times \mathbf{B}$. It is essential for computation, and it comes out beautifully. The square roots in $|\mathbf{A}||\mathbf{B}||\sin \theta|$ will disappear in formula (6) for $\mathbf{A} \times \mathbf{B}$. (The square roots also disappeared in $\mathbf{A} \cdot \mathbf{B}$, which is $|\mathbf{A}||\mathbf{B}|\cos \theta$. But $|\mathbf{A}||\mathbf{B}|\tan \theta$ would be terrible.) Since $\mathbf{A} \times \mathbf{B}$ is a vector we need to find *three components*.

Start with the two-dimensional case. The vectors $a_1\mathbf{i} + a_2\mathbf{j}$ and $b_1\mathbf{i} + b_2\mathbf{j}$ are in the xy plane. Their cross product must go in the z direction. Therefore $\mathbf{A} \times \mathbf{B} = \underline{\quad ? \quad}\mathbf{k}$ and there is only one nonzero component. It must be $|\mathbf{A}||\mathbf{B}|\sin \theta$ (with the correct sign), but we want a better formula. There are two clean ways to compute $\mathbf{A} \times \mathbf{B}$, either by algebra (*a*) or by a bridge (*b*) to the dot product and geometry:

$$(a) \quad (a_1\mathbf{i} + a_2\mathbf{j}) \times (b_1\mathbf{i} + b_2\mathbf{j}) = a_1b_1\mathbf{i} \times \mathbf{i} + a_1b_2\mathbf{i} \times \mathbf{j} + a_2b_1\mathbf{j} \times \mathbf{i} + a_2b_2\mathbf{j} \times \mathbf{j}. \quad (4)$$

On the right are $\mathbf{0}$, $a_1b_2\mathbf{k}$, $-a_2b_1\mathbf{k}$ and $\mathbf{0}$. *The cross product is* $(a_1b_2 - a_2b_1)\mathbf{k}$.

(*b*) Rotate $\mathbf{B} = b_1\mathbf{i} + b_2\mathbf{j}$ clockwise through $90°$ into $\mathbf{B}^* = b_2\mathbf{i} - b_1\mathbf{j}$. Its length is unchanged (and $\mathbf{B} \cdot \mathbf{B}^* = 0$). Then $|\mathbf{A}||\mathbf{B}^*|\sin \theta$ equals $|\mathbf{A}||\mathbf{B}^*|\cos \theta$, which is $\mathbf{A} \cdot \mathbf{B}^*$:

$$|\mathbf{A}||\mathbf{B}|\sin \theta = \mathbf{A} \cdot \mathbf{B}^* = \begin{bmatrix} a_1 \\ a_2 \end{bmatrix} \cdot \begin{bmatrix} b_2 \\ -b_1 \end{bmatrix} = a_1b_2 - a_2b_1. \quad (5)$$

11F In the xy plane, $\mathbf{A} \times \mathbf{B}$ equals $(a_1b_2 - a_2b_1)\mathbf{k}$. The parallelogram with sides $\mathbf{A}$ and $\mathbf{B}$ has area $|a_1b_2 - a_2b_1|$. The triangle OAB has area $\frac{1}{2}|a_1b_2 - a_2b_1|$.

EXAMPLE 5   For $\mathbf{A} = \mathbf{i} + 2\mathbf{j}$ and $\mathbf{B} = 4\mathbf{i} + 5\mathbf{j}$ the cross product is $(1 \cdot 5 - 2 \cdot 4)\mathbf{k} = -3\mathbf{k}$. Area of parallelogram $= 3$, area of triangle $= 3/2$. The minus sign in $\mathbf{A} \times \mathbf{B} = -3\mathbf{k}$ is absent in the areas.

Note   Splitting $\mathbf{A} \times \mathbf{B}$ into four separate cross products is correct, but it does not follow easily from $|\mathbf{A}||\mathbf{B}|\sin \theta$. Method (*a*) is not justified until Remark 1 below. An algebraist would change the definition of $\mathbf{A} \times \mathbf{B}$ to start with the distributive law (splitting rule) and the anticommutative law:

$$\mathbf{A} \times (\mathbf{B} + \mathbf{C}) = (\mathbf{A} \times \mathbf{B}) + (\mathbf{A} \times \mathbf{C}) \quad \text{and} \quad \mathbf{A} \times \mathbf{B} = -(\mathbf{B} \times \mathbf{A}).$$

THE CROSS PRODUCT FORMULA (3 COMPONENTS)

We move to three dimensions. The goal is to compute all three components of $\mathbf{A} \times \mathbf{B}$ (not just the length). Method (*a*) splits each vector into its $\mathbf{i}, \mathbf{j}, \mathbf{k}$ components, making nine separate cross products:

$$(a_1\mathbf{i} + a_2\mathbf{j} + a_3\mathbf{k}) \times (b_1\mathbf{i} + b_2\mathbf{j} + b_3\mathbf{k}) = a_1b_2(\mathbf{i} \times \mathbf{i}) + a_1b_2(\mathbf{i} \times \mathbf{j}) + \text{ seven more terms.}$$

Remember $\mathbf{i}\times\mathbf{i}=\mathbf{j}\times\mathbf{j}=\mathbf{k}\times\mathbf{k}=\mathbf{0}$. Those three terms disappear. The other six terms come in pairs, and *please notice the cyclic pattern*:

FORMULA $\mathbf{A}\times\mathbf{B}=(a_2b_3-a_3b_2)\mathbf{i}+(a_3b_1-a_1b_3)\mathbf{j}+(a_1b_2-a_2b_1)\mathbf{k}.$ (6)

The $\mathbf{k}$ component is the 2×2 answer, when $a_3=b_3=0$. The $\mathbf{i}$ component involves indices 2 and 3, $\mathbf{j}$ involves 3 and 1, $\mathbf{k}$ involves 1 and 2. The cross product formula is written as a "determinant" in equation (14) below—many people use that form to compute $\mathbf{A}\times\mathbf{B}$.

EXAMPLE 6 $(\mathbf{i}+2\mathbf{j}+3\mathbf{k})\times(4\mathbf{i}+5\mathbf{j}+6\mathbf{k})=(2\cdot6-3\cdot5)\mathbf{i}+(3\cdot4-1\cdot6)\mathbf{j}+(1\cdot5-2\cdot4)\mathbf{k}$. The $\mathbf{i},\mathbf{j},\mathbf{k}$ components give $\mathbf{A}\times\mathbf{B}=-3\mathbf{i}+6\mathbf{j}-3\mathbf{k}$. Never add the $-3,6$, and -3.

Remark 1   The three-dimensional formula (6) is still to be matched with $\mathbf{A}\times\mathbf{B}$ from geometry. One way is to rotate $\mathbf{B}$ into $\mathbf{B}^*$ as before, staying in the plane of $\mathbf{A}$ and $\mathbf{B}$. Fortunately there is an easier test. The vector in equation (6) satisfies all four geometric requirements on $\mathbf{A}\times\mathbf{B}$: *perpendicular to $\mathbf{A}$, perpendicular to $\mathbf{B}$, correct length, right-hand rule*. The length is checked in Problem 16—here is the zero dot product with $\mathbf{A}$:

$$\mathbf{A}\cdot(\mathbf{A}\times\mathbf{B})=a_1(a_2b_3-a_3b_2)+a_2(a_3b_1-a_1b_3)+a_3(a_1b_2-a_2b_1)=0.$$ (7)

Remark 2   (Optional) There is a wonderful extension of the Pythagoras formula $a^2+b^2=c^2$. Instead of sides of a triangle, we go to *areas of projections* on the yz, xz, and xy planes. $3^2+6^2+3^2$ is the square of the parallelogram area in Example 6.

For triangles these areas are cut in half. Figure 11.15a shows three projected triangles of area $\frac{1}{2}$. Its Pythagoras formula is $\left(\frac{1}{2}\right)^2+\left(\frac{1}{2}\right)^2+\left(\frac{1}{2}\right)^2=(\text{area of } PQR)^2$.

EXAMPLE 7 $P=(1,0,0), Q=(0,1,0), R=(0,0,1)$ lie in a plane. Find its equation.

Idea for any P,Q,R: Find vectors $\mathbf{A}$ and $\mathbf{B}$ in the plane. Compute the normal $\mathbf{N}=\mathbf{A}\times\mathbf{B}$.

Solution   The vector from P to Q has components $-1,1,0$. It is $\mathbf{A}=\mathbf{j}-\mathbf{i}$ (subtract to go from P to Q). Similarly the vector from P to R is $\mathbf{B}=\mathbf{k}-\mathbf{i}$. Since $\mathbf{A}$ and $\mathbf{B}$ are in the plane of Figure 11.15, $\mathbf{N}=\mathbf{A}\times\mathbf{B}$ is perpendicular:

$$(\mathbf{j}-\mathbf{i})\times(\mathbf{k}-\mathbf{i})=(\mathbf{j}\times\mathbf{k})-(\mathbf{i}\times\mathbf{k})-(\mathbf{j}\times\mathbf{i})+(\mathbf{i}\times\mathbf{i})=\mathbf{i}+\mathbf{j}+\mathbf{k}.$$ (8)

The normal vector is $\mathbf{N}=\mathbf{i}+\mathbf{j}+\mathbf{k}$. The equation of the plane is $1x+1y+1z=d$.

With the right choice $d=1$, this plane contains P, Q, R. The equation is $x+y+z=1$.

EXAMPLE 8   What is the area of this same triangle PQR?

Solution   The area is half of the cross-product length $|\mathbf{A}\times\mathbf{B}|=|\mathbf{i}+\mathbf{j}+\mathbf{k}|=\sqrt{3}$.

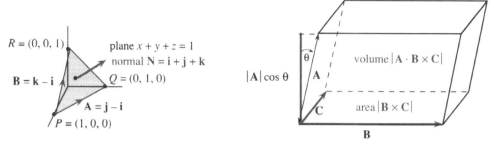

Fig. 11.15   Area of PQR is $\sqrt{3}/2$. **N** is $PQ \times PR$. **Volume of box is $|\mathbf{A} \cdot (\mathbf{B} \times \mathbf{C})|$.**

DETERMINANTS AND VOLUMES

We are close to good algebra. The two plane vectors $a_1\mathbf{i} + a_2\mathbf{j}$ and $b_1\mathbf{i} + b_2\mathbf{j}$ are the sides of a parallelogram. Its area is $a_1 b_2 - a_2 b_1$, possibly with a sign change. There is a special way to write these four numbers—in a "*square matrix*." There is also a name for the combination that leads to area. It is the "*determinant of the matrix*":

$$\textit{The matrix is } \begin{bmatrix} a_1 & a_2 \\ b_1 & b_2 \end{bmatrix}, \textit{ its determinant is } \begin{vmatrix} a_1 & a_2 \\ b_1 & b_2 \end{vmatrix} = a_1 b_2 - a_2 b_1.$$

This is a 2 by 2 matrix (notice brackets) and a 2 by 2 determinant (notice vertical bars). The matrix is an array of four numbers and the determinant is one number:

$$\text{Examples of determinants:} \quad \begin{vmatrix} 2 & 1 \\ 4 & 3 \end{vmatrix} = 6 - 4 = 2, \quad \begin{vmatrix} 2 & 1 \\ 2 & 1 \end{vmatrix} = 0, \quad \begin{vmatrix} 1 & 0 \\ 0 & 1 \end{vmatrix} = 1.$$

The second has no area because $\mathbf{A} = \mathbf{B}$. The third is a unit square ($\mathbf{A} = \mathbf{i}, \mathbf{B} = \mathbf{j}$).

Now move to three dimensions, where determinants are most useful. The parallelogram becomes a parallelepiped. The word "box" is much shorter, and we will use it, but remember that *the box is squashed*. (Like a rectangle squashed to a parallelogram, the angles are generally not 90°.) The three edges from the origin are $\mathbf{A} = (a_1, a_2, a_3), \mathbf{B} = (b_1, b_2, b_3), \mathbf{C} = (c_1, c_2, c_3)$. Those edges are at right angles only when $\mathbf{A} \cdot \mathbf{B} = \mathbf{A} \cdot \mathbf{C} = \mathbf{B} \cdot \mathbf{C} = 0$.

Question: **What is the volume of the box**? The right-angle case is easy—it is length times width times height. The volume is $|\mathbf{A}|$ times $|\mathbf{B}|$ times $|\mathbf{C}|$, when the angles are 90°. For a squashed box (Figure 11.15) we need the perpendicular height, not the sloping height.

There is a beautiful formula for volume. $\mathbf{B}$ and $\mathbf{C}$ give a parallelogram in the base, and $|\mathbf{B} \times \mathbf{C}|$ is the base area. This cross product points straight up. The third vector $\mathbf{A}$ points up at an angle—its perpendicular height is $|\mathbf{A}| \cos\theta$. Thus the volume is area $|\mathbf{B} \times \mathbf{C}|$ times $|\mathbf{A}|$ times $\cos\theta$. **The volume is the dot product of $\mathbf{A}$ with $\mathbf{B} \times \mathbf{C}$.**

> **11G** The **triple scalar product** is $\mathbf{A} \cdot (\mathbf{B} \times \mathbf{C})$. **Volume of box $= |\mathbf{A} \cdot (\mathbf{B} \times \mathbf{C})|$.**

Important: $\mathbf{A} \cdot (\mathbf{B} \times \mathbf{C})$ is a number, not a vector. This volume is zero when $\mathbf{A}$ is in the same plane as $\mathbf{B}$ and $\mathbf{C}$ (the box is totally flattened). Then $\mathbf{B} \times \mathbf{C}$ is perpendicular to $\mathbf{A}$ and their dot product is zero.

Useful facts: $\mathbf{A} \cdot (\mathbf{B} \times \mathbf{C}) = (\mathbf{A} \times \mathbf{B}) \cdot \mathbf{C} = \mathbf{C} \cdot (\mathbf{A} \times \mathbf{B}) = \mathbf{B} \cdot (\mathbf{C} \times \mathbf{A}).$

Remember $\mathbf{i} \times \mathbf{i} = \mathbf{j} \times \mathbf{j} = \mathbf{k} \times \mathbf{k} = \mathbf{0}$. Those three terms disappear. The other six terms come in pairs, and *please notice the cyclic pattern*:

FORMULA $\quad \mathbf{A} \times \mathbf{B} = (a_2 b_3 - a_3 b_2)\mathbf{i} + (a_3 b_1 - a_1 b_3)\mathbf{j} + (a_1 b_2 - a_2 b_1)\mathbf{k}.$ $\qquad$ (6)

The $\mathbf{k}$ component is the 2×2 answer, when $a_3 = b_3 = 0$. The $\mathbf{i}$ component involves indices 2 and 3, $\mathbf{j}$ involves 3 and 1, $\mathbf{k}$ involves 1 and 2. The cross product formula is written as a "determinant" in equation (14) below—many people use that form to compute $\mathbf{A} \times \mathbf{B}$.

EXAMPLE 6 $\quad (\mathbf{i} + 2\mathbf{j} + 3\mathbf{k}) \times (4\mathbf{i} + 5\mathbf{j} + 6\mathbf{k}) = (2 \cdot 6 - 3 \cdot 5)\mathbf{i} + (3 \cdot 4 - 1 \cdot 6)\mathbf{j} + (1 \cdot 5 - 2 \cdot 4)\mathbf{k}$. The $\mathbf{i}, \mathbf{j}, \mathbf{k}$ components give $\mathbf{A} \times \mathbf{B} = -3\mathbf{i} + 6\mathbf{j} - 3\mathbf{k}$. Never add the $-3, 6,$ and -3.

Remark 1 The three-dimensional formula (6) is still to be matched with $\mathbf{A} \times \mathbf{B}$ from geometry. One way is to rotate $\mathbf{B}$ into $\mathbf{B}^*$ as before, staying in the plane of $\mathbf{A}$ and $\mathbf{B}$. Fortunately there is an easier test. The vector in equation (6) satisfies all four geometric requirements on $\mathbf{A} \times \mathbf{B}$: *perpendicular to* $\mathbf{A}$, *perpendicular to* $\mathbf{B}$, *correct length, right-hand rule*. The length is checked in Problem 16—here is the zero dot product with $\mathbf{A}$:

$$\mathbf{A} \cdot (\mathbf{A} \times \mathbf{B}) = a_1(a_2 b_3 - a_3 b_2) + a_2(a_3 b_1 - a_1 b_3) + a_3(a_1 b_2 - a_2 b_1) = 0. \qquad (7)$$

Remark 2 (Optional) There is a wonderful extension of the Pythagoras formula $a^2 + b^2 = c^2$. Instead of sides of a triangle, we go to *areas of projections* on the yz, xz, and xy planes. $3^2 + 6^2 + 3^2$ is the square of the parallelogram area in Example 6.

For triangles these areas are cut in half. Figure 11.15a shows three projected triangles of area $\frac{1}{2}$. Its Pythagoras formula is $\left(\frac{1}{2}\right)^2 + \left(\frac{1}{2}\right)^2 + \left(\frac{1}{2}\right)^2 = (\text{area of } PQR)^2$.

EXAMPLE 7 $\quad P = (1, 0, 0), Q = (0, 1, 0), R = (0, 0, 1)$ lie in a plane. Find its equation.

Idea for any P, Q, R: *Find vectors* $\mathbf{A}$ *and* $\mathbf{B}$ *in the plane. Compute the normal* $\mathbf{N} = \mathbf{A} \times \mathbf{B}$.

Solution $\quad$ The vector from P to Q has components $-1, 1, 0$. It is $\mathbf{A} = \mathbf{j} - \mathbf{i}$ (subtract to go from P to Q). Similarly the vector from P to R is $\mathbf{B} = \mathbf{k} - \mathbf{i}$. Since $\mathbf{A}$ and $\mathbf{B}$ are in the plane of Figure 11.15, $\mathbf{N} = \mathbf{A} \times \mathbf{B}$ is perpendicular:

$$(\mathbf{j} - \mathbf{i}) \times (\mathbf{k} - \mathbf{i}) = (\mathbf{j} \times \mathbf{k}) - (\mathbf{i} \times \mathbf{k}) - (\mathbf{j} \times \mathbf{i}) + (\mathbf{i} \times \mathbf{i}) = \mathbf{i} + \mathbf{j} + \mathbf{k}. \qquad (8)$$

The normal vector is $\mathbf{N} = \mathbf{i} + \mathbf{j} + \mathbf{k}$. *The equation of the plane is* $1x + 1y + 1z = d$.

With the right choice $d = 1$, this plane contains P, Q, R. The equation is $x + y + z = 1$.

EXAMPLE 8 $\quad$ What is the area of this same triangle PQR?

Solution $\quad$ The area is half of the cross-product length $|\mathbf{A} \times \mathbf{B}| = |\mathbf{i} + \mathbf{j} + \mathbf{k}| = \sqrt{3}$.

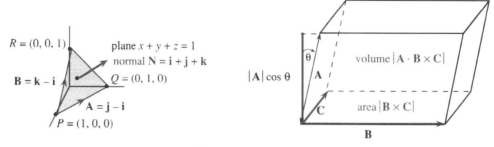

Fig. 11.15   Area of PQR is $\sqrt{3}/2$. **N** is $PQ \times PR$. **Volume of box is $|\mathbf{A} \cdot (\mathbf{B} \times \mathbf{C})|$.**

DETERMINANTS AND VOLUMES

We are close to good algebra. The two plane vectors $a_1\mathbf{i}+a_2\mathbf{j}$ and $b_1\mathbf{i}+b_2\mathbf{j}$ are the sides of a parallelogram. Its area is $a_1b_2 - a_2b_1$, possibly with a sign change. There is a special way to write these four numbers—in a "*square matrix*." There is also a name for the combination that leads to area. It is the "*determinant of the matrix*":

$$\textit{The matrix is } \begin{bmatrix} a_1 & a_2 \\ b_1 & b_2 \end{bmatrix}, \textit{ its determinant is } \begin{vmatrix} a_1 & a_2 \\ b_1 & b_2 \end{vmatrix} = a_1b_2 - a_2b_1.$$

This is a 2 by 2 matrix (notice brackets) and a 2 by 2 determinant (notice vertical bars). The matrix is an array of four numbers and the determinant is one number:

$$\text{Examples of determinants:} \quad \begin{vmatrix} 2 & 1 \\ 4 & 3 \end{vmatrix} = 6 - 4 = 2, \quad \begin{vmatrix} 2 & 1 \\ 2 & 1 \end{vmatrix} = 0, \quad \begin{vmatrix} 1 & 0 \\ 0 & 1 \end{vmatrix} = 1.$$

The second has no area because $\mathbf{A} = \mathbf{B}$. The third is a unit square ($\mathbf{A} = \mathbf{i}, \mathbf{B} = \mathbf{j}$).

Now move to three dimensions, where determinants are most useful. The parallelogram becomes a parallelepiped. The word "box" is much shorter, and we will use it, but remember that *the box is squashed.* (Like a rectangle squashed to a parallelogram, the angles are generally not $90°$.) The three edges from the origin are $\mathbf{A} = (a_1, a_2, a_3), \mathbf{B} = (b_1, b_2, b_3), \mathbf{C} = (c_1, c_2, c_3)$. Those edges are at right angles only when $\mathbf{A} \cdot \mathbf{B} = \mathbf{A} \cdot \mathbf{C} = \mathbf{B} \cdot \mathbf{C} = 0$.

Question: *What is the volume of the box*? The right-angle case is easy—it is length times width times height. The volume is $|\mathbf{A}|$ times $|\mathbf{B}|$ times $|\mathbf{C}|$, when the angles are $90°$. For a squashed box (Figure 11.15) we need the perpendicular height, not the sloping height.

There is a beautiful formula for volume. $\mathbf{B}$ and $\mathbf{C}$ give a parallelogram in the base, and $|\mathbf{B} \times \mathbf{C}|$ is the base area. This cross product points straight up. The third vector $\mathbf{A}$ points up at an angle—its perpendicular height is $|\mathbf{A}| \cos \theta$. Thus the volume is area $|\mathbf{B} \times \mathbf{C}|$ times $|\mathbf{A}|$ times $\cos \theta$. **The volume is the dot product of $\mathbf{A}$ with $\mathbf{B} \times \mathbf{C}$.**

***

11G The *triple scalar product* is $\mathbf{A} \cdot (\mathbf{B} \times \mathbf{C})$. **Volume of box $= |\mathbf{A} \cdot (\mathbf{B} \times \mathbf{C})|$.**

***

Important: $\mathbf{A} \cdot (\mathbf{B} \times \mathbf{C})$ is a number, not a vector. This volume is zero when $\mathbf{A}$ is in the same plane as $\mathbf{B}$ and $\mathbf{C}$ (the box is totally flattened). Then $\mathbf{B} \times \mathbf{C}$ is perpendicular to $\mathbf{A}$ and their dot product is zero.

Useful facts: $\mathbf{A} \cdot (\mathbf{B} \times \mathbf{C}) = (\mathbf{A} \times \mathbf{B}) \cdot \mathbf{C} = \mathbf{C} \cdot (\mathbf{A} \times \mathbf{B}) = \mathbf{B} \cdot (\mathbf{C} \times \mathbf{A}).$

All those come from the same box, with different sides chosen as base—but no change in volume. Figure 11.15 has **B** and **C** in the base but it can be **A** and **B** or **A** and **C**. *The triple product* $\mathbf{A} \cdot (\mathbf{C} \times \mathbf{B})$ *has opposite sign*, since $\mathbf{C} \times \mathbf{B} = -(\mathbf{B} \times \mathbf{C})$. This order **ACB** is not cyclic like **ABC** and **CAB** and **BCA**.

To compute this triple product $\mathbf{A} \cdot (\mathbf{B} \times \mathbf{C})$, we take $\mathbf{B} \times \mathbf{C}$ from equation (6):

$$\mathbf{A} \cdot (\mathbf{B} \times \mathbf{C}) = a_1(b_2c_3 - b_3c_2) + a_2(b_3c_1 - b_1c_3) + a_3(b_1c_2 - b_2c_1). \tag{9}$$

The numbers a_1, a_2, a_3 multiply 2 by 2 determinants to give a 3 by 3 determinant! There are three terms with plus signs (like $a_1b_2c_3$). The other three have minus signs (like $-a_1b_3c_2$). The plus terms have indices $123, 231, 312$ in cyclic order. The minus terms have anticyclic indices $132, 213, 321$. Again there is a special way to write the nine components of $\mathbf{A}, \mathbf{B}, \mathbf{C}$—as a "3 by 3 matrix." The combination in (9), which gives volume, is a "3 by 3 determinant:"

$$\textbf{matrix} = \begin{bmatrix} a_1 & a_2 & a_3 \\ b_1 & b_2 & b_3 \\ c_1 & c_2 & c_3 \end{bmatrix}, \quad \textbf{determinant} = \mathbf{A} \cdot (\mathbf{B} \times \mathbf{C}) = \begin{vmatrix} a_1 & a_2 & a_3 \\ b_1 & b_2 & b_3 \\ c_1 & c_2 & c_3 \end{vmatrix}.$$

A single number is produced out of nine numbers, by formula (9). The nine numbers are multiplied three at a time, as in $a_1b_1c_2$—except this product is not allowed. *Each row and column must be represented once*. This gives the six terms in the determinant:

$$\begin{vmatrix} a_1 & a_2 & a_3 \\ b_1 & b_2 & b_3 \\ c_1 & c_2 & c_3 \end{vmatrix} = \begin{array}{l} a_1b_2c_3 + a_2b_3c_1 + a_3b_1c_2 \\ -a_1b_3c_2 - a_2b_1c_3 - a_3b_2c_1 \end{array} \tag{10}$$

The trick is in the $\pm$ signs. Products down to the right are "*plus*":

$$\begin{vmatrix} 2 & 1 & 1 \\ 1 & 2 & 1 \\ 1 & 1 & 2 \end{vmatrix} = \begin{array}{c} 2 \cdot 2 \cdot 2 + 1 \cdot 1 \cdot 1 + 1 \cdot 1 \cdot 1 \\ -2 \cdot 1 \cdot 1 - 1 \cdot 1 \cdot 2 - 1 \cdot 2 \cdot 1 \end{array} = \begin{array}{c} 8 + 1 + 1 \\ -2 - 2 - 2 \end{array} = 4.$$

With practice the six products like $2 \cdot 2 \cdot 2$ are done in your head. Write down only $8 + 1 + 1 - 2 - 2 - 2 = 4$. This is the determinant and the volume.

Note the special case when the vectors are $\mathbf{i}, \mathbf{j}, \mathbf{k}$. The box is a unit cube:

$$\textbf{volume of cube} = \begin{vmatrix} 1 & 0 & 0 \\ 0 & 1 & 0 \\ 0 & 0 & 1 \end{vmatrix} = \begin{array}{c} 1 + 0 + 0 \\ -0 - 0 - 0 \end{array} = 1.$$

If $\mathbf{A}, \mathbf{B}, \mathbf{C}$ *lie in the same plane, the volume is zero*. A zero determinant is the test to see whether three vectors lie in a plane. Here row $\mathbf{A} = $ row $\mathbf{B} - $ row $\mathbf{C}$:

$$\begin{vmatrix} 0 & 1 & -1 \\ -1 & 1 & 0 \\ -1 & 0 & 1 \end{vmatrix} = \begin{array}{c} 0 \cdot 1 \cdot 1 + 1 \cdot 0 \cdot (-1) + (-1) \cdot (-1) \cdot 0 \\ -0 \cdot 0 \cdot 0 - 1 \cdot (-1) \cdot 1 - (-1) \cdot 1 \cdot (-1) \end{array} = 0. \tag{11}$$

Zeros in the matrix simplify the calculation. All three products with plus signs—down to the right—are zero. The only two nonzero products cancel each other.

If the three -1's are changed to $+1$'s, the determinant is -2. The determinant can be negative when all nine entries are positive! A negative determinant only means that the rows $\mathbf{A}, \mathbf{B}, \mathbf{C}$ form a "left-handed triple." This extra information from the sign—right-handed vs. left-handed—is free and useful, but the volume is the absolute value.

The determinant yields the volume also in higher dimensions. In physics, four dimensions give space-time. Ten dimensions give superstrings. Mathematics uses all dimensions. The 64 numbers in an 8 by 8 matrix give the volume of an eight-dimensional box—with $8! = 40,320$ terms instead of $3! = 6$. Under pressure from my class I omit the formula.

Question When is the point (x, y, z) on the plane through the origin containing $\mathbf{B}$ and $\mathbf{C}$? For the vector $\mathbf{A} = x\mathbf{i} + y\mathbf{j} + z\mathbf{k}$ to lie in that plane, the volume $\mathbf{A} \cdot (\mathbf{B} \times \mathbf{C})$ must be zero. The equation of the plane is ***determinant = zero***.

Follow this example for $\mathbf{B} = \mathbf{j} - \mathbf{i}$ and $\mathbf{C} = \mathbf{k} - \mathbf{i}$ to find the plane parallel to $\mathbf{B}$ and $\mathbf{C}$:

$$\begin{vmatrix} x & y & z \\ -1 & 1 & 0 \\ -1 & 0 & 1 \end{vmatrix} = \begin{matrix} x \cdot 1 \cdot 1 + y \cdot 0 \cdot (-1) + z \cdot 0 \cdot (-1) \\ -x \cdot 0 \cdot 0 - y \cdot 1 \cdot (-1) - z \cdot 1 \cdot (-1) \end{matrix} = 0. \tag{12}$$

This equation is $x + y + z = 0$. The normal vector $\mathbf{N} = \mathbf{B} \times \mathbf{C}$ has components $1, 1, 1$.

THE CROSS PRODUCT AS A DETERMINANT

There is a connection between 3 by 3 and 2 by 2 determinants that you have to see. The numbers in the top row multiply determinants from the other rows:

$$\begin{vmatrix} \underline{a_1} & a_2 & a_3 \\ b_1 & \underline{b_2} & \underline{b_3} \\ c_1 & \underline{c_2} & \underline{c_3} \end{vmatrix} = \underline{a_1} \begin{vmatrix} \underline{b_2} & \underline{b_3} \\ \underline{c_2} & \underline{c_3} \end{vmatrix} - a_2 \begin{vmatrix} b_1 & b_3 \\ c_1 & c_3 \end{vmatrix} + a_3 \begin{vmatrix} b_1 & b_2 \\ c_1 & c_2 \end{vmatrix}. \tag{13}$$

The highlighted product $a_1(b_2c_3 - b_3c_2)$ gives two of the six terms. ***All six products contain an a and b and c from different columns***. There are $3! = 6$ different orderings of columns $1, 2, 3$. Note how a_3 multiplies a determinant from columns 1 and 2.

Equation (13) is identical with equations (9) and (10). We are meeting the same six terms in different ways. The new feature is the minus sign in front of a_2—and the common mistake is to forget that sign. In a 4 by 4 determinant, $a_1, -a_2, a_3, -a_4$ would multiply 3 by 3 determinants.

Now comes a key step. We write $\mathbf{A} \times \mathbf{B}$ as a determinant. The vectors $\mathbf{i}, \mathbf{j}, \mathbf{k}$ go in the top row, the components of $\mathbf{A}$ and $\mathbf{B}$ go in the other rows. ***The "determinant" is exactly*** $\mathbf{A} \times \mathbf{B}$:

$$\mathbf{A} \times \mathbf{B} = \begin{vmatrix} \mathbf{i} & \mathbf{j} & \mathbf{k} \\ \underline{a_1} & a_2 & \underline{a_3} \\ \underline{b_1} & b_2 & \underline{b_3} \end{vmatrix} = \mathbf{i} \begin{vmatrix} a_2 & a_3 \\ b_2 & b_3 \end{vmatrix} - \mathbf{j} \begin{vmatrix} \underline{a_1} & \underline{a_3} \\ \underline{b_1} & \underline{b_3} \end{vmatrix} + \mathbf{k} \begin{vmatrix} a_1 & a_2 \\ b_1 & b_2 \end{vmatrix}. \tag{14}$$

This time we highlighted the $\mathbf{j}$ component with its minus sign. There is no great mathematics in formula (14)—it is probably illegal to mix $\mathbf{i}, \mathbf{j}, \mathbf{k}$ with six numbers but it works. This is the good way to remember and compute $\mathbf{A} \times \mathbf{B}$. In the example $(\mathbf{j}-\mathbf{i}) \times (\mathbf{k}-\mathbf{i})$ from equation (8), those two vectors go into the last two rows:

$$\begin{vmatrix} \mathbf{i} & \mathbf{j} & \mathbf{k} \\ -1 & 1 & 0 \\ -1 & 0 & 1 \end{vmatrix} = \mathbf{i} \begin{vmatrix} 1 & 0 \\ 0 & 1 \end{vmatrix} - \mathbf{j} \begin{vmatrix} -1 & 0 \\ -1 & 1 \end{vmatrix} + \mathbf{k} \begin{vmatrix} -1 & 1 \\ -1 & 0 \end{vmatrix} = \mathbf{i}+\mathbf{j}+\mathbf{k}.$$

The $\mathbf{k}$ component is highlighted, to see $a_1 b_2 - a_2 b_1$ again. Note the change from equation (11), which had $0,1,-1$ in the top row. That triple product was a number (zero). This cross product is a vector $\mathbf{i}+\mathbf{j}+\mathbf{k}$.

Review question 1 With the $\mathbf{i}, \mathbf{j}, \mathbf{k}$ row changed to $3, 4, 5$, what is the determinant?
Answer $\quad 3 \cdot 1 + 4 \cdot 1 + 5 \cdot 1 = 12$. That triple product is the volume of a box.

Review question 2 When is $\mathbf{A} \times \mathbf{B} = \mathbf{0}$ and when is $\mathbf{A} \cdot (\mathbf{B} \times \mathbf{C}) = 0$? Zero vector, zero number.
Answer $\quad$ When $\mathbf{A}$ and $\mathbf{B}$ are on the same line. When $\mathbf{A}, \mathbf{B}, \mathbf{C}$ are in the same plane.

Review question 3 Does the parallelogram area $|\mathbf{A} \times \mathbf{B}|$ equal a 2 by 2 determinant?
Answer $\quad$ If $\mathbf{A}$ and $\mathbf{B}$ lie in the xy plane, *yes*. Generally *no*.

Review question 4 What are the *vector* triple products $(\mathbf{A} \times \mathbf{B}) \times \mathbf{C}$ and $\mathbf{A} \times (\mathbf{B} \times \mathbf{C})$?
Answer $\quad$ Not computed yet. These are two new vectors in Problem 47.

Review question 5 Find the plane through the origin containing $\mathbf{A} = \mathbf{i}+\mathbf{j}+2\mathbf{k}$ and $\mathbf{B} = \mathbf{i}+\mathbf{k}$. Find the cross product of those same vectors $\mathbf{A}$ and $\mathbf{B}$.
Answer $\quad$ The position vector $\mathbf{P} = x\mathbf{i}+y\mathbf{j}+z\mathbf{k}$ is perpendicular to $\mathbf{N} = \mathbf{A} \times \mathbf{B}$:

$$\mathbf{P} \cdot (\mathbf{A} \times \mathbf{B}) = \begin{vmatrix} x & y & z \\ 1 & 1 & 2 \\ 1 & 0 & 1 \end{vmatrix} = x+y-z = 0. \quad \mathbf{A} \times \mathbf{B} = \begin{vmatrix} \mathbf{i} & \mathbf{j} & \mathbf{k} \\ 1 & 1 & 2 \\ 1 & 0 & 1 \end{vmatrix} = \mathbf{i}+\mathbf{j}-\mathbf{k}.$$

11.3 EXERCISES

Read-through questions

The cross product $\mathbf{A} \times \mathbf{B}$ is a __a__ whose length is __b__. Its direction is __c__ to $\mathbf{A}$ and $\mathbf{B}$. That length is the area of a __d__, whose base is $|\mathbf{A}|$ and whose height is __e__. When $\mathbf{A} = a_1\mathbf{i}+a_2\mathbf{j}$ and $\mathbf{B} = b_1\mathbf{i}+b_2\mathbf{j}$, the area is __f__. This equals a 2 by 2 __g__. In general $|\mathbf{A} \cdot \mathbf{B}|^2 + |\mathbf{A} \times \mathbf{B}|^2 = $ __h__.

The rules for cross product are $\mathbf{A} \times \mathbf{A} = $ __i__ and $\mathbf{A} \times \mathbf{B} = -($ __j__ $)$ and $\mathbf{A} \times (\mathbf{B}+\mathbf{C}) = \mathbf{A} \times \mathbf{B}+$ __k__. In particular $\mathbf{A} \times \mathbf{B}$ needs the __l__-hand rule to decide its direction. If the fingers curl from $\mathbf{A}$ towards $\mathbf{B}$ (not more than $180°$), then __m__ points __n__. By this rule $\mathbf{i} \times \mathbf{j} = $ __o__ and $\mathbf{i} \times \mathbf{k} = $ __p__ and $\mathbf{j} \times \mathbf{k} = $ __q__.

The vectors $a_1\mathbf{i}+a_2\mathbf{j}+a_3\mathbf{k}$ and $b_1\mathbf{i}+b_2\mathbf{j}+b_3\mathbf{k}$ have cross product __r__ $\mathbf{i}+$ __s__ $\mathbf{j}+$ __t__ $\mathbf{k}$. The vectors $\mathbf{A} = \mathbf{i}+\mathbf{j}+\mathbf{k}$ and $\mathbf{B} = \mathbf{i}+\mathbf{j}$ have $\mathbf{A} \times \mathbf{B} = $ __u__. (*This is also the* 3 *by* 3 *determinant* __v__.) Perpendicular to the plane containing $(0,0,0),(1,1,1),(1,1,0)$ is the normal vector $\mathbf{N} = $ __w__. The area of the triangle with those three vertices is __x__, which is half the area of the parallelogram with fourth vertex at __y__.

Vectors, $\mathbf{A}, \mathbf{B}, \mathbf{C}$ from the origin determine a __z__. Its volume $|\mathbf{A} \cdot ($ __A__ $)|$ comes from a 3 by 3 __B__. There are six terms, __C__ with a plus sign and __D__ with minus. In every term each row and __E__ is represented once. The rows $(1,0,0),(0,0,1),$

and $(0,1,0)$ have determinant $=$ __F__ . That box is a __G__ , but its sides form a __H__-handed triple in the order given.

If $\mathbf{A},\mathbf{B},\mathbf{C}$ lie in the same plane then $\mathbf{A}\cdot(\mathbf{B}\times\mathbf{C})$ is __I__ . For $\mathbf{A}=x\mathbf{i}+y\mathbf{j}+z\mathbf{k}$ the first row contains the letters __J__ . So the plane containing $\mathbf{B}$ and $\mathbf{C}$ has the equation __K__ $=0$. When $\mathbf{B}=\mathbf{i}+\mathbf{j}$ and $\mathbf{C}=\mathbf{k}$ that equation is __L__ . $\mathbf{B}\times\mathbf{C}$ is __M__ .

A 3 by 3 determinant splits into __N__ 2 by 2 determinants. They come from rows 2 and 3, and are multiplied by the entries in row 1. With $\mathbf{i},\mathbf{j},\mathbf{k}$ in row 1, this determinant equals the __O__ product. Its $\mathbf{j}$ component is __P__ , including the __Q__ sign which is easy to forget.

Compute the cross products 1–8 from formula (6) or the determinant (14). Do one example both ways.

1 $(\mathbf{i}\times\mathbf{j})\times\mathbf{k}$ 2 $(\mathbf{i}\times\mathbf{j})\times\mathbf{i}$

3 $(2\mathbf{i}+3\mathbf{j})\times(\mathbf{i}+\mathbf{k})$ 4 $(2\mathbf{i}+3\mathbf{j}+\mathbf{k})\times(2\mathbf{i}+3\mathbf{j}-\mathbf{k})$

5 $(2\mathbf{i}+3\mathbf{j}+\mathbf{k})\times(\mathbf{i}-\mathbf{j}-\mathbf{k})$ 6 $(\mathbf{i}+\mathbf{j}-\mathbf{k})\times(\mathbf{i}-\mathbf{j}+\mathbf{k})$

7 $(\mathbf{i}+2\mathbf{j}+3\mathbf{k})\times(4\mathbf{i}-9\mathbf{j})$

8 $(\mathbf{i}\cos\theta+\mathbf{j}\sin\theta)\times(\mathbf{i}\sin\theta-\mathbf{j}\cos\theta)$

9 When are $|\mathbf{A}\times\mathbf{B}|=|\mathbf{A}||\mathbf{B}|$ and $|\mathbf{A}\cdot(\mathbf{B}\times\mathbf{C})|=|\mathbf{A}||\mathbf{B}||\mathbf{C}|$?

10 **True or false**:
   (a) $\mathbf{A}\times\mathbf{B}$ never equals $\mathbf{A}\cdot\mathbf{B}$.
   (b) If $\mathbf{A}\times\mathbf{B}=\mathbf{0}$ and $\mathbf{A}\cdot\mathbf{B}=0$, then either $\mathbf{A}=\mathbf{0}$ or $\mathbf{B}=\mathbf{0}$.
   (c) If $\mathbf{A}\times\mathbf{B}=\mathbf{A}\times\mathbf{C}$ and $\mathbf{A}\neq\mathbf{0}$, then $\mathbf{B}=\mathbf{C}$.

In 11–16 find $|\mathbf{A}\times\mathbf{B}|$ by equation (1) and then by computing $\mathbf{A}\times\mathbf{B}$ and its length.

11 $\mathbf{A}=\mathbf{i}+\mathbf{j}+\mathbf{k},\ \mathbf{B}=\mathbf{i}$ 12 $\mathbf{A}=\mathbf{i}+\mathbf{j},\ \mathbf{B}=\mathbf{i}-\mathbf{j}$

13 $\mathbf{A}=-\mathbf{B}$ 14 $\mathbf{A}=\mathbf{i}+\mathbf{j},\ \mathbf{B}=\mathbf{j}+\mathbf{k}$

15 $\mathbf{A}=a_1\mathbf{i}+a_2\mathbf{j},\ \mathbf{B}=b_1\mathbf{i}+b_2\mathbf{j}$

16 $\mathbf{A}=(a_1,a_2,a_3),\ \mathbf{B}=(b_1,b_2,b_3)$

In Problem 16 (the general case), equation (1) proves that the length from equation (6) is correct.

17 **True or false**, by testing on $\mathbf{A}=\mathbf{i},\ \mathbf{B}=\mathbf{j},\ \mathbf{C}=\mathbf{k}$:
   (a) $\mathbf{A}\times(\mathbf{A}\times\mathbf{B})=\mathbf{0}$ (b) $\mathbf{A}\cdot(\mathbf{B}\times\mathbf{C})=(\mathbf{A}\times\mathbf{B})\cdot\mathbf{C}$
   (c) $\mathbf{A}\cdot(\mathbf{B}\times\mathbf{C})=\mathbf{C}\cdot(\mathbf{B}\times\mathbf{A})$
   (d) $(\mathbf{A}-\mathbf{B})\times(\mathbf{A}+\mathbf{B})=2(\mathbf{A}\times\mathbf{B})$.

18 (a) From $\mathbf{A}\times\mathbf{B}=-(\mathbf{B}\times\mathbf{A})$ deduce that $\mathbf{A}\times\mathbf{A}=\mathbf{0}$.
   (b) Split $(\mathbf{A}+\mathbf{B})\times(\mathbf{A}+\mathbf{B})$ into four terms, to deduce that $(\mathbf{A}\times\mathbf{B})=-(\mathbf{B}\times\mathbf{A})$.

What are the normal vectors to the planes 19–22 ?

19 $(2,1,0)\cdot(x,y,z)=4$ 20 $3x+4z=5$

21 $\begin{vmatrix} x & y & z \\ 1 & 1 & 0 \\ 0 & 1 & 1 \end{vmatrix}=2$ 22 $\begin{vmatrix} x & y & z \\ 1 & 1 & 1 \\ 1 & 1 & 2 \end{vmatrix}=0$

Find N and the equation of the plane described in 23–29.

23 Contains the points $(2,1,1)$, $(1,2,1)$, $(1,1,2)$

24 Contains the points $(0,1,2)$, $(1,2,3)$, $(2,3,4)$

25 Through $(0,0,0)$, $(1,1,1)$, (a,b,c) [What if $a=b=c$?]

26 Parallel to $\mathbf{i}+\mathbf{j}$ and $\mathbf{k}$

27 $\mathbf{N}$ makes a $45°$ angle with $\mathbf{i}$ and $\mathbf{j}$

28 $\mathbf{N}$ makes a $60°$ angle with $\mathbf{i}$ and $\mathbf{j}$

29 $\mathbf{N}$ makes a $90°$ angle with $\mathbf{i}$ and $\mathbf{j}$

30 The triangle with sides $\mathbf{i}$ and $\mathbf{j}$ is _____ as large as the parallelogram with those sides. The tetrahedron with edges $\mathbf{i},\mathbf{j},\mathbf{k}$ is _____ as large as the box with those edges. Extra credit: In four dimensions the "simplex" with edges $\mathbf{i},\mathbf{j},\mathbf{k},\mathbf{l}$ has volume $=$ _____ .

31 If the points (x,y,z), $(1,1,0)$, and $(1,2,1)$ lie on a plane through the origin, what determinant is zero? What equation does this give for the plane ?

32 Give an example of a right-hand triple and left-hand triple. Use vectors other than just $\mathbf{i},\mathbf{j},\mathbf{k}$.

33 When $\mathbf{B}=3\mathbf{i}+\mathbf{j}$ is rotated $90°$ clockwise in the xy plane it becomes $\mathbf{B}^*=$ _____ . When rotated $90°$ counterclockwise it is _____ . When rotated $180°$ it is _____ .

34 From formula (6) verify that $\mathbf{B}\cdot(\mathbf{A}\times\mathbf{B})=0$.

35 Compute

$$\begin{vmatrix} 1 & 2 & 3 \\ 2 & 3 & 4 \\ 3 & 4 & 6 \end{vmatrix},\ \begin{vmatrix} 2 & 1 & 0 \\ 1 & 2 & 1 \\ 0 & 1 & 2 \end{vmatrix},\ \begin{vmatrix} 1 & 0 & 2 \\ 0 & 3 & 0 \\ 2 & 0 & 1 \end{vmatrix}.$$

36 Which of the following are equal to $\mathbf{A}\times\mathbf{B}$?
$(\mathbf{A}+\mathbf{B})\times\mathbf{B},\ (-\mathbf{B})\times(-\mathbf{A}),\ |\mathbf{A}||\mathbf{B}||\sin\theta|,\ (\mathbf{A}+\mathbf{C})\times(\mathbf{B}-\mathbf{C})$, $\frac{1}{2}(\mathbf{A}-\mathbf{B})\times(\mathbf{A}+\mathbf{B})$.

37 Compare the six terms on both sides to prove that

$$\begin{vmatrix} a_1 & b_1 & c_1 \\ a_2 & b_2 & c_2 \\ a_3 & b_3 & c_3 \end{vmatrix}=\begin{vmatrix} a_1 & a_2 & a_3 \\ b_1 & b_2 & b_3 \\ c_1 & c_2 & c_3 \end{vmatrix}.$$

The matrix is "*transposed*"—same determinant.

38 Compare the six terms to prove that

$$\begin{vmatrix} a_1 & a_2 & a_3 \\ b_1 & b_2 & b_3 \\ c_1 & c_2 & c_3 \end{vmatrix}=-b_1\begin{vmatrix} a_2 & a_3 \\ c_2 & c_3 \end{vmatrix}+b_2\begin{vmatrix} a_1 & a_3 \\ c_1 & c_3 \end{vmatrix}-b_3\begin{vmatrix} a_1 & a_2 \\ c_1 & c_2 \end{vmatrix}.$$

This is an "expansion on row 2." Note minus signs.

39 Choose the signs and 2 by 2 determinants in

$$\begin{vmatrix} a_1 & a_2 & a_3 \\ b_1 & b_2 & b_3 \\ c_1 & c_2 & c_3 \end{vmatrix} = \pm c_1 \begin{vmatrix} a_2 & a_3 \\ b_2 & b_3 \end{vmatrix} \pm c_2 \underline{\hspace{1cm}} \pm c_3 \underline{\hspace{1cm}}.$$

40 Show that $(\mathbf{A} \times \mathbf{B}) + (\mathbf{B} \times \mathbf{C}) + (\mathbf{C} \times \mathbf{A})$ is perpendicular to $\mathbf{B} - \mathbf{A}$ and $\mathbf{C} - \mathbf{B}$ and $\mathbf{A} - \mathbf{C}$.

Problems 41–44 compute the areas of triangles.

41 The triangle PQR in Example 7 has squared area $(\sqrt{3}/2)^2 = (\frac{1}{2})^2 + (\frac{1}{2})^2 + (\frac{1}{2})^2$, from the 3D version of Pythagoras in Remark 2. Find the area of PQR when $P = (a,0,0)$, $Q = (0,b,0)$, and $R = (0,0,c)$. Check with $\frac{1}{2}|\mathbf{A} \times \mathbf{B}|$.

42 A triangle in the xy plane has corners at $(a_1,b_1), (a_2,b_2)$ and (a_3,b_3). Its area A is half the area of a parallelogram. Find two sides of the parallelogram and explain why

$$A = \tfrac{1}{2}|(a_2 - a_1)(b_3 - b_1) - (a_3 - a_1)(b_2 - b_1)|.$$

43 By Problem 42 find the area A of the triangle with corners $(2,1)$ and $(4,2)$ and $(1,2)$. Where is a fourth corner to make a parallelogram?

44 Lifting the triangle of Problem 42 up to the plane $z = 1$ gives corners $(a_1,b_1,1), (a_2,b_2,1), (a_3,b_3,1)$. The area of the triangle times $\frac{1}{3}$ is the volume of the upside-down pyramid from $(0,0,0)$ to these corners. This pyramid volume is $\frac{1}{6}$ the box volume, so $\frac{1}{3}$ (area of triangle) $= \frac{1}{6}$(volume of box):

$$area\ of\ triangle = \frac{1}{2} \begin{vmatrix} a_1 & b_1 & 1 \\ a_2 & b_2 & 1 \\ a_3 & b_3 & 1 \end{vmatrix}.$$

Find the area A in Problem 43 from this determinant.

45 (1) The projections of $\mathbf{A} = a_1\mathbf{i} + a_2\mathbf{j} + a_3\mathbf{k}$ and $\mathbf{B} = b_1\mathbf{i} + b_2\mathbf{j} + b_3\mathbf{k}$ onto the xy plane are \underline{\hspace{1cm}}.

(2) The parallelogram with sides $\mathbf{A}$ and $\mathbf{B}$ projects to a parallelogram with area \underline{\hspace{1cm}}.

(3) General fact: The projection onto the plane normal to the unit vector $\mathbf{n}$ has area $(\mathbf{A} \times \mathbf{B}) \cdot \mathbf{n}$. Verify for $\mathbf{n} = \mathbf{k}$.

46 (a) For $\mathbf{A} = \mathbf{i} + \mathbf{j} - 4\mathbf{k}$ and $\mathbf{B} = -\mathbf{i} + \mathbf{j}$, compute $(\mathbf{A} \times \mathbf{B}) \cdot \mathbf{i}$ and $(\mathbf{A} \times \mathbf{B}) \cdot \mathbf{j}$ and $(\mathbf{A} \times \mathbf{B}) \cdot \mathbf{k}$. By Problem 45 those are the areas of projections onto the yz and xz and xy planes.

(b) Square and add those areas to find $|\mathbf{A} \times \mathbf{B}|^2$. This is the Pythagoras formula in space (Remark 2).

47 (a) The triple cross product $(\mathbf{A} \times \mathbf{B}) \times \mathbf{C}$ is in the plane of $\mathbf{A}$ and $\mathbf{B}$, because it is perpendicular to the cross product \underline{\hspace{1cm}}.

(b) Compute $(\mathbf{A} \times \mathbf{B}) \times \mathbf{C}$ when $\mathbf{A} = a_1\mathbf{i} + a_2\mathbf{j} + a_3\mathbf{k}, \mathbf{B} = b_1\mathbf{i} + b_2\mathbf{j} + b_3\mathbf{k}, \mathbf{C} = \mathbf{i}$.

(c) Compute $(\mathbf{A} \cdot \mathbf{C})\mathbf{B} - (\mathbf{B} \cdot \mathbf{C})\mathbf{A}$ when $\mathbf{C} = \mathbf{i}$. The answers in (b) and (c) should agree. This is also true if $\mathbf{C} = \mathbf{j}$ or $\mathbf{C} = \mathbf{k}$ or $\mathbf{C} = c_1\mathbf{i} + c_2\mathbf{j} + c_3\mathbf{k}$. That proves the tricky formula

$$(\mathbf{A} \times \mathbf{B}) \times \mathbf{C} = (\mathbf{A} \cdot \mathbf{C})\mathbf{B} - (\mathbf{B} \cdot \mathbf{C})\mathbf{A}. \tag{$*$}$$

48 Take the dot product of equation ($*$) with $\mathbf{D}$ to prove

$$(\mathbf{A} \times \mathbf{B}) \cdot (\mathbf{C} \times \mathbf{D}) = (\mathbf{A} \cdot \mathbf{C})(\mathbf{B} \cdot \mathbf{D}) - (\mathbf{B} \cdot \mathbf{C})(\mathbf{A} \cdot \mathbf{D}).$$

49 The plane containing $P = (0,1,1)$ and $Q = (1,0,1)$ and $R = (1,1,0)$ is perpendicular to the cross product $\mathbf{N} = $ \underline{\hspace{1cm}}. Find the equation of the plane and the area of triangle PQR.

50 Let $P = (1,0,-1), Q = (1,1,1), R = (2,2,1)$. Choose S so that $PQRS$ is a parallelogram and compute its area. Choose T, U, V so that $OPQRSTUV$ is a box (parallelepiped) and compute its volume.

11.4 Matrices and Linear Equations

We are moving from geometry to algebra. Eventually we get back to calculus, where functions are nonlinear—but linear equations come first. In Chapter 1, $y = mx + b$ produced a line. Two equations produce two lines. If they cross, the intersection point solves both equations—and we want to find it.

Three equations in three variables x, y, z produce three planes. Again they go through one point (*usually*). Again the problem is to find that intersection point —which solves the three equations.

The ultimate problem is to solve n equations in n unknowns. There are n hyperplanes in n-dimensional space, which meet at the solution. We need a test to be sure they meet. We also want the solution. These are the objectives of **linear algebra**, which joins with calculus at the center of pure and applied mathematics.†

Like every subject, linear algebra requires a good notation. To state the equations and solve them, we introduce a "matrix." ***The problem will be*** $A\mathbf{u} = \mathbf{d}$. ***The solution will be*** $\mathbf{u} = A^{-1}\mathbf{d}$. It remains to understand where the equations come from, where the answer comes from, and what the matrices A and A^{-1} stand for.

TWO EQUATIONS IN TWO UNKNOWNS

Linear algebra has no reason to choose one variable as special. The equation $y - y_0 = m(x - x_0)$ separates y from x. A better equation for a line is $ax + by = d$. (A vertical line like $x = 5$ appears when $b = 0$. The first form did not allow slope $m = \infty$.) This section studies two lines:

$$a_1 x + b_1 y = d_1$$
$$a_2 x + b_2 y = d_2. \tag{1}$$

By solving both equations at once, we are asking (x, y) to lie on both lines. The practical question is: Where do the lines cross? The mathematician's question is: Does a solution exist and is it unique?

To understand everything is not possible. There are parts of life where you never know what is going on (until too late). But two equations in two unknowns can have no mysteries. There are three ways to write the system—by **rows**, by **columns**, and by **matrices**. Please look at all three, since setting up a problem is generally harder and more important than solving it. After that comes the concession to the real world: we compute x and y.

EXAMPLE 1   How do you invest \$5000 to earn \$400 a year interest, if a money market account pays 5% and a deposit account pays 10%?

Set up equations by rows: With x dollars at 5% the interest is $.05x$. With y dollars at 10% the interest is $.10y$. One row for principal, another row for interest:

$$x + \quad y = 5000$$
$$.05x + .10y = \quad 400. \tag{2}$$

---

†Linear algebra dominates some applications while calculus governs others. Both are essential. A fuller treatment is presented in the author's book *Linear Algebra and Its Applications* (Harcourt Brace Jovanovich, 3rd edition 1988), and in many other texts.

Same equations by columns: The left side of (2) contains x times one vector plus y times another vector. The right side is a third vector. The equation by columns is

$$x \begin{bmatrix} 1 \\ .05 \end{bmatrix} + y \begin{bmatrix} 1 \\ .10 \end{bmatrix} = \begin{bmatrix} 5000 \\ 400 \end{bmatrix}. \tag{3}$$

Same equations by matrices: Look again at the left side. There are two unknowns x and y, which go into a vector $\mathbf{u}$. They are multiplied by the four numbers 1, .05, 1, and .10, which go into a *two by two matrix* A. The left side becomes *a matrix times a vector*:

$$A\mathbf{u} = \begin{bmatrix} 1 & 1 \\ .05 & .10 \end{bmatrix} \begin{bmatrix} x \\ y \end{bmatrix} = \begin{bmatrix} 5000 \\ 400 \end{bmatrix}. \tag{4}$$

Now you see where the "rows" and "columns" came from. They are the rows and columns of a matrix. The rows entered the separate equations (2). The columns entered the vector equation (3). The matrix-vector multiplication $A\mathbf{u}$ is defined so that all these equations are the same:

$$A\mathbf{u} \text{ by rows:} \quad \begin{bmatrix} a_1 & b_1 \\ a_2 & b_2 \end{bmatrix} \begin{bmatrix} x \\ y \end{bmatrix} = \begin{bmatrix} a_1 x + b_1 y \\ a_2 x + b_2 y \end{bmatrix} \quad \text{(each row is a dot product)}$$

$$A\mathbf{u} \text{ by columns:} \quad \begin{bmatrix} a_1 & b_1 \\ a_2 & b_2 \end{bmatrix} \begin{bmatrix} x \\ y \end{bmatrix} = x \begin{bmatrix} a_1 \\ a_2 \end{bmatrix} + y \begin{bmatrix} b_1 \\ b_2 \end{bmatrix} \quad \text{(combination of column vectors)}$$

A is the *coefficient matrix*. The unknown vector is $\mathbf{u}$. The known vector on the right side, with components 5000 and 400, is $\mathbf{d}$. The matrix equation is $A\mathbf{u} = \mathbf{d}$.

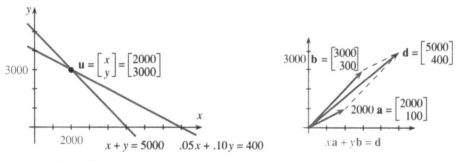

Fig. 11.16   Each row of $A\mathbf{u} = \mathbf{d}$ gives a line. Each column gives a vector.

This notation $A\mathbf{u} = \mathbf{d}$ continues to apply when there are more equations and more unknowns. The matrix A has a *row for each equation* (usually m rows). It has a *column for each unknown* (usually n columns). For 2 equations in 3 unknowns it is a 2 by 3 matrix (therefore rectangular). For 6 equations in 6 unknowns the matrix is 6 by 6 (therefore square). The best way to get familiar with matrices is to work with them. Note also the pronunciation: "matri*sees*" and never "matrixes."

Answer to the practical question The solution is $x = 2000$, $y = 3000$. That is the intersection point in the row picture (Figure 11.16). It is also the correct combination in the column picture. The matrix equation checks both at once, because matrices are multiplied by rows *or* by columns. The product either way is $\mathbf{d}$:

$$\begin{bmatrix} 1 & 1 \\ .05 & .10 \end{bmatrix} \begin{bmatrix} 2000 \\ 3000 \end{bmatrix} = \begin{bmatrix} 2000 + 3000 \\ (.05)2000 + (.10)3000 \end{bmatrix} = \begin{bmatrix} 5000 \\ 400 \end{bmatrix} = \mathbf{d}.$$

Singular case In the row picture, the lines cross at the solution. But there is a case that gives trouble. *When the lines are parallel*, they never cross and there is *no* solution. When the lines are the same, there is an *infinity* of solutions:

$$\textit{parallel lines}\quad \begin{array}{l} 2x + y = 0 \\ 2x + y = 1 \end{array} \quad \textit{same line}\quad \begin{array}{l} 2x + \ y = 0 \\ 4x + 2y = 0 \end{array} \tag{5}$$

This trouble also appears in the column picture. The columns are vectors **a** and **b**. The equation $A\mathbf{u} = \mathbf{d}$ is the same as $x\mathbf{a} + y\mathbf{b} = \mathbf{d}$. We are asked to find the combination of **a** and **b** (with coefficients x and y) that produces **d**. In the singular case **a** and **b** lie along the same line (Figure 11.17). No combination can produce **d**, unless it happens to lie on this line.

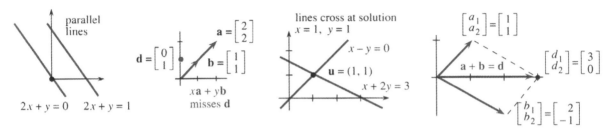

Fig. 11.17   Row and column pictures: *singular* (no solution) and *nonsingular* ($x = y = 1$).

The investment problem is *non*singular, and $2000\,\mathbf{a} + 3000\,\mathbf{b}$ equals **d**. We also drew **EXAMPLE 2**: The matrix A multiplies $\mathbf{u} = (1, 1)$ to solve $x + 2y = 3$ and $x - y = 0$:

$$A\mathbf{u} = \begin{bmatrix} 1 & 2 \\ 1 & -1 \end{bmatrix}\begin{bmatrix} 1 \\ 1 \end{bmatrix} = \begin{bmatrix} 1+2 \\ 1-1 \end{bmatrix} = \begin{bmatrix} 3 \\ 0 \end{bmatrix}. \quad \text{By columns}\ \begin{bmatrix} 1 \\ 1 \end{bmatrix} + \begin{bmatrix} 2 \\ -1 \end{bmatrix} = \begin{bmatrix} 3 \\ 0 \end{bmatrix}.$$

The crossing point is $(1, 1)$ in the row picture. The solution is $x = 1, y = 1$ in the column picture (Figure 11.17b). Then 1 times **a** plus 1 times **b** equals the right side **d**.

SOLUTION BY DETERMINANTS

Up to now we just wrote down the answer. The real problem is to find x and y when they are unknown. We solve two equations with letters not numbers:

$$a_1 x + b_1 y = d_1$$
$$a_2 x + b_2 y = d_2.$$

The key is to eliminate x. Multiply the first equation by a_2 and the second equation by a_1. Subtract the first from the second and the x's disappear:

$$(a_1 b_2 - a_2 b_1) y = (a_1 d_2 - a_2 d_1). \tag{6}$$

To eliminate y, subtract b_1, times the second equation from b_2 times the first:

$$(b_2 a_1 - b_1 a_2) x = (b_2 d_1 - b_1 d_2). \tag{7}$$

What you see in those parentheses are 2 by 2 determinants! Remember from Section 11.3:

The determinant of $\begin{bmatrix} a_1 & b_1 \\ a_2 & b_2 \end{bmatrix}$ *is the number* $\begin{vmatrix} a_1 & b_1 \\ a_2 & b_2 \end{vmatrix} = a_1 b_2 - a_2 b_1.$

This number appears on the left side of (6) and (7). The right side of (7) is also a determinant—but it has d's in place of a's. The right side of (6) has d's in place of b's. So x and y are *ratios of determinants*, given by Cramer's Rule:

11H *Cramer's Rule*   The solution is $x = \dfrac{\begin{vmatrix} d_1 & b_1 \\ d_2 & b_2 \end{vmatrix}}{\begin{vmatrix} a_1 & b_1 \\ a_2 & b_2 \end{vmatrix}}, \quad y = \dfrac{\begin{vmatrix} a_1 & d_1 \\ a_2 & d_2 \end{vmatrix}}{\begin{vmatrix} a_1 & b_1 \\ a_2 & b_2 \end{vmatrix}}.$

The investment example is solved by three determinants from the three columns:

$$\begin{vmatrix} 1 & 1 \\ .05 & .10 \end{vmatrix} = .05 \qquad \begin{vmatrix} 5000 & 1 \\ 400 & .10 \end{vmatrix} = 100 \qquad \begin{vmatrix} 1 & 5000 \\ .05 & 400 \end{vmatrix} = 150.$$

Cramer's Rule has $x = 100/.05 = 2000$ and $y = 150/.05 = 3000$. This is the solution. The singular case is when *the determinant of A is zero*—and we can't divide by it.

11I Cramer's Rule breaks down when det $A = 0$—which is the singular case. Then the lines in the row picture are parallel, and one column is a multiple of the other column.

EXAMPLE 3   The lines $2x + y = 0$, $2x + y = 1$ are parallel. The determinant is zero:

$$\begin{bmatrix} 2 & 1 \\ 2 & 1 \end{bmatrix} \begin{bmatrix} x \\ y \end{bmatrix} = \begin{bmatrix} 0 \\ 1 \end{bmatrix} \text{ has det } A = \begin{vmatrix} 2 & 1 \\ 2 & 1 \end{vmatrix} = 0.$$

The lines in Figure 11.17a don't meet. Notice the columns: $\begin{bmatrix} 2 \\ 2 \end{bmatrix}$ is a multiple of $\begin{bmatrix} 1 \\ 1 \end{bmatrix}$.

One final comment on 2 by 2 systems. They are small enough so that all solution methods apply. Cramer's Rule uses *determinants*. Larger systems use *elimination* (3 by 3 matrices are on the borderline). A third solution (the same solution!) comes from the *inverse matrix* A^{-1}, to be described next. But the inverse is more a symbol for the answer than a new way of computing it, because to find A^{-1} we still use determinants or elimination.

THE INVERSE OF A MATRIX

The symbol A^{-1} is pronounced "A inverse." It stands for a matrix—the one that solves $A\mathbf{u} = \mathbf{d}$. I think of A as a matrix that takes $\mathbf{u}$ to $\mathbf{d}$. Then A^{-1} is a matrix that takes $\mathbf{d}$ back to $\mathbf{u}$. If $A\mathbf{u} = \mathbf{d}$ then $\mathbf{u} = A^{-1}\mathbf{d}$ (provided the inverse exists). This is exactly like functions and inverse functions: $g(x) = y$ and $x = g^{-1}(y)$. Our goal is to find A^{-1} when we know A.

The first approach will be very direct. Cramer's Rule gave formulas for x and y, the components of $\mathbf{u}$. From that rule we can read off A^{-1}, *assuming that* $D = a_1b_2 - a_2b_1$ *is not zero*. D is det A and we divide by it:

$$\textit{Cramer}: \mathbf{u} = \frac{1}{D}\begin{bmatrix} b_2d_1 - b_1d_2 \\ -a_2d_1 + a_1d_2 \end{bmatrix} \quad \textbf{This is } A^{-1}\mathbf{d} = \frac{1}{D}\begin{bmatrix} b_2 & -b_1 \\ -a_2 & a_1 \end{bmatrix}\begin{bmatrix} d_1 \\ d_2 \end{bmatrix} \quad (8)$$

The matrix on the right (including $1/D$ in all four entries) is A^{-1}. Notice the sign pattern and the subscript pattern. The inverse exists if D is not zero—this is important. Then the solution comes from a matrix-vector multiplication, A^{-1} times $\mathbf{d}$. We repeat the rules for that multiplication:

DEFINITION   A matrix M times a vector $\mathbf{v}$ equals a vector of dot products:

$$M\mathbf{v} = \begin{bmatrix} \text{row 1} \\ \text{row 2} \end{bmatrix}\begin{bmatrix} \mathbf{v} \end{bmatrix} = \begin{bmatrix} (\text{row 1}) \cdot \mathbf{v} \\ (\text{row 2}) \cdot \mathbf{v} \end{bmatrix}. \tag{9}$$

Equation (8) follows this rule with $M = A^{-1}$ and $\mathbf{v} = \mathbf{d}$. Look at Example 1:

$$A = \begin{bmatrix} 1 & 1 \\ .05 & .10 \end{bmatrix}, \quad \det A = .05, \quad A^{-1} = \frac{1}{.05}\begin{bmatrix} .10 & -1 \\ -.05 & 1 \end{bmatrix} = \begin{bmatrix} 2 & -20 \\ -1 & 20 \end{bmatrix}.$$

There stands the inverse matrix. It multiplies $\mathbf{d}$ to give the solution $\mathbf{u}$:

$$A^{-1}\mathbf{d} = \begin{bmatrix} 2 & -20 \\ -1 & 20 \end{bmatrix}\begin{bmatrix} 5000 \\ 400 \end{bmatrix} = \begin{bmatrix} (2)(5000) - (20)(400) \\ (-1)(5000) + (20)(400) \end{bmatrix} = \begin{bmatrix} 2000 \\ 3000 \end{bmatrix}.$$

The formulas work perfectly, but you have to see a direct way to reach $A^{-1}\mathbf{d}$. *Multiply both sides of* $A\mathbf{u} = \mathbf{d}$ *by* A^{-1}. The multiplication "cancels" A on the left side, and leaves $\mathbf{u} = A^{-1}\mathbf{d}$. This approach comes next.

MATRIX MULTIPLICATION

To understand the power of matrices, we must multiply them. The product of A^{-1} with $A\mathbf{u}$ is a matrix times a vector. But that multiplication can be done another way. First A^{-1} multiplies A, a matrix times a matrix. The product $A^{-1}A$ is another matrix (a very special matrix). Then this new matrix multiplies $\mathbf{u}$.

The matrix-matrix rule comes directly from the matrix-vector rule. Effectively, a vector $\mathbf{v}$ is a matrix V with only one column. When there are more columns, M times V splits into separate matrix-vector multiplications, side by side:

DEFINITION   A matrix M times a matrix V equals a matrix of dot products:

$$MV = \begin{bmatrix} \text{row 1} \\ \text{row 2} \end{bmatrix}\begin{bmatrix} \mathbf{v}_1 & \mathbf{v}_2 \end{bmatrix} = \begin{bmatrix} (\text{row 1}) \cdot \mathbf{v}_1 & (\text{row 1}) \cdot \mathbf{v}_2 \\ (\text{row 2}) \cdot \mathbf{v}_1 & (\text{row 2}) \cdot \mathbf{v}_2 \end{bmatrix}. \tag{10}$$

EXAMPLE 4 $\begin{bmatrix} 1 & 2 \\ 3 & 4 \end{bmatrix}\begin{bmatrix} 5 & 6 \\ 7 & 8 \end{bmatrix} = \begin{bmatrix} 1 \cdot 5 + 2 \cdot 7 & 1 \cdot 6 + 2 \cdot 8 \\ 3 \cdot 5 + 4 \cdot 7 & 3 \cdot 6 + 4 \cdot 8 \end{bmatrix} = \begin{bmatrix} 19 & 22 \\ 43 & 50 \end{bmatrix}.$

EXAMPLE 5   Multiplying A^{-1} times A produces the "*identity matrix*" $\begin{bmatrix} 1 & 0 \\ 0 & 1 \end{bmatrix}$:

$$A^{-1}A = \frac{\begin{bmatrix} b_2 & -b_1 \\ -a_2 & a_1 \end{bmatrix}}{D} \begin{bmatrix} a_1 & b_1 \\ a_2 & b_2 \end{bmatrix} = \frac{\begin{bmatrix} a_1 b_2 - a_2 b_1 & 0 \\ 0 & -a_2 b_1 + a_1 b_2 \end{bmatrix}}{D} = \begin{bmatrix} 1 & 0 \\ 0 & 1 \end{bmatrix}. \quad (11)$$

This identity matrix is denoted by I. It has 1's on the diagonal and 0's off the diagonal. It acts like the number 1. *Every vector satisfies* $I\mathbf{u} = \mathbf{u}$.

11J (*Inverse matrix and identity matrix*) $AA^{-1} = I$ and $A^{-1}A = I$ and $I\mathbf{u} = \mathbf{u}$:

$$A = \begin{bmatrix} a & b \\ c & d \end{bmatrix} \qquad A^{-1} = \frac{1}{D} \begin{bmatrix} d & -b \\ -c & a \end{bmatrix} \qquad \begin{bmatrix} 1 & 0 \\ 0 & 1 \end{bmatrix} \begin{bmatrix} x \\ y \end{bmatrix} = \begin{bmatrix} x \\ y \end{bmatrix}. \quad (12)$$

Note the placement of a, b, c, d. With these letters D is $ad - bc$.

The next section moves to three equations. The algebra gets more complicated (and 4 by 4 is worse). It is not easy to write out A^{-1}. So we stay longer with the 2 by 2 formulas, where each step can be checked. Multiplying $A\mathbf{u} = \mathbf{d}$ by the inverse matrix gives $A^{-1}A\mathbf{u} = A^{-1}\mathbf{d}$—and the left side is $I\mathbf{u} = \mathbf{u}$.

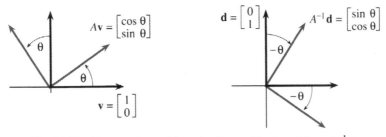

Fig. 11.18   Rotate $\mathbf{v}$ forward into $A\mathbf{v}$. Rotate $\mathbf{d}$ backward into $A^{-1}\mathbf{d}$.

EXAMPLE 6 $A = \begin{bmatrix} \cos\theta & -\sin\theta \\ \sin\theta & \cos\theta \end{bmatrix}$ rotates every $\mathbf{v}$ to $A\mathbf{v}$, through the angle θ.

Question 1   Where is the vector $\mathbf{v} = \begin{bmatrix} 1 \\ 0 \end{bmatrix}$ rotated to?

Question 2   What is A^{-1}?

Question 3   Which vector $\mathbf{u}$ is rotated into $\mathbf{d} = \begin{bmatrix} 0 \\ 1 \end{bmatrix}$?

Solution 1 $\mathbf{v}$ rotates into $A\mathbf{v} = \begin{bmatrix} \cos\theta & -\sin\theta \\ \sin\theta & \cos\theta \end{bmatrix} \begin{bmatrix} 1 \\ 0 \end{bmatrix} = \begin{bmatrix} \cos\theta \\ \sin\theta \end{bmatrix}.$

Solution 2　　$\det A = 1$ so $A^{-1} = \begin{bmatrix} \cos\theta & \sin\theta \\ -\sin\theta & \cos\theta \end{bmatrix} = $ rotation through $-\theta$.

Solution 3　　If $A\mathbf{u} = \mathbf{d}$ then $\mathbf{u} = A^{-1}\mathbf{d} = \begin{bmatrix} \cos\theta & \sin\theta \\ -\sin\theta & \cos\theta \end{bmatrix} \begin{bmatrix} 0 \\ 1 \end{bmatrix} = \begin{bmatrix} \sin\theta \\ \cos\theta \end{bmatrix}$.

Historical note　I was amazed to learn that it was Leibniz (again!) who proposed the notation we use for matrices. ***The entry in row*** i ***and column*** j ***is*** a_{ij}. The identity matrix has $a_{11} = a_{22} = 1$ and $a_{12} = a_{21} = 0$. This is in a linear algebra book by Charles Dodgson—better known to the world as Lewis Carroll, the author of *Alice in Wonderland*. I regret to say that he preferred his own notation $i \, \rfloor \, j$ instead of a_{ij}. "I have turned the symbol toward the left, to avoid all chance of confusion with $\int$." It drove his typesetter mad.

PROJECTION ONTO A PLANE = LEAST SQUARES FITTING BY A LINE

We close with a genuine application. It starts with three-dimensional vectors $\mathbf{a}, \mathbf{b}, \mathbf{d}$ and leads to a 2 by 2 system. One good feature: $\mathbf{a}, \mathbf{b}, \mathbf{d}$ can be n-dimensional with no change in the algebra. In practice that happens. Second good feature: There is a calculus problem in the background. The example is ***to fit points by a straight line***.

There are three ways to state the problem, and they look different:

1. Solve $x\mathbf{a} + y\mathbf{b} = \mathbf{d}$ as well as possible (three equations, two unknowns x and y).
2. Project the vector $\mathbf{d}$ onto the plane of the vectors $\mathbf{a}$ and $\mathbf{b}$.
3. Find the closest straight line ("***least squares***") to three given points.

Figure 11.19 shows a three-dimensional vector $\mathbf{d}$ above the plane of $\mathbf{a}$ and $\mathbf{b}$. Its projection onto the plane is $\mathbf{p} = x\mathbf{a} + y\mathbf{b}$. The numbers x and y are unknown, and our goal is to find them. The calculation will use the dot product, which is always the key to right angles.

The difference $\mathbf{d} - \mathbf{p}$ is the "*error*." There has to be an error, because no combination of $\mathbf{a}$ and $\mathbf{b}$ can produce $\mathbf{d}$ exactly. (Otherwise $\mathbf{d}$ is in the plane.) The projection $\mathbf{p}$ is the closest point to $\mathbf{d}$, and it is governed by one fundamental law: ***The error is perpendicular to the plane***. That makes the error perpendicular to both vectors $\mathbf{a}$ and $\mathbf{b}$:

$$\mathbf{a} \cdot (x\mathbf{a} + y\mathbf{b} - \mathbf{d}) = 0 \qquad \text{and} \qquad \mathbf{b} \cdot (x\mathbf{a} + y\mathbf{b} - \mathbf{d}) = 0. \tag{13}$$

Rewrite those as two equations for the two unknown numbers x and y:

$$\begin{aligned} (\mathbf{a} \cdot \mathbf{a})x + (\mathbf{a} \cdot \mathbf{b})y &= \mathbf{a} \cdot \mathbf{d} \\ (\mathbf{b} \cdot \mathbf{a})x + (\mathbf{b} \cdot \mathbf{b})y &= \mathbf{b} \cdot \mathbf{d}. \end{aligned} \tag{14}$$

These are the famous ***normal equations*** in statistics, to compute x and y and $\mathbf{p}$.

EXAMPLE 7　For $\mathbf{a} = (1,1,1)$ and $\mathbf{b} = (1,2,3)$ and $\mathbf{d} = (0,5,4)$, solve equation (14):

$$\begin{aligned} 3x + 6y &= 9 \\ 6x + 14y &= 22 \end{aligned} \quad \text{gives} \quad \begin{aligned} x &= -1 \\ y &= 2 \end{aligned} \quad \text{so} \quad \mathbf{p} = -\mathbf{a} + 2\mathbf{b} = (1,3,5) = \textit{projection.}$$

Notice the three equations that we are not solving (we can't): $x\mathbf{a} + y\mathbf{b} = \mathbf{d}$ is

$$\begin{aligned} x + y &= 0 \\ x + 2y &= 5 \\ x + 3y &= 4 \end{aligned} \quad \text{with the 3 by 2 matrix } A = \begin{bmatrix} 1 & 1 \\ 1 & 2 \\ 1 & 3 \end{bmatrix}. \tag{15}$$

For $\mathbf{d} = (0, 5, 4)$ there is no solution; $\mathbf{d}$ is not in the plane of $\mathbf{a}$ and $\mathbf{b}$. For $\mathbf{p} = (1, 3, 5)$ there is a solution, $x = -1$ and $y = 2$. The vector $\mathbf{p}$ is in the plane. The error $\mathbf{d} - \mathbf{p}$ is $(-1, 2, -1)$. This error is perpendicular to the columns $(1, 1, 1)$ and $(1, 2, 3)$, so it is perpendicular to their plane.

SAME EXAMPLE (*written as a line-fitting problem*) Fit the points $(1, 0)$ and $(2, 5)$ and $(3, 4)$ as closely as possible ("least squares") by a straight line.

Two points determine a line. The example asks the line $f = x + yt$ to go through *three* points. That gives the three equations in (15), which can't be solved with two unknowns. We have to settle for the closest line, drawn in Figure 11.19b. This line is computed again below, by calculus.

Notice that the closest line has heights $1, 3, 5$ where the data points have heights $0, 5, 4$. Those are the numbers in $\mathbf{p}$ and $\mathbf{d}$! The heights $1, 3, 5$ fit onto a line; the heights $0, 5, 4$ do not. In the first figure, $\mathbf{p} = (1, 3, 5)$ is in the plane and $\mathbf{d} = (0, 5, 4)$ is not. Vectors in the plane lead to heights that lie on a line.

Notice another coincidence. The coefficients $x = -1$ and $y = 2$ give the projection $-\mathbf{a} + 2\mathbf{b}$. They also give the closest line $f = -1 + 2t$. All numbers appear in both figures.

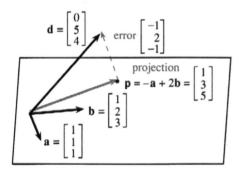 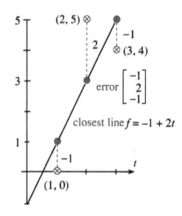

Fig. 11.19 Projection onto plane is $(1, 3, 5)$ with coefficients $-1, 2$. Closest line has heights $1, 3, 5$ with coefficients $-1, 2$. Error in both pictures is $-1, 2, -1$.

Remark Finding the closest line is a *calculus problem*: **Minimize a sum of squares.** The numbers x and y that minimize E give the least squares solution:

$$E(x, y) = (x + y - 0)^2 + (x + 2y - 5)^2 + (x + 3y - 4)^2. \qquad (16)$$

Those are the three errors in equation (15), squared and added. They are also the three errors in the straight line fit, between the line and the data points. The projection minimizes the error (by geometry), the normal equations (14) minimize the error (by algebra), and now calculus minimizes the error by setting the derivatives of E to zero.

The new feature is this: E depends on two variables x and y. *Therefore E has two derivatives.* They both have to be zero at the minimum. That gives two equations for x and y:

x derivative of E is zero: $2(x + y) + 2(x + 2y - 5) \quad + 2(x + 3y - 4) \quad = 0$

y derivative of E is zero: $2(x + y) + 2(x + 2y - 5)(2) + 2(x + 3y - 4)(3) = 0.$

When we divide by 2, those are the normal equations $3x + 6y = 9$ and $6x + 14y = 22$. The minimizing x and y from calculus are the same numbers -1 and 2.

The x derivative treats y as a constant. The y derivative treats x as a constant. These are *partial derivatives*. This calculus approach to least squares is in Chapter 13, as an important application of partial derivatives.

We now summarize the *least squares problem*—to find the closest line to n data points. In practice n may be 1000 instead of 3. The points have horizontal coordinates $b_1, b_2, \ldots, b_n$. The vertical coordinates are $d_1, d_2, \ldots, d_n$. These vectors $\mathbf{b}$ and $\mathbf{d}$, together with $\mathbf{a} = (1, 1, \ldots, 1)$, determine a projection—the combination $\mathbf{p} = x\mathbf{a} + y\mathbf{b}$ that is closest to $\mathbf{d}$. This problem is the same in n dimensions—the error $\mathbf{d} - \mathbf{p}$ is perpendicular to $\mathbf{a}$ and $\mathbf{b}$. That is still tested by dot products, $\mathbf{p} \cdot \mathbf{a} = \mathbf{d} \cdot \mathbf{a}$ and $\mathbf{p} \cdot \mathbf{b} = \mathbf{d} \cdot \mathbf{b}$, which give the normal equations for x and y:

$$
\begin{aligned}
(\mathbf{a} \cdot \mathbf{a})x + (\mathbf{a} \cdot \mathbf{b})y &= \mathbf{a} \cdot \mathbf{d} \\
(\mathbf{b} \cdot \mathbf{a})x + (\mathbf{b} \cdot \mathbf{b})y &= \mathbf{b} \cdot \mathbf{d}
\end{aligned}
\quad \text{or} \quad
\begin{aligned}
(n) \quad x + (\Sigma b_i)y &= \Sigma d_i \\
(\Sigma b_i)x + (\Sigma b_i^2)y &= \Sigma b_i d_i .
\end{aligned}
\tag{17}
$$

> **11K** The least squares problem projects $\mathbf{d}$ onto the plane of $\mathbf{a}$ and $\mathbf{b}$. The projection is $\mathbf{p} = x\mathbf{a} + y\mathbf{b}$, in n dimensions. The closest line $f = x + yt$, in two dimensions. The normal equations (17) give the best x and y.

11.4 EXERCISES

Read-through questions

The equations $3x + y = 8$ and $x + y = 6$ combine into the vector equation $x\underline{\quad \text{a} \quad} + y\underline{\quad \text{b} \quad} = \underline{\quad \text{c} \quad} = \mathbf{d}$. The left side is $A\mathbf{u}$, with coefficient matrix $A = \underline{\quad \text{d} \quad}$ and unknown vector $\mathbf{u} = \underline{\quad \text{e} \quad}$. The determinant of A is $\underline{\quad \text{f} \quad}$, so this problem is not $\underline{\quad \text{g} \quad}$. The row picture shows two intersecting $\underline{\quad \text{h} \quad}$. The column picture shows $x\mathbf{a} + y\mathbf{b} = \mathbf{d}$, where $\mathbf{a} = \underline{\quad \text{i} \quad}$ and $\mathbf{b} = \underline{\quad \text{j} \quad}$. The inverse matrix is $A^{-1} = \underline{\quad \text{k} \quad}$. The solution is $\mathbf{u} = A^{-1}\mathbf{d} = \underline{\quad \text{l} \quad}$.

A matrix-vector multiplication produces a vector of dot $\underline{\quad \text{m} \quad}$ from the rows, and also a combination of the $\underline{\quad \text{n} \quad}$:

$$
\begin{bmatrix} \mathbf{A} \\ \mathbf{B} \end{bmatrix} \begin{bmatrix} \mathbf{u} \end{bmatrix} = \begin{bmatrix} - \\ - \end{bmatrix}, \quad \begin{bmatrix} \mathbf{a} & \mathbf{b} \end{bmatrix} \begin{bmatrix} x \\ y \end{bmatrix} = \begin{bmatrix} - \\ - \end{bmatrix}, \quad \begin{bmatrix} 3 & 1 \\ 1 & 1 \end{bmatrix} \begin{bmatrix} 1 \\ 5 \end{bmatrix} = \begin{bmatrix} - \\ - \end{bmatrix}.
$$

If the entries are a, b, c, d, the determinant is $D = \underline{\quad \text{o} \quad}$. A^{-1} is $[\underline{\quad \text{p} \quad}]$ divided by D. Cramer's Rule shows components of $\mathbf{u} = A^{-1}\mathbf{d}$ as ratios of determinants: $x = \underline{\quad \text{q} \quad}/D$ and $y = \underline{\quad \text{r} \quad}/D$.

A matrix-matrix multiplication MV yields a matrix of dot products, from the rows of $\underline{\quad \text{s} \quad}$ and the columns of $\underline{\quad \text{t} \quad}$:

$$
\begin{bmatrix} \mathbf{A} \\ \mathbf{B} \end{bmatrix} \begin{bmatrix} \mathbf{v}_1 & \mathbf{v}_2 \end{bmatrix} = \begin{bmatrix} - & - \\ - & - \end{bmatrix}, \quad \begin{bmatrix} 3 & 1 \\ 1 & 1 \end{bmatrix} \begin{bmatrix} 1 & 2 \\ 5 & 6 \end{bmatrix} = \begin{bmatrix} - & - \\ - & - \end{bmatrix}.
$$

$$
\begin{bmatrix} 3 & 1 \\ 1 & 1 \end{bmatrix} \begin{bmatrix} 1/2 & -1/2 \\ -1/2 & 3/2 \end{bmatrix} = \begin{bmatrix} - & - \\ - & - \end{bmatrix} \quad \begin{bmatrix} 1 & 0 \\ 0 & 1 \end{bmatrix} \begin{bmatrix} A \end{bmatrix} = \begin{bmatrix} - & - \\ - & - \end{bmatrix}.
$$

The last line contains the $\underline{\quad \text{u} \quad}$ matrix, denoted by I. It has the property that $IA = AI = \underline{\quad \text{v} \quad}$ for every matrix A, and $I\mathbf{u} = \underline{\quad \text{w} \quad}$ for every vector $\mathbf{u}$. The inverse matrix satisfies $A^{-1}A = \underline{\quad \text{x} \quad}$. Then $A\mathbf{u} = \mathbf{d}$ is solved by multiplying both sides by $\underline{\quad \text{y} \quad}$, to give $\mathbf{u} = \underline{\quad \text{z} \quad}$. There is no inverse matrix when $\underline{\quad \text{A} \quad}$.

The combination $x\mathbf{a} + y\mathbf{b}$ is the projection of $\mathbf{d}$ when the error $\underline{\quad \text{B} \quad}$ is perpendicular to $\underline{\quad \text{C} \quad}$ and $\underline{\quad \text{D} \quad}$. If $\mathbf{a} = (1, 1, 1)$, $\mathbf{b} = (1, 2, 3)$, and $\mathbf{d} = (0, 8, 4)$, the equations for x and y are $\underline{\quad \text{E} \quad}$. Solving them also gives the closest $\underline{\quad \text{F} \quad}$ to the data points $(1, 0)$, $\underline{\quad \text{G} \quad}$, and $(3, 4)$. The solution is $x = 0$, $y = 2$, which means the best line is $\underline{\quad \text{H} \quad}$. The projection is $0\mathbf{a} + 2\mathbf{b} = \underline{\quad \text{I} \quad}$. The three error components are $\underline{\quad \text{J} \quad}$. Check perpendicularity: $\underline{\quad \text{K} \quad} = 0$ and $\underline{\quad \text{L} \quad} = 0$. Applying calculus to this problem, x and y minimize the sum of squares $E = \underline{\quad \text{M} \quad}$.

In 1–8 find the point (x, y) where the two lines intersect (if they do). Also show how the right side is a combination of the columns on the left side (if it is). Also find the determinant D.

1 $x + y = 7$
   $x - y = 3$

2 $2x + y = 11$
   $x + y = 6$

3 $3x - y = 8$
$\quad\ x - 3y = 0$

4 $x + 2y = 3$
$\quad 2x + 4y = 7$

5 $2x - 4y = 0$
$\quad\ x - 2y = 0$

6 $10x + y = 1$
$\quad\ x + y = 1$

7 $ax + \ by = 0$
$\quad 2ax + 2by = 2$

8 $ax + by = 1$
$\quad cx + dy = 1$

9 Solve Problem 3 by Cramer's Rule.

10 Try to solve Problem 4 by Cramer's Rule.

11 What are the ratios for Cramer's Rule in Problem 5?

12 If $A = I$ show how Cramer's Rule solves $A\mathbf{u} = \mathbf{d}$.

13 Draw the row picture and column picture for Problem 1.

14 Draw the row and column pictures for Problem 6.

15 Find A^{-1} in Problem 1.

16 Find A^{-1} in Problem 8 if $ad - bc = 1$.

17 A 2 by 2 system is *singular* when the two lines in the row picture _____ . This system is still solvable if one equation is a _____ of the other equation. In that case the two lines are _____ and the number of solution is _____ .

18 Try Cramer's Rule when there is no solution or infinitely many:

$$3x + \ y = 0 \qquad\qquad 3x + \ y = 1$$
$$\qquad\qquad\ \text{or}$$
$$6x + 2y = 2 \qquad\qquad 6x + 2y = 2.$$

19 $A\mathbf{u} = \mathbf{d}$ is singular when the columns of A are _____ . A solution exists if the right side $\mathbf{d}$ is _____ . In this solvable case the number of solutions is _____ .

20 The equations $x - y = d_1$ and $9x - 9y = d_2$ can be solved if _____ .

21 Suppose $x = \frac{1}{4}$ billion people live in the U.S. and $y = 5$ billion live outside. If 4 per cent of those inside move out and 2 per cent of those outside move in, find the populations d_1 inside and d_2 outside after the move. Express this as a matrix multiplication $A\mathbf{u} = \mathbf{d}$ (and find the matrix).

22 In Problem 21 what is special about $a_1 + a_2$ and $b_1 + b_2$ (the sums down the columns of A)? Explain why $d_1 + d_2$ equals $x + y$.

23 With the same percentages moving, suppose $d_1 = 0.58$ billion are inside and $d_2 = 4.92$ billion are outside *at the end*. Set up and solve two equations for the original populations x and y.

24 What is the determinant of A in Problems 21–23? What is A^{-1}? Check that $A^{-1}A = I$.

25 The equations $ax + y = 0$, $x + ay = 0$ have the solution $x = y = 0$. For which two values of a are there other solutions (and what are the other solutions)?

26 The equations $ax + by = 0$, $cx + dy = 0$ have the solution $x = y = 0$. There are other solutions if the two lines are _____ . This happens if a, b, c, d satisfy _____ .

27 Find the determinant and inverse of $A = \begin{bmatrix} 2 & 4 \\ 3 & 5 \end{bmatrix}$. Do the same for $2A, A^{-1}, -A$, and I.

28 Show that the determinant of A^{-1} is $1/\det A$:

$$A^{-1} = \begin{bmatrix} d/(ad - bc) & -b/(ad - bc) \\ -c/(ad - bc) & a/(ad - bc) \end{bmatrix}$$

29 Compute AB and BA and also BC and CB:

$$A = \begin{bmatrix} 1 & 4 \\ 2 & -1 \end{bmatrix} \quad B = \begin{bmatrix} 3 & 1 \\ 1 & 1 \end{bmatrix} \quad C = \begin{bmatrix} 1 & 1 \\ 0 & 2 \end{bmatrix}.$$

Verify the *associative law*: AB times C equals A times BC.

30 (a) Find the determinants of A, B, AB, and BA above.

    (b) Propose a law for the determinant of BC and test it.

31 For $A = \begin{bmatrix} a & b \\ c & d \end{bmatrix}$ and $B = \begin{bmatrix} e & f \\ g & h \end{bmatrix}$ write out AB and factor its determinant into $(ad - bc)(eh - fg)$. Therefore $\det(AB) = (\det A)(\det B)$.

32 Usually $\det(A + B)$ does *not* equal $\det A + \det B$. Find examples of inequality and equality.

33 Find the inverses, and check $A^{-1}A = I$ and $BB^{-1} = I$, for

$$A = \begin{bmatrix} 1 & 4 \\ 0 & 2 \end{bmatrix} \quad \text{and} \quad B = \begin{bmatrix} 2 & 2 \\ 0 & 1 \end{bmatrix}.$$

34 In Problem 33 compute AB and the inverse of AB. Check that this inverse equals B^{-1} times A^{-1}.

35 The matrix product $ABB^{-1}A^{-1}$ equals the _____ matrix. Therefore the inverse of AB is _____ . *Important*: The associative law in Problem 29 allows you to multiply BB^{-1} first.

36 The matrix multiplication $C^{-1}B^{-1}A^{-1}ABC$ yields the _____ matrix. Therefore the inverse of ABC is _____ .

37 The equations $x + 2y + 3z$ and $4x + 5y + cz = 0$ always have a nonzero solution. The vector $\mathbf{u} = (x, y, z)$ is required to be _____ to $\mathbf{v} = (1, 2, 3)$ and $\mathbf{w} = (4, 5, c)$. So choose $\mathbf{u} = $ _____ .

38 Find the combination $\mathbf{p} = x\mathbf{a} + y\mathbf{b}$ of the vectors $\mathbf{a} = (1, 1, 1)$ and $\mathbf{b} = (-1, 0, 1)$ that comes closest to $\mathbf{d} = (2, 6, 4)$. (a) Solve the normal equations (14) for x and y. (b) Check that the error $\mathbf{d} - \mathbf{p}$ is perpendicular to $\mathbf{a}$ and $\mathbf{b}$.

39 Plot the three data points $(-1, 2), (0, 6), (1, 4)$ in a plane. Draw the straight line $x + yt$ with the same x and y as in Problem 38. Locate the three errors up or down from the data points and compare with Problem 38.

40 Solve equation (14) to find the combination $x\mathbf{a} + y\mathbf{b}$ of $\mathbf{a} = (1, 1, 1)$ and $\mathbf{b} = (-1, 1, 2)$ that is closest to $\mathbf{d} = (1, 1, 3)$. Draw the corresponding straight line for the data points $(-1, 1), (1, 1)$, and $(2, 3)$. What is the vector of three errors and what is it perpendicular to?

41 Under what condition on d_1, d_2, d_3 do the three points $(0, d_1), (1, d_2), (2, d_3)$ lie on a line ?

42 Find the matrices that reverse x and y and project:

$$M \begin{bmatrix} x \\ y \end{bmatrix} = \begin{bmatrix} y \\ x \end{bmatrix} \quad \text{and} \quad P \begin{bmatrix} x \\ y \end{bmatrix} = \begin{bmatrix} x \\ 0 \end{bmatrix}.$$

43 Multiplying by $P = \begin{bmatrix} .5 & .5 \\ .5 & .5 \end{bmatrix}$ projects $\mathbf{u}$ onto the $45°$ line.

(a) Find the projection $P\mathbf{u}$ of $\mathbf{u} = \begin{bmatrix} 1 \\ 0 \end{bmatrix}$.

(b) Why does P times P equal P ?

(c) Does P^{-1} exist ? What vectors give $P\mathbf{u} = \mathbf{0}$?

44 Suppose $\mathbf{u}$ is not the zero vector but $A\mathbf{u} = \mathbf{0}$. Then A^{-1} can't exist: It would multiply _____ and produce $\mathbf{u}$.

11.5 Linear Algebra

This section moves from two to three dimensions. There are three unknowns x, y, z and also three equations. This is at the crossover point between formulas and algorithms—it is real linear algebra. The formulas give a direct solution using determinants. The algorithms use elimination and the numbers x, y, z appear at the end. In practice that end result comes quickly. *Computers solve linear equations by elimination.*

The situation for a nonlinear equation is similar. Quadratic equations $ax^2 + bx + c = 0$ are solved by a formula. Cubic equations are solved by Newton's method (even though a formula exists). For equations involving x^5 or x^{10}, algorithms take over completely.

Since we are at the crossover point, we look both ways. This section has a lot to do, in mixing geometry, determinants, and 3 by 3 matrices:

1. The row picture: three planes intersect at the solution
2. The column picture: a vector equation combines the columns
3. The formulas: determinants and Cramer's Rule
4. Matrix multiplication and A^{-1}
5. The algorithm: Gaussian elimination.

Part of our goal is three-dimensional calculus. Another part is n-dimensional algebra. And a third possibility is that you may not take mathematics next year. If that happens, I hope you will *use* mathematics. Linear equations are so basic and important, in such a variety of applications, that the effort in this section is worth making.

An example is needed. It is convenient and realistic if the matrix contains zeros. Most equations in practice are fairly simple—a thousand equations each with 990 zeros would be very reasonable. Here are three equations in three unknowns:

$$
\begin{aligned}
x + y \quad\;\;\; &= \;\; 1 \\
x \quad\;\; + 2z &= \;\; 0 \\
-2y + 2z &= -4.
\end{aligned}
\tag{1}
$$

In matrix-vector form, the unknown $\mathbf{u}$ has components x, y, z. The right sides $1, 0, -4$ go into $\mathbf{d}$. The nine coefficients, including three zeros, enter the matrix A:

$$
\begin{bmatrix} 1 & 1 & 0 \\ 1 & 0 & 2 \\ 0 & -2 & 2 \end{bmatrix}
\begin{bmatrix} x \\ y \\ z \end{bmatrix}
=
\begin{bmatrix} 1 \\ 0 \\ -4 \end{bmatrix}
\quad \text{or} \quad A\mathbf{u} = \mathbf{d}.
\tag{2}
$$

The goal is to understand that system geometrically, and then solve it.

THE ROW PICTURE: INTERSECTING PLANES

Start with the first equation $x + y = 1$. In the xy plane that produces a line. In three dimensions it is a *plane*. It has the usual form $ax + by + cz = d$, except that c happens to be zero. The plane is easy to visualize (Figure 11.20a), because it cuts straight down through the line. The equation $x + y = 1$ allows z to have any value, so the graph includes all points above and below the line.

The second equation $x + 2z = 0$ gives a second plane, which goes through the origin. *When the right side is zero, the point* $(0,0,0)$ *satisfies the equation.* This time y is absent from the equation, so the plane contains the whole y axis. All points $(0, y, 0)$

meet the requirement $x + 2z = 0$. *The normal vector to the plane is* $\mathbf{N} = \mathbf{i} + 2\mathbf{k}$. The plane cuts across, rather than down, in 11.20b.

Before the third equation we combine the first two. ***The intersection of two planes is a line***. In three-dimensional space, two equations (not one) describe a line. The points on the line have to satisfy $x + y = 1$ and also $x + 2z = 0$. A convenient point is $P = (0, 1, 0)$. Another point is $Q = (-1, 2, \frac{1}{2})$. The line through P and Q extends out in both directions.

The solution is on that line. The third plane decides where.

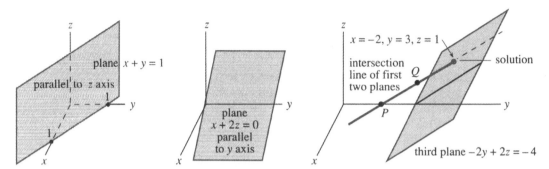

Fig. 11.20   First plane, second plane, interesection line meets third plane at solution.

The third equation $-2y + 2z = -4$ gives the third plane—which misses the origin because the right side is not zero. What is important is *the point where the three planes meet*. The intersection line of the first two planes crosses the third plane. We used determinants (but elimination is better) to find $x = -2$, $y = 3$, $z = 1$. This solution satisfies the three equations and lies on the three planes.

A brief comment on 4 by 4 systems. The first equation might be $x + y + z - t = 0$. It represents a three-dimensional "hyperplane" in four-dimensional space. (In physics this is space-time.) The second equation gives a second hyperplane, and its intersection with the first one is two-dimensional. The third equation (third hyperplane) reduces the intersection to a line. The fourth hyperplane meets that line at a point, which is the solution. It satisfies the four equations and lies on the four hyperplanes. In this course three dimensions are enough.

COLUMN PICTURE: COMBINATION OF COLUMN VECTORS

There is an extremely important way to rewrite our three equations. In (1) they were separate, in (2) they went into a matrix. Now they become a vector equation:

$$x \begin{bmatrix} 1 \\ 1 \\ 0 \end{bmatrix} + y \begin{bmatrix} 1 \\ 0 \\ -2 \end{bmatrix} + z \begin{bmatrix} 0 \\ 2 \\ 2 \end{bmatrix} = \begin{bmatrix} 1 \\ 0 \\ -4 \end{bmatrix}. \tag{3}$$

The columns of the matrix are multiplied by x, y, z. That is a special way to see matrix-vector multiplication: $A\mathbf{u}$ ***is a combination of the columns of*** A. We are looking for the numbers x, y, z so that the combination produces the right side $\mathbf{d}$.

The column vectors $\mathbf{a}, \mathbf{b}, \mathbf{c}$ are shown in Figure 11.21a. The vector equation is $x\mathbf{a} + y\mathbf{b} + z\mathbf{c} = \mathbf{d}$. The combination that solves this equation must again be $x = -2$, $y = 3, z = 1$. That agrees with the intersection point of the three planes in the row picture.

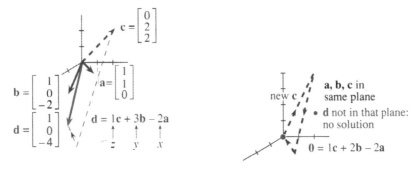

Fig. 11.21 Columns combine to give **d**. Columns combine to give **zero** (singular case).

THE DETERMINANT AND THE INVERSE MATRIX

For a 3 by 3 determinant, the section on cross products gave two formulas. One was the triple product $\mathbf{a} \cdot (\mathbf{b} \times \mathbf{c})$. The other wrote out the six terms:

$$\det A = \mathbf{a} \cdot (\mathbf{b} \times \mathbf{c}) = a_1(b_2 c_3 - b_3 c_2) + a_2(b_3 c_1 - b_1 c_3) + a_3(b_1 c_2 - b_2 c_1).$$

Geometrically this is *the volume of a box*. The columns $\mathbf{a}, \mathbf{b}, \mathbf{c}$ are the edges going out from the origin. In our example the determinant and volume are 2:

$$\begin{vmatrix} a_1 & b_1 & c_1 \\ a_2 & b_2 & c_2 \\ a_3 & b_3 & c_3 \end{vmatrix} = \begin{vmatrix} 1 & 1 & 0 \\ 1 & 0 & 2 \\ 0 & -2 & 2 \end{vmatrix} = \begin{matrix} (1)(0)(2) - (1)(-2)(2) + (1)(-2)(0) \\ -(1)(1)(2) + (0)(1)(2) - (0)(0)(0) \end{matrix} = 2.$$

A slight dishonesty is present in that calculation, and will be admitted now. In Section 11.3 the vectors $\mathbf{A}, \mathbf{B}, \mathbf{C}$ were *rows*. In this section $\mathbf{a}, \mathbf{b}, \mathbf{c}$ are *columns*. It doesn't matter, because the determinant is the same either way. Any matrix can be "transposed"—exchanging rows for columns—without altering the determinant. The six terms ($a_1 b_2 c_3$ is the first) may come in a different order, but they are the same six terms. Here four of those terms are zero, because of the zeros in the matrix. The sum of all six terms is $D = \det A = 2$.

Since D is not zero, the equations can be solved. The three planes meet at a point. The column vectors $\mathbf{a}, \mathbf{b}, \mathbf{c}$ produce a genuine box, and are not flattened into the same plane (with zero volume). The solution involves **dividing by** D—which is only possible if $D = \det A$ is not zero.

11L When the determinant D is not zero, A bas an inverse: $AA^{-1} = A^{-1} A = I$. Then the equations $A\mathbf{u} = \mathbf{d}$ have one and only one solution $\mathbf{u} = A^{-1}\mathbf{d}$.

The 3 by 3 identity matrix I is at the end of equation (5). Always $I\mathbf{u} = \mathbf{u}$.

We now compute A^{-1}, first with letters and then with numbers. The neatest formula uses cross products of the columns of A—it is special for 3 by 3 matrices.

Every entry is divided by D: *The inverse matrix is* $A^{-1} = \dfrac{1}{D} \begin{bmatrix} \mathbf{b} \times \mathbf{c} \\ \mathbf{c} \times \mathbf{a} \\ \mathbf{a} \times \mathbf{b} \end{bmatrix}.$ (4)

To test this formula, multiply by A. **Matrix multiplication produces a matrix of dot products**—from the rows of the first matrix and the columns of the second, $A^{-1}A = I$:

$$\frac{1}{D}\begin{bmatrix} \mathbf{b}\times\mathbf{c} \\ \mathbf{c}\times\mathbf{a} \\ \mathbf{a}\times\mathbf{b} \end{bmatrix}\begin{bmatrix} \mathbf{a} & \mathbf{b} & \mathbf{c} \end{bmatrix} = \frac{1}{D}\begin{bmatrix} \mathbf{a}\cdot(\mathbf{b}\times\mathbf{c}) & \mathbf{b}\cdot(\mathbf{b}\times\mathbf{c}) & \mathbf{c}\cdot(\mathbf{b}\times\mathbf{c}) \\ \mathbf{a}\cdot(\mathbf{c}\times\mathbf{a}) & \mathbf{b}\cdot(\mathbf{c}\times\mathbf{a}) & \mathbf{c}\cdot(\mathbf{c}\times\mathbf{a}) \\ \mathbf{a}\cdot(\mathbf{a}\times\mathbf{b}) & \mathbf{b}\cdot(\mathbf{a}\times\mathbf{b}) & \mathbf{c}\cdot(\mathbf{a}\times\mathbf{b}) \end{bmatrix} = \begin{bmatrix} 1 & 0 & 0 \\ 0 & 1 & 0 \\ 0 & 0 & 1 \end{bmatrix}. \quad (5)$$

On the right side, six of the triple products are zero. They are the off-diagonals like $\mathbf{b}\cdot(\mathbf{b}\times\mathbf{c})$, which contain the same vector twice. Since $\mathbf{b}\times\mathbf{c}$ is perpendicular to $\mathbf{b}$, this triple product is zero. The same is true of the others, like $\mathbf{a}\cdot(\mathbf{a}\times\mathbf{b}) = 0$. That is the volume of a box with two identical sides. The six off-diagonal zeros are the volumes of completely flattened boxes.

On the main diagonal the triple products equal D. The order of vectors can be **abc** or **bca** or **cab**, and the volume of the box stays the same. Dividing by this number D, which is placed outside for that purpose, gives the 1's in the identity matrix I.

Now we change to numbers. The goal is to find A^{-1} and to test it.

EXAMPLE 1   *The inverse of* $A = \begin{bmatrix} 1 & 1 & 0 \\ 1 & 0 & 2 \\ 0 & -2 & 2 \end{bmatrix}$ *is* $A^{-1} = \frac{1}{2}\begin{bmatrix} 4 & -2 & 2 \\ -2 & 2 & -2 \\ -2 & 2 & -1 \end{bmatrix}$.

That comes from the formula, and it absolutely has to be checked. Do not fail to multiply A^{-1} times A (or A times A^{-1}). Matrix multiplication is much easier than the formula for A^{-1}. We highlight row 3 times column 1, with dot product zero:

$$\frac{1}{2}\begin{bmatrix} 4 & -2 & 2 \\ -2 & 2 & -2 \\ -2 & 2 & -1 \end{bmatrix}\begin{bmatrix} 1 & 1 & 0 \\ 1 & 0 & 2 \\ 0 & -2 & 2 \end{bmatrix} = \frac{1}{2}\begin{bmatrix} 4-2 & 4-4 & -4+4 \\ -2+2 & -2+4 & 4-4 \\ -2+2 & -2+2 & 4-2 \end{bmatrix} = \begin{bmatrix} 1 & 0 & 0 \\ 0 & 1 & 0 \\ 0 & 0 & 1 \end{bmatrix}.$$

Remark on A^{-1} Inverting a matrix requires $D \neq 0$. We divide by $D = \det A$. The cross products $\mathbf{b}\times\mathbf{c}$ and $\mathbf{c}\times\mathbf{a}$ and $\mathbf{a}\times\mathbf{b}$ give A^{-1} in a neat form, but errors are easy. We prefer to avoid writing $\mathbf{i},\mathbf{j},\mathbf{k}$. There are nine 2 by 2 determinants to be calculated, and here is A^{-1} in full—containing the nine "*cofactors*" divided by D:

$$A^{-1} = \frac{1}{D}\begin{bmatrix} b_2c_3 - b_3c_2 & b_3c_1 - b_1c_3 & b_1c_2 - b_2c_1 \\ c_2a_3 - c_3a_2 & c_3a_1 - c_1a_3 & c_1a_2 - c_2a_1 \\ a_2b_3 - a_3b_2 & a_3b_1 - a_1b_3 & a_1b_2 - a_2b_1 \end{bmatrix}. \quad (6)$$

Important: The first row of A^{-1} does not use the first column of A, except in $1/D$. In other words, $\mathbf{b}\times\mathbf{c}$ does not involve $\mathbf{a}$. Here are the 2 by 2 determinants that produce $4, -2, 2$—which is divided by $D = 2$ in the top row of A^{-1}:

$$\begin{bmatrix} 1 & 1 & 0 \\ 1 & 0 & 2 \\ 0 & -2 & 2 \end{bmatrix}\begin{bmatrix} 1 & 1 & 0 \\ 1 & 0 & 2 \\ 0 & -2 & 2 \end{bmatrix}\begin{bmatrix} 1 & 1 & 0 \\ 1 & 0 & 2 \\ 0 & -2 & 2 \end{bmatrix}\quad \begin{bmatrix} + & - & + \\ - & + & - \\ + & - & + \end{bmatrix}. \quad (7)$$

The second highlighted determinant looks like $+2$ not -2. But the **sign matrix** on the right assigns a minus to that position in A^{-1}. We reverse the sign of $b_1c_3 - b_3c_1$, to find the cofactor $b_3c_1 - b_1c_3$ in the top row of (6).

To repeat: **For a row of** A^{-1}, **cross out the corresponding column of** A. **Find the three** 2 **by** 2 **determinants, use the sign matrix, and divide by** D.

EXAMPLE 2 $B = \begin{bmatrix} 1 & 1 & 1 \\ 0 & 1 & 1 \\ 0 & 0 & 1 \end{bmatrix}$ has $D = 1$ and $B^{-1} = \begin{bmatrix} 1 & -1 & 0 \\ 0 & 1 & -1 \\ 0 & 0 & 1 \end{bmatrix}$. (8)

The multiplication $BB^{-1} = I$ checks the arithmetic. Notice how $\begin{smallmatrix} 1 & 1 \\ 1 & 1 \end{smallmatrix}$ in B leads to a zero in the top row of B^{-1}. To find row 1, column 3 of B^{-1} we ignore column 1 and row 3 of B. (Also: the inverse of a triangular matrix is triangular.) The minus signs come from the sign matrix.

THE SOLUTION $\mathbf{u} = A^{-1}\mathbf{d}$

The purpose of A^{-1} is to solve the equation $A\mathbf{u} = \mathbf{d}$. Multiplying by A^{-1} produces $I\mathbf{u} = A^{-1}\mathbf{d}$. The matrix becomes the identity, $I\mathbf{u}$ equals $\mathbf{u}$, and the solution is immediate:

$$\mathbf{u} = A^{-1}\mathbf{d} = \frac{1}{D}\begin{bmatrix} \mathbf{b}\times\mathbf{c} \\ \mathbf{c}\times\mathbf{a} \\ \mathbf{a}\times\mathbf{b} \end{bmatrix}\begin{bmatrix} \\ \mathbf{d} \\ \\ \end{bmatrix} = \frac{1}{D}\begin{bmatrix} \mathbf{d}\cdot(\mathbf{b}\times\mathbf{c}) \\ \mathbf{d}\cdot(\mathbf{c}\times\mathbf{a}) \\ \mathbf{d}\cdot(\mathbf{a}\times\mathbf{b}) \end{bmatrix}. \qquad (9)$$

By writing those components x, y, z as *ratios of determinants*, we have Cramer's Rule:

11M (*Cramer's Rule*)

The solution is $x = \dfrac{\begin{vmatrix} \mathbf{d} & \mathbf{b} & \mathbf{c} \end{vmatrix}}{\begin{vmatrix} \mathbf{a} & \mathbf{b} & \mathbf{c} \end{vmatrix}}$, $\quad y = \dfrac{\begin{vmatrix} \mathbf{a} & \mathbf{d} & \mathbf{c} \end{vmatrix}}{\begin{vmatrix} \mathbf{a} & \mathbf{b} & \mathbf{c} \end{vmatrix}}$, $\quad z = \dfrac{\begin{vmatrix} \mathbf{a} & \mathbf{b} & \mathbf{d} \end{vmatrix}}{\begin{vmatrix} \mathbf{a} & \mathbf{b} & \mathbf{c} \end{vmatrix}}$. (10)

The right side $\mathbf{d}$ replaces, in turn, columns $\mathbf{a}$ and $\mathbf{b}$ and $\mathbf{c}$. All denominators are $D = \mathbf{a}\cdot(\mathbf{b}\times\mathbf{c})$. The numerator of x is the determinant $\mathbf{d}\cdot(\mathbf{b}\times\mathbf{c})$ in (9). The second numerator agrees with the second component $\mathbf{d}\cdot(\mathbf{c}\times\mathbf{a})$, because the cyclic order is correct. The third determinant with columns **abd** equals the triple product $\mathbf{d}\cdot(\mathbf{a}\times\mathbf{b})$ in $A^{-1}\mathbf{u}$. Thus (10) is the same as (9).

EXAMPLE A: Multiply by A^{-1} to find the known solution $x = -2, y = 3, z = 1$:

$$\mathbf{u} = A^{-1}\mathbf{d} = \frac{1}{2}\begin{bmatrix} 4 & -2 & 2 \\ -2 & 2 & -2 \\ -2 & 2 & -1 \end{bmatrix}\begin{bmatrix} 1 \\ 0 \\ -4 \end{bmatrix} = \frac{1}{2}\begin{bmatrix} 4-8 \\ -2+8 \\ -2+4 \end{bmatrix} = \begin{bmatrix} -2 \\ 3 \\ 1 \end{bmatrix}.$$

EXAMPLE B: Multiply by B^{-1} to solve $B\mathbf{u} = \mathbf{d}$ when $\mathbf{d}$ is the column $(6, 5, 4)$:

$$\mathbf{u} = B^{-1}\mathbf{d} = \begin{bmatrix} 1 & -1 & 0 \\ 0 & 1 & -1 \\ 0 & 0 & 1 \end{bmatrix}\begin{bmatrix} 6 \\ 5 \\ 4 \end{bmatrix} = \begin{bmatrix} 1 \\ 1 \\ 4 \end{bmatrix}. \quad Check\ B\mathbf{u} = \begin{bmatrix} 1 & 1 & 1 \\ 0 & 1 & 1 \\ 0 & 0 & 1 \end{bmatrix}\begin{bmatrix} 1 \\ 1 \\ 4 \end{bmatrix} = \begin{bmatrix} 6 \\ 5 \\ 4 \end{bmatrix}.$$

EXAMPLE C: Put $\mathbf{d} = (6, 5, 4)$ in each column of B. Cramer's Rule gives $\mathbf{u} = (1, 1, 4)$:

$$\begin{vmatrix} 6 & 1 & 1 \\ 5 & 1 & 1 \\ 4 & 0 & 1 \end{vmatrix} = 1 \quad \begin{vmatrix} 1 & 6 & 1 \\ 0 & 5 & 1 \\ 0 & 4 & 1 \end{vmatrix} = 1 \quad \begin{vmatrix} 1 & 1 & 6 \\ 0 & 1 & 5 \\ 0 & 0 & 4 \end{vmatrix} = 4 \quad \text{all divided by } D = \begin{vmatrix} 1 & 1 & 1 \\ 0 & 1 & 1 \\ 0 & 0 & 1 \end{vmatrix} = 1.$$

This rule fills the page with determinants. Those are good ones to check by eye, without writing down the six terms (three $+$ and three $-$).

The formulas for A^{-1} are honored chiefly in their absence. They are not used by the computer, even though the algebra is in some ways beautiful. In big calculations, the computer never finds A^{-1}—just the solution.

We now look at the singular case $D = 0$. Geometry-algebra-algorithm must all break down. After that is the algorithm: Gaussian elimination.

THE SINGULAR CASE

Changing one entry of a matrix can make the determinant zero. The triple product $\mathbf{a} \cdot (\mathbf{b} \times \mathbf{c})$, which is also the volume, becomes $D = 0$. The box is flattened and the matrix is singular. That happens in our example when the lower right entry is changed from 2 to 4:

$$S = \begin{bmatrix} 1 & 1 & 0 \\ 1 & 0 & 2 \\ 0 & -2 & 4 \end{bmatrix} \text{ has determinant } D = 0.$$

This does more than change the inverse. It *destroys* the inverse. We can no longer divide by D. There is no S^{-1}.

What happens to the row picture and column picture? For 2 by 2 systems, the singular case had two parallel lines. Now the row picture has three planes, which need not be parallel. Here the planes are *not parallel*. Their normal vectors are the rows of S, which go in different directions. But somehow the planes fail to go through a common point.

What happens is more subtle. The intersection line from two planes misses the third plane. The line is parallel to the plane and stays above it (Figure 11.22)a. When all three planes are drawn, they form an open tunnel. The picture tells more than the numbers, about how three planes can fail to meet. The third figure shows an end view, where the planes go directly into the page. Each pair meets in a line, but those lines don't meet in a point.

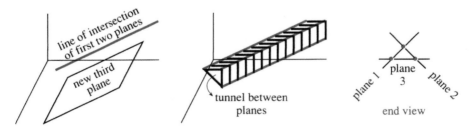

Fig. 11.22   The row picture in the singular case: no interestion point, no solutions.

When two planes are parallel, the determinant is again zero. One row of the matrix is a multiple of another row. The extreme case has all three planes parallel—as in a matrix with nine 1's.

The column picture must also break down. In the 2 by 2 failure (previous section), the columns were on the same line. *Now the three columns are in the same plane*. The combinations of those columns produce **d** only if it happens to lie in that particular plane. Most vectors **d** will be outside the plane, so most singular systems have no solution.

When the determinant is zero, $A\mathbf{u} = \mathbf{d}$ has no solution or infinitely many.

THE ELIMINATION ALGORITHM

Go back to the 3 by 3 example $A\mathbf{u} = \mathbf{d}$. If you were given those equations, you would never think of determinants. You would—*quite correctly*—start with the first equation. It gives $x = 1 - y$, which goes into the next equation to eliminate x:

$$
\begin{array}{rl}
x + y & = 1 \\
x \quad + 2z & = 0 \\
-2y + 2z & = -4
\end{array}
\quad \xrightarrow{x = 1-y} \quad
\begin{array}{rl}
1 - y + 2z & = 0 \\
-2y + 2z & = -4.
\end{array}
$$

Stop there for a minute. On the right is a 2 by 2 system for y and z. The first equation and first unknown are eliminated—exactly what we want. But that step was not organized in the best way, because a "1" ended up on the left side. Constants should stay on the right side—the pattern should be preserved. It is better to take the same step by *subtracting the first equation from the second*:

$$
\begin{array}{rl}
x + y & = 1 \\
x \quad + 2z & = 0 \\
-2y + 2z & = -4
\end{array}
\quad \longrightarrow \quad
\begin{array}{rl}
- y + 2z & = -1 \\
-2y + 2z & = -4.
\end{array}
\tag{11}
$$

Same equations, better organization. Now look at the corner term $-y$. Its coefficient -1 is the *second pivot*. (The first pivot was $+1$, the coefficient of x in the first corner.) We are ready for the next elimination step:

> *Plan*: Subtract a multiple of the "pivot equation" from the equation below it.
> *Goal*: To produce a zero below the pivot, so y is eliminated.
> *Method*: Subtract 2 times the pivot equation to cancel $-2y$.

$$
\begin{array}{rl}
- y + 2z & = -1 \\
-2y + 2z & = -4
\end{array}
\quad \rightarrow \quad
-2z = -2.
\tag{12}
$$

The answer comes by *back substitution*. Equation (12) gives $z = 1$. Then equation (11) gives $y = 3$. Then the first equation gives $x = -2$. This is much quicker than determinants. You may ask: *Why use Cramer's Rule*? Good question.

With numbers elimination is better. It is faster and also safer. (To check against error, substitute $-2, 3, 1$ into the original equations.) The algorithm reaches the answer *without the determinant and without the inverse*. Calculations with letters use $\det A$ and A^{-1}.

Here are the steps in a definite order (top to bottom):

> Subtract a multiple of equation 1 to produce $0x$ in equation 2
> Subtract a multiple of equation 1 to produce $0x$ in equation 3
> Subtract a multiple of equation 2 (new) to produce $0y$ in equation 3.

EXAMPLE (notice the zeros appearing under the pivots):

$$\begin{matrix} x+ & y+ & z= & 1 \\ 2x+ & 5y+ & 3z= & 7 \\ 4x+ & 7y+ & 6z= & 11 \end{matrix} \quad \rightarrow \quad \begin{matrix} x+ & y+ & z=1 \\ & 3y+ & z=5 \\ & 3y+ & 2z=7 \end{matrix} \quad \rightarrow \quad \begin{matrix} x+ & y+z=1 \\ & 3y+z=5 \\ & z=2. \end{matrix}$$

Elimination leads to a ***triangular system***. The coefficients below the diagonal are zero. First $z=2$, then $y=1$, then $x=-2$. *Back substitution solves triangular systems* (fast).

As a final example, try the singular case $S\mathbf{u}=\mathbf{d}$ when the corner entry is changed from 2 to 4. With $D=0$, there is no inverse matrix S^{-1}. Elimination also fails, by reaching an impossible equation $0=-2$:

$$\begin{matrix} x+ & y & =1 \\ x & +2z= & 0 \\ & -2y+4z= & -4 \end{matrix} \quad \rightarrow \quad \begin{matrix} x+ & y & =1 \\ & -y+2z= & -1 \\ & -2y+4z= & -4 \end{matrix} \quad \rightarrow \quad \begin{matrix} x+ & y & =1 \\ & -y+2z= & -1 \\ & 0= & -2 \end{matrix}$$

The three planes do not meet at a point—a fact that was not obvious at the start. Algebra discovers this fact from $D=0$. Elimination discovers it from $0=-2$. The chapter is ending at the point where my linear algebra book begins.

One final comment. In actual computing, you will use a code written by professionals. The steps will be the same as above. A multiple of equation 1 is subtracted from each equation below it, to eliminate the first unknown x. With one fewer unknown and equation, elimination starts again. (A parallel computer executes many steps at once.) Extra instructions are included to reduce roundoff error. You only see the result! But it is more satisfying to know what the computer is doing.

In the end, solving linear equations is the key step in solving nonlinear equations. The central idea of differential calculus is to *linearize* near a point.

11.5 EXERCISES

Read-through questions

Three equations in three unknowns can be written as $A\mathbf{u}=\mathbf{d}$. The __a__ $\mathbf{u}$ has components x,y,z and A is a __b__. The row picture has a __c__ for each equation. The first two planes intersect in a __d__, and all three planes intersect in a __e__, which is __f__. The column picture starts with vectors $\mathbf{a,b,c}$ from the columns of __g__ and combines them to produce __h__. The vector equation is __i__ $=\mathbf{d}$.

The determinant of A is the triple product __j__. This is the volume of a box, whose edges from the origin are __k__. If $\det A=$ __l__ then the system is __m__. Otherwise there is an __n__ matrix such that $A^{-1}A=$ __o__ (the __p__ matrix). In this case the solution to $A\mathbf{u}=\mathbf{d}$ is $\mathbf{u}=$ __q__.

The rows of A^{-1} are the cross products $\mathbf{b}\times\mathbf{c}$, __r__, __s__, divided by D. The entries of A^{-1} are 2 by 2 __t__, divided by D.

The upper left entry equals __u__. The 2 by 2 determinants needed for a row of A^{-1} do not use the corresponding __v__ of A.

The solution is $\mathbf{u}=A^{-1}\mathbf{d}$. Its first component x is a ratio of determinants, $|\mathbf{d\,b\,c}|$ divided by __w__. Cramer's Rule breaks down when $\det A=$ __x__. Then the columns $\mathbf{a,b,c}$ lie in the same __y__. There is no solution to $x\mathbf{a}+y\mathbf{b}+z\mathbf{c}=\mathbf{d}$, if $\mathbf{d}$ is not on that __z__. In a singular row picture, the intersection of planes 1 and 2 is __A__ to the third plane.

In practice $\mathbf{u}$ is computed by __B__. The algorithm starts by subtracting a multiple of row 1 to eliminate x from __c__. If the first two equations are $x-y=1$ and $3x+z=7$, this elimination step leaves __D__. Similarly x is eliminated from the third equation, and then __E__ is eliminated. The equations are solved by back __F__. When the system has no solution, we reach an impossible equation like __G__. The example $x - y = 1, 3x + z = 7$ has no solution if the third equation is __H__.

Rewrite 1–4 as matrix equations $Au = d$ (do not solve).

1 $d = (0, 0, 8)$ is a combination of $a = (1, 2, 0)$ and $b = (2, 3, 2)$ and $c = (2, 5, 2)$.

2 The planes $x + y = 0, x + y + z = 1$, and $y + z = 0$ intersect at $u = (x, y, z)$.

3 The point $u = (x, y, z)$ is on the planes $x = y$, $y = z$, $x - z = 1$.

4 A combination of $a = (1, 0, 0)$ and $b = (0, 2, 0)$ and $c = (0, 0, 3)$ equals $d = (5, 2, 0)$.

5 Show that Problem 3 has no solution in two ways: find the determinant of A, and combine the equations to produce $0 = 1$.

6 Solve Problem 2 in two ways: by inspection and Cramer's Rule.

7 Solve Problem 4 in two ways: by inspection and by computing the determinant and inverse of the *diagonal matrix*

$$A = \begin{bmatrix} 1 & 0 & 0 \\ 0 & 2 & 0 \\ 0 & 0 & 3 \end{bmatrix}.$$

8 Solve the three equations of Problem 1 by elimination.

9 The vectors b and c lie in a plane which is perpendicular to the vector _____. In case the vector a also lies in that plane, it is also perpendicular and $a \cdot$ _____ $= 0$. The _____ of the matrix with columns in a plane is _____.

10 The plane $a_1 x + b_1 y + c_1 z = d_1$ is perpendicular to its normal vector $N_1 =$ _____. The plane $a_2 x + b_2 y + c_2 z = d_2$ is perpendicular to $N_2 =$ _____. The planes meet in a line that is perpendicular to both vectors, so the line is parallel to their _____ product. If this line is also parallel to the third plane and perpendicular to N_3, the system is _____. The matrix has no _____, which happens when $(N_1 \times N_2) \cdot N_3 = 0$.

Problems 11–24 use the matrices A, B, C.

$$A = \begin{bmatrix} 1 & 4 & 0 \\ 0 & 2 & 6 \\ 0 & 0 & 3 \end{bmatrix} \quad B = \begin{bmatrix} 0 & 0 & 1 \\ 2 & 1 & 0 \\ 6 & 4 & 0 \end{bmatrix} \quad C = \begin{bmatrix} 1 & -1 & -3 \\ -1 & 2 & 0 \\ 0 & -1 & 3 \end{bmatrix}.$$

11 Find the determinants $|A|, |B|, |C|$. Since A is triangular, its determinant is the product _____.

12 Compute the cross products of each pair of columns in B (three cross products).

13 Compute the inverses of A and B above. Check that $A^{-1} A = I$ and $B^{-1} B = I$.

14 Solve $Au = \begin{bmatrix} 1 \\ 0 \\ 0 \end{bmatrix}$ and $Bu = \begin{bmatrix} 1 \\ 0 \\ 0 \end{bmatrix}$. With this right side d, why is u the first column of the inverse?

15 Suppose all three columns of a matrix add to zero, as in C above. The dot product of each column with $v = (1, 1, 1)$ is _____. All three columns lie in the same _____. The determinant of C must be _____.

16 Find a nonzero solution to $Cu = 0$. Find all solutions to $Cu = 0$.

17 Choose any right side d that is perpendicular to $v = (1, 1, 1)$ and solve $Cu = d$. Then find a second solution.

18 Choose any right side d that is not perpendicular to $v = (1, 1, 1)$. Show by elimination (reach an impossible equation) that $Cu = d$ has no solution.

19 Compute the matrix product AB and then its determinant. How is $\det AB$ related to $\det A$ and $\det B$?

20 Compute the matrix products BC and CB. All columns of CB add to _____, and its determinant is _____.

21 Add A and C by adding each entry of A to the corresponding entry of C. Check whether the determinant of $A + C$ equals $\det A + \det C$.

22 Compute $2A$ by multiplying each entry of A by 2. The determinant of $2A$ equals _____ times the determinant of A.

23 Which four entries of A give the upper left corner entry p of A^{-1}, after dividing by $D = \det A$? Which four entries of A give the entry q in row 1, column 2 of A^{-1}? Find p and q.

24 The 2 by 2 determinants from the first two rows of B are -1 (from columns 2, 3) and -2 (from columns 1, 3) and _____ (from columns 1, 2). These numbers go into the third _____ of B^{-1}, after dividing by _____ and changing the sign of _____.

25 Why does every inverse matrix A^{-1} have an inverse?

26 From the multiplication $ABB^{-1}A^{-1} = I$ it follows that the inverse of AB is _____. The separate inverses come in _____ order. If you put on socks and then shoes, the inverse begins by taking off _____.

27 Find the determinants of these four *permutation matrices*:

$$P = \begin{bmatrix} 0 & 1 & 0 \\ 1 & 0 & 0 \\ 0 & 0 & 1 \end{bmatrix} \quad Q = \begin{bmatrix} 0 & 0 & 1 \\ 0 & 1 & 0 \\ 1 & 0 & 0 \end{bmatrix} \quad PQ = \begin{bmatrix} 0 & 1 & 0 \\ 0 & 0 & 1 \\ 1 & 0 & 0 \end{bmatrix}$$

and $QP =$ _____. Multiply $u = (x, y, z)$ by each permutation to find Pu, Qu, PQu, and QPu.

28 Find all six of the 3 by 3 permutation matrices (including I), with a single 1 in each row and column. Which of them are "even" (determinant 1) and which are "odd" (determinant -1)?

29 How many 2 by 2 permutation matrices are there, including I? How many 4 by 4?

30 Multiply any matrix A by the permutation matrix P and explain how PA is related to A. In the opposite order explain how AP is related to A.

31 Eliminate x from the last two equations by subtracting the first equation. Then eliminate y from the new third equation by using the new second equation:

(a)
$$\begin{aligned} x + y + z &= 2 \\ x + 3y + 3z &= 0 \\ x + 3y + 7z &= 2 \end{aligned}$$

(b)
$$\begin{aligned} x + y \quad\ \ &= 1 \\ x + \quad z &= 3 \\ y + z &= 5. \end{aligned}$$

After elimination solve for z, y, x (back substitution).

32 By elimination and back substitution solve

(a)
$$\begin{aligned} x + 2y + 2z &= 0 \\ 2x + 3y + 5z &= 0 \\ 2y + 2z &= 8 \end{aligned}$$

(b)
$$\begin{aligned} x - y \quad\ \ &= 1 \\ x \quad - z &= 4 \\ y - z &= 7. \end{aligned}$$

33 Eliminate x from equation 2 by using equation 1:

$$\begin{aligned} x + 2y + 2z &= 0 \\ 2x + 4y + 5z &= 0 \\ 2y + 2z &= 8. \end{aligned}$$

Why can't the new second equation eliminate y from the third equation? Is there a solution or is the system singular?

Note: If elimination creates a zero in the "pivot position," try to exchange that pivot equation with an equation below it. Elimination succeeds when there is a full set of pivots.

34 The pivots in Problem 32a are $1, -1$, and 4. Circle those as they appear along the diagonal in elimination. Check that the product of the pivots equals the determinant. (This is how determinants are computed.)

35 Find the pivots and determinants in Problem 31.

36 Find the inverse of $A = \begin{bmatrix} 1 & 1 & 0 \\ 0 & 1 & 1 \\ 0 & 0 & 1 \end{bmatrix}$ and also of $B = A^2$.

37 The symbol a_{ij} stands for the entry in row i, column j. Find a_{12} and a_{21} in Problem 36. The formula $\Sigma a_{ij} b_{jk}$ gives the entry in which row and column of the matrix product AB ?

38 Write down a 3 by 3 singular matrix S in which no two rows are parallel. Find a combination of rows 1 and 2 that is parallel to row 3. Find a combination of columns 1 and 2 that is parallel to column 3. Find a nonzero solution to $Su = 0$.

39 Compute these determinants. The 2 by 2 matrix is invertible if _____ . The 3 by 3 matrix (is)(is not) invertible.

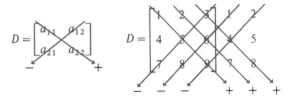

CHAPTER 12

Motion Along a Curve

This chapter is about "vector functions." The vector $2\mathbf{i}+4\mathbf{j}+8\mathbf{k}$ is constant. The vector $\mathbf{R}(t)=t\mathbf{i}+t^2\mathbf{j}+t^3\mathbf{k}$ is moving. It is a function of the parameter t, which often represents time. At each time t, the position vector $\mathbf{R}(t)$ locates the moving body:

$$\textit{position vector} = \mathbf{R}(t) = x(t)\mathbf{i}+ y(t)\mathbf{j}+z(t)\mathbf{k}. \tag{1}$$

Our example has $x = t$, $y = t^2$, $z = t^3$. As t varies, these points trace out a ***curve in space***. The parameter t tells when the body passes each point on the curve. The constant vector $2\mathbf{i}+4\mathbf{j}+8\mathbf{k}$ is the position vector $\mathbf{R}(2)$ at the instant $t = 2$.

What are the questions to be asked? Every student of calculus knows the first question: *Find the derivative.* If something moves, the Navy salutes it and we differentiate it. At each instant, the body moving along the curve has a speed and a direction. This information is contained in another vector function—the velocity vector $\mathbf{v}(t)$ which is the derivative of $\mathbf{R}(t)$:

$$\mathbf{v}(t) = \frac{d\mathbf{R}}{dt} = \frac{dx}{dt}\mathbf{i}+ \frac{dy}{dt}\mathbf{j}+ \frac{dz}{dt}\mathbf{k}. \tag{2}$$

Since $\mathbf{i},\mathbf{j},\mathbf{k}$ are fixed vectors, their derivatives are zero. In polar coordinates $\mathbf{i}$ and $\mathbf{j}$ are replaced by moving vectors. Then the velocity $\mathbf{v}$ has more terms from the product rule (Section 12.4).

Two important cases are uniform motion ***along a line and around a circle***. We study those motions in detail ($\mathbf{v}$ = constant on line, $\mathbf{v}$ = tangent to circle). This section also finds the speed and distance and acceleration for any motion $\mathbf{R}(t)$.

Equation (2) is the computing rule for the velocity $d\mathbf{R}/dt$. It is not the *definition* of $d\mathbf{R}/dt$, which goes back to basics and does not depend on coordinates:

$$\frac{d\mathbf{R}}{dt} = \lim_{\Delta t\to 0} \frac{\Delta\mathbf{R}}{\Delta t} = \lim_{\Delta t\to 0} \frac{\mathbf{R}(t +\Delta t) -\mathbf{R}(t)}{\Delta t}.$$

We repeat: $\mathbf{R}$ is a vector so $\Delta\mathbf{R}$ is a vector so $d\mathbf{R}/dt$ is a vector. All three vectors are in Figure 12.1 (t is not a vector!). This figure reveals the key fact about the geometry: *The velocity $\mathbf{v} = d\mathbf{R}/dt$ is tangent to the curve.*

517

The vector $\Delta\mathbf{R}$ goes from one point on the curve to a nearby point. Dividing by Δt changes its length, not its direction. That direction lines up with the tangent to the curve, as the points come closer.

EXAMPLE 1 $\mathbf{R}(t) = t\mathbf{i} + t^2\mathbf{j} + t^3\mathbf{k}$ $\mathbf{v}(t) = \mathbf{i} + 2t\mathbf{j} + 3t^2\mathbf{k}$

This curve swings upward as t increases. When $t = 0$ the velocity is $\mathbf{v} = \mathbf{i}$. The tangent is along the x axis, since the $\mathbf{j}$ and $\mathbf{k}$ components are zero. When $t = 1$ the velocity is $\mathbf{i} + 2\mathbf{j} + 3\mathbf{k}$, and the curve is climbing.

For the shadow on the xy plane, drop the $\mathbf{k}$ component. Position on the shadow is $t\mathbf{i} + t^2\mathbf{j}$. Velocity along the shadow is $\mathbf{i} + 2t\mathbf{j}$. The shadow is a plane curve.

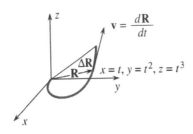

Fig. 12.1   Position vector $\mathbf{R}$, change $\Delta\mathbf{R}$, velocity $d\mathbf{R}/dt$.

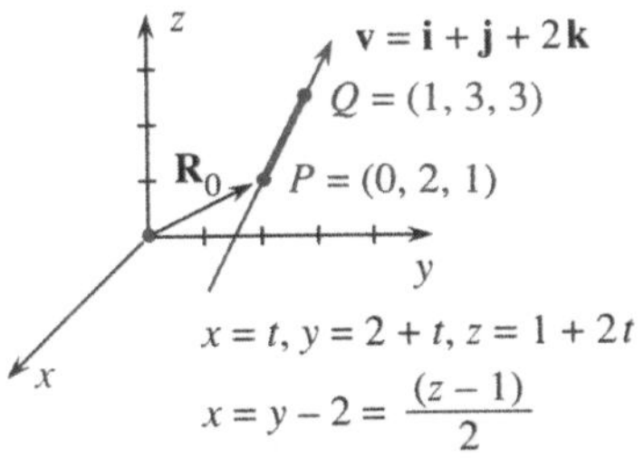

Fig. 12.2   Equations of a line, with and without the parameter t.

EXAMPLE 2   Uniform motion in a straight line: *the velocity vector $\mathbf{v}$ is constant*.

The speed and direction don't change. The position vector moves with $d\mathbf{R}/dt = \mathbf{v}$:

$$\mathbf{R}(t) = \mathbf{R}_0 + t\mathbf{v} \quad (\mathbf{R}_0 \text{ fixed, } \mathbf{v} \text{ fixed, } t \text{ varying}) \qquad (3)$$

That is the **equation of a line** in vector form. Certainly $d\mathbf{R}/dt = \mathbf{v}$. The starting point $\mathbf{R}_0 = x_0\mathbf{i} + y_0\mathbf{j} + z_0\mathbf{k}$ is given. The velocity $\mathbf{v} = v_1\mathbf{i} + v_2\mathbf{j} + v_3\mathbf{k}$ is also given. Separating the x, y and z components, equation (3) for a line is

line with parameter: $x = x_0 + tv_1, \quad y = y_0 + tv_2, \quad z = z_0 + tv_3.$ (4)

The speed along the line is $|\mathbf{v}| = \sqrt{v_1^2 + v_2^2 + v_3^2}$. The direction of the line is the unit vector $\mathbf{v}/|\mathbf{v}|$. We have three equations for x, y, z, and eliminating t leaves two equations. The parameter t equals $(x - x_0)/v_1$ from equation (4). It also equals $(y - y_0)/v_2$ and $(z - z_0)/v_3$. So these ratios equal each other, and t is gone:

line without parameter: $\dfrac{x - x_0}{v_1} = \dfrac{y - y_0}{v_2} = \dfrac{z - z_0}{v_3}.$ (5)

An example is $x = y/2 = z/3$. In this case $(x_0, y_0, z_0) = (0, 0, 0)$—the line goes through the origin. Another point on the line is $(x, y, z) = (2, 4, 6)$. Because t is gone, we cannot say when we reach that point and how fast we are going. The equations $x/4 = y/8 = z/12$ give the same line. Without t we can't know the velocity $\mathbf{v} = d\mathbf{R}/dt$.

EXAMPLE 3   Find an equation for the line through $P = (0, 2, 1)$ and $Q = (1, 3, 3)$.

Solution   We have choices! $\mathbf{R}_0$ can go to *any point* on the line. The velocity $\mathbf{v}$ can be *any multiple* of the vector from P to Q. The decision on $\mathbf{R}_0$ controls where we start, and $\mathbf{v}$ controls our speed.

The vector from P to Q is $\mathbf{i} + \mathbf{j} + 2\mathbf{k}$. Those numbers $1, 1, 2$ come from subtracting $0, 2, 1$ from $1, 3, 3$. We choose this vector $\mathbf{i} + \mathbf{j} + 2\mathbf{k}$ as a first $\mathbf{v}$, and double it for a second **v**. We choose the vector $\mathbf{R}_0 = \mathbf{P}$ as a first start and $\mathbf{R}_0 = \mathbf{Q}$ as a second start. Here are two different expressions for the same line—they are $\mathbf{P} + t\mathbf{v}$ and $\mathbf{Q} + t(2\mathbf{v})$:

$$\mathbf{R}(t) = (2\mathbf{j} + \mathbf{k}) + t(\mathbf{i} + \mathbf{j} + 2\mathbf{k}) \qquad \mathbf{R}^*(t) = (\mathbf{i} + 3\mathbf{j} + 3\mathbf{k}) + t(2\mathbf{i} + 2\mathbf{j} + 4\mathbf{k}).$$

The vector $\mathbf{R}(t)$ gives $x = t$, $y = 2 + t$, $z = 1 + 2t$. The vector $\mathbf{R}^*$ is at a different point on the same line at the same time: $x^* = 1 + 2t$, $y^* = 3 + 2t$, $z^* = 3 + 4t$.

If I pick $t = 1$ in $\mathbf{R}$ and $t = 0$ in $\mathbf{R}^*$, the point is $(1, 3, 3)$. We arrive there at different times. You are seeing how parameters work, to tell "where" and also "when." If t goes from $-\infty$ to $+\infty$, all points on one line are also on the other line. The path is the same, but the "twins" are going at different speeds.

Question 1   When do these twins meet? When does $\mathbf{R}(t) = \mathbf{R}^*(t)$?
Answer   They meet at $t = -1$, when $\mathbf{R} = \mathbf{R}^* = -\mathbf{i} + \mathbf{j} - \mathbf{k}$.

Question 2   What is an equation for the segment between P and Q (not beyond)?
Answer   In the equation for $\mathbf{R}(t)$, let t go from 0 to 1 (not beyond):

$$x = t \quad y = 2 + t \quad z = 1 + 2t \quad [0 \leqslant t \leqslant 1 \text{ for segment}]. \tag{6}$$

At $t = 0$ we start from $P = (0, 2, 1)$. At $t = 1$ we reach $Q = (1, 3, 3)$.

Question 3   What is an equation for the line without the parameter t?
Answer   Solve equations (6) for t or use (5): $x/1 = (y - 2)/1 = (z - 1)/2$.

Question 4   Which point on the line is closest to the origin?
Answer   The derivative of $x^2 + y^2 + z^2 = t^2 + (2 + t)^2 + (1 + 2t)^2$ is $8 + 8t$. This derivative is zero at $t = -1$. So the closest point is $(-1, 1, -1)$.

Question 5   Where does the line meet the plane $x + y + z = 11$?
Answer   Equation (6) gives $x + y + z = 3 + 4t = 11$. So $t = 2$. The meeting point is $x = t = 2, y = t + 2 = 4, z = 1 + 2t = 5$.

Question 6   What line goes through $(3, 1, 1)$ perpendicular to the plane $x - y - z = 1$?
Answer   The normal vector to the plane is $\mathbf{N} = \mathbf{i} - \mathbf{j} - \mathbf{k}$. *That is* **v**. The position vector to $(3, 1, 1)$ is $\mathbf{R}_0 = 3\mathbf{i} + \mathbf{j} + \mathbf{k}$. Then $\mathbf{R} = \mathbf{R}_0 + t\mathbf{v}$.

COMPARING LINES AND PLANES

A line has one parameter or two equations. We give the starting point and velocity: $(x, y, z) = (x_0, y_0, z_0) + t(v_1, v_2, v_3)$. That tells directly which points are on the line. Or we eliminate t to find the two equations in (5).

A plane has one equation or two parameters! The equation is $ax + by + cz = d$. That tells us *in*directly which points are on the plane. (Instead of knowing x, y, z, we know the equation they satisfy. Instead of directions **v** and **w** in the plane, we are told the perpendicular direction $\mathbf{N} = (a, b, c)$.) With parameters, the line contains $\mathbf{R}_0 + t\mathbf{v}$ and the plane contains $\mathbf{R}_0 + t\mathbf{v} + s\mathbf{w}$. A plane looks worse with parameters (t and s), a line looks better.

Questions 5 and 6 connected lines to planes. Here are two more. See Problems 41 − 44:

Question 7   When is the line $\mathbf{R}_0 + t\mathbf{v}$ parallel to the plane? When is it perpendicular?
Answer   The test is $\mathbf{v} \cdot \mathbf{N} = 0$. The test is $\mathbf{v} \times \mathbf{N} = 0$.

EXAMPLE 4   Find the plane containing $P_0 = (1, 2, 1)$ and the line of points $(1, 0, 0) + t(2, 0, -1)$. That vector $\mathbf{v}$ will be in the plane.

Solution   The vector $\mathbf{v} = 2\mathbf{i} - \mathbf{k}$ goes along the line. The vector $\mathbf{w} = 2\mathbf{j} + \mathbf{k}$ goes from $(1, 0, 0)$ to $(1, 2, 1)$. Their cross product is

$$\mathbf{N} = \mathbf{v} \times \mathbf{w} = \begin{vmatrix} \mathbf{i} & \mathbf{j} & \mathbf{k} \\ 2 & 0 & -1 \\ 0 & 2 & 1 \end{vmatrix} = 2\mathbf{i} - 2\mathbf{j} + 4\mathbf{k}.$$

The plane $2x - 2y + 4z = 2$ has this normal $\mathbf{N}$ and contains the point $(1, 2, 1)$.

SPEED, DIRECTION, DISTANCE, ACCELERATION

We go back to the curve traced out by $\mathbf{R}(t)$. The derivative $\mathbf{v}(t) = d\mathbf{R}/dt$ is the velocity vector along that curve. The **speed** is the magnitude of $\mathbf{v}$:

$$\text{speed} = |\mathbf{v}| = \sqrt{(dx/dt)^2 + (dy/dt)^2 + (dz/dt)^2}. \tag{7}$$

The **direction** of the velocity vector is $\mathbf{v}/|\mathbf{v}|$. This is a unit vector, since $\mathbf{v}$ is divided by its length. **The unit tangent vector $\mathbf{v}/|\mathbf{v}|$ is denoted by $\mathbf{T}$.**

The tangent vector is constant for lines. It changes direction for curves.

EXAMPLE 5   (important)   Find $\mathbf{v}$ and $|\mathbf{v}|$ and $\mathbf{T}$ for steady motion around a circle:

$$x = r \cos \omega t, \quad y = r \sin \omega t, \quad z = 0.$$

Solution   The position vector is $\mathbf{R} = r \cos \omega t\, \mathbf{i} + r \sin \omega t\, \mathbf{j}$. The velocity is

$$\mathbf{v} = d\mathbf{R}/dt = -\omega r \sin \omega t\, \mathbf{i} + \omega r \cos \omega t\, \mathbf{j} \quad \text{(tangent, not unit tangent)}$$

The speed is the radius r times the angular velocity ω:

$$|\mathbf{v}| = \sqrt{(-\omega r \sin \omega t)^2 + (\omega r \cos \omega t)^2} = \omega r.$$

The unit tangent vector is $\mathbf{v}$ divided by $|\mathbf{v}|$:

$$\mathbf{T} = -\sin \omega t\, \mathbf{i} + \cos \omega t\, \mathbf{j} \quad \text{(length 1 since } \sin^2 \omega t + \cos^2 \omega t = 1\text{)}.$$

Think next about the *distance traveled*. Distance along a curve is always denoted by s (called **arc length**). I don't know why we use s—certainly not as the initial for speed. In fact speed is distance divided by time. The ratio s/t gives average speed; ds/dt is instantaneous speed. We are back to Chapter 1 and Section 8.3, the relation of speed to distance:

$$\text{speed } |\mathbf{v}| = ds/dt \qquad \text{distance } s = \int (ds/dt)\, dt = \int |\mathbf{v}(t)|\, dt.$$

Notice that $|\mathbf{v}|$ and s and t are scalars. The direction vector is $\mathbf{T}$:

$$\mathbf{T} = \frac{\mathbf{v}}{|\mathbf{v}|} = \frac{d\mathbf{R}/dt}{ds/dt} = \frac{d\mathbf{R}}{ds} = \textit{unit tangent vector.} \tag{8}$$

In Figure 12.3, the chord length (straight) is $|\Delta\mathbf{R}|$. The arc length (curved) is Δs. As $\Delta\mathbf{R}$ and Δs approach zero, the ratio $|\Delta\mathbf{R}/\Delta s|$ approaches $|\mathbf{T}| = 1$.

Think finally about the *acceleration vector* **a**(t). It is the rate of change of velocity (not the rate of change of speed):

$$\mathbf{a} = \frac{d\mathbf{v}}{dt} = \frac{d^2\mathbf{R}}{dt^2} = \frac{d^2x}{dt^2}\mathbf{i} + \frac{d^2y}{dt^2}\mathbf{j} + \frac{d^2z}{dt^2}\mathbf{k}. \qquad (9)$$

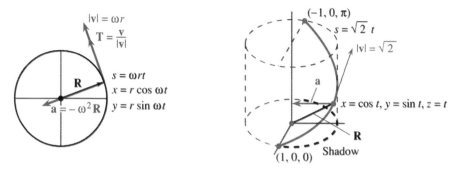

Fig. 12.3 Steady motion around a circle. Half turn up a helix.

For steady motion along a line, as in $x = t, y = 2 + t, z = 1 + 2t$, there is no acceleration. The second derivatives are all zero. For steady motion around a circle, there *is* acceleration. In driving a car, you accelerate with the gas pedal or the brake. *You also accelerate by turning the wheel*. It is the velocity vector that changes, not the speed.

EXAMPLE 6 Find the distance $s(t)$ and acceleration **a**(t) for circular motion.

Solution   The speed in Example 5 is $ds/dt = \omega r$. After integrating, the distance is $s = \omega r t$. At time t we have gone through an angle of ωt. The radius is r, so the distance traveled agrees with ωt times r. Note that the dimension of ω is $1/\text{time}$. (Angles are dimensionless.) At time $t = 2\pi/\omega$ we have gone once around the circle—to $s = 2\pi r$ not back to $s = 0$.

The acceleration is $\mathbf{a} = d^2\mathbf{R}/dt^2$. Remember $\mathbf{R} = r\cos\omega t\,\mathbf{i} + r\sin\omega t\,\mathbf{j}$:

$$\mathbf{a}(t) = -\omega^2 r\cos\omega t\,\mathbf{i} - \omega^2 r\sin\omega t\,\mathbf{j}. \qquad (10)$$

That direction is opposite to **R**. This is a special motion, with no action on the gas pedal or the brake. All the acceleration is from turning. The magnitude is $|\mathbf{a}| = \omega^2 r$, with the correct dimension of $\text{distance}/(\text{time})^2$.

EXAMPLE 7 Find **v** and s and **a** around the helix $\mathbf{R} = \cos t\,\mathbf{i} + \sin t\,\mathbf{j} + t\,\mathbf{k}$.

Solution   The velocity is $\mathbf{v} = -\sin t\,\mathbf{i} + \cos t\,\mathbf{j} + \mathbf{k}$. The speed is

$$ds/dt = |\mathbf{v}| = \sqrt{\sin^2 t + \cos^2 t + 1} = \sqrt{2} \text{ (constant)}.$$

Then distance is $s = \sqrt{2}t$. At time $t = \pi$, a half turn is complete. The distance along the shadow is π (a half circle). The distance along the helix is $\sqrt{2}\pi$, because of its $45°$ slope.

The unit tangent vector is velocity/speed, and the acceleration is $d\mathbf{v}/dt$:

$$\mathbf{T} = (-\sin t\,\mathbf{i} + \cos t\,\mathbf{j} + \mathbf{k})/\sqrt{2} \qquad \mathbf{a} = -\cos t\,\mathbf{i} - \sin t\,\mathbf{j}.$$

EXAMPLE 8   Find $\mathbf{v}$ and s and $\mathbf{a}$ around the ellipse $x = \cos t, y = 2 \sin t, z = 0$.

Solution   Take derivatives: $\mathbf{v} = -\sin t\,\mathbf{i} + 2 \cos t\,\mathbf{j}$ and $|\mathbf{v}| = \sqrt{\sin^2 t + 4\cos^2 t}$. This is the speed ds/dt. For the distance s, something bad happens (or something normal). The speed is not simplified by $\sin^2 t + \cos^2 t = 1$. We cannot integrate ds/dt to find a formula for s. The square root defeats us.

The acceleration $-\cos t\,\mathbf{i} - 2 \sin t\,\mathbf{j}$ still points to the center. This is *not* the Earth going around the sun. The path is an ellipse but the speed is wrong. See Section 12.4 (the pound note) for a terrible error in the position of the sun.

12A The basic formulas for motion along a curve are

$$\mathbf{v} = \frac{d\mathbf{R}}{dt} \qquad \mathbf{a} = \frac{d\mathbf{v}}{dt} \qquad |\mathbf{v}| = \frac{ds}{dt} \qquad \mathbf{T} = \frac{\mathbf{v}}{|\mathbf{v}|} = \frac{d\mathbf{R}/dt}{ds/dt} = \frac{d\mathbf{R}}{ds}.$$

Suppose we know the acceleration $\mathbf{a}(t)$ and the initial velocity $\mathbf{v}_0$ and position $\mathbf{R}_0$. Then $\mathbf{v}(t)$ and $\mathbf{R}(t)$ are also known. We integrate each component:

$$\mathbf{a}(t) = \text{constant} \Rightarrow \mathbf{v}(t) = \mathbf{v}_0 + \mathbf{a}t \qquad \Rightarrow \mathbf{R}(t) = \mathbf{R}_0 + \mathbf{v}_0 t + \tfrac{1}{2}\mathbf{a}t^2$$

$$\mathbf{a}(t) = \cos t\,\mathbf{k} \ \Rightarrow \mathbf{v}(t) = \mathbf{v}_0 + \sin t\,\mathbf{k} \Rightarrow \mathbf{R}(t) = \mathbf{R}_0 + \mathbf{v}_0 t - \cos t\,\mathbf{k}.$$

THE CURVE OF A BASEBALL

There is a nice discussion of curve balls in the calculus book by Edwards and Penney. We summarize it here (optionally). The ball leaves the pitcher's hand five feet off the ground: $\mathbf{R}_0 = 0\mathbf{i} + 0\mathbf{j} + 5\mathbf{k}$. The initial velocity is $\mathbf{v}_0 = 120\mathbf{i} - 2\mathbf{j} + 2\mathbf{k}$ (120 ft/sec is more than 80 miles per hour). The acceleration is $-32\mathbf{k}$ from gravity, plus a new term from **spin**. If the spin is around the z axis, and the ball goes along the x axis, then this acceleration is in the y direction. (It comes from the cross product $\mathbf{k} \times \mathbf{i}$—there is a pressure difference on the sides of the ball.) A good pitcher can achieve $\mathbf{a} = 16\mathbf{j} - 32\mathbf{k}$. The batter integrates as fast as he can:

$$\mathbf{v}(t) = \mathbf{v}_0 + \mathbf{a}t = 120\mathbf{i} + (-2 + 16t)\mathbf{j} + (2 - 32t)\mathbf{k}$$

$$\mathbf{R}(t) = \mathbf{R}_0 + \mathbf{v}_0 t + \tfrac{1}{2}\mathbf{a}t^2 = 120t\,\mathbf{i} + (-2t + 8t^2)\mathbf{j} + (5 + 2t - 16t^2)\mathbf{k}.$$

Notice the t^2. The effect of spin is small at first, then suddenly bigger (as every batter knows). So is the effect of gravity—the ball starts to dive. At $t = \tfrac{1}{2}$, the $\mathbf{i}$ component is 60 feet and the ball reaches the batter. The $\mathbf{j}$ component is 1 foot and the $\mathbf{k}$ component is 2 feet—the curve goes low over the outside corner.

At $t = \tfrac{1}{4}$, when the batter saw the ball halfway, the $\mathbf{j}$ component was zero. It looked as if it was coming right over the plate.

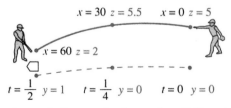

Fig. 12.4   A curve ball approaches home plate. Halfway it is on line.

Read-through questions

The position vector __a__ along the curve changes with the parameter t. The velocity is __b__. The acceleration is __c__. If the position is $\mathbf{i}+t\mathbf{j}+t^2\mathbf{k}$, then $\mathbf{v} =$ __d__ and $\mathbf{a} =$ __e__. In that example the speed is $|\mathbf{v}| =$ __f__. This equals ds/dt, where s measures the __g__. Then $s = \int$ __h__. The tangent vector is in the same direction as the __i__, but $\mathbf{T}$ is a __j__ vector. In general $\mathbf{T} =$ __k__ and in the example $\mathbf{T} =$ __l__.

Steady motion along a line has $\mathbf{a} =$ __m__. If the line is $x = y = z$, the unit tangent vector is $\mathbf{T} =$ __n__. If the speed is $|\mathbf{v}| = \sqrt{3}$, the velocity vector is $\mathbf{v} =$ __o__. If the initial position is $(1,0,0)$, the position vector is $\mathbf{R}(t) =$ __p__. The general equation of a line is $x = x_0 + tv_1$, $y =$ __q__, $z =$ __r__. In vector notation this is $\mathbf{R}(t) =$ __s__. Eliminating t leaves the equations $(x - x_0)/v_1 = (y - y_0)/v_2 =$ __t__. A line in space needs __u__ equations where a plane needs __v__. A line has one parameter where a plane has __w__. The line from $\mathbf{R}_0 = (1,0,0)$ to $(2,2,2)$ with $|\mathbf{v}| = 3$ is $\mathbf{R}(t) =$ __x__.

Steady motion around a circle (radius r, angular velocity ω) has $x =$ __y__, $y =$ __z__, $z = 0$. The velocity is $\mathbf{v} =$ __A__. The speed is $|\mathbf{v}| =$ __B__. The acceleration is $\mathbf{a} =$ __C__, which has magnitude __D__ and direction __E__. Combining upward motion $\mathbf{R} = t\mathbf{k}$ with this circular motion produces around a __F__. Then $\mathbf{v} =$ __G__ and $|\mathbf{v}| =$ __H__.

1 Sketch the curve with parametric equations $x = t$, $y = t^3$. Find the velocity vector and the speed at $t = 1$.

2 Sketch the path with parametric equations $x = 1 + t$, $y = 1 - t$. Find the xy equation of the path and the speed along it.

3 On the circle $x = \cos t$, $y = \sin t$ explain by the chain rule and then by geometry why $dy/dx = -\cot t$.

4 Locate the highest point on the curve $x = 6t$, $y = 6t - t^2$. This curve is a _____. What is the acceleration $\mathbf{a}$?

5 Find the velocity vector and the xy equation of the tangent line to $x = e^t$, $y = e^{-t}$ at $t = 0$. What is the xy equation of the curve?

6 Describe the shapes of these curves: (a) $x = 2^t$, $y = 4^t$; (b) $x = 4^t$, $y = 8^t$; (c) $x = 4^t$, $y = 4t$.

Note: *To find "parametric equations" is to find $x(t)$, $y(t)$, and possibly $z(t)$.*

7 Find parametric equations for the line through $P = (1,2,4)$ and $Q = (5,5,4)$. Probably your speed is 5; change the equations so the speed is 10. Probably your $\mathbf{R}_0$ is P; change the start to Q.

8 Find an equation for any one plane that is perpendicular to the line in Problem 7. Also find equations for any one line that is perpendicular.

9 On a straight line from $(2,3,4)$ with velocity $\mathbf{v} = \mathbf{i} - \mathbf{k}$, the position vector is $\mathbf{R}(t) =$ _____. If the velocity vector is changed to $t\mathbf{i} - t\mathbf{k}$, then $\mathbf{R}(t) =$ _____. The path is still _____.

10 Find parametric equations for steady motion from $P = (3,1,-2)$ at $t = 0$ on a line to $Q = (0,0,0)$ at $t = 3$. What is the speed? Change parameters so the speed is e^t.

11 The equations $x - 1 = \frac{1}{2}(y - 2) = \frac{1}{3}(z - 2)$ describe a _____. The same path is given parametrically by $x = 1 + t$, $y =$ _____, $z =$ _____. The same path is also given by $x = 1 + 2t$, $y =$ _____, $z =$ _____.

12 Find parametric equations to go around the unit circle with speed e^t starting from $x = 1$, $y = 0$. When is the circle completed?

13 The path $x = 2y = 3z = 6t$ is a _____ traveled with speed _____. If t is restricted by $t \geqslant 1$ the path starts at _____. If t is restricted by $0 \leqslant t \leqslant 1$ the path is a _____.

14 Find the closest point to the origin on the line $x = 1 + t$, $y = 2 - t$. When and where does it cross the $45°$ line through the origin? Find the equation of a line it never crosses.

15 (a) How far apart are the two parallel lines $x = y$ and $x = y + 1$? (b) How far is the point $x = t$, $y = t$ from the point $x = t$, $y = t + 1$? (c) What is the closest distance if their speeds are different: $x = t$, $y = t$ and $x = 2t$, $y = 2t + 1$?

16 Which vectors follow the same path as $\mathbf{R} = t\mathbf{i} + t^2\mathbf{j}$? The speed along the path may be different.

(a) $2t\mathbf{i} + 2t^2\mathbf{j}$ (b) $2t\mathbf{i} + 4t^2\mathbf{j}$ (c) $-t\mathbf{i} + t^2\mathbf{j}$ (d) $t^3\mathbf{i} + t^6\mathbf{j}$

17 Find a parametric form for the straight line $y = mx + b$.

18 The line $x = 1 + v_1 t$, $y = 2 + v_2 t$ passes through the origin provided _____ $v_1 +$ _____ $v_2 = 0$. This line crosses the $45°$ line $y = x$ unless _____ $v_1 +$ _____ $v_2 = 0$.

19 Find the velocity $\mathbf{v}$ and speed $|\mathbf{v}|$ and tangent vector $\mathbf{T}$ for these motions: (a) $\mathbf{R} = t\mathbf{i} + t^{-1}\mathbf{j}$ (b) $\mathbf{R} = t \cos t\,\mathbf{i} + t \sin t\,\mathbf{j}$ (c) $\mathbf{R} = (t + 1)\mathbf{i} + (2t + 1)\mathbf{j} + (2t + 2)\mathbf{k}$.

20 If the velocity $dx/dt\,\mathbf{i} + dy/dt\,\mathbf{j}$ is always perpendicular to the position vector $x\mathbf{i} + y\mathbf{j}$, show from their dot product that $x^2 + y^2$ is constant. The point stays on a circle.

21 Find two paths $\mathbf{R}(t)$ with the same $\mathbf{v} = \cos t\,\mathbf{i} + \sin t\,\mathbf{j}$. Find a third path with a different $\mathbf{v}$ but the same acceleration.

22 If the acceleration is a constant vector, the path must be _____. If the path is a straight line, the acceleration vector must be _____.

23 Find the minimum and maximum speed if $x = t + \cos t$, $y = t - \sin t$. Show that $|\mathbf{a}|$ is constant but not $\mathbf{a}$. The point is going around a circle while the center is moving on what line?

24 Find $x(t), y(t)$ so that the point goes around the circle $(x-1)^2 + (y-3)^2 = 4$ with speed 1.

25 A ball that is circling with $x = \cos 2t, y = \sin 2t$ flies off on a tangent at $t = \pi/8$. Find its departure point and its position at a later time t (linear motion; compute its constant velocity $\mathbf{v}$).

26 Why is $|\mathbf{a}|$ generally different from $d^2 s/dt^2$? Give an example of the difference, and an example where they are equal.

27 Change t so that the speed along the helix $\mathbf{R} = \cos t \mathbf{i} + \sin t \mathbf{j} + t \mathbf{k}$ is 1 instead of $\sqrt{2}$. Call the new parameter s.

28 Find the speed ds/dt on the line $x = 1 + 6t, y = 2 + 3t$, $z = 2t$. Integrate to find the length s from $(1,2,0)$ to $(13,8,4)$. Check by using $12^2 + 6^2 + 4^2$.

29 Find $\mathbf{v}$ and $|\mathbf{v}|$ and $\mathbf{a}$ for the curve $x = \tan t, y = \sec t$. What is this curve? At what time does it go to infinity, and along what line?

30 Construct parametric equations for travel on a helix with speed t.

31 Suppose the unit tangent vector $\mathbf{T}(t)$ is the derivative of $\mathbf{R}(t)$. What does that say about the speed? Give a noncircular example.

32 For travel on the path $y = f(x)$, with no parameter, it is impossible to find the _____ but still possible to find the _____ at each point of the path.

Find $x(t)$ and $y(t)$ for paths 33–36.

33 Around the square bounded by $x = 0, x = 1, y = 0, y = 1$, with speed 2. The formulas have four parts.

34 Around the unit circle with speed e^{-t}. Do you get all the way around?

35 Around a circle of radius 4 with acceleration $|\mathbf{a}| = 1$.

36 Up and down the y axis with constant acceleration $-\mathbf{j}$, returning to $(0,0)$ at $t = 10$.

37 True (with reason) or false (with example):

(a) If $|\mathbf{R}| = 1$ for all t then $|\mathbf{v}| = $ constant.

(b) If $\mathbf{a} = 0$ then $\mathbf{R} = $ constant.

(c) If $\mathbf{v} \cdot \mathbf{v} = $ constant then $\mathbf{v} \cdot \mathbf{a} = 0$.

(d) If $\mathbf{v} \cdot \mathbf{R} = 0$ then $\mathbf{R} \cdot \mathbf{R} = $ constant.

(e) There is no path with $\mathbf{v} = \mathbf{a}$.

38 Find the position vector to the shadow of $t \mathbf{i} + t^2 \mathbf{j} + t^3 \mathbf{k}$ on the xz plane. Is the curve ever parallel to the line $x = y = z$?

39 On the ellipse $x = a \cos t, y = b \sin t$, the angle θ from the center is not the same as t because _____.

40 Two particles are racing from $(1,0)$ to $(0,1)$. One follows $x = \cos t, y = \sin t$, the other follows $x = 1 + v_1 t, y = v_2 t$. Choose v_1 and v_2 so that the second particle goes slower but wins.

41 Two lines in space are given by $\mathbf{R}(t) = \mathbf{P} + t\mathbf{v}$ and $\mathbf{R}(t) = \mathbf{Q} + t\mathbf{w}$. Four possibilities: The lines are parallel or the same or intersecting or skew. Decide which is which based on the vectors $\mathbf{v}$ and $\mathbf{w}$ and $\mathbf{u} = \mathbf{Q} - \mathbf{P}$ (which goes between the lines):

(a) The lines are parallel if _____ are parallel.

(b) The lines are the same if _____ are parallel.

(c) The lines intersect if _____ are not parallel but _____ lie in the same plane.

(d) The lines are skew if the triple product $\mathbf{u} \cdot (\mathbf{v} \times \mathbf{w})$ is _____.

42 If the lines are skew (not in the same plane), find a formula based on $\mathbf{u}, \mathbf{v}, \mathbf{w}$ for the distance between them. The vector $\mathbf{u}$ may not be perpendicular to the two lines, so project it onto a vector that is.

43 The distance from $\mathbf{Q}$ to the line $\mathbf{P} + t\mathbf{v}$ is the projection of $\mathbf{u} = \mathbf{Q} - \mathbf{P}$ perpendicular to $\mathbf{v}$. How far is $\mathbf{Q} = (9,4,5)$ from the line $x = 1 + t, y = 1 + 2t, z = 3 + 2t$?

44 Solve Problem 43 by calculus: substitute for x, y, z in $(x-9)^2 + (y-4)^2 + (z-5)^2$ and minimize. Which (x, y, z) on the line is closest to $(9,4,5)$?

45 Practice with parameters, starting from $x = F(t), y = G(t)$.

(a) The mirror image across the $45°$ line is $x = $ _____, $y = $ _____.

(b) Write the curve $x = t^3, y = t^2$ as $y = f(x)$.

(c) Why can't $x = t^2, y = t^3$ be written as $y = f(x)$?

(d) If F is invertible then $t = F^{-1}(x)$ and $y = $ _____ (x).

46 From 12:00 to 1:00 a snail crawls steadily out the minute hand (one meter in one hour). Find its position at time t starting from $(0,0)$.

12.2 Plane Motion: Projectiles and Cycloids

The previous section started with $\mathbf{R}(t)$. From this position vector we computed $\mathbf{v}$ and $\mathbf{a}$. Now we find $\mathbf{R}(t)$ itself, from more basic information. The laws of physics govern projectiles, and the motion of a wheel produces a cycloid (which enters problems in robotics). The projectiles fly without friction, so the only force is gravity.

These motions occur in a plane. The two components of position will be x (across) and y (up). A projectile moves as t changes, so we look for $x(t)$ and $y(t)$. We are shooting a basketball or firing a gun or peacefully watering the lawn, and we have to aim in the right direction (not directly at the target). If the hose delivers water at 10 meters/second, can you reach the car 12 meters away?

The usual initial position is $(0,0)$. Some flights start higher, at $(0,h)$. The initial velocity is $(v_0 \cos \alpha, v_0 \sin \alpha)$, where v_0 is the speed and α is the angle with the horizontal. The acceleration from gravity is purely vertical: $d^2 y/dt^2 = -g$. So the horizontal velocity stays at its initial value. The upward velocity decreases by $-gt$:

$$dx/dt = v_0 \cos \alpha, \; dy/dt = v_0 \sin \alpha - gt.$$

The horizontal distance $x(t)$ is steadily increasing. The height $y(t)$ increases and then decreases. To find the position, integrate the velocities (for a high start add h to y):

The projectile path is $x(t) = (v_0 \cos \alpha)t, \; y(t) = (v_0 \sin \alpha)t - \frac{1}{2}gt^2.$ (1)

This path is a *parabola*. But it is not written as $y = ax^2 + bx + c$. It could be, if we eliminated t. Then we would lose track of time. The parabola is $y(x)$, with no parameter, where we have $x(t)$ and $y(t)$.

Basic question: *Where does the projectile hit the ground*? For the parabola, we solve $y(x) = 0$. That gives the position x. *For the projectile we solve $y(t) = 0$.* That gives the *time* it hits the ground, not the place. If that time is T, then $x(T)$ gives the place.

The information is there. It takes two steps instead of one, but we learn more.

EXAMPLE 1 Water leaves the hose at 10 meters/second (this is v_0). It starts up at the angle α. Find the time T when y is zero again, and find where the projectile lands.

Solution   The flight ends when $y = (10 \sin \alpha)T - \frac{1}{2}gT^2 = 0$. The flight time is $T = (20 \sin \alpha)/g$. At that time, the horizontal distance is

$$x(T) = (10 \cos \alpha)T = (200 \cos \alpha \sin \alpha)/g. \text{ This is the } \textbf{\textit{range }} R.$$

The projectile (or water from the hose) hits the ground at $x = R$. To simplify, replace $200 \cos \alpha \sin \alpha$ by $100 \sin 2\alpha$. Since $g = 9.8$ meters/sec^2, *we can't reach the car*:

The range $R = (100 \sin 2\alpha)/9.8$ is at most $100/9.8$. This is less than 12.

The range is greatest when $\sin 2\alpha = 1 (\alpha$ is $45°)$. To reach 12 meters we could stand on a ladder (Problem 14). To hit a baseball against air resistance, the best angle is nearer to $35°$. Figure 12.5 shows symmetric parabolas (no air resistance) and unsymmetric flight paths that drop more steeply.

12B The flight time T and the horizontal range $R = x(T)$ are reached when $y = 0$, which means $(v_0 \sin \alpha)T = \frac{1}{2}gT^2$:

$$T = (2v_0 \sin \alpha)/g \text{ and } R = (v_0 \cos \alpha)T = (v_0^2 \sin 2\alpha)/g.$$

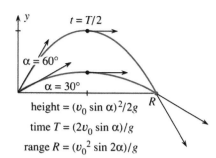

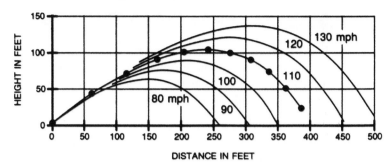

Fig. 12.5   Equal range R, different times T. Baseballs hit at 35° with increasing v_0. The dots are at half-seconds (from *The Physics of Baseball* by Robert Adair: Harper and Row 1990).

EXAMPLE 2   What are the correct angles α for a given range R and given v_0 ?

Solution   The range is $R = (v_0^2 \sin 2\alpha)/g$. This determines the sine of 2α—*but two angles can have the same sine*. We might find $2\alpha = 60°$ or $120°$. The starting angles $\alpha = 30°$ and $\alpha = 60°$ in Figure 12.5 give the same $\sin 2\alpha$ and the same range R. The flight times contain $\sin \alpha$ and are different.

By calculus, the maximum height occurs when $dy/dt = 0$. Then $v_0 \sin \alpha = gt$, which means that $t = (v_0 \sin \alpha)/g$. This is half of the total flight time T—the time going up equals the time coming down. The value of y at this halfway time $t = \frac{1}{2}T$ is

$$y_{\max} = (v_0 \sin \alpha)(v_0 \sin \alpha)/g - \tfrac{1}{2}g(v_0 \sin \alpha/g)^2 = (v_0 \sin \alpha)^2/2g. \qquad (2)$$

EXAMPLE 3   If a ski jumper goes 90 meters down a 30° slope, after taking off at 28 meters/second, find equations for the flight time and the ramp angle α.

Solution   The jumper lands at the point $x = 90 \cos 30°$, $y = -90 \sin 30°$ (minus sign for obvious reasons). The basic equation (2) is $x = (28 \cos \alpha)t$, $y = (28 \sin \alpha)t - \frac{1}{2}gt^2$. Those are two equations for α and t. Note that t is not T, the flight time to $y = 0$.

Conclusion   The position of a projectile involves three parameters v_0, α, and t. ***Three pieces of information determine the flight*** (almost). The reason for the word *almost* is the presence of $\sin \alpha$ and $\cos \alpha$. Some flight requirements cannot be met (reaching a car at 12 meters). Other requirements can be met in two ways (when the car is close). The equation $\sin \alpha = c$ is more likely to have no solution or two solutions than exactly one solution.

Watch for the three pieces of information in each problem. When a football starts at $v_0 = 20$ meters/second and hits the ground at $x = 40$ meters, the third fact is _____. This is like a lawyer who is asked the fee and says \$1000 for three questions. "Isn't that steep?" says the client. "Yes," says the lawyer, "now what's your last question?"

CYCLOIDS

A projectile's path is a parabola. To compute it, eliminate t from the equations for x and y. Problem 5 finds $y = ax^2 + bx$, a parabola through the origin. The path of a point on a wheel seems equally simple, but eliminating t is virtually impossible. The cycloid is a curve that really needs and uses a parameter.

To trace out a cycloid, ***roll a circle of radius*** a ***along the*** x ***axis***. Watch the point that starts at the bottom of the circle. It comes back to the bottom at $x = 2\pi a$, after a complete turn of the circle. The path in between is shown in Figure 12.6. After a century of looking for the xy equation, a series of great scientists (Galileo, Christopher Wren, Huygens, Bernoulli, even Newton and l'Hôpital) found the right way to study a cycloid—by introducing a parameter. We will call it θ; it could also be t.

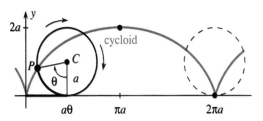
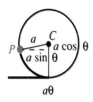
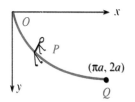

Fig. 12.6   Path of P on a rolling circle is a cycloid. Fastest slide to Q.

The parameter is the angle θ through which the circle turns. (This angle is not at the origin, like θ in polar coordinates.) The circle rolls a distance $a\theta$, radius times angle, along the x axis. So the center of the circle is at $x = a\theta$, $y = a$. To account for the segment CP, subtract $a \sin \theta$ from x and $a \cos \theta$ from y:

$$\textbf{\textit{The point } P \textbf{ \textit{has} }} x = a(\theta - \sin \theta) \textbf{ \textit{and} } y = a(1 - \cos \theta). \tag{3}$$

At $\theta = 0$ the position is $(0,0)$. At $\theta = 2\pi$ the position is $(2\pi a, 0)$. In between, the slope of the cycloid comes from the chain rule:

$$\frac{dy}{dx} = \frac{dy/d\theta}{dx/d\theta} = \frac{a \sin \theta}{a(1 - \cos \theta)}. \tag{4}$$

This is infinite at $\theta = 0$. The point on the circle starts straight upward and the cycloid has a *cusp*. Note how all calculations use the parameter θ. We go quickly:

Question 1   Find the area under one arch of the cycloid ($\theta = 0$ to $\theta = 2\pi$).
Answer   The area is $\int y \, dx = \int_0^{2\pi} a(1 - \cos \theta)a(1 - \cos \theta)d\theta$. This equals $3\pi a^2$.

Question 2   Find the length of the arch, using $ds = \sqrt{(dx/d\theta)^2 - (dy/d\theta)^2} \, d\theta$.
Answer $\int ds = \int_0^{2\pi} a\sqrt{(1 - \cos \theta)^2 - (\sin \theta)^2} \, d\theta = \int_0^{2\pi} a\sqrt{2 - 2 \cos \theta} \, d\theta$.
Now substitute $1 - \cos \theta = 2 \sin^2 \frac{1}{2}\theta$. The square root is $2 \sin \frac{1}{2}\theta$. The length is $8a$.

Question 3   If the cycloid is turned over (y is downward), find the time to slide to the bottom. The slider starts with $v = 0$ at $y = 0$.
Answer   Kinetic plus potential energy is $\frac{1}{2}mv^2 - mgy = 0$ (it starts from zero and can't change). So the speed is $v = \sqrt{2gy}$. This is ds/dt and we know ds:

$$\text{sliding time} = \int dt = \int \frac{ds}{\sqrt{2gy}} = \int_0^{\pi} \frac{a\sqrt{2 - 2 \cos \theta} \, d\theta}{\sqrt{2ga(1 - \cos \theta)}} = \pi \sqrt{a/g}.$$

Check dimensions: $a = $ distance, $g = $ distance$/(\text{times})^2$, $\pi \sqrt{a/g} = $ time. ***That is the shortest sliding time for any curve***. The cycloid solves the "brachistochrone problem," which minimizes the time down curves from O to Q (Figure 12.6). You might think a straight path would be quicker—it is certainly shorter. A straight line has the equation $x = \pi y/2$, so the sliding time is

$$\int dt = \int ds/\sqrt{2gy} = \int_0^{2a} \sqrt{(\pi/2)^2 + 1} \, dy/\sqrt{2gy} = \sqrt{\pi^2 + 4} \sqrt{a/g}. \tag{5}$$

This is larger than the cycloid time $\pi \sqrt{a/g}$. It is better to start out vertically and pick up speed early, even if the path is longer.

Instead of publishing his solution, John Bernoulli turned this problem into an international challenge: *Prove that the cycloid gives the fastest slide.* Most mathematicians couldn't do it. The problem reached Isaac Newton (this was later in his life). As you would expect, Newton solved it. For some reason he sent back his proof with no name. But when Bernoulli received the answer, he was not fooled for a moment: "I recognize the lion by his claws."

What is also amazing is a further property of the cycloid: ***The time to Q is the same if you begin anywhere along the path***. Starting from rest at P instead of O, the bottom is reached at the same time. This time Bernoulli got carried away: "You will be petrified with astonishment when I say...".

There are other beautiful curves, closely related to the cycloid. For an ***epicycloid***, the circle rolls around the outside of another circle. For a ***hypocycloid***, the rolling circle is inside the fixed circle. The *astroid* is the special case with radii in the ratio 1 to 4. It is the curved star in Problem 34, where $x = a \cos^3 \theta$ and $y = a \sin^3 \theta$.

The cycloid even solves the old puzzle: *What point moves backward when a train starts forward?* The train wheels have a flange that extends below the track, and $dx/dt < 0$ at the bottom of the flange.

12.2 EXERCISES

Read-through questions

A projectile starts with speed v_0 and angle α. At time t its velocity is $dx/dt =$ __a__, $dy/dt =$ __b__ (the downward acceleration is g). Starting from $(0,0)$, the position at time t is $x =$ __c__, $y =$ __d__. The flight time back to $y = 0$ is $T =$ __e__. At that time the horizontal range is $R =$ __f__. The flight path is a __g__.

The three quantities v_0, __h__, __i__ determine the projectile's motion. Knowing v_0 and the position of the target, we (can) (cannot) solve for α. Knowing α and the position of the target, we (can) (cannot) solve for v_0.

A __j__ is traced out by a point on a rolling circle. If the radius is a and the turning angle is θ, the center of the circle is at $x =$ __k__, $y =$ __l__. The point is at $x =$ __m__, $y =$ __n__, starting from $(0,0)$. It travels a distance __o__ in a full turn of the circle. The curve has a __p__ at the end of every turn. An upside-down cycloid gives the __q__ slide between two points.

Problems 1–18 and 41 are about projectiles

1 Find the time of flight T, the range R, and the maximum height Y of a projectile with $v_0 = 16$ ft/sec and

   (a) $\alpha = 30°$ (b) $\alpha = 60°$ (c) $\alpha = 90°$.

2 If $v_0 = 32$ ft/sec and the projectile returns to the ground at $T = 1$, find the angle α and the range R.

3 A ball is thrown at $60°$ with $v_0 = 20$ meters/sec to clear a wall 2 meters high. How far away is the wall?

4 If $\mathbf{v}(0) = 3\mathbf{i} + 3\mathbf{j}$ find $\mathbf{v}(t), \mathbf{v}(1), \mathbf{v}(2)$ and $\mathbf{R}(t), \mathbf{R}(1), \mathbf{R}(2)$.

5 (a) Eliminate t from $x = t, y = t - \frac{1}{2}t^2$ to find the xy equation of the path. At what x is $y = 0$?

   (b) Do the same for any v_0 and α.

6 Find the angle α for a ball kicked at 30 meters/second if it clears 6 meters traveling horizontally.

7 How far out does a stone hit the water h feet below, starting with velocity v_0 at angle $\alpha = 0$?

8 How far out does the same stone go, starting at angle α? Find an equation for the angle that maximizes the range.

9 A ball starting from $(0,0)$ passes through $(5,2)$ after 2 seconds. Find v_0 and α. (The units are meters.)

*10 With x and y from equation (1), show that

$$v_0^2 \geqslant (gx/v_0)^2 + 2gy.$$

If a fire is at height H and the water velocity is v_0, how far can the fireman put the hose back from the fire? (The parabola in this problem is the "envelope" enclosing all possible paths.)

11 Estimate the initial speed of a 100-meter golf shot hit at $\alpha = 45°$. Is the true v_0 larger or smaller, when air friction is included?

12 $T = 2v_0(\sin\alpha)/g$ is in seconds and $R = (v_0^2 \sin 2\alpha)/g$ is in meters if v_0 and g are in _____.

13 (a) What is the greatest height a ball can be thrown? Aim straight up with $v_0 = 28$ meters/sec.

14 If a baseball goes 100 miles per hour for 60 feet, how long does it take (in seconds) and how far does it fall from gravity (in feet)? Use $\frac{1}{2}gt^2$.

15 If you double v_0, what happens to the range and maximum height? If you change the angle by $d\alpha$, what happens to those numbers?

16 At what point on the path is the speed of the projectile (a) least (b) greatest?

17 If the hose with $v_0 = 10$m/sec is at a $45°$ angle, x reaches 12 meters when $t =$ _____ and $y =$ _____. From a ladder of height _____ the water will reach the car (12 meters).

18 Describe the two trajectories a golf ball can take to land right in the hole, if it starts with a large known velocity v_0. In reality (with air resistance) which of those shots would fall closer?

Problems 19–34 are about cycloids and related curves

19 Find the unit tangent vector **T** to the cycloid. Also find the speed at $\theta = 0$ and $\theta = \pi$, if the wheel turns at $d\theta/dt = 1$.

20 The slope of the cycloid is infinite at $\theta = 0$:
$$\frac{dy}{dx} = \frac{dy/d\theta}{dx/x\theta} = \frac{\sin\theta}{1 - \cos\theta}.$$
By whose rule? Estimate the slope at $\theta = \frac{1}{10}$ and $\theta = -\frac{1}{10}$. Where does the slope equal one?

21 Show that the tangent to the cycloid at P in Figure 12.6a goes through $x = a\theta$, $y = 2a$. Where is this point on the rolling circle?

22 For a **trochoid**, the point P is a distance d from the center of the rolling circle. Redraw Figure 12.6b to find $x = a\theta - d\sin\theta$ and $y =$ _____ .

23 If a circle of radius a rolls inside a circle of radius $2a$, show that one point on the small circle goes across on a straight line.

24 Find d^2y/dx^2 for the cycloid, which is concave _____ .

25 If $d\theta/dt = c$, find the velocities dx/dt and dy/dt along the cycloid. Where is dx/dt greatest and where is dy/dt greatest?

26 Experiment with graphs of $x = a\cos\theta + b\sin\theta$, $y = c\cos\theta + d\sin\theta$ using a computer. What kind of curves are they? Why are they closed?

27 A stone in a bicycle tire goes along a cycloid. Find equations for the stone's path if it flies off at the top (a projectile).

28 Draw curves on a computer with $x = a\cos\theta + b\cos 3\theta$ and $y = c\sin\theta + d\sin 3\theta$. Is there a limit to the number of loops?

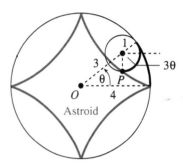

Astroid

35 Find the area inside the astroid.

36 Explain why $x = 2a\cot\theta$ and $y = 2a\sin^2\theta$ for the point P on the **witch of Agnesi**. Eliminate θ to find the xy equation.

Note: Maria Agnesi wrote the first three-semester calculus text (l'Hôpital didn't do integral calculus). The word "witch" is a total mistranslation, nothing to do with her or the curve.

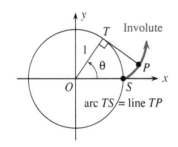

arc TS = line TP

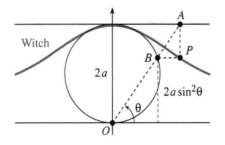

Witch

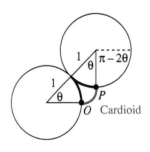

Cardioid

29 When a penny rolls completely around another penny, the head makes _____ turns. When it rolls inside a circle four times larger (for the astroid), the head makes _____ turns.

30 Display the cycloid family with computer graphics:

(a) **cycloid**

(b) **epicycloid** $x = C\cos\theta - \cos C\theta, y = C\sin\theta + \sin C\theta$

(c) **hypocycloid** $x = c\cos\theta + \cos c\theta, y = c\sin\theta - \sin c\theta$

(d) **astroid** ($c = 3$)

(e) **deltoid** ($c = 2$).

31 If one arch of the cycloid is revolved around the x axis, find the surface area and volume.

32 For a hypocycloid the fixed circle has radius $c + 1$ and the circle rolling inside has radius 1. There are $c + 1$ cusps if c is an integer. How many cusps (use computer graphics if possible) for $c = 1/2$? $c = 3/2$? $c = \sqrt{2}$? What curve for $c = 1$?

33 When a string is unwound from a circle find $x(\theta)$ and $y(\theta)$ for point P. Its path is the "**involute**" of the circle.

34 For the point P on the **astroid**, explain why $x = 3\cos\theta + \cos 3\theta$ and $y = 3\sin\theta - \sin 3\theta$. The angle in the figure is 3θ

because both circular arcs have length _____. Convert to $x = 4\cos^3\theta, y = 4\sin^3\theta$ by triple-angle formulas.

37 For a **cardioid** the radius $C-1$ of the fixed circle equals the radius 1 of the circle rolling outside (epicycloid with $C=2$). (a) The coordinates of P are $x = -1 + 2\cos\theta - \cos 2\theta$, $y = $ _____. (b) The double-angle formulas yield $x = 2\cos\theta (1 - \cos\theta), y = $ _____. (c)$x^2 + y^2 = $ _____ so its square root is $r = $ _____.

38 Explain the last two steps in equation (5) for the sliding time down a straight path.

39 On an upside-down cycloid the slider takes the same time T to reach bottom *wherever it starts*. Starting at $\theta = \alpha$, write $1 - \cos\theta = 2\sin^2\theta/2$ and $1 - \cos\alpha = 2\sin^2\alpha/2$ to show that

$$T = \int_\alpha^\pi \frac{\sqrt{2a^2(1 - \cos\theta)}d\theta}{\sqrt{2ag(\cos\alpha - \cos\theta)}} = \pi\sqrt{\frac{a}{g}}.$$

40 Suppose a heavy weight is attached to the top of the rolling circle. What is the path of the weight?

41 The wall in Fenway Park is 37 feet high and 315 feet from home plate. A baseball hit 3 feet above the ground at $\alpha = 22.5°$ will just go over if $v_0 = $ _____. The time to reach the wall is _____.

12.3 Curvature and Normal Vector

A driver produces acceleration three ways—by the gas pedal, the brake, and steering wheel. The first two change the speed. Turning the wheel changes the direction. *All three change the velocity* (they give acceleration). For steady motion around a circle, the change is from steering—the acceleration $d\mathbf{v}/dt$ points to the center. We now look at motion along other curves, to separate change in the speed $|\mathbf{v}|$ from change in the direction $\mathbf{T}$.

The direction of motion is $\mathbf{T} = \mathbf{v}/|\mathbf{v}|$. It depends on the path but not the speed (because we divide by $|\mathbf{v}|$). For turning we measure two things:

    1. How fast T turns: this will be the *curvature* κ (kappa).
    2. Which direction T turns: this will be the *normal vector* N.

κ and $\mathbf{N}$ depend, like s and $\mathbf{T}$, only on the shape of the curve. Replacing t by $2t$ or t^2 leaves them unchanged. For a circle we give the answers in advance. The normal vector $\mathbf{N}$ points to the center. The curvature κ is $1/\text{radius}$.

A smaller turning circle means a larger curvature κ: more bending.

The curvature κ is *change in direction* $|d\mathbf{T}|$ *divided by change in position* $|ds|$. There are three formulas for κ—a direct one for graphs $y(x)$, a brutal but valuable one for any parametric curve $(x(t), y(t))$, and a neat formula that uses the vectors $\mathbf{v}$ and $\mathbf{a}$. We begin with the definition and the neat formula.

        DEFINITION $\quad \kappa = |d\mathbf{T}/ds| \qquad$ **FORMULA** $\quad \kappa = |\mathbf{v} \times \mathbf{a}|/|\mathbf{v}|^3 \qquad$ (1)

The definition does not involve the parameter t—but the calculations do. The position vector $\mathbf{R}(t)$ yields $\mathbf{v} = d\mathbf{R}/dt$ and $\mathbf{a} = d\mathbf{v}/dt$. If t is changed to $2t$, the velocity $\mathbf{v}$ is doubled and $\mathbf{a}$ is multiplied by 4. Then $|\mathbf{v} \times \mathbf{a}|$ and $|\mathbf{v}|^3$ are multiplied by 8, and their ratio κ is unchanged.

Proof of formula (1) Start from $\mathbf{v} = |\mathbf{v}|\mathbf{T}$ and compute its derivative $\mathbf{a}$:

$$\mathbf{a} = \frac{d|\mathbf{v}|}{dt}\mathbf{T} + |\mathbf{v}|\frac{d\mathbf{T}}{dt} \text{ by the product rule.}$$

Now take the cross product with $\mathbf{v} = |\mathbf{v}|\mathbf{T}$. Remember that $\mathbf{T} \times \mathbf{T} = \mathbf{0}$:

$$\mathbf{v} \times \mathbf{a} = |\mathbf{v}|\mathbf{T} \times |\mathbf{v}|\frac{d\mathbf{T}}{dt}. \qquad (2)$$

We know that $|\mathbf{T}| = 1$. Equation (4) will show that $\mathbf{T}$ is perpendicular to $d\mathbf{T}/dt$. So $|\mathbf{v} \times \mathbf{a}|$ is the first length $|\mathbf{v}|$ times the second length $|\mathbf{v}||d\mathbf{T}/dt|$. The factor $\sin\theta$ in the length of a cross product is 1 from the $90°$ angle. In other words

$$\left|\frac{d\mathbf{T}}{dt}\right| = \frac{|\mathbf{v} \times \mathbf{a}|}{|\mathbf{v}|^2} \quad \text{and} \quad \kappa = \left|\frac{d\mathbf{T}}{ds}\right| = \left|\frac{d\mathbf{T}/dt}{ds/dt}\right| = \frac{|\mathbf{v} \times \mathbf{a}|}{|\mathbf{v}|^3}. \qquad (3)$$

The chain rule brings the extra $|ds/dt| = |\mathbf{v}|$ into the denominator.

Before any examples, we show that $d\mathbf{T}/dt$ is perpendicular to $\mathbf{T}$. The reason is that $\mathbf{T}$ is a unit vector. Differentiate both sides of $\mathbf{T} \cdot \mathbf{T} = 1$:

$$\frac{d\mathbf{T}}{dt} \cdot \mathbf{T} + \mathbf{T} \cdot \frac{d\mathbf{T}}{dt} = 0 \qquad or \qquad 2\mathbf{T} \cdot \frac{d\mathbf{T}}{dt} = 0. \qquad (4)$$

That proof used the product rule $\mathbf{U}' \cdot \mathbf{V} + \mathbf{U} \cdot \mathbf{V}'$ for the derivative of $\mathbf{U} \cdot \mathbf{V}$ (Problem 23, with $\mathbf{U} = \mathbf{V} = \mathbf{T}$). Think of the vector $\mathbf{T}$ moving around the unit sphere.

To keep a constant length $(\mathbf{T}+d\mathbf{T})\cdot(\mathbf{T}+d\mathbf{T})=1$, we need $2\mathbf{T}\cdot d\mathbf{T}=0$. Movement $d\mathbf{T}$ is perpendicular to radius vector $\mathbf{T}$.

Our first examples will be **plane curves**. The position vector $\mathbf{R}(t)$ has components $x(t)$ and $y(t)$ but no $z(t)$. Look at the components of $\mathbf{v}$ and $\mathbf{a}$ and $\mathbf{v}\times\mathbf{a}$ (x' means dx/dt):

$\mathbf{R}$	$x(t)$	$y(t)$	0
$\mathbf{v}$	$x'(t)$	$y'(t)$	0
$\mathbf{a}$	$x''(t)$	$y''(t)$	0
$\mathbf{v}\times\mathbf{a}$	0	0	$x'y''-y'y''$

$$|\mathbf{v}| = \sqrt{|x'|^2+|y'|^2}$$

$$\kappa = \frac{|x'y''-y'x''|}{((x')^2+(y')^2)^{3/2}} \tag{5}$$

Equation (5) is the brutal but valuable formula for κ. Apply it to movement around a circle. We should find $\kappa = 1/\text{radius } a$:

EXAMPLE 1 When $x = a\cos\omega t$ and $y = a\sin\omega t$ we substitute x', y', x'', y'' into (5):

$$\kappa = \frac{(-\omega a\sin\omega t)(-\omega^2 a\sin\omega t)-(\omega a\cos\omega t)(-\omega^2 a\cos\omega t)}{[(\omega a\sin\omega t)^2+(\omega a\cos\omega t)^2]^{3/2}} = \frac{\omega^2 a^2}{[\omega^2 a^2]^{3/2}}.$$

This is $\omega^3 a^2/\omega^3 a^3$ and ω cancels. The speed makes no difference to $\kappa = 1/a$.

The third formula for κ applies to an ordinary plane curve given by $y(x)$. The parameter t is x! You see the square root in the speed $|\mathbf{v}| = ds/dx$:

$\mathbf{R}$	x	$y(x)$	0
$\mathbf{v}$	1	dy/dx	0
$\mathbf{a}$	0	d^2y/dx^2	0
$\mathbf{v}\times\mathbf{a}$	0	0	d^2y/dx^2

$$|\mathbf{v}| = \sqrt{1+(dy/dx)^2}$$

$$\kappa = \frac{|d^2y/dx^2|}{(1+(dy/dx)^2)^{3/2}} \tag{6}$$

In practice this is the most popular formula for κ. The most popular approximation is $|d^2y/dx^2|$. (The denominator is omitted.) For the bending of a beam, the nonlinear equation uses κ and the linear equation uses d^2y/dx^2. We can see the difference for a parabola:

EXAMPLE 2   The curvature of $y = \frac{1}{2}x^2$ is $\kappa = |y''|/(1+(y')^2)^{3/2} = 1/(1+x^2)^{3/2}$.

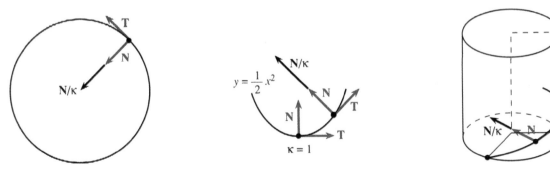

Fig. 12.7   Normal $\mathbf{N}$ divided by curvature κ for circle and parabola and unit helix.

The approximation is $y'' = 1$. This agrees with κ at $x = 0$, where the parabola turns the corner. But for large x, the curvature approaches zero. Far out on the parabola, we go a long way for a small change in direction.

The parabola $y = -\frac{1}{2}x^2$, opening down, has the same κ. Now try a space curve.

EXAMPLE 3   Find the curvature of the unit helix $\mathbf{R} = \cos t\,\mathbf{i} + \sin t\,\mathbf{j} + t\mathbf{k}$.

Take the cross product of $\mathbf{v} = -\sin t\,\mathbf{i} + \cos t\,\mathbf{j} + \mathbf{k}$ and $\mathbf{a} = -\cos t\,\mathbf{i} - \sin t\,\mathbf{j}$:

$$\mathbf{v} \times \mathbf{a} = \begin{vmatrix} \mathbf{i} & \mathbf{j} & \mathbf{k} \\ -\sin t & \cos t & 1 \\ -\cos t & -\sin t & 0 \end{vmatrix} = \sin t\,\mathbf{i} - \cos t\,\mathbf{j} + \mathbf{k}.$$

This cross product has length $\sqrt{2}$. Also the speed is $|\mathbf{v}| = \sqrt{\sin^2 t + \cos^2 t + 1} = \sqrt{2}$:

$$\kappa = |\mathbf{v} \times \mathbf{a}|/|\mathbf{v}|^3 = \sqrt{2}/(\sqrt{2})^3 = \tfrac{1}{2}.$$

Compare with a unit circle. Without the climbing term $t\mathbf{k}$, the curvature would be 1. Because of climbing, each turn of the helix is longer and $\kappa = \tfrac{1}{2}$.

That makes one think: Is the helix twice as long as the circle? No. The length of a turn is only increased by $|\mathbf{v}| = \sqrt{2}$. The other $\sqrt{2}$ is because the tangent $\mathbf{T}$ slopes upward. The shadow in the base turns a full $360°$, but $\mathbf{T}$ turns less.

THE NORMAL VECTOR N

The discussion is bringing us to an important vector. Where κ measures the *rate* of turning, the unit vector $\mathbf{N}$ gives the *direction* of turning. $\mathbf{N}$ is perpendicular to $\mathbf{T}$, and in the plane that leaves practically no choice. Turn left or right. For a space curve, follow $d\mathbf{T}$. Remember equation (4), which makes $d\mathbf{T}$ perpendicular to $\mathbf{T}$.

The normal vector $\mathbf{N}$ is a unit vector along $d\mathbf{T}/dt$. It is perpendicular to $\mathbf{T}$:

DEFINITION $\mathbf{N} = \dfrac{d\mathbf{T}/ds}{|d\mathbf{T}/ds|} = \dfrac{1}{\kappa}\dfrac{d\mathbf{T}}{ds}$ **FORMULA** $\mathbf{N} = \dfrac{d\mathbf{T}/dt}{|d\mathbf{T}/dt|}.$ (7)

EXAMPLE 4   Find the normal vector $\mathbf{N}$ for the same helix $\mathbf{R} = \cos t\,\mathbf{i} + \sin t\,\mathbf{j} + t\mathbf{k}$.

Solution   Copy $\mathbf{v}$ from Example 3, divide by $|\mathbf{v}|$, and compute $d\mathbf{T}/dt$:

$$\mathbf{T} = \mathbf{v}/|\mathbf{v}| = (-\sin t\,\mathbf{i} + \cos t\,\mathbf{j} + \mathbf{k})/\sqrt{2} \quad \text{and} \quad d\mathbf{T}/dt = (-\cos t\,\mathbf{i} - \sin t\,\mathbf{j})/\sqrt{2}.$$

To change $d\mathbf{T}/dt$ into a unit vector, cancel the $\sqrt{2}$. *The normal vector is* $\mathbf{N} = -\cos t\,\mathbf{i} - \sin t\,\mathbf{j}$. It is perpendicular to $\mathbf{T}$. Since the $\mathbf{k}$ component is zero, $\mathbf{N}$ is horizontal. The tangent $\mathbf{T}$ slopes up at $45°$—it goes around the circle at that latitude. The normal $\mathbf{N}$ is tangent to this circle ($\mathbf{N}$ is tangent to the path of the tangent!). So $\mathbf{N}$ stays horizontal as the helix climbs.

There is also a third direction, perpendicular to $\mathbf{T}$ and $\mathbf{N}$. It is the *binormal* vector $\mathbf{B} = \mathbf{T} \times \mathbf{N}$, computed in Problems 25 – 30. The unit vectors $\mathbf{T}, \mathbf{N}, \mathbf{B}$ provide the natural coordinate system for the path—along the curve, in the plane of the curve, and out of that plane. The theory is beautiful but the computations are not often done—we stop here.

TANGENTIAL AND NORMAL COMPONENTS OF ACCELERATION

May I return a last time to the gas pedal and brake and steering wheel? The first two give acceleration along $\mathbf{T}$. Turning gives acceleration along $\mathbf{N}$. The rate of turning (curvature κ) and the direction $\mathbf{N}$ are established. We now ask about the *force* required.

Newton's Law is $\mathbf{F} = m\mathbf{a}$, so we need the acceleration $\mathbf{a}$—especially its component along $\mathbf{T}$ and its component along $\mathbf{N}$.

$$\text{The acceleration is } \mathbf{a} = \frac{d^2s}{dt^2}\mathbf{T} + \kappa\left[\frac{ds}{dt}\right]^2\mathbf{N}. \tag{8}$$

For a straight path, d^2s/dt^2 is the only acceleration—the ordinary second derivative. The term $\kappa(ds/dt)^2$ is the acceleration in turning. Both have the dimension of length/(time)^2.

The force to steer around a corner depends on curvature and speed—as all drivers know. Acceleration is the derivative of $\mathbf{v} = |\mathbf{v}|\mathbf{T} = (ds/dt)\mathbf{T}$:

$$\mathbf{a} = \frac{d^2s}{dt^2}\mathbf{T} + \frac{ds}{dt}\frac{d\mathbf{T}}{dt} = \frac{d^2s}{dt^2}\mathbf{T} + \frac{ds}{dt}\frac{d\mathbf{T}}{ds}\frac{ds}{dt}. \tag{9}$$

That last term is $\kappa(ds/dt)^2\mathbf{N}$, since $d\mathbf{T}/ds = \kappa\mathbf{N}$ by formula (7). So (8) is proved.

EXAMPLE 5 A fixed speed $ds/dt = 1$ gives $d^2s/dt^2 = 0$. The only acceleration is $\kappa\mathbf{N}$.

EXAMPLE 6 Find the components of $\mathbf{a}$ for circular speed-up $\mathbf{R}(t) = \cos t^2\,\mathbf{i} + \sin t^2\,\mathbf{j}$.

Without stopping to think, compute $d\mathbf{R}/dt = \mathbf{v}$ and $ds/dt = |\mathbf{v}|$ and $\mathbf{v}/|\mathbf{v}| = \mathbf{T}$:

$$\mathbf{v} = -2t\sin t^2\,\mathbf{i} + 2t\cos t^2\,\mathbf{j}, \ |\mathbf{v}| = 2t, \ \mathbf{T} = -\sin t^2\,\mathbf{i} + \cos t^2\,\mathbf{j}.$$

The derivative of $ds/dt = |\mathbf{v}|$ is $d^2s/dt^2 = 2$. The derivative of $\mathbf{v}$ is $\mathbf{a}$:

$$\mathbf{a} = -2\sin t^2\,\mathbf{i} + 2\cos t^2\,\mathbf{j} - 4t^2\cos t^2\,\mathbf{i} - 4t^2\sin t^2\,\mathbf{j}.$$

In the first terms of $\mathbf{a}$ we see $2\mathbf{T}$. In the last terms we must be seeing $\kappa|\mathbf{v}|^2\mathbf{N}$. Certainly $|\mathbf{v}|^2 = 4t^2$ and $\kappa = 1$, because the circle has radius 1. Thus $\mathbf{a} = 2\mathbf{T} + 4t^2\mathbf{N}$ has the tangential component 2 and normal component $4t^2$—acceleration along the circle and in to the center.

Table of Formulas

$\mathbf{v} = d\mathbf{R}/dt \quad \mathbf{a} = d\mathbf{v}/dt$

$|\mathbf{v}| = ds/st \quad \mathbf{T} = \mathbf{v}/|\mathbf{v}| = |d\mathbf{R}/ds|$

Curvature $\kappa = |d\mathbf{T}/ds| = |\mathbf{v}\times\mathbf{a}|/|\mathbf{v}|^3$

Plane curves $\kappa = \dfrac{|x'y'' - y'x''|}{((x')^2+(y')^2)^{3/2}} = \dfrac{|d^2y/dx^2|}{(1+(dy/dx)^2)^{3/2}}$

Normal vector $\mathbf{N} = \dfrac{1}{\kappa}\dfrac{d\mathbf{T}}{ds} = \dfrac{d\mathbf{T}/dt}{|d\mathbf{T}/dt|}$

Acceleration $\mathbf{a} = (d^2s/dt^2)\mathbf{T} + \kappa|\mathbf{v}|^2\mathbf{N}$

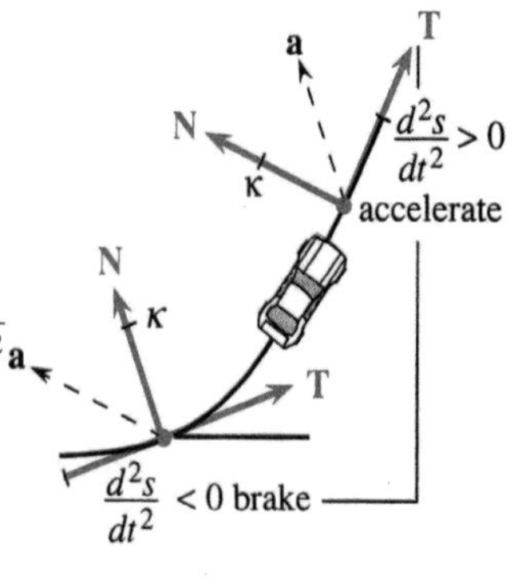

Fig. 12.8   Components of $\mathbf{a}$ as car turns corner.

Read-through questions

The curvature tells how fast the curve __a__ . For a circle of radius a, the direction changes by 2π in a distance __b__ , so $\kappa =$ __c__ . For a plane curve $y = f(x)$ the formula is $\kappa = |y''|/$ __d__ . The curvature of $y = \sin x$ is __e__ . At a point where $y'' = 0$ (an __f__ point) the curve is momentarily straight and $\kappa =$ __g__ . For a space curve $\kappa = |\mathbf{v} \times \mathbf{a}|/$ __h__ .

The normal vector $\mathbf{N}$ is perpendicular to __i__ . It is a __j__ vector along the derivative of $\mathbf{T}$, so $\mathbf{N} =$ __k__ . For motion around a circle $\mathbf{N}$ points __l__ . Up a helix $\mathbf{N}$ also points __m__ . Moving at unit speed on any curve, the time t is the same as the __n__ s. Then $|\mathbf{v}| =$ __o__ and $d^2 s/dt^2 =$ __p__ and $\mathbf{a}$ is in the direction of __q__ .

Acceleration equals __r__ $\mathbf{T} +$ __s__ $\mathbf{N}$. At unit speed around a unit circle, those components are __t__ . An astronaut who spins once a second in a radius of one meter has $|\mathbf{a}| =$ __u__ meters/sec^2, which is about __v__ g.

Compute the curvature κ in Problems 1–8.

1 $y = e^x$

2 $y = \ln x$ (where is κ largest ?)

3 $x = 2 \cos t, y = 2 \sin t$

4 $x = \cos t^2, y = \sin t^2$

5 $x = 1 + t^2, y = 3t^2$ (the path is a _____).

6 $x = \cos^3 t, y = \sin^3 t$

7 $r = \theta = t$ (so $x = t \cos t, y =$ _____)

8 $x = t, y = \ln \cos t$

9 Find $\mathbf{T}$ and $\mathbf{N}$ in Problem 4.

10 Show that $\mathbf{N} = \sin t\, \mathbf{i} + \cos t\, \mathbf{j}$ in Problem 6.

11 Compute $\mathbf{T}$ and $\mathbf{N}$ in Problem 8.

12 Find the speed $|\mathbf{v}|$ and curvature κ of a projectile:

$$x = (v_0 \cos \alpha)t, \ y = (v_0 \sin \alpha)t - \tfrac{1}{2}gt^2.$$

13 Find $\mathbf{T}$ and $|\mathbf{v}|$ and κ for the helix $\mathbf{R} = 3 \cos t\, \mathbf{i} + 3 \sin t\, \mathbf{j} + 4t\, \mathbf{k}$. How much longer is a turn of the helix than the corresponding circle ? What is the upward slope of $\mathbf{T}$?

14 When $\kappa = 0$ the path is a _____ . This happens when $\mathbf{v}$ and $\mathbf{a}$ are _____ . Then $\mathbf{v} \times \mathbf{a} =$ _____ .

15 Find the curvature of a cycloid $x = a(t - \sin t)$, $y = a(1 - \cos t)$.

16 If all points of a curve are moved twice as far from the origin $(x \to 2x, y \to 2y)$, what happens to κ ? What happens to $\mathbf{N}$?

17 Find κ and $\mathbf{N}$ at $\theta = \pi$ for the hypocycloid $x = 4 \cos \theta + \cos 4\theta$, $y = 4 \sin \theta - \sin 4\theta$.

18 From $\mathbf{v} = |\mathbf{v}|\mathbf{T}$ and $\mathbf{a}$ in equation (8), derive $\kappa = |\mathbf{v} \times \mathbf{a}|/|\mathbf{v}|^3$.

19 From a point on the curve, go along the vector $\mathbf{N}/\kappa$ to find the *center of curvature*. Locate this center for the point $(1,0)$ on the circle $x = \cos t, y = \sin t$ and the ellipse $x = \cos t, y = 2 \sin t$ and the parabola $y = \tfrac{1}{2}(x^2 - 1)$. The path of the center of curvature is the *"evolute"* of the curve.

20 Which of these depend only on the shape of the curve, and which depend also on the speed ? $\mathbf{v}, \mathbf{T}, |\mathbf{v}|, s, \kappa, \mathbf{a}, \mathbf{N}, \mathbf{B}$.

21 A plane curve through $(0,0)$ and $(2,0)$ with constant curvature κ is the circular arc _____ . For which κ is there no such curve ?

22 Sketch a smooth curve going through $(0,0), (1,-1)$, and $(2,0)$. Somewhere $d^2 y/dx^2$ is at least _____ . Somewhere the curvature is at least _____ . (Proof is for instructors only.)

23 For plane vectors, the ordinary product rule applied to $U_1 V_1 + U_2 V_2$ shows that $(\mathbf{U} \cdot \mathbf{V})' = \mathbf{U}' \cdot \mathbf{V} +$ _____ .

24 If $\mathbf{v}$ is perpendicular to $\mathbf{a}$, prove that the speed is constant. True or false: The path is a circle.

Problems 25–30 work with the T-N-B system—along the curve, in the plane of the curve, perpendicular to that plane.

25 Compute $\mathbf{B} = \mathbf{T} \times \mathbf{N}$ for the helix $\mathbf{R} = \cos t\, \mathbf{i} + \sin t\, \mathbf{j} + t\, \mathbf{k}$ in Examples $3 - 4$.

26 Using Problem 23, differentiate $\mathbf{B} \cdot \mathbf{T} = 0$ and $\mathbf{B} \cdot \mathbf{B} = 1$ to show that $\mathbf{B}'$ is perpendicular to $\mathbf{T}$ and $\mathbf{B}$. So $d\mathbf{B}/ds = -\tau \mathbf{N}$ for some number τ called the *torsion*.

27 Compute the torsion $\tau = |d\mathbf{B}/ds|$ for the helix in Problem 25.

28 Find $\mathbf{B} = \mathbf{TN}$ for the curve $x = 1, y = t, z = t^2$.

29 A circle lies in the xy plane. Its normal $\mathbf{N}$ lies _____ and $\mathbf{B} =$ and $\tau = |d\mathbf{B}/ds| =$ _____ .

30 The Serret-Frenet formulas are $d\mathbf{T}/ds = \kappa \mathbf{N}$, $d\mathbf{N}/ds = -k\mathbf{T} + \tau\mathbf{B}$, $d\mathbf{B}/ds = -\tau\mathbf{N}$. We know the first and third. Differentiate $\mathbf{N} = -\mathbf{T} \times \mathbf{B}$ to find the second.

31 The angle θ from the x axis to the tangent line is $\theta = \tan^{-1}(dy/dx)$, when dy/dx is the slope of the curve.

(a) Compute $d\theta/dx$.

(b) Divide by $ds/dx = (1 + (dy/dx)^2)^{1/2}$ to show that $|d\theta/ds|$ is κ in equation (5). Curvature is change in direction $|d\theta|$ divided by change in position $|ds|$.

32 If the tangent direction is at angle θ then $\mathbf{T} = \cos \theta\, \mathbf{i} + \sin \theta\, \mathbf{j}$. In Problem 31 $|d\theta/ds|$ agreed with $\kappa = |d\mathbf{T}/ds|$ because $|d\mathbf{T}/d\theta| =$ _____ .

In 33–37 find the T and N components of acceleration.

33 $x = 5 \cos \omega t, y = 5 \sin \omega t, z = 0$ (circle)

34 $x = 1 + t, y = 1 + 2t, z = 1 + 3t$ (line)

35 $x = t \cos t, y = t \sin t, z = 0$ (spiral)

36 $x = e^t \cos t, y = e^t \sin t, z = 0$ (spiral)

37 $x = 1, y = t, z = t^2$.

38 For the spiral in 36, show that the angle between **R** and **a** (position and acceleration) is constant. Find the angle.

39 Find the curvature of a polar curve $r = F(\theta)$.

12.4 Polar Coordinates and Planetary Motion

This section has a general purpose—to do vector calculus in *polar coordinates*. It also has a specific purpose—to study *central forces* and the *motion of planets*. The main gravitational force on a planet is from the sun. It is a central force, because it comes from the sun at the center. Polar coordinates are natural, so the two purposes go together.

You may feel that the planets are too old for this course. But Kepler's laws are more than theorems, they are something special in the history of mankind—"the greatest scientific discovery of all time." If we can recapture that glory we should do it. Part of the greatness is in the difficulty—Kepler was working sixty years before Newton discovered calculus. From pages of observations, and some terrific guesses, a theory was born. We will try to preserve the greatness without the difficulty, and show how elliptic orbits come from calculus. The first conclusion is quick.

Motion in a central force field always stays in a plane.

$\mathbf{F}$ is a multiple of the vector $\mathbf{R}$ from the origin (central force). $\mathbf{F}$ also equals $m\mathbf{a}$ (Newton's Law). Therefore $\mathbf{R}$ and $\mathbf{a}$ are in the same direction and $\mathbf{R} \times \mathbf{a} = \mathbf{0}$. Then $\mathbf{R} \times \mathbf{v}$ has zero derivative and is constant:

$$\text{by the product rule:} \quad \frac{d}{dt}(\mathbf{R} \times \mathbf{v}) = \mathbf{v} \times \mathbf{v} + \mathbf{R} \times \mathbf{a} = \mathbf{0} + \mathbf{0}. \tag{1}$$

$\mathbf{R} \times \mathbf{v}$ is a constant vector $\mathbf{H}$. So $\mathbf{R}$ stays in the plane perpendicular to $\mathbf{H}$.

How does a planet move in that plane? We turn to polar coordinates. At each point except the origin (where the sun is), $\mathbf{u}_r$ *is the unit vector pointing outward*. It is the position vector $\mathbf{R}$ divided by its length r (which is $\sqrt{x^2 + y^2}$):

$$\mathbf{u}_r = \mathbf{R}/r = (x\mathbf{i} + y\mathbf{j})/r = \cos\theta\,\mathbf{i} + \sin\theta\,\mathbf{j}. \tag{2}$$

That is a unit vector because $\cos^2\theta + \sin^2\theta = 1$. It goes out from the center. Figure 12.9 shows $\mathbf{u}_r$ and the second unit vector $\mathbf{u}_\theta$ at a $90°$ angle:

$$\mathbf{u}_\theta = -\sin\theta\,\mathbf{i} + \cos\theta\,\mathbf{j}. \tag{3}$$

The dot product is $\mathbf{u}_r \cdot \mathbf{u}_\theta = 0$. The subscripts r and θ indicate direction (not derivative).

Question 1: How do $\mathbf{u}_r$ and $\mathbf{u}_\theta$ change as r changes (out a ray)? *They don't.*

Question 2: How do $\mathbf{u}_r$ and $\mathbf{u}_\theta$ change as θ changes? *Take the derivative:*

$$\begin{aligned} d\mathbf{u}_r/d\theta &= -\sin\theta\,\mathbf{i} + \cos\theta\,\mathbf{j} = \mathbf{u}_\theta \\ d\mathbf{u}_\theta/d\theta &= -\cos\theta\,\mathbf{i} + \sin\theta\,\mathbf{j} = -\mathbf{u}_r. \end{aligned} \tag{4}$$

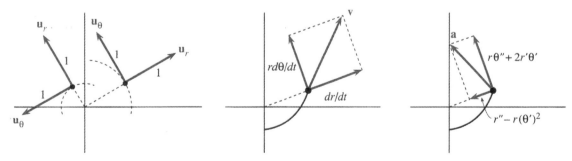

Fig. 12.9 $\mathbf{u}_r$ is outward, $\mathbf{u}_\theta$ is around the center. Components of $\mathbf{v}$ and $\mathbf{a}$ in this directions.

Since $\mathbf{u}_r = \mathbf{R}/r$, one formula is simple: **The position vector is** $\mathbf{R} = r\mathbf{u}_r$. For its derivative $\mathbf{v} = d\mathbf{R}/dt$, use the chain rule $d\mathbf{u}_r/dt = (d\mathbf{u}_r/d\theta)(d\theta/dt) = (d\theta/dt)\mathbf{u}_\theta$:

$$\textit{The velocity is } \mathbf{v} = \frac{d}{dt}(r\mathbf{u}_r) = \frac{dr}{dt}\mathbf{u}_r + r\frac{d\theta}{dt}\mathbf{u}_\theta. \tag{5}$$

The outward speed is dr/dt. The circular speed is $r\,d\theta/dt$. The sum of squares is $|\mathbf{v}|^2$.

Return one more time to steady motion around a circle, say $r = 3$ and $\theta = 2t$. The velocity is $\mathbf{v} = 6\mathbf{u}_\theta$, all circular. The acceleration is $-12\mathbf{u}_r$, all inward. For circles $\mathbf{u}_\theta$ is the tangent vector $\mathbf{T}$. But the unit vector $\mathbf{u}_r$ points outward and $\mathbf{N}$ points inward—the way the curve turns.

Now we tackle acceleration for any motion in polar coordinates. There can be speedup in r and speedup in θ (also change of direction). Differentiate $\mathbf{v}$ in (5) by the product rule:

$$\frac{d\mathbf{v}}{dt} = \frac{d^2r}{dt^2}\mathbf{u}_r + \frac{dr}{dt}\frac{d\mathbf{u}_r}{dt} + \frac{dr}{dt}\frac{d\theta}{dt}\mathbf{u}_\theta + r\frac{d^2\theta}{dt^2}\mathbf{u}_\theta + r\frac{d\theta}{dt}\frac{d\mathbf{u}_\theta}{dt}.$$

For $d\mathbf{u}_r/dt$ and $d\mathbf{u}_\theta/dt$, multiply equation (4) by $d\theta/dt$. Then all terms contain $\mathbf{u}_r$ or $\mathbf{u}_\theta$. The formula for $\mathbf{a}$ is famous but not popular (except it got us to the moon):

$$\mathbf{a} = \frac{d\mathbf{v}}{dt} = \left(\frac{d^2r}{dt^2} - r\left(\frac{d\theta}{dt}\right)^2\right)\mathbf{u}_r + \left(r\frac{d^2\theta}{dt^2} + 2\frac{dr}{dt}\frac{d\theta}{dt}\right)\mathbf{u}_\theta. \tag{6}$$

In the steady motion with $r = 3$ and $\theta = 2t$, only one acceleration term is nonzero: $\mathbf{a} = -12\mathbf{u}_r$. Formula (6) can be memorized (maybe). Problem 14 gives a new way to reach it, using $re^{i\theta}$.

EXAMPLE 1   Find $\mathbf{R}$ and $\mathbf{v}$ and $\mathbf{a}$ for speedup $\theta = t^2$ around the circle $r = 1$.

Solution   The position vector is $\mathbf{R} = \mathbf{u}_r$. Then $\mathbf{v}$ and $\mathbf{a}$ come from $(5-6)$:

$$\mathbf{v} = (r\,d\theta/dt)\mathbf{u}_\theta = 2t\mathbf{u}_\theta \qquad \mathbf{a} = -(2t)^2\mathbf{u}_r + 2\mathbf{u}_\theta.$$

This question and answer were also in Example 6 of the previous section. The acceleration was $2\mathbf{T} + 4t^2\mathbf{N}$. Notice again that $\mathbf{T} = \mathbf{u}_\theta$ and $\mathbf{N} = -\mathbf{u}_r$, going round the circle.

EXAMPLE 2   Find $\mathbf{R}$ and $\mathbf{v}$ and $|\mathbf{v}|$ and $\mathbf{a}$ for the spiral motion $r = 3t$, $\theta = 2t$.

Solution   The position vector is $\mathbf{R} = 3t\,\mathbf{u}_r$. Equation (5) gives velocity and speed:

$$\mathbf{v} = 3\mathbf{u}_r + = 6t\mathbf{u}_\theta \quad \text{and} \quad |\mathbf{v}| = \sqrt{(3)^2 + (6t)^2}.$$

The motion goes *out* and also *around*. From (6) the acceleration is $-12t\,\mathbf{u}_r + 12\mathbf{u}_\theta$. The same answers would come more slowly from $\mathbf{R} = 3t\cos 2t\,\mathbf{i} + 3t\sin 2t\,\mathbf{j}$.

This example uses polar coordinates, but *the motion is not circular*. One of Kepler's inspirations, after many struggles, was to get away from circles.

KEPLER'S LAWS

You may know that before Newton and Leibniz and calculus and polar coordinates, Johannes Kepler discovered three laws of planetary motion. He was the court mathematician to the Holy Roman Emperor, who mostly wanted predictions of wars. Kepler also determined the date of every Easter—no small problem. His triumph was to discover patterns in the observations made by astronomers (especially by Tycho Brahe). Galileo and Copernicus expected circles, but Kepler found ellipses.

Law 1:   Each planet travels in an ellipse with one focus at the sun.

Law 2:   The vector from sun to planet sweeps out area at a steady rate: $dA/dt =$ constant.

Law 3:   The length of the planet's year is $T = ka^{3/2}$, where a = maximum distance from the center (not the sun) and $k = 2\pi/\sqrt{GM}$ is the same for all planets.

With calculus the proof of these laws is a thousand times quicker. But Law 2 is the only easy one. The sun exerts a central force. Equation (1) gave $\mathbf{R} \times \mathbf{v} = \mathbf{H} =$ constant for central forces. Replace $\mathbf{R}$ by $r\mathbf{u}_r$ and replace $\mathbf{v}$ by equation (5):

$$\mathbf{H} = r\mathbf{u}_r \times \left(\frac{dr}{dt}\mathbf{u}_r + r\frac{d\theta}{dt}\mathbf{u}_\theta\right) = r^2\frac{d\theta}{dt}(\mathbf{u}_r \times \mathbf{u}_\theta). \qquad (7)$$

This vector $\mathbf{H}$ is constant, so *its length $h = r^2 d\theta/dt$ is constant*. In polar coordinates, the area is $dA = \frac{1}{2}r^2 d\theta$. This area dA is swept out by the planet (Figure 12.10), and we have proved Law 2:

$$dA/dt = \tfrac{1}{2}r^2 d\theta/dt = \tfrac{1}{2}h = constant. \qquad (8)$$

Near the sun r is small. So $d\theta/dt$ is big and planets go around faster.

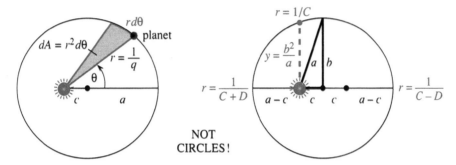

Fig. 12.10   The planet is on an ellipse with the sun at a focus. Note a, b, c, q.

Now for Law 1, about ellipses. We are aiming for $1/r = C - D\cos\theta$, which is *the polar coordinate equation of an ellipse*. It is easier to write q than $1/r$, and find an equation for q. The equation we will reach is $d^2q/d\theta^2 + q = C$. The desired $q = C - D\cos\theta$ solves that equation (check this), and gives us Kepler's ellipse.

The first step is to connect dr/dt to $dq/d\theta$ by the chain rule:

$$\frac{dr}{dt} = \frac{d}{dt}\left(\frac{1}{q}\right) = \frac{-1}{q^2}\frac{dq}{dt} = \frac{-1}{q^2}\frac{dq}{d\theta}\frac{d\theta}{dt} = -h\frac{dq}{d\theta}. \qquad (9)$$

Notice especially $d\theta/dt = h/r^2 = hq^2$. What we really want are second derivatives:

$$\frac{d^2r}{dt^2} = -h\frac{d}{dt}\left(\frac{dq}{d\theta}\right) = -h\frac{d}{d\theta}\left(\frac{dq}{d\theta}\right)\frac{d\theta}{dt} = -h^2q^2\frac{d^2q}{d\theta^2}. \qquad (10)$$

After this trick of introducing q, we are ready for physics. The planet obeys Newton's Law $\mathbf{F} = m\mathbf{a}$, and the central force $\mathbf{F}$ is the sun's gravity:

$$\frac{\mathbf{F}}{m} = \mathbf{a} \quad \text{is} \quad -\frac{GM}{r^2} = \frac{d^2r}{dt^2} - r\left(\frac{d\theta}{dt}\right)^2. \qquad (11)$$

That right side is the $\mathbf{u}_r$ component of $\mathbf{a}$ in (6). Change r to $1/q$ and change $d\theta/dt$ to hq^2. The preparation in(10) allows us to rewrite d^2r/dt^2 in equation (11). That equation becomes

$$-GM\ q^2 = -h^2q^2\frac{d^2q}{d\theta^2} - \frac{1}{q}(hq^2)^2.$$

Dividing by $-h^2q^2$ gives what we hoped for—the simple equation for q:

$$d^2q/d\theta^2 + q = GM/h^2 = C \ \textit{(a constant)}. \tag{12}$$

The solution is $q = C - D\cos\theta$. Section 9.3 gave this polar equation for an ellipse or parabola or hyperbola. To be sure it is an ellipse, an astronomer computes C and D from the sun's mass M and the constant G and the earth's position and velocity. *The main point is that $C > D$.* Then q is never zero and r is never infinite. Hyperbolas and parabolas are ruled out, and the orbit in Figure 12.10 must be an ellipse.†

Astronomy is really impressive. You should visit the Greenwich Observatory in London, to see how Halley watched his comet. He amazed the world by predicting the day it would return. Also the discovery of Neptune was pure mathematics—the path of Uranus was not accounted for by the sun and known planets. LeVerrier computed a point in the sky and asked a Berlin astronomer to look. Sure enough Neptune was there.

Recently one more problem was solved—to explain the gap in the asteroids around Jupiter. The reason is "*chaos*"—the three-body problem goes unstable and an asteroid won't stay in that orbit. We have come a long way from circles.

Department of Royal Mistakes The last pound note issued by the Royal Mint showed Newton looking up from his great book *Principia Mathematica*. He is not smiling and we can see why. The artist put the sun at the center! Newton has just proved it is at the focus. True, the focus is marked S and the planet is P. But those rays at the center brought untold headaches to the Mint—the note is out of circulation. I gave an antique dealer three pounds for it (in coins).

Kepler's third law gives the time T to go around the ellipse—the planet's year. What is special in the formula is $a^{3/2}$—and for Kepler himself, the 15th of May 1618 was unforgettable: "the right ratio outfought the darkness of my mind, by the great proof afforded by my labor of seventeen years on Brahe's observations." The second law $dA/dt = \frac{1}{2}h$ is the key, plus two facts about an ellipse—its area πab and the height b^2/a above the sun:

1. The area $A = \int_0^T \frac{dA}{dt}dt = \frac{1}{2}hT$ must equal πab, so $T = \dfrac{2\pi ab}{h}$

2. The distance $r = 1/C$ at $\theta = \pi/2$ must equal b^2/a, so $b = \sqrt{a/C}$.

The height b^2/a is in Figure 12.10 and Problems $25-26$. The constant $C = GM/h^2$ is in equation (12). Put them together to find the period:

$$T = \frac{2\pi ab}{h} = \frac{2\pi a}{h}\sqrt{\frac{a}{C}} = \frac{2\pi}{\sqrt{GM}}a^{3/2}. \tag{13}$$

---

† An amateur sees the planet come around again, and votes for an ellipse.

To think of Kepler guessing $a^{3/2}$ is amazing. To think of Newton proving Kepler's laws by calculus is also wonderful—because we can do it too.

EXAMPLE 3 When a satellite goes around in a circle, find the time T.

Let r be the radius and ω be the angular velocity. The time for a complete circle (angle 2π) is $T = 2\pi/\omega$. The acceleration is GM/r^2 from gravity, and it is also $r\omega^2$ for circular motion. Therefore Kepler is proved right:

$$r\omega^2 = GM/r^2 \quad \Rightarrow \quad \omega = \sqrt{GM/r^3} \quad \Rightarrow \quad T = 2\pi/\omega = 2\pi r^{3/2}/\sqrt{GM}.$$

12.4 EXERCISES

Read-through questions

A central force points toward ___a___. Then $\mathbf{R} \times d^2\mathbf{R}/dt^2 = 0$ because ___b___. Therefore $\mathbf{R} \times d\mathbf{R}/dt$ is a ___c___ (called $\mathbf{H}$).

In polar coordinates, the outward unit vector is $\mathbf{u}_r = \cos\theta\,\mathbf{i} + $___d___. Rotated by $90°$ this becomes $\mathbf{u}_\theta = $___e___. The position vector $\mathbf{R}$ is the distance r times ___f___. The velocity $\mathbf{v} = d\mathbf{R}/dt$ is ___g___ $\mathbf{u}_r + $___h___ $\mathbf{u}_\theta$. For steady motion around the circle $r = 5$ with $\theta = 4t$, $\mathbf{v}$ is ___i___ and $|\mathbf{v}|$ is ___j___ and $\mathbf{a}$ is ___k___.

For motion under a circular force, r^2 times ___l___ is constant. Dividing by 2 gives Kepler's second law $dA/dt = $___m___. The first law says that the orbit is an ___n___ with the sun at ___o___. The polar equation for a conic section is ___p___ $= C - D\cos\theta$. Using $\mathbf{F} = m\mathbf{a}$ we found $q_{\theta\theta} + $___q___ $= C$. So the path is a conic section; it must be an ellipse because ___r___. The properties of an ellipse lead to the period $T = $___s___, which is Kepler's third law.

1 Find the unit vectors $\mathbf{u}_r$ and $\mathbf{u}_\theta$ at the point $(0, 2)$. The $\mathbf{u}_r$ and $\mathbf{u}_\theta$ components of $\mathbf{v} = \mathbf{i} + \mathbf{j}$ at that point are _____.

2 Find $\mathbf{u}_r$ and $\mathbf{u}_\theta$ at $(3, 3)$. If $\mathbf{v} = \mathbf{i} + \mathbf{j}$ then $\mathbf{v} = $_____ $\mathbf{u}_r$. Equation (5) gives $dr/dt = $_____ and $d\theta/dt = $_____.

3 At the point $(1, 2)$, velocities in the direction _____ will give $dr/dt = 0$. Velocities in the direction _____ will give $d\theta/dt = 0$.

4 Traveling on the cardioid $r = 1 - \cos\theta$ with $d\theta/dt = 2$, what is $\mathbf{v}$? How long to go around the cardioid (no integration involved)?

5 If $r = e^\theta$ and $\theta = 3t$, find $\mathbf{v}$ and $\mathbf{a}$ when $t = 1$.

6 If $r = 1$ and $\theta = \sin t$, describe the path and find $\mathbf{v}$ and $\mathbf{a}$ from equations $(5 - 6)$. Where is the velocity zero?

7 (important) $\mathbf{R} = 4\cos 5t\,\mathbf{i} + 4\sin 5t\,\mathbf{j} = 4\mathbf{u}_r$ travels on a circle of radius 4 with $\theta = 5t$ and speed 20. Find the components of $\mathbf{v}$ and $\mathbf{a}$ in three systems: $\mathbf{i}$ and $\mathbf{j}$, $\mathbf{T}$ and $\mathbf{N}$, u_r and $\mathbf{u}_\theta$.

8 When is the circle $r = 4$ completed, if the speed is $8t$? Find $\mathbf{v}$ and $\mathbf{a}$ at the return to the starting point $(4, 0)$.

9 The $\mathbf{u}_\theta$ component of acceleration is _____ $= 0$ for a central force, which is in the direction of _____. Then $r^2 d\theta/dt$ is constant (new proof) because its derivative is r times _____.

10 If $r^2 d\theta/dt = 2$ for travel up the line $x = 1$, draw a triangle to show that $r = \sec \theta$ and integrate to find the time to reach $(1,1)$.

11 A satellite is $r = 10,000$ km from the center of the Earth, traveling perpendicular to the radius vector at $4\,\text{km/sec}$. Find $d\theta/dt$ and h.

12 From $|\mathbf{u}_r| = 1$, it follows that $d\mathbf{u}_r/dr$ and $d\mathbf{u}_r/d\theta$ are _____ to $\mathbf{u}_r$ (Section 12.3). In fact $d\mathbf{u}_r/dr$ is _____ and $d\mathbf{u}_r/d\theta$ is _____.

13 Momentum is $m\mathbf{v}$ and its derivative is $m\mathbf{a} =$ force. Angular momentum is $m\mathbf{H} = m\mathbf{R} \times \mathbf{v}$ and its derivative is _____ $=$ torque. Angular momentum is constant under a central force because the _____ is zero.

14 To find (and remember) $\mathbf{v}$ and $\mathbf{a}$ in polar coordinates, start with the complex number $re^{i\theta}$ and take its derivatives:

$$\mathbf{R} = re^{i\theta} \qquad \frac{d\mathbf{R}}{dt} = \frac{dr}{dt}e^{i\theta} + ir\frac{d\theta}{dt}e^{i\theta}$$

$$\frac{d^2\mathbf{R}}{dt^2} = \underline{\quad} + \underline{\quad} + \underline{\quad} + \underline{\quad} + \underline{\quad}.$$

Key idea: The coefficients of $e^{i\theta}$ and $ie^{i\theta}$ are the $\mathbf{u}_r$ and $\mathbf{u}_\theta$ components of $\mathbf{R}, \mathbf{v}, \mathbf{a}$:

$$\mathbf{R} = r\mathbf{u}_r + 0\mathbf{u}_\theta \qquad \mathbf{v} = \frac{dr}{dt}\mathbf{u}_r + r\frac{d\theta}{dt}\mathbf{u}_\theta \qquad \mathbf{a} = \underline{\quad}.$$

(a)   Fill in the five terms from the derivative of $d\mathbf{R}/dt$

(b)   Convert $e^{i\theta}$ to $\mathbf{u}_r$ and $ie^{i\theta}$ to $\mathbf{u}_\theta$ to find $\mathbf{a}$

(c)   Compare $\mathbf{R}, \mathbf{v}, \mathbf{a}$ with formulas (5–6)

(d)   (for instructors only) Why does this method work?

Note how $e^{i\theta} = \cos\theta + i\sin\theta$ corresponds to $\mathbf{u}_r = \cos\theta\,\mathbf{i} + \sin\theta\,\mathbf{j}$. This is one place where electrical engineers are allowed to write j instead of i for $\sqrt{-1}$.

15 If the period is T find from (13) a formula for the distance a.

16 To stay above New York what should be the period of a satellite? What should be its distance a from the center of the Earth?

17 From T and a find a formula for the mass M.

18 If the moon has a period of 28 days at an average distance of $a = 380,000$ km, estimate the mass of the _____.

19 The Earth takes $365\frac{1}{4}$ days to go around the sun at a distance $a \approx 93$ million miles ≈ 150 million kilometers. Find the mass of the sun.

20   True or false:

(a)   The paths of all comets are ellipses.

(b)   A planet in a circular orbit has constant speed.

(c)   Orbits in central force fields are conic sections.

21 $\sqrt{GM} \approx 2 \cdot 10^7$ in what units, based on the Earth's mass $M = 6 \cdot 10^{24}$ kg and the constant $G - 6.67 \cdot 10^{-11}\,\text{Nm}^2/\text{kg}^2$? A force of one $\text{kg} \cdot \text{meter/sec}^2$ is a Newton N.

22 If a satellite circles the Earth at 9000 km from the center, estimate its period T in seconds.

23 The Viking 2 orbiter around Mars had a period of about $10,000$ seconds. If the mass of Mars is $M = 6.4 \cdot 10^{23}$ kg, what was the value of a?

24 Convert $1/r = C - D\cos\theta$, or $1 = Cr - Dx$, into the xy equation of an ellipse.

25 The distances a and c on the ellipse give the constants in $r = 1/(C - D\cos\theta)$. Substitute $\theta = 0$ and $\theta = \pi$ as in Figure 12.10 to find $D = c/(a^2 - c^2)$ and $C = a/(a^2 - c^2) = a/b^2$.

26 Show that $x = -c$, $y = b^2/a$ lies on the ellipse $x^2/a^2 + y^2/b^2 = 1$. Thus y is the height $1/C$ above the sun in Figure 12.10. The distance from the sun to the center has $c^2 = a^2 - b^2$.

27 The point $x = a\cos 2\pi t/T$, $y = b\sin 2\pi t/T$ travels around an ellipse centered at $(0,0)$ and returns at time T. By symmetry it sweeps out area at the same rate at both ends of the major axis. Why does this *break* Kepler's second law?

28 If a central force is $F = -ma(r)\mathbf{u}_r$, explain why $d^2r/dt^2 - r(d\theta/dt)^2 = -a(r)$. What is $a(r)$ for gravity? Equation (12) for $q = 1/r$ leads to $q_{\theta\theta} + q = r^2 a(r)$.

29 When $F = 0$ the body should travel in a straight _____.

The equation $q_{\theta\theta} + q = 0$ allows $q = \cos\theta$, in which case the path $1/r = \cos\theta$ is _____. Extra credit: Mark off equal distances on a line, connect them to the sun, and explain why the triangles have equal area. So dA/dt is still constant.

30 The strong nuclear force increases with distance, $a(r) = r$. It binds quarks so tightly that up to now no top quarks have been seen (reliably). Problem 28 gives $q_{\theta\theta} + q = 1/q^3$.

(a)   Multiply by q_θ and integrate to find $\frac{1}{2}q_\theta^2 + \frac{1}{2}q^2 =$ _____ $+ C$.

*(b)   Integrate again (with tables) after setting $u = q^2$, $u_\theta = 2qq_\theta$.

31 The path of a quark in 30(b) can be written as $r^2(A + B\cos 2\theta) = 1$. Show that this is the same as the ellipse $(A + B)x^2 + (A - B)y^2 = 1$ with the origin at the *center*. The nucleus is not at a focus, and the pound note is correct for Newton watching quarks. (Quantum mechanics not accounted for.)

32 When will Halley's comet appear again? It disappeared in 1986 and its mean distance to the sun (average of $a + c$ and $a - c$) is $a = 1.6 \cdot 10^9$ kilometers.

33 You are walking at 2 feet/second toward the center of a merry-go-round that turns once every ten seconds. Starting from $r = 20$, $\theta = 0$ find $r(t)$, $\theta(t)$, $\mathbf{v}(t)$, $\mathbf{a}(t)$ and the length of your path to the center.

34 From Kepler's laws $r = 1/(C - D \cos \theta)$ and $r^2 d\theta/dt = h$, show that

    1. $dr/dt = -Dh \sin \theta$ **2.** $d^2r/dt^2 = \left(\dfrac{1}{r} - C\right)h^2/r^2$

    3. $d^2r/dt^2 - r(d\theta/dt) = -Ch^2/r^2$.

When Newton reached **3**, he knew that Kepler's laws required a central force of Ch^2/r^2. This is his *inverse square law*. Then he went backwards, in our equations (8–12), to show that this force yields Kepler's laws.

35 How long is our year? The Earth's orbit has $a = 149.57 \cdot 10^6$ kilometers.

CHAPTER 13

Partial Derivatives

This chapter is at the center of multidimensional calculus. Other chapters and other topics may be optional; this chapter and these topics are required. We are back to the basic idea of calculus—*the derivative*. There is a function f, the variables move a little bit, and f moves. The question is how much f moves and how fast. Chapters 1–4 answered this question for $f(x)$, a function of one variable. Now we have $f(x, y)$ or $f(x, y, z)$—with two or three or more variables that move independently. As x and y change, f changes. The fundamental problem of differential calculus is to connect Δx and Δy to Δf. Calculus solves that problem in the limit. *It connects dx and dy to df.* In using this language I am building on the work already done. You know that df/dx is the limit of $\Delta f/\Delta x$. Calculus computes the rate of change— which is the slope of the tangent line. The goal is to extend those ideas to

$$f(x, y) = x^2 - y^2 \quad \text{or} \quad f(x, y) = \sqrt{x^2 + y^2} \quad \text{or} \quad f(x, y, z) = 2x + 3y + 4z.$$

These functions have graphs, they have derivatives, and they must have tangents.

The heart of this chapter is summarized in six lines. The subject is *differential calculus*—small changes in a short time. Still to come is *integral* calculus—adding up those small changes. We give the words and symbols for $f(x, y)$, matched with the words and symbols for $f(x)$. Please use this summary as a guide, to know where calculus is going.

Curve $y = f(x)$ vs.   ***Surface*** $z = f(x, y)$

$\dfrac{df}{dx}$ becomes two partial derivatives $\dfrac{\partial f}{\partial x}$ and $\dfrac{\partial f}{\partial y}$

$\dfrac{d^2 f}{dx^2}$ becomes four second derivatives $\dfrac{\partial^2 f}{\partial x^2}, \dfrac{\partial^2 f}{\partial x \partial y}, \dfrac{\partial^2 f}{\partial y \partial x}, \dfrac{\partial^2 f}{\partial y^2}$

$\Delta f \approx \dfrac{df}{dx} \Delta x$ becomes the linear approximation $\Delta f \approx \dfrac{\partial f}{\partial x} \Delta x + \dfrac{\partial f}{\partial y} \Delta y$

tangent line   becomes the tangent plane $z - z_0 = \dfrac{\partial f}{\partial x}(x - x_0) + \dfrac{\partial f}{\partial y}(y - y_0)$

$\dfrac{dy}{dt} = \dfrac{dy}{dx}\dfrac{dx}{dt}$ becomes the chain rule $\dfrac{dz}{dt} = \dfrac{\partial z}{\partial x}\dfrac{dx}{dt} + \dfrac{\partial z}{\partial y}\dfrac{dy}{dt}$

$\dfrac{df}{dx} = 0$ becomes two maximum-minimum equations $\dfrac{\partial f}{\partial x} = 0$ and $\dfrac{\partial f}{\partial y} = 0.$

13.1 Surfaces and Level Curves

The graph of $y = f(x)$ is a curve in the xy plane. There are two variables—x is independent and free, y is dependent on x. Above x on the base line is the point (x, y) on the curve. The curve can be displayed on a two-dimensional printed page. The graph of $z = f(x, y)$ is a **surface in xyz space**. There are *three variables*—x and y are independent, z is dependent. Above (x, y) in the base plane is the point (x, y, z) on the surface (Figure 13.1). Since the printed page remains two-dimensional, we shade or color or project the surface. The eyes are extremely good at converting two-dimensional images into three-dimensional understanding—they get a lot of practice. The mathematical part of our brain also has something new to work on—*two partial derivatives*.

This section uses examples and figures to illustrate surfaces and their level curves. The next section is also short. Then the work begins.

EXAMPLE 1   Describe the **surface** and the **level curves** for $z = f(x, y) = \sqrt{x^2 + y^2}$.

The surface is a cone. Reason: $\sqrt{x^2 + y^2}$ is the distance in the base plane from $(0, 0)$ to (x, y). When we go out a distance 5 in the base plane, we go up the same distance 5 to the surface. The cone climbs with slope 1. The distance out to (x, y) equals the distance up to z (this is a 45° cone).

The level curves are circles. At height 5, the cone contains a circle of points—all at the same "level" on the surface. The plane $z = 5$ meets the surface $z = \sqrt{x^2 + y^2}$ at those points (Figure 13.1b). The circle below them (in the base plane) is the level curve.

DEFINITION   A *level curve* or *contour line* of $z = f(x, y)$ contains all points (x, y) that share the *same value* $f(x, y) = c$. Above those points, the surface is at the height $z = c$.

There are different level curves for different c. To see the curve for $c = 2$, cut through the surface with the horizontal plane $z = 2$. The plane meets the surface above the points where $f(x, y) = 2$. **The level curve in the base plane has the equation** $f(x, y) = 2$. Above it are all the points at "level 2" or "level c" on the surface.

Every curve $f(x, y) = c$ is labeled by its constant c. This produces a **contour map** (the base plane is full of curves). For the cone, the level curves are given by $\sqrt{x^2 + y^2} = c$, and the contour map consists of circles of radius c.

Question   What are the level curves of $z = f(x, y) = x^2 + y^2$?
Answer   *Still circles.* But the surface is not a cone (it bends up like a parabola). The circle of radius 3 is the level curve $x^2 + y^2 = 9$. On the surface above, the height is 9.

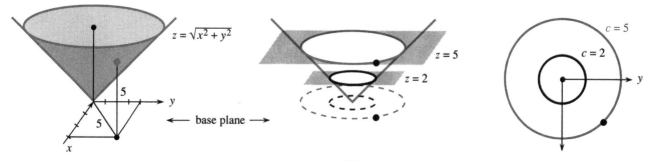

Fig. 13.1   The surface for $z = f(x, y) = \sqrt{x^2 + y^2}$ is a cone. The level curves are circles.

EXAMPLE 2   For the *linear function* $f(x,y) = 2x + y$, the surface is a plane. Its level curves are straight lines. ***The surface*** $z = 2x + y$ ***meets the plane*** $z = c$ ***in the line*** $2x + y = c$. That line is above the base plane when c is positive, and below when c is negative. The contour lines are *in* the base plane. Figure 13.2b labels these parallel lines according to their height in the surface.

Question If the level curves are all straight lines, must they be parallel?
Answer   No. The surface $z = y/x$ has level curves $y/x = c$. Those lines $y = cx$ swing around the origin, as the surface climbs like a spiral playground slide.

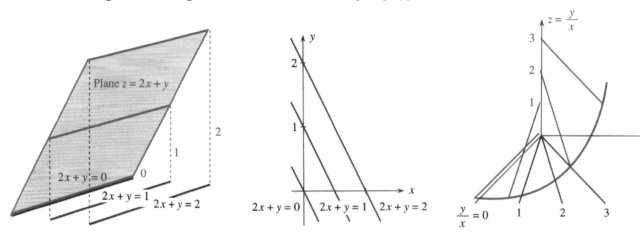

Fig. 13.2   A plane has parallel level lines. The spiral slide $z = y/x$ has lines $y/x = c$.

EXAMPLE 3   The weather map shows contour lines of the *temperature function*. Each level curve connects points at a constant temperature. One line runs from Seattle to Omaha to Cincinnati to Washington. In winter it is painful even to think about the line through L.A. and Texas and Florida. *USA Today* separates the contours by color, which is better. We had never seen a map of universities.

Question From a contour map, how do you find the highest point?
Answer   The level curves form *loops* around the maximum point. As c increases the loops become tighter. Similarly the curves squeeze to the lowest point as c decreases.

EXAMPLE 4   A contour map of a mountain may be the best example of all. Normally the level curves are separated by 100 feet in height. On a steep trail those curves are bunched together—the trail climbs quickly. In a flat region the contour lines are far apart. Water runs perpendicular to the level curves. On my map of New Hampshire that is true of creeks but looks doubtful for rivers.

Question Which direction in the base plane is uphill on the surface?
Answer   The steepest direction is perpendicular to the level curves. This is important. Proof to come.

EXAMPLE 5   In economics $x^2 y$ is a *utility function* and $x^2 y = c$ is an *indifference curve*.

The utility function $x^2 y$ gives the value of x hours awake and y hours asleep. Two hours awake and fifteen minutes asleep have the value $f = (2^2)(\frac{1}{4})$. This is the same as one hour of each: $f = (1^2)(1)$. Those lie on the same level curve in Figure 13.4a. We are indifferent, and willing to exchange any two points on a level curve.

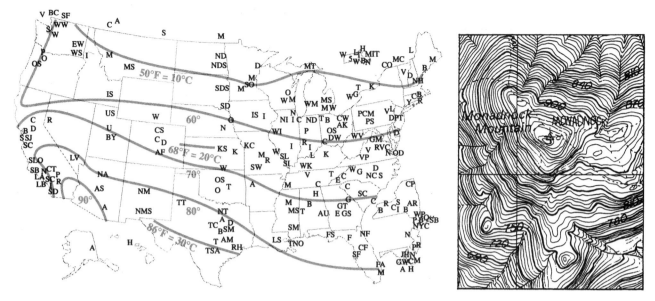

Fig. 13.3 The temperature at many U.S. and Canadian universities. Mt. Monadnock in New Hampshire is said to be the most climbed mountain (except Fuji?) at $125,000$/year. Contour lines every 6 meters.

The indifference curve is "*convex*." ***We prefer the average of any two points***. The line between two points is up on higher level curves.

Figure 13.4b shows an extreme case. The level curves are straight lines $4x + y = c$. Four quarters are freely substituted for one dollar. The value is $f = 4x + y$ dollars.

Figure 13.4c shows the other extreme. Extra left shoes or extra right shoes are useless. The value (or utility) is the *smaller* of x and y. That counts *pairs* of shoes.

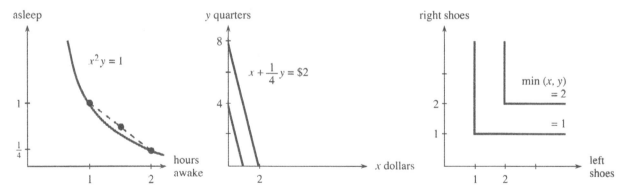

Fig. 13.4 Utility functions $x^2 y$, $4x + y$, $\min(x, y)$. Convex, straight substitution, complements.

Read-through questions

The graph of $z = f(x, y)$ is a __a__ in __b__-dimensional space. The __c__ curve $f(x, y) = 7$ lies down in the base plane. Above this level curve are all points at height __d__ in the surface. The __e__ $z = 7$ cuts through the surface at those points. The level curves $f(x, y) = $__f__ are drawn in the xy plane and labeled by __g__. The family of labeled curves is a __h__ map.

For $z = f(x, y) = x^2 - y^2$, the equation for a level curve is __i__. This curve is a __j__. For $z = x - y$ the curves are __k__. *Level curves never cross because* __l__. They crowd together when the surface is __m__. The curves tighten to a point when __n__. The steepest direction on a mountain is __o__ to the __p__.

1 Draw the surface $z = f(x, y)$ for these four functions:

$$f_1 = \sqrt{4 - x^2 - y^2} \qquad f_2 = 2 - \sqrt{x^2 + y^2}$$
$$f_3 = 2 - \tfrac{1}{2}(x^2 + y^2) \qquad f_4 = 1 + e^{-x^2 - y^2}$$

2 The level curves of all four functions are _____. They enclose the maximum at _____. Draw the four curves $f(x, y) = 1$ and rank them by increasing radius.

3 Set $y = 0$ and compute the x derivative of each function at $x = 2$. Which mountain is flattest and which is steepest at that point?

4 Set $y = 1$ and compute the x derivative of each function at $x = 1$.

For f_5 to f_{10} draw the level curves $f = 0, 1, 2$. Also $f = -4$.

5 $f_5 = x - y$

6 $f_6 = (x + y)^2$

7 $f_7 = xe^{-y}$

8 $f_8 = \sin(x - y)$

9 $f_9 = y - x^2$

10 $f_{10} = y/x^2$

11 Suppose the level curves are parallel straight lines. Does the surface have to be a plane?

12 Construct a function whose level curve $f = 0$ is in two separate pieces.

13 Construct a function for which $f = 0$ is a circle and $f = 1$ is not.

14 Find a function for which $f = 0$ has infinitely many pieces.

15 Draw the contour map for $f = xy$ with level curves $f = -2, -1, 0, 1, 2$. Describe the surface.

16 Find a function $f(x, y)$ whose level curve $f = 0$ consists of a circle and *all points inside it*.

Draw two level curves in 17–20. Are they ellipses, parabolas, or hyperbolas? Write $\sqrt{} - 2x = c$ as $\sqrt{} = c + 2x$ before squaring both sides.

17 $f = \sqrt{4x^2 + y^2}$

18 $f = \sqrt{4x^2 + y^2 - 2x}$

19 $f = \sqrt{5x^2 + y^2 - 2x}$

20 $f = \sqrt{3x^2 + y^2 - 2x}$

21 The level curves of $f = (y - 2)/(x - 1)$ are _____ through the point $(1, 2)$ except that this point is not _____.

22 Sketch a map of the US with lines of constant temperature (isotherms) based on today's paper.

23 (a) The contour lines of $z = x^2 + y^2 - 2x - 2y$ are circles around the point _____, where z is a minimum.

(b) The contour lines of $f = $ _____ are the circles $x^2 + y^2 = c + 1$ on which $f = c$.

24 Draw a contour map of any state or country (lines of constant height above sea level). Florida may be too flat.

25 The graph of $w = F(x, y, z)$ is a _____-dimensional surface in $xyzw$ space. Its level sets $F(x, y, z) = c$ are _____-dimensional surfaces in xyz space. For $w = x - 2y + z$ those level sets are _____. For $w = x^2 + y^2 + z^2$ those level sets are _____.

26 The surface $x^2 + y^2 - z^2 = -1$ is in Figure 13.8. There is empty space when z^2 is smaller than 1 because _____.

27 The level sets of $F = x^2 + y^2 + qz^2$ look like footballs when q is _____, like basketballs when q is _____, and like frisbees when q is _____.

28 Let $T(x, y)$ be the driving time from your home at $(0, 0)$ to nearby towns at (x, y). Draw the level curves.

29 (a) The level curves of $f(x, y) = \sin(x - y)$ are _____.

(b) The level curves of $g(x, y) = \sin(x^2 - y^2)$ are _____.

(c) The level curves of $h(x, y) = \sin(x - y^2)$ are _____.

30 Prove that if $x_1 y_1 = 1$ and $x_2 y_2 = 1$ then their average $x = \tfrac{1}{2}(x_1 + x_2)$, $y = \tfrac{1}{2}(y_1 + y_2)$ has $xy \geqslant 1$. The function $f = xy$ has convex level curves (hyperbolas).

31 The hours in a day are limited by $x + y = 24$. Write $x^2 y$ as $x^2(24 - x)$ and maximize to find the optimal number of hours to stay awake.

32 Near $x = 16$ draw the level curve $x^2 y = 2048$ and the line $x + y = 24$. Show that the curve is convex and the line is tangent.

33 The surface $z = 4x + y$ is a _____. The surface $z = \min(x, y)$ is formed from two _____. We are willing to exchange 6 left and 2 right shoes for 2 left and 4 right shoes but better is the average _____.

34 Draw a contour map of the top of your shoe.

13.2 Partial Derivatives

The central idea of differential calculus is the derivative. A change in x produces a change in f. The ratio $\Delta f/\Delta x$ approaches the derivative, or slope, or rate of change. What to do if f depends on both x and y?

The new idea is to vary x and y one at a time. First, only x moves. If the function is $x + xy$, then Δf is $\Delta x + y\Delta x$. The ratio $\Delta f/\Delta x$ is $1 + y$. The "x derivative" of $x + xy$ is $1 + y$. For all functions the method is the same: **Keep y constant, change x, take the limit of $\Delta f/\Delta x$:**

DEFINITION $\qquad \dfrac{\partial f}{\partial x}(x, y) = \lim\limits_{\Delta x \to 0} \dfrac{\Delta f}{\Delta x} = \lim\limits_{\Delta x \to 0} \dfrac{f(x + \Delta x, y) - f(x, y)}{\Delta x}.$ $\qquad$ (1)

On the left is a new symbol $\partial f/\partial x$. It signals that only x is allowed to vary—$\partial f/\partial x$ is a **partial derivative**. The different form ∂ of the same letter (still say "d") is a reminder that x is not the only variable. Another variable y is present but not moving.

EXAMPLE 1 $\quad f(x, y) = x^2 y^2 + xy + y \qquad \dfrac{\partial f}{\partial x}(x, y) = 2xy^2 + y + 0.$

Do not treat y as zero! Treat it as a constant, like 6. Its x derivative is zero. If $f(x) = \sin 6x$ then $df/dx = 6\cos 6x$. If $f(x, y) = \sin xy$ then $\partial f/\partial x = y\cos xy$.

Spoken aloud, $\partial f/\partial x$ is still "$df\,dx$." It is a function of x and y. When more is needed, call it "the partial of f with respect to x." The symbol f' is no longer available, since it gives no special indication about x. Its replacement f_x is pronounced "$f\,x$" or "f sub x," which is shorter than $\partial f/\partial x$ and means the same thing.

We may also want to indicate the point (x_0, y_0) where the derivative is computed:

$$\frac{\partial f}{\partial x}(x_0, y_0) \quad \text{or} \quad f_x(x_0, y_0) \quad \text{or} \quad \left.\frac{\partial f}{\partial x}\right|_{(x_0, y_0)} \quad \text{or just} \quad \left(\frac{\partial f}{\partial x}\right)_0.$$

EXAMPLE 2 $\quad f(x, y) = \sin 2x \cos y \qquad f_x = 2\cos 2x \cos y \quad (\cos y \text{ is constant for } \partial/\partial x)$

The particular point (x_0, y_0) is $(0,0)$. The height of the surface is $f(0,0) = 0$. The slope in the x direction is $f_x = 2$. At a different point $x_0 = \pi$, $y_0 = \pi$ we find $f_x(\pi, \pi) = -2$.

Now keep x constant and vary y. The ratio $\Delta f/\Delta y$ approaches $\partial f/\partial y$:

$$f_y(x, y) = \lim\limits_{\Delta y \to 0} \frac{\Delta f}{\Delta y} = \lim\limits_{\Delta y \to 0} \frac{f(x, y + \Delta y) - f(x, y)}{\Delta y}. \qquad (2)$$

This is the slope in the y direction. Please realize that a surface can go up in the x direction and down in the y direction. The plane $f(x, y) = 3x - 4y$ has $f_x = 3$ (up) and $f_y = -4$ (down). We will soon ask what happens in the $45°$ direction.

EXAMPLE 3 $\quad f(x, y) = \sqrt{x^2 + y^2} \qquad \dfrac{\partial f}{\partial x} = \dfrac{x}{\sqrt{x^2 + y^2}} = \dfrac{x}{f} \qquad \dfrac{\partial f}{\partial y} = \dfrac{y}{\sqrt{x^2 + y^2}} = \dfrac{y}{f}.$

The x derivative of $\sqrt{x^2 + y^2}$ is really one-variable calculus, because y is constant. The exponent drops from $\frac{1}{2}$ to $-\frac{1}{2}$, and there is $2x$ from the chain rule. **This distance function has the curious derivative** $\partial f/\partial x = x/f$.

The graph is a cone. Above the point $(0,2)$ the height is $\sqrt{0^2+2^2}=2$. The partial derivatives are $f_x=0/2$ and $f_y=2/2$. At that point, Figure 13.5 climbs in the y direction. It is level in the x direction. An actual step Δx will increase 0^2+2^2 to $(\Delta x)^2+2^2$. But this change is of order $(\Delta x)^2$ and the x derivative is zero.

Figure 13.5 is rather important. It shows how $\partial f/\partial x$ and $\partial f/\partial y$ are the ordinary derivatives of $f(x,y_0)$ and $f(x_0,y)$. It is natural to call these **partial functions**. The first has y fixed at y_0 while x varies. The second has x fixed at x_0 while y varies. Their graphs are **cross sections down the surface** —cut out by the vertical planes $y=y_0$ and $x=x_0$. Remember that the level curve is cut out by the horizontal plane $z=c$.

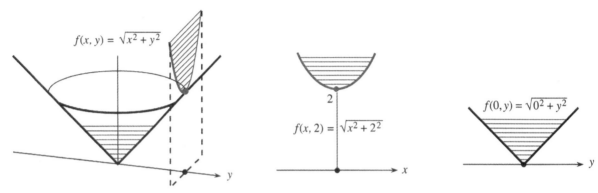

Fig. 13.5   Partial functions $\sqrt{x^2+2^2}$ and $\sqrt{0^2+y^2}$ of the distance function $f=\sqrt{x^2+y^2}$.

The limits of $\Delta f/\Delta x$ and $\Delta f/\Delta y$ are computed as always. With partial functions we are back to a single variable. **The partial derivative is the ordinary derivative of a partial function** (constant y or constant x). For the cone, $\partial f/\partial y$ exists at all points except $(0,0)$. The figure shows how the cross section down the middle of the cone produces the absolute value function: $f(0,y)=|y|$. It has one-sided derivatives but not a two-sided derivative.

Similarly $\partial f/\partial x$ will not exist at the sharp point of the cone. We develop the idea of a *continuous function* $f(x,y)$ as needed (the definition is in the exercises). Each partial derivative involves one direction, but limits and continuity involve all directions. The distance function is continuous at $(0,0)$, where it is not differentiable.

EXAMPLE 4 $f(x,y)=y^2-x^2$ $\partial f/\partial x=-2x$ $\partial f/\partial y=2y$

Move in the x direction from $(1,3)$. Then y^2-x^2 has the partial function $9-x^2$. With y fixed at 3, a parabola opens downward. In the y direction (along $x=1$) the partial function y^2-1 opens upward. The surface in Figure 13.6 is called a *hyperbolic paraboloid*, because the level curves $y^2-x^2=c$ are hyperbolas. Most people call it a saddle, and the special point at the origin is a **saddle point**. The origin is special for y^2-x^2 because both derivatives are zero. **The bottom of the y parabola at $(0,0)$ is the top of the x parabola**. The surface is momentarily flat in all directions. It is the top of a hill and the bottom of a mountain range at the same time. A saddle point is neither a maximum nor a minimum, although both derivatives are zero.

Note   Do not think that $f(x,y)$ must contain y^2 and x^2 to have a saddle point. The function $2xy$ does just as well. The level curves $2xy=c$ are still hyperbolas. The partial functions $2xy_0$ and $2x_0y$ now give straight lines—which is remarkable. Along the $45°$ line $x=y$, the function is $2x^2$ and climbing. Along the $-45°$ line $x=-y$, the function is $-2x^2$ and falling. The graph of $2xy$ is Figure 13.6 rotated by $45°$.

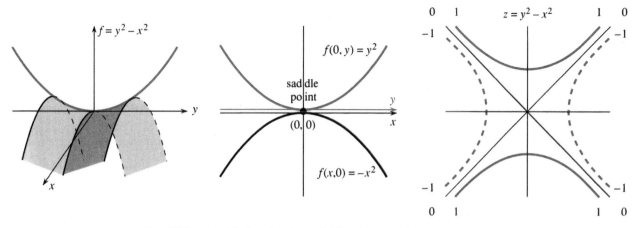

Fig. 13.6 A saddle function, its partial functions, and its level curves.

EXAMPLES 5–6 $\quad f(x,y,z) = x^2 + y^2 + z^2 \quad P(T,V) = nRT/V$

Example 5 shows more variables. Example 6 shows that the variables may not be named x and y. Also, the function may not be named f! Pressure and temperature and volume are P and T and V. The letters change but nothing else:

$$\partial P/\partial T = nR/V \quad \partial P/\partial V = -nRT/V^2 \quad \text{(note the derivative of } 1/V\text{).}$$

There is no $\partial P/\partial R$ because R is a constant from chemistry—not a variable.

Physics produces six variables for a moving body—the coordinates x, y, z and the momenta p_x, p_y, p_z. Economics and the social sciences do better than that. If there are 26 products there are 26 variables—sometimes 52, to show prices as well as amounts. The profit can be a complicated function of these variables. *The partial derivatives are the marginal profits*, as one of the 52 variables is changed. A spreadsheet shows the 52 values and the effect of a change. An infinitesimal spreadsheet shows the derivative.

SECOND DERIVATIVE

Genius is not essential, to move to second derivatives. The only difficulty is that *two* first derivatives f_x and f_y lead to *four* second derivatives f_{xx} and f_{xy} and f_{yx} and f_{yy}. (Two subscripts: f_{xx} is the x derivative of the x derivative. Other notations are $\partial^2 f/\partial x^2$ and $\partial^2 f/\partial x\partial y$ and $\partial^2 f/\partial y\partial x$ and $\partial^2 f/\partial y^2$.) Fortunately f_{xy} equals f_{yx}, as we see first by example.

EXAMPLE 7 $\quad f = x/y$ has $f_x = 1/y$, which has $f_{xx} = 0$ and $f_{xy} = -1/y^2$.

The function x/y is linear in x (which explains $f_{xx} = 0$). Its y derivative is $f_y = -x/y^2$. This has the x derivative $f_{yx} = -1/y^2$. *The mixed derivatives f_{xy} and f_{yx} are equal*.

In the pure y direction, the second derivative is $f_{yy} = 2x/y^3$. One-variable calculus is sufficient for all these derivatives, because only one variable is moving.

EXAMPLE 8 $f = 4x^2 + 3xy + y^2$ has $f_x = 8x + 3y$ and $f_y = 3x + 2y$.

Both "*cross derivatives*" f_{xy} and f_{yx} equal 3. The second derivative in the x direction is $\partial^2 f/\partial x^2 = 8$ or $f_{xx} = 8$. Thus "$f\,xx$" is "d second $f\,d\,x$ squared." Similarly $\partial^2 f/\partial y^2 = 2$. The only change is from d to ∂.

If $f(x, y)$ has continuous second derivatives then $f_{xy} = f_{yx}$. Problem 43 sketches a proof based on the Mean Value Theorem. For third derivatives almost any example shows that $f_{xxy} = f_{xyx} = f_{yxx}$ is different from $f_{yyx} = f_{yxy} = f_{xyy}$.

Question *How do you plot a space curve $x(t), y(t), z(t)$ in a plane?* One way is to look parallel to the direction $(1,1,1)$. On your XY screen, plot $X = (y - x)/\sqrt{2}$ and $Y = (2z - x - y)/\sqrt{6}$. The line $x = y = z$ goes to the point $(0,0)$!

How do you graph a surface $z = f(x, y)$? Use the same X and Y. Fix x and let y vary, for curves one way in the surface. Then fix y and vary x, for the other partial function. For a parametric surface like $x = (2 + v\sin\frac{1}{2}u)\cos u$, $y = (2 + v\sin\frac{1}{2}u)\sin u$, $z = v\cos\frac{1}{2}u$, vary u and then v. Dick Williamson showed how this draws a one-sided "Möbius strip."

13.2 EXERCISES

Read-through questions

The __a__ derivative $\partial f/\partial y$ comes from fixing __b__ and moving __c__. It is the limit of __d__. If $f = e^{2x}\sin y$ then $\partial f/\partial x =$ __e__ and $\partial f/\partial y =$ __f__. If $f = (x^2 + y^2)^{1/2}$ then $f_x =$ __g__ and $f_y =$ __h__. At (x_0, y_0) the partial derivative f_x is the ordinary derivative of the __i__ function $f(x, y_0)$. Similarly f_y comes from $f($ __j__ $)$. Those functions are cut out by vertical planes $x = x_0$ and __k__, while the level curves are cut out by __l__ planes.

The four second derivatives are f_{xx}, __m__, __n__, __o__. For $f = xy$ they are __p__. For $f = \cos 2x \cos 3y$ they are __q__. In those examples the derivatives __r__ and __s__ are the same. That is always true when the second derivatives are __t__. At the origin, $\cos 2x \cos 3y$ is curving __u__ in the x and y directions, while xy goes __v__ in the 45° direction and __w__ in the −45° direction.

Find $\partial f/\partial x$ and $\partial f/\partial y$ for the functions in 1–12.

1 $3x - y + x^2 y^2$ 2 $\sin(3x - y) + y$
3 $x^3 y^2 - x^2 - e^y$ 4 xe^{x+4}
5 $(x + y)/(x - y)$ 6 $1/\sqrt{x^2 + y^2}$
7 $(x^2 + y^2)^{-1}$ 8 $\ln(x + 2y)$
9 $\ln\sqrt{x^2 + y^2}$ 10 y^x
11 $\tan^{-1}(y/x)$ 12 $\ln(xy)$

Compute $f_{xx}, f_{xy} = f_{yx}$, and f_{yy} for the functions in 13–20.

13 $x^2 + 3xy + 2y^2$ 14 $(x + 3y)^2$
15 $(x + iy)^3$ 16 e^{ax+by}
17 $1/\sqrt{x^2 + y^2}$ 18 $(x + y)^n$
19 $\cos ax \cos by$ 20 $1/(x + iy)$

Find the domain and range (all inputs and outputs) for the functions 21–26. Then compute f_x, f_y, f_z, f_t.

21 $1/(x - y)^2$ 22 $\sqrt{x^2 + y^2 - t^2}$
23 $(y - x)/(z - t)$ 24 $\ln(x + t)$
25 $x^{\ln t}$ *Why does this equal $t^{\ln x}$?*   26 $\cos x \cos^{-1} y$

27 Verify $f_{xy} = f_{yx}$ for $f = x^m y^n$. If $f_{xy} = 0$ then f_x does not depend on ____ and f_y is independent of ____. The function must have the form $f(x, y) = G(x) +$ ____.

28 In terms of v, compute f_x and f_y for $f(x, y) = \int_x^y v(t)\,dt$. First vary x. Then vary y.

29 Compute $\partial f/\partial x$ for $f = \int_0^{xy} v(t)\,dt$. Keep y constant.

30 What is $f(x, y) = \int_x^y dt/t$ and what are f_x and f_y?

31 Calculate all eight third derivatives $f_{xxx}, f_{xxy}, \dots$ of $f = x^3 y^3$. How many are different?

In 32–35, choose $g(y)$ so that $f(x, y) = e^{cx} g(y)$ satisfies the equation.

32 $f_x + f_y = 0$ 33 $f_x = 7f_y$
34 $f_y = f_{xx}$ 35 $f_{xx} = 4f_{yy}$

36 Show that $t^{-1/2}e^{-x^2/4t}$ satisfies the **heat equation** $f_t = f_{xx}$. This $f(x,t)$ is the temperature at position x and time t due to a point source of heat at $x = 0, t = 0$.

37 The equation for heat flow in the xy plane is $f_t = f_{xx} + f_{yy}$. Show that $f(x,y,t) = e^{-2t}\sin x \sin y$ is a solution. What exponent in $f = e^{\underline{\quad}}\sin 2x \sin 3y$ gives a solution?

38 Find solutions $f(x,y) = e^{\underline{\quad}}\sin mx \cos ny$ of the heat equation $f_t = f_{xx} + f_{yy}$. Show that $t^{-1}e^{-x^2/4t}e^{-y^2/4t}$ is also a solution.

39 The basic **wave equation** is $f_{tt} = f_{xx}$. Verify that $f(x,t) = \sin(x+t)$ and $f(x,t) = \sin(x-t)$ are solutions. Draw both graphs at $t = \pi/4$. Which wave moved to the left and which moved to the right?

40 Continuing 39, the peaks of the waves moved a distance $\Delta x = \underline{\quad}$ in the time step $\Delta t = \pi/4$. The wave velocity is $\Delta x/\Delta t = \underline{\quad}$.

41 Which of these satisfy the wave equation $f_{tt} = c^2 f_{xx}$?

$$\sin(x-ct), \quad \cos(x+ct), \quad e^{x-ct}, \quad e^x - e^{ct}, \quad e^x \cos ct.$$

42 Suppose $\partial f/\partial t = \partial f/\partial x$. Show that $\partial^2 f/\partial t^2 = \partial^2 f/\partial x^2$.

43 The proof of $f_{xy} = f_{yx}$ studies $f(x,y)$ in a small rectangle. The top-bottom difference is $g(x) = f(x,B) - f(x,A)$. The difference at the corners 1, 2, 3, 4 is:

$$
\begin{aligned}
Q &= [f_4 - f_3] - [f_2 - f_1] \\
&= g(b) - g(a) \quad \text{(definition of } g\text{)} \\
&= (b-a)g_x(c) \quad \text{(Mean Value Theorem)} \\
&= (b-a)[f_x(c,B) - f_x(c,A)] \quad \text{(compute } g_x\text{)} \\
&= (b-a)(B-A)f_{xy}(c,C) \quad \text{(MVT again)}
\end{aligned}
$$

(a) The right-left difference is $h(y) = f(b,y) - f(a,y)$. The same Q is $h(B) - h(A)$. Change the steps to reach $Q = (B-A)(b-a)f_{yx}(c^*,C^*)$.

(b) The two forms of Q make f_{xy} at (c,C) equal to f_{yx} at (c^*,C^*). Shrink the rectangle toward (a,A). What assumption yields $f_{xy} = f_{yx}$ at that typical point?

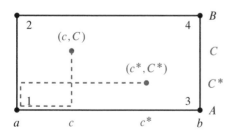

44 Find $\partial f/\partial x$ and $\partial f/\partial y$ where they exist, based on equations (1) and (2).

(a) $f = |xy|$ (b) $f = x^2 + y^2$ if $x \neq 0$, $f = 0$ if $x = 0$

Questions 45–52 are about limits in two dimensions.

45 Complete these *four correct definitions of limit*: **1** The points (x_n, y_n) approach the point (a,b) if x_n converges to a and $\underline{\quad}$. **2** For any circle around (a,b), the points (x_n, y_n) eventually go $\underline{\quad}$ the circle and stay $\underline{\quad}$. **3** The distance from (x_n, y_n) to (a,b) is $\underline{\quad}$ and it approaches $\underline{\quad}$. **4** For any $\varepsilon > 0$ there is an N such that the distance $\underline{\quad} < \varepsilon$ for all $n > \underline{\quad}$.

46 Find (x_2, y_2) and (x_4, y_4) and the limit (a,b) if it exists. Start from $(x_0, y_0) = (1,0)$.

(a) $(x_n, y_n) = (1/(n+1), n/(n+1))$
(b) $(x_n, y_n) = (x_{n-1}, y_{n-1})$
(c) $(x_n, y_n) = (y_{n-1}, x_{n-1})$
(d) $(x_n, y_n) = (x_{n-1} + y_{n-1}, x_{n-1} - y_{n-1})$

47 (*Limit of $f(x,y)$*) **1** Informal definition: the numbers $f(x_n, y_n)$ approach L when the points (x_n, y_n) approach (a,b). **2** *Epsilon-delta definition* : For each $\varepsilon > 0$ there is a $\delta > 0$ such that $|f(x,y) - L|$ is less than $\underline{\quad}$ when the distance from (x,y) to (a,b) is $\underline{\quad}$. The value of f at (a,b) is not involved.

48 Write down the limit L as $(x,y) \to (a,b)$. At which points (a,b) does $f(x,y)$ have no limit?

(a) $f(x,y) = \sqrt{x^2 + y^2}$ (b) $f(x,y) = x/y$

(c) $f(x,y) = 1/(x+y)$ (d) $f(x,y) = xy/(x^2 + y^2)$

In (d) find the limit at $(0,0)$ along the line $y = mx$. The limit changes with m, so L does not exist at $(0,0)$. Same for x/y.

49 *Definition of continuity*: $f(x,y)$ is continuous at (a,b) if $f(a,b)$ is defined and $f(x,y)$ approaches the limit $\underline{\quad}$ as (x,y) approaches (a,b). Construct a function that is *not* continuous at $(1,2)$.

50 Show that $x^2y/(x^4 + y^2) \to 0$ along every straight line $y = mx$ to the origin. But traveling down the parabola $y = x^2$, the ratio equals $\underline{\quad}$.

51 Can you define $f(0,0)$ so that $f(x,y)$ is continuous at $(0,0)$?
(a) $f = |x| + |y - 1|$ (b) $f = (1+x)^y$ (c) $f = x^{1+y}$.

52 Which functions approach zero as $(x,y) \to (0,0)$ and why ?

(a) $\dfrac{xy^2}{x^2 + y^2}$ (b) $\dfrac{x^2y^2}{x^4 + y^4}$ (c) $\dfrac{x^m y^n}{x^m + y^n}$.

13.3 Tangent Planes and Linear Approximations

Over a short range, a smooth curve $y = f(x)$ is almost straight. The curve changes direction, but the tangent line $y - y_0 = f'(x_0)(x - x_0)$ keeps the same slope forever. The tangent line immediately gives the linear approximation to $y = f(x)$: $y \approx y_0 + f'(x_0)(x - x_0)$.

What happens with two variables? The function is $z = f(x, y)$, and its graph is a **surface**. We are at a point on that surface, and we are near-sighted. We don't see far away. The surface may curve out of sight at the horizon, or it may be a bowl or a saddle. To our myopic vision, the surface looks flat. We believe we are on a plane (not necessarily horizontal), and we want the equation of this **tangent plane**.

Notation The basepoint has coordinates x_0 and y_0. The height on the surface is $z_0 = f(x_0, y_0)$. Other letters are possible: the point can be (a, b) with height w. The subscript $_0$ indicates the value of x or y or z or $\partial f / \partial x$ or $\partial f / \partial y$ *at the point.*

With one variable the tangent line has slope df / dx. With two variables there are two derivatives $\partial f / \partial x$ and $\partial f / \partial y$. At the particular point, they are $(\partial f / \partial x)_0$ and $(\partial f / \partial y)_0$. **Those are the slopes of the tangent plane** . Its equation is the key to this chapter:

13A The tangent plane at (x_0, y_0, z_0) has the same slopes as the surface $z = f(x, y)$. The equation of the tangent plane (a linear equation) is

$$z - z_0 = \left(\frac{\partial f}{\partial x}\right)_0 (x - x_0) + \left(\frac{\partial f}{\partial y}\right)_0 (y - y_0). \tag{1}$$

The normal vector $\mathbf{N}$ to that plane has components $(\partial f / \partial x)_0$, $(\partial f / \partial y)_0$, -1.

EXAMPLE 1   Find the tangent plane to $z = 14 - x^2 - y^2$ at $(x_0, y_0, z_0) = (1, 2, 9)$.

Solution   The derivatives are $\partial f / \partial x = -2x$ and $\partial f / \partial y = -2y$. When $x = 1$ and $y = 2$ those are $(\partial f / \partial x)_0 = -2$ and $(\partial f / \partial y)_0 = -4$. The equation of the tangent plane is

$$z - 9 = -2(x - 1) - 4(y - 2) \qquad \text{or} \qquad z + 2x + 4y = 19. \tag{2}$$

This $z(x, y)$ has derivatives -2 and -4, just like the surface. So the plane is tangent.

The normal vector $\mathbf{N}$ has components $-2, -4, -1$. **The equation of the normal line is** $(x, y, z) = (1, 2, 9) + t(-2, -4, -1)$. Starting from $(1, 2, 9)$ the line goes out along $\mathbf{N}$—perpendicular to the plane and the surface.

Figure 13.7 shows more detail about the tangent plane. The dotted lines are the x and y tangent lines. They lie in the plane. All tangent lines lie in the tangent plane! These particular lines are tangent to the "partial functions"—where y is fixed at $y_0 = 2$ or x is fixed at $x_0 = 1$. The plane is balancing on the surface and touching at the tangent point.

More is true. In the surface, *every curve through the point is tangent to the plane.* Geometrically, the curve goes up to the point and "kisses" the plane.† The tangent $\mathbf{T}$ to the curve and the normal $\mathbf{N}$ to the surface are perpendicular: $\mathbf{T} \cdot \mathbf{N} = 0$.

EXAMPLE 2   Find the tangent plane to the sphere $z^2 = 14 - x^2 - y^2$ at $(1, 2, 3)$.

---
†A safer word is "osculate." At saddle points the plane is kissed from both sides.

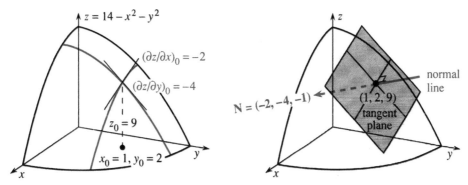

Fig. 13.7   The tangent plane contains the x and y tangent lines, perpendicular to **N**.

Solution   Instead of $z = 14 - x^2 - y^2$ we have $z = \sqrt{14 - x^2 - y^2}$. At $x_0 = 1$, $y_0 = 2$ the height is now $z_0 = 3$. The surface is a sphere with radius $\sqrt{14}$. The only trouble from the square root is its derivatives:

$$\frac{\partial z}{\partial x} = \frac{\partial}{\partial x}\sqrt{14 - x^2 - y^2} = \frac{\frac{1}{2}(-2x)}{\sqrt{14 - x^2 - y^2}} \quad \text{and} \quad \frac{\partial z}{\partial y} = \frac{\frac{1}{2}(-2y)}{\sqrt{14 - x^2 - y^2}} \qquad (3)$$

At $(1,2)$ those slopes are $-\frac{1}{3}$ and $-\frac{2}{3}$. The equation of the tangent plane is linear: $z - 3 = -\frac{1}{3}(x - 1) - \frac{2}{3}(y - 2)$. I cannot resist improving the equation, by multiplying through by 3 and moving all terms to the left side:

$$\textit{tangent plane to sphere}: \quad 1(x - 1) + 2(y - 2) + 3(z - 3) = 0. \qquad (4)$$

If mathematics is the "science of patterns," equation (4) is a prime candidate for study. The numbers $1, 2, 3$ appear twice. The coordinates are $(x_0, y_0, z_0) = (1, 2, 3)$. The normal vector is $1\mathbf{i} + 2\mathbf{j} + 3\mathbf{k}$. The tangent equation is $1x + 2y + 3z = 14$. None of this can be an accident, but the square root of $14 - x^2 - y^2$ made a simple pattern look complicated.

  This square root is not necessary. Calculus offers a direct way to find dz/dx—*implicit differentiation*. Just differentiate every term as it stands:

$$x^2 + y^2 + z^2 = 14 \quad \text{leads to} \quad 2x + 2z\,\partial z/\partial x = 0 \quad \text{and} \quad 2y + 2z\,\partial z/\partial y = 0. \quad (5)$$

Canceling the 2's, the derivatives on a sphere are $-x/z$ and $-y/z$. Those are the same as in (3). The equation for the tangent plane has an extremely symmetric form:

$$z - z_0 = -\frac{x_0}{z_0}(x - x_0) - \frac{y_0}{z_0}(y - y_0) \quad \text{or} \quad x_0(x - x_0) + y_0(y - y_0) + z_0(z - z_0) = 0. \quad (6)$$

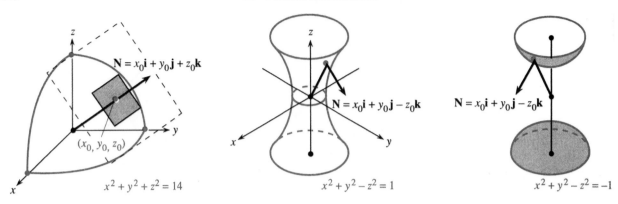

Fig. 13.8   Tangent plane and normal **N** for a sphere. Hyperboloids of 1 and 2 sheets.

Reading off $\mathbf{N} = x_0\mathbf{i} + y_0\mathbf{j} + z_0\mathbf{k}$ from the last equation, calculus proves something we already knew: **The normal vector to a sphere points outward along the radius**.

THE TANGENT PLANE TO $F(x, y, z) = c$

The sphere suggests a question that is important for other surfaces. Suppose the equation is $F(x, y, z) = c$ instead of $z = f(x, y)$. Can the partial derivatives and tangent plane be found directly from F?

The answer is *yes*. It is not necessary to solve first for z. The derivatives of F, computed at (x_0, y_0, z_0), give a second formula for the tangent plane and normal vector.

13B The tangent plane to the surface $F(x, y, z) = c$ has the linear equation

$$\left(\frac{\partial F}{\partial x}\right)_0 (x - x_0) + \left(\frac{\partial F}{\partial y}\right)_0 (y - y_0) + \left(\frac{\partial F}{\partial z}\right)_0 (z - z_0) = 0. \qquad (7)$$

The normal vector is $\mathbf{N} = \left(\dfrac{\partial F}{\partial x}\right)_0 \mathbf{i} + \left(\dfrac{\partial F}{\partial y}\right)_0 \mathbf{j} + \left(\dfrac{\partial F}{\partial z}\right)_0 \mathbf{k}$.

Notice how this includes the original case $z = f(x, y)$. **The function F becomes** $f(x, y) - z$. Its partial derivatives are $\partial f / \partial x$ and $\partial f / \partial y$ and -1. (The -1 is from the derivative of $-z$.) Then equation (7) is the same as our original tangent equation (1).

EXAMPLE 3   The surface $F = x^2 + y^2 - z^2 = c$ is a **hyperboloid**. Find its tangent plane.

Solution   The partial derivatives are $F_x = 2x, F_y = 2y, F_z = -2z$. Equation (7) is

$$\textbf{\textit{tangent plane}}: \qquad 2x_0(x - x_0) + 2y_0(y - y_0) - 2z_0(z - z_0) = 0. \qquad (8)$$

We can cancel the 2's. The normal vector is $\mathbf{N} = x_0\mathbf{i} + y_0\mathbf{j} - z_0\mathbf{k}$. For $c > 0$ *this hyperboloid has one sheet* (Figure 13.8). For $c = 0$ it is a cone and for $c < 0$ it breaks into two sheets (Problem 13.1.26).

DIFFERENTIALS

Come back to the linear equation $z - z_0 = (\partial z/\partial x)_0(x - x_0) + (\partial z/\partial y)_0(y - y_0)$ for the tangent plane. That may be the most important formula in this chapter. Move along the tangent plane instead of the curved surface. Movements in the plane are dx and dy and dz—while Δx and Δy and Δz are movements in the surface. The d's are governed by the tangent equation—the Δ's are governed by $z = f(x, y)$. In Chapter 2 the d's were **differentials** along the tangent line:

$$dy = (dy/dx)dx \quad \text{(straight line)} \quad \text{and} \quad \Delta y \approx (dy/dx)\Delta x \quad \text{(on the curve).} \qquad (9)$$

Now y is independent like x. The dependent variable is z. The idea is the same. The distances $x - x_0$ and $y - y_0$ and $z - z_0$ (on the tangent plane) are dx and dy and dz. *The equation of the plane is*

$$dz = (\partial z/\partial x)_0 dx + (\partial z/\partial y)_0 dy \quad \text{or} \quad df = f_x dx + f_y dy. \qquad (10)$$

This is the **total differential**. All letters dz and df and dw can be used, but ∂z and ∂f are not used. Differentials suggest small movements in x and y; then dz is the resulting movement in z. On the tangent plane, equation (10) holds exactly.

A "centering transform" has put x_0, y_0, z_0 at the center of coordinates. Then the "zoom transform" stretches the surface into its tangent plane.

EXAMPLE 4   The area of a triangle is $A = \frac{1}{2}ab \sin\theta$. Find the total differential dA.

Solution   The base has length b and the sloping side has length a. The angle between them is θ. You may prefer $A = \frac{1}{2}bh$, where h is the perpendicular height $a \sin\theta$. Either way we need the partial derivatives. If $A = \frac{1}{2}ab\sin\theta$, then

$$\frac{\partial A}{\partial a} = \frac{1}{2}b\sin\theta \qquad \frac{\partial A}{\partial b} = \frac{1}{2}a\sin\theta \qquad \frac{\partial A}{\partial\theta} = \frac{1}{2}ab\cos\theta. \qquad (11)$$

These lead immediately to the total differential dA (like a product rule):

$$dA = \left(\frac{\partial A}{\partial a}\right)da + \left(\frac{\partial A}{\partial b}\right)db + \left(\frac{\partial A}{\partial\theta}\right)d\theta = \frac{1}{2}b\sin\theta\, da + \frac{1}{2}a\sin\theta\, db + \frac{1}{2}ab\cos\theta\, d\theta.$$

EXAMPLE 5   The volume of a cylinder is $V = \pi r^2 h$. Decide whether V is more sensitive to a change from $r = 1.0$ to $r = 1.1$ or from $h = 1.0$ to $h = 1.1$.

Solution   The partial derivatives are $\partial V/\partial r = 2\pi rh$ and $\partial V/\partial h = \pi r^2$. **They measure the sensitivity to change**. Physically, they are the side area and base area of the cylinder. The volume differential dV comes from a shell around the side plus a layer on top:

$$dV = \text{shell} + \text{layer} = 2\pi rh\, dr + \pi r^2 dh. \qquad (12)$$

Starting from $r = h = 1$, that differential is $dV = 2\pi dr + \pi dh$. With $dr = dh = .1$, the shell volume is $.2\pi$ and the layer volume is only $.1\pi$. So V is sensitive to dr.

For a short cylinder like a penny, the layer has greater volume. V is more sensitive to dh. In our case $V = \pi r^2 h$ increases from $\pi(1)^3$ to $\pi(1.1)^3$. **Compare ΔV to dV:**

$$\Delta V = \pi(1.1)^3 - \pi(1)^3 = .331\pi \qquad \text{and} \qquad dV = 2\pi(.1) + \pi(.1) = .300\pi.$$

The difference is $\Delta V - dV = .031\pi$. The shell and layer missed a small volume in Figure 13.9, just above the shell and around the layer. The mistake is of order $(dr)^2 + (dh)^2$. For $V = \pi r^2 h$, the differential $dV = 2\pi rh\, dr + \pi r^2 dh$ is a **linear approximation** to the true change ΔV. We now explain that properly.

LINEAR APPROXIMATION

Tangents lead immediately to linear approximations. That is true of tangent planes as it was of tangent lines. The plane stays close to the surface, as the line stayed close to the curve. Linear functions are simpler than $f(x)$ or $f(x, y)$ or $F(x, y, z)$. All we need are first derivatives *at the point*. Then the approximation is good *near the point*.

This key idea of calculus is already present in differentials. On the plane, df equals $f_x dx + f_y dy$. On the curved surface that is a linear approximation to Δf:

13C The linear approximation to $f(x, y)$ near the point (x_0, y_0) is

$$f(x, y) \approx f(x_0, y_0) + \left(\frac{\partial f}{\partial x}\right)_0 (x - x_0) + \left(\frac{\partial f}{\partial y}\right)_0 (y - y_0). \qquad (13)$$

In other words $\Delta f \approx f_x \Delta x + f_y \Delta y$, as proved in Problem 24. The right side of (13) is a linear function $f_L(x, y)$. At (x_0, y_0), the functions f and f_L have the same slopes. Then $f(x, y)$ curves away from f_L with an error of "second order:"

$$|f(x, y) - f_L(x, y)| \leqslant M[(x - x_0)^2 + (y - y_0)^2]. \qquad (14)$$

This assumes that f_{xx}, f_{xy}, and f_{yy} are continuous and bounded by M along the line from (x_0, y_0) to (x, y). Example 3 of Section 13.5 shows that $|f_{tt}| \leqslant 2M$ along that line. A factor $\frac{1}{2}$ comes from equation 3.8.12, for the error $f - f_L$ with one variable.

For the volume of a cylinder, r and h went from 1.0 to 1.1. The second derivatives of $V = \pi r^2 h$ are $V_{rr} = 2\pi h$ and $V_{rh} = 2\pi r$ and $V_{hh} = 0$. They are below $M = 2.2\pi$. Then (14) gives the error bound $2.2\pi(.1^2 + .1^2) = .044\pi$, not far above the actual error $.031\pi$. The main point is that *the error in linear approximation comes from the quadratic terms*—those are the first terms to be ignored by f_L.

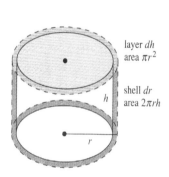

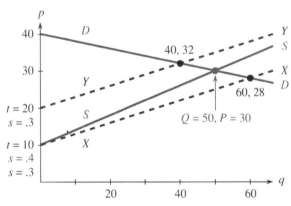

Fig. 13.9   Shell plus layer gives $dV = .300\pi$. Including top ring gives $\Delta V = .331\pi$.

Fig. 13.10   Quantity Q and price P move with the lines.

EXAMPLE 6   Find a linear approximation to the distance function $r = \sqrt{x^2 + y^2}$.

Solution   The partial derivatives are x/r and y/r. Then $\Delta r \approx (x/r)\Delta x + (y/r)\Delta y$.

For (x, y, r) near $(1, 2, \sqrt{5})$: $\sqrt{x^2 + y^2} \approx \sqrt{1^2 + 2^2} + (x - 1)/\sqrt{5} + 2(y - 2)/\sqrt{5}$.

If y is fixed at 2, this is a one-variable approximation to $\sqrt{x^2 + 2^2}$. If x is fixed at 1, it is a linear approximation in y. Moving both variables, you might think dr would involve dx and dy in a square root. It doesn't. Distance involves x and y in a square root, but: *change of distance* is linear in Δx and Δy—to a first approximation.

There is a rough point at $x = 0, y = 0$. Any movement from $(0,0)$ gives $\Delta r = \sqrt{(\Delta x)^2 + (\Delta y)^2}$. The square root has returned. The reason is that **the partial derivatives x/r and y/r are not continuous at** $(0,0)$. The cone has a sharp point with no tangent plane. *Linear approximation breaks down.*

The next example shows how to approximate Δz from Δx and Δy, when the equation is $F(x, y, z) = c$. We use the implicit derivatives in (7) instead of the explicit derivatives in (1). The idea is the same: Look at the tangent equation as a way to find Δz, instead of an equation for z. Here is Example 6 with new letters.

EXAMPLE 7 From $F = -x^2 - y^2 + z^2 = 0$ find a linear approximation to Δz.

Solution (implicit derivatives) Use the derivatives of F: $-2x\Delta x - 2y\Delta y + 2z\Delta z \approx 0$. Then solve for Δz, which gives $\Delta z \approx (x/z)\Delta x + (y/z)\Delta y$—the same as Example 6.

EXAMPLE 8 How does the equilibrium price change when the supply curve changes?

The equilibrium price is at the intersection of the supply and demand curves (**supply = demand**). As the price p rises, the demand q drops (the slope is $-.2$):

$$\text{demand line } DD : p = -.2q + 40. \tag{15}$$

The supply (also q) goes *up* with the price. The slope s is positive (here $s = .4$):

$$\text{supply line } SS : p = sq + t = .4q + 10.$$

Those lines are in Figure 13.10. They meet at the **equilibrium price** $P = \$30$. The quantity $Q = 50$ is available at P (on SS) and demanded at P (on DD). So it is sold.

Where do partial derivatives come in? The reality is that those lines DD and SS are not fixed for all time. Technology changes, and competition changes, and the value of money changes. Therefore the lines move. Therefore the crossing point (Q, P) also moves. Please recognize that derivatives are hiding in those sentences.

Main point: **The equilibrium price P is a function of s and t.** Reducing s by better technology lowers the supply line to $p = .3q + 10$. The demand line has not changed. The customer is as eager or stingy as ever. But the price P and quantity Q are different. The new equilibrium is at $Q = 60$ and $P = \$28$, where the new line XX crosses DD.

If the technology is expensive, the supplier will raise t when reducing s. Line YY is $p = .3q + 20$. That gives a higher equilibrium $P = \$32$ at a lower quantity $Q = 40$—the demand was too weak for the technology.

Calculus question Find $\partial P/\partial s$ and $\partial P/\partial t$. The difficulty is that P is not given as a function of s and t. So take implicit derivatives of the supply = demand equations:

$$
\begin{aligned}
\text{supply} = \text{demand}: && P &= -.2Q + 40 = sQ + t & &(16) \\
s \text{ derivative}: && P_s &= -.2Q_s = sQ_s + Q & &(\text{note } t_s = 0) \\
t \text{ derivative}: && P_t &= -.2Q_t = sQ_t + 1 & &(\text{note } t_t = 1)
\end{aligned}
$$

Now substitute $s = .4, t = 10, P = 30, Q = 50$. That is the starting point, around which we are finding a linear approximation. The last two equations give $P_s = 50/3$ and

$P_t = 1/3$ (Problem 25). The linear approximation is

$$P = 30 + 50(s - .4)/3 + (t - 10)/3 \tag{17}$$

Comment This example turned out to be subtle (so is economics) . I hesitated before including it. The equations are linear and their derivatives are easy, but something in the problem is hard—there is no explicit formula for P. The function $P(s,t)$ is not known. Instead of a point on a surface, we are following the intersection of two lines. *The solution changes as the equation changes.***The derivative of the solution comes from the derivative of the equation**.

Summary The foundation of this section is equation (1) for the tangent plane. Every thing builds on that—total differential, linear approximation, sensitivity to small change. Later sections go on to the chain rule and "directional derivatives" and "gradients." The central idea of differential calculus is $\Delta f \approx f_x \Delta x + f_y \Delta y$.

NEWTON'S METHOD FOR TWO EQUATIONS

Linear approximation is used *to solve equations*. To find out where a function is zero, look first to see where its approximation is zero. To find out where a graph crosses the xy plane, look to see where its tangent plane crosses.

Remember Newton's method for $f(x) = 0$. The current guess is x_n. Around that point, $f(x)$ is close to $f(x_n) + (x - x_n)f'(x_n)$. This is zero at the next guess $x_{n+1} = x_n - f(x_n)/f'(x_n)$. That is where the tangent line crosses the x axis.

With two variables the idea is the same—but two unknowns x and y require two equations. We solve $g(x, y) = 0$ and $h(x, y) = 0$. Both functions have linear approximations that start from the current point (x_n, y_n)—where derivatives are computed:

$$g(x, y) \approx g(x_n, y_n) + (\partial g/\partial x)(x - x_n) + (\partial g/\partial y)(y - y_n)$$
$$h(x, y) \approx h(x_n, y_n) + (\partial h/\partial x)(x - x_n) + (\partial h/\partial y)(y - y_n). \tag{18}$$

The natural idea is to *set these approximations to zero*. That gives linear equations for $x - x_n$ and $y - y_n$. Those are the steps Δx and Δy that take us to the next guess in Newton's method:

13D Newton's method to solve $g(x, y) = 0$ and $h(x, y) = 0$ has linear equations for the steps Δx and Δy that go from (x_n, y_n) to (x_{n+1}, y_{n+1}):

$$\left(\frac{\partial g}{\partial x}\right)\Delta x + \left(\frac{\partial g}{\partial y}\right)\Delta y = -g(x_n, y_n) \text{ and } \left(\frac{\partial h}{\partial x}\right)\Delta x + \left(\frac{\partial h}{\partial y}\right)\Delta y = -h(x_n, y_n). \tag{19}$$

EXAMPLE 9 $g = x^3 - y = 0$ and $h = y^3 - x = 0$ have 3 solutions $(1, 1), (0, 0), (-1, -1)$.

I will start at different points (x_0, y_0). The next guess is $x_1 = x_0 + \Delta x, y_1 = y_0 + \Delta y$. It is of extreme interest to know which solution Newton's method will choose—if it converges at all. I made three small experiments.

1. Suppose $(x_0, y_0) = (2, 1)$. At that point $g = 2^3 - 1 = 7$ and $h = 1^3 - 2 = -1$. The derivatives are $g_x = 3x^2 = 12$, $g_y = -1$, $h_x = -1$, $h_y = 3y^2 = 3$. The steps Δx and Δy come from solving (19):

$$\begin{array}{llll} 12\Delta x - \Delta y = -7 & & \Delta x = -4/7 & & x_1 = x_0 + \Delta x = 10/7 \\ -\Delta x + 3\Delta y = +1 & \Rightarrow & \Delta y = +1/7 & \Rightarrow & y_1 = y_0 + \Delta y = 8/7. \end{array}$$

This new point $(10/7, 8/7)$ is closer to the solution at $(1, 1)$. The next point is $(1.1, 1.05)$ and convergence is clear. Soon convergence is fast.

2. Start at $(x_0, y_0) = (\frac{1}{2}, 0)$. There we find $g = 1/8$ and $h = -1/2$:

$$\begin{array}{llll} (3/4)\Delta x - \Delta y = -1/8 & & \Delta x = -1/2 & & x_1 = x_0 + \Delta x = 0 \\ -\Delta x + 0\Delta y = +1/2 & \Rightarrow & \Delta y = +1/4 & \Rightarrow & y_1 = y_0 + \Delta y = -1/4. \end{array}$$

Newton has jumped from $\left(\frac{1}{2}, 0\right)$ on the x axis to $\left(0, -\frac{1}{4}\right)$ on the y axis. The next step goes to $(1/32, 0)$, back on the x axis. We are in the "basin of attraction" of $(0, 0)$.

3. Now start further out the axis at $(1, 0)$, where $g = 1$ and $h = -1$:

$$\begin{array}{llll} 3\Delta x - \Delta y = -1 & & \Delta x = -1 & & x_1 = x_0 + \Delta x = 0 \\ -\Delta x + 0\Delta y = +1 & \Rightarrow & \Delta y = -2 & \Rightarrow & y_1 = y_0 + \Delta y = -2. \end{array}$$

Newton moves from $(1, 0)$ to $(0, -2)$ to $(16, 0)$. Convergence breaks down—the method blows up. This danger is ever-present, when we start far from a solution.

Please recognize that even a small computer will uncover amazing patterns. It can start from hundreds of points (x_0, y_0), and follow Newton's method. Each solution has a **basin of attraction**, containing all (x_0, y_0) leading to that solution. There is also a basin leading to infinity. The basins in Figure 13.11 are completely mixed together—a color figure shows them as **fractals**. The most extreme behavior is on the borderline between basins, when Newton can't decide which way to go. Frequently we see chaos.

Chaos is irregular movement that follows a definite rule. Newton's method determines an *iteration* from each point (x_n, y_n) to the next. In scientific problems it normally converges to the solution we want. (We start close enough.) But the computer makes it possible to study iterations from faraway points. This has created a new part of mathematics—so new that any experiments you do are likely to be original.

Section 3.7 found chaos when trying to solve $x^2 + 1 = 0$. But don't think Newton's method is a failure. On the contrary, it is the best method to solve nonlinear equations. The error is squared as the algorithm converges, because linear approximations have errors of order $(\Delta x)^2 + (\Delta y)^2$. Each step doubles the number of correct digits, *near the solution*. The example shows why it is important to be near.

Fig. 13.11   The basins of attraction to $(1,1)$, $(0,0)$, $(-1,-1)$, and infinity.

13.3 EXERCISES

Read-through questions

The tangent line to $y = f(x)$ is $y - y_0 = \underline{\ \ a\ \ }$. The tangent plane to $w = f(x, y)$ is $w - w_0 = \underline{\ \ b\ \ }$. The normal vector is $\mathbf{N} = \underline{\ \ c\ \ }$. For $w = x^3 + y^3$ the tangent equation at $(1, 1, 2)$ is $\underline{\ \ d\ \ }$. The normal vector is $\mathbf{N} = \underline{\ \ e\ \ }$. For a sphere, the direction of $\mathbf{N}$ is $\underline{\ \ f\ \ }$.

The surface given implicitly by $F(x, y, z) = c$ has tangent equation $(\partial F / \partial x)_0 (x - x_0) + \underline{\ \ g\ \ }$. For $xyz = 6$ at $(1, 2, 3)$ the tangent plane is $\underline{\ \ h\ \ }$. On that plane the differentials satisfy $\underline{\ \ i\ \ }\ dx + \underline{\ \ j\ \ }\ dy + \underline{\ \ k\ \ }\ dz = 0$. The differential of $z = f(x, y)$ is $dz = \underline{\ \ l\ \ }$. This holds exactly on the tangent plane, while $\Delta z \approx \underline{\ \ m\ \ }$ holds approximately on the $\underline{\ \ n\ \ }$. The height $z = 3x + 7y$ is more sensitive to a change in $\underline{\ \ o\ \ }$ than in x, because the partial derivative $\underline{\ \ p\ \ }$ is larger than $\underline{\ \ q\ \ }$.

The linear approximation to $f(x, y)$ is $f(x_0, y_0) + \underline{\ \ r\ \ }$. This is the same as $\Delta f \approx \underline{\ \ s\ \ }\ \Delta x + \underline{\ \ t\ \ }\ \Delta y$. The error is of order $\underline{\ \ u\ \ }$. For $f = \sin xy$ the linear approximation around $(0, 0)$ is $f_L = \underline{\ \ v\ \ }$. We are moving along the $\underline{\ \ w\ \ }$ instead of the $\underline{\ \ x\ \ }$. When the equation is given as $F(x, y, z) = c$, the linear approximation is $\underline{\ \ y\ \ }\ \Delta x + \underline{\ \ z\ \ }\ \Delta y + \underline{\ \ A\ \ }\ \Delta z = 0$.

Newton's method solves $g(x, y) = 0$ and $h(x, y) = 0$ by a $\underline{\ \ B\ \ }$ approximation. Starting from x_n, y_n the equations are replaced by $\underline{\ \ C\ \ }$ and $\underline{\ \ D\ \ }$. The steps Δx and Δy go to the next point $\underline{\ \ E\ \ }$. Each solution has a basin of $\underline{\ \ F\ \ }$. Those basins are likely to be $\underline{\ \ G\ \ }$.

In 1–8 find the tangent plane and the normal vector at P.

1 $z = \sqrt{x^2 + y^2}, P = (0, 1, 1)$

2 $x + y + z = 17, P = (3, 4, 10)$

3 $z = x/y, P = (6, 3, 2)$

4 $z = e^{x+2y}, P = (0, 0, 1)$

5 $x^2 + y^2 + z^2 = 6, P = (1, 2, 1)$

6 $x^2 + y^2 + 2z^2 = 7, P = (1, 2, 1)$

7 $z = x^y, P = (1, 1, 1)$

8 $V = \pi r^2 h, P = (2, 2, 8\pi)$.

9 Show that the tangent plane to $z^2 - x^2 - y^2 = 0$ goes through the origin and makes a $45°$ angle with the z axis.

10 The planes $z = x + 4y$ and $z = 2x + 3y$ meet at $(1, 1, 5)$. The whole line of intersection is $(x, y, z) = (1, 1, 5) + \mathbf{v}t$. Find $\mathbf{v} = \mathbf{N}_1 \times \mathbf{N}_2$.

11 If $z = 3x - 2y$ find dz from dx and dy. If $z = x^3/y^2$ find dz from dx and dy at $x_0 = 1, y_0 = 1$. If x moves to 1.02 and y moves to 1.03, find the approximate dz and exact Δz for both functions. The first surface is the _____ to the second surface.

12 The surfaces $z = x^2 + 4y$ and $z = 2x + 3y^2$ meet at $(1, 1, 5)$. Find the normals N_1 and N_2 and also $v = N_1 \times N_2$. The line in this direction $\mathbf{v}$ is tangent to what curve?

13 The normal $\mathbf{N}$ to the surface $F(x, y, z) = 0$ has components F_x, F_y, F_z. The *normal line* has $x = x_0 + F_x t, y = y_0 + F_y t, z = $ _____. For the surface $xyz - 24 = 0$, find the tangent plane and normal line at $(4, 2, 3)$.

14 For the surface $x^2 y^2 - z = 0$, the normal line at $(1, 2, 4)$ has $x = $ _____ , $y = $ _____ , $z = $ _____ .

15 For the sphere $x^2 + y^2 + z^2 = 9$, find the equation of the tangent plane through $(2, 1, 2)$. Also find the equation of the normal line and show that it goes through $(0, 0, 0)$.

16 If the normal line at every point on $F(x,y,z)=0$ goes through $(0,0,0)$, show that $F_x=cx, F_y=cy, F_z=cz$. The surface must be a sphere.

17 For $w=xy$ near (x_0,y_0), the linear approximation is $dw=$ _____. This looks like the _____ rule for derivatives. The difference between $\Delta w=xy-x_0y_0$ and this approximation is _____.

18 If $f=xyz$ (3 independent variables) what is df?

19 You invest $P=\$4000$ at $R=8\%$ to make $I=\$320$ per year. If the numbers change by dP and dR what is dI? If the rate drops by $dR=.002$ (to 7.8%) what change dP keeps $dI=0$? Find the exact interest I after those changes in R and P.

20 Resistances R_1 and R_2 have parallel resistance R, where $1/R=1/R_1+1/R_2$. Is R more sensitive to ΔR_1 or ΔR_2 if $R_1=1$ and $R_2=2$?

21 (a) If your batting average is $A=(25\text{ hits})/(100\text{ at bats})=.250$, compute the increase (to 26/101) with a hit and the decrease (to 25/101) with an out.
(b) If $A=x/y$ then $dA=$ _____ $dx+$ _____ dy. A hit $(dx=dy=1)$ gives $dA=(1-A)/y$. An out $(dy=1)$ gives $dA=-A/y$. So at $A=.250$ a hit has _____ times the effect of an out.

22 (a) 2 hits and 3 outs $(dx=2,dy=5)$ will raise your average $(dA>0)$ provided A is less than _____.
(b) A player batting $A=.500$ with $y=400$ at bats needs $dx=$ _____ hits to raise his average to .505.

23 If x and y change by Δx and Δy, find the approximate change $\Delta\theta$ in the angle $\theta=\tan^{-1}(y/x)$.

24 The *Fundamental Lemma* behind equation (13) writes $\Delta f=a\Delta x+b\Delta y$. The Lemma says that $a\to f_x(x_0,y_0)$ and $b\to f_y(x_0,y_0)$ when $\Delta x\to0$ and $\Delta y\to0$. The proof takes Δx first and then Δy:

(1) $f(x_0+\Delta x,y_0)-f(x_0,y_0)=\Delta x f_x(c,y_0)$ where c is between _____ and _____ (by which theorem?)
(2) $f(x_0+\Delta x,y_0+\Delta y)-f(x_0+\Delta x,y_0)=\Delta y f_y(x_0+\Delta x,C)$ where C is between _____ and _____.
(3) $a=f_x(c,y_0)\to f_x(x_0,y_0)$ provided f_x is _____.
(4) $b=f_y(x_0+\Delta x,C)\to f_y(x_0,y_0)$ provided f_y is _____.

25 If the supplier reduces s, Figure 13.10 shows that P decreases and Q _____.

(a) Find $P_s=50/3$ and $P_t=1/3$ in the economics equation (17) by solving the equations above it for Q_s and Q_t.
(b) What is the linear approximation to Q around $s=.4$, $t=10, P=30, Q=50$?

26 Solve the equations $P=-.2Q+40$ and $P=sQ+t$ for P and Q. Then find $\partial P/\partial s$ and $\partial P/\partial t$ explicitly. At the same s,t,P,Q check 50/3 and 1/3.

27 If the supply=demand equation (16) changes to $P=sQ+t=-Q+50$, find P_s and P_t at $s=1, t=10$.

28 To find out how the roots of $x^2+bx+c=0$ vary with b, take partial derivatives of the equation with respect to _____. Compare $\partial x/\partial b$ with $\partial x/\partial c$ to show that a root at $x=2$ is more sensitive to b.

29 Find the tangent planes to $z=xy$ and $z=x^2-y^2$ at $x=2, y=1$. Find the Newton point where those planes meet the xy plane (set $z=0$ in the tangent equations).

30 (a) To solve $g(x,y)=0$ and $h(x,y)=0$ is to find the meeting point of three surfaces: $z=g(x,y)$ and $z=h(x,y)$ and _____.
(b) Newton finds the meeting point of three planes: the tangent plane to the graph of g, _____, and _____.

Problems 31–36 go further with Newton's method for $g=x^3-y$ and $h=y^3-x$. This is Example 9 with solutions $(1,1),(0,0),(-1,-1)$.

31 Start from $x_0=1, y_0=1$ and find Δx and Δy. Where are x_1 and y_1, and what line is Newton's method moving on?

32 Start from $(\frac{1}{2},\frac{1}{2})$ and find the next point. This is in the basin of attraction of which solution?

33 Starting from $(a,-a)$ find Δy which is also $-\Delta x$. Newton goes toward $(0,0)$. But can you find the sharp point in Figure 13.11 where the lemon meets the spade?

34 Starting from $(a,0)$ show that Newton's method goes to $(0,-2a^3)$ and find the next point (x_2,y_2). Which numbers a lead to convergence? Which special number a leads to a cycle, in which (x_2,y_2) is the same as the starting point $(a,0)$?

35 Show that $x^3=y, y^3=x$ has exactly three solutions.

36 Locate a point from which Newton's method diverges.

37 Apply Newton's method to a linear problem: $g=x+2y-5=0, h=3x-3=0$. From any starting point show that (x_1,y_1) is the exact solution (convergence in *one step*).

38 The complex equation $(x+iy)^3=1$ contains two real equations, $x^3-3xy^2=1$ from the real part and $3x^2y-y^3=0$ from the imaginary part. Search by computer for the basins of attraction of the three solutions $(1,0),(-1/2,\sqrt{3}/2)$, and $(-1/2,-\sqrt{3}/2)$—which give the cube roots of 1.

39 In Newton's method the new guess comes from (x_n,y_n) by an *iteration*: $x_{n+1}=G(x_n,y_n)$ and $y_{n+1}=H(x_n,y_n)$. What are G and H for $g=x^2-y=0, h=x-y=0$? First find Δx and Δy; then $x_n+\Delta x$ gives G and $y_n+\Delta y$ gives H.

40 In Problem 39 find the basins of attraction of the solution $(0,0)$ and $(1,1)$.

41 The matrix in Newton's method is the **Jacobian**:

$$J=\begin{bmatrix}\partial g/\partial x & \partial g/\partial y\\ \partial h/\partial x & \partial h/\partial y\end{bmatrix}\quad\text{and}\quad J\begin{bmatrix}\Delta x\\ \Delta y\end{bmatrix}=\begin{bmatrix}-g_n\\ -h_n\end{bmatrix}.$$

Find J and Δx and Δy for $g=e^x-1, h=e^y+x$.

42 Find the Jacobian matrix at $(1,1)$ when $g=x^2+y^2$ and $h=xy$. This matrix is _____ and Newton's method fails. The graphs of g and h have _____ tangent planes.

43 Solve $g = x^2 - y^2 + 1 = 0$ and $h = 2xy = 0$ by Newton's method from three starting points: $(0,2)$ and $(-1,1)$ and $(2,0)$. Take ten steps by computer or one by hand. The solution $(0,1)$ attracts when $y_0 > 0$. If $y_0 = 0$ you should find the chaos iteration $x_{n+1} = \frac{1}{2}(x_n - x_n^{-1})$.

13.4 Directional Derivatives and Gradients

As x changes, we know how $f(x, y)$ changes. The partial derivative $\partial f/\partial x$ treats y as constant. Similarly $\partial f/\partial y$ keeps x constant, and gives the slope in the y direction. But east-west and north-south are not the only directions to move. We could go along a $45°$ line, where $\Delta x = \Delta y$. In principle, before we draw axes, no direction is preferred. The graph is a surface with slopes in *all* directions.

On that surface, calculus looks for the rate of change (or the slope). There is a *directional derivative*, whatever the direction. In the $45°$ case we are inclined to divide Δf by Δx, but we would be wrong. Let me state the problem. We are given $f(x, y)$ around a point $P = (x_0, y_0)$. We are also given a direction $\mathbf{u}$ (a unit vector). There must be a natural definition of $D_{\mathbf{u}} f$—*the derivative of f in the direction $\mathbf{u}$*. To compute this slope at P, we need a formula. Preferably the formula is based on $\partial f/\partial x$ and $\partial f/\partial y$, which we already know.

Note that the $45°$ direction has $\mathbf{u} = \mathbf{i}/2 + \mathbf{j}/2$. The square root of 2 is going to enter the derivative. This shows that dividing Δf by Δx is wrong. We should divide by the step length Δs.

EXAMPLE 1   Stay on the surface $z = xy$. When (x, y) moves a distance Δs in the $45°$ direction from $(1, 1)$, what is $\Delta z/\Delta s$?

Solution   The step is Δs times the unit vector $\mathbf{u}$. Starting from $x = y = 1$ the step ends at $x = y = 1 + \Delta s/\sqrt{2}$. (The components of $\mathbf{u}\Delta s$ are $\Delta s/\sqrt{2}$.) Then $z = xy$ is

$$z = (1 + \Delta s/\sqrt{2})^2 = 1 + \sqrt{2}\Delta s + \tfrac{1}{2}(\Delta s)^2, \text{ which means } \Delta z = \sqrt{2}\Delta s + \tfrac{1}{2}(\Delta s)^2.$$

The ratio $\Delta z/\Delta s$ approaches $\sqrt{2}$ as $\Delta s \to 0$. That is the slope in the $45°$ direction.

DEFINITION   *The derivative of f in the direction $\mathbf{u}$ at the point P is $D_{\mathbf{u}} f(P)$:*

$$D_{\mathbf{u}} f(P) = \lim_{\Delta s \to 0} \frac{\Delta f}{\Delta s} = \lim_{\Delta s \to 0} \frac{f(P + \mathbf{u}\Delta s) - f(P)}{\Delta s}. \tag{1}$$

The step from $P = (x_0, y_0)$ has length Δs. It takes us to $(x_0 + u_1\Delta s, y_0 + u_2\Delta s)$. We compute the change Δf and divide by Δs. But formula (2) below saves time.

The x direction is $\mathbf{u} = (1, 0)$. Then $\mathbf{u}\Delta s$ is $(\Delta s, 0)$ and we recover $\partial f/\partial x$:

$$\frac{\Delta f}{\Delta s} = \frac{f(x_0 + \Delta s, y_0) - f(x_0, y_0)}{\Delta s} \text{ approaches } D_{(1,0)} f = \frac{\partial f}{\partial x}.$$

Similarly $D_{\mathbf{u}} f = \partial f/\partial y$, when $\mathbf{u} = (0, 1)$ is in the y direction. *What is $D_{\mathbf{u}} f$ when $\mathbf{u} = (0, -1)$*? That is the negative y direction, so $D_{\mathbf{u}} f = -\partial f/\partial y$.

CALCULATING THE DIRECTIONAL DERIVATIVE

$D_{\mathbf{u}} f$ is the slope of the surface $z = f(x, y)$ in the direction $\mathbf{u}$. How do you compute it? From $\partial f/\partial x$ and $\partial f/\partial y$, in two special directions, there is a quick way to find $D_{\mathbf{u}} f$ in all directions. **Remember that $\mathbf{u}$ is a unit vector.**

13E   The *directional derivative $D_{\mathbf{u}} f$* in the direction $\mathbf{u} = (u_1, u_2)$ equals

$$D_{\mathbf{u}} f = \frac{\partial f}{\partial x} u_1 + \frac{\partial f}{\partial y} u_2. \tag{2}$$

The reasoning goes back to the linear approximation of Δf:

$$\Delta f \approx \frac{\partial f}{\partial x} \Delta x + \frac{\partial f}{\partial y} \Delta y = \frac{\partial f}{\partial x} u_1 \Delta s + \frac{\partial f}{\partial y} u_2 \Delta s.$$

Divide by Δs and let Δs approach zero. Formula (2) is the limit of $\Delta f / \Delta s$, as the approximation becomes exact. A more careful argument guarantees this limit provided f_x and f_y are continuous at the basepoint (x_0, y_0).

Main point: *Slopes in all directions are known from slopes in two directions*.

EXAMPLE 1 (repeated) $f = xy$ and $P = (1, 1)$ and $\mathbf{u} = (1/\sqrt{2}, 1/\sqrt{2})$. Find $D_\mathbf{u} f(P)$.

The derivatives $f_x = y$ and $f_y = x$ equal 1 at P. The $45°$ derivative is

$$D_\mathbf{u} f(P) = f_x u_1 + f_y u_2 = 1(1/\sqrt{2}) + 1(1/\sqrt{2}) = \sqrt{2} \text{ as before}.$$

EXAMPLE 2   The linear function $f = 3x + y + 1$ has slope $D_\mathbf{u} f = 3u_1 + u_2$.

The x direction is $u = (1, 0)$, and $D_\mathbf{u} f = 3$. That is $\partial f / \partial x$. In the y direction $D_\mathbf{u} f = 1$. *Two other directions are special—along the level lines of $f(x, y)$ and perpendicular*:

Level direction: $D_\mathbf{u} f$ is zero because f is constant

Steepest direction: $D_\mathbf{u} f$ is as large as possible (with $u_1^2 + u_2^2 = 1$).

To find those directions, look at $D_\mathbf{u} f = 3u_1 + u_2$. The level direction has $3u_1 + u_2 = 0$. Then u is proportional to $(1, -3)$. Changing x by 1 and y by -3 produces no change in $f = 3x + y + 1$.

In the steepest direction $\mathbf{u}$ is proportional to $(3, 1)$. Note the partial derivatives $f_x = 3$ and $f_y = 1$. The dot product of $(3, 1)$ and $(1, -3)$ is zero—*steepest direction is perpendicular to level direction*. To make $(3, 1)$ a unit vector, divide by $\sqrt{10}$.

Steepest climb: $D_\mathbf{u} f = 3(3/\sqrt{10}) + 1(1/\sqrt{10}) = 10/\sqrt{10} = \sqrt{10}$

Steepest descent: Reverse to $\mathbf{u} = (-3/\sqrt{10}, -1/\sqrt{10})$ and $D_\mathbf{u} f = -\sqrt{10}$.

The contour lines around a mountain follow $D_\mathbf{u} f = 0$. The creeks are perpendicular. On a plane like $f = 3x + y + 1$, those directions stay the same at all points (Figure 13.12). On a mountain the steepest direction changes as the slopes change.

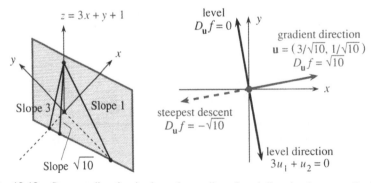

Fig. 13.12   Steepest direction is along the gradient. Level direction is perpendicular.

THE GRADIENT VECTOR

Look again at $f_x u_1 + f_y u_2$, which is the directional derivative $D_{\mathbf{u}} f$. *This is the dot product of two vectors*. One vector is $\mathbf{u} = (u_1, u_2)$, which sets the direction. The other vector is (f_x, f_y), which comes from the function. This second vector is the **gradient**.

DEFINITION   The **gradient** of $f(x, y)$ is the vector whose components are $\dfrac{\partial f}{\partial x}$ and $\dfrac{\partial f}{\partial y}$:

$$\operatorname{grad} f = \nabla f = \frac{\partial f}{\partial x}\mathbf{i} + \frac{\partial f}{\partial y}\mathbf{j} \quad \left(\text{add } \frac{\partial f}{\partial z}\mathbf{k} \text{ in three dimensions}\right).$$

The space-saving symbol ∇ is read as "grad." In Chapter 15 it becomes "del."

For the linear function $3x + y + 1$, the gradient is the constant vector $(3, 1)$. It is the way to climb the plane. For the nonlinear function $x^2 + xy$, the gradient is the non-constant vector $(2x + y, x)$. Notice that grad f shares the two derivatives in $\mathbf{N} = (f_x, f_y, -1)$. But the gradient is not the normal vector. $\mathbf{N}$ is in three dimensions, pointing away from the surface $z = f(x, y)$. *The gradient vector is in the xy plane*! The gradient tells which way on the surface is up, but it does that from down in the base.

The level curve is also in the xy plane, perpendicular to the gradient. The contour map is a projection on the base plane of what the hiker sees on the mountain. The vector grad f tells the **direction** of climb, and its length $|\operatorname{grad} f|$ gives the **steepness**.

13F The directional derivative is $D_{\mathbf{u}} f = (\operatorname{grad} f) \cdot \mathbf{u}$. The level direction is perpendicular to grad f, since $D_{\mathbf{u}} f = 0$. *The slope $D_{\mathbf{u}} f$ is largest when $\mathbf{u}$ is parallel to* grad f. That maximum slope is the length $|\operatorname{grad} f| = \sqrt{f_x^2 + f_y^2}$:

$$\text{for} \quad \mathbf{u} = \frac{\operatorname{grad} f}{|\operatorname{grad} f|} \quad \text{the slope is} \quad (\operatorname{grad} f) \cdot \mathbf{u} = \frac{|\operatorname{grad} f|^2}{|\operatorname{grad} f|} = |\operatorname{grad} f|.$$

The example $f = 3x + y + 1$ had grad $f = (3, 1)$. Its steepest slope was in the direction $\mathbf{u} = (3, 1)/\sqrt{10}$. The maximum slope was $\sqrt{10}$. That is $|\operatorname{grad} f| = \sqrt{9 + 1}$.

Important point: *The maximum of $(\operatorname{grad} f) \cdot \mathbf{u}$ is the length $|\operatorname{grad} f|$.* In nonlinear examples, the gradient and steepest direction and slope will vary. But look at one particular point in Figure 13.13. Near that point, and near any point, the linear picture takes over.

On the graph of f, the special vectors are the level direction $\mathbf{L} = (f_y, -f_x, 0)$ and the uphill direction $\mathbf{U} = (f_x, f_y, f_x^2 + f_y^2)$ and the normal $\mathbf{N} = (f_x, f_y, -1)$. Problem 18 checks that those are perpendicular.

EXAMPLE 3   The gradient of $f(x, y) = (14 - x^2 - y^2)/3$ is $\nabla f = (-2x/3, -2y/3)$.

On the surface, the normal vector is $\mathbf{N} = (-2x/3, -2y/3, -1)$. At the point $(1, 2, 3)$, this perpendicular is $\mathbf{N} = (-2/3, -4/3, -1)$. At the point $(1, 2)$ down in the base, the gradient is $(-2/3, -4/3)$. The length of grad f is the slope $\sqrt{20}/3$.

Probably a hiker does not go straight up. A "grade" of $\sqrt{20}/3$ is fairly steep (almost 150%). To estimate the slope in other directions, measure the distance along the path between two contour lines. If $\Delta f = 1$ in a distance $\Delta s = 3$ the slope is about $1/3$. This calculation is not exact until the limit of $\Delta f/\Delta s$, which is $D_{\mathbf{u}} f$.

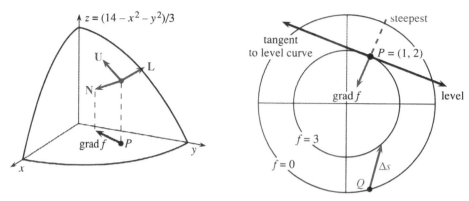

Fig. 13.13   **N** perpendicular to surface and grad f perpendicular to level line (in the base).

EXAMPLE 4   The gradient of $f(x, y, z) = xy + yz + xz$ has three components.

The pattern extends from $f(x, y)$ to $f(x, y, z)$. The gradient is now the three-dimensional vector (f_x, f_y, f_z). For this function grad f is $(y + z, x + z, x + y)$. To draw the graph of $w = f(x, y, z)$ would require a four-dimensional picture, with axes in the $xyzw$ directions.

Notice the dimensions. The graph is a 3-dimensional "surface" in 4-dimensional space. The gradient is down below in the 3-dimensional base. The level sets of f come from $xy + yz + zx = c$—they are 2-dimensional. The gradient is perpendicular to that level set (still down in 3 dimensions). The gradient is not **N**! The normal vector is $(f_x, f_y, f_z, -1)$, perpendicular to the surface up in 4-dimensional space.

EXAMPLE 5   Find grad z when $z(x, y)$ is given implicitly: $F(x, y, z) = x^2 + y^2 - z^2 = 0$.

In this case we find $z = \pm\sqrt{x^2 + y^2}$. The derivatives are $\pm x/\sqrt{x^2 + y^2}$ and $\pm y/\sqrt{x^2 + y^2}$, which go into grad z. But the point is this: To find that gradient faster, differentiate $F(x, y, z)$ as it stands. Then divide by F_z:

$$F_x dx + F_y dy + F_z dz = 0 \qquad \text{or} \qquad dz = (-F_x dx - F_y dy)/F_z. \qquad (3)$$

The gradient is $(-F_x/F_z, -F_y/F_z)$. Those derivatives are evaluated at (x_0, y_0). The computation does not need the explicit function $z = f(x, y)$:

$$F = x^2 + y^2 - z^2 \Rightarrow F_x = 2x, F_y = 2y, F_z = -2z \Rightarrow \text{grad } z = (x/z, y/z).$$

To go uphill on the cone, move in the direction $(x/z, y/z)$. That gradient direction goes radially outward. The steepness of the cone is the length of the gradient vector:

$$|\text{grad } z| = \sqrt{(x/z)^2 + (y/z)^2} = 1 \text{ because } z^2 = x^2 + y^2 \text{ on the cone.}$$

DERIVATIVES ALONG CURVED PATHS

On a straight path the derivative of f is $D_{\mathbf{u}} f = (\text{grad } f) \cdot \mathbf{u}$. What is the derivative on a curved path? ***The path direction*** $\mathbf{u}$ ***is the tangent vector*** $\mathbf{T}$. So replace $\mathbf{u}$ by $\mathbf{T}$, which gives the "direction" of the curve.

The path is given by the position vector $\mathbf{R}(t) = x(t)\mathbf{i} + y(t)\mathbf{j}$. The velocity is $\mathbf{v} = (dx/dt)\mathbf{i} + (dy/dt)\mathbf{j}$. The tangent vector is $\mathbf{T} = \mathbf{v}/|\mathbf{v}|$. Notice the choice—to move at any speed (with **v**) or to go at unit speed (with **T**.) There is the same choice for the derivative of $f(x, y)$ along this curve:

$$\textit{rate of change} \quad \frac{df}{dt} = (\text{grad } f) \cdot \mathbf{v} = \frac{\partial f}{\partial x}\frac{dx}{dt} + \frac{\partial f}{\partial y}\frac{dy}{dt} \tag{4}$$

$$\textit{slope} \quad \frac{df}{ds} = (\text{grad } f) \cdot \mathbf{T} = \frac{\partial f}{\partial x}\frac{dx}{ds} + \frac{\partial f}{\partial y}\frac{dy}{ds} \tag{5}$$

The first involves *time*. If we move faster, df/dt increases. The second involves *distance*. If we move a distance ds, at any speed, the function changes by df. So the slope in that direction is df/ds. Chapter 1 introduced velocity as df/dt and slope as dy/dx and mixed them up. Finally we see the difference. Uniform motion on a straight line has $\mathbf{R} = \mathbf{R}_0 + \mathbf{v}t$. The velocity **v** is constant. The direction $\mathbf{T} = \mathbf{u} = \mathbf{v}/|\mathbf{v}|$ is also constant. The directional derivative is $(\text{grad } f) \cdot \mathbf{u}$, but the rate of change is $(\text{grad } f) \cdot \mathbf{v}$.

Equations (4) and (5) look like chain rules. They are chain rules. The next section extends $df/dt = (df/dx)(dx/dt)$ to more variables, proving (4) and (5). Here we focus on the meaning: *df/ds is the derivative of f in the direction $\mathbf{u} = \mathbf{T}$ along the curve.*

EXAMPLE 7 Find df/dt and df/ds for $f = r$. The curve is $x = t^2, y = t$ in Figure 13.14a.

Solution The velocity along the curve is $\mathbf{v} = 2t\mathbf{i} + \mathbf{j}$. At the typical point $t = 1$ it is $\mathbf{v} = 2\mathbf{i} + \mathbf{j}$. The unit tangent is $\mathbf{T} = \mathbf{v}/\sqrt{5}$. The gradient is a unit vector $\mathbf{i}/\sqrt{2} + \mathbf{j}/\sqrt{2}$ pointing outward, when $f(x, y)$ is the distance r from the center. The dot product with **v** is $df/dt = 3/\sqrt{2}$. The dot product with **T** is $df/ds = 3/\sqrt{10}$.

When we slow down to speed 1 (with **T**), the changes in $f(x, y)$ slow down too.

EXAMPLE 8 Find df/ds for $f = xy$ along the circular path $x = \cos t, y = \sin t$.

First take a direct approach. On the circle, xy equals $(\cos t)(\sin t)$. Its derivative comes from the product rule: $df/dt = \cos^2 t - \sin^2 t$. Normally this is different from df/ds, because the time t need not equal the arc length s. There is a speed factor ds/dt to divide by—but here the speed is 1. (A circle of length $s = 2\pi$ is completed at $t = 2\pi$.) Thus the slope df/ds along the roller-coaster in Figure 13.14 is $\cos^2 t - \sin^2 t$.

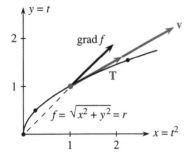
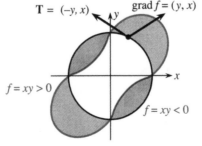
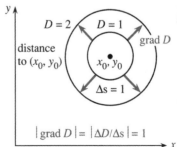

Fig. 13.14 The distance $f = r$ changes along the curve. The slope of the roller-coaster is $(\text{grad } f) \cdot \mathbf{T}$. The distance D from (x_0, y_0) has grad D = unit vector.

The second approach uses the vectors grad f and **T**. The gradient of $f = xy$ is $(y, x) = (\sin t, \cos t)$. The unit tangent vector to the path is $\mathbf{T} = (-\sin t, \cos t)$. Their dot product is the same df/ds:

$$\textbf{\textit{slope along path}} = (\text{grad } f) \cdot \mathbf{T} = -\sin^2 t + \cos^2 t.$$

GRADIENTS WITHOUT COORDINATES

This section ends with a little "philosophy." What is the *coordinate-free definition* of the gradient? Up to now, grad $f = (f_x, f_y)$ depended totally on the choice of x and y axes. But the steepness of a surface is independent of the axes. Those are added later, to help us compute.

The steepness df/ds involves only f and the direction, nothing else. The gradient should be a "tensor"—its meaning does not depend on the coordinate system. The gradient has different formulas in different systems (xy or $r\theta$ or ...), but the direction and length of grad f are determined by df/ds—without any axes:

The **direction** of grad f is the one in which df/ds is largest.
The **length** $|\text{grad } f|$ is that largest slope.

The key equation is $(\textbf{\textit{change in }} f) \approx (\textbf{\textit{gradient of }} f) \cdot (\textbf{\textit{change in position}})$. That is another way to write $\Delta f \approx f_x \Delta x + f_y \Delta y$. It is the multivariable form—we used two variables—of the basic linear approximation $\Delta y \approx (dy/dx)\Delta x$.

EXAMPLE 9 $D(x, y) = $ distance from (x, y) to (x_0, y_0). Without derivatives prove $|\text{grad } D| = 1$. The graph of $D(x, y)$ is a cone with slope 1 and sharp point (x_0, y_0).

First question   In which direction does the distance $D(x, y)$ increase fastest?
Answer   Going directly away from (x_0, y_0). Therefore this is the direction of grad D.

Second question   How quickly does D increase in that steepest direction?
Answer   A step of length Δs increases D by Δs. Therefore $|\text{grad } D| = \Delta s/\Delta s = 1$.

Conclusion   grad D **is a unit vector**. The derivatives of D in Problem 48 are $(x - x_0)/D$ and $(y - y_0)/D$. The sum of their squares is 1, because $(x - x_0)^2 + (y - y_0)^2$ equals D^2.

13.4   EXERCISES

Read-through questions

$D_{\mathbf{u}} f$ gives the rate of change of __a__ in the direction __b__. It can be computed from the two derivatives __c__ in the special directions __d__. In terms of u_1, u_2 the formula is $D_{\mathbf{u}} f = $ __e__. This is a __f__ product of $\mathbf{u}$ with the vector __g__, which is called the __h__. For the linear function $f = ax + by$, the gradient is grad $f = $ __i__ and the directional derivative is $D_{\mathbf{u}} f = $ __j__ $\cdot$ __k__.

The gradient $\nabla f = (f_x, f_y)$ is not a vector in __l__ dimensions, it is a vector in the __m__. It is perpendicular to the __n__ lines. It points in the direction of __o__ climb. Its

magnitude $|\text{grad } f|$ is __p__. For $f = x^2 + y^2$ the gradient points __q__ and the slope in that steepest direction is __r__.

The gradient of $f(x, y, z)$ is __s__. This is different from the gradient on the surface $F(x, y, z) = 0$, which is $-(F_x/F_z)\mathbf{i} + $ __t__. Traveling with velocity $\mathbf{v}$ on a curved path, the rate of change of f is $df/dt = $ __u__. When the tangent direction is $\mathbf{T}$, the slope of f is $df/ds = $ __v__. In a straight direction $\mathbf{u}$, df/ds is the same as __w__.

Compute grad f, **then** $D_{\mathbf{u}} f = (\text{grad} f) \cdot \mathbf{u}$, **then** $D_{\mathbf{u}} f$ **at** P.

1   $f(x, y) = x^2 - y^2$ $\mathbf{u} = (\sqrt{3}/2, 1/2) \; P = (1, 0)$

2   $f(x, y) = 3x + 4y + 7$ $\mathbf{u} = (3/5, 4/5) \quad P = (0, \pi/2)$

3   $f(x, y) = e^x \cos y$ $\mathbf{u} = (0, 1) \qquad P = (0, \pi/2)$

4   $f(x, y) = y^{10}$ $\mathbf{u} = (0, -1) \qquad P = (1, -1)$

5   $f(x, y) = $ distance to $(0, 3) \; \mathbf{u} = (1, 0) \qquad P = (1, 1)$

Find grad $f = (f_x, f_y, f_z)$ **for the functions 6–8 from physics.**

6   $1/\sqrt{x^2 + y^2 + z^2}$ (point source at the origin)

7   $\ln(x^2 + y^2)$ (line source along z axis)

8   $1/\sqrt{(x-1)^2 + y^2 + z^2} - 1/\sqrt{(x+1)^2 + y^2 + z^2}$ (dipole)

9   For $f = 3x^2 + 2y^2$ find the steepest direction and the level direction at $(1, 2)$. Compute $D_{\mathbf{u}} f$ in those directions.

10   Example 2 claimed that $f = 3x + y + 1$ has steepest slope $\sqrt{10}$. Maximize $D_{\mathbf{u}} f = 3u_1 + u_2 = 3u_1 + \sqrt{1 - u_1^2}$.

11   **True or false**, when $f(x, y)$ is any smooth function:

    (a)   There is a direction $\mathbf{u}$ at P in which $D_{\mathbf{u}} f = 0$.

    (b)   There is a direction $\mathbf{u}$ in which $D_{\mathbf{u}} f = \text{grad} \, f$.

    (c)   There is a direction $\mathbf{u}$ in which $D_{\mathbf{u}} f = 1$.

    (d)   The gradient of $f(x)g(x)$ equals $g \,\text{grad}\, f + f \,\text{grad}\, g$.

12   What is the gradient of $f(x)$? (One component only.) What are the two possible directions $\mathbf{u}$ and the derivatives $D_{\mathbf{u}} f$? What is the normal vector $\mathbf{N}$ to the curve $y = f(x)$? (Two components.)

In 13–16 find the direction u in which f **increases fastest at** $P = (1, 2)$. **How fast?**

13   $f(x, y) = ax + by$ 14 $f(x, y) = $ smaller of $2x$ and y

15   $f(x, y) = e^{x-y}$ 16 $f(x, y) = \sqrt{5 - x^2 - y^2}$ (careful)

17   (Looking ahead) At a point where $f(x, y)$ is a maximum, what is grad f? Describe the level curve containing the maximum point (x, y).

18   (a)   Check by dot products that the normal and uphill and level directions on the graph are perpendicular: $\mathbf{N} = (f_x, f_y, -1), \mathbf{U} = (f_x, f_y, f_x^2 + f_y^2), \mathbf{L} = (f_y, -f_x, 0)$.

    (b) $\mathbf{N}$ is _____ to the tangent plane, $\mathbf{U}$ and $\mathbf{L}$ are _____ to the tangent plane.

    (c)   The gradient is the xy projection of _____ and also of _____ . The projection of $\mathbf{L}$ points along the _____ .

19   Compute the $\mathbf{N}, \mathbf{U}, \mathbf{L}$ vectors for $f = 1 - x + y$ and draw them at a point on the flat surface.

20   Compute the $\mathbf{N}, \mathbf{U}, \mathbf{L}$ for $x^2 + y^2 - z^2 = 0$ and draw them at a typical point on the cone.

With gravity in the negative z **direction, in what direction** $-\mathbf{U}$ **will water flow down the roofs 21–24?**

21   $z = 2x$ (flat roof)   22 $z = 4x - 3y$ (flat roof)

23   $z = \sqrt{1 - x^2 - y^2}$ (sphere)   24 $z = -\sqrt{x^2 + y^2}$ (cone)

25   Choose two functions $f(x, y)$ that depend only on $x + 2y$. Their gradients at $(1, 1)$ are in the direction _____ . Their level curves are _____ .

26   The level curve of $f = y/x$ through $(1, 1)$ is _____ . The direction of the gradient must be _____ . Check grad f.

27   Grad f is perpendicular to $2\mathbf{i} + \mathbf{j}$ with length 1, and grad g is parallel to $2\mathbf{i} + \mathbf{j}$ with length 5. Find grad f, grad g, f, and g.

28   **True or false**:

    (a)   If we know grad f, we know f.

    (b)   The line $x = y = -z$ is perpendicular to the plane $z = x + y$.

    (c)   The gradient of $z = x + y$ lies along that line.

29   Write down the level direction $\mathbf{u}$ for $\theta = \tan^{-1}(y/x)$ at the point $(3, 4)$. Then compute grad θ and check $D_{\mathbf{u}} \theta = 0$.

30   On a circle around the origin, distance is $\Delta s = r\Delta\theta$. Then $d\theta/ds = 1/r$. Verify by computing grad θ and $\mathbf{T}$ and $(\text{grad } \theta) \cdot \mathbf{T}$.

31   At the point $(2, 1, 6)$ on the mountain $z = 9 - x - y^2$, which way is up? On the roof $z = x + 2y + 2$, which way is down? The roof is _____ to the mountain.

32   Around the point $(1, -2)$ the temperature $T = e^{-x^2 - y^2}$ has $\Delta T \approx $ _____ $\Delta x + $ _____ Δy. In what direction $\mathbf{u}$ does it get hot fastest?

33   Figure A shows level curves of $z = f(x, y)$.

    (a)   Estimate the direction and length of grad f at P, Q, R.

    (b)   Locate two points where grad f is parallel to $\mathbf{i} + \mathbf{j}$.

    (c)   Where is $|\text{grad } f|$ largest? Where is it smallest?

    (d)   What is your estimate of $z_{\max}$ on this figure?

    (e)   On the straight line from P to R, describe z and estimate its maximum.

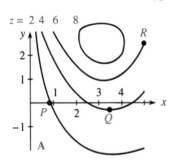

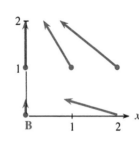

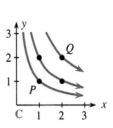

34 A quadratic function $ax^2 + by^2 + cx + dy$ has the gradients shown in Figure B. Estimate a, b, c, d and sketch two level curves.

35 The level curves of $f(x, y)$ are circles around $(1, 1)$. The curve $f = c$ has radius $2c$. What is f? What is grad f at $(0, 0)$?

36 Suppose grad f is tangent to the hyperbolas $xy = $ constant in Figure C. Draw three level curves of $f(x, y)$. Is $|\text{grad } f|$ larger at P or Q? Is $|\text{grad } f|$ constant along the hyperbolas? Choose a function that could be $f: x^2 + y^2, x^2 - y^2, xy, x^2y^2$.

37 Repeat Problem 36, if grad f is *perpendicular* to the hyperbolas in Figure C.

38 If $f = 0, 1, 2$ at the points $(0, 1), (1, 0), (2, 1)$, estimate grad f by assuming $f = Ax + By + C$.

39 What functions have the following gradients?

(a) $(2x + y, x)$ (b) $(e^{x-y}, -e^{x-y})$ (c) $(y, -x)$ (careful)

40 Draw level curves of $f(x, y)$ if grad $f = (y, x)$.

In 41–46 find the velocity v and the tangent vector T. Then compute the rate of change $df/dt = \text{grad } f \cdot \mathbf{v}$ and the slope $df/ds = \text{grad } f \cdot \mathbf{T}$.

41 $f = x^2 + y^2$ $x = t$ $y = t^2$

42 $f = x$ $x = \cos 2t$ $y = \sin 2t$

43 $f = x^2 - y^2$ $x = x_0 + 2t$ $y = y_0 + 3t$

44 $f = xy$ $x = t^2 + 1$ $y = 3$

45 $f = \ln xyz$ $x = e^t$ $y = e^{2t}$ $z = e^{-t}$

46 $f = 2x^2 + 3y^2 + z^2$ $x = t$ $y = t^2$ $z = t^3$

47 (a) Find df/ds and df/dt for the roller-coaster $f = xy$ along the path $x = \cos 2t, y = \sin 2t$. (b) Change to $f = x^2 + y^2$ and explain why the slope is zero.

48 The distance D from (x, y) to $(1, 2)$ has $D^2 = (x-1)^2 + (y-2)^2$. Show that $\partial D/\partial x = (x-1)/D$ and $\partial D/\partial y = (y-2)/D$ and $|\text{grad } D| = 1$. The graph of $D(x, y)$ is a _____ with its vertex at _____.

49 If $f = 1$ and grad $f = (2, 3)$ at the point $(4, 5)$, find the tangent plane at $(4, 5)$. If f is a linear function, find $f(x, y)$.

50 Define the derivative of $f(x, y)$ in the direction $\mathbf{u} = (u_1, u_2)$ at the point $P = (x_0, y_0)$. What is Δf (approximately)? What is $D_u f$ (exactly)?

51 The slope of f along a level curve is $df/ds = $ _____ $= 0$. This says that grad f is perpendicular to the vector _____ in the level direction.

13.5 The Chain Rule

Calculus goes back and forth between solving problems and getting ready for harder problems. The first is "*application*," the second looks like "*theory*." If we minimize f to save time or money or energy, that is an application. If we don't take derivatives to find the minimum—maybe because f is a function of other functions, and we don't have a chain rule—then it is time for more theory. The chain rule is a fundamental working tool, because $f(g(x))$ appears all the time in applications. So do $f(g(x,y))$ and $f(x(t),y(t))$ and worse. We have to know their derivatives. Otherwise calculus can't continue with the applications.

You may instinctively say: Don't bother with the theory, just teach me the formulas. That is not possible. You now regard the derivative of $\sin 2x$ as a trivial problem, unworthy of an answer. That was not always so. Before the chain rule, the slopes of $\sin 2x$ and $\sin x^2$ and $\sin^2 x^2$ were hard to compute from $\Delta f/\Delta x$. We are now at the same point for $f(x,y)$. We know the *meaning* of $\partial f/\partial x$, but if $f = r\tan\theta$ and $x = r\cos\theta$ and $y = r\sin\theta$, we need a way to *compute $\partial f/\partial x$*. A little theory is unavoidable, if the problem-solving part of calculus is to keep going.

To repeat: ***The chain rule applies to a function of a function***. In one variable that was $f(g(x))$. With two variables there are more possibilities:

1. $f(z)$ with $z = g(x,y)$ Find $\partial f/\partial x$ and $\partial f/\partial y$
2. $f(x,y)$ with $x = x(t), y = y(t)$ Find df/dt
3. $f(x,y)$ with $x = x(t,u), y = y(t,u)$ Find $\partial f/\partial t$ and $\partial f/\partial u$

All derivatives are assumed continuous. More exactly, the *input* derivatives like $\partial g/\partial x$ and dx/dt and $\partial x/\partial u$ are continuous. Then the output derivatives like $\partial f/\partial x$ and df/dt and $\partial f/\partial u$ will be continuous from the chain rule. We avoid points like $r = 0$ in polar coordinates—where $\partial r/\partial x = x/r$ has a division by zero.

A Typical Problem Start with a function of x and y, for example x times y. Thus $f(x,y) = xy$. Change x to $r\cos\theta$ and y to $r\sin\theta$. The function becomes $(r\cos\theta)$ times $(r\sin\theta)$. We want its derivatives with respect to r and θ. First we have to decide on its *name*.

To be correct, we should not reuse the letter f. The new function can be F:

$$f(x,y) = xy \quad f(r\cos\theta, r\sin\theta) = (r\cos\theta)(r\sin\theta) = F(r,\theta).$$

Why not call it $f(r,\theta)$? Because strictly speaking that is r times θ! If we follow the rules, then $f(x,y)$ is xy and $f(r,\theta)$ should be $r\theta$. The new function F does the right thing—it multiplies $(r\cos\theta)(r\sin\theta)$. But in many cases, the rules get bent and the letter F is changed back to f.

This crime has already occurred. The end of the last page ought to say $\partial F/\partial t$. Instead the printer put $\partial f/\partial t$. The purpose of the chain rule is to find derivatives in the new variables t and u (or r and θ). In our example we want ***the derivative of F with respect to r***. Here is the chain rule:

$$\frac{\partial F}{\partial r} = \frac{\partial f}{\partial x}\frac{\partial x}{\partial r} + \frac{\partial f}{\partial y}\frac{\partial y}{\partial r} = (y)(\cos\theta) + (x)(\sin\theta) = 2r\sin\theta\cos\theta.$$

I substituted $r\sin\theta$ and $r\cos\theta$ for y and x. You immediately check the answer: $F(r,\theta) = r^2\cos\theta\sin\theta$ does lead to $\partial F/\partial r = 2r\cos\theta\sin\theta$. The derivative is correct. The only incorrect thing—but we do it anyway—is to write f instead of F.

Question What is $\dfrac{\partial f}{\partial\theta}$?   Answer   It is $\dfrac{\partial f}{\partial x}\dfrac{\partial x}{\partial\theta} + \dfrac{\partial f}{\partial y}\dfrac{\partial y}{\partial\theta}$.

THE DERIVATIVES OF $f(g(x,y))$

Here g depends on x and y, and f depends on g. Suppose x moves by dx, while y stays constant. Then g moves by $dg = (\partial g/\partial x)dx$. When g changes, f also changes: $df = (df/dg)dg$. Now substitute for dg to make the chain: $df = (df/dg)(\partial g/\partial x)dx$. This is the first rule:

$$\boxed{\textbf{13G} \quad \textbf{\textit{Chain rule for }} f(g(x,y)): \frac{\partial f}{\partial x} = \frac{df}{dg}\frac{\partial g}{\partial x} \quad \text{and} \quad \frac{\partial f}{\partial y} = \frac{df}{dg}\frac{\partial g}{\partial y}. \quad (1)}$$

EXAMPLE 1   Every $f(x+cy)$ satisfies the 1-way wave equation $\partial f/\partial y = c\,\partial f/\partial x$.

The inside function is $g = x + cy$. The outside function can be anything, g^2 or $\sin g$ or e^g. The composite function is $(x+cy)^2$ or $\sin(x+cy)$ or e^{x+cy}. In each separate case we could check that $\partial f/\partial y = c\,\partial f/\partial x$. The chain rule produces this equation in all cases at once, from $\partial g/\partial x = 1$ and $\partial g/\partial y = c$:

$$\frac{\partial f}{\partial x} = \frac{df}{dg}\frac{\partial g}{\partial x} = 1\frac{df}{dg} \quad \text{and} \quad \frac{\partial f}{\partial y} = \frac{df}{dg}\frac{\partial g}{\partial y} = c\frac{df}{dg} \quad \text{so} \quad \frac{\partial f}{\partial y} = c\frac{\partial f}{\partial x}. \quad (2)$$

This is important: $\partial f/\partial y = c\,\partial f/\partial x$ is our first example of a ***partial differential equation***. The unknown $f(x,y)$ has two variables. Two partial derivatives enter the equation.

Up to now we have worked with dy/dt and ***ordinary differential equations***. The independent variable was time *or* space (and only one dimension in space). For partial differential equations the variables are time *and* space (possibly several dimensions in space). The great equations of mathematical physics—heat equation, wave equation, Laplace's equation—are partial differential equations.

Notice how the chain rule applies to $f = \sin xy$. Its x derivative is $y\cos xy$. A patient reader would check that f is $\sin g$ and g is xy and f_x is $f_g g_x$. Probably you are not so patient—you know the derivative of $\sin xy$. Therefore we pass quickly to the next chain rule. Its outside function depends on *two* inside functions, and each of those depends on t. We want df/dt.

THE DERIVATIVE OF $f(x(t),y(t))$

Before the formula, here is the idea. Suppose t changes by Δt. That affects x and y; they change by Δx and Δy. There is a domino effect on f; it changes by Δf. Tracing backwards,

$$\Delta f \approx \frac{\partial f}{\partial x}\Delta x + \frac{\partial f}{\partial y}\Delta y \quad \text{and} \quad \Delta x \approx \frac{dx}{dt}\Delta t \quad \text{and} \quad \Delta y \approx \frac{dy}{dt}\Delta t.$$

Substitute the last two into the first, connecting Δf to Δt. Then let $\Delta t \to 0$:

$$\boxed{\textbf{13H} \quad \textbf{\textit{Chain rule for }} f(x(t),y(t)): \frac{df}{dt} = \frac{\partial f}{\partial x}\frac{dx}{dt} + \frac{\partial f}{\partial y}\frac{dy}{dt}. \quad (3)}$$

This is close to the one-variable rule $dz/dx = (dz/dy)(dy/dx)$. There we could "cancel" dy. (We actually canceled Δy in $(\Delta z/\Delta y)(\Delta y/\Delta x)$, and then approached the limit.) Now Δt affects Δf in two ways, through x and through y. *The chain rule has two terms.* If we cancel in $(\partial f/\partial x)(dx/dt)$ we only get one of the terms!

We mention again that the true name for $f(x(t), y(t))$ is $F(t)$ not $f(t)$. For $f(x, y, z)$ the rule has three terms: $f_x x_t + f_y y_t + f_z z_t$ is f_t (or better dF/dt.)

EXAMPLE 2 How quickly does the temperature change when you drive to Florida?

Suppose the Midwest is at $30°F$ and Florida is at $80°F$. Going 1000 miles south increases the temperature $f(x, y)$ by $50°$, or .05 degrees per mile. Driving straight south at 70 miles per hour, the rate of increase is $(.05)(70) = 3.5$ degrees per hour. *Note how* (**degrees/mile**) *times* (**miles/hour**) *equals* (**degrees/hour**). That is the ordinary chain rule $(df/dx)(dx/dt) = (df/dt)$—there is no y variable going south.

If the road goes southeast, the temperature is $f = 30 + .05x + .01y$. Now $x(t)$ is distance south and $y(t)$ is distance east. What is df/dt if the speed is still 70?

Solution $\quad \dfrac{df}{dt} = \dfrac{\partial f}{\partial x}\dfrac{dx}{dt} + \dfrac{\partial f}{\partial y}\dfrac{dy}{dt} = .05\dfrac{70}{\sqrt{2}} + .01\dfrac{70}{\sqrt{2}} \approx 3$ degrees/hour.

In reality there is another term. The temperature also depends *directly on* t, because of night and day. The factor $\cos(2\pi t/24)$ has a period of 24 hours, and it brings an extra term into the chain rule:

For $f(x, y, t)$ the chain rule is $\quad \dfrac{df}{dt} = \dfrac{\partial f}{\partial x}\dfrac{dx}{dt} + \dfrac{\partial f}{\partial y}\dfrac{dy}{dt} + \dfrac{\partial f}{\partial t}. \qquad (4)$

This is the **total** derivative df/dt, from all causes. Changes in x, y, t all affect f. The partial derivative $\partial f/\partial t$ is only one part of df/dt. (Note that $dt/dt = 1$.) If night and day add $12\cos(2\pi t/24)$ to f, the extra term is $\partial f/\partial t = -\pi \sin(2\pi t/24)$. At nightfall that is $-\pi$ degrees per hour. You have to drive faster than 70 mph to get warm.

SECOND DERIVATIVES

What is $d^2 f/dt^2$? We need the derivative of (4), which is painful. It is like acceleration in Chapter 12, with many terms. So start with movement in a straight line.

Suppose $x = x_0 + t\cos\theta$ and $y = y_0 + t\sin\theta$. We are moving at the fixed angle θ, with speed 1. The derivatives are $x_t = \cos\theta$ and $y_t = \sin\theta$ and $\cos^2\theta + \sin^2\theta = 1$. Then df/dt is immediate from the chain rule:

$$f_t = f_x x_t + f_y y_t = f_x \cos\theta + f_y \sin\theta. \qquad (5)$$

For the second derivative f_{tt}, apply this rule to f_t. Then f_{tt} is

$$(f_t)_x \cos\theta + (f_t)_y \sin\theta = (f_{xx}\cos\theta + f_{yx}\sin\theta)\cos\theta + (f_{xy}\cos\theta + f_{yy}\sin\theta)\sin\theta.$$

Collect terms: $\quad f_{tt} = f_{xx}\cos^2\theta + 2f_{xy}\cos\theta\sin\theta + f_{yy}\sin^2\theta. \qquad (6)$

In polar coordinates change t to r. When we move in the r direction, θ is fixed. Equation (6) gives f_{rr} from f_{xx}, f_{xy}, f_{yy}. Second derivatives on curved paths (with new terms from the curving) are saved for the exercises.

EXAMPLE 3 If f_{xx}, f_{xy}, f_{yy} are all continuous and bounded by M, find a bound on f_{tt}. This is the second derivative along any line.

Solution Equation (6) gives $|f_{tt}| \leqslant M\cos^2\theta + M\sin 2\theta + M\sin^2\theta \leqslant 2M$. This upper bound $2M$ was needed in equation 13.3.14, for the error in linear approximation.

THE DERIVATIVES OF $f(x(t,u), y(t,u))$

Suppose there are two inside functions x and y, each depending on t and u. When t moves, x and y both move: $dx = x_t dt$ and $dy = y_t dt$. Then dx and dy force a change in f: $df = f_x dx + f_y dy$. The chain rule for $\partial f/\partial t$ is no surprise:

13I Chain rule for $f(x(t,u), y(t,u))$: $\dfrac{\partial f}{\partial t} = \dfrac{\partial f}{\partial x}\dfrac{\partial x}{\partial t} + \dfrac{\partial f}{\partial y}\dfrac{\partial y}{\partial t}.$ $\qquad$ (7)

This rule has $\partial/\partial t$ instead of d/dt, because of the extra variable u. The symbols remind us that u is constant. Similarly t is constant while u moves, and there is a second chain rule for $\partial f/\partial u$: $f_u = f_x x_u + f_y y_u$.

EXAMPLE 4 In polar coordinates find f_θ and $f_{\theta\theta}$. Start from $f(x, y) = f(r\cos\theta, r\sin\theta)$.

The chain rule uses the θ derivatives of x and y:

$$\frac{\partial f}{\partial \theta} = \frac{\partial f}{\partial x}\frac{\partial x}{\partial \theta} + \frac{\partial f}{\partial y}\frac{\partial y}{\partial \theta} = \left(\frac{\partial f}{\partial x}\right)(-r\sin\theta) + \left(\frac{\partial f}{\partial y}\right)(r\cos\theta). \qquad (8)$$

The second θ derivative is harder, because (8) has four terms that depend on θ. Apply the chain rule to the first term $\partial f/\partial x$. It is a function of x and y, and x and y are functions of θ:

$$\frac{\partial}{\partial \theta}\left(\frac{\partial f}{\partial x}\right) = \frac{\partial}{\partial x}\left(\frac{\partial f}{\partial x}\right)\frac{\partial x}{\partial \theta} + \frac{\partial}{\partial y}\left(\frac{\partial f}{\partial x}\right)\frac{\partial y}{\partial \theta} = f_{xx}(-r\sin\theta) + f_{xy}(r\cos\theta).$$

The θ derivative of $\partial f/\partial y$ is similar. So apply the product rule to (8):

$$f_{\theta\theta} = [f_{xx}(-r\sin\theta) + f_{xy}(r\cos\theta)](-r\sin\theta) + f_x(-r\cos\theta)$$

$$+ [f_{yx}(-r\sin\theta) + f_{yy}(r\cos\theta)](r\cos\theta) + f_y(-r\sin\theta). \qquad (9)$$

This formula is not attractive. In mathematics, a messy formula is almost always a signal of asking the wrong question. In fact the combination $f_{xx} + f_{yy}$ is much more special than the separate derivatives. We might hope the same for $f_{rr} + f_{\theta\theta}$, but dimensionally that is impossible—since r is a length and θ is an angle. The dimensions of f_{xx} and f_{yy} are matched by f_{rr} and f_r/r and $f_{\theta\theta}/r^2$. We could even hope that

$$f_{xx} + f_{yy} = f_{rr} + \frac{1}{r}f_r + \frac{1}{r^2}f_{\theta\theta}. \qquad (10)$$

This equation is true. Add (5) + (6) + (9) with t changed to r. **Laplace's equation** $f_{xx} + f_{yy} = 0$ **is now expressed in polar coordinates**: $f_{rr} + f_r/r + f_{\theta\theta}/r^2 = 0$.

A PARADOX

Before leaving polar coordinates there is one more question. It goes back to $\partial r/\partial x$, which was practically the first example of partial derivatives:

$$\frac{\partial r}{\partial x} = \frac{\partial}{\partial x}\sqrt{x^2 + y^2} = x/\sqrt{x^2 + y^2} = x/r. \qquad (11)$$

My problem is this. We know that x is $r\cos\theta$. So x/r on the right side is $\cos\theta$. On the other hand r is $x/\cos\theta$. So $\partial r/\partial x$ is also $1/\cos\theta$. **How can $\partial r/\partial x$ lead to** $\cos\theta$ **one way and** $1/\cos\theta$ **the other way**?

I will admit that this cost me a sleepless night. There must be an explanation—we cannot end with $\cos\theta = 1/\cos\theta$. This paradox brings a new respect for partial derivatives. May I tell you what I finally noticed? You could cover up the next paragraph and think about the puzzle first.

The key to partial derivatives is to ask: **Which variable is held constant?** In equation (11), y is constant. But when $r = x/\cos\theta$ gave $\partial r/\partial x = 1/\cos\theta$, θ was constant. In both cases we change x and look at the effect on r. The movement is on a horizontal line (constant y) or on a radial line (constant θ). Figure 13.15 shows the difference.

Remark This example shows that $\partial r/\partial x$ is different from $1/(\partial x/\partial r)$. The neat formula $(\partial r/\partial x)(\partial x/\partial r) = 1$ is not generally true. May I tell you what takes its place? We have to include $(\partial r/\partial y)(\partial y/\partial r)$. With two variables xy and two variables $r\theta$, we need 2 by 2 matrices! Section 14.4 gives the details:

$$\begin{bmatrix} \partial r/\partial x & \partial r/\partial y \\ \partial\theta/\partial x & \partial\theta/\partial y \end{bmatrix} \begin{bmatrix} \partial x/\partial r & \partial x/\partial\theta \\ \partial y/\partial r & \partial y\partial\theta \end{bmatrix} = \begin{bmatrix} 1 & 0 \\ 0 & 1 \end{bmatrix}.$$

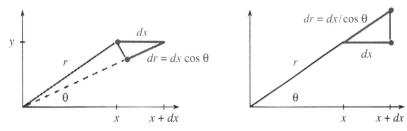

Fig. 13.15 $dr = dx\cos\theta$ when y is constant, $dr = dx/\cos\theta$ when θ is constant.

NON-INDEPENDENT VARIABLES

This paradox points to a serious problem. In computing partial derivatives of $f(x, y, z)$, we assumed that x, y, z were independent. Up to now, x could move while y and z were fixed. In physics and chemistry and economics that may not be possible. If there is a relation between x, y, z, then x can't move by itself.

EXAMPLE 5 The gas law $PV = nRT$ relates pressure to volume and temperature. P, V, T *are not independent.* What is the meaning of $\partial V/\partial P$? Does it equal $1/(\partial P/\partial V)$?

Those questions have no answers, until we say what is held constant. In the paradox, $\partial r/\partial x$ had one meaning for fixed y and another meaning for fixed θ. **To indicate what is held constant, use an extra subscript** (not denoting a derivative):

$$(\partial r/\partial x)_y = \cos\theta \qquad (\partial r/\partial x)_\theta = 1/\cos\theta. \tag{12}$$

$(\partial f/\partial P)_V$ has constant volume and $(\partial f/\partial P)_T$ has constant temperature. The usual $\partial f/\partial P$ has both V and T constant. But then the gas law won't let us change P.

EXAMPLE 6 Let $f = 3x + 2y + z$. Compute $\partial f/\partial x$ on the plane $z = 4x + y$.

Solution 1   Think of x and y as independent. Replace z by $4x + y$:

$$f = 3x + 2y + (4x + y) \quad \text{so} \quad (\partial f/\partial x)_y = 7.$$

Solution 2   Keep x and y independent. Deal with z by the chain rule:

$$(\partial f/\partial x)_y = \partial f/\partial x + (\partial f/\partial z)(\partial z/\partial x) = 3 + (1)(4) = 7.$$

Solution 3   (*different*) Make x and z independent. Then $y = z - 4x$:

$$(\partial f/\partial x)_z = \partial f/\partial x + (\partial f/\partial y)(\partial y/\partial x) = 3 + (2)(-4) = -5.$$

Without a subscript, $\partial f/\partial x$ means: Take the x derivative the usual way. The answer is $\partial f/\partial x = 3$, when y and z don't move. But on the plane $z = 4x + y$, one of them must move! 3 is only part of the total answer, which is $(\partial f/\partial x)_y = 7$ or $(\partial f/\partial x)_z = -5$.

Here is the geometrical meaning. We are on the plane $z = 4x + y$. The derivative $(\partial f/\partial x)_y$ moves x but not y. To stay on the plane, dz is $4dx$. The change in $f = 3x + 2y + z$ is $df = 3dx + 0 + dz = 7dx$.

EXAMPLE 7   On the world line $x^2 + y^2 + z^2 = t^2$ find $(\partial f/\partial y)_{x,z}$ for $f = xyzt$.

The subscripts x, z mean that x and z are fixed. The chain rule skips $\partial f/\partial x$ and $\partial f/\partial z$:

$$(\partial f/\partial y)_{x,z} = \partial f/\partial y + (\partial f/\partial t)(\partial t/\partial y) = xzt + (xyz)(y/t). \text{ Why } y/t?$$

EXAMPLE 8   From the law $PV = T$, compute the product $(\partial P/\partial V)_T (\partial V/\partial T)_P (\partial T/\partial P)_V$.

Any intelligent person cancels ∂V's, ∂T's, and ∂P's to get 1. The right answer is -1:

$$(\partial P/\partial V)_T = -T/V^2 \qquad (\partial V/\partial T)_P = 1/P \qquad (\partial T/\partial P)_V = V.$$

The product is $-T/PV$. **This is** -1 **not** $+1$! The chain rule is tricky (Problem 42).

EXAMPLE 9   Implicit differentiation was used in Chapter 4. The chain rule explains it:

$$\text{If } F(x, y) = 0 \text{ then } F_x + F_y y_x = 0 \text{ so } dy/dx = -F_x/F_y. \tag{13}$$

13.5 EXERCISES

Read-through questions

The chain rule applies to a function of a ___a___. The x derivative of $f(g(x, y))$ is $\partial f/\partial x = $ ___b___. The y derivative is $\partial f/\partial y = $ ___c___. The example $f = (x + y)^n$ has $g = $ ___d___. Because $\partial g/\partial x = \partial g/\partial y$ we know that ___e___ $=$ ___f___. This ___g___ differential equation is satisfied by any function of $x + y$.

Along a path, the derivative of $f(x(t), y(t))$ is $df/dt = $ ___h___. The derivative of $f(x(t), y(t), z(t))$ is ___i___. If $f = xy$ then the chain rule gives $df/dt = $ ___j___. That is the same as the ___k___ rule! When $x = u_1 t$ and $y = u_2 t$ the path is ___l___. The chain rule

for $f(x, y)$ gives $df/dt = $ ___m___. That is the ___n___ derivative $D_{\mathbf{u}} f$.

The chain rule for $f(x(t, u), y(t, u))$ is $\partial f/\partial t = $ ___o___. We don't write df/dt because ___p___. If $x = r \cos \theta$ and $y = r \sin \theta$, the variables t, u change to ___q___. In this case $\partial f/\partial r = r$ and $\partial f/\partial \theta = $ ___s___. That connects the derivatives in ___t___ and ___u___ coordinates. The difference between $\partial r/\partial x = x/r$ and $\partial r/\partial x = 1/\cos \theta$ is because ___v___ is constant in the first and ___w___ is constant in the second.

With a relation like $xyz = 1$, the three variables are ___x___ independent. The derivatives $(\partial f/\partial x)_y$ and $(\partial f/\partial x)_z$ and $(\partial f/\partial x)$ mean ___y___ and ___z___ and ___A___. For $f = x^2 + y^2 + z^2$ with $xyz = 1$, we compute $(\partial f/\partial x)_z$ from the chain rule ___B___. In that rule $\partial z/\partial x = $ ___C___ from the relation $xyz = 1$.

Find f_x and f_y in Problems 1–4. What equation connects them?

1 $f(x,y) = \sin(x+cy)$ **2** $f(x,y) = (ax+by)^{10}$

3 $f(x,y) = e^{x+7y}$ **4** $f(x,y) = \ln(x+7y)$

5 Find both terms in the t derivative of $(g(x(t), y(t)))^3$.

6 If $f(x,y) = xy$ and $x = u(t)$ and $y = v(t)$, what is df/dt? Probably all other rules for derivatives follow from the chain rule.

7 The step function $f(x)$ is zero for $x < 0$ and one for $x > 0$. Graph $f(x)$ and $g(x) = f(x+2)$ and $h(x) = f(x+4)$. If $f(x+2t)$ represents a wall of water (a tidal wave), which way is it moving and how fast?

8 The wave equation is $f_{tt} = c^2 f_{xx}$. (a) Show that $(x+ct)^n$ is a solution. (b) Find C different from c so that $(x+Ct)^n$ is also a solution.

9 If $f = \sin(x-t)$, draw two lines in the xt plane along which $f = 0$. Between those lines sketch a sine wave. Skiing on top of the sine wave, what is your speed dx/dt?

10 If you float at $x = 0$ in Problem 9, do you go up first or down first? At time $t = 4$ what is your height and upward velocity?

11 *Laplace's equation* is $f_{xx} + f_{yy} = 0$. Show from the chain rule that any function $f(x+iy)$ satisfies this equation if $i^2 = -1$. Check that $f = (x+iy)^2$ and its real part _____ and its imaginary part _____ all satisfy Laplace's equation.

12 Equation (10) gave the polar form $f_{rr} + f_r/r + f_{\theta\theta}/r^2 = 0$ of Laplace's equation. (a) Check that $f = r^2 e^{2i\theta}$ and its real part $r^2 \cos 2\theta$ and its imaginary part $r^2 \sin 2\theta$ all satisfy Laplace's equation. (b) Show from the chain rule that any function $f(re^{i\theta})$ satisfies this equation if $i^2 = -1$.

In Problems 13–18 find df/dt from the chain rule (3).

13 $f = x^2 + y^2, x = t, y = t^2$

14 $f = \sqrt{x^2+y^2}, x = t, y = t^2$

15 $f = xy, x = 1-\sqrt{t}, y = 1+\sqrt{t}$

16 $f = x/y, x = e^t, y = 2e^t$

17 $f = \ln(x+y), x = e^t, y = e^t$

18 $f = x^4, x = t, y = t$

19 If a cone grows in height by $dh/dt = 1$ and in radius by $dr/dt = 2$, starting from zero, how fast is its volume growing at $t = 3$?

20 If a rocket has speed $dx/dt = 6$ down range and $dy/dt = 2t$ upward, how fast is it moving away from the launch point at $(0,0)$? How fast is the angle θ changing, if $\tan\theta = y/x$?

21 If a train approaches a crossing at 60 mph and a car approaches (at right angles) at 45 mph, how fast are they coming together? (a) Assume they are both 90 miles from the crossing. (b) Assume they are going to hit.

22 In Example 2 does the temperature increase faster if you drive due south at 70 mph or southeast at 80 mph?

23 On the line $x = u_1 t, y = u_2 t, z = u_3 t$, what combination of f_x, f_y, f_z gives df/dt? This is the *directional derivative* in 3D.

24 On the same line $x = u_1 t, y = u_2 t, z = u_3 t$, find a formula for $d^2 f/dt^2$. Apply it to $f = xyz$.

25 For $f(x,y,t) = x+y+t$ find $\partial f/\partial t$ and df/dt when $x = 2t$ and $y = 3t$. Explain the difference.

26 If $z = (x+y)^2$ then $x = \sqrt{z} - y$. Does $(\partial z/\partial x)(\partial x/\partial z) = 1$?

27 Suppose $x_t = t$ and $y_t = 2t$, not constant as in (5–6). For $f(x,y)$ find f_t and f_{tt}. The answer involves $f_x, f_y, f_{xx}, f_{xy}, f_{yy}$.

28 Suppose $x_t = t$ and $y_t = t^2$. For $f = (x+y)^3$ find f_t and then f_{tt} from the chain rule.

29 Derive $\partial f/\partial r = (\partial f/\partial x)\cos\theta + (\partial f/\partial y)\sin\theta$ from the chain rule. Why do we take $\partial x/\partial r$ as $\cos\theta$ and not $1/\cos\theta$?

30 Compute f_{xx} for $f(x,y) = (ax+by+c)^{10}$. If $x = t$ and $y = t$ compute f_{tt}. True or false: $(\partial f/\partial x)(\partial x/\partial t) = \partial f/\partial t$.

31 Show that $\partial^2 r/\partial x^2 = y^2/r^3$ in two ways:
(1) Find the x derivative of $\partial r/\partial x = x/\sqrt{x^2+y^2}$
(2) Find the x derivative of $\partial r/\partial x = x/r$ by the chain rule.

32 Reversing x and y in Problem 31 gives $r_{yy} = x^2/r^3$. But show that $r_{xy} = -xy/r^3$.

33 If $\sin z = x+y$ find $(\partial z/\partial x)_y$ in two ways:
(1) Write $z = \sin^{-1}(x+y)$ and compute its derivative.
(2) Take x derivatives of $\sin z = x+y$. Verify that these answers, explicit and implicit, are equal.

34 By direct computation find f_x and f_{xx} and f_{xy} for $f = \sqrt{x^2+y^2}$.

35 Find a formula for $\partial^2 f/\partial r\partial\theta$ in terms of the x and y derivatives of $f(x,y)$.

36 Suppose $z = f(x,y)$ is solved for x to give $x = g(y,z)$. Is it true that $\partial z/\partial x = 1/(\partial x/\partial z)$? Test on examples.

37 Suppose $z = e^{xy}$ and therefore $x = (\ln z)/y$. Is it true or not that $(\partial z/\partial x) = 1/(\partial x/\partial z)$?

38 If $x = x(t,u,v)$ and $y = y(t,u,v)$ and $z = z(t,u,v)$, find the t derivative of $f(x,y,z)$.

39 The t derivative of $f(x(t,u), y(t,u))$ is in equation (7). What is f_{tt}?

40 (a) For $f = x^2 + y^2 + z^2$ compute $\partial f/\partial x$ (no subscript, x, y, z all independent).
(b) When there is a further relation $z = x^2 + y^2$, use it to remove z and compute $(\partial f/\partial x)_y$.

(c) Compute $(\partial f/\partial x)_y$ using the chain rule $(\partial f/\partial x) + (\partial f/\partial z)(\partial z/\partial x)$.

(d) Why doesn't that chain rule contain $(\partial f/\partial y)(\partial y/\partial x)$?

41 For $f = ax + by$ on the plane $z = 3x + 5y$, find $(\partial f/\partial x)_z$ and $(\partial f/\partial x)_y$ and $(\partial f/\partial z)_x$.

42 The gas law in physics is $PV = nRT$ or a more general relation $F(P, V, T) = 0$. Show that the three derivatives in Example 8 still multiply to give -1. First find $(\partial P/\partial V)_T$ from $\partial F/\partial V + (\partial F/\partial P)(\partial P/\partial V)_T = 0$.

43 If Problem 42 changes to four variables related by $F(x, y, z, t) = 0$, what is the corresponding product of four derivatives?

44 Suppose $x = t + u$ and $y = tu$. Find the t and u derivatives of $f(x, y)$. Check when $f(x, y) = x^2 - 2y$.

45 (a) For $f = r^2 \sin^2 \theta$ find f_x and f_y.

(b) For $f = x^2 + y^2$ find f_r and f_θ.

46 On the curve $\sin x + \sin y = 0$, find dy/dx and $d^2 y/dx^2$ by implicit differentiation.

47 (horrible) Suppose $f_{xx} + f_{yy} = 0$. If $x = u + v$ and $y = u - v$ and $f(x, y) = g(u, v)$, find g_u and g_v. Show that $g_{uu} + g_{vv} = 0$.

48 A function has *constant returns to scale* if $f(cx, cy) = cf(x, y)$ When x and y are doubled so are $f = \sqrt{x^2 + y^2}$ and $f = \sqrt{xy}$. In economics , input/output is constant. In mathematics f is *homogeneous* of degree one.

Prove that $x \partial f/\partial x + y \partial f/\partial y = f(x, y)$, by computing the c derivative at $c = 1$. Test this equation on the two examples and find a third example.

49 ***True or false***: The directional derivative of $f(r, \theta)$ in the direction of $\mathbf{u}_\theta$ is $\partial f/\partial \theta$.

13.6 Maxima, Minima, and Saddle Points

The outstanding equation of differential calculus is also the simplest: $df/dx = 0$. The slope is zero and the tangent line is horizontal. Most likely we are at the top or bottom of the graph—a maximum or a minimum. This is the point that the engineer or manager or scientist or investor is looking for—maximum stress or production or velocity or profit. With more variables in $f(x, y)$ and $f(x, y, z)$, the problem becomes more realistic. The question still is: ***How to locate the maximum and minimum?***

The answer is in the *partial derivatives*. When the graph is level, they are zero. Deriving the equations $f_x = 0$ and $f_y = 0$ is pure mathematics and pure pleasure. Applying them is the serious part. We watch out for saddle points, and also for a minimum at a boundary point—this section takes extra time. Remember the steps for $f(x)$ in one-variable calculus:

1. The leading candidates are ***stationary*** points (where $df/dx = 0$).
2. The other candidates are ***rough points*** (no derivative) and ***endpoints*** (a or b).
3. Maximum vs. minimum is decided by the sign of the ***second derivative***.

In two dimensions, a stationary point requires $\partial f/\partial x = 0$ and $\partial f/\partial y = 0$. The tangent line becomes a tangent plane. The endpoints a and b are replaced by a ***boundary curve***. In practice boundaries contain about 40% of the minima and 80% of the work.

Finally there are three second derivatives f_{xx}, f_{xy}, and f_{yy}. They tell how the graph bends away from the tangent plane—up at a *minimum*, down at a *maximum*, both ways at a *saddle point*. This will be determined by comparing $(f_{xx})(f_{yy})$ with $(f_{xy})^2$.

STATIONARY POINT → HORIZONTAL TANGENT → ZERO DERIVATIVES

Suppose f has a minimum at the point (x_0, y_0). This may be an ***absolute minimum*** or only a ***local minimum***. In both cases $f(x_0, y_0) \leqslant f(x, y)$ near the point. For an absolute minimum, this inequality holds wherever f is defined. For a local minimum, the inequality can fail far away from (x_0, y_0). The bottom of your foot is an absolute minimum, the end of your finger is a local minimum. We assume for now that (x_0, y_0) is an *interior point* of the domain of f. At a boundary point, we cannot expect a horizontal tangent and zero derivatives.

Main conclusion: At a minimum or maximum (absolute or local) a nonzero derivative is impossible. The tangent plane would tilt. In some direction f would decrease. Note that the minimum *point* is (x_0, y_0), and the minimum *value* is $f(x_0, y_0)$.

13J If derivatives exist at an interior minimum or maximum, they are zero:

$$\partial f/\partial x = 0 \quad \text{and} \quad \partial f/\partial y = 0 \quad (\textbf{\textit{together this is}} \text{ grad } f = 0). \qquad (1)$$

For a function $f(x, y, z)$ of three variables, add the third equation $\partial f/\partial z = 0$.

The reasoning goes back to the one-variable case. That is because we look along the lines $x = x_0$ and $y = y_0$. The minimum of $f(x, y)$ is at the point where the lines meet. So this is also the minimum ***along each line separately***.

Moving in the x direction along $y = y_0$, we find $\partial f/\partial x = 0$. Moving in the y direction, $\partial f/\partial y = 0$ at the same point. ***The slope in every direction is zero***. In other words grad $f = \mathbf{0}$.

Graphically, (x_0, y_0) is the low point of the surface $z = f(x, y)$. Both cross sections in Figure 13.16 touch bottom. The phrase "if derivatives exist" rules out the vertex of a cone, which is a **rough point**. The absolute value $f = |x|$ has a minimum without $df/dx = 0$, and so does the distance $f = r$. The rough point is $(0, 0)$.

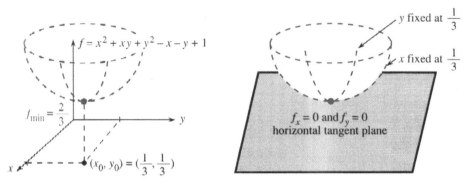

Fig. 13.16 $\partial f/\partial x = 0$ and $\partial f/\partial y = 0$ at the minimum. Quadratic f has linear derivatives.

EXAMPLE 1   Minimize the quadratic $f(x, y) = x^2 + xy + y^2 - x - y + 1$.

To locate the minimum (or maximum), set $f_x = 0$ and $f_y = 0$:

$$f_x = 2x + y - 1 = 0 \quad \text{and} \quad f_y = x + 2y - 1 = 0. \tag{2}$$

Notice what's important: ***There are two equations for two unknowns*** x ***and*** y. Since f is quadratic, the equations are linear. Their solution is $x_0 = \frac{1}{3}$, $y_0 = \frac{1}{3}$ (the stationary point). This is actually a minimum, but to prove that you need to read further.

The constant 1 affects the minimum value $f = \frac{2}{3}$—but not the minimum point. The graph shifts up by 1. The linear terms $-x - y$ affect f_x and f_y. They move the minimum away from $(0, 0)$. The quadratic part $x^2 + xy + y^2$ makes the surface curve upwards. Without that curving part, a plane has its minimum and maximum at boundary points.

EXAMPLE 2   *(Steiner's problem)* *Find the point that is nearest to three given points.*

This example is worth your attention. We are locating an airport close to three cities. Or we are choosing a house close to three jobs. The problem is to get as near as possible to the corners of a triangle. The best point depends on the meaning of "*near.*"

The distance to the first corner (x_1, y_1) is $d_1 = \sqrt{(x - x_1)^2 + (y - y_1)^2}$. The distances to the other corners (x_2, y_2) and (x_3, y_3) are d_2 and d_3. Depending on whether cost equals (*distance*) or (*distance*)^2 or (*distance*)^p, our problem will be:

$$\textit{Minimize} \quad d_1 + d_2 + d_3 \quad \textit{or} \quad d_1^2 + d_2^2 + d_3^2 \quad \textit{or even} \quad d_1^p + d_2^p + d_3^p.$$

The second problem is the easiest, when d_1^2 and d_2^2 and d_3^2 are quadratics:

$$f(x, y) = (x - x_1)^2 + (y - y_1)^2 + (x - x_2)^2 + (y - y_2)^2 + (x - x_3)^2 + (y - y_3)^2$$

$$\partial f/\partial x = 2[x - x_1 + x - x_2 + x - x_3] = 0 \quad \partial f/\partial y = 2[y - y_1 + y - y_2 + y - y_3] = 0.$$

Solving $\partial f/\partial x = 0$ gives $x = \frac{1}{3}(x_1 + x_2 + x_3)$. Then $\partial f/\partial y = 0$ gives $y = \frac{1}{3}(y_1 + y_2 + y_3)$. The best point is the **centroid of the triangle** (Figure 13.17a). It is the nearest point to the corners when the cost is (distance)^2. Note how squaring makes the derivatives linear. *Least squares* dominates an enormous part of applied mathematics.

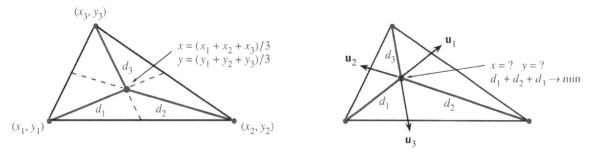

Fig. 13.17   The centroid minimizes $d_1^2 + d_2^2 + d_3^2$. The Steiner point minimizes $d_1 + d_2 + d_3$.

The real "*Steiner problem*" is to minimize $f(x, y) = d_1 + d_2 + d_3$. We are laying down roads from the corners, with cost proportional to length. The equations $f_x = 0$ and $f_y = 0$ look complicated because of square roots. But the nearest point in Figure 13.17b has a remarkable property, which you will appreciate.

Calculus takes derivatives of $d_1^2 = (x - x_1)^2 + (y - y_1)^2$. The x derivative leaves $2d_1(\partial d_1 / \partial x) = 2(x - x_1)$. Divide both sides by $2d_1$:

$$\frac{\partial d_1}{\partial x} = \frac{x - x_1}{d_1} \quad \text{and} \quad \frac{\partial d_1}{\partial y} = \frac{y - y_1}{d_1} \quad \text{so} \quad \text{grad } d_1 = \left(\frac{x - x_1}{d_1}, \frac{y - y_1}{d_1} \right). \quad (3)$$

This gradient is a unit vector. The sum of $(x - x_1)^2 / d_1^2$ and $(y - y_1)^2 / d_1^2$ is $d_1^2 / d_1^2 = 1$. This was already in Section 13.4: Distance increases with slope 1 away from the center. The gradient of d_1 (call it $\mathbf{u}_1$) is a unit vector from the center point (x_1, y_1).

Similarly the gradients of d_2 and d_3 are unit vectors $\mathbf{u}_2$ and $\mathbf{u}_3$. They point directly away from the other corners of the triangle. The total cost is $f(x, y) = d_1 + d_2 + d_3$, so we add the gradients. The equations $f_x = 0$ and $f_y = 0$ combine into the vector equation

$$\text{grad } f = \mathbf{u}_1 + \mathbf{u}_2 + \mathbf{u}_3 = \mathbf{0} \textbf{ at the minimum}.$$

The three unit vectors add to zero! Moving away from one corner brings us closer to another. The nearest point to the three corners is where those movements cancel. This is the meaning of "grad $f = \mathbf{0}$ at the minimum."

It is unusual for three unit vectors to add to zero—this can only happen in one way. *The three directions must form angles of* 120°. The best point has this property, which is repeated in Figure 13.18a. The unit vectors cancel each other. At the "Steiner point," the roads to the corners make 120° angles. This optimal point solves the problem, except for one more possibility.

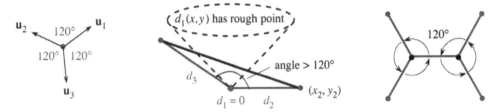

Fig. 13.18 Gradients $\mathbf{u}_1 + \mathbf{u}_2 + \mathbf{u}_3 = \mathbf{0}$ for $120°$ angles. Corner wins at wide angle. *Four corners.* In this case two branchpoints are better—still $120°$.

The other possibility is a minimum at a ***rough point***. The graph of the distance function $d_1(x, y)$ is a cone. It has a sharp point at the center (x_1, y_1). All three corners of the triangle are rough points for $d_1 + d_2 + d_3$, so all of them are possible minimizers.

Suppose the angle at a corner exceeds $120°$. Then there is no Steiner point. Inside the triangle, the angle would become even wider. The best point must be a rough point—one of the corners. The winner is the corner with the wide angle. In the figure that means $d_1 = 0$. Then the sum $d_2 + d_3$ comes from the two shortest edges.

Summary   The solution is at a $120°$ point or a wide-angle corner. That is the theory. The real problem is to compute the Steiner point—which I hope you will do.

Remark 1   Steiner's problem for *four points* is surprising. We don't minimize $d_1 + d_2 + d_3 + d_4$—there is a better problem. Connect the four points with roads, minimizing their total length, ***and allow the roads to branch***. A typical solution is in Figure 13.18c. The angles at the branch points are $120°$. There are at most two branch points (two less than the number of corners).

Remark 2 For other powers p, the cost is $(d_1)^p + (d_2)^p + (d_3)^p$. The x derivative is

$$\partial f / \partial x = p(d_1)^{p-2}(x - x_1) + p(d_2)^{p-2}(x - x_2) + p(d_1)^{p-2}(x - x_3). \qquad (4)$$

The key equations are still $\partial f / \partial x = 0$ and $\partial f / \partial y = 0$. Solving them requires a computer and an algorithm. To share the work fairly, I will supply the algorithm (Newton's method) if you supply the computer. Seriously, this is a terrific example. It is typical of real problems—we know $\partial f / \partial x$ and $\partial f / \partial y$ but not the point where they are zero. You can calculate that nearest point, which changes as p changes. You can also discover new mathematics, about how that point moves. I will collect all replies I receive to Problems 38 and 39.

MINIMUM OR MAXIMUM ON THE BOUNDARY

Steiner's problem had no boundaries. The roads could go anywhere. But most applications have restrictions on x and y, like $x \geqslant 0$ or $y \leqslant 0$ or $x^2 + y^2 \geqslant 1$. The minimum with these restrictions is probably higher than the absolute minimum. There are three possibilities:

(1) *stationary point* $f_x = 0$, $f_y = 0$ (2) *rough point*   (3) *boundary point*

That third possibility requires us to maximize or minimize $f(x, y)$ along the boundary.

EXAMPLE 3 Minimize $f(x,y) = x^2 + xy + y^2 - x - y + 1$ in the *half-plane $x \geq 0$*.

The minimum in Example 1 was $\frac{2}{3}$. It occurred at $x_0 = \frac{1}{3}, y_0 = \frac{1}{3}$. *This point is still allowed.* It satisfies the restriction $x \geq 0$. So the minimum is not moved.

EXAMPLE 4   Minimize the same $f(x,y)$ restricted to the *lower half-plane $y \leq 0$*.

Now the absolute minimum at $(\frac{1}{3}, \frac{1}{3})$ is not allowed. There are no rough points. We look for a minimum on the boundary line $y = 0$ in Figure 13.19a. Set $y = 0$, so f depends only on x. Then choose the best x:

$$f(x,0) = x^2 + 0 - x - 0 + 1 \quad \text{and} \quad f_x = 2x - 1 = 0.$$

The minimum is at $x = \frac{1}{2}$ and $y = 0$, where $f = \frac{3}{4}$. This is up from $\frac{2}{3}$.

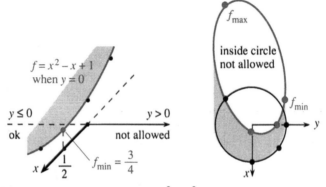

Fig. 13.19   The boundaries $y = 0$ and $x^2 + y^2 = 1$ contain the minimum points.

EXAMPLE 5   *Minimize the same $f(x,y)$ on or outside the circle $x^2 + y^2 = 1$.*

One possibility is $f_x = 0$ and $f_y = 0$. But this is at $(\frac{1}{3}, \frac{1}{3})$, inside the circle. The other possibility is a minimum at a boundary point, *on the circle*.

To follow this boundary we can set $y = \sqrt{1 - x^2}$. The function f gets complicated, and df/dx is worse. There is a way to avoid square roots: Set $x = \cos t$ and $y = \sin t$. Then $f = x^2 + xy + y^2 - x - y + 1$ is a function of the angle t:

$$f(t) = 1 + \cos t \sin t - \cos t - \sin t + 1$$

$$df/dt = \cos^2 t - \sin^2 t + \sin t - \cos t = (\cos t - \sin t)(\cos t + \sin t - 1).$$

Now $df/dt = 0$ locates a minimum or maximum along the boundary. The first factor $(\cos t - \sin t)$ is zero when $x = y$. The second factor is zero when $\cos t + \sin t = 1$, or $x + y = 1$. Those points *on the circle* are the candidates. Problem 24 sorts them out, and Section 13.7 finds the minimum in a new way—using "*Lagrange multipliers*." Minimization on a boundary is a serious problem—it gets difficult quickly—and multipliers are ultimately the best solution.

MAXIMUM VS. MINIMUM VS. SADDLE POINT

How to separate the maximum from the minimum? When possible, try all candidates and decide. Compute f at every stationary point and other critical point (maybe also out at infinity), and compare. Calculus offers another approach, based on *second derivatives*.

With one variable the second derivative test was simple: $f_{xx} > 0$ at a minimum, $f_{xx} = 0$ at an inflection point, $f_{xx} < 0$ at a maximum. This is a local test, which may not give a global answer. But it decides whether the slope is increasing (bottom of the graph) or decreasing (top of the graph). We now find a similar test for $f(x, y)$.

The new test involves all three second derivatives. It applies where $f_x = 0$ and $f_y = 0$. The tangent plane is horizontal. **We ask whether the graph of f goes above or below that plane.** The tests $f_{xx} > 0$ and $f_{yy} > 0$ guarantee a minimum in the x and y directions, but there are other directions.

EXAMPLE 6 $f(x, y) = x^2 + 10xy + y^2$ has $f_{xx} = 2, f_{xy} = 10, f_{yy} = 2$ (*minimum or not?*)

All second derivatives are positive—but wait and see. The stationary point is $(0,0)$, where $\partial f/\partial x$ and $\partial f/\partial y$ are both zero. Our function is the sum of $x^2 + y^2$, which goes upward, and $10xy$ which has a saddle. The second derivatives must decide whether $x^2 + y^2$ or $10xy$ is stronger.

Along the x axis, where $y = 0$ and $f = x^2$, our point is at the bottom. The minimum in the x direction is at $(0,0)$. Similarly for the y direction. But $(0,0)$ is *not a minimum point* for the whole function, because of $10xy$.

Try $x = 1, y = -1$. Then $f = 1 - 10 + 1$, which is negative. The graph goes *below* the xy plane in that direction. The stationary point at $x = y = 0$ is a *saddle point*.

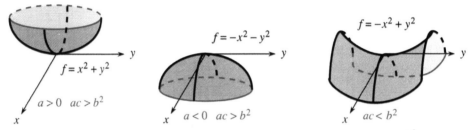

Fig. 13.20   Minimum, maximum, saddle point based on the signs of a and $ac - b^2$.

EXAMPLE 7 $f(x, y) = x^2 + xy + y^2$ has $f_{xx} = 2, f_{xy} = 1, f_{yy} = 2$ (*minimum or not?*)

The second derivatives $2, 1, 2$ are again positive. The graph curves up in the x and y directions. But there is a big difference from Example 6: f_{xy} is reduced from 10 to 1. **It is the size of f_{xy} (*not its sign!*) that makes the difference.** The extra terms $-x - y + 4$ in Example 1 moved the stationary point to $(\frac{1}{3}, \frac{1}{3})$. The second derivatives are still $2, 1, 2$, and they pass the test for a minimum:

13K At $(0,0)$ the quadratic function $f(x, y) = ax^2 + 2bxy + cy^2$ has a

$$\text{minimum if } \genfrac{}{}{0pt}{}{a>0}{ac>b^2} \qquad \text{maximum if } \genfrac{}{}{0pt}{}{a<0}{ac>b^2} \qquad \textit{saddle point} \text{ if } ac < b^2.$$

For a direct proof, split $f(x, y)$ into two parts by "completing the square:"

$$ax^2 + 2bxy + cy^2 = a\left(x + \frac{b}{a}y\right)^2 + \frac{ac - b^2}{a}y^2.$$

That algebra can be checked (notice the $2b$). It is the conclusion that's important:

if $a > 0$ and $ac > b^2$, both parts are positive: ***minimum*** at $(0,0)$

if $a < 0$ and $ac > b^2$, both parts are negative: ***maximum*** at $(0,0)$

if $ac < b^2$, the parts have opposite signs: ***saddle point*** at $(0,0)$.

Since the test involves the *square* of b, its sign has no importance. Example 6 had $b = 5$ and a saddle point. Example 7 had $b = \frac{1}{2}$ and a minimum. Reversing to $-x^2 - xy - y^2$ yields a maximum. So does $-x^2 + xy - y^2$.

Now comes the final step, from $ax^2 + 2bxy + cy^2$ to a general function $f(x, y)$. For all functions, quadratics or not, it is the ***second order terms*** that we test.

EXAMPLE 8 $f(x, y) = e^x - x - \cos y$ has a stationary point at $x = 0, y = 0$.

The first derivatives are $e^x - 1$ and $\sin y$, both zero. The second derivatives are $f_{xx} = e^x = 1$ and $f_{yy} = \cos y = 1$ and $f_{xy} = 0$. We only use the derivatives *at the stationary point*. The first derivatives are zero, so the second order terms come to the front in the series for $e^x - x - \cos y$:

$$(1 + x + \tfrac{1}{2}x^2 + \cdots) - x - (1 - \tfrac{1}{2}y^2 + \cdots) = \tfrac{1}{2}x^2 + \tfrac{1}{2}y^2 + \text{higher order terms.} \quad (7)$$

There is a *minimum* at the origin. The quadratic part $\frac{1}{2}x^2 + \frac{1}{2}y^2$ goes upward. The x^3 and y^4 terms are too small to protest. Eventually those terms get large, but near a stationary point it is the quadratic that counts. We didn't need the whole series, because from $f_{xx} = f_{yy} = 1$ and $f_{xy} = 0$ we knew it would start with $\frac{1}{2}x^2 + \frac{1}{2}y^2$.

13L The test in **13K** applies to the second derivatives $a = f_{xx}, b = f_{xy}, c = f_{yy}$ of any $f(x, y)$ at any stationary point. At all points the test decides whether the graph is concave up, concave down, or "indefinite."

EXAMPLE 9 $f(x, y) = e^{xy}$ has $f_x = ye^{xy}$ and $f_y = xe^{xy}$. The stationary point is $(0,0)$.

The second derivatives at that point are $a = f_{xx} = 0, b = f_{xy} = 1$, and $c = f_{yy} = 0$. The test $b^2 > ac$ makes this a saddle point. Look at the infinite series:

$$e^{xy} = 1 + xy + \tfrac{1}{2}x^2y^2 + \cdots.$$

No linear term because $f_x = f_y = 0$: The origin is a ***stationary point***. No x^2 or y^2 term (only xy): The stationary point is a ***saddle point***.

At $x = 2, y = -2$ we find $f_{xx}f_{yy} > (f_{xy})^2$. The graph is concave up at that point—but it's not a minimum since the first derivatives are not zero.

The series begins with the constant term—not important. Then come the linear terms—extremely important. Those terms are decided by *first* derivatives, and they give the tangent plane. It is only at stationary points—when the linear part disappears and the tangent plane is horizontal—that second derivatives take over. Around any basepoint, ***these constant-linear-quadratic terms are the start of the Taylor series***.

THE TAYLOR SERIES

We now put together the whole infinite series. It is a "Taylor series"—which means it is ***a power series that matches all derivatives of*** f (at the basepoint). For one variable, the powers were x^n when the basepoint was 0. For two variables, the powers are x^n times y^m when the basepoint is $(0,0)$. Chapter 10 multiplied the nth derivative $d^n f/dx^n$ by $x^n/n!$ *Now every mixed derivative* $(\partial/\partial x)^n (\partial/\partial y)^m f(x,y)$ *is computed at the basepoint* (subscript $_0$).

We multiply those numbers by $x^n y^m/n!m!$ to match each derivative of $f(x,y)$:

13M When the basepoint is $(0,0)$, the Taylor series is a double sum $\Sigma\Sigma a_{nm}x^n y^m$. The term $a_{nm}x^n y^m$ has the same mixed derivative at $(0,0)$ as $f(x,y)$. The series is

$$f(0,0)+x\left(\frac{\partial f}{\partial x}\right)_0 + y\left(\frac{\partial f}{\partial y}\right)_0 + \frac{x^2}{2}\left(\frac{\partial^2 f}{\partial x^2}\right)_0 + xy\left(\frac{\partial^2 f}{\partial x\partial y}\right)_0$$

$$+\frac{y^2}{2}\left(\frac{\partial^2 f}{\partial y^2}\right)_0 + \sum\sum_{n+m>2}\frac{x^n y^m}{n!m!}\left(\frac{\partial^{n+m} f}{\partial x^n\partial^m}\right)_0 .$$

The derivatives of this series agree with the derivatives of $f(x,y)$ at the basepoint.

The first three terms are the linear approximation to $f(x,y)$. They give the tangent plane at the basepoint. The x^2 term has $n=2$ and $m=0$, so $n!m!=2$. The xy term has $n=m=1$, and $n!m!=1$. *The quadratic part* $\frac{1}{2}(ax^2+2bxy+cy^2)$ *is in control when the linear part is zero*.

EXAMPLE 10   All derivatives of e^{x+y} equal one at the origin. The Taylor series is

$$e^{x+y} = 1+x+y+\frac{x^2}{2}+xy+\frac{y^2}{2}+\cdots = \sum\sum\frac{x^n y^m}{n!m!}$$

This happens to have $ac=b^2$, the special case that was omitted in **13M** and **13N**. *It is the two-dimensional version of an inflection point*. The second derivatives fail to decide the concavity. When $f_{xx}f_{yy}=(f_{xy})^2$, the decision is passed up to the higher derivatives. But in ordinary practice, the Taylor series is stopped after the quadratics.

If the basepoint moves to (x_0,y_0), the powers become $(x-x_0)^n(y-y_0)^m$—and all derivatives are computed at this new basepoint.

Final question: ***How would you compute a minimum numerically?*** One good way is to solve $f_x=0$ and $f_y=0$. These are the functions g and h of Newton's method (Section 13.3). At the current point (x_n,y_n), the derivatives of $g=f_x$ and $h=f_y$ give linear equations for the steps Δx and Δy. Then the next point $x_{n+1}=x_n+\Delta x, y_{n+1}=y_n+\Delta y$ comes from those steps. The input is (x_n,y_n), the output is the new point, and the linear equations are

$$\begin{array}{ll}(g_x)\Delta x+(g_y)\Delta y=-g(x_n,y_n) & (f_{xx})\Delta x+(f_{xy})\Delta y=-f_x(x_n,y_n)\\ (h_x)\Delta x+(h_y)\Delta y=-h(x_n,y_n) & (f_{xy})\Delta x+(f_{yy})\Delta y=-f_y(x_n,y_n).\end{array} \qquad (5)$$

$$\text{or}$$

When the second derivatives of f are available, use Newton's method.

When the problem is too complicated to go beyond first derivatives, here is an alternative—*steepest descent*. The goal is to move down the graph of $f(x,y)$, like a boulder rolling down a mountain. The steepest direction at any point is given by the *gradient*, with a minus sign to go down instead of up. So move in the direction $\Delta x = -s\,\partial f/\partial x$ and $\Delta y = -s\,\partial f/\partial y$.

The question is: How far to move? Like a boulder, a steep start may not aim directly toward the minimum. The stepsize s is monitored, to end the step when the function

f starts upward again (Problem 54). At the end of each step, compute first derivatives and start again in the new steepest direction.

13.6 EXERCISES

Read-through questions

A minimum occurs at a __a__ point (where $f_x = f_y = 0$) or a __b__ point (no derivative) or a __c__ point. Since $f = x^2 - xy + 2y$ has $f_x = $__d__ and $f_y = $__e__, the stationary point is $x = $__f__, $y = $__g__. This is not a minimum, because f decreases when __h__.

The minimum of $d^2 = (x - x_1)^2 + (y - y_1)^2$ occurs at the rough point __i__. The graph of d is a __j__ and grad d is a __k__ vector that points __l__. The graph of $f = |xy|$ touches bottom along the lines __m__. Those are "rough lines" because the derivative __n__. The maximum of d and f must occur on the __o__ of the allowed region because it doesn't occur __p__.

When the boundary curve is $x = x(t), y = y(t)$, the derivative of $f(x, y)$ along the boundary is __q__ (chain rule). If $f = x^2 + 2y^2$ and the boundary is $x = \cos t, y = \sin t$, then $df/dt = $__r__. It is zero at the points __s__. The maximum is at __t__ and the minimum is at __u__. Inside the circle f has an absolute minimum at __v__.

To separate maximum from minimum from __w__, compute the __x__ derivatives at a __y__ point. The tests for a minimum are __z__. The tests for a maximum are __A__. In case $ac < $__B__ or $f_{xx} f_{yy} < $__C__, we have a __D__. At all points these tests decide between concave up and __E__ and "indefinite." For $f = 8x^2 - 6xy + y^2$, the origin is a __F__. The signs of f at $(1,0)$ and $(1,3)$ are __G__.

The Taylor series for $f(x, y)$ begins with the six terms __H__. The coefficient of $x^n y^m$ is __I__. To find a stationary point numerically, use __J__ or __K__.

Find all stationary points ($f_x = f_y = 0$) in 1–16. Separate minimum from maximum from saddle point. Test 13K applies to $a = f_{xx}, b = f_{xy}, c = f_{yy}$.

1 $x^2 + 2xy + 3y^2$

2 $xy - x + y$

3 $x^2 + 4xy + 3y^2 - 6x - 12y$

4 $x^2 - y^2 + 4y$

5 $x^2 y^2 - x$

6 $xe^y - e^x$

7 $-x^2 + 2xy - 3y^2$

8 $(x + y)^2 + (x + 2y - 6)^2$

9 $x^2 + y^2 + z^2 - 4z$

10 $(x + y) + (x + 2y - 6)$

11 $(x - y)^2$

12 $(1 + x^2)/(1 + y^2)$

13 $(x + y)^2 - (x + 2y)^2$

14 $\sin x - \cos y$

15 $x^3 + y^3 - 3x^2 + 3y^2$

16 $8xy - x^4 - y^4$

17 A rectangle has sides on the x and y axes and a corner on the line $x + 3y = 12$. Find its maximum area.

18 A box has a corner at $(0,0,0)$ and all edges parallel to the axes. If the opposite corner (x, y, z) is on the plane $3x + 2y + z = 1$, what position gives maximum volume? Show first that the problem maximizes $xy - 3x^2 y - 2xy^2$.

19 (Straight line fit, Section 11.4) Find x and y to minimize the error
$$E = (x + y)^2 + (x + 2y - 5)^2 + (x + 3y - 4)^2.$$
Show that this gives a minimum not a saddle point.

20 (*Least squares*) What numbers x, y come closest to satisfying the three equations $x - y = 1, 2x + y = -1, x + 2y = 1$? Square and add the errors, $(x - y - 1)^2 + $_____ + _____. Then minimize.

21 Minimize $f = x^2 + xy + y^2 - x - y$ restricted by
(a) $x \leqslant 0$ (b) $y \geqslant 1$ (c) $x \leqslant 0$ and $y \geqslant 1$.

22 Minimize $f = x^2 + y^2 + 2x + 4y$ in the regions
(a) all x, y (b) $y \geqslant 0$ (c) $x \geqslant 0, y \geqslant 0$

23 Maximize and minimize $f = x + \sqrt{3} y$ on the circle $x = \cos t, y = \sin t$.

24 Example 5 followed $f = x^2 + xy + y^2 - x - y + 1$ around the circle $x^2 + y^2 = 1$. The four stationary points have $x = y$ or $x + y = 1$. Compute f at those points and locate the minimum.

25 (a) Maximize $f = ax + by$ on the circle $x^2 + y^2 = 1$.
(b) Minimize $x^2 + y^2$ on the line $ax + by = 1$.

26 For $f(x, y) = \frac{1}{4}x^4 - xy + \frac{1}{4}y^4$, what are the equations $f_x = 0$ and $f_y = 0$? What are their solutions? What is $f_{\min}$?

27 Choose $c > 0$ so that $f = x^2 + xy + cy^2$ has a saddle point at $(0,0)$. Note that $f > 0$ on the lines $x = 0$ and $y = 0$ and $y = x$ and $y = -x$, so checking four directions does not confirm a minimum.

Problems 28–42 minimize the Steiner distance $f = d_1 + d_2 + d_3$ and related functions. A computer is needed for 33 and 36–39.

28 Draw the triangle with corners at $(0,0), (1,1)$, and $(1,-1)$. By symmetry the Steiner point will be on the x axis. Write down the distances d_1, d_2, d_3 to $(x,0)$ and find the x that minimizes $d_1 + d_2 + d_3$. Check the $120°$ angles.

29 Suppose three unit vectors add to zero. Prove that the angles between them must be $120°$.

30 In three dimensions, Steiner minimizes the total distance $f(x,y,z) = d_1 + d_2 + d_3 + d_4$ from four points. Show that grad d_1 is still a unit vector (in which direction?) At what angles do four unit vectors add to zero?

31 With four points in a plane, the Steiner problem allows branches (Figure 13.18c). Find the shortest network connecting the corners of a rectangle, if the side lengths are (a) 1 and 2 (b) 1 and 1 (two solutions for a square) (c) 1 and 0.1.

32 Show that a Steiner point ($120°$ angles) can never be outside the triangle.

33 Write a program to minimize $f(x,y) = d_1 + d_2 + d_3$ by Newton's method in equation (5). Fix two corners at $(0,0), (3,0)$, vary the third from $(1,1)$ to $(2,1)$ to $(3,1)$ to $(4,1)$, and compute Steiner points.

34 Suppose one side of the triangle goes from $(-1,0)$ to $(1,0)$. Above that side are points from which the lines to $(-1,0)$ and $(1,0)$ meet at a $120°$ angle. Those points lie on a circular arc—draw it and find its center and its radius.

35 Continuing Problem 34, there are circular arcs for all three sides of the triangle. On the arcs, every point sees one side of the triangle at a $120°$ angle. Where is the Steiner point? (Sketch three sides with their arcs.)

36 Invent an algorithm to converge to the Steiner point based on Problem 35. Test it on the triangles of Problem 33.

37 Write a code to minimize $f = d_1^4 + d_2^4 + d_3^4$ by solving $f_x = 0$ and $f_y = 0$. Use Newton's method in equation (5).

38 Extend the code to allow all powers $p \geq 1$, not only $p = 4$. Follow the minimizing point from the centroid at $p = 2$ to the Steiner point at $p = 1$ (try $p = 1.8, 1.6, 1.4, 1.2$).

39 Follow the minimizing point with your code as p increases: $p = 2, p = 4, p = 8, p = 16$. Guess the limit at $p = \infty$ and test whether it is equally distant from the three corners.

40 At $p = \infty$ we are making the largest of the distances d_1, d_2, d_3 as small as possible. The best point for a $1, 1, \sqrt{2}$ right triangle is _____.

41 Suppose the road from corner 1 is wider than the others, and the total cost is $f(x,y) = \sqrt{2} d_1 + d_2 + d_3$. Find the gradient of f and the angles at which the best roads meet.

42 Solve Steiner's problem for *two* points. Where is $d_1 + d_2$ a minimum? Solve also for three points if only the three corners are allowed.

Find all derivatives at $(0,0)$. Construct the Taylor series:

43 $f(x,y) = (x+y)^3$

44 $f(x,y) = xe^y$

45 $f(x,y) = \ln(1-xy)$

Find $f_x, f_y, f_{xx}, f_{xy}, f_{yy}$ **at the basepoint. Write the quadratic approximation to** $f(x,y)$—**the Taylor series through second-order terms**:

46 $f = e^{x+y}$ at $(0,0)$

47 $f = e^{x+y}$ at $(1,1)$

48 $f = \sin x \cos y$ at $(0,0)$

49 $f = x^2 + y^2$ at $(1,-1)$

50 The Taylor series around (x,y) is also written with steps h and $k: f(x+h,y+k) = f(x,y) + h$ _____ $+ k$ _____ $+ \frac{1}{2}h^2$ _____ $+ hk$ _____ $+ \cdots$. Fill in those four blanks.

51 Find lines along which $f(x,y)$ is constant (these functions have $f_{xx}f_{yy} = f_{xy}^2$ or $ac = b^2$):

    (a) $f = x^2 - 4xy + 4y^2$ (b) $f = e^x e^y$

52 For $f(x,y,z)$ the first three terms after $f(0,0,0)$ in the Taylor series are _____ . The next six terms are _____ .

53 (a) For the error $f - f_L$ in linear approximation, the Taylor series at $(0,0)$ starts with the quadratic terms _____ .
    (b) The graph of f goes up from its tangent plane (and $f > f_L$) if _____ . Then f is concave upward.
    (c) For $(0,0)$ to be a minimum we also need _____

54 The gradient of $x^2 + 2y^2$ at the point $(1,1)$ is $(2,4)$. Steepest descent is along the line $x = 1 - 2s, y = 1 - 4s$ (minus sign to go downward). Minimize $x^2 + 2y^2$ with respect to the stepsize s. That locates the next point _____ , where steepest descent begins again.

55 Newton's method minimizes $x^2 + 2y^2$ in one step. Starting at $(x_0, y_0) = (1,1)$, find Δx and Δy from equation (5).

56 If $f_{xx} + f_{yy} = 0$, show that $f(x,y)$ cannot have an interior maximum or minimum (only saddle points).

57 The value of x theorems and y exercises is $f = x^2 y$ (maybe). The most that a student or author can deal with is $4x + y = 12$. Substitute $y = 12 - 4x$ and maximize f. Show that the line $4x + y = 12$ is tangent to the level curve $x^2 y = f_{max}$.

58 The desirability of x houses and y yachts is $f(x,y)$. The constraint $px + qy = k$ limits the money available. The cost of a house is _____ , the cost of a yacht is _____ . Substitute $y = (k - px)/q$ into $f(x,y) = F(x)$ and use the chain rule for dF/dx. Show that the slope $-f_x/f_y$ at the best x is $-p/q$.

59 At the farthest point in a baseball field, explain why the fence is perpendicular to the line from home plate. Assume it is not a rough point (corner) or endpoint (foul line).

13.7 Constraints and Lagrange Multipliers

This section faces up to a practical problem. We often minimize one function $f(x, y)$ while another function $g(x, y)$ is fixed. There is a **constraint** on x and y, given by $g(x, y) = k$. This restricts the material available or the funds available or the energy available. With this constraint, the problem is to do the best possible (f_{max} or f_{min}).

At the absolute minimum of $f(x, y)$, the requirement $g(x, y) = k$ is probably violated. In that case the minimum point is not allowed. We cannot use $f_x = 0$ and $f_y = 0$—those equations don't account for g.

Step 1   Find equations for the **constrained minimum** or **constrained maximum**. They will involve f_x and f_y and also g_x and g_y, which give local information about f and g. To see the equations, look at two examples.

EXAMPLE 1   Minimize $f = x^2 + y^2$ subject to the constraint $g = 2x + y = k$.

Trial runs The constraint allows $x = 0$, $y = k$, where $f = k^2$. Also $(\frac{1}{2}k, 0)$ satisfies the constraint, and $f = \frac{1}{4}k^2$ is smaller. Also $x = y = \frac{1}{3}k$ gives $f = \frac{2}{9}k^2$ (best so far).

Idea of solution Look at the level curves of $f(x, y)$ in Figure 13.21. They are circles $x^2 + y^2 = c$. When c is small, the circles do not touch the line $2x + y = k$. There are no points that satisfy the constraint, when c is too small. *Now increase c.*

Eventually the growing circles $x^2 + y^2 = c$ will *just touch the line $x + 2y = k$*. The point where they touch is the winner. It gives the smallest value of c that can be achieved on the line. The touching point is (x_{min}, y_{min}), and the value of c is f_{min}.

What equation describes that point? When the circle touches the line, they are *tangent*. They have the same slope. *The perpendiculars to the circle and the line go in the same direction.* That is the key fact, which you see in Figure 13.21a. The direction perpendicular to $f = c$ is given by grad $f = (f_x, f_y)$. The direction perpendicular to $g = k$ is given by grad $g = (g_x, g_y)$. The key equation says that those two vectors are parallel. One gradient vector is a multiple of the other gradient vector, with a multiplier λ (called lambda) that is unknown:

13N At the minimum of $f(x, y)$ subject to $g(x, y) = k$, the gradient of f is parallel to the gradient of g—with an unknown number λ as the multiplier:

$$\text{grad } f = \lambda \text{ grad } g \quad \text{so} \quad \frac{\partial f}{\partial x} = \lambda \frac{\partial g}{\partial x} \quad \text{and} \quad \frac{\partial f}{\partial y} = \lambda \frac{\partial g}{\partial y}. \quad (1)$$

Step 2   There are now three unknowns x, y, λ. There are also *three equations*:

$$\partial f / \partial x = \lambda \partial g / \partial x \text{ is} \qquad\qquad 2x = 2\lambda$$
$$\partial f / \partial y = \lambda \partial g / \partial y \text{ is} \qquad\qquad 2y = \lambda \qquad\qquad (2)$$
$$g(x, y) = k \text{ is} \qquad\qquad 2x + y = k.$$

In the third equation, substitute 2λ for $2x$ and $\frac{1}{2}\lambda$ for y. Then $2x + y$ equals $\frac{5}{2}\lambda$ equals k. Knowing $\lambda = \frac{2}{5}k$, go back to the first two equations for x, y, and f_{min}:

$$x = \lambda = \frac{2}{5}k, \quad y = \frac{1}{2}\lambda = \frac{1}{5}k, \quad f_{min} = \left(\frac{2}{5}k\right)^2 + \left(\frac{1}{5}k\right)^2 = \frac{5}{25}k^2 = \frac{1}{5}k^2.$$

The winning point (x_{min}, y_{min}) is $(\frac{2}{5}k, \frac{1}{5}k)$. It minimizes the "distance squared," $f = x^2 + y^2 = \frac{1}{5}k^2$, from the origin to the line.

Question *What is the meaning of the Lagrange multiplier λ?*

Mysterious answer ***The derivative of $\frac{1}{5}k^2$ is $\frac{2}{5}k$, which equals λ. The multiplier λ is the derivative of f_{min} with respect to k.*** Move the line by Δk, and f_{min} changes by about $\lambda \Delta k$. Thus the Lagrange multiplier measures the ***sensitivity*** to k.

Pronounce his name "Lagronge" or better "Lagrongh" as if you are French.

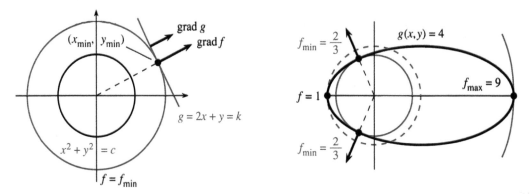

Fig. 13.21   Circles $f = c$ tangent to line $g = k$ and ellipse $g = 4$: parallel gradients

EXAMPLE 2   Maximize and minimize $f = x^2 + y^2$ on the ellipse $g = (x-1)^2 + 4y^2 = 4$.

Idea and equations The circles $x^2 + y^2 = c$ grow until they touch the ellipse. The touching point is (x_{min}, y_{min}) and that smallest value of c is f_{min}. As the circles grow they cut through the ellipse. Finally there is a point (x_{max}, y_{max}) where the last circle touches. That largest value of c is f_{max}.

The minimum and maximum are described by the same rule: the circle is tangent to the ellipse (Figure 13.21b). ***The perpendiculars go in the same direction.*** Therefore (f_x, f_y) is a multiple of (g_x, g_y), and the unknown multiplier is λ:

$$\begin{aligned}
f_x &= \lambda g_x : & 2x &= \lambda 2(x-1) \\
f_y &= \lambda g_y : & 2y &= \lambda 8y & (3) \\
g &= k : & (x-1)^2 + 4y^2 &= 4.
\end{aligned}$$

Solution   The second equation allows two possibilities: $y = 0$ or $\lambda = \frac{1}{4}$. Following up $y = 0$, the last equation gives $(x-1)^2 = 4$. Thus $x = 3$ or $x = -1$. Then the first equation gives $\lambda = 3/2$ or $\lambda = 1/2$. The values of f are $x^2 + y^2 = 3^2 + 0^2 = 9$ and $x^2 + y^2 = (-1)^2 + 0^2 = 1$.

Now follow $\lambda = 1/4$. The first equation yields $x = -1/3$. Then the last equation requires $y^2 = 5/9$. Since $x^2 = 1/9$ we find $x^2 + y^2 = 6/9 = 2/3$. This is f_{min}.

Conclusion The equations (3) have four solutions, at which the circle and ellipse are tangent. The four points are $(3,0), (-1,0), (-1/3, \sqrt{5}/3)$, and $(-1/3, -\sqrt{5}/3)$. The four values of f are $9, 1, \frac{2}{3}, \frac{2}{3}$.

Summary   The three equations are $f_x = \lambda g_x$ and $f_y = \lambda g_y$ and $g = k$. The unknowns are x, y, and λ. There is no absolute system for solving the equations (unless they are linear; then use elimination or Cramer's Rule). Often the first two equations yield x and y in terms of λ, and substituting into $g = k$ gives an equation for λ.

At the minimum, the level curve $f(x, y) = c$ is tangent to the constraint curve $g(x, y) = k$. If that constraint curve is given parametrically by $x(t)$ and $y(t)$, then minimizing $f(x(t), y(t))$ uses the chain rule:

$$\frac{df}{dt} = \frac{\partial f}{\partial x} \frac{dx}{dt} + \frac{\partial f}{\partial y} \frac{dy}{dt} = 0 \quad \text{or} \quad (\text{grad } f) \cdot (\text{tangent to curve}) = 0.$$

This is the calculus proof that grad f is perpendicular to the curve. *Thus grad f is parallel to grad g.* This means $(f_x, f_y) = \lambda(g_x, g_y)$.

We have lost $f_x = 0$ and $f_y = 0$. But a new function L has *three* zero derivatives:

13O The Lagrange function is $L(x, y, \lambda) = f(x, y) - \lambda(g(x, y) - k)$. Its three derivatives are $L_x = L_y = L_\lambda = 0$ at the solution:

$$\frac{\partial L}{\partial x} = \frac{\partial f}{\partial x} - \lambda \frac{\partial g}{\partial x} = 0 \quad \frac{\partial L}{\partial y} = \frac{\partial f}{\partial y} - \lambda \frac{\partial g}{\partial y} = 0 \quad \frac{\partial L}{\partial \lambda} = -g + k = 0. \quad (4)$$

Note that $\partial L / \partial \lambda = 0$ automatically produces $g = k$. The constraint is "***built in***" to L. *Lagrange has included a term $\lambda(g - k)$,* which is destined to be zero—but its derivatives are absolutely needed in the equations! At the solution, $g = k$ and $L = f$ and $\partial L / \partial k = \lambda$.

What is important is $f_x = \lambda g_x$ and $f_y = \lambda g_y$, coming from $L_x = L_y = 0$. In words: The constraint $g = k$ forces $dg = g_x dx + g_y dy = 0$. *This restricts the movements dx and dy.* They must keep to the curve. The equations say that $df = f_x dx + f_y dy$ is equal to λdg. Thus df is zero *in the allowed direction*—which is the key point.

MAXIMUM AND MINIMUM WITH TWO CONSTRAINTS

The whole subject of min(max)imization is called ***optimization***. Its applications to business decisions make up *operations research*. The special case of linear functions is always important—in this part of mathematics it is called ***linear programming***. A book about those subjects won't fit inside a calculus book, but we can take one more step—to allow a second constraint.

The function to minimize or maximize is now $f(x, y, z)$. The constraints are $g(x, y, z) = k_1$ and $h(x, y, z) = k_2$. The multipliers are λ_1 and λ_2. We need at least three variables x, y, z because two constraints would completely determine x and y.

13P To minimize $f(x, y, z)$ subject to $g(x, y, z) = k_1$, and $h(x, y, z) = k_2$, solve five equations for $x, y, z, \lambda_1, \lambda_2$. Combine $g = k_1$ and $h = k_2$ with

$$\frac{\partial f}{\partial x} = \lambda_1 \frac{\partial g}{\partial x} + \lambda_2 \frac{\partial h}{\partial x}, \quad \frac{\partial f}{\partial y} = \lambda_1 \frac{\partial g}{\partial y} + \lambda_2 \frac{\partial h}{\partial y}, \quad \frac{\partial f}{\partial z} = \lambda_1 \frac{\partial g}{\partial z} + \lambda_2 \frac{\partial h}{\partial z}. \quad (5)$$

Figure 13.22a shows the geometry behind these equations. For convenience f is $x^2 + y^2 + z^2$, so we are minimizing distance (squared). The constraints $g = x + y + z = 9$ and $h = x + 2y + 3z = 20$ are linear—their graphs are planes. The constraints keep (x, y, z) on both planes—and therefore on the line where they meet. We are finding *the squared distance from* $(0, 0, 0)$ *to a line.*

What equation do we solve? The level surfaces $x^2 + y^2 + z^2 = c$ are spheres. They grow as c increases. The first sphere to touch the line is tangent to it. That touching point gives the solution (the smallest c). **All three vectors** grad f, grad g, grad h **are perpendicular to the line**:

line tangent to sphere $\Rightarrow$ grad f perpendicular to line

line in both planes $\Rightarrow$ grad g and grad h perpendicular to line.

Thus grad f, grad g, grad h are *in the same plane*—perpendicular to the line. With three vectors in a plane, grad f is a combination of grad g and grad h:

$$(f_x, f_y, f_z) = \lambda_1(g_x, g_y, g_z) + \lambda_2(h_x, h_y, h_z). \tag{6}$$

This is the key equation (5). It applies to curved surfaces as well as planes.

EXAMPLE 3   Minimize $x^2 + y^2 + z^2$ when $x + y + z = 9$ and $x + 2y + 3z = 20$.

In (Figure 13.22b), the normals to those planes are grad $g = (1, 1, 1)$ and grad $h = (1, 2, 3)$. The gradient of $f = x^2 + y^2 + z^2$ is $(2x, 2y, 2z)$. The equations (5)–(6) are

$$2x = \lambda_1 + \lambda_2, \quad 2y = \lambda_1 + 2\lambda_2, \quad 2z = \lambda_1 + 3\lambda_2.$$

Substitute these x, y, z into the other two equations $g = x + y + z = 9$ and $h = 20$:

$$\frac{\lambda_1 + \lambda_2}{2} + \frac{\lambda_1 + 2\lambda_2}{2} + \frac{\lambda_1 + 3\lambda_2}{2} = 9 \quad \text{and} \quad \frac{\lambda_1 + \lambda_2}{2} + 2\frac{\lambda_1 + 2\lambda_2}{2} + 3\frac{\lambda_1 + 3\lambda_2}{2} = 20.$$

After multiplying by 2, these simplify to $3\lambda_1 + 6\lambda_2 = 18$ and $6\lambda_1 + 14\lambda_2 = 40$. The solutions are $\lambda_1 = 2$ and $\lambda_2 = 2$. Now the previous equations give $(x, y, z) = (2, 3, 4)$.

The Lagrange function with two constraints is $L(x, y, z, \lambda_1, \lambda_2) = f - \lambda_1(g - k_1) - \lambda_2(h - k_2)$. Its five derivatives are zero—those are our five equations. Lagrange has increased the number of unknowns from 3 to 5, by adding λ_1 and λ_2. The best point $(2, 3, 4)$ gives $f_{\min} = 29$. The λ's give $\partial f / \partial k$—the sensitivity to changes in 9 and 20.

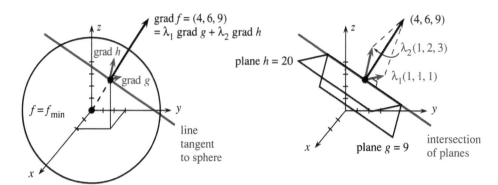

Fig. 13.22   Perpendicular vector grad f is a combination λ_1 grad $g + \lambda_2$ grad h.

INEQUALITY CONSTRAINTS

In practice, applications involve *inequalities* as well as equations. The constraints might be $g \leqslant k$ and $h \geqslant 0$. The first means: It is not required to use the whole resource k, but you cannot use more. The second means: h measures a quantity that cannot be negative. *At the minimum point, the multipliers must satisfy the same inequalities*: $\lambda_1 \leqslant 0$ and $\lambda_2 \geqslant 0$. There are inequalities on the λ's when there are inequalities in the constraints.

Brief reasoning: With $g \leqslant k$ the minimum can be *on or inside* the constraint curve. Inside the curve, where $g < k$, we are free to move in all directions. The constraint is not really constraining. This brings back $f_x = 0$ and $f_y = 0$ and $\lambda = 0$—an ordinary minimum. On the curve, where $g = k$ constrains the minimum from going lower, we have $\lambda < 0$. We don't know in advance which to expect.

For 100 constraints $g_i \leqslant k_i$, there are $100\,\lambda$'s. Some λ's are zero (when $g_i < k_i$) and some are nonzero (when $g_i = k_i$). It is those 2^{100} possibilities that make optimization interesting. In *linear programming* with two variables, the constraints are $x \geqslant 0, y \geqslant 0$:

EXAMPLE 4 Minimixe $f = 5x + 6y$ with $g = x + y = 4$ and $h = x \geqslant 0$ and $H = y \geqslant 0$.

The constraint $g = 4$ is an equation, h and H yield inequalities. Each has its own Lagrange multiplier—and the inequalities require $\lambda_2 \geqslant 0$ and $\lambda_3 \geqslant 0$. The derivatives of f, g, h, H are no problem to compute:

$$\frac{\partial f}{\partial x} = \lambda_1 \frac{\partial g}{\partial x} + \lambda_2 \frac{\partial h}{\partial x} + \lambda_3 \frac{\partial H}{\partial x} \quad \text{yields} \quad 5 = \lambda_1 + \lambda_2$$

$$\frac{\partial f}{\partial y} = \lambda_1 \frac{\partial g}{\partial y} + \lambda_2 \frac{\partial h}{\partial y} + \lambda_3 \frac{\partial H}{\partial y} \quad \text{yields} \quad 6 = \lambda_1 + \lambda_3. \tag{7}$$

Those equations make λ_3 larger than λ_2. Therefore $\lambda_3 > 0$, which means that the constraint on H must be an equation. (Inequality for the multiplier means equality for the constraint.) In other words $H = y = 0$. Then $x + y = 4$ leads to $x = 4$. The solution is at $(x_{\min}, y_{\min}) = (4, 0)$, where $f_{\min} = 20$.

At this minimum, $h = x = 4$ is above zero. The multiplier for the constraint $h \geqslant 0$ must be $\lambda_2 = 0$. Then the first equation gives $\lambda_1 = 5$. As always, the multiplier measures sensitivity. When $g = 4$ is increased by Δk, the cost $f_{\min} = 20$ is increased by $5\Delta k$. In economics $\lambda_1 = 5$ is called a *shadow price*—it is the cost of increasing the constraint.

Behind this example is a nice problem in geometry. The constraint curve $x + y = 4$ is a line. The inequalities $x \geqslant 0$ and $y \geqslant 0$ leave a piece of that line—from P to Q in Figure 13.23. The level curves $f = 5x + 6y = c$ move out as c increases, until they touch the line. *The first touching point is $Q = (4, 0)$, which is the solution. It is always an endpoint*—or a corner of the triangle PQR. It gives the smallest cost $f_{\min}$, which is $c = 20$.

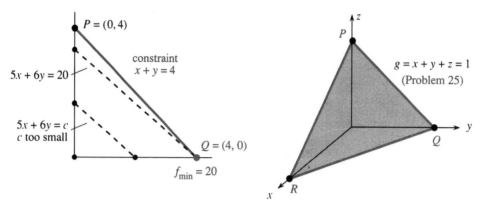

Fig. 13.23   Linear programming: f and g are linear, inequalities cut off x and y.

13.7 EXERCISES

Read-through questions

A restriction $g(x, y) = k$ is called a __a__ . The minimizing equations for $f(x, y)$ subject to $g = k$ are __b__ . The number λ is the Lagrange __c__ . Geometrically, grad f is __d__ to grad g at the minimum. That is because the __e__ curve $f = f_{min}$ is __f__ to the constraint curve $g = k$. The number λ turns out to be the derivative of __g__ with respect to __h__ . The Lagrange function is $L = $ __i__ and the three equations for x, y, λ are __j__ and __k__ and __l__ .

To minimize $f = x^2 - y$ subject to $g = x - y = 0$, the three equations for x, y, λ are __m__ . The solution is __n__ . In this example the curve $f(x, y) = f_{min} = $ __o__ is a __p__ which is __q__ to the line $g = 0$ at (x_{min}, y_{min}).

With two constraints $g(x, y, z) = k_1$ and $h(x, y, z) = k_2$ there are __r__ multipliers. The five unknowns are __s__ . The five equations are __t__ . The level surface $f = f_{min}$ is __u__ to the curve where $g = k_1$ and $h = k_2$. Then grad f is __v__ to this curve, and so are grad g and __w__ . Thus __x__ is a combination of grad g and __y__ . With nine variables and six constraints, there will be __z__ multipliers and eventually __A__ equations. If a constraint is an __B__ $g \leqslant k$, then its multiplier must satisfy $\lambda \leqslant 0$ at a minimum.

1 Example 1 minimized $f = x^2 + y^2$ subject to $2x + y = k$. Solve the constraint equation for $y = k - 2x$, substitute into f, and minimize this function of x. The minimum is at $(x, y) = $ _____ , where $f = $ _____ .

Note: This direct approach *reduces* to one unknown x. Lagrange *increases* to x, y, λ. But Lagrange is better when the first step of solving for y is difficult or impossible.

Minimize and maximize $f(x, y)$ **in 2–6. Find** x, y, **and** λ.

2 $f = x^2 y$ with $g = x^2 + y^2 = 1$

3 $f = x + y$ with $g = \dfrac{1}{x} + \dfrac{1}{y} = 1$

4 $f = 3x + y$ with $g = x^2 + 9y^2 = 1$

5 $f = x^2 + y^2$ with $g = x^6 + y^6 = 2$.

6 $f = x + y$ with $g = x^{1/3} y^{2/3} = k$. With $x = $ capital and $y = $ labor, g is a Cobb-Douglas function in economics. Draw two of its level curves.

7 Find the point on the circle $x^2 + y^2 = 13$ where $f = 2x - 3y$ is a maximum. Explain the answer.

8 Maximize $ax + by + cz$ subject to $x^2 + y^2 + z^2 = k^2$. Write your answer as the Schwarz inequality for dot products: $(a, b, c) \cdot (x, y, z) \leqslant $ _____ k.

9 Find the plane $z = ax + by + c$ that best fits the points $(x, y, z) = (0, 0, 1), (1, 0, 0), (1, 1, 2), (0, 1, 2)$. The answer a, b, c minimizes the sum of $(z - ax - by - c)^2$ at the four points.

10 The base of a triangle is the top of a rectangle (5 sides, combined area $= 1$). What dimensions minimize the distance around?

11 Draw the hyperbola $xy = -1$ touching the circle $g = x^2 + y^2 = 2$. The minimum of $f = xy$ on the circle is reached at the points _____ . The equations $f_x = \lambda g_x$ and $f_y = \lambda g_y$ are satisfied at those points with $\lambda = $ _____ .

12 Find the maximum of $f = xy$ on the circle $g = x^2 + y^2 = 2$ by solving $f_x = \lambda g_x$ and $f_y = \lambda g_y$ and substituting x and y into f. Draw the level curve $f = f_{max}$ that touches the circle.

13 Draw the level curves of $f = x^2 + y^2$ with a closed curve C across them to represent $g(x, y) = k$. Mark a point where C crosses a level curve. Why is that point not a minimum of f on

C? Mark a point where C is *tangent* to a level curve. Is that the minimum of f on C?

14 On the circle $g = x^2 + y^2 = 1$, Example 5 of 13.6 minimized $f = xy - x - y$. (a) Set up the three Lagrange equations for x, y, λ. (b) The first two equations give $x = y =$ _____. (c) There is another solution for the special value $\lambda = -\frac{1}{2}$, when the equations become _____. This is easy to miss but it gives $f_{min} = -1$ at the point _____.

Problems 15–18 develop the theory of Lagrange multipliers.

15 (*Sensitivity*) Certainly $L = f - \lambda(g - k)$ has $\partial L/\partial k = \lambda$. Since $L = f_{min}$ and $g = k$ at the minimum point, this seems to prove the key formula $df_{min}/dk = \lambda$. But x_{min}, y_{min}, λ, and f_{min} all change with k. We need the *total* derivative of $L(x, y, \lambda, k)$:

$$\frac{dL}{dk} = \frac{\partial L}{\partial x}\frac{dx}{dk} + \frac{\partial L}{\partial y}\frac{dy}{dk} + \frac{\partial L}{\partial \lambda}\frac{d\lambda}{dk} + \frac{\partial L}{\partial k}\frac{dk}{dk}.$$

Equation (1) at the minimum point should now yield the sensitivity formula $df_{min}/dk = \lambda$.

16 (*Theory behind* λ) When $g(x, y) = k$ is solved for y, it gives a curve $y = R(x)$. Then minimizing $f(x, y)$ along this curve yields

$$\frac{\partial f}{\partial x} + \frac{\partial f}{\partial y}\frac{dR}{dx} = 0, \frac{\partial g}{\partial x} + \frac{\partial g}{\partial y}\frac{dR}{dx} = 0.$$

Those come from the _____ rule: $df/dx = 0$ at the minimum and $dg/dx = 0$ along the curve because $g =$ _____. Multiplying the second equation by $\lambda = (\partial f/\partial y)/(\partial g/\partial y)$ and subtracting from the first gives _____ $= 0$. Also $\partial f/\partial y = \lambda \partial g/\partial y$. These are the equations (1) for x, y, λ.

17 (*Example of failure*) $\lambda = f_y/g_y$ breaks down if $g_y = 0$ at the minimum point.

(a) $g = x^2 - y^3 = 0$ does not allow negative y because _____.

(b) When $g = 0$ the minimum of $f = x^2 + y$ is at the point _____.

(c) At that point $f_y = \lambda g_y$ becomes _____ which is impossible.

(d) Draw the pointed curve $g = 0$ to see why it is not tangent to a level curve of f.

18 (*No maximum*) Find a point on the line $g = x + y = 1$ where $f(x, y) = 2x + y$ is greater than 100 (or 1000). Write out grad $f = \lambda$ grad g to see that there is no solution.

19 Find the minimum of $f = x^2 + 2y^2 + z^2$ if (x, y, z) is restricted to the planes $g = x + y + z = 0$ and $h = x - z = 1$.

20 (a) Find by Lagrange multipliers the volume $V = xyz$ of the largest box with sides adding up to $x + y + z = k$. (b) Check that $\lambda = dV_{max}/dk$. (c) United Airlines accepts baggage with $x + y + z = 108''$. If it changes to $111''$, approximately how much (by λ) and exactly how much does V_{max} increase?

21 The planes $x = 0$ and $y = 0$ intersect in the line $x = y = 0$, which is the z axis. Write down a vector perpendicular to the plane $x = 0$ and a vector perpendicular to the plane $y = 0$. Find λ_1 times the first vector plus λ_2 times the second. This combination is perpendicular to the line _____.

22 Minimize $f = x^2 + y^2 + z^2$ on the plane $ax + by + cz = d$—one constraint and one multiplier. Compare f_{min} with the distance formula $|d|/\sqrt{a^2 + b^2 + c^2}$ in Section 11.2.

23 At the absolute minimum of $f(x, y)$, the derivatives _____ are zero. If this point happens to fall on the curve $g(x, y) = k$ then the equations $f_x = \lambda g_x$ and $f_y = \lambda g_y$ hold with $\lambda =$ _____.

Problems 24–33 allow inequality constraints, optional but good.

24 Find the minimum of $f = 3x + 5y$ with the constraints $g = x + 2y = 4$ and $h = x \geqslant 0$ and $H = y \geqslant 0$, using equations like (7). Which multiplier is zero?

25 Figure 13.23 shows the constraint plane $g = x + y + z = 1$ chopped off by the inequalities $x \geqslant 0, y \geqslant 0, z \geqslant 0$. What are the three "endpoints" of this triangle? Find the minimum and maximum of $f = 4x - 2y + 5z$ on the triangle, by testing f at the endpoints.

26 With an inequality constraint $g \leqslant k$, the multiplier at the minimum satisfies $\lambda \leqslant 0$. If k is increased, f_{min} goes down (since $\lambda = df_{min}/dk$). *Explain the reasoning*: By increasing k, (more) (fewer) points satisfy the constraints. Therefore (more) (fewer) points are available to minimize f. Therefore f_{min} goes (up) (down).

27 With an inequality constraint $g \leqslant k$, the multiplier at a *maximum* point satisfies $\lambda \geqslant 0$. Change the reasoning in **26**.

28 When the constraint $h \geqslant k$ is a strict inequality $h > k$ at the minimum, the multiplier is $\lambda = 0$. *Explain the reasoning*: For a small increase in k, the same minimizer is still available (since $h > k$ leaves room to move). Therefore f_{min} is (changed) (unchanged), and $\lambda = df_{min}/dk$ is _____.

29 Minimize $f = x^2 + y^2$ subject to the inequality constraint $x + y \leqslant 4$. The minimum is obviously at _____, where f_x and f_y are zero. The multiplier is $\lambda =$ _____. A small change from 4 will leave $f_{min} =$ _____ so the sensitivity df_{min}/dk still equals λ.

30 Minimize $f = x^2 + y^2$ subject to the inequality constraint $x + y \geqslant 4$. Now the minimum is at _____ and the multiplier is $\lambda =$ _____ and $f_{min} =$ _____. A small change to $4 + dk$ changes f_{min} by what multiple of dk?

31 Minimize $f = 5x + 6y$ with $g = x + y = 4$ and $h = x \geqslant 0$ and $H = y \leqslant 0$. Now $\lambda_3 \leqslant 0$ and the sign change destroys Example 4. Show that equation (7) has no solution, and choose x, y to make $5x + 6y < -1000$.

32 Minimize $f = 2x + 3y + 4z$ subject to $g = x + y + z = 1$ and $x, y, z \geqslant 0$. These constraints have multipliers $\lambda_2 \geqslant 0$, $\lambda_3 \geqslant 0$, $\lambda_4 \geqslant 0$. The equations are $2 = \lambda_1 + \lambda_2$, _____, and $4 = \lambda_1 + \lambda_4$. Explain why $\lambda_3 > 0$ and $\lambda_4 > 0$ and $f_{min} = 2$.

33 A wire $40''$ long is used to enclose *one or two* squares (side x and side y). Maximize the total area $x^2 + y^2$ subject to $x \geqslant 0, y \geqslant 0, 4x + 4y = 40$.

CHAPTER 14

Multiple Integrals

14.1 Double Integrals

This chapter shows how to integrate functions of two or more variables. First, a double integral is defined as the limit of sums. Second, we find a fast way to compute it. *The key idea is to replace a double integral by two ordinary "single" integrals*.

The double integral $\iint f(x, y)dy\,dx$ starts with $\int f(x, y)dy$. For each fixed x we integrate with respect to y. The answer depends on x. Now integrate again, this time with respect to x. The limits of integration need care and attention! Frequently those limits on y and x are the hardest part.

Why bother with sums and limits in the first place? Two reasons. There has to be a definition and a computation to fall back on, when the single integrals are difficult or impossible. And also—this we emphasize—*multiple integrals represent more than area and volume*. Those words and the pictures that go with them are the easiest to understand. You can almost see the volume as a "sum of slices" or a "double sum of thin sticks." The true applications are mostly to other things, but the central idea is always the same: *Add up small pieces and take limits*.

We begin with the area of R and the volume of V, by double integrals.

A LIMIT OF SUMS

The graph of $z = f(x, y)$ is a curved surface above the xy plane. At the point (x, y) in the plane, the height of the surface is z. (The surface is *above* the xy plane only when z is positive. Volumes below the plane come with minus signs, like areas below the x axis.) We begin by choosing a positive function—for example $z = 1 + x^2 + y^2$.

The base of our solid is a region R in the xy plane. That region will be chopped into small rectangles (sides Δx and Δy). When R itself is the rectangle $0 \leqslant x \leqslant 1$, $0 \leqslant y \leqslant 2$, the small pieces fit perfectly. For a triangle or a circle, the rectangles miss part of R. But they do fit in the limit, and any region with a piecewise smooth boundary will be acceptable.

Question What is the volume above R and below the graph of $z = f(x, y)$?
Answer It is a double integral—*the integral of $f(x, y)$ over R*. To reach it we begin with a sum, as suggested by Figure 14.1.

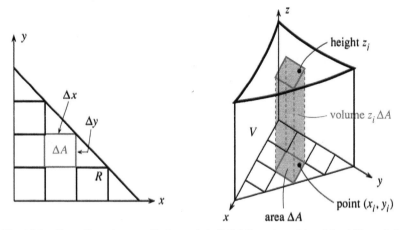

Fig. 14.1 Base R cut into small pieces ΔA. Solid V cut into thin sticks $\Delta V = z \Delta A$.

For single integrals, the interval $[a,b]$ is divided into short pieces of length Δx. For double integrals, R is divided into small rectangles of area $\Delta A = (\Delta x)(\Delta y)$. Above the ith rectangle is a "thin stick" with small volume. That volume is the base area ΔA times the height above it—except that this height $z = f(x,y)$ varies from point to point. Therefore we select a point (x_i, y_i) in the ith rectangle, and compute the volume from the height above that point:

volume of one stick $= f(x_i, y_i) \Delta A$ *volume of all sticks* $= \sum f(x_i, y_i) \Delta A.$

This is the crucial step for any integral—to see it as a sum of small pieces.

Now take limits: $\Delta x \to 0$ and $\Delta y \to 0$. The height $z = f(x,y)$ is nearly constant over each rectangle. (We assume that f is a continuous function.) The sum approaches a limit, which depends only on the base R and the surface above it. The limit is the volume of the solid, and it is the **double integral** of $f(x,y)$ over R:

$$\iint_R f(x,y)\, dA = \lim_{\substack{\Delta x \to 0 \\ \Delta y \to 0}} \sum f(x_i, y_i) \Delta A. \tag{1}$$

To repeat: The limit is the same for all choices of the rectangles and the points (x_i, y_i). The rectangles will not fit exactly into R, if that base area is curved. The heights are not exact, if the surface $z = f(x,y)$ is also curved. But the errors on the sides and top, where the pieces don't fit and the heights are wrong, approach zero. Those errors are the volume of the "icing" around the solid, which gets thinner as $\Delta x \to 0$ and $\Delta y \to 0$. A careful proof takes more space than we are willing to give. But the properties of the integral need and deserve attention:

1. Linearity: $\iint (f + g) dA = \iint f\, dA + \iint g\, dA$
2. Constant comes outside: $\iint c f(x,y) dA = c \iint f(x,y) dA$
3. R splits into S and T (not overlapping): $\iint_R f dA = \iint_S f dA + \iint_T f\, dA.$

In **1** the volume under $f + g$ has two parts. The "thin sticks" of height $f + g$ split into thin sticks under f and under g. In **2** the whole volume is stretched upward by c. In **3** the volumes are side by side. As with single integrals, these properties help in computations.

By writing dA, we allow shapes other than rectangles. Polar coordinates have an extra factor r in $dA = r\, dr\, d\theta$. By writing $dx\, dy$, we choose rectangular coordinates and prepare for the splitting that comes now.

SPLITTING A DOUBLE INTEGRAL INTO TWO SINGLE INTEGRALS

The double integral $\iint f(x, y)dy\, dx$ will now be reduced to single integrals in y and then x. (Or vice versa. Our first integral could equally well be $\int f(x, y)dx$.) Chapter 8 described the same idea for solids of revolution. First came the area of a slice, which is a single integral. Then came a second integral to add up the slices. For solids formed by revolving a curve, all slices are circular disks—now we expect other shapes.

Figure 14.2 shows a slice of area $A(x)$. It cuts through the solid at a fixed value of x. The cut starts at $y = c$ on one side of R, and ends at $y = d$ on the other side. This particular example goes from $y = 0$ to $y = 2$ (R is a rectangle). The area of a slice is the y integral of $f(x, y)$. *Remember that x is fixed and y goes from c to d:*

$$A(x) = \textbf{\textit{area of slice}} = \int_c^d f(x, y)dy \quad \text{(the answer is a function of } x\text{)}.$$

EXAMPLE 1 $\quad A = \int_{y=0}^{2} (1 + x^2 + y^2)dy = \left[y + x^2 y + \frac{y^3}{3} \right]_{y=0}^{y=2} = 2 + 2x^2 + \frac{8}{3}.$

This is the reverse of a partial derivative! The integral of $x^2 dy$, with x constant, is $x^2 y$. This "partial integral" is actually called an ***inner integral***. After substituting the limits $y = 2$ and $y = 0$ and subtracting, we have the area $A(x) = 2 + 2x^2 + \frac{8}{3}$. Now the ***outer integral*** adds slices to find the volume $\int A(x)\, dx$. The answer is a *number*:

$$\text{volume} = \int_{x=0}^{1} \left(2 + 2x^2 + \frac{8}{3} \right) dx = \left[2x + \frac{2x^3}{3} + \frac{8}{3}x \right]_0^1 = 2 + \frac{2}{3} + \frac{8}{3} = \frac{16}{3}.$$

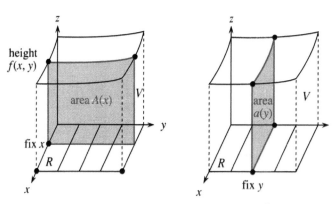

Fig. 14.2 A slice of V at a fixed x has area $A(x) = \int f(x, y)dy$.

To complete this example, check the volume when the x integral comes first:

$$\text{inner integral} = \int_{x=0}^{1} (1 + x^2 + y^2)dx = \left[x + \frac{1}{3}x^3 + y^2 x \right]_{x=0}^{x=1} = \frac{4}{3} + y^2$$

$$\text{outer integral} = \int_{y=0}^{2} \left(\frac{4}{3} + y^2 \right) dy = \left[\frac{4}{3}y + \frac{1}{3}y^3 \right]_{y=0}^{y=2} = \frac{8}{3} + \frac{8}{3} = \frac{16}{3}.$$

The fact that double integrals can be split into single integrals is *Fubini's Theorem*.

14A if $f(x, y)$ is continuous on the rectangle R, then

$$\iint_R f(x, y) dA = \int_a^b \left[\int_c^d f(x, y) dy \right] dx = \int_c^d \left[\int_a^b f(x, y) dx \right] dy. \quad (2)$$

The inner integrals are the cross-sectional areas $A(x)$ and $a(y)$ of the slices. The outer integrals add up the volumes $A(x)dx$ and $a(y)dy$. Notice the reversing of limits.

Normally the brackets in (2) are omitted. When the y integral is first, dy is written inside dx. *The limits on y are inside too*. I strongly recommend that you compute the inner integral on one line and the outer integral on a *separate line*.

EXAMPLE 2   Find the volume below the plane $z = x - 2y$ and above the base triangle R.

The triangle R has sides on the x and y axes and the line $x + y = 1$. The strips in the y direction have varying lengths. (So do the strips in the x direction.) This is the main point of the example—the base is not a rectangle. The upper limit on the inner integral changes as x changes. *The top of the triangle is at $y = 1 - x$.*

Figure 14.3 shows the strips. The region should always be drawn (except for rectangles). Without a figure the limits are hard to find. A sketch of R makes it easy:

$$y \text{ goes from } c = 0 \text{ to } d = 1 - x. \text{ Then } x \text{ goes from } a = 0 \text{ to } b = 1.$$

The inner integral has *variable limits* and the outer integral has *constant limits*:

$$\text{inner: } \int_{y=0}^{y=1-x} (x - 2y) dy = \left[xy - y^2 \right]_{y=0}^{y=1-x} = x(1-x) - (1-x)^2 = -1 + 3x - 2x^2$$

$$\text{outer: } \int_{x=0}^1 (-1 + 3x - 2x^2) dx = \left[-x + \frac{3}{2}x^2 - \frac{2}{3}x^3 \right]_0^1 = -1 + \frac{3}{2} - \frac{2}{3} = -\frac{1}{6}.$$

The volume is negative. Most of the solid is below the xy plane. To check the answer $-\frac{1}{6}$, do the x integral first: x *goes from 0 to $1 - y$. Then y goes from 0 to 1.*

$$\text{inner: } \int_{x=0}^{1-y} (x - 2y) dx = \left[\frac{1}{2}x^2 - 2xy \right]_0^{1-y} = \frac{1}{2}(1-y)^2 - 2(1-y)y = \frac{1}{2} - 3y + \frac{5}{2}y^2$$

$$\text{outer: } \int_{y=0}^1 \left(\frac{1}{2} - 3y + \frac{5}{2}y^2 \right) dy = \left[\frac{1}{2}y - \frac{3}{2}y^2 + \frac{5}{6}y^3 \right]_0^1 = \frac{1}{2} - \frac{3}{1} + \frac{5}{6} = -\frac{1}{6}.$$

Same answer, very probably right. The next example computes $\iint 1 \, dx \, dy = $ area of R.

EXAMPLE 3   The area of R is $\int_{x=0}^1 \int_{y=0}^{1-x} dy \, dx$ and also $\int_{y=0}^1 \int_{x=0}^{1-y} dx \, dy$.

The first has vertical strips. The inner integral equals $1 - x$. Then the outer integral (of $1 - x$) has limits 0 and 1, and the area is $\frac{1}{2}$. It is like an indefinite integral inside a definite integral.

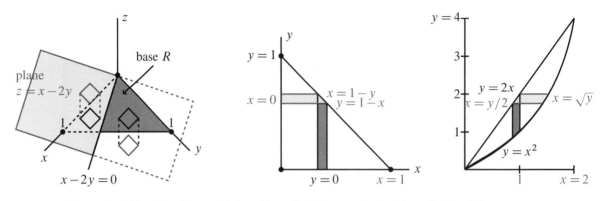

Fig. 14.3   Thin sticks above and below (Example 2). Reversed order (Examples 3 and 4).

EXAMPLE 4   Reverse the order of integration in $\int_{x=0}^{2}\int_{y=x^2}^{2x} x^3 dy\, dx$.

Solution   Draw a figure! The inner integral goes from the parabola $y = x^2$ up to the straight line $y = 2x$. This gives vertical strips. The strips sit side by side between $x = 0$ and $x = 2$. They stop where $2x$ equals x^2, and the line meets the parabola.

The problem is to put the x integral first. It goes along horizontal strips. On each line $y =$ constant, we need the *entry value* of x and the *exit value* of x. From the figure, x goes from $\frac{1}{2}y$ to $\sqrt{y}$. Those are the inner limits. Pay attention also to the outer limits, because they now apply to y. The region starts at $y = 0$ and ends at $y = 4$. *No change in the integrand x^3*—that is the height of the solid:

$$\int_{x=0}^{2}\int_{y=x^2}^{2x} x^3 dy\, dx \quad \text{\textit{is reversed to}} \quad \int_{y=0}^{4}\int_{x=\frac{1}{2}y}^{\sqrt{y}} x^3 dx\, dy. \tag{3}$$

EXAMPLE 5   Find the volume bounded by the planes $x = 0$, $y = 0$, $z = 0$, and $2x + y + z = 4$.

Solution   The solid is a tetrahedron (four sides). It goes from $z = 0$ (the xy plane) up to the plane $2x + y + z = 4$. On that plane $z = 4 - 2x - y$. This is the height function $f(x, y)$ to be integrated.

Figure 14.4 shows the base R. To find its sides, set $z = 0$. The sides of R are the lines $x = 0$ and $y = 0$ and $2x + y = 4$. Taking vertical strips, dy is inner:

$$\text{inner:}\quad \int_{y=0}^{4-2x}(4-2x-y)dy = \left[(4-2x)y - \frac{1}{2}y^2\right]_{0}^{4-2x} = \frac{1}{2}(4-2x)^2$$

$$\text{outer:}\quad \int_{x=0}^{2}\frac{1}{2}(4-2x)^2 dx = \left[-\frac{(4-2x)^3}{2\cdot3\cdot2}\right]_{0}^{2} = \frac{4^3}{2\cdot3\cdot2} = \frac{16}{3}.$$

Question   What is the meaning of the inner integral $\frac{1}{2}(4-2x)^2\left(\text{and also}\frac{16}{3}\right)$?

Answer   The first is $A(x)$, the area of the slice. $\frac{16}{3}$ is the solid volume.

Question   What if the inner integral $\int f(x, y)dy$ has limits that depend on y ?
Answer   It can't. Those limits must be wrong. Find them again.

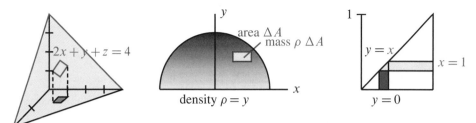

Fig. 14.4 Tetrahedron in Example 5, semicircle in Example 6, triangle in Example 7.

EXAMPLE 6 Find the mass in a semicircle $0 \leqslant y \leqslant \sqrt{1-x^2}$ if the density is $\rho = y$.

This is a new application of double integrals. The total mass is a sum of small masses (ρ times ΔA) in rectangles of area ΔA. The rectangles don't fit perfectly inside the semicircle R, and the density is not constant in each rectangle—but those problems disappear in the limit. We are left with a double integral:

$$\text{\textit{total mass }} M = \iint_R \rho \, dA = \iint_R \rho(x, y) dx \, dy. \tag{4}$$

Set $\rho = y$. Figure 14.4 shows the limits on x and y (try both $dy \, dx$ and $dx \, dy$):

$$\text{mass } M = \int_{x=-1}^{1} \int_{y=0}^{\sqrt{1-x^2}} y \, dy \, dx \quad \text{and also} \quad M = \int_{y=0}^{1} \int_{-\sqrt{1-y^2}}^{\sqrt{1-y^2}} y \, dx \, dy.$$

The first inner integral is $\frac{1}{2}y^2$. Substituting the limits gives $\frac{1}{2}(1-x^2)$. The outer integral of $\frac{1}{2}(1-x^2)$ yields the total mass $M = \frac{2}{3}$.

The second inner integral is xy. Substituting the limits on x gives _____. Then the outer integral is $-\frac{2}{3}(1-y^2)^{3/2}$. Substituting $y = 1$ and $y = 0$ yields $M = $ _____ .

Remark This same calculation also produces the *moment* around the x axis, when the density is $\rho = 1$. The factor y is the distance to the x axis. *The moment is* $M_x = \iint y \, dA = \frac{2}{3}$. Dividing by the area of the semicircle (which is $\pi/2$) locates the centroid: $\overline{x} = 0$ by symmetry and

$$\overline{y} = \text{height of centroid} = \frac{\text{moment}}{\text{area}} = \frac{2/3}{\pi/2} = \frac{4}{3\pi}. \tag{5}$$

This is the "average height" of points inside the semicircle, found earlier in 8.5.

EXAMPLE 7 Integrate $\int_{y=0}^{y=1}\int_{x=y}^{x=1} \cos x^2 dx \, dy$ avoiding the impossible $\int \cos x^2 \, dx$.

This is a famous example where reversing the order makes the calculation possible. The base R is the triangle in Figure 14.4 (note that x goes from y to 1). *In the opposite order y goes from* 0 *to* x. Then $\int \cos x^2 dy = x \cos x^2$ contains the factor x that we need:

$$\text{outer integral: } \int_0^1 x \cos x^2 dx = \left[\tfrac{1}{2}\sin x^2\right]_0^1 = \tfrac{1}{2}\sin 1.$$

Read-through questions

The double integral $\iint_R f(x,y)dA$ gives the volume between R and __a__. The base is first cut into small __b__ of area ΔA. The volume above the ith piece is approximately __c__. The limit of the sum __d__ is the volume integral. Three properties of double integrals are __e__ (linearity) and __f__ and __g__.

If R is the rectangle $0 \leqslant x \leqslant 4, 4 \leqslant y \leqslant 6$, the integral $\iint x\, dA$ can be computed two ways. One is $\iint x\, dy\, dx$, when the inner integral is __h__ $]_4^6 =$ __i__. The outer integral gives __j__ $]_0^4 =$ __k__. When the x integral comes first it equals $\int x\, dx =$ __l__ $]_0^4 =$ __m__. Then the y integral equals __n__. This is the volume between __o__ (describe V).

The area of R is $\iint$ __p__ $dy\, dx$. When R is the triangle between $x = 0, y = 2x$, and $y = 1$, the inner limits on y are __q__. This is the length of a __r__ strip. The (outer) limits on x are __s__. The area is __t__. In the opposite order, the (inner) limits on x are __u__. Now the strip is __v__ and the outer integral is __w__. When the density is $\rho(x,y)$, the total mass in the region R is $\iint$ __x__. The moments are $M_y =$ __y__ and $M_x =$ __z__. The centroid has $\bar{x} = M_y/M$.

Compute the double integrals 1–4 by two integrations.

1. $\int_{y=0}^{1} \int_{x=0}^{2} x^2 dx\, dy$ and $\int_{y=0}^{1} \int_{x=0}^{2} y^2 dx\, dy$

2. $\int_{y=2}^{2e} \int_{x=1}^{e} 2xy\, dx\, dy$ and $\int_{y=2}^{2e} \int_{x=1}^{e} dx\, dy/xy$

3. $\int_{0}^{\pi/2} \int_{0}^{\pi/4} \sin(x+y)\, dx\, dy$ and $\int_{1}^{2} \int_{0}^{2} dy\, dx/(x+y)^2$

4. $\int_{0}^{1} \int_{1}^{2} ye^{xy} dx\, dy$ and $\int_{-1}^{1} \int_{0}^{3} dy\, dx/\sqrt{3+2x+y}$

In 5–10, draw the region and compute the area.

5. $\int_{x=1}^{2} \int_{y=1}^{2x} dy\, dx$

6. $\int_{0}^{1} \int_{x^3}^{x} dy\, dx$

7. $\int_{0}^{\infty} \int_{e^{-2x}}^{e^{-x}} dy\, dx$

8. $\int_{-1}^{1} \int_{x^2-1}^{1-x^2} dy\, dx$

9. $\int_{-1}^{1} \int_{y^2}^{1} dx\, dy$

10. $\int_{-1}^{1} \int_{x=y}^{|y|} dx\, dy$

In 11–16 reverse the order of integration (and find the new limits) in 5–10 respectively.

In 17–24 find the limits on $\iint dy\, dx$ and $\iint dx\, dy$. Draw R and compute its area.

17. $R =$ triangle inside the lines $x = 0, y = 1, y = 2x$.

18. $R =$ triangle inside the lines $x = -1, y = 0, x + y = 0$.

19. $R =$ triangle inside the lines $y = x, y = -x, y = 3$.

20. $R =$ triangle inside the lines $y = x, y = 2x, y = 4$.

21. $R =$ triangle with vertices $(0,0), (4,4), (4,8)$.

22. $R =$ triangle with vertices $(0,0), (-2,-1), (1,-2)$.

23. $R =$ triangle with vertices $(0,0), (2,0), (1,b)$. Here $b > 0$.

*24. $R =$ triangle with vertices $(0,0), (a,b), (c,d)$. The sides are $y = bx/a, y = dx/c$, and $y = b + (x-a)(d-b)/(c-a)$. Find $A = \iint dy\, dx$ when $0 < a < c, 0 < d < b$.

25. Evaluate $\int_{0}^{b} \int_{0}^{a} \partial^2 f/\partial x\partial y\, dx\, dy$.

26. Evaluate $\int_{0}^{b} \int_{0}^{a} \partial f/\partial x\, dx\, dy$.

In 27–28, divide the unit square R into triangles S and T and verify $\iint_R f\, dA = \iint_S f\, dA + \iint_T f\, dA$.

27. $f(x,y) = 2x - 3y + 1$

28. $f(x,y) = xe^y - ye^x$

29. The area under $y = f(x)$ is a single integral from a to b or a double integral (*find the limits*):

$$\int_{a}^{b} f(x)\, dx = \iint 1\, dy\, dx.$$

30. Find the limits and the area under $y = 1 - x^2$:

$$\int (1-x^2)\, dx \text{ and } \iint 1\, dx\, dy \text{ (reversed from 29).}$$

31. A city inside the circle $x^2 + y^2 = 100$ has population density $\rho(x,y) = 10(100 - x^2 - y^2)$. Integrate to find its population.

32. Find the volume bounded by the planes $x = 0, y = 0, z = 0$, and $ax + by + cz = 1$.

In 33–34 the rectangle with corners $(1,1), (1,3), (2,1), (2,3)$ has density $\rho(x,y) = x^2$. The moments are $M_y = \iint x\rho\, dA$ and $M_x = \iint y\rho\, dA$.

33. Find the mass.   34. Find the center of mass.

In 35–36 the region is a circular wedge of radius 1 between the lines $y = x$ and $y = -x$.

35. Find the area.   36. Find the centroid $(\bar{x}, \bar{y})$.

37. Write a program to compute $\int_0^1 \int_0^1 f(x,y)dx\, dy$ by the midpoint rule (*midpoints of n^2 small squares*). Which $f(x,y)$ are integrated exactly by your program?

38. Apply the midpoint code to integrate x^2 and xy and y^2. The errors decrease like what power of $\Delta x = \Delta y = 1/n$?

Use the program to compute the volume under $f(x,y)$ in 39–42. Check by integrating exactly or doubling n.

39. $f(x,y) = 3x + 4y + 5$

40. $f(x,y) = 1/\sqrt{x^2 + y^2}$

41. $f(x,y) = x^y$

42. $f(x,y) = e^x \sin \pi y$

43 In which order is $\iint x^y dx\, dy = \iint x^y dy\, dx$ easier to integrate over the square $0 \leqslant x \leqslant 1$, $0 \leqslant y \leqslant 1$? By reversing order, integrate $(x-1)/\ln x$ from 0 to 1—its antiderivative is unknown.

44 Explain in your own words the definition of the double integral of $f(x, y)$ over the region R.

45 $\sum y_i \Delta A$ might not approach $\iint y\, dA$ if we only know that $\Delta A \to 0$. In the square $0 \leqslant x, y \leqslant 1$, take rectangles of sides Δx and 1 (not Δx and Δy). If (x_i, y_i) is a point in the rectangle where $y_i = 1$, then $\sum y_i \Delta A =$ _____ . But $\iint y\, dA =$ _____ .

14.2 Change to Better Coordinates

You don't go far with double integrals before wanting to *change variables*. Many regions simply do not fit with the x and y axes. Two examples are in Figure 14.5, a tilted square and a ring. Those are excellent shapes—*in the right coordinates*.

We have to be able to answer basic questions like these:

$$\text{Find the area } \iint dA \text{ and moment } \iint x\, dA \text{ and moment of inertia } \{\}$$

The problem is: *What is dA?* We are leaving the xy variables where $dA = dx\, dy$.

The reason for changing is this: The limits of integration in the y direction are miserable. I don't know them and I don't want to know them. For every x we would need the entry point P of the line $x = $ constant, and the exit point Q. The heights of P and Q are the limits on $\int dy$, the inner integral. The geometry of the square and ring are totally missed, if we stick rigidly to x and y.

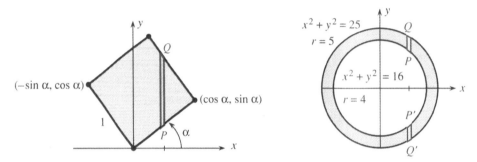

Fig. 14.5   Unit square turned through angle α. Ring with radii 4 and 5.

Which coordinates are better? Any sensible person agrees that the area of the tilted square is 1. "Just turn it and the area is obvious." But that sensible person may not know the moment or the center of gravity or the moment of inertia. So we actually have to do the turning.

The new coordinates u and v are in Figure 14.6a. The limits of integration on v are 0 and 1. So are the limits on u. *But when you change variables, you don't just change limits.* Two other changes come with new variables:

1. The small area $dA = dx\, dy$ becomes $dA = $ _____ $du\, dv$.
2. The integral of x becomes the integral of _____ .

Substituting $u = \sqrt{x}$ in a single integral, we make the same changes. Limits $x = 0$ and $x = 4$ become $u = 0$ and $u = 2$. Since x is u^2, dx is $2u\, du$. The purpose of the change is to find an antiderivative. For double integrals, the usual purpose is to improve the limits—but we have to accept the whole package.

To turn the square, there are formulas connecting x and y to u and v. The geometry is clear—*rotate axes by α*—but it has to be converted into algebra:

$$u = x\cos\alpha + y\sin\alpha \qquad\qquad x = u\cos\alpha - v\sin\alpha$$
$$\text{and in reverse} \qquad\qquad\qquad\qquad (1)$$
$$v = -x\sin\alpha + y\cos\alpha \qquad\qquad y = u\sin\alpha + v\cos\alpha.$$

Figure 14.6 shows the rotation. As points move, the whole square turns. A good way to remember equation (1) is to follow the corners as they become $(1,0)$ and $(0,1)$.

The change from $\iint x\,dA$ to $\iint \underline{\hspace{1cm}} du\,dv$ is partly decided by equation (1). It gives x as a function of u and v. We also need dA. For a pure rotation the first guess is correct: **The area $dx\,dy$ equals the area $du\,dv$. For most changes of variable this is false.** The general formula for dA comes after the examples.

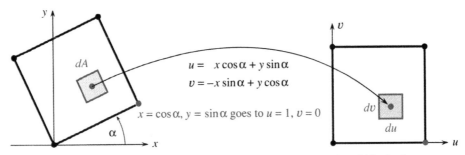

$$u = x\cos\alpha + y\sin\alpha$$
$$v = -x\sin\alpha + y\cos\alpha$$

$x = \cos\alpha, y = \sin\alpha$ goes to $u = 1, v = 0$

Fig. 14.6 Change of coordinates—axes turned by α. For rotation dA is $du\,dv$.

EXAMPLE 1 Find $\iint dA$ and $\iint x\,dA$ and $\overline{x}$ and also $\iint x^2\,dA$ for the tilted square.

Solution The area of the square is $\int_0^1\int_0^1 du\,dv = 1$. Notice the good limits. Then

$$\iint x\,dA = \int_0^1\int_0^1(u\cos\alpha - v\sin\alpha)du\,dv = \tfrac{1}{2}\cos\alpha - \tfrac{1}{2}\sin\alpha. \qquad (2)$$

This is the *moment around the y axis*. The factors $\tfrac{1}{2}$ come from $\tfrac{1}{2}u^2$ and $\tfrac{1}{2}v^2$. The x coordinate of the center of gravity is

$$\overline{x} = \iint x\,dA \Big/ \iint dA = \left(\tfrac{1}{2}\cos\alpha - \tfrac{1}{2}\sin\alpha\right)/1.$$

Similarly the integral of y leads to $\overline{y}$. The answer is no mystery—the point $(\overline{x}, \overline{y})$ is at the center of the square! Substituting $x = u\cos\alpha - v\sin\alpha$ made $x\,dA$ look worse, but the limits 0 and 1 are much better.

The moment of inertia I_y around the y axis is also simplified:

$$\iint x^2\,dA = \int_0^1\int_0^1(u\cos\alpha - v\sin\alpha)^2 du\,dv = \frac{\cos^2\alpha}{3} - \frac{\cos\alpha\sin\alpha}{2} + \frac{\sin^2\alpha}{3}. \qquad (3)$$

You know this next fact but I will write it anyway: *The answers don't contain u or v.* Those are dummy variables like x and y. The answers do contain α, because the square has turned. (The area is fixed at 1.) The moment of inertia $I_x = \iint y^2\,dA$ is the same as equation (3) but with all plus signs.

Question The sum $I_x + I_y$ simplifies to $\tfrac{2}{3}$ (a constant). Why no dependence on α?
Answer $I_x + I_y$ equals I_0. This moment of inertia around $(0,0)$ is unchanged by rotation. We are turning the square around one of its corners.

CHANGE TO POLAR COORDINATES

The next change is to r and θ. A small area becomes $dA = r\,dr\,d\theta$ (definitely not $dr\,d\theta$). Area always comes from multiplying two lengths, and $d\theta$ is not a length. Figure 14.7 shows the crucial region—a "polar rectangle" cut out by rays and circles. Its area ΔA is found in two ways, both leading to $r\,dr\,d\theta$:

(*Approximate*) The straight sides have length Δr. The circular arcs are *close to* $r\Delta\theta$. The angles are $90°$. So ΔA is close to $(\Delta r)(r\Delta\theta)$.

(*Exact*) A wedge has area $\frac{1}{2}r^2\Delta\theta$. The difference between wedges is ΔA:

$$\Delta A = \frac{1}{2}\left(r + \frac{\Delta r}{2}\right)^2 \Delta\theta - \frac{1}{2}\left(r - \frac{\Delta r}{2}\right)^2 \Delta\theta = r\,\Delta r\,\Delta\theta.$$

The exact method places r dead center (see figure). The approximation says: Forget the change in $r\Delta\theta$ as you move outward. Keep only the first-order terms.

A third method is coming, which requires no picture and no geometry. Calculus always has a third method! The change of variables $x = r\cos\theta, y = r\sin\theta$ will go into a general formula for dA, and out will come the area $r\,dr\,d\theta$.

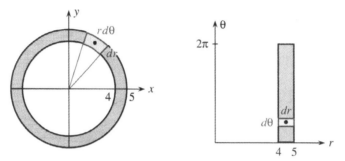

Fig. 14.7   Ring and polar rectangle in xy and $r\theta$, with stretching factor $r = 4.5$.

EXAMPLE 2   Find the area and center of gravity of the ring. Also find $\iint x^2 dA$.

Solution   The limits on r are 4 and 5. The limits on θ are 0 and 2π. *Polar coordinates are perfect for a ring.* Compared with limits like $x = \sqrt{25 - y^2}$, the change to $r\,dr\,d\theta$ is a small price to pay:

$$\text{area} = \int_0^{2\pi}\int_4^5 r\,dr\,d\theta = 2\pi\left[\frac{1}{2}r^2\right]_4^5 = \pi 5^2 - \pi 4^2 = 9\pi.$$

The θ integral is 2π (full circle). Actually the ring is a giant polar rectangle. We could have used the exact formula $r\,\Delta r\,\Delta\theta$, with $\Delta\theta = 2\pi$ and $\Delta r = 5 - 4$. When the radius r is centered at 4.5, the product $r\,\Delta r\,\Delta\theta$ is $(4.5)(1)(2\pi) = 9\pi$ as above.

Since the ring is symmetric around $(0, 0)$, the integral of $x\,dA$ must be *zero*:

$$\iint_R x\,dA = \int_0^{2\pi}\int_4^5 (r\cos\theta)r\,dr\,d\theta = \left[\frac{1}{3}r^3\right]_4^5 \left[\sin\theta\right]_0^{2\pi} = 0.$$

Notice $r\cos\theta$ from x—the other r is from dA. The moment of inertia is

$$\iint_R x^2 dA = \int_0^{2\pi}\int_4^5 r^2\cos^2\theta\,r\,dr\,d\theta = \left[\frac{1}{4}r^4\right]_4^5 \int_0^{2\pi}\cos^2\theta\,d\theta = \frac{1}{4}(5^4 - 4^4)\pi.$$

This θ integral is π not 2π, because the average of $\cos^2\theta$ is $\frac{1}{2}$ not 1.

For reference here are the moments of inertia when the density is $\rho(x, y)$:

$$I_y = \iint x^2\rho\,dA \quad I_x = \iint y^2\rho\,dA \quad I_0 = \iint r^2\rho\,dA = \textbf{\textit{polar moment}} = I_x + I_y. \quad (4)$$

EXAMPLE 3 Find masses and moments for semicircular plates: $\rho = 1$ and
$\rho = 1 - r$.

Solution   The semicircles in Figure 14.8 have $r = 1$. The angle goes from 0 to π
(the upper half-circle). Polar coordinates are best. ***The mass is the integral of the
density ρ:***

$$M = \int_0^\pi \int_0^1 r \, dr \, d\theta = (\tfrac{1}{2})(\pi) \quad \text{and} \quad M = \int_0^\pi \int_0^1 (1-r)r \, dr \, d\theta = (\tfrac{1}{6})(\pi).$$

The first mass $\pi/2$ equals the area (because $\rho = 1$). The second mass $\pi/6$ is smaller
(because $\rho < 1$). Integrating $\rho = 1$ is the same as finding a volume when the height is
$z = 1$ (part of a cylinder). Integrating $\rho = 1 - r$ is the same as finding a volume when
the height is $z = 1 - r$ (part of a cone). Volumes of cones have the extra factor $\tfrac{1}{3}$. The
center of gravity involves the moment $M_x = \iint y\rho \, dA$. ***The distance from the x axis
is y, the mass of a small piece is $\rho \, dA$, integrate to add mass times distance.*** Polar
coordinates are still best, with $y = r \sin\theta$. Again $\rho = 1$ and $\rho = 1 - r$:

$$\iint y \, dA = \int_0^\pi \int_0^1 r \sin\theta \, r \, dr \, d\theta = \tfrac{2}{3} \qquad \iint y(1-r) \, dA = \int_0^\pi \int_0^1 r \sin\theta(1-r)r \, dr \, d\theta = \tfrac{1}{6}.$$

The height of the center of gravity is $\overline{y} = M_x/M = $ ***moment divided by mass***:

$$\overline{y} = \frac{2/3}{\pi/2} = \frac{4}{3\pi} \quad \text{when } \rho = 1 \qquad \overline{y} = \frac{1/6}{\pi/6} = \frac{1}{\pi} \quad \text{when } \rho = 1 - r.$$

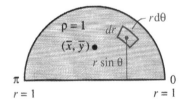  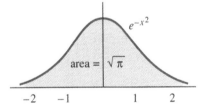

Fig. 14.8   Semicircles with density piled above them.   **Fig. 14.9**   Bell-shaped curve.

Question   Compare $\overline{y}$ for $\rho = 1$ and $\rho = $ other positive constants and $\rho = 1 - r$.
Answer Any constant ρ gives $\overline{y} = 4/3\pi$. Since $1 - r$ is dense at $r = 0$, $\overline{y}$ drops to $1/\pi$.

Question   How is $\overline{y} = 4/3\pi$ related to the "average" of y in the semicircle ?
Answer   They are identical. This is the point of $\overline{y}$. Divide the integral by the area:

$$\textit{The average value of a function is} \quad \iint f(x,y) dA \Big/ \iint dA. \qquad (5)$$

The integral of f is divided by the integral of 1 (the area). In one dimension $\int_a^b v(x) \, dx$
was divided by $\int_a^b 1 \, dx$ (the length $b - a$). That gave the average value of $v(x)$ in
Section 5.6. Equation (5) is the same idea for $f(x, y)$.

EXAMPLE 4 Compute $A = \int_{-\infty}^{\infty} e^{-x^2} dx = \sqrt{\pi}$ from $A^2 = \int_{-\infty}^{\infty} e^{-x^2} dx \int_{-\infty}^{\infty} e^{-y^2} dy = \pi$.

A is the area under a "bell-shaped curve"—see Figure 14.9. This is the most important definite integral in the study of probability. It is difficult because a factor $2x$ is not present. Integrating $2xe^{-x^2}$ gives $-e^{-x^2}$, but integrating e^{-x^2} is impossible—except approximately by a computer. How can we hope to show that A is exactly $\sqrt{\pi}$? The trick is to go from an area integral A to a volume integral A^2. This is unusual (and hard to like), but the end justifies the means:

$$A^2 = \int_{x=-\infty}^{\infty} \int_{y=-\infty}^{\infty} e^{-x^2} e^{-y^2} dy\, dx = \int_{\theta=0}^{2\pi} \int_{r=0}^{\infty} e^{-r^2} r\, dr\, d\theta. \qquad (6)$$

The double integrals cover the whole plane. The r^2 comes from $x^2 + y^2$, and the key factor r appears in polar coordinates. It is now possible to substitute $u = r^2$. The r integral is $\frac{1}{2}\int_0^{\infty} e^{-u} du = \frac{1}{2}$. The θ integral is 2π. The double integral is $(\frac{1}{2})(2\pi)$. Therefore $A^2 = \pi$ and the single integral is $A = \sqrt{\pi}$.

EXAMPLE 5 Apply Example 4 to the **"normal distribution"** $p(x) = e^{-x^2/2}/\sqrt{2\pi}$.

Section 8.4 discussed probability. It emphasized the importance of this particular $p(x)$. At that time we could not verify that $\int p(x) dx = 1$. Now we can:

$$x = \sqrt{2}y \quad \text{yields} \quad \frac{1}{\sqrt{2\pi}} \int_{-\infty}^{\infty} e^{-x^2/2} dx = \frac{1}{\sqrt{\pi}} \int_{-\infty}^{\infty} e^{-y^2} dy = 1. \qquad (7)$$

Question Why include the 2's in $p(x)$? The integral of $e^{-x^2}/\sqrt{\pi}$ also equals 1.
Answer With the 2's the "**variance**" is $\int x^2 p(x) dx = 1$. This is a convenient number.

CHANGE TO OTHER COORDINATES

A third method was promised, to find $r\, dr\, d\theta$ without a picture and without geometry. The method works directly from $x = r \cos \theta$ and $y = r \sin \theta$. It also finds the 1 in $du\, dv$, after a rotation of axes. Most important, this new method finds the factor J in the area $dA = J\, du\, dv$, for any change of variables. The change is from xy to uv.

For single integrals, the "**stretching factor**" J between the original dx and the new du is (not surprisingly) the ratio dx/du. Where we have dx, we write $(dx/du)du$. Where we have $(du/dx)dx$, we write du. That was the idea of substitutions—the main way to simplify integrals.

For double integrals the stretching factor appears in the area: $dx\, dy$ becomes $|J|\, du\, dv$. **The old and new variables are related by** $x = x(u, v)$ **and** $y = y(u, v)$. The point with coordinates u and v comes from the point with coordinates x and y. A whole region S, full of points in the uv plane, comes from the region R full of corresponding points in the xy plane. A small piece with area $|J|\, du\, dv$ comes from a small piece with area $dx\, dy$. **The formula for** J **is a two-dimensional version of** dx/du.

> **14B** The stretching factor for area is the 2 by 2 *Jacobian determinant* $J(u,v)$:
>
> $$J = \begin{vmatrix} \partial x/\partial u & \partial x/\partial v \\ \partial y/\partial u & \partial y/\partial v \end{vmatrix} = \frac{\partial x}{\partial u}\frac{\partial y}{\partial v} - \frac{\partial x}{\partial v}\frac{\partial y}{\partial u}. \tag{8}$$
>
> An integral over R in the xy plane becomes an integral over S in the uv plane:
>
> $$\iint_R f(x,y)\,dx\,dy = \iint_S f(x(u,v), y(u,v))|J|\,du\,dv. \tag{9}$$

The determinant J is often written $\partial(x,y)/\partial(u,v)$, as a reminder that this stretching factor is like dx/du. We require $J \neq 0$. That keeps the stretching and shrinking under control.

You naturally ask: Why take the absolute value $|J|$ in equation (9)? Good question—it wasn't done for single integrals. The reason is in the limits of integration. The single integral $\int_0^1 dx$ is $\int_0^{-1}(-du)$ after changing x to $-u$. *We keep the minus sign and allow single integrals to run backward.* Double integrals could too, but normally they go left to right and down to up. We use the absolute value $|J|$ and run forward.

EXAMPLE 6 Polar coordinates have $x = u\cos v = r\cos\theta$ and $y = u\sin v = r\sin\theta$.

With no geometry:
$$J = \begin{vmatrix} \partial x/\partial r & \partial x/\partial\theta \\ \partial y/\partial r & \partial y/\partial\theta \end{vmatrix} = \begin{vmatrix} \cos\theta & -r\sin\theta \\ \sin\theta & r\cos\theta \end{vmatrix} = r. \tag{10}$$

EXAMPLE 7 Find J for the *linear change* to $x = au + bv$ and $y = cu + dv$.

Ordinary determinant:
$$J = \begin{vmatrix} \partial x/\partial u & \partial x/\partial v \\ \partial y/\partial u & \partial y/\partial v \end{vmatrix} = \begin{vmatrix} a & b \\ c & d \end{vmatrix} = ad - bc. \tag{11}$$

Why make this simple change, in which a, b, c, d are all constant? It straightens parallelograms into squares (and rotates those squares). Figure 14.10 is typical.

Common sense indicated $J = 1$ for pure rotation—no change in area. Now $J = 1$ comes from equations (1) and (11), because $ad - bc$ is $\cos^2\alpha + \sin^2\alpha$.

In practice, xy rectangles generally go into uv rectangles. The sides can be curved (as in polar rectangles) but the angles are often 90°. The change is "*orthogonal.*" The next example has angles that are not 90°, and J still gives the answer.

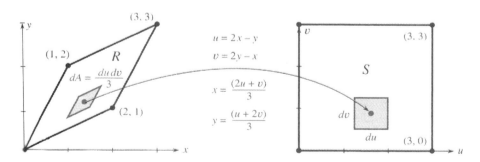

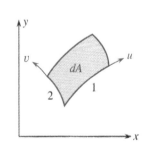

Fig. 14.10 Change from xy to uv has $J = \frac{1}{3}$.

Fig. 14.11 Curved areas are also $dA = |J|\,du\,dv$.

EXAMPLE 8   Find the area of R in Figure 14.10. Also compute $\iint\limits_{R} e^x dx\, dy$.

Solution   The figure shows $x = \frac{2}{3}u + \frac{1}{3}v$ and $y = \frac{1}{3}u + \frac{2}{3}v$. The determinant is

$$J = \begin{vmatrix} \partial x/\partial u & \partial x/\partial v \\ \partial y/\partial u & \partial y/\partial v \end{vmatrix} = \begin{vmatrix} 2/3 & 1/3 \\ 1/3 & 2/3 \end{vmatrix} = \frac{4}{9} - \frac{1}{9} = \frac{1}{3}.$$

The area of the xy parallelogram becomes an integral over the uv square:

$$\iint\limits_{R} dx\, dy = \iint\limits_{S} |J|\, du\, dv = \int_0^3 \int_0^3 \frac{1}{3}\, du\, dv = \frac{1}{3}\cdot 3\cdot 3 = 3.$$

The square has area 9, the parallelogram has area 3. I don't know if $J = \frac{1}{3}$ is a stretching factor or a shrinking factor. The other integral $\iint e^x dx\, dy$ is

$$\int_0^3 \int_0^3 e^{2u/3+v/3}\frac{1}{3} du\, dv = \left[\frac{3}{2} e^{2u/3}\right]_0^3 \left[3e^{v/3}\right]_0^3 \frac{1}{3} = \frac{3}{2}(e^2 - 1)(e - 1).$$

Main point: The change to u and v makes the limits easy (just 0 and 3).

Why is the stretching factor J a determinant? With straight sides, this goes back to Section 11.3 on vectors. *The area of a parallelogram is a determinant.* Here the sides are curved, but that only produces $(du)^2$ and $(dv)^2$, which we ignore.

A change du gives one side of Figure 14.11—it is $(\partial x/\partial u\, \mathbf{i} + \partial y/\partial u\, \mathbf{j})du$. Side 2 is $(\partial x/\partial v\, \mathbf{i} + \partial y/\partial v\, \mathbf{j})dv$. The curving comes from second derivatives. The area (the cross product of the sides) is $|J|\, du\, dv$.

Final remark I can't resist looking at the change in the reverse direction. Now the rectangle is in xy and the parallelogram is in uv. In all formulas, exchange x for u and y for v:

$$new\ J = \begin{vmatrix} \partial u/\partial x & \partial u/\partial y \\ \partial v/\partial x & \partial v/\partial y \end{vmatrix} = \frac{\partial(u,v)}{\partial(x,y)} = \frac{1}{old\ J}. \qquad (12)$$

This is exactly like $du/dx = 1/(dx/du)$. It is the derivative of the inverse function. The product of slopes is 1—stretch out, shrink back. From xy to uv we have 2 by 2 matrices, and the identity matrix I takes the place of 1:

$$\frac{dx}{du}\frac{du}{dx} = 1 \quad becomes \quad \begin{bmatrix} \partial x/\partial u & \partial x/\partial v \\ \partial y/\partial u & \partial y/\partial v \end{bmatrix} \begin{bmatrix} \partial u/\partial x & \partial u/\partial y \\ \partial v/\partial x & \partial v/\partial y \end{bmatrix} = \begin{bmatrix} 1 & 0 \\ 0 & 1 \end{bmatrix}. \qquad (13)$$

The first row times the first column is $(\partial x/\partial u)(\partial u/\partial x) + (\partial x/\partial v)(\partial v/\partial x) = \partial x/\partial x = 1$. The first row times the second column is $(\partial x/\partial u)(\partial u/\partial y) + (\partial x/\partial v)(\partial v/\partial y) = \partial x/\partial y = 0$. *The matrices are inverses of each other.* The determinants of a matrix and its inverse obey our rule: old J times new $J = 1$. Those J's cannot be zero, just as dx/du and du/dx were not zero. (Inverse functions increase steadily or decrease steadily.)

In two dimensions, an area $dx\, dy$ goes to $J\, du\, dv$ and comes back to $dx\, dy$.

Read-through questions

We change variables to improve the __a__ of integration. The disk $x^2 + y^2 \leqslant 9$ becomes the rectangle $0 \leqslant r \leqslant$ __b__, $0 \leqslant \theta \leqslant$ __c__. The inner limits on $\iint dy\, dx$ are $y = \pm$ __d__. In polar coordinates this area integral becomes $\iint$ __e__ = __f__.

A polar rectangle has sides dr and __g__. Two sides are not __h__ but the angles are still __i__. The area between the circles $r = 1$ and $r = 3$ and the rays $\theta = 0$ and $\theta = \pi/4$ is __j__. The integral $\iint x\, dy\, dx$ changes to $\iint$ __k__. This is the __l__ around the __m__ axis. Then $\bar{x}$ is the ratio __n__. This is the x coordinate of the __o__, and it is the __p__ value of x.

In a rotation through α, the point that reaches (u, v) starts at $x = u \cos \alpha - v \sin \alpha$, $y =$ __q__. A rectangle in the uv plane comes from a __r__ in xy. The areas are __s__ so the stretching factor is $J =$ __t__. This is the determinant of the matrix __u__ containing $\cos \alpha$ and $\sin \alpha$. The moment of inertia $\iint x^2 dx\, dy$ changes to $\iint$ __v__ $du\, dv$.

For single integrals dx changes to __w__ du. For double integrals $dx\, dy$ changes to $J\, du\, dv$ with $J =$ __x__. The stretching factor J is the determinant of the 2 by 2 matrix __y__. The functions $x(u, v)$ and $y(u, v)$ connect an xy region R to a uv region S, and $\iint_R dx\, dy = \iint_S$ __z__ = area of __A__. For polar coordinates $x =$ __B__, $y =$ __C__. For $x = u$, $y = u + 4v$ the 2 by 2 determinant is $J =$ __D__. A square in the uv plane comes from a __E__ in xy. In the opposite direction the change has $u = x$ and $v = \frac{1}{4}(y - x)$ and a new $J =$ __F__. This J is constant because this change of variables is __G__.

In 1–12 R is a pie-shaped wedge: $0 \leqslant r \leqslant 1$ and $\pi/4 \leqslant \theta \leqslant 3\pi/4$.

1 What is the area of R? Check by integration in polar coordinates.

2 Find limits on $\iint dy\, dx$ to yield the area of R, and integrate. *Extra credit*: Find limits on $\iint dx\, dy$.

3 Equation (1) with $\alpha = \pi/4$ rotates R into the uv region $S = $ _____. Find limits on $\iint du\, dv$.

4 Compute the centroid height $\bar{y}$ of R by changing $\iint y\, dx\, dy$ to polar coordinates. Divide by the area of R.

5 The region R has $\bar{x} = 0$ because _____. After rotation through $\alpha = \pi/4$, the centroid $(\bar{x}, \bar{y})$ of R becomes the centroid _____ of S.

6 Find the centroid of any wedge $0 \leqslant r \leqslant a$, $0 \leqslant \theta \leqslant b$.

7 Suppose R^* is the wedge R moved up so that the sharp point is at $x = 0$, $y = 1$.
  (a) Find limits on $\iint dy\, dx$ to integrate over R^*.
  (b) With $x^* = x$ and $y^* = y - 1$, the xy region R^* corresponds to what region in the $x^* y^*$ plane?
  (c) After that change $dx\, dy$ equals _____ $dx^* dy^*$.

8 Find limits on $\iint r\, dr\, d\theta$ to integrate over R^* in Problem 7.

9 The right coordinates for R^* are r^* and θ^*, with $x = r^* \cos \theta^*$ and $y = r^* \sin \theta^* + 1$.
  (a) Show that $J = r^*$ so $dA = r^* dr^* d\theta^*$.
  (b) Find limits on $\iint r^* dr^* d\theta^*$ to integrate over R^*.

10 If the centroid of R is $(0, \bar{y})$, the centroid of R^* is _____. The centroid of the circle with radius 3 and center $(1, 2)$ is _____. The centroid of the upper half of that circle is _____.

11 The moments of inertia I_x, I_y, I_0 of the original wedge R are _____.

12 The moments of inertia I_x, I_y, I_0 of the shifted wedge R^* are _____.

Problems 13–16 change four-sided regions to squares.

13 R has straight sides $y = 2x$, $x = 1$, $y = 1 + 2x$, $x = 0$. Locate its four corners and draw R. Find its area by geometry.

14 Choose a, b, c, d so that the change $x = au + bv$, $y = cu + dv$ takes the previous R into S, the unit square $0 \leqslant u \leqslant 1$, $0 \leqslant v \leqslant 1$. From the stretching factor $J = ad - bc$ find the area of R.

15 The region R has straight sides $x = 0$, $x = 1$, $y = 0$, $y = 2x + 3$. Choose a, b, c so that $x = u$ and $y = au + bv + cuv$ change R to the unit square S.

16 A nonlinear term uv was needed in Problem 15. Which regions R could change to the square S with a linear $x = au + bv$, $y = cu + dv$?

Draw the xy region R that corresponds in 17–22 to the uv square S with corners $(0, 0)$, $(1, 0)$, $(0, 1)$, $(1, 1)$. Locate the corners of R and then its sides (like a jigsaw puzzle).

17 $x = 2u + v$, $y = u + 2v$

18 $x = 3u + 2v$, $y = u + v$

19 $x = e^{2u+v}$, $y = e^{u+2v}$

20 $x = uv$, $y = v^2 - u^2$

21 $x = u$, $y = v(1 + u^2)$

22 $x = u \cos v$, $y = u \sin v$ (only three corners)

23 In Problems 17 and 19, compute J from equation (8). Then find the area of R from $\iint_S |J| du\, dv$.

24 In 18 and 20, find $J = \partial(x, y)/\partial(u, v)$ and the area of R.

25 If R lies between $x = 0$ and $x = 1$ under the graph of $y = f(x) > 0$, then $x = u$, $y = vf(u)$ takes R to the unit square S. Locate the corners of R and the point corresponding to $u = \frac{1}{2}$, $v = 1$. Compute J to prove what we know:

$$\text{area of } R = \int_0^1 f(x) dx = \int_0^1 \int_0^1 J\, du\, dv.$$

26 From $r = \sqrt{x^2 + y^2}$ and $\theta = \tan^{-1}(y/x)$, compute $\partial r/\partial x$, $\partial r/\partial y$, $\partial \theta/\partial x$, $\partial \theta/\partial y$, and the determinant $J = \partial(r,\theta)/\partial(x,y)$. How is this J related to the factor $r = \partial(x,y)/\partial(r,\theta)$ that enters $r \, dr \, d\theta$?

27 Example 4 integrated e^{-x^2} from 0 to ∞ (answer $\sqrt{\pi}$). Also $B = \int_0^1 e^{-x^2} dx$ leads to $B^2 = \int_0^1 e^{-x^2} dx \int_0^1 e^{-y^2} dy$. Change this double integral over the unit square to r and θ—and find the limits on r that make exact integration impossible.

28 Integrate by parts to prove that the standard normal distribution $p(x) = e^{-x^2/2}/\sqrt{2\pi}$ has $\sigma^2 = \int_{-\infty}^{\infty} x^2 p(x) dx = 1$.

29 Find the average distance from a point on a circle to the points inside. Suggestion: Let $(0,0)$ be the point and let $0 \leq r \leq 2a\cos\theta$, $0 \leq \theta \leq \pi$ be the circle (radius a). The distance is r, so the average distance is $\bar{r} = \iint \underline{\quad} / \iint \underline{\quad}$.

30 Draw the region $R: 0 \leq x \leq 1$, $0 \leq y \leq \infty$ and describe it with polar coordinates (limits on r and θ). Integrate $\iint_R (x^2 + y^2)^{-3/2} dx \, dy$ in polar coordinates.

31 Using polar coordinates, find the volume under $z = x^2 + y^2$ above the unit disk $x^2 + y^2 \leq 1$.

32 The end of Example 1 stated the moment of inertia $\iint y^2 dA$. Check that integration.

33 In the square $-1 \leq x \leq 2$, $-2 \leq y \leq 1$, where could you distribute a unit mass (with $\iint \rho \, dx \, dy = 1$) to maximize

(a) $\iint x^2 \rho \, dA$ (b) $\iint y^2 \rho \, dA$ (c) $\iint r^2 \rho \, dA$?

34 *True or false*, with a reason:

(a) If the uv region S corresponds to the xy region R, then area of S = area of R.

(b) $\iint x \, dA \leq \iint x^2 \, dA$

(c) The average value of $f(x,y)$ is $\iint f(x,y) dA$

(d) $\int_{-\infty}^{\infty} x e^{-x^2} dx = 0$

(e) A polar rectangle has the same area as a straight-sided region with the same corners.

35 Find the mass of the tilted square in Example 1 if the density is $\rho = xy$.

36 Find the mass of the ring in Example 2 if the density is $\rho = x^2 + y^2$. This is the same as which moment of inertia with which density ?

37 Find the polar moment of inertia I_0 of the ring in Example 2 if the density is $\rho = x^2 + y^2$.

38 Give the following statement an appropriate name: $\iint_R f(x,y) dA = f(P)$ times (area of R), where P is a point in R. Which point P makes this correct for $f = x$ and $f = y$?

39 Find the xy coordinates of the top point in Figure 14.6a and check that it goes to $(u,v) = (1,1)$.

14.3 Triple Integrals

At this point in the book, I feel I can speak to you directly. You can guess what triple integrals are like. Instead of a small interval or a small rectangle, there is a small box. Instead of length dx or area $dx\,dy$, the box has volume $dV = dx\,dy\,dz$. That is length times width times height. The goal is to put small boxes together (by integration). The main problem will be to discover the correct limits on x, y, z.

We could dream up more and more complicated regions in three-dimensional space. But I don't think you can see the method clearly without seeing the region clearly. In practice six shapes are the most important:

$$\textit{box} \quad \textit{prism} \quad \textit{cylinder} \quad \textit{cone} \quad \textit{tetrahedron} \quad \textit{sphere}.$$

The box is easiest and the sphere may be the hardest (but no problem in spherical coordinates). Circular cylinders and cones fall in the middle, where xyz coordinates are possible but $r\theta z$ are the best. I start with the box and prism and xyz.

EXAMPLE 1 By triple integrals find the volume of a box and a prism (Figure 14.12).

$$\iiint_{\text{box}} dV = \int_{z=0}^{1} \int_{y=0}^{3} \int_{x=0}^{2} dx\,dy\,dz \quad \text{and} \quad \iiint_{\text{prism}} dV = \int_{z=0}^{1} \int_{y=0}^{3-3z} \int_{x=0}^{2} dx\,dy\,dz$$

The inner integral for both is $\int dx = 2$. Lines in the x direction have length 2, cutting through the box and the prism. The middle integrals show the limits on y (since dy comes second):

$$\int_{y=0}^{3} 2\,dy = 6 \quad \text{and} \quad \int_{y=0}^{3-3z} 2\,dy = 6 - 6z.$$

After two integrations these are *areas*. The first area 6 is for a plane section through the box. The second area $6 - 6z$ is cut through the prism. The shaded rectangle goes from $y = 0$ to $y = 3 - 3z$—we needed and used the equation $y + 3z = 3$ for the boundary of the prism. *At this point z is still constant!* But the area depends on z, because the prism gets thinner going upwards. The base area is $6 - 6z = 6$, the top area is $6 - 6z = 0$.

The outer integral multiplies those areas by dz, to give the volume of slices. They are horizontal slices because z came last. Integration adds up the slices to find the total volume:

$$\text{box volume} = \int_{z=0}^{1} 6\,dz = 6 \qquad \text{prism volume} = \int_{z=0}^{1} (6 - 6z)\,dz = \left[6z - 3z^2\right]_0^1 = 3.$$

The box volume $2 \cdot 3 \cdot 1$ didn't need calculus. The prism is half of the box, so its volume was sure to be 3—but it is satisfying to see how $6z - 3z^2$ gives the answer. Our purpose is to see how a triple integral works.

Question Find the prism volume in the order $dz\,dy\,dx$ (*six orders are possible*).

Answer $\displaystyle \int_0^2 \int_0^3 \int_0^{(3-y)/3} dz\,dy\,dx = \int_0^2 \int_0^3 \left(\frac{3-y}{3}\right) dy\,dx = \int_0^2 \frac{3}{2}\,dx = 3.$

To find those limits on the z integral, follow a line in the z direction. It enters the prism at $z = 0$ and exits at the sloping face $y + 3z = 3$. That gives the upper limit $z = (3 - y)/3$. It is the height of a thin stick as in Section 14.1. This section writes out $\int dz$ for the height, but a quicker solution starts at the double integral.

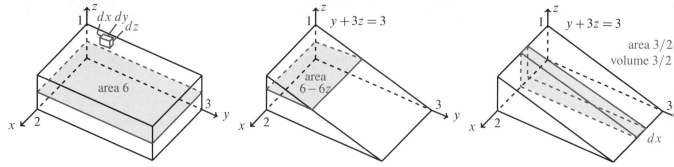

Fig. 14.12 Box with sides $2,3,1$. The prism is half of the box: volume $\int (6-6z)dz$ or $\int \frac{3}{2}dx$.

What is the number $\frac{3}{2}$ in the last integral ? It is the ***area of a vertical slice***, cut by a plane $x = $ constant. The outer integral adds up slices.

$$\iiint f(x,y,z)\,dV \text{ is computed from three single integrals } \int\left[\int\left[\int f\,dx\,\right]dy\,\right]dz.$$

That step cannot be taken in silence—some basic calculus is involved. The triple integral is the limit of $\sum f_i \Delta V$, a sum over small boxes of volume ΔV. Here f_i is any value of $f(x,y,z)$ in the ith box. (In the limit, the boxes fit a curved region.) Now take those boxes *in a certain order*. Put them into lines in the x direction and put the lines of boxes into planes. The lines lead to the inner x integral, whose answer depends on y and z. The y integral combines the lines into planes. Finally the outer integral accounts for all planes and all boxes.

Example 2 is important because it displays more possibilities than a box or prism.

EXAMPLE 2   Find the volume of a tetrahedron (4-sided pyramid). Locate $(\overline{x},\overline{y},\overline{z})$.

Solution A tetrahedron has four flat faces, all triangles. The fourth face in Figure 14.13 is on the plane $x+y+z=1$. A line in the x direction enters at $x = 0$ and exits at $x = 1-y-z$. (The length depends on y and z. The equation of the boundary plane gives x.) Then those lines are put into plane slices by the y integral:

$$\int_{y=0}^{1-z}\int_{x=0}^{1-y-z} dx\,dy = \int_{y=0}^{1-z}(1-y-z)dy = \left[y-\tfrac{1}{2}y^2-zy\right]_0^{1-z} = \tfrac{1}{2}(1-z)^2.$$

What is this number $\frac{1}{2}(1-z)^2$? ***It is the area at height*** z. The plane at that height slices out a right triangle, whose legs have length $1-z$. The area is correct, but look at the limits of integration. ***If*** x ***goes to*** $1-y-z$, ***why does*** y ***go to*** $1-z$? Reason: We are assembling lines, not points. The figure shows a line at every y up to $1-z$.

Adding the slices gives the volume: $\int_0^1 \frac{1}{2}(1-z)^2 dz = \left[\frac{1}{6}(z-1)^3\right]_0^1 = \frac{1}{6}$. This agrees with $\frac{1}{3}$(base times height), the volume of a pyramid.

The height $\overline{z}$ of the centroid is "z_{average}." We compute $\iiint z\,dV$ and divide by the volume. Each horizontal slice is multiplied by its height z, and the limits of integration don't change:

$$\iiint z\,dV = \int_0^1\int_0^{1-y}\int_0^{1-y-z} z\,dx\,dy\,dz = \int_0^1 \frac{z(1-z)^2}{2}\,dz = \frac{1}{24}.$$

This is quick because z is constant in the x and y integrals. Each triangular slice contributes z times its area $\frac{1}{2}(1-z)^2$ times dz. Then the z integral gives the moment

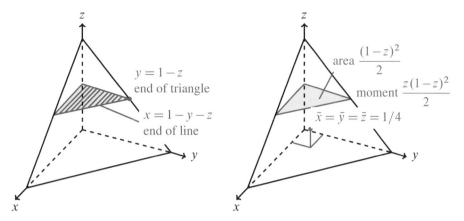

Fig. 14.13 Lines end at plane $x + y + z = 1$. Triangles end at edge $y + z = 1$. The average height is $\bar{z} = \iiint z \, dV / \iiint dV$.

1/24. To find the *average* height, divide $1/24$ by the volume:

$$\bar{z} = \text{ height of centroid} = \frac{\iiint z \, dV}{\iiint dV} = \frac{1/24}{1/6} = \frac{1}{4}.$$

By symmetry $\bar{x} = \frac{1}{4}$ and $\bar{y} = \frac{1}{4}$. The centroid is the point $(\frac{1}{4}, \frac{1}{4}, \frac{1}{4})$. Compare that with $(\frac{1}{3}, \frac{1}{3})$, the centroid of the standard right triangle. Compare also with $\frac{1}{2}$, the center of the unit interval. There must be a five-sided region in four dimensions centered at $(\frac{1}{5}, \frac{1}{5}, \frac{1}{5}, \frac{1}{5})$.

For area and volume we meet another pattern. Length of standard interval is 1, area of standard triangle is $\frac{1}{2}$, volume of standard tetrahedron is $\frac{1}{6}$, hypervolume in four dimensions must be _____ . The interval reaches the point $x = 1$, the triangle reaches the line $x + y = 1$, the tetrahedron reaches the plane $x + y + z = 1$. The four-dimensional region stops at the hyperplane _____ $= 1$.

EXAMPLE 3 Find the volume $\iiint dx \, dy \, dz$ inside the unit sphere $x^2 + y^2 + z^2 = 1$.

First question: What are the limits on x ? If a needle goes through the sphere in the x direction, where does it enter and leave ? Moving in the x direction, the numbers y and z stay constant. The inner integral deals only with x. The smallest and largest x are at the boundary where $x^2 + y^2 + z^2 = 1$. This equation does the work—we solve it for x. Look at the limits on the x integral:

$$\text{volume of sphere} = \int_{?}^{?} \int_{?}^{?} \int_{-\sqrt{1-y^2-z^2}}^{\sqrt{1-y^2-z^2}} dx \, dy \, dz = \int_{?}^{?} \int_{?}^{?} 2\sqrt{1-y^2-z^2} \, dy \, dz. \quad (1)$$

The limits on y are $-\sqrt{1-z^2}$ and $+\sqrt{1-z^2}$. You can use algebra on the boundary equation $x^2 + y^2 + z^2 = 1$. But notice that x is gone! We want the smallest and largest y, for each z. It helps very much to draw the plane at height z, slicing through the sphere in Figure 14.14. The slice is a circle of radius $r = \sqrt{1-z^2}$. So the area is πr^2, which must come from the y integral:

$$\int 2\sqrt{1-y^2-z^2} \, dy = \text{ area of slice } = \pi(1-z^2). \quad (2)$$

I admit that I didn't integrate. Is it cheating to use the formula πr^2 ? I don't think so. Mathematics is hard enough, and we don't have to work blindfolded. The goal is understanding, and if you know the area then use it. Of course the integral of $\sqrt{1-y^2-z^2}$ can be done if necessary—use Section 7.2.

The triple integral is down to a single integral. We went from one needle to a circle of needles and now to a sphere of needles. The volume is a sum of slices of area $\pi(1-z^2)$. The South Pole is at $z=-1$, the North Pole is at $z=+1$, and the integral is the volume $4\pi/3$ inside the unit sphere:

$$\int_{-1}^{1} \pi(1-z^2)dz = \pi\left(z-\frac{1}{3}z^3\right)\Big]_{-1}^{1} = \frac{2}{3}\pi - \left(-\frac{2}{3}\pi\right) = \frac{4}{3}\pi. \qquad (3)$$

Question 1   A cone also has circular slices. How is the last integral changed?

Answer   The slices of a cone have radius $1-z$. Integrate $(1-z)^2$ not $\sqrt{1-z^2}$.

Question 2   How does this compare with a circular cylinder (height 1, radius 1)?

Answer   Now all slices have radius 1. Above $z=0$, a cylinder has volume π and a half-sphere has volume $\frac{2}{3}\pi$ and a cone has volume $\frac{1}{3}\pi$.
For solids with equal surface area, the sphere has largest volume.

Question 3   What is the average height $\bar{z}$ in the cone and half-sphere and cylinder?

Answer $$\bar{z} = \frac{\int z(\text{slice area})dz}{\int(\text{slice area})dz} = \frac{1}{4} \quad \text{and} \quad \frac{3}{8} \quad \text{and} \quad \frac{1}{2}.$$

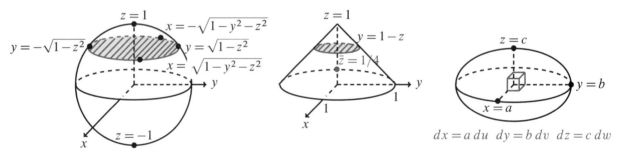

Fig. 14.14 $\int dx = $ length of needle, $\iint dx\,dy = $ area of slice. Ellipsoid is a stretched sphere.

EXAMPLE 4   Find the volume $\iiint dx\,dy\,dz$ inside the ellipsoid $x^2/a^2 + y^2/b^2 + z^2/c^2 = 1$.

The limits on x are now $\pm\sqrt{1-y^2/b^2-z^2/c^2}$. The algebra looks terrible. The geometry is better—all slices are ellipses. A *change of variable* is absolutely the best.

Introduce $u=x/a$ and $v=y/b$ and $w=z/c$. Then the outer boundary becomes $u^2+v^2+w^2=1$. In these new variables the shape is a sphere. The triple integral for a sphere is $\iiint du\,dv\,dw = 4\pi/3$. But what volume dV in xyz space corresponds to a small box with sides du and dv and dw?

Every uvw box comes from an xyz box. The box is stretched with no bending or twisting. Since u is x/a, the length dx is $a\,du$. Similarly $dy=b\,dv$ and $dz=c\,dw$. The volume of the xyz box (Figure 14.14) is $dx\,dy\,dz=(abc)\,du\,dv\,dw$.

The **stretching factor** $J = abc$ is a constant, and the volume of the ellipsoid is

$$\underset{\substack{\text{bad limits}\\ \text{ellipsoid}}}{\iiint} dx\, dy\, dz = \underset{\substack{\text{better limits}\\ \text{sphere}}}{\iiint} (abc)\, du\, dv\, dw = \frac{4\pi}{3} abc. \qquad (4)$$

You realize that this is special—other volumes are much more complicated. The sphere and ellipsoid are curved, but the small xyz boxes are straight. The next section introduces spherical coordinates, and we can finally write "*good limits*." But then we need a different J.

14.3 EXERCISES

Read-through questions

Six important solid shapes are __a__. The integral $\iiint dx\, dy\, dz$ adds the volume __b__ of small __c__. For computation it becomes __d__ single integrals. The inner integral $\int dx$ is the __e__ of a line through the solid. The variables __f__ and __g__ are held constant. The double integral $\iint dx\, dy$ is the __h__ of a slice, with __i__ held constant. Then the z integral adds up the volumes of __j__.

If the solid region V is bounded by the planes $x = 0$, $y = 0$, $z = 0$, and $x + 2y + 3z = 1$, the limits on the inner x integral are __k__. The limits on y are __l__. The limits on z are __m__. In the new variables $u = x$, $v = 2y$, $w = 3z$, the equation of the outer boundary is __n__. The volume of the tetrahedron in uvw space is __o__. From $dx = du$ and $dy = dv/2$ and $dz = $__p__, the volume of an xyz box is $dx\, dy\, dz = $__q__ $du\, dv\, dw$. So the volume of V is __r__.

To find the average height $\bar{z}$ in V we compute __s__ / __t__. To find the total mass in V if the density is $\rho = e^z$ we compute the integral __u__. To find the average density we compute __v__ / __w__. In the order $\iiint dz\, dx\, dy$ the limits on the inner integral can depend on __x__. The limits on the middle integral can depend on __y__. The outer limits for the ellipsoid $x^2 + 2y^2 + 3z^2 \leqslant 8$ are __z__.

1 For the solid region $0 \leqslant x \leqslant y \leqslant z \leqslant 1$, find the limits in $\iiint dx\, dy\, dz$ and compute the volume.

2 Reverse the order in Problem 1 to $\iiint dz\, dy\, dx$ and find the limits of integration. The four faces of this tetrahedron are the planes $x = 0$ and $y = x$ and _____.

3 This tetrahedron and five others like it fill the unit cube. Change the inequalities in Problem 1 to describe the other five.

4 Find the centroid $(\bar{x}, \bar{y}, \bar{z})$ in Problem 1.

Find the limits of integration in $\iiint dx\, dy\, dz$ and the volume of solids 5–16. Draw a very rough picture.

5 A cube with sides of length 2, centered at $(0, 0, 0)$.

6 Half of that cube, the box above the xy plane.

7 Part of the same cube, the prism above the plane $z = y$.

8 Part of the same cube, above $z = y$ and $z = 0$.

9 Part of the same cube, above $z = x$ and below $z = y$.

10 Part of the same cube, where $x \leqslant y \leqslant z$. What shape is this?

11 The tetrahedron bounded by planes $x = 0$, $y = 0$, $z = 0$, and $x + y + 2z = 2$.

12 The tetrahedron with corners $(0,0,0)$, $(2,0,0)$, $(0,4,0)$, $(0,0,4)$. First find the plane through the last three corners.

13 The part of the tetrahedron in Problem 11 below $z = \frac{1}{2}$.

14 The tetrahedron in Problem 12 with its top sliced off by the plane $z = 1$.

15 The volume above $z = 0$ below the cone $\sqrt{x^2 + y^2} = 1 - z$.

*16 The tetrahedron in Problem 12, after it falls across the x axis onto the xy plane.

In 17–20 find the limits in $\iiint dx\, dy\, dz$ or $\iiint dz\, dy\, dx$. Compute the volume.

17 A circular cylinder with height 6 and base $x^2 + y^2 \leqslant 1$.

18 The part of that cylinder below the plane $z = x$. Watch the base. Draw a picture.

19 The volume shared by the cube (Problem 5) and cylinder.

20 The same cylinder lying along the x axis.

21 A cube is inscribed in a sphere: radius 1, both centers at $(0,0,0)$. What is the volume of the cube?

22 Find the volume and the centroid of the region bounded by $x = 0$, $y = 0$, $z = 0$, and $x/a + y/b + z/c = 1$.

23 Find the volume and centroid of the solid $0 \leqslant z \leqslant 4 - x^2 - y^2$.

24 Based on the text, what is the volume inside $x^2 + 4y^2 + 9z^2 = 16$? What is the "hypervolume" of the 4-dimensional pyramid that stops at $x + y + z + w = 1$?

25 Find the partial derivatives $\partial I/\partial x$, $\partial I/\partial y$, $\partial^2 I/\partial y\,\partial z$ of

$$I = \int_0^z \int_0^y dx\,dy \quad \text{and} \quad I = \int_0^z \int_0^y \int_0^x f(x,y,z)\,dx\,dy\,dz.$$

26 Define the average value of $f(x,y,z)$ in a solid V.

27 Find the moment of inertia $\iiint l^2 dV$ of the cube $|x| \leqslant 1$, $|y| \leqslant 1$, $|z| \leqslant 1$ when l is the distance to
(a) the x axis (b) the edge $y = z = 1$ (c) the diagonal $x = y = z$.

28 Add upper limits to produce the volume of a unit cube from small cubes: $V = \sum\limits_{i=1}^{} \sum\limits_{j=1}^{} \sum\limits_{k=1}^{} (\Delta x)^3 = 1$.

*29 Find the limit as $\Delta x \to 0$ of $\sum\limits_{i=1}^{3/\Delta x} \sum\limits_{j=1}^{2/\Delta x} \sum\limits_{k=1}^{j} (\Delta x)^3$.

30 The midpoint rule for an integral over the unit cube chooses the center value $f(\frac{1}{2}, \frac{1}{2}, \frac{1}{2})$. Which functions $f = x^m y^n z^p$ are integrated correctly ?

31 The trapezoidal rule estimates $\int_0^1 \int_0^1 \int_0^1 f(x,y,z)\,dx\,dy\,dz$ as $\frac{1}{8}$ times the sum of $f(x,y,z)$ at 8 corners. This correctly integrates $x^m y^n z^p$ for which m, n, p ?

32 Propose a 27-point "Simpson's Rule" for integration over a cube. If many small cubes fill a large box, why are there only 8 new points per cube ?

14.4 Cylindrical and Spherical Coordinates

Cylindrical coordinates are good for describing solids that are *symmetric around an axis*. The solid is three-dimensional, so there are three coordinates r, θ, z:

$$r: out \text{ from the axis} \qquad \theta: around \text{ the axis} \qquad z: along \text{ the axis.}$$

This is a mixture of polar coordinates $r\theta$ in a plane, plus z upward. You will not find $r\theta z$ difficult to work with. Start with a cylinder centered on the z axis:

$$solid \, cylinder: 0 \leqslant r \leqslant 1 \quad flat \, bottom \, and \, top: 0 \leqslant z \leqslant 3 \quad half\text{-}cylinder: 0 \leqslant \theta \leqslant \pi$$

Integration over this half-cylinder is $\int_0^3 \int_0^\pi \int_0^1 \underline{\quad ? \quad} \, dr \, d\theta \, dz$. These limits on r, θ, z are especially simple. Two other axially symmetric solids are almost as convenient:

$$\boldsymbol{cone}: integrate \, to \, r + z = 1 \qquad \boldsymbol{sphere}: integrate \, to \, r^2 + z^2 = R^2$$

I would not use cylindrical coordinates for a box. Or a tetrahedron.

The integral needs one thing more—the volume dV. The movements dr and $d\theta$ and dz give a "curved box" in xyz space, drawn in Figure 14.15c. The base is a polar rectangle, with area $r \, dr \, d\theta$. The new part is the height dz. **The volume of the curved box is $r \, dr \, d\theta \, dz$.** Then r goes in the blank space in the triple integral—the stretching factor is $J = r$. There are six orders of integration (we give two):

$$\text{volume} = \int_z \int_\theta \int_r r \, dr \, d\theta \, dz = \int_\theta \int_z \int_r r \, dr \, dz \, d\theta. \qquad (1)$$

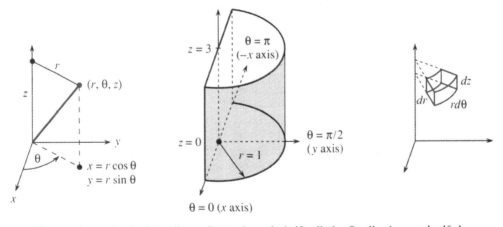

Fig. 14.15 Cylindrical coordinates for a point and a half-cylinder. Small volume $r \, dr \, d\theta \, dz$.

EXAMPLE 1 (Volume of the half-cylinder). The integral of $r \, dr$ from 0 to 1 is $\frac{1}{2}$. The θ integral is π and the z integral is 3. The volume is $3\pi/2$.

EXAMPLE 2 The surface $r = 1 - z$ encloses the cone in Figure 14.16. Find its volume.

First solution   Since r goes out to $1 - z$, the integral of $r \, dr$ is $\frac{1}{2}(1 - z)^2$. The θ integral is 2π (a full rotation). Stop there for a moment.

We have reached $\int\int r \, dr \, d\theta = \frac{1}{2}(1 - z)^2 \, 2\pi$. This is the *area of a slice at height z*. The slice is a circle, its radius is $1 - z$, its area is $\pi(1 - z)^2$. The z integral adds those slices to give $\pi/3$. That is correct, but it is not the only way to compute the volume.

Second solution   Do the z and θ integrals first. Since z goes up to $1-r$, and θ goes around to 2π, those integrals produce $\iint r \, dz \, d\theta = r(1-r)2\pi$. Stop again—this must be the area of something.

After the z and θ integrals we have a **shell at radius** r. The height is $1-r$ (the outer shells are shorter). This height times $2\pi r$ gives the area around the shell. The choice between shells and slices is exactly as in Chapter 8. **Different orders of integration give different ways to cut up the solid.**

The volume of the shell is area times thickness dr. The volume of the complete cone is the integral of shell volumes: $\int_0^1 r(1-r)2\pi \, dr = \pi/3$.

Third solution   Do the r and z integrals first: $\iint r \, dr \, dz = \frac{1}{6}$. Then the θ integral is $\int \frac{1}{6} \, d\theta$, which gives $\frac{1}{6}$ times 2π. This is the volume $\pi/3$—but what is $\frac{1}{6} \, d\theta$?

The third cone is cut into wedges. The volume of a wedge is $\frac{1}{6} \, d\theta$. It is quite common to do the θ integral last, especially when it just multiplies by 2π. It is not so common to think of wedges.

Question Is the volume $\frac{1}{6} \, d\theta$ equal to an area $\frac{1}{6}$ times a thickness $d\theta$?

Answer   No! The triangle in the third cone has area $\frac{1}{2}$ not $\frac{1}{6}$. *Thickness is never $d\theta$.*

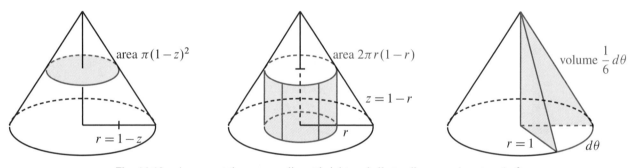

Fig. 14.16   A cone cut three ways: slice at height z, shell at radius r, wedge at angle θ.

This cone is typical of a **solid of revolution**. The axis is in the z direction. The θ integral yields 2π, whether it comes first, second, or third. The r integral goes out to a radius $f(z)$, which is 1 for the cylinder and $1-z$ for the cone. The integral $\iint r \, dr \, d\theta$ is $\pi(f(z))^2 =$ area of circular slice. This leaves the z integral $\int \pi(f(z))^2 dz$. That is our old volume formula $\int \pi(f(x))^2 dx$ from Chapter 8, where the slices were cut through the x axis.

EXAMPLE 3   The **moment of inertia** around the z axis is $\iiint r^3 dr \, d\theta \, dz$. The extra r^2 is (distance to axis)^2. For the cone this triple integral is $\pi/10$.

EXAMPLE 4   The **moment** around the z axis is $\iiint r^2 \, dr \, d\theta \, dz$. For the cone this is $\pi/6$. The *average distance* $\overline{r}$ is (moment)/(volume) $= (\pi/6)/(\pi/3) = \frac{1}{2}$.

EXAMPLE 5   A sphere of radius R has the boundary $r^2 + z^2 = R^2$, in cylindrical coordinates. The outer limit on the r integral is $\sqrt{R^2 - z^2}$. That is not acceptable in difficult problems. To avoid it we now change to coordinates that are natural for a sphere.

SPHERICAL COORDINATES

The Earth is a solid sphere (or near enough). On its surface we use two coordinates—latitude and longitude. To dig inward or fly outward, there is a third coordinate—*the distance ρ from the center*. This Greek letter *rho* replaces r to avoid confusion with cylindrical coordinates. Where r is measured from the z axis, ρ is measured from the origin. Thus $r^2 = x^2 + y^2$ and $\rho^2 = x^2 + y^2 + z^2$.

The angle θ is the same as before. It goes from 0 to 2π. It is the longitude, which increases as you travel east around the Equator.

The angle ϕ is new. It equals 0 at the North Pole and π (not 2π) at the South Pole. It is the ***polar angle***, measured down from the z axis. The Equator has a latitude of 0 but a polar angle of $\pi/2$ (halfway down). Here are some typical shapes:

solid sphere (or ball): $0 \leqslant \rho \leqslant R$ $\qquad$ surface of sphere: $\rho = R$

upper half-sphere: $0 \leqslant \phi \leqslant \pi/2$ $\qquad$ eastern half-sphere: $0 \leqslant \theta \leqslant \pi$

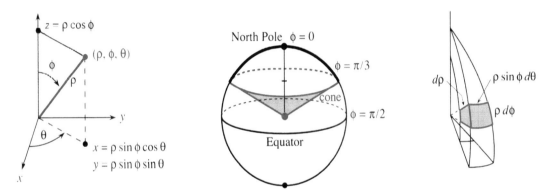

Fig. 14.17 Spherical coordinates $\rho\phi\theta$. The volume $dV = \rho^2 \sin \phi \, d\rho \, d\phi \, d\theta$ of a spherical box.

The angle ϕ is constant on a cone from the origin. It cuts the surface in a circle (Figure 14.17b), but not a great circle. The angle θ is constant along a half-circle from pole to pole. The distance ρ is constant on each inner sphere, starting at the center $\rho = 0$ and moving out to $\rho = R$.

In spherical coordinates the volume integral is $\iiint \rho^2 \sin \phi \, d\rho \, d\phi \, d\theta$. To explain that surprising factor $J = \rho^2 \sin \phi$, start with $x = r \cos \theta$ and $y = r \sin \theta$. In spherical coordinates r is $\rho \sin \phi$ and z is $\rho \cos \phi$—see the triangle in the figure. So substitute $\rho \sin \phi$ for r:

$$x = \rho \sin \phi \cos \theta, \ y = \rho \sin \phi \sin \theta, \ z = \rho \cos \phi. \tag{1}$$

Remember those two steps, $\rho\phi\theta$ to $r\theta z$ to xyz. We check that $x^2 + y^2 + z^2 = \rho^2$:

$$\rho^2(\sin^2 \phi \, \cos^2 \theta + \sin^2 \phi \, \sin^2 \theta + \cos^2 \phi) = \rho^2(\sin^2 \phi + \cos^2 \phi) = \rho^2.$$

The volume integral is explained by Figure 14.17c. That shows a "*spherical box*" with right angles and curved edges. Two edges are $d\rho$ and $\rho d\phi$. The third edge is horizontal. The usual $rd\theta$ becomes $\rho \sin \phi \, d\theta$. Multiplying those lengths gives dV.

The volume of the box is $dV = \rho^2 \sin \phi \, d\rho \, d\phi \, d\theta$. This is a *distance cubed*, from $\rho^2 d\rho$.

EXAMPLE 6   A solid ball of radius R has known volume $V = \frac{4}{3}\pi R^3$. Notice the limits:

$$\int_0^{2\pi} \int_0^{\pi} \int_0^R \rho^2 \sin \phi \, d\rho \, d\phi \, d\theta = \left[\tfrac{1}{3}\rho^3\right]_0^R \left[-\cos \phi\right]_0^{\pi} \left[\theta\right]_0^{2\pi} = (\tfrac{1}{3}R^3)(2)(2\pi).$$

Question   What is the volume above the cone in Figure 14.17?
Answer   The ϕ integral stops at $[-\cos \phi]_0^{\pi/3} = \frac{1}{2}$. The volume is $(\frac{1}{3}R^3)(\frac{1}{2})(2\pi)$.

EXAMPLE 7   The *surface area* of a sphere is $A = 4\pi R^2$. Forget the ρ integral:

$$A = \int_0^{2\pi} \int_0^{\pi} R^2 \sin \phi \, d\phi \, d\theta = R^2 \left[-\cos \phi\right]_0^{\pi} \left[\theta\right]_0^{2\pi} = R^2(2)(2\pi).$$

After those examples from geometry, here is the real thing from science. I want to compute one of the most important triple integrals in physics—"the gravitational attraction of a solid sphere." For some reason Isaac Newton had trouble with this integral. He refused to publish his masterpiece on astronomy until he had solved it. I think he didn't use spherical coordinates—and the integral is not easy even now.

The answer that Newton finally found is beautiful. *The sphere acts as if all its mass were concentrated at the center*. At an outside point $(0, 0, D)$, the force of gravity is proportional to $1/D^2$. The force from a uniform solid sphere equals the force from a point mass, at every outside point P. That is exactly what Newton wanted and needed, to explain the solar system and to prove Kepler's laws.

Here is the difficulty. Some parts of the sphere are closer than D, some parts are farther away. The actual distance q, from the outside point P to a typical inside point, is shown in Figure 14.18. The *average* distance $\bar{q}$ to all points in the sphere is not D. But what Newton needed was a different average, and by good luck or some divine calculus it works perfectly: *The average of* $1/q$ *is* $1/D$. This gives the potential energy:

$$\textit{potential at point } P = \iiint_{\text{sphere}} \frac{1}{q} dV = \frac{\text{Volume of sphere}}{D}. \tag{2}$$

A small volume dV at the distance q contributes dV/q to the potential (Section 8.6, with density 1). The integral adds the contributions from the whole sphere. Equation (2) says that the potential at $r = D$ is not changed when the sphere is squeezed to the center. The potential equals the whole volume divided by the single distance D.

Important point: The average of $1/q$ is $1/D$ and not $1/\bar{q}$. The average of $\frac{1}{2}$ and $\frac{1}{4}$ is not $\frac{1}{3}$. Smaller point: I wrote "sphere" where I should have written "ball." The sphere is solid: $0 \leqslant \rho \leqslant R, 0 \leqslant \phi \leqslant \pi, 0 \leqslant \theta \leqslant 2\pi$. What about the force? For the small volume it is proportional to dV/q^2 (this is the inverse square law). But *force is a vector*, pulling the outside point toward dV—not toward the center of the sphere. The figure shows the geometry and the symmetry. *We want the z component of the force.* (By symmetry the overall x and y components are zero.) The angle between the force vector and the z axis is α, so for the z component we multiply by $\cos \alpha$. The total force comes from the integral that Newton discovered:

$$\textbf{\textit{force at point }} P = \iiint\limits_{\text{sphere}} \frac{\cos \alpha}{q^2} dV = \frac{\text{volume of sphere}}{D^2}. \tag{3}$$

I will compute the integral (2) and leave you the privilege of solving (3). I mean that word seriously. If you have come this far, you deserve the pleasure of doing what at one time only Isaac Newton could do. Problem 26 offers a suggestion (just the law of cosines) but the integral is yours.

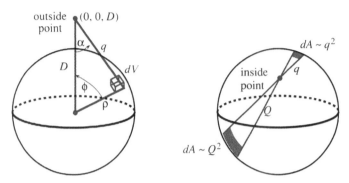

Fig. 14.18 Distance q from outside point to inside point. Distances q and Q to surface.

The law of cosines also helps with (2). For the triangle in the figure it gives $q^2 = D^2 - 2\rho D \cos \phi + \rho^2$. Call this whole quantity u. We do the surface integral first ($d\phi$ and $d\theta$ with ρ fixed). Then $q^2 = u$ and $q = \sqrt{u}$ and $du = 2\rho D \sin \phi \, d\phi$:

$$\int_0^{2\pi} \int_0^\pi \frac{\rho^2 \sin \phi \, d\phi \, d\theta}{q} = \int \frac{2\pi\rho^2}{2\rho D} \frac{du}{\sqrt{u}} = \left[\frac{2\pi\rho}{D} \sqrt{u} \right]_{\phi=0}^{\phi=\pi}. \tag{4}$$

2π came from the θ integral. The integral of $du/\sqrt{u}$ is $2\sqrt{u}$. Since $\cos \phi = -1$ at the upper limit, u is $D^2 + 2\rho D + \rho^2$. *The square root of u is $D + \rho$.* At the lower limit $\cos \phi = +1$ and $u = D^2 - 2\rho D + \rho^2$. This is another perfect square—its square root is $D - \rho$. The surface integral (4) with fixed ρ is

$$\iint \frac{dA}{q} = \frac{2\pi\rho}{D} [(D+\rho) - (D-\rho)] = \frac{4\pi\rho^2}{D}. \tag{5}$$

Last comes the ρ integral: $\int_0^R 4\pi\rho^2 d\rho / D = \frac{4}{3}\pi R^3 / D$. This proves formula (2): *potential equals volume of the sphere divided by D.*

Note 1 Physicists are also happy about equation (5). The average of $1/q$ is $1/D$ not only over the solid sphere but over each spherical shell of area $4\pi\rho^2$. The shells can have different densities, as they do in the Earth, and still Newton is correct. This also applies to the force integral (3)—*each separate shell acts as if its mass were concentrated at the center.* Then the final ρ integral yields this property for the solid sphere.

Note 2 Physicists also know that force is minus the derivative of potential. The derivative of (2) with respect to D produces the force integral (3). Problem 27 explains this shortcut to Equation (3).

EXAMPLE 8   *Everywhere inside a hollow sphere the force of gravity is zero.*

When D is smaller than ρ, the lower limit $\sqrt{u}$ in the integral (4) changes from $D - \rho$ to $\rho - D$. That way the square root stays positive. This changes the answer in (5) to $4\pi\rho^2/\rho$, so the potential no longer depends on D. *The potential is constant inside the hollow shell.* Since the force comes from its derivative, the force is zero.

A more intuitive proof is in the second figure. The infinitesimal areas on the surface are proportional to q^2 and Q^2. But the distances to those areas are q and Q, so the forces involve $1/q^2$ and $1/Q^2$ (the inverse square law). Therefore the two areas exert equal and opposite forces on the inside point, and they cancel each other. The total force from the shell is zero.

I believe this zero integral is the reason that the inside of a car is safe from lightning. Of course a car is not a sphere. But electric charge distributes itself to keep the surface at constant potential. The potential stays constant inside—therefore no force. The tires help to prevent conduction of current (and electrocution of driver).

P.S. Don't just step out of the car. Let a metal chain conduct the charge to the ground. Otherwise you could be the conductor.

CHANGE OF COORDINATES—STRETCHING FACTOR J

Once more we look to calculus for a formula. We need the volume of a small curved box in any uvw coordinate system. The $r\theta z$ box and the $\rho\phi\theta$ box have right angles, and their volumes were read off from the geometry (stretching factors $J = r$ and $J = \rho^2 \sin\phi$ in Figures 14.15 and 14.17). Now we change from xyz to other coordinates uvw—which are chosen to fit the problem.

Going from xy to uv, the area $dA = J\, du\, dv$ was a 2 by 2 determinant. In three dimensions the determinant is 3 by 3. The matrix is always the "Jacobian matrix," containing first derivatives. There were four derivatives from xy to uv, now there are nine from xyz to uvw.

14C Suppose x, y, z are given in terms of u, v, w. Then a small box in uvw space (sides du, dv, dw) comes from a volume $dV = J\, du\, dv\, dw$ in xyz space:

$$J = \begin{vmatrix} \partial x/\partial u & \partial x/\partial v & \partial x/\partial w \\ \partial y/\partial u & \partial y/\partial v & \partial y/\partial w \\ \partial z/\partial u & \partial z/\partial v & \partial z/\partial w \end{vmatrix} = \text{stretching factor} \frac{\partial(x, y, z)}{\partial(u, v, w)}. \tag{6}$$

The volume integral $\iiint dx\, dy\, dz$ becomes $\iiint |J|\, du\, dv\, dw$, with limits on uvw.

Remember that a 3 by 3 determinant is the sum of six terms (Section 11.5). One term in J is $(\partial x/\partial u)(\partial y/\partial v)(\partial z/\partial w)$, along the main diagonal. This comes from pure stretching, and the other five terms allow for rotation. The best way to exhibit the formula is for spherical coordinates—where the nine derivatives are easy but the determinant is not:

EXAMPLE 9 Find the factor J for $x = \rho \sin \phi \cos \theta$, $y = \rho \sin \phi \sin \theta$, $z = \rho \cos \phi$.

$$J = \frac{\partial(x,y,z)}{\partial(\rho,\phi,\theta)} = \begin{vmatrix} \sin\phi\cos\theta & \rho\cos\phi\cos\theta & -\rho\sin\phi\sin\theta \\ \sin\phi\sin\theta & \rho\cos\phi\sin\theta & \rho\sin\phi\cos\theta \\ \cos\phi & -\rho\sin\phi & 0 \end{vmatrix}.$$

The determinant has six terms, but two are zero—because of the zero in the corner. The other four terms are $\rho^2 \sin\phi \cos^2\phi \sin^2\theta$ and $\rho^2 \sin\phi \cos^2\phi \cos^2\theta$ and $\rho^2 \sin^3\phi \sin^2\theta$ and $\rho^2 \sin^3\phi \cos^2\theta$. Add the first two (note $\sin^2\theta + \cos^2\theta$) and separately add the second two. Then add the sums to reach $J = \rho^2 \sin\phi$.

Geometry already gave this answer. For most uvw variables, use the determinant.

14.4 EXERCISES

Read-through questions

The three __a__ coordinates are $r\theta z$. The point at $x = y = z = 1$ has $r =$ __b__, $\theta =$ __c__, $z =$ __d__. The volume integral is $\iiint$ __e__. The solid region $1 \leqslant r \leqslant 2, 0 \leqslant \theta \leqslant 2\pi, 0 \leqslant z \leqslant 4$ is a __f__. Its volume is __g__. From the r and θ integrals the area of a __h__ equals __i__. From the z and θ integrals the area of a __j__ equals __k__. In $r\theta z$ coordinates the shapes of __l__ are convenient, while __m__ are not.

The three __n__ coordinates are $\rho\phi\theta$. The point at $x = y = z = 1$ has $\rho =$ __o__, $\phi =$ __p__, $\theta =$ __q__. The angle ϕ is measured from __r__. θ is measured from __s__. ρ is the distance to __t__, where r was the distance to __u__. If $\rho\phi\theta$ are known then $x =$ __v__, $y =$ __w__, $z =$ __x__. The stretching factor J is a 3 by 3 __y__, and volume is $\iiint$ __z__.

The solid region $1 \leqslant \rho \leqslant 2, 0 \leqslant \phi \leqslant \pi, 0 \leqslant \theta \leqslant 2\pi$ is a __A__. Its volume is __B__. From the ϕ and θ integrals the area of a __C__ at radius ρ equals __D__. Newton discovered that the outside gravitational attraction of a __E__ is the same as for an equal mass located at __F__.

Convert the xyz coordinates in 1–4 to $r\theta z$ and $\rho\phi\theta$.

1 $(D,0,0)$

2 $(0,-D,0)$

3 $(0,0,D)$(watch θ)

4 $(3,4,5)$

Convert the spherical coordinates in 5–7 to xyz and $r\theta z$.

5 $\rho = 4$, $\phi = \pi/4$, $\theta = -\pi/4$

6 $\rho = 2$, $\phi = \pi/3$, $\theta = \pi/6$

7 $\rho = 1$, $\phi = \pi$, $\theta =$ anything.

8 Where does $x = r$ and $y = \theta$?

9 Find the polar angle ϕ for the point with cylindrical coordinates $r\theta z$.

10 What are $x(t), y(t), z(t)$ on the great circle from $\rho = 1$, $\phi = \pi/2, \theta = 0$ with speed 1 to $\rho = 1, \phi = \pi/4, \theta = \pi/2$?

From the limits of integration describe each region in 11–20 and find its volume. The inner integral has the inner limits.

11 $\int_{\theta=0}^{2\pi} \int_{r=0}^{1/\sqrt{2}} \int_{z=r}^{\sqrt{1-r^2}} r \, dz \, dr \, d\theta$

12 $\int_0^\pi \int_0^1 \int_0^{1+r^2} r \, dz \, dr \, d\theta$

13 $\int_{\theta=0}^{2\pi} \int_{z=0}^{1} \int_{r=0}^{2-z} r \, dr \, dz \, d\theta$

14 $\int_0^\pi \int_0^\pi \int_0^\pi r \, d\theta \, dr \, dz$

15 $\int_0^{\pi/2} \int_0^{\pi/2} \int_0^1 \rho^2 \sin\phi \, d\rho \, d\phi \, d\theta$

16 $\int_0^{2\pi} \int_0^{\pi/3} \int_{\sec\phi}^{2} \rho^2 \sin\phi \, d\rho \, d\phi \, d\theta$

17 $\int_0^\pi \int_0^\pi \int_0^{\sin\phi} \rho^2 \sin\phi \, d\rho \, d\phi \, d\theta$

18 $\int_0^{2\pi} \int_0^{\pi/4} \int_1^3 \rho^2 \sin\phi \, d\rho \, d\phi \, d\theta$

19 $\int_0^\pi \int_0^\pi \int_0^\pi \rho^2 \sin\phi \, d\rho \, d\phi \, d\theta$

20 $\int_0^1 \int_0^1 \int_0^1 \rho^2 \sin\phi \, d\rho \, d\phi \, d\theta$

21 Example 5 gave the volume integral for a sphere in $r\theta z$ coordinates. What is the area of the circular slice at height z ? What is the area of the cylindrical shell at radius r ? Integrate over slices (dz) and over shells (dr) to reach $4\pi R^3/3$.

22 Describe the solid with $0 \leqslant \rho \leqslant 1 - \cos\phi$ and find its volume.

23 A cylindrical tree has radius a. A saw cuts horizontally, ending halfway in at the x axis. Then it cuts on a sloping plane (angle α with the horizontal), also ending at the x axis. What is the volume of the wedge that falls out?

24 Find the mass of a planet of radius R, if its density at each radius ρ is $\delta = (\rho+1)/\rho$. Notice the infinite density at the center, but finite mass $M = \iiint \delta \, dV$. Here ρ is radius, not density.

25 For the cone out to $r = 1-z$, the average distance from the z axis is $\bar{r} = \frac{1}{2}$. For the triangle out to $r = 1-z$ the average is $\bar{r} = \frac{1}{3}$. How can they be different when rotating the triangle produces the cone?

Problems 26–32, on the attraction of a sphere, use Figure 14.18 and the law of cosines $q^2 = D^2 - 2\rho D \cos\phi + \rho^2 = u$.

26 *Newton's achievement* Show that $\iiint (\cos\alpha) dV/q^2$ equals volume/D^2. *One hint only:* Find $\cos\alpha$ from a second law of cosines $\rho^2 = D^2 - 2qD\cos\alpha + q^2$. The ϕ integral should involve $1/q$ and $1/q^3$. Equation (2) integrates $1/q$, leaving $\iiint dV/q^3$ still to do.

27 Compute $\partial q/\partial D$ in the first cosine law and show from Figure 14.18 that it equals $\cos\alpha$. Then the derivative of equation (2) with respect to D is a shortcut to Newton's equation (3).

28 The lines of length D and q meet at the angle α. Move the meeting point up by ΔD. Explain why the other line stretches by $\Delta q \approx \Delta D \cos\alpha$. So $\partial q/\partial D = \cos\alpha$ as before.

29 Show that the average distance is $\bar{q} = 4R/3$, from the North Pole $(D = R)$ to points on the Earth's surface $(\rho = R)$. To compute: $\bar{q} = \iint qR^2 \sin\phi \, d\phi \, d\theta/(\text{area } 4\pi R^2)$. Use the same substitution u.

30 Show as in Problem 29 that the average distance is $\bar{q} = D + \frac{1}{3}\rho^2/D$, from the outside point $(0,0,D)$ to points on the shell of radius ρ. Then integrate $\iiint q \, dV$ and divide by $4\pi R^3/3$ to find $\bar{q}$ for the solid sphere.

31 In Figure 14.18b, it is not true that the areas on the surface are exactly proportional to q^2 and Q^2. Why not? What happens to the second proof in Example 8?

32 For two solid spheres attracting each other (sun and planet), can we concentrate *both* spheres into point masses at their centers?

*33 Compute $\iiint \cos\alpha \, dV/q^3$ to find the force of gravity at $(0,0,D)$ from a *cylinder* $x^2 + y^2 \leqslant a^2, 0 \leqslant z \leqslant h$. Show from a figure why $q^2 = r^2 + (D-z)^2$ and $\cos\alpha = (D-z)/q$.

34 A *linear* change of variables has $x = au + bv + cw, y = du + ev + fw$, and $z = gu + hv + iw$. Write down the six terms in the determinant J. Three terms have minus signs.

35 A pure stretching has $x = au, y = bv$, and $z = cw$. Find the 3 by 3 matrix and its determinant J. What is special about the xyz box in this case?

36 (a) The matrix in Example 9 has three columns. Find the lengths of those three vectors (sum of squares, then square root). Compare with the edges of the box in Figure 14.17.

(b) Take the dot product of every column in J with every other column. Zero dot products mean right angles in the box. So J is the product of the column lengths.

37 Find the stretching factor J for cylindrical coordinates from the matrix of first derivatives.

38 Follow Problem 36 for cylindrical coordinates—find the length of each column in J and compare with the box in Figure 14.15.

39 Find the moment of inertia around the z axis of a spherical shell (radius ρ, density 1). The distance from the axis to a point on the shell is $r =$ _____. Substitute for r to find

$$I(\rho) = \int_0^{2\pi} \int_0^\pi r^2 \rho^2 \sin\phi \, d\phi \, d\theta.$$

Divide by mr^2 (which is $4\pi\rho^4$) to compute the number J for a hollow ball in the rolling experiment of Section 8.5.

40 The moment of inertia of a solid sphere (radius R, density 1) adds up the hollow spheres of Problem 39: $I = \int_0^R I(\rho) d\rho =$ _____. Divide by mR^2 (which is $\frac{4}{3}\pi R^5$) to find J in the rolling experiment. A solid ball rolls faster than a hollow ball because _____.

41 Inside the Earth, the force of gravity is proportional to the distance ρ from the center. Reason: The inner ball of radius ρ has mass proportional to _____ (assume constant density). The force is proportional to that mass divided by _____. The rest of the Earth (sphere with hole) exerts no force because _____.

42 Dig a tunnel through the center to Australia. Drop a ball in the tunnel at $y = R$; Australia is $y = -R$. The force of gravity is $-cy$ by Problem 41. Newton's law is $my'' = -cy$. What does the ball do when it reaches Australia?

CHAPTER 15

Vector Calculus

Chapter 14 introduced double and triple integrals. We went from $\int dx$ to $\iint dx\,dy$ and $\iiint dx\,dy\,dz$. All those integrals add up small pieces, and the limit gives area or volume or mass. What could be more natural than that? I regret to say, after the success of those multiple integrals, that something is missing. It is even more regrettable that we didn't notice it. The missing piece is nothing less than the Fundamental Theorem of Calculus.

The double integral $\iint dx\,dy$ equals the area. To compute it, we did not use an antiderivative of 1. At least not consciously. The method was almost trial and error, and the hard part was to find the limits of integration. This chapter goes deeper, to show how the step from a double integral to a single integral is really a new form of the Fundamental Theorem—when it is done right.

Two new ideas are needed early, one pleasant and one not. You will like *vector fields*. You may not think so highly of *line integrals*. Those are ordinary single integrals like $\int v(x)dx$, but they go along curves instead of straight lines. The nice step dx becomes the confusing step ds. Where $\int dx$ equals the length of the interval, $\int ds$ is the length of the curve. The point is that regions are enclosed by curves, and we have to integrate along them. The Fundamental Theorem in its two-dimensional form (Green's Theorem) connects *a double integral over the region* to *a single integral along its boundary curve*.

The great applications are in science and engineering, where vector fields are so natural. But there are changes in the language. Instead of an antiderivative, we speak about a *potential function*. Instead of the derivative, we take the "*divergence*" and "*curl*." Instead of area, we compute *flux* and *circulation* and *work*. Examples come first.

15.1 Vector Fields

For an ordinary scalar function, the input is a number x and the output is a number $f(x)$. For a vector field (or vector function), the input is a point (x, y) and the output is a two-dimensional vector $\mathbf{F}(x, y)$. There is a "field" of vectors, one at every point. In three dimensions the input point is (x, y, z) and the output vector $\mathbf{F}$ has three components.

DEFINITION   Let R be a region in the xy plane. A **vector field** $\mathbf{F}$ assigns to every point (x, y) in R a vector $\mathbf{F}(x, y)$ with two components:

$$\mathbf{F}(x, y) = M(x, y)\mathbf{i} + N(x, y)\mathbf{j}. \tag{1}$$

This plane vector field involves *two functions of two variables*. They are the components M and N, which vary from point to point. A vector has fixed components, a vector field has varying components.

A three-dimensional vector field has components $M(x, y, z)$ and $N(x, y, z)$ and $P(x, y, z)$. Then the vectors are $\mathbf{F} = M\mathbf{i} + N\mathbf{j} + P\mathbf{k}$.

EXAMPLE 1   The **position vector** at (x, y) is $\mathbf{R} = x\mathbf{i} + y\mathbf{j}$. Its components are $M = x$ and $N = y$. The vectors grow larger as we leave the origin (Figure 15.1a). Their direction is outward and their length is $|\mathbf{R}| = \sqrt{x^2 + y^2} = r$. The vector $\mathbf{R}$ is boldface, the number r is lightface.

EXAMPLE 2   The vector field $\mathbf{R}/r$ consists of **unit vectors** $\mathbf{u}_r$, pointing outward. We divide $\mathbf{R} = x\mathbf{i} + y\mathbf{j}$ by its length, at every point except the origin. The components of $\mathbf{R}/r$ are $M = x/r$ and $N = y/r$. Figure 15.1 shows a third field $\mathbf{R}/r^2$, whose length is $1/r$.

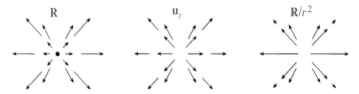

Fig. 15.1   The vector fields $\mathbf{R}$ and $\mathbf{R}/r$ and $\mathbf{R}/r^2$ are radial. Lengths r and 1 and $1/r$.

EXAMPLE 3   The **spin field** or rotation field or turning field goes around the origin instead of away from it. The field is $\mathbf{S}$. Its components are $M = -y$ and $N = x$:

$$\mathbf{S} = -y\mathbf{i} + x\mathbf{j} \text{ also has length } |\mathbf{S}| = \sqrt{(-y)^2 + x^2} = r. \tag{2}$$

$\mathbf{S}$ is perpendicular to $\mathbf{R}$—their dot product is zero: $\mathbf{S} \cdot \mathbf{R} = (-y)(x) + (x)(y) = 0$. The spin fields $\mathbf{S}/r$ and $\mathbf{S}/r^2$ have lengths 1 and $1/r$:

$$\frac{\mathbf{S}}{r} = -\frac{y}{r}\mathbf{i} + \frac{x}{r}\mathbf{j} \text{ has } \left|\frac{\mathbf{S}}{r}\right| = 1 \qquad \frac{\mathbf{S}}{r^2} = -\frac{y}{x^2 + y^2}\mathbf{i} + \frac{x}{x^2 + y^2}\mathbf{j} \text{ has } \left|\frac{\mathbf{S}}{r^2}\right| = \frac{1}{r}.$$

The unit vector $\mathbf{S}/r$ is $\mathbf{u}_\theta$. Notice the blank at $(0, 0)$, where this field is not defined.

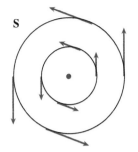 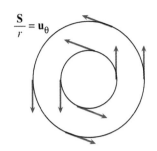 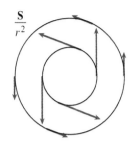

Fig. 15.2   The spin fields $\mathbf{S}$ and $\mathbf{S}/r$ and $\mathbf{S}/r^2$ go around the origin. Lengths r and 1 and $1/r$.

EXAMPLE 4   A *gradient field* starts with an ordinary function $f(x, y)$. The components M and N are the partial derivatives $\partial f/\partial x$ and $\partial f/\partial y$. Then the field $\mathbf{F}$ is the gradient of f:

$$\mathbf{F} = \text{grad } f = \nabla f = \partial f/\partial x\,\mathbf{i} + \partial f/\partial y\,\mathbf{j}. \tag{3}$$

This vector field grad f *is everywhere perpendicular to the level curves* $f(x, y) = c$. The length $|\text{grad } f|$ tells how fast f is changing (in the direction it changes fastest). Invent a function like $f = x^2 y$, and you immediately have its gradient field $\mathbf{F} = 2xy\mathbf{i} + x^2\mathbf{j}$. To repeat, M is $\partial f/\partial x$ and N is $\partial f/\partial y$.

For every vector field you should ask two questions: *Is it a gradient field? If so, what is* f? Here are answers for the radial fields and spin fields:

15A   The radial fields $\mathbf{R}$ and $\mathbf{R}/r$ and $\mathbf{R}/r^2$ are all gradient fields.
 The spin fields $\mathbf{S}$ and $\mathbf{S}/r$ are not gradients of any $f(x, y)$.
 The spin field $\mathbf{S}/r^2$ is the gradient of the polar angle $\theta = \tan^{-1}(y/x)$.

The derivatives of $f = \frac{1}{2}(x^2 + y^2)$ are x and y. Thus $\mathbf{R}$ is a gradient field. The gradient of $f = r$ is the unit vector $\mathbf{R}/r$ pointing outwards. Both fields are perpendicular to circles around the origin. Those are the level curves of $f = \frac{1}{2}r^2$ and $f = r$.

Question   Is every $\mathbf{R}/r^n$ a gradient field?
Answer   *Yes.* But among the spin fields, the only gradient is $\mathbf{S}/r^2$.

A major goal of this chapter is to recognize gradient fields by a simple test. The rejection of $\mathbf{S}$ and $\mathbf{S}/r$ will be interesting. For some reason $-y\mathbf{i} + x\mathbf{j}$ is rejected and $y\mathbf{i} + x\mathbf{j}$ is accepted. (It is the gradient of _____.) The acceptance of $\mathbf{S}/r^2$ as the gradient of $f = \theta$ contains a surprise at the origin (Section 15.3).

Gradient fields are called ***conservative***. The function f is the ***potential function***. These words, and the next examples, come from physics and engineering.

EXAMPLE 5   The *velocity field* is $\mathbf{V}$ and the *flow field* is $\rho\mathbf{V}$.

Suppose fluid moves steadily down a pipe. Or a river flows smoothly (no waterfall). Or the air circulates in a fixed pattern. The velocity can be different at different points, but there is no change with time. The velocity vector $\mathbf{V}$ gives the *direction of flow* and *speed of flow* at every point.

In reality the velocity field is $\mathbf{V}(x, y, z)$, with three components M, N, P. Those are the velocities v_1, v_2, v_3 in the x, y, z directions. The speed $|\mathbf{V}|$ is the length: $|\mathbf{V}|^2 = v_1^2 + v_2^2 + v_3^2$. In a "plane flow" the $\mathbf{k}$ component is zero, and the velocity field is $v_1 \mathbf{i} + v_2 \mathbf{j} = M \mathbf{i} + N \mathbf{j}$.

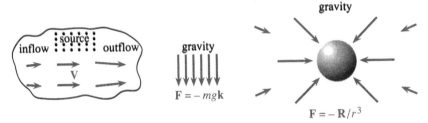

Fig. 15.3   A steady velocity field $\mathbf{V}$ and two force fields $\mathbf{F}$.

For a compact disc or a turning wheel, $\mathbf{V}$ is a spin field ($\mathbf{V} = \omega \mathbf{S}$, $\omega =$ angular velocity). A tornado might be closer to $\mathbf{V} = \mathbf{S}/r^2$ (except for a dead spot at the center). An explosion could have $\mathbf{V} = \mathbf{R}/r^2$. A quieter example is flow in and out of a lake with steady rain as a source term.

The *flow field* $\rho \mathbf{V}$ is the density ρ times the velocity field. While $\mathbf{V}$ gives the rate of movement, $\rho \mathbf{V}$ gives the ***rate of movement of mass***. A greater density means a greater rate $|\rho \mathbf{V}|$ of "mass transport." It is like the number of passengers on a bus times the speed of the bus.

EXAMPLE 6   Force fields from gravity: $\mathbf{F}$ is downward in the classroom, $\mathbf{F}$ is radial in space.

When gravity pulls downward, it has only one nonzero component: $\mathbf{F} = -mg\mathbf{k}$. This assumes that vectors to the center of the Earth are parallel—almost true in a classroom. Then $\mathbf{F}$ is the gradient of $-mgz$ (note $\partial f / \partial z = -mg$).
In physics the usual potential is not $-mgz$ but $+mgz$. The force field is *minus* grad f also in electrical engineering. Electrons flow from high potential to low potential. The mathematics is the same, but the sign is reversed.

In space, the force is radial inwards: $\mathbf{F} = -mMG\mathbf{R}/r^3$. Its magnitude is proportional to $1/r^2$ (Newton's inverse square law). The masses are m and M, and the gravitational constant is $G = 6.672 \times 10^{-11}$—with distance in meters, mass in kilograms, and time in seconds. The dimensions of G are (force)(distance)^2/(mass)^2. This is different from the acceleration $g = 9.8 \text{m/sec}^2$, which already accounts for the mass and radius of the Earth.

Like all radial fields, ***gravity is a gradient field***. It comes from a potential f:

$$f = \frac{mMG}{r} \text{ and } \frac{\partial f}{\partial x} = -\frac{mMGx}{r^3} \text{ and } \frac{\partial f}{\partial y} = -\frac{mMGy}{r^3} \text{ and } \frac{\partial f}{\partial z} = -\frac{mMGz}{r^3}. \quad (4)$$

EXAMPLE 7   (a short example) Current in a wire produces a ***magnetic field*** $\mathbf{B}$. It is the spin field $\mathbf{S}/r^2$ around the wire, times the strength of the current.

STREAMLINES AND LINES OF FORCE

Drawing a vector field is not always easy. Even the spin field looks messy when the vectors are too long (they go in circles and fall across each other). *The circles give a clearer picture than the vectors.* In any field, the vectors are tangent to "*field lines*"—which in the spin case are circles.

DEFINITION C is a *field line* or *integral curve* if the vectors $\mathbf{F}(x, y)$ are tangent to C. The slope dy/dx of the curve C equals the slope N/M of the vector $\mathbf{F} = M\mathbf{i} + N\mathbf{j}$:

$$\frac{dy}{dx} = \frac{N(x,y)}{M(x,y)} \quad \left(= -\frac{x}{y} \text{ for the spin field}\right). \tag{6}$$

We are still drawing the field of vectors, but now they are infinitesimally short. They are connected into curves! What is lost is their length, because $\mathbf{S}$ and $\mathbf{S}/r$ and $\mathbf{S}/r^2$ all have the same field lines (circles). For the position field $\mathbf{R}$ and gravity field $\mathbf{R}/r^3$, the field lines are rays from the origin. In this case the "curves" are actually straight.

EXAMPLE 8 Show that the field lines for the velocity field $\mathbf{V} = y\mathbf{i} + x\mathbf{j}$ are hyperbolas.

$$\frac{dy}{dx} = \frac{N}{M} = \frac{x}{y} \quad \Rightarrow \quad y\,dy = x\,dx \quad \Rightarrow \quad \tfrac{1}{2}y^2 - \tfrac{1}{2}x^2 = \text{constant}.$$

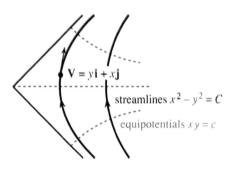

streamlines $x^2 - y^2 = C$

equipotentials $xy = c$

Fig. 15.4 Velocity fields are tangent to streamlines. Gradient fields also have equipotentials.

At every point these hyperbolas line up with the velocity $\mathbf{V}$. *Each particle of fluid travels on a field line.* In fluid flow those hyperbolas are called *streamlines*. Drop a leaf into a river, and it follows a streamline. Figure 15.4 shows the streamlines for a river going around a bend.

Don't forget the essential question about each vector field. Is it a gradient field? For $\mathbf{V} = y\mathbf{i} + x\mathbf{j}$ the answer is *yes*, and the potential is $f = xy$:

$$\text{the gradient of } xy \text{ is } (\partial f/\partial x)\mathbf{i} + (\partial f/\partial y)\mathbf{j} = y\mathbf{i} + x\mathbf{j}. \tag{7}$$

When there is a potential, it has level curves. They connect points of equal potential, so the curves $f(x, y) = c$ are called *equipotentials*. Here they are the curves $xy = c$—also hyperbolas. Since gradients are perpendicular to level curves, *the streamlines are perpendicular to the equipotentials*. Figure 15.4 is sliced one way by streamlines and the other way by equipotentials.

A gradient field $\mathbf{F} = \partial f/\partial x\,\mathbf{i} + \partial f/\partial y\,\mathbf{j}$ is tangent to the field lines (streamlines) and perpendicular to the equipotentials (level curves of f).

In the gradient direction f changes fastest. In the level direction f doesn't change at all. The chain rule along $f(x, y) = c$ proves these directions to be perpendicular:

$$\frac{\partial f}{\partial x}\frac{dx}{dt} + \frac{\partial f}{\partial y}\frac{dy}{dt} = 0 \quad \text{or} \quad (\text{grad } f)\cdot(\text{tangent to level curve}) = 0.$$

EXAMPLE 9   The streamlines of $\mathbf{S}/r^2$ are circles around $(0,0)$. The equipotentials are rays $\theta = c$. Add rays to Figure 15.2 for the gradient field $\mathbf{S}/r^2$.

For the gravity field those are reversed. A body is pulled in along the field lines (rays). The equipotentials are the circles where $f = 1/r$ is constant. The plane is crisscrossed by "orthogonal trajectories"—curves that meet everywhere at right angles.

   If you bring a magnet near a pile of iron filings, a little shake will display the field lines. In a force field, they are "lines of force." *Here are the other new words.*

Vector field $\mathbf{F}(x, y, z) = M\mathbf{i} + N\mathbf{j} + P\mathbf{k}$ Plane field $\mathbf{F} = M(x, y)\mathbf{i} + N(x, y)\mathbf{j}$

Radial field: multiple of $\mathbf{R} = x\mathbf{i} + y\mathbf{j} + z\mathbf{k}$ Spin field: multiple of $\mathbf{S} = -y\mathbf{i} + x\mathbf{j}$

Gradient field = conservative field: $M = \partial f/\partial x, N = \partial f/\partial y, P = \partial f/\partial z$

Potential $f(x, y)$ (not a vector)   Equipotential curves $f(x, y) = c$

Streamline = field line = integral curve: a curve that has $\mathbf{F}(x, y)$ as its tangent vectors.

15.1 EXERCISES

Read-through questions

A vector field assigns a __a__ to each point (x, y) or (x, y, z). In two dimensions $\mathbf{F}(x, y) =$ __b__ $\mathbf{i} +$ __c__ $\mathbf{j}$. An example is the position field $\mathbf{R} =$ __d__. Its magnitude is $|\mathbf{R}| =$ __e__ and its direction is __f__. It is the gradient field for $f =$ __g__. The level curves are __h__, and they are __i__ to the vectors $\mathbf{R}$.

   Reversing this picture, the spin field is $\mathbf{S} =$ __j__. Its magnitude is $|\mathbf{S}| =$ __k__ and its direction is __l__. It is not a gradient field, because no function has $\partial f/\partial x =$ __m__ and $\partial f/\partial y =$ __n__. $\mathbf{S}$ is the velocity field for flow going __o__. The streamlines or __p__ lines or integral __q__ are __r__. The flow field $\rho\mathbf{V}$ gives the rate at which __s__ is moved by the flow.

   A gravity field from the origin is proportional to $\mathbf{F} =$ __t__ which has $|\mathbf{F}| =$ __u__. This is Newton's __v__ square law. It is a gradient field, with potential $f =$ __w__. The equipotential curves $f(x, y) = c$ are __x__. They are __y__ to the field lines which are __z__. This illustrates that the __A__ of a function $f(x, y)$ is __B__ to its level curves.

The velocity field $y\mathbf{i} + x\mathbf{j}$ is the gradient of $f =$ __C__. Its streamlines are __D__. The slope dy/dx of a streamline equals the ratio __E__ of velocity components. The field is __F__ to the streamlines. Drop a leaf onto the flow, and it goes along __G__.

Find a potential $f(x, y)$ for the gradient fields 1–8. Draw the streamlines perpendicular to the equipotentials $f(x, y) = c$.

1 $\mathbf{F} = \mathbf{i} + 2\mathbf{j}$ (constant field)   **2** $\mathbf{F} = x\mathbf{i} + \mathbf{j}$

3 $\mathbf{F} = \cos(x + y)\mathbf{i} + \cos(x + y)\mathbf{j}$ **4** $\mathbf{F} = (1/y)\mathbf{i} - (x/y^2)\mathbf{j}$

5 $\mathbf{F} = (2x\mathbf{i} + 2y\mathbf{j})/(x^2 + y^2)$ **6** $\mathbf{F} = x^2\mathbf{i} + y^2\mathbf{j}$

7 $\mathbf{F} = xy\mathbf{i} +$ _____ $\mathbf{j}$ **8** $\mathbf{F} = \sqrt{y}\mathbf{i} +$ _____ $\mathbf{j}$

9 Draw the shear field $\mathbf{F} = x\mathbf{j}$. Check that it is not a gradient field: If $\partial f/\partial x = 0$ then $\partial f/\partial y = x$ is impossible. What are the streamlines (field lines) in the direction of $\mathbf{F}$?

10 Find all functions that satisfy $\partial f/\partial x = -y$ and show that none of them satisfy $\partial f/\partial y = x$. Then the spin field $\mathbf{S} = -y\mathbf{i} + x\mathbf{j}$ is not a gradient field.

Compute $\partial f/\partial x$ and $\partial f/\partial y$ in 11–18. Draw the gradient field $F = \text{grad } f$ and the equipotentials $f(x, y) = c$:

11 $f = 3x + y$ **12** $f = x - 3y$

13 $f = x + y^2$ **14** $f = (x - 1)^2 + y^2$

15 $f = x^2 - y^2$ **16** $f = e^x \cos y$

17 $f = e^{x-y}$ **18** $f = y/x$

Find equations for the streamlines in 19–24 by solving $dy/dx = N/M$ (including a constant C). Draw the streamlines.

19 $\mathbf{F} = \mathbf{i} - \mathbf{j}$ **20** $\mathbf{F} = \mathbf{i} + x\mathbf{j}$

21 $\mathbf{F} = \mathbf{S}$ (spin field)   **22** $\mathbf{F} = \mathbf{S}/r$ (spin field)

23 $\mathbf{F} = \text{grad } (x/y)$ **24** $\mathbf{F} = \text{grad } (2x + y)$.

25 The Earth's gravity field is radial, but in a room the field lines seem to go straight down into the floor. This is because nearby field lines always look _____ .

26 A line of charges produces the electrostatic force field $\mathbf{F} = \mathbf{R}/r^2 = (x\mathbf{i} + y\mathbf{j})/(x^2 + y^2)$. Find the potential $f(x, y)$. ($\mathbf{F}$ is also the gravity field for a line of masses.)

In 27–32 write down the vector fields $M\mathbf{i} + N\mathbf{j}$.

27 $\mathbf{F}$ points radially away from the origin with magnitude 5.

28 The velocity is perpendicular to the curves $x^3 + y^3 = c$ and the speed is 1.

29 The gravitational force $\mathbf{F}$ comes from two unit masses at $(0, 0)$ and $(1, 0)$.

30 The streamlines are in the $45°$ direction and the speed is 4.

31 The streamlines are circles clockwise around the origin and the speed is 1.

32 The equipotentials are the parabolas $y = x^2 + c$ and $\mathbf{F}$ is a gradient field.

33 Show directly that the hyperbolas $xy = 2$ and $x^2 - y^2 = 3$ are perpendicular at the point $(2, 1)$, by computing both slopes dy/dx and multiplying to get -1.

34 The derivative of $f(x, y) = c$ is $f_x + f_y(dy/dx) = 0$. Show that the slope of this level curve is $dy/dx = -M/N$. It is perpendicular to streamlines because $(-M/N)(N/M) = $ _____ .

35 The x and y derivatives of $f(r)$ are $\partial f/\partial x = $ _____ and $\partial f/\partial y = $ _____ by the chain rule. (Test $f = r^2$.) The equipotentials are _____ .

36 $\mathbf{F} = (ax + by)\mathbf{i} + (bx + cy)\mathbf{j}$ is a gradient field. Find the potential f and describe the equipotentials.

37 *True or false*:

1. The constant field $\mathbf{i} + 2\mathbf{k}$ is a gradient field.

2. For non-gradient fields, equipotentials meet streamlines at non-right angles.

3. In three dimensions the equipotentials are surfaces instead of curves.

4. $\mathbf{F} = x^2\mathbf{i} + y^2\mathbf{j} + z^2\mathbf{k}$ points outward from $(0, 0, 0)$— a radial field.

38 Create and draw f and $\mathbf{F}$ and your own equipotentials and streamlines.

39 How can different vector fields have the same streamlines? Can they have the same equipotentials? Can they have the same f?

40 Draw arrows at six or eight points to show the direction and magnitude of each field:

    (a) $\mathbf{R} + \mathbf{S}$ (b) $\mathbf{R}/r - \mathbf{S}/r$ (c) $x^2\mathbf{i} + x^2\mathbf{j}$ (d) $y\mathbf{i}$.

15.2 Line Integrals

A line integral is *an integral along a curve*. It can equal an area, but that is a special case and not typical. Instead of area, here are two important line integrals in physics and engineering:

$$\textit{Work along a curve} = \int_c \mathbf{F} \cdot \mathbf{T} \, ds \qquad \textit{Flow across a curve} = \int_c \mathbf{F} \cdot \mathbf{n} \, ds.$$

In the first integral, $\mathbf{F}$ is a *force field*. In the second integral, $\mathbf{F}$ is a *flow field*. Work is done in the direction of movement, so we integrate $\mathbf{F} \cdot \mathbf{T}$. Flow is measured through the curve C, so we integrate $\mathbf{F} \cdot \mathbf{n}$. Here $\mathbf{T}$ is the unit *tangent* vector, and $\mathbf{F} \cdot \mathbf{T}$ is the force component along the curve. Similarly $\mathbf{n}$ is the unit *normal* vector, at right angles with $\mathbf{T}$. Then $\mathbf{F} \cdot \mathbf{n}$ is the component of flow perpendicular to the curve.

We will write those integrals in several forms. They may never be as comfortable as $\int y(x) dx$, but eventually we get them under control. I mention these applications early, so you can see where we are going. This section concentrates on work, and flow comes later. (It is also called *flux*—the Latin word for flow.) You recognize ds as the step along the curve, corresponding to dx on the x axis. Where $\int dx$ gives the length of an interval (it equals $b - a$), $\int ds$ is the length of the curve.

EXAMPLE 1   Flight from Atlanta to Los Angeles on a straight line and a semicircle.

According to Delta Airlines, the distance straight west is 2000 miles. Atlanta is at $(1000, 0)$ and Los Angeles is at $(-1000, 0)$, with the origin halfway between. The semicircle route C has radius 1000. *This is not a great circle route.* It is more of a "flat circle," which goes north past Chicago. No plane could fly it (it probably goes into space).

The equation for the semicircle is $x^2 + y^2 = 1000^2$. Parametrically this path is $x = 1000 \cos t, y = 1000 \sin t$. For a line integral the parameter is better. The plane leaves Atlanta at $t = 0$ and reaches L.A. at $t = \pi$, more than three hours later. On the straight 2000-mile path, Delta could almost do it. Around the semicircle C, the distance is 1000π miles and the speed has to be 1000 miles per hour. Remember that speed is distance ds divided by time dt:

$$ds/dt = \sqrt{(dx/dt)^2 + (dy/dt)^2} = 1000\sqrt{(-\sin t)^2 + (-\cos t)^2} = 1000. \quad (1)$$

The tangent vector to C is proportional to $(dx/dt, dy/dt) = (-1000 \sin t, 1000 \cos t)$. But $\mathbf{T}$ is a unit vector, so we divide by 1000—which is the speed.

Suppose the wind blows due east with force $\mathbf{F} = M\mathbf{i}$. The components are M and zero. For $M =$ constant, compute the dot product $\mathbf{F} \cdot \mathbf{T}$ and the work $-2000 M$:

$$\mathbf{F} \cdot \mathbf{T} = M\mathbf{i} \cdot (-\sin t \, \mathbf{i} + \cos t \, \mathbf{j}) = M(-\sin t) + 0(\cos t) = -M \sin t$$

$$\int_c \mathbf{F} \cdot \mathbf{T} \, ds = \int_{t=0}^\pi (-M \sin t) \left(\frac{ds}{dt} dt \right) = \int_0^\pi -1000 M \sin t \, dt = -2000 M.$$

Work is force times distance moved. It is negative, because the wind acts *against* the movement. You may point out that the work could have been found more simply—go 2000 miles and multiply by $-M$. I would object that *this straight route is a different path*. But you claim that *the path doesn't matter*—the work of the wind is $-2000M$ on every path. I concede that this time you are right (but not always).

Most line integrals depend on the path. Those that don't are crucially important. For a *gradient field*, we only need to know the starting point P and the finish Q.

> **15B** When **F** is the gradient of a potential function $f(x, y)$, the work $\int_c \mathbf{F} \cdot \mathbf{T} \, ds$ depends only on the endpoints P and Q. *The work is the change in* f:
>
> $$\text{If} \quad \mathbf{F} = \partial f/\partial x \, \mathbf{i} + \partial f/\partial y \, \mathbf{j} \quad \text{then} \quad \int_c \mathbf{F} \cdot \mathbf{T} \, ds = f(Q) - f(P). \quad (2)$$

When $\mathbf{F} = M\mathbf{i}$, its components M and zero are the partial derivatives of $f = Mx$. To compute the line integral, just evaluate f at the endpoints. Atlanta has $x = 1000$, Los Angeles has $x = -1000$, and *the potential function* $f = Mx$ *is like an antiderivative*:

$$\text{work} = f(Q) - f(P) = M(-1000) - M(1000) = -2000M. \quad (3)$$

Fig. 15.5 Force $M\mathbf{i}$, work $-2000M$ on all paths. Force $My\mathbf{i}$, no work on straight path.

May I give a rough explanation of the work integral $\int \mathbf{F} \cdot \mathbf{T} \, ds$? It becomes clearer when the small movement $\mathbf{T} \, ds$ is written as $dx \, \mathbf{i} + dy \, \mathbf{j}$. The work is the dot product with **F**:

$$\mathbf{F} \cdot \mathbf{T} \, ds = \left(\frac{\partial f}{\partial x} \mathbf{i} + \frac{\partial f}{\partial y} \mathbf{j} \right) \cdot (dx \, \mathbf{i} + dy \, \mathbf{j}) = \frac{\partial f}{\partial x} dx + \frac{\partial f}{\partial y} dy = df. \quad (4)$$

The infinitesimal work is df. The total work is $\int df = f(Q) - f(P)$. This is the ***Fundamental Theorem for a line integral***. Only one warning: When **F** is not the gradient of any f (Example 2), the Theorem does not apply.

EXAMPLE 2 Fly these paths against the non-constant force field $\mathbf{F} = My\mathbf{i}$. Compute the work.

There is no force on the straight path where $y = 0$. Along the x axis the wind does no work. But the semicircle goes up where $y = 1000 \sin t$ and the wind is strong:

$$\mathbf{F} \cdot \mathbf{T} \ (My\mathbf{i}) \cdot (-\sin t \, \mathbf{i} + \cos t \, \mathbf{j}) = -My \sin t = -1000M \sin^2 t$$

$$\int_c \mathbf{F} \cdot \mathbf{T} \, ds = \int_0^\pi (-1000M \sin^2 t) \frac{ds}{dt} dt = \int_0^\pi -10^6 M \sin^2 t \, dt = -\frac{\pi}{2} 10^6 M.$$

This work is enormous (and unrealistic). But the calculations make an important point—everything is converted to the parameter t. The second point is that $\mathbf{F} = My\mathbf{i}$ is not a gradient field. *First reason*: The work was zero on the straight path and nonzero on the semicircle. *Second reason*: No function has $\partial f/\partial x = My$ and $\partial f/\partial y = 0$. (The first makes f depend on y and the second forbids it. This **F** is called a *shear force*.) Without a potential we cannot substitute P and Q—and the work depends on the path.

THE DEFINITION OF LINE INTEGRALS

We go back to the start, to define $\int \mathbf{F} \cdot \mathbf{T} \, ds$. We can think of $\mathbf{F} \cdot \mathbf{T}$ as a function $g(x, y)$ along the path, and define its integral as a limit of sums:

$$\int_C g(x, y) \, ds = \text{limit of} \sum_{i=1}^{N} g(x_i, y_i) \Delta s_i \quad \text{as} \quad (\Delta s)_{\max} \to 0. \tag{5}$$

The points (x_i, y_i) lie on the curve C. The last point Q is (x_N, y_N); the first point P is (x_0, y_0). The step Δs_i is the distance to (x_i, y_i) from the previous point. As the steps get small ($\Delta s \to 0$) the straight pieces follow the curve. Exactly as in Section 8.2, the special case $g = 1$ gives the arc length. As long as $g(x, y)$ is piecewise continuous (jumps allowed) and the path is piecewise smooth (corners allowed), the limit exists and defines the line integral.

When g is the density of a wire, the line integral is the total mass. When g is $\mathbf{F} \cdot \mathbf{T}$, the integral is the work. But nobody does the calculation by formula (5). We now introduce a parameter t—which could be the time, or the arc length s, or the distance x along the base.

The differential ds becomes $(ds/dt)dt$. Everything changes over to t:

$$\int g(x, y) ds = \int_{t=a}^{t=b} g(x(t), y(t)) \sqrt{(dx/dt)^2 + (dy/dt)^2} \, dt. \tag{6}$$

The curve starts when $t = a$, runs through the points $(x(t), y(t))$, and ends when $t = b$. The square root in the integral is the speed ds/dt. In three dimensions the points on C are $(x(t), y(t), z(t))$ and $(dz/dt)^2$ is in the square root.

EXAMPLE 3   The points on a coil spring are $(x, y, z) = (\cos t, \sin t, t)$. Find the mass of two complete turns (from $t = 0$ to $t = 4\pi$) if the density is $\rho = 4$.

Solution   The key is $(dx/dt)^2 + (dy/dt)^2 + (dz/dt)^2 = \sin^2 t + \cos^2 t + 1 = 2$. Thus $ds/dt = \sqrt{2}$. To find the mass, integrate the mass per unit length which is $g = \rho = 4$:

$$\text{mass} = \int_0^{4\pi} \rho \frac{ds}{dt} dt = \int_0^{4\pi} 4\sqrt{2} \, dt = 16\sqrt{2}\,\pi.$$

That is a line integral in three-dimensional space. It shows how to introduce t. But it misses the main point of this section, because it contains no vector field $\mathbf{F}$. This section is about *work*, not just mass.

DIFFERENT FORMS OF THE WORK INTEGRAL

The work integral $\int \mathbf{F} \cdot \mathbf{T} \, ds$ can be written in a better way. The force is $\mathbf{F} = M\mathbf{i} + N\mathbf{j}$. A small step along the curve is $dx\,\mathbf{i} + dy\,\mathbf{j}$. Work is force times distance, but it is only the force component *along the path* that counts. The dot product $\mathbf{F} \cdot \mathbf{T} \, ds$ finds that component automatically.

> **15C** The vector to a point on C is $\mathbf{R} = x\mathbf{i} + y\mathbf{j}$. Then $d\mathbf{R} = \mathbf{T}\,ds = dx\,\mathbf{i} + dy\,\mathbf{j}$:
>
> $$\mathbf{work} = \int_c \mathbf{F} \cdot d\mathbf{R} = \int_c M\,dx + N\,dy. \qquad (7)$$
>
> Along a space curve the work is $\int \mathbf{F} \cdot \mathbf{T}\,ds = \int \mathbf{F} \cdot d\mathbf{R} = \int M\,dx + N\,dy + P\,dz.$

The product $M\,dx$ is (force in x direction)(movement in x direction). This is zero if either factor is zero. When the only force is gravity, pushing a piano takes no work. It is friction that hurts. Carrying the piano up the stairs brings in $P\,dz$, and the total work is the piano weight P times the change in z.

To connect the new $\int \mathbf{F} \cdot d\mathbf{R}$ with the old $\int \mathbf{F} \cdot \mathbf{T}\,ds$, remember the tangent vector **T**. It is $d\mathbf{R}/ds$. *Therefore* $\mathbf{T}\,ds$ *is* $d\mathbf{R}$. The best for computations is $d\mathbf{R}$, because the unit vector **T** has a division by $ds/dt = \sqrt{(dx/dt)^2 + (dy/dt)^2}$. Later we multiply by this square root, in converting ds to $(ds/dt)dt$. It makes no sense to compute the square root, divide by it, and then multiply by it. That is avoided in the improved form $\int M\,dx + N\,dy$.

EXAMPLE 4   Vector field $\mathbf{F} = -y\mathbf{i} + x\mathbf{j}$, path from $(1,0)$ to $(0,1)$: Find the work.

Note 1   This **F** is the spin field **S**. It goes *around* the origin, while $\mathbf{R} = x\mathbf{i} + y\mathbf{j}$ goes outward. Their dot product is $\mathbf{F} \cdot \mathbf{R} = -yx + xy = 0$. This does not mean that $\mathbf{F} \cdot d\mathbf{R} = 0$. The force is perpendicular to $\mathbf{R}$, but not to the *change* in $\mathbf{R}$. The work to move from $(1,0)$ to $(0,1)$, x axis to y axis, is not zero.

Note 2   We have not described the path C. That must be done. The spin field is not a gradient field, and the work along a straight line does not equal the work on a quarter-circle:

$$\text{straight line } x = 1 - t,\, y = t \qquad \text{quarter-circle } x = \cos t,\, y = \sin t.$$

Calculation of work   Change $\mathbf{F} \cdot d\mathbf{R} = M\,dx + N\,dy$ to the parameter t:

Straight line: $\displaystyle\int -y\,dx + x\,dy = \int_0^1 -t(-dt) + (1-t)dt = 1$

Quarter-circle: $\displaystyle\int -y\,dx + x\,dy = \int_0^{\pi/2} -\sin t(-\sin t\,dt) + \cos t(\cos t\,dt) = \frac{\pi}{2}.$

General method   The path is given by $x(t)$ and $y(t)$. Substitute those into $M(x,y)$ and $N(x,y)$—then **F** is a function of t. Also find dx/dt and dy/dt. Integrate $M\,dx/dt + N\,dy/dt$ from the starting time t to the finish.

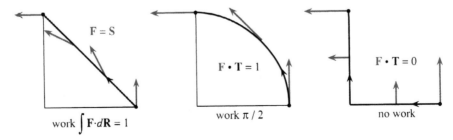

work $\int \mathbf{F} \cdot d\mathbf{R} = 1$ work $\pi/2$ no work

Fig. 15.6   Three paths for $\int \mathbf{F} \cdot d\mathbf{R} = \int -y\,dx + x\,dy = 1, \pi/2, 0.$

For practice, take the path down the x axis to the origin ($x = 1 - t, y = 0$). Then go up the y axis ($x = 0, y = t - 1$). The starting time at $(1, 0)$ is $t = 0$. The turning time at the origin is $t = 1$. The finishing time at $(0, 1)$ is $t = 2$. The integral has two parts because this new path has two parts:

$$\text{Bent path: } \int -y \, dx + x \, dy = 0 + 0 (y = 0 \text{ on one part, then } x = 0).$$

Note 3 The answer depended on the path, for this spin field $\mathbf{F} = \mathbf{S}$. The answer did *not* depend on the choice of parameter. If we follow the same path at a different speed, the work is the same. We can choose another parameter τ, since $(ds/dt) dt$ and $(ds/d\tau) d\tau$ both equal ds. Traveling twice as fast on the straight path ($x = 1 - 2\tau$, $y = 2\tau$) we finish at $\tau = \frac{1}{2}$ instead of $t = 1$. The work is still 1:

$$\int -y \, dx + x \, dy = \int_0^{1/2} (-2\tau)(-2 d\tau) + (1 - 2\tau)(2 d\tau) = \int_0^{1/2} 2 \, d\tau = 1.$$

CONSERVATION OF TOTAL ENERGY (KINETIC + POTENTIAL)

When a force field does work on a mass m, it normally gives that mass a new velocity. Newton's Law is $\mathbf{F} = m\mathbf{a} = m \, d\mathbf{v}/dt$. (It is a vector law. Why write out three components?) The work $\int \mathbf{F} \cdot d\mathbf{R}$ is

$$\int (m \, d\mathbf{v}/dt) \cdot (\mathbf{v} \, dt) = \tfrac{1}{2} m \mathbf{v} \cdot \mathbf{v} \Big]_P^Q = \tfrac{1}{2} m |\mathbf{v}(Q)|^2 - \tfrac{1}{2} m |\mathbf{v}(P)|^2. \tag{8}$$

The work equals the change in the kinetic energy $\frac{1}{2} m |\mathbf{v}|^2$. But for a gradient field the work is also the *change in potential*—with a minus sign from physics:

$$work = \int \mathbf{F} \cdot d\mathbf{R} = -\int df = f(P) - f(Q). \tag{9}$$

Comparing (8) with (9), the combination $\frac{1}{2} m |\mathbf{v}|^2 + f$ is the same at P and Q. *The total energy, kinetic plus potential, is conserved.*

INDEPENDENCE OF PATH: GRADIENT FIELDS

The work of the spin field $\mathbf{S}$ depends on the path. Example 4 took three paths—straight line, quarter-circle, bent line. The work was 1, $\pi/2$, and 0, different on each path. This happens for more than 99.99% of all vector fields. It does not happen for the most important fields. Mathematics and physics concentrate on very special fields—for which the work depends only on the endpoints. We now explain what happens, *when the integral is independent of the path.*

Suppose you integrate from P to Q on one path, and back to P on another path. Combined, that is a *closed path* from P to P (Figure 15.7). But a backward integral is the negative of a forward integral, since $d\mathbf{R}$ switches sign. *If the integrals from P to Q are equal, the integral around the closed path is zero*:

$$\oint_P^P \mathbf{F} \cdot d\mathbf{R} = \int_P^Q \mathbf{F} \cdot d\mathbf{R} + \int_Q^P \mathbf{F} \cdot d\mathbf{R} = \int_P^Q \mathbf{F} \cdot d\mathbf{R} - \int_P^Q \mathbf{F} \cdot d\mathbf{R} = 0. \tag{10}$$

$$\text{closed} \qquad \text{path 1} \qquad \text{back path 2} \qquad \text{path 1} \qquad \text{path 2}$$

The circle on the first integral indicates a closed path. Later we will drop the $P's$.

Not all closed path integrals are zero! For most fields **F**, different paths yield different work. For "*conservative*" fields, all paths yield the same work. Then zero work around a closed path conserves energy. The big question is: *How to decide which fields are conservative, without trying all paths*? Here is the crucial information about conservative fields, in a plane region R *with no holes*:

15D $\mathbf{F} = M(x,y)\mathbf{i} + N(x,y)\mathbf{j}$ is a conservative field if it has these properties:

A. The work $\int \mathbf{F} \cdot d\mathbf{R}$ around every closed path is zero.

B. The work $\int_P^Q \mathbf{F} \cdot d\mathbf{R}$ depends only on P and Q, not on the path.

C. **F** is a *gradient field*: $M = \partial f / \partial x$ and $N = \partial f / \partial y$ for some potential $f(x,y)$.

D. The components satisfy $\partial M / \partial y = \partial N / \partial x$.

A field with one of these properties has them all. **D** is the quick test.

These statements **A** – **D** bring everything together for conservative fields (alias gradient fields). A closed path goes one way to Q and back the other way to P. The work cancels, and statements **A** and **B** are equivalent. We now connect them to **C**. *Note*: Test **D** says that the "*curl*" of **F** is zero. That can wait for Green's Theorem in the next section—the full discussion of the curl comes in 15.6.

First, *a gradient field* **F** = grad f *is conservative*. The work is $f(Q) - f(P)$, by the fundamental theorem for line integrals. It depends only on the endpoints and not the path. Therefore statement **C** leads back to **B**.

Our job is in the other direction, to show that conservative fields $M\mathbf{i} + N\mathbf{j}$ are gradients. Assume that the work integral depends only on the endpoints. We must construct a potential f, so that **F** is its gradient. In other words, $\partial f / \partial x$ must be M and $\partial f / \partial y$ must be N.

> *Fix the point P. Define $f(Q)$ as the work to reach Q. Then **F** equals* grad f.

Check the reasoning. At the starting point P, f is zero. At every other point Q, f is the work $\int M\,dx + N\,dy$ to reach that point. *All paths from P to Q give the same $f(Q)$*, because the field is assumed conservative. After two examples we prove that grad f agrees with **F**—the construction succeeds.

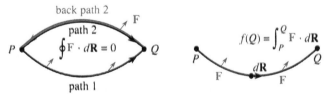

Fig. 15.7 Conservative fields: $\oint \mathbf{F} \cdot d\mathbf{R} = 0$ and $\int_P^Q \mathbf{F} \cdot d\mathbf{R} = f(Q) - f(P)$. Here $f(P) = 0$.

EXAMPLE 5 Find $f(x,y)$ when $\mathbf{F} = M\mathbf{i} + N\mathbf{j} = 2xy\mathbf{i} + x^2\mathbf{j}$. We want $\partial f / \partial x = 2xy$ and $\partial f / \partial y = x^2$.

Solution 1 Choose $P = (0,0)$. Integrate $M\,dx + N\,dy$ along to $(x,0)$ and up to (x,y):

$$\int_{(0,0)}^{(x,0)} 2xy\,dy = 0 \text{ (since } y = 0) \qquad \int_{(x,0)}^{(x,y)} x^2\,dy = x^2 y \text{ (which is } f).$$

Certainly $f = x^2 y$ meets the requirements: $f_x = 2xy$ and $f_y = x^2$. Thus $\mathbf{F} = \text{grad } f$. Note that $dy = 0$ in the first integral (on the x axis). Then $dx = 0$ in the second integral (x is fixed). The integrals add to $f = x^2 y$.

Solution 2   Integrate $2xy\,dx + x^2\,dy$ on the straight line (xt, yt) from $t = 0$ to $t = 1$:

$$\int_0^1 2(xt)(yt)(x\,dt) + (xt)^2(y\,dt) = \int_0^1 3x^2 yt^2 dt = x^2 yt^3]_0^1 = x^2 y.$$

Most authors use Solution 1. I use Solution 2. *Most students use Solution 3*:

Solution 3   Directly solve $\partial f/\partial x = M = 2xy$ and then fix up $df/dy = N = x^2$:

$$\partial f/\partial x = 2xy \quad \text{gives} \quad f = x^2 y \ (\textit{plus any function of } y).$$

In this example $x^2 y$ already has the correct derivative $\partial f/\partial y = x^2$. No additional function of y is necessary. When we integrate with respect to x, *the constant of integration* (usually C) *becomes a function $C(y)$.*

You will get practice in finding f. This is only possible for conservative fields! I tested $M = 2xy$ and $N = x^2$ in advance (using $\mathbf{D}$) to be sure that $\partial M/\partial y = \partial N/\partial x$.

EXAMPLE 6   Look for $f(x, y)$ when $M\mathbf{i} + N\mathbf{j}$ is the spin field $-y\mathbf{i} + x\mathbf{j}$.

Attempted solution 1   Integrate $-y\,dx + x\,dy$ from $(0,0)$ to $(x,0)$ to (x,y):

$$\int_{(0,0)}^{(x,0)} -y\,dx = 0 \quad \text{and} \quad \int_{(x,0)}^{(x,y)} x\,dy = xy \text{ (which seems like } f).$$

Attempted solution 2   Integrate $-y\,dx + x\,dy$ on the line (xt, yt) from $t = 0$ to 1:

$$\int_0^1 -(yt)(x\,dt) + (xt)(y\,dt) = 0 \text{ (a different } f, \text{also wrong)}.$$

Attempted solution 3   Directly solve $\partial f/\partial x = -y$ and try to fix up $\partial f/\partial y = x$:

$$\partial f/\partial x = -y \quad \text{gives} \quad f = -xy \text{ (plus any function } C(y)).$$

The y derivative of this f is $-x + dC/dy$. That does not agree with the required $\partial f/\partial y = x$. **Conclusion: The spin field $-y\mathbf{i} + x\mathbf{j}$ is not conservative.** There is no f. Test $\mathbf{D}$ gives $\partial M/\partial y = -1$ and $\partial N/\partial x = +1$.

To finish this section, we move from examples to a proof. The potential $f(Q)$ is defined as the work to reach Q. We must show that its partial derivatives are M and N. This seems reasonable from the formula $f(Q) = \int M\,dx + N\,dy$, but we have to think it through.

Remember statement **A**, that all paths give the same $f(Q)$. Take a path that goes from P to the left of Q. It comes in to Q on a line $y = $ constant (so $dy = 0$). As the path reaches Q, we are only integrating $M\,dx$. The derivative of this integral, at Q, is $\partial f/\partial x = M$. That is the Fundamental Theorem of Calculus.

To show that $\partial f/\partial y = N$, take a different path. Go from P to a point below Q. The path comes up to Q on a vertical line (so $dx = 0$). Near Q we are only integrating $N\,dy$, so $\partial f/\partial y = N$.

The requirement that the region must have no holes will be critical for test **D**.

EXAMPLE 7   Find $f(x,y) = \int_{(0,0)}^{(x,0)} x\,dx + y\,dy$. Test **D** is passed: $\partial N/\partial x = 0 = \partial M/\partial y$.

Solution 1 $\int_{(0,0)}^{(x,0)} x\,dx = \frac{1}{2}x^2$ is added to $\int_{(0,0)}^{(x,0)} y\,dy = \frac{1}{2}y^2$.

Solution 2 $\int_0^1 (xt)(x\,dt) + (yt)(y\,dt) = \int_0^1 (x^2+y^2)t\,dt = \frac{1}{2}(x^2+y^2)$.

Solution 3 $\partial f/\partial x = x$ gives $f = \frac{1}{2}x^2 + C(y)$. Then $\partial f/\partial y = y$ needs $C(y) = \frac{1}{2}y^2$.

15.2   EXERCISES

Read-through questions

Work is the __a__ of $\mathbf{F} \cdot d\mathbf{R}$. Here $\mathbf{F}$ is the __b__ and $\mathbf{R}$ is the __c__. The __d__ product finds the component of __e__ in the direction of movement $d\mathbf{R} = dx\mathbf{i} + dy\mathbf{j}$. The straight path $(x,y) = (t,2t)$ goes from __f__ at $t=0$ to __g__ at $t=1$ with $d\mathbf{R} = dt\mathbf{i} + $ __h__. The work of $\mathbf{F} = 3\mathbf{i} + \mathbf{j}$ is $\int \mathbf{F} \cdot d\mathbf{R} = \int$ __i__ $dt = $ __j__.

Another form of $d\mathbf{R}$ is $\mathbf{T}ds$, where $\mathbf{T}$ is the __k__ vector to the path and $ds = \sqrt{\underline{\quad l \quad}}$. For the path $(t,2t)$, the unit vector $\mathbf{T}$ is __m__ and $ds = $ __n__ dt. For $\mathbf{F} = 3\mathbf{i} + \mathbf{j}$, $\mathbf{F} \cdot \mathbf{T}ds$ is still __o__ dt. This $\mathbf{F}$ is the gradient of $f = $ __p__. The change in $f = 3x + y$ from $(0,0)$ to $(1,2)$ is __q__.

When $\mathbf{F} = \text{grad } f$, the dot product $\mathbf{F} \cdot d\mathbf{R}$ is $(\partial f/\partial x)dx + $ __r__ $= df$. The work integral from P to Q is $\int df = $ __s__. In this case the work depends on the __t__ but not on the __u__. Around a closed path the work is __v__. The field is called __w__. $\mathbf{F} = (1+y)\mathbf{i} + x\mathbf{j}$ is the gradient of $f = $ __x__. The work from $(0,0)$ to $(1,2)$ is __y__, the change in potential.

For the spin field $\mathbf{S} = $ __z__, the work (does)(does not) depend on the path. The path $(x,y) = (3\cos t, 3\sin t)$ is a circle with $\mathbf{S} \cdot d\mathbf{R} = $ __A__. The work is __B__ around the complete circle. Formally $\int g(x,y)ds$ is the limit of the sum __C__.

The four equivalent properties of a conservative field $\mathbf{F} = M\mathbf{i} + N\mathbf{j}$ are **A**: __D__, **B**: __E__, **C**: __F__, and **D**: __G__. Test **D** is (passed)(not passed) by $\mathbf{F} = (y+1)\mathbf{i} + x\mathbf{j}$. The work $\int \mathbf{F} \cdot d\mathbf{R}$ around the circle $(\cos t, \sin t)$ is __H__. The work on the upper semicircle equals the work on __I__. This field is the gradient of $f = $ __J__, so the work to $(-1,0)$ is __K__.

Compute the line integrals in 1–6.

1   $\int_c ds$ and $\int_c dy$: $x=t$, $y=2t$, $0 \le t \le 1$.

2   $\int_c x\,ds$ and $\int_c xy\,ds$: $x=\cos t$, $y=\sin t$, $0 \le t \le \pi/2$.

3   $\int_c xy\,ds$: bent line from $(0,0)$ to $(1,1)$ to $(1,0)$.

4   $\int_c y\,dx - x\,dy$: any square path, sides of length 3.

5   $\int_c dx$ and $\int_c y\,dx$: any closed circle of radius 3.

6   $\int_c (ds/dt)\,dt$: any path of length 5.

7   Does $\int_P^Q xy\,dy$ equal $\frac{1}{2}xy^2]_P^Q$?

8   Does $\int_P^Q x\,dx$ equal $\frac{1}{2}x^2]_P^Q$?

9   Does $(\int_c ds)^2 = (\int_c dx)^2 + (\int_c dy)^2$?

10   Does $\int_c (ds)^2$ make sense?

In 11–16 find the work in moving from $(1,0)$ to $(0,1)$. When F is conservative, construct f. choose your own path when F is not conservative.

11   $\mathbf{F} = \mathbf{i} + y\mathbf{j}$ 12 $\mathbf{F} = y\mathbf{i} + \mathbf{j}$

13   $\mathbf{F} = xy^2\mathbf{i} + yx^2\mathbf{j}$ 14 $\mathbf{F} = e^y\mathbf{i} + xe^y\mathbf{j}$

15   $\mathbf{F} = (x/r)\mathbf{i} + (y/r)\mathbf{j}$ 16 $\mathbf{F} = -y^2\mathbf{i} + x^2\mathbf{j}$

17   For which powers n is $\mathbf{S}/r^n$ a gradient by test **D**?

18   For which powers n is $\mathbf{R}/r^n$ a gradient by test **D**?

19   A wire hoop around a vertical circle $x^2 + z^2 = a^2$ has density $\rho = a + z$. Find its mass $M = \int \rho\,ds$.

20   A wire of constant density ρ lies on the semicircle $x^2 + y^2 = a^2$, $y > 0$. Find its mass M and also its moment $M_x = \int \rho y\,ds$. Where is its center of mass $\bar{x} = M_y/M$, $\bar{y} = M_x/M$?

21   If the density around the circle $x^2 + y^2 = a^2$ is $\rho = x^2$, what is the mass and where is the center of mass?

22   Find $\int \mathbf{F} \cdot d\mathbf{R}$ along the space curve $x=t$, $y=t^2$, $z=t^3$, $0 \le t \le 1$.

   (a) $\mathbf{F} = \text{grad}(xy + xz)$ (b) $\mathbf{F} = y\mathbf{i} - x\mathbf{j} + z\mathbf{k}$

23   (a)   Find the unit tangent vector **T** and the speed ds/dt along the path $\mathbf{R} = 2t\mathbf{i} + t^2\mathbf{j}$.

   (b)   For $\mathbf{F} = 3x\mathbf{i} + 4\mathbf{j}$, find $\mathbf{F} \cdot \mathbf{T}ds$ using (a) and $\mathbf{F} \cdot d\mathbf{R}$ directly.

   (c)   What is the work from $(2,1)$ to $(4,4)$?

24 If $M(x,y,z)\mathbf{i}+N(x,y,z)\mathbf{j}$ is the gradient of $f(x,y,z)$, show that none of these functions can depend on z.

25 Find all gradient fields of the form $M(y)\mathbf{i}+N(x)\mathbf{j}$.

26 Compute the work $W(x,y)=\int M\,dx+N\,dy$ on the straight line path (xt,yt) from $t=0$ to $t=1$. Test to see if $\partial W/\partial x=M$ and $\partial W/\partial y=N$.

    (a) $M=y^3, N=3xy^2$ (b) $M=x^3, N=3yx^2$

    (c) $M=x/y, N=y/x$ (d) $M=e^{x+y}, N=e^{x+y}$

27 Find a field $\mathbf{F}$ whose work around the unit square ($y=0$ then $x=1$ then $y=1$ then $x=0$) equals 4.

28 Find a nonconservative $\mathbf{F}$ whose work around the unit circle $x^2+y^2=1$ is zero.

In 29–34 compute $\int \mathbf{F}\cdot d\mathbf{R}$ along the straight line $\mathbf{R}=t\mathbf{i}+t\mathbf{j}$ and the parabola $\mathbf{R}=t\mathbf{i}+t^2\mathbf{j}$, from (0,0) to (1,1). When F is a gradient field, use its potential $f(x,y)$.

29 $\mathbf{F}=\mathbf{i}-2\mathbf{j}$ **30** $\mathbf{F}=x^2\mathbf{j}$

31 $\mathbf{F}=2xy^2\mathbf{i}+2yx^2\mathbf{j}$ **32** $\mathbf{F}=x^2y\mathbf{i}+xy^2\mathbf{j}$

33 $\mathbf{F}=y\mathbf{i}-x\mathbf{j}$ **34** $\mathbf{F}=(x\mathbf{i}+y\mathbf{j})/(x^2+y^2+1)$

35 For which numbers a and b is $\mathbf{F}=axy\mathbf{i}+(x^2+by)\mathbf{j}$ a gradient field?

36 Compute $\int -y\,dx+x\,dy$ from $(1,0)$ to $(0,1)$ on the line $x=1-t^2, y=t^2$ and the quarter-circle $x=\cos 2t, y=\sin 2t$. Example 4 found 1 and $\pi/2$ with different parameters.

Apply the test $N_x=M_y$ to 37–42. Find f when test D is passed.

37 $\mathbf{F}=y^2e^{-x}\mathbf{i}-2ye^{-x}\mathbf{j}$ **38** $\mathbf{F}=ye^x\mathbf{i}-2ye^x\mathbf{j}$

39 $\mathbf{F}=\dfrac{x\mathbf{i}+y\mathbf{j}}{|x\mathbf{i}+y\mathbf{j}|}$ **40** $\mathbf{F}=\dfrac{\text{grad }xy}{|\text{grad }xy|}$

41 $\mathbf{F}=\mathbf{R}+\mathbf{S}$ **42** $\mathbf{F}=(ax+by)\mathbf{i}+(cx+dy)\mathbf{j}$

43 Around the unit circle find $\oint ds$ and $\oint dx$ and $\oint x\,ds$.

44 *True or false*, with reason:

    (a) When $\mathbf{F}=y\mathbf{i}$ the line integral $\int \mathbf{F}\cdot d\mathbf{R}$ along a curve from P to Q equals the usual area under the curve.

    (b) That line integral depends only on P and Q, not on the curve.

    (c) That line integral around the unit circle equals π.

15.3 Green's Theorem

This section contains the Fundamental Theorem of Calculus, extended to two dimensions. That sounds important and it is. The formula was discovered 150 years after Newton and Leibniz, by an ordinary mortal named George Green. His theorem connects a ***double integral over a region*** R to a ***line integral along its boundary*** C.

The integral of df/dx equals $f(b) - f(a)$. This connects a one-dimensional integral to a zero-dimensional integral. The boundary only contains two points a and b! The answer $f(b) - f(a)$ is some kind of a "point integral." It is this absolutely crucial idea—to integrate a derivative from information *at the boundary*—that Green's Theorem extends into two dimensions.

There are two important integrals around C. The ***work*** is $\int \mathbf{F} \cdot \mathbf{T} \, ds = \int M \, dx + N \, dy$. The ***flux*** is $\int \mathbf{F} \cdot \mathbf{n} \, ds = \int M \, dy - N \, dx$ (notice the switch). The first is for a force field, the second is for a flow field. The tangent vector $\mathbf{T}$ turns $90°$ clockwise to become the normal vector $\mathbf{n}$. Green's Theorem handles both, in two dimensions. In three dimensions they split into the Divergence Theorem (15.5) and Stokes' Theorem (15.6).

Green's Theorem applies to "smooth" functions $M(x, y)$ and $N(x, y)$, with continuous first derivatives in a region slightly bigger than R. Then all integrals are well defined. M and N will have a definite and specific meaning in each application—to electricity or magnetism or fluid flow or mechanics. The purpose of a *theorem* is to capture the central ideas once and for all. We do that now, and the applications follow.

15E *Green's Theorem* Suppose the region R is bounded by the simple closed piecewise smooth curve C. Then an integral over R equals a line integral around C:

$$\oint_C M \, dx + N \, dy = \iint_R \left(\frac{\partial N}{\partial x} - \frac{\partial M}{\partial y} \right) dx \, dy. \qquad (1)$$

A curve is "simple" if it doesn't cross itself (figure 8's are excluded). It is "closed" if its endpoint Q is the same as its starting point P. This is indicated by the closed circle on the integral sign. The curve is "smooth" if its tangent $\mathbf{T}$ changes continuously—the word "piecewise" allows a finite number of corners. Fractals are not allowed, but all reasonable curves are acceptable (later we discuss figure 8's and rings). First comes an understanding of the formula, by testing it on special cases.

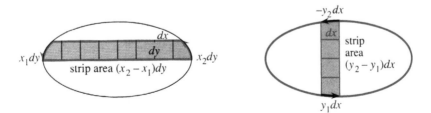

Fig. 15.8   Area of R adds up strips: $\oint x \, dy = \iint dx \, dy$ and $\oint y \, dx = -\iint dy \, dx$.

Special case 1: $M = 0$ and $N = x$. Green's Theorem with $\partial N / \partial x = 1$ becomes

$$\oint_C x \, dy = \iint_R 1 \, dx \, dy \quad \text{(which is the area of } R\text{)}. \qquad (2)$$

The integrals look equal, because the inner integral of dx is x. Then both integrals have $x\,dy$—but we need to go carefully. The area of a layer of R is dy times the difference in x (the length of the strip). The line integral in Figure 15.8 agrees. It has an upward dy times x (at the right) plus a downward $-dy$ times x (at the left). The integrals add up the strips, to give the total area.

Special case 2: $M = y$ and $N = 0$ and $\oint_C y\,dx = \iint_R (-1)dx\,dy = -(\text{area of } R)$.

Now Green's formula has a minus sign, because the line integral is *counterclockwise*. The top of each slice has $dx < 0$ (going left) and the bottom has $dx > 0$ (going right). Then $y\,dx$ at the top and bottom combine to give *minus* the area of the slice in Figure 15.8b.

Special case 3: $\oint 1\,dx = 0$. The dx's to the right cancel the dx's to the left (the curve is closed). With $M = 1$ and $N = 0$, Green's Theorem is $0 = 0$.

Most important case: $M\mathbf{i} + N\mathbf{j}$ is a *gradient field*. It has a potential function $f(x, y)$. Green's Theorem is $0 = 0$, because $\partial M / \partial y = \partial N / \partial x$. This is test **D**:

$$\frac{\partial M}{\partial y} = \frac{\partial}{\partial y}\left(\frac{\partial f}{\partial x}\right) \textit{ is the same as } \frac{\partial N}{\partial x} = \frac{\partial}{\partial x}\left(\frac{\partial f}{\partial y}\right). \tag{3}$$

The cross derivatives always satisfy $f_{yx} = f_{xy}$. That is why gradient fields pass test **D**.

When the double integral is zero, the line integral is also zero: $\oint_C M\,dx + N\,dy = 0$. The work is zero. ***The field is conservative***! This last step in $\mathbf{A} \Rightarrow \mathbf{B} \Rightarrow \mathbf{C} \Rightarrow \mathbf{D} \Rightarrow \mathbf{A}$ will be complete when Green's Theorem is proved.

Conservative examples are $\oint x\,dx = 0$ and $\oint y\,dy = 0$. Area is not involved.

Remark   The special cases $\oint x\,dy$ and $-\oint y\,dx$ led to the area of R. As long as $1 = \partial N / \partial x - \partial M / \partial y$, the double integral becomes $\iint 1\,dx\,dy$. This gives a way to compute area by a line integral.

$$\textit{The area of } R \textit{ is } \oint_C x\,dy = -\oint_C y\,dx = \frac{1}{2}\oint_C (x\,dy - y\,dx). \tag{4}$$

EXAMPLE 1   The area of the triangle in Figure 15.9 is 2. Check Green's Theorem.

The last area formula in (4) uses $\frac{1}{2}\mathbf{S}$, half the spin field. $N = \frac{1}{2}x$ and $M = -\frac{1}{2}y$ yield $N_x - M_y = \frac{1}{2} + \frac{1}{2} = 1$. On one side of Green's Theorem is $\iint 1\,dx\,dy = $ area of triangle. On the other side, the line integral has three pieces.

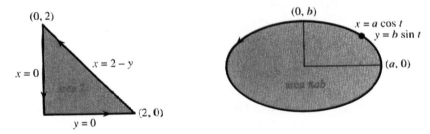

Fig. 15.9   Green's Theorem: Line integral around triangle, area integral for ellipse.

Two pieces are zero: $x\,dy - y\,dx = 0$ on the sides where $x = 0$ and $y = 0$. The sloping side $x = 2 - y$ has $dx = -dy$. The line integral agrees with the area, confirming Green's Theorem:

$$\frac{1}{2}\oint_C x\,dy - y\,dx = \frac{1}{2}\int_{y=0}^{2}(2-y)dy + y\,dy = \frac{1}{2}\int_0^2 2dy = 2.$$

EXAMPLE 2 *The area of an ellipse is πab when the semiaxes have lengths a and b.*

This is a classical example, which all authors like. The points on the ellipse are $x = a\cos t, y = b\sin t$, as t goes from 0 to 2π. (The ellipse has $(x/a)^2 + (y/b)^2 = 1$.) By computing the boundary integral, we discover the area inside. Note that the differential $x\,dy - y\,dx$ is just $ab\,dt$:

$$(a\cos t)(b\cos t\,dt) - (b\sin t)(-a\sin t\,dt) = ab(\cos^2 t + \sin^2 t)dt = ab\,dt.$$

The line integral is $\frac{1}{2}\int_0^{2\pi} ab\,dt = \pi ab$. This area πab is πr^2, for a circle with $a = b = r$.

Proof of Green's Theorem: In our special cases, the two sides of the formula were equal. We now show that they are always equal. The proof uses the Fundamental Theorem to integrate $(\partial N/\partial x)dx$ and $(\partial M/\partial y)dy$. Frankly speaking, this one-dimensional theorem is all we have to work with—since we don't know M and N.

The proof is a step up in mathematics, to work with symbols M and N instead of specific functions. The integral in (6) below has no numbers. The idea is to deal with M and N in two separate parts, which added together give Green's Theorem:

$$\oint_C M\,dx = \iint_R -\frac{\partial M}{\partial y}dx\,dy \quad \text{and separately} \quad \oint_C N\,dy = \iint_R \frac{\partial N}{\partial x}dx\,dy. \quad (5)$$

Start with a "very simple" region (Figure 15.10a). Its top is given by $y = f(x)$ and its bottom by $y = g(x)$. In the double integral, integrate $-\partial M/\partial y$ first with respect to y. The inner integral is

$$\int_{g(x)}^{f(x)} -\frac{\partial M}{\partial y}\,dy = -M(x,y)\Big]_{g(x)}^{f(x)} = -M(x, f(x)) + M(x, g(x)). \quad (6)$$

The Fundamental Theorem (in the y variable) gives this answer that depends on x. If we knew M and f and g, we could do the outer integral—from $x = a$ to $x = b$. But we have to leave it and go to the other side of Green's Theorem—the line integral:

$$\oint M\,dx = \int_{\text{top}} M(x,y)dx + \int_{\text{bottom}} M(x,y)dx = \int_b^a M(x, f(x))dx + \int_a^b M(x, g(x))dx. \quad (7)$$

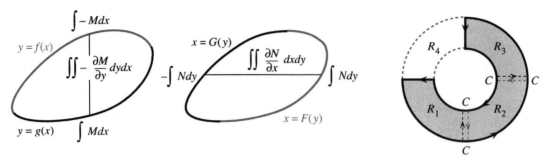

Fig. 15.10　Very simple region (a−b). Simple region (c) is a union of very simple regions.

Compare (7) with (6). The integral of $M(x, g(x))$ is the same for both. The integral of $M(x, f(x))$ has a minus sign from (6). In (7) it has a plus sign but the integral is from b to a. So life is good.

The part for N uses the same idea. Now the x integral comes first, because $(\partial N/\partial x)dx$ is practically asking to be integrated—from $x = G(y)$ at the left to $x = F(y)$ at the right. We reach $N(F(y), y) - N(G(y), y)$. Then the y integral matches $\oint N \, dy$ and completes (5). Adding the two parts of (5) proves Green's Theorem.

Finally we discuss the shape of R. The broken ring in Figure 15.10 is not "very simple," because horizontal lines go in and out and in and out. Vertical lines do the same. The x and y strips break into pieces. Our reasoning assumed no break between $y = f(x)$ at the top and $y = g(x)$ at the bottom.

There is a nice idea that saves Green's Theorem. Separate the broken ring into three very simple regions R_1, R_2, R_3. The three double integrals equal the three line integrals around the R's. Now *add these separate results*, to produce the double integral over all of R. When we add the line integrals, *the crosscuts CC are covered twice and they cancel*. The cut between R_1 and R_2 is covered upward (around R_1) and downward (around R_2). That leaves the integral around the boundary equal to the double integral inside—which is Green's Theorem.

When R is a complete ring, including the piece R_4, the theorem is still true. The integral around the outside is still counterclockwise. But the integral is *clockwise* around the inner circle. ***Keep the region R to your left as you go around C***. The complete ring is "doubly" connected, not "simply" connected. Green's Theorem allows any finite number of regions R_i and crosscuts CC and holes.

EXAMPLE 3　The area under a curve is $\int_a^b y \, dx$, as we always believed.

In computing area we never noticed the whole boundary. The true area is a line integral $-\oint y \, dx$ around the ***closed curve*** in Figure 15.11a. But $y = 0$ on the x axis. Also $dx = 0$ on the vertical lines (up and down at b and a). Those parts contribute zero to the integral of $y \, dx$. The only nonzero part is back along the curve—which is the area $-\int_b^a y \, dx$ or $\int_a^b y \, dx$ that we know well.

What about signs, when the curve dips below the x axis? That area has been counted as negative since Chapter 1. I saved the proof for Chapter 15. The reason lies in the arrows on C.

The line integral around that part *goes the other way*. The arrows are clockwise, the region is on the *right*, and the area counts as negative. With the correct rules, a figure 8 is allowed after all.

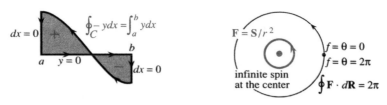

Fig. 15.11 Closed path gives the sign of the area. Nonconservative field because of hole.

CONSERVATIVE FIELDS

We never leave gradients alone! They give conservative fields—the work around a closed path is $f(P) - f(P) = 0$. But a potential function $f(x, y)$ is only available when test **D** is passed: *If $\partial f/\partial x = M$ and $\partial f/\partial y = N$ then $\partial M/\partial y = \partial N/\partial x$.* The reason is that $f_{xy} = f_{yx}$.

Some applications prefer the language of "differentials." Instead of looking for $f(x, y)$, we look for df:

DEFINITION The expression $M(x, y)dx + N(x, y)dy$ is a ***differential form***. When it agrees with the differential $df = (\partial f/\partial x)dx + (\partial f/\partial y)dy$ of some function, the form is called ***exact***. *The test for an exact differential is* **D**: $\partial N/\partial x = \partial M/\partial y$.

Nothing is new but the language. Is $y\,dx$ an exact differential? *No*, because $M_y = 1$ and $N_x = 0$. Is $y\,dx + x\,dy$ an exact differential? *Yes*, it is the differential of $f = xy$. That is the product rule! Now comes an important example, to show why R should be ***simply connected*** (a region with no holes).

EXAMPLE 4 The spin field $\mathbf{S}/r^2 = (-y\mathbf{i} + x\mathbf{j})/(x^2 + y^2)$ *almost* passes test **D**.

$$N_x = \frac{\partial}{\partial x}\left(\frac{x}{x^2 + y^2}\right) = \frac{x^2 + y^2 - x(2x)}{(x^2 + y^2)^2} = M_y = \frac{\partial}{\partial y}\left(\frac{-y}{x^2 + y^2}\right) = \frac{-(x^2 + y^2) + y(2y)}{(x^2 + y^2)^2}. \quad (8)$$

Both numerators are $y^2 - x^2$. Test **D** looks good. To find f, integrate $M = \partial f/\partial x$:

$$f(x, y) = \int -y\,dx/(x^2 + y^2) = \tan^{-1}(y/x) + C(y).$$

The extra part $C(y)$ can be zero—the y derivative of $\tan^{-1}(y/x)$ gives N with no help from $C(y)$. ***The potential f is the angle*** θ in the usual x, y, r right triangle.

Test **D** is passed and **F** is grad θ. What am I worried about? It is only this, that ***Green's Theorem on a circle seems to give*** $2\pi = 0$. The double integral is $\iint(N_x - M_y)dx\,dy$. According to (8) this is the integral of zero. But the line integral is 2π:

$$\oint \mathbf{F} \cdot d\mathbf{R} = \oint(-y\,dx + x\,dy)/(x^2 + y^2) = 2(\text{area of circle})/a^2 = 2\pi a^2/a^2 = 2\pi. \quad (9)$$

With $x = a\cos t$ and $y = a\sin t$ we would find the same answer. ***The work is*** 2π (not zero!) ***when the path goes around the origin***.

We have a paradox. If Green's Theorem is wrong, calculus is in deep trouble. Some requirement must be violated to reach $2\pi = 0$. Looking at $\mathbf{S}/r^2$, the problem is at the origin. The field is not defined when $r = 0$ (it blows up). The derivatives in (8) are not continuous. Test **D** does not apply at the origin, and was not passed. *We could remove* $(0,0)$, *but then the region where test* **D** *is passed would have a hole.*

It is amazing how one point can change everything. When the path circles the origin, the line integral is not zero. ***The potential function*** $f = \theta$ ***increases by*** 2π. That agrees with $\int \mathbf{F} \cdot d\mathbf{R} = 2\pi$ from (9). It disagrees with $\iint 0\, dx\, dy$. The 2π is right, the zero is wrong. $N_x - M_y$ must be a "*delta function of strength* 2π."

The double integral is 2π from an infinite spike over the origin—even though $N_x = M_y$ everywhere else. In fluid flow the delta function is a "vortex."

FLOW ACROSS A CURVE: GREEN'S THEOREM TURNED BY 90°

A flow field is easier to visualize than a force field, because something is really there and it moves. Instead of gravity in empty space, water has velocity $M(x,y)\mathbf{i} + N(x,y)\mathbf{j}$. At the boundary C it can flow in or out. The new form of Green's Theorem is a fundamental "balance equation" of applied mathematics:

> ***Flow through*** C (out minus in) $=$ ***replacement in*** R (source minus sink).

The flow is *steady*. Whatever goes out through C must be replaced in R. When there are no sources or sinks (negative sources), the total flow through C must be zero. This balance law is Green's Theorem in its "normal form" (for $\mathbf{n}$) instead of its "tangential form" (for $\mathbf{T}$):

15F For a steady flow field $\mathbf{F} = M(x,y)\mathbf{i} + N(x,y)\mathbf{j}$, the flux $\int \mathbf{F} \cdot \mathbf{n}\, ds$ through the boundary C balances the replacement of fluid inside R:

$$\oint_C M\, dy - N\, dx = \iint_R \left(\frac{\partial M}{\partial x} + \frac{\partial N}{\partial y} \right) dx\, dy. \tag{10}$$

Figure 15.12 shows the 90° turn. $\mathbf{T}$ becomes $\mathbf{n}$ and "circulation" along C becomes flux through C. In the original form of Green's Theorem, change N and M to M and $-N$ to obtain the flux form:

$$\oint M\, dx + N\, dy \rightarrow \oint -N\, dx + M\, dy \quad \iint (N_x - M_y)dx\, dy \rightarrow \iint (M_x + N_y)dx\, dy. \tag{11}$$

Playing with letters has proved a new theorem! The two left sides in (11) are equal, so the right sides are equal—which is Green's Theorem (10) for the flux. The components M and N can be chosen freely and named freely.

The change takes $M\mathbf{i} + N\mathbf{j}$ into its perpendicular field $-N\mathbf{i} + M\mathbf{j}$. The field is turned at every point (we are not just turning the plane by 90°). The spin field $\mathbf{S} = -y\mathbf{i} + x\mathbf{j}$ changes to the position field $\mathbf{R} = x\mathbf{i} + y\mathbf{j}$. The position field $\mathbf{R}$ changes to $-\mathbf{S}$. Streamlines of one field are equipotentials of the other field. The new form (10) of Green's Theorem is just as important as the old one—in fact I like it better. It is easier to visualize flow across a curve than circulation along it.

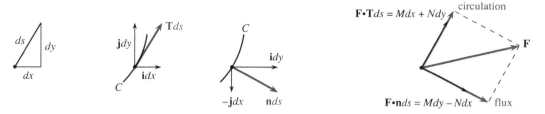

Fig. 15.12 The perpendicular component $\mathbf{F} \cdot \mathbf{n}$ flows through C. Note $\mathbf{n}\,ds = dy\,\mathbf{i} - dx\,\mathbf{j}$.

The change of letters was just for the proof. From now on $\mathbf{F} = M\mathbf{i} + N\mathbf{j}$.

EXAMPLE 5 Compute both sides of the new form (10) for $\mathbf{F} = 2x\mathbf{i} + 3y\mathbf{j}$. The region R is a rectangle with sides a and b.

Solution   This field has $\partial M/\partial x + \partial N/\partial y = 2 + 3$. The integral over R is $\iint_R 5\,dx\,dy = 5ab$. The line integral has four parts, because R has four sides. Between the left and right sides, $M = 2x$ increases by $2a$. Down the left and up the right, $\int M\,dy = 2ab$ (those sides have length b). Similarly $N = 3y$ changes by $3b$ between the bottom and top. Those sides have length a, so they contribute $3ab$ to a total line integral of $5ab$.

Important: **The "divergence" of a flow field is $\partial M/\partial x + \partial N/\partial y$.** The example has divergence $= 5$. To maintain this flow we must replace 5 units continually—not just at the origin but everywhere. (A one-point source is in example 7.) The divergence is the source strength, because it equals the outflow. **To understand Green's Theorem for any vector field $M\mathbf{i} + N\mathbf{j}$, look at a tiny rectangle** (sides dx and dy):

Flow out the right side minus flow in the left side $=$ (change in M) times dy

Flow out the top minus flow in the bottom $=$ (change in N) times dx

Total flow out of rectangle: $dM\,dy + dN\,dx = (\partial M/\partial x + \partial N/\partial y)dx\,dy$.

The divergence times the area $dx\,dy$ equals the total flow out. Section 15.5 gives more detail with more care in three dimensions. The divergence is $M_x + N_y + P_z$.

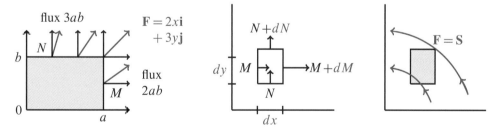

Fig. 15.13 $M_x + N_y = 2 + 3 = 5$ yields flux $= 5(\text{area}) = 5ab$. The flux is $dM\,dy + dN\,dx = (M_x + N_y)dx\,dy$. The spin field has no flux.

EXAMPLE 6 Find the flux through a closed curve C of the spin field $\mathbf{S} = -y\mathbf{i} + x\mathbf{j}$.

Solution   The field has $M = -y$ and $N = x$ and $M_x + N_y = 0$. *The double integral is zero.* Therefore the total flow (out minus in) is also zero—through any closed curve. Figure 15.13 shows flow entering and leaving a square. No fluid is added or removed. There is no rain and no evaporation. **When the divergence $M_x + N_y$ is zero, there is no source or sink.**

FLOW FIELDS WITHOUT SOURCES

This is really quite important. Remember that conservative fields do no work around C, they have a potential f, and they have "zero curl." Now turn those statements through $90°$, to find their twins. *Source-free fields have no flux through* C, *they have stream functions* g, *and they have "zero divergence."* The new statements **E–F–G–H** describe fields without sources.

15G The field $\mathbf{F} = M(x, y)\mathbf{i} + N(x, y)\mathbf{j}$ is source-free if it has these properties:

E   The total flux $\oint \mathbf{F} \cdot \mathbf{n}\, ds$ through every closed curve is zero.

F   Across all curves from P to Q, the flux $\int_P^Q \mathbf{F} \cdot \mathbf{n}\, ds$ is the same.

G   There is a ***stream function*** $g(x, y)$, for which $M = \partial g / \partial y$ and $N = -\partial g / \partial x$.

H   The components satisfy $\partial M / \partial x + \partial N / \partial y = 0$ (***the divergence is zero***).

A field with one of these properties has them all. **H** is the quick test.

The spin field $-y\mathbf{i} + x\mathbf{j}$ passed this test (Example 6 was source-free). The field $2x\mathbf{i} + 3y\mathbf{j}$ does not pass (Example 5 had $M_x + N_y = 5$). ***Example 7 almost passes***.

EXAMPLE 7   The radial field $\mathbf{R}/r^2 = (x\mathbf{i} + y\mathbf{j})/(x^2 + y^2)$ has a ***point source*** at $(0, 0)$.

The new test **H** is *divergence* $= \partial M / \partial x + \partial N / \partial y = 0$. Those two derivatives are

$$\frac{\partial}{\partial x}\left(\frac{x}{x^2 + y^2}\right) = \frac{x^2 + y^2 - x(2x)}{(x^2 + y^2)^2} \quad \text{and} \quad \frac{\partial}{\partial y}\left(\frac{y}{x^2 + y^2}\right) = \frac{x^2 + y^2 - y(2y)}{(x^2 + y^2)^2}. \quad (12)$$

They add to zero. There seems to be no source (if the calculation is correct). The flow through a circle $x^2 + y^2 = a^2$ should be zero. But it's not:

$$\oint M\, dy - N\, dx = \oint (x\, dy - y\, dx)/(x^2 + y^2) = 2(\text{area of circle})/a^2 = 2\pi. \quad (13)$$

A source is hidden somewhere. Looking at $\mathbf{R}/r^2$, the problem is at $(0, 0)$. The field is not defined when $r = 0$ (it blows up). The derivatives in (12) are not continuous. Test **H** does not apply, and was not passed. The divergence $M_x + N_y$ must be a "delta function" of strength 2π. There is a ***point source*** sending flow out through all circles.

I hope you see the analogy with Example 4. The field $\mathbf{S}/r^2$ is curl-free except at $r = 0$. The field $\mathbf{R}/r^2$ is divergence-free except at $r = 0$. The mathematics is parallel and the fields are perpendicular. A potential f and a stream function g require a region without holes.

THE BEST FIELDS: CONSERVATIVE AND SOURCE-FREE

*What if **F** is conservative and also source-free*? Those are outstandingly important fields. The curl is zero and the divergence is zero. Because the field is conservative, it comes from a potential. Because it is source-free, there is a stream function:

$$M = \frac{\partial f}{\partial x} = \frac{\partial g}{\partial y} \quad \text{and} \quad N = \frac{\partial f}{\partial y} = -\frac{\partial g}{\partial x}. \tag{14}$$

Those are the ***Cauchy-Riemann equations***, named after a great mathematician of his time and one of the greatest of all time. I can't end without an example.

EXAMPLE 8 Show that $y\mathbf{i} + x\mathbf{j}$ is both conservative and source-free. Find f and g.

Solution   With $M = y$ and $N = x$, check first that $\partial M/\partial y = 1 = \partial N/\partial x$. There must be a potential function. It is $f = xy$, which achieves $\partial f/\partial x = y$ and $\partial f/\partial y = x$. Note that $f_{xx} + f_{yy} = 0$.

Check next that $\partial M/\partial x + \partial N/\partial y = 0 + 0$. There must be a stream function. It is $g = \frac{1}{2}(y^2 - x^2)$, which achieves $\partial g/\partial y = y$ and $\partial g/\partial x = -x$. Note that $g_{xx} + g_{yy} = 0$.

The curves $f = $ constant are the equipotentials. The curves $g = $ constant are the streamlines (Figure 15.4). These are the twin properties—a conservative field with a potential and a source-free field with a stream function. They come together into the fundamental partial differential equation of equilibrium—*Laplace's equation* $f_{xx} + f_{yy} = 0$.

15H There is a potential and stream function when $M_y = N_x$ and $M_x = -N_y$. They satisfy ***Laplace's equation***:

$$f_{xx} + f_{yy} = M_x + N_y = 0 \quad \text{and} \quad g_{xx} + g_{yy} = -N_x + M_y = 0. \tag{15}$$

If we have f without g, as in $f = x^2 + y^2$ and $M = 2x$ and $N = 2y$, we don't have Laplace's equation: $f_{xx} + f_{yy} = 4$. This is a gradient field that needs a source. If we have g without f, as in $g = x^2 + y^2$ and $M = 2y$ and $N = -2x$, we don't have Laplace's equation. The field is source-free but it has spin. The first field is $2\mathbf{R}$ and the second field is $2\mathbf{S}$.

With no source and no spin, we are with Laplace at the center of mathematics and science.

Green's Theorem: Tangential form $\oint \mathbf{F} \cdot \mathbf{T} \, ds$ and normal form $\oint \mathbf{F} \cdot \mathbf{n} \, ds$

$$\underset{\text{work}}{\oint_C M \, dx + N \, dy} = \underset{\text{curl}}{\iint_R (N_x - M_y) dx \, dy} \qquad \underset{\text{flux}}{\oint_C M \, dy - N \, dx} = \underset{\text{divergence}}{\iint_R (M_x + N_y) dx \, dy}$$

Conservative: work $=$ zero, $N_x = M_y$, gradient of a potential: $M = f_x$ and $N = f_y$
Source-free: flux $=$ zero, $M_x = -N_y$, has a stream function: $M = g_y$ and $N = -g_x$
Conservative $+$ source-free: Cauchy-Riemann $+$ Laplace equations for f and g.

Read-through questions

The work integral $\oint M\,dx + N\,dy$ equals the double integral __a__ by __b__'s Theorem. For $\mathbf{F} = 3\mathbf{i} + 4\mathbf{j}$ the work is __c__. For $\mathbf{F} = $ __d__ and __e__, the work equals the area of R. When $M = \partial f/\partial x$ and $N = \partial f/\partial y$, the double integral is zero because __f__. The line integral is zero because __g__. An example is $\mathbf{F} = $ __h__. The direction on C is __i__ around the outside and __j__ around the boundary of a hole. If R is broken into very simple pieces with crosscuts between them, the integrals of __k__ cancel along the crosscuts.

Test **D** for gradient fields is __l__. A field that passes this test has $\oint \mathbf{F} \cdot d\mathbf{R} = $ __m__. There is a solution to $f_x = $ __n__ and $f_y = $ __o__. Then $df = M\,dx + N\,dy$ is an __p__ differential. The spin field $\mathbf{S}/r^2$ passes test **D** except at __q__. Its potential $f = $ __r__ increases by __s__ going around the origin. The integral $\iint (N_x - M_y)dx\,dy$ is not zero but __t__.

The flow form of Green's Theorem is __u__ = __v__. The normal vector in $\mathbf{F} \cdot \mathbf{n}\,ds$ points __w__ and $|\mathbf{n}| = $ __x__ and $\mathbf{n}\,ds$ equals $dy\,\mathbf{i} - dx\,\mathbf{j}$. The divergence of $M\mathbf{i} + N\mathbf{j}$ is __y__. For $\mathbf{F} = x\mathbf{i}$ the double integral is __z__. There (is)(is not) a source. For $\mathbf{F} = y\mathbf{i}$ the divergence is __A__. The divergence of $\mathbf{R}/r^2$ is zero except at __B__. This field has a __C__ source.

A field with no source has properties $\mathbf{E} = $ __D__, $\mathbf{F} = $ __E__, $\mathbf{G} = $ __F__, $\mathbf{H} = $ zero divergence. The stream function g satisfies the equations __G__. Then $\partial M/\partial x + \partial N/\partial y = 0$ because $\partial^2 g/\partial x\partial y = $ __H__. The example $\mathbf{F} = y\mathbf{i}$ has $g = $ __I__. There (is)(is not) a potential function. The example $\mathbf{F} = x\mathbf{i} - y\mathbf{j}$ has $g = $ __J__ and also $f = $ __K__. This f satisfies Laplace's equation __L__, because the field $\mathbf{F}$ is both __M__ and __N__. The functions f and g are connected by the __O__ equations $\partial f/\partial x = \partial g/\partial y$ and __P__.

Compute the line integrals 1–6 and (separately) the double integrals in Green's Theorem (1). The circle has $x = a\cos t$, $y = a\sin t$. The triangle has sides $x = 0$, $y = 0$, $x + y = 1$.

1 $\oint x\,dy$ along the circle

2 $\oint x^2 y\,dy$ along the circle

3 $\oint x\,dx$ along the triangle

4 $\oint y\,dx$ along the triangle

5 $\oint x^2 y\,dx$ along the circle

6 $\oint x^2 y\,dx$ along the triangle

7 Compute both sides of Green's Theorem in the form (10):

(a) $\mathbf{F} = x\mathbf{i} + y\mathbf{j}$, $R = $ upper half of the disk $x^2 + y^2 \leqslant 1$.

(b) $\mathbf{F} = x^2\mathbf{i} + xy\mathbf{j}$, $C = $ square with sides $y = 0$, $x = 1$, $y = 1$, $x = 0$.

8 Show that $\oint_C (x^2 y + 2x)dy + xy^2 dx$ depends only on the area of R. Does it equal the area?

9 Find the area inside the hypocycloid $x = \cos^3 t$, $y = \sin^3 t$ from $\frac{1}{2}\oint x\,dy - y\,dx$.

10 For constants b and c, how is $\oint by\,dx + cx\,dy$ related to the area inside C? If $b = 7$, which c makes the integral zero?

11 For $\mathbf{F} = \operatorname{grad}\sqrt{x^2 + y^2}$, show in three ways that $\oint \mathbf{F} \cdot d\mathbf{R} = 0$ around $x = \cos t$, $y = \sin t$.

(a) $\mathbf{F}$ is a gradient field so _____.

(b) Compute $\mathbf{F}$ and directly integrate $\mathbf{F} \cdot d\mathbf{R}$.

(c) Compute the double integral in Green's Theorem.

12 Devise a way to find the one-dimensional theorem $\int_a^b (df/dx)dx = f(b) - f(a)$ as a special case of Green's Theorem when R is a square.

13 (a) Choose $x(t)$ and $y(t)$ so that the path goes from $(1,0)$ to $(1,0)$ after circling the origin *twice*.

(b) Compute $\oint y\,dx$ and compare with the area inside your path.

(c) Compute $\oint (y\,dx - x\,dy)/(x^2 + y^2)$ and compare with 2π in Example 7.

14 In Example 4 of the previous section, the work $\int \mathbf{S} \cdot d\mathbf{R}$ between $(1,0)$ and $(0,1)$ was 1 for the straight path and $\pi/2$ for the quarter-circle path. Show that the work is always twice the area between the path and the axes.

Compute both sides of $\oint \mathbf{F} \cdot \mathbf{n}\,ds = \iint (M_x + N_y)dx\,dy$ in 15–20.

15 $\mathbf{F} = y\mathbf{i} + x\mathbf{j}$ in the unit circle

16 $\mathbf{F} = xy\mathbf{i}$ in the unit square $0 \leqslant x$, $y \leqslant 1$

17 $\mathbf{F} = \mathbf{R}/r$ in the unit circle

18 $\mathbf{F} = \mathbf{S}/r$ in the unit square

19 $\mathbf{F} = x^2 y\mathbf{j}$ in the unit triangle (sides $x = 0$, $y = 0$, $x + y = 1$)

20 $\mathbf{F} = \operatorname{grad} r$ in the top half of the unit circle.

21 Suppose div $\mathbf{F} = 0$ except at the origin. Then the flux $\oint \mathbf{F} \cdot \mathbf{n}\,ds$ is the same through any two circles around the origin, because _____. (What is $\iint (M_x + N_y)dx\,dy$ between the circles?)

22 Example 7 has div $\mathbf{F} = 0$ except at the origin. The flux through every circle $x^2 + y^2 = a^2$ is 2π. The flux through a square around the origin is also 2π because _____. (Compare Problem 21.)

23 Evaluate $\oint a(x,y)dx + b(x,y)dy$ by both forms of Green's Theorem. The choice $M = a$, $N = b$ in the work form gives the double integral _____. The choice $M = b$, $N = -a$ in the flux form gives the double integral _____. There was only one Green.

24 Evaluate $\oint \cos^3 y\,dy - \sin^3 x\,dx$ by Green's Theorem.

25 The field $\mathbf{R}/r^2$ in Example 7 has zero divergence except at $r = 0$. Solve $\partial g/\partial y = x/(x^2 + y^2)$ to find an attempted stream function g. Does g have trouble at the origin?

26 Show that $\mathbf{S}/r^2$ has zero divergence (except at $r = 0$). Find a stream function by solving $\partial g/\partial y = y/(x^2 + y^2)$. Does g have trouble at the origin?

27 Which differentials are exact: $y\,dx - x\,dy$, $x^2 dx + y^2 dy$, $y^2 dx + x^2 dy$?

28 If $M_x + N_y = 0$ then the equations $\partial g/\partial y =$ _____ and $\partial g/\partial x =$ _____ yield a stream function. If also $N_x = M_y$, show that g satisfies Laplace's equation.

Compute the divergence of each field in 29–36 and solve $g_y = M$ and $g_x = -N$ for a stream function (if possible).

29 $2xy\mathbf{i} - y^2\mathbf{j}$

30 $3xy^2\mathbf{i} - y^3\mathbf{j}$

31 $x^2\mathbf{i} + y^2\mathbf{j}$

32 $y^2\mathbf{i} + x^2\mathbf{j}$

33 $e^x \cos y\mathbf{i} - e^x \sin y\mathbf{j}$

34 $e^{x+y}(\mathbf{i} - \mathbf{j})$

35 $2y\mathbf{i}/x + y^2\mathbf{j}/x^2$

36 $xy\mathbf{i} - xy\mathbf{j}$

37 Compute $N_x - M_y$ for each field in 29–36 and find a potential function f when possible.

38 The potential $f(Q)$ is the work $\int_P^Q \mathbf{F} \cdot \mathbf{T}\,ds$ to reach Q from a fixed point P (Section 15.2). In the same way, the stream function $g(Q)$ can be constructed from the integral _____. Then $g(Q) - g(P)$ represents *the flux across the path from P to Q*. Why do all paths give the same answer?

39 The real part of $(x + iy)^3 = x^3 + 3ix^2 y - 3xy^2 - iy^3$ is $f = x^3 - 3xy^2$. Its gradient field is $\mathbf{F} = \text{grad } f =$ _____. The divergence of $\mathbf{F}$ is _____. Therefore f satisfies Laplace's equation $f_{xx} + f_{yy} = 0$ (check that it does).

40 Since div $\mathbf{F} = 0$ in Problem 39, we can solve $\partial g/\partial y =$ _____ and $\partial g/\partial x =$ _____. The stream function is $g =$ _____. It is the imaginary part of the same $(x + iy)^3$. Check that f and g satisfy the Cauchy–Riemann equations.

41 The real part f and imaginary part g of $(x + iy)^n$ satisfy the Laplace and Cauchy-Riemann equations for $n = 1, 2, \dots$. (They give all the polynomial solutions.) Compute f and g for $n = 4$.

42 When is $M\,dy - N\,dx$ an exact differential dg?

43 The potential $f = e^x \cos y$ satisfies Laplace's equation. There must be a g. Find the field $\mathbf{F} = \text{grad } f$ and the stream function $g(x, y)$.

44 Show that the spin field $\mathbf{S}$ does work around every simple closed curve.

45 For $\mathbf{F} = f(x)\mathbf{j}$ and $R =$ unit square $0 \leqslant x \leqslant 1$, $0 \leqslant y \leqslant 1$, integrate both sides of Green's Theorem (1). What formula is required from one-variable calculus?

46 A region R is "*simply connected*" when every closed curve inside R can be squeezed to a point without leaving R. Test these regions:

1. xy plane without $(0,0)$
2. xyz space without $(0,0,0)$
3. sphere $x^2 + y^2 + z^2 = 1$
4. a torus (or doughnut)
5. a sweater
6. a human body
7. the region between two spheres
8. xyz space with circle removed.

15.4 Surface Integrals

The double integral in Green's Theorem is over a flat surface R. Now the region moves out of the plane. It becomes a **curved surface** S, part of a sphere or cylinder or cone. When the surface has only one z for each (x, y), it is the graph of a function $z(x, y)$. In other cases S can twist and close up—a sphere has an upper z and a lower z. In all cases we want to compute area and flux. This is a necessary step (it is our last step) before moving Green's Theorem to three dimensions.

First a quick review. The basic integrals are $\int dx$ and $\iint dx\, dy$ and $\iiint dx\, dy\, dz$. The one that didn't fit was $\int ds$—the length of a curve. When we go from curves to surfaces, ds becomes dS. **Area is** $\iint dS$ **and flux is** $\iint \mathbf{F} \cdot \mathbf{n}\, dS$, with double integrals because the surfaces are two-dimensional. The main difficulty is in dS.

All formulas are summarized in a table at the end of the section.

There are two ways to deal with ds (along curves). The same methods apply to dS (on surfaces). The first is in xyz coordinates; the second uses parameters. Before this subject gets complicated, I will explain those two methods.

> *Method* **1** *is for the graph of a function: curve* $y(x)$ *or surface* $z(x, y)$.

A small piece of the curve is almost straight. It goes across by dx and up by dy:

$$\text{length}\, ds = \sqrt{(dx)^2 + (dy)^2} = \sqrt{1 + (dy/dx)^2}\, dx. \tag{1}$$

A small piece of the surface is practically flat. Think of a tiny sloping rectangle. One side goes across by dx and up by $(\partial z/\partial x)dx$. The neighboring side goes along by dy and up by $(\partial z/\partial y)dy$. Computing the area is a linear problem (from Chapter 11), because the flat piece is in a plane.

Two vectors $\mathbf{A}$ and $\mathbf{B}$ form a parallelogram. **The length of their cross product is the area**. In the present case, the vectors are $\mathbf{A} = \mathbf{i} + (\partial z/\partial x)\mathbf{k}$ and $\mathbf{B} = \mathbf{j} + (\partial z/\partial y)\mathbf{k}$. Then $\mathbf{A}dx$ and $\mathbf{B}dy$ are the sides of the small piece, and we compute $\mathbf{A} \times \mathbf{B}$:

$$\mathbf{A} \times \mathbf{B} = \begin{vmatrix} \mathbf{i} & \mathbf{j} & \mathbf{k} \\ 1 & 0 & \partial z/\partial x \\ 0 & 1 & \partial z/\partial y \end{vmatrix} = -\partial z/\partial x\, \mathbf{i} - \partial z/\partial y\, \mathbf{j} + \mathbf{k}. \tag{2}$$

This is exactly the **normal vector** $\mathbf{N}$ to the tangent plane and the surface, from Chapter 13. Please note: The small flat piece is actually a parallelogram (not always a rectangle). Its area dS is much like ds, but the length of $\mathbf{N} = \mathbf{A} \times \mathbf{B}$ involves two derivatives:

$$\text{area}\, dS = |\mathbf{A}dx \times \mathbf{B}dy| = |\mathbf{N}|dx\, dy = \sqrt{1 + (\partial z/\partial x)^2 + (\partial z/\partial y)^2}\, dx\, dy. \tag{3}$$

EXAMPLE 1   Find the area on the plane $z = x + 2y$ above a base area A.

This is the example to visualize. The area down in the xy plane is A. The area up on the sloping plane is greater than A. A roof has more area than the room underneath it. If the roof goes up at a $45°$ angle, the ratio is $\sqrt{2}$. Formula (3) yields the correct ratio for any surface—including our plane $z = x + 2y$.

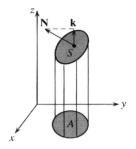
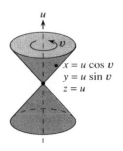
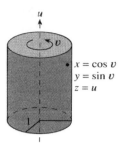

Fig. 15.14 Roof area = base area times $|\mathbf{N}|$. Cone and cylinder with parameters u and v.

The derivatives are $\partial z / \partial x = 1$ and $\partial z / \partial y = 2$. They are constant (planes are easy). The square root in (3) contains $1 + 1^2 + 2^2 = 6$. Therefore $dS = \sqrt{6}\, dx\, dy$. An area in the xy plane is multiplied by $\sqrt{6}$ up in the surface (Figure 15.14a). The vectors $\mathbf{A}$ and $\mathbf{B}$ are no longer needed—their work was done when we reached formula (3)—but here they are:

$$\mathbf{A} = \mathbf{i} + (\partial z / \partial x)\mathbf{k} = \mathbf{i} + \mathbf{k} \quad \mathbf{B} = \mathbf{j} + (\partial z / \partial y)\mathbf{k} = \mathbf{j} + 2\mathbf{k} \quad \mathbf{N} = -\mathbf{i} - 2\mathbf{j} + \mathbf{k}.$$

The length of $\mathbf{N} = \mathbf{A} \times \mathbf{B}$ is $\sqrt{6}$. The angle between $\mathbf{k}$ and $\mathbf{N}$ has $\cos\theta = 1/\sqrt{6}$. ***That is the angle between base plane and sloping plane***. Therefore the sloping area is $\sqrt{6}$ times the base area. For curved surfaces the idea is the same, except that the square root in $|\mathbf{N}| = 1/\cos\theta$ changes as we move around the surface.

Method 2 is for curves $x(t)$, $y(t)$ and surfaces $x(u, v)$, $y(u, v)$, $z(u, v)$ with parameters.

A curve has one parameter t. A surface has two parameters u and v (it is two-dimensional). One advantage of parameters is that x, y, z get equal treatment, instead of picking out z as $f(x, y)$. Here are the first two examples:

$$\textit{cone } x = u \cos v, \ y = u \sin v, \ z = u \quad \textit{cylinder } x = \cos v, \ y = \sin v, \ z = u. \quad (4)$$

Each choice of u and v gives a point on the surface. By making all choices, we get the complete surface. Notice that a parameter can equal a coordinate, as in $z = u$. Sometimes both parameters are coordinates, as in $x = u$ and $y = v$ and $z = f(u, v)$. That is just $z = f(x, y)$ in disguise—the surface without parameters. In other cases *we find the xyz equation by eliminating u and v*:

cone $\quad (u \cos v)^2 + (u \sin v)^2 = u^2 \quad$ or $\quad x^2 + y^2 = z^2 \quad$ or $\quad z = \sqrt{x^2 + y^2}$

cylinder $\quad (\cos v)^2 + (\sin v)^2 = 1 \quad$ or $\quad x^2 + y^2 = 1.$

The cone is the graph of $f = \sqrt{x^2 + y^2}$. The cylinder is *not* the graph of any function. There is a line of z's through each point on the circle $x^2 + y^2 = 1$. That is what $z = u$ tells us: Give u all values, and you get the whole line. Give u and v all values, and you get the whole cylinder. Parameters allow a surface to close up and even go through itself—which the graph of $f(x, y)$ can never do.

Actually $z = \sqrt{x^2 + y^2}$ gives only the top half of the cone. (A function produces only one z.) The parametric form gives the bottom half also. Similarly $y = \sqrt{1 - x^2}$ gives only the top of a circle, while $x = \cos t$, $y = \sin t$ goes all the way around.

Now we find dS, using parameters. Small movements give a piece of the surface, practically flat. One side comes from the change du, the neighboring side comes from dv. The two sides are given by small vectors $\mathbf{A}\,du$ and $\mathbf{B}\,dv$:

$$\mathbf{A} = \frac{\partial x}{\partial u}\mathbf{i} + \frac{\partial y}{\partial u}\mathbf{j} + \frac{\partial z}{\partial u}\mathbf{k} \quad \text{and} \quad \mathbf{B} = \frac{\partial x}{\partial v}\mathbf{i} + \frac{\partial y}{\partial v}\mathbf{j} + \frac{\partial z}{\partial v}\mathbf{k}. \tag{5}$$

To find the area dS of the parallelogram, start with the cross product $\mathbf{N} = \mathbf{A} \times \mathbf{B}$:

$$\mathbf{N} = \begin{vmatrix} \mathbf{i} & \mathbf{j} & \mathbf{k} \\ x_u & y_u & z_u \\ x_v & y_v & z_v \end{vmatrix} = \left(\frac{\partial y}{\partial u}\frac{\partial z}{\partial v} - \frac{\partial z}{\partial u}\frac{\partial y}{\partial v}\right)\mathbf{i} + \left(\frac{\partial z}{\partial u}\frac{\partial x}{\partial v} - \frac{\partial x}{\partial u}\frac{\partial z}{\partial v}\right)\mathbf{j} + \left(\frac{\partial x}{\partial u}\frac{\partial y}{\partial v} - \frac{\partial y}{\partial u}\frac{\partial x}{\partial v}\right)\mathbf{k}. \tag{6}$$

Admittedly this looks complicated—actual examples are often fairly simple. The area dS of the small piece of surface is $|\mathbf{N}|\,du\,dv$. The length $|\mathbf{N}|$ is a square root:

$$dS = \sqrt{\left(\frac{\partial y}{\partial u}\frac{\partial z}{\partial v} - \frac{\partial z}{\partial u}\frac{\partial y}{\partial v}\right)^2 + \left(\frac{\partial z}{\partial u}\frac{\partial x}{\partial v} - \frac{\partial x}{\partial u}\frac{\partial z}{\partial v}\right)^2 + \left(\frac{\partial x}{\partial u}\frac{\partial y}{\partial v} - \frac{\partial y}{\partial u}\frac{\partial x}{\partial v}\right)^2}\; du\,dv. \tag{7}$$

EXAMPLE 2   Find $\mathbf{A}$ and $\mathbf{B}$ and $\mathbf{N} = \mathbf{A} \times \mathbf{B}$ and dS for the cone and cylinder.

The cone has $x = u\cos v$, $y = u\sin v$, $z = u$. The u derivatives produce $\mathbf{A} = \partial\mathbf{R}/\partial u = \cos v\,\mathbf{i} + \sin v\,\mathbf{j} + \mathbf{k}$. The v derivatives produce the other tangent vector $\mathbf{B} = \partial\mathbf{R}/\partial v = -u\sin v\,\mathbf{i} + u\cos v\,\mathbf{j}$. The normal vector is $\mathbf{A} \times \mathbf{B} = -u\cos v\,\mathbf{i} - u\sin v\,\mathbf{j} + u\,\mathbf{k}$. Its length gives dS:

$$dS = |\mathbf{A} \times \mathbf{B}|\,du\,dv = \sqrt{(u\cos v)^2 + (u\sin v)^2 + u^2}\; du\,dv = \sqrt{2}\,u\,du\,dv.$$

The cylinder is even simpler: $dS = du\,dv$. In these and many other examples, $\mathbf{A}$ is perpendicular to $\mathbf{B}$. *The small piece is a rectangle*. Its sides have length $|\mathbf{A}|\,du$ and $|\mathbf{B}|\,dv$. (The cone has $|\mathbf{A}| = u$ and $|\mathbf{B}| = \sqrt{2}$, the cylinder has $|\mathbf{A}| = |\mathbf{B}| = 1$). The cross product is hardly needed for area, when we can just multiply $|\mathbf{A}|\,du$ times $|\mathbf{B}|\,dv$.

Remark on the two methods Method 1 also used parameters, but a very special choice—u is x and v is y. The parametric equations are $x = x$, $y = y$, $z = f(x, y)$. If you go through the long square root in (7), changing u to x and v to y, it simplifies to the square root in (3). (The terms $\partial y/\partial x$ and $\partial x/\partial y$ are zero; $\partial x/\partial x$ and $\partial y/\partial y$ are 1.) Still it pays to remember the shorter formula from Method 1.

Don't forget that after computing dS, you have to integrate it. Many times the good is with polar coordinates. Surfaces are often symmetric around an axis or a point. Those are the *surfaces of revolution*—which we saw in Chapter 8 and will come back to.

Strictly speaking, the integral starts with ΔS (not dS). A flat piece has area $|\mathbf{A} \times \mathbf{B}|\Delta x\Delta y$ or $|\mathbf{A} \times \mathbf{B}|\Delta u\Delta v$. The area of a curved surface is properly defined as a limit. The key step of calculus, from sums of ΔS to the integral of dS, is safe for smooth surfaces. In examples, the hard part is computing the double integral and substituting the limits on x, y or u, v.

EXAMPLE 3 Find the surface area of the cone $z = \sqrt{x^2 + y^2}$ up to the height $z = a$. We use Method 1 (no parameters). The derivatives of z are computed, squared, and added:

$$\frac{\partial z}{\partial x} = \frac{x}{\sqrt{x^2 + y^2}} \qquad \frac{\partial z}{\partial y} = \frac{y}{\sqrt{x^2 + y^2}} \qquad |\mathbf{N}|^2 = 1 + \frac{x^2}{x^2 + y^2} + \frac{y^2}{x^2 + y^2} = 2.$$

Conclusion: $|\mathbf{N}| = \sqrt{2}$ and $dS = \sqrt{2}\,dx\,dy$. The cone is on a 45° slope, so the area $dx\,dy$ in the base is multiplied by $\sqrt{2}$ in the surface above it (Figure 15.15). The square root in dS accounts for the extra area due to slope. A horizontal surface has $dS = \sqrt{1}\,dx\,dy$, as we have known all year.

Now for a key point. ***The integration is down in the base plane***. The limits on x and y are given by the "*shadow*" of the cone. To locate that shadow set $z = \sqrt{x^2 + y^2}$ equal to $z = a$. The plane cuts the cone at the circle $x^2 + y^2 = a^2$. We integrate over the inside of that circle (where the shadow is):

$$\text{surface area of cone} = \iint\limits_{\text{shadow}} \sqrt{2}\,dx\,dy = \sqrt{2}\,\pi a^2.$$

EXAMPLE 4 Find the same area using $dS = \sqrt{2}\,u\,du\,dv$ from Example 2.

With parameters, dS looks different and the shadow in the base looks different. The circle $x^2 + y^2 = a^2$ becomes $u^2 \cos^2 v + u^2 \sin^2 v = a^2$. In other words $u = a$. (The cone has $z = u$, the plane has $z = a$, they meet when $u = a$.) The angle parameter v goes from 0 to 2π. The effect of these parameters is to switch us "automatically" to polar coordinates, where area is $r\,dr\,d\theta$:

$$\text{surface area of cone} = \iint dS = \int\limits_{0}^{2\pi} \int\limits_{0}^{a} \sqrt{2}\,u\,du\,dv = \sqrt{2}\,\pi a^2.$$

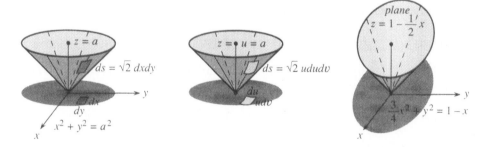

Fig. 15.15 Cone cut by plane leaves shadow in the base. Integrate over the shadow.

EXAMPLE 5 Find the area of the same cone up to the sloping plane $z = 1 - \frac{1}{2}x$.

Solution The cone still has $dS = \sqrt{2}\,dx\,dy$, but the limits of integration are changed. The plane cuts the cone in an ellipse. Its shadow down in the xy plane is another ellipse (Figure 15.15c). ***To find the edge of the shadow, set $z = \sqrt{x^2 + y^2}$ equal to $z = 1 - \frac{1}{2}x$***. We square both sides:

$$x^2 + y^2 = 1 - x + \frac{1}{4}x^2 \qquad \text{or} \qquad \frac{3}{4}(x + \tfrac{2}{3})^2 + y^2 = \tfrac{4}{3}.$$

This is the ellipse in the base—where height makes no difference and z is gone. The area of an ellipse is πab, when the equation is in the form $(x/a)^2 + (y/b)^2 = 1$. After multiplying by 3/4 we find $a = 4/3$ and $b = \sqrt{4/3}$. Then $\iint \sqrt{2}\, dx\, dy = \sqrt{2}\,\pi ab$ is the surface area of the cone.

The hard part was finding the shadow ellipse (I went quickly). Its area πab came from Example 15.3.2. The new part is $\sqrt{2}$ from the slope.

EXAMPLE 6   Find the surface area of a sphere of radius a (known to be $4\pi a^2$).

This is a good example, because both methods almost work. The equation of the sphere is $x^2 + y^2 + z^2 = a^2$. Method 1 writes $z = \sqrt{a^2 - x^2 - y^2}$. The x and y derivatives are $-x/z$ and $-y/z$:

$$1 + \left(\frac{\partial z}{\partial x}\right)^2 + \left(\frac{\partial z}{\partial y}\right)^2 = \frac{z^2}{z^2} + \frac{x^2}{z^2} + \frac{y^2}{z^2} = \frac{a^2}{z^2} = \frac{a^2}{a^2 - x^2 - y^2}.$$

The square root gives $dS = a\, dx\, dy/\sqrt{a^2 - x^2 - y^2}$. Notice that z is gone (as it should be). Now integrate dS over the shadow of the sphere, which is a circle. Instead of $dx\, dy$, switch to polar coordinates and $r\, dr\, d\theta$:

$$\iint_{\text{shadow}} dS = \int_0^{2\pi} \int_0^a \frac{a\, r\, dr\, d\theta}{\sqrt{a^2 - r^2}} = -2\pi a \sqrt{a^2 - r^2}\Big]_0^a = 2\pi a^2. \qquad (8)$$

This calculation is successful but wrong. $2\pi a^2$ is the area of the *half-sphere* above the xy plane. The lower half takes the negative square root of $z^2 = a^2 - x^2 - y^2$. This shows the danger of Method 1, when the surface is not the graph of a function.

EXAMPLE 7   (same sphere by Method 2: *use parameters*) The natural choice is spherical coordinates. Every point has an angle $u = \phi$ down from the North Pole and an angle $v = \theta$ around the equator. The xyz coordinates from Section 14.4 are $x = a \sin\phi \cos\theta$, $y = a \sin\phi \sin\theta$, $z = a \cos\phi$. The radius $\rho = a$ is fixed (not a parameter). Compute the first term in equation (6), noting $\partial z/\partial\theta = 0$:

$$(\partial y/\partial\phi)(\partial z/\partial\theta) - (\partial z/\partial\phi)(\partial y/\partial\theta) = -(-a\sin\phi)(a\sin\phi\cos\theta) = a^2\sin^2\phi\cos\theta.$$

The other terms in (6) are $a^2 \sin^2\phi \sin\theta$ and $a^2 \sin\phi\cos\phi$. Then dS in equation (7) squares these three components and adds. We factor out a^4 and simplify:

$$a^4(\sin^4\phi\cos^2\theta + \sin^4\phi\sin^2\theta + \sin^2\phi\cos^2\phi) = a^4(\sin^4\phi + \sin^2\phi\cos^2\phi) = a^4\sin^2\phi.$$

Conclusion: $dS = a^2 \sin\phi\, d\phi\, d\theta$. A spherical person will recognize this immediately. It is the volume element $dV = \rho^2 \sin\phi\, d\rho\, d\phi\, d\theta$, except $d\rho$ is missing. The small box has area dS and thickness $d\rho$ and volume dV. Here we only want dS:

$$\text{area of sphere} = \iint dS = \int_0^{2\pi} \int_0^{\pi} a^2 \sin\phi\, d\phi\, d\theta = 4\pi a^2. \qquad (9)$$

Figure 15.16a shows a small surface with sides $a\, d\phi$ and $a\, \sin\phi\, d\theta$. Their product is dS. Figure 15.16b goes back to Method 1, where equation (8) gave $dS = (a/z)\, dx\, dy$.

I doubt that you will like Figure 15.16c—and you don't need it. With parameters ϕ and θ, the shadow of the sphere is a rectangle. The equator is the line down the middle, where $\phi = \pi/2$. The height is $z = a\cos\phi$. The area $d\phi\, d\theta$ in the base is the shadow of $dS = a^2\sin\phi\, d\phi\, d\theta$ up in the sphere. Maybe this figure shows what we don't halve to know about parameters.

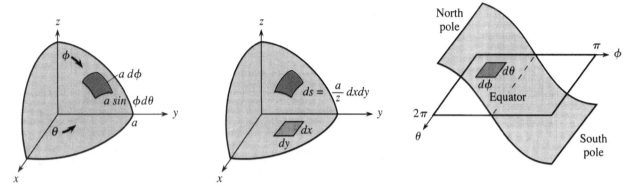

Fig. 15.16   Surface area on a sphere: (a) spherical coordinates (b) xyz coordinates (c) $\phi\theta$ space.

EXAMPLE 8 Rotate $y = x^2$ around the x axis. Find the surface area using parameters. The first parameter is x (from a to b). The second parameter is the rotation angle θ (from 0 to 2π). The points on the surface in Figure 15.17 are $x = x$, $y = x^2\cos\theta$, $z = x^2\sin\theta$. Equation (7) leads after much calculation to $dS = x^2\sqrt{1 + 4x^2}\, dx\, d\theta$.

   Main point: dS agrees with Section 8.3, where the area was $\int 2\pi y \sqrt{1 + (dy/dx)^2}\, dx$. The 2π comes from the θ integral and y is x^2. Parameters give this formula automatically.

VECTOR FIELDS AND THE INTEGRAL OF $\mathbf{F}\cdot\mathbf{n}$

Formulas for surface area are dominated by square roots. There is a square root in dS, as there was in ds. Areas are like arc lengths, one dimension up. The good point about line integrals $\int \mathbf{F}\cdot\mathbf{n}\, ds$ is that the square root disappears. It is in the denominator of $\mathbf{n}$, where ds cancels it: $\mathbf{F}\cdot\mathbf{n}\, ds = M\, dy - N\, dx$. The same good thing will now happen for surface integrals $\iint \mathbf{F}\cdot\mathbf{n}\, dS$.

15I Through the surface $z = f(x, y)$, the vector field $\mathbf{F}(x, y, z) = M\mathbf{i} + N\mathbf{j} + P\mathbf{k}$ has

$$\textbf{flux} = \underset{\text{surface}}{\iint}\ \mathbf{F}\cdot\mathbf{n}\, dS = \underset{\text{shadow}}{\iint} \left(-M\frac{\partial f}{\partial x} - N\frac{\partial f}{\partial y} + P\right) dx\, dy. \qquad (10)$$

This formula tells what to integrate, given the surface and the vector field (f and $\mathbf{F}$). The xy limits come from the shadow. Formula (10) takes the normal vector from Method 1:

$$\mathbf{N} = -\partial f/\partial x\, \mathbf{i} - \partial f/\partial y\, \mathbf{j} + \mathbf{k} \text{ and } |\mathbf{N}| = \sqrt{1 + (\partial f/\partial x)^2 + (\partial f/\partial y)^2}.$$

For the *unit* normal vector $\mathbf{n}$, divide $\mathbf{N}$ by its length: $\mathbf{n} = \mathbf{N}/|\mathbf{N}|$. The square root is in the denominator, and the same square root is in dS. See equation (3):

$$\mathbf{F} \cdot \mathbf{n} \, dS = \frac{\mathbf{F} \cdot \mathbf{N}}{\sqrt{\quad}} \sqrt{\quad} \, dx \, dy = \left(-M \frac{\partial f}{\partial x} - N \frac{\partial f}{\partial y} + P \right) dx \, dy. \quad (11)$$

That is formula (10), with cancellation of square roots. The expression $\mathbf{F} \cdot \mathbf{n} \, dS$ is often written as $\mathbf{F} \cdot d\mathbf{S}$, again relying on boldface to make $d\mathbf{S}$ a vector. Then $d\mathbf{S}$ equals $\mathbf{n} \, dS$, with direction $\mathbf{n}$ and magnitude dS.

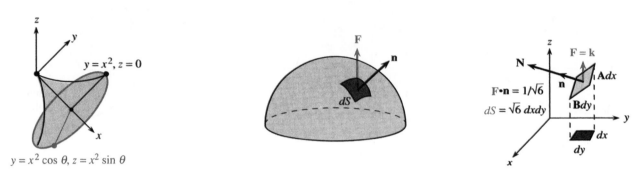

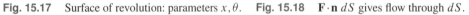

Fig. 15.17 Surface of revolution: parameters x, θ. **Fig. 15.18** $\mathbf{F} \cdot \mathbf{n} \, dS$ gives flow through dS.

EXAMPLE 9 Find $\mathbf{n} \, dS$ for the plane $z = x + 2y$. Then find $\mathbf{F} \cdot \mathbf{n} \, dS$ for $\mathbf{F} = \mathbf{k}$. This plane produced $\sqrt{6}$ in Example 1 (for area). For flux the $\sqrt{6}$ disappears:

$$\mathbf{n} \, dS = \frac{\mathbf{N}}{|\mathbf{N}|} \, dS = \frac{-\mathbf{i} - 2\mathbf{j} + \mathbf{k}}{\sqrt{6}} \sqrt{6} \, dx \, dy = (-\mathbf{i} - 2\mathbf{j} + \mathbf{k}) \, dx \, dy.$$

For the flow field $\mathbf{F} = \mathbf{k}$, the dot product $\mathbf{k} \cdot \mathbf{n} \, dS$ reduces to $1 \, dx \, dy$. The slope of the plane makes no difference! *The flow through the base also flows through the plane.* The areas are different, but flux is like rain. Whether it hits a tent or the ground below, it is the same rain (Figure 15.18). In this case $\iint \mathbf{F} \cdot \mathbf{n} \, dS = \iint dx \, dy =$ shadow area in the base.

EXAMPLE 10 Find the flux of $\mathbf{F} = x\mathbf{i} + y\mathbf{j} + z\mathbf{k}$ through the cone $z = \sqrt{x^2 + y^2}$.

Solution $$\mathbf{F} \cdot \mathbf{n} \, dS = \left[-x \left(\frac{x}{\sqrt{x^2 + y^2}} \right) - y \left(\frac{y}{\sqrt{x^2 + y^2}} \right) + \sqrt{x^2 + y^2} \right] dx \, dy = 0.$$

The zero comes as a surprise, but it shouldn't. The cone goes straight out from the origin, and so does $\mathbf{F}$. The vector $\mathbf{n}$ that is perpendicular to the cone is also perpendicular to $\mathbf{F}$. There is no flow *through* the cone, because $\mathbf{F} \cdot \mathbf{n} = 0$. The flow travels out along rays.

$\iint \mathbf{F} \cdot \mathbf{n} \, dS$ FOR A SURFACE WITH PARAMETERS

In Example 10 the cone was $z = f(x, y) = \sqrt{x^2 + y^2}$. We found dS by Method 1. Parameters were not needed (more exactly, they were x and y). For surfaces that fold and twist, the formulas with u and v look complicated but the actual calculations can be simpler. This was certainly the case for $dS = du \, dv$ on the cylinder.

A small piece of surface has area $dS = |\mathbf{A} \times \mathbf{B}| \, du \, dv$. The vectors along the sides are $\mathbf{A} = x_u \mathbf{i} + y_u \mathbf{j} + z_u \mathbf{k}$ and $\mathbf{B} = x_v \mathbf{i} + y_v \mathbf{j} + z_v \mathbf{k}$. They are tangent to the surface.

Now we put their cross product $\mathbf{N} = \mathbf{A} \times \mathbf{B}$ to another use, because $\mathbf{F} \cdot \mathbf{n} dS$ involves not only area but *direction*. We need the unit vector $\mathbf{n}$ to see how much flow goes through.

The direction vector is $\mathbf{n} = \mathbf{N}/|\mathbf{N}|$. Equation (7) is $dS = |\mathbf{N}| du \, dv$, so the square root $|\mathbf{N}|$ cancels in $\mathbf{n} dS$. This leaves a nice formula for the "normal component" of flow:

15J Through a surface with parameters u and v, the field $\mathbf{F} = M\mathbf{i} + N\mathbf{j} + P\mathbf{k}$ has

$$\text{flux} = \iint \mathbf{F} \cdot \mathbf{n} dS = \iint \mathbf{F} \cdot \mathbf{N} \, du \, dv = \iint \mathbf{F} \cdot (\mathbf{A} \times \mathbf{B}) \, du \, dv. \qquad (12)$$

EXAMPLE 11 Find the flux of $\mathbf{F} = x\mathbf{i} + y\mathbf{j} + z\mathbf{k}$ through the cylinder $x^2 + y^2 = 1$, $0 \leqslant z \leqslant b$.

Solution The surface of the cylinder is $x = \cos u$, $y = \sin u$, $z = v$. The tangent vectors from (5) are $\mathbf{A} = (-\sin u)\mathbf{i} + (\cos u)\mathbf{j}$ and $\mathbf{B} = \mathbf{k}$. The normal vector in Figure 15.19 goes straight out through the cylinder:

$$\mathbf{N} = \mathbf{A} \times \mathbf{B} = \cos u \, \mathbf{i} + \sin u \, \mathbf{j} \qquad (\text{check } \mathbf{A} \cdot \mathbf{N} = 0 \text{ and } \mathbf{B} \cdot \mathbf{N} = 0).$$

To find $\mathbf{F} \cdot \mathbf{N}$, switch $\mathbf{F} = x\mathbf{i} + y\mathbf{j} + z\mathbf{k}$ to the parameters u and v. Then $\mathbf{F} \cdot \mathbf{N} = 1$:

$$\mathbf{F} \cdot \mathbf{N} = (\cos u \, \mathbf{i} + \sin u \, \mathbf{j} + v \, \mathbf{k}) \cdot (\cos u \, \mathbf{i} + \sin u \, \mathbf{j}) = \cos^2 u + \sin^2 u.$$

For the flux, integrate $\mathbf{F} \cdot \mathbf{N} = 1$ and apply the limits on $u = \theta$ and $v = z$:

$$\text{flux} = \int_0^b \int_0^{2\pi} 1 \, du \, dv = 2\pi b = \text{surface area of the cylinder}.$$

Note that the top and bottom were not included! We can find those fluxes too. The outward direction is $\mathbf{n} = \mathbf{k}$ at the top and $\mathbf{n} = -\mathbf{k}$ down through the bottom. Then $\mathbf{F} \cdot \mathbf{n}$ is $+z = b$ at the top and $-z = 0$ at the bottom. The bottom flux is zero, the top flux is b times the area (or πb). The total flux is $2\pi b + \pi b = 3\pi b$. Hold that answer for the next section.

Apology: I made u the angle and v the height. Then $\mathbf{N}$ goes outward not inward.

EXAMPLE 12 Find the flux of $\mathbf{F} = \mathbf{k}$ out the top half of the sphere $x^2 + y^2 + z^2 = a^2$.

Solution Use spherical coordinates. Example 7 had $u = \phi$ and $v = \theta$. We found

$$\mathbf{N} = \mathbf{A} \times \mathbf{B} = a^2 \sin^2 \phi \cos \theta \, \mathbf{i} + a^2 \sin^2 \phi \sin \theta \, \mathbf{j} + a^2 \sin^2 \phi \cos \phi \, \mathbf{k}.$$

The dot product with $\mathbf{F} = \mathbf{k}$ is $\mathbf{F} \cdot \mathbf{N} = a^2 \sin \phi \cos \phi$. The integral goes from the pole to the equator, $\phi = 0$ to $\phi = \pi/2$, and around from $\theta = 0$ to $\theta = 2\pi$:

$$\text{flux} = \int_0^{2\pi} \int_0^{\pi/2} a^2 \sin \phi \, \cos \phi \, d\phi \, d\theta = 2\pi a^2 \frac{\sin^2 \phi}{2} \Big]_0^{\pi/2} = \pi a^2.$$

The next section will show that the flux remains at πa^2 through *any surface* (!) that is bounded by the equator. A special case is a flat surface—the disk of radius a at the equator. Figure 15.18 shows $\mathbf{n} = \mathbf{k}$ pointing directly up, so $\mathbf{F} \cdot \mathbf{n} = \mathbf{k} \cdot \mathbf{k} = 1$. The flux is $\iint 1\,dS=$ area of disk $=\pi a^2$. **All fluid goes past the equator and out through the sphere**.

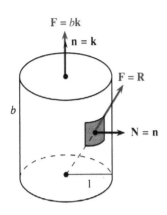

Fig. 15.19   Flow through cylinder.

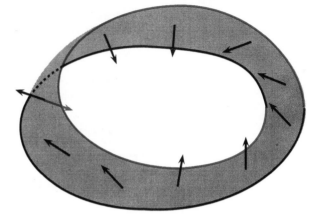

Fig. 15.20   Möbius strip (no way to choose **n**).

I have to mention one more problem. It might not occur to a reasonable person, but sometimes a surface has only *one side*. The famous example is the **Möbius strip**, for which you take a strip of paper, twist it once, and tape the ends together. Its special property appears when you run a pen along the "inside." The pen in Figure 15.20 suddenly goes "outside." After another round trip it goes back "inside." Those words are in quotation marks, because on a Möbius strip they have no meaning.

Suppose the pen represents the normal vector. On a sphere **n** points outward. Alternatively **n** could point inward; we are free to choose. But the Möbius strip makes the choice impossible. After moving the pen continuously, it comes back in the opposite direction. ***This surface is not orientable***. We cannot integrate $\mathbf{F}\cdot\mathbf{n}$ to compute the flux, because we cannot decide the direction of **n**.

A surface is *oriented* when we can and do choose **n**. This uses the final property of cross products, that they have length and direction and also a *right-hand rule*. We can tell $\mathbf{A}\times\mathbf{B}$ from $\mathbf{B}\times\mathbf{A}$. Those give the two orientations of **n**. For an open surface (like a wastebasket) you can select either one. For a closed surface (like a sphere) it is conventional for **n** to be outward. By making that decision once and for all, the sign of the flux is established: *outward flux is positive*.

FORMULAS FOR SURFACE INTEGRALS

Method 1: Parameters x,y	*Method* 2: Parameters u,v
Coordinates $x,y,z(x,y)$	$x(u,v),y(u,v),z(u,v)$ on surface
$\mathbf{A}=\mathbf{i}+\partial z/\partial x\,\mathbf{k}\quad \mathbf{N}=\mathbf{A}\times\mathbf{B}$	$\mathbf{A}=\partial x/\partial u\,\mathbf{i}+\partial y/\partial u\,\mathbf{j}+\partial z/\partial u\,\mathbf{k}$
$\mathbf{B}=\mathbf{j}+\partial z/\partial y\,\mathbf{k}\quad \mathbf{n}=\mathbf{N}/|\mathbf{N}|$	$\mathbf{B}=\partial x/\partial v\,\mathbf{i}+\partial y/\partial v\,\mathbf{j}+\partial z/\partial v\,\mathbf{k}$
$dS=|\mathbf{N}|dx\,dy=\sqrt{1+z_x^2+z_y^2}\,dx\,dy$	$dS=|\mathbf{N}|du\,dv$
$\mathbf{n}dS=\mathbf{N}dx\,dy=(-\partial z/\partial x\,\mathbf{i}-\partial z/\partial y\,\mathbf{j}+\mathbf{k})dx\,dy$	$\mathbf{n}dS=\mathbf{N}\,du\,dv$

15.4 EXERCISES

Read-through questions

A small piece of the surface $z = f(x, y)$ is nearly __a__. When we go across by dx, we go up by __b__. That movement is $\mathbf{A} dx$, where the vector $\mathbf{A}$ is $\mathbf{i} +$ __c__. The other side of the piece is $\mathbf{B} dy$, where $\mathbf{B} = \mathbf{j} +$ __d__. The cross product $\mathbf{A} \times \mathbf{B}$ is $\mathbf{N} =$ __e__. The area of the piece is $dS = |\mathbf{N}| dx\, dy$. For the surface $z = xy$, the vectors are $\mathbf{A} =$ __f__ and $\mathbf{B} =$ __g__ and $\mathbf{N} =$ __h__. The area integral is $\iint dS =$ __i__ $dx\, dy$.

With parameters u and v, a typical point on a $45°$ cone is $x = u \cos v$, $y =$ __j__, $z =$ __k__. A change in u moves that point by $\mathbf{A} du = (\cos v\, \mathbf{i} +$ __l__$)du$. A change in v moves the point by $\mathbf{B} dv =$ __m__. The normal vector is $\mathbf{N} = \mathbf{A} \times \mathbf{B} =$ __n__. The area is $dS =$ __o__ $du\, dv$. In this example $\mathbf{A} \cdot \mathbf{B} =$ __p__ so the small piece is a __q__ and $dS = |\mathbf{A}||\mathbf{B}| du\, dv$.

For flux we need $\mathbf{n} dS$. The __r__ vector $\mathbf{n}$ is $\mathbf{N} = \mathbf{A} \times \mathbf{B}$ divided by __s__. For a surface $z = f(x, y)$, the product $\mathbf{n} dS$ is the vector __t__ (to memorize from table). The particular surface $z = xy$ has $\mathbf{n} dS =$ __u__ $dx\, dy$. For $\mathbf{F} = x\mathbf{i} + y\mathbf{j} + z\mathbf{k}$ the flux through $z = xy$ is $\mathbf{F} \cdot \mathbf{n} dS =$ __v__ $dx\, dy$.

On a $30°$ cone the points are $x = 2u \cos v$, $y = 2u \sin v$, $z = u$. The tangent vectors are $\mathbf{A} =$ __w__ and $\mathbf{B} =$ __x__. This cone has $\mathbf{n} dS = \mathbf{A} \times \mathbf{B}\, du\, dv =$ __y__. For $\mathbf{F} = x\mathbf{i} + y\mathbf{j} + z\mathbf{k}$, the flux element through the cone is $\mathbf{F} \cdot \mathbf{n} dS =$ __z__. The reason for this answer is __A__. The reason we don't compute flux through a Möbius strip is __B__.

In 1–14 find N and $dS = |\mathbf{N}| dx\, dy$ and the surface area $\iint dS$. Integrate over the xy shadow which ends where the z's are equal ($x^2 + y^2 = 4$ in Problem 1).

1 Paraboloid $z = x^2 + y^2$ below the plane $z = 4$.

2 Paraboloid $z = x^2 + y^2$ between $z = 4$ and $z = 8$.

3 Plane $z = x - y$ inside the cylinder $x^2 + y^2 = 1$.

4 Plane $z = 3x + 4y$ above the square $0 \leqslant x \leqslant 1$, $0 \leqslant y \leqslant 1$.

5 Spherical cap $x^2 + y^2 + z^2 = 1$ above $z = 1/\sqrt{2}$.

6 Spherical band $x^2 + y^2 + z^2 = 1$ between $z = 0$ and $1/\sqrt{2}$.

7 Plane $z = 7y$ above a triangle of area A.

8 Cone $z^2 = x^2 + y^2$ between planes $z = a$ and $z = b$.

9 The monkey saddle $z = \frac{1}{3}x^3 - xy^2$ inside $x^2 + y^2 = 1$.

10 $z = x + y$ above triangle with vertices $(0,0)$, $(2,2)$, $(0,2)$.

11 Plane $z = 1 - 2x - 2y$ inside $x \geqslant 0$, $y \geqslant 0$, $z \geqslant 0$.

12 Cylinder $x^2 + z^2 = a^2$ inside $x^2 + y^2 = a^2$. Only set up $\iint dS$.

13 Right circular cone of radius a and height h. Choose $z = f(x, y)$ or parameters u and v.

14 Gutter $z = x^2$ below $z = 9$ and between $y = \pm 2$.

In 15–18 compute the surface integrals $\iint g(x, y, z) dS$.

15 $g = xy$ over the triangle $x + y + z = 1$, $x, y, z \geqslant 0$.

16 $g = x^2 + y^2$ over the top half of $x^2 + y^2 + z^2 = 1$ (use ϕ, θ).

17 $g = xyz$ on $x^2 + y^2 + z^2 = 1$ above $z^2 = x^2 + y^2$ (use ϕ, θ).

18 $g = x$ on the cylinder $x^2 + y^2 = 4$ between $z = 0$ and $z = 3$.

In 19–22 calculate A, B, N, and dS.

19 $x = u$, $y = v + u$, $z = v + 2u + 1$.

20 $x = uv$, $y = u + v$, $z = u - v$.

21 $x = (3 + \cos u) \cos v$, $y = (3 + \cos u) \sin v$, $z = \sin u$.

22 $x = u \cos v$, $y = u \sin v$, $z = v$ (not $z = u$).

23–26 In Problems 1–4 respectively find the flux $\iint \mathbf{F} \cdot \mathbf{n} dS$ for $\mathbf{F} = x\mathbf{i} + y\mathbf{j} + z\mathbf{k}$.

27–28 In Problems 19–20 respectively compute $\iint \mathbf{F} \cdot \mathbf{n} dS$ for $\mathbf{F} = y\mathbf{i} - x\mathbf{j}$ through the region $u^2 + v^2 \leqslant 1$.

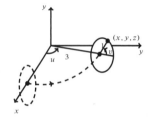

29 A unit circle is rotated around the z axis to give a torus (see figure). The center of the circle stays a distance 3 from the z axis. Show that Problem 21 gives a typical point (x, y, z) on the torus and find the surface area $\iint dS = \iint |\mathbf{N}| du\, dv$.

30 The surface $x = r \cos \theta$, $y = r \sin \theta$, $z = a^2 - r^2$ is bounded by the equator $(r = a)$. Find $\mathbf{N}$ and the flux $\iint \mathbf{k} \cdot \mathbf{n} dS$, and compare with Example 12.

31 Make a "double Möbius strip" from a strip of paper by twisting it twice and taping the ends. Does a normal vector (use a pen) have the same direction after a round trip?

32 Make a "triple Möbius strip" with three twists. Is it orientable—does the normal vector come back in the same or opposite direction?

33 If a very wavy surface stays close to a smooth surface, are their areas close?

34 Give the equation of a plane with roof area $dS = 3$ times base area $dx\, dy$.

35 The points $(x, f(x) \cos \theta, f(x) \sin \theta)$ are on the surface of revolution: $y = f(x)$ revolved around the x axis, parameters $u = x$ and $v = \theta$. Find $\mathbf{N}$ and compare $dS = |\mathbf{N}| dx\, d\theta$ with Example 8 and Section 8.3.

15.5 The Divergence Theorem

This section returns to the fundamental law (*flow out*) − (*flow in*) = (*source*). In two dimensions, the flow was in and out through a closed curve C. The plane region inside was R. In three dimensions, the flow enters and leaves through a closed surface S. The solid region inside is V. Green's Theorem in its normal form (for the flux of a smooth vector field) now becomes the great three-dimensional balance equation— the *Divergence Theorem*:

> **15K**   The flux of $\mathbf{F} = M\mathbf{i} + N\mathbf{j} + P\mathbf{k}$ through the boundary surface S equals the integral of the divergence of $\mathbf{F}$ inside V. **The Divergence Theorem is**
> $$\iint \mathbf{F} \cdot \mathbf{n}\, dS = \iiint \operatorname{div} \mathbf{F}\, dV = \iiint \left(\frac{\partial M}{\partial x} + \frac{\partial N}{\partial y} + \frac{\partial P}{\partial z} \right) dx\, dy\, dz. \quad (1)$$

In Green's Theorem the divergence was $\partial M/\partial x + \partial N/\partial y$. The new term $\partial P/\partial z$ accounts for upward flow. Notice that a constant upward component P adds nothing to the divergence (its derivative is zero). It also adds nothing to the flux (flow up through the top equals flow up through the bottom). When the whole field $\mathbf{F}$ is constant, the theorem becomes $0 = 0$.

There are other vector fields with div $\mathbf{F} = 0$. They are of the greatest importance. The Divergence Theorem for those fields is again $0 = 0$, and there is conservation of fluid. **When div $\mathbf{F} = 0$, *flow in equals flow out*.** We begin with examples of these "*divergence-free*" fields.

EXAMPLE 1   The spin fields $-y\mathbf{i} + x\mathbf{j} + 0\mathbf{k}$ and $0\mathbf{i} - z\mathbf{j} + y\mathbf{k}$ have zero divergence.

The first is an old friend, spinning around the z axis. The second is new, spinning around the x axis. Three-dimensional flow has a great variety of spin fields. The separate terms $\partial M/\partial x$, $\partial N/\partial y$, $\partial P/\partial z$ are all zero, so div $\mathbf{F} = 0$. The flow goes around in circles, and whatever goes out through S comes back in. (We might have put a circle on $\iint_s$ as we did on $\oint_c$, to emphasize that S is closed.)

EXAMPLE 2   The position field $\mathbf{R} = x\mathbf{i} + y\mathbf{j} + z\mathbf{k}$ has div $\mathbf{R} = 1 + 1 + 1 = 3$.

This is radial flow, straight out from the origin. Mass has to be added *at every point* to keep the flow going. On the right side of the divergence theorem is $\iiint 3\, dV$. Therefore the flux is three times the volume.

Example 11 in Section 15.4 found the flux of $\mathbf{R}$ through a cylinder. The answer was $3\pi b$. Now we also get $3\pi b$ from the Divergence Theorem, since the volume is πb. This is one of many cases in which the triple integral is easier than the double integral.

EXAMPLE 3   An *electrostatic field* $\mathbf{R}/\rho^3$ or *gravity field* $-\mathbf{R}/\rho^3$ almost has div $\mathbf{F} = 0$.

The vector $\mathbf{R} = x\mathbf{i} + y\mathbf{j} + z\mathbf{k}$ has length $\sqrt{x^2 + y^2 + z^2} = \rho$. Then $\mathbf{F}$ has length ρ/ρ^3 (inverse square law). Gravity from a point mass pulls *inward* (minus sign). The electric field from a point charge repels *outward*. The three steps almost show that div $\mathbf{F} = 0$:

Step 1. $\partial\rho/\partial x = x/\rho, \partial\rho/\partial y = y/\rho, \partial\rho/\partial z = z/\rho$—but do not add those three. $\mathbf{F}$ is not ρ or $1/\rho^2$ (these are scalars). The vector field is $\mathbf{R}/\rho^3$. We need $\partial M/\partial x, \partial N/\partial y, \partial P/\partial z$.

Step 2. $\partial M/\partial x = \partial/\partial x (x/\rho^3)$ is equal to $1/\rho^3 - (3x\,\partial\rho/\partial x)/\rho^4 = 1/\rho^3 - 3x^2/\rho^5$. For $\partial N/\partial y$ and $\partial P/\partial z$, replace $3x^2$ by $3y^2$ and $3z^2$. Now add those three.

Step 3. div $\mathbf{F} = 3/\rho^3 - 3(x^2 + y^2 + z^2)/\rho^5 = 3/\rho^3 - 3/\rho^3 = 0$.

The calculation div $\mathbf{F} = 0$ leaves a puzzle. One side of the Divergence Theorem seems to give $\iiint 0\,dV = 0$. Then the other side should be $\iint \mathbf{F} \cdot \mathbf{n}\,dS = 0$. But the flux is *not* zero when all flow is outward:

> The unit normal vector to the sphere $\rho =$ constant is $\mathbf{n} = \mathbf{R}/\rho$.
> The outward flow $\mathbf{F} \cdot \mathbf{n} = (\mathbf{R}/\rho^3) \cdot (\mathbf{R}/\rho) = \rho^2/\rho^4$ is always positive.
> Then $\iiint \mathbf{F} \cdot \mathbf{n}\,dS = \iint dS/\rho^2 = 4\pi\rho^2/\rho^2 = 4\pi$. *We have reached* $4\pi = 0$.

This paradox in three dimensions is the same as for $\mathbf{R}/r^2$ in two dimensions. Section 15.3 reached $2\pi = 0$, and the explanation was a point source at the origin. Same explanation here: M, N, P are infinite when $\rho = 0$. The divergence is a "delta function" times 4π, from the point source. The Divergence Theorem does not apply (unless we allow delta functions). That single point makes all the difference.

Every surface enclosing the origin has flux $= 4\pi$. Our calculation was for a sphere. The surface integral is much harder when S is twisted (Figure 15.21a). But the Divergence Theorem takes care of everything, because div $\mathbf{F} = 0$ in the volume V between these surfaces. Therefore $\iint \mathbf{F} \cdot \mathbf{n}\,dS = 0$ for the two surfaces together. The flux $\iint \mathbf{F} \cdot \mathbf{n}\,dS = -4\pi$ into the sphere must be balanced by $\iint \mathbf{F} \cdot \mathbf{n}\,dS = 4\pi$ out of the twisted surface.

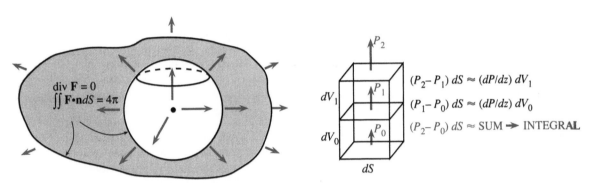

Fig. 15.21 Point source: flux 4π through all enclosing surfaces. Net flux upward $= \iiint (\partial P/\partial z)\,dV$.

Instead of a paradox $4\pi = 0$, this example leads to Gauss's Law. A mass M at the origin produces a gravity field $\mathbf{F} = -GM\mathbf{R}/\rho^3$. A charge q at the origin produces an electric field $\mathbf{E} = (q/4\pi\varepsilon_0)\mathbf{R}/\rho^3$. The physical constants are G and ε_0, the mathematical constant is the relation between divergence and flux. Equation (1) yields equation (2), in which the mass densities $M(x, y, z)$ and charge densities $q(x, y, z)$ need not be concentrated at the origin:

15L   Gauss's law in differential form: div $\mathbf{F} = -4\pi GM$ and div $\mathbf{E} = q/\varepsilon_0$.
Gauss's law in integral form: Flux is proportional to total mass or charge:

$$\iint \mathbf{F} \cdot \mathbf{n}\,dS = -\iiint 4\pi GM\,dV \quad \text{and} \quad \iint \mathbf{E} \cdot \mathbf{n}\,dS = \iiint q\,dV/\varepsilon_0. \qquad (2)$$

THE REASONING BEHIND THE DIVERGENCE THEOREM

The general principle is clear: Flow out minus flow in equals source. Our goal is to see why *the divergence of* **F** *measures the source*. In a small box around each point, we show that div **F** dV balances **F** $\cdot$ **n** dS through the six sides.

So consider a small box. Its center is at (x, y, z). Its edges have length $\Delta x, \Delta y, \Delta z$. Out of the top and bottom, the normal vectors are **k** and $-$**k**. The dot product with **F** $= M$**i** $+ N$**j** $+ P$**k** is $+P$ or $-P$. The area ΔS is $\Delta x \Delta y$. So the two fluxes are close to $P(x, y, z + \frac{1}{2}\Delta z)\Delta x \Delta y$ and $-P(x, y, z - \frac{1}{2}\Delta z)\Delta x \Delta y$. When the top is combined with the bottom, the difference of those P's is ΔP:

$$\textit{net flux upward} \ \approx \Delta P \Delta x \Delta y = (\Delta P / \Delta z)\Delta x \Delta y \Delta z \approx (\partial P / \partial z)\Delta V. \qquad (3)$$

Similarly, the combined flux on two side faces is approximately $(\partial N / \partial y)\Delta V$. On the front and back it is $(\partial M / \partial x)\Delta V$. Adding the six faces, we reach the key point:

$$\textit{flux out of the box} \ \approx (\partial M / \partial x + \partial N / \partial y + \partial P / \partial z)\Delta V. \qquad (4)$$

This is (div **F**)ΔV. For a constant field both sides are zero—the flow goes straight through. For **F** $= x$**i** $+ y$**j** $+ z$**k**, a little more goes out than comes in. The divergence is 3, so $3\Delta V$ is created inside the box. By the balance equation the flux is also $3\Delta V$.

The approximation symbol $\approx$ means that the leading term is correct (probably not the next term). The ratio $\Delta P / \Delta z$ is not exactly $\partial P / \partial z$. The difference is of order Δz, so the error in (3) is of higher order $\Delta V \Delta z$. Added over many boxes (about $1/\Delta V$ boxes), this error disappears as $\Delta z \to 0$.

The sum of (div **F**)ΔV over all the boxes approaches $\iiint$(div **F**)dV. On the other side of the equation is a sum of fluxes. There is **F** $\cdot$ **n** ΔS out of the top of one box, plus **F** $\cdot$ **n** ΔS out of the bottom of the box above. The first has **n** $=$ **k** and the second has **n** $= -$**k**. *They cancel each other—the flow goes from box to box*. This happens every time two boxes meet. The only fluxes that survive (because nothing cancels them) are at the outer surface S. The final step, as $\Delta x, \Delta y, \Delta z \to 0$, is that those outside terms approach $\iint$ **F** $\cdot$ **n** dS. Then the local divergence theorem (4) becomes the global Divergence Theorem (1).

Remark on the proof   That "final step" is not easy, because the box surfaces don't line up with the outer surface S. A formal proof of the Divergence Theorem would imitate the proof of Green's Theorem. On a very simple region $\iiint(\partial P / \partial z)dx\,dy\,dz$ equals $\iint P\,dx\,dy$ over the top minus $\iint P\,dx\,dy$ over the bottom. After checking the orientation this is $\iint P$**k** $\cdot$ **n** dS. Similarly the volume integrals of $\partial M / \partial x$ and $\partial N / \partial y$ are the surface integrals $\iint M$**i** $\cdot$ **n** dS and $\iint N$**j** $\cdot$ **n** dS. Adding the three integrals gives the Divergence Theorem. Since Green's Theorem was already proved in this way, the reasoning behind (4) is more helpful than repeating a detailed proof.

The discoverer of the Divergence Theorem was probably Gauss. His notebooks only contain the outline of a proof—but after all, this is Gauss. Green and Ostrogradsky both published proofs in 1828, one in England and the other in St. Petersburg (now Leningrad). As the theorem was studied, the requirements came to light (smoothness of **F** and S, avoidance of one-sided Möbius strips).

New applications are discovered all the time—***when a scientist writes down a balance equation in a small box***. The source is known. The equation is div **F** $=$source. After Example 5 we explain **F**.

EXAMPLE 4   If the temperature inside the sun is $T = \ln 1/\rho$, find the heat flow **F** $= -$grad T and the source div **F** and the flux $\iint$ **F** $\cdot$ **n** dS. The sun is a ball of radius ρ_0.

Solution $\mathbf{F}$ is $-\text{grad}\,\ln 1/\rho = +\text{grad}\,\ln\rho$. Derivatives of $\ln\rho$ bring division by ρ:

$$\mathbf{F} = (\partial\rho/\partial x\,\mathbf{i} + \partial\rho/\partial y\,\mathbf{j} + \partial\rho/\partial z\,\mathbf{k})/\rho = (x\mathbf{i} + y\mathbf{j} + z\mathbf{k})/\rho^2.$$

This flow is radially outward, of magnitude $1/\rho$. The normal vector $\mathbf{n}$ is also radially outward, of magnitude 1. The dot product on the sun's surface is $1/\rho_0$:

$$\iint \mathbf{F}\cdot\mathbf{n}\,dS = \iint dS/\rho_0 = (\text{surface area})/\rho_0 = 4\pi\rho_0^2/\rho_0 = 4\pi\rho_0. \tag{5}$$

Check that answer by the Divergence Theorem. Example 5 will find div $\mathbf{F} = 1/\rho^2$. Integrate over the sun. In spherical coordinates we integrate $d\rho$, $\sin\phi\,d\phi$, and $d\theta$:

$$\iiint_{\text{sun}} \text{div}\,\mathbf{F}\,dV = \int_0^{2\pi}\int_0^{\pi}\int_0^{\rho_0} \rho^2\,\sin\phi\,d\rho\,d\phi\,d\theta/\rho^2 = (\rho_0)(2)(2\pi)\ \text{as in (5)}.$$

This example illustrates *the basic framework of equilibrium*. The pattern appears everywhere in applied mathematics—electromagnetism, heat flow, elasticity, even relativity. There is usually a constant c that depends on the material (the example has $c = 1$). The names change, but we always take *the divergence of the gradient*:

---

potential $f \rightarrow$ force field $-c\,\text{grad}\,f$. Then $\text{div}(-c\,\text{grad}\,f) = $ electric charge

temperature $T \rightarrow$ flow field $-c\,\text{grad}\,T$. Then $\text{div}(-c\,\text{grad}\,T) = $ heat source

displacement $u \rightarrow$ stress field $+c\,\text{grad}\,u$. Then $\text{div}(-c\,\text{grad}\,u) = $ outside force.

You are studying calculus, not physics or thermodynamics or elasticity. But please notice the main point. The equation to solve is $\text{div}(-c\,\text{grad}\,f) = $ known source. The divergence and gradient are exactly what the applications need. Calculus teaches the right things.

This framework is developed in many books, including my own text *Introduction to Applied Mathematics* (Wellesley-Cambridge Press). It governs equilibrium, in matrix equations and differential equations.

PRODUCT RULE FOR VECTORS: INTEGRATION BY PARTS

May I go back to basic facts about the divergence? First the definition:

$$\mathbf{F}(x, y, z) = M\mathbf{i} + N\mathbf{j} + P\mathbf{k}\ \text{has div}\ \mathbf{F} = \nabla\cdot\mathbf{F} = \partial M/\partial x + \partial N/\partial y + \partial P/\partial z.$$

The divergence is a scalar (not a vector). At each point div $\mathbf{F}$ is a number. In fluid flow, it is the rate at which mass leaves—the "flux per unit volume" or "flux density."

The symbol ∇ stands for a vector whose components are *operations not numbers*:

$$\nabla = \text{``del''} = \mathbf{i}\,\partial/\partial x + \mathbf{j}\,\partial/\partial y + \mathbf{k}\,\partial/\partial z. \tag{6}$$

This vector is illegal but very useful. First, apply it to an ordinary function $f(x, y, z)$:

$$\nabla f = \text{``del}\,f\text{''} = \mathbf{i}\,\partial f/\partial x + \mathbf{j}\,\partial f/\partial y + \mathbf{k}\,\partial f/\partial z = \textbf{\textit{gradient of }} f. \tag{7}$$

Second, take the dot product $\nabla\cdot\mathbf{F}$ with a vector function $\mathbf{F}(x, y, z) = M\mathbf{i} + N\mathbf{j} + P\mathbf{k}$:

$$\nabla\cdot\mathbf{F} = \text{``del dot }\mathbf{F}\text{''} = \partial M/\partial x + \partial N/\partial y + \partial P/\partial z = \textbf{\textit{divergence of }}\mathbf{F}. \tag{8}$$

Third, take the cross product $\nabla \times \mathbf{F}$. This produces the vector curl $\mathbf{F}$ (next section):

$$\nabla \times \mathbf{F} = \text{"del cross } \mathbf{F}\text{"} = \ldots \text{(to be defined)} \ldots = \mathbf{\textit{curl of}} \ \mathbf{F}. \tag{9}$$

The gradient and divergence and curl are ∇ and $\nabla\cdot$ and $\nabla\times$. The three great operations of vector calculus use a single notation! You are free to write ∇ or not—to make equations shorter or to help the memory. Notice that Laplace's equation shrinks to

$$\nabla \cdot \nabla f = \frac{\partial}{\partial x}\left(\frac{\partial f}{\partial x}\right) + \frac{\partial}{\partial y}\left(\frac{\partial f}{\partial y}\right) + \frac{\partial}{\partial z}\left(\frac{\partial f}{\partial z}\right) = 0. \tag{10}$$

Equation (10) gives the potential when the source is zero (very common). $\mathbf{F} = \text{grad } f$ combines with div $\mathbf{F} = 0$ into Laplace's equation div grad $f = 0$. This equation is so important that it shrinks further to $\nabla^2 f = 0$ and even to $\Delta f = 0$. Of course $\Delta f = f_{xx} + f_{yy} + f_{zz}$ has nothing to do with $\Delta f = f(x + \Delta x) - f(x)$. Above all, remember that f is a scalar and $\mathbf{F}$ is a vector: **gradient of scalar is vector** and **divergence of vector is scalar**.

Underlying this chapter is the idea of extending calculus to vectors. So far we have emphasized the Fundamental Theorem. The integral of df/dx is now the integral of div $\mathbf{F}$. Instead of endpoints a and b, we have a curve C or surface S. But it is the *rules* for derivatives and integrals that make calculus work, and we need them now for vectors. Remember the derivative of u times v and the integral (by parts) of $u\, dv/dx$:

15M   Scalar functions $u(x, y, z)$ and vector fields $\mathbf{V}(x, y, z)$ obey the **product rule**:

$$\text{div}(u\mathbf{V}) = u \text{ div } \mathbf{V} + \mathbf{V} \cdot (\text{grad } u). \tag{11}$$

The reverse of the product rule is integration by parts (**Gauss's Formula**):

$$\iiint u \text{ div } \mathbf{V} \, dx\, dy\, dz = -\iiint \mathbf{V} \cdot (\text{grad } u)\, dx\, dy\, dz + \iint u\, \mathbf{V} \cdot \mathbf{n}\, dS. \tag{12}$$

For a plane field this is **Green's Formula** (and $u = 1$ gives Green's Theorem):

$$\iint u\left(\frac{\partial M}{\partial x} + \frac{\partial N}{\partial y}\right) dx\, dy = -\iint \left(M\frac{\partial u}{\partial x} + N\frac{\partial u}{\partial y}\right) dx\, dy + \int u(M\mathbf{i} + N\mathbf{j}) \cdot \mathbf{n}\, ds. \tag{13}$$

Those look like heavy formulas. They are too much to memorize, unless you use them often. The important point is to connect vector calculus with "scalar calculus," which is not heavy. Every product rule yields two terms:

$$(uM)_x = u\, \partial M/\partial x + M\, \partial u/\partial x \quad (uN)_y = u\, \partial N/\partial y + N\, \partial u/\partial y \quad (uP)_z = u\, \partial P/\partial z + P\, \partial u/\partial z.$$

Add those ordinary rules and you have the vector rule (11) for the divergence of $u\mathbf{V}$.

Integrating the two parts of $\text{div}(u\mathbf{V})$ gives $\iint u\mathbf{V} \cdot \mathbf{n}\, dS$ by the Divergence Theorem. Then one part moves to the other side, producing the minus signs in (12) and (13). **Integration by parts leaves a boundary term**, in three and two dimensions as it did in one dimension: $\int uv' dx = -\int u'v dx + [uv]_a^b$.

EXAMPLE 5   Find the divergence of $\mathbf{F} = \mathbf{R}/\rho^2$, starting from grad $\rho = \mathbf{R}/\rho$.

Solution   Take $\mathbf{V} = \mathbf{R}$ and $u = 1/\rho^2$ in the product rule (11). Then div $\mathbf{F} = (\text{div } \mathbf{R})/\rho^2 + \mathbf{R} \cdot (\text{grad } 1/\rho^2)$. The divergence of $\mathbf{R} = x\mathbf{i} + y\mathbf{j} + z\mathbf{k}$ is 3. For grad $1/\rho^2$ apply the chain rule:

$$\mathbf{R} \cdot (\text{grad } 1/\rho^2) = -2\mathbf{R} \cdot (\text{grad } \rho)/\rho^3 = -2\mathbf{R} \cdot \mathbf{R}/\rho^4 = -2/\rho^2.$$

The two parts of div $\mathbf{F}$ combine into $3/\rho^2 - 2/\rho^2 = 1/\rho^2$—as claimed in Example 4.

EXAMPLE 6 Find the balance equation for flow with velocity $\mathbf{V}$ and fluid density ρ.

$\mathbf{V}$ is the rate of movement of fluid, while $\rho\mathbf{V}$ is the rate of movement of *mass*. Comparing the ocean to the atmosphere shows the difference. Air has a greater velocity than water, but a much lower density. So normally $\mathbf{F} = \rho\mathbf{V}$ is larger for the ocean. (Don't confuse the density ρ with the radial distance ρ. The Greeks only used 24 letters.)

There is another difference between water and air. Water is virtually incompressible (meaning $\rho = $ constant). Air is certainly compressible (its density varies). The balance equation is a fundamental law—the conservation of mass or the "*continuity equation*" for fluids. This is a mathematical statement about a physical flow without sources or sinks:

$$\textit{Continuity Equation}: \text{div}(\rho\mathbf{V}) + \partial\rho/\partial t = 0. \qquad (14)$$

Explanation: The mass in a region is $\iiint \rho \, dV$. Its rate of decrease is $-\iiint \partial p/\partial t \, dV$. The decrease comes from flow out through the surface (normal vector $\mathbf{n}$). The dot product $\mathbf{F} \cdot \mathbf{n} = \rho\mathbf{V} \cdot \mathbf{n}$ is the rate of mass flow through the surface. So the integral $\iint \mathbf{F} \cdot \mathbf{n} \, dS$ is the total rate that mass goes out. By the Divergence Theorem this is $\iiint \text{div } \mathbf{F} \, dV$.

To balance $-\iiint \partial\rho/\partial t \, dV$ in every region, div $\mathbf{F}$ must equal $-\partial\rho/\partial t$ at every point. The figure shows this continuity equation (14) for flow in the x direction.

mass in		extra mass out	mass loss
$\rightarrow$	$\boxed{\text{mass } \rho dS \, dx}$ $\rightarrow$		$=$
$\rho\mathbf{V} \, dS \, dt$		$d(\rho\mathbf{V}) \, dS \, dt$	$-d\rho \, dS \, dx$

Fig. 15.22   Conservation of mass during time dt : $d(\rho\mathbf{V})/dx + d\rho/dt = 0$.

15.5 EXERCISES

Read-through questions

In words, the basic balance law is __a__. The flux of $\mathbf{F}$ through a closed surface S is the double integral __b__. The divergence of $M\mathbf{i} + N\mathbf{j} + P\mathbf{k}$ is __c__, and it measures __d__. The total source is the triple integral __e__. That equals the flux by the __f__ Theorem.

For $\mathbf{F} = 5z\mathbf{k}$ the divergence is __g__. If V is a cube of side a then the triple integral equals __h__. The top surface where $z = a$ has $\mathbf{n} = $__i__ and $\mathbf{F} \cdot \mathbf{n} = $__j__. The bottom and sides have $\mathbf{F} \cdot \mathbf{n} = $__k__. The integral $\iint \mathbf{F} \cdot \mathbf{n} \, dS$ equals __l__.

The field $\mathbf{F} = \mathbf{R}/\rho^3$ has div $\mathbf{F} = 0$ except __m__. $\iint \mathbf{F} \cdot \mathbf{n} \, dS$ equals __n__ over any surface around the origin. This illustrates Gauss's Law __o__. The field $\mathbf{F} = x\mathbf{i} + y\mathbf{j} - 2z\mathbf{k}$ has div $\mathbf{F} = $__p__ and $\iint \mathbf{F} \cdot \mathbf{n} \, dS = $__q__. For this $\mathbf{F}$, the flux out through a pyramid and in through its base are __r__.

The symbol ∇ stands for __s__. In this notation div $\mathbf{F}$ is __t__. The gradient of f is __u__. The divergence of grad f is __v__. The equation div grad $f = 0$ is __w__'s equation.

The divergence of a product is div $(u\mathbf{V}) = $__x__. Integration by parts is $\iiint u \text{ div } \mathbf{V} \, dx \, dy \, dz = $__y__ $+$__z__. In two dimensions this becomes __A__. In one dimension it becomes __B__. For steady fluid flow the continuity equation is div $\rho\mathbf{V} = $__C__.

In 1–10 compute the flux $\iint \mathbf{F} \cdot \mathbf{n} \, dS$ by the Divergence Theorem.

1 $\mathbf{F} = x\mathbf{i} + x\mathbf{j} + x\mathbf{k}$, S: unit sphere $x^2 + y^2 + z^2 = 1$.

2 $\mathbf{F} = -y\mathbf{i} + x\mathbf{j}$, V: unit cube $0 \leqslant x \leqslant 1$, $0 \leqslant y \leqslant 1$, $0 \leqslant z \leqslant 1$.

3 $\mathbf{F} = x^2\mathbf{i} + y^2\mathbf{j} + z^2\mathbf{k}$, S: unit sphere

4 $\mathbf{F} = x^2\mathbf{i} + 8y^2\mathbf{j} + z^2\mathbf{k}$, V: unit cube.

5 $\mathbf{F} = x\mathbf{i} + 2y\mathbf{j}$, S: sides $x = 0$, $y = 0$, $z = 0$, $x + y + z = 1$.

6 $\mathbf{F} = \mathbf{u}_r = (x\mathbf{i} + y\mathbf{j} + z\mathbf{k})/\rho$, S: sphere $\rho = a$.

7 $\mathbf{F} = \rho(x\mathbf{i} + y\mathbf{j} + z\mathbf{k})$, S: sphere $\rho = a$

8 $\mathbf{F} = x^3\mathbf{i} + y^3\mathbf{j} + z^3\mathbf{k}$, S: sphere $\rho = a$.

9 $\mathbf{F} = z^2\mathbf{k}$, V: upper half of ball $\rho \leqslant a$.

10 $\mathbf{F} = \text{grad} \, (xe^y \sin z)$, S: sphere $\rho = a$.

11 Find $\iiint \text{div} \, (x^2\mathbf{i} + y\mathbf{j} + 2\mathbf{k}) \, dV$ in the cube $0 \leqslant x, y, z \leqslant a$. Also compute $\mathbf{n}$ and $\iint \mathbf{F} \cdot \mathbf{n} \, dS$ for all six faces and add.

12 When a is small in problem 11, the answer is close to ca^3. Find the number c. At what point does div $\mathbf{F} = c$?

13 (a) Integrate the divergence of $\mathbf{F} = \rho\mathbf{i}$ in the ball $\rho \leqslant a$.

(b) Compute $\iint \mathbf{F} \cdot \mathbf{n} \, dS$ over the spherical surface $\rho = a$.

14 Integrate $\iint \mathbf{R} \cdot \mathbf{n} \, dS$ over the faces of the box $0 \leqslant x \leqslant 1$, $0 \leqslant y \leqslant 2$, $0 \leqslant z \leqslant 3$ and check by the Divergence Theorem.

15 Evaluate $\iint \mathbf{F} \cdot \mathbf{n} \, dS$ when $\mathbf{F} = x\mathbf{i} + z^2\mathbf{j} + y^2\mathbf{k}$ and:

(a) S is the cone $z^2 = x^2 + y^2$ bounded above by the plane $z = 1$.

(b) S is the pyramid with corners $(0,0,0)$, $(1,0,0)$, $(0,1,0)$, $(0,0,1)$.

16 Compute all integrals in the Divergence Theorem when $\mathbf{F} = x(\mathbf{i} + \mathbf{j} - \mathbf{k})$ and V is the unit cube $0 \leqslant x, y, z \leqslant 1$.

17 Following Example 5, compute the divergence of $(x\mathbf{i} + y\mathbf{j} + z\mathbf{k})/\rho^7$.

18 $(\text{grad} \, f) \cdot \mathbf{n}$ is the _____ derivative of f in the direction _____. It is also written $\partial f/\partial n$. If $f_{xx} + f_{yy} + f_{zz} = 0$ in V, derive $\iint \partial f/\partial n \, dS = 0$ from the Divergence Theorem.

19 Describe the closed surface S and outward normal $\mathbf{n}$:

(a) $V = $ hollow ball $1 \leqslant x^2 + y^2 + z^2 \leqslant 9$.

(b) $V = $ solid cylinder $x^2 + y^2 \leqslant 1$, $|z| \leqslant 7$.

(c) $V = $ pyramid $x \geqslant 0$, $y \geqslant 0$, $z \geqslant 0$, $x + 2y + 3z \leqslant 1$.

(d) $V = $ solid cone $x^2 + y^2 \leqslant z^2 \leqslant 1$.

20 Give an example where $\iint \mathbf{F} \cdot \mathbf{n} \, dS$ is easier than $\iiint \text{div} \, \mathbf{F} \, dV$.

21 Suppose $\mathbf{F} = M(x, y)\mathbf{i} + N(x, y)\mathbf{j}$, R is a region in the xy plane, and (x, y, z) is in V if (x, y) is in R and $|z| \geqslant 1$.

(a) Describe V and reduce $\iiint \text{div} \, \mathbf{F} \, dV$ to a double integral.

(b) Reduce $\iint \mathbf{F} \cdot \mathbf{n} \, dS$ to a line integral (check top, bottom, side).

(c) Whose theorem says that the double integral equals the line integral?

22 Is it possible to have $\mathbf{F} \cdot \mathbf{n} = 0$ at all points of S and also div $\mathbf{F} = 0$ at all points in V? $\mathbf{F} = \mathbf{0}$ is not allowed.

23 Inside a solid ball (radius a, density 1, mass $M = 4\pi a^3/3$) the gravity field is $\mathbf{F} = -GM\mathbf{R}/a^3$.

(a) Check div $\mathbf{F} = -4\pi G$ in Gauss's Law.

(b) The force at the surface is the same as if the whole mass M were _____ .

(c) Find a gradient field with div $\mathbf{F} = 6$ in the ball $\rho \leqslant a$ and div $\mathbf{F} = 0$ outside.

24 The outward field $\mathbf{F} = \mathbf{R}/\rho^3$ has magnitude $|\mathbf{F}| = 1/\rho^2$. Through an area A on a sphere of radius ρ, the flux is _____ . A spherical box has faces at ρ_1 and ρ_2 with $A = \rho_1^2 \sin \phi \, d\phi \, d\theta$ and $A = \rho_2^2 \sin \phi \, d\phi \, d\theta$. Deduce that the flux out of the box is zero, which confirms div $\mathbf{F} = 0$.

25 In Gauss's Law, what charge distribution $q(x, y, z)$ gives the unit field $\mathbf{E} = \mathbf{u}_r$? What is the flux through the unit sphere?

26 If a fluid with velocity $\mathbf{V}$ is incompressible (constant density ρ), then its continuity equation reduces to _____ . If it is irrotational then $\mathbf{F} = \text{grad} \, f$. If it is both then f satisfies _____ equation.

27 **True or false**, with a good reason.

(a) If $\iint \mathbf{F} \cdot \mathbf{n} \, dS = 0$ for every closed surface, $\mathbf{F}$ is constant.

(b) If $\mathbf{F} = \text{grad} \, f$ then div $\mathbf{F} = 0$.

(c) If $|\mathbf{F}| \leqslant 1$ at all points then $\iiint \text{div} \, \mathbf{F} \, dV \leqslant$ area of the surface S.

(d) If $|\mathbf{F}| \leqslant 1$ at all points then $|\text{div} \, \mathbf{F}| \leqslant 1$ at all points.

28 Write down statements $\mathbf{E} - \mathbf{F} - \mathbf{G} - \mathbf{H}$ for source-free fields $\mathbf{F}(x, y, z)$ in three dimensions. In statement $\mathbf{F}$, paths sharing the same endpoint become surfaces sharing the same boundary curve. In $\mathbf{G}$, the stream function becomes a *vector field* such that $\mathbf{F} = \text{curl} \, \mathbf{g}$.

29 Describe two different surfaces bounded by the circle $x^2 + y^2 = 1$, $z = 0$. The field $\mathbf{F}$ automatically has the same flux through both if _____ .

30 The boundary of a bounded region R has no boundary. Draw a plane region and explain what that means. What does it mean for a solid ball?

15.6 Stokes' Theorem and the Curl of F

For the Divergence Theorem, the surface was closed. S was the boundary of V. Now the surface is not closed and S has its own boundary—a curve called C. We are back near the original setting for Green's Theorem (region bounded by curve, double integral equal to work integral). But Stokes' Theorem, also called Stokes's Theorem, is in three-dimensional space. There is a **curved surface** S bounded by a **space curve** C. This is our first integral around a space curve.

The move to three dimensions brings a change in the vector field. The plane field $\mathbf{F}(x, y) = M\mathbf{i} + N\mathbf{j}$ becomes a space field $\mathbf{F}(x, y, z) = M\mathbf{i} + N\mathbf{j} + P\mathbf{k}$. The work $M\,dx + N\,dy$ now includes $P\,dz$. The critical quantity in the double integral (it was $\partial N / \partial x - \partial M / \partial y$) must change too. We called this scalar quantity "curl $\mathbf{F}$," but in reality it is only the third component of a vector. Stokes' Theorem needs all three components of that vector—which is curl $\mathbf{F}$.

DEFINITION   The curl of a vector field $\mathbf{F}(x, y, z) = M\mathbf{i} + N\mathbf{j} + P\mathbf{k}$ is the vector field

$$\text{curl } \mathbf{F} = \left(\frac{\partial P}{\partial y} - \frac{\partial N}{\partial z} \right)\mathbf{i} + \left(\frac{\partial M}{\partial z} - \frac{\partial P}{\partial x} \right)\mathbf{j} + \left(\frac{\partial N}{\partial x} - \frac{\partial M}{\partial y} \right)\mathbf{k}. \qquad (1)$$

The symbol $\nabla \times \mathbf{F}$ stands for a determinant that yields those six derivatives:

$$\text{curl } \mathbf{F} = \nabla \times \mathbf{F} = \begin{vmatrix} \mathbf{i} & \mathbf{j} & \mathbf{k} \\ \partial/\partial x & \partial/\partial y & \partial/\partial z \\ M & N & P \end{vmatrix}. \qquad (2)$$

The three products $\mathbf{i}\,\partial/\partial y\,P$ and $\mathbf{j}\,\partial/\partial z\,M$ and $\mathbf{k}\,\partial/\partial x\,N$ have plus signs. The three products like $\mathbf{k}\,\partial/\partial y\,M$, down to the left, have minus signs. There is a cyclic symmetry. This determinant helps the memory, even if it looks and is illegal. A determinant is not supposed to have a row of vectors, a row of operators, and a row of functions.

EXAMPLE 1   The plane field $M(x, y)\mathbf{i} + N(x, y)\mathbf{j}$ has $P = 0$ and $\partial M / \partial z = 0$ and $\partial N / \partial z = 0$. Only two terms survive: curl $\mathbf{F} = (\partial N / \partial x - \partial M / \partial y)\mathbf{k}$. Back to Green.

EXAMPLE 2   The cross product $\mathbf{a} \times \mathbf{R}$ is a **spin field** $\mathbf{S}$. Its axis is the fixed vector $\mathbf{a} = a_1\mathbf{i} + a_2\mathbf{j} + a_3\mathbf{k}$. The flow in Figure 15.23 turns around $\mathbf{a}$, and its components are

$$\mathbf{S} = \mathbf{a} \times \mathbf{R} = \begin{vmatrix} \mathbf{i} & \mathbf{j} & \mathbf{k} \\ a_1 & a_2 & a_3 \\ x & y & z \end{vmatrix} = (a_2 z - a_3 y)\mathbf{i} + (a_3 x - a_1 z)\mathbf{j} + (a_1 y - a_2 x)\mathbf{k}. \qquad (3)$$

Our favorite spin field $-y\mathbf{i} + x\mathbf{j}$ has $(a_1, a_2, a_3) = (0, 0, 1)$ and its axis is $\mathbf{a} = \mathbf{k}$.

The divergence of a spin field is $M_x + N_y + P_z = 0 + 0 + 0$. Note how the divergence uses M_x while the curl uses N_x and P_x. **The curl of S is the vector** $2\mathbf{a}$:

$$\left(\frac{\partial P}{\partial y} - \frac{\partial N}{\partial z} \right)\mathbf{i} + \left(\frac{\partial M}{\partial z} - \frac{\partial P}{\partial x} \right)\mathbf{j} + \left(\frac{\partial N}{\partial x} - \frac{\partial M}{\partial y} \right)\mathbf{k} = 2a_1\mathbf{i} + 2a_2\mathbf{j} + 2a_3\mathbf{k} = 2\mathbf{a}.$$

This example begins to reveal the meaning of the curl. It measures the spin! The direction of curl $\mathbf{F}$ is the **axis of rotation**—in this case along $\mathbf{a}$. The magnitude of curl

F is *twice the speed of rotation*. In this case $|\text{curl } \mathbf{F}| = 2|\mathbf{a}|$ and the angular velocity is $|\mathbf{a}|$.

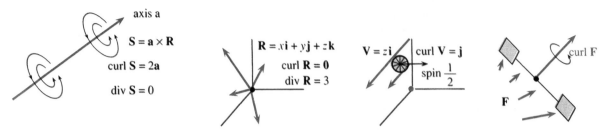

Fig. 15.23 Spin field $\mathbf{S} = \mathbf{a} \times \mathbf{R}$, position field $\mathbf{R}$, velocity field (shear field)$\mathbf{V} = z\mathbf{i}$, any field $\mathbf{F}$.

EXAMPLE 3 (!!)   *Every gradient field* $\mathbf{F} = \partial f/\partial x\, \mathbf{i} + \partial f/\partial y\, \mathbf{j} + \partial f/\partial z\, \mathbf{k}$ *has* curl $\mathbf{F} = \mathbf{0}$:

$$\text{curl } \mathbf{F} = \left(\frac{\partial}{\partial y} \frac{\partial f}{\partial z} - \frac{\partial}{\partial z} \frac{\partial f}{\partial y} \right) \mathbf{i} + \left(\frac{\partial}{\partial z} \frac{\partial f}{\partial x} - \frac{\partial}{\partial x} \frac{\partial f}{\partial z} \right) \mathbf{j} + \left(\frac{\partial}{\partial x} \frac{\partial f}{\partial y} - \frac{\partial}{\partial y} \frac{\partial f}{\partial x} \right) \mathbf{k} = \mathbf{0}. \quad (4)$$

Always f_{yz} equals f_{zy}. They cancel. Also $f_{xz} = f_{zx}$ and $f_{yx} = f_{xy}$. So curl grad $f = \mathbf{0}$.

EXAMPLE 4   (twin of Example 3) The divergence of curl **F** is also automatically zero:

$$\text{div curl } \mathbf{F} = \frac{\partial}{\partial x} \left(\frac{\partial P}{\partial y} - \frac{\partial N}{\partial z} \right) + \frac{\partial}{\partial y} \left(\frac{\partial M}{\partial z} - \frac{\partial P}{\partial x} \right) + \frac{\partial}{\partial z} \left(\frac{\partial N}{\partial x} - \frac{\partial M}{\partial y} \right) = 0. \quad (5)$$

Again the mixed derivatives give $P_{xy} = P_{yx}$ and $N_{xz} = N_{zx}$ and $M_{zy} = M_{yz}$. The terms cancel in pairs. In "curl grad" and "div curl", everything is arranged to give zero.

15N The curl of the gradient of every $f(x, y, z)$ is curl grad $f = \nabla \times \nabla f = \mathbf{0}$.
The divergence of the curl of every $\mathbf{F}(x, y, z)$ is div curl $\mathbf{F} = \nabla \cdot \nabla \times \mathbf{F} = 0$.

The spin field **S** has no divergence. The position field **R** has no curl. **R** is the gradient of $f = \frac{1}{2}(x^2 + y^2 + z^2)$. **S** is the curl of a suitable **F**. Then div **S** $=$ div curl **F** and curl **R** $=$ curl grad f are automatically zero.

You correctly believe that curl **F** measures the "spin" of the field. You may expect that curl $(\mathbf{F} + \mathbf{G})$ is curl $\mathbf{F} +$ curl **G**. Also correct. Finally you may think that a field of parallel vectors has no spin. That is wrong. Example 5 has parallel vectors, but their different lengths produce spin.

EXAMPLE 5   The field $\mathbf{V} = z\mathbf{i}$ in the x direction has curl $\mathbf{V} = \mathbf{j}$ in the y direction.

If you put a wheel in the xz plane, *this field will turn it*. The velocity $z\mathbf{i}$ at the top of the wheel is greater than $z\mathbf{i}$ at the bottom (Figure 15.23c). So the top goes faster and the wheel rotates. The axis of rotation is curl $\mathbf{V} = \mathbf{j}$. The turning speed is $\frac{1}{2}$, because this curl has magnitude 1.

Another velocity field $\mathbf{v} = -x\mathbf{k}$ produces the same spin: curl $\mathbf{v} = \mathbf{j}$. The flow is in the z direction, it varies in the x direction, and the spin is in the y direction. Also interesting is $\mathbf{V} + \mathbf{v}$. The two "shear fields" add to a perfect spin field $\mathbf{S} = z\mathbf{i} - x\mathbf{k}$, whose curl is $2\mathbf{j}$.

THE MEANING OF CURL F

Example 5 put a paddlewheel into the flow. This is possible for any vector field **F**, and it gives insight into curl **F**. The turning of the wheel (if it turns) depends on its location (x, y, z). The turning also depends on the *orientation* of the wheel. We could put it into a spin field, and if the wheel axis **n** is perpendicular to the spin axis **a**, the wheel won't turn! The general rule for turning speed is this: ***the angular velocity of the wheel is*** $\frac{1}{2}$(curl **F**) · **n**. This is the "*directional spin*," just as (grad f) · **u** was the "*directional derivative*"—and **n** is a unit vector like **u**.

There is no spin anywhere in a gradient field. It is ***irrotational***: curl grad $f = \mathbf{0}$.

The pure spin field $\mathbf{a} \times \mathbf{R}$ has curl $\mathbf{F} = 2\mathbf{a}$. The angular velocity is $\mathbf{a} \cdot \mathbf{n}$ (note that $\frac{1}{2}$ cancels 2). This turning is everywhere, ***not just at the origin***. If you put a penny on a compact disk, it turns once when the disk rotates once. That spin is "*around itself*," and it is the same whether the penny is at the center or not.

The turning speed is greatest when the wheel axis **n** lines up with the spin axis **a**. Then $\mathbf{a} \cdot \mathbf{n}$ is the full length $|\mathbf{a}|$. The gradient gives the direction of fastest growth, and the curl gives the direction of fastest turning:

maximum growth rate of f is $|\text{grad } f|$ in the direction of grad f

maximum rotation rate of **F** is $\frac{1}{2}|\text{curl } \mathbf{F}|$ in the direction of curl **F**.

STOKES' THEOREM

Finally we come to the big theorem. It will be like Green's Theorem—a line integral equals a surface integral. The line integral is still the work $\oint \mathbf{F} \cdot d\mathbf{R}$ around a curve. The surface integral in Green's Theorem is $\iint (N_x - M_y) dx \, dy$. The surface is flat (in the xy plane). Its normal direction is **k**, and we now recognize $N_x - M_y$ as the **k** component of the curl. Green's Theorem uses only this component because the normal direction is always **k**. For Stokes' Theorem on a curved surface, we need all three components of curl **F**.

Figure 15.24 shows a hat-shaped surface S and its boundary C (a closed curve). Walking in the positive direction around C, with your head pointing in the direction of **n**, the surface is *on your left*. You may be standing straight up ($\mathbf{n} = \mathbf{k}$ in Green's Theorem). You may even be upside down ($\mathbf{n} = -\mathbf{k}$ is allowed). In that case you must go the other way around C, to keep the two sides of equation (6) equal. The surface is still on the left. A Möbius strip is not allowed, because its normal direction cannot be established. The unit vector **n** determines the "counterclockwise direction" along C.

15O (*Stokes' Theorem*) $\qquad \oint_C \mathbf{F} \cdot d\mathbf{R} = \iint_S (\text{curl } \mathbf{F}) \cdot \mathbf{n} \, dS.$ (6)

The right side adds up small spins in the surface. The left side is the total circulation (or work) around C. That is not easy to visualize—this may be the hardest theorem in the book—but notice one simple conclusion. *If* curl $\mathbf{F} = 0$ *then* $\oint \mathbf{F} \cdot d\mathbf{R} = 0$. ***This applies above all to gradient fields***—as we know.

A gradient field has no curl, by (4). A gradient field does no work, by (6). In three dimensions as in two dimensions, ***gradient fields are conservative fields***. They will be the focus of this section, after we outline a proof (or two proofs) of Stokes' Theorem.

The first proof shows *why* the theorem is true. The second proof shows that it really is true (and how to compute). You may prefer the first.

First proof Figure 15.24 has a triangle ABC attached to a triangle ACD. Later there can be more triangles. S will be *piecewise flat*, close to a curved surface. Two triangles are enough to make the point. In the plane of each triangle (they have different **n**'s) Green's Theorem is known:

$$\oint_{AB+BC+CA} \mathbf{F}\cdot d\mathbf{R} = \iint_{ABC} \text{curl } \mathbf{F}\cdot\mathbf{n}dS \qquad \oint_{AC+CD+DA} \mathbf{F}\cdot d\mathbf{R} = \iint_{ACD} \text{curl } \mathbf{F}\cdot\mathbf{n}dS.$$

Now add. The right sides give $\iint \text{curl } \mathbf{F}\cdot\mathbf{n}dS$ over the two triangles. On the left, *the integral over CA cancels the integral over AC*. The "crosscut" disappears. That leaves $AB + BC + CD + DA$. This line integral goes around the outer boundary C—which is the left side of Stokes' Theorem.

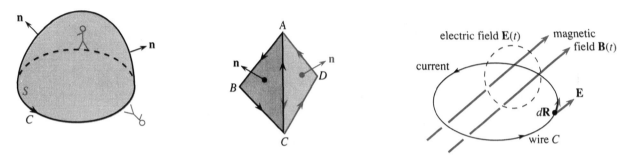

Fig. 15.24   Surfaces S and boundary curves C. Change in $\mathbf{B} \rightarrow$ curl $\mathbf{E} \rightarrow$ current in C.

Second proof Now the surface can be curved. A new proof may seem excessive, but it brings formulas you could compute with. From $z = f(x, y)$ we have

$$dz = \partial f/\partial x\, dx + \partial f/\partial y\, dy \quad \text{and} \quad \mathbf{n}dS = (-\partial f/\partial x\, \mathbf{i} - \partial f/\partial y\, \mathbf{j} + \mathbf{k})dx\, dy.$$

For $\mathbf{n}dS$, see equation 15.4.11. With this dz, the line integral in Stokes' Theorem is

$$\oint_C \mathbf{F}\cdot d\mathbf{R} = \oint_{\text{shadow of C}} M\, dx + N\, dy + P(\partial f/\partial x\, dx + \partial f/\partial y\, dy). \tag{7}$$

The dot product of curl $\mathbf{F}$ and $\mathbf{n}dS$ gives the surface integral $\iint_S \text{curl } \mathbf{F}\cdot\mathbf{n}dS$:

$$\iint_{\text{shadow of S}} [(P_y - N_z)(-\partial f/\partial x) + (M_z - P_x)(-\partial f/\partial y) + (N_x - M_y)]dx\, dy. \tag{8}$$

To prove $(7) = (8)$, change M in Green's Theorem to $M + P\partial f/\partial x$. Also change N to $N + P\partial f/\partial y$. Then $(7) = (8)$ is Green's Theorem down on the shadow (Problem 47). This proves Stokes' Theorem up on S. Notice how Green's Theorem (flat surface) was the key to both proofs of Stokes' Theorem (curved surface).

EXAMPLE 6   Stokes' Theorem in electricity and magnetism yields Faraday's Law.

Stokes' Theorem is not heavily used for calculations—equation (8) shows why. But the spin or curl or *vorticity* of a flow is absolutely basic in fluid mechanics. The other important application, coming now, is to electric fields. Faraday's Law is to Gauss's Law as Stokes' Theorem is to the Divergence Theorem.

Suppose the curve C is an actual wire. We can produce current along C by varying the magnetic field $\mathbf{B}(t)$. The flux $\varphi = \iint \mathbf{B} \cdot \mathbf{n} \, dS$, passing within C and changing in time, creates an electric field $\mathbf{E}$ that does work:

$$\textit{Faraday's Law} \text{ (integral form)} : \text{work} = \oint_C \mathbf{E} \cdot d\mathbf{R} = -d\varphi/dt.$$

That is physics. It may be true, it may be an approximation. Now comes mathematics (surely true), which turns this integral form into a differential equation. Information at points is more convenient than information around curves. Stokes converts the line integral of $\mathbf{E}$ into the surface integral of curl $\mathbf{E}$:

$$\oint_C \mathbf{E} \cdot d\mathbf{R} = \iint_S \text{curl } \mathbf{E} \cdot \mathbf{n} \, dS \text{ and also } -\partial \varphi/\partial t = \iint_S -(\partial \mathbf{B}/\partial t) \cdot \mathbf{n} \, dS.$$

These are equal for any curve C, however small. So the right sides are equal for any surface S. We squeeze to a point. The right hand sides give one of Maxwell's equations:

$$\textit{Faraday's Law} \text{(differential form): curl } \mathbf{E} = -\partial \mathbf{B}/\partial t.$$

CONSERVATIVE FIELDS AND POTENTIAL FUNCTIONS

The chapter ends with our constant and important question: Which fields do no work around closed curves? Remember test **D** for plane curves and plane vector fields:

$$\text{if } \partial M/\partial y = \partial N/\partial x \text{ then } \mathbf{F} \text{ is conservative and } \mathbf{F} = \text{ grad } f \text{ and } \oint \mathbf{F} \cdot d\mathbf{R} = 0.$$

Now allow a three-dimensional field like $\mathbf{F} = 2xy \, \mathbf{i} + (x^2 + z)\mathbf{j} + y\mathbf{k}$. Does it do work around a space curve? *Or is it a gradient field?* That will require $\partial f/\partial x = 2xy$ and $\partial f/\partial y = x^2 + z$ and $\partial f/\partial z = y$. *We have three equations for one function* $f(x, y, z)$. Normally they can't be solved. When test **D** is passed (now it is the three-dimensional test: curl $\mathbf{F} = \mathbf{0}$) they *can* be solved. This example passes test **D**, and f is $x^2 y + yz$.

15P $\mathbf{F}(x, y, z) = M\mathbf{i} + N\mathbf{j} + P\mathbf{k}$ is a conservative field if it has these properties:

  A. The work $\oint \mathbf{F} \cdot d\mathbf{R}$ around every closed path in space is zero.

  B. The work $\int_P^Q \mathbf{F} \cdot d\mathbf{R}$ depends on P and Q, not on the path in space.

  C. $\mathbf{F}$ is a *gradient field*: $M = \partial f/\partial x$ and $N = \partial f/\partial y$ and $P = \partial f/\partial z$.

  D. The components satisfy $M_y = N_x$, $M_z = P_x$, and $N_z = P_y$ (curl $\mathbf{F}$ *is zero*).

A field with one of these properties has them all. **D** is the quick test.

A detailed proof of $\mathbf{A} \Rightarrow \mathbf{B} \Rightarrow \mathbf{C} \Rightarrow \mathbf{D} \Rightarrow \mathbf{A}$ is not needed. Only notice how $\mathbf{C} \Rightarrow \mathbf{D}$: curl grad $\mathbf{F}$ is always zero. The newest part is $\mathbf{D} \Rightarrow \mathbf{A}$. *If* curl $\mathbf{F} = \mathbf{0}$ *then* $\oint \mathbf{F} \cdot d\mathbf{R} = 0$. But that is not news. It is Stokes' Theorem.

   The interesting problem is to solve the three equations for f, when test **D** is passed. The example above had

$$\partial f/\partial x = 2xy \Rightarrow f = \int 2xy \, dx = x^2 y \text{ plus any function } C(y, z)$$

$$\partial f/\partial y = x^2 + z = x^2 + \partial C/\partial y \Rightarrow C = yz \text{ plus any function } c(z)$$

$$\partial f/\partial z = y = y + dc/dz \Rightarrow c(z) \text{ can be zero.}$$

The first step leaves an arbitrary $C(y,z)$ to fix the second step. The second step leaves an arbitrary $c(z)$ to fix the third step (not needed here). Assembling the three steps, $f = x^2 y + C = x^2 y + yz + c = x^2 y + yz$. Please recognize that the "fix-up" is only possible when curl $\mathbf{F} = \mathbf{0}$. Test **D** must be passed.

EXAMPLE 7   Is $\mathbf{F} = (z-y)\mathbf{i} + (x-z)\mathbf{j} + (y-x)\mathbf{k}$ the gradient of any f?

Test **D** says *no*. This $\mathbf{F}$ is a spin field $\mathbf{a} \times \mathbf{R}$. Its curl is $2\mathbf{a} = (2,2,2)$, which is not zero. A search for f is bound to fail, but we can try. To match $\partial f / \partial x = z - y$, we must have $f = zx - yx + C(y,z)$. The y derivative is $-x + \partial C / \partial y$. That never matches $N = x - z$, so f can't exist.

EXAMPLE 8   What choice of P makes $\mathbf{F} = yz^2\mathbf{i} + xz^2\mathbf{j} + P\mathbf{k}$ conservative? Find f.

Solution   We need curl $\mathbf{F} = 0$, by test **D**. First check $\partial M / \partial y = z^2 = \partial N / \partial x$. Also

$$\partial P / \partial x = \partial M / \partial z = 2yz \quad \text{and} \quad \partial P / \partial y = \partial N / \partial z = 2xz.$$

$P = 2xyz$ passes all tests. To find f we can solve the three equations, or notice that $f = xyz^2$ is successful. Its gradient is $\mathbf{F}$.

A third method defines $f(x,y,z)$ as *the work to reach* (x,y,z) *from* $(0,0,0)$. The path doesn't matter. For practice we integrate $\mathbf{F} \cdot d\mathbf{R} = M\,dx + N\,dy + P\,dz$ along the straight line (xt, yt, zt):

$$f(x,y,z) = \int_0^1 (yt)(zt)^2(x\,dt) + (xt)(zt)^2(y\,dt) + 2(xt)(yt)(zt)(z\,dt) = xyz^2.$$

EXAMPLE 9   Why is div curl grad f automatically zero (in two ways)?

Solution   First, curl grad f is zero (always). Second, div curl $\mathbf{F}$ is zero (always). Those are the key identities of vector calculus. We end with a review.

$$\textit{Green's Theorem}: \quad \oint \mathbf{F} \cdot d\mathbf{R} = \iint (\partial N / \partial x - \partial M / \partial y)\,dx\,dy$$

$$\oint \mathbf{F} \cdot \mathbf{n}\,ds = \iint (\partial M / \partial x + \partial N / \partial y)\,dx\,dy$$

$$\textit{Divergence Theorem}: \quad \iint \mathbf{F} \cdot \mathbf{n}\,dS = \iiint (\partial M / \partial x + \partial N / \partial y + \partial P / \partial z)\,dx\,dy\,dz$$

$$\textit{Stokes' Theorem}: \quad \oint \mathbf{F} \cdot d\mathbf{R} = \iint \text{curl } \mathbf{F} \cdot \mathbf{n}\,dS.$$

The first form of Green's Theorem leads to Stokes' Theorem. The second form becomes the Divergence Theorem. You may ask, *why not go to three dimensions in the first place*? The last two theorems contain the first two (take $P = 0$ and a flat surface). We could have reduced this chapter to two theorems, not four. I admit that, but a fundamental principle is involved: "It is easier to generalize than to specialize."

For the same reason df/dx came before partial derivatives and the gradient.

Read-through questions

The curl of $M\mathbf{i}+N\mathbf{j}+P\mathbf{k}$ is the vector __a__. It equals the 3 by 3 determinant __b__. The curl of $x^2\mathbf{i}+z^2\mathbf{k}$ is __c__. For $\mathbf{S}=y\mathbf{i}-(x+z)\mathbf{j}+y\mathbf{k}$ the curl is __d__. This $\mathbf{S}$ is a __e__ field $\mathbf{a}\times\mathbf{R}=\frac{1}{2}(\text{curl }\mathbf{F})\times\mathbf{R}$, with axis vector $\mathbf{a}=$ __f__. For any gradient field $f_x\mathbf{i}+f_y\mathbf{j}+f_z\mathbf{k}$ the curl is __g__. That is the important identity curl grad $f=$ __h__. It is based on $f_{xy}=f_{yx}$ and __i__ and __j__. The twin identity is __k__.

The curl measures the __l__ of a vector field. A paddlewheel in the field with its axis along $\mathbf{n}$ has turning speed __m__. The spin is greatest when $\mathbf{n}$ is in the direction of __n__. Then the angular velocity is __o__.

Stokes' Theorem is __p__ = __q__. The curve C is the __r__ of the __s__ S. This is __t__ Theorem extended to __u__ dimensions. Both sides are zero when $\mathbf{F}$ is a gradient field because __v__.

The four properties of a conservative field are $\mathbf{A}=$ __w__, $\mathbf{B}=$ __x__, $\mathbf{C}=$ __y__, $\mathbf{D}=$ __z__. The field $y^2z^2\mathbf{i}+2xy^2z\mathbf{k}$ (passes)(fails) test $\mathbf{D}$. This field is the gradient of $f=$ __A__. The work $\int\mathbf{F}\cdot d\mathbf{R}$ from $(0,0,0)$ to $(1,1,1)$ is __B__ (on which path?). For every field $\mathbf{F}$, $\iint$ curl $\mathbf{F}\cdot\mathbf{n}ds$ is the same out through a pyramid and up through its base because __C__.

In Problems 1–6 find curl F.

1 $\mathbf{F}=z\mathbf{i}+x\mathbf{j}+y\mathbf{k}$

2 $\mathbf{F}=$ grad $(xe^y\sin z)$

3 $\mathbf{F}=(x+y+z)(\mathbf{i}+\mathbf{j}+\mathbf{k})$

4 $\mathbf{F}=(x+y)\mathbf{i}-(x+y)\mathbf{k}$

5 $\mathbf{F}=\rho^n(x\mathbf{i}+y\mathbf{j}+z\mathbf{k})$

6 $\mathbf{F}=(\mathbf{i}+\mathbf{j})\times\mathbf{R}$

7 Find a potential f for the field in Problem 3.

8 Find a potential f for the field in Problem 5.

9 When do the fields $x^m\mathbf{i}$ and $x^n\mathbf{j}$ have zero curl?

10 When does $(a_1x+a_2y+a_3z)\mathbf{k}$ have zero curl?

In 11–14, compute curl F and find $\oint_c \mathbf{F}\cdot d\mathbf{R}$ by Stokes' Theorem.

11 $\mathbf{F}=x^2\mathbf{i}+y^2\mathbf{k}, C=$ circle $x^2+z^2=1, y=0$.

12 $\mathbf{F}=\mathbf{i}\times\mathbf{R}, C=$ circle $x^2+z^2=1, y=0$.

13 $\mathbf{F}=(\mathbf{i}+\mathbf{j})\times\mathbf{R}, C=$ circle $y^2+z^2=1, x=0$.

14 $\mathbf{F}=(y\mathbf{i}-x\mathbf{j})\times(x\mathbf{i}+y\mathbf{j}), C=$ circle $x^2+y^2=1, z=0$.

15 (important) Suppose two surfaces S and T have the same boundary C, and the direction around C is the same.

(a) Prove $\iint_S$ curl $\mathbf{F}\cdot\mathbf{n}dS=\iint_T$ curl $\mathbf{F}\cdot\mathbf{n}dS$.

(b) Second proof: The difference between those integrals is $\iiint$ div(curl $\mathbf{F})dV$. By what Theorem? What region is V? Why is this integral zero?

16 In 15, suppose S is the top half of the earth ($\mathbf{n}$ goes out) and T is the bottom half ($\mathbf{n}$ comes in). What are C and V? Show by example that $\iint_S \mathbf{F}\cdot\mathbf{n}dS=\iint_T \mathbf{F}\cdot\mathbf{n}dS$ is *not* generally true.

17 Explain why $\iint$ curl $\mathbf{F}\cdot\mathbf{n}dS=0$ over the closed boundary of any solid V.

18 Suppose curl $\mathbf{F}=\mathbf{0}$ and div $\mathbf{F}=0$. (a) Why is $\mathbf{F}$ the gradient of a potential? (b) Why does the potential satisfy Laplace's equation $f_{xx}+f_{yy}+f_{zz}=0$?

In 19–22, find a potential f if it exists.

19 $\mathbf{F}=z\mathbf{i}+\mathbf{j}+x\mathbf{k}$.

20 $\mathbf{F}=2xyz\mathbf{i}+x^2z\mathbf{j}+x^2y\mathbf{k}$

21 $\mathbf{F}=e^{x-z}\mathbf{i}-e^{x-z}\mathbf{k}$

22 $\mathbf{F}=yz\mathbf{i}+xz\mathbf{j}+(xy+z^2)\mathbf{k}$

23 Find a field with curl $\mathbf{F}=(1,0,0)$.

24 Find all fields with curl $\mathbf{F}=(1,0,0)$.

25 $\mathbf{S}=\mathbf{a}\times\mathbf{R}$ is a spin field. Compute $\mathbf{F}=\mathbf{b}\times\mathbf{S}$ (constant vector $\mathbf{b}$) and find its curl.

26 How fast is a paddlewheel turned by the field $\mathbf{F}=y\mathbf{i}-x\mathbf{k}$ (a) if its axis direction is $\mathbf{n}=\mathbf{j}$? (b) if its axis is lined up with curl $\mathbf{F}$? (c) if its axis is perpendicular to curl $\mathbf{F}$?

27 How is curl $\mathbf{F}$ related to the angular velocity ω in the spin field $\mathbf{F}=\omega(-y\mathbf{i}+x\mathbf{j})$? How fast does a wheel spin, if it is in the plane $x+y+z=1$?

28 Find a vector field $\mathbf{F}$ whose curl is $\mathbf{S}=y\mathbf{i}-x\mathbf{j}$.

29 Find a vector field $\mathbf{F}$ whose curl is $\mathbf{S}=\mathbf{a}\times\mathbf{R}$.

30 *True or false*: when two vector fields have the same curl at all points: (a) their difference is a constant field (b) their difference is a gradient field (c) they have the same divergence.

In 31–34, compute $\iint$ curl $\mathbf{F}\cdot\mathbf{n}dS$ over the top half of the sphere $x^2+y^2+z^2=1$ and (separately) $\oint\mathbf{F}\cdot d\mathbf{R}$ around the equator.

31 $\mathbf{F}=y\mathbf{i}-x\mathbf{j}$

32 $\mathbf{F}=\mathbf{R}/\rho^2$

33 $\mathbf{F}=\mathbf{a}\times\mathbf{R}$

34 $\mathbf{F}=(\mathbf{a}\times\mathbf{R})\times\mathbf{R}$

35 The circle C in the plane $x+y+z=6$ has radius r and center at $(1,2,3)$. The field $\mathbf{F}$ is $3z\mathbf{j}+2y\mathbf{k}$. Compute $\oint\mathbf{F}\cdot d\mathbf{R}$ around C.

36 S is the top half of the unit sphere and $\mathbf{F}=z\mathbf{i}+x\mathbf{j}+xyz\mathbf{k}$. Find $\iint$ curl $\mathbf{F}\cdot\mathbf{n}dS$.

37 Find $g(x,y)$ so that curl $g\mathbf{k}=y\mathbf{i}+x^2\mathbf{j}$. What is the name for g in Section 15.3? It exists because $y\mathbf{i}+x^2\mathbf{j}$ has zero _____.

38 Construct $\mathbf{F}$ so that curl $\mathbf{F}=2x\mathbf{i}+3y\mathbf{j}-5z\mathbf{k}$ (which has zero divergence).

39 Split the field $\mathbf{F}=xy\mathbf{i}$ into $\mathbf{V}+\mathbf{W}$ with curl $\mathbf{V}=\mathbf{0}$ and div $\mathbf{W}=0$.

40 Ampère's law for a steady magnetic field $\mathbf{B}$ is curl $\mathbf{B} = \mu\mathbf{J}$ ($\mathbf{J}$ = current density, μ = constant). Find the work done by $\mathbf{B}$ around a space curve C from the current passing through it.

Maxwell allows varying currents which brings in the electric field.

41 For $\mathbf{F} = (x^2 + y^2)\mathbf{i}$, compute curl (curl $\mathbf{F}$) and grad (div $\mathbf{F}$) and $\mathbf{F}_{xx} + \mathbf{F}_{yy} + \mathbf{F}_{zz}$.

42 For $\mathbf{F} = v(x, y, z)\mathbf{i}$, prove these useful identities:

(a) curl(curl $\mathbf{F}$) = grad (div $\mathbf{F}$) − ($\mathbf{F}_{xx} + \mathbf{F}_{yy} + \mathbf{F}_{zz}$).

(b) curl($f\mathbf{F}$) = f curl $\mathbf{F}$ + (grad f) × $\mathbf{F}$.

43 If $\mathbf{B} = \mathbf{a}\cos t$ (constant direction $\mathbf{a}$), find curl $\mathbf{E}$ from Faraday's Law. Then find the alternating spin field $\mathbf{E}$.

44 With $\mathbf{G}(x, y, z) = m\mathbf{i} + n\mathbf{j} + p\mathbf{k}$, write out $\mathbf{F} \times \mathbf{G}$ and take its divergence. Match the answer with $\mathbf{G} \cdot$ curl $\mathbf{F} - \mathbf{F} \cdot$ curl $\mathbf{G}$.

45 Write down Green's Theorem in the xz plane from Stokes' Theorem.

46 *True or false*: $\nabla \times \mathbf{F}$ is perpendicular to $\mathbf{F}$.

47 (a) The second proof of Stokes' Theorem took $M^* = M(x, y, f(x, y)) + P(x, y, f(x, y))\partial f/\partial x$ as the M in Green's Theorem. Compute $\partial M^*/\partial y$ from the chain rule and product rule (there are five terms).

(b) Similarly $N^* = N(x, y, f) + P(x, y, f)\partial f/\partial y$ has the x derivative $N_x + N_z f_x + P_x f_y + P_z f_x f_y + P f_{yx}$. Check that $N_x^* - M_y^*$ matches the right side of equation (8), as needed in the proof.

48 "The shadow of the boundary is the boundary of the shadow." This fact was used in the second proof of Stokes' Theorem, going to Green's Theorem on the shadow. Give two examples of S and C and their shadows.

49 Which integrals are equal when C = boundary of S or S = boundary of V?

$$\oint \mathbf{F} \cdot d\mathbf{R} \qquad \oint (\text{curl } \mathbf{F}) \cdot d\mathbf{R} \qquad \oint (\text{curl } \mathbf{F}) \cdot \mathbf{n}\, ds \qquad \iint \mathbf{F} \cdot \mathbf{n}\, dS$$

$$\iint \text{div } \mathbf{F}\, dS \quad \iint (\text{curl } \mathbf{F}) \cdot \mathbf{n}\, dS \quad \iint (\text{grad div } \mathbf{F}) \cdot \mathbf{n}\, dS \quad \iiint \text{div } \mathbf{F}\, dV$$

50 Draw the field $\mathbf{V} = -x\mathbf{k}$ spinning a wheel in the xz plane. What wheels would *not* spin?

CHAPTER 16

Mathematics after Calculus

I would like this book to do more than help you pass calculus. (I hope it does that too.) After calculus you will have choices—**Which mathematics course to take next**?— and these pages aim to serve as a guide. Part of the answer depends on where you are going—toward engineering or management or teaching or science or another career where mathematics plays a part. The rest of the answer depends on where the courses are going. This chapter can be a useful reference, to give a clearer idea than course titles can do:

Linear Algebra   Differential Equations   Discrete Mathematics

Advanced Calculus (with Fourier Series)   Numerical Methods   Statistics

Pure mathematics is often divided into analysis and algebra and geometry. Those parts come together in the "mathematical way of thinking"—a mixture of logic and ideas. It is a deep and creative subject—here we make a start.

Two main courses after calculus are **linear algebra** and **differential equations**. I hope you can take both. To help you later, Sections 16.1 and 16.2 organize them by examples. First a few words to compare and contrast those two subjects.

Linear algebra is about *systems* of equations. There are n variables to solve for. A change in one affects the others. They can be prices or velocities or currents or concentrations—outputs from any model with interconnected parts.

Linear algebra makes only one assumption—*the model must be linear*. A change in one variable produces proportional changes in all variables. Practically every subject begins that way. (When it becomes nonlinear, we solve by a sequence of linear equations. Linear programming is nonlinear because we require $x \geqslant 0$.) Elsewhere I wrote that "Linear algebra has become as basic and as applicable as calculus, and fortunately it is easier." I recommend taking it.

A differential equation is *continuous* (from calculus), where a matrix equation is *discrete* (from algebra). The rate dy/dt is determined by the present state y—which changes by following that rule. Section 16.2 solves $y' = cy + s(t)$ for economics and life sciences, and $y'' + by' + cy = f(t)$ for physics and engineering. Please keep it and refer to it.

A third key direction is ***discrete mathematics***. Matrices are a part, networks and algorithms are a bigger part. Derivatives are not a part—this is closer to algebra. It is needed in computer science. Some people have a knack for counting the ways a computer can send ten messages in parallel—and for finding the fastest way.

Typical question: ***Can*** 25 ***states be matched with*** 25 ***neighbors, so one state in each pair has an even number of letters***? New York can pair with New Jersey, Texas with Oklahoma, California with Arizona. We need rules for Hawaii and Alaska. This matching question doesn't sound mathematical, but it is.

Section 16.3 selects four topics from discrete mathematics, so you can decide if you want more.

Go back for a moment to calculus and differential equations. A completely realistic problem is seldom easy, but we can solve models. (Developing a good model is a skill in itself.) One method of solution involves complex numbers:

$$\text{any function}\quad u(x+iy)\quad \text{solves}\quad u_{xx}+u_{yy}=0\quad \text{(Laplace equation)}$$

$$\text{any function}\quad e^{ik(x+ct)}\quad \text{solves}\quad u_{tt}-c^2u_{xx}=0\quad \text{(wave equation)}.$$

From those building blocks we assemble solutions. For the wave equation, a signal starts at $t=0$. It is a combination of pure oscillations e^{ikx}. The coefficients in that combination make up the ***Fourier transform***—to tell how much of each frequency is in the signal. A lot of engineers and scientists would rather know those Fourier coefficients than $f(x)$.

A Fourier series breaks the signal into $\Sigma\, a_k \cos kx$ or $\Sigma\, b_k \sin kx$ or $\Sigma\, c_k e^{ikx}$. These sums can be infinite (like power series). Instead of values of $f(x)$, or derivatives at the basepoint, the function is described by a_k, b_k, c_k. Everything is computed by the "***Fast Fourier Transform***." This is the greatest algorithm since Newton's method.

A radio signal is near one frequency. A step function has many frequencies. A delta function has every frequency in the same amount: $\delta(x)=\Sigma \cos kx$. Channel 4 can't broadcast a perfect step function. You wouldn't want to hear a delta function.

We mentioned ***computing***. For nonlinear equations this means Newton's method. For $Ax=b$ it means elimination—***algorithms take the place of formulas***. Exact solutions are gone—speed and accuracy and stability become essential. It seems right to make scientific computing a part of applied mathematics, and teach the algorithms with the theory. My text *Introduction to Applied Mathematics* is one step in this direction, trying to present advanced calculus as it is actually used.

We cannot discuss applications and forget ***statistics***. Our society produces oceans of data—somebody has to draw conclusions. To decide if a new drug works, and if oil spills are common or rare, and how often to have a checkup, we can't just guess. I am astounded that the connection between smoking and health was hidden for centuries. It was in the data! Eventually the statisticians uncovered it. Professionals can find patterns, and the rest of us can understand (with a little mathematics) what has been found.

One purpose in studying mathematics is to know more about your own life. Calculus lights up a key idea: ***Functions***. Shapes and populations and heart signals and profits and growth rates, all are given by functions. They change in time. They have integrals and derivatives. To understand and use them is a challenge—mathematics takes effort. A lot of people have contributed, in whatever way they could—as you and I are doing. We may not be Newton or Leibniz or Gauss or Einstein, but we can share some part of what they created.

16.1 Vector Spaces and Linear Algebra

You have met the idea of a *matrix*. An m by n matrix A has m rows and n columns (it is square if $m = n$). It multiplies a vector x that has n components. The result is a vector Ax with m components. The central problem of linear algebra is to go backward: ***From $Ax = b$ find*** x. That is possible when A is square and invertible. Otherwise there is no solution x—or there are infinitely many.

The crucial property of matrix multiplication is linearity. If $Ax = b$ and $AX = B$ then A times $x + X$ is $b + B$. Also A times $2x$ is $2b$. In general A times cx is cb. In particular A times 0 is 0 (one vector has n zeros, the other vector has m zeros). The whole subject develops from linearity. Derivatives and integrals obey linearity too.

Question 1   What are the solutions to $Ax = 0$*?* One solution is $x = 0$. There may be other solutions and they fill up the "***nullspace***":

$$\begin{bmatrix} 1 & 2 \\ 0 & 3 \end{bmatrix} \begin{bmatrix} x \\ y \end{bmatrix} = \begin{bmatrix} 0 \\ 0 \end{bmatrix} \text{ requires } \begin{matrix} x = 0 \\ y = 0 \end{matrix} \qquad \begin{bmatrix} 1 & 2 & 0 \\ 0 & 3 & 1 \end{bmatrix} \begin{bmatrix} x \\ y \\ z \end{bmatrix} = \begin{bmatrix} 0 \\ 0 \end{bmatrix} \text{ also allows } \begin{matrix} x = & 2 \\ y = & -1 \\ z = & 3 \end{matrix}$$

When there are more unknowns than equations—when A has more columns than rows—***the system*** $Ax = 0$ ***has many solutions***. They are not scattered randomly around! Another solution is $X = 4, Y = -2, Z = 6$. This lies on the same line as $(2, -1, 3)$ and $(0, 0, 0)$. Always the solutions to $Ax = 0$ form a "*space*" of vectors—which brings us to a central idea of linear algebra .

Note   These pages are not concentrating on the mechanics of multiplying or inverting matrices. Those are explained in all courses. My own teaching emphasizes that *Ax is a combination of the columns of A.* The solution $x = A^{-1}b$ is computed by elimination. Here we explain the deeper idea of a ***vector space***—and especially the particular spaces that control $Ax = b$. I cannot go into the same detail as in my book on *Linear Algebra and Its Applications*, where examples and exercises develop the new ideas. Still these pages can be a useful support.

All vectors with n components lie in *n-dimensional space*. You can add them and subtract them and multiply them by any c. (Don't multiply two vectors and never write $1/x$ or $1/A$). The results $x + X$ and $x - X$ and cx are still vectors in the space. Here is the important point:

The line of solutions to $Ax = 0$ *is a* "*subspace*"—a vector space in its own right. The sum $x + X$ has components $6, -3, 9$—which is another solution. The difference $x - X$ is a solution, and so is $4x$. These operations leave us in the subspace.

The nullspace consists of all solutions to $Ax = 0$. It may contain only the zero vector (as in the first example). It may contain a line of vectors (as in the second example). It may contain a whole plane of vectors (Problem 5). In every case $x + X$ and $x - X$ and cx are also in the nullspace. We are assigning a new word to an old idea—the equation $x - 2y = 0$ has always been represented by a line (its nullspace). Now we have 6-dimensional subspaces of an 8-dimensional vector space.

Notice that $x^2 - y = 0$ does not produce a subspace (a parabola instead). Even the x and y axes together, from $xy = 0$, do not form a subspace. We go off the axes when we add $(1, 0)$ to $(0, 1)$. You might expect the straight line $x - 2y = 1$ to be a subspace, but again it is not so. When x and y are doubled, we have $X - 2Y = 2$. Then (X, Y) is on a different line. Only $Ax = 0$ is guaranteed to produce a subspace.

Figure 16.1 shows the nullspace and "row space." **Check dot products** (both zero).

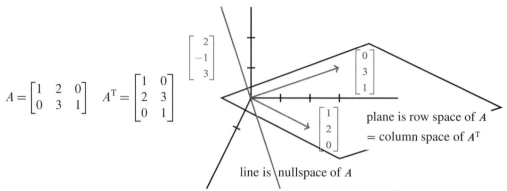

Fig. 16.1   The nullspace is perpendicular to the rows of A (the columns of A^T).

Question 2   When A multiplies a vector x, **what subspace does Ax lie in**? The product Ax is a combination of the columns of A—hence the name "**column space**":

$$\begin{bmatrix} 2 & 1 \\ 0 & 1 \\ 0 & 0 \end{bmatrix} \begin{bmatrix} 3 \\ 2 \end{bmatrix} = \begin{matrix} 3(\text{column 1}) \\ + \\ 2(\text{column 2}) \end{matrix} = \begin{bmatrix} 8 \\ 2 \\ 0 \end{bmatrix} \quad \text{This is in the column space;} \quad \begin{bmatrix} 0 \\ 0 \\ 1 \end{bmatrix} \text{ is not.}$$

No choice of x can produce $Ax = (0,0,1)$. For this A, all combinations of the columns end in a _____ . The column space is like the xy plane within xyz space. It is a subspace of m-dimensional space, containing every vector b that is a combination of the columns:

> *The system $Ax = b$ has a solution exactly when b is in the column space.*

When A has an inverse, the column space is the whole n-dimensional space. The nullspace contains only $x = 0$. There is exactly one solution $x = A^{-1}b$. This is the good case—and we outline four more key topics in linear algebra.

1. *Basis and dimension of a subspace*. A one-dimensional subspace is a line. A plane has dimension two. The nullspace above contained all multiples of $(2, -1, 3)$—by knowing that "basis vector" we know the whole line. The column space was a plane containing column 1 and column 2. Again those vectors are a "***basis***"—by knowing the columns we know the whole column space.

Our 2 by 3 matrix has three columns: $(1,0)$ and $(2,3)$ and $(0,1)$. Those are *not* a basis for the column space! This space is only a plane, and three vectors are too many. ***The dimension is two***. By combining $(1,0)$ and $(0,1)$ we can produce the other vector $(2,3)$. There are only two ***independent*** columns, and they form a basis for this column space.

In general: When a subspace contains r independent vectors, and no more, those vectors are a basis and the dimension is r. "Independent" means that no vector is a combination of the others. In the example, $(1,0)$ and $(2,3)$ are also a basis. A subspace has many bases, just as a plane has many axes.

2. *Least squares*. If $Ax = b$ has no solution, we look for the x that comes closest. Section 11.4 found the straight line nearest to a set of points. We make the length of $Ax - b$ as small as possible, when zero length is not possible. No vector solves

$Ax = b$, when b is not in the column space. So b is projected onto that space. This leads to the "normal equations" that produce the best x:

$$A^T Ax = A^T b. \tag{1}$$

When a rectangular matrix appears in applications, its **transpose** generally comes too. The columns of A are the rows of A^T. The rows of A are the columns of A^T. Then $A^T A$ is square and symmetric—equal to its transpose and vital for applied mathematics.

3. **Eigenvalues (for square matrices only)**. Normally Ax points in a direction different from x. For certain special **eigenvectors**, Ax is parallel to x. Here is a 2 by 2 matrix with two eigenvectors—in one case $Ax = 5x$ and in the other $Ax = 2x$;

$$Ax = \lambda x : \begin{bmatrix} 3 & 2 \\ 1 & 4 \end{bmatrix} \begin{bmatrix} 1 \\ 1 \end{bmatrix} = \begin{bmatrix} 5 \\ 5 \end{bmatrix} = 5 \begin{bmatrix} 1 \\ 1 \end{bmatrix} \text{ and } \begin{bmatrix} 3 & 2 \\ 1 & 4 \end{bmatrix} \begin{bmatrix} 2 \\ -1 \end{bmatrix} = \begin{bmatrix} 4 \\ -2 \end{bmatrix} = 2 \begin{bmatrix} 2 \\ -1 \end{bmatrix}.$$

The multipliers 5 and 2 are the **eigenvalues** of A. An 8 by 8 matrix has eight eigenvalues, which tell what the matrix is doing (to the eigenvectors). The eigenvectors are uncoupled, and they go their own way. A system of equations $dy/dt = Ay$ acts like one equation—*when y is an eigenvector*:

$$\begin{aligned} dy_1/dt &= 3y_1 + 2y_2 \\ dy_2/dt &= y_1 + 4y_2 \end{aligned} \quad \text{has the solution} \quad \begin{aligned} y_1 &= e^{5t} \\ y_2 &= e^{5t} \end{aligned} \quad \text{which is} \quad e^{5t} \begin{bmatrix} 1 \\ 1 \end{bmatrix}.$$

The eigenvector is $(1,1)$. The eigenvalue $\lambda = 5$ is in the exponent. When you substitute y_1 and y_2 the differential equations become $5e^{5t} = 5e^{5t}$. The fundamental principle for $dy/dt = cy$ still works for the system $dy/dt = Ay$: **Look for pure exponential solutions**. The eigenvalue "lambda" is the growth rate in the exponent.

I have to add: Find the eigenvectors also. The second eigenvector $(2, -1)$ has eigenvalue $\lambda = 2$. A second solution is $y_1 = 2e^{2t}, y_2 = -e^{2t}$. Substitute those into the equation—they are even better at displaying the general rule:

If $Ax = \lambda x$ then $d/dt(e^{\lambda t}x) = A(e^{\lambda t}x)$. **The pure exponentials are $y = e^{\lambda t}x$.**

The four entries of A pull together for the eigenvector. So do the 64 entries of an 8 by 8 matrix—again $e^{\lambda t}x$ solves the equation. Growth or decay is decided by $\lambda > 0$ or $\lambda < 0$. When $\lambda = k + i\omega$ is a complex number, growth and oscillation combine in $e^{\lambda t} = e^{kt}e^{i\omega t} = e^{kt}(\cos \omega t + i \sin \omega t)$.

Subspaces govern static problems $Ax = b$. Eigenvalues and eigenvectors govern dynamic problems $dy/dt = Ay$. Look for exponentials $y = e^{\lambda t}x$.

4. **Determinants and inverse matrices** . A 2 by 2 matrix has determinant $D = ad - bc$. This matrix has no inverse if $D = 0$. Reason: A^{-1} divides by D:

$$A^{-1} = \frac{1}{D} \begin{bmatrix} d & -b \\ -c & a \end{bmatrix} \quad \text{times} \quad A = \begin{bmatrix} a & b \\ c & d \end{bmatrix} \quad \text{equals} \quad I = \begin{bmatrix} 1 & 0 \\ 0 & 1 \end{bmatrix}.$$

This pattern extends to n by n matrices, but D and A^{-1} become more complicated.

For 3 by 3 matrices D has six terms. Section 11.5 identified D as a triple product $a \cdot (b \times c)$ of the columns. Three events come together in the singular case: D **is zero and** A **has no inverse and the columns lie in a plane**. The opposite events produce the "nonsingular" case: D is nonzero and A^{-1} exists. Then $Ax = b$ is solved by $x = A^{-1}b$.

D is also the product of the pivots and the product of the eigenvalues. The *pivots* arise in elimination—the practical way to solve $Ax = b$ without A^{-1}. To find eigenvalues we turn $Ax = \lambda x$ into $(A - \lambda I)x = 0$. By a nice twist of fate, ***this matrix*** $A - \lambda I$ ***has*** $D = 0$. Go back to the example:

$$\begin{bmatrix} 3 & 2 \\ 1 & 4 \end{bmatrix} - \lambda \begin{bmatrix} 1 & 0 \\ 0 & 1 \end{bmatrix} = \begin{bmatrix} 3-\lambda & 2 \\ 1 & 4-\lambda \end{bmatrix} \quad \text{has} \quad D = (3-\lambda)(4-\lambda) - 2 = \lambda^2 - 7\lambda + 10.$$

The equation $\lambda^2 - 7\lambda + 10 = 0$ gives $\lambda = 5$ and $\lambda = 2$. The eigenvalues come first, to make $D = 0$. Then $(A - 5I)x = 0$ and $(A - 2I)x = 0$ yield the eigenvectors. These x's go into $y = e^{\lambda t}x$ to solve differential equations—which come next.

16.1 EXERCISES

Read-through questions

If $Ax = b$ and $AX = B$, then A times $2x + 3X$ equals __a__. If $Ax = 0$ and $AX = 0$ then A times $2x + 3X$ equals __b__. In this care x and X are in the __c__ of A, and so is the combination __d__. The nullspace contains all solutions to __e__. It is a subspace, which means __f__. If $x = (1, 1, 1)$ is in the nullspace then the columns add to __g__, so they are (independent)(dependent).

Another subspace is the __h__ space of A, containing all combinations of the columns. The system $Ax = b$ can be solved when b is __i__. Otherwise the best solution comes from $A^T Ax =$ __j__. Here A^T is the __k__ matrix, whose rows are __l__. The nullspace of A^T contains all solutions to __m__. The __n__ space of A^T (row space of A) is the fourth fundamental subspace. Each subspace has a basis containing as many __o__ vectors as possible. The number of vectors in the basis is the __p__ of the subspace.

When $Ax = \lambda x$, the number λ is an __q__ and x is an __r__. The equation $dy/dt = Ay$ has the exponential solution $y =$ __s__. A 7 by 7 matrix has __t__ eigenvalues, whose product is the __u__ D. If D is nonzero the matrix A has an __v__. Then $Ax = b$ is solved by $x =$ __w__. The formula for D contains $7! = 5040$ terms, so x is better computed by __x__. On the other hand $Ax = \lambda x$ means that $A - \lambda I$ has determinant __y__. The eigenvalue is computed before the __z__.

Find the nullspace in 1–6. Along with x go all cx.

1 $A = \begin{bmatrix} 1 & 2 \\ 2 & 4 \end{bmatrix}$ (solve $Ax = 0$)

2 $B = \begin{bmatrix} 12 & -6 \\ -6 & 3 \end{bmatrix}$

3 $C = \begin{bmatrix} 1 & 0 \\ 0 & 1 \\ 1 & 2 \end{bmatrix}$ (solve $Cx = 0$)

4 $C^T = \begin{bmatrix} 1 & 0 & 1 \\ 0 & 1 & 2 \end{bmatrix}$

5 $E = \begin{bmatrix} 1 & 1 & 1 \\ 1 & 1 & 1 \\ 1 & 1 & 1 \end{bmatrix}$

6 $F = \begin{bmatrix} 2 & 1 \\ 1 & 2 \end{bmatrix}$

7 Change Problem 1 to $Ax = \begin{bmatrix} 3 \\ 6 \end{bmatrix}$ (a) Find any particular solution x_P. (b) Add any x_0 from the nullspace and show that $x_P + x_0$ is also a solution.

8 Change Problem 1 to $Ax = \begin{bmatrix} 1 \\ 0 \end{bmatrix}$ and find all solutions. Graph the lines $x_1 + 2x_2 = 1$ and $2x_1 + 4x_2 = 0$ in a plane.

9 Suppose $Ax_P = b$ and $Ax_0 = 0$. Then by linearity $A(x_P + x_0) =$ _____. Conclusion: The sum of a particular solution x_P and any nullvector x_0 is _____.

10 Suppose $Ax = b$ and $Ax_P = b$. Then by linearity $A(x - x_P) =$ _____. The difference between solutions is a vector in _____. Conclusion: Every solution has the form $x = x_P + x_0$, *one particular solution plus a vector in the nullspace.*

11 Find three vectors b in the column space of E. Find all vectors b for which $Ex = b$ can be solved.

12 If $Ax = 0$ then the rows of A are perpendicular to x. Draw the row space and nullspace (lines in a plane) for A above.

13 Compute CC^T and $C^T C$. Why not C^2?

14 Show that $Cx = b$ has no solution, if $b = (-1, 1, 1)$. Find the best solution form $C^T Cx = C^T b$.

15 C^T has three columns. How many are independent? Which ones?

16 Find two independent vectors that are in the column space of C but are not columns of C.

17 For which of the matrices $ABCEF$ are the columns a basis for the column space ?

18 Explain the reasoning: If the columns of a matrix A are independent, the only solution to $Ax = 0$ is $x = 0$.

19 Which of the matrices $ABCEF$ have nonzero determinants ?

20 Find a basis for the full three-dimensional space using only vectors with positive components.

21 Find the matrix F^{-1} for which $FF^{-1} = I = \begin{bmatrix} 1 & 0 \\ 0 & 1 \end{bmatrix}$.

22 Verify that (determinant of F)^2 = (determinant of F^2).

23 (Important) Write down $F - \lambda I$ and compute its determinant. Find the two numbers λ that make this determinant zero. For those two numbers find eigenvectors x such that $Fx = \lambda x$.

24 Compute $G = F^2$. Find the determinant of $G - \lambda I$ and the two λ's that make it zero. For those λ's find eigenvectors x such that $Gx = \lambda x$. Conclusion: if $Fx = \lambda x$ then $F^2 x = \lambda^2 x$.

25 From Problem 23 find two exponential solutions to the equation $dy/dt = Fy$. Then find a combination of those solutions that starts from $y_0 = (1, 0)$ at $t = 0$.

26 From Problem 24 find two solutions to $dy/dt = Gy$. Then find the solution that starts from $y_0 = (2, 1)$.

27 Compute the determinant of $E - \lambda I$. Find all λ's that make this determinant zero. Which eigenvalue is repeated ?

28 Which previous problem found eigenvectors for $Ex = 0x$? Find an eigenvector for $Ex = 3x$.

29 Find the eigenvalues and eigenvectors of A.

30 Explain the reasoning: A matrix has a zero eigenvalue if and only if its determinant is zero.

31 Find the matrix H whose eigenvalues are 0 and 4 with eigenvectors $(1, 1)$ and $(1, -1)$.

32 If $Fx = \lambda x$ then multiplying both sides by _____ gives $F^{-1}x = \lambda^{-1}x$. If F has eigenvalues 1 and 3 then F^{-1} has eigenvalues _____ . The determinants of F and F^{-1} are _____ .

33 *True or false*, with a reason or an example.

  (a) The solutions to $Ax = b$ form a subspace.

  (b) $\begin{bmatrix} 0 & 2 \\ 0 & 0 \end{bmatrix}$ has $\begin{bmatrix} 1 \\ 0 \end{bmatrix}$ in its null space and column space.

  (c) $A^T A$ has the same entry in its upper right and lower left corners.

  (d) If $Ax = \lambda x$ then $y = e^{\lambda t}$ solves $dy/dt = Ay$.

  (e) If the columns of A are not independent, their combinations still form a subspace.

16.2 Differential Equations

We just solved differential equations by linear algebra. Those were special systems $dy/dt = Ay$, linear with constant coefficients. The solutions were exponentials, involving $e^{\lambda t}$. The eigenvalues of A were the "growth factors" λ. This section solves other equations—by no means all. We concentrate on a few that have important applications.

Return for a moment to the beginning—when direct integration was king:

$$\textbf{1.}\ \ dy/dt = s(t) \quad \textbf{2.}\ \ dy/dt = cy \quad \textbf{3.}\ \ dy/dt = u(y)c(t).$$

In **1**, $y(t)$ is the integral of $s(t)$. In **2**, $y(t)$ is the integral of $cy(t)$. That sounds circular—it only made sense after the discovery of $y = e^{ct}$. This exponential has the correct derivative cy. To find it by integration instead of inventing it, separate y from t:

Separation and integration also solve 3: $\int dy/u(y) = \int c(t)dt$. The model logistic equation has $u = y - y^2 = quadratic$. Equation **2** has $u = y = linear$. Equation **1** is also a special case with $u = 1 = constant$. But **2** and **1** are very different, for the following reason.

The compound interest equation $y' = cy$ is growing from *inside*. The equation $y' = s(t)$ is growing from *outside*. Where c is a "growth rate," s is a "source." They don't have the same meaning, and they don't have the same units. The combination $y' = cy + s$ was solved in Chapter 6, provided c and s are constant—but applications force us to go further.

In three examples we introduce non-constant source terms.

EXAMPLE 1　Solve $dy/dt = cy + s$ with the new source term $s = e^{kt}$.

Method　Substitute $y = Be^{kt}$, with an "undetermined coefficient" B to make it right:

$$kBe^{kt} = cBe^{kt} + e^{kt} \quad \text{yields} \quad B = 1/(k-c).$$

The source e^{kt} is the ***driving term***. The solution Be^{kt} is the ***response***. The exponent is the same! The key idea is to expect e^{kt} in the response.

Initial condition　To match y_0 at $t = 0$, the solution needs another exponential. It is the ***free response*** Ae^{ct}, which satisfies $dy/dt = cy$ with *no source*. To make $y = Ae^{ct} + Be^{kt}$ agree with y_0, choose $A = y_0 - B$:

Final solution　$y = (y_0 - B)e^{ct} + Be^{kt} = y_0e^{ct} + (e^{kt} - e^{ct})/(k-c).$ 　　　(1)

Exceptional case　$B = 1/(k-c)$ grows larger as k approaches c. When $k = c$ the method breaks down—the response Be^{kt} is no longer correct. The solution (1) approaches $0/0$, and in the limit we get a derivative. It has an extra factor t:

$$\frac{e^{kt} - e^{ct}}{k - c} = \frac{\text{change in } e^{ct}}{\text{change in } c} \rightarrow \frac{d}{dc}(e^{ct}) = te^{ct}. \qquad (2)$$

The correct response is te^{ct} when $k = c$. This is the form to substitute, when the driving rate k equals the natural rate c (called *resonance*).

Add the free response y_0e^{ct} to match the initial condition.

EXAMPLE 2   Solve $dy/dt = cy + s$ with the new source term $s = \cos kt$.

Substitute $y = B \sin kt + D \cos kt$. This has *two* undetermined coefficients B and D:

$$kB \cos kt - kD \sin kt = c(B \sin kt + D \cos kt) + \cos kt. \tag{3}$$

Matching cosines gives $kB = cD + 1$. The sines give $-kD = cB$. Algebra gives B, D, y:

$$B = \frac{c}{k^2 + c^2} \qquad D = \frac{k}{k^2 + c^2} \qquad y = \frac{c \sin kt + k \cos kt}{k^2 + c^2}. \tag{4}$$

Question Why do we need both $B \sin kt$ and $D \cos kt$ in the response to $\cos kt$?

First Answer   Equation (3) is impossible if we leave out B or D.

Second Answer $\cos kt$ is $\frac{1}{2}e^{ikt} + \frac{1}{2}e^{-ikt}$. So e^{ikt} and e^{-ikt} are both in the response.

EXAMPLE 3   Solve $dy/dt = cy + s$ with the new source term $s = te^{kt}$.

Method Look for $y = Be^{kt} + Dte^{kt}$. Problem 13 determines B and D. Add Ae^{ct} as needed, to match the initial value y_0.

SECOND-ORDER EQUATIONS

The equation $dy/dt = cy$ is *first-order*. The equation $d^2y/dt^2 = -cy$ is *second-order*. The first is typical of problems in life sciences and economics—the rate dy/dt depends on the present situation y. The second is typical of engineering and physical sciences—the acceleration d^2y/dt^2 enters the equation.

If you put money in a bank, it starts growing immediately. If you turn the wheels of a car, it changes direction *gradually*. The path is a curve, not a sharp corner. Newton's law is $F = ma$, not $F = mv$.

A mathematician compares a straight line to a parabola. The straight line crosses the x axis no more than once. The parabola can cross twice. The equation $ax^2 + bx + c = 0$ has *two solutions*, provided we allow them to be complex or equal. These are exactly the possibilities we face below: *two real solutions, two complex solutions, or one solution that counts twice*. The quadratic could be $x^2 - 1$ or $x^2 + 1$ or x^2. The roots are 1 and $-1, i$ and $-i, 0$ and 0.

In solving differential equations the roots appear in the exponent, and are called λ.

     EXAMPLE 4 $y'' = +y$:   solutions $y = e^t$ and $y = e^{-t}$ $\lambda = 1, -1$

     EXAMPLE 5 $y'' = 0y$:   solutions $y = 1$ and $y = t$ $\lambda = 0, 0$

     EXAMPLE 6 $y'' = -y$:   solutions $y = \cos t$ and $y = \sin t$ $\lambda = i, -i$

Where are the complex solutions? They are hidden in Example 6, which could be written $y = e^{it}$ and $y = e^{-it}$. These satisfy $y'' = -y$ since $i^2 = -1$. The use of sines and cosines avoids the imaginary number i, but it breaks the pattern of $e^{\lambda t}$.

Example 5 also seems to break the pattern—again $e^{\lambda t}$ is hidden. The solution $y = 1$ is e^{0t}. The other solution $y = t$ is te^{0t}. *The zero exponent is repeated*—another exceptional case that needs an extra factor t.

Exponentials solve every equation with constant coefficients and zero right hand side:

       To solve $ay'' + by' + cy = 0$ *substitute* $y = e^{\lambda t}$ *and find* λ.

This method has three steps, leading to the right exponents $\lambda = r$ and $\lambda = s$:

1. With $y = e^{\lambda t}$ the equation is $a\lambda^2 e^{\lambda t} + b\lambda e^{\lambda t} + ce^{\lambda t} = 0$. Cancel $e^{\lambda t}$.
2. Solve $a\lambda^2 + b\lambda + c = 0$. Factor or use the formula $\lambda = (-b \pm \sqrt{b^2 - 4ac})/2a$.
3. Call those roots $\lambda = r$ and $\lambda = s$. The complete solution is $y = Ae^{rt} + Be^{st}$.

The pure exponentials are $y = e^{rt}$ ***and*** $y = e^{st}$. Depending on r and s, they grow or decay or oscillate. They are combined with constants A and B to match the *two* conditions at $t = 0$. The initial state y_0 equals $A + B$. *The initial velocity* y_0' *equals* $rA + sB$ (the derivative at $t = 0$).

EXAMPLE 7　Solve $y'' - 3y' + 2y = 0$ with $y_0 = 5$ and $y_0' = 4$.

Step 1　substitutes $y = e^{\lambda t}$. The equation becomes $\lambda^2 e^{\lambda t} - 3\lambda e^{\lambda t} + 2e^{\lambda t} = 0$. Cancel $e^{\lambda t}$.

Step 2　solves $\lambda^2 - 3\lambda + 2 = 0$. Factor into $(\lambda - 1)(\lambda - 2) = 0$. The exponents r, s are $1, 2$.

Step 3　produces $y = Ae^t + Be^{2t}$. The initial conditions give $A + B = 5$ and $1A + 2B = 4$.

The constants are $A = 6$ and $B = -1$. ***The solution*** is $y = 6e^t - e^{2t}$.

This solution grows because there is a positive λ. The equation is "unstable." It becomes stable when the middle term $-3y'$ is changed to $+3y'$. ***When the damping is positive the solution decays. The*** λ***'s are negative***:

EXAMPLE 8　$(\lambda^2 + 3\lambda + 2)$ factors into $(\lambda + 1)(\lambda + 2)$. The exponents are -1 and -2. The solution is $y = Ae^{-t} + Be^{-2t}$. It decays to zero for any initial condition.

EXAMPLES 9–10　Solve $y'' + 2y' + 2y = 0$ and $y'' + 2y' + y = 0$. How do they differ?

Key difference　$\lambda^2 + 2\lambda + 2$ has ***complex roots***, $\lambda^2 + 2\lambda + 1$ has a ***repeated root***:

$$\lambda^2 + 2\lambda + 2 = 0 \quad \text{gives} \quad \lambda = -1 \pm i \quad (\lambda + 1)^2 = 0 \quad \text{gives} \quad \lambda = -1, -1.$$

The -1 in all these λ's means decay. The i means oscillation. The first exponential is $e^{(-1+i)t}$, which splits into e^{-t} (*decay*) times e^{it} (*oscillation*). Even better, change e^{it} and e^{-it} into cosines and sines:

$$y = Ae^{(-1+i)t} + Be^{(-1-i)t} = e^{-t}(a\cos t + b\sin t). \tag{5}$$

At $t = 0$ this produces $y_0 = a$. Then matching y_0' leads to b.

　　Example 10 has $r = s = -1$ (repeated root). One solution is e^{-t} as usual. The second solution cannot be another e^{-t}. Problem 21 shows that it is te^{-t}—again the exceptional case multiplies by t! The general solution is $y = Ae^{-t} + Bte^{-t}$.

　　Without the damping term $2y'$, these examples are $y'' + 2y = 0$ or $y'' + y = 0$— pure oscillation. A small amount of damping mixes oscillation and decay. Large damping gives pure decay. The borderline is when λ is repeated ($r = s$). That occurs when $b^2 - 4ac$ in the square root is zero. ***The borderline between two real roots and two complex roots is two repeated roots***.

　　The method of solution comes down to one idea: *Substitute* $y = e^{\lambda t}$. The equations apply to mechanical vibrations and electrical circuits (also other things, but those two are of prime importance). While describing these applications I will collect the information that comes from λ.

SPRINGS AND CIRCUITS: MECHANICAL AND ELECTRICAL ENGINEERING

A mass is hanging from a spring. We pull it down an extra distance y_0 and give it a starting velocity y_0'. The mass moves up or down, obeying Newton's law: *mass times acceleration equals spring force plus damping force*:

$$my'' = -ky - dy' \quad \text{or} \quad my'' + dy' + ky = 0. \tag{6}$$

This is *free oscillation*. The spring force $-ky$ is proportional to the stretching y (Hooke's law). The damping acts like a shock absorber or air resistance—it takes out energy. Whether the system goes directly toward zero or swings back and forth is decided by the three numbers m, d, k. They were previously called a, b, c.

16A The solutions $e^{\lambda t}$ to $my'' + dy' + ky = 0$ are controlled by the roots of $m\lambda^2 + d\lambda + k = 0$. With $d > 0$ there is damping and decay. From $\sqrt{d^2 - 4mk}$ there may be oscillation:

    overdamping: $d^2 > 4mk$ gives real roots and pure decay (Example 8)

    underdamping: $d^2 < 4mk$ gives complex roots and oscillation (Example 9)

    critical damping: $d^2 = 4mk$ gives a real repeated root $-d/2m$ (Example 10)

We are using letters when the examples had numbers, but the results are the same:

$$m\lambda^2 + d\lambda + k = 0 \quad \text{has roots} \quad r, s = -\frac{d}{2m} \pm \frac{1}{2m}\sqrt{d^2 - 4mk}.$$

Overdamping has no imaginary parts or oscillations: $y = Ae^{rt} + Be^{st}$. Critical damping has $r = s$ and an exceptional solution with an extra $t : y = Ae^{rt} + Bte^{rt}$. (*This is only a solution when $r = s$.*) Underdamping has decay from $-d/2m$ and oscillation from the imaginary part. An undamped spring ($d = 0$) has pure oscillation at the natural frequency $\omega_0 = \sqrt{k/m}$.

All these possibilities are in Figure 16.2, created by Alar Toomre. At the top is pure oscillation ($d = 0$ and $y = \cos 2t$). The equation is $y'' + dy' + 4y = 0$ and d starts to grow. When d reaches 4, the quadratic is $\lambda^2 + 4\lambda + 4$ or $(\lambda + 2)^2$. The repeated root yields e^{-2t} and te^{-2t}. After that the oscillation is gone. There is a smooth transition from one case to the next—as complex roots join in the repeated root and split into real roots.

At the bottom right, the final value $y(2\pi)$ increases with large damping. This was a surprise. At $d = 5$ the roots are -1 and -4. At $d = 8.5$ the roots are $-\frac{1}{2}$ and -8. The small root gives slow decay (like molasses). *As $d \to \infty$ the solution approaches $y = 1$.*

If we are serious about using mathematics, we should take advantage of anything that helps. For second-order equations, the formulas look clumsy but the examples are quite neat. The idea of $e^{\lambda t}$ is absolutely basic. The good thing is that electrical circuits satisfy the same equation. There is a beautiful analogy between springs and circuits:

$$\text{mass } m \leftrightarrow \text{inductance } L$$

$$\text{damping constant } d \leftrightarrow \text{resistance } R$$

$$\text{elastic constant } k \leftrightarrow 1/(\text{capacitance } C)$$

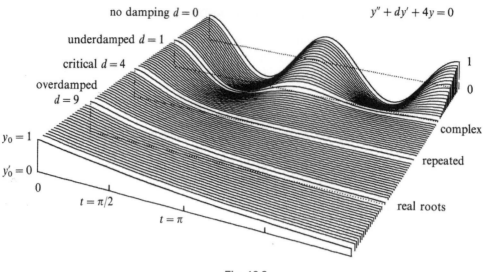

Fig. 16.2

The resistor takes out energy as the shock absorber did—converting into heat by friction. Without resistance we have pure oscillation. Electric charge is stored in the capacitor (like potential energy). It is released as current (like kinetic energy). It is stored up again (like a stretched spring). This continues at a frequency $\omega_0 = 1/\sqrt{LC}$ (like the spring's natural frequency $\sqrt{k/m}$). These analogies turn mechanical engineers into electrical engineers and vice versa.

The equation for the current $y(t)$ now includes a driving term on the right:

$$L\frac{dy}{dt} + R\,y + \frac{1}{C}\int y\,dt = \text{applied voltage} = V \sin \omega t. \tag{7}$$

To match networks with springs, differentiate both sides of (7):

$$Ly'' + Ry' + y/C = V\omega \cos \omega t. \tag{8}$$

The oscillations are *free* when $V = 0$ and *forced* when $V \neq 0$. The free oscillations $e^{\lambda t}$ are controlled by $L\lambda^2 + R\lambda + 1/C = 0$. Notice the undamped case $R = 0$ when $\lambda = \pm i/\sqrt{LC}$. This shows the natural frequency $\omega_0 = 1/\sqrt{LC}$. Damped free oscillations are in the exercises—what is new and important is the forcing from the right hand side. Our last step is to solve equation (8).

PARTICULAR SOLUTIONS—THE METHOD OF UNDETERMINED COEFFICIENTS

The forcing term is a multiple of $\cos \omega t$. **The "particular solution" is a multiple of $\cos \omega t$ plus a multiple of $\sin \omega t$.** To discover the undetermined coefficients in $y = a \cos \omega t + b \sin \omega t$, substitute into the differential equation (8):

$$-L\omega^2(a \cos \omega t + b \sin \omega t) + R\omega(-a \sin \omega t + b \cos \omega t)$$
$$+(a \cos \omega t + b \sin \omega t)/C = V\omega \cos \omega t.$$

The terms in $\cos \omega t$ and the terms in $\sin \omega t$ give two equations for a and b:

$$-aL\omega^2 + bR\omega + a/C = V\omega \quad \text{and} \quad -bL\omega^2 - aR\omega + b/C = 0. \tag{9}$$

EXAMPLE 11   Solve $y'' + y = \cos \omega t$. The oscillations are *forced* at frequency ω. The oscillations are *free* ($y'' + y = 0$) at frequency 1. The solution contains both.

Particular solution   Set $y = a \cos \omega t + b \sin \omega t$ at the driving frequency ω, and (9) becomes

$$-a\omega^2 + 0 + a = 1 \quad \text{and} \quad -b\omega^2 - 0 + b = 0.$$

The second equation gives $b = 0$. No sines are needed because the problem has no dy/dt. The first equation gives $a = 1/(1 - \omega^2)$, which multiplies the cosine:

$$y = (\cos \omega t)/(1 - \omega^2) \quad \text{solves} \quad y'' + y = \cos \omega t. \tag{10}$$

General solution   Add to this particular solution any solution to $y'' + y = 0$:

$$y = y_{\text{particular}} + y_{\text{homogeneous}} = (\cos \omega t)/(1 - \omega^2) + C \cos t + D \sin t. \tag{11}$$

Problem of resonance   When the driving frequency is $\omega = 1$, the solution (11) becomes meaningless—its denominator is zero. *Reason*: The natural frequency in $\cos t$ and $\sin t$ is also 1. A new particular solution comes from $t \cos t$ and $t \sin t$.

The key to success is to know the form for y. The table displays four right hand sides and the correct y's for any constant-coefficient equation:

Right hand side	**Particular solution**
e^{kt}	$y = Be^{kt}$ (same exponent)
$\cos \omega t$ or $\sin \omega t$	$y = a \cos \omega t + b \sin \omega t$ (include both)
polynomial in t	$y = $ polynomial of the same degree
$e^{kt} \cos \omega t$ or $e^{kt} \sin \omega t$	$y = ae^{kt} \cos \omega t + be^{kt} \sin \omega t$

Exception   If one of the roots λ for free oscillation equals k or $i\omega$ or 0 or $k + i\omega$, the corresponding y in the table is *wrong*. The proposed solution would give zero on the right hand side. The correct form for y includes an extra t. All particular solutions are computed by substituting into the differential equation.

Apology   Constant-coefficient equations hardly use calculus (only $e^{\lambda t}$). They reduce directly to algebra (substitute y, solve for λ and a and b). I find the **S**-curve from the logistic equation much more remarkable. The nonlinearity of epidemics or heartbeats or earthquakes demands all the calculus we know. The solution is not so predictable. The extreme of unpredictability came when Lorenz studied weather prediction and discovered chaos.

NUMERICAL METHODS

Those four pages explained how to solve linear equations with constant coefficients: *Substitute* $y = e^{\lambda t}$. The list of special solutions becomes longer in a course on differential equations. But for most nonlinear problems we enter another world— where solutions are numerical and approximate, not exact.

In actual practice, numerical methods for $dy/dt = f(t, y)$ divide in two groups:

1. Single-step methods like Euler and Runge-Kutta
2. Multistep methods like Adams-Bashforth

The unknown y and the right side f can be vectors with n components. The notation stays the same: the step is $\Delta t = h$, the time t_n is nh, and y_n is the approximation to the true y at that time. We test the first step, to find y_1 from $y_0 = 1$. The equation is $dy/dt = y$, so the right side is $f = y$ and the true solution is $y = e^t$.

Notice how the first value of f (in this case 1) is used inside the second f:

Improved Euler $y_{n+1} = y_n + \frac{1}{2}h[f(t_n, y_n) + f(t_{n+1}, y_n + hf(y_n, t_n))]$

TEST $y_1 = 1 + \frac{1}{2}h[1 + (1+h)] = 1 + h + \frac{1}{2}h^2$

At time h the true solution equals e^h. Its infinite series is correct through h^2 for Improved Euler (a second-order method). The ordinary Euler method $y_{n+1} = y_n + hf(t_n, y_n)$ is first-order. TEST: $y_1 = 1 + h$. Now try Runge-Kutta (a fourth-order method):

Runge-Kutta $y_{n+1} = y_n + \frac{1}{6}[k_1 + 2k_2 + 2k_3 + k_4]$ with $k_1 = hf(t_n, y_n)$

$k_2 = hf(t_n + \frac{1}{2}h, y_n + \frac{1}{2}k_1)$ $k_3 = hf(t_n + \frac{1}{2}h, y_n + \frac{1}{2}k_2)$ $k_4 = hf(t_n + h, y_n + k_3)$

Now the first value of f is used in the second (for k_2), the second is used in the third, and then k_3 is used in k_4. The programming is easy. Check the accuracy with another test on $dy/dt = y$:

$$\text{TEST } y_1 = 1 + \frac{1}{6}\left[h + 2h\left(1 + \frac{h}{2}\right) + 2h\left(1 + \frac{h}{2}\left(1 + \frac{h}{2}\right)\right) + h\left(1 + h\left(1 + \frac{h}{2}\left(1 + \frac{h}{2}\right)\right)\right)\right]$$

$$= 1 + h + \frac{h^2}{2} + \frac{h^3}{6} + \frac{h^4}{24}. \qquad \text{This answer agrees with } e^h \text{ through } h^4.$$

These formulas are included in the book so that you can apply them directly— for example to see the **S**-shape from the logistic equation with $f = cy - by^2$.

Multistep formulas are simpler and quicker, but they need a single-step method to get started. Here is y_4 in a fourth-order formula that needs y_0, y_1, y_2, y_3. Just shift all indices for y_5, y_6, and y_{n+1}:

Multistep $y_4 = y_3 + \frac{h}{24}[55y_3' - 59y_2' + 37y_1' - 9y_0']$.

The advantage is that each step needs only *one* new evaluation of $y_n' = f(t_n, y_n)$. Runge-Kutta needs four evaluations for the same accuracy .

Stability is the key requirement for any method. Now the good test is $y' = -y$. The solution should decay and not blow up. Section 6.6 showed how a large time step makes Euler's method unstable—the same will happen for more accurate formulas. The price of total stability is an "implicit method" like $y_1 = y_0 + \frac{1}{2}h(y_n' + y_1')$, where the unknown y_1 appears also in y_1'. There is an equation to be solved at every step. Calculus is ending as it started—with the methods of Isaac Newton.

16.2 EXERCISES

Read-through questions

The solution to $y' - 5y = 10$ is $y = Ae^{5t} + B$. The homogeneous part Ae^{5t} satifies $y' - 5y = \underline{\quad a \quad}$. The particular solution B equals $\underline{\quad b \quad}$. The initial condition y_0 is matched by $A = \underline{\quad c \quad}$. For $y' - 5y = e^{kt}$ the right form is $y = Ae^{5t} + \underline{\quad d \quad}$. For $y' - 5y = \cos t$ the form is $y = Ae^{5t} + \underline{\quad e \quad} + \underline{\quad f \quad}$.

The equation $y'' + 4y' + 5y = 0$ is second-order because $\underline{\quad g \quad}$. The pure exponential solutions come from the roots of $\underline{\quad h \quad}$, which are $r = \underline{\quad i \quad}$ and $s = \underline{\quad j \quad}$. The general solution is $y = \underline{\quad k \quad}$. Changing $4y'$ to $\underline{\quad l \quad}$ yields pure oscillation. Changing to $2y'$ yields $\lambda = -1 \pm 2i$, when the solutions become $y = \underline{\quad m \quad}$. This oscillation is (over)(under)(critically) damped. A spring with $m = 1$, $d = 2$, $k = 5$ goes (back and forth)(directly to zero). An electrical network with $L = 1$, $R = 2$, $C = \frac{1}{5}$ also $\underline{\quad n \quad}$.

One particular solution of $y'' + 4y = e^t$ is e^t times $\underline{\quad o \quad}$. If the right side is $\cos t$, the form of y_p is $\underline{\quad p \quad}$. If the right side is 1 then $y_p = \underline{\quad q \quad}$. If the right side is $\underline{\quad r \quad}$ we have resonance and y_p contains an extra factor $\underline{\quad s \quad}$.

Problems 1–14 are about first-order linear equations.

1 Substitute $y = Be^{3t}$ into $y' - y = 8e^{3t}$ to find a particular solution.

2 Substitute $y = a\cos 2t + b\sin 2t$ into $y' + y = 4\sin 2t$ to find a particular solution.

3 Substitute $y = a + bt + ct^2$ into $y' + y = 1 + t^2$ to find a particular solution.

4 Substitute $y = ae^t \cos t + be^t \sin t$ into $y' = 2e^t \cos t$ to find a particular solution.

5 In Problem 1 we can add Ae^t because this solves the equation $\underline{\quad}$. Choose A so that $y(0) = 7$.

6 In Problem 2 we can add Ae^{-t}, which solves $\underline{\quad}$. Choose A to match $y(0) = 0$.

7 In Problem 3 we add $\underline{\quad}$ to match $y(0) = 2$.

8 In Problem 4 we can add $y = A$. Why ?

9 Starting from $y_0 = 0$ solve $y' = e^{kt}$ and also solve $y' = 1$. Show that the first solution approaches the second as $k \to 0$.

10 Solve $y' - y = e^{kt}$ starting from $y_0 = 0$. What happens to your formula as $k \to 1$? By l'Hôpital's rule show that y approaches te^t as $k \to 1$.

11 Solve $y' - y = e^t + \cos t$. What form do you assume for y with two terms on the right side ?

12 Solve $y' + y = e^t + t$. What form to assume for y ?

13 Solve $y' = cy + te^t$ following Example 3 ($c \neq 1$).

14 Solve $y' = y + t$ following Example 3 ($c = 1$ *and* $k = 0$).

Problems 15–28 are about second-order linear equations.

15 Substitute $y = e^{\lambda t}$ into $y'' + 6y' + 5y = 0$. (a) Find all λ's. (b) The solution decays because $\underline{\quad}$. (c) The general solution with constants A and B is $\underline{\quad}$.

16 Substitute $y = e^{\lambda t}$ into $y'' + 9y = 0$. (a) Find all λ's. (b) The solution oscillates because $\underline{\quad}$. (c) The general solution with constants a and b is $\underline{\quad}$.

17 Substitute $y = e^{\lambda t}$ into $y'' + 2y' + 3y = 0$. Find both λ's. The solution oscillates as it decays because $\underline{\quad}$. The general solution with A and B and $e^{\lambda t}$ is $\underline{\quad}$. The general solution with e^{-t} times sine and cosine is $\underline{\quad}$.

18 Substitute $y = e^{\lambda t}$ into $y'' + 6y' + 9y = 0$. (a) Find all λ's. (b) The general solution with $e^{\lambda t}$ and $te^{\lambda t}$ is $\underline{\quad}$.

19 For $y'' + dy' + y = 0$ find the type of damping at $d = 0, 1, 2, 3$.

20 For $y'' + 2y' + ky = 0$ find the type of damping at $k = 0, 1, 2$.

21 If $\lambda^2 + b\lambda + c = 0$ has a repeated root prove it is $\lambda = -b/2$. In this case compute $y'' + by' + cy$ when $y = te^{\lambda t}$.

22 $\lambda^2 + 3\lambda + 2 = 0$ has roots -1 and -2(not repeated). Show that te^{-t} does *not* solve $y'' + 3y' + 2y = 0$.

23 Find $y = a\cos t + b\sin t$ to solve $y'' + y' + y = \cos t$.

24 Find $y = a\cos \omega t + b\sin \omega t$ to solve $y'' + y' + y = \sin \omega t$.

25 Solve $y'' + 9y = \cos 5t$ with $y_0 = 0$ and $y'_0 = 0$. The solution contains $\cos 3t$ and $\cos 5t$.

26 The difference $\cos 5t - \cos 3t$ equals $2\sin 4t \sin t$. Graph it to see fast oscillations inside slow oscillation (beats).

27 The solution to $y'' + \omega_0^2 y = \cos \omega t$ with $y_0 = 0$ and $y'_0 = 0$ is what multiple of $\cos \omega t - \cos \omega_0 t$? The formula breaks down when $\omega = \underline{\quad}$.

28 Substitute $y = Ae^{i\omega t}$ into the circuit equation $Ly' + Ry + \int y \, dt/C = Ve^{i\omega t}$. Cancel $e^{i\omega t}$ to find A. Its denominator is the *impedance*.

Problems 29–32 have the four right sides in the table (end of section). Find $y_{particular}$ by using the correct form.

29 $y'' + 3y = e^{5t}$

30 $y'' + 3y = \sin t$

31 $y'' + 2y = 1 + t$

32 $y'' + 2y = e^t \cos t$.

33 Find the coefficients of y in Problems 29–31 for which the forms in the table are wrong. Why are they wrong ? What new forms are correct ?

34 The magic factor t entered equation (2). The series for $e^{kt} - e^{ct}$ starts with $1 + kt + \frac{1}{2}k^2t^2$ minus $1 + ct + \frac{1}{2}c^2t^2$. Divide by $k - c$ and set $k = c$ to start the series for te^{ct}.

35 Find four exponentials $y = e^{\lambda t}$ for $d^4y/dt^4 - y = 0$.

36 Find a particular solution to $d^4y/dt^4 + y = e^t$.

37 The solution is $y = Ae^{-2t} + Bte^{-2t}$ when $d = 4$ in Figure 16.2. Choose A and B to match $y_0 = 1$ and $y_0' = 0$. How large is $y(2\pi)$?

38 When d reaches 5 the quadratic for Figure 16.2 is $\lambda^2 + 5\lambda + 4 = (\lambda + 1)(\lambda + 4)$. Match $y = Ae^{-t} + Be^{-4t}$ to $y_0 = 1$ and $y_0' = 0$. How large is $y(2\pi)$?

39 When the quadractic for Figure 16.2 has roots $-r$ and $-4/r$, the solution is $y = Ae^{-rt} + Be^{-4t/r}$.

   (a) Match the initial conditions $y_0 = 1$ and $y_0' = 0$.

   (b) Show that y approaches 1 as $r \to 0$.

40 In one sentence tell why $y'' = 6y$ has exponential solutions but $y'' = 6y^2$ does not. What power $y = x^n$ solves this equation ?

41 The solution to $dy/dt = f(t)$, with no y on the right side, is $y = \int f(t)\, dt$. Show that the Runge-Kutta method becomes Simpson's Rule.

42 Test all methods on the logistic equation $y' = y - y^2$ to see which gives $y_\infty = 1$ most accurately. Start at the inflection point $y_0 = \frac{1}{2}$ with $h = \frac{1}{10}$. Begin the multistep method with exact values of $y = (1 + e^{-t})^{-1}$.

43 Extend the tests of Improved Euler and Runge-Kutta to $y' = -y$ with $y_0 = 1$. They are *stable* if $|y_1| \leqslant 1$. How large can h be ?

44 Apply Runge-Kutta to $y' = -100y + 100\sin t$ with $y_0 = 0$ and $h = .02$. Increase h to .03 to see that instability is no joke.

16.3 Discrete Mathematics: Algorithms

Discrete mathematics is not like calculus. *Everything is finite.* I can start with the 50 states of the U.S. I ask if Maine is connected to California, by a path through neighboring states. You say *yes*. I ask for the **shortest path** (fewest states on the way). You get a map and try all possibilities (not really all—but your answer is right). Then I close all boundaries between states like Illinois and Indiana, because one has an even number of letters and the other has an odd number. *Is New York still connected to Washington*? You ask what kind of game this is—but I hope you will read on.

Far from being dumb, or easy, or useless, discrete mathematics asks good questions. It is important to know the fastest way across the country. It is more important to know the fastest way through a phone network. When you call long distance, a quick connection has to be found. Some lines are tied up, like Illinois to Indiana, and there is no way to try every route.

The example connects New York to New Jersey (7 letters and 9). Washington is connected to Oregon (10 letters and 6). As you read those words, your mind jumps to this fact—there is no path from New York with 7 letters to Washington with 10. Somewhere you must get stuck. There might be a path between all states with an odd number of letters—I doubt it. Graph theory gives a way to find out.

GRAPHS

A model for a large part of finite mathematics is a **graph**. It is not the graph of $y = f(x)$. The word "graph" is used in a totally different way, for a collection of **nodes** and **edges** . The nodes are like the 50 states. *The edges go between two nodes—* the neighboring states. A network of computers fits this model. So do the airline connections between cities. A pair of cities may or may not have an edge between them—depending on flight schedules. The model is determined by V and E.

DEFINITION   A **graph** is a set V of *nodes* (or *vertices*) and a set E of *edges*.

EXAMPLE 1   How many edges are possible with n nodes, in a **complete graph** ?

The first node has edges to the $n - 1$ other nodes. (*An edge to itself is not allowed.*) The second node has $n - 2$ new edges. The third node has another $n - 3$. The total count of edges, when none are missing, is the sum from Section 5.3:

$$1 + 2 + \ldots + (n - 1) = \tfrac{1}{2}n(n - 1) \quad \text{edges in a complete graph.}$$

Fifty states have $25 \cdot 49 = 1225$ possible edges. The "neighboring states graph" has less than 200. A line of 6 nodes has 5 edges, out of $\tfrac{1}{2} \cdot 6 \cdot 5 = 15$ possible.

EXAMPLE 2   Which states with an odd number of letters are reachable from New York ? Boundaries to states like Pennsylvania (12 letters) are closed.

Method of solution   Start from New York (7). There is an edge to Connecticut (11). That touches Massachusetts (13), which is a neighbor of Vermont (7). But we missed Rhode Island, and how do we get back ? *The order depends on our search method—* and two methods are specially important.

Depth fist search (DFS) "From the current state, go to *one* new state if possible." But what do we do from Vermont, when New Hampshire (12) is not allowed ? The answer is: *backtrack to Massachusetts*. That becomes the next current state.

We label every state as we reach it, to show which state we came from. Then VT has the label MA, and we easily cross back. From MA we go to RI. Then backtrack to MA and CT and NY. At every step I searched for a new state with no success. From NY we see NJ (9). Finally we are in a corner.

The depth first search is ended, by a barrier of even states. Unless we allow Ontario and keep going to Minnesota.

Breadth first search (**BFS**) "From the current state, add *all* possible new states to the bottom of the list. But take the next current state from the top of the list." There is no need to backtrack.

From NY we reach VT and MA and CT and NJ. What comes next?

Where **DFS** moves from the *last* possible state, breadth first search moves from the *first* possible state. No move from VT is possible—so we "scan" from Massachusetts. We see Rhode Island (barely). That ends **BFS**.

The same six states are reached both ways. Only the order is different. **DFS** is *last in-first out*. **BFS** is *first in-first out*. You have the same choice in drawing a family tree—follow a path as far as it goes and backtrack, or list all brothers and sisters before their children. The **BFS** graph in Figure 16.3 is a tree. So is the DFS graph, using forward edges only.

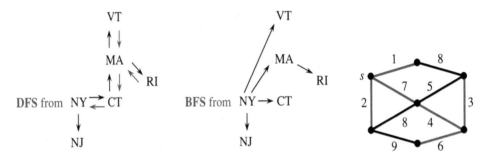

Fig. 16.3   Search trees from New York. The minimum spanning tree.

DEFINITION   A *tree* is a connected graph with *no loops*. Its N nodes are connected by $N - 1$ edges. If $N = n$, so every node is in the tree, it is a *spanning tree*.

The path from VA to KY to TN to NC to VA is a *loop* (or *cycle*). If one of those four edges is removed, we have a tree. If two edges are removed, we have two trees (a small forest).

EXAMPLE 3   Allow an edge between neighboring states only when *one state is even and the other is odd*. Are the lower 48 states connected?

Start anywhere—say California. Apply either type of search—maybe DFS. Go to Arizona (7) then Utah (4) then WY (9) then CO (8) then NM then OK then TX. (I am writing this on an airplane, looking at the map.) We will never get to Florida! It is blocked by Alabama and Georgia.

The search creates a tree, but not a spanning tree. This graph is not connected. An odd-to-even graph is special and important. It is called "*bipartite*," meaning *two parts*. The odd states are in one part, the even states are in the other. All edges go *between* parts. No edges are within a part.†

---

†Exactly half the states have an even number of letters (a real trivia question). This is the little-known reason for admitting Alaska and Hawaii.

EXAMPLE 4   Is there a "***complete matching***" between 25 even and 25 odd states ? This requires neighboring states to be paired off (with no repetition).

Method 1   Start pairing them off: $CA-AZ$, $UT-WY$, $NV-ID$, $NE-SD$, $WA-MT$. What about Oregon ? Maybe it should have been paired with Idaho. Then Nevada could pair with Arizona. Trial and error goes nowhere fast.

Method 2   Think first. The four states $CA-OR-WA-NV$ are even. This whole group is only connected to three odd states (AZ, ID, MT). The matching is impossible.

16B A complete matching is impossible if a group of nodes in one part is connected to fewer nodes in the other part. If every group has enough connections, the matching can be achieved.

This is Hall's Theorem . In a course on graphs, it would be proved. Our purpose here is to see the ideas and questions in discrete mathematics, more than the proofs.

THE GREEDY ALGORITHM

Put back all edges between neighboring states. The nodes could be provinces of Canada or states of Australia. If they are countries of Europe–Asia–Africa (or the Americas), we need a new map. The essential thing is the new problem.

   In a network each edge has a "length." A positive number c_{ij} is assigned to the edge from node i to node j. In an economics problem, c_{ij} is the *cost*. In a flow problem it is the *capacity*, in an electrical circuit it is the *conductance*. We look for paths that minimize these "lengths."

PROBLEM   *Find the minimum spanning tree*. Connect all nodes by a tree with the smallest possible total length.

The six cheapest highways connecting seven cities form a minimum spanning tree. It is cheapest to build, not cheapest to drive—you have to follow the tree. Where there is no edge we set $c_{ij} = \infty$ (or an extremely large value, in an actual code). Then the algorithm works with a complete network—all $n(n-1)/2$ edges are allowed. How does it find the minimum spanning tree in Figure 16.3c ?

Method 1   *Always add the shortest edge that goes out from the current tree.*

Starting from node s, this rule chooses edges of length $1, 2, 7, 4, 3$. Now it skips 5, which would close a loop. It chooses 6, for total length 23.

Method 2   *Add edges in order, from shortest to longest.* Reject an edge that closes a loop. Several trees grow together (a forest). At the end we have a minimum spanning tree.

This variation chooses edge lengths in the order 1, 2, 3, 4, 6 (*rejecting* 5), 7. In our network both methods produce the same tree. When many edges have equal length, there can be many shortest trees.

   These methods are examples of the ***Greedy Algorithm***: ***Do the best thing at every step***. Don't look ahead. Stick to a decision once it is made. In most network problems the Greedy Algorithm is not optimal—in this spanning tree problem it is.

   Method 2 looks faster than Method 1. Sort the edges by length, and go down the list. Just avoid loops. But sorting takes time! It is a fascinating problem in itself—bubble sort or insertion sort or heapsort. We go on to a final example of discrete mathematics and its algorithms.

PROBLEM   *Find the **shortest path** from the source node s to each other node.*

The shortest path may not go along the minimum spanning tree. In the figure, the best path going east has length $1 + 8$. There is a new *shortest path tree*, in which the source plays a special role as the "root."

How do we find shortest paths? Listing all possibilities is more or less insane. A good algorithm builds out from the source, selecting one new edge at every step. After k steps we know the distances $d_1, ..., d_k$ to the k nearest nodes.

     Algorithm: Minimize $d_i + c_{ij}$ over all settled nodes i and all remaining nodes j.

The best new node j is a distance c_{ij} from a settled node, which is a distance d_i from the source. In the example network, the first edges are $1, 2, 7$. *Next is* 8. The northeast node is closest to the source at this step. The final tree does not use edges $3, 5, 6$—even though they are short.

These pages were written to show you the algorithmic part of discrete mathematics. The other part is *algebra*—permutations, partitions, groups, counting problems, generating functions. There is no calculus, but that's fair. The rest of the book was written to show what calculus can do—***I hope very much that you enjoyed it***.

Thank you for reading, and thinking, and working.

16.3 EXERCISES

Read-through questions

A graph is a set V of __a__ and a set E of __b__. With 6 nodes, a complete graph has __c__ edges. A spanning tree has only __d__. A tree is defined as __e__, and it is spanning if __f__. It has __g__ path between each pair of nodes.

To find a path from node i to node j, two search methods are __h__. As nodes are reached, DFS looks out from the __i__ node for a new one. BFS looks out from __j__. DFS must be prepared to __k__ to earlier nodes. In case of fire, BFS locates all doors from the room you are in before __l__.

In a bipartite graph, all edges go from one part to __m__. A matching is impossible if k nodes in one part are connected to __n__ nodes in the other part. The edges in a *network* have __o__ c_{ij}. A minimum spanning tree is __p__. It can be found by the __q__ algorithm, which accepts the shortest edge to a new node without worrying about __r__.

1 Start form one node of a hexagon (six nodes, six edges). Number the other nodes by (a) breadth first search (b) depth first search.

2 Draw two squares with one node in common (7-node graph). From that node number all others by **DFS** and **BFS**. Indicate backtracks.

3 How many spanning trees in the hexagon graph?

4 Draw a spanning tree in the two-square graph. How many spanning trees does it have?

5 Define a connected graph. If a graph has 7 edges and 9 nodes, prove that it is not connected.

6 Define a loop. If a connected graph has 8 edges and 9 nodes, prove that it has no loops.

7 Find the shortest path (minimum number of edges) from Maine to California.

8 Which state is farthest (how many edges are needed) from the state you are in? Why would it come last in **BFS**?

9 List the steps of **BFS** from your state to Georgia or Colorado or New Jersey. (*There are edges Hawaii–California and Alaska–Washington.*)

10 With edges between odd neighboring states and between even neighbors, what is the largest connected set of states? Map required.

11 With edges only form odd to even neighbors, how many states can be matched? (Answer unknown to author—please advise.)

12 A matching is a forest of two-node trees. Give another description.

13 Find the minimum spanning tree for network **A**.

14 Find the shortest path tree from the center of network **A**.

15 Is there a complete matching between left and right nodes in graph **B**? If not, which group of nodes has too few connections?

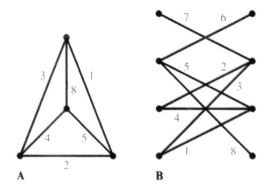

A   **B**

16 Find the loop in network **B**. Then find a minimum spanning tree by Method 1 and Method 2.

17 How many spanning trees in graph **B**? It has one loop.

18 Show that a graph cannot have 0, 1, 2, 3, and 4 edges going into its five nodes.

19 If the only edges into a node have lengths 6 and 8, can they both be in a minimum spanning tree?

20 In Problem 19, prove that a minimum spanning tree contains edge (6) if it contains edge (8).

21 *True or false,* with reason or example.

(a) In a complete network, the minimum spanning tree contains the $n-1$ shortest edges.

(b) If a graph has 9 nodes and 9 edges, it has a loop.

(c) A graph with a complete matching must be connected.

22 Draw a tree that is perfect for (a) DFS; (b) BFS.

23 The *adjacency matrix* has $a_{ij} = 1$ if there is an edge from node i to node j. Write down this matrix for graphs A and B.

24 In a complete network start with $d_{ij} = c_{ij}$. Show that the d_{ij} at the end of this program are shortest distances:

$$\text{for} \quad i = 1 \text{ to } n \text{ do}$$
$$\text{for} \quad j = 1 \text{ to } n \text{ do}$$
$$\text{for} \quad k = 1 \text{ to } n \text{ do}$$
$$d_{ij} = \max(d_{ij}, d_{ik} + d_{kj})$$

25 How many spanning trees in graph **A**?

26 A maximum spanning tree has greatest possible length. Give an algorithm to find it.

27 Write a code that will find a spanning tree (or stop), given a list of edges like $(1,2), (1,3), (4,7), \ldots$.

STUDENT STUDY GUIDE
CHAPTER 1

This first chapter of the Study Guide is included with the book, so it will help immediately at the start of calculus. The Guide is correlated chapter by chapter with the text. It contains four components which experience has shown are most helpful:

1. Model problems with comments and complete solutions

2. Extra drill problems included with exercises for chapter review

3. Read-through questions with the blanks filled in

4. Solutions to selected even-numbered problems

The complete Student Study Guide can be obtained directly from Wellesley-Cambridge Press, Box 812060, Wellesley MA 02482. Whether you read this first chapter or the whole Guide, we very much hope that you will find it useful.

1.1 Velocity and Distance

The first step in calculus (in my opinion) is to begin working with *functions*. The functions can be described by formulas or by graphs. The first graphs to study are straight lines, and the functions that go with them are linear: $y = mx + b$ or (using other letters) $f = vt + C$. It takes practice to connect the formula $y =$ to the straight line with slope 7 going through the point $(2, 4)$. You need to be able to rewrite that formula as $y = 7t - 10$. You also need to be able to find the formula from the graph. Here is an example: *Find the equations for these straight lines.*

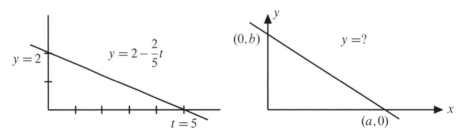

The first line goes down 2 and across 5. The slope is $\frac{-2}{5} = -.4$. The minus sign is because y decreases as t increases – the slope of the graph is negative. The "starting point" is $y = 2$ when $t = 0$. The equation is $y = 2 - .4t$. Check that this gives $y = 0$ when $t = 5$, so the formula correctly predicts that second point.

Problem 1: *Find the equation for the second line.* Comment: The points $(0, b)$ and $(a, 0)$ are given by letters not numbers. This is typical of mathematics, which is not so much about numbers as most people think – it is really about patterns. Sometimes numbers help, other times they get in the way. Many exercises in mathematics books are really asking you to find the pattern by solving a general problem (with letters) instead of a special problem (with numbers). But don't forget: You can always substitute 5 and 2 for the letters a and b.

Solution to Problem 1: The slope is $-\frac{b}{a}$. The equation is $y = b - \frac{b}{a}x$. Compare with $y = 2 - \frac{2}{5}t$.

Pairs of functions. Calculus deals with two functions at once. We have to understand both functions, and *how they are related*. In writing the book I asked myself: Where do we find two functions? What example can we start with? The example should be familiar and it should be real – not just made up. The best examples build on what we already know. The first pair of functions is distance $f(t)$ and velocity $v(t)$.

Comment: This example is taken from *life*, not from physics. Don't be put off by symbols like $f(t)$ and $v(t)$. The velocity is certainly connected to the distance traveled. This connection is clearest when the velocity is constant. If you go 120 miles in 2 hours at steady speed, then $v = \frac{120\,\text{miles}}{2\,\text{hours}} = 60$ miles per hour. That example converts to straight line graphs: The distance graph goes up with slope 60. Starting from $f(0) = 0$, the equation for the line is $f(t) = 60t$.

Most people see how $f = 60t$ leads to $v = 60$. Now look at the connection between $f(t)$ and $v(t)$ in the next graphs.

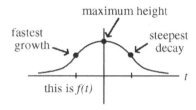
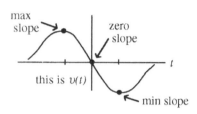

What do you see? The slope of f is zero at the center. Thus $v = 0$ where f is a maximum. The maximum of v is where the graph of f is steepest (upwards). Then v is largest (positive). When the graph of f flattens out, v drops back toward zero.

With formulas I could specify these functions exactly. The distance might be $f(t) = \frac{1}{1+t^2}$. Then Chapter 2 will find $\frac{-2t}{(1+t^2)^2}$ for the velocity $v(t)$. Very often calculus is swept up by formulas, and the ideas get lost. You need to know the rules for computing $v(t)$, and exams ask for them, but it is not right for calculus to turn into pure manipulations. Our goal from the start is to *see the ideas*.

Again comes the question: Where to go after $f = 60t$ and $v = 60$? The next step is to allow two velocities (two slopes). The velocity jumps. The distance graph switches from one straight line to another straight line – with a different slope. It takes practice to write down the formulas and draw the graphs. Here are six questions about straight line graphs that change slopes. The figure shows $f(t)$, the questions are about $v(t)$.

Answers are given for the first graph. Test yourself on the second graph.

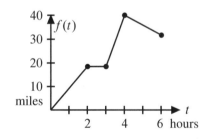
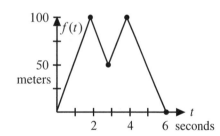

2 During the first two hours, the car travels _____ miles. The velocity during this time is _____ miles per hour. The slope of the distance graph from $t = 0$ to $t = 2$ is _____.
Answer: Miles traveled $= 20$, velocity $= 10$ miles per hour, slope $= \frac{20}{2} = 10$.

3 From $t = 2$ to $t = 3$ the car travels _____ miles. The velocity during this time is _____ miles per hour. The slope of this segment is _____. Answer: $0, 0, 0$.

4 From $t = 3$ to $t = 4$ the car travels _____ miles. The velocity during this time is _____ mph. The slope of the f-graph from $t = 3$ to $t = 4$ is _____. Answer: $20, 20, \frac{40-20}{4-3} = 20$.

5 From $t = 4$ to $t = 6$, the car travels *backwards* _____ miles. The change in f is _____ miles. The velocity is _____ miles per hour. The slope of this segment of the f-graph is _____. Answer: Backwards 10 miles, change in f is -10, velocity -5 miles per hour, slope $\frac{30-40}{6-4} = -5$.

6 *Draw graphs of v(t) from those two graphs of f(t).*

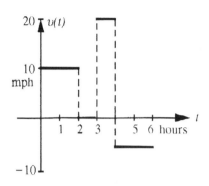 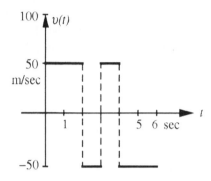

7 Give a 4-part formula for $f(t)$. Each piece is a line segment described by a linear function $f(t) = vt + C$. The first piece is just $f(t) = vt$ because it starts from $f(0) = 0$.

- Use the form $vt + C$ if you know the slope v and the height C when $t = 0$. With different letters this form is $y = mx + b$. Use $f(t) = f(3) + v(t - 3)$ if you know v and you also know the distance $f(3)$ at the particular time $t = 3$. Section 2.3 on *tangent lines* uses both forms. The second form becomes $y = f(a) + f'(a)(x - a)$ and the slope is denoted by $f'(a)$.

From $t = 3$ to $t = 4$, our slope is 20 and the starting point is $(3, 20)$. The formula for this segment is $f(t) = 20 + 20(t - 3), 3 \le t \le 4$. This simplifies to $f(t) = 20t - 40$. The complete solution is:

$$f(t) = \begin{cases} 10t + 0, 0 \le t \le 2 \\ 20,\ 2 \le t \le 3 \\ 20 + 20(t - 3),\ 3 \le t \le 4 \quad \text{or} \quad 20t - 40 \\ 40 - 5(t - 4),\ 4 \le t \le 6 \quad \text{or} - 5t + 60 \end{cases}$$

IMPORTANT If we are given $v(t)$ we can discover $f(t)$. But we do need to know the distance at one particular time like $t = 0$. The table of f's below was started from $f(0) = 0$. Questions **8–11** allow you to fill in the missing parts of the table. Then the piecewise constant $v(t)$ produces a piecewise linear $f(t)$.

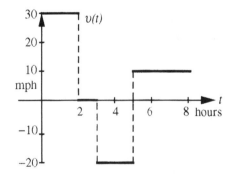

t	$f(t)$
0	0
2	60
3	60 + 0
5	60 + 0 + _____
8	60 + 0 + (−40)
	+ _____ = _____

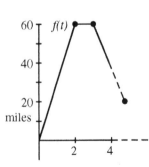

8 From $t = 0$ to $t = 2$, the velocity is _____ mph. The distance traveled during this time is _____ miles. The area of this rectangular portion of the graph is _____ × _____ = _____ . Answer: 30, 60, 30 × 2 = 60.

9 From $t = 2$ to $t = 3$, the velocity is _____ miles per hour. The "area" of this portion of the graph (a rectangle with no height) is _____ × _____ = _____ . Answer: 0, 1 × 0 = 0.

10 From $t = 3$ to $t = 6$, the velocity is _____ mph. During this time the vehicle travels *backward* _____ miles. The area of this rectangular portion is _____ × _____ = _____ . Keep in mind that area below the horizontal axis counts as *negative*.
Answer: -20, 40, $-20 \times 2 = -40$.

11 From $t = 5$ to $t = 8$ the velocity is _____ mph. During this time the vehicle travels forward _____ miles. The area of this rectangular portion is _____ × _____ = _____ .
Answer: 10, 30, $10 \times 3 = 30$.

12 Using the information in **8-11**, finish the table and complete the distance graph out to time $t = 8$. Connect the points $(t, f(t))$ in the table with straight line segments.

13 What is the distance traveled from $t = 0$ to $t = 4$? From $t = 0$ to $t = 7$?
Answer: The points $(4, 40)$ and $(7, 40)$ can be read directly from the graph. The distance in each case is 40 miles.

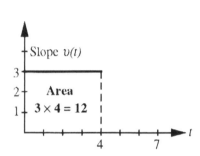

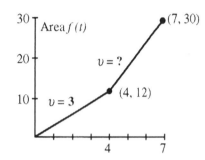

14 This is Problem 1.1.32 in the text. Suppose $v = 3$ up to time $t = 4$. What constant velocity will lead to $f(7) = 30$ if $f(0) = 0$? Give graphs and formulas for $v(t)$ and $f(t)$.

- We know that $f = 12$ when $t = 4$ because distance = rate × time = $3 \times 4 = 12$. Note that the slope of the f-graph from $t = 0$ to $t = 4$ is 3, the velocity. Since we want $f(7) = 30$, we must go $30 - 12 = 18$ more miles. The slope of the second piece of the f-graph should be $\frac{30-12}{7-4} = \frac{18}{3} = 6$. The desired velocity is $v = 6$. The equation of

  this segment is $f(t) = 12 + 6(t - 4)$. At the breakpoint $t = 4$, the velocity $v(4)$ is *not* defined, but the distance $f(4)$ *is* defined. The symbol in $v(t)$ is $<$ while the symbol in $f(t)$ is $\leq$.

$$v(t) = \begin{cases} 3, & 0 < t < 4 \\ 6, & 4 < t < 7 \end{cases} \qquad f(t) = \begin{cases} 3t + 0, & 0 \leq t \leq 4 \\ 12 + 6(t - 4), & 4 \leq t \leq 7. \end{cases}$$

Here are solutions to the read-through questions and selected even-numbered exercises for Section 1.1.

Starting from $f(0) = 0$ at constant velocity v, the distance function is $f(t) = \mathbf{vt}$. When $f(t) = \mathbf{55t}$ the velocity is $v = \mathbf{55}$. When $f(t) = 55(t) + 1000$ the velocity is still **55** and the starting value is $f(0) = \mathbf{1000}$. In each case v is the slope of the graph of f. When $\mathbf{v(t)}$ is negative, the graph of $\mathbf{f(t)}$ goes downward. In that case area in the v-graph counts as **negative**.

Forward motion from $f(0) = 0$ to $f(2) = 10$ has $v = \mathbf{5}$. Then backward motion to $f(4) = 0$ has $v = -\mathbf{5}$. The distance function is $f(t) = 5t$ for $0 \leq t \leq 2$ and then $f(t)$ equals $\mathbf{5(4 - t)}$ (not $-5t$). The slopes are **5** and $-\mathbf{5}$. The distance $f(3) = \mathbf{5}$. The area under the v-graph up to time 1.5 is **7.5**. The domain of f is the time interval $\mathbf{0 \leq t \leq 4}$, and the range is the distance interval $\mathbf{0 \leq f \leq 10}$. The range of $v(t)$ is only **5 and** $-\mathbf{5}$.

The value of $f(t) = 3t + 1$ at $t = 2$ is $f(2) = \mathbf{7}$. The value 19 equals $f(6)$. The difference $f(4) - f(1) = \mathbf{9}$. That is the change in distance, when $4 - 1$ is the change in time. The ratio of those changes equals **3**, which is the slope of the graph. The formula for $f(t) + 2$ is $3t + 3$ whereas $f(t + 2)$ equals $\mathbf{3t + 7}$. Those functions have the same slope as f: the graph of $f(t) + 2$ is shifted **up** and $f(t + 2)$ is shifted **to the left**. The formula for $f(5t)$ is $\mathbf{15t + 1}$. The formula for $5f(t)$ is $\mathbf{15t + 5}$. The slope has jumped from 3 to **15**.

The set of inputs to a function is its **domain**. The set of outputs is its range. The functions $f(t) = 7 + 3(t - 2)$ and $f(t) = vt + C$ are **linear**. Their graphs are **straight lines with slopes equal** to 3 and **v**. They are the same function, if $v = \mathbf{3}$ and $C = \mathbf{1}$.

10 $v(t)$ is negative-zero-positive; $v(t)$ is above 55 then equal to 55; $v(t)$ increases in jumps; $v(t)$ is zero then positive. All with corresponding $f(t)$.

26 The function increases by 2 in one time unit so the slope (velocity) is 2; $f(t) = \mathbf{2t + C}$ with constant $C = f(0)$.

36 At $t = 0$ the reading was $.061 + 10(.015) = \mathbf{.211}$. A drop of $.061 - .04 = .021$ would take $\mathbf{.021/.015}$ hours. This was the Exxon Valdes accident in Alaska.

1.2 Calculus Without Limits

This section was rewritten for the second printing of the book, in order to bring out the central ideas. One idea starts with numbers like $f = 3, 5, 9$. Their differences are 2 and 4. The sum of those differences is $2 + 4 = 6$. This equals $9 - 3$, or $f_{last} - f_{first}$. (It is like a card trick, where you think of numbers and only tell me the differences. Then I tell you the last number minus the first number.) The middle number 5 cancels out:

$$(5 - 3) + (9 - 5) = 9 - 3 \quad \text{or} \quad (f_1 - f_0) + (f_2 - f_1) = f_2 - f_0.$$

Taking sums is the "inverse" *of taking differences*. This is an early pointer to calculus.

- The "*derivative*" of $f(t)$ is the slope $v(t)$. This is like differences.
- The "*integral*" of $v(t)$ is the area $f(t)$. This is like sums.

Teaching question: *Why mention these ideas so early?* They are not proved. They can't even be explained in complete detail. Here is my answer: You the student should know where the course is going. I believe in seeing an idea several times, instead of "catch it once or lose it forever". Here was a case where a few numbers give a preview of the Fundamental Theorem of Calculus.

As the idea unfolds and develops, so does the notation. At first you have specific numbers 3, 5, 9. Then you have general numbers f_0, f_1, f_2. Then you have a piecewise linear function. Then you have a general function $f(t)$.

Similarly: At first you have differences $5 - 3$ and $9 - 5$. Then you have $v_1 = f_1 - f_0$ and $v_j = f_j - f_{j-1}$. Then you have a piecewise constant function. Then you have the derivative $v(t) = \frac{df}{dt}$.

The *equation for a straight line* is another idea that unfolds and develops. It was already in some of the problems for Section 1.1. Now Section 1.2 brings it out again, especially the income tax example. The tax in the second income bracket used to be

$$f(x) = \text{tax on } \$20,350 + (\text{tax rate } .28)(\text{income over } \$20,350).$$

This is so typical of what comes later: $f(x) = f(a) + (\text{slope})(x - a)$. Do those letters x and a make the equation less clear? For many students I believe they do, at least at first. The number $20,350$ is definite where the letter a is vague. The number $.28$ is specific where the slope $f'(a)$ will be general. A teacher is always caught by this dilemma. Mathematics expresses the pattern by symbols, but it is understood first for numbers.

The best way is to see the same idea several times, which we do for straight lines. Section 2.3 finds the tangent line to a curve. Section 3.1 uses the slope for a linear approximation. Section

3.7 explains Newton's method for solving $f(x) = 0$ – replace the curve by its tangent line. Will you see that the underlying idea is the same? I hope so, because it is the key idea of differential calculus: *Lines stay close to curves* (at least for a while).

1 Suppose $v = 3$ for $t < T$ and $v = 1$ for $t > T$. If $f(0) = 0$, find formulas for $f(t)$.

- The equation of the first segment is $f(t) = 0 + 3t$. This continues to $t = T$ when $f = 3T$. Using the fact that $(T, 3T)$ starts the second segment and $v = 1$ is the slope, we find $f(t) = 3T + 1(t - T)$ for $T \leq t$.

2 Suppose f_0, f_1, f_2, f_3 are the distances $0, 8, 12, 14$ at times $t = 0, 1, 2, 3$. Find velocities v_1, v_2, v_3 and a formula that fits v_j. Graph $v(t)$ and $f(t)$ with constant and linear pieces.

- The differences are $v_1 = 8, v_2 = 4$, and $v_3 = 2$. The velocities are halved at each step: $v_1 = \frac{1}{2}(16), v_2 = \frac{1}{4}(16), v_3 = \frac{1}{8}(16)$. The formula $v_j = \left(\frac{1}{2}\right)^j (16)$ or $v_j = 16/2^j$ fits all velocities.

3 Using the differences $v_1, v_2, \cdots, v_j$ and the starting value f_0, give a formula for f_j.

- $f_j = f_0 + v_1 + v_2 + \cdots + v_j$. *You should notice this formula*.

4 Suppose the tax rates are increased by 2 percentage points to .17, .30, and .33. At the same time, a tax credit of \$500 is allowed. If your original income was x, does your tax $f(x)$ go up or down after the changes? What if the \$500 is only a deduction from income?

- The answer depends on x. At $x = \$10,000$ the rate increase costs only 2% of \$10,000, which is \$200. You win. At $x = \$25,000$ the 2% rate increase balances the \$500 credit.

  If the \$500 is *deducted from income* of \$10,000, then you pay 17% of \$9,500 instead of 15% of \$10,000. This time you lose. To find the income at which deduction balances the 2% rate increase, solve $.15x = .17(x - 500)$ to find $x = \$4,250$.

5 (Algebra) If the distances are $f_j = j^2$, show that the velocities are $v_j = 2j - 1$.

- Since the difference v_j is $f_j - f_{j-1}$, you have to subtract $(j-1)^2$ from j^2:

$$j^2 - (j-1)^2 = j^2 - (j^2 - 2j + 1) = 2j - 1.$$

Please notice the summary at the end of Section 1.2, and then go over these read-through problems. Here are the blanks filled in, followed by even-numbered solutions.

Start with the numbers $f = 1, 6, 2, 5$. Their differences are $v = $ **5, −4, 3**. The sum of those differences is **4**. This is equal to f_{last} minus $\mathbf{f}_{\text{first}}$. The numbers 6 and 2 have no effect on this answer because in $(6-1) + (2-6) + (5-2)$ the numbers 6 and 2 **cancel**. The slope of the line between $f(0) = 1$ and $f(1) = 6$ is **5**. The equation of that line is $f(t) = 1 + \mathbf{5t}$.

With distances $1, 5, 25$ at unit times, the velocities are **4 and 20**. These are the **slopes** of the f-graph. The slope of the tax graph is the tax **rate**. If $f(t)$ is the postage cost for t ounces or t grams, the slope is the **cost per ounce (or per gram)**. For distances 0, 1, 4, 9 the velocities are **1, 3, 5**. The sum of the first j odd numbers is $f_j = \mathbf{j^2}$. Then f_{10} is **100** and the velocity v_{10} is **19**.

The piecewise linear sine has slopes $1, 0, -1, -1, 0, 1$. Those form a piecewise constant cosine. Both functions have **period** equal to 6, which means that $f(t + 6) = f(t)$ for every t. The velocities $v = 1, 2, 4, 8, \ldots$ have $v_j = \mathbf{2^{j-1}}$. In that case $f_0 = 1$ and $f_j = \mathbf{2^j}$. The sum of $1, 2, 4, 8, 16$ is **31**. The difference $2^j - 2^{j-1}$ equals $\mathbf{2^{j-1}}$. After a burst of speed V to time T, the distance is **VT**. If $f(T) = 1$ and V increases, the burst lasts only to $T = \mathbf{1/V}$. When V approaches infinity, $f(t)$ approaches a **step** function. The velocities approach a **delta** function, which is concentrated at $t = 0$ but has area **1** under its graph. The slope of a step function is **zero or Infinity**.

8 $f(t) = 1 + \mathbf{10t}$ for $0 \leq t \leq \frac{1}{10}$, $f(t) = \mathbf{2}$ for $t \geq \frac{1}{10}$

10 $f(3) = \mathbf{12}; g(f(3)) = g(12) = \mathbf{25}; g(f(t)) = g(4t) = \mathbf{8t + 1}$. When t is changed to $4t$, distance increases **four** times as fast and the velocity is multiplied by 4.

28 The second difference $f_{j+1} - 2f_j + f_{j-1}$ equals $(f_{j+1} - f_j) - (f_j - f_{j-1}) = v_{j+1} - v_j = 4$.

32 The period of $v + w$ is **30**, the smallest multiple of both 6 and 10. Then v completes five cycles and w completes three. An example for functions is $v = \sin \frac{\pi x}{3}$ and $w = \sin \frac{\pi x}{5}$.

42 The ratios $\frac{\Delta f}{\Delta t}$ are $\frac{e - 1}{1} = 1.718$ and $\frac{e^{.1} - 1}{.1} = 1.052$ and $\frac{e^{.01} - 1}{.01} = 1.005$. They are approaching 1.

1.3 The Velocity at an Instant

Worked examples 1-4 bring out the key ideas of Section 1.3. You are computing the slope of a curve – genuine calculus! To compute this slope at a particular time $t = 1$, find the change in distance $f(1+h) - f(1)$. To compute the velocity at a general time t, find the change $f(t+h) - f(t)$. In both cases, divide by h for average slope. Let $h \to 0$ for instantaneous velocity = slope at point.

1 For $f(t) = t^2 - t$ find the *average* speed between (a) $t = 1$ and $t = 2$ (b) $t = 0.9$ and $t = 1.0$ (c) $t = 1$ and $t = 1 + h$. (d) Use part (c) as $h \to 0$ to find the *instantaneous* speed at $t = 1$. (e) What is the *average* speed from t to $t + h$? (f) What is the formula for $v(t)$?

- (a) $f(2) = 2^2 - 2 = 2$ and $f(1) = 1^2 - 1 = 0$. Average speed $= \frac{2 - 0}{2 - 1} = 2$

- (b) $f(0.9) = 0.81 - 0.9 = -0.09$. Average speed is $\frac{f(1) - f(0.9)}{1 - (0.9)} = \frac{.09}{0.1} = 0.9$

- (c) $\frac{f(1+h) - f(1)}{(1+h) - (1)} = \frac{[(1+h)^2 - (1+h)] - [1^2 - 1]}{(1+h) - 1} = \frac{[(1 + 2h + h^2) - (1+h)] - 0}{h} = \frac{h^2 + h}{h} = h + 1.$

- (d) Let h get extremely small. Then $h + 1$ is near 1. The speed at $t = 1$ is $v = 1$.

- (e) $\frac{f(t+h) - f(t)}{(t+h) - t} = \frac{[(t+h)^2 - (t+h)] - [t^2 - t]}{h} = \frac{t^2 + 2ht + h^2 - t - h - t^2 + t}{h}.$
  After cancelling this is

$$\frac{h^2 + 2ht - h}{h} = h + 2t - 1 = average \ speed.$$

- (f) Let h go to zero. The average $h + 2t - 1$ goes to $0 + 2t - 1$. Thus $v(t) = 2t - 1$. *Check $v = 1$ at $t = 1$.*

2 If $v(t) = 6 - 2t$, find a formula for $f(t)$ and graph it.

- $f(t)$ is the area under the v-graph from 0 to t. For small t, this is the area of a *trapezoid* with heights 6 and $6 - 2t$. The base is t. Since area = average height times base, $f(t) = \frac{1}{2}(6 + 6 - 2t)t = (6 - t)t = 6t - t^2$. At $t = 3$ we reach $f(3) = 18 - 9 = 9$. This is the area of the triangle.

  Does this area formula $f(t) = 6t - t^2$ continue to apply after $t = 3$, when some area is negative? *Yes it does.* The triangle beyond $t = 3$ has base $t - 3$ and height (or depth) $6 - 2t$. Its area is $\frac{1}{2}bh = \frac{1}{2}(t - 3)(6 - 2t) = -9 + 6t - t^2$. After adding the positive area 9 of the first triangle, we still have $6t - t^2$.

  The negative height automatically gave negative area. The graph is a parabola that turns down at $t = 3$, when new area is negative.

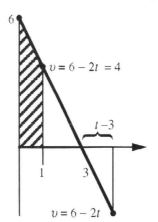
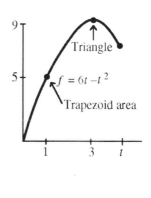

3 If $f(t) = 4t^2$, find the slope at time t. Start with the average from t to $t + h$.

- Step 1: The average slope is

$$\frac{f(t+h) - f(t)}{(t+h) - t} = \frac{4(t+h)^2 - 4t^2}{h} = \frac{(4t^2 + 8th + 4h^2) - 4t^2}{h}$$

This simplifies to $\frac{8th + 4h^2}{h} = \mathbf{8t + 4h}$.

- Step 2: Decide what happens to the average slope as h approaches zero. The average $8t + 4h$ approaches the slope at a point (the instantaneous velocity) which is $8t$.

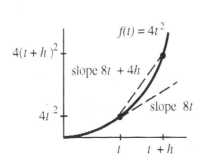
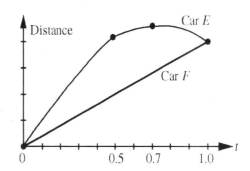

4 The graph shows distances for two cars. E and F start and end at the same place (why?) but their trips are different. Car F travels at a steady velocity. Describe what is happening to car E at these times:

(a) $t = 0$ to $t = .5$ (b) $t = .5$ (c) $t = .5$ to $t = .7$ (d) $t = .7$ (e) $t = .7$ to $t = 1$.

- (a) Car E covers more ground than F. It is going faster than F.

- (b) Since the slope of the graph *at this instant* is the same for E and F, their speed is the same.

- (c) E gains less ground than F, so E is going slower than F.

- (d) *At this instant E has speed zero.* The distance graph is level. *Maximum distance, zero speed.*

- (e) Car E loses ground, the slope of the distance graph is negative: E is going backward.

Here are the read-throughs and selected solutions to even-numbered problems.

Between the distances $f(2) = 100$ and $f(6) = 200$, the average velocity is $\frac{100}{4} = \mathbf{25}$. If $f(t) = \frac{1}{4}t^2$ then $f(6) = \mathbf{9}$ and $f(8) = \mathbf{16}$. The average velocity in between is $\mathbf{3.5}$. The instantaneous velocities at $t = 6$ and $t = 8$ are $\mathbf{3}$ and $\mathbf{4}$.

The average velocity is computed from $f(t)$ and $f(t+h)$ by $v_{\text{ave}} = \frac{1}{h}(\mathbf{f(t+h) - f(t)})$. If $f(t) = t^2$ then $v_{\text{ave}} = \mathbf{2t + h}$. From $t = 1$ to $t = 1.1$ the average is $\mathbf{2.1}$. The instantaneous velocity is the **limit** of v_{ave}. If the distance is $f(t) = \frac{1}{2}at^2$ then the velocity is $v(t) = \mathbf{at}$ and the acceleration is $\mathbf{a}$.

On the graph of $f(t)$, the average velocity between A and B is the slope of **the secant line**. The velocity at A is found by **letting B approach A**. The velocity at B is found by **letting A approach B**. When the velocity is positive, the distance is **increasing**. When the velocity is increasing, the car is **accelerating**.

2 (c) $\frac{\frac{1}{2}a(t^2 + 2th + h^2) - \frac{1}{2}at^2}{h} = at + \frac{1}{2}ah$. The limit at $h = 0$ is $v = at :=$ (acceleration) $\times$ (time).

(e) $\frac{6-6}{h} = 0$ (limit is 0); (f) the limit is $v(t) = 2t$ (and $f(t) = t^2$ gives $\frac{(t+h)^2 - t^2}{h} = 2t + h$).

14 True $\left(\text{the slope is } \frac{\Delta f}{\Delta t}\right)$; False (the curve is partly steeper and partly flatter than the secant line which gives the average slope); True (because $\Delta f = \Delta F$); False (V could be larger in between).

18 The graph is a parabola $f(t) = \frac{1}{2}t^2$ out to $f = 2$ at $t = 2$. After that the slope of f stays constant at 2.

20 Area to $t = 1$ is $\frac{1}{2}$; to $t = 2$ is $\frac{3}{2}$; to $t = 3$ is 2; to $t = 4$ is $\frac{3}{2}$; to $t = 5$ is $\frac{1}{2}$; area from $t = 0$ to $t = 6$ is zero. The graph of $f(t)$ through these points is parabola-line-parabola (symmetric)-line-parabola to zero.

26 $f(t) = t - t^2$ has $v(t) = 1 - 2t$ and $f(3t) = 3t - 9t^2$. The slope of $f(3t)$ is $3 - 18t$. This is $\mathbf{3v(3t)}$.

28 To find $f(t)$ multiply the time t by the average velocity. This is because $v_{\text{ave}} = \frac{f(t) - f(0)}{t - 0} = \frac{f(t)}{t}$.

1.4 Circular Motion

This section is important but optional (if that combination is possible). It is important because it introduces sines and cosines, and it also introduces the idea of a "parameter" t. No other example could do this better. These are *periodic* functions, which repeat every time you go around the circle. For any class studying physics, circular motion is essential. For calculus the section is optional, because we return in Section 2.4 to compute the derivatives of $\sin t$ and $\cos t$ from $\Delta f / \Delta t$.

It is certainly possible to study *circular motion* and omit up-and-down *harmonic motion*. In the circular case the ball is at $x = \cos t$, $y = \sin t$. In the up-and-down case the shadow is still at height $y = \sin t$, but now $x = 0$. My idea was that this up-and-down motion is made easy by its connection to circular motion. When we know the velocity of the ball on the circle, we know the derivatives of $\sin t$ and $\cos t$.

The standard motion has speed 1. The velocity vector is tangent to the unit circle of radius 1. The derivative of $\sin t$ is $\cos t$, and the derivative of $\cos t$ is $-\sin t$. Every student is going to learn those rules – by seeing the circle, you see what they mean. Calculus concentrates so much on a few particular functions (amazingly few), and sines and cosines are on that short list. I am not in favor of "late trigonometry," because seeing a function several times is the way to understand it.

Conclusion: This section *introduces* $\sin t$ and $\cos t$. *Don't study them to death*. Just begin to see how they work.

Questions **1-4** refer to a ball going counterclockwise around the unit circle, centered at $(0,0)$ with radius 1. The speed is 1 (say 1 meter/sec to have units). Assume the ball starts at angle zero. Then $x = \cos t$ and $y = \sin t$.

1 How long does it take for 10 revolutions? This means 10 times around the circle.

- The distance around a circle is $2\pi r$, which is 2π meters since $r = 1$. One revolution at speed 1 m/sec takes 2π seconds. So 10 revolutions take 20π seconds.

2 At time $t = \pi$, where is the ball?

- The ball is halfway around the circle at $x = -1, y = 0$.

3 Where is the ball (approximately) at $t = 14$ seconds?

- $\frac{14}{2\pi} \approx 2.23$, so the ball has made about $2\frac{1}{4}$ revolutions. Its position after that $\frac{1}{4}$ revolution is near the top of the circle where $x = 0, y = 1$. More accurately $y = \sin 14 = .99$.

4 Suppose that string is cut at $t = \frac{5\pi}{3}$. When and where would the ball hit the x axis?

- You need to know some trigonometry, which is shown in the figure. The ball is at $B = (\cos \frac{5\pi}{3}, \sin \frac{5\pi}{3}) = (\frac{1}{2}, -\frac{\sqrt{3}}{2})$. When the string is cut, the ball travels along BP at one meter per second. The tangent of the angle $POB = \frac{\pi}{3}$ is $\frac{PB}{OB} = \frac{\sqrt{3}}{1}$. Therefore the ball goes from B to P in $\sqrt{3}$ seconds. The time of arrival is $\frac{5\pi}{3} + \sqrt{3}$.
  The right triangle has $OB^2 + BP^2 = 1^2 + (\sqrt{3})^2 = 4$, so the hypotenuse is $OP = 2$.

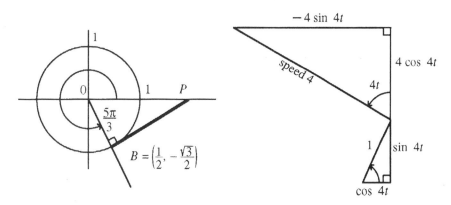

5 Another ball travels the unit circle counterclockwise with speed 4 m/sec starting at angle 0.

(a) What angle does it reach at time t? Answer: On the unit circle, a distance of 1 on the circumference corresponds to a central angle of 1 radian. Since the speed is 4 m/sec the central angle is increasing at 4 radians/sec. *At time t the angle is $4t$ radians.*

(b) What are the ball's x and y coordinates at time t? Answer: $(x, y) = (\cos 4t, \sin 4t)$.

(c) What are the ball's x and y velocities at time t? Answer: The velocity (the tangent vector) is drawn with length 4 because the speed is 4. The triangle inside the circle is similar to the triangle outside the circle. The sides of the larger triangle are four times longer. Thus the vertical velocity is $4\cos 4t$ and the horizontal velocity is $-4\sin 4t$.

  - Summary: Motion around the unit circle with speed k (counterclockwise) gives $x = \cos kt$ and $y = \sin kt$. The velocity is $-k \sin kt$ (horizontal) and $k \cos kt$ (vertical).

6 What are the components of velocity if a ball travels at 1 radian per second around a circle of radius R feet (starting at angle 0)?

- The ball's position at time t is $(R \cos t, R \sin t)$. Since the ball travels R feet along the circumference for each radian, the speed is R ft/sec. The horizontal velocity is $-R \sin t$ and the vertical velocity is $R \cos t$.

Motion around a circle of radius R feet at k radians per second gives $x = R \cos kt$ and $y = R \sin kt$. The speed is kR ft/sec. The horizontal velocity is $-Rk \sin kt$. The vertical velocity is $Rk \cos kt$.

Here are read-throughs and selected solutions.

A ball at angle t on the unit circle has coordinates $x = \mathbf{cos\ t}$ and $y = \mathbf{sin\ t}$. It completes a full circle at $t = \mathbf{2\pi}$. Its speed is $\mathbf{1}$. Its velocity points in the direction of the **tangent**, which is **perpendicular** to the radius coming out from the center. The upward velocity is $\mathbf{cos\ t}$ and the horizontal velocity is $\mathbf{-sin\ t}$.

A mass going up and down level with the ball has height $f(t) = \mathbf{sin\ t}$. This is called simple **harmonic** motion. The velocity is $v(t) = \mathbf{cos\ t}$. When $t = \pi/2$ the height is $f = \mathbf{1}$ and the velocity is $v = \mathbf{0}$. If a speeded-up mass reaches $f = \sin 2t$ at time t, its velocity is $v = \mathbf{2\cos 2t}$. A shadow traveling *under* the ball has $f = \cos t$ and $v = -\mathbf{sin\ t}$. When f is distance = area = integral, v is **velocity = slope = derivative**.

14 The ball goes halfway around the circle in time π. For the mass to fall a distance 2 in time π we need $2 = \frac{1}{2}a\pi^2$ so $a = 4/\pi^2$.

16 The area is $f(t) = \sin t$, and $\sin \frac{\pi}{6} - \sin 0 = \frac{1}{2}$.

18 The area is still $f(t) = \sin t$, and $\sin \frac{3\pi}{2} - \sin \frac{\pi}{2} = -1 - 1 = -2$.

20 The radius is 2 and time is speeded up by 3 so the speed is 6. There is a minus sign because the cosine starts downward (ball moving to left).

26 Counterclockwise with radius 3 starting at $(3,0)$ with speed 12.

1.5 Review of Trigonometry

Right triangles show the usual picture of trigonometry. The sine and cosine and tangent are ratios of the sides. The cosecant and secant and cotangent are the same ratios turned upside down.

A circle shows a better picture of trigonometry. The angle goes all the way to $360°$ and beyond. Of course we change $360°$ to 2π radians. Then distance around the circle is $r\theta$ = radius times angle. The derivative of $\sin t$ is $\cos t$, when angles are in radians. Otherwise we would have factors $\frac{2\pi}{360}$ which nobody wants. This section is helpful as a quick reference to the laws of trigonometry. *Don't forget*:

$$\sin 2t = 2 \sin t \cos t \quad \text{and} \quad \cos 2t = 2\cos^2 t - 1 = 1 - 2\sin^2 t.$$

Three more to remember: $\mathbf{sin^2}\theta + \mathbf{cos^2}\theta = 1$ and $1 + \mathbf{tan^2}\theta = \mathbf{sec^2}\theta$ and $1 + \mathbf{cot^2}\theta = \mathbf{csc^2}\theta$. Those all come from $x^2 + y^2 = r^2$, when you divide by r^2 then x^2 then y^2.

The distance from $(2,5)$ to $(3,3)$ is $\sqrt{1^2 + 2^2} = \sqrt{5}$. The book proves the addition formulas $\cos(s-t) = \mathbf{cos\ s\ cos\ t} + \mathbf{sin\ s\ sin\ t}$ and after sign change $\cos(s+t) = \mathbf{cos\ s\ cos\ t} - \mathbf{sin\ s\ sin\ t}$.

When $s = t$ you get $2t$, the "double-angle": $\cos 2t = \mathbf{cos^2 t} - \mathbf{sin^2 t}$ or $\mathbf{2\ cos^2 t} - \mathbf{1}$. Therefore $\frac{1}{2}(1 + \cos 2t) = \mathbf{cos^2 t}$, a formula needed in calculus.

1 (Problem 1.5.8) Find the distance from $(1,0)$ to $(0,1)$ along (a) a straight line (b) a quarter-circle and (c) a semicircle centered at $(\frac{1}{2}, \frac{1}{2})$.

- (a) Straight distance $\sqrt{1^2 + 1^2} = \sqrt{2}$. (b) The central angle is $\frac{\pi}{2}$, so the distance is $r\theta = \frac{\pi}{2}$. (c) The central angle is π and the radius is $\frac{\sqrt{2}}{2}$, half the length in part (a). Distance $\frac{\sqrt{2}}{2}\pi$.

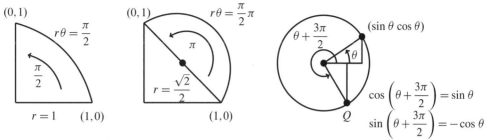

2 Simplify $\sin(\theta + \frac{3\pi}{2})$. Use formula (a), then use the circle.

- $\sin\theta\cos\frac{3\pi}{2} + \cos\theta\sin\frac{3\pi}{2}$ equals $(\sin\theta)(0) + (\cos\theta)(-1) = -\cos\theta$.

- Graphical solution: $\theta + \frac{3\pi}{2}$ is $\frac{3}{4}$ of a rotation ahead of θ. From the picture you see that $\sin(\theta + \frac{3\pi}{2})$ is the y coordinate of Q. This is $-\cos\theta$.

3 Find every angle θ that satisfies $\cos\theta = -1$.

- We know that $\cos\pi = -1$. Since the cosine repeats when θ is increased or decreased by 2π, the answer is $\theta = \pi \pm n(2\pi)$ for $n = 0, 1, 2, \cdots$. This gives the set $\theta = \pm\pi, \pm 3\pi, \pm 5\pi, \cdots$.

Read-throughs and selected even-numbered solutions :

Starting with a right triangle, the six basic functions are the ratios of the sides. Two ratios (the cosine x/r and the sine y/r) are below 1. Two ratios (the secant r/x and the cosecant **r/y**) are above 1. Two ratios (the **tangent** and the **cotangent**) can take any value. The six functions are defined for all angles θ, by changing from a triangle to a **circle**.

The angle θ is measured in **radians**. A full circle is $\theta = 2\pi$, when the distance around is $2\pi r$. The distance to angle θ is θr. All six functions have period **2π**. Going clockwise changes the sign of θ and **$\sin\theta$** and **$\tan\theta$**. Since $\cos(-\theta) = \cos\theta$, the cosine is **unchanged (or even)**.

4 $\cos 2(\theta + \pi)$ is the same as $\cos(2\theta + 2\pi)$ which is $\cos 2\theta$. Since $\cos^2\theta = \frac{1}{2} + \frac{1}{2}\cos 2\theta$, this also has period π.

14 $\sin 3t = \sin(2t + t) = \sin 2t \cos t + \cos 2t \sin t$. This equals $(2\sin t \cos t)\cos t + (\cos^2 t - \sin^2 t)\sin t$ or $3\sin t - 4\sin^3 t$.

26 $\sin\theta = \theta$ at $\theta = $ **0 and never again**. Reason: The right side θ has slope 1 and the left side has slope $\cos\theta < 1$. (So the graphs of $\sin\theta$ and θ can't meet a second time.)

30 $A\sin(x + \phi)$ equals $A\sin x\cos\phi + A\cos x\sin\phi$. That expression must match with $a\sin x + b\cos x$. Thus $a = A\cos\phi$ and $b = A\sin\phi$. Then $a^2 + b^2 = A^2\cos^2\phi + A^2\sin^2\phi = A^2$. Therefore $\mathbf{A} = \sqrt{\mathbf{a^2 + b^2}}$ and $\tan\phi = \frac{A\sin\phi}{A\cos\phi} = \frac{b}{a}$.

34 The amplitude and period of $2\sin\pi x$ are both **2**.

1.6 A Thousand Points of Light

There are many "point-graphs" that you could experiment with. Try $y = \sin cn$ for different choices of c and different ranges of n. The graphs are surprising.

The **Mathematica** command is ListPlot [Table[Sin [cn], n,N]]. Put in c and N.

1.7 Computing in Calculus (This has been updated as Section 0.5)

Computer labs and computer assignments have become an accepted part of calculus courses. Projects take 1-2 weeks, activities take 1-2 days. An essential part of each project is a report. Work in groups is demonstrated as very successful – the problem of equal work for equal grade is not severe in actual practice, and the advantages are great.

My favorite in-class activity is this *Water Tank Problem*, which can come early to emphasize slopes and graphs. It was created by Peter Taylor, adapted at Ithaca College, and is slightly extended here. Teachers and students are invited to work through it together, on the board and *in the class*.

At time $t = 0$, water begins to flow into an empty tank at the rate of 40 liters/minute. This flow rate is held constant for two minutes. From $t = 2$ to $t = 4$, the flow rate is gradually reduced to 5 liters/minute. This rate is constant for the final two minutes. At $t = 6$ the tank contains 120 liters.

A Draw a graph of the volume $V(t)$ of water in the tank for $0 \le t \le 6$.

B What is the average rate of flow into the tank over the entire six-minute period? Show how this can be interpreted on your graph.

C Starting at time $t = 2$, water is pumped out at the constant rate of 15 liters/minute. Now $V(t)$ above represents the total amount of water that has flowed *in*, and we let $W(t)$ represent the total amount pumped *out*. Plot $W(t)$ on the same axes. Show how to interpret the volume of water in the tank at any time.

D (*This is calculus*) Find on your graph the point at which the water level in the tank is a *maximum*. What is the instantaneous flow rate from the hose into the tank at this time?

E Graph the *rates* of flow in and out, for $0 \le t \le 6$. Mark the point where the water level is a maximum. Extra question: Can the flow rate into the tank be linear from $t = 2$ to $t = 4$?

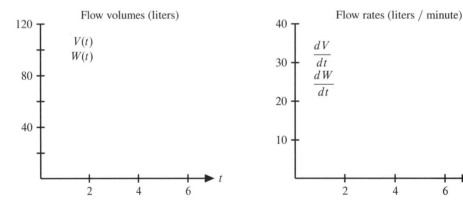

CHAPTER REVIEW PROBLEMS

Graph Problems These questions are asking for sketches by hand, not for works of art. It is surprising how much of calculus shows up in these pictures.

G1 Draw $f(x) = \sin x$ between $x = 0$ and $x = 4\pi$. Mark all maximum points by M and all minimum points by m. Mark the points where the slope is steepest (going up or down) by S.

Under that graph draw $v(x) = \cos x$ between $x = 0$ and $x = 4\pi$. *Line it up correctly.* Repeat the letters M, m, S down on the cosine graph. Notice that M and m are at points where the cosine is _____ , and S is at points where the cosine is _____ .

G2 Draw $y = 2 - x^2$ between $x = -1$ and $x = 1$. Mark the maximum point by M and the steepest points by S. Directly under that graph draw $v = -2x$. Move the letters M and S

down to this slope graph. When y has a maximum its slope is _____ . When y is steepest its slope has a _____ .

G3 Draw a smooth curve that goes up then down then up again. A specific choice would be the graph of $y = x^3 - x$ from $x = -2$ to $x = 2$. Mark the maximum, minimum, and steepest points by M, m, S. Under your curve draw a graph of its slope. (The slope of $x^3 - x$ is $3x^2 - 1$.) If the graphs are lined up, then M and m are at points where the slope is zero.

New question: What is different about the M point and the m point? How can you tell a maximum from a minimum, looking at the "zero crossing" on the slope graph?

Answer: At a maximum of $y(x)$, its slope crosses from _____ to _____ .

G4 **The triangles of trigonometry**. Draw a right triangle with sides marked $\cos\theta, \sin\theta$, and 1. Enlarge it to a triangle with sides marked $1, \tan\theta$, and $\sec\theta$. (You multiplied the sides of the first triangle by $\sec\theta$, which is greater than 1.) From the second triangle read off the identity $1 + \tan^2\theta =$ _____ . Now draw a third triangle with sides $\cot\theta, 1$, and $\csc\theta$. You have multiplied the first triangle by $\csc\theta$. This is $1/\sin\theta$ and again it is greater than 1. From this third triangle read off the identity $1 + \cot^2\theta =$ _____ .

Question: Is the third triangle always larger than the second triangle? *Answer: No.* To get the third triangle from the second, multiply by $\frac{\cos\theta}{\sin\theta}$. This is larger than 1 when θ is larger than _____ . The triangles have $0 < \theta < 90°$. At $\theta = 45°$, which two triangles are the same?

Computing Problems Use a computer or calculator. You do not need a supercomputer.

C1 Find the tax on incomes of $x_1 = \$1,000,000$ and $x_2 = \$2,000,000$. Tax rates for a single person are in Section 1.2. How large is the tax difference $f(x_2) - f(x_1)$?

C2 What income x would leave you with $\$1,000,000$ after paying tax in full?

C3 For the function $f(x) = x^2 + x^3 + x^4$ compute $f(1.1)$, $f(1.01)$, and $f(1.001)$. Subtract $f(1)$ from each of those and divide by $.1$, $.01$, and $.001$.

C4 Starting from $f_0 = 0$ with differences $v = 1, \frac{1}{2}, \frac{1}{3}, \frac{1}{4}, \frac{1}{5}, \cdots$ compute $f_{100}, f_{200}, f_{400}, f_{800}$.

Review Problems

R1 Under what condition on y_2 will the line through $(1, 5)$ and $(4, y_2)$ have a positive slope?

R2 Under what condition on x_2 will the line through $(1, 5)$ and $(x_2, 3)$ have a positive slope?

R3 Find the slope of the line through (x_1, y_1) and (x_2, y_2). When is the slope infinite?

R4 Explain the basic idea of *"Calculus without limits"* in Section 1.2.

R5 Explain how the average velocity $= \dfrac{\text{change in } f}{\text{change in } t}$ leads to an instantaneous velocity.

R6 For $f(t) = t^3$ find the average velocity between $t = 0$ and $t = h$. What is the instantaneous velocity at $t = 0$?

R7 For $f(t) = t^3$ find the average velocity between $t = 1$ and $t = 1 + h$. As $h \to 0$ find $v(1)$.

Drill Problems

D1 Find the equation of the line through $x = 0, y = 3$ with slope 5.

D2 Find the equation of the line through $(0, 3)$ and $(2, 7)$.

D3 Find the distance function $f(t)$ if $f(0) = 3$ and $f(2) = 7$ and the velocity is constant.

D4 If $f(0) = 6$ and $f(10) = 26$ find the average velocity between $t = 0$ and $t = 10$. *True or false*: The actual velocity is above and below that average for equal times (5 above and 5 below).

D5 For $f(t) = 7t - 5$ write down $f(t + h)$, subtract $f(t)$, and divide by h.

D6 For $f(t) = 7t^2 - 5$ write down $f(t + h)$, subtract $f(t)$, and divide by h.

D7 Find the differences v for $f = 1, 2, 4, 2, 1$. The sum of differences is _____ .

D8 Find f_3 starting from $f_0 = 10$ if the differences v are $2, -6, 7$.

D9 Express $\cos 2t$ in terms of $\cos^2 t$ and then in terms of $\sin^2 t$.

CHAPTER 1 INTRODUCTION TO CALCULUS

Section 1.1 Velocity and Distance   (page 6)

1 $v = 30, 0, -30$; $v = -10, 20$ **3** $v(t) = \begin{cases} 2 & \text{for} \;\; 0 < t < 10 \\ 1 & \text{for} \;\; 10 < t < 20 \\ -3 & \text{for} \;\; 20 < t < 30 \end{cases}$ $v(t) = \begin{cases} 0 & \text{for} \;\; 0 < t < T \\ \frac{1}{T} & \text{for} \;\; T < t < 2T \\ 0 & \text{for} \;\; 2T < t < 3T \end{cases}$

5 25; 22; $t + 10$ **7** 6; -30 **9** $v(t) = \begin{cases} 20 & \text{for} \;\; t < .2 \\ 0 & \text{for} \;\; t > .2 \end{cases}$ $f(t) = \begin{cases} 20t & \text{for} \;\; t \le .2 \\ 4 & \text{for} \;\; t \ge .2 \end{cases}$ **11** 10%; $12\frac{1}{2}\%$

13 $f(t) = 0, 30(t-1), 30$; $f(t) = -30t, -60, 30(t-6)$ **15** Average $8, 20$ **17** $40t - 80$ for $1 \le t \le 2.5$

21 $0 \le t \le 3, -40 \le f \le 20$; $0 \le t \le 3T, 0 \le f \le 60T$ **23** $3 - 7t$ **25** $6t - 2$ **27** $3t + 7$

29 Slope -2; $1 \le f \le 9$ **31** $v(t) = \begin{cases} 8 & \text{for} \;\; 0 < t < T \\ -2 & \text{for} \;\; T < t < 5T \end{cases}$ $f(t) = \begin{cases} 8t & \text{for} \;\; 0 \le t \le T \\ 10T - 2t & \text{for} \;\; T \le t \le 5T \end{cases}$

33 $\frac{9}{5}C + 32$; slope $\frac{9}{5}$ **35** $f(w) = \frac{w}{1000}$; slope = conversion factor   **37** $1 \le t \le 5, 0 \le f \le 2$

39 $0 \le t \le 5, 0 \le f \le 4$ **41** $0 \le t \le 5, 1 \le t \le 32$ **43** $\frac{1}{2}t + 4$; $\frac{1}{2}t + \frac{7}{2}$; $2t + 12$; $2t + 3$

45 Domains $-1 \le t \le 1$: ranges $0 \le 2t + 2 \le 4$, $-3 \le t - 2 \le -1$, $-2 \le -f(t) \le 0, 0 \le f(-t) \le 2$

47 $\frac{3}{2}V$; $\frac{3}{2}V$ **49**   input * input $\to A$ input * input $\to A$ $B*B \to C$ input$+1 \to A$
                       input $+A \to$ output   input $+A \to B$ $B + C \to$ output $A * A \to B$
                                                              $A + B \to$ output

51 $3t + 5, 3t + 1, 6t - 2, 6t - 1, -3t - 1, 9t - 4$; slopes $3, 3, 6, 6, -3, 9$

53 The graph goes up and down *twice*. $f(f(t)) = \begin{cases} 2(2t) & 0 \le t \le 1.5 & 12 - 2(12 - 2t) & 3 \le t \le 4.5 \\ 12 - 4t & 1.5 \le t \le 3 & 2(12 - 2t) & 4.5 \le t \le 6 \end{cases}$

Section 1.2 Calculus Without Limits   (page 14)

1 $2 + 5 + 3 = 10$; $f = 1, 3, 8, 11$; 10 **3** $f = 3, 4, 6, 7, 7, 6$; max f at $v = 0$ or at break from $v = 1$ to -1

5 $1.1, -2, 5$; $f(6) = 6.6, -11, 4$; $f(7) = 7.7, -13, 9$ **7** $f(t) = 2t$ for $t \le 5$, $10 + 3(t - 5)$ for $t \ge 5$; $f(10) = 25$

9 $7, 28, 8t + 4$; multiply slopes   **11** $f(8) = 8.8, -15, 14$; $\frac{\Delta f}{\Delta t} = 1.1, -2, 5$

13 $f(x) = 3052.50 + .28(x - 20, 350)$; then $11, 580.50$ is $f(49, 300)$ **15** $19\frac{1}{4}\%$

17 Credit subtracts $1,000$, deduction only subtracts 15% of 1000 **19** All $v_j = 2$; $v_j = (-1)^{j-1}$; $v_j = \left(\frac{1}{2}\right)^j$

21 L's have area $1, 3, 5, 7$ **23** $f_j = j$; sum $j^2 + j$; sum $\frac{j^2}{2} + \frac{j}{2}$ **25** $(101^2 - 99^2)/2 = \frac{400}{2}$ **27** $v_j = 2j$ **29** $f_{31} = 5$

31 $a_j = -f_j$ **33** 0; 1; $.1$ **35** $v = 2, 6, 18, 54$; $2 \cdot 3^{j-1}$ **37** $\frac{\Delta f}{\Delta t} = 1, .7177, .6956, .6934 \to \ln 2 = .6931$ in Chapter 6

39 $v_j = -\left(\frac{1}{2}\right)^j$ **41** $v_j = 2(-1)^j$, sum is $f_j - 1$ **45** $v = 1000, t = 10/V$

47 M,N   **51** $\sqrt{9} < 2 \cdot 9 < 9^2 < 2^9$; $\left(\frac{1}{9}\right)^2 < 2\left(\frac{1}{9}\right) < \sqrt{1/9} < 2^{1/9}$

Section 1.3 The Velocity at an Instant   (page 21)

1 $6, 6, \frac{13}{2}a, -12, 0, 13$ **3** $4, 3.1, 3 + h, 2.9$ **5** Velocity at $t = 1$ is 3 **7** Area $f = t + t^2$, slope of f is $1 + 2t$

9 F; F; F; T   **11** 2; $2t$ **13** $12 + 10t^2$; $2 + 10t^2$ **15** Time 2, height 1, stays above $\frac{3}{4}$ from $t = \frac{1}{2}$ to $\frac{3}{2}$

17 $f(6) = 18$ **21** $v(t) = -2t$ then $2t$ **23** Average to $t = 5$ is 2; $v(5) = 7$ **25** $4v(4t)$ **27** $v_{\text{ave}} = t, v(t) = 2t$

Section 1.4 Circular Motion   (page 28)

1 10π, $(0, -1)$, $(-1, 0)$ **3** $(4\cos t, 4\sin t)$; 4 and $4t$; $4\cos t$ and $-4\sin t$

5 $3t$; $(\cos 3t, \sin 3t)$; $-3\sin 3t$ and $3\cos 3t$ **7** $x = \cos t$; $\sqrt{2}/2$; $-\sqrt{2}/2$ **9** $2\pi/3$; 1; 2π

11 Clockwise starting at $(1, 0)$ **13** Speed $\frac{2}{\pi}$ **15** Area 2 **17** Area 0

19 4 from speed, 4 from angle **21** $\frac{1}{4}$ from radius times 4 from angle gives 1 in velocity

23 Slope $\frac{1}{2}$; average $(1 - \frac{\sqrt{3}}{2})/(\pi/6) = \frac{3(2-\sqrt{3})}{\pi} = .256$ **25** Clockwise with radius 1 from $(1,0)$, speed 3

27 Clockwise with radius 5 from $(0,5)$, speed 10 **29** Counterclockwise with radius 1 from $(\cos 1, \sin 1)$, speed 1

31 Left and right from $(1,0)$ to $(-1,0)$, $v = -\sin t$ **33** Up and down between 2 and -2; start $2\sin\theta$, $v = 2\cos(t+\theta)$

35 Up and down from $(0,-2)$ to $(0,2)$; $v = \sin\frac{1}{2}t$ **37** $x = \cos\frac{2\pi t}{360}$, $y = \sin\frac{2\pi t}{360}$, speed $\frac{2\pi}{360}$, $v_{up} = \cos\frac{2\pi t}{360}$

Section 1.5 A Review of Trigonometry (page 33)

1 Connect corner to midpoint of opposite side, producing $30°$ angle **3** π **7** $\frac{\theta}{2\pi} \to$ area $\frac{1}{2}r^2\theta$

9 $d = 1$, distance around hexagon $<$ distance around circle **11** T; T; T; F

13 $\cos(2t+t) = \cos 2t \cos t - \sin 2t \sin t = 4\cos^3 t - 3\cos t$

15 $\frac{1}{2}\cos(s-t) + \frac{1}{2}\cos(s+t)$; $\frac{1}{2}\cos(s-t) - \frac{1}{2}\cos(s+t)$ **17** $\cos\theta = \sec\theta = \pm 1$ at $\theta = n\pi$

19 Use $\cos\left(\frac{\pi}{2} - s - t\right) = \cos\left(\frac{\pi}{2} - s\right)\cos t + \sin\left(\frac{\pi}{2} - s\right)\sin t$ **23** $\theta = \frac{3\pi}{2} +$ multiple of 2π

25 $\theta = \frac{\pi}{4} +$ multiple of π **27** No θ **29** $\phi = \frac{\pi}{4}$ **31** $|OP| = a$, $|OQ| = b$

CHAPTER 2 DERIVATIVES

Section 2.1 The Derivative of a Function (page 49)

1 (b) and (c) **3** $12 + 3h$; $13 + 3h$; 3; 3 **5** $f(x) + 1$ **7** -6 **9** $2x + \Delta x + 1$; $2x + 1$

11 $\frac{4}{t+\Delta t} - \frac{4}{t} = \frac{-4}{t(t+\Delta t)} \to \frac{-4}{t^2}$ **13** 7; 9; corner **15** $A = 1$, $B = -1$ **17** F; F; T; F

19 $b = B$; m and M; m or undefined **21** Average $x_2 + x_1 \to 2x_1$

25 $\frac{1}{2}$; no limit (one-sided limits $1, -1$); 1; 1 if $t \neq 0$, -1 if $t = 0$ **27** $f'(3)$; $f(4) - f(3)$

29 $2x^4(4x^3) = 8x^7$ **31** $\frac{du}{dx} = \frac{1}{2u} = \frac{1}{2\sqrt{x}}$ **33** $\frac{\Delta f}{\Delta x} = -\frac{1}{2}$; $f'(2)$ doesn't exist **35** $2f\frac{df}{dx} = 4u^3\frac{du}{dx}$

Section 2.2 Powers and Polynomials (page 56)

1 $6x^5$; $30x^4$; $f''''''' = 720 = 6!$ **3** $2x + 7$ **5** $1 + 2x + 3x^2 + 4x^3$ **7** $nx^{n-1} - nx^{-n-1}$

9 $1 + x + \frac{1}{2}x^2 + \frac{1}{6}x^3$ **11** $-\frac{1}{x}$, $\left(-\frac{1}{x}\right) + 5$ **13** $x^{-2/3}$; $x^{-4/3}$; $-\frac{1}{9}x^{-4/3}$

15 $3x^2 - 1 = 0$ at $x = \frac{1}{\sqrt{3}}$ and $\frac{-1}{\sqrt{3}}$ **17** 8 ft/sec; -8 ft/sec; 0 **19** Decreases for $-1 < x < \frac{1}{3}$

21 $\frac{(x+h)-x}{h(\sqrt{x+h}-\sqrt{x})} \to \frac{1}{2\sqrt{x}}$ **23** 1 5 10 10 5 1 adds to $(1+1)^5$ $(x = h = 1)$

25 $3x^2$; $2h$ is difference of x's **27** $\frac{\Delta f}{\Delta x} = 2x + \Delta x + 3x^2 + 3x\Delta x + (\Delta x)^2 \to 2x + 3x^2 =$ sum of separate derivatives

29 $7x^6$; $7(x+1)^6$ **31** $\frac{1}{24}x^4$ plus any cubic **33** $x + \frac{1}{2}x^2 + \frac{1}{3}x^3 + \frac{1}{4}x^4 + C$ **35** $\frac{1}{24}x^4$, $\frac{1}{120}x^5$

37 F; F; F; T; T **39** $\frac{y}{x} = .12$ so $\frac{\Delta y}{\Delta x} = \frac{1}{2}(.12)$; six cents **41** $\frac{\Delta y}{\Delta x} = \frac{1}{\Delta x}\left(\frac{c}{x+\Delta x} - \frac{c}{x}\right)$, $\frac{dy}{dx} = -\frac{c}{x^2}$

43 $E = \frac{2x}{2x+3}$ **45** t to $\sqrt[3]{2t}$ **47** $\frac{1}{10}x^{10}$; $\frac{1}{n+1}x^{n+1}$; divide by $n+1 = 0$

49 $.7913, -3.7913, 1.618, -.618; 0, 1.266, -2.766$

Section 2.3 The Slope and the Tangent Line (page 63)

1 $\frac{-12}{x^2}$; $y - 6 = -3(x-2)$; $y - 6 = \frac{1}{3}(x-2)$; $y - 6 = -\frac{3}{2}(x-2)$ **3** $y + 1 = 3(x-1)$; $y = 3x - 4$

5 $y = x$; $(3,3)$ **7** $y - a^2 = (c+a)(x-a)$; $y - a^2 = 2a(x-a)$ **9** $y = \frac{1}{5}x^2 + 2$; $y - 7 = -\frac{1}{2}(x-5)$

11 $y = 1$; $x = \frac{\pi}{2}$ **13** $y - \frac{1}{a} = -\frac{1}{a^2}(x-a)$; $y = \frac{2}{a}$, $x = 2a$; 2 **15** $c = 4$, tangent at $x = 2$

17 $(-3, 19)$ and $\left(\frac{1}{3}, \frac{13}{27}\right)$ **19** $c = 4$, $y = 3 - x$ tangent at $x = 1$

21 $(1+h)^3$; $3h + 3h^2 + h^3$; $3 + 3h + h^2$; 3 **23** Tangents parallel, *same* normal

25 $y = 2ax - a^2$, $Q = (0, -a^2)$; distance $a^2 + \frac{1}{4}$; angle of incidence = angle of reflection

27 $x = 2p$; focus has $y = \frac{x^2}{4p} = p$ **29** $y - \frac{1}{\sqrt{2}} = x + \frac{1}{\sqrt{2}}$; $x = -\frac{2}{\sqrt{2}} = -\sqrt{2}$

31 $y - a^2 = -\frac{1}{2a}(x-a)$; $y = a^2 + \frac{1}{2}$; $a = \frac{\sqrt{3}}{2}$ **33** $\left(\frac{1}{x^2}\right)(1000) = 10$ at $x = 10$ hours   **35** $a = 2$

37 1.01004512; $1 + 10(.001) = 1.01$ **39** $(2+\Delta x)^3 - (8 + 6\Delta x) = 6(\Delta x)^2 + (\Delta x)^3$ **41** $x_1 = \frac{5}{4}$; $x_2 = \frac{41}{40}$

43 $T = 8$ sec; $f(T) = 96$ meters   **45** $a > \frac{4}{5}$ meters/sec^2

Section 2.4   The Derivative of the Sine and Cosine   (page 70)

1 (a) and (b)   **3** $0; 1; 5; \frac{1}{5}$ **5** $\sin(x + 2\pi)$; $(\sin h)/h \to 1$; 2π **7** $\cos^2 \theta \approx 1 - \theta^2 + \frac{1}{4}\theta^4$; $\frac{1}{4}\theta^4$ is small

9 $\sin\frac{1}{2}\theta \approx \frac{1}{2}\theta$ **11** $\frac{3}{2}$; 4 **13** $PS = \sin h$; area $OPR = \frac{1}{2}\sin h <$ curved area $\frac{1}{2}h$

15 $\cos x = 1 - \frac{x^2}{2\cdot 1} + \frac{x^4}{4\cdot 3\cdot 2\cdot 1} - \cdots$ **17** $\frac{1}{2h}(\cos(x+h) - \cos(x-h)) = \frac{1}{h}(-\sin x \sin h) \to -\sin x$

19 $y' = \cos x - \sin x = 0$ at $x = \frac{\pi}{4} + n\pi$ **21** $(\tan h)/h = \sin h / h \cos h < \frac{1}{\cos h} \to 1$

23 Slope $\frac{1}{2}\cos\frac{1}{2}x = \frac{1}{2}, 0, -\frac{1}{2}, \frac{1}{2}$; no   **25** $y = 2\cos x + \sin x$; $y'' = -y$ **27** $y = -\frac{1}{3}\cos 3x$; $y = \frac{1}{3}\sin 3x$

29 In degrees $(\sin h)/h \to 2\pi/360 = .01745$ **31** $2\sin x \cos x + 2\cos x(-\sin x) = 0$

Section 2.5   The Product and Quotient and Power Rules   (page 77)

1 $2x$ **3** $\frac{-1}{(1+x)^2} - \frac{\cos x}{(1+\sin x)^2}$ **5** $(x-2)(x-3) + (x-1)(x-3) + (x-1)(x-2)$

7 $-x^2 \sin x + 4x\cos x + 2\sin x$ **9** $2x - 1 - \frac{1}{\sin^2 x}$ **11** $2\sqrt{x}\sin x \cos x + \frac{1}{2}x^{-1/2}\sin^2 x + \frac{1}{2}(\sin x)^{-1/2}\cos x$

13 $4x^3 \cos x - x^4 \sin x + \cos^4 x - 4x\cos^3 x \sin x$ **15** $\frac{1}{2}x^2 \cos x + 2x \sin x$ **17** 0 **19** $-\frac{8}{3}(x-5)^{-5/3} + \frac{8}{3}(5-x)^{-5/3}(=0?)$

21 $3(\sin x \cos x)^2(\cos^2 x - \sin^2 x) + 2\cos 2x$ **23** $u'vwz + v'uwz + w'uvz + z'uvw$ **25** $-\csc^2 x - \sec^2 x$

27 $V = \frac{t\cos t}{1+t}$, $V' = \frac{\cos t - t\sin t - t^2 \sin t}{(1+t)^2}$ $A = 2\left(\frac{t}{t+1} + t\cos t + \frac{\cos t}{t+1}\right)$ $A' = 2(\cos t - t\sin t + \frac{1-\cos t}{(t+1)^2} - \frac{\sin t}{t+1})$

29 $10t$ for $t < 10$, $\frac{50}{\sqrt{t-10}}$ for $t > 10$ **31** $\frac{2t^3 + 3t^2}{(1+t)^2}$; $\frac{2t^3 + 6t^2 + 6t}{(1+t)^3}$

33 $u''v + 2u'v' + uv''$; $u'''v + 3u''v' + 3u'v'' + v'''$ **35** $\frac{1}{2}\sin^2 t$; $\frac{1}{2}\tan^2 t$; $\frac{2}{3}[(1+t)^{3/2} - 1]$

39 T; F; F; T; F   **41** degree $2n - 1$/degree $2n$ **43** $v(t) = \cos t - t\sin t$ $(t \le \frac{\pi}{2})$; $v(t) = -\frac{\pi}{2}(t \ge \frac{\pi}{2})$

45 $y = \frac{2hx^3}{L^3} + \frac{3hx^2}{L^2}$ has $\frac{dy}{dx} = 0$ at $x = 0$ (no crash) and at $x = -L$ (no dive). Then $\frac{dy}{dx} = \frac{6Vh}{L}\left(\frac{x^2}{L^2} + \frac{x}{L}\right)$ and $\frac{d^2 y}{dx^2} = \frac{6V^2 h}{L^2}\left(\frac{2x}{L} + 1\right)$.

Section 2.6   Limits   (page 84)

1 $\frac{1}{4}$, $L = 0$, after $N = 10$; $\frac{25}{24}$, ∞, no N; $\frac{1}{4}$, 0, after 5; 1.1111, $\frac{10}{9}$, all n; $\sqrt{2}$, 1, after 38; $\sqrt{20} - 4$, $\frac{1}{2}$, all n; $\frac{625}{256}$, $e = 2.718\cdots$, after $N = 12$.

3 (c) and (d)   **5** Outside any interval around zero there are only a finite number of a's   **7** $\frac{5}{2}$ **9** $\frac{f(h) - f(0)}{h}$ **11** 1

13 1 **15** $\sin 1$ **17** No limit   **19** $\frac{1}{2}$ **21** Zero if $f(x)$ is continuous at a **23** 2

25 $.001, .0001, .005, .1$ **27** $|f(x) - L|$; $\frac{4x}{1+x}$ **29** 0; $X = 100$ **33** 4; ∞; 7; 7 **35** 3; no limit; 0; 1

37 $\frac{1}{1-r}$ if $|r| < 1$; no limit if $|r| \ge 1$ **39** $.0001$; after $N = 7$ (or 8?)   **41** $\frac{1}{2}$

43 9; $8\frac{1}{2}$; $a_n - 8 = \frac{1}{2}(a_{n-1} - 8) \to 0$

45 $a_n - L \le b_n - L \le c_n - L$ so $|b_n - L| < \varepsilon$ if $|a_n - L| < \varepsilon$ and $|c_n - L| < \varepsilon$

Section 2.7 Continuous Functions (page 89)

1 $c = \sin 1$; no c **3** Any c; $c = 0$ **5** $c = 0$ or 1; no c **7** $c = 1$; no c **9** no c; no c

11 $c = \frac{1}{64}$; $c = \frac{1}{64}$ **13** $c = -1$; $c = -1$ **15** $c = 1$; $c = 1$ **17** $c = -1$; $c = -1$

19 $c = 2, 1, 0, -1, \cdots$; same c **21** $f(x) = 0$ except at $x = 1$ **23** $\sqrt{x-1}$ **25** $-\frac{2x}{|x|}$ **27** $\frac{5}{x-1}$

29 One; two; two **31** No; yes; no **33** $xf(x), (f(x))^2, x, f(x), 2(f(x) - x), f(x) + 2x$ **35** F; F; F; T

37 Step; $f(x) = \sin\frac{1}{x}$ with $f(0) = 0$ **39** Yes; no; no; yes ($f_4(0) = 1$)

41 $g\left(\frac{1}{2}\right) = f(1) - f\left(\frac{1}{2}\right) = f(0) - f\left(\frac{1}{2}\right) = -g(0)$; zero is an intermediate value between $g(0)$ and $g\left(\frac{1}{2}\right)$

43 $f(x) - x$ is ≥ 0 at $x = 0$ and ≤ 0 at $x = 1$

CHAPTER 3 APPLICATIONS OF THE DERIVATIVE

Section 3.1 Linear Approximation (page 95)

1 $Y = x$ **3** $Y = 1 + 2(x - \frac{\pi}{4})$ **5** $Y = 2\pi(x - 2\pi)$ **7** $2^6 + 6 \cdot 2^5 \cdot .001$ **9** 1

11 $1 - 1(-.02) = 1.02$ **13** Error .000301 vs. $\frac{1}{2}(.0001)6$ **15** $.0001 - \frac{1}{3}10^{-8}$ vs. $\frac{1}{2}(.0001)(2)$

17 Error .59 vs. $\frac{1}{2}(.01)(90)$ **19** $\frac{d}{dx}\sqrt{1-x} = \frac{-1}{2\sqrt{1-x}} = -\frac{1}{2}$ at $x = 0$

21 $\frac{d}{du}\sqrt{c^2+u} = \frac{1}{2\sqrt{c^2+u}} = \frac{1}{2c}$ at $u = 0$, $c + \frac{u}{2c} = c + \frac{x^2}{2c}$ **23** $dV = 3(10)^2(.1)$

25 $A = 4\pi r^2$, $dA = 8\pi r\, dr$ **27** $V = \pi r^2 h$, $dV = 2\pi rh\, dr$ (plus $\pi r^2\, dh$) **29** $1 + \frac{1}{2}x$ **31** 32nd root

Section 3.2 Maximum and Minimum Problems (page 103)

1 $x = -2$: abs min **3** $x = -1$: rel max, $x = 0$: abs min, $x = 4$: abs max

5 $x = -1$: abs max, $x = 0, 1$: abs min, $x = \frac{1}{2}$: rel max **7** $x = -3$: abs min, $x = 0$: rel max, $x = 1$: rel min

9 $x = 1, 9$: abs min, $x = 5$: abs max **11** $x = \frac{1}{3}$: rel max, $x = 1$: rel min, $x = 0$: stationary (not min or max)

13 $x = 0, 1, 2, \cdots$: abs min, $x = \frac{1}{2}, \frac{3}{2}, \frac{5}{2}, \cdots$: abs max **15** $|x| \leq 1$: all min, $x = -3$ abs max, $x = 2$ rel max

17 $x = 0$: rel min, $x = \frac{1}{3}$: abs max, $x = 4$: abs min

19 $x = 0$: abs min, $x = \pi$: stationary (not min or max), $x = 2\pi$: abs max

21 $\theta = 0$: rel min, $\tan\theta = -\frac{4}{3}(\sin\theta = \frac{4}{5}$ and $\cos\theta = -\frac{3}{5}$ abs max, $\sin\theta = -\frac{4}{5}$ and $\cos\theta = \frac{3}{5}$ abs min), $\theta = 2\pi$: rel max

23 $h = \frac{1}{3}(62''$ or 158 cm); cube **25** $\frac{v}{av^2+b}$; $2\sqrt{ab}$ gallons/mile, $\frac{1}{2\sqrt{ab}}$ miles/gallon at $v = \sqrt{\frac{b}{a}}$

27 (b) $\theta = \frac{3\pi}{8} = 67.5°$ **29** $x = \frac{a}{\sqrt{3}}$; compare Example 7; $\frac{a}{b} = \sqrt{3}$

31 $R(x) - C(x)$; $\frac{R(x) - C(x)}{x}$; $\frac{dR}{dx} - \frac{dC}{dx}$; profit **33** $x = \frac{d-a}{2(b-e)}$; zero **35** $x = 2$

37 $V = x(6 - \frac{3x}{2})(12 - 2x)$; $x \approx 1.6$ **39** $A = \pi r^2 + x^2$, $x = \frac{1}{4}(4 - 2\pi r)$; $r_{\min} = \frac{2}{2+\pi}$

41 max area 2500 vs $\frac{10000}{\pi} = 3185$ **43** $x = 2, y = 3$ **45** $P(x) = 12 - x$; thin rectangle up y axis

47 $h = \frac{H}{3}, r = \frac{2R}{3}, V = \frac{4\pi R^2 H}{27} = \frac{4}{9}$ of cone volume

49 $r = \frac{HR}{2(H-R)}$; best cylinder has *no* height, area $2\pi R^2$ from top and bottom(?)

51 $r = 2, h = 4$ **53** 25 and 0 **55** 8 and $-\infty$

57 $\sqrt{r^2+x^2} + \sqrt{q^2+(s-x)^2}$; $\frac{df}{dx} = \frac{x}{\sqrt{r^2+x^2}} - \frac{s-x}{\sqrt{q^2+(s-x)^2}} = 0$ when $\sin a = \sin c$

59 $y = x^2 = \frac{3}{2}$ **61** $(1, -1), (\frac{13}{5}, -\frac{1}{5})$ **63** $m = 1$ gives nearest line **65** $m = \frac{1}{3}$ **67** equal; $x = \frac{1}{2}$

69 $\frac{1}{x}x^2$ **71** True (use sign change of f'')

73 Radius R, swim $2R\cos\theta$, run $2R\theta$, time $\frac{2R\cos\theta}{v} + \frac{2R\theta}{10v}$; *max* when $\sin\theta = \frac{1}{10}$; *min* all run

Section 3.3 Second Derivatives: Bending and Acceleration (page 110)

3 $y = -1 - x^2$; no $\cdots$ **5** False **7** True **9** True (f' has 8 zeros, f'' has 7)

11 $x = 3$ is min: $f''(3) = 2$ **13** $x = 0$ not max or min; $x = \frac{9}{2}$ is min: $f''(\frac{9}{2}) = 81$

15 $x = \frac{3\pi}{4}$ is max: $f''(\frac{3\pi}{4}) = -\sqrt{2}$; $x = \frac{7\pi}{4}$ is min: $f''(\frac{7\pi}{4}) = \sqrt{2}$

17 Concave down for $x > \frac{1}{3}$ (inflection point)

19 $x = 3$ is max: $f''(3) = -4$; $x = 2, 4$ are min but $f'' = 0$ **21** $f(\Delta x) = f(-\Delta x)$ **23** $1 + x - \frac{x^2}{2}$

25 $1 - \frac{x^2}{6}$ **27** $1 - \frac{1}{2}x - \frac{1}{8}x^2$ **29** Error $\frac{1}{2}f''(x)\Delta x$ **31** Error $0\Delta x + \frac{1}{3}f'''(x)(\Delta x)^2$

37 $\frac{1}{.99} = 1.010101$; $\frac{1}{1.1} = .90909$ **39** Inflection **41** 18 vs. 17 **43** Concave up; below

Section 3.4 Graphs (page 119)

1 120; 150; $\frac{60}{x}$ **3** Odd; $x = 0, y = x$ **5** Even; $x = 1, x = -1, y = 0$ **7** Even; $y = 1$ **9** Even

11 Even; $x = 1, x = -1, y = 0$ **13** $x = 0, x = -1, y = 0$ **15** $x = 1, y = 1$ **17** Odd **19** $\frac{2x}{x-1}$

21 $x + \frac{1}{x-4}$ **23** $\sqrt{x^2 + 1}$ **25** Of the same degree **27** Have degree $P <$ degree Q; none

29 $x = 1$ and $y = 3x + C$ if f is the polynomial; but $f(x) = (x-1)^{1/3} + 3x$ has no asymptote $x = 1$

31 $(x - 3)^2$ **39** $x = \sqrt{2}, x = -\sqrt{2}, y = x$ **41** $Y = 100 \sin \frac{2\pi X}{360}$ **45** $c = 3, d = 10$; $c = 4, d = 20$

47 $x^* = \sqrt{5} = 2.236$ **49** $y = x - 2$; $Y = X$; $\mathbf{y} = 2\mathbf{x}$ **51** $x_{max} = .281, x_{min} = 6.339$; $x_{infl} = 4.724$

53 $x_{min} = .393, x_{max} = 1.53, x_{min} = 3.33$; $x_{infl} = .896, 2.604$

55 $x_{min} = -.7398, x_{max} = .8135$; $x_{infl} = .04738$; $x_{blowup} = \pm 2.38$ **57** 8 digits

Section 3.5 Parabolas, Ellipses, and Hyperbolas (page 128)

1 $dy/dx = 0$ at $\frac{-b}{2a}$ **3** $V = (1, -4), F = (1, -3.75)$ **5** $V = (0, 0), F = (0, -1)$ **7** $F = (1, 1)$

9 $V = (0, \pm 3); F = (0, \pm \sqrt{8})$ **11** $V = (0, \pm 1); F = (0, \pm \sqrt{\frac{5}{4}})$ **13** Two lines, $a = b = c = 0; V = F = (0, 0)$

15 $y = 5x^2 - 4x$ **17** $y + p = \sqrt{x^2 + (y-p)^2} \rightarrow 4py = x^2; F = (0, \frac{1}{12}), y = -\frac{1}{12}; (\pm \frac{\sqrt{11}}{6}, \frac{11}{12})$

19 $x = ay^2$ with $a > 0; y = \frac{(x+p)^2}{4p}; y = -ax^2 + ax$ with $a > 0$

21 $\frac{x^2}{4} + y^2 = 1; \frac{(x-1)^2}{4} + (y-1)^2 = 1$ **23** $\frac{x^2}{25} + \frac{y^2}{9} = 1; \frac{(x-3)^2}{36} + \frac{(y-1)^2}{32} = 1; x^2 + y^2 = 25$

25 Circle, hyperbola, ellipse, parabola **27** $\frac{dy}{dx} = -\frac{4}{5}; y = -\frac{4}{5}x + 5$ **29** $\frac{5}{4}; \frac{9}{40} = \frac{1}{2}(\frac{5}{4} - \frac{4}{5})$

31 Circle; $(3, 1); 2; X = \frac{x-3}{2}, Y = \frac{y-1}{2}$ **33** $3x'^2 + y'^2 = 2$ **35** $y^2 - \frac{1}{3}x^2 = 1; \frac{y^2}{9} - \frac{4x^2}{9} = 1; y^2 - x^2 = 5$

37 $\frac{x^2}{25} - \frac{y^2}{39} = 1$ **39** $y^2 - 4y + 4, 2x^2 + 12x + 18; -14, (-3, 2)$, right-left

41 $F = (\pm \frac{\sqrt{5}}{2}, 0); y = \pm \frac{x}{2}$ **43** $(x + y + 1)^2 = 0$

45 $(a^2 - 1)x^2 + 2abxy + (b^2 - 1)y^2 + 2acx + 2bcy + c^2 = 0; 4(a^2 + b^2 - 1)$; if $a^2 + b^2 < 1$ then $B^2 - 4AC < 0$

Section 3.6 Iterations $x_{n+1} = F(x_n)$ (page 136)

1 $-.366; \infty$ **3** $1; 1$ **5** $\frac{2}{3}; \pm\infty$ **7** $-2; -2$

9 $\frac{1-\sqrt{3}}{2}$ attracts, $\frac{1+\sqrt{3}}{2}$ repels; $\frac{1}{2}$ attracts, 0 repels; 1 attracts, 0 repels; 1 attracts; $\frac{2}{3}$ attracts; 0 repels; $\pm\sqrt{2}$ repel

11 Negative **13** .900 **15** .679 **17** $|a| < 1$ **19** Unstable $|F'| > 1$ **21** $x^* = \frac{s}{1-a}; |a| < 1$

23 $2000; $2000   **25** $x_0, b/x_0, x_0, b/x_0, \cdots$ **27** $F' = -\frac{\sqrt{2}}{2}x^{-3/2} = -\frac{1}{2}$ at x^*

29 $F' = 1 - 2cx = 1 - 4c$ at $x^* = 2; 0 < c < \frac{1}{2}$ succeeds

31 $F' = 1 - 9c(x-2)^8 = 1 - 9c$ at $x^* = 3; 0 < c < \frac{2}{9}$ succeeds

33 $x_{n+1} = x_n - \frac{x_n^3 - 2}{3x_n^2}; x_{n+1} = x_n - \frac{\sin x_n - \frac{1}{2}}{\cos x_n}$ **35** $x^* = 4$ if $x_0 > 2.5; x^* = 1$ if $x_0 < 2.5$

37 $m = 1 + c$ at $x^* = 0, m = 1 - c$ at $x^* = 1$ (converges if $0 < c < 2$)   **39** 0   **43** $F' = 1$ at $x^* = 0$

Section 3.7 Newton's Method and Chaos   (page 145)

1 $x_{n+1} = x_n - \frac{x_n^3 - b}{3x_n^2} = \frac{2x_n}{3} + \frac{b}{3x_n^2}$ **5** $x_1 = x_0; x_1$ is not defined (∞)   **7** $x^* = 1$ or 5 from $x_0 < 3, x_0 > 3$

11 $x_0 < \frac{1}{2}$ to $x^* = 0; x_0 > \frac{1}{2}$ to $x^* = 1$ **21** $x_{n+1} = x_n - \frac{x_n^k - 7}{k x_n^{k-1}}$ **23** $x_4 = \cot \pi = \infty; x_3 = \cot \frac{8\pi}{7} = \cot \frac{\pi}{7}$

25 π is not a fraction   **27** $= \frac{1}{4}x_n^2 + \frac{1}{2} + \frac{1}{4x_n^2} = \frac{(x_n^2 + 1)^2}{4x_n^2} = \frac{y_n^2}{4(y_n - 1)}$ **29** $16z - 80z^2 + 128z^3 - 64z^4; 4; 2$

31 $|x_0| < 1$ **33** $\Delta x = 1$, one-step convergence for quadratics   **35** $\frac{\Delta f}{\Delta x} = \frac{5.25}{1.5}; x_2 = 1.86$

37 $1.75 < x^* < 2.5; 1.75 < x^* < 2.125$ **39** $8; 3 < x^* < 4$ **41** Increases by 1; doubles for Newton

45 $x_1 = x_0 + \cot x_0 = x_0 + \pi$ gives $x_2 = x_1 + \cot x_1 = x_1 + \pi$ **49** $a = 2, Y'$s approach $\frac{1}{2}$

Section 3.8 The Mean Value Theorem and l'Hôpital's Rule   (page 152)

1 $c = \sqrt{\frac{4}{3}}$ **3** No c **5** $c = 1$ **7** Corner at $\frac{1}{2}$ **9** Cusp at 0

11 $\sec^2 x - \tan^2 x = $ constant   **13** 6   **15** -2 **17** -1 **19** n **21** $-\frac{1}{2}$ **23** Not $\frac{0}{0}$

25 -1 **27** 1; $\frac{1 - \sin x}{1 + \cos x}$ has no limit   **29** $f'(c) = \frac{4^3 - 1^3}{4 - 1}; c = \sqrt{7}$

31 $0 = x^* - x_{n+1} + \frac{f''(c)}{2f'(x_n)}(x^* - x_n)^2$ gives $M \approx \frac{f''(x^*)}{2f'(x^*)}$ **33** $f'(0); \frac{f'(x)}{1}$; singularity   **35** $\frac{f(x)}{g(x)} \to \frac{3}{4}$ **37** 1

CHAPTER 4   DERIVATIVES BY THE CHAIN RULE

Section 4.1 The Chain Rule   (page 158)

1 $z = y^3, y = x^2 - 3, z' = 6x(x^2 - 3)^2$ **3** $z = \cos y, y = x^3, z' = -3x^2 \sin x^3$

5 $z = \sqrt{y}, y = \sin x, z' = \cos x / 2\sqrt{\sin x}$ **7** $z = \tan y + (1/\tan x), y = 1/x, z' = \left(\frac{-1}{x^2}\right) \sec^2 \left(\frac{1}{x}\right) - (\tan x)^{-2} \sec^2 x$

9 $z = \cos y, y = x^2 + x + 1, z' = -(2x+1)\sin(x^2 + x + 1)$ **11** $17 \cos 17x$ **13** $\sin(\cos x) \sin x$

15 $x^2 \cos x + 2x \sin x$ **17** $(\cos \sqrt{x+1})\frac{1}{2}(x+1)^{-1/2}$ **19** $\frac{1}{2}(1 + \sin x)^{-1/2}(\cos x)$ **21** $\cos \left(\frac{1}{\sin x}\right)\left(\frac{-\cos x}{\sin^2 x}\right)$

23 $8x^7 = 2(x^2)^2(2x^2)(2x)$ **25** $2(x+1) + \cos(x + \pi) = 2x + 2 - \cos x$

27 $(x^2 + 1)^2 + 1; \sin U$ from 0 to $\sin 1; U(\sin x)$ is 1 and 0 with period $2\pi; R$ from 0 to $x; R(\sin x)$ is half-waves.

29 $g(x) = x + 2, h(x) = x^2 + 2, k(x) = 3$ **31** $f'(f(x))f'(x)$; no; $(-1/(1/x)^2)(-1/x^2) = 1$ and $f(f(x)) = x$

33 $\frac{1}{2}\left(\frac{1}{2}x + 8\right) + 8; \frac{1}{8}x + 14; \frac{1}{16}$ **35** $f(g(x)) = x, g(f(y)) = y$

37 $f(g(x)) = \frac{1}{1-x}, g(f(x)) = 1 - \frac{1}{x}, f(f(x)) = x = g(g(x)), g(f(g(x))) = \frac{x}{x-1} = f(g(f(x)))$

39 $f(y) = y - 1, g(x) = 1$ **43** $2\cos(x^2 + 1) - 4x^2 \sin(x^2 + 1); -(x^2 - 1)^{-3/2}; -(\cos \sqrt{x})/4x + (\sin \sqrt{x})/4x^{3/2}$

45 $f'(u(t))u'(t)$ **47** $(\cos^2 u(x) - \sin^2 u(x))\frac{du}{dx}$ **49** $2xu(x) + x^2\frac{du}{dx}$ **51** $1/4\sqrt{1 - \sqrt{1-x}}\sqrt{1-x}$

53 df/dt **55** $f'(g(x))g'(x) = 4(x^3)^3 3x^2 = 12x^{11}$ **57** $3600; \frac{1}{2}; 18$ **59** $3; \frac{1}{3}$

Section 4.2 Implicit Differentiation and Related Rates (page 163)

1 $-x^{n-1}/y^{n-1}$ 3 $\frac{dy}{dx}=1$ 5 $\frac{dy}{dx}=\frac{1}{F'(y)}$ 7 $(y^2-2xy)/(x^2-2xy)$ or 1   9 $\frac{1}{\sec^2 y}$ or $\frac{1}{1+x^2}$

11 First $\frac{dy}{dx}=-\frac{y}{x}$, second $\frac{dy}{dx}=\frac{x}{y}$ 13 Faster, faster   15 $2zz'=2yy'\to z'=\frac{y}{x}y'=y'\sin\theta$

17 $\sec^2\theta=\frac{c}{200\pi}$ 19 $500\frac{df}{dx};500\sqrt{1+\left(\frac{df}{dx}\right)^2}$ 21 $\frac{dy}{dt}=-\frac{8}{3};\frac{dy}{dt}=-2\sqrt{3};\infty$ then 0

23 $V=\pi r^2 h;\frac{dh}{dt}=\frac{1}{4\pi}\frac{dV}{dt}=-\frac{1}{4\pi}$ in/sec   25 $A=\frac{1}{2}ab\sin\theta,\frac{dA}{dt}=7$ 27 1.6 m/sec; 9 m/sec; 12.8 m/sec

29 $-\frac{7}{5}$ 31 $\frac{dz}{dt}=\frac{\sqrt{2}}{2}\frac{dy}{dt};\frac{d\theta}{dt}=\frac{1}{10}\cos^2\theta\frac{d\theta}{dt};\theta''=\frac{\cos\theta}{10}y''-\frac{1}{50}\cos^3\theta\sin\theta(y')^2$

Section 4.3 Inverse Functions and Their Derivatives (page 170)

1 $x=\frac{y+6}{3}$ 3 $x=\sqrt{y+1}$ (x unrestricted $\to$ no inverse)   5 $x=\frac{1}{y-1}$ 7 $x=(1+y)^{1/3}$

9 (x unrestricted $\to$ no inverse)   11 $y=\frac{1}{x-a}$ 13 $2<f^{-1}(x)<3$ 15 f goes up and down

17 $f(x)g(x)$ and $\frac{1}{f(x)}$ 19 $m\neq 0;m\geq 0;|m|\geq 1$ 21 $\frac{dy}{dx}=5x^4,\frac{dy}{dx}=\frac{1}{5}y^{-4/5}$

23 $\frac{dy}{dx}=3x^2;\frac{dx}{dy}=\frac{1}{3}(1+y)^{-2/3}$ 25 $\frac{dy}{dx}=\frac{-1}{(x-1)^2},\frac{dx}{dy}=\frac{-1}{(y-1)^2}$ 27 $y;\frac{1}{2}y^2+C$

29 $f(g(x))=-1/3x^3;g^{-1}(y)=\frac{-1}{y};g(g^{-1}(x))=x$ 39 $2/\sqrt{3}$ 41 $1/6\cos 9$

43 Decreasing; $\frac{dx}{dy}=\frac{1}{dy/dx}<0$ 45 F; T; F   47 $g(x)=x^m,f(y)=y^n,x=(z^{1/n})^{1/m}$

49 $g(x)=x^3,f(y)=y+6,x=(z-6)^{1/3}$ 51 $g(x)=10^x,f(y)=\log y,x=\log(10^y)=y$

53 $y=x^3,y''=6x,d^2x/dy^2=-\frac{2}{9}y^{-5/3};\text{m/sec}^2,\text{sec/m}^2$ 55 $p=\frac{1}{\sqrt{y}}-1;0<y\leq 1$

57 $\max=G=\frac{3}{8}y^{4/3},G'=\frac{1}{2}y^{1/3}$ 59 $y^2/100$

Section 4.4 Inverses of Trigonometric Functions (page 175)

1 $0,\frac{\pi}{2},0$ 3 $\frac{\pi}{2},0,\frac{\pi}{4}$ 5 π is outside $[-\frac{\pi}{2},\frac{\pi}{2}]$ 7 $y=-\sqrt{3}/2$ and $\sqrt{3}/2$

9 $\sin x=\sqrt{1-y^2};\sqrt{1-y^2}$ and 1   11 $\frac{d(\sin^{-1}y)}{dy}\cos x=1\to\frac{d(\sin^{-1}y)}{dy}=\frac{1}{\cos x}=\frac{1}{\sqrt{1-y^2}}$

13 $y=0:1,-1,1;y=1:0,0,\frac{1}{2}$ 15 F; F; T; T; F; F   17 $\frac{du}{dx}=\frac{1}{\sqrt{1-x^2}}$ 19 $\frac{dz}{dx}=3$

21 $\frac{dz}{dx}=\frac{2\sin^{-1}x}{\sqrt{1-x^2}}$ 23 $1-\frac{y\sin^{-1}y}{\sqrt{1-y^2}}$ 25 $\frac{dx}{dy}=\frac{1}{|y+1|\sqrt{y^2+2y}}$ 27 $u=1$ so $\frac{du}{dy}=0$ 31 $\sec x=\sqrt{y^2+1}$

33 $\frac{1}{10},1,\frac{1}{2}$ 35 $-y/\sqrt{1-y^2}$ 37 $\frac{1}{2}\sec\frac{x}{2}\tan\frac{x}{2}$ 39 $\frac{nx^{n-1}}{|x^n|\sqrt{x^{2n}-1}}$ 41 $\frac{dy}{dx}=\frac{1}{1+x^2}$

43 $\frac{dy}{dx}=\pm\frac{1}{1+x^2}$ 47 $u=4\sin^{-1}y$ 49 π 51 $-\pi/4$

CHAPTER 5 INTEGRALS

Section 5.1 The Idea of the Integral (page 181)

1 $1,3,7,15,127$ 3 $-\frac{1}{2}-\frac{1}{4}-\frac{1}{8}=\frac{1}{8}-1$ 5 $f_j-f_0=\frac{r^j-1}{r-1}$ 7 $3x$ for $x\leq 7,7x-4$ for $x\geq 1$

9 $\frac{1}{52}\frac{1}{\sqrt{52}},\frac{2}{52},\frac{1}{52}\sqrt{\frac{j}{52}}$ 11 Lower by 2   13 Up, down; rectangle   15 $\sqrt{x+\Delta x}-\sqrt{x};\Delta x;\frac{df}{dx};\sqrt{x}$

17 6; 18; triangle   19 18 rectangles   21 $6x-\frac{1}{2}x^2-10;6-x$ 23 $\frac{14}{27}$ 25 $x^2;x^2;\frac{1}{3}x^3$

Section 5.2 Antiderivatives (page 186)

1 $x^5 + \frac{2}{3}x^6; \frac{5}{3}$ 3 $2\sqrt{x}; 2$ 5 $\frac{3}{4}x^{4/3}(1+2^{1/3}); \frac{3}{4}(1+2^{1/3})$ 7 $-2\cos x - \frac{1}{2}\cos 2x; \frac{5}{2} - 2\cos 1 - \frac{1}{2}\cos 2$

9 $x\sin x + \cos x; \sin 1 + \cos 1 - 1$ 11 $\frac{1}{2}\sin^2 x; \frac{1}{2}\sin^2 1$ 13 $f = C; 0$ 15 $f(b) - f(a); f_7 - f_2$

17 $8 + \frac{8}{N}$ 19 $\frac{\pi}{3}(1+\sqrt{3}); \frac{\pi}{6}(3+\sqrt{3}); 2$ 21 $\frac{5}{2}; \frac{205}{36}; \infty$ 23 $f(x) = 2\sqrt{x}$ 25 $\frac{1}{2}$, below $-1; \frac{1}{4}, \frac{5}{4}$

27 Increase - decrease; increase - decrease - increase

29 Area under B – area under D; time when $B = D$; time when $B - D$ is largest   33 T; F; F; T; F

Section 5.3 Summation Versus Integration (page 194)

1 $\frac{25}{12}; 16$ 3 $127; 2^{n+1} - 1$ 5 $\sum_{j=1}^{50} 2j = 2550; \sum_{i=1}^{100}(2j-1) = 10,000; \sum_{k=1}^{4}(-1)^{k+1}/k = \frac{7}{12}$

7 $\sum_{k=0}^{n} a_k x^k; \sum_{j=1}^{n} \sin\frac{2\pi j}{n}$ 9 $5.18738; 7.48547$ 11 $2(a_i^2 + b_i^2)$ 13 $2^n - 1; \frac{1}{11} - \frac{1}{1}$ 15 F; T

17 $\frac{df}{dx} + C; f_9 - f_8 - f_1 + f_0$ 19 $f_1 = 1; n^2 + (2n+1) = (n+1)^2$

21 $a+b+c = 1, 2a+4b+8c = 5, 3a+9b+27c = 14$; sum of squares   23 $S_{400} = 80200; E_{400} = .0025 = \frac{1}{n}$

25 $S_{100,1/3} \approx 350, E_{100,1/3} \approx .00587; S_{100,3} = 25502500, E_{100,3} = .0201$ 27 v_1 and v_2 have the same sign

29 $v_1 = 9, v_2 = 12, \Sigma\Sigma = 21$ 31 At $N = 1, 2^{N-2}$ is not 1   33 $0; \frac{1}{n}(v_1 + \cdots + v_n)$

35 $\Delta x \sum_{j=1}^{n} v(j\Delta x)$ 37 $f(1) - f(0) = \int_0^1 \frac{df}{dx} dx$

Section 5.4 Indefinite Integrals and Substitutions (page 200)

1 $\frac{2}{3}(2+x)^{3/2} + C$ 3 $(x+1)^{n+1}/(n+1) + C (n \neq -1)$ 5 $\frac{1}{12}(x^2+1)^6 + C$ 7 $-\frac{1}{4}\cos^4 x + C$

9 $-\frac{1}{8}\cos^4 2x + C$ 11 $\sin^{-1} t + C$ 13 $\frac{1}{3}(1+t^2)^{3/2} - (1+t^2)^{1/2} + C$ 15 $2\sqrt{x} + x + C$

17 $\sec x + C$ 19 $-\cos x + C$ 21 $\frac{1}{3}x^3 + \frac{2}{3}x^{3/2}$ 23 $-\frac{1}{3}(1-2x)^{3/2}$ 25 $y = \sqrt{2x}$

27 $\frac{1}{2}x^2$ 29 $a\sin x + b\cos x$ 31 $\frac{4}{15}x^{5/2}$ 33 F; F; F; F   35 $f(x-1); 2f\left(\frac{\pi}{2}\right)$

37 $x - \tan^{-1} x$ 39 $\int \frac{1}{u} du$ 41 $4.9t^2 + C_1 t + C_2$ 43 $f(t+3); f(t) + 3t; 3f(t); \frac{1}{3}f(3t)$

Section 5.5 The Definite Integral (page 205)

1 $C = -f(2)$ 3 $C = f(3)$ 5 $f(t)$ is wrong   7 $C = 0$ 9 $C = f(-a) - f(-b)$

11 $u = x^2 + 1; \int_1^2 u^{10}\frac{du}{2} = \frac{u^{11}}{22}\Big]_1^2 = \frac{2^{11}-1}{22}$ 13 $u = \tan x; \int_0^1 u\, du = \frac{1}{2}$

15 $u = \sec x; \int_1^{\sqrt{2}} u\, du = \frac{1}{2}$ (same as 13)   17 $u = \frac{1}{x}, x = \frac{1}{u}, dx = \frac{-du}{u^2}; \int_1^{1/2} \frac{-du}{u}$

19 $S = \frac{1}{2}\left(\frac{1}{4}+1\right)^4 + \frac{1}{2}(1+1)^4; s = \frac{1}{2}(0) + \frac{1}{2}\left(\frac{1}{4}+1\right)^4$

21 $S = \frac{1}{2}\left[\left(\frac{1}{2}\right)^3 + 1^3 + \left(\frac{3}{2}\right)^3 + 2^3\right]; s = \frac{1}{2}\left[0^3 + \left(\frac{1}{2}\right)^3 + 1^3 + \left(\frac{3}{2}\right)^3\right]$

23 $S = \frac{1}{4}\left[\left(\frac{17}{16}\right)^4 + \left(\frac{5}{4}\right)^4 + \left(\frac{25}{16}\right)^4 + 2^4\right]$ 25 Last rectangle minus first rectangle

27 $S = .07$ since 7 intervals have points where $W = 1$. The integral of $W(x)$ exists and equals zero.

29 M is increasing so Problem 25 gives $S - s = \Delta x(1-0)$; area from graph up to $y = 1$ is $\frac{1}{2} \cdot 1 + \frac{1}{4} \cdot \frac{1}{2} + \cdots =$
$\frac{1}{2}\left(1 + \frac{1}{4} + \frac{1}{16} + \cdots\right) = \frac{\frac{1}{2}}{1-\frac{1}{4}} = \frac{2}{3}$; area under graph is $\frac{1}{3}$.

31 $f(x) = 3 + \int_0^x v(x)dx; f(x) = \int_3^x v(x)dx$ 33 T; F; T; F; T; F; T

Section 5.6 Properties of the Integral and Average Value   (page 212)

1 $\bar{v} = \frac{1}{2}\int_{-1}^{1} x^4 dx = \frac{1}{5}$ equals c^4 at $c = \pm(\frac{1}{5})^{1/4}$ **3** $\bar{v} = \frac{1}{\pi}\int_0^\pi \cos^2 x\, dx = \frac{1}{2}$ equals $\cos^2 c$ at $c = \frac{\pi}{4}$ and $\frac{3\pi}{4}$

5 $\bar{v} = \int_1^2 \frac{dx}{x^2} = \frac{1}{2}$ equals $\frac{1}{c^2}$ at $c = \sqrt{2}$ **7** $\int_3^5 v(x)dx$ **9** False, take $v(x) < 0$

11 True; $\frac{1}{3}\int_0^1 v(x)dx + \frac{2}{3} \cdot \frac{1}{2}\int_1^3 v(x)dx = \frac{1}{3}\int_0^3 v(x)dx$ **13** False; when $v(x) = x^2$ the function $x^2 - \frac{1}{3}$ is even

15 False; take $v(x) = 1$; factor $\frac{1}{2}$ is missing   **17** $\bar{v} = \frac{1}{b-a}\int_a^b v(x)dx$ **19** 0 and $\frac{2}{\pi}$

21 $v(x) = Cx^2; v(x) = C$. This is "constant elasticity" in economics (Section 2.2)   **23** $\overline{V} \to 0; \overline{V} \to 1$

25 $\frac{1}{2}\int_0^2 (a-x)dx = a+1$ if $a > 2$; $\frac{1}{2}\int_0^2 |a-x|dx = \frac{1}{2}$ area$= \frac{a^2}{2} - a + 1$ if $a < 2$; distance = absolute value

27 Small interval where $y = \sin \theta$ has probability $\frac{d\theta}{\pi}$; the average y is $\int_0^\pi \frac{\sin \theta\, d\theta}{\pi} = \frac{2}{\pi}$

29 Area under $\cos \theta$ is 1. Rectangle $0 \le \theta \le \frac{\pi}{2}, 0 \le y \le 1$ has area $\frac{\pi}{2}$. Chance of falling across a crack is $\frac{1}{\pi/2} = \frac{2}{\pi}$.

31 $\frac{1}{6^3}, \frac{3}{6^3}, \cdots, \frac{1}{6^3}; 10.5$ **33** $\frac{1}{t}\int_0^t 220 \cos \frac{2\pi t}{60} dt = \frac{1}{t} \cdot 220 \cdot \frac{60}{2\pi} \sin \frac{2\pi t}{60} = V_{\text{ave}}$

35 Any $v(x) = v_{\text{even}}(x) + v_{\text{odd}}(x); (x+1)^3 = (3x^2+1) + (x^3+3x); \frac{1}{x+1} = \frac{1}{1-x^2} - \frac{x}{1-x^2}$

37 16 per class; $\frac{6}{64}; E(x) = \frac{1800}{64} = \frac{225}{8}$ **39** F; F; T; T

41 $f(x) = \begin{cases} \frac{1}{2}(x-2)^2 & x \ge 2 \\ -\frac{1}{2}(x-2)^2 & x \le 2 \end{cases} + C; f(5) - f(0) = \frac{9}{2} + \frac{4}{2} = \frac{13}{2}$

Section 5.7 The Fundamental Theorem and Its Applications   (page 219)

1 $\cos^2 x$ **3** 0   **5** $(x^2)^3(2x) = 2x^7$ **7** $v(x+1) - v(x)$ **9** $\frac{\sin^2 x}{x} - \frac{1}{x^2}\int_0^x \sin^2 t\, dt$

11 $\int_0^x v(u)du$ **13** 0   **15** $2\sin x^2$ **17** $u(x)v(x)$ **19** $\sin^{-1}(\sin x) \cos x = x \cos x$

21 F; F; F; T   **23** Taking derivatives $v(x) = (x\cos x)' = \cos x - x \sin x$

25 Taking derivatives $-v(-x)(-1) = v(x)$ so v is even   **27** F; T; T; F

29 $\int_1^x v(t)dt = \int_0^x v(t)dt - \int_0^1 v(t)dt = \frac{x}{x+2} - \frac{1}{1+2}$ (in revised printing)

31 $V = s^3; A = 3s^2$; half of hollow cube; $\Delta V \approx 3s^2 dS; 3s^2$ (which is A)

33 $dH/dr = 2\pi^2 r^3$ **35** Wedge has length $r \approx$ height of triangle; $\int_0^{\pi/2} \frac{1}{2}r^2 d\theta = \frac{\pi r^2}{4}$

37 $r = \frac{1}{\cos \theta}; \frac{d\theta}{2\cos^2 \theta}; \int_0^{\pi/4} \frac{d\theta}{2\cos^2 \theta} = \frac{\tan \theta}{2}\big]_0^{\pi/4} = \frac{1}{2}$

39 $x = y^2; \int_0^2 y^2 dy = \frac{y^3}{3}\big]_0^2 = \frac{8}{3}$; vertical strips have length $2 - \sqrt{x}$

41 Length $\sqrt{2}a$; width $\frac{da}{\sqrt{2}}; \int_0^1 a\, da = \frac{1}{2}$ **43** The differences of the sums $f_j = v_1 + v_2 + \cdots + v_j$ are $f_j - f_{j-1} = v_j$

45 No, $\int_0^x a(t)dt = \frac{df}{dx}(x) - \frac{df}{dx}(0)$ and $\int_0^1 \left(\int_0^x a(t)dt\right) dx = f(1) - f(0) - \frac{df}{dx}(0)$

Section 5.8 Numerical Integration   (page 226)

1 $\frac{1}{2}\Delta x(v_0 - v_n)$ **3** $1, .5625, .3025; 0, .0625, .2025$ **5** $L_8 \approx .1427, T_8 \approx .2052, S_8 \approx .2000$

7 $p = 2$: for $y = x^2, \frac{1}{4} \cdot 0^2 + \frac{1}{2} \cdot (\frac{1}{2})^2 + \frac{1}{4} \cdot 1^2 \ne \frac{1}{3}$ **9** For $y = x^2$, error $\frac{1}{6}(\Delta x)^2$ from $\frac{1}{2} - \frac{1}{3}, y_1' = 2\Delta x$

13 8 intervals give $\frac{(\Delta x)^2}{12}\left[-\frac{1}{b^2} + \frac{1}{a^2}\right] = \frac{1}{1024} < .001$ **15** $f''(c)$ is $y'(c)$ **17** $\infty; .683, .749, .772 \to \frac{\pi}{4}$

19 $A + B + C = 1, \frac{1}{2}B + C = \frac{1}{2}, \frac{1}{4}B + C = \frac{1}{3}$; Simpson

21 $y = 1$ and x on $[0,1]: L_n = 1$ and $\frac{1}{2} - \frac{1}{2n}, R_n = 1$ and $\frac{1}{2} + \frac{1}{2n}$, so only $\frac{1}{2}L_n + \frac{1}{2}R_n$ gives 1 and $\frac{1}{2}$

23 $T_{10} \approx 500,000,000; T_{100} \approx 50,000,000; 25,000\pi$

25 $a = 4, b = 2, c = 1; \int_0^1 (4x^2 + 2x + 1)dx = \frac{10}{3}$; Simpson fits parabola   **27** $c = \frac{1}{4320}$

CHAPTER 6 EXPONENTIALS AND LOGARITHMS

Section 6.1 An Overview (page 234)

1 $5; -5; -1; \frac{1}{5}; \frac{3}{2}; 2$ 5 $1; -10; 80; 1; 4; -1$ 7 $n \log_b x$ 9 $\frac{10}{3}; \frac{3}{10}$ 13 10^5

15 $0; I_{SF} = 10^7 I_0; 8.3 + \log_{10} 4$ 17 $A = 7, b = 2.5$ 19 $A = 4, k = 1.5$

21 $\frac{1}{cx}; \frac{2}{cx}; \log 2$ 23 $y - 1 = cx; y - 10 = c(x - 1)$ 25 $(.1^{-h} - 1)/(-h) = (10^h - 1)/(-h)$

27 $y'' = c^2 b^x; x'' = -1/cy^2$ 29 Logarithm

Section 6.2 The Exponential e^x (page 241)

1 $49e^{7x}$ 3 $8e^{8x}$ 5 $3^x \ln 3$ 7 $\left(\frac{2}{3}\right)^x \ln \frac{2}{3}$ 9 $\frac{-e^x}{(1+e^x)^2}$ 11 2   13 xe^x 15 $\frac{4}{(e^x + e^{-x})^2}$

17 $e^{\sin x} \cos x + e^x \cos e^x$ 19 $.1246, .0135, .0014$ are close to $\frac{e}{2n}$ 21 $\frac{1}{e}; \frac{1}{e}$

23 $Y(h) = 1 + \frac{1}{10}; Y(1) = \left(1 + \frac{1}{10}\right)^{10} = 2.59$ 25 $\left(1 + \frac{1}{x}\right)^x < e < e^x < e^{3x/2} < e^{2x} < 10^x < x^x$

27 $\frac{e^{3x}}{3} + \frac{e^{7x}}{7}$ 29 $x + \frac{2^x}{\ln 2} + \frac{3^x}{\ln 3}$ 31 $\frac{(2e)^x}{\ln(2e)} + 2e^x$ 33 $\frac{e^{x^2}}{2} - \frac{e^{-x^2}}{2}$

35 $2e^{x/2} + \frac{e^{2x}}{2}$ 37 e^{-x} drops faster at $x = 0$ (slope -1); meet at $x = 1; e^{-x^2}/e^{-x} < e^{-9}/e^{-3} < \frac{1}{100}$ for $x > 3$

39 $y - e^a = e^a(x - a)$; need $-e^a = -ae^a$ or $a = 1$

41 $y' = x^x(\ln x + 1) = 0$ at $x_{\min} = \frac{1}{e}; y'' = x^x[(\ln x + 1)^2 + \frac{1}{x}] > 0$

43 $\frac{d}{dx}(e^{-x} y) = e^{-x} \frac{dy}{dx} - e^{-x} y = 0$ so $e^{-x} y = $ Constant or $y = Ce^x$

45 $\frac{e^{2x}}{2}]_0^1 = \frac{e^2 - 1}{2}$ 47 $\frac{2^x}{\ln 2}]_{-1}^1 = \frac{2 - \frac{1}{2}}{\ln 2} = \frac{3}{2\ln 2}$ 49 $-e^{-x}]_0^\infty = 1$ 51 $e^{1+x}]_0^1 = e^2 - e$ 53 $\frac{2^{\sin x}}{\ln 2}]_0^\pi = 0$

55 $\int \frac{du/dx}{e^u} dx = -e^{-u} + C; \int (e^u)^2 \frac{du}{dx} dx = \frac{1}{2} e^{2u} + C$ 57 $yy' = 1$ gives $\frac{1}{2} y^2 = x + C$ or $y = \sqrt{2x + 2C}$

59 $\frac{dF}{dx} = (n - x)x^{n-1}/e^x < 0$ for $x > n; F(2x) < \frac{\text{constant}}{e^x} \to 0$ 61 $\frac{6!}{\sqrt{12\pi}} \approx 117; \left(\frac{6}{e}\right)^6 \approx 116; 7$ digits

Section 6.3 Growth and Decay in Science and Economics (page 250)

1 $t^2 + y_0$ 3 $y_0 e^{2t}$ 5 $10 e^{4t}; t = \frac{\ln 10}{4}$ 7 $\frac{1}{4} e^{4t} + 9.75; t = \frac{\ln 361}{4}$ 11 $c = \frac{\ln 2}{2}; t = \frac{\ln 10}{c}$

13 $\frac{5568}{-.7} \ln \left(\frac{1}{5}\right)$ 15 $c = \frac{\ln 2}{20}; t = \frac{1}{c} \ln \left(\frac{8}{5}\right)$ 17 $t = \frac{\ln(1/240)}{\ln(.98)}$ 19 $e^c = 3$ so $y_0 = e^{-3c} 1000 = \frac{1000}{27}$

21 $p = 1013 e^{ch}; 50 = 1013 e^{20c}; c = \frac{1}{20} \ln \left(\frac{50}{1013}\right); p(10) = 1013 e^{10c} = 1013 \sqrt{\frac{50}{1013}} = \sqrt{(1013)(50)}$

23 $c = \frac{\ln 2}{3}; \left(\frac{1}{2}\right)^3 = \frac{1}{8}$ 25 $y = y_0 - at$ reaches y_1 at $t = \frac{y_0 - y_1}{a}$; then $y = Ae^{-at/y_1}$ 27 F; F; T; T

29 $A = \frac{1}{3}, B = -\frac{1}{3}$ 31 $e^t - 1$ 33 $1 - e^{-t}$ 35 $6; 6 + Ae^{-2t}; 6 - 6e^{-2t}, 6 + 4e^{-2t}; 6$

37 $4; 4 - \frac{1}{e}; 4$ 39 $ye^{-t}; y(t) = te^t$ 41 $A = 1, B = -1, C = -1$ 43 $e^{.0725} > .075$ 45 $s(e - 1); \frac{s(e-1)}{e}$

47 $(1.02)(1.03) \to 5.06\%; 5\%$ by Problem 27   49 $20,000 \, e^{(20-T)(.05)} = 34,400$ (it grows for $20 - T$ years)

51 $s = -cy_0 e^{ct}/(e^{ct} - 1) = -(.01)(1000)e^{.60}/(e^{.60} - 1)$ 53 $y_0 = \frac{100}{.005}\left(1 - e^{-.005(48)}\right)$

55 $e^{4c} = 1.20$ so $c = \frac{\ln 1.20}{4}$ 57 $24e^{36.5} = ?$ 59 To $-\infty$; constant; to $+\infty$

61 $\frac{dY}{dT} = 60cY; \frac{dY}{dT} = 60(-Y + 5)$; still $Y_\infty = 5$

63 $y = 60e^{ct} + 20, 60 = 60e^{12c} + 20, c = \frac{1}{12} \ln \left(\frac{40}{60}\right); 100 = 60e^{ct} + 20$ at $t = \frac{1}{c} \ln \left(\frac{80}{60}\right)$ 65 0

Section 6.4 Logarithms   (page 258)

1 $\frac{1}{x}$ **3** $\frac{-1}{x(\ln x)^2}$ **5** $\ln x$ **7** $\frac{\cos x}{\sin x} = \cot x$ **9** $\frac{7}{x}$ **11** $\frac{1}{3}\ln t + C$ **13** $\ln\frac{4}{3}$

15 $\frac{1}{2}\ln 5$ **17** $-\ln(\ln 2)$ **19** $\ln(\sin x) + C$ **21** $-\frac{1}{3}\ln(\cos 3x) + C$ **23** $\frac{1}{3}(\ln x)^3 + C$

27 $\ln y = \frac{1}{2}\ln(x^2 + 1); \frac{dy}{dx} = \frac{x}{\sqrt{x^2+1}}$ **29** $\frac{dy}{dx} = e^{\sin x}\cos x$

31 $\frac{dy}{dx} = e^x e^{e^x}$ **33** $\ln y = e^x \ln x; \frac{dy}{dx} = y e^x \left(\ln x + \frac{1}{x}\right)$ **35** $\ln y = -1$ so $y = \frac{1}{e}, \frac{dy}{dx} = 0$ **37** 0

39 $-\frac{1}{x}$ **41** $\sec x$ **47** $.1; .095; .095310179$ **49** $-.01; -.01005; -.010050335$

51 l'Hôpital: 1   **53** $\frac{1}{\ln b}$ **55** $3 - 2\ln 2$ **57** Rectangular area $\frac{1}{2} + \cdots + \frac{1}{n} < \int_1^n \frac{dt}{t} = \ln n$

59 Maximum at e **61** 0   **63** $\log_{10} e$ or $\frac{1}{\ln 10}$ **65** $1 - x; 1 + x\ln 2$

67 Fraction is $y = 1$ when $\ln(T + 2) - \ln 2 = 1$ or $T = 2e - 2$ **69** $y' = \frac{2}{(t+2)^2} \to y = 1 - \frac{2}{t+2}$ never equals 1

71 $\ln p = x\ln 2;$ **LD** $2^x \ln 2;$ **ED** $p = e^{x\ln 2}, p' = \ln 2\, e^{x\ln 2}$

75 $2^4 = 4^2; y\ln x = x\ln y \to \frac{\ln x}{x} = \frac{\ln y}{y}; \frac{\ln x}{x}$ decreases after $x = e$, and the only the integers before e are 1 and 2.

Section 6.5 Separable Equations Including the Logistic Equation   (page 266)

1 $7e^t - 5$ **3** $\left(\frac{3}{2}x^2 + 1\right)^{1/3}$ **5** x **7** $e^{1-\cos t}$ **9** $\left(\frac{ct}{2} + \sqrt{y_0}\right)^2$ **11** $y_\infty = 0; t = \frac{1}{by_0}$

15 $z = 1 + e^{-t}, y$ is in **13**   **17** $ct = \ln 3, ct = \ln 9$

19 $b = 10^{-9}, c = 13\cdot 10^{-3}; y_\infty = 13\cdot 10^6;$ at $y = \frac{c}{2b}$ (10) gives $\ln\frac{1}{b} = ct + \ln\frac{10^6}{c - 10^6 b}$ so $t = 1900 + \frac{\ln 12}{c} = 2091$

21 y^2 dips down and up (a valley)   **23** $sc = 1 = sbr$ so $s = \frac{1}{c}, r = \frac{c}{b}$

25 $y = \frac{N}{1 + e^{-Nt}(N-1)}; T = \frac{\ln(N-1)}{N} \to 0$ **27** Dividing cy by $y + K > 1$ slows down y'

29 $\frac{dR}{dy} = \frac{cK}{(y+K)^2} > 0, \frac{cy}{y+K} \to c$

31 $\frac{dY}{dT} = \frac{-Y}{Y+1};$ multiply $e^{y/K}\frac{y}{K} = e^{-ct/K}e^{y_0/K}\left(\frac{y_0}{K}\right)$ by K and take the Kth power to reach (19)

33 $y' = (3 - y)^2; \frac{1}{3-y} = t + \frac{1}{3}; y = 2$ at $t = \frac{2}{3}$

35 $Ae^t + D = Ae^t + B + Dt + t \to D = -1, B = -1; y_0 = A + B$ gives $A = 1$

37 $y \to 1$ from $y_0 > 0, y \to -\infty$ from $y_0 < 0; y \to 1$ from $y_0 > 0, y \to -1$ from $y_0 < 0$

39 $\int \frac{\cos y\, dy}{\sin y} = \int dt \to \ln(\sin y) = t + C = t + \ln\frac{1}{2}$. Then $\sin y = \frac{1}{2}e^t$ stops at 1 when $t = \ln 2$

Section 6.6 Powers Instead of Exponentials   (page 276)

1 $1 - x + \frac{x^2}{2} - \frac{x^3}{6} + \cdots$ **3** $1 \pm x + \frac{x^2}{2} \pm \frac{x^3}{6} + \cdots$ **5** $1050.62; 1050.95; 1051.25$

7 $1 + n\left(\frac{-1}{n}\right) + \frac{n(n+1)}{2}\left(\frac{-1}{n}\right)^2 \to 1 - 1 + \frac{1}{2}$ **9** square of $\left(1 + \frac{1}{n}\right)^n$; set $N = 2n$

11 Increases; $\ln\left(1 + \frac{1}{x}\right) - \frac{1}{x+1} > 0$ **13** $y(3) = 8$ **15** $y(t) = 4(3^t)$ **17** $y(t) = t$

19 $y(t) = \frac{1}{2}(3^t - 1)$ **21** $s\left(\frac{a^t - 1}{a-1}\right)$ if $a \neq 1; st$ if $a = 1$ **23** $y_0 = 6$ **25** $y_0 = 3$

27 $-2, -10, -26 \to -\infty; -5, -\frac{17}{2}, -\frac{41}{4} \to -12$ **29** $P = \frac{b}{c+d}$ **31** 10.38% **33** $100(1.1)^{20} = \$673$

35 $\frac{100,000(.1/12)}{1 - (1 + .1/12)^{-240}} = 965$ **37** $\frac{1000}{.1}(1.1^{20} - 1) = 57,275$ **39** $y_\infty = 1500$ **41** $2; \left(\frac{53}{52}\right)^{52} = 2.69; e$

43 $1.0142^{12} = 1.184 \to$ Visa charges 18.4%

Section 6.7 Hyperbolic Functions   (page 280)

1 $e^x, e^{-x}, \frac{e^{2x} - e^{-2x}}{4} = \frac{1}{2}\sinh 2x$ **7** $\sinh nx$ **9** $3\sinh(3x + 1)$ **11** $\frac{-\sinh x}{\cosh^2 x} = -\tanh x \operatorname{sech} x$

13 $4\cosh x \sinh x$ **15** $\frac{x}{\sqrt{x^2+1}}\left(\operatorname{sech}\sqrt{x^2+1}\right)^2$ **17** $6\sinh^5 x \cosh x$

19 $\cosh(\ln x) = \frac{1}{2}\left(x + \frac{1}{x}\right) = 1$ at $x = 1$ **21** $\frac{5}{13}, \frac{13}{5}, -\frac{12}{5}, -\frac{13}{12}, -\frac{5}{12}$ **23** $0, 0, 1, \infty, \infty$

25 $\frac{1}{2}\sinh(2x + 1)$ **27** $\frac{1}{3}\cosh^3 x$ **29** $\ln(1 + \cosh x)$ **31** e^x

33 $\int y\,dx = \int \sinh t(\sinh t\,dt)$; $A = \frac{1}{2}\sinh t\cosh t - \int y\,dx$; $A' = \frac{1}{2}$; $A = 0$ at $t = 0$ so $A = \frac{1}{2}t$.

41 $e^y = x + \sqrt{x^2+1}$, $y = \ln[x + \sqrt{x^2+1}]$ **47** $\frac{1}{4}\ln|\frac{2+x}{2-x}|$ **49** $\sinh^{-1}x$ (see 41)   **51** $-\operatorname{sech}^{-1}x$

53 $\frac{1}{2}\ln 3; \infty$ **55** $y(x) = \frac{1}{c}\cosh cx$; $\frac{1}{c}\cosh cL - \frac{1}{c}$

57 $y'' = y - 3y^2$; $\frac{1}{2}(y')^2 = \frac{1}{2}y^2 - y^3$ is satisfied by $y = \frac{1}{2}\operatorname{sech}^2\frac{x}{2}$

CHAPTER 7 TECHNIQUES OF INTEGRATION

Section 7.1 Integration by Parts   (page 287)

1 $-x\cos x + \sin x + C$ **3** $-xe^{-x} - e^{-x} + C$ **5** $x^2\sin x + 2x\cos x - 2\sin x + C$

7 $\frac{1}{2}(2x+1)\ln(2x+1) - x + C$ **9** $\frac{1}{2}e^x(\sin x - \cos x) + C$ **11** $\frac{e^{ax}}{a^2+b^2}(a\sin bx - b\cos bx) + C$

13 $\frac{x}{2}(\sin(\ln x) - \cos(\ln x)) + C$ **15** $x(\ln x)^2 - 2x\ln x + 2x + C$ **17** $x\sin^{-1}x + \sqrt{1-x^2} + C$

19 $\frac{1}{2}(x^2+1)\tan^{-1}x - \frac{x}{2} + C$ **21** $x^3\sin x + 3x^2\cos x - 6x\sin x - 6\cos x + C$

23 $e^x(x^3 - 3x^2 + 6x - 6) + C$ **25** $x\tan x + \ln(\cos x) + C$ **27** -1 **29** $-\frac{3}{4}e^{-2} + \frac{1}{4}$ **31** -2

33 $3\ln 10 - 6 + 2\tan^{-1}3$ **35** $u = x^n, v = e^x$ **37** $u = x^n, v = \sin x$ **39** $u = (\ln x)^n, v = x$

41 $u = x\sin x, v = e^x \to \int e^x\sin x\,dx$ in **9** and $-\int x\cos xe^x dx$. Then $u = -x\cos x, v = e^x \to \int e^x\cos x\,dx$ in **10** and $-\int x\sin xe^x dx$
(move to left side): $\frac{e^x}{2}(x\sin x - x\cos x + \cos x)$. Also try $u = xe^x, v = -\cos x$.

43 $\int \frac{1}{2}u\sin u\,du = \frac{1}{2}(\sin u - u\cos u) = \frac{1}{2}(\sin x^2 - x^2\cos x^2)$; odd

45 $3\cdot$ step function; $3e^x\cdot$ step function   **49** $0; x\delta(x)] - \int \delta(x)dx = -1; v(x)\delta(x)] - \int v(x)\delta(x)dx$

51 $v(x) = \int_x^1 f(x)dx$

53 $u(x) = \frac{1}{k}\int_0^x v(x)dx; \frac{1}{k}(\frac{x}{2} - \frac{x^3}{6}); \frac{x}{k}$ for $x \le \frac{1}{2}, \frac{1}{k}(2x - x^2 - \frac{1}{4})$ for $x \ge \frac{1}{2}; \frac{x}{k}$ for $x \le \frac{1}{2}, \frac{1}{2k}$ for $x \ge \frac{1}{2}$.

55 $u = x^2, v = -\cos x \to -x^2\cos x + (2x)\sin x - \int 2\sin x\,dx$ **57** Compare **23**

59 $uw']_0^1 - \int_0^1 u'w' - u'w]_0^1 + \int_0^1 u'w' = [uw' - u'w]_0^1$

61 No mistake: $e^x\cosh x - e^x\sin hx = 1$ is part of the constant C

Section 7.2 Trigonometric Integrals   (page 293)

1 $\int(1 - \cos^2 x)\sin x\,dx = -\cos x + \frac{1}{3}\cos^3 x + C$ **3** $\frac{1}{2}\sin^2 x + C$

5 $\int(1 - u^2)^2 u^2(-du) = -\frac{1}{3}\cos^3 x + \frac{2}{5}\cos^5 x - \frac{1}{7}\cos^7 x + C$ **7** $\frac{2}{3}(\sin x)^{3/2} + C$

9 $\frac{1}{8}\int \sin^3 2x\,dx = \frac{1}{16}(-\cos 2x + \frac{1}{3}\cos^3 2x) + C$ **11** $\frac{\pi}{2}$ **13** $\frac{1}{3}(\frac{3x}{2} + \frac{\sin 6x}{4}) + C$

15 $x + C$ **17** $\frac{1}{6}\cos^5 x\sin x + \frac{5}{6}\int \cos^4 x\,dx$; use equation (5)

19 $\int_0^{\pi/2}\cos^n x\,dx = \frac{n-1}{n}\int_0^{\pi/2}\cos^{n-2}x\,dx = \cdots = \frac{n-1}{n}\frac{n-3}{n-2}\cdots\frac{1}{2}\int_0^{\pi/2}dx$

21 $I = -\sin^{n-1}x\cos x + (n-1)\int \sin^{n-2}x\cos^2 x\,dx = -\sin^{n-1}x\cos x + (n-1)\int \sin^{n-2}x\,dx - (n-1)I$.
So $nI = -\sin^{n-1}x\cos x + (n-1)\int \sin^{n-2}x\,dx$.

23 $0, +, 0, 0, 0, -$ **25** $-\frac{2}{3}\cos^3 x, 0$ **27** $-\frac{1}{2}(\frac{\cos 2x}{2} + \frac{\cos 200x}{200}), 0$ **29** $\frac{1}{2}(\frac{\sin 200x}{200} + \frac{\sin 2x}{2}), 0$

31 $-\frac{1}{2}\cos x, 0$ **33** $\int_0^\pi x\sin x\,dx = \int_0^\pi A\sin^2 x\,dx \to A = 2$ **35** Sum = zero = $\frac{1}{2}$ (left+right)

37 p is even   **39** $p - q$ is even   **41** $\sec x + C$ **43** $\frac{1}{3}\tan^3 x + C$ **45** $\frac{1}{3}\sec^3 x + C$

47 $\frac{1}{3}\tan^3 x - \tan x + x + C$ **49** $\ln|\sin x| + C$ **51** $\frac{1}{2\cos^2 x} + C$ **53** $A = \sqrt{2}, -\sqrt{2}\sin(x + \frac{\pi}{4})$

55 $4\sqrt{2}$ **57** $\frac{1000}{\sqrt{3}}$ **59** $\frac{1-\cos x+\sin x}{1+\cos x+\sin x} + C$ **61** p and q are 10 and 1

Section 7.3 Trigonometric Substitutions (page 299)

1 $x = 2\sin\theta; \int d\theta = \sin^{-1}\frac{x}{2} + C$ 3 $x = 2\sin\theta; \int 4\cos^2\theta\, d\theta = 2\sin^{-1}\frac{x}{2} + x\sqrt{1-\frac{x^2}{4}} + C$

5 $x = \sin\theta; \int \sin^2\theta\, d\theta = \frac{1}{2}\sin^{-1}x - \frac{1}{2}x\sqrt{1-x^2} + C$

7 $x = \tan\theta; \int \cos^2\theta\, d\theta = \frac{1}{2}\tan^{-1}x + \frac{1}{2}\frac{x}{1+x^2} + C$

9 $x = 5\sec\theta; \int 5(\sec^2\theta - 1)d\theta = \sqrt{x^2 - 25} - 5\sec^{-1}\frac{x}{5} + C$

11 $x = \sec\theta; \int \cos\theta\, d\theta = \frac{\sqrt{x^2-1}}{x} + C$ 13 $x = \tan\theta; \int \cos\theta\, d\theta = \frac{x}{\sqrt{1+x^2}} + C$

15 $x = 3\sec\theta; \int \frac{\cos\theta\, d\theta}{9\sin^2\theta} = \frac{-1}{9\sin\theta} + C = \frac{-x}{9\sqrt{x^2-9}} + C$

17 $x = \sec\theta; \int \sec^3\theta\, d\theta = \frac{1}{2}\sec\theta\tan\theta + \frac{1}{2}\ln(\sec\theta + \tan\theta) + C = \frac{1}{2}x\sqrt{x^2-1} + \frac{1}{2}\ln(x + \sqrt{x^2-1}) + C$

19 $x = \tan\theta; \int \frac{\cos\theta\, d\theta}{\sin^2\theta} = -\frac{1}{\sin\theta} + C = \frac{-\sqrt{x^2+1}}{x} + C$

21 $\int \frac{-\sin\theta\, d\theta}{\sin\theta} = -\theta + C = -\cos^{-1}x + C$; with $C = \frac{\pi}{2}$ this is $\sin^{-1}x$

23 $\int \frac{\tan\theta\sec^2\theta\, d\theta}{\sec^2\theta} = -\ln(\cos\theta) + C = \ln\sqrt{x^2+1} + C$ which is $\frac{1}{2}\ln(x^2+1) + C$

25 $x = a\sin\theta; \int_{-\pi/2}^{\pi/2} a^2\cos^2\theta\, d\theta = \frac{a^2\pi}{2}$ = area of semicircle   27 $\sin^{-1}x]_{.5}^{1} = \frac{\pi}{2} - \frac{\pi}{6} = \frac{\pi}{3}$

30 Like Example 6: $x = \sin\theta$ with $\theta = \frac{\pi}{2}$ when $x = \infty, \theta = \frac{\pi}{3}$ when $x = 2, \int_{\pi/3}^{\pi/2} \frac{\cos\theta\, d\theta}{\sin^2\theta} = -1 + \frac{2}{\sqrt{3}}$

31 $x = 3\tan\theta; \int_{-\pi/2}^{\pi/2} \frac{3\sec^2\theta\, d\theta}{9\sec^2\theta} = \frac{\theta}{3}]_{-\pi/2}^{\pi/2} = \frac{\pi}{3}$ 33 $\int \frac{x^{n+1} + x^{n-1}}{x^2+1}dx = \int x^{n-1}dx = \frac{x^n}{n}$

35 $x = \sec\theta; \frac{1}{2}(e^f + e^{-f}) = \frac{1}{2}(x + \sqrt{x^2-1} + \frac{1}{x+\sqrt{x^2-1}}) = \frac{1}{2}(x + \sqrt{x^2-1} + x - \sqrt{x^2-1}) = x$

37 $x = \cosh\theta; \int d\theta = \cosh^{-1}x + C$

39 $x = \cosh\theta; \int \sinh^2\theta\, d\theta = \frac{1}{2}(\sinh\theta\cosh\theta - \theta) + C = \frac{1}{2}x\sqrt{x^2-1} - \frac{1}{2}\ln(x + \sqrt{x^2-1}) + C$

41 $x = \tanh\theta; \int d\theta = \tanh^{-1}x + C$ 43 $(x-2)^2 + 4$ 45 $(x-3)^2 - 9$ 47 $(x+1)^2$

49 $u = x - 2, \int \frac{du}{u^2+4} = \frac{1}{2}\tan^{-1}\frac{u}{2} = \frac{1}{2}\tan^{-1}(\frac{x-2}{2}) + C; u = x - 3, \int \frac{du}{u^2-9} = \frac{1}{6}\ln\frac{u-3}{u+3} = \frac{1}{6}\ln\frac{x-6}{x} + C;$

   $u = x + 1, \int \frac{du}{u^2} = \frac{-1}{u} = \frac{-1}{x+1} + C$

51 $u = x + b; \int \frac{du}{u^2 - b^2 + c}$ uses $u = a\sec\theta$ if $b^2 > c, u = a\tan\theta$ if $b^2 < c$, equals $-\frac{1}{u} = \frac{-1}{x+b}$ if $b^2 = c$

53 $\cos\theta$ is negative $(-\sqrt{1-x^2})$ from $\frac{\pi}{2}$ to $\frac{3\pi}{2}$; then $\int_0^1 - \int_1^{-1} + \int_{-1}^0 \sqrt{1-x^2}dx = \pi$ = area of unit circle

55 Divide y by 4, multiply dx by 4, same $\int y\, dx$

57 No $\sin^{-1}x$ for $x > 1$; the square root is imaginary. All correct with complex numbers.

Section 7.4 Partial Fractions   (page 304)

1 $A = -1, B = 1, -\ln x + \ln(x-1) + C$ 3 $\frac{1}{x-3} - \frac{1}{x-2}$ 5 $\frac{1}{2x} - \frac{2}{x+1} + \frac{5/2}{x+2}$

7 $\frac{3}{x} + \frac{1}{x^2}$ 9 $3 - \frac{3}{x^2+1}$ 11 $-\frac{1}{x} - \frac{1}{x^2} + \frac{1}{x-1}$ 13 $-\frac{1/6}{x} + \frac{1/2}{x-1} - \frac{1/2}{x-2} + \frac{1/6}{x-3}$

15 $\frac{A}{x+1} + \frac{B}{x-1} + \frac{Cx+D}{x^2+1}; A = -\frac{1}{4}, B = \frac{1}{4}, C = 0, D = -\frac{1}{2}$

17 Coefficients of $y: 0 = -Ab + B$; match constants $1 = Ac; A = \frac{1}{c}, B = \frac{b}{c}$

19 $A = 1$, then $B = 2$ and $C = 1; \int \frac{dx}{x-1} + \int \frac{(2x+1)dx}{x^2+x+1} =$

   $\ln(x-1) + \ln(x^2+x+1) = \ln(x-1)(x^2+x+1) = \ln(x^3-1)$

21 $u = e^x; \int \frac{du}{u^2-u} = \int \frac{du}{u-1} - \int \frac{du}{u} = \ln(\frac{u-1}{u}) + C = \ln(\frac{e^x-1}{e^x}) + C$

23 $u = \cos\theta; \int \frac{-du}{1-u^2} = -\frac{1}{2}\int \frac{du}{1-u} - \frac{1}{2}\int \frac{du}{1+u} = \frac{1}{2}\ln(1-u) - \frac{1}{2}\ln(1+u) = \frac{1}{2}\ln\frac{1-\cos\theta}{1+\cos\theta} + C$. We can reach

   $\frac{1}{2}\ln\frac{(1-\cos\theta)^2}{1-\cos^2\theta} = \ln\frac{1-\cos\theta}{\sin\theta} = \ln(\csc\theta - \cot\theta)$ or a different way $\frac{1}{2}\ln\frac{1-\cos^2\theta}{(1+\cos\theta)^2} = \ln\frac{\sin\theta}{1+\cos\theta} = -\ln\frac{1+\cos\theta}{\sin\theta} = -\ln(\csc\theta + \cot\theta)$

25 $u = e^x; du = e^x dx = u\, dx; \int \frac{1+u}{(1-u)u}du = \int \frac{2du}{1-u} + \int \frac{du}{u} = -2\ln(1-e^x) + \ln e^x + C = -2\ln(1-e^x) + x + C$

27 $x+1=u^2, dx=2u\,du; \int \frac{2u\,du}{1+u} = \int \left[2-\frac{2}{1+u}\right]du = 2u-2\ln(1+u)+C = 2\sqrt{x+1}-2\ln(1+\sqrt{x+1})+C$

29 Note $Q(a)=0$. Then $\frac{x-a}{Q(x)} = \frac{x-a}{Q(x)-Q(a)} \to \frac{1}{Q'(a)}$ by definition of derivative. At a double root $Q'(a)=0$.

Section 7.5 Improper Integrals   (page 309)

1 $\frac{x^{1-e}}{1-e}\Big]_1^\infty = \frac{1}{e-1}$ **3** $-2(1-x)^{1/2}\Big]_0^1 = 2$ **5** $\tan^{-1} x\Big]_{-\pi/2}^0 = \frac{\pi}{2}$ **7** $\frac{1}{2}(\ln x)^2\Big]_0^1 = -\infty$

9 $x\ln x - x\Big]_0^e = 0$ **11** $\ln(\ln(\ln x))\Big]_{100}^\infty = \infty$ **13** $\frac{1}{2}(x+\sin x\cos x)\Big]_0^\infty = \infty$

15 $\frac{x^{1-p}}{1-p}\Big]_0^\infty$ diverges for every p!   **17** Less than $\int_1^\infty \frac{dx}{x^6} = \frac{1}{5}$

19 Less than $\int_0^1 \frac{dx}{x^2+1} + \int_1^\infty \frac{\sqrt{x}\,dx}{x^2} = \tan^{-1} x\Big]_0^1 - \frac{2}{\sqrt{x}}\Big]_1^\infty = \frac{\pi}{4}+2$

21 Less than $\int_1^\infty e^{-x}dx = \frac{1}{e}$, greater than $-\frac{1}{e}$

23 Less than $\int_0^1 e^2 dx + e\int_1^\infty e^{-(x-1)^2}dx = e^2 + e\int_1^\infty e^{-u^2}du = e^2 + \frac{e}{\sqrt{\pi}}$

25 $\int_0^1 \frac{\sin^2 x\,dx}{x^2} + \int_1^\infty \frac{\sin^2 x\,dx}{x^2}$ less than $1+\int_1^\infty \frac{dx}{x^2} = 2$ **27** $p! = p$ times $(p-1)!$; $1 = 1$ times $0!$

29 $u=x, dv = xe^{-x^2}dx: -x\frac{e^{-x^2}}{2}\Big]_0^\infty + \int_0^\infty \frac{e^{-x^2}}{2}dx = \frac{1}{4}\sqrt{\pi}$ **31** $\int_0^\infty 1000e^{-.1t}dt = -10{,}000e^{-.1t}\Big]_0^\infty = \$10{,}000$

33 $W = \frac{-GMm}{x}\Big]_R^\infty = \frac{GMm}{R} = \frac{1}{2}mv_0^2$ if $v_0 = \sqrt{\frac{2GM}{R}}$

35 $\int_0^\infty \frac{dx}{2^x} = \int_0^\infty e^{-x\ln 2}dx = \frac{e^{-x\ln 2}}{-\ln 2}\Big]_0^\infty = \frac{1}{\ln 2}$

37 $\int_0^{\pi/2}(\sec x - \tan x)dx = \left[\ln(\sec x + \tan x) + \ln(\cos x)\right]_0^{\pi/2} = \left[\ln(1+\sin x)\right]_0^{\pi/2} = \ln 2$.
The areas under $\sec x$ and $\tan x$ separately are infinite   **39** Only $p=0$

CHAPTER 8   APPLICATIONS OF THE INTEGRAL

Section 8.1 Areas and Volumes by Slices   (page 318)

1 $x^2-3=1$ gives $x=\pm 2; \int_{-2}^2 \left[(1-(x^2-3)\right]dx = \frac{32}{3}$

3 $y^2=x=9$ gives $y=\pm 3; \int_{-3}^3 [9-y^2]dy = 36$

5 $x^4-2x^2=2x^2$ gives $x=\pm 2$ (or $x=0$); $\int_{-2}^2 [2x^2-(x^4-2x^2)]dx = \frac{128}{15}$

7 $y=x^2=-x^2+18x$ gives $x=0,9; \int_0^9 [(-x^2+18x)-x^2]dx = 243$

9 $y=\cos x = \cos^2 x$ when $\cos x = 1$ or $0, x=0$ or $\frac{\pi}{2}$ or $\cdots \int_0^{\pi/2}(\cos x - \cos^2 x)dx = 1-\frac{\pi}{4}$

11 $e^x = e^{2x-1}$ gives $x=1; \int_0^1 [e^x - e^{2x-1}]dx = (e-1) - \left(\frac{e-e^{-1}}{2}\right)$

13 Intersections $(0,0),(1,3),(2,2); \int_0^1 [3x-x]dx + \int_1^2 [4-x-x]dx = 2$

16 Inside, since $1-x^2 < \sqrt{1-x^2}; \int_{-1}^1 [\sqrt{1-x^2}-(1-x^2)]dx = \frac{\pi}{2} - \frac{4}{3}$

17 $V = \int_{-a}^a \pi y^2 dx = \int_{-a}^a \pi b^2\left(1-\frac{x^2}{a^2}\right)dx = \frac{4\pi b^2 a}{3}$; around y axis $V = \frac{4\pi a^2 b}{3}$; rotating
$x=2, y=0$ around y axis gives a circle not in the first football

19 $V = \int_0^\pi 2\pi x\sin x\,dx = 2\pi^2$ **21** $\int_0^8 \pi(8-x)^2 dx = \frac{512\pi}{3}; \int_0^8 2\pi x(8-x)dx = \frac{512\pi}{.3}$ (same cone tipped over)

23 $\int_0^1 \pi\cdot 1^2 dx - \int_0^1 \pi(x^4)^2 dx = \frac{8\pi}{9}; \int_0^1 2\pi(1-x^4)x\,dx = \frac{2\pi}{3}$

25 $\int_{1/3}^2 \pi(3^2)dx - \int_{1/3}^2 \pi(\frac{1}{x})^2 dx = \frac{25\pi}{2}; \int_{1/3}^2 2\pi x(3-\frac{1}{x})dx = \frac{25\pi}{3}$

27 $\int_0^1 \pi[(x^{2/3})^2 - (x^{3/2})^2]dx = \frac{5\pi}{28}; \int_0^1 2\pi x(x^{2/3}-x^{3/2})dx = \frac{5\pi}{28}$ (notice xy symmetry)

29 $x^2 = R^2 - y^2, V = \int_{R-h}^R \pi(R^2-y^2)dy = \pi\left(Rh^2 - \frac{h^3}{3}\right)$

31 $\int_{-a}^a (2\sqrt{a^2-x^2})^2 dx = \frac{16}{3}a^3$ **33** $\int_0^1 (2\sqrt{1-y})^2 dy = 2$ **37** $\int A(x)dx$ or in this case $\int a(y)dy$

39 Ellipse; $\sqrt{1-x^2}\tan\theta; \frac{1}{2}(1-x^2)\tan\theta; \frac{2}{3}\tan\theta$

41 Half of $\pi r^2 h$; rectangles   **43** $\int_1^3 \pi(5^2-2^2)dx = 42\pi$ **45** $\int_1^3 \pi(4^2-1^2)dx = 30\pi$

47 $\int_0^{b-a} \pi((b-y)^2 - a^2)dy = \frac{\pi}{3}(b^3 - 3a^2b + 2a^3)$ **49** $\int_0^2 \pi(3-x)^2 dx; \int_0^1 2\pi y(2)dy + \int_1^3 2\pi y(3-y)dy$

51 $\int_a^b \pi \left(\frac{y}{m}\right)^2 dy = \frac{\pi(b^3-a^3)}{3m^2}$ **53** $960\,\pi$ cm   **55** $\frac{\pi}{2}$ **57** $\frac{2\pi}{3}$

59 2π **61** $\int_0^4 2\pi y(2 - \sqrt{y})dy = \frac{32\pi}{5}$ **63** $3\pi e$ **65** Height 1; $\int_0^a 2\pi x\, dx = \pi a^2$; cylinder

67 Length of hole is $2\sqrt{b^2 - a^2} = 2$, so $b^2 - a^2 = 1$ and volume is $\frac{4\pi}{3}$ **69** F; T(?); F; T

Section 8.2 Length of a Plane Curve   (page 324)

1 $\int_0^1 \sqrt{1 + \left(\frac{3}{2}x^{1/2}\right)^2}\,dx = \frac{8}{27}\left[\left(\frac{13}{4}\right)^{3/2} - 1\right] = \frac{13\sqrt{13}-8}{27}$ **3** $\int_0^1 \sqrt{1 + x^2(x^2 + 2)}\,dx = \int_0^1 (1 + x^2)dx = \frac{4}{3}$

5 $\int_1^3 \sqrt{1 + \left(x^2 - \frac{1}{4x^2}\right)^2}\,dx = \int_1^3 \left(x^2 + \frac{1}{4x^2}\right)dx = \frac{53}{6}$

7 $\int_1^4 \sqrt{1 + \left(x^{1/2} - \frac{1}{4}x^{-1/2}\right)^2}\,dx = \int_1^4 \left(x^{1/2} + \frac{1}{4}x^{-1/2}\right)dx = \frac{31}{6}$

9 $\int_0^{\pi/2} \sqrt{9\cos^4 t \sin^2 t + 9\sin^4 t \cos^2 t}\,dt = \int_0^{\pi/2} 3\cos t \sin t\,dt = \frac{3}{2}$

11 $\int_0^{\pi/2} \sqrt{\sin^2 t + (1 - \cos t)^2}\,dt = \int_0^{\pi/2} \sqrt{2 - 2\cos t}\,dt = \int_0^{\pi/2} 2\sin\frac{t}{2}\,dt = 4 - 2\sqrt{2}$

13 $\int_0^1 \sqrt{t^2 + 2t + 1}\,dt = \int_0^1 (t+1)dt = \frac{3}{2}$ **15** $\int_0^\pi \sqrt{1 + \cos^2 x}\,dx = 3.820$ **17** $\int_1^e \sqrt{1 + \frac{1}{x^2}}\,dx = 2.003$

19 Graphs are flat toward $(1,0)$ then steep up to $(1,1)$; limiting length is 2

21 $\frac{ds}{dt} = \sqrt{36\sin^2 3t + 36\cos^2 3t} = 6$ **23** $\int_0^1 \sqrt{26}\,dy = \sqrt{26}$

25 $\int_{-1}^1 \sqrt{\frac{1}{4}(e^y - e^{-y})^2 + 1}\,dy = \int_{-1}^1 \frac{1}{2}(e^y + e^{-y})dy = \frac{1}{2}(e^y - e^{-y})\big]_{-1}^1 = e - \frac{1}{e}$.

   Using $x = \cosh y$ this is $\int \sqrt{1 + \sinh^2 y}\,dy = \int \cosh y\,dy = \sinh y\big]_{-1}^1 = 2\sinh 1$

27 Ellipse; two y's for the same x **29** Carpet length $2 \neq$ straight distance $\sqrt{2}$

31 $(ds)^2 = (dx)^2 + (dy)^2 + (dz)^2; ds = \sqrt{\left(\frac{dx}{dt}\right)^2 + \left(\frac{dy}{dt}\right)^2 + \left(\frac{dz}{dt}\right)^2}\,dt;$
   $ds = \sqrt{\sin^2 t + \cos^2 t + 1}\,dt = \sqrt{2}\,dt; 2\pi\sqrt{2};$ curve $=$ helix, shadow $=$ circle

33 $L = \int_0^1 \sqrt{1 + 4x^2}\,dx; \int_0^2 \sqrt{1 + x^2}\,dx = \int_0^1 \sqrt{1 + 4u^2}\,2\,du = 2L;$ stretch xy plane by $2\left(y = x^2 \text{ becomes } \frac{y}{2} = \left(\frac{x}{2}\right)^2\right)$

Section 8.3 Area of a Surface of Revolution   (page 327)

1 $\int_2^6 2\pi \sqrt{x}\sqrt{1 + \left(\frac{1}{2\sqrt{x}}\right)^2}\,dx = \int_2^6 2\pi\sqrt{x + \frac{1}{4}}\,dx = \frac{49\pi}{3}$ **3** $2\int_0^1 2\pi(7x)\sqrt{50}\,dx = 14\pi\sqrt{50}$

5 $\int_{-1}^1 2\pi \sqrt{4 - x^2}\sqrt{1 + \frac{x^2}{4 - x^2}}\,dx = \int_{-1}^1 4\pi\,dx = 8\pi$ **7** $\int_0^2 2\pi x\sqrt{1 + (2x)^2}\,dx = \frac{\pi}{6}(1 + 4x^2)^{3/2}\big]_0^2 = \frac{\pi}{6}[17^{3/2} - 1]$

9 $\int_0^3 2\pi x\sqrt{2}\,dx = 9\pi\sqrt{2}$ **11** Figure shows radius s times angle $\theta = $ arc $2\pi R$

13 $2\pi r\Delta s = \pi(R + R')(s - s') = \pi Rs - \pi R's'$ because $R's - Rs' = 0$

15 Radius a, center at $(0,b); \left(\frac{dx}{dt}\right)^2 + \left(\frac{dy}{dt}\right)^2 = a^2$, surface area $\int_0^{2\pi} 2\pi(b + a\sin t)a\,dt = 4\pi^2 ab$

17 $\int_1^2 2\pi x\sqrt{1 + \frac{(1-x)^2}{2x - x^2}}\,dx = \int_1^2 \frac{2\pi x\,dx}{\sqrt{2x - x^2}} = \pi^2 + 2\pi$ (write $2x - x^2 = 1 - (x-1)^2$ and set $x - 1 = \sin\theta$)

19 $\int_{1/2}^1 2\pi x\sqrt{1 + \frac{1}{x^4}}\,dx$ (can be done)

21 Surface area $= \int_1^\infty 2\pi\frac{1}{x}\sqrt{1 + \frac{1}{x^4}}\,dx > \int_1^\infty \frac{2\pi\,dx}{x} = 2\pi\ln x\big]_1^\infty = \infty$ but volume $= \int_1^\infty \pi\left(\frac{1}{x}\right)^2 dx = \pi$

23 $\int_0^\pi 2\pi\sin t\sqrt{2\sin^2 t + \cos^2 t}\,dt = \int_0^\pi 2\pi\sin t\sqrt{2 - \cos^2 t}\,dt = \int_{-1}^1 2\pi\sqrt{2 - u^2}\,du =$
   $\pi u\sqrt{2 - u^2} + 2\pi\sin^{-1}\frac{u}{\sqrt{2}}\big]_{-1}^1 = 2\pi + \pi^2$

Section 8.4 Probability and Calculus   (page 334)

1 $P(X < 4) = \frac{7}{8}, P(X = 4) = \frac{1}{16}, P(X > 4) = \frac{1}{16}$ **3** $\int_0^\infty p(x)dx$ is not $1; p(x)$ is negative for large x

5 $\int_2^\infty e^{-x}dx = \frac{1}{e^2}; \int_1^{1.01} e^{-x}dx \approx (.01)\frac{1}{e}$ **7** $p(x) = \frac{1}{\pi}; F(x) = \frac{x}{\pi}$ for $0 \le x \le \pi$ ($F = 1$ for $x > \pi$)

9 $\mu = \frac{1}{7} \cdot 1 + \frac{1}{7} \cdot 2 + \cdots + \frac{1}{7} \cdot 7 = 4$ 11 $\int_0^\infty \frac{2x\,dx}{\pi(1+x^2)} = \frac{1}{\pi} \ln(1+x^2)]_0^\infty = +\infty$

13 $\int_0^\infty axe^{-ax}\,dx = \left[-xe^{-ax} \right]_0^\infty + \int_0^\infty e^{-ax}\,dx = \frac{1}{a}$

15 $\int_0^x \frac{2\,dx}{\pi(1+x^2)} = \frac{2}{\pi} \tan^{-1} x; \int_0^x e^{-x}\,dx = 1 - e^{-x}; \int_0^x ae^{-ax}\,dx = 1 - e^{-ax}$ 17 $\int_{10}^\infty \frac{1}{10} e^{-x/10}\,dx = -e^{-x/10}]_{10}^\infty = \frac{1}{e}$

19 *Exponential* better than Poisson: 60 years $\to \int_0^{60} .01e^{-.01x}\,dx = 1 - e^{-.6} = .45$

21 $y = \frac{x-\mu}{\sigma}$; three areas $\approx \frac{1}{3}$ each because $\mu - \sigma$ to μ is the same as μ to $\mu + \sigma$ and areas add to 1

23 $-2\mu \int xp(x)\,dx + \mu^2 \int p(x)\,dx = -2\mu \cdot \mu + \mu^2 \cdot 1 = -\mu^2$

25 $\mu = 0 \cdot \frac{1}{3} + 1 \cdot \frac{1}{3} + 2 \cdot \frac{1}{3} = 1; \sigma^2 = (0-1)^2 \cdot \frac{1}{3} + (1-1)^2 \cdot \frac{1}{3} + (2-1)^2 \cdot \frac{1}{3} = \frac{2}{3}$.
Also $\sum n^2 p_n - \mu^2 = 0 \cdot \frac{1}{3} + 1 \cdot \frac{1}{3} + 4 \cdot \frac{1}{3} - 1 = \frac{2}{3}$

27 $\mu = \int_0^\infty \frac{xe^{-x/2}\,dx}{2} = 2; 1 - \int_0^4 \frac{e^{-x/2}\,dx}{2} = 1 + \left[e^{-x/2} \right]_0^4 = e^{-2}$

29 Standard deviation (yes – no poll) $\leq \frac{1}{2\sqrt{N}} = \frac{1}{2\sqrt{900}} = \frac{1}{60}$ Poll showed $\frac{870}{900} = \frac{29}{30}$ peaceful.
95% confidence interval is from $\frac{29}{30} - \frac{2}{60}$ to $\frac{29}{30} + \frac{2}{60}$, or 93% to 100% peaceful.

31 95% confidence of unfair if more than $\frac{2\sigma}{\sqrt{N}} = \frac{1}{\sqrt{2500}} = 2\%$ away from 50% heads.
2% of $2500 = 50$. So unfair if more than 1300 or less than 1200.

33 55 is 1.5σ below the mean, and the area up to $\mu - 1.5\sigma$ is about 8% so 24 students fail.
A grade of 57 is 1.3σ below the mean and the area up to $\mu - 1.3\sigma$ is about 10%.

35 .999; $.999^{1000} = \left(1 - \frac{1}{1000}\right)^{1000} \approx \frac{1}{e}$ because $\left(1 - \frac{1}{n}\right)^n \to \frac{1}{e}$.

Section 8.5 Masses and Moments (page 340)

1 $\bar{x} = \frac{10}{6}$ 3 $\bar{x} = \frac{4}{4}$ 5 $\bar{x} = \frac{3.5}{3}$ 7 $\bar{x} = \frac{2}{3} = \bar{y}$ 9 $\bar{x} = \frac{7/2}{7} = \bar{y}$ 11 $\bar{x} = \frac{1/3}{\pi/4} = \bar{y}$ 13 $\bar{x} = \frac{1/4}{1/2}, \bar{y} = \frac{1/8}{1/2}$

15 $\bar{x} = \frac{0}{3\pi} = \bar{y}$ 21 $I = \int x^2 \rho\,dx - 2t \int x\rho\,dx + t^2 \int \rho\,dx; \frac{dI}{dt} = -2 \int x\rho\,dx + 2t \int \rho\,dx = 0$ for $t = \bar{x}$

23 South Dakota 25 $2\pi^2 a^2 b$ 27 $M_x = 0, M_y = \frac{\pi}{2}$ 29 $\frac{2}{\pi}$ 31 Moment

33 $I = \sum m_n r_n^2; \frac{1}{2} \sum m_n r_n^2 \omega_n^2; 0$ 35 $14\pi\ell \frac{r^2}{2}; 14\pi\ell \frac{r^4}{4}; \frac{1}{2}$

37 $\frac{2}{3}$; solid ball, solid cylinder, hallow ball, hollow cylinder 39 No

41 $T \approx \sqrt{1+J}$ by problem **40** so $T \approx \sqrt{1.4}, \sqrt{1.5}, \sqrt{5/3}, \sqrt{2}$

Section 8.6 Force, Work, and Energy (page 346)

1 2.4 ft lb; 2.424... ft lb 3 24000 lb/ft; $83\frac{1}{3}$ ft lb 5 $10x$ ft lb; $10x$ ft lb 7 25000 ft lb; 20000 ft lb

9 864,000 Nkm 11 $5.6 \cdot 10^7$ Nkm 13 $k = 10$ lb/ft; $W = 25$ ft lb 15 $\int 60wh\,dh = 48000w, 12000w$

17 $\frac{1}{2}wAH^2; \frac{3}{8}wAH^2$ 19 $9600w$ 21 $\left(1 - \frac{v^2}{c^2}\right)^{-3/2}$ 23 $(800)(9800)$ kg 25 $\pm$ force

CHAPTER 9 POLAR COORDINATES AND COMPLEX NUMBERS

Section 9.1 Polar Coordinates (page 350)

1 $\left(1, \frac{\pi}{2}\right)$ 3 $\left(2, \frac{\pi}{4}\right)$ 5 $\left(\sqrt{2}, \frac{5\pi}{4}\right)$ 7 $(0, 2)$ 9 $\left(\sqrt{10}, \sqrt{10}\right)$ 11 $\left(\sqrt{3}, -1\right)$ 13 $2\sqrt{2}$

15 $\sqrt{r^2 + R^2 - 2rR \cos(\theta - \phi)}$

17 $0 < r < \infty, -\frac{\pi}{2} < \theta < \frac{\pi}{2}; 0 \leq r < \infty, \pi \leq \theta \leq 2\pi; \sqrt{4} < r < \sqrt{5}, 0 \leq \theta < 2\pi; 0 \leq r < \infty, -\frac{\pi}{4} \leq \theta \leq \frac{\pi}{4}$

19 $y = x \tan \theta, r = x \sec \theta$ 21 $\theta = \frac{\pi}{4}$, all $r; r = \frac{1}{\sin\theta+\cos\theta}; r = \cos\theta + \sin\theta$

23 $x^2 + y^2 = y$ 25 $x = r \sin\theta \cos\theta, y = r \sin^2\theta, x^2 + y^2 = y$

27 $x^2 + y^2 = x + y, \left(x - \frac{1}{2}\right)^2 + \left(y - \frac{1}{2}\right)^2 = \left(\frac{\sqrt{2}}{2}\right)^2$ 29 $x = \frac{\cos\theta}{\cos\theta+\sin\theta}, y = \frac{\sin\theta}{\cos\theta+\sin\theta}$ 31 $\left(x^2 + y^2\right)^3 = x^4$

Section 9.2 Polar Equations and Graphs   (page 355)

1 Line $y = 1$ 3 Circle $x^2 + y^2 = 2x$ 5 Ellipse $3x^2 + 4y^2 = 1 - 2x$ 7 x, y, r symmetries

9 x symmetry only   11 No symmetry   13 x, y, r symmetries!

15 $x^2 + y^2 = 6y + 8x \rightarrow (x-4)^2 + (y-3)^2 = 5^2$, center $(4,3)$ 17 $(2,0), (0,0)$

19 $r = 1 - \frac{\sqrt{2}}{2}, \theta = \frac{3\pi}{4}; r = 1 + \frac{\sqrt{2}}{2}, \theta = \frac{7\pi}{4}; (0,0)$ 21 $r = 2, \theta = \pm\frac{\pi}{12}, \pm\frac{5\pi}{12}, \pm\frac{7\pi}{12}, \pm\frac{11\pi}{12}$

23 $(x, y) = (1, 1)$ 25 $r = \cos 5\theta$ has 5 petals   27 $(x^2 + y^2 - x)^2 = x^2 + y^2$

29 $(x^2 + y^2)^3 = (x^2 - y^2)^2$ 31 $\cos\theta = -\frac{1}{2}, \sin\theta = \frac{\sqrt{3}}{2} \rightarrow y = \frac{2\sqrt{3}}{2}, x = -\frac{2}{3}$ 33 $x = \frac{4}{3}, r = -\frac{5}{3}$ 35 .967

Section 9.3 Slope, Length, and Area for Polar Curves   (page 359)

1 Area $\frac{3\pi}{2}$ 3 Area $\frac{9\pi}{2}$ 5 Area $\frac{\pi}{8}$ 7 Area $\frac{\pi}{8} - \frac{1}{4}$ 9 $\int_{-\pi/3}^{\pi/3} \left(\frac{9}{2}\cos^2\theta - \frac{(1+\cos\theta)^2}{2}\right)d\theta = \pi$

11 Area 8π 13 Only allow $r^2 > 0$, then $4\int_0^{\pi/4} \frac{1}{2}\cos 2\theta \, d\theta = 1$ 15 $2 + \frac{\pi}{4}$

17 $\theta = 0$; left points $r = \frac{1}{2}, \theta = \pm\frac{2\pi}{3}, x = -\frac{1}{4}, y = \pm\frac{\sqrt{3}}{4}$

19 $\frac{r^2}{2c}\big]_6^{14} = 40,000$; $\frac{1}{2c}\left[r\sqrt{r^2+c^2} + c^2\ln\left(r + \sqrt{r^2+c^2}\right)\right]_6^{14} = 40,000.001$

21 $\tan\psi = \tan\theta$ 23 $x = 0, y = 1$ is on limacon but not circle   25 $\frac{1}{2}\ln(2\pi + \sqrt{1+4\pi^2}) + \pi\sqrt{1+4\pi^2}$

27 $\frac{3\pi}{2}$ 29 $\frac{1}{2}$(base)(height)$\approx \frac{1}{2}(r\Delta\theta)r$ 31 $\frac{4\pi}{5}\sqrt{2}$ 33 $2\pi(2-\sqrt{2})$ 35 $\frac{8\pi}{3}$ 39 $\sec\theta$

Section 9.4 Complex Numbers   (page 364)

1 Sum $= 4$, product $= 5$ 5 Angles $\frac{3\pi}{4}, \frac{3\pi}{2}, \frac{9\pi}{4}$ 7 Real axis; imaginary axis; $\frac{1}{2}$ axis $x \geq 0$; unit circle

9 $cd = 5 + 10i, \frac{c}{d} = \frac{11-2i}{25}$ 11 $2\cos\theta, 1; -1, 1$ 13 Sum $= 0$, product $= -1$ 15 $r^4 e^{4i\theta}, \frac{1}{r}e^{-i\theta}, \frac{1}{r^4}e^{-4i\theta}$

17 Evenly spaced on circle around origin   19 e^{it}, e^{-it} 21 e^t, e^{-t}, e^0 23 $\cos 7t, \sin 7t$

29 $t = -\frac{2\pi}{\sqrt{3}}, y = -e^{\pi/\sqrt{3}}$ 31 F; T; at most 2; Re $c < 0$ 33 $\frac{1}{r}e^{-i\theta}, x = \frac{1}{r}\cos\theta, y = -\frac{1}{r}\sin\theta; \pm\frac{1}{\sqrt{r}}e^{-i\theta/2}$

CHAPTER 10   INFINITE SERIES

Section 10.1 Geometric Series   (page 373)

1 Subtraction leaves $G - xG = 1$ or $G = \frac{1}{1-x}$ 3 $\frac{1}{2}; \frac{4}{5}; \frac{100}{11}; 3\frac{4}{99}$ 5 $2\cdot 1 + 3\cdot 2x + 4\cdot 3x^2 + \cdots = \frac{2}{(1-x)^3}$

7 .142857 repeats because the next step divides 7 into 1 again

9 If q (prime, not 2 or 5) divides $10^N - 10^M$ then it divides $10^{N-M} - 1$ 11 This decimal does not repeat

13 $\frac{87}{99}; \frac{123}{999}$ 15 $\frac{x}{1-x^2}$ 17 $\frac{x^3}{1-x^3}$ 19 $\frac{\ln x}{1-\ln x}$ 21 $\frac{1}{x-1}$ 23 $\tan^{-1}(\tan x) = x$

26 $(1 + x + x^2 + x^3 \cdots)(1 - x + x^2 - x^3 \cdots) = 1 + x^2 + x^4 + \cdots$

27 $2(.1234\ldots)$ is $2\cdot\frac{1}{10}\cdot\frac{1}{\left(1-\frac{1}{10}\right)^2} = \frac{20}{81}; 1 - .0123\ldots$ is $1 - \frac{1}{100}\frac{1}{\left(1-\frac{1}{10}\right)^2} = \frac{80}{81}$ 29 $\frac{2}{3}\frac{1}{1-\frac{1}{3}} = 1$

31 $-\ln(1-.1) = -\ln.9$ 33 $\frac{1}{2}\ln\frac{1.1}{.9}$ 35 $(n+1)!$ 37 $y = \frac{b}{1-bx}$

39 All products like $a_1 b_2$ are missed; $(1+1)(1+1) \neq 1 + 1$ 41 Take $x = \frac{1}{2}$ in (13): $\ln 3 = 1.0986$

43 In 3 seconds the ball goes 78 feet   45 $\tan z = \frac{2}{3}$; (18)is slower with $x = \frac{2}{3}$

Section 10.2 Convergence Tests: Positive Series (page 380)

1 $\frac{1}{2} + \frac{1}{4} + \cdots$ is smaller than $1 + \frac{1}{3} + \cdots$

3 $a_n = s_n - s_{n-1} = \frac{1}{n^2 - n}, s = 1; a_n = 4, s = \infty; a_n = \ln \frac{2n}{n+1} - \ln \frac{2(n-1)}{n} = \ln \frac{n^2}{n-1}, s = \ln 2$

5 No decision on $\sum b_n$ 7 Diverges: $\frac{1}{100}\left(1 + \frac{1}{2} + \cdots\right)$ 9 $\sum \frac{1}{100 + n^2}$ converges: $\sum \frac{1}{n^2}$ is larger

11 Converges: $\sum \frac{1}{n^2}$ is larger   13 Diverges: $\sum \frac{1}{2n}$ is smaller   15 Diverges: $\sum \frac{1}{2n}$ is smaller

17 Converges: $\sum \frac{2}{2^n}$ is larger   19 Converges: $\sum \frac{3}{3^n}$ is larger   21 $L = 0$ 23 $L = 0$ 25 $L = \frac{1}{2}$

27 root $\left(\frac{n-1}{n}\right)^n \to L = \frac{1}{e}$ 29 $s = 1$ (only survivor)   31 If y decreases, $\sum_2^n y(i) \le \int_1^n y(x)dx \le \sum_1^{n-1} y(i)$

33 $\sum_1^\infty e^{-n} \le \int_0^\infty e^{-x}dx = 1; \frac{1}{e} + \frac{1}{e^2} + \cdots = \frac{1}{e-1}$ 35 Converges faster than $\int \frac{dx}{x^2+1}$

37 Diverges because $\int_0^\infty \frac{x\,dx}{x^2+1} = \frac{1}{2}\ln(x^2+1)\big]_0^\infty = \infty$ 39 Diverges because $\int_1^\infty x e^{-\pi}dx = \frac{x^{e-\pi+1}}{e-\pi+1}\big]_0^\infty = \infty$

41 Converges (geometric) because $\int_1^\infty \left(\frac{e}{\pi}\right)^x dx < \infty$ 43 (b) $\int_n^{n+1} \frac{dx}{x} > $ (base 1) $\left(\text{height } \frac{1}{n+1}\right)$

45 After adding we have $1 + \frac{1}{2} + \cdots + \frac{1}{2n}$ (close to $\ln 2n$); thus originally close to $\ln 2n - \ln n = \ln \frac{2n}{n} = \ln 2$

47 $\int_{100}^{1000} \frac{dx}{x^2} = \frac{1}{100} - \frac{1}{1000} = .009$ 49 Comparison test: $\sin a_n < a_n$; if $a_n = \pi_n$ then $\sin a_n = 0$ but $\sum a_n = \infty$

51 $a_n = n^{-5/2}$ 53 $a_n = \frac{2^n}{n^n}$ 55 Ratios are $1, \frac{1}{2}, 1, \frac{1}{2}, \cdots$ (no limit L); $\left(\frac{1}{2^k}\right)^{1/2k} = \frac{1}{\sqrt{2}}$; yes

57 Root test $\frac{1}{\ln n} \to L = 0$ 59 Root test $L = \frac{1}{10}$ 61 Divergence: N terms add to $\ln \frac{N+2}{2} \to \infty$

63 Diverge $\left(\text{compare } \sum \frac{1}{n}\right)$ 65 Root test $L = \frac{3}{4}$ 67 Beyond some point $\frac{a_n}{b_n} < 1$ or $a_n < b_n$

Section 10.3 Convergence Tests: All Series (page 384)

1 Terms don't approach zero   3 Absolutely   5 Conditionally not absolutely   7 No convergence

9 Absolutely   11 No convergence   13 By comparison with $\sum |a_n|$

15 Even sums $\frac{1}{2} + \frac{1}{4} + \frac{1}{6} + \cdots$ diverge; a_n's are not decreasing   17 (b) If $a_n > 0$ then s_n is too large so $s - s_n < 0$

19 $s = 1 - \frac{1}{e}$; below by less than $\frac{1}{5!}$

21 Subtract $2\left(\frac{1}{2^2} + \frac{1}{4^2} + \cdots\right) = \frac{2}{4}\left(\frac{1}{1^2} + \frac{1}{2^2} + \cdots\right) = \frac{\pi^2}{12}$ from positive series to get alternating series

23 Text proves: If $\sum |a_n|$ converges so does $\sum a_n$

25 New series$= \left(\frac{1}{2}\right) - \frac{1}{4} + \left(\frac{1}{6}\right) - \frac{1}{8} \cdots = \frac{1}{2}\left(1 - \frac{1}{2} + \frac{1}{3} - \frac{1}{4} \cdots\right)$ 27 $\frac{3}{2}\ln 2$: add $\ln 2$ series to $\frac{1}{2}$ ($\ln 2$ series)

29 Terms alternate and decrease to zero; partial sums are $1 + \frac{1}{2} + \cdots + \frac{1}{n} - \ln n \to \gamma$

31 .5403?   33 Hint + comparison test   35 Partial sums $a_n - a_0$; sum $-a_0$ if $a_n \to 0$

37 $\frac{1}{1 - \frac{1}{2}} \frac{1}{1 - \frac{1}{3}} = 3$ but product is not $1 + \frac{2}{3} + \cdots$

39 Write x to base 2, as in 1.0010 which keeps $1 + \frac{1}{8}$ and deletes $\frac{1}{2}, \frac{1}{4}, \cdots$

41 $\frac{1}{9} + \frac{1}{27} + \cdots$ adds to $\frac{1/9}{1 - 1/3} = \frac{1}{6}$ and can't cancel $\frac{1}{3}$

43 $\frac{\sin 1}{1 - \cos 1} = \cot \frac{1}{2}$ (trig identity) $= \tan\left(\frac{\pi}{2} - \frac{1}{2}\right); s = \sum \frac{e^{in}}{n} = -\log(1 - e^i)$ by 10a in Section 10.1; take imaginary part

Section 10.4 The Taylor Series for e^x, $\sin x$, and $\cos x$ (page 390)

1 $1 + 2x + \frac{(2x)^2}{2!} + \cdots$; derivatives $2^n; 1 + 2 + \frac{2^2}{2!} + \cdots$ 3 Derivatives $i^n; 1 + ix + \cdots$

5 Derivatives $2^n n!; 1 + 2x + 4x^2 + \cdots$ 7 Derivatives $-(n-1)!; -x - \frac{x^2}{2} - \frac{x^3}{3} - \cdots$

9 $y = 2 - e^x = 1 - x - \frac{x^2}{2!} - \cdots$ 11 $y = x - \frac{x^3}{6} + \cdots = \sin x$ 13 $y = xe^x = x + x^2 + \frac{x^3}{2!} + \cdots$

15 $1 + 2x + x^2; 4 + 4(x-1) + (x-1)^2$ 17 $-(x-1)^5$ 19 $1 - (x-1) + (x-1)^2 - \cdots$

21 $(x-1) - \frac{(x-1)^2}{2} + \frac{(x-1)^3}{3} - \cdots = \ln(1 + (x-1))$ 23 $e^{-1}e^{1-x} = e^{-1}\left(1 - (x-1) + \frac{(x-1)^2}{2!} - \cdots\right)$

25 $x + 2x^2 + 2x^3$ 27 $\frac{1}{2} - \frac{x^2}{24} + \frac{x^4}{720}$ 29 $x - \frac{x^3}{18} + \frac{x^5}{600}$ 31 $1 + x^2 + \frac{x^4}{2}$ 33 $1 + x - \frac{x^3}{2}$

35 ∞ slope; $1 + \frac{1}{2}(x-1)$ **37** $x - \frac{x^3}{3} + \frac{x^5}{5}$ **39** $x + \frac{x^3}{3} + \frac{2x^5}{15}$ **41** $1 + x + \frac{x^2}{2}$ **43** $1 + 0x - x^2$

45 $\cos\theta = \frac{e^{i\theta} + e^{-i\theta}}{2}, \sin\theta = \frac{e^{i\theta} - e^{-i\theta}}{2i}$ **47** 99th powers $-1, -i, e^{3\pi i/4}, -i$

49 $e^{-i\pi/3}$ and -1; sum zero, product -1 **53** $i\frac{\pi}{2}, i\frac{\pi}{2} + 2\pi i$ **55** $2e^x$

Section 10.5 Power Series   (page 395)

1 $1 + 4x + (4x)^2 + \cdots; r = \frac{1}{4}; x = \frac{1}{4}$ **3** $e(1 - x + \frac{x^2}{2!} - \cdots); r = \infty$

5 $\ln e + \ln(1 + \frac{x}{e}) = 1 + \frac{x}{e} - \frac{1}{2}(\frac{x}{e})^2 + \cdots; r = e; x = -e$

7 $|\frac{x-1}{2}| < 1$ or $(-1,3); \frac{2}{3-x}$ **9** $|x-a| < 1; -\ln(1-(x-a))$

11 $1 + \frac{x}{2!} + \frac{x^2}{3!} + \cdots$; add to 1 at $x = 0$ **13** $a_1, a_3, \cdots$ are all zero   **15** $\frac{1 - (1 - \frac{1}{2}x^2 \cdots)}{x^2} \to \frac{1}{2}$

17 $f^{(8)}(c) = \cos c < 1$; alternating terms might not decrease (as required)

19 $f = \frac{1}{1-x}, |R_n| \le \frac{x^{n+1}}{(1-c)^{n+2}}; R_n = \frac{x^{n+1}}{1-x}; (1-c)^4 = 1 - \frac{1}{2}$

21 $f^{(n+1)}(x) = \frac{n!}{(1-x)^{n+1}}, |R_n| \le \frac{x^{n+1}}{(1-c)^{n+1}}(\frac{1}{n+1}) \to 0$ when $x = \frac{1}{2}$ and $1 - c > \frac{1}{2}$

23 $R_2 = f(x) - f(a) - f'(a)(x-a) - \frac{1}{2}f''(a)(x-a)^2$ so $R_2 = R_2' = R_2'' = 0$ at $x = a, R_2''' = f'''$;
Generalized Mean Value Theorem in 3.8 gives $a < c < c_2 < c_1 < x$

25 $1 + \frac{1}{2}x^2 + \frac{3}{8}(x^2)^2$ **27** $(-1)^n; (-1)^n(n+1)$

29 (a) one friend k times, the other $n - k$ times, $0 \le k \le n; 21$ **33** $(16-1)^{1/4} \approx 1.968$

35 $(1 + .1)^{1.1} = 1(1.1)(.1) + \frac{(1.1)(.1)}{2}(.1)^2 \approx 1.1105$ **37** $1 + \frac{x^2}{2} + \frac{5x^4}{24}; r = \frac{\pi}{2}$ **41** $x + x^2 + \frac{5}{6}x^3 + \frac{5}{6}x^4$

43 $x^2 - \frac{1}{3}x^4 + \frac{2}{45}x^6$ **45** $1 + \frac{x}{2} + \frac{3x}{8} + \frac{5x}{16}$ **47** .2727   **49** $-\frac{1}{6} - \frac{1}{3} = -\frac{1}{2}$ **51** $r = 1, r = \frac{\pi}{2} - 1$

CHAPTER 11 VECTORS AND MATRICES

Section 11.1 Vectors and Dot Products   (page 405)

1 $(0,0,0); (5,5,5); 3; -3; \cos\theta = -1$ **3** $2\mathbf{i} - \mathbf{j} - \mathbf{k}; -\mathbf{i} - 7\mathbf{j} + 8\mathbf{k}; 6; -3; \cos\theta = -\frac{1}{2}$

5 $(v_2, -v_1); (v_2, -v_1, 0), (v_3, 0, -v_1)$ **7** $(0,0); (0,0,0)$ **9** Cosine of θ; projection of $\mathbf{w}$ on $\mathbf{v}$

11 F;T;F   **13** Zero; sum = 10 o'clock vector; sum = 8 o'clock vector times $\frac{1+\sqrt{3}}{2}$

15 $45°$ **17** Circle $x^2 + y^2 = 4; (x-1)^2 + y^2 = 4$; vertical line $x = 2$; half-line $x \ge 0$

19 $\mathbf{v} = -3\mathbf{i} + 2\mathbf{j}, \mathbf{w} = 2\mathbf{i} - \mathbf{j}; \mathbf{i} = 4\mathbf{v} - \mathbf{w}$ **21** $d = -6; C = \mathbf{i} - 2\mathbf{j} + \mathbf{k}$

23 $\cos\theta = \frac{1}{\sqrt{3}}; \cos\theta = \frac{2}{\sqrt{6}}; \cos\theta = \frac{1}{3}$ **25** $\mathbf{A} \cdot (\mathbf{A} + \mathbf{B}) = 1 + \mathbf{A} \cdot \mathbf{B} = 1 + \mathbf{B} \cdot \mathbf{A} = \mathbf{B} \cdot (\mathbf{A} + \mathbf{B})$; equilateral, $60°$

27 $a = \mathbf{A} \cdot \mathbf{I}, b = \mathbf{A} \cdot \mathbf{J}$ **29** $(\cos t, \sin t)$ and $(-\sin t, \cos t); (\cos 2t, \sin 2t)$ and $(-2\sin 2t, 2\cos 2t)$

31 $\mathbf{C} = \mathbf{A} + \mathbf{B}, \mathbf{D} = \mathbf{A} - \mathbf{B}; \mathbf{C} \cdot \mathbf{D} = \mathbf{A} \cdot \mathbf{A} + \mathbf{B} \cdot \mathbf{A} - \mathbf{A} \cdot \mathbf{B} - \mathbf{B} \cdot \mathbf{B} = r^2 - r^2 = 0$

33 $\mathbf{U} + \mathbf{V} - \mathbf{W} = (2,5,8), \mathbf{U} - \mathbf{V} + \mathbf{W} = (0,-1,-2), -\mathbf{U} + \mathbf{V} + \mathbf{W} = (4,3,6)$

35 c and $\sqrt{a^2 + b^2}; b/a$ and $\sqrt{a^2 + b^2 + c^2}$

37 $\mathbf{M}_1 = \frac{1}{2}\mathbf{A} + \mathbf{C}, \mathbf{M}_2 = \mathbf{A} + \frac{1}{2}\mathbf{B}, \mathbf{M}_3 = \mathbf{B} + \frac{1}{2}\mathbf{C}; \mathbf{M}_1 + \mathbf{M}_2 + \mathbf{M}_3 = \frac{3}{2}(\mathbf{A} + \mathbf{B} + \mathbf{C}) = \mathbf{0}$

39 $8 \le 3 \cdot 3; 2\sqrt{xy} \le x + y$ **41** Cancel a^2c^2 and b^2d^2; then $b^2c^2 + a^2d^2 \ge 2abcd$ because $(bc - ad)^2 \ge 0$

43 F; T; T; F   **45** all $2\sqrt{2}; \cos\theta = -\frac{1}{3}$

Section 11.2 Planes and Projections   (page 414)

1 $(0,0,0)$ and $(2,-1,0); \mathbf{N} = (1,2,3)$ **3** $(0,5,6)$ and $(0,6,7); \mathbf{N} = (1,0,0)$

5 $(1,1,1)$ and $(1,2,2); \mathbf{N} = (1,1,-1)$ **7** $x + y = 3$ **9** $x + 2y + z = 2$

11 Parallel if $\mathbf{N} \cdot \mathbf{V} = 0$; perpendicular if $\mathbf{V} = $ multiple of $\mathbf{N}$

13 $\mathbf{i} + \mathbf{j} + \mathbf{k}$ (vector between points) is not perpendicular to $\mathbf{N}$; $\mathbf{V} \cdot \mathbf{N}$ is not zero; plane through first three is $x + y + z = 1$; $x + y - z = 3$ succeeds; right side must be zero

15 $ax + by + cz = 0$; $a(x - x_0) + b(y - y_0) + c(z - z_0) = 0$ 17 $\cos \theta = \frac{\sqrt{5}}{3}, \frac{\sqrt{5}}{3}, \frac{1}{3}$

19 $\frac{2}{36}\mathbf{A}$ has length $\frac{1}{3}$ 21 $\mathbf{P} = \frac{1}{2}\mathbf{A}$ has length $\frac{1}{2}|\mathbf{A}|$ 23 $\mathbf{P} = -\mathbf{A}$ has length $|\mathbf{A}|$ 25 $\mathbf{P} = \mathbf{O}$

27 Projection on $\mathbf{A} = (1, 2, 2)$ has length $\frac{5}{3}$; force down is 4; mass moves in the direction of $\mathbf{F}$

29 $|\mathbf{P}|_{\min} = \frac{5}{|\mathbf{N}|} = $ distance from plane to origin   31 Distances $\frac{1}{\sqrt{3}}$ and $\frac{2}{\sqrt{3}}$ both reached at $\left(\frac{1}{3}, \frac{1}{3}, -\frac{1}{3}\right)$

33 $\mathbf{i} + \mathbf{j} + \mathbf{k}$; $t = -\frac{4}{3}$; $\left(\frac{2}{3}, -\frac{1}{3}, -\frac{1}{3}\right)$; $\frac{4}{\sqrt{3}}$

35 Same $\mathbf{N} = (2, -2, 1)$; for example $\mathbf{Q} = (0, 0, 1)$; then $\mathbf{Q} + \frac{2}{9}\mathbf{N} = \left(\frac{4}{9}, -\frac{4}{9}, \frac{11}{9}\right)$ is on second plane; $\frac{2}{9}|\mathbf{N}| = \frac{2}{3}$

37 $3\mathbf{i} + 4\mathbf{j}$; $(3t, 4t)$ is on the line if $3(3t) + 4(4t) = 10$ or $t = \frac{10}{25}$; $P = \left(\frac{30}{25}, \frac{40}{25}\right)$, $|P| = 2$

39 $2x + 2\left(\frac{10}{4} - \frac{3}{4}x\right)\left(-\frac{3}{4}\right) = 0$ so $x = \frac{30}{25} = \frac{6}{5}$; $3x + 4y = 10$ gives $y = \frac{8}{5}$

41 Use equations (8) and (9) with $\mathbf{N} = (a, b)$ and $\mathbf{Q} = (x_1, y_1)$ 43 $t = \frac{\mathbf{A} \cdot \mathbf{B}}{|\mathbf{A}|^2}$; $\mathbf{B}$ onto $\mathbf{A}$

45 $aV\mathbf{L} = \frac{1}{2}\mathbf{L}_I - \frac{1}{2}\mathbf{L}_{III}$; $aV\mathbf{F} = \frac{1}{2}\mathbf{L}_{II} + \frac{1}{2}\mathbf{L}_{III}$

47 $\mathbf{V} \cdot \mathbf{L}_I = 2 - 1$; $\mathbf{V} \cdot \mathbf{L}_{II} = -3 - 1$, $\mathbf{V} \cdot \mathbf{L}_{III} = -3 - 2$; thus $\mathbf{V} \cdot 2\mathbf{i} = 1$, $\mathbf{V} \cdot (\mathbf{i} - \sqrt{3}\mathbf{j}) = -4$, and $\mathbf{V} = \frac{1}{2}\mathbf{i} + \frac{3\sqrt{3}}{2}\mathbf{j}$

Section 11.3 Cross Products and Determinants   (page 423)

1 $\mathbf{O}$ 3 $3\mathbf{i} - 2\mathbf{j} - 3\mathbf{k}$ 5 $-2\mathbf{i} + 3\mathbf{j} - 5\mathbf{k}$ 7 $27\mathbf{i} + 12\mathbf{j} - 17\mathbf{k}$

9 $\mathbf{A}$ perpendicular to $\mathbf{B}$; $\mathbf{A}, \mathbf{B}, \mathbf{C}$ mutually perpendicular   11 $|\mathbf{A} \times \mathbf{B}| = \sqrt{2}$, $\mathbf{A} \times \mathbf{B} = \mathbf{j} - \mathbf{k}$ 13 $\mathbf{A} \times \mathbf{B} = \mathbf{O}$

15 $|\mathbf{A} \times \mathbf{B}|^2 = (a_1^2 + a_2^2)(b_1^2 + b_2^2) - (a_1 b_1 + a_2 b_2)^2 = (a_1 b_2 - a_2 b_1)^2$; $\mathbf{A} \times \mathbf{B} = (a_1 b_2 - a_2 b_1)\mathbf{k}$

17 F;T;F;T   19 $\mathbf{N} = (2, 1, 0)$ or $2\mathbf{i} + \mathbf{j}$ 21 $x - y + z = 2$ so $\mathbf{N} = \mathbf{i} - \mathbf{j} + \mathbf{k}$

23 $[(1, 2, 1) - (2, 1, 1)] \times [(1, 1, 2) - (2, 1, 1)] = \mathbf{N} = \mathbf{i} + \mathbf{j} + \mathbf{k}$; $x + y + z = 4$

25 $(1, 1, 1) \times (a, b, c) = \mathbf{N} = (c - b)\mathbf{i} + (a - c)\mathbf{j} + (b - a)\mathbf{k}$; points on a line if $a = b = c$ (many planes)

27 $\mathbf{N} = \mathbf{i} + \mathbf{j}$, plane $x + y = $ constant   29 $\mathbf{N} = \mathbf{k}$, plane $z = $ constant

31 $\begin{vmatrix} x & y & z \\ 1 & 1 & 0 \\ 1 & 2 & 1 \end{vmatrix} = x - y + z = 0$ 33 $\mathbf{i} - 3\mathbf{j}$; $-\mathbf{i} + 3\mathbf{j}$; $-3\mathbf{i} - \mathbf{j}$ 35 $-1, 4, -9$

39 $+c_1 \begin{vmatrix} a_2 & a_3 \\ b_2 & b_3 \end{vmatrix} - c_2 \begin{vmatrix} a_1 & a_3 \\ b_1 & b_3 \end{vmatrix} + c_3 \begin{vmatrix} a_1 & a_2 \\ b_1 & b_2 \end{vmatrix}$

41 area$^2 = \left(\frac{1}{2}ab\right)^2 + \left(\frac{1}{2}ac\right)^2 + \left(\frac{1}{2}bc\right)^2 = \left(\frac{1}{2}|\mathbf{A} \times \mathbf{B}|\right)^2$ when $\mathbf{A} = a\mathbf{i} - b\mathbf{j}$, $\mathbf{B} = a\mathbf{i} - c\mathbf{k}$

43 $A = \frac{1}{2}(2 \cdot 1 - (-1)1) = \frac{3}{2}$; fourth corner can be $(3, 3)$

45 $a_1\mathbf{i} + a_2\mathbf{j}$ and $b_1\mathbf{i} + b_2\mathbf{j}$; $|a_1 b_2 - a_2 b_1|$; $\mathbf{A} \times \mathbf{B} = \cdots + (a_1 b_2 - a_2 b_1)\mathbf{k}$

47 $\mathbf{A} \times \mathbf{B}$; from Eq. (6), $(\mathbf{A} \times \mathbf{B}) \times \mathbf{i} = -(a_3 b_1 - a_1 b_3)\mathbf{k} + (a_1 b_2 - a_2 b_1)\mathbf{j}$; $(\mathbf{A} \cdot \mathbf{i})\mathbf{B} - (\mathbf{B} \cdot \mathbf{i})\mathbf{A} = a_1(b_1\mathbf{i} + b_2\mathbf{j} + b_3\mathbf{k}) - b_1(a_1\mathbf{j} + a_2\mathbf{j} + a_3\mathbf{k})$

49 $\mathbf{N} = (Q - P) \times (R - P) = \mathbf{i} + \mathbf{j} + \mathbf{k}$; area $\frac{1}{2}\sqrt{3}$; $x + y + z = 2$

Section 11.4 Matrices and Linear Equations   (page 433)

1 $x = 5, y = 2, D = -2$, $\begin{bmatrix} 7 \\ 3 \end{bmatrix} = 5\begin{bmatrix} 1 \\ 1 \end{bmatrix} + 2\begin{bmatrix} 1 \\ -1 \end{bmatrix}$ 3 $x = 3, y = 1$, $\begin{bmatrix} 8 \\ 0 \end{bmatrix} = 3\begin{bmatrix} 3 \\ 1 \end{bmatrix} + \begin{bmatrix} -1 \\ -3 \end{bmatrix}$, $D = -8$

5 $x = 2y, y = $ anything, $D = 0, 2y\begin{bmatrix} 2 \\ 1 \end{bmatrix} + y\begin{bmatrix} -4 \\ -2 \end{bmatrix} = \begin{bmatrix} 0 \\ 0 \end{bmatrix}$ 7 no solution, $D = 0$

9 $x = \frac{1}{D}\begin{vmatrix} 8 & -1 \\ 0 & -3 \end{vmatrix} = \frac{-24}{-8} = 3, y = \frac{1}{D}\begin{bmatrix} 3 & 8 \\ 1 & 0 \end{bmatrix} = \frac{-8}{-8} = 1$ 11 $\frac{0}{0}$

15 $ad - bc = -2$ so $A^{-1} = \begin{bmatrix} .5 & .5 \\ .5 & -.5 \end{bmatrix}$ **17** Are parallel; multiple; the same; infinite

19 Multiples of each other; in the same direction as the columns; infinite

21 $d_1 = .34, d_2 = 4.91$ **23** $.96x + .02y = .58, .04x + .98y = 4.92; D = .94, x = .5, y = 5$

25 $a = 1$ gives any $x = -y; a = -1$ gives any $x = y$

27 $D = -2, A^{-1} = -\frac{1}{2} \begin{bmatrix} 5 & -4 \\ -3 & 2 \end{bmatrix}; D = -8; (2A)^{-1} = \frac{1}{2}A^{-1}; D = \frac{1}{-2}, (A^{-1})^{-1} = $ original A;

    $D = -2 (\text{not} +2), (-A)^{-1} = -A^{-1}; D = 1, I^{-1} = I$

29 $AB = \begin{bmatrix} 7 & 5 \\ 5 & 1 \end{bmatrix}, BA = \begin{bmatrix} 5 & 11 \\ 3 & 3 \end{bmatrix}, BC = \begin{bmatrix} 3 & 5 \\ 1 & 3 \end{bmatrix}, CB = \begin{bmatrix} 4 & 2 \\ 2 & 2 \end{bmatrix}$

31 $AB = \begin{bmatrix} ae + bg & af + bh \\ ce + dg & cf + dh \end{bmatrix}, \begin{matrix} aecf + aedh + bgcf + bgdh \\ -afce - afdg - bhce - bhdg \end{matrix} = (ad - bc)(eh - fg)$

33 $A^{-1} = \begin{bmatrix} 1 & -2 \\ 0 & \frac{1}{2} \end{bmatrix}, B^{-1} = \begin{bmatrix} \frac{1}{2} & -1 \\ 0 & 1 \end{bmatrix}$ **35** Identity; $B^{-1}A^{-1}$ **37** Perpendicular; $\mathbf{u} = \mathbf{v} \times \mathbf{w}$

39 Line $4 + t$, errors $-1, 2, -1$ **41** $d_1 - 2d_2 + d_3 = 0$ **43** A^{-1} can't multiply $\mathbf{O}$ and produce $\mathbf{u}$

Section 11.5 Linear Algebra   (page 443)

1 $\begin{bmatrix} 1 & 2 & 2 \\ 2 & 3 & 5 \\ 0 & 2 & 2 \end{bmatrix} \begin{bmatrix} x \\ y \\ z \end{bmatrix} = \begin{bmatrix} 0 \\ 0 \\ 8 \end{bmatrix}$ **3** $\begin{bmatrix} 1 & -1 & 0 \\ 0 & 1 & -1 \\ 1 & 0 & -1 \end{bmatrix} \begin{bmatrix} x \\ y \\ z \end{bmatrix} = \begin{bmatrix} 0 \\ 0 \\ 1 \end{bmatrix}$

5 $\det A = 0$, add 3 equations $\to 0 = 1$ **7** $5\mathbf{a} + 1\mathbf{b} + 0\mathbf{c} = \mathbf{d}, A^{-1} = \begin{bmatrix} 1 & 0 & 0 \\ 0 & \frac{1}{2} & 0 \\ 0 & 0 & \frac{1}{3} \end{bmatrix}$

9 $\mathbf{b} \times \mathbf{c}; \mathbf{a} \cdot \mathbf{b} \times \mathbf{c} = 0$; determinant is zero   **11** $6, 2, 0$; product of diagonal entries

13 $A^{-1} = \begin{bmatrix} 1 & -2 & 4 \\ 0 & \frac{1}{2} & -1 \\ 0 & 0 & \frac{1}{3} \end{bmatrix}, B^{-1} = \begin{bmatrix} 0 & 2 & -\frac{1}{2} \\ 0 & -3 & 1 \\ 1 & 0 & 0 \end{bmatrix}$ **15** Zero; same plane; D is zero

17 $\mathbf{d} = (1, -1, 0); \mathbf{u} = (1, 0, 0)$ or $(7, 3, 1)$ **19** $AB = \begin{bmatrix} 8 & 4 & 1 \\ 40 & 26 & 0 \\ 18 & 12 & 0 \end{bmatrix}$, $\det AB = 12 = (\det A)$ times $(\det B)$

21 $A + C = \begin{bmatrix} 2 & 3 & -3 \\ -1 & 4 & 6 \\ 0 & -1 & 6 \end{bmatrix}$, $\det(A + C)$ is not $\det A + \det C$

23 $p = \frac{(2)(3) - (0)(6)}{6} = 1, q = \frac{-(4)(3) + (0)(0)}{6} = -2$ **25** $(A^{-1})^{-1}$ is always A

27 $-1, -1, 1, 1, ; (y, x, z), (z, y, x), (y, z, x), (z, x, y)$ **29** $2! = 2, 4! = 24$

31 $z = \frac{1}{2}, y = -\frac{3}{2}, x = 3; z = \frac{7}{2}, y = \frac{3}{2}, x = -\frac{1}{2}$

33 New second equation $3z = 0$ doesn't contain y; exchange with third equation; there is a solution

35 Pivots $1, 2, 4, D = 8$; pivots $1, -1, 2, D = -2$ **37** $a_{12} = 1, a_{21} = 0, \sum a_{ij}b_{jk} = $ row i, column k in AB

39 $a_{11}a_{22} - a_{12}a_{21} \neq 0; D = 0$

CHAPTER 12   MOTION ALONG A CURVE

Section 12.1 The Position Vector   (page 452)

1 $\mathbf{v}(1) = \mathbf{i} + 3\mathbf{j}$; speed $\sqrt{10}$;   **3** $\frac{dy}{dx} = \frac{dy/dt}{dx/dt} = \frac{\cos t}{-\sin t}$; tangent to circle is perpendicular to $\frac{x}{y} = \frac{\cos t}{\sin t}$

5 $\mathbf{v} = e^t \mathbf{i} - e^{-t}\mathbf{j} = \mathbf{i} - \mathbf{j}; y - 1 = -(x - 1); xy = 1$

7 $\mathbf{R}=(1,2,4)+(4,3,0)t$; $\mathbf{R}=(1,2,4)+(8,6,0)t$; $\mathbf{R}=(5,5,4)+(8,6,0)t$

9 $\mathbf{R}=(2+t,3,4-t)$; $\mathbf{R}=(2+\frac{t^2}{2},3,4-\frac{t^2}{2})$; the same line

11 Line; $y=2+2t,z=2+3t$; $y=2+4t,z=2+6t$

13 Line; $\sqrt{36+9+4}=7$; $(6,3,2)$; line segment   15 $\frac{\sqrt{2}}{2};1;\frac{\sqrt{2}}{2}$ 17 $x=t,y=mt+b$

19 $\mathbf{v}=\mathbf{i}-\frac{1}{t^2}\mathbf{j},|\mathbf{v}|=\sqrt{1+t^{-4}},\mathbf{T}=\mathbf{v}/|\mathbf{v}|$; $\mathbf{v}=(\cos t-t\sin t)\mathbf{i}+(\sin t+t\cos t)\mathbf{j};|\mathbf{v}|=\sqrt{1+t^2}$;
$\mathbf{T}=\mathbf{v}/|\mathbf{v}|$; $\mathbf{v}=\mathbf{i}+2\mathbf{j}+2\mathbf{k},|\mathbf{v}|=3,\mathbf{T}=\frac{1}{3}\mathbf{v}$

21 $\mathbf{R}=-\sin t\,\mathbf{i}+\cos t\,\mathbf{j}+$ any $\mathbf{R}_0$; same $\mathbf{R}$ plus any $\mathbf{w}t$

23 $\mathbf{v}=(1-\sin t)\mathbf{i}+(1-\cos t)\mathbf{j};|\mathbf{v}|=\sqrt{3-2\sin t-2\cos t},|\mathbf{v}|_{\min}=\sqrt{3-2\sqrt{2}},|\mathbf{v}|_{\max}=\sqrt{3+2\sqrt{2}}$;
$\mathbf{a}=-\cos t\,\mathbf{i}+\sin t\,\mathbf{j},|\mathbf{a}|=1$; center is on $x=t,y=t$

25 Leaves at $(\frac{\sqrt{2}}{2},\frac{\sqrt{2}}{2})$; $\mathbf{v}=(-\sqrt{2},\sqrt{2})$; $\mathbf{R}=(\frac{\sqrt{2}}{2},\frac{\sqrt{2}}{2})+v(t-\frac{\pi}{8})$

27 $\mathbf{R}=\cos\frac{s}{\sqrt{2}}\mathbf{i}+\sin\frac{s}{\sqrt{2}}\mathbf{j}+\frac{1}{\sqrt{2}}\mathbf{k}$

29 $\mathbf{v}=\sec^2 t\,\mathbf{i}+\sec t\tan t\,\mathbf{j};|\mathbf{v}|=\sec^2 t\sqrt{1+\sin^2 t};\mathbf{a}=2\sec^2 t\tan t\,\mathbf{i}+(\sec^3 t+\sec t\tan^2 t)\mathbf{j}$;
curve is $y^2-x^2=1$; hyperbola has asymptote $y=x$

31 If $\mathbf{T}=\mathbf{v}$ then $|\mathbf{v}|=1$; line $\mathbf{R}=t\mathbf{i}$ or helix in Problem 27

33 $(x(t),y(t))=\begin{matrix}(2t,0) & 0\le t\le\frac{1}{2} & (3-2t,1) & 1\le t\le\frac{3}{2}\\ (1,2t-1) & \frac{1}{2}\le t\le 1 & (0,4-2t) & \frac{3}{2}\le t\le 2\end{matrix}$

35 $x(t)=4\cos\frac{t}{2},y(t)=4\sin\frac{t}{2}$ 37 F; F; T; T; F   39 $\frac{y}{x}=\tan\theta$ but $\frac{y}{x}\ne\tan t$

41 $\mathbf{v}$ and $\mathbf{w}$; $\mathbf{v}$ and $\mathbf{w}$ and $\mathbf{u}$; $\mathbf{v}$ and $\mathbf{w},\mathbf{v}$ and $\mathbf{w}$ and $\mathbf{u}$; not zero

43 $\mathbf{u}=(8,3,2)$; projection perpendicular to $\mathbf{v}=(1,2,2)$ is $(6,-1,-2)$ which has length $\sqrt{41}$

45 $x=G(t),y=F(t)$; $y=x^{2/3}$; $t=1$ and $t=-1$ give the same x so they would give the same y; $y-G(F^{-1}(x))$

Section 12.2 Plane Motion: Projectiles and Cycloids   (page 457)

1 (a) $T=16/g$ sec, $R=128\sqrt{3}/g$ ft, $Y=32/g$ ft   (b) $\frac{16\sqrt{2}}{g};\frac{128\sqrt{3}}{g},\frac{96}{g}$ (c) $\frac{32}{g},0,\frac{128}{g}$ 3 $x=1.2$ or 33.5

5 $y=x-\frac{1}{2}x^2=0$ at $x=2$; $y=x\tan x-\frac{g}{2}\left(\frac{x}{v_0\cos\alpha}\right)^2=0$ at $x=R$ 7 $x=v_0\sqrt{\frac{2h}{g}}$

9 $v_0\approx 11.2,\tan\alpha\approx 4.32$ 11 $v_0=\sqrt{gR}=\sqrt{980}$ m/sec; larger   13 $v_0^2/2g=40$ meters

15 Multiply R and H by 4; $dR=2v_0^2\cos 2\alpha\,d\alpha/g,dH=v_0^2\sin\alpha\cos\alpha\,d\alpha/g$

17 $t=\frac{12\sqrt{2}}{10}$ sec; $y=12-\frac{144g}{100}\approx-2.1$m; $+2.1$m   19 $\mathbf{T}=\frac{(1-\cos\theta)\mathbf{i}+\sin\theta\mathbf{j}}{\sqrt{2-2\cos\theta}}$

21 Top of circle   25 $ca(1-\cos\theta),ca\sin\theta;\theta=\pi,\frac{\pi}{2}$ 27 After $\theta=\pi:x=\pi a+v_0 t$ and $y=2a-\frac{1}{2}gt^2$ 29 2;3

31 $\frac{64\pi a^2}{3};5\pi^2 a^3$ 33 $x=\cos\theta+\theta\sin\theta,y=\sin\theta-\theta\cos\theta$ 35 $(a=4)6\pi$

37 $y=2\sin\theta-\sin 2\theta=2\sin\theta(1-\cos\theta);x^2+y^2=4(1-\cos\theta)^2;r=2(1-\cos\theta)$

Section 12.3 Curvature and Normal Vector   (page 463)

1 $\frac{e^x}{(1+e^{2x})^{3/2}}$ 3 $\frac{1}{2}$ 5 0 (line)   7 $\frac{2+t^2}{(1+t^2)^{3/2}}$ 9 $(-\sin t^2,\cos t^2);(-\cos t^2,-\sin t^2)$

11 $(\cos t,\sin t);(-\sin t,-\cos t)$ 13 $(-\frac{3}{5}\sin t,\frac{3}{5}\cos t,\frac{4}{5});|\mathbf{v}|=5,k=\frac{3}{25};\frac{5}{3}$ longer; $\tan\theta=\frac{4}{3}$

15 $\frac{1}{2\sqrt{2}a\sqrt{1-\cos\theta}}$ 17 $k=\frac{3}{16},\mathbf{N}=\mathbf{i}$ 19 $(0,0);(-3,0)$ with $\frac{1}{k}=4;(-1,2)$ with $\frac{1}{k}=2\sqrt{2}$

21 Radius $\frac{1}{k}$, center $(1,\pm\sqrt{\frac{1}{k^2}-1})$ for $k\le 1$ 23 $\mathbf{U}\cdot\mathbf{V}'$ 25 $\frac{1}{\sqrt{2}}(\sin t\,\mathbf{i}-\cos t\,\mathbf{j}+\mathbf{k})$ 27 $\frac{1}{2}$

29 $\mathbf{N}$ in the plane, $\mathbf{B}=\mathbf{k},\tau=0$ 31 $\frac{d^2 y/dx^2}{1+(dy/dx)^2}$ 33 $\mathbf{a}=0$ $\mathbf{T}+5\omega^2\mathbf{N}$ 35 $\mathbf{a}=\frac{t}{\sqrt{1+t^2}}\mathbf{T}+\frac{2+t^2}{\sqrt{1+t^2}}=\mathbf{N}$

37 $\mathbf{a}=\frac{4t}{\sqrt{1+4t^2}}\mathbf{T}+\frac{2}{\sqrt{1+4t^2}}\mathbf{N}$ 39 $|F^2+2(F')^2-FF''|/(F^2+F'^2)^{3/2}$

Section 12.4 Polar Coordinates and Planetary Motion (page 468)

1 $\mathbf{j}, -\mathbf{i}; \mathbf{i}+\mathbf{j} = \mathbf{u}_r - \mathbf{u}_\theta$ **3** $(2,-1); (1,2)$ **5** $\mathbf{v} = 3e^3(\mathbf{u}_r + \mathbf{u}_\theta) = 3e^3(\cos 3 - \sin 3)\mathbf{i} + 3e^3(\sin 3 + \cos 3)\mathbf{j}$

7 $\mathbf{v} = -20\sin 5t\,\mathbf{i} + 20\cos 5t\,\mathbf{j} = 20\,\mathbf{T} = 20\mathbf{u}_\theta; \mathbf{a} = -100\cos 5t\,\mathbf{i} - 100\sin 5t\,\mathbf{j} = 100\,\mathbf{N} = -100\,\mathbf{u}_r$

9 $r\frac{d^2\theta}{dt^2} + 2\frac{dr}{dt}\frac{d\theta}{dt} = 0 = \frac{1}{r}\frac{d}{dt}\left(r^2\frac{d\theta}{dt}\right)$ **11** $\frac{d\theta}{dt} = .0004$ radians/sec; $h = r^2\frac{d\theta}{dt} = 40,000$

13 $m\mathbf{R}\times\mathbf{a}$; torque **15** $T^{2/3}(GM/4\pi^2)^{1/3}$ **17** $4\pi^2 a^3/T^2 G$ **19** $\frac{4\pi^2(150)^3 10^{27}}{(365\frac{1}{4})^2(24)^2(3600)^2(6.67)10^{-11}}$ kg

23 Use Problem **15** **25** $a+c = \frac{1}{C-D}, a-c = \frac{1}{C+D}$, solve for C, D

27 Kepler measures area from focus (sun) **29** Line; $x = 1$

33 $r = 20 - 2t, \theta = \frac{2\pi t}{10}, \mathbf{v} = -2\mathbf{u}_r + (20-2t)\frac{2\pi}{10}\mathbf{u}_\theta; \mathbf{a} = (2t-20)\left(\frac{2\pi}{10}\right)^2\mathbf{u}_r - 4\left(\frac{2\pi}{10}\right)\mathbf{u}_\theta; \int_0^{10}|\mathbf{v}|dt$

CHAPTER 13 PARTIAL DERIVATIVES

Section 13.1 Surfaces and Level Curves (page 475)

3 x derivatives $\infty, -1, -2, -4e^{-4}$ (flattest) **5** Straight lines **7** Logarithm curves

9 Parabolas **11** No: $f = (x+y)^n$ or $(ax+by)^n$ or any function of $ax+by$ **13** $f(x,y) = 1 - x^2 - y^2$

15 Saddle **17** Ellipses $4x^2 + y^2 = c^2$ **19** Ellipses $5x^2 + y^2 = c^2 + 4cx + x^2$

21 Straight lines not reaching $(1,2)$ **23** Center $(1,1); f = x^2 + y^2 - 1$ **25** Four, three, planes, spheres

27 Less than 1, equal to 1, greater than 1 **29** Parallel lines, hyperbolas, parabolas

31 $\frac{d}{dx}: 48x - 3x^2 = 0, x = 16$ hours **33** Plane; planes; 4 left and 3 right (3 pairs)

Section 13.2 Partial Derivatives (page 479)

1 $3 + 2xy^2; -1 + 2yx^2$ **3** $3x^2 y^2 - 2x; 2x^3 y - e^y$ **5** $\frac{-2y}{(x-y)^2}; \frac{2x}{(x-y)^2}$ **7** $\frac{-2x}{(x^2+y^2)^2}; \frac{-2y}{(x^2+y^2)^2}$

9 $\frac{x}{x^2+y^2}; \frac{y}{x^2+y^2}$ **11** $\frac{-y}{x^2+y^2}; \frac{x}{x^2+y^2}$ **13** $2,3,4$ **15** $6(x+iy), 6i(x+iy), -6(x+iy)$

17 $\left(f = \frac{1}{r}\right) f_{xx} = \frac{2x^2 - y^2}{r^5}; f_{xy} = \frac{3xy}{r^5}; f_{yy} = \frac{2y^2 - x^2}{r^5}$ **19** $-a^2 \cos ax \cos by, ab\sin ax\sin by, -b^2 \cos ax\cos by$

21 Omit line $x = y$; all positive numbers; $f_x = -2(x-y)^{-3}, f_y = 2(x-y)^{-3}$

23 Omit $z = t$; all numbers; $\frac{-1}{z-t}, \frac{1}{z-t}, \frac{(x-y)}{(z-t)^2}, \frac{(y-x)}{(z-t)^2}$

25 $x > 0, t > 0$ and $x = 0, t > 1$ and $x = -1, -2, \cdots, t = e, e^2, \cdots; f_x = (\ln t)x^{\ln t - 1}, f_t = (\ln x)t^{\ln x - 1}$

27 $y, x; f = G(x) + H(y)$ **29** $\frac{\partial f}{\partial x} = \frac{\partial(xy)}{\partial x}v(xy) = yv(xy)$

31 $f_{xxx} = 6y^3, f_{yyy} = 6x^3, f_{xxy} = f_{xyx} = f_{yxx} = 18xy^2, f_{yyx} = f_{yxy} = f_{xyy} = 18x^2 y$

33 $g(y) = Ae^{cy/7}$ **35** $g(y) = Ae^{cy/2} + Be^{-cy/2}$

37 $f_t = -2f, f_{xx} = f_{yy} = -e^{-2t}\sin x\sin y; e^{-13t}\sin 2x\sin 3y$

39 $\sin(x+t)$ moves left **41** $\sin(x-ct), \cos(x+ct), e^{x-ct}$

43 $(B-A)h_y(C^*) = (B-A)\left[f_y(b,C^*) - f_y(a,C^*)\right] = (B-A)(b-a)f_{yx}(c^*,C^*)$; continuous f_{xy} and f_{yx}

45 y converges to b; inside and stay inside; $d_n = \sqrt{(x_n - a)^2 + (y_n - b)^2} \to$ zero; $d_n < \epsilon$ for $n > N$

47 ϵ, less than δ **49** $f(a,b); \frac{1}{x-1}$ or $\frac{1}{(x-1)(y-2)}$ **51** $f(0,0) = 1; f(0,0) = 1$; not defined for $x < 0$

Section 13.3 Tangent Planes and Linear Approximations (page 488)

1 $z - 1 = y - 1; \mathbf{N} = \mathbf{j} - \mathbf{k}$ 3 $z - 2 = \frac{1}{3}(x - 6) - \frac{2}{3}(y - 3); \mathbf{N} = \frac{1}{3}\mathbf{i} - \frac{2}{3}\mathbf{j} - \mathbf{k}$

5 $2(x - 1) + 4(y - 2) + 2(z - 1) = 0; \mathbf{N} = 2\mathbf{i} + 4\mathbf{j} + 2\mathbf{k}$ 7 $z - 1 = x - 1; \mathbf{N} = \mathbf{i} - \mathbf{k}$

9 Tangent plane $2z_0(z - z_0) - 2x_0(x - x_0) - 2y_0(y - y_0) = 0; (0, 0, 0)$ satisfies this equation because

$z_0^2 - x_0^2 - y_0^2 = 0$ on the surface; $\cos\theta = \dfrac{\mathbf{N} \cdot \mathbf{k}}{|\mathbf{N}||\mathbf{k}|} = \dfrac{-z_0}{\sqrt{x_0^2 + y_0^2 + z_0^2}} = \dfrac{-1}{\sqrt{2}}$ (surface is the $45°$ cone)

11 $dz = 3dx - 2dy$ for both; $dz = 0$ for both; $\Delta z = 0$ for $3x - 2y, \Delta z = .00029$ for x^3/y^2; tangent plane

13 $z = z_0 + F_z t$; plane $6(x - 4) + 12(y - 2) + 8(z - 3) = 0$; normal line $x = 4 + 6t, y = 2 + 12t, z = 3 + 8t$

15 Tangent plane $4(x - 2) + 2(y - 1) + 4(z - 2) = 0$; normal line $x = 2 + 4t, y = 1 + 2t, z = 2 + 4t; (0, 0, 0)$ at $t = -\frac{1}{2}$

17 $dw = y_0 dx + x_0 dy$; product rule; $\Delta w - dw = (x - x_0)(y - y_0)$

19 $dI = 4000dR + .08dP; dP = \$100; I = (.78)(4100) = \$319.80$

21 Increase $= \frac{26}{101} - \frac{25}{100} = \frac{3}{404}$, decrease $= \frac{25}{100} - \frac{25}{101} = \frac{1}{404}; dA = \frac{1}{y}dx - \frac{x}{y^2}dy; 3$ 23 $\Delta\theta \approx \dfrac{-y\Delta x + x\Delta y}{\sqrt{x^2 + y^2}}$

25 Q increases; $Q_s = -\frac{250}{3}, Q_t = \frac{-5}{3}, P_s = -.2Q_s = \frac{50}{3}, P_t = -.2Q_t = \frac{1}{3}; Q = 50 - \frac{250}{3}(s - .4) - \frac{5}{3}(t - 10)$

27 $s = 1, t = 10$ gives $Q = 40$: $\begin{array}{l} P_s = -Q_s = sQ_s + Q = Q_s + 40 \\ P_t = -Q_t = sQ_t + 1 = Q_t + 1 \end{array}; Q_s = -20, Q_t = -\frac{1}{2}, P_s = 20, P_t = \frac{1}{2}$

29 $z - 2 = x - 2 + 2(y - 1)$ and $z - 3 = 4(x - 2) - 2(y - 1); x = 1, y = \frac{1}{2}, z = 0$

31 $\Delta x = -\frac{1}{2}, \Delta y = \frac{1}{2}; x_1 = \frac{1}{2}; y_1 = -\frac{1}{2};$ line $x + y = 0$

33 $3a^2\Delta x - \Delta y = -a - a^3$ gives $\Delta y = -\Delta x = \dfrac{a + a^3}{1 + 3a^2}$; lemon starts at $(1/\sqrt{3}, -1/\sqrt{3})$

$-\Delta x + 3a^2\Delta y = a + a^3$

35 If $x^3 = y$ then $y^3 = x^9$. Then $x^9 = x$ only if $x = 0$ or 1 or -1 (or complex number)

37 $\Delta x = -x_0 + 1, \Delta y = -y_0 + 2, (x_1, y_1) = (1, 2) =$ solution

39 $G = H = \frac{x_n^2}{2x_n - 1}$ 41 $J = \begin{bmatrix} e^x & 0 \\ 1 & e^y \end{bmatrix}, \Delta x = -1 + e^{-x_n}, \Delta y = -1 - (x_n - 1 + e^{-x_n})e^{-y_n}$

43 $(x_1, y_1) = \left(0, \frac{5}{4}\right), \left(-\frac{3}{4}, \frac{5}{4}\right), \left(\frac{3}{4}, 0\right)$

Section 13.4 Directional Derivatives and Gradients (page 495)

1 grad $f = 2x\mathbf{i} - 2y\mathbf{j}, D_\mathbf{u}f = \sqrt{3}x - y, D_\mathbf{u}f(P) = \sqrt{3}$

3 grad $f = e^x\cos y\,\mathbf{i} - e^x\sin y\mathbf{j}, D_\mathbf{u}f = -e^x\sin y, D_\mathbf{u}f(P) = -1$

5 $f = \sqrt{x^2 + (y - 3)^2}$, grad $f = \frac{x}{f}\mathbf{i} + \frac{y - 3}{f}\mathbf{j}, D_\mathbf{u}f = \frac{x}{f}, D_\mathbf{u}f(P) = \frac{1}{\sqrt{5}}$ 7 grad $f = \frac{2x}{x^2 + y^2}\mathbf{i} + \frac{2y}{x^2 + y^2}\mathbf{j}$

9 grad $f = 6x\mathbf{i} + 4y\mathbf{j} = 6\mathbf{i} + 8\mathbf{j} =$ steepest direction at P; level direction $-8\mathbf{i} + 6\mathbf{j}$ is perpendicular; $10, 0$

11 T; F (grad f is a vector); F; T   13 $\mathbf{u} = \left(\frac{a}{\sqrt{a^2 + b^2}}, \frac{b}{\sqrt{a^2 + b^2}}\right), D_\mathbf{u}f = \sqrt{a^2 + b^2}$

15 grad $f = (e^{x - y}, -e^{x - y}) = (e^{-1}, -e^{-1})$ at $P; \mathbf{u} = \left(\frac{1}{\sqrt{2}}, \frac{-1}{\sqrt{2}}\right), D_\mathbf{u}f = \sqrt{2}e^{-1}$

17 grad $f = \mathbf{O}$ at maximum; level curve is one point   19 $\mathbf{N} = (-1, 1, -1), \mathbf{U} = (-1, 1, 2), \mathbf{L} = (1, 1, 0)$

21 Direction $-\mathbf{U} = (-2, 0, -4)$ 23 $-\mathbf{U} = \left(\frac{x}{\sqrt{1 - x^2 - y^2}}, \frac{y}{\sqrt{1 - x^2 - y^2}}, \frac{-x^2 - y^2}{1 - x^2 - y^2}\right)$

25 $f = (x + 2y)$ and $(x + 2y)^2; \mathbf{i} + 2\mathbf{j}$; straight lines $x + 2y =$ constant (perpendicular to $\mathbf{i} + 2\mathbf{j}$)

27 grad $f = \pm\left(\frac{1}{\sqrt{5}}, \frac{-2}{\sqrt{5}}\right)$; grad $g = \pm(2\sqrt{5}, \sqrt{5}), f = \pm\left(\frac{x}{\sqrt{5}} - \frac{2y}{\sqrt{5}}\right) + C, g = \pm(2\sqrt{5}x + \sqrt{5}y) + C$

29 $\theta =$ constant along ray in direction $\mathbf{u} = \dfrac{3\mathbf{i} + 4\mathbf{j}}{5}$; grad $\theta = \dfrac{-y\mathbf{i} + x\mathbf{j}}{x^2 + y^2} = \dfrac{-4\mathbf{i} + 3\mathbf{j}}{25}; \mathbf{u} \cdot$ grad $\theta = 0$

31 $\mathbf{U} = (f_x, f_y, f_x^2 + f_y^2) = (-1, -2.5); -\mathbf{U} = (-1, -2, 5)$; tangent at the point $(2, 1, 6)$

33 grad f toward $2\mathbf{i} + \mathbf{j}$ at $P, \mathbf{j}$ at $Q, -2\mathbf{i} + \mathbf{j}$ at $R; (2, \frac{1}{2})$ and $(2\frac{1}{2}, 2)$; largest upper left, smallest lower right;

$z_{max} > 9; z$ goes from 2 to 8 and back to 6

35 $f = \frac{1}{2}\sqrt{(x-1)^2+(y-1)^2}; \left(\frac{\partial f}{\partial x}, \frac{\partial f}{\partial y}\right)_{0,0} = \left(\frac{-1}{2\sqrt{2}}, \frac{-1}{2\sqrt{2}}\right)$

37 Figure C now shows level curves; $|\text{grad } f|$ is varying; f could be xy

39 $x^2+xy; e^{x-y}$; no function has $\frac{\partial f}{\partial x} = y$ and $\frac{\partial f}{\partial y} = -x$ because then $f_{xy} \neq f_{yx}$

41 $\mathbf{v} = (1,2t); \mathbf{T} = \mathbf{v}/\sqrt{1+4t^2}; \frac{df}{dt} = \mathbf{v}\cdot(2t,2t^2) = 2t+4t^3; \frac{df}{ds} = (2t+4t^3)/\sqrt{1+4t^2}$

43 $\mathbf{v} = (2,3); \mathbf{T} = \frac{\mathbf{v}}{\sqrt{13}}; \frac{df}{dt} = \mathbf{v}\cdot(2x_0+4t, -2y_0-6t) = 4x_0-6y_0-10t; \frac{df}{ds} = \frac{df/dt}{\sqrt{13}}$

45 $\mathbf{v} = (e^t, 2e^{2t}, -e^{-t}); \mathbf{T} = \frac{\mathbf{v}}{|\mathbf{v}|}; \text{grad } f = \left(\frac{1}{x}, \frac{1}{y}, \frac{1}{z}\right) = (e^{-t}, e^{-2t}, e^t), \frac{df}{dt} = 1+2-1, \frac{df}{ds} = \frac{2}{|\mathbf{v}|}$

47 $\mathbf{v} = (-2\sin 2t, 2\cos 2t), \mathbf{T} = (-\sin 2t, \cos 2t); \text{grad } f = (y,x), \frac{df}{ds} = -2\sin^2 2t + 2\cos^2 2t, \frac{df}{dt} = \frac{1}{2}\frac{df}{ds};$
zero slope because $f = 1$ on this path

49 $z-1 = 2(x-4)+3(y-5); f = 1+2(x-4)+3(y-5)$ **51** $\text{grad } f \cdot \mathbf{T} = 0; \mathbf{T}$

Section 13.5 The Chain Rule   (page 503)

1 $f_y = cf_x = c\cos(x+cy)$ **3** $f_y = 7f_x = 7e^{x+7y}$ **5** $3g^2\frac{\partial g}{\partial x}\frac{dx}{dt} + 3g^2\frac{\partial g}{\partial y}\frac{dy}{dt}$ **7** Moves left at speed 2

9 $\frac{dx}{dt} = 1$ (wave moves at speed 1)

11 $\frac{\partial^2}{\partial x^2}f(x+iy) = f''(x+iy), \frac{\partial^2}{\partial y^2}f(x+iy) = i^2 f''(x+iy)$
so $f_{xx} + f_{yy} = 0; (x+iy)^2 = (x^2-y^2)+i(2xy)$

13 $\frac{df}{dt} = 2x(1)+2y(2t) = 2t+4t^3$ **15** $\frac{df}{dt} = y\frac{dx}{dt} + x\frac{dy}{dt} = -1$ **17** $\frac{df}{dt} = \frac{1}{x+y}\frac{dx}{dt} + \frac{1}{x+y}\frac{dy}{dt} = 1$

19 $V = \frac{1}{3}\pi r^2 h, \frac{dV}{dt} = \frac{2\pi rh}{3}\frac{dr}{dt} + \frac{\pi r^2}{3}\frac{dh}{dt} = 36\pi$

21 $\frac{dD}{dt} = \frac{90}{\sqrt{90^2+90^2}}(60) + \frac{90}{\sqrt{90^2+90^2}}(45) = \frac{105}{\sqrt{2}}$ mph; $\frac{dD}{dt} = \frac{60}{\sqrt{45^2+60^2}}(60) + \frac{45}{\sqrt{45^2+60^2}}(45) \approx 74$ mph

23 $\frac{df}{dt} = u_1\frac{\partial f}{\partial x} + u_2\frac{\partial f}{\partial y} + u_3\frac{\partial f}{\partial z}$ **25** $\frac{df}{dt} = 1$ with x and y fixed; $\frac{df}{dt} = 6$

27 $f_t = f_x t + f_y(2t); f_{tt} = f_{xt}t + f_x + 2f_{yt}t + 2f_y = (f_{xx}t + f_{yx}(2t))t + f_x + 2(f_{xy}t + f_{yy}(2t))t + 2f_y$

29 $\frac{\partial f}{\partial r} = \frac{\partial f}{\partial x}\frac{\partial x}{\partial r} + \frac{\partial f}{\partial y}\frac{\partial y}{\partial r} = \frac{\partial f}{\partial x}\cos\theta + \frac{\partial f}{\partial y}\sin\theta, \theta$ is fixed

31 $r_{xx} = \frac{1}{\sqrt{x^2+y^2}} - \frac{x^2}{(x^2+y^2)^{3/2}} = \frac{y^2}{(x^2+y^2)^{3/2}}; \frac{\partial}{\partial x}\left(\frac{x}{r}\right) = \frac{1}{r} - xr^{-2}\frac{\partial r}{\partial x} = \frac{1}{r} - \frac{x^2}{r^3} = \frac{y^2}{r^3}$

33 $\left(\frac{\partial z}{\partial x}\right)_y = \frac{1}{\sqrt{1-(x+y)^2}}; (\cos z)\left(\frac{\partial z}{\partial x}\right)_y = 1$; first answer is also $\frac{1}{\sqrt{1-\sin^2 z}} = \frac{1}{\cos z}$

35 $f_r = f_x\cos\theta + f_y\sin\theta, f_{r\theta} = -f_x\sin\theta + f_y\cos\theta + f_{xx}(-r\sin\theta\cos\theta) + f_{xy}(-r\sin^2\theta + r\cos^2\theta) + f_{yy}(r\cos\theta\sin\theta)$

37 Yes (with y constant): $\frac{\partial z}{\partial x} = ye^{xy}, \frac{\partial x}{\partial z} = \frac{1}{zy} = \frac{1}{ye^{xy}}$ **39** $f_t = f_x x_t + f_y y_t; f_{tt} = f_{xx}x_t^2 + 2f_{xy}x_t y_t + f_{yy}y_t^2$

41 $\left(\frac{\partial f}{\partial x}\right)_z = \frac{\partial f}{\partial x} + \frac{\partial f}{\partial y}\frac{\partial y}{\partial x} = a - \frac{3}{2}b; \left(\frac{\partial f}{\partial x}\right)_y = a; \left(\frac{\partial f}{\partial z}\right)_x = \frac{\partial f}{\partial y}\frac{\partial y}{\partial z} = \frac{1}{5}b$

43 1   **45** $f = y^2$ so $f_x = 0, f_y = 2y = 2r\sin\theta; f = r^2$ so $f_r = 2r = 2\sqrt{x^2+y^2}, f_\theta = 0$

47 $g_u = f_x x_u + f_y y_u = f_x + f_y; g_v = f_x x_v + f_y y_v = f_x - f_y; g_{uu} = f_{xx}x_u + f_{xy}y_u + f_{yx}x_u + f_{yy}y_u$
$= f_{xx} + 2f_{xy} + f_{yy}; g_{vv} = f_{xx}x_v + f_{xy}y_v - f_{yx}x_v - f_{yy}y_v = f_{xx} - 2f_{xy} + f_{yy}$. Add $g_{uu} + g_{vv}$ **49** False

Section 13.6 Maxima, Minima, and Saddle Points   (page 512)

1 $(0,0)$ is a minimum   **3** $(3,0)$ is a saddle point   **5** No stationary points   **7** $(0,0)$ is a maximum

9 $(0,0,2)$ is a minimum   **11** All points on the line $x = y$ are minima   **13** $(0,0)$ is a saddle point

15 $(0,0)$ is a saddle point; $(2,0)$ is a minimum; $(0,-2)$ is a maximum; $(2,-2)$ is a saddle point

17 Maximum of area $(12-3y)y$ is 12

19 $\begin{array}{l} 2(x+y)+2(x+2y-5)+2(x+3y-4) = 0 \\ 2(x+y)+4(x+2y-5)+6(x+3y-4) = 0 \end{array}$ gives $\begin{array}{l} x = 2; \\ y = -1 \end{array}$ min because $E_{xx}E_{yy} = (6)(28) > E_{xy}^2 = 12^2$

21 Minimum at $\left(0,\frac{1}{2}\right); (0,1); (0,1)$

23 $\frac{df}{dt} = 0$ when $\tan t = \sqrt{3}$; $f_{max} = 2$ at $\left(\frac{1}{2}, \frac{\sqrt{3}}{2}\right)$, $f_{min} = -2$ at $\left(-\frac{1}{2}, -\frac{\sqrt{3}}{2}\right)$

25 $(ax+by)_{max} = \sqrt{a^2+b^2}$; $(x^2+y^2)_{min} = \frac{1}{a^2+b^2}$ 27 $0 < c < \frac{1}{4}$

29 The vectors head-to-tail form a 60-60-60 triangle. The outer angle is $120°$ 31 $2+\sqrt{3}$; $1+\sqrt{3}$; $1+\frac{\sqrt{3}}{10}$

35 Steiner point where the arcs meet   39 Best point for $p = \infty$ is equidistant from corners

41 grad $f = \left(\sqrt{2}\frac{x-x_1}{d_1} + \frac{x-x_2}{d_2} + \frac{x-x_3}{d_3}, \sqrt{2}\frac{y-y_1}{d_1} + \frac{y-y_2}{d_2} + \frac{y-y_3}{d_3}\right)$; angles are 90-135-135

43 Third derivatives all 6; $f = \frac{6}{3!}x^3 + \frac{6}{2!}x^2y + \frac{6}{2!}xy^2 + \frac{6}{3!}y^3$

45 $\left(\frac{\partial}{\partial x}\right)^n \left(\frac{\partial}{\partial y}\right)^m \ln(1-xy)]_{0,0} = n!(n-1)!$ for $m = n > 0$, other derivatives zero; $f = -xy - \frac{x^2y^2}{2} - \frac{x^3y^3}{3} - \cdots$

47 All derivatives are e^2 at $(1,1)$; $f \approx e^2\left[1 + (x-1) + (y-1) + \frac{1}{2}(x-1)^2 + (x-1)(y-1) + \frac{1}{2}(y-1)^2\right]$

49 $x = 1, y = -1$: $f_x = 2, f_y = -2, f_{xx} = 2, f_{xy} = 0, f_{yy} = 2$; series must recover $x^2 + y^2$

51 Line $x - 2y = $ constant; $x + y = $ constant

53 $\frac{x^2}{2}f_{xx} + xyf_{xy} + \frac{y^2}{2}f_{yy}]_{0,0}$; $f_{xx} > 0$ and $f_{xx}f_{yy} > f_{xy}^2$ at $(0,0)$; $f_x = f_y = 0$ 55 $\Delta x = -1, \Delta y = -1$

57 $f = x^2(12 - 4x)$ has $f_{max} = 16$ at $(2,4)$; line has slope -4, $y = \frac{16}{x^2}$ has slope $\frac{-32}{8} = -4$

59 If the fence were not perpendicular, a point to the left or right would be closer

Section 13.7 Constraints and Lagrange Multipliers   (page 519)

1 $f = x^2 + (k-2x)^2$; $\frac{df}{dx} = 2x - 4(k-2x) = 0$; $\left(\frac{2k}{5}, \frac{k}{5}\right), \frac{k^2}{5}$ 3 $\lambda = -4, x_{min} = 2, y_{min} = 2$

5 $\lambda = \frac{1}{3(4)^{1/3}}$: $(x,y) = (\pm 2^{1/6}, 0)$ or $(0, \pm 2^{1/6})$, $f_{min} = 2^{1/3}$; $\lambda = \frac{1}{3}$: $(x,y) = (\pm 1, \pm 1)$, $f_{max} = 2$

7 $\lambda = \frac{1}{2}, (x,y) = (2, -3)$; tangent line is $2x - 3y = 13$

9 $(1-c)^2 + (-a-c)^2 + (2-a-b-c)^2 + (2-b-c)^2$ is minimized at $a = -\frac{1}{2}, b = \frac{3}{2}, c = \frac{3}{4}$

11 $(1, -1)$ and $(-1, 1)$; $\lambda = -\frac{1}{2}$

13 f is not a minimum when C crosses to lower curve; stationary point when C is tangent to level curve

15 Substituting $\frac{\partial L}{\partial x} = \frac{\partial L}{\partial y} = \frac{\partial L}{\partial \lambda} = 0$ and $L = f_{min}$ leaves $\frac{df_{min}}{dk} = \lambda$

17 x^2 is never negative; $(0,0)$; $1 = \lambda(-3y^2)$ but $y = 0$; $g = 0$ has a cusp at $(0,0)$

19 $2x = \lambda_1 + \lambda_2, 4y = \lambda_1, 2z = \lambda_1 - \lambda_2, x + y + z = 0, x - z = 1$ gives $\lambda_1 = 0, \lambda_2 = 1, f_{min} = \frac{1}{2}$ at $\left(\frac{1}{2}, 0, -\frac{1}{2}\right)$

21 $(1,0,0); (0,1,0); (\lambda_1, \lambda_2, 0); x = y = 0$ 23 $\frac{\partial f}{\partial x}$ and $\frac{\partial f}{\partial y}$; $\lambda = 0$

25 $(1,0,0), (0,1,0), (0,0,1)$; at these points $f = 4$ and -2 (min) and 5(max)

27 By increasing k, more points are available so f_{max} goes up. Then $\lambda = \frac{df_{min}}{dk} \geq 0$

29 $(0,0); \lambda = 0$; f_{min} stays at 0

31 $5 = \lambda_1 + \lambda_2, 6 = \lambda_1 + \lambda_3, \lambda_2 \geq 0, \lambda_3 \leq 0$; subtraction $5 - 6 = \lambda_2 - \lambda_3$ or $-1 \geq 0$ (impossible); $x = 2004, y = -2000$ gives $5x + 6y = -1980$

33 $2x = 4\lambda_1 + \lambda_2, 2y = 4\lambda_1 + \lambda_3, \lambda_2 \geq 0, \lambda_3 \geq 0, 4x + 4y = 40$; max area 100 at $(10,0)(0,10)$; min 25 at $(5,5)$

CHAPTER 14 MULTIPLE INTEGRALS

Section 14.1 Double Integrals   (page 526)

1 $\frac{8}{3}; \frac{2}{3}$ 3 $1; \ln\frac{3}{2}$ 5 2 7 $\frac{1}{2}$ 9 $\frac{4}{3}$ 11 $\int_{y=1}^2 \int_{x=1}^2 dx\,dy + \int_{y=2}^4 \int_{y/2}^2 dx\,dy$

13 $\int_{y=0}^1 \int_{x=-\frac{1}{2}\ln y}^{-\ln y} dx\,dy$ 15 $\int_{x=0}^1 \int_{y=-\sqrt{x}}^{\sqrt{x}} dy\,dx$ 17 $\int_0^1 \int_0^{y/2} dx\,dy = \int_0^{1/2} \int_{2x}^1 dy\,dx = \frac{1}{4}$

19 $\int_0^3 \int_{-y}^y dx\,dy = \int_{-1}^0 \int_{-x}^3 dy\,dx + \int_0^1 \int_x^3 dy\,dx = 9$ 21 $\int_0^4 \int_{y/2}^y dx\,dy + \int_4^8 \int_{y/2}^4 dx\,dy = \int_0^4 \int_x^{2x} dy\,dx = 8$

23 $\int_0^1 \int_0^{bx} dy\,dx + \int_1^2 \int_0^{b(2-x)} dy\,dx = \int_0^b \int_{y/b}^{2-(y/b)} dx\,dy = b$ 25 $f(a,b) - f(a,0) - f(0,b) + f(0,0)$

27 $\int_0^1 \int_0^1 (2x - 3y + 1)dx\,dy = \frac{1}{2}$ 29 $\int_a^b f(x)dx = \int_a^b \int_0^{f(x)} 1\,dy\,dx$ 31 $50{,}000\pi$

33 $\int_1^3 \int_1^2 x^2 dx\, dy = \frac{14}{3}$ **35** $2\int_0^{1/\sqrt{2}} \int_0^{\sqrt{1-y^2}} 1 dx\, dy = \frac{\pi}{4}$

37 $\frac{1}{n^2} \sum_{j=1}^n \sum_{i=1}^n f\left(\frac{i-\frac{1}{2}}{n}, \frac{j-\frac{1}{2}}{n}\right)$ is exact for $f = 1, x, y, xy$ **39** Volume 8.5 **41** Volumes $\ln 2, 2\ln(1+\sqrt{2})$

43 $\int_0^1 \int_0^1 x^y dx\, dy = \int_0^1 \frac{1}{y+1} dy = \ln 2; \int_0^1 \int_0^1 x^y dy\, dx = \int_0^1 \frac{x-1}{\ln x} dx = \ln 2$

45 With long rectangles $\sum y_i \Delta A = \sum \Delta A = 1$ but $\int \int y\, dA = \frac{1}{2}$

Section 14.2 Change to Better Coordinates   (page 534)

1 $\int_{\pi/4}^{3\pi/4} \int_0^1 r\, dr\, d\theta = \frac{\pi}{4}$ **3** S = quarter-circle with $u \geq 0$ and $v \geq 0$; $\int_0^1 \int_0^{\sqrt{1-v^2}} du\, dv$

5 R is symmetric across the y axis; $\int_0^1 \int_0^{\sqrt{1-v^2}} u\, du\, dv = \frac{1}{3}$ divided by area gives $(\bar{u}, \bar{v}) = (4/3\pi, 4/3\pi)$

7 $2\int_0^{1/\sqrt{2}} \int_{1+x}^{1+\sqrt{1-x^2}} dy\, dx$; xy region R^* becomes R in the x^*y^* plane; $dx\, dy = dx^*dy^*$ when region moves

9 $J = \begin{vmatrix} \partial x/\partial r^* & \partial x/\partial \theta^* \\ \partial y/\partial r^* & \partial y/\partial \theta^* \end{vmatrix} = \begin{vmatrix} \cos\theta^* & -r^*\sin\theta^* \\ \sin\theta^* & r^*\cos\theta^* \end{vmatrix} = r^*; \int_{\pi/4}^{3\pi/4} \int_0^1 r^*\, dr^*\, d\theta^*$

11 $I_y = \int\int_R x^2 dx\, dy = \int_{\pi/4}^{3\pi/4} \int_0^1 r^2\cos^2\theta\, r\, dr\, d\theta = \frac{\pi}{16} - \frac{1}{8}; I_x = \frac{\pi}{16} + \frac{1}{8}; I_0 = \frac{\pi}{8}$

13 $(0,0), (1,2), (1,3), (0,1)$; area of parallelogram is 1

15 $x = u, y = u + 3v + uv$; then $(u,v) = (1,0), (1,1), (0,1)$ give corners $(x,y) = (1,0), (1,5), (0,3)$

17 Corners $(0,0), (2,1), (3,3), (1,2)$; sides $y = \frac{1}{2}x, y = 2x - 3, y = \frac{1}{2}x + \frac{3}{2}, y = 2x$

19 Corners $(1,1), (e^2, e), (e^3, e^3), (e, e^2)$; sides $x = y^2, y = x^2/e^3, x = y^2/e^3, y = x^2$

21 Corners $(0,0), (1,0), (1,2), (0,1)$; sides $y = 0, x = 1, y = 1 + x^2, x = 0$

23 $J = \begin{vmatrix} 2 & 1 \\ 1 & 2 \end{vmatrix} = 3$, area $\int_0^1 \int_0^1 3du\, dv = 3; J = \begin{vmatrix} 2e^{2u+v} & e^{2u+v} \\ e^{u+2v} & 2e^{u+2v} \end{vmatrix} = 3e^{3u+3v}, \int_0^1 \int_0^1 3e^{3u+3v} du\, dv =$
$\int_0^1 (e^{3+3v} - e^{3v}) dv = \frac{1}{3}(e^6 - 2e^3 + 1)$

25 Corners $(x,y) = (0,0), (1,0), (1, f(1)), (0, f(0)); (\frac{1}{2}, 1)$ gives $x = \frac{1}{2}, y = f(\frac{1}{2}); J = \begin{vmatrix} 1 & 0 \\ vf'(u) & f(u) \end{vmatrix} = f(u)$

27 $B^2 = 2\int_0^{\pi/4} \int_0^{1/\sin\theta} e^{-r^2} r\, dr\, d\theta = \int_0^{\pi/4}(e^{-1/\sin^2\theta} - 1)d\theta$

29 $\bar{r} = \int\int r^2 dr\, d\theta / \int\int r\, dr\, d\theta = \int_0^\pi \frac{8}{3}a^3\cos^3\theta\, d\theta / \pi a^2 = \frac{32a}{9\pi}$ **31** $\int_0^{2\pi} \int_0^1 r^2 r\, dr\, d\theta = \frac{\pi}{2}$

33 Along the right side; along the bottom; at the bottom right corner

35 $\int\int xy\, dx\, dy = \int_0^1 \int_0^1 (u\cos\alpha - v\sin\alpha)(u\sin\alpha + v\cos\alpha)du\, dv = \frac{1}{4}(\cos^2\alpha - \sin^2\alpha)$

37 $\int_0^{2\pi} \int_4^5 r^2 r^2 r\, dr\, d\theta = \frac{2\pi}{6}(5^6 - 4^6)$ **39** $x = \cos\alpha - \sin\alpha, y = \sin\alpha + \cos\alpha$ goes to $u = 1, v = 1$

Section 14.3 Triple Integrals   (page 540)

1 $\int_0^1 \int_0^z \int_0^y dx\, dy\, dz = \frac{1}{6}$

3 $0 \leq y \leq x \leq z \leq 1$ and all other orders xzy, yzx, zxy, zyx; all six contain $(0,0,0)$; to contain $(1,0,1)$

5 $\int_{-1}^1 \int_{-1}^1 \int_{-1}^1 dx\, dy\, dz = 8$ **7** $\int_{-1}^1 \int_{-1}^z \int_{-1}^1 dx\, dy\, dz = 4$ **9** $\int_{-1}^1 \int_z^1 \int_1^z dx\, dy\, dz = \frac{4}{3}$

11 $\int_0^1 \int_0^{2-2z} \int_0^{2-y-2z} dx\, dy\, dz = \frac{2}{3}$ **13** $\int_0^{1/2} \int_0^{2-2z} \int_0^{2-y-2z} dx\, dy\, dz = \frac{7}{12}$

15 $\int_0^1 \int_0^{1-z} \int_0^{\sqrt{(1-z)^2-y^2}} dx\, dy\, dz = \frac{\pi}{3}$ **17** $\int_0^6 \int_0^1 \int_0^{\sqrt{1-y^2}} dx\, dy\, dz = 6\pi$ **19** $\int_0^1 \int_0^1 \int_0^{\sqrt{1-y^2}} dx\, dy\, dz = \pi$

21 Corner of cube at $\left(\frac{1}{\sqrt{3}}, \frac{1}{\sqrt{3}}, \frac{1}{\sqrt{3}}\right)$; sides $\frac{2}{\sqrt{3}}$; area $\frac{8}{3\sqrt{3}}$

23 Horizontal slices are circles of area $\pi r^2 = \pi(4-z)$; volume $= \int_0^4 \pi(4-z)dz = 8\pi$; centroid
    has $\bar{x} = 0, \bar{y} = 0, \bar{z} = \int_0^4 z\pi(4-z)dz/8\pi = \frac{4}{3}$

25 $I = \frac{z^2}{2}$ gives zeros; $\frac{\partial I}{\partial x} = \int_0^z \int_0^y f \, dy \, dz, \frac{\partial I}{\partial y} = \int_0^z \int_0^x f \, dx \, dz, \frac{\partial^2 I}{\partial y \, \partial z} = \int_0^x f \, dx$

27 $\int_{-1}^1 \int_{-1}^1 \int_{-1}^1 (y^2 + x^2) dx \, dy \, dz = \frac{16}{3}; \int \int \int x^2 dV = \frac{8}{3}; 3 \int \int \int (x - \frac{x+y+z}{3})^2 dV = \frac{16}{3}$

29 $\int_0^3 \int_0^2 \int_0^y dx \, dy \, dz = 6$ **31** Trapezoidal rule is second-order; correct for $1, x, y, z, xy, xz, yz, xyz$

Section 14.4 Cylindrical and Spherical Coordinates (page 547)

1 $(r, \theta, z) = (D, 0, 0); (\rho, \phi, \theta) = (D, \frac{\pi}{2}, 0)$ **3** $(r, \theta, z) = (0, \text{any angle}, D); (\rho, \phi, \theta) = (D, 0, \text{any angle})$

5 $(x, y, z) = (2, -2, 2\sqrt{2}); (r, \theta, z) = (2\sqrt{2}, -\frac{\pi}{4}, 2\sqrt{2})$ **7** $(x, y, z) = (0, 0, -1); (r, \theta, z) = (0, \text{any angle}, -1)$

9 $\phi = \tan^{-1}(\frac{r}{z})$ **11** 45° cone in unit sphere: $\frac{2\pi}{3}(1 - \frac{1}{\sqrt{2}})$ **13** cone without top: $\frac{7\pi}{3}$

15 $\frac{1}{4}$ hemisphere: $\frac{\pi}{6}$ **17** $\frac{\pi^2}{8}$ **19** Hemisphere of radius π : $\frac{2}{3}\pi^4$ **21** $\pi(R^2 - z^2); 4\pi r \sqrt{R^2 - r^2}$

23 $\frac{2}{3}a^3 \tan \alpha$ (see 8.1.39) **27** $\frac{\partial q}{\partial D} = \frac{\rho - D \cos \phi}{q} = \frac{\text{near side}}{\text{hypotenuse}} = \cos \alpha$

31 Wedges are not exactly similar; the error is higher order $\Rightarrow$ proof is correct

33 Proportional to $1 + \frac{1}{h}(\sqrt{a^2 + (D-h)^2} - \sqrt{a^2 + D^2})$

35 $J = \begin{vmatrix} a & & \\ & b & \\ & & c \end{vmatrix} = abc$; straight edges at right angles **37** $\begin{vmatrix} \cos \theta & -r \sin \theta & 0 \\ \sin \theta & r \cos \theta & 0 \\ 0 & 0 & 1 \end{vmatrix} = r$

39 $\frac{8\pi\rho^4}{3}; \frac{2}{3}$ **41** $\rho^3; \rho^2$; force $= 0$ inside hollow sphere

CHAPTER 15 VECTOR CALCULUS

Section 15.1 Vector Fields (page 554)

1 $f(x, y) = x + 2y$ **3** $f(x, y) = \sin(x + y)$ **5** $f(x, y) = \ln(x^2 + y^2) = 2\ln r$

7 $\mathbf{F} = xy\mathbf{i} + \frac{x^2}{2}\mathbf{j}, f(x, y) = \frac{x^2 y}{2}$ **9** $\frac{\partial f}{\partial x} = 0$ so f cannot depend on x; streamlines are vertical ($y = $ constant)

11 $\mathbf{F} = 3\mathbf{i} + \mathbf{j}$ **13** $\mathbf{F} = \mathbf{i} + 2y\mathbf{j}$ **15** $\mathbf{F} = 2x\mathbf{i} - 2y\mathbf{j}$ **17** $\mathbf{F} = e^{x-y}\mathbf{i} - e^{x-y}\mathbf{j}$

19 $\frac{dy}{dx} = -1; y = -x + C$ **21** $\frac{dy}{dx} = -\frac{x}{y}; x^2 + y^2 = C$ **23** $\frac{dy}{dx} = \frac{-x/y^2}{1/y} = \frac{-x}{y}; x^2 + y^2 = C$ **25** parallel

27 $\mathbf{F} = \frac{5x}{r}\mathbf{i} + \frac{5y}{r}\mathbf{j}$ **29** $\mathbf{F} = \frac{-mMG}{r^3}(x\mathbf{i} + y\mathbf{j}) - \frac{mMG}{((x-1)^2 + y^2)^{3/2}}((x-1)\mathbf{i} + y\mathbf{j})$

31 $\mathbf{F} = \frac{\sqrt{2}}{2}y\mathbf{i} - \frac{\sqrt{2}}{2}x\mathbf{j}$ **33** $\frac{dy}{dx} = \frac{-2}{x^2} = -\frac{1}{2}; \frac{dy}{dx} = \frac{x}{\sqrt{x^2-3}} = 2$

35 $\frac{\partial f}{\partial x} = \frac{\partial f}{\partial r}\frac{\partial r}{\partial x} = \frac{\partial f}{\partial r}\frac{x}{r}; \frac{\partial f}{\partial y} = \frac{\partial f}{\partial r}\frac{y}{r}; f(r) = C$ gives circles

37 T; F (no equipotentials); T; F (not multiple of $x\mathbf{i} + y\mathbf{j} + z\mathbf{k}$)

39 $\mathbf{F}$ and $\mathbf{F} + \mathbf{i}$ and $2\mathbf{F}$ have the same streamlines (different velocities) and equipotentials (different potentials). But if f is given, $\mathbf{F}$ must be grad f.

Section 15.2 Line Integrals (page 562)

1 $\int_0^1 \sqrt{1^2 + 2^2} dt = \sqrt{5}; \int_0^1 2 \, dt = 2$ **3** $\int_0^1 t^2 \sqrt{2} dt + \int_1^2 1 \cdot (2 - t) dt = \frac{\sqrt{2}}{3} + \frac{1}{2}$

5 $\int_0^{2\pi}(-3\sin t)dt = 0$ (gradient field); $\int_0^{2\pi} -9\sin^2 t \, dt = -9\pi = -$area

7 No, $xy\mathbf{j}$ is not a gradient field; take line $x = t, y = t$ from $(0,0)$ to $(1,1)$ and $\int t^2 \, dt \neq \frac{1}{2}$

9 No, for a circle $(2\pi r)^2 \neq 0^2 + 0^2$ **11** $f = x + \frac{1}{2}y^2; f(0,1) - f(1,0) = -\frac{1}{2}$

13 $f = \frac{1}{2}x^2 y^2; f(0,1) - f(1,0) = 0$ **15** $f = r = \sqrt{x^2 + y^2}; f(0,1) - f(1,0) = 0$

17 Gradient for $n = 2$; after calculation $\frac{\partial M}{\partial y} - \frac{\partial N}{\partial x} = \frac{n-2}{r^n}$

19 $x = a \cos t, z = a \sin t, ds = a \, dt, M = \int_0^{2\pi}(a + a \sin t)a \, dt = 2\pi a^2$

21 $x = a\cos t, y = a\sin t, ds = a\,dt, M = \int_0^{2\pi} a^3 \cos^2 t\,dt = \pi a^3, (\bar{x}, \bar{y}) = (0,0)$ by symmetry

23 $\mathbf{T} = \frac{2\mathbf{i}+2t\mathbf{j}}{\sqrt{4+4t^2}} = \frac{\mathbf{i}+t\mathbf{j}}{\sqrt{1+t^2}}; \mathbf{F} = 3x\mathbf{i}+4\mathbf{j} = 6t\mathbf{i}+4\mathbf{j}, ds = 2\sqrt{1+t^2}dt, \mathbf{F}\cdot\mathbf{T}ds = (6t\mathbf{i}+4\mathbf{j})\cdot(\frac{\mathbf{i}+t\mathbf{j}}{\sqrt{1+t^2}})2\sqrt{1+t^2}dt = 20t\,dt; \mathbf{F}\cdot d\mathbf{R} =$
$(6t\mathbf{i}+4\mathbf{j})\cdot(2\,dt\mathbf{i}+2t\,dt\,\mathbf{j}) = 20t\,dt$; work $= \int_1^2 20t\,dt = 30$

25 If $\frac{\partial M(y)}{\partial y} = \frac{\partial N(x)}{\partial x}$ then $M = ay+b, N = ax+c$, constants a, b, c

27 $\mathbf{F} = 4x\mathbf{j}$ (work $= 4$ from $(1,0)$ up to $(1,1)$)   29 $f = [x-2y]_{(0,0)}^{(1,1)} = -1$ 31 $f = [xy^2]_{(0,0)}^{(1,1)} = 1$

33 Not conservative; $\int_0^1 (t\mathbf{i}-t\mathbf{j})\cdot(dt\,\mathbf{i}+dt\,\mathbf{j}) = \int 0\,dt = 0; \int_0^1 (t^2\mathbf{i}-t\mathbf{j})\cdot(dt\,\mathbf{i}+2t\,dt\,\mathbf{j}) = \int_0^1 -t^2 dt = -\frac{1}{3}$

35 $\frac{\partial M}{\partial y} = ax, \frac{\partial N}{\partial x} = 2x+b$, so $a = 2, b$ is arbitrary   37 $\frac{\partial M}{\partial y} = 2ye^{-x} = \frac{\partial N}{\partial x}; f = -y^2 e^{-x}$

39 $\frac{\partial M}{\partial y} = \frac{-xy}{r^3} = \frac{\partial N}{\partial x}; f = r = \sqrt{x^2+y^2} = |x\mathbf{i}+y\mathbf{j}|$

41 $\mathbf{F} = (x-y)\mathbf{i}+(x+y)\mathbf{j}$ has $\frac{\partial M}{\partial y} = -1, \frac{\partial N}{\partial x} = 1$, no f 43 $2\pi; 0; 0$

Section 15.3 Green's Theorem   (page 571)

1 $\int_0^{2\pi} (a\cos t)a\cos t\,dt = \pi a^2; N_x - M_y = 1, \iint dx\,dy = $ area πa^2

3 $\int_0^1 x\,dx + \int_1^0 x\,dx = 0, N_x - M_y = 0, \iint 0\,dx\,dy = 0$

5 $\int x^2 y\,dx = \int_0^{2\pi} (a\cos t)^2(a\sin t)(-a\sin t\,dt) = -\frac{a^4}{4}\int_0^{2\pi}(\sin 2t)^2 dt = -\frac{\pi a^4}{4};$
$N_x - M_y = -x^2, \iint(-x^2)dx\,dy = \int_0^{2\pi}\int_0^a -r^2\cos^2\theta\, r\,dr\,d\theta = -\frac{\pi a^4}{4}$

7 $\int x\,dy - y\,dx = \int_0^\pi (\cos^2 t + \sin^2 t)\,dt = \pi; \iint(1+1)dx\,dy = 2$ (area) $= \pi; \int x^2 dy - xy\,dx = \frac{1}{2}+1; \int_0^1\int_0^1(2x+x)dx\,dy = \frac{3}{2}$

9 $\frac{1}{2}\int_0^{2\pi}(3\cos^4 t\sin^2 t + 3\sin^4 t\cos^2 t)dt = \frac{1}{2}\int_0^{2\pi} 3\cos^2 t\sin^2 t\,dt = \frac{3}{2}\frac{\pi}{4}$ (see Answer 5)

11 $\int \mathbf{F}\cdot d\mathbf{R} = 0$ around any loop; $\mathbf{F} = \frac{x}{r}\mathbf{i}+\frac{y}{r}\mathbf{j}$ and $\int \mathbf{F}\cdot d\mathbf{R} = \int_0^{2\pi}[-\sin t\cos t + \sin t\cos t]dt = 0; \frac{\partial M}{\partial y} = \frac{\partial N}{\partial x}$ gives $\iint 0\,dx\,dy$

13 $x = \cos 2t, y = \sin 2t, t$ from 0 to $2\pi; \int_0^{2\pi} -2\sin^2 2t\,dt = -2\pi = -2$ (area);
$\int_0^{2\pi} -2dt = -4\pi = -2$ times Example 7

15 $\int M\,dy - N\,dx = \int_0^{2\pi} 2\sin t\cos t\,dt = 0; \iint(M_x + N_y)dx\,dy = \iint 0\,dx\,dy = 0$

17 $M = \frac{x}{r}, N = \frac{y}{r}, \int M\,dy - N\,dx = \int_0^{2\pi}(\cos^2 t + \sin^2 t)dt = 2\pi; \iint(M_x + N_y)dx\,dy = \iint(\frac{1}{r}-\frac{x^3}{r^3}+\frac{1}{r}-\frac{y^2}{r^3})dx\,dy =$
$\iint \frac{1}{r}dx\,dy = \iint dr\,d\theta = 2\pi$

19 $\int M\,dy - N\,dx = \int -x^2 y\,dx = \int_1^0 -x^2(1-x)dx = \frac{1}{12}; \int_0^1\int_0^{1-y} x^2 dx\,dy = \frac{1}{12}$

21 $\iint(M_x + N_y)dx\,dy = \iint \text{div}\,\mathbf{F}\,dx\,dy = 0$ between the circles

23 Work: $\int a\,dx + b\,dy = \iint(\frac{\partial b}{\partial x} - \frac{\partial a}{\partial y})dx\,dy$; Flux: *same integral*

25 $g = \tan^{-1}(\frac{y}{x}) = \theta$ is undefined at $(0,0)$ 27 Test $M_y = N_x : x^2 dx + y^2 dy$ is exact $= d(\frac{1}{3}x^3 + \frac{1}{3}y^3)$

29 div $\mathbf{F} = 2y - 2y = 0; g = xy^2$ 31 div $\mathbf{F} = 2x + 2y$; no g 33 div $\mathbf{F} = 0; g = e^x \sin y$

35 div $\mathbf{F} = 0; g = \frac{y^2}{x}$

37 $N_x - M_y = -2x, -6xy, 0, 2x-2y, 0, -2e^{x+y}$; in **31** and **33** $f = \frac{1}{3}(x^3 + y^3)$ and $f = e^x \cos y$

39 $\mathbf{F} = (3x^2 - 3y^2)\mathbf{i} - 6xy\,\mathbf{j}$; div $\mathbf{F} = 0$ 41 $f = x^4 - 6x^2 y^2 + y^4; g = 4x^3 y - 4xy^3$

43 $\mathbf{F} = e^x \cos y\,\mathbf{i} - e^x \sin y\,\mathbf{j}; g = e^x \sin y$

45 $N = f(x), \int M\,dx + N\,dy = \int_0^1 f(1)dy + \int_1^0 f(0)dy = f(1) - f(0); \int \int(N_x - M_y)dx\,dy =$
$\iint \frac{\partial f}{\partial x}dx\,dy = \int_0^1 \frac{\partial f}{\partial x}dx$ (Fundamental Theorem of Calculus)

Section 15.4 Surface Integrals (page 581)

1 $N = -2x\mathbf{i} - 2y\mathbf{j} + \mathbf{k};\ dS = \sqrt{1 + 4x^2 + 4y^2}\ dx\ dy;\ \int_0^{2\pi} \int_0^2 \sqrt{1 + 4r^2}\ r\ dr\ d\theta = \frac{\pi}{6}(17^{3/2} - 1)$

3 $N = -\mathbf{i} + \mathbf{j} + \mathbf{k};\ dS = \sqrt{3}\ dx\ dy;\ \text{area } \sqrt{3}\pi$

5 $N = \frac{-x\mathbf{i} - y\mathbf{j}}{\sqrt{1 - x^2 - y^2}} + \mathbf{k};\ dS = \frac{dx\ dy}{\sqrt{1 - x^2 - y^2}};\ \int_0^{2\pi} \int_0^{1/\sqrt{2}} \frac{r\ dr\ d\theta}{\sqrt{1 - r^2}} = \pi(2 - \sqrt{2})$

7 $N = -7\mathbf{j} + \mathbf{k}; dS = 5\sqrt{2}\ dx\ dy;\ \text{area } 5\sqrt{2}A$

9 $N = (y^2 - x^2)\mathbf{i} - 2xy\mathbf{j} + \mathbf{k}; dS = \sqrt{1 + (y^2 - x^2)^2 + 4x^2y^2}\ dx\ dy = \sqrt{1 + (y^2 - x^2)^2}\ dx\ dy;$
$\int_0^{2\pi} \int_0^1 \sqrt{1 + r^4}\ r\ dr\ d\theta = \frac{\pi}{\sqrt{2}} + \frac{\pi \ln(1 + \sqrt{2})}{2}$

11 $N = 2\mathbf{i} + 2\mathbf{j} + \mathbf{k}; dS = 3dx\ dy\ ;\ 3\ (\text{area of triangle with } 2x + 2y \le 1) = \frac{3}{8}$

13 $\pi a\sqrt{a^2 + h^2}$ 15 $\int_0^1 \int_0^{1-y} xy(\sqrt{3}\ dx\ dy) = \frac{\sqrt{3}}{24}$

17 $\int_0^{2\pi} \int_0^{\pi/4} \sin^2\phi\ \cos\phi\ \sin\theta\ \cos\theta(\sin\phi\ d\phi\ d\theta) = 0$ 19 $A = \mathbf{i} + \mathbf{j} + 2\mathbf{k}; B = \mathbf{j} + \mathbf{k}; N = -\mathbf{i} - \mathbf{j} + \mathbf{k}; dS = \sqrt{3}\ du\ dv$

21 $A = -\sin u(\cos v\,\mathbf{i} + \sin v\,\mathbf{j}) + \cos u\,\mathbf{k}; B = -(3 + \cos u)\sin v\,\mathbf{i} + (3 + \cos u)\cos v\,\mathbf{j};$
$N = -(3 + \cos u)(\cos u \cos v\,\mathbf{i} + \cos u \sin v\,\mathbf{j} + \sin u\,\mathbf{k}); dS = (3 + \cos u)du\ dv$

23 $\iint(-M\frac{\partial f}{\partial x} - N\frac{\partial f}{\partial y} + P)dx\ dy = \iint(-2x^2 - 2y^2 + z)dx\ dy = \iint -r^2(r\ dr\ d\theta) = -8\pi$

25 $\mathbf{F} \cdot \mathbf{N} = -x + y + z = 0$ on plane

27 $N = -\mathbf{i} - \mathbf{j} + \mathbf{k}, \mathbf{F} = (v + u)\mathbf{i} - u\mathbf{j}, \iint \mathbf{F} \cdot \mathbf{N} dS = \iint -v\ du\ dv = 0$

29 $\iint dS = \int_0^{2\pi} \int_0^{2\pi}(3 + \cos u)du\ dv = 12\pi^2$ 31 Yes   33 No

35 $A = \mathbf{i} + f'\cos\theta\,\mathbf{j} + f'\sin\theta\,\mathbf{k}; B = -f\sin\theta\,\mathbf{j} + f\cos\theta\,\mathbf{k}; N = ff'\mathbf{i} - f\cos\theta\,\mathbf{j} - f\sin\theta\,\mathbf{k}; dS = |N|dx\ d\theta = f(x)\sqrt{1 + f'^2}\ dx\ d\theta$

Section 15.5 The Divergence Theorem (page 588)

1 $\text{div } \mathbf{F} = 1, \iiint dV = \frac{4\pi}{3}$ 3 $\text{div } \mathbf{F} = 2x + 2y + 2z, \iiint \text{div } \mathbf{F}\ dV = 0$ 5 $\text{div } \mathbf{F} = 3, \iint 3dV = \frac{3}{6} = \frac{1}{2}$

7 $\mathbf{F} \cdot \mathbf{N} = \rho^2, \iint_{\rho=a} \rho^2 dS = 4\pi a^4$ 9 $\text{div } \mathbf{F} = 2z, \int_0^{2\pi} \int_0^{\pi/2} \int_0^a 2\rho\cos\phi(\rho^2 \sin\phi\ dp\ d\phi\ d\theta) = \frac{1}{2}\pi a^4$

11 $\int_0^a \int_0^a \int_0^a (2x + 1)dx\ dy\ dz = a^4 + a^3; -2a^2 + 2a^2 + 0 + a^4 + 0 + a^3$

13 $\text{div } \mathbf{F} = \frac{x}{\rho}, \iiint \frac{x}{\rho} dV = 0; \mathbf{F} \cdot \mathbf{n} = x, \iint x dS = 0$ 15 $\text{div } \mathbf{F} = 1; \iiint 1 dV = \frac{\pi}{3}; \iiint 1\ dV = \frac{1}{6}$

17 $\text{div}\left(\frac{\mathbf{R}}{\rho^7}\right) = \frac{\text{div } \mathbf{R}}{\rho^7} + \mathbf{R} \cdot \text{grad } \frac{1}{\rho^7} = \frac{3}{\rho^7} - \frac{7}{\rho^8} \mathbf{R} \cdot \text{grad } \rho$

19 Two spheres, $\mathbf{n}$ radial out, $\mathbf{n}$ radial in, $\mathbf{n} = \mathbf{k}$ on top, $\mathbf{n} = -\mathbf{k}$ on bottom, $\mathbf{n} = \frac{x\mathbf{i} + y\mathbf{j}}{\sqrt{x^2 + y^2}}$ on side;
$\mathbf{n} = -\mathbf{i}, -\mathbf{j}, -\mathbf{k}, \mathbf{i} + 2\mathbf{j} + 3\mathbf{k}$ on 4 faces; $\mathbf{n} = \mathbf{k}$ on top, $\mathbf{n} = \frac{1}{\sqrt{2}}\left(\frac{x}{r}\mathbf{i} + \frac{y}{r}\mathbf{j} - \mathbf{k}\right)$ on cone

21 $V = \text{cylinder}, \iiint \text{div } \mathbf{F}\ dV = \iint\left(\frac{\partial M}{\partial x} + \frac{\partial N}{\partial y}\right)dx\ dy\ (z \text{ integral} = 1); \iint \mathbf{F} \cdot \mathbf{n} dS = \int M dy - N dx, z \text{ integral} = 1$ on side, $\mathbf{F} \cdot \mathbf{n} = 0$ top and bottom; Green's flux theorem.

23 $\text{div } \mathbf{F} = \frac{-3GM}{a^3} = -4\pi G$; at the center; $\mathbf{F} = 2\mathbf{R}$ inside, $\mathbf{F} = 2\left(\frac{a}{\rho}\right)^3 \mathbf{R}$ outside

25 $\text{div } \mathbf{u}_r = \frac{2}{\rho}, q = \frac{2\epsilon_0}{\rho}, \iint \mathbf{E} \cdot \mathbf{n}\ dS = \iint 1 dS = 4\pi$ 27 F $(\text{div } \mathbf{F} = 0); \text{F}; \text{T } (\mathbf{F} \cdot \mathbf{n} \le 1);$ F

29 Plane circle; top half of sphere; $\text{div } \mathbf{F} = 0$

Section 15.6 Stokes' Theorem and the Curl of F (page 595)

1 $\text{curl } \mathbf{F} = \mathbf{i} + \mathbf{j} + \mathbf{k}$ 3 $\text{curl } \mathbf{F} = \mathbf{0}$ 5 $\text{curl } \mathbf{F} = \mathbf{0}$ 7 $f = \frac{1}{2}(x + y + z)^2$

9 $\text{curl } x^m\mathbf{i} = \mathbf{0}; x^n\mathbf{j}$ has zero curl if $n = 0$ 11 $\text{curl } \mathbf{F} = 2y\mathbf{i}; \mathbf{n} = \mathbf{j}$ on circle so $\iint \mathbf{F} \cdot \mathbf{n} dS = 0$

13 $\text{curl } \mathbf{F} = 2\mathbf{i} + 2\mathbf{j}, \mathbf{n} = \mathbf{i}, \iint \text{curl } \mathbf{F} \cdot \mathbf{n} dS = \iint 2 dS = 2\pi$

15 Both integrals equal $\int \mathbf{F} \cdot d\mathbf{R}$; Divergence Theorem, V = region between S and T, always div curl $\mathbf{F} = 0$

17 Always div curl $\mathbf{F} = 0$ **19** $f = xz + y$ **21** $f = e^{x-z}$ **23** $\mathbf{F} = y\mathbf{k}$

25 curl $\mathbf{F} = (a_3 b_2 - a_2 b_3)\mathbf{i} + (a_1 b_3 - a_3 b_1)\mathbf{j} + (a_2 b_1 - a_1 b_2)\mathbf{k}$ **27** curl $\mathbf{F} = 2\omega \mathbf{k}$; curl $\mathbf{F} \cdot \frac{\mathbf{i}+\mathbf{j}+\mathbf{k}}{\sqrt{3}} = 2\omega/\sqrt{3}$

29 $\mathbf{F} = x(a_3 z + a_2 y)\mathbf{i} + y(a_1 x + a_3 z)\mathbf{j} + z(a_1 x + a_2 y)\mathbf{k}$

31 curl $\mathbf{F} = -2\mathbf{k}, \int\int -2\mathbf{k} \cdot \mathbf{R}dS = \int_0^{2\pi} \int_0^{\pi/2} -2\cos\phi(\sin\phi\, d\phi\, d\theta) = -2\pi; \int y\, dx - x\, dy =$
$\int_0^{2\pi} (-\sin^2 t - \cos^2 t)dt = -2\pi$

33 curl $\mathbf{F} = 2\mathbf{a}, 2\int\int (a_1 x + a_2 y + a_3 z)dS = 0 + 0 + 2a_3 \int_0^{2\pi} \int_0^{\pi/2} \cos\phi \, \sin\phi\, d\phi\, d\theta = 2\pi a_3$

35 curl $\mathbf{F} = -\mathbf{i}, \mathbf{n} = \frac{\mathbf{i}+\mathbf{j}+\mathbf{k}}{\sqrt{3}}, \int\int \mathbf{F} \cdot \mathbf{n}dS = -\frac{1}{\sqrt{3}}\pi r^2$

37 $g = \frac{y^2}{2} - \frac{x^3}{3} =$ stream function; zero divergence

39 div $\mathbf{F} =$ div $(\mathbf{V} + \mathbf{W}) =$ div $\mathbf{V}$ so $y =$ div $\mathbf{V}$ so $\mathbf{V} = \frac{y^2}{2}\mathbf{j}$ (has zero curl). Then $\mathbf{W} = \mathbf{F} - \mathbf{V} = xy\mathbf{i} - \frac{y^2}{2}\mathbf{j}$

41 curl (curl $\mathbf{F}$) = curl $(-2y\mathbf{k}) = -2\mathbf{i}$; grad (div $\mathbf{F}$) = grad $2x = 2\mathbf{i}$; $\mathbf{F}_{xx} + \mathbf{F}_{yy} + \mathbf{F}_{zz} = 4\mathbf{i}$

43 curl $\mathbf{E} = -\frac{\partial \mathbf{B}}{\partial t} = \mathbf{a}\sin t$ so $\mathbf{E} = \frac{1}{2}(\mathbf{a} \times \mathbf{R})\sin t$

45 $\mathbf{n} = \mathbf{j}$ so $\int M\, dx + P\, dz = \int\int \left(\frac{\partial M}{\partial z} - \frac{\partial P}{\partial x}\right)dx\, dz$ **47** $M_y^* = M_y + M_z f_y + P_y f_x + P_z f_y f_x + P f_{xy}$

49 $\int \mathbf{F} \cdot d\mathbf{R} = \int\int$ curl $\mathbf{F} \cdot \mathbf{n}\, dS; \int\int \mathbf{F} \cdot \mathbf{n}dS = \int\int\int$ div $\mathbf{F}\, dV$

CHAPTER 16   MATHEMATICS AFTER CALCULUS

Section 16.1 Linear Algebra   (page 602)

1 All vectors $c\begin{bmatrix} 2 \\ -1 \end{bmatrix}$ **3** Only $x = 0$ **5** Plane of vectors with $x_1 + x_2 + x_3 = 0$

7 $x_p = \begin{bmatrix} 3 \\ 0 \end{bmatrix}, A(x_p + x_0) = \begin{bmatrix} 3 \\ 6 \end{bmatrix} + \begin{bmatrix} 0 \\ 0 \end{bmatrix}$ **9** $A(x_p + x_0) = b + 0 = b$; another solution

11 $\begin{bmatrix} 1 \\ 1 \\ 1 \end{bmatrix}, \begin{bmatrix} 0 \\ 0 \\ 0 \end{bmatrix}, \begin{bmatrix} 2 \\ 2 \\ 2 \end{bmatrix}; b = \begin{bmatrix} c \\ c \\ c \end{bmatrix}$

13 $CC^T = \begin{bmatrix} 1 & 0 & 1 \\ 0 & 1 & 2 \\ 1 & 2 & 5 \end{bmatrix}; C^T C = \begin{bmatrix} 2 & 2 \\ 2 & 5 \end{bmatrix}$; (2 by 3)(2 by 3) is impossible

15 Any two are independent   **17** C and F have independent columns

19 det $F = 3$ **21** $F^{-1} = \frac{1}{3}\begin{bmatrix} 2 & -1 \\ -1 & 2 \end{bmatrix}$

23 det $(F - \lambda I) =$ det $\begin{bmatrix} 2-\lambda & 1 \\ 1 & 2-\lambda \end{bmatrix} = (2-\lambda)^2 - 1 = 3 - 4\lambda + \lambda^2 = 0$ if $\lambda = 1$ or $\lambda = 3$;

$F\begin{bmatrix} 1 \\ -1 \end{bmatrix} = 1\begin{bmatrix} 1 \\ -1 \end{bmatrix}, F\begin{bmatrix} 1 \\ 1 \end{bmatrix} = 3\begin{bmatrix} 1 \\ 1 \end{bmatrix}$

25 $y = e^t \begin{bmatrix} 1 \\ -1 \end{bmatrix}, y = e^{3t}\begin{bmatrix} 1 \\ 1 \end{bmatrix}, y = \frac{e^t}{2}\begin{bmatrix} 1 \\ -1 \end{bmatrix} + \frac{e^{3t}}{2}\begin{bmatrix} 1 \\ 1 \end{bmatrix}$

27 det $\begin{bmatrix} 1-\lambda & 1 & 1 \\ 1 & 1-\lambda & 1 \\ 1 & 1 & 1-\lambda \end{bmatrix} = (1-\lambda)^3 - 3(1-\lambda) + 2 = \lambda^3 - 3\lambda^2 = 0$ if $\lambda = 3$ or $\lambda = 0$ (repeated)

29 $\det \begin{bmatrix} 1-\lambda & 2 \\ 2 & 4-\lambda \end{bmatrix} = \lambda^2 - 5\lambda = 0$ if $\lambda = 0$ or $\lambda = 5$; $A\begin{bmatrix} 2 \\ -1 \end{bmatrix} = 0\begin{bmatrix} 2 \\ -1 \end{bmatrix}, A\begin{bmatrix} 1 \\ 2 \end{bmatrix} = 5\begin{bmatrix} 1 \\ 2 \end{bmatrix}$

31 $H = \begin{bmatrix} 2 & -2 \\ -2 & 2 \end{bmatrix}$ **33** F if $b \neq 0$; T; T; F ($e^{\lambda t}$ is not a vector); T

Section 16.2 Differential Equations   (page 610)

1 $3Be^{3t} - Be^{3t} = 8e^{3t}$ gives $B = 4$: $y = 4e^{3t}$ **3** $y = 3 - 2t + t^2$ **5** $Ae^t + 4e^{3t} = 7$ at $t = 0$ if $A = 3$

7 Add $y = Ae^{-t}$ because $y' + y = 0$; choose $A = -1$ so $-e^{-t} + 3 - 2t + t^2 = 2$ at $t = 0$

9 $y = \frac{e^{kt}-1}{k}$; $y = t$; by l'Hôpital $\lim_{k \to 0} \frac{e^{kt}-1}{k} = \lim_{k \to 0} \frac{te^{kt}}{1} = t$

11 Substitute $y = Ae^t + Bte^t + C\cos t + D\sin t$ in equation: $B = 1, C = \frac{1}{2}, D = -\frac{1}{2}$ any A

13 Particular solution $y = Ate^t + Be^t$; $y' = Ate^t + (A+B)e^t = c(Ate^t + Be^t) + te^t$
gives $A = cA + 1, A + B = cB, A = \frac{1}{1-c}, B = \frac{-1}{(1-c)^2}$

15 $\lambda^2 e^{\lambda t} + 6\lambda e^{\lambda t} + 5e^{\lambda t} = 0$ gives $\lambda^2 + 6\lambda + 5 = 0, (\lambda+5)(\lambda+1) = 0, \lambda = -1$ or -5
(both negative so decay); $y = Ae^{-t} + Be^{-5t}$

17 $(\lambda^2 + 2\lambda + 3)e^{\lambda t} = 0, \lambda = -1 \pm \sqrt{-2}$ has imaginary part and negative real part;
$y = Ae^{(-1+\sqrt{2}i)t} + Be^{(-1-\sqrt{2}i)t}$; $y = Ce^{-t}\cos\sqrt{2}t + De^{-t}\sin\sqrt{2}t$

19 $d = 0$ no damping; $d = 1$ underdamping; $d = 2$ critical damping; $d = 3$ overdamping

21 $\lambda = -\frac{b}{2} \pm \frac{\sqrt{b^2-4c}}{2}$ is repeated when $b^2 = 4c$ and $\lambda = -\frac{b}{2}$; $(t\lambda^2 + 2\lambda)e^{\lambda t} + b(t\lambda+1)e^{\lambda t} + cte^{\lambda t} = 0$
when $\lambda^2 + b\lambda + c = 0$ and $2\lambda + b = 0$

23 $-a\cos t - b\sin t - a\sin t + b\cos t + a\cos t + b\sin t = \cos t$ if $a = 0, b = 1$, $y = \sin t$

25 $y = A\cos 3t + B\cos 5t$; $y'' + 9y = -25B\cos 5t + 9B\cos 5t = \cos 5t$ gives $B = \frac{-1}{16}$;
$y_0 = 0$ gives $A = \frac{1}{16}$

27 $y = A(\cos\omega t - \cos\omega_0 t), y'' = -A\omega^2\cos\omega t + A\omega_0^2\cos\omega_0 t, y'' + \omega_0^2 y = \cos\omega t$ gives $A(-\omega^2 + \omega_0^2) = 1$;
breaks down when $\omega^2 = \omega_0^2$

29 $y = Be^{5t}$; $25B + 3B = 1, B = \frac{1}{28}$ **31** $y = A + Bt = \frac{1}{2} + \frac{1}{2}t$

33 $y'' - 25y = e^{5t}$; $y'' + y = \sin t$; $y'' = 1 + t$; right side solves homogeneous equation so particular
solution needs extra factor t

35 $e^t, e^{-t}, e^{it}, e^{-it}$ **37** $y = e^{-2t} + 2te^{-2t}$; $y(2\pi) = (1 + 4\pi)e^{-4\pi} \approx 0$

39 $y = (4e^{-rt} - r^2 e^{-4t/r})/(4 - r^2) \to 1$ as $r \to 0$ **43** $h \leq 2$; $h \leq 2.8$

Section 16.3 Discrete Mathematics   (page 615)

1 Two then two then last one; go around hexagon   **3** Six (each deletes one edge)

5 Connected: there is a path between any two nodes; connecting each new node requires an edge

13 Edge lengths $1, 2, 4$

15 No; $1, 3, 4$ on left connect only to $2, 3$ on right; $1, 3$ on right connect only to 2 on left   **17** 4

19 Yes   **21** F (may loop); T   **25** 16

Table of Integrals

1 $\int u^n \, dx = \frac{u^{n+1}}{a(n+1)}$ except for $\int \frac{dx}{u} = \frac{\ln |u|}{a}$ All the integrals 1 - 17 involve $u = ax + b$

2 $\int xu^n \, dx = \frac{u^{n+2}}{a^2(n+2)} - \frac{bu^{n+1}}{a^2(n+1)}$ except for $\int \frac{x\,dx}{u} = \frac{x}{a} - \frac{b\ln|u|}{a^2}$ and $\int \frac{x\,dx}{u^2} = \frac{b}{a^2 u} + \frac{\ln|u|}{a^2}$

3 $\int \frac{x^2\,dx}{u} = \frac{1}{a^3}\left(\frac{u^2}{2} - 2bu + b^2 \ln |u|\right)$ **4** $\int \frac{x^2\,dx}{u^2} = \frac{1}{a^3}\left(u - 2b\ln|u| - \frac{b^2}{u}\right)$ **5** $\int \frac{x^2\,dx}{u^3} = \frac{1}{a^3}\left(\ln|u| + \frac{2b}{u} - \frac{b^2}{2u^2}\right)$

6 $\int \frac{dx}{xu} = \frac{1}{b}\ln\left|\frac{x}{u}\right|$ **7** $\int \frac{dx}{x^2 u} = -\frac{1}{bx} + \frac{a}{b^2}\ln\left|\frac{u}{x}\right|$ **8** $\int \frac{dx}{xu^2} = \frac{1}{bu} - \frac{1}{b^2}\ln\left|\frac{u}{x}\right|$ **9** $\int \frac{dx}{x^2 u^2} = -\frac{b + 2ax}{b^2 xu} + \frac{2a}{b^3}\ln\left|\frac{u}{x}\right|$

10 $\int \sqrt{u}\,dx = \frac{2}{3a} u^{3/2}$ **11** $\int x\sqrt{u}\,dx = \frac{2(3ax - 2b)}{15a^2} u^{3/2}$ **12** $\int x^2 \sqrt{u}\,dx = \frac{2(15a^2 x^2 - 12abx + 8b^2)}{105a^3} u^{3/2}$

13 $\int \frac{\sqrt{u}}{x}\,dx = 2\sqrt{u} + b\int \frac{dx}{x\sqrt{u}}$ **14** $\int \frac{x\,dx}{\sqrt{u}} = \frac{2(ax - 2b)}{3a^2}\sqrt{u}$ **15** $\int \frac{x^2\,dx}{\sqrt{u}} = \frac{2(3a^2 x^2 - 4abx + 8b^2)}{15a^3}\sqrt{u}$

16 $\int \frac{dx}{x\sqrt{u}} = \frac{1}{\sqrt{b}}\ln\left|\frac{\sqrt{u}-\sqrt{b}}{\sqrt{u}+\sqrt{b}}\right|$ $(b>0)$ or $\frac{2}{\sqrt{-b}}\tan^{-1}\frac{\sqrt{u}}{\sqrt{-b}}$ $(b<0)$ **17** $\int \frac{\sqrt{u}}{x^2}\,dx = -\frac{\sqrt{u}}{x} + \frac{a}{2}\int \frac{dx}{x\sqrt{u}}$

18 $\int \frac{dx}{(ax+b)(cx+d)} = \frac{1}{bc-ad}\ln\left|\frac{cx+d}{ax+b}\right|$ **19** $\int \frac{x\,dx}{(ax+b)(cx+d)} = \frac{1}{bc-ad}\left(\frac{b}{a}\ln|ax+b| - \frac{d}{c}\ln|cx+d|\right)$

20 $\int \sqrt{x^2 \pm a^2}\,dx = \frac{x}{2}\sqrt{x^2 \pm a^2} \pm \frac{a^2}{2}\ln|x + \sqrt{x^2 \pm a^2}|$ **21** $\int \frac{dx}{\sqrt{x^2 \pm a^2}} = \ln|x + \sqrt{x^2 \pm a^2}|$

22 $\int \frac{\sqrt{x^2 + a^2}}{x}\,dx = \sqrt{x^2 + a^2} - a\ln\left(\frac{a + \sqrt{x^2 + a^2}}{x}\right)$ **23** $\int \frac{\sqrt{x^2 - a^2}}{x}\,dx = \sqrt{x^2 - a^2} - a\sec^{-1}\frac{x}{a}$

24 $\int \frac{x^2\,dx}{\sqrt{x^2 \pm a^2}} = \frac{x}{2}\sqrt{x^2 \pm a^2} \mp \frac{a^2}{2}\ln|x + \sqrt{x^2 \pm a^2}|$ **25** $\int \frac{\sqrt{x^2 \pm a^2}}{x^2}\,dx = -\frac{\sqrt{x^2 \pm a^2}}{x} + \ln|x + \sqrt{x^2 \pm a^2}|$

26 $\int x^2 \sqrt{x^2 \pm a^2}\,dx = \frac{x}{8}(2x^2 \pm a^2)\sqrt{x^2 \pm a^2} - \frac{a^4}{8}\ln|x + \sqrt{x^2 \pm a^2}|$ **27** $\int \frac{dx}{x^2\sqrt{x^2 \pm a^2}} = \mp\frac{\sqrt{x^2 \pm a^2}}{a^2 x}$

28 $\int (x^2 \pm a^2)^{3/2}\,dx = \frac{x}{8}(2x^2 \pm 5a^2)\sqrt{x^2 \pm a^2} + \frac{3a^4}{8}\ln|x + \sqrt{x^2 \pm a^2}|$ **29** $\int \frac{dx}{(x^2 \pm a^2)^{3/2}} = \frac{\pm x}{a^2\sqrt{x^2 \pm a^2}}$

30 $\int \sqrt{a^2 - x^2}\,dx = \frac{x}{2}\sqrt{a^2 - x^2} + \frac{a^2}{2}\sin^{-1}\frac{x}{a}$ **31** $\int \frac{dx}{\sqrt{a^2 - x^2}} = \sin^{-1}\frac{x}{a}$ **32** $\int \frac{dx}{(a^2 - x^2)^{3/2}} = \frac{x}{a^2\sqrt{a^2 - x^2}}$

33 $\int \frac{\sqrt{a^2 - x^2}}{x}\,dx = \sqrt{a^2 - x^2} - a\ln\left|\frac{a + \sqrt{a^2 - x^2}}{x}\right|$ **34** $\int x^2\sqrt{a^2 - x^2}\,dx = \frac{x}{8}(2x^2 - a^2)\sqrt{a^2 - x^2} + \frac{a^4}{8}\sin^{-1}\frac{x}{a}$

35 $\int \frac{dx}{x^2\sqrt{a^2 - x^2}} = -\frac{\sqrt{a^2 - x^2}}{a^2 x}$ **36** $\int \frac{\sqrt{a^2 - x^2}}{x^2}\,dx = -\frac{\sqrt{a^2 - x^2}}{x} - \sin^{-1}\frac{x}{a}$ **37** $\int \frac{dx}{x\sqrt{a^2 - x^2}} = -\frac{1}{a}\ln\left|\frac{a + \sqrt{a^2 - x^2}}{x}\right|$

38 $\int \frac{x^2\,dx}{\sqrt{a^2 - x^2}} = -\frac{x}{2}\sqrt{a^2 - x^2} + \frac{a^2}{2}\sin^{-1}\frac{x}{a}$ **39** $\int (a^2 - x^2)^{3/2}\,dx = \frac{x}{8}(5a^2 - 2x^2)\sqrt{a^2 - x^2} + \frac{3a^4}{8}\sin^{-1}\frac{x}{a}$

40 $\int \frac{dx}{b + c\sin ax} = \frac{-2}{a\sqrt{b^2 - c^2}}\tan^{-1}\left[\sqrt{\frac{b-c}{b+c}}\tan\left(\frac{\pi}{4} - \frac{ax}{2}\right)\right]$, $b^2 > c^2$ **41** $\int \frac{dx}{1 + \sin ax} = -\frac{1}{a}\tan\left(\frac{\pi}{4} - \frac{ax}{2}\right)$

42 $\int \frac{dx}{b + c\sin ax} = \frac{-1}{a\sqrt{c^2 - b^2}}\ln\left|\frac{c + b\sin ax + \sqrt{c^2 - b^2}\cos ax}{b + c\sin ax}\right|$, $b^2 < c^2$ **43** $\int \frac{dx}{1 - \sin ax} = \frac{1}{a}\tan\left(\frac{\pi}{4} + \frac{ax}{2}\right)$

44 $\int \frac{dx}{b + c\cos ax} = \frac{2}{a\sqrt{b^2 - c^2}}\tan^{-1}\left[\sqrt{\frac{b-c}{b+c}}\tan\frac{ax}{2}\right]$, $b^2 > c^2$ **45** $\int \frac{dx}{1 + \cos ax} = \frac{1}{a}\tan\frac{ax}{2}$

46 $\int \frac{dx}{b + c\cos ax} = \frac{1}{a\sqrt{c^2 - b^2}}\ln\left|\frac{c + b\cos ax + \sqrt{c^2 - b^2}\sin ax}{b + c\cos ax}\right|$, $b^2 < c^2$ **47** $\int \frac{dx}{1 - \cos ax} = -\frac{1}{a}\cot\frac{ax}{2}$

48 $\int \sin^{-1} ax\,dx = x\sin^{-1} ax + \frac{1}{a}\sqrt{1 - a^2 x^2}$ **49** $\int x^n \sin^{-1} ax\,dx = \frac{x^{n+1}}{n+1}\sin^{-1} ax - \frac{a}{n+1}\int \frac{x^{n+1}\,dx}{\sqrt{1 - a^2 x^2}}$

50 $\int \tan^{-1} ax\,dx = x\tan^{-1} ax - \frac{1}{2a}\ln(1 + a^2 x^2)$ **51** $\int x^n \tan^{-1} ax\,dx = \frac{x^{n+1}}{n+1}\tan^{-1} ax - \frac{a}{n+1}\int \frac{x^{n+1}\,dx}{1 + a^2 x^2}$

52 $\int e^{ax}\,dx = \frac{e^{ax}}{a}$ **53** $\int xe^{ax}\,dx = \frac{e^{ax}}{a^2}(ax - 1)$ **54** $\int x^2 e^{ax}\,dx = \frac{e^{ax}}{a^3}(a^2 x^2 - 2ax + 2)$ $(b^{ax}$ is $e^{a(\ln b)x})$

55 $\int \frac{dx}{x\ln ax} = \ln|\ln ax|$ Not elementary: $\int e^{x^2}\,dx$, $\int e^x \ln x\,dx$, $\int \frac{dx}{\ln x}$, $\int \frac{e^x}{x}\,dx$, $\int \frac{\sin x}{x}\,dx$, $\int \frac{\sin^{-1} x}{x}\,dx$

Exponentials and Logarithms

$y = b^x \leftrightarrow x = \log_b y \quad y = e^x \leftrightarrow x = \ln y$

$e = \lim(1 + \frac{1}{n})^n = \sum_{n=0}^{\infty} \frac{1}{n!} = 2.71828\cdots$

$e^x = \lim(1 + \frac{x}{n})^n = \sum_{n=0}^{\infty} \frac{x^n}{n!}$

$\ln y = \int_1^y \frac{dx}{x} \qquad \ln 1 = 0 \quad \ln e = 1$

$\ln xy = \ln x + \ln y \quad \ln x^n = n \ln x$

$\log_a y = (\log_a b)(\log_b y) \quad \log_a b = 1/\log_b a$

$e^{x+y} = e^x e^y \quad b^x = e^{x \ln b} \quad e^{\ln y} = y$

Vectors and Determinants

$\mathbf{A} = a_1 \mathbf{i} + a_2 \mathbf{j} + a_3 \mathbf{k}$

$|\mathbf{A}|^2 = \mathbf{A} \cdot \mathbf{A} = a_1^2 + a_2^2 + a_3^2$ (length squared)

$\mathbf{A} \cdot \mathbf{B} = a_1 b_1 + a_2 b_2 + a_3 b_3 = |\mathbf{A}||\mathbf{B}| \cos \theta$

$|\mathbf{A} \cdot \mathbf{B}| \leq |\mathbf{A}||\mathbf{B}|$ (Schwarz inequality: $|\cos \theta| \leq 1$)

$|\mathbf{A} + \mathbf{B}| \leq |\mathbf{A}| + |\mathbf{B}|$ (triangle inequality)

$|\mathbf{A} \times \mathbf{B}| = |\mathbf{A}||\mathbf{B}||\sin \theta|$ (cross product)

$\mathbf{A} \times \mathbf{B} = \begin{vmatrix} \mathbf{i} & \mathbf{j} & \mathbf{k} \\ a_1 & a_2 & a_3 \\ b_1 & b_2 & b_3 \end{vmatrix} = \begin{array}{l} \mathbf{i}(a_2 b_3 - a_3 b_2) \\ +\mathbf{j}(a_3 b_1 - a_1 b_3) \\ +\mathbf{k}(a_1 b_2 - a_2 b_1) \end{array}$

Right hand rule $\mathbf{i} \times \mathbf{j} = \mathbf{k}, \mathbf{j} \times \mathbf{k} = \mathbf{i}, \mathbf{k} \times \mathbf{i} = \mathbf{j}$

Parallelogram area $= |a_1 b_2 - a_2 b_1| = |\text{Det}|$

Box volume $= |\mathbf{A} \cdot (\mathbf{B} \times \mathbf{C})| = |\text{Determinant}|$

SI Units   Symbols

length	meter	m
mass	kilogram	kg
time	second	s
current	ampere	A
frequency	hertz	Hz $\sim$ 1/s
force	newton	N $\sim$ kg·m/s^2
pressure	pascal	Pa $\sim$ N/m^2
energy, work	joule	J $\sim$ N·m
power	watt	W $\sim$ J/s
charge	coulomb	C $\sim$ A·s
temperature	kelvin	K

Speed of light $c = 2.9979 \times 10^8$ m/s

Gravity $G = 6.6720 \times 10^{-11}$ Nm^2/kg^2

Equations and Their Solutions

$y' = cy$	$y_0 e^{ct}$
$y' = cy + s$	$y_0 e^{ct} + \frac{s}{c}(e^{ct} - 1)$
$y' = cy - by^2$	$\frac{c}{b + de^{-ct}} \quad d = \frac{c - by_0}{y_0}$
$y'' = -\lambda^2 y$	$\cos \lambda t$ and $\sin \lambda t$
$my'' + dy' + ky = 0$	$e^{\lambda_1 t}$ and $e^{\lambda_2 t}$ or $te^{\lambda_1 t}$
$y_{n+1} = ay_n + s$	$a^n y_0 + s\frac{a^n - 1}{a - 1}$

Matrices and Inverses

$Ax = $ combination of columns $= b$

Solutions $x = A^{-1}b$ if $A^{-1}A = I$

Least squares $A^T A\overline{x} = A^T b$

$Ax = \lambda x \quad$ (λ is an eigenvalue)

$\begin{bmatrix} a & b \\ c & d \end{bmatrix}^{-1} = \frac{1}{ad - bc} \begin{bmatrix} d & -b \\ -c & a \end{bmatrix}$

$(AB)^{-1} = B^{-1}A^{-1}, (AB)^T = B^T A^T$

$\begin{bmatrix} \mathbf{a} & \mathbf{b} & \mathbf{c} \end{bmatrix}^{-1} = \frac{1}{D} \begin{bmatrix} \mathbf{b} \times \mathbf{c} \\ \mathbf{c} \times \mathbf{a} \\ \mathbf{a} \times \mathbf{b} \end{bmatrix}$

$\begin{vmatrix} a_1 & a_2 & a_3 \\ b_1 & b_2 & b_3 \\ c_1 & c_2 & c_3 \end{vmatrix} = \begin{array}{l} + a_1 b_2 c_3 + a_2 b_3 c_1 + a_3 b_1 c_2 \\ - a_1 b_3 c_2 - a_2 b_1 c_3 - a_3 b_2 c_1 \end{array}$

From	To	Multiply by
degrees	radians	.01745
calories	joules	4.1868
BTU	joules	1055.1
foot-pounds	joules	1.3558
feet	meters	.3048
miles	km	1.609
feet/sec	km/hr	1.0973
pounds	kg	.45359
ounces	kg	.02835
gallons	liters	3.785
horsepower	watts	745.7

Radius at Equator $R = 6378$ km $= 3964$ miles

Acceleration $g = 9.8067$ m/s$^2 = 32.174$ ft/s^2